J. Gregory

LIGNINS

KARL FREUDENBERG

Professor Freudenberg and his school in Heidelberg, Germany have made an unparalleled series of contributions to the understanding of the biogenesis and the structure of lignins.

LIGNINS
OCCURRENCE, FORMATION, STRUCTURE AND REACTIONS

Edited by

K.V. SARKANEN

College of Forest Resources
University of Washington
Seattle, Washington

C.H. LUDWIG

Georgia Pacific Corporation
Bellingham Division
Bellingham, Washington

WILEY–INTERSCIENCE

a division of John Wiley & Sons, Inc., New York – London – Sydney – Toronto

Copyright © 1971, by John Wiley & Sons, Inc.

All rights reserved. Published simultaneously in Canada.

No part of this book may be reproduced by any means, nor transmitted, nor translated into a machine language without the written permission of the publisher.

Library of Congress Catalogue Card Number: 79-148456

ISBN 0-471-75422-6

Printed in the United States of America.

10 9 8 7 6 5 4 3 2 1

FOREWORD

Knowledge of lignins has evolved over a period of more than one hundred years and their importance is widely recognized. It is now known that lignins exist as polymeric cell wall constituents in almost all dry-land plants, and among the natural polymers, lignins are second only to carbohydrates in natural abundance.

The complex structure of these polymers, apparently the result of a unique principle of polymerization, has delayed progress in this area of chemical research. Thus the evolution of satisfactory structural concepts has lagged behind the rapid advances in the understanding of other natural as well as synthetic polymers. In recent years, however, applications of modern methods of chemical analysis and of modern concepts of the biogenesis of natural compounds have significantly advanced the field of lignin chemistry toward the realm of a well-organized discipline.

In the economy of Man, lignins play an important although not yet very conspicuous role. Wood seems to have essentially the nature of a lignin-hemicellulose plastic reinforced by cellulosic fibers and, based upon this nature, seemingly will continue to be an outstandingly important structural material and the primary source of paper and board products. The vast forest areas both in semi-arctic and tropical regions, form currently an enormous reserve for expanding the production of useful products from wood. Another virtually untapped reserve is found in residues of agriculture.

The demands of the expanding world population for the foods and raw materials are generating serious concerns. The Malthusian preachers of gloom, however, seem to have been quite uninformed about the vast available lignocellulosic reserve. After all, our supplies of edible materials can be substantially augmented by sugars obtainable through the hydrolysis of the carbohydrates present in woody materials provided that we are prepared to pay the price. Likewise, the use of lignocellulose as a raw material for the chemical industry is technologically possible, but does not occur primarily because of the low cost of petroleum and its processing. While the future industrial prospects of lignocellulose have not yet been developed extensively, increasing interest is currently being directed toward mitigation of the hazards of pollution arising from lignin containing effluents from pulp mills, and this attention, hopefully, will encourage broader utilization of the lignocellulose and industrially available by-products.

Because of these important considerations, an ever growing number of scientists, technologists, engineers, economists and others will welcome the appearance of the book, "Lignins."

Professor K. V. Sarkanen and Dr. C. H. Ludwig, one a colleague and the other a prior doctoral student with me at the University of Washington, along with the distinguished authors who have written chapters, have produced, as the first collective endeavor by a number of widely known authorities, a well balanced and clear presentation of this complex field. Comprehensive monographs of lignin chemistry have been published in the past, and yet the field of lignin chemistry has still remained bewildering to the uninitiated because of the often contradictory opinions maintained by investigators in this area.

It is hoped that the critical evaluation of available information by the authors whose writings appear in "Lignins" will help to eliminate earlier contradictions and to stimulate further advances in this important field.

JOSEPH L. McCARTHY

University of Washington
Seattle, Washington

PREFACE

The first extensive English language monograph on lignins, "The Chemistry of Lignin" by F. E. Brauns, appeared in 1952, followed by a "Supplement Volume" by F. E. Brauns and D. A. Brauns in 1960. Monographs by the same title have been published by Y. Hachihama and S. Jyodai (in Japanese) in 1946 and by I. A. Pearl in 1967. Specific aspects of lignins have received coverage in "Biochemistry of Lignin" by W. J. Schubert and most recently, in "Biogenesis and Constitution of Lignin" published jointly by K. Freudenberg and A. C. Neish in 1969.

As in many other fields, it is now almost impossible for a single person to be a master of all aspects of lignin chemistry, biochemistry and utilization. A patent solution for such a situation is to resort to multiple authorship, which in the present volume has, for the first time, been applied to a lignin monograph. As editors we feel pleased to have been able to recruit some of the best talent in the field to this endeavor. The contributors were asked at the onset to be selective rather than comprehensive in their choice of material. As a consequence, the reader will note some divergence of emphasis and opinions in different chapters.

The editors are much indebted to the authors of individual chapters for their splendid cooperation and unrelenting patience during the admittedly lengthy preparation of this book. In its various stages of evolution, the book was also inflicted upon unwary colleagues and captive graduate students alike. Among these, we wish to express our deepest gratitude to Vincent F. Felicetta, Jack R. Kelley, W. Scott Briggs and Terry E. McEwen for their comprehensive criticism of several manuscripts and to Yuan-Zong Lai, Tapio Mattila and Hannu Makkonen for their extensive help in the final editing of the chapters. We are also grateful for the assistance and advice given to us by E. G. King, A. E. Markham, E. V. White, Carl Adolphson, Sea-Bong Chang and June R. P. Ross and wish to acknowledge the help of Hanna E. Sarkanen, Joyce S. Krell, Robert F. Frick, John Q. Murphy and the Institute of Forest Products at the University of Washington in the preparation of the final reproduction copy. Our respective wives, Hanna E. Sarkanen and Mildred H. Ludwig, endured a period of Spartan life during the preparation of the book, and we wish to affectionately acknowledge their decisive role in its realization.   K. V. S. & C. H. L.

CONTRIBUTORS

G. GRAHAM ALLAN
College of Forest Resources
University of Washington
Seattle, Washington 98105

HOU-MIN CHANG
School of Forest Resources
North Carolina State University
Raleigh, North Carolina 27607

RUSSELL F. CHRISTMAN
Department of Civil Engineering
University of Washington
Seattle, Washington 98105

CARLTON W. DENCE
College of Forest Resources
University of Syracuse
Syracuse, New York 13210

DOUGLAS W. GLENNIE
Hooker Chemical Corporation
1940 Ward Street
Niagara Falls, New York 14302

DAVID W. GOHEEN
Central Research Division
Crown Zellerbach Corporation
Camas, Washington 98607

OTTO GOLDSCHMID
Olympic Research Division
ITT Rayonier
Shelton, Washington 98584

D. A. I. GORING
Physical Chemistry Division
Pulp & Paper Research Inst. of Canada
Montreal, Canada

HERBERT L. HERGERT
ITT Rayonier
161 East 42nd Street
New York, New York 10017

CHARLES H. HOYT
Chemical Products Division
Crown Zellerbach Corporation
Camas, Washington 98607

BJORN F. HRUTFIORD
College of Forest Resources
University of Washington
Seattle, Washington 98105

YUAN-ZONG LAI
Department of Chemistry
University of Montana
Missoula, Montana 59801

CHARLES H. LUDWIG
Research Laboratory
Bellingham Division
Georgia-Pacific Corporation
Bellingham, Washington 98225

JOSEPH MARTON
Westvaco, Research Laboratory
Johns Hopkins Road
Laurel, Maryland 20810

TAPIO MATTILA
Research Center
Neste Oy, Kullo
Finland

RAY T. OGLESBY
Department of Conservation
Cornell University
Ithaca, New York 14850

K. V. SARKANEN
College of Forest Resources
University of Washington
Seattle, Washington 98105

ADRIAN F. A. WALLIS
Division of Forest Products
Commonwealth Scientific and Industrial
    Research Organization, Australia
P. O. Box 310
South Melbourne, Victoria 3205
Australia

A. B. WARDROP
School of Biological Sciences
La Trobe University
Victoria, Melbourne
Australia

CONTENTS

Frontispiece	ii
Foreword	v
Preface	vii
Contributors	ix

PART ONE
INTRODUCTION

1. DEFINITION AND NOMENCLATURE
   K. V. Sarkanen and C. H. Ludwig

I.	Introduction	1
II.	Criteria for the Presence of Lignin in Plant Tissues	6
III.	Nomenclature in Lignin Chemistry	10
	A. Ambiguities in the Term Lignin	10
	B. Notation for Carbon Atoms in Phenylpropane Units	11
	C. Inconsistencies in Compound Names	12
	D. Abbreviated Names and Structural Formulae	14
References		17

PART TWO
LIGNINS IN THE PLANT KINGDOM

2. OCCURRENCE AND FORMATION IN PLANTS 19
   A. B. Wardrop

I.	The Occurrence of Lignin in Plants	19
II.	The Texture and Organization of the Cell Wall in Fibres and Tracheids	22
III.	Cell Wall Organization	23
IV.	Lignification	25

V.	Biochemical Aspects of Lignification	28
VI.	Physiological Aspects of Lignification	33
VII.	The Influence of Lignification on Cell Wall Properties	36
VII.	Summary	38
	References	38

3. CLASSIFICATION AND DISTRIBUTION
K. V. Sarkanen and H. L. Hergert

- I. Major Classes of Lignins — 43
- II. Guaiacyl Lignins — 44
    - A. Lignins in Conifer Woods — 44
        1. Gross Distribution; 2. Variation of Lignin Content within Growth Rings; 3. Distribution of Lignin in the Cell Wall of Conifer Tracheids and Parenchyma; 4. Compositional Variation in Normal Conifer Lignins; 5. Compression Wood Lignins; 6. Tissue Culture Lignins.
    - B. Cycad Lignins — 61
    - C. Lignins in Vascular Cryptogams — 61
    - D. General Characteristics of Guaiacyl Lignins — 63
- III. Guaiacyl-Syringyl Lignins — 63
    - A. Exceptional Conifers — 63
    - B. Angiosperm Wood Lignins — 65
        1. Distribution; 2. Distribution in Single Cells; 3. Relative Amounts of Guaiacyl- and Syringylpropane Units in Hardwood Lignins; 4. Within-Tree Variation in Hardwood Lignins.
    - C. Lignins in Herbaceous Dicotyledons — 74
    - D. Lignins in Monocotyledons — 74
        1. Lignin Contents of Mature Plants; 2. Chemical Composition; 3. Changes During Maturation.
    - E. Guaiacyl-Syringyl Lignins in Cryptogams — 79
- IV. The Problem of First Appearance of Lignins in the Plant Kingdom — 79
- V. Lignins and Phenolic Polymers in Tree Barks — 81
    - A. Histological Evidence for Lignin in Bark — 82
    - B. Bark Phenolic Acids — 83
    - C. Bark Lignin — 85
    - D. Proximate Analysis of Bark Lignin — 88
- References — 89

PART THREE
DEHYDROGENATIVE POLYMERIZATION AND STRUCTURE OF LIGNINS

4. PRECURSORS AND THEIR POLYMERIZATION
K. V. Sarkanen

I.	Biogenesis of Lignin Precursors		95
	A.	Application of Tracer Methods	95
	B.	Shikimic Acid Pathway in Higher Plants	97
	C.	Shikimic Acid Pathway in Lignin Formation	99
	D.	Conversion of Phenylalanine and Tyrosine to Lignin	104
	E.	Conversion of Cinnamic Acid Intermediates to Lignins	105
	F.	The Role of Precursor Glucosides	109
	G.	Enzymatic Aspects of Polymerization to Lignin	111
	H.	Lignification in Maturing Plants	112
II.	Dehydrogenative Polymerization to Lignin		116
	A.	Introduction	116
	B.	Oxidative Coupling of Simple Phenols	117
	C.	Dehydrogenation of 2,4,6-Trisubstituted Phenols	120
	D.	Displacement Reactions	122
	E.	Coupling Reactions at Unsubstituted Sites of Phenoxyradicals	125
	F.	Dehydrogenation of p-Propenylphenol Derivatives	132
	G.	Dehydrogenations of Coniferyl and Sinapyl Alcohols	138
	H.	Cross-Couplings Between Mono- and Dilignols	145
	I.	Coupling Reactions of Di- and Oligolignols	148
	J.	Structural Variance in Dehydrogenation Polymers	150
References			155

5. ISOLATION AND STRUCTURAL STUDIES
Y. Z. Lai and K. V. Sarkanen

I.	Introduction		165
II.	Milled Wood Lignins		166
	A.	Method of Preparation	166
	B.	Physical and Chemical Changes during the Milling Process	170
III.	Brauns Lignins and "Enzymicly Liberated Lignins"		175
	A.	Characteristics of Brauns Lignins	175
	B.	"Enzymicly Liberated Lignins"	179
	C.	The Possible Nature of Brauns and Enzymicly Liberated Lignins	179
IV.	Periodate Lignins		181

V.	Isolation Methods Based on Acidic Hydrolysis of Polysaccharides	182
	A. Condensation Reactions Caused by Mineral Acids	182
	B. Lignin Isolated by Means of Mineral Acids	185
	C. Cuoxam Lignins	186
	D. Dioxane Acidolysis Lignins	187
VI.	Alkali Lignins	188
VII.	Lignin Thioglycolic Acids	189
VIII.	Limitations of Lignin Preparations	189
IX.	Quantitative Determination of Lignin	190
	A. Applicability of Klason Lignin Method to Wood Samples	190
	B. Lignins in Bark Tissues	193
	C. Herbaceous Plants Including Grasses	193
	D. Lignin in Pulps	194
X.	Structure of Lignins	195
	A. End Groups with Unattached Side-Chains	195
	B. Units with Free Phenolic Hydroxyl Groups	201
	C. β-O-4 Lignol (Arylglycerol-β-Ether) Structures	205
	D. β-1 Lignol Structures	208
	E. 4-O-1 and 5-1 Linked Structures	211
	F. β-5 Lignol Structures	211
	G. β-β Lignol Structures	213
	H. 5-5 and 5-O-4 Lignol Structures	216
	I. Minor Linkage Types in Conifer Lignins	217
	J. Possible Covalent Bonds Between Lignin and Polysaccharides	218
	K. Isolated Lignin-Carbohydrate Complexes	220
	L. Structural Implications of Compositions and Functional Group Analysis of Milled Wood Lignins	224
	M. Relationship Between Lignins and Extractives	228
	References	230

PART FOUR
SPECTROSCOPIC CHARACTERIZATION OF LIGNINS

6. ULTRAVIOLET SPECTRA
Otto Goldschmid

I.	Introduction	241
II.	Origin of Ultraviolet Absorption Spectra of Aromatic Compounds	241
III.	Ultraviolet Spectra of Lignins	245
	A. Effect of Constituent Groups	246
	1. Phenolic Hydroxyl Groups; 2. Carbonyl Groups; 3. α-β Double Bonds; 4. Biphenyl Groups; 5. Summary.	

B.	Effect of Species	256
IV.	Spectrophotometric Lignin Determinations	258
A.	Determination in Wood Cellulose	258
B.	Determination in Pulping Liquor	260
C.	Determination in Receiving Waters	262
References		263

7. INFRARED SPECTRA
Herbert L. Hergert

I.	Introduction	267
II.	Methods of Measurement	268
	A. Effect of State upon Spectral Measurement	268
	B. Near and Far Infrared Spectroscopy	269
III.	Interpretation of Spectra	270
	A. Lignin Absorption Band Assignment	270
	B. Role of Model Compounds and Lignin Derivatives in Band Assignment	272
	1. Hydroxyl Groups; 2. Methyl and Methylene Groups; 3. Carbonyl Groups; 4. Ethylenic Double Bond; 5. Aromatic Skeletal Bands.	
	C. Problems and Pitfalls in the Use of Lignin Infrared Spectra	280
IV.	Application of Spectra	281
	A. Polymers Related to Lignin	281
	B. Whole Wood Lignins	284
	C. Commercial Lignins	288
	D. Bark Lignin and Humic Acids	290
	E. Chemotaxonomic Studies	291
	F. Quantitative Analyses in Wood and Pulp	292
References		293

8. MAGNETIC RESONANCE SPECTRA
C. H. Ludwig

I.	Introduction	299
II.	Nuclear Magnetic Resonance	299
	A. Introduction	299
	1. Proton Magnetic Resonance Spectroscopy; 2. Special Problems with Polymers; 3. Selection of Solvents for Lignin Models and Lignins.	
	B. Lignin Models and Lignin Degradation Products	301
	1. Spectra of Dimeric and Oligomeric Lignin Models; 2. Ranges of Chemical Shifts for Protons in Model Acetates; 3. The identifi-	

cation of Lignin Degradation Products;
4. Reactivity Studies on Lignin Model
Compounds.
  C. Polymeric Lignin Derivatives 313
    1. Interpretation of Spectra; 2. Semi-
Quantitative Functional Group Analyses;
3. Broad Line PMR Studies.
  D. Future Potential in Lignin Research 326
 III. Electron Paramagnetic Resonance 326
  A. Introduction 326
  B. The Measurement of Free Radical Content of Lignins
and Related Materials 327
    1. Lignocellulose in Wood; 2. Isolated Lignins
  C. Structural and Reactivity Studies of Lignins and Models 332
 References 340

PART FIVE
REACTIONS OF LIGNINS

9. SOLVOLYSIS BY ACIDS AND BASES
A. F. A. Wallis

 I. Introduction 345
 II. Acid Solvolysis 345
  A. Mild Acid Hydrolysis 345
  B. Hydroxyl Displacement Reactions 350
    1. Introduction of Alkoxy Groups;
2. Mercaptolysis; 3. Water Elimination.
  C. Solvolytic Ether Cleavages 354
    1. α-Ether Cleavage; 2. β-Ether Cleavage
  D. Solvolytic C-C Cleavages 359
    1. β-γ Cleavage; 2. α-β Cleavage
  E. Side-Chain Rearrangements 360
 III. Basic Solvolysis 361
  A. Solvolytic Ether Cleavages 361
    1. α-Ether Cleavage; 2. β-Ether Cleavage;
3. Diaryl Ether Cleavage; 4. Methoxyl
Group Cleavage.
  B. Solvolytic C-C Cleavages 365
    1. β-γ Cleavage; 2. α-β Cleavage; 3. α-1
Cleavage; 4. Cleavages at High Temperatures.
 References 369

10. HALOGENATION AND NITRATION
C. W. Dence

I.	Introduction	373
II.	Halogenation and Nitration Systems	375
	A. Halogenating Agents	375
	B. Nitrating Agents	379
III.	Halogenation and Nitration Reactions of Lignin Model Compounds	381
	A. Reactions of the Aromatic Moiety	381
	1. Electrophilic Substitution; 2. Electrophilic Displacement of Ring Substituents; 3. Cleavage of Alkyl Aryl Ether Linkages; 4. Oxidation and Other Reactions.	
	B. Reactions of the Side-Chain Moiety	398
IV.	Halolignins	404
	A. Isolation and Purification	404
	B. Physical Properties	405
	C. Chemical Constitution	406
	D. Reactions	416
V.	Nitrolignin	418
	A. Isolation	418
	B. Physical Properties	418
	C. Chemical Constitution	418
	D. Reactions of Nitrolignin	421
References		422

11. OXIDATION
H.-M. Chang and G. G. Allan

I.	Introduction	433
II.	Oxidations Degrading Lignin to Aromatic Carbonyl-Containing Compounds	434
	A. Alkaline Nitrobenzene Oxidation	434
	B. Metal Oxide Oxidations	444
	1. Probable Mechanism of Metal Oxide Oxidation; 2. Cupric Oxide Oxidation; 3. Other Metal Oxide Oxidations.	
	C. Molecular Oxygen and Alkali	449
	D. Permanganate Oxidation	452
III.	Oxidative Processes Degrading Aromatic Nuclei	457
	A. Acid-Catalyzed Peracetic Acid Oxidation	457
	1. Peracetic Acid Oxidation of Model Compounds; 2. Peracetic Acid Oxidation of Lignins and Wood.	

       B. Sodium Hypochlorite, Chlorine Dioxide and Sodium Chlorite   463
       C. Acid-Catalyzed Reaction of Hydrogen Peroxide   468
       D. Other Oxidation Processes Degrading Aromatic Rings   469
  IV. Oxidation Processes Limited to Specific Groups   472
       A. Periodate Oxidation   472
       B. Fremy's Salt   472
       C. Dichlorodicyano-p-quinone   473
       D. Hydrogen and Sodium Peroxides in Alkaline Media   473
  V. Photoxidation of Lignin   477
References   478

12. REDUCTION AND HYDROGENOLYSIS
B. F. Hrutfiord

  I. Introduction   487
       A. The Impact of Hydrogenolysis Studies on the Elucidation of Structure   487
       B. The Concurrence of Hydrolysis and Hydrogenation   488
  II. Non-Degradative Reduction of Lignin   488
  III. Mild Catalytic Hydrogenolysis of Lignin   489
       A. Hydrogenolysis   489
       B. Characterization of Hydrogenolysis Lignin   490
       C. Monomeric Degradation Products from Hydrogenolysis Lignins   490
       D. The Isolation of Dimers from Hydrogenolysis Lignin   492
       E. Polymeric Hydrogenolysis Lignins   494
       F. Mechanism of Formation of Hydrogenolysis Lignin   496
       G. Hydrogenolysis of Milled Wood Lignin   500
       H. The Isolation of a C_6C_3 Biphenyl Dimer   502
       I. Other Hydrogenolysis Studies Producing Aromatic Monomers   503
  IV. Hydrogenation of Lignins to Cyclohexane Derivatives   503
       A. Studies on Isolated Lignins   503
  V. Hydrogenolysis of Lignin by Potassium and Sodium in Liquid Ammonia   504
       A. Isolation and Characterization of Reaction Products   504
       B. Mechanism   505
References   507

13. MODIFICATION REACTIONS
G. Graham Allan

  I. Introduction   511
  II. Reactive Sites Available in Lignins   511

	A. Ether Linkages	511
	B. Hydroxyl Groups	519
	C. Carbonyl Groups	531
	D. Carboxyl and Ester Functions	537
	E. Ethylenic Moieties	537
	F. Sulfur-Containing Groups	537
III.	Modifications at Structural Sites	539
	A. On Side-Chains	539
	B. Reactive Sites on Aromatic Rings	545
References	559

14. HIGH ENERGY DEGRADATION
G. Graham Allan and Tapio Mattila

I.	Introduction	575
II.	Thermal Decomposition of Lignin	575
	A. The Nature of the Pyrolysis Residue	575
	B. The Composition of the Aqueous Distillate	576
	C. The Constituents of the Thermolysis Tar	577
	D. Gaseous Pyrolysis Products	579
	E. Effect of Additives on the Pyrolysis	579
	F. Development of Technical Thermolytic Processes	580
	G. Thermogravimetric Analysis	581
	H. Fast Cracking	583
III.	Alkali Fusion of Lignin	586
	A. Formation of Protocatechuic Acid, Catechol and Oxalic Acid	586
	B. Formation of "Lignin Acids"	587
	C. Fusion of Various Lignin Preparations	588
References	592

PART SIX
DELIGNIFICATION IN PULPING PROCESSES

15. REACTIONS IN SULFITE PULPING
D. W. Glennie

I.	Introduction	597
II.	Early Research on Lignin Sulfonation	597
III.	Methods of Sulfonating Lignin	598
	A. Technical Processes	598
	B. Experimental Processes	600
IV.	Methods of Isolating Lignin Sulfonates	601

V.	Mechanism of Sulfonation and Delignification		602
	A. Rate Studies and Reactive Groups in Lignin		602
	1. Wood Lignins; 2. Kinetics of Sulfonation of Model Compounds.		
	B. Topochemistry of Sulfonation and Delignification		612
	C. Use of Tracers in Sulfonation and Delignification Studies		614
VI.	Functional Groups in Lignin Sulfonates		
	A. Relationship Between Functional Groups and Molecular Weight		614
	B. Phenolic Hydroxyl Groups		618
	1. Content in Lignin Sulfonates from Bisulfite and Acid Sulfite Treatments; 2. Phenolic Hydroxyl in Lignin Sulfonates from Neutral Sulfite Pulping.		
	C. The Sulfonate Group		620
	1. Relationship of Sulfonate Group Content to Sulfonation Conditions and Polydispersity of Lignin Sulfonates; 2. Presence of Non-Sulfonic Acid Sulfur in Lignin Sulfonates; 3. Location of the Sulfonate Groups; 4. Alkaline Hydrolysis of Lignin Sulfonates.		
	D. The Carbonyl Group		625
	1. Behavior and Formation of Carbonyl Groups During Sulfonation; 2. Effect of Reduction on Sulfonatability.		
	E. Carboxyl Groups in Lignin Sulfonates		629
	F. Catechol Groups Formed by Demethylation		630
VII.	Concluding Remarks		631
	References		631

16. REACTIONS IN ALKALINE PULPING
Joseph Marton

I.	Morphological and Topochemical Considerations		639
	A. Introduction		639
	B. Pulping Chemicals		640
	C. Penetration of Pulping Liquor		642
	D. Distribution of Lignin in the Wood and Pulp Fibers		643
	E. Changes at Fiber-Liquor Interface		645
	F. Kinetics of Delignification, Reaction Mechanisms		645
II.	Base Catalyzed Fragmentation of Lignin		650
	A. Reactive Structures		650
	B. Hydrolysis of α-Ethers		651
	C. Reactions of β-O-4 Alkyl Aryl Ethers		653

CONTENTS

III.	Condensation Reactions	658
	A. The Extent of Condensation in Delignification	658
	B. Condensation Reactions Involving Ring-Position 5	659
	C. Reactions of Kraft Lignin with Formaldehyde	662
	D. Condensations Between Side-Chain Positions	664
IV.	The Role of Sulfide in Kraft Pulping	665
	A. Differences Between Soda and Kraft Pulping	665
	B. Intermediate Thiolignins and Their Conversion to Kraft Lignins	666
	C. Cleavage of Methoxyl Groups	669
V.	Structure and Properties of Kraft Lignin	670
	A. Isolation	670
	B. Analytical Composition	671
	C. Functional Groups	673
	1. Hydroxyl Groups; 2. Carboxylic Acid, Benzyl Alcohol and -Ether Groups; 3. Unsaturated Structures; 4. Carbonyl Groups; 5. The Aromatic Ring.	
	D. Formulation for Kraft Lignin	679
	E. Spectral Properties	681
	F. Macromolecular Properties	684
References		689

PART SEVEN
MACROMOLECULAR ASPECTS

17. POLYMER PROPERTIES OF LIGNIN AND LIGNIN DERIVATIVES
D. A. I. Goring

I.	Introduction	695
II.	Molecular Weight	698
	A. Molecular Weight and Molecular Weight Distributions of Lignin Derivatives	698
	B. Reasons for the Polydispersity of Soluble Lignins	701
III.	Molecular Shape	703
	A. Conformity of Viscosity, Sedimentation and Diffusion with Behaviour Expected for Spherical Microgel	703
	B. Shape of Lignin Macromolecules as Determined by Electron Microscopy	706
IV.	Polyelectrolyte Behaviour	708
	A. Relationship of Viscometric Behaviour to Ionic Strength	708
	B. Effect of Polyelectrolyte Swelling on Gel Filtration of Lignosulphonates	710

	C.	Electrophoresis	711
V.	Colloidal Behaviour		713
	A.	The Dispersing Properties of Lignosulphonates and Alkali Lignins; Theory	713
	B.	The Gelling Reaction of Lignosulphonate with Dichromate	716
VI.	Molecular Weight Determination		719
	A.	Number Average Molecular Weights; Methods	719
	B.	The Light Scattering Method; Limitations	719
	C.	Ultracentrifuge Methods	720
	D.	Combination of Viscosity and Diffusion Methods	723
	E.	Gel Filtration	727
VII.	Glass Transitions		729
	A.	Softening Temperatures, Effects of Molecular Weight and Water Content	729
	B.	Glass Transitions and Adhesion Properties	733
	C.	The Significance of Glass Transitions in Paper and Board Manufacture	734
	D.	Thermal Expansion and Second Order Transition Phenomena	740
VIII.	Solubility, Accessibility and Sorption		741
	A.	Hildebrand's Solubility Parameter and Hydrogen Bonding Capacity	741
	B.	Relation Between Accessibility and Solubility	744
	C.	Sorption of Vapours	745
	D.	Sorption of Ions	746
IX.	The Lignin Polymer in Wood		748
	A.	Network Theory	748
	B.	Small-Molecule Theory	748
	C.	Lignin-Carbohydrate Bonds and Possible "Snake Cage" Effects	749
	D.	Aggregation by Secondary Bonds	750
	E.	Heterogeneity in Original Lignin	751
	F.	Topochemistry of Delignification	753
	G.	Sol-Gel Concept of Lignin in Wood	758
	H.	Gel Degradation Theory of Delignification	758
	I.	Concluding Remarks	761
References			761

PART EIGHT
BIOLOGICAL TRANSFORMATIONS OF LIGNINS

18. MICROBIOLOGICAL DEGRADATION AND THE FORMATION OF HUMUS
R. F. Christman and R. T. Oglesby

I.	Introduction	769
II.	Biological Agents	769
III.	Microbiological Degradation of Lignin	771
	A. Initial Microbiological Attack	771
	B. Phenylpropanoid Intermediates	774
	C. Benzyl Derivatives	777
IV.	Conversion to Humus	
	A. The Nature of Humus	783
	B. Characteristics of Lignin Humification	785
References		793

PART NINE
UTILIZATION OF LIGNINS

19. LOW MOLECULAR WEIGHT CHEMICALS
D. W. Goheen

I.	Introduction	797
II.	Vanillin and Related Products	798
	A. Early Studies	798
	B. Vanillin Syntheses	798
	C. Vanillin from Lignins	800
	D. Purification	804
	E. Properties, Uses and Derivatives	804
III.	Organic Sulfur Chemicals from Lignins	806
	A. Production of Dimethyl Sulfide and Methyl Mercaptan	806
	B. Mechanism of Dimethyl Sulfide and Methyl Mercaptan Formation	807
	C. Properties and Reactions of Dimethyl Sulfide	809
	1. Properties; 2. Halogenation; 3. Reactions Involving Unshared Electron Pairs; 4. Oxidations.	
	D. Dimethyl Sulfoxide	812
	1. As a Reactant; 2. As a Catalyzing Reaction Solvent; 3. As a Solubilizing Reaction Solvent; 4. In Biological Systems.	

		E.	Dimethyl Sulfone	817
		F.	Methyl Mercaptan	817
	IV.	Potential Phenolic Products by High Temperature Sodium Sulfide Treatments		817
	V.	Hydrogenolysis of Lignins		818
	VI.	Oxidation of Lignins		821
	VII.	Carbonization of Lignins		821
	VIII.	Future Prospects for Low Molecular Weight Chemicals from Lignins		823
	References			824

20. POLYMERIC PRODUCTS
C. H. Hoyt and D. W. Goheen

	I.	Introduction		833
		A.	Lignin in Wood	833
		B.	Available Lignin Products	833
	II.	Preparation of Lignin Products		837
		A.	Lignin Sulfonates	837
		B.	Kraft Lignins	838
			1. Isolation; 2. Properties.	
		C.	Hydrolysis Lignin	840
	III.	Modification of Lignin Products		840
		A.	Modifications of Sulfite Lignins	840
			1. Removal of Sugars and Other Non-Lignin Components; 2. Cation Selection; 3. Desulfonated Products from Lignin Sulfonates.	
		B.	Modifications of Kraft Lignin	845
			1. Chemical Modifications.	
	IV.	Utilization of Lignin Products		846
		A.	Oil Well Drilling Muds	846
		B.	Cement and Concrete Additives	848
			1. Grinding Aid for Portland Cement; 2. Air Entraining Agents; 3. Concrete Additives; 4. Oil Well Cementing.	
		C.	Dispersants	849
		D.	Ore Flotation	851
		E.	Emulsifiers and Stabilizers	851
		F.	Grinding Aids	851
		G.	Electrolytic Refining	852
		H.	Binders and Adhesives	852
		I.	Resin Ingredients	854
		J.	Rubber Additives	855
		K.	Protein Precipitants	856

		L. Tanning Agents	856

       L. Tanning Agents   856
       M. Sequestering Agents   856
       N. Miscellaneous   857
            1. Storage Battery Plates; 2. Lime Plaster;
            3. Crystal Growth Inhibitor; 4. Ingot Mold
            Wash; 5. Flame Retardant.
       O. Liquor Disposal   858
  V.   Utilization of Hydrolysis Lignin   858
  VI. Concluding Remarks   859
References   859

Index   867

LIGNINS

1

DEFINITION AND NOMENCLATURE
K. V. Sarkanen and C. H. Ludwig

I. <u>Introduction</u>. The word "lignin" is derived from the Latin word "<u>lignum</u>" meaning wood and, indeed, lignins form an essential component of the woody stems of arborescent gymnosperms and angiosperms in which their amounts range from 15% to 36%. Lignins are not, however, restricted to arborescent plants, but are found as integral cell wall constituents in all vascular plants including the herbaceous varieties. Their presence has been demonstrated in tissues associated with stems as well as in foliage and roots.

As cell wall constituents, lignins do not merely act as "encrusting materials" of some secondary nature, as is often misleadingly stated. Rather, they perform a multiple function that is essential to the life of the plant. By decreasing the permeation of water across the cell walls in the conducting xylem tissues, lignins play an important role in the intricate internal transport of water, nutrients and metabolites. Secondly, lignins impart rigidity to the cell walls and, in woody parts, act as permanent bonding agents between cells generating a composite structure outstandingly resistant towards impact, compression and bending. Finally, lignified tissues effectively resist attacks by microorganisms by impeding penetration of destructive enzymes into the cell wall.

The giant redwoods (<u>Sequoia</u>) of California illustrate the strength and permanence of lignified plant tissues in an impressive fashion. The gigantic trunks of these trees may reach the height of over 300 feet and support crown structures several tons in weight. Yet, the lifespan of redwoods is probably unparalleled among living organisms and may extend to four to five thousand years. During this time period, which in length is not much shorter than the total time of human civilization, these towering giants have successfully resisted countless attacks by storms, fires, insects and microorganisms!

The exact definition and the differentiation of lignins from other polyphenolic plant constituents has remained a matter of debate until very recently. As the result of gradual clarification of ideas, a definition of lignins has finally emerged to which most of the investigators in the field subscribe today.

According to this definition, we understand lignins to be polymeric natural products arising from an enzyme-initiated dehydrogenative polymerization of three primary precursors: trans-coniferyl <u>1</u>, trans-sinapyl <u>2</u> and trans-p-coumaryl <u>3</u> alcohols.

$$\underset{1}{\text{HO}\underset{\text{OMe}}{\bigcirc}\text{CH=CHCH}_2\text{OH}} \quad \underset{2}{\text{HO}\underset{\text{OMe}}{\overset{\text{OMe}}{\bigcirc}}\text{CH=CHCH}_2\text{OH}} \quad \underset{3}{\text{HO}\bigcirc\text{CH=CHCH}_2\text{OH}}$$

Strictly speaking, the above definition pertains to what S. A. Brown (1) has quite appropriately called the "lignin core." Lignins, as they exist in the cell wall, are always associated with hemicelluloses, not only in intimate physical admixture but also anchored to the latter by actual covalent bonds. Secondly, most lignins contain varying amounts of certain aromatic carboxylic acids in esterlike combination. These acids are most probably not generated from the three primary precursors in the dehydrogenative polymerization process.

In the dehydrogenative polymerization that results in formation of the lignin macromolecule <u>in vivo</u>, coupling processes between phenoxy radicals constitute the main <u>principle</u> of molecular growth. Lignins are not the only natural products formed through this type of growth mechanism but share the dehydrogenative polymerization mode with a wealth of other naturally occurring phenolic materials including a number of alkaloids, tannins, depsidones, xanthones, usnic acid, aphid pigments, terpenoids, lignans etc. A recently compiled monograph (2) summarizes the accumulated information in these areas in an outstanding fashion.

Interestingly enough, the idea of enzymatic dehydrogenation as the governing principle of lignin formation in nature was explicitly expressed at a very early stage of lignin investigations by H. Erdtman (3) who also made meaningful predictions about the nature of chemical bonds involved. This concept, a prime example of imaginative intuition as it was, did not win universal acceptance at that time for the lack of supporting experimental information. The confirmatory evidence, as well as the detailed elaboration of the dehydrogenative polymerization principle, stems essentially from monumental studies carried out by K. Freudenberg and his school at the University of Heidelberg (4). These developments, the main part of which took place during the post-World War II period, received outstanding contributions from structural discoveries in other laboratories, notably those of Adler (5) and Kratzl (6). The rapid concurrent progress in the clarification of biogenetic pathways leading to the formation of primary lignin precursors further helped to consolidate the picture of the lignin polymerization process and will be

DEFINITION AND NOMENCLATURE

discussed in detail in Chapter 4.

The theory of dehydrogenative polymerization provides the essential basis for understanding the peculiar structure of the unique class of lignin polymers. The need for a unifying structural principle is particularly desirable in the characterization of lignins, which are notorious for their resistance to detailed chemical characterization. It is well known that already the isolation of representative lignin preparations presents formidable difficulties, and the chemical characterization of these preparations is frustrated by their tendency to undergo self-condensation reactions of unknown nature. Finally, all known methods of chemical degradation yield identifiable small molecular weight products in modest yields only.

Under these circumstances, the task of devising a satisfactory structural picture of lignin macromolecules is very difficult and may be likened to an attempt to compose a picture-puzzle with an incomplete number of pieces. What is contained in the missing pieces can only be guessed at, and there may be more than one way to assemble the available pieces. This means that structural representations that are proposed for lignins at the present stage of knowledge by necessity contain uncertain elements and are therefore subject to future amendments.

In 1964, Freudenberg (7) made the first serious attempt to unify the available information on the dehydrogenative polymerization of coniferyl alcohol together with the combined analytical and reactivity data (8) on spruce lignin. Modifications to the originally proposed structure were first introduced by Freudenberg and Harkin (9), and somewhat later, by Harkin (10), to accommodate newly acquired information. The last mentioned formula is reproduced in Figure 1.1. It represents an average fragment of a larger lignin molecule and contains altogether twenty monomeric units. The majority of these are of the guaiacylpropane type while two (units 5 and 12) are assumed to be of the p-hydroxyphenylpropane type, and one (unit 10), of the syringylpropane type. Structures present in minor quantities are shown as alternatives to units 6, 13 and 14.

To an organic chemist, the formula in Figure 1.1 conveys a clear idea of the difficulties facing the investigator in structural research. It may be noted that most of the monomeric units are linked together by bonds of extraordinary stability. These include carbon-to-carbon linkages, either of the biphenyl type, such as between units 9 and 10, or of the alkyl-aryl type (units 17 and 18 as well as 16 and 20). Even the ether linkages, with the exception of α-aryl ether bonds (units 3 and 4) are quite resistant towards hydrolysis. On top of the resistivity against degradation to simple molecular species comes the tendency of lignin molecules to undergo self-condensation reactions, particularly in acidic media. Structures possessing this tendency include units of the cinnamyl alcohol type (unit 14c) and those containing benzylic hydroxyl groups (unit 6).

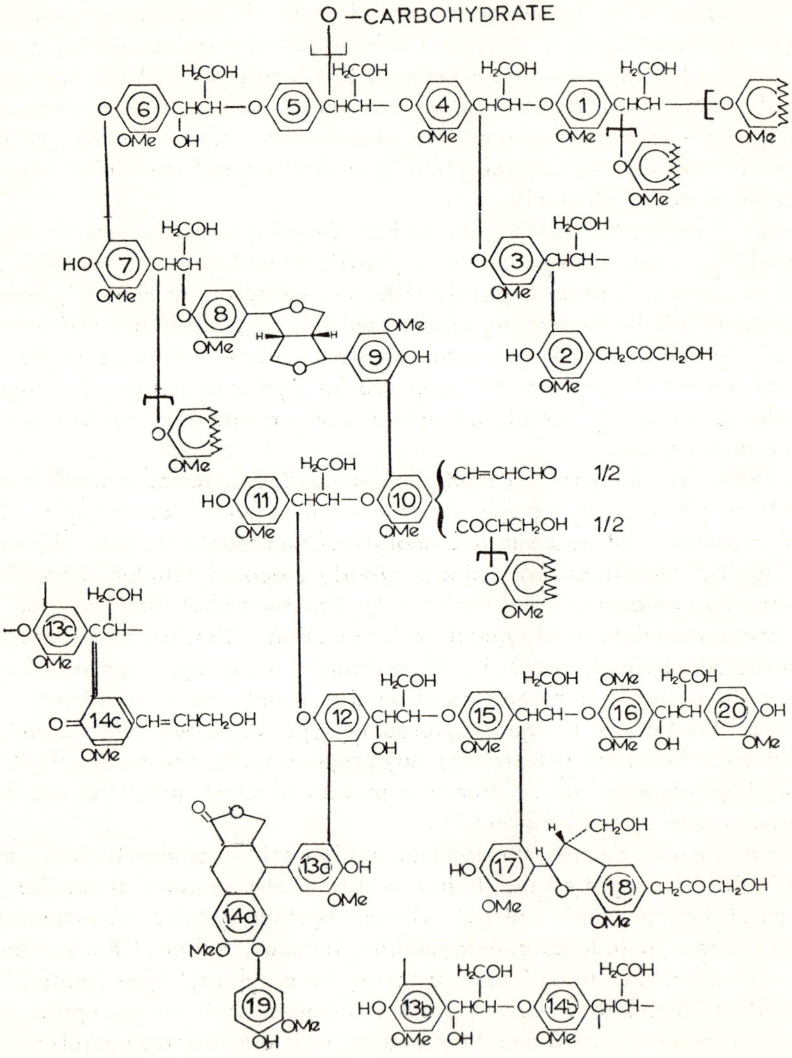

Figure 1.1. Schematic formula for spruce wood lignin by Freudenberg (8), with modifications by Harkin (10).

DEFINITION AND NOMENCLATURE

The structure proposed by Freudenberg and coworkers for spruce lignin has substantial merits in providing a reasonable picture of the molecular architecture of lignins, and in forming a basis for mapping the course of various chemical reactions. For example, Ludwig, Nist and McCarthy (11) have drawn attention to the opportunity for extensive chelation between the numerous hydroxylic and ether functions present, and such internal hydrogen bonds can be anticipated to impart a degree of rigidity to lignin molecules.

The Freudenbergian concept of lignin structure is clearly more realistic than the structural sketches promoted recently by Forss and coworkers (12). While future developments in all likelihood will not require profound departures from the Freudenberg formulation, the most recent results in dehydrogenative polymerization and structural studies are already suggestive of some needed modifications. Such modifications are discussed in detail in Chapters 4 and 5, and are incorporated, in a tentative manner only, to the amended structure shown in Figure 1.2.

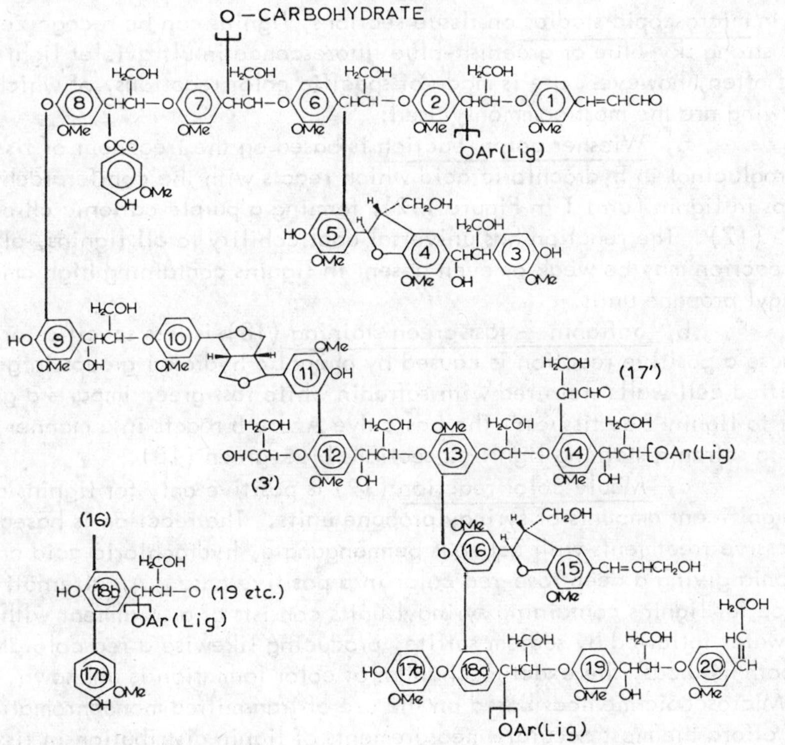

Figure 1.2. Tentative structure for normal conifer wood lignin, based on modifications of the spruce lignin formulae by Freudenberg et al. (8, 9).

The structure of spruce lignin is probably representative of gymnosperm wood lignins in general, and there is good reason to believe that more or less analogous structural patterns are present in all plant lignins. On the other hand, detailed characterization studies have been, perhaps somewhat one-sidedly, concentrated in the area of gymnosperm lignins, and the clarification of detailed structural features of other plant lignins must await future studies.

II. Criteria for the Presence of Lignin in Plant Tissues. The identification of lignin in plant tissues is commonly not a difficult task, except when the tissues also contain certain insoluble polyphenols, often related to tannins or flavonols. These constituents have certain characteristics common with lignins and, as a consequence, have been frequently confused with the latter. Such phenolic materials are found, for example, in the barks of trees ("bark phenolic acids," see Chapter 3, Section V), in Eucalyptus wood as so-called "kino" (13), in primitive plant species such as mosses (14), and in the cortex of tropical tree-ferns (15).

In microscopic studies on tissue sections, lignins can be recognized by their strong sky-blue or greenish-blue fluorescence in ultraviolet light (16). More often, however, use is made of specific color reactions, of which the following are the most commonly used:

a, Wiesner color reaction is based on the treatment of tissue with phloroglucinol in hydrochloric acid which reacts with the coniferaldehyde groups in lignin (unit 1 in Figure 1.2), forming a purple cationic chromophore (17). The reaction has universal applicability to all lignins, although the reaction may be weak or even absent in lignins containing high amounts of syringyl propane units.

b, Safranin — fast green staining (18) is less specific for lignin, because a positive reaction is caused by phenolic hydroxyl groups in general. Lignified cell walls stain red with safranin while fast green imparts a green color to lignin-free tissues. The basic dye Azure B reacts in a manner analogous to safranin staining lignified sections blue-green (18).

c, Mäule color reaction (19) is positive only for lignins containing significant amounts of syringylpropane units. The reaction is based on successive treatments with aqueous permanganate, hydrochloric acid and ammonia giving a deep rose-red color in a positive case. An alternative method for lignins containing syringyl units consists of a treatment with chlorine water followed by sodium sulfite, producing likewise a red coloration. For both methods, the exact mechanism of color formation is unknown.

Microscopic methods based on the use of transmitted monochromatic UV light afford the most accurate measurements of lignin distribution in tissue sections (20).

Pellets of finely pulverized plant material in KCl or KBr pellets may be used to measure the UV (21) and IR spectra (22) of the lignin component.

DEFINITION AND NOMENCLATURE

The UV spectrum displays a characteristic broad maximum at about 280 mµ while polysaccharides are totally transparent in this region. The infrared spectra of lignins are quite characteristic, and the carbohydrate absorption is sufficiently weak in the region 1200 to 1680 cm^{-1} to reveal the absorption maxima due to lignin. The spectrum of western hemlock wood meal in a KBr pellet is shown as an example in Figure 1.3. The strong maxima at 1500 and 1600 cm^{-1} are especially characteristic of lignins. It is pertinent to note that the spectral pattern of non-lignin polyphenols is different in this region, with notably weak and diffuse absorption at the 1500 cm^{-1} region.

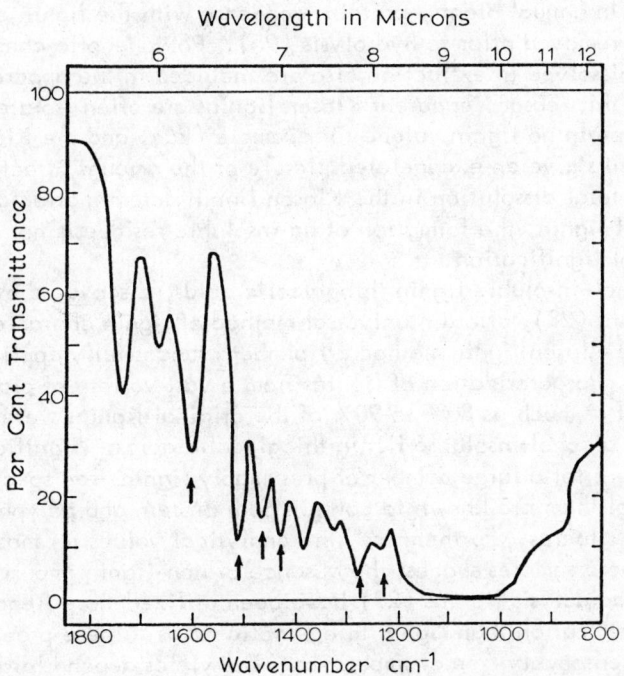

Figure 1.3. Infrared spectrum of western hemlock wood meal in KBr pellet. Absorption maxima marked by arrows are caused by lignin (22).

Chemical methods of identifying and characterizing lignins in extracted plant tissues include the following:
1, Isolation of lignins as an insoluble residue (Klason lignin) after hydrolysis of carbohydrate constituents with sulfuric acid.
2, Conversion of lignins to isolatable lignin thioglycolic acids and and dioxane-acidolysis lignins.

3, Isolation of so-called "Hibbert's ketones" from the ethanolysis products of lignin.
4, Nitrobenzene oxidation to characteristic aldehydes.
5, Hydrogenation to monomeric phenylpropane derivatives of lignin.

Of these methods, the isolation of Klason lignin is based on total hydrolysis of polysaccharides by 72% sulfuric acid (23). The method works outstandingly well for conifer woods, giving quantitative yields of heavily condensed lignin. Less than quantitative yields are generally obtained for hardwoods (24). In annual plants, proteins condense with the lignin component unless totally removed prior to hydrolysis (25). Polyphenolic constituents remaining undissolved in extraction also are included in the apparent "Klason lignin." For this reason, apparent Klason lignins are often isolated from plants that contain no lignin, algae for example (26), and the Klason lignin contents of barks give an exaggerated picture of the amount of actual lignin (27). While total dissolution in the Klason lignin determination demonstrates the absence of lignin, the formation of an insoluble residue is not a sufficient proof of actual lignification.

The alcohol-insoluble lignin thioglycolic acids, discovered by Holmberg at an early date (28), afford a universal method of lignin characterization superior to the Klason lignin method. Holmberg successfully applied this method to the characterization of lignins from a vast variety of plant sources. It appears that as much as 80% to 90% of the original lignins were recovered in the form of alcohol insoluble lignin thioglycolic acids. Significantly, the yields were zero for a large number of presumably lignin-free species, although many of them are known to contain both protein and polyphenolic cell-wall constituents. Furthermore, the analytical values for most lignin thioglycolic acid samples suggest the absence of non-lignin impurities.

Dioxane acidolysis lignins (29) have been utilized more frequently for lignin characterization than lignin thioglycolic acids. These preparations appear to be reasonably free of impurities. The yields depend largely on the conditions and length of hydrolysis, and both depolymerization and condensation reactions are known to occur during hydrolysis.

The most reliable criterion for the presence of lignin may be obtained by isolating so-called "Hibbert's ketones" from the ethanolysis products of the specimen. The exclusive sources of these ketones are the arylglycerol-β-ether structures $\underline{4}$ of the original lignin (such as units 1, 4, 5, 6, and 11 in Figure 1.1). They are formed through a complex depolymerization process (30). Guaiacyl-, syringyl- and p-hydroxyphenyl glycerol-β-ether structures each generate a separate group of Hibbert's ketones commonly converted to the corresponding α,β-dicarbonyl compounds $\underline{5}$ for identification (31), equation 1.

DEFINITION AND NOMENCLATURE

"Hibbert's ketones"

[Structures 4 and 5, with reaction scheme showing conversion via EtOH/HCl to Hibbert's ketones, followed by H_2O_2 treatment] (1)

Ar = guaiacyl, syringyl, or p-hydroxyphenyl groups

Hibbert's ketones may also be used to determine the participation of guaiacyl-, syringyl- and p-hydroxyphenylpropane units in the make-up of the core lignin.

Alternatively, arylglycerol-β-ether structures can be converted to arylpropane derivatives 6 and 7 by hydrogenolysis (32), and similar monomeric products are obtained in treatment with sodium in liquid ammonia (see Chapter 12) (33). The exclusive formation of compounds 6 and 7 from arylglycerol-β-ether structures is, however, not as conclusive as that of Hibbert's ketones.

[Structures 6 and 7: propylguaiacol derivatives]

[Structures 8 (vanillin), 9 (syringaldehyde), 10 (p-hydroxybenzaldehyde)]

The oxidation of plant lignin with nitrobenzene and alkali (34) to vanillin 8, syringaldehyde 9 and p-hydroxybenzaldehyde 10 is the most effective way of obtaining large yields of identifiable degradation products from lignin. Consequently, this method has been extensively used as a diagnostic test.

The combined yields of the three aldehydes may amount to 65% of some lignins (35). All structural units in lignins, except those containing carbon-to-carbon or ether linkages attached to positions 5 or 6 are converted to these aldehydes in nitrobenzene oxidation, albeit in varying yields. Consequently, both the total yields of the three aldehydes and their relative ratios provide important information for the structural characterization of lignins in nature as well as for the characterization of various lignin derivatives.

As a diagnostic test for the presence of lignin, the identification of any of the three aromatic aldehydes 8, 9, or 10 among the nitrobenzene oxidation products is not as convincing as the isolation of Hibbert's ketones. In Chapter 3, examples are given of the isolation of these aldehydes from plants in which the occurrence of lignin is in doubt (Section IV). In addition to lignin units, vanillin 8 and p-hydroxybenzaldehyde 10 may stem from ferulic and p-coumaric acid ester groups (36) which may not always be even associated with the lignin component.

Most of the methods of lignin characterization discussed in the preceding paragraphs are commonly applied to extractive-free cell wall material rather than to isolated, pure lignins. This is done for the simple reason that isolated lignin preparations rarely possess exactly the same chemical characteristics as the original lignins and furthermore, represent only a part of the total lignin. As a matter of fact, the unavailability of representative lignin samples has always been a formidable roadblock preventing systematic progress in this area.

The problem of lignin isolation has not as yet been solved in a totally satisfactory manner. Currently "Milled Wood Lignins," first isolated by Bjorkman (37), are considered to be, in many respects, almost identical with the original lignin. These preparations are extracted with aqueous dioxane from plant material that has been milled to particles of colloidal size. Usually, less than one-half of lignin can be isolated in this way, and hence the need for a more effective method is apparent. Specific enzymatic hydrolysis of the cell wall polysaccharides would obviously afford the most promising approach to solving the problem, and efforts in this direction by Pew (38) show considerable promise, although the incomplete accessibility of the polysaccharidic components prevents the total removal of these components (see Chapter 5, Section II).

III. Nomenclature in Lignin Chemistry. The current nomenclature in the lignin field is the product of gradual evolution. Unlike in many other fields, no systematic revisions of nomenclature have been carried out, and as a consequence, the usage of certain terms is not always consistent with the meanings accepted for these terms elsewhere. In a few instances, dual meanings for the same term are in common usage, forming a source of confusion to the unfamiliar reader.

A. Ambiguities in the Term Lignin. Usage of the term "lignin" is

DEFINITION AND NOMENCLATURE 11

itself not free of ambiguities. In current literature, the word lignin is associated with altogether four slightly different meanings:

1, In its narrowest meaning, "lignin" is used to refer to the lignin component in gymnosperm woods, or more specifically, to that present in the so-called "normal wood" (see Chapter 3, Section II). There is no reason why this type of lignin should not be explicitly referred to as "normal gymnosperm wood lignin" or "normal softwood lignin."

2, In a wider sense, "lignin" refers to the lignin components of various plants which differ from each other according to species and location in the plant tissue. The separation of lignins from their natural association in the cell wall involves, at least, rupturing lignin-polysaccharide bonds and a reduction in molecular weight. Consequently, it is often desirable to call lignins associated with the plant tissue "protolignins" or "lignins in situ." The use of the term "native lignin" should perhaps be discouraged in this connection, because this term for a period of time was associated with a soluble lignin component isolated from wood extractives by Brauns (39). More recently the term "Brauns Lignin" (BL) has become commonly accepted for this preparation.

3, When the word "lignin" refers to isolated lignin preparations, the varying degrees of chemical change occurring during the isolation form a source of confusion, accentuated by the natural tendency of investigators to underestimate these changes. For example, it is highly optimistic, to say the least, to call lignins extracted from decayed wood "enzymicly liberated lignins" (40), or to assume lignins to be unreactive towards periodate (41). Isolation methods based on the acidic hydrolysis of polysaccharides cause variable degrees of changes, depending on the severity of the treatment. Some hydrochloric acid ("Willstätter") lignins retain many of the properties of lignins in situ, while sulfuric acid ("Klason") lignins are generally heavily condensed (see Chapter 5, Section V for further details).

4, Finally, lignin derivatives may quite appropriately be called lignins, as long as the name clearly indicates that we are dealing with a derivative, not an isolated lignin preparation. Accordingly, ethanolysis lignin is a better name than "ethanol lignin," hydrogenolysis lignin preferable to "hydrol lignin" and dioxane acidolysis lignin to "dioxane lignin." Many reactions, on the other hand, change the polymeric phenylpropane structure characteristic of lignins, and it seems somewhat excessive to include such products in the category of lignin derivatives. Here, a revision of nomenclature is clearly warranted. For example, it is somewhat misleading to call the lignin products from the Kraft pulping process "Kraft lignins" (see Chapter 16), and even more misleading to put the extensively degraded chlorination and nitration products in the category of "chlorolignins" or "nitrolignins" (see Chapter 10).

B. Notation for Carbon Atoms in Phenylpropane Units. A special discussion on this topic would otherwise be trivial, except for the circumstance that three separate systems are in current use to denote side-chain carbon atoms in phenylpropane (C_6C_3-) units:

```
  (OMe)
    |           |   |   |
 -O-⟨ ⟩-   -C - C - C-
    |           |   |   |
  (OMe)       α   β   γ     Common notation
              γ   β   α     Revision proposed by Freudenberg (42)
             (3) (2) (1)    Notation used by Brown (43)
```

The revision proposed by Freudenberg is based on the argument that the same notation system should be applied to the phenylpropane units in lignin and to the cinnamyl alcohol precursors of lignin. In the latter, the α-notation must be assigned to the carbinol carbon, and therefore, Freudenberg argues (42), the earlier side-chain notation in lignin units should be reversed. The revised notation is used by Freudenberg in his later publications. On the other hand the notation used by Brown conforms with that applied to a number of naturally occurring C_6C_3-compounds (43).

Actually, there seems to be hardly sufficient reason for revising the original notation. In phenylpropane itself, the carbon adjacent to the aromatic ring is the α-carbon, and a notation based on phenylpropane is no worse than another based on cinnamyl alcohols.

C. Inconsistencies in Compound Names. In lignin chemistry, the radical name guaiacyl commonly means a 3-methoxy-4-hydroxyphenyl group (e.g. in guaiacylpropane). On the other hand, according to Chemical Abstracts (44), "guaiacyl" is the proper name only for o-methoxyphenyl groups. The latter meaning appears occasionally in lignin model compounds. For example, the widely accepted name for compound 11 is "guaiacylglycerol-β-guaiacyl ether." In this name, the first guaiacyl means 3-methoxy-4-hydroxyphenyl and the second, o-methoxyphenyl. The logic of such nomenclature has not greatly disturbed the lignin chemists, who have become accustomed to other dual meanings as well, calling compound 12 "veratryl glycerol" and compound 13 "veratryl alcohol." Veratryl, in the former case has the meaning 3,4-dimethoxyphenyl and in the latter name, 3,4-dimethoxybenzyl. Finally, the radical name syringyl means 3,5-dimethoxy-4-hydroxyphenyl in syringylpropane and 3,5-dimethoxy-4-hydroxybenzyl in syringyl alcohol.

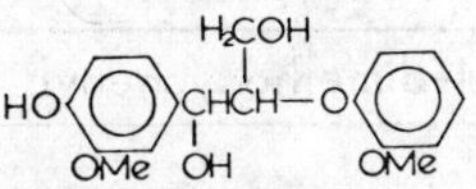

DEFINITION AND NOMENCLATURE

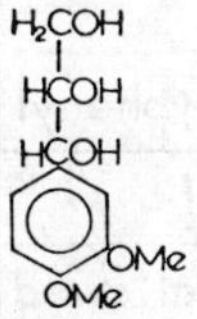

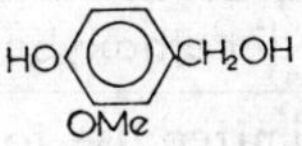

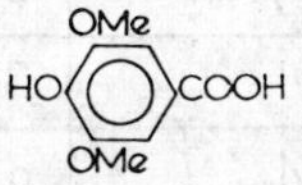

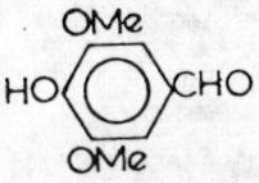

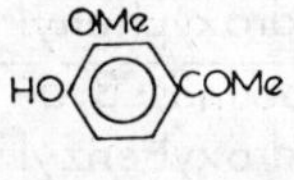

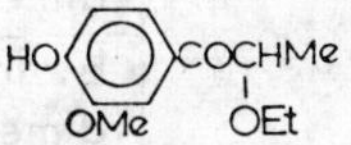

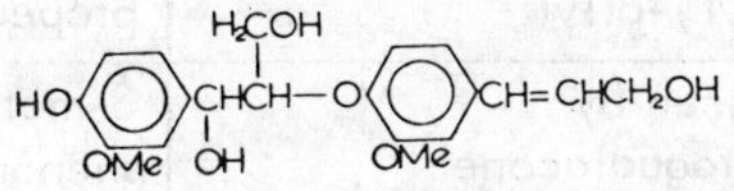

11 HO-C6H3(OMe)(OH)-CHCH(CH2OH)-O-C6H4(OMe)

12 C6H4(OMe)-CH(CH2OH)(HCOH)(HCOH)

13 HO-C6H3(OMe)-CH2OH

14 HO-C6H2(OMe)2-COOH

15 HO-C6H3(OMe)-CHO (with OMe)

16 HO-C6H3(OMe)-COMe

17 HO-C6H3(OMe)-COCHMe-OEt

18 HO-C6H3(OMe)(OH)-CHCH(CH2OH)-O-C6H3(OMe)-CH=CHCH2OH

The IUPAC nomenclature offers little help for deciding which ones of the dual meanings ought to be eliminated. The radical name guaiacyl is too widely used in the meaning of 3-methoxy-4-hydroxyphenyl to be changed to denote o-methoxyphenyl only. IUPAC nomenclature accepts veratryl in the meaning of 3,4-dimethoxybenzyl but does not even recognize "syringyl" as a radical name. The use of veratryl in the meaning of 3,4-dimethoxyphenyl can be discouraged without great difficulty, but the situation is more difficult with the term syringyl. Inasmuch as the name syringyl-propane has become deeply ingrained in common usage in the lignin area, the meaning 3,5-dimethoxy-4-hydroxyphenyl for syringyl groups can hardly be eliminated, although it contradicts the meaning associated with this group in compound names syringyl alcohol, syringaldehyde 15 and syringic acid 14. Until an internationally accepted nomenclature in lignin chemistry is established, dual meanings for the radical syringyl must be tolerated.

Table 1.1 summarizes the preferred meanings for radical names discussed above, listing also examples of other inconsistencies in compound names. These include the incorrect name "acetovanillone" for compound 16, "α-ethoxypropiovanillone" for compound 17 and the inappropriate use of radical name coniferyl in "guaiacylglycerol-β-coniferyl ether" 18.

Table 1.1 Examples of Dual Meanings in Lignin Nomenclature

	Meanings in Common Use	
	Preferable	Less Desirable
Guaiacyl	3-methoxy-4-hydroxyphenyl	o-methoxyphenyl
Veratryl	3,4-dimethoxybenzyl	3,4-dimethoxyphenyl
Syringyl	a. Limited use to be allowed in reference to lignins and lignin models as 3,5-dimethoxy-4-hydroxyphenyl b. In other compounds: 3,5-dimethoxy-4-hydroxybenzyl	---
Coniferyl	3-(3-methoxy-4-hydroxy-phenyl)-allyl	2-methoxy-4-(3-hydroxy-propenyl)-phenyl
Acetovanillone	Replaced by "acetoguaiacone"	3-methoxy-4-hydroxy-aceto-phenone
α-Ethoxypropio-vanillone	Replaced by "α-ethoxypropioguaiacone"	3-methoxy-4-hydroxy-α-ethoxy-propiophenone
Coniferyl sulfonic acid	Should be replaced by Isoeugenol γ-sulfonic acid	---
Vanillyl sulfonic acid	Should be replaced by Creosol α-sulfonic acid	---

D. **Abbreviated Names and Structural Formulae.** As our knowledge of lignin structure becomes more detailed, it obviously becomes desirable to expand the current nomenclature. For example, Freudenberg (42) recently made the excellent proposition of calling low-molecular weight intermediates in lignin formation "lignols." Thus monomeric C_6C_3- precursors, in addition to primary precursors which include other participating monomers such as coniferyl aldehyde and ferulic acid, may be called "monolignols," dimeric intermediates "dilignols" etc. This system, a counterpart of that used in the carbohydrate field, will no doubt be accepted for general use. By adding a notation expressing the mode of coupling between the monomeric units, the system may be made more specific. As an example, dimeric intermediate 18 could be called a β-O-4 dilignol."

DEFINITION AND NOMENCLATURE

Abbreviated formulae have been developed in the fields of both protein and polysaccharide chemistry, to express structures for complex molecular aggregates. So far, similar systems have not evolved for complex lignin structures, although the need for an abbreviated formula system becomes obvious by inspecting the Freudenberg lignin formula in Figure 1.1, for example.

Devising a workable abbreviation system for structures present in lignin is obviously not as straight-forward a task because of the complexity of interunit bonding. A reasonably convenient system may, however, be set up starting with the following conventions:

1. Ⓖ, Ⓢ and Ⓗ are symbols for the aromatic groups, representing guaiacyl, syringyl and p-hydroxyphenyl groups, respectively.

2. The symbol "p" represents a generalized side-chain structure, including both the allyl alcohol side-chains in primary precursor molecules as well as structures resulting from free radical coupling and quinone methide

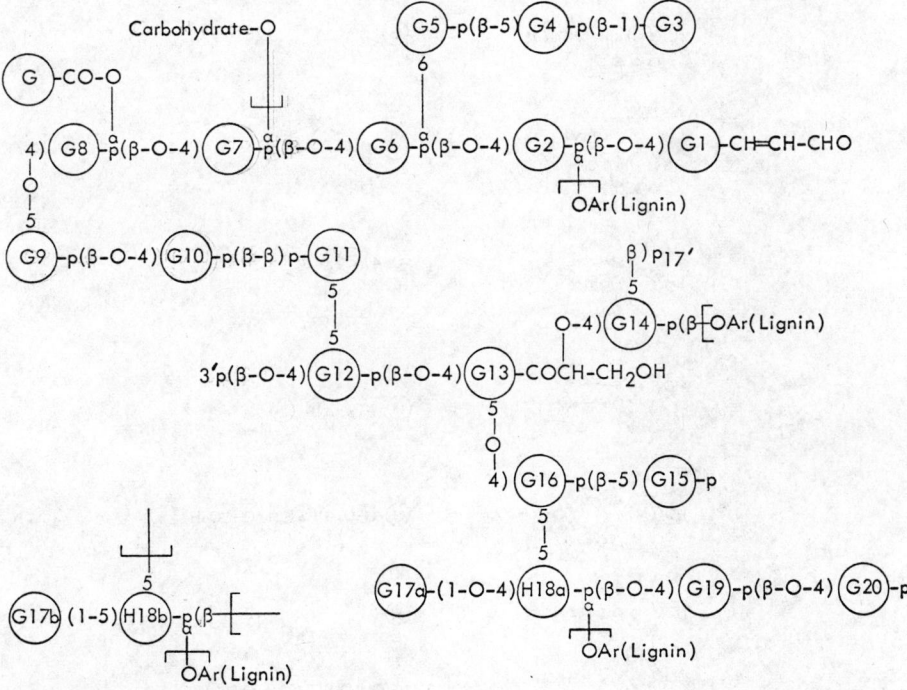

Figure 1.4. Normal gymnosperm lignin structure from Figure 1.2, expressed in abbreviated notation.

addition reactions, but excluding side-chains such as those of the coniferyl aldehyde or the ferulic acid type, or allyl, propenyl or propyl side-chains, all of which should be explicitly expressed. The symbol p is also appropriate for denoting displaced side-chain structures such as the glyceraldehyde group attached to the 4-position of unit 12 in Figure 1.2.

3. The bonding between units is expressed by making use of the conventional notation for carbon atoms in phenylpropane units (α, β, γ for side-chain carbons and 1 to 6 for ring carbons). Ether linkages formed through intramolecular hydroxyl addition to quinone methide intermediates can usually be omitted from the formulae without sacrificing clarity.

For illustration, an abbreviated version of the gymnosperm lignin structure in Figure 1.2 is reproduced in Figure 1.4. Table 1.2 provides further examples of structures that can be conveniently expressed by the abbreviated formula system.

Table 1.2. Examples of the Use of the Abbreviated Formula System

DEFINITION AND NOMENCLATURE 17

1. S. A. Brown, Ann. Rev. Plant Physiol., 17, 238 (1966).
2. Oxidative Coupling of Phenols, (W. I. Taylor and A. R. Battersby, Eds.), Marcel Dekker, Inc. New York, 1967.
3. H. Erdtmann, Biochem. Z., 258, 172 (1933); Liebigs Ann. Chem., 503, 283 (1933).
4. K. Freudenberg, Adv. in Chem. Ser., 59, 1 (1966); Science, 148, 595 (1965).
5. E. Adler, Adv. in Chem. Ser., 59, 22 (1966); Paperi Puu, 43, 670 (1961).
6. K. Kratzl, in Cellular Ultrastructure of Woody Plants (W. A. Côté, Jr., Ed.), Syracuse Univ. Press, 1965, p. 157; Holz Roh- und Werkstoff, 19, 218 (1961); in Biochemistry of Wood (K. Kratzl and E. Billek, Eds.), Pergamon Press, 1959, p. 247.
7. K. Freudenberg, Holzforschung, 18, 3 (1964).
8. K. Freudenberg, Science, 148, 595 (1965).
9. K. Freudenberg and J. M. Harkin, Holzforschung, 18, 166 (1964).
10. J. M. Harkin, Reference 2, p. 274.
11. C. H. Ludwig, B. J. Nist and J. L. McCarthy, J. Am. Chem. Soc., 86, 1196 (1964).
12. K. Forss and K. E. Fremer, Paperi Puu, 47, 443 (1965).
13. W. E. Cohen, Australian Council Sci. Ind. Res., Div. For. Products, Technical Paper No. 20 (1936).
14. E. Nilsson and O. Tottmar, Acta Chem. Scand., 21, 1558 (1967).
15. A. M. Latif and K. V. Sarkanen, unpublished results.
16. S. M. Manskaya and M. N. Kochneva, Doklady Akad. Nauk S.S.S.R. 6, 505 (1948).
17. E. Adler, K. J. Björkquist and S. Häggroth, Acta Chem. Scand., 2, 93 (1948); E. Adler and L. R. Ellmer, ibid., 2, 839 (1948).
18. W. A. Jensen, Botanical Histochemistry, W. H. Freeman & Co., San Francisco, 1962.
19. G. H. N. Towers and R. D. Gibbs, Nature, 172, 25 (1953).
20. P. W. Lange, Svensk Papperstidn., 57, 525, 533 (1954).
21. H. I. Bolker and N. G. Somerville, Tappi, 45, 826 (1962).
22. K. V. Sarkanen, H.-M. Chang and G. G. Allan, ibid., 50, 583 (1967).
23. Tappi Stardard Method T13m.
24. N. Migita and M. Kawamura, J. Agric. Chem. Soc. Japan, 20, 348 507 (1944); I. A. Pearl and L. R. Busche, Tappi, 43, 961 (1960).
25. F. E. Brauns and D. A. Brauns, The Chemistry of Lignin, Supplement Vol., 1960, pp. 128-136.
26. S. M. Manskaya, in Biochemistry of Wood, (K. Kratzl and G. Billek, Eds.), Pergamon Press, London, 1959.
27. T. Higuchi, Y. Ito, M. Shimada and I. Kawamura, Cell. Chem. and Techn., 1, 585 (1967).

28. B. Holmberg, Svensk Papperstidn., 33, 679 (1930); Ing. Vetenskaps Akad. Handl., No. 131 (1934).
29. J. M. Pepper and P. D. S. Wood, Can. J. Chem., 40, 1026 (1962).
30. E. Adler, Chim. Biochim. Lignine, Cellulose, Hemicelluloses, Grenoble, 1964, p. 73.
31. K. Kratzl, G. Billek, E. Klein and K. Buchtela, Monatsh. Chem., 88, 721 (1957).
32. S. A. Brown and A. C. Neish, J. Am. Chem. Soc., 81, 2419 (1959).
33. A. F. Semechkina and N. N. Shorygina, Zhur. Obshchei Khim., 23, 1593 (1953); Faserforsch. U. Textil Tech., 5, 79 (1954).
34. Reference 25, pp. 531-546.
35. R. H. J. Creighton, J. L. McCarthy and H. Hibbert, J. Am. Chem. Soc., 63, 3049 (1941).
36. T. Higuchi, Y. Ito and I. Kawamura, Phytochem., 6, 875 (1967).
37. A. Björkman, Svensk Papperstidn., 59, 477 (1956); 60, 243, 329 (1957); A. Björkman and B. Person, ibid., 60, 158, 285 (1957).
38. J. C. Pew, Tappi, 40, 553 (1957).
39. F. E. Brauns, J. Am. Chem. Soc., 61, 2120 (1939); J. Org. Chem., 10, 211 (1945).
40. G. de Stevens and F. F. Nord, Proc. Natl. Acad. Sci., U.S., 39, 80 (1953).
41. W. J. Wold, P. F. Ritchie and C. B. Purves, J. Am. Chem. Soc., 69, 1371 (1947).
42. K. Freudenberg, Chim. Biochim. Lignine, Cellulose, Hemicelluloses, Grenoble, 1964, p. 39.
43. S. A. Brown, in Biochemistry of Phenolic Compounds,(J. B. Harborne, Ed.), Academic Press, New York, 1964, pp. 373, 382.
44. Chem. Abstr., 51, Subject Index Issue (1)(1957).

2

OCCURRENCE AND FORMATION IN PLANTS

A. B. Wardrop

1. The Occurrence of Lignin in Plants. Investigations of the chemistry of lignin have shown it to be a substance of great structural complexity (1). Its structure is based on molecular units of the phenylpropane type. The complexity of the lignin molecule arises in part from the manner in which the C_6-C_3 units are linked to each other and in part from the fact that these units are not chemically identical.

In anatomical studies it is extremely useful that certain groupings in the lignin molecule can be recognised by colour reactions, Figure 2.1, so that surveys for the presence of these groupings in different families of plants can be readily carried out.

Studies of this kind have been made, notably by Manskaya and Kochneva (2) Towers and Gibbs (3) and Srivastava (4). It was shown that the most primitive land plants—in the sense of those longest recorded in the geological record—have lignin in which the guaiacyl nuclei predominate; whereas in plants of more recent origin, such as the angiosperms, both guaiacyl and syringyl nuclei are present. However there are many exceptions to this generalization (3, 4).

An additional fact to be considered in relation to these studies is that different kinds of lignin may occur within the same plant. Thus in Eucalyptus elaeophora it was shown (5) that the primary xylem stained brown with chlorine water-sodium sulphite, resembling the reaction with the xylem of conifers, whereas the secondary xylem stained red. In Phormium tenax (New Zealand flax) the sclerenchyma caps showed almost no reaction with phloroglucinol-hydrochloric acid but stained red with chlorine water-sodium sulphite and showed some ultra-violet absorption. In an unpublished investigation of the lignified collenchyma of Eryngium by Wardrop it has also been observed that with chlorine water-sodium sulphite the xylem stained brown and the collenchyma red, but with phloroglucinol-hydrochloric acid the xylem stained deeply red and collenchyma gave little or no reaction. Both the xylem and collenchyma showed ultra-violet absorption. In an investigation of Phleum pratense (timothy), Stafford (6) showed that the xylem stained with phloroglucinol-hydrochloric acid at an earlier stage than with chlorine water-sodium sulphite, and that the sclerenchyma of the upper internodes stained with chlorine water-sodium sulphite at an earlier stage than with phloroglucinol-hydrochloric acid.

Reagent	Color Reaction	Grouping Indicated
Phloroglucinol	red	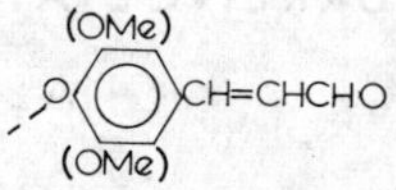 (OMe) / (OMe) CH=CHCHO (Cinnamaldehyde end group)
Chlorine water-sodium sulphite	red	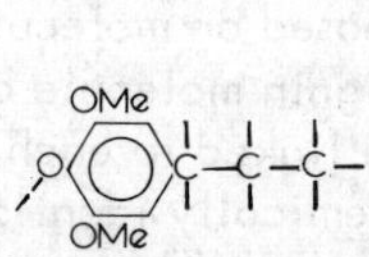 OMe / OMe —C—C—C— (Syringyl nucleus)
	brown	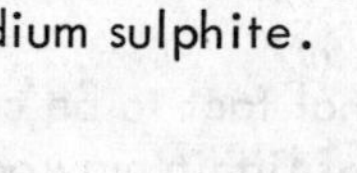 OMe —C—C—C— (Guaiacyl nucleus)

Figure 2.1. The colour reactions given by various molecular groups in lignins with phloroglucinol and with chlorine water-sodium sulphite.

Chemical studies complementing studies of the above kind have been carried out by Bland (7). Thus, as pointed out above, the primary xylem gave different staining reactions with chlorine water-sodium sulphite from that given by the secondary xylem. In the leaf petioles and leaf midribs primary xylem predominates. Chemical analysis of the primary xylem lignin by Bland showed that it was predominantly of the guaiacyl type whereas that in the secondary xylem was of the guaiacyl-syringyl type. Thus both the staining and chemical reactions are indicative of the presence of lignins, differing in the nature of the C_6-C_3 units composing their structure, in different parts of the same plant.

It should also be appreciated that lignification of the cell wall is not a characteristic of all cells, but is confined to cell groups (tissues) of specialized functions such as that of the conduction of solutes or of mechanical support. Lignin is absent from the algae and its presence is doubtful in the mosses (Musci), only some members of which contain cells specialized for water conduction and mechanical support. It is only in the true vascular plants in cells specialized for conduction that lignin can be detected unequivocally in quantity. The most primitive of such plants which contain recognisable tracheids are the Psilopsida, represented by the rootless Psilotum and Tmesipteris. Thus from the botanical standpoint the phenomenon of lignification in tissues is essentially associated with development of a specialized conduction (vascular) system in the plant body. Because lignin is a substance unique to vascular plants and because of the rigidity it imparts to the cell wall, Barghoorn (8) has regarded the development of the capacity of cells to lignify as conferring on the vascular plants, properties on which forces of natural selection could operate in the process of adaption of plants to life on land.

In a discussion of the same subject Neish (9) has considered the evolutionary significance of lignin as a secondary growth substance in terms of the ideas advanced by Frankel (10). Neish considers that the development of the lignification mechanism of cell walls has made possible the development of mechanical and vascular tissues which in turn have influenced the development of arborescent forms which have competitive advantage for light over herbaceous forms. It is perhaps relevant that apple trees suffering a virus infection which results in the formation of non-lignified fibres, have xylem of greatly reduced rigidity so that the branches and even the stems tend to bend downwards (11).

The need to establish excretory mechanisms may have influenced the development of the lignification process. This idea has been discussed by Steward (12, 13), by Reznik (14) and by Neish (9). In their view unicellular organisms can dispose of excretory products to the external medium in which they grow and into the cell vacuole. With increasing structural complexity however, the problem of excretion would become more acute so that excretory products might be deposited in the cell wall. In the event of a mutation producing a substance conferring functionally advantageous properties to the cell wall, such as lignin, it might be supposed that the evolutionary advantage would permit the development of more massive forms, at the same time providing a mechanism for the disposal of metabolic products not required by the organism.

While the above generalizations are of broad botanical interest, appreciation of the possible biological functions of lignin and of its technological significance requires consideration of the location of lignin in the cell wall and the manner of its deposition. This can best be done by reference to studies of the lignification of tracheids and angiosperm fibres, on which the most extensive studies have been carried out. Because the amount of lignin in different parts of the cell wall is variable and the closeness of its association with the various wall constituents differs, it is necessary to consider first the general nature of the wall in these cells.

II. The Texture and Organization of the Cell Wall in Fibres and Tracheids. The major components of the cell wall of fibres and tracheids may be conveniently classified into three categories. These are: the framework components represented in higher plants by cellulose, the matrix components consisting of non-cellulosic polysaccharides and their derivatives; and of encrusting components, of which lignin is the major representative in these cells, and, as pointed out above, differs from the other wall components in its phenolic nature.

The cellulose framework is largely crystalline. According to Mühlethaler (15) the cellulose molecules are aggregated to form "elementary fibrils" about 35 Å in width and these may be variously aggregated and fasciated into larger units approximately 100 Å in width and of indefinite length—the microfibrils. This view thus envisages the "elementary fibril" as the basic structural unit. An alternative view first proposed by Frey-Wyssling (16) is that the microfibril is the basic structural unit and that within it are crystalline regions which correspond to the "micelles" of earlier workers. In either view, however, the microfibril as seen in the electron microscope can be envisaged as containing within it regions of greater and lesser degrees of perfection of molecular order. Furthermore x-ray diffraction studies of the degree of lateral order of the crystalline regions of the cellulose show that this quantity (as measured by the diffraction line breadth) increases with hydrolysis. Such a change is most readily interpreted in terms of the concept that crystallization of cellulose accompanies hydrolysis and leads to the view that the crystalline regions are surrounded by molecules oriented in the direction of those in crystalline regions but not in other planes, i. e. there exists a paracrystalline phase of cellulose molecules around the crystalline phase (17). This would imply the presence of a paracrystalline phase surrounding the elementary fibrils or around the crystalline regions (depending on the model of microfibril structure adopted) and this in turn implies that the surface of the microfibril consists of a paracrystalline phase of cellulose molecules.

In contrast to this ordered physical state of the cellulose, the matrix components are largely amorphous, especially those with a branched molecular form. However from x-ray diffraction studies, and from studies of infrared absorption spectra it is reasonable to conclude that some of the matrix components such as xylan are oriented in the direction of the cellulose microfibrils and may in fact be associated with the paracrystalline phase (18).

Like the matrix components the lignin lies between the microfibrils. From x-ray diffraction studies it appears to be amorphous, and although it shows ultra-violet dichroism this is considered to be wholly form dichroism. At least it can be said that the amount of oriented lignin in the cell wall is extremely small (19).

In view of its location between the cellulose microfibrils it is not surprising that chemical evidence shows, that insofar as lignin is associated with other wall components, the association is with matrix constituents. This is supported by the fact that lignin isolated mechanically is associated with matrix components (20). The nature of the matrix-lignin bond or bonds has not been established with certainty. However, Freudenberg (21) considers that an ether linkage between the polysaccharides and the benzyl alcohol group of the lignin is involved. Freudenberg also considers that some direct linkage to cellulose exists. If this is so, it is reasonably certain that the linkage would be to the paracrystalline phase of cellulose. Certainly this phase is intimately associated with some of the lignin since even the most careful delignification results in a decrease in the width of the diffraction lines corresponding to planes parallel to the cellulose molecules (002). This suggests that some crystallization of cellulose occurs as a result of delignification, although it is to be expected that some hydrolysis also occurs (22).

III. Cell Wall Organization. The mature cell wall of fibres and tracheids consists of two structures. The primary wall is the structure first formed and is the cell envelope enclosing the protoplasm during the growth of the cell as it differentiates from the cambium. The secondary wall is formed after surface growth of the cell has ceased.

In the primary wall the microfibrils are approximately transverse on the inner surface of the wall, but on the outer surface they differ from this orientation in a manner dependent on the extent and polarity of the growth which takes place during differentiation. The microfibrils appear to be interwoven so that there is a gradual transition in the microfibril orientation through the thickness of the cell wall.

In contrast to the primary wall, the much thicker secondary wall is conspicuously layered. Each layer is composed of a number of lamellae each consisting of a sheet of microfibrils. In the first and third layers of the secondary wall successive lamellae differ in the direction of microfibrillar orientation whereas in the second layer the orientation of microfibrils in all lamellae is approximately the same, Figure 2.2. Analysis of differentiating cells (23) and of cell wall fragments isolated by micromanipulation (24) show that the cellulose content of the primary wall is 20-28 per cent compared with that of 45-55 per cent in the secondary wall. The volume occupied by the cellulose framework in the two structures would, because of density considerations, be expected to be rather less than these values. Thus, it can be seen that in the primary wall the matrix-encrusting components are predominant whereas in the secondary wall the structural component is predominant.

Early studies, designed to establish the distribution of lignin in the cell wall using the techniques of staining and differential solubility, established that the percentage of lignin was highest in the middle lamella and primary walls (25). More recently attempts at the quantitative estimation of lignin distribution have been made using the techniques of ultra-violet microspectrophotometry (26), interference microscopy (27) and ultra-violet fluorescence microscopy (28). From microspectrophotometric measurements Lange concluded that the concentration of lignin in the region external to the layer S2 in Picea (spruce) was about 73 per cent and for Betula (birch) the corresponding value somewhat greater. The properties of the total cell wall lignin in the different wall layers have been discussed by Berlyn and Mark (28a).

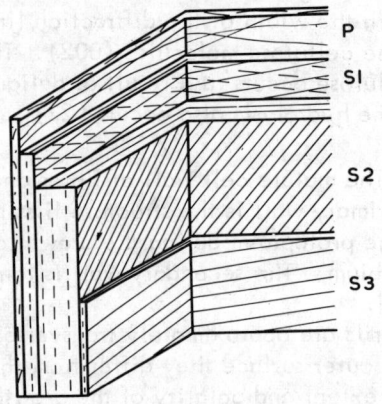

Figure 2.2. Diagrammatic representation of the orientation of the microfibrils of the cellulose framework in the cell wall of a conifer tracheid.

P = primary wall
S = secondary wall
  1: first layer
  2: second layer
  3: third layer

Studies by Ruch and Hentgartner (28) using the technique of ultra-violet fluorescence have shown for jute that the distribution of lignin is almost uniform through the secondary wall and increases greatly in amount in the primary wall and middle lamella. A similar conclusion was reached for Picea. These observations are in agreement with electron microscopic observations using potassium permanganate as an electron stain (29), Figure 2.3. The difference in pattern of lignin distribution between the results of Lange, using ultra-violet microspectrophotometry, and those of Ruch and Hentgartner using ultra-violet fluorescence has been attributed by Frey-Wyssling (19) to diffraction effects at the primary-secondary wall interface (see also Chapter 3, Section II-A-3). The level of lignification in the middle lamella is variable around the circumference of the cell, being greatest at the cell corners, and greater on the radial as compared with the tangential walls (30). This observation is of interest in relation to the sequence of lignification.

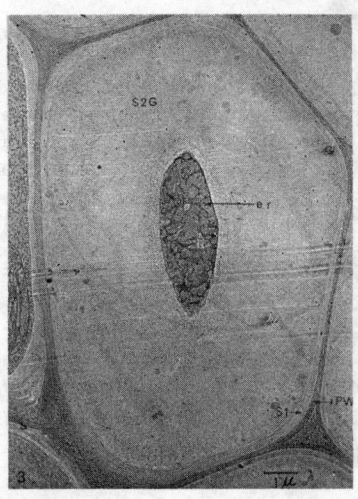

Figure 2.3. An electron micrograph of a transverse section of the reaction wood (tension wood) of Eucalyptus elaeophora fixed in potassium permanganate showing the localization of lignin in the primary wall and middle lamella (29). (A. B. Wardrop, in Cellular Ultrastructure of Woody Plants, copyright Syracuse University Press, Syracuse, New York).

IV. Lignification. In the developing secondary xylem of woody plants, the fibre or tracheid initials, formed by division in the cambium first undergo a phase of surface growth during which they increase in both length and radial breadth. This phase of surface growth ceases first near the middle of the cells and is followed by a phase of wall thickening (secondary wall formation) (31). About the time of the onset of wall thickening the first indication of lignification can be seen. In a study using ultra-violet microscopy it was shown by Wardrop (5) that the first deposition was at the cell corners within or just inside the primary wall. The process then extended rapidly to the middle lamellae at the cell corners. This pattern for the initiation of lignification has been confirmed by "feeding" experiments using tritiated lignin precursors by Saleh et al. (32). These observations also agree with subsequent electron microscopy observations (29), Figure 2.4. Following this initial deposition at the cell corners, lignification then extends along the middle lamella and into the secondary wall. Frequently the tangential walls appear to become lignified somewhat before the adjacent radial walls. From a study combining ultra-violet and polarization microscopy it was shown that wall thickening always proceeds somewhat in advance of lignification. Furthermore, although lignification of the middle lamella begins at an early stage, measurements of ultra-violet absorption showed that lignification continued in these regions simultaneously with the subsequently initiated lignification of the secondary walls. This is illustrated in Figure 2.5 (33). It can be seen that in tracheids

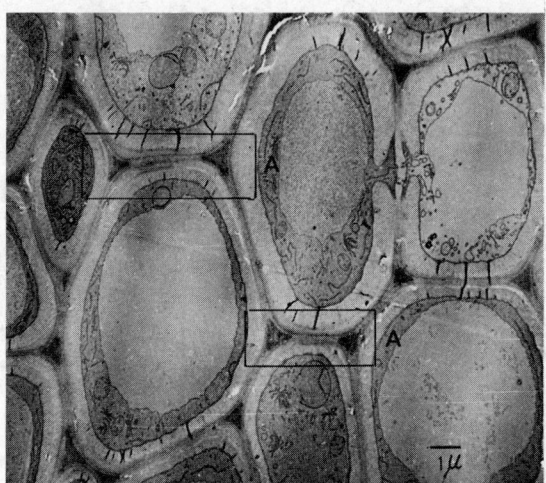

Figure 2.4. A transverse section of differentiating xylem of Eucalyptus elaeophora fixed in potassium permanganate showing the layer S2 of the secondary wall. In the areas, A, note the lignification in the middle lamella and the outer regions of the layer S1 (29). (A. B. Wardrop, in Cellular Ultrastructure of Woody Plants, copyright Syracuse University Press, Syracuse New York).

of Pinus radiata (Monterey pine), the middle lamellae between cells 38 and 39 from the cambium (in which formation of layer S3 had just begun) showed some absorption but the secondary wall showed very little absorption. However in the middle lamellae between cells 44 and 45 absorption was greater than that between cells 38 and 39, and that in the secondary wall showed considerable absorption. Thus the lignification process must be envisaged as proceeding continuously through the wall thickness during cell differentiation.

Figure 2.5. Microphotometer traces through ultra-violet photomicrographs showing greater absorption in the middle lamella between tracheids of Pinus radiata of increasing maturity (see text) (33). (A. B. Wardrop and D. E. Bland in Proc. Intern. Congr. Biochem., Vienna, 1958).

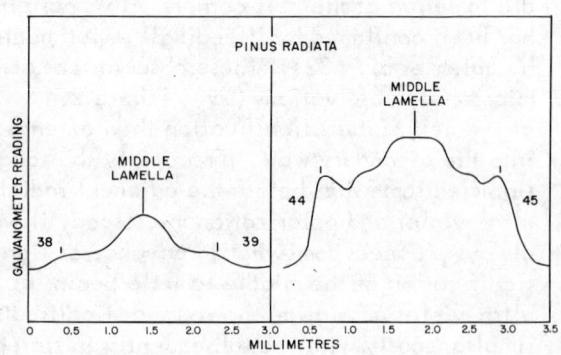

It would seem important that although in the later stages of lignification the process appears to proceed continuously through the differentiating xylem, its initiation takes place cell by cell. It also appears (5) that the cessation of lignification takes place in a similar way. These observations thus suggest that lignification is essentially a process determined by individual cells, as indeed might be anticipated from the study of other tissues where lignified cells occur surrounded by unlignified tissue. This is of relevance in relation to consideration of the possible mechanisms of lignification which have been advanced.

In microscopic studies of lignification other changes have been observed to accompany the process. Thus the cell walls stain strongly with ruthenium red before the inception of lignification and the intensity of staining progressively decreases as the process proceeds. Because ruthenium red is considered to stain substances rich in carboxyl groups it has been suggested that when lignification begins, phenylpropane units are attached through the carboxyl group probably as a uronic acid ester linkage (34). It was noted further by Allen (35) that the intensity of staining with ruthenium red was greater on the radial as compared with the tangential walls. This observation may be related to the apparently greater lignification of the radial walls noted by Wardrop, et al. (30).

Electron microscopic studies of lignification are not extensive. However this approach has been facilitated by the fact that potassium permanganate which has been used as a cytological fixative also acts as an electron stain for lignin. Thus it was shown by Crocker (36) that potassium permanganate is reduced by lignin to manganese dioxide. The precipitation of this electron dense substance in regions where lignin is present serves as a means of locating its presence.

When differentiating xylem is fixed with potassium permanganate the cambial region shows little or no stain in the cell wall. In more mature tissue however there is intense staining in the middle lamellae of the cell corners but at this stage the primary wall appears unstained, Figure 2.4. In some instances the outer part of the layer S1 shows some staining. Results similar to those obtained using ultra-violet microscopy showing the initiation of lignification at the cell corners before lignification occurred in the middle lamellae were not obtained in the electron-microscopic studies. However it is clear from results such as that in Figure 2.4 that at an early stage lignification occurs in the middle lamella and outer part of the layer S1. Subsequently the process extends along the middle lamella, through the primary wall, and ultimately to the secondary wall as observed in the ultra-violet absorption studies.

From the above anatomical observations it would seem reasonable to conclude that lignification proceeds cell by cell but because of the concentration of lignin in the middle lamella, the cellular nature of the process, especially in secondary xylem, tends to be obscured. It is of interest to consider these

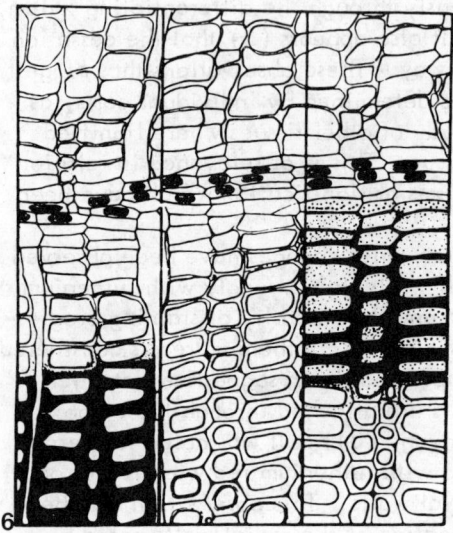

Figure 2.6. The relative distribution of lignin (left) and β-glucosidase (right) in differentiating xylem. Data by Freudenberg et al. (37).

observations in relation to the biochemical hypothesis of the nature of lignification advanced by Freudenberg et al. (37), see also Chapter 4.

V. Biochemical Aspects of Lignification. The hypothesis of Freudenberg is based on the recognition of substances which are possibly precursors of lignin in the zone of differentiating xylem. The first of these substances to be recognized was coniferin, which was isolated by Tiemann and Haarmann (38). Subsequently other similar glycosides such as syringin and p-glucocoumaryl alcohol have been isolated (39).

Freudenberg showed, in addition, that the distribution of lignin in the zone of differentiation was complementary to that of β-glucosidase. The presence of this enzyme was demonstrated by the precipitation of indoxyl which took place when a solution of indican was applied to the tissue. These observations are summarized in Figure 2.6. In addition it was noted by Freudenberg et al. (37) and by Siegel (40) that two phenol dehydrogenases—laccase and peroxidase—were present in the zone of differentiation and that the amount of these enzymes decreased in the more heavily lignified cells.

On the basis of these observations it was proposed that the lignin precursors such as coniferin were produced in the cambium or were translocated from another region of the plant; were hydrolysed by the glycosidase present; and in conifers, the aglycone-coniferyl alcohol was oxidized and polymerized to form lignin in the differentiating cell walls.

This hypothesis is consistent with the earlier observation of Griffoen (41) who attempted to demonstrate a gradient of redox potential in zone of xylem differentiation using redox indicators. It was concluded that the region of highest reducing activity was in the cambium and that the cytological environment became progressively more oxidizing in the cells of increasing degree of lignification. Higuchi (42) working with bamboo observed the presence of a reduced form of glutathione and ascorbic acid in the unlignified regions of the shoot and that these substances inhibited the dehydrogenation-polymerization of coniferyl alcohol. Furthermore, Manskaya (43) showed that during the annual cycle of growth in conifers the amount of peroxidase present is inversely related to the amount of coniferin produced. Finally, it was shown by Freudenberg (44) and by Kratzl (45) that a C^{14} labelled, possible lignin precursor such as coniferin could be injected into the differentiating xylem and was incorporated into the lignin which was formed. From the foregoing it can be seen that the hypothesis of Freudenberg is consistent with observations of the general histological pattern of lignification as observed in the secondary xylem.

Although it is clear that the work of Freudenberg and his school has established a body of evidence which justifies the concept proposed as being more than a working hypothesis, it will be appreciated that the concept is essentially biochemical in nature and does not specify the manner in which the individual cells control the lignification of their walls.

Evidence that the lignification of the cell wall is controlled by the individual cell is contained in many anatomical observations. Thus, although the cell wall of fibres and tracheids is lignified the membranes of the pits usually remain unlignified (46). More strikingly the fibres of tension wood (the reaction wood of angiosperms) in the initial stages of their differentiation frequently develop in a manner which cannot be distinguished from that of normal fibres with a well developed lignified secondary wall. However, in addition a thick unlignified layer may then be formed, Figure 2.7. What appears to be the converse of this situation can be seen in the lignified collenchyma of many Umbelliferae, discussed by Duchaigne (47) and exemplified in the leaf collenchyma of Eryngium, Figure 2.8. In these cells, initially an unlignified collenchyma wall is formed but as the cell matures a thick lignified layer is deposited. In the primary xylem of Pinus radiata it was observed (5) that the lignified helical thickenings of the cell wall are deposited on the primary wall which does not lignify.

None of these observations preclude the extra-cellular origin of the lignin precursor as suggested by Freudenberg, but clearly the protoplast must control a mechanism which governs some vital step in lignin synthesis. Because the deposition of cellulose and matrix components always proceeds in advance of lignification, in the case of differentiating secondary xylem, Figure 2.5, this means that the surface of the protoplast is spatially removed from the region of most intensive lignification in the wall. If then, as the

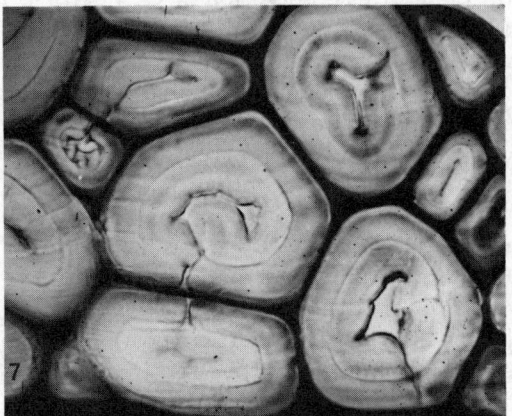

Figure 2.7. An ultra-violet photomicrograph of a transverse section of the reaction wood (tension wood) of Fagus sylvatica showing the strong ultra-violet absorption in the middle lamella primary wall and first layer of the secondary wall compared with the later formed "gelatinous" secondary wall. Micrograph by Wardrop and Dadswell, Aust. J. Bot., 3, 177 (1955).

anatomical evidence suggests, the protoplast controls the lignification, it is reasonable to suppose that the precursor undergoes metabolic modification in the protoplast before being "fed" to the developing wall. Alternatively, the chemical nature of the matrix could be controlled by the protoplast in a manner such that it constitutes a suitable substrate upon which the extra-cellular precursor of lignin can subsequently be deposited. That the nature of the substrate may influence the amount of lignin deposited on it was suggested by the experiments of Siegel (48) using model systems of filter paper of cellulose derivatives impregnated with peroxidase, in the presence of hydrogen peroxide and eugenol.

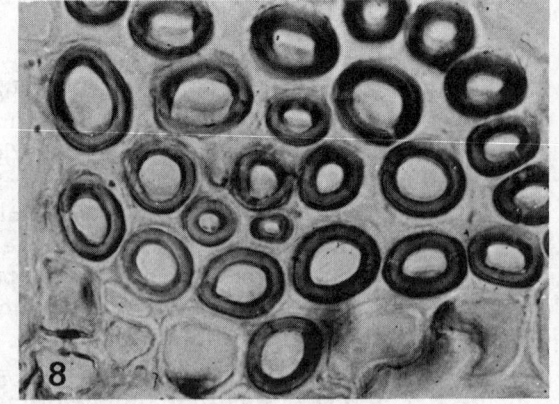

Figure 2.8. An ultra-violet photomicrograph of a transverse section of the sclerozed collenchyma of Eryngium vesiculosum showing the strong absorption in the secondary wall compared with the thickened primary wall. Micrograph by A. Dughaigne, Ann. Des. Sc. Nat. Bot., 11, 456 (1955).

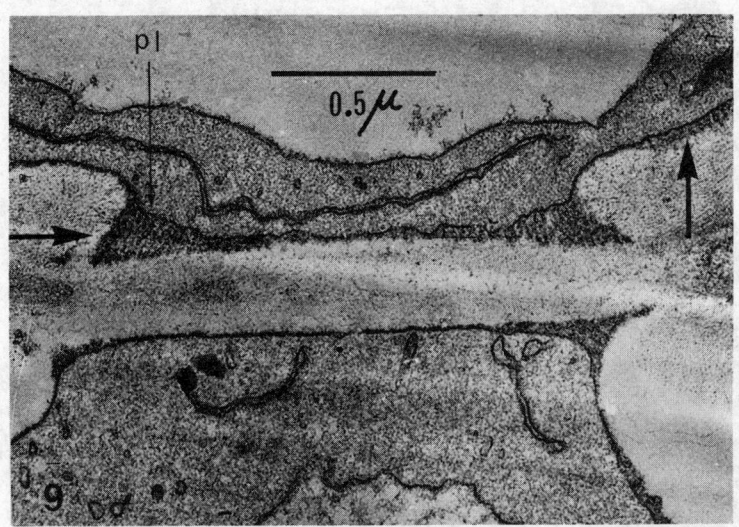

Figure 2.9. An electron micrograph of a transverse section through a bordered pit between adjacent xylem fibres of Eucalyptus elaeophora showing vesicle accumulation in the region between the plasmalemma (pl) and the cell wall (29). (A. B. Wardrop, in Cellular Ultrastructure of Woody Plants, copyright Syracuse University Press, Syracuse, New York).

Barskaya (49) has suggested that there are two components involved in lignification such that one is formed in the cytoplasm and the other arises in the cambium zone as proposed by Freudenberg. This conclusion was based on the observation that cells in which the whole wall showed positive staining using the Mäule reaction, were stained only in the middle lamella, when the phloroglucinol-hydrochloric acid reaction was employed. It has however been pointed out by Bland (7) that although the phloroglucinol reaction is more sensitive than the Mäule reaction, there are few aldehyde groups present in lignin so that it may be that a positive phloroglucinol reaction is obtained only when the concentration of lignin becomes sufficiently high in the middle lamella for the reaction to be observed.

As already indicated, studies of lignification of the cell wall with the electron microscope using potassium permanganate as a lignin stain are in broad agreement with the results obtained from earlier optical studies. There is little evidence however relating the onset of lignification to the cytological characteristics of the cell. Nevertheless it was observed (29) that in cells undergoing lignification small vesicles were accumulated external to the plasmalemma and usually at the edges of the pit chambers, Figure 2.9, although

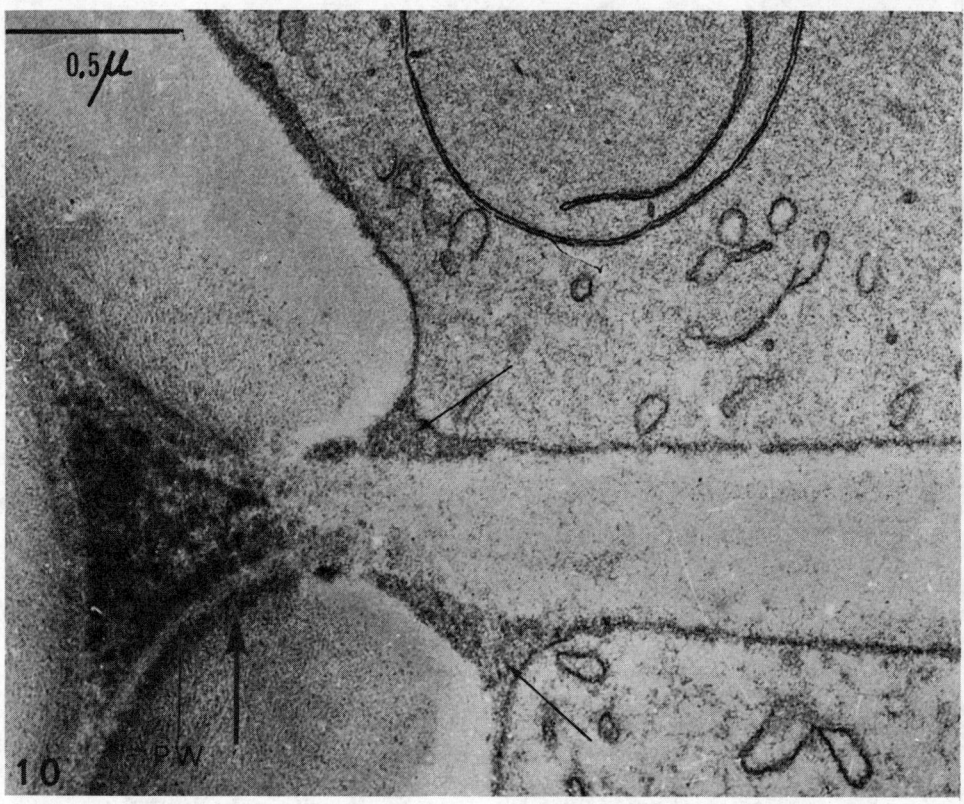

Figure 2.10. Similar to Figure 2.9, showing the apparent tangential continuity of vesicles secreted in the region of the pit border with the heavily straining region of the cell wall (heavy arrow) internal to the primary wall (29). (A. B. Wardrop, in Cellular Ultrastructure of Woody Plants, copyright Syracuse University Press, Syracuse, New York).

they were observed at other points around the periphery of the cell. In some instances there appeared to be a fairly sharp line of demarcation between the accumulated vesicles and the cell wall but more frequently this was not so and they appeared to merge with the developing wall. Although there is no evidence of the nature of the contents of these vesicles, it was pointed out that they could contain lignin precursors, enzymes or mineral substances (29). The possibility that they may in some way be involved in lignification was suggested by the observation that frequently the accumulated vesicles appeared, when observed in transverse sections, to be tangentially continuous with lignifying regions of the cell wall, Figure 2.10.

In more recent studies it has been shown by Pickett-Heaps (50) that when tritiated cinnamic acid was fed to the differentiating xylem of wheat roots, it was incorporated into the lignin of the wall thickenings and could also be observed to be associated with vesicles derived from the Golgi apparatus and the endoplasmic reticulum. It has been suggested by Wardrop (51) that β-glucosidase is released into the cell wall from the cytoplasm during lignification and that the release of this enzyme through its function of releasing the aglycone from the precursor glucoside may control the lignification process. These results are, however, only of an exploratory nature and further studies along similar lines can be expected to add further to our understanding of the process.

VI. Physiological Aspects of Lignification. In woody plants the annual cycle of growth is reflected in the types of cells which are formed and in their dimensions and degree of lignification. These differences are apparent microscopically, and, macroscopically, as the growth rings of stems and roots. Analysis of early wood and late wood show considerable variation in the lignin content (see also Chapter 3, Section II-A-2); that of late wood being less than, equal to, or greater than that of the early wood depending on the location of the specimen in the tree. Such analyses however are not useful in attempting to assess the lignin content per cell, 1, because a given weight of material analysed would represent fewer cells of late wood than of early wood, 2, because of the difference in cell wall thickness; and, 3, in angiosperms, because of the additional factor of the varying proportions of cell types in early wood and late wood. However, the work of Manskaya (43), Figure 2.11, has demonstrated that in the annual cycle of growth the differentiation of early wood and late wood is paralleled by changes in the amount of lignin precursors and of oxidizing enzymes in the zone of differentiation.

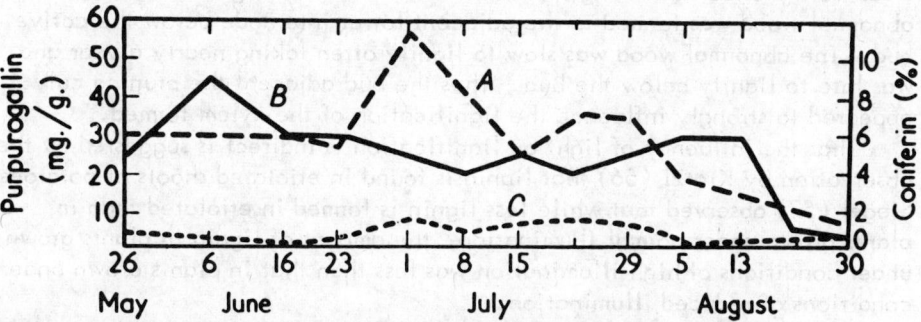

Figure 2.11. The seasonal variation of coniferin (B) and of peroxidase activity in the cambium near the xylem (A) and in the cambial sap (C). Data by S. M. Manskaya (43).

That genetic factors may influence lignification has been demonstrated by the investigations of Kuc and Nelson (52). Using the brown midrib-1 type (bm_1) of maize described originally by Jorgensen (53), it was shown that nitrobenzene oxidation of the lignin from this type yielded less aromatic aldehydes and a greater fraction resistant to oxidation, and was lower in methoxyl content compared with lignin from normal maize (Bm_1).

It was shown further that, when normal plants and the mutant were fed with uniformly labelled L-phenylalanine or L-tyrosine, the incorporation of either compound into the mutant lignin was considerably less than into the normal lignin. These results were interpreted by Kuc and Nelson (52) to mean that the gene of the normal (Bm_1) plants controls, at least in part, the manner in which the lignin monomers are incorporated into the lignin and that in the mutant (bm_1) the lignin has fewer sites at which p-coumaric acid can be esterified. These conclusions are thus at variance with the concept of Freudenberg in that an ordered assembly of the lignin monomers is implied.

Relatively little work has been carried out on the influence of environmental factors on lignification. Phillips (54) working on Fraxinus (ash) demonstrated a direct or indirect effect of light on lignification of the xylem. As in many deciduous species cambial activity is initiated before the opening of the buds—the plant presumably drawing on nutrient reserves during this phase of its development. In one experiment Phillips shaded a shoot for two weeks when the leaves were 2 inches long (May). The shoot was then allowed to grow for 1 month under normal conditions and then shaded for an additional three weeks. Transverse sections of the shoot showed alternate zones of unlignified and lignified fibres in the growth ring corresponding to the year in which the experiment was performed. It cannot be concluded with certainty that this experiment demonstrates the formation of a lignin precursor in the leaves since lignification in the stem is known to be influenced by auxin, the formation of which is influenced by the physiological state of the leaves. Thus in the early work of Wray (55) it was observed that when apple trees were pruned in the English summer just after radial growth had ceased, abnormal wood was formed in the adjacent lower internode below the active bud. The abnormal wood was slow to lignify often taking nearly a year and was late to lignify below the bud. Thus the bud adjacent the pruning cut appeared to strongly influence the lignification of the xylem formed.

That the influence of light on lignification is indirect is suggested by the observation by Kratzl (56) that lignin is found in etiolated shoots of potatoes. Siegel (57) observed that while less lignin is formed in etiolated than in plants grown under normal illumination, the amount of lignin in plants grown under conditions of high illumination was less than that in plants grown under conditions of reduced illumination.

The observations of Phillips (54) and of Wray (55) indicate the possible influence on lignification of correlative factors such as auxins which are associated with the activity of the buds. In agreement with this contention are numerous observations on the anatomical changes which accompany geotropic responses in woody plants. This subject has been reviewed by Wardrop (58). In leaning stems or branches of gymnosperms abnormal xylem tracheids, characterized by their high level of lignification, are formed on the lower sides of the stems and branches. In angiosperms however, in similar situations abnormal xylem fibres are formed on the upper side of leaning stems and branches, and are characterized by their low level of lignification compared with xylem opposite to it, and by their extremely thick cell walls. The reaction wood has been termed "compression wood" in gymnosperms and "tension wood" in angiosperms.

The distribution of auxins in stems bent horizontally has been studied experimentally by Nečesány (59) in Pinus sylvestris (Scots pine) and Populus alba (white poplar). Ether extracts were made of the differentiating xylem from the upper and lower sides of the bent stems and the effect of chromatographically separated components on the extension and inhibition of growth of coleoptile segments of Triticum aestivum (wheat) was measured. It was observed that substances stimulating the growth of coleoptile sections were present mainly on the lower side in both species, and of these 3-indole acetic acid (IAA) appeared to be the most important of the substances stimulating elongation. Thus in both instances the IAA concentration was greatest in the region of most intense lignification. It is also of interest that kinetin has been found to enhance lignification in carrot (60) and tobacco tissue cultures (61). These observations are consistent with the earlier observation of Wershing and Bailey (62) that the application of hyperphysiological concentrations of IAA induce the formation of compression wood in conifer stems. It is also consistent with the recent observation of Kennedy and Farrar (63) that the anti-auxin, tri-iodobenzoic acid, causes the formation of tension wood in angiosperms. Similarly, Koblitz (64) observed that IAA in concentrations greater than 1 p.p.m stimulated lignification in carrot tissue cultures.

These results do not agree with the observation of Siegel (40) that IAA inhibited "lignification" in the model system eugenol-H_2O_2-peroxidase using stems of Elodea (waterweed) as the matrix for the reaction. It was also observed that IAA inhibited intracellular oxidations governed by peroxidase (65), but this may be due to some concentration effect of the IAA, since Koblitz (64) has observed that lignification was inhibited by IAA at concentrations of less than 1 p.p.m in the presence of coconut milk.

However, the evidence from natural systems is that lignification in general appears to proceed under relatively high physiological concentrations of auxin but may be inhibited by anti-auxins as in the case of tension wood formation. The mechanism by which auxin influences lignification has not been elucidated but it may be noted that peroxidase which is involved in the oxidation phase of lignification also functions as an IAA oxidase which converts IAA to an oxindole derivative (66). It has further been pointed out by Thimann (67) that peroxidase is inhibited by diphenols and polyphenols, that diphenols and polyphenols are also oxidized by peroxidase to form polymerization products and that this reaction is inhibited by IAA. Thus the balance between IAA, peroxidase, and phenols appears to be a factor involved in the lignification process although the nature of the mutual interaction between these substances remains to be elucidated. The investigations of Siegel (48) indicate, as pointed out above, that the nature of the polysaccharide matrix on which the lignin is formed is also a significant factor in the process of lignification.

Another way in which lignification may be influenced is indicated by the studies of Lipetz et al. (68, 69) on the effect of calcium ions on the progress of lignification. It has been shown by Davis (70) that plants growing under low calcium levels tended to be more heavily lignified than when the calcium level was high. By investigating this effect using tissue cultures of crown gall-tumor tissue, Lipetz (68) observed similar effects. In addition however, it was shown that calcium and other ions affected the peroxidase level in the cell wall and that the efficiency of various ions in effecting the release of wall peroxidase was in the order $Ca^{++} > Sr^{++} > Ba^{++} > Mg^{++} > NH_4^+$. On the basis of these observations it was concluded that ions such as Ca^{++} may influence the level of lignification of the cell wall through their control of the peroxidase level within it. On the other hand Koblitz (71) showed that boron stimulates lignification in young roots of maize. It was shown subsequently by Dulta and McIlrath (72) and by McIlrath and Skok (73) that in callus and root cultures of Helianthus (sunflower) the peroxidase level was less in boron deficient cultures than in normal cultures.

At the cytological level it would seem on the basis of the work of Pickett-Heaps (50) that vesicles derived from the Golgi apparatus or endoplasmic reticulum may serve to carry phenylpropane precursors from the cytoplasm to the cell wall.

It will be clear, however, that, although lignification may be influenced by the ionic concentration in the cell wall, there is no indication as to the mechanism by which intra-cellular lignification may be controlled. At this stage it would seem that intense investigation of the nature of the wall-cytoplasm relation in lignifying and non-lignifying cells, complemented by appropriate histochemical investigations are required if further understanding of the mechanism of selective lignification in cells is to be obtained.

VII. The Influence of Lignification on Cell Wall Properties. The progressive deposition of lignin in association with the matrix between the cellulose microfibrils and its close association with the paracrystalline phase surrounding them, results in changes in optical, staining and mechanical properties of the cell wall. Thus lignification is accompanied by increasing ultraviolet fluorescence of the wall and by an increasing UV absorption. The slight UV dichroism of the wall observed by Lange (74) and Ruch (75) is considered to be entirely form dichroism resulting from elongated form of lignin aggregates between the microfibrils (19).

Changes in staining reactions which accompany lignification reflect chemical or purely physical changes resulting from lignin deposition. Thus it was observed by Allen (35) and later by Wardrop (5) that the intensity of staining of the cell wall by ruthenium red decreased as lignification progressed. This suggests that groups stained in the unlignified wall are blocked by the deposition of lignin. This is consistent with the previously mentioned suggestion (34) that lignin is linked to the non-cellulose carbohydrates through uronic acid-ester linkages. This suggestion also agrees with the observation of Allen that staining by ruthenium red was enhanced if the tissue was previously hydrolysed.

It has also been observed that unlignified cell walls stained with chlorine-zinc iodide (76) or congo red (77) show intense dichroism whereas lignified walls are monochroic and less intensely stained. This change in optical properties of the stained walls can be understood, since the "micelles" of the iodine or the congo red can pass into the inter-microfibrillar regions of the unlignified walls but are too large to penetrate these regions in the lignified walls. The different staining reactions of lignified and unlignified walls using metachromatic dyes such as oxamine blue (41) can be explained in similar terms.

Because of its location between the microfibrils the deposition of lignin in the wall can cause a degree of swelling. This has been observed by Frey (76) and by Preston and Middlebrook (78). For the same reason lignified walls tend to shrink less than unlignified ones. It is of interest that Siegel has shown that cell walls in slices of pea roots artificially "lignified" by the system: eugenol-peroxide-peroxidase, showed considerably less shrinkage compared with untreated tissue; were less wettable with water and acid; and did not swell as much as controls when placed in dilute alkali.

So far as is known, lignification has no influence on tensile properties of the wall but is considered to increase resistance of the wall to compressive forces. This has been discussed by Frey-Wyssling (79). Thus if a column e.g. a microfibril, is considered to be loaded axially in compression, then at the initiation of failure it tends to buckle laterally. If lignin is packed between the microfibrils then the tendency to buckle is reduced and the load in compression necessary to cause buckling is increased and so in this way lignin is considered to contribute to the compressive strength of the cell wall.

It is of interest that early workers found that the lignin content of wood was greatest in regions of greatest mechanical (compressive) stress. This is accentuated, for example, in branches of conifers when reaction wood is formed and the lignin content on the lower side of the growth ring is increased. More recently Siegel (57) has shown for a Poinsettia stem that from the apex to the base a correlation exists between the lignin content of the xylem and the increasing pressure exerted by the plant itself (Figure 2.12).

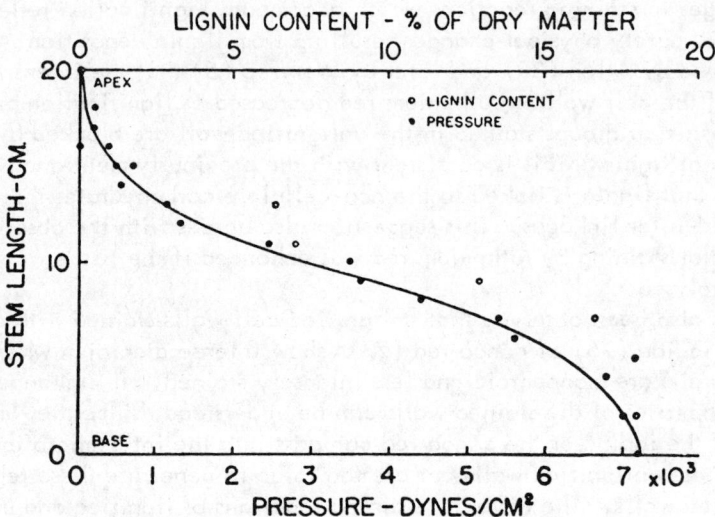

Figure 2.12. The variation of lignin content at different heights in a stem of Poinsettia. Data by S. M. Siegel (57).

VIII. Summary. The foregoing survey indicates that the phenomenon of lignification is associated with the evolution of plants with a developed vascular system and it is reasonable to suppose that it is significant as a factor associated with adaptation to a terrestrial habit. Its location in the cell wall and the manner of its association with other wall constituents and the time of its development during cell ontogeny are consistent with such a view. The observations made to this stage indicate that further insight into the mechanism of the process of lignification may well derive from detailed cytological studies directed especially to the elucidation of the problem of the manner of selective lignification of the cell wall.

REFERENCES

1. F. E. Brauns and D. A. Brauns, The Chemistry of Lignin (Supplement Volume) Academic Press, N. Y., London, 1965.
2. S. M. Manskaya and M. N. Kochneva, Doklady Akad. Nauk S.S.S.R. 6, 505 (1948).
3. G. H. N. Towers and R. D. Gibbs, Nature, 172, 25 (1953).
4. L. M. Srivastava, Tappi, 49, 173 (1966).
5. A. B. Wardrop, Tappi, 40, 225 (1957).

6. H. A. Stafford, Plant Physiol., 37, 643 (1962).
7. D. E. Bland, Holzforschung, 20, 12 (1966).
8. E. S. Barghoorn in The Formation of Wood in Forest Trees, (M. H. Zimmermann, ed.), p. 3, Academic Press, N. Y., 1964.
9. A. C. Neish in The Formation of Wood in Forest Trees, (M. H. Zimmermann, ed.), Academic Press, N. Y., 1964.
10. G. S. Frankel, Science, 129, 1466 (1959).
11. A. B. Beakbane and E. C. Thompson, Nature, 156, 145 (1945).
12. C. M. Stewart, Nature, 186, 374 (1960); Science, 153, 1068 (1966).
13. C. M. Stewart, The Chemistry of Secondary Growth in Trees, C.S.I.R.O. (Australia) Division of Forest Products. Tech. Pap., 43, 1966.
14. H. Reznik, Ergeb. Biol., 23, 14 (1960).
15. K. Mühlethaler in Cellular Ultrastructure of Woody Plants (W.A. Côté, ed.), p. 191, University of Syracuse Press, 1965.
16. A. Frey-Wyssling, Science, 119, 80 (1954).
17. A. B. Wardrop, Nature, 164, 366 (1949).
18. R. D. Preston in The Formation of Wood in Forest Trees (M. H. Zimmermann, ed.) Academic Press, N. Y., 1964.
19. A. Frey-Wyssling in The Formation of Wood in Forest Trees (M. H. Zimmermann, ed.) Academic Press, N. Y., 1964.
20. A. Bjorkman, Svensk Papperstid., 59, 477 (1956).
21. K. Freudenberg in The Formation of Wood in Forest Trees, p.151 (M. H. Zimmermann, ed.) Academic Press, N. Y., 1964.
22. A. B. Wardrop and H. E. Dadswell, Aust. J. Bot., 3, 177 (1955).
23. A. Allsop and P. Misra, Biochem. J., 34, 1078 (1940).
24. H. Meier, J. Polymer Sci., 51, 11 (1961).
25. T. Kerr and I. W. Bailey, J. Arnold Arb., 15, 327 (1934).
26. P. W. Lange, Svensk Papperstid., 53, 749 (1950).
27. P. W. Lange and A. Kjaer, Norsk Skogindustri, 11, 425 (1957).
28. F. Ruch and H. Hentgartner, Beih. Z. Schweiz.Forstver.,30, 75 (1960).
28a. G. P. Berlyn and R. E. Mark, For. Prod. J., 140 (1965).
29. A. B. Wardrop, in Cellular Ultrastructure of Woody Plants (W. A. Côté, ed.) Syracuse University Press, 1965.
30. A. B. Wardrop, H. E. Dadswell and G. W. Davies, Appita, 14, 185 (1961).
31. A. B. Wardrop, in The Formation of Wood in Forest Trees (M. H. Zimmermann, ed.) Academic Press, N. Y., 1964, p.87.
32. T. M. Saleh, L. Leney and K. V. Sarkanen, Holzforschung, 21, 116 (1967).
33. A. B. Wardrop and D. E. Bland, Proc. Intern. Congr. Biochem., Fourth Congr., Vienna, 1958, 2, 92 (Published 1959).
34. C. M. Stewart, J. F. Kottek, H. W. Dadswell and A. J. Watson, Tappi, 44, 798 (1961).

35. C. E. Allen, Bot. Gaz., 32, 1-34 (1901).
36. E. C. Crocker, Ind. Eng. Chem., 13, 625 (1921).
37. K. Freudenberg, H. Reznik, W. Fuchs and M. Reichert, Naturwissenschaften, 42, 29 (1955).
38. F. Tiemann and W. Haarmann, Chem. Ber., 7, 608 (1874).
39. K. Freudenberg and J. M. Harkin, Phytochemistry, 2, 189 (1963).
40. S. M. Siegel, Physiologia Plantarum, 7, 41 (1954).
41. K. Griffoen, Rec. Trav. Bot. Neerl., 35, 322 (1938).
42. T. Higuchi, in The Biochemistry of Wood, (K. Kratzl and G. Billek, ed.) Pergamon Press (1959).
43. S. M. Manskaya, Doklady Akad. Nauk. S.S.S.R. Pok., 62, 369 (1948).
44. K. Freudenberg, in Biochemistry of Wood, (K. Kratzl and G. Billek, ed.) Pergamon Press, London, p. 121, 1963.
45. K. Kratzl, Holz Roh- u. Werkstoff, 19, 219 (1961).
46. R. K. Bamber, Nature, 191, 409 (1961).
47. A. Duchaigne, Ann. Des. Sc. Nat. Bot., 11, 456 (1955).
48. S. M. Siegel, J. Amer. Chem. Soc., 79, 628 (1957).
49. E. I. Barskaya, Fiziologiia Rastennii, 9(2), 210 (1962).
50. J. D. Pickett-Heaps, Protoplasma, 65, 181 (1968).
51. A. B. Wardrop, unpubl. results.
52. J. Kuc and O. E. Nelson, Arch. Biochem. Biophys., 105, 103 (1964).
53. L. R. Jorgensen, J. Amer. Soc. Agron., 23, 549 (1931).
54. E. W. J. Phillips, Nature, 174, 85 (1954).
55. Wray, Proc. Leeds. Phil. Lit. Soc., 2, 560 (1934).
56. K. Kratzl, Experientia, 4, 110 (1948).
57. S. M. Siegel, Quart. Rev. Biol., 31, 1 (1956).
58. A. B. Wardrop, The Reaction Anatomy of Arborescent Angiosperms, in The Formation of Wood in Forest Trees (M. H. Zimmermann, ed.) Academic Press, N. Y., 1964.
59. V. Nečesány, Phyton (Buenos Aires), 11, 117 (1958).
60. H. Koblitz, Faserforsch.Textiltech., 13, 270 (1962).
61. L. Bergmann, Planta, 62, 221 (1964).
62. H. T. Wershing and I. W. Bailey, J. Forestry, 40, 411 (1942).
63. R. W. Kennedy and J. L. Farrar, Nature, 208, 406 (1965).
64. H. Koblitz, Flora, 154, 511 (1964).
65. S. M. Siegel, P. Frost and F. Porto, Plant Physiol., 35, 163 (1960).
66. P. M. Ray, Arch. Biochem. Biophys., 64, 175 (1960).
67. K. V. Thimann, in the Formation of Wood in Forest Trees (M. H. Zimmermann, ed.) p. 452, Academic Press, N. Y., 1964.
68. J. Lipetz, Amer. J. Bot., 49, 460 (1962).
69. J. Lipetz and A. J. Garrow, J. Cell. Biol., 25, 109 (1965).
70. D. E. Davis, Amer. J. Bot., 36, 276 (1949).
71. H. Koblitz, Z. Botan., 43, 45 (1955).

72. T. R. Dulta and W. J. McIlrath, Botan. Gaz., 125, 89 (1964).
73. W. J. McIlrath and J. Skok, Botan. Gaz., 125, 268 (1964).
74. P. W. Lange, Svensk Papperstid., 57, 235 (1954).
75. F. Ruch, Exper. Cell. Research, 11, 680 (1951).
76. A. Frey, Jahrb. Wiss. Bot., 65, 210 (1926).
77. O. Walchli, Schweiz.Arch. Angew. Wiss. Tech., 11, 129 (1945).
78. R. D. Preston and M. J. Middlebrook, Text. Inst., 40, 1715 (1949).
79. A. Frey-Wyssling, Die Pflanzliche Zellwand, Springer-Verlag, Berlin, 1959.

3

CLASSIFICATION AND DISTRIBUTION
K. V. Sarkanen and H. L. Hergert

1. Major Classes of Lignins. The classification of lignins has so far not received extensive attention. Up to the recent times, most investigators have been satisfied with the broad division of better known lignins to (a) gymnosperm or softwood lignins, (b) angiosperm wood or hardwood lignins and (c) grass lignins. These three types of lignins are best differentiated on the basis of nitrobenzene oxidation products. While gymnosperm lignins yield mainly vanillin with some p-hydroxybenzaldehyde (1), both syringaldehyde and vanillin are obtained from angiosperm wood lignins (2) and significant amounts of all three aldehydes are obtained from grass lignins (3).

This classification is unsatisfactory in leaving out lignins of herbaceous angiosperms as well as those of Pteridophyta. Furthermore, some exceptional gymnosperms contain lignin of the angiosperm type (4). Gibbs (4) has consequently introduced the division of lignins in the plant kingdom into two major classes only, namely "guaiacyl lignins" and "guaiacyl-syringyl lignins." The former lignins include those present in the majority of gymnosperms, while all angiosperm lignins, including grass lignins, belong to "guaiacyl-syringyl lignins." Guaiacyl-syringyl lignins may be recognized by their positive Mäule reaction and always yield significant amounts of syringaldehyde in nitrobenzene oxidation. In contrast, guaiacyl lignins invariably display a negative Mäule reaction and only traces of syringaldehyde are formed in nitrobenzene oxidation.

The division of plant lignins into two major classes is a logical and convenient one, and has also been accepted by Higuchi and Kawamura (5). In the classification set up by these authors, guaiacyl lignins are called "Type N lignins" and guaiacyl-syringyl lignins, "Type L lignins." For general usage, the nomenclature of Gibbs is preferable for its simplicity. Higuchi and Kawamura also attempted to divide both classes to subgroups using as criteria the IR spectra of isolated milled wood lignins and the composition of nitrobenzene oxidation products. Their proposed classification is as follows:

(A) "Type N lignins" (Guaiacyl lignins).

1, "Standard" guaiacyl lignins (Type Ns), present in the

majority of gymnosperm woods.
2, Lignins in Pteridophyta (Type Np), including representatives of ferns, tree-ferns and clubmosses.
3, Lignins in Cycadales (Type Nc).
(B) "Type L lignins" (Guaiacyl-syringyl lignins).
1, "Standard" guaiacyl-syringyl lignins (Type Ls) present in hardwood species of the temperate zone and in herbaceous angiosperms, with the exception of monocotyledons.
2, Lignins in coniferous Gnetales (Type Lg).
3, Tropical hardwood lignins (Type Lt).
4, Grass lignins (Type Lm).
5, Lignins in monocotyledons other than grasses (Type Lm').

The classification of Higuchi and Kawamura is a purely tentative one. Infrared spectroscopy and nitrobenzene oxidation reveal only minor differences between the three subclasses of guaiacyl lignins, and the distinction between Ls and Lm' lignins is likewise not very clear. Furthermore, the nature of lignin even within the same plant may vary, resulting sometimes in the coexistence of guaiacyl lignins and guaiacyl-syringyl lignins in different tissues. Any classification of lignins on a taxonomic basis must therefore remain an approximate one only.

II. Guaiacyl Lignins
   A. Lignins in Conifer Woods.
      1. Gross Distribution. The lignin contents of conifer woods vary in the general range from 24 to 33% (6). For some species, the reported lignin contents are quite constant. For Norway spruce (Picea excelsa), for example, the determined values differ little from an average of 28.6% (6). Douglas-fir (Pseudotsuga Menziesii), on the other hand, offers a good example of substantial variation in lignin contents. For example, Erickson and Arima (7) found the average lignin content to be at the 24.5% level, in good conformity with the values by Kennedy and Javorsky (8) for the same species. However, most of the recorded lignin contents for Douglas-fir lie between 27 and 29% (9) and occasionally very high values, such as 33.5% (10) and 34.5% (11) have been obtained. Although a part of the variation in recorded lignin contents may be caused by such factors as the presence of compression wood in the samples, or differences in the lignin determination methods, in the main the variance in the lignin content, and probably in the overall chemical composition, within certain conifer species appears to be real. As a consequence, comparisons between species can only be possible on the basis of extensive statistical studies, generally unavailable at the present time. For a number of species, such as spruces and firs (Picea, Abies), pines (Pinus) and larches (Larix) the reported lignin contents are generally in the range 24 to 29% (6). Tentatively, at least, these species could be classified as "low lignin conifers." Persistently high lignin contents (31 to

CLASSIFICATION AND DISTRIBUTION

33%) have been recorded for hemlocks (Tsuga), cedars (Thuja, Chamaecyparis, Juniperus), redwoods (Sequoia) and Cryptomeria, among others (6).

The overall compositions of six North-American conifers have been determined by Timell (12) and are shown in Table 3.1. Since the cellulose contents of conifers as well as angiosperm woods are generally 42±2%, low lignin contents are usually associated with high hemicellulose and vice versa. No correlation appears to exist between the lignin content and the composition of hemicellulose.

Table 3.1. Compositions of woods from five conifers (12)

Species	Percent of extracted wood		
	Klason Lignin %	Cellulose %	Hemicellulose %
White spruce Picea glauca	27.1	44	29
Jack pine Pinus banksiana	28.6	41	30
Tamarack Larix laricina	28.6	45	28
Balsam fir Abies balsamea	29.4	44	27
Eastern white cedar Thuja occidentalis	30.7	44	25
Eastern hemlock Tsuga canadensis	32.5	42	26

The varying lignin contents of wood samples deriving from the same species are at least in part due to the somewhat uneven distribution of lignin in the woody xylem of a single tree. Erickson and Arima (7) have determined the radial variation in the lignin content of Douglas-fir, using eleven 30-year old trees for the study. Some of the average lignin contents, all obtained on wood samples collected from the breast-height level of the trunks, are presented in a graphical form in Figure 3.1.

It can be seen that the lignin content gradually decreases during the juvenile growth period and approaches a constant level in mature wood, in good conformity with the increasing trend in the values for alpha cellulose. However, the total increment in alpha cellulose is more than twice as large as the decrease in the lignin content. Both of these changes are concurrent with an increase in late wood percentage with age. This basic pattern can, however, be modified by a change in growth conditions. When fertilization and thinning was applied to a group of 19-year old Douglas-fir trees, the resulting

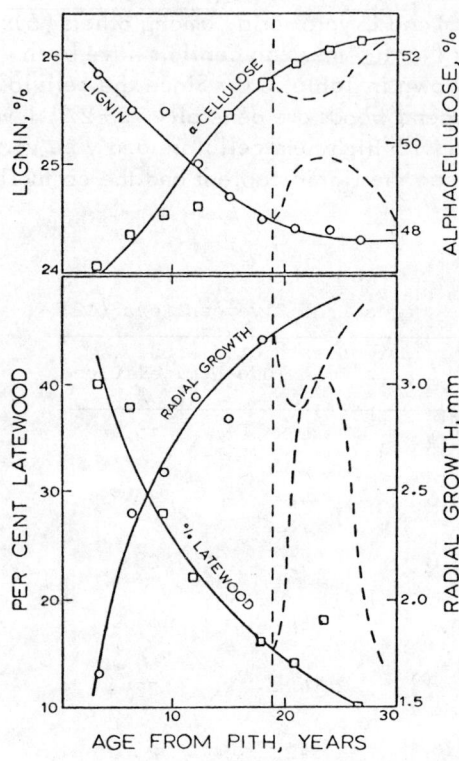

Figure 3.1. Variation in Douglas-fir wood (<u>Pseudotsuga Menziesii</u>) as a function of age (7). Percent lignin and alpha cellulose, radial growth and percent late wood as functions of age. Dotted curves are for trees which received fertilization treatment during the 19th growth year.

acceleration in growth was accompanied by an increase in lignin and decrease in alpha cellulose content (dotted lines, Figure 3.1).

Although a negative gradient of lignin content in the radial direction has also been established for <u>Cedrus deodora</u> and <u>Cupressus torulosa</u> (13), it can not be generalized to all conifers. The opposite trend, increasing lignin content from the pith outwards, has been claimed for Japanese red pine (<u>Pinus densiflora</u>) by Hata (14), and is also apparent in the data by Anderson and Pigman for loblolly pine (15). In the latter case, both early wood and late wood were found to have lower lignin contents in heartwood than in sapwood. In view of the fact that the alpha cellulose content has been found to increase radially for many pine species, including <u>Pinus radiata</u> (16), and <u>P. taeda</u> (17), the reported trend in the lignin content of the two pine species appears unexpected. In the vertical direction upwards, an increase in lignin content was reported for <u>Pinus densiflora</u> (14), while the opposite trend was found for <u>Cedrus deodora</u> and <u>Cupressus torulosa</u> (13).

The highest lignin contents are found in the so-called "compression" or "reaction wood," distinguishable from "normal wood" in morphological, physical and chemical characteristics (18). Straight, vertical stems are generally free of compression wood, with the exception of narrow areas around knots. In leaning stems, the lower parts of the trunk consist of compression wood, and its amount is the larger the more the stem deviates from the vertical direction. The wood opposite to the compression wood zone ("opposite wood") as well as the so-called "side-wood" appear to possess essentially normal wood characteristics. Likewise, a compression wood zone invariably exists in the lower sides of branches. To the naked eye, compression wood usually appears darker and can be recognized easily by inspecting thin discs in transmitted light. The transition from normal to compression wood zones is usually abrupt, although different degrees of compression wood nature have been observed. In "mild compression wood," the early wood and late wood zones can be differentiated, while in pronounced compression wood these two zones are almost uniform.

The lignin content of pronounced compression wood is 34 to 41% higher than in normal wood and significant differences are apparent in the polysaccharide composition, as shown by Timell et al. (19, 20) (Table 3.2). The methoxyl content of compression wood lignin is lower than that of normal wood lignin (21).

Table 3.2. Overall Composition of Normal and Compression Woods in Balsam Fir (Abies balsamea) (19) and Tamarack (Larix Laricina) (20), on Extractive-free Basis

Component	Balsam fir wood		Tamarack wood	
	Normal	Compression	Normal	Compression
Lignin	29.9	40.1	27	38
Uronic anhydride	5.5	3.8	2.8	3.1
Residues of				
Galactose	1.4	7.6	3.6	17.7
Glucose	44.6	31.6	68.3	59.0
Mannose	10.0	6.3	18.2	9.2
Arabinose	1.6	1.8	2.1	2.1
Xylose	5.4	7.5	7.8	11.0

The lignin content of 1 to 3-year old trees has been found to be lower than in more mature trees and these "juvenile" lignins contain less than normal amounts of methoxyl groups, as shown in Table 3.3. (22, 23).

Table 3.3. Lignin in Immature Conifer Xylem

Species	Age of Plant Years	Lignin Content %	MeO in Lignin %
Pinus thunbergi (22)	1	24.4[a]	12.5
	2	27.3[a]	13.4
	3	27.6[a]	14.3
	Mature wood	28.3[a]	15.0
Cryptomeria (22)	1	32.7[a]	14.8
	Mature wood	34.8[a]	14.7
Sequoia sempervirens (23)	120 days	18.6[b]	10.0[c]
	Mature wood	31.6[b]	13.5[d]

[a] Klason lignin
[b] Yields of lignin thioglycolic acids
[c] MeO in purified dioxane acidolysis lignin
[d] MeO in purified alkali lignin

2. Variation of Lignin Content within Growth Rings. Wood in horizontal cross-sections of conifer trunks consists of macroscopically visible, concentric annual rings, each divided into the so-called early wood and late wood zones. The cells in the early wood zones are thin-walled and often hexagonal in shape, whereas the late wood cells are rectangular, with thick cell walls and short radial diameter. The transition from early to late wood is usually abrupt.

Ritter and Fleck (24) were the first to find the highest lignin content for early wood in three conifer species, and conversely, the highest alpha cellulose content was found in the late wood. This relationship was confirmed for a number of other conifer species by other investigators (14, 25-32).

The most detailed studies on lignin and alpha cellulose variation across growth rings have been carried out by Wilson and coworkers (30-32). The growth rings were separated into six zones, and the lignin contents determined for each zone individually. The change in lignin content was found to vary across the growth ring along a sigmoidal curve, with a maximum in the early wood, and a minimum usually, but not always, in the late wood section. The change in lignin content was found to be much more gradual than the rather abrupt morphological transition from early to late wood which is commonly accompanied with a steep increase in specific gravity from 0.1 - 0.2 to 0.5 - 0.6 (33). While the lignin distribution did not suggest a good correlation with cell morphology, it showed a reasonable inverse relationship with the alpha cellulose content, the sum of both components remaining at the approximately constant level of 72 to 75% across the growth ring.

CLASSIFICATION AND DISTRIBUTION

The maximum and minimum lignin contents in successive growth rings are quoted in Table 3.4, and it can be noted that the average differences between these values range from 1.0% for hemlock to 3.5% for Douglas-fir. The difference between maximum and minimum values might be species-dependent, because Anderson and Pigman, in a less detailed study (15), found a difference of 3.4% for the lignin contents of early and late wood of Douglas-fir, while the same difference for two pine species was only 1.0 to 2.6%.

Table 3.4. Maximum and Minimum Lignin Contents for Three Successive Annual Increments (31, 32)

Species	Years	Klason lignin variation in successive rings						Average difference %
		1		2		3		
		Max	Min	Max	Min	Max	Min	
Western hemlock *Tsuga heterophylla*	56-58	33.1	32.6	34.0	32.4	33.3	32.5	1.0
Western red cedar *Thuja plicata*	72-74	32.1	29.9	32.6	29.8	32.5	31.3	2.1
Pacific silver fir *Abies amabilis*	78-80	29.0	26.2	29.0	26.3	29.2	26.3	2.8
Douglas fir *Pseudotsuga menziesii*	64-66 40	27.7 30.8	25.3 25.9	28.8	25.4	28.6	24.0	3.5
Sitka spruce *Picea sitchensis*	69-71	26.6	25.0	26.6	24.9	27.0	25.0	1.6
Black spruce *Picea mariana*	34	30.4	28.4					

However, in exceptional cases the lignin content of late wood can be equal to or even higher than that of early wood. Irregularities of this nature have been reported for redwood (Sequoia sempervirens) (25) and Norway spruce (Picea excelsa) (28). In the detailed study by Jansons (28) altogether 124 comparisons were made on samples taken from four Norway spruce stems. Of these, only one stem showed "irregular" properties and the lignin content of late wood was found occasionally to contain up to 3.1% more lignin than early wood. The lignin content in late wood varied as much as from 22.3 to 34%, while that in early wood was more constant, 25.2 to 29.5%.

3. **Distribution of Lignin in the Cell Wall of Conifer Tracheids and Parenchyma.** Electron micrographs of thin sections of wood from which the polysaccharidic components have been removed by means of 80% hydrofluoric acid or periodate (34) or by fungal degradation (35) reveal the "lignin

skeleton" of individual cells. An example of such lignin micrographs is given in Figure 3.4. The distribution of lignin in individual cells can also be accentuated in electron micrographs without removing the polysaccharides, by treating the sections with either p-(acetoxymercuri)aniline (34) or potassium permanganate (see Figure 2.3 in Chapter 2). Outstanding as these methods are in revealing minute details of cell wall structure, they do not lend themselves for the quantitative estimation of lignin distribution across the cell wall. For the latter purpose, methods based on UV microscopy of thin wood sections are of best service. This approach was initiated by Lange (36) who estimated the middle lamellae of Norway spruce tracheids to contain approximately 73% lignin. This value was in outstanding accordance with the 72% lignin content obtained earlier by Bailey (26) for segregated middle lamellae of Douglas-fir by direct analytical means. Lange's method has been applied by Wardrop and Bland (37) to the characterization of lignin deposition in the cambial layer tracheids of Pinus radiata (See Figure 2.5 in Chapter 2). Goring and coworkers (38-40) have further refined the UV-microscopy method by reducing the section thickness to 0.5µ and thus avoiding errors caused by non-parallel illumination. A UV-micrograph obtained by Goring and Fergus (40) for early wood cells of black spruce (Picea mariana) is shown in Figure 3.2. The uniformity of lignin concentration in the secondary wall as well as the sudden increase in lignin at the boundary of the middle lamella-primary wall region are discernible in the micrograph. Goring and

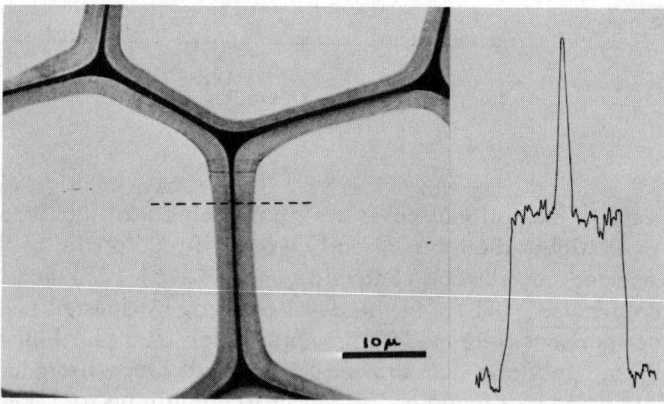

Figure 3.2. Cross section of Epon-embedded tracheids of black spruce early wood photographed in ultraviolet light of wavelength 240 mµ. The densitometer tracing was taken across the tracheid wall on the dotted line (40).

CLASSIFICATION AND DISTRIBUTION

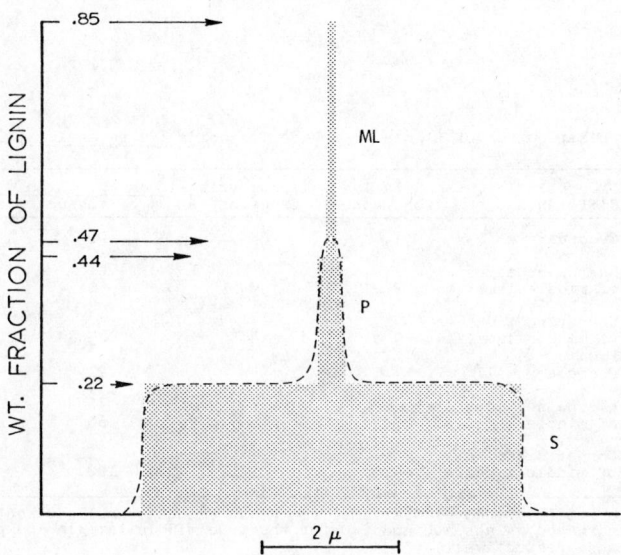

Figure 3.3. Corrected distribution of lignin in early wood tracheids of black spruce (40).

Fergus (40) used the improved method to determine the distribution of lignin in the early wood and latewood of black spruce. Higher estimates were obtained for the corner sections of middle lamellae than for the thin radial and tangential parts. This difference is, in all likelihood, only apparent and due to the extreme thinness ($\sim 0.1\mu$) of the true middle lamella outside of the corner section. Using oblique sectioning, Goring and Fergus succeeded in estimating separately the lignin concentrations in the true middle lamella and the primary wall areas. The corrected lignin distribution for early wood cells is shown in Figure 3.3.

The results by Goring and Fergus are summarized in Table 3.5. It may be noted that the lignin content of the middle lamella-primary wall regions exceeds earlier estimates, especially in late wood, where these regions appear to contain little if any material other than lignin. The lignin content of secondary wall (22%) is slightly higher than the earlier estimate of Lange (16%) for Norway spruce wood (36), and appears to be the same in early and late wood cells. The volume fraction of the middle lamella-primary wall is quite small, 13% for early and 6% for late wood. Therefore, middle lamella lignin accounts only for 28 and 19% of the lignin in early and late wood, respectively, and the frequently quoted percentage for middle lamella lignin,

Table 3.5. Distribution of Lignin in Spruce Tracheids (Picea mariana) (39)

Wood	Morphological differentiation	Relative absorbance	Tissue volume fraction, %	Lignin % of tissue	Lignin % of total
Early wood	Secondary wall	1.0	87	22	72
	Narrow sections of middle lamella	2.2[1]	9	50	16 ⎫ 28
	Corner sections of middle lamella	3.8	4	85	12 ⎭
Late wood	Secondary wall	1.0	94	22	81
	Narrow sections of middle lamella	2.7[1]	4	60	10 ⎫ 19
	Corner sections of middle lamella	4.5	2	100	9 ⎭

1) Measured for the compound middle lamella. Measurements on oblique sections suggest absorbance values 3.8 and 2.0 for the true middle lamella and primary wall sections, respectively.

72% of total lignin, is totally erroneous, as pointed out by Berlyn and Mark (41).

Côté, Piccard and Timell demonstrated that the distribution of lignin in compression wood cells presents a totally different picture. As shown in Figure 3.4   empty intercellular spaces are often present in the corner sections of the middle lamellae, and the percentage of middle lamella lignin must therefore be less than in normal wood. Three distinct zones of lignification are present in the secondary wall. The outermost, extraordinarily thick S_1 to S_2 is very heavily lignified, whereas the remainder of the secondary wall has a lignin content intermediate between S_1 and the transition layer.

The ray parenchyma cells, which have been estimated to contain 4.0 to 4.7% of the tissue material in black spruce wood (40), are heavily lignified as shown in Figure 3.5. No differences in lignification have been observed in ray cells associated with early and late wood. Bailey (26) determined analytically the lignin content of 41% for segregated rays of Douglas-fir, and Harada and Wardrop estimated the percentage of lignin in the ray parenchyma of Cryptomeria japonica to be 44% (42). In most conifer species, rays lignify concurrently with the normal tracheid cells, but differ from the latter in retaining their cytoplasm and living functions after completed lignification and die finally at the sapwood-heartwood boundary (43). Interestingly, certain hard pines, Pinus banksiana for example, form an exception from this

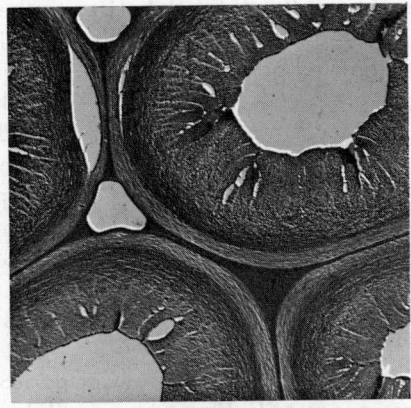

Figure 3.4. Transverse section of balsam fir compression wood after removal of polysaccharides with hydrofluoric acid. The outer part of S_2 is most heavily lignified, the S_1 is least lignified. Electron micrograph by Côté, Piccard and Timell (Tappi 50, 352 (1967)).

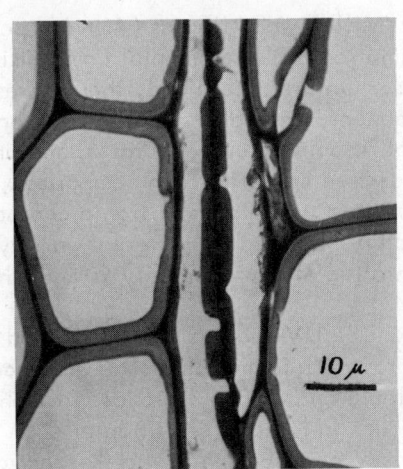

Figure 3.5. Cross section of an Epon-embedded ray parenchyma cell. Note the high UV absorption of the ray cell wall ($\lambda=240m\mu$) (40).

rule, as shown by Balatinecz and Kennedy (43). In these species, most ray parenchyma cells remain unlignified in sapwood, and secondary wall thickening and lignification occur at the sapwood-heartwood boundary.

In the primary xylem tissue, present near growing tips and the piths of mature conifer trees, lignification is limited to secondary walls and the intercellular space remains lignin-free, as shown by Wardrop (44).

It appears that various environmental influences can either prematurely terminate the lignification of single cells or, conversely, extend the process beyond its normal level. For example, Wardrop (44) studied the growth of a ring-barked Pinus radiata, in which the phloem flow of metabolites to the tissues below the debarked zone is totally disrupted. He found that the tissue formed in this zone contained totally unlignified parenchyma cells. Tracheids, when formed, displayed widely varying degrees of lignification. Kennedy and Farrar (45) found that when Jack pine seedlings were suddenly tilted from vertical position, normal tracheids in the differentiating xylem on the lower side of the stem became heavily lignified. Conversely, round compression wood tracheids with relatively low lignin contents were produced when tilted seedlings were brought to vertical position. Sometimes single cells differ radically in their lignification pattern from the surrounding cells. Balatinecz and Kennedy (43) located heavily lignified parenchyma cells in the midst of totally unlignified cells in the rays of hard pines.

4. Compositional Variation in Normal Conifer Lignins. Comparisons between lignins from different sources are necessarily limited by available methodology which presently may not be sufficiently developed to reveal all significant differences. The following characteristics of lignins have been generally used in comparative studies:

i. Analytical Composition. Milled wood lignins (46) are perhaps the most useful preparations for the estimation of the elemental composition and functional group contents of lignins. On the other hand, they represent only 30 to 50% of the total lignin component, generally contain small and somewhat variable amounts (2 to 4%) of associated carbohydrates, and are not always readily dried to completely water-free condition. In comparison, lignin thioglycolic acids (47) possess certain advantages in lignin characterization. They are readily and reproducibly prepared from a great variety of plant materials and usually are representative of approximately 90% of the total lignin component.

ii. Spectral Characteristics. The UV-spectra of milled wood lignins and lignin thioglycolic acids can provide information on the presence of syringylpropane units, conjugated ethylenic and carbonyl groups, free phenolic hydroxyl groups and certain ester groups in lignins (see Chapter 6). Syringylpropane units make a characteristic contribution to the IR-spectra of these preparations and information is also obtained on the nature of carbonyl and ester groups present (see Chapter 7).

iii. Nitrobenzene Oxidation and Ethanolysis Products. The total yield of aromatic aldehydes in the nitrobenzene oxidation of extracted wood serves as an index of the presence of "uncondensed" units in lignin. The molar ratios syringaldehyde/vanillin (S/V) and p-hydroxybenzaldehyde/vanillin (H/V) in the aldehyde mixtures may be used for the comparison of the three basic units in lignins. Likewise, the total yield of Hibbert's ketones is dependent on the amount of β-O-4 lignol structures in lignin. The molar ratios syringoyl acetyl/vanilloyl acetyl (SA/VA) and p-hydroxybenzoyl

Table 3.6. Compositions of Milled Wood Lignins from Conifer Woods

Wood species	Ref.	C %	H %	O %	OMe %	Groups/ C₉-unit OMe	Phenol-OH
Normal wood lignins:							
Norway spruce, Picea excelsa	(46)	63.84	6.04	29.68	15.75	0.95	0.29
Black spruce, Picea mariana	(46)	63.66	6.29	29.39	15.41	0.93	0.28
Scots pine, Pinus sylvestris	(46)	63.96	6.14	29.82	15.74	0.95	0.27
Western hemlock, Tsuga heterophylla	(46)	63.35	6.30	29.76	15.70	0.96	0.29
	(49)	62.97	5.61	31.42	14.29	0.87	-
Western red cedar, Thuja plicata	(49)	63.82	6.11	30.09	16.12	0.98	-
		61.78	5.79	32.43	15.73	0.98	-
Western larch, Larix occidentalis	(49)	61.64	5.84	32.52	16.16	1.03	-
Douglas fir, Pseudotsuga menziesii	(49)	63.37	6.07	30.56	15.07	0.91	-
		60.08	5.75	29.24	13.67[a]	0.90	-
Himalayan pine, Griffithi excelsa (1 yr. old stem)	(49)	63.13	6.03	30.84	14.37	0.89	-
Ponderosa pine, Pinus ponderosa	(49)	62.45	6.03	31.52	15.05	0.93	-
					Average	0.94	
Compression wood lignins:							
Douglas fir, Pseudotsuga menziesii	(49)	58.38	5.76	29.13	10.93[b]	0.71	-

a) 96.3% Klason lignin
b) 94.2% Klason lignin

acetyl/vanilloyl acetyl (HA/VA) are indicative of the nature of units participating in β-O-4 lignol structures.

Inspecting first the analytical values for milled wood lignins of various conifers, shown in Table 3.6, it should be considered that the percentages of hydrogen and oxygen are affected by moisture content and carbohydrate impurities, and the variation in these figures is not necessarily due to intrinsic differences. The average number of methoxyl groups per phenylpropanoid unit can be computed from carbon and methoxyl contents, and varies from 0.87 to 1.0, with an average of 0.94. The variation appears to be real, but not necessarily species-dependent. The amount of units with free phenolic groups is remarkably constant.

Table 3.7 lists the analytical compositions determined by Holmberg (47) for alcohol-insoluble lignin thioglycolic acids of various conifers. Only a small variation can be noted in sulfur contents, suggesting an approximate constancy of functional groups reactive towards thioglycolic acid. The average number of methoxyl groups in phenylpropane units may be computed by correcting for carbon introduced in the reaction with thioglycolic acid. When this is done, essentially the same range, 0.87 to 1.01, is obtained as was derived from the compositions of milled wood lignins. White fir lignin thioglycolic acid with the exceptionally low value of 0.80 forms the single exception. It is not impossible, however, that the low methoxyl content is caused by the presence of compression wood in the original sample, and this assumption is actually supported by the unusually high yield of lignin thioglycolic acid in this case.

Table 3.7. Analytical Compositions of Thioglycolic Acid Derivatives of Guaiacyl Lignins (47)

Species	Part of plant	Yield %	Analytical compositions				
			C %	H %	MeO %	S %	MeO/C$_9$
Conifers:							
Norway spruce Picea excelsa	Trunkwood	30	55.4	5.2	11.3	10.1	.90
Scots pine Pinus sylvestris	Heartwood	28	55.5	5.2	12.5	9.5	1.01
	Sapwood	29	55.4	5.2	11.8	10.2	.94
Eastern hemlock Tsuga canadensis	Trunkwood	33	54.8	5.4	11.4	10.1	.93
White fir Abies concolor	Trunkwood	34.5	55.3	5.3	10.1	10.1	.80
Larch Larix decidua	Trunkwood	30.5	54.4	5.4	11.6	9.8	.95
Juniper Juniperus communis	Trunkwood	33	54.6	5.4	11.0	9.3	.88
Gingko Gingko biloba	Branch wood	35	55.5	5.0	10.9	9.9	.87
						Average	.93
Cycads:							
Cycas revoluta	Leaf stems	18	54.7	5.4	10.7	9.3	.85
	Vessel bundles	16.5	54.4	5.2	10.3	8.0	.81
	Pith	2	41.3	4.9	-	.5	-
Cryptogams:							
Club moss Lycopodium annoticum	Stems	29	55.5	5.3	11.9	9.7	.95
Fern Pteris aquilina	Leaf stems	22	54.8	5.2	10.0	10.0	.80
	Rhizomes	4.6	53.7	5.2	3.2	6.8	.24
Fern Dryopteris filixmas	Leaf stems	15	53.9	4.7	7.5	7.8	.56
	Rhizomes	18	54.1	5.2	7.3	7.8	.57
	Pith	6.4	41.8	5.1	.1	.9	.09

The average number of methoxyl groups in phenylpropane units reflects the relative frequencies of the three basic units in the conifer lignin molecule. Since syringaldehyde is either totally absent among the nitrobenzene oxidation products of conifer woods or only present in minute quantitities (1), and the isolated Hibbert's ketones are exclusively of the guaiacyl and p-hydroxyphenyl types (48), the amount of syringylpropane units, in all likelihood, is small enough in conifer lignins to be justifiably ignored for practical purposes. Consequently, it is tempting to ascribe the variance in the MeO/C9-values to the presence of varying amounts of units of the p-hydroxyphenyl propane type, ranging perhaps from zero to 13 mole % in normal conifer lignins.

There exists a general uniformity in the spectra of conifer lignins. The ultraviolet spectra of milled wood lignins have been found to be essentially superimposable (46, 50) ruling out major differences in the nature and amount

Table 3.8. Nitrobenzene Oxidation Products of Conifers, according to Creighton, Gibbs and Hibbert (2)

Species	Klason lignin %	Total aldehydes, % of K.L.	Vanillin % of K.L.
Picea glauca (White spruce)	28.6	23.7	23.5
Tsuga canadensis (Hemlock)	31.1	23.2	22.1
Pinus strobus (White pine)	34.9	20.1	18.5
Thuja plicata (Western red cedar)	33.9	24.6	24.0
Sequoia sempervirens (Redwood)	31.8	24.8	23.5
Taxus canadensis (Ground hemlock)	32.5	22.0	20.7
Agathis australis	30.8	19.9	18.7
Podocarpus acutifolium	34.9	20.2	19.8
P. macrophyllus v. Maki	39.9 39.0	17.4 17.0	15.1 14.9
Callitris rhomboides	34.8	19.7	18.8

of aromatic moieties, conjugated ethylenic and carbonyl groups and other more extended chromophores, such as coniferyl aldehyde groups. Likewise, the infrared spectra are essentially identical in the region from 1700 to 800 cm^{-1} (49).

Nitrobenzene oxidation was first systematically applied to the characterization of lignins by Creighton, Gibbs and Hibbert (2) and their results on guaiacyl lignins are shown in Table 3.8. A similar study was later conducted by Leopold and Malmström (1) (Table 3.9). The yields of total aldehydes, consisting of over 90% of vanillin varied from 17 to 25% of Klason lignin in the former study, while somewhat larger values (21.4 to 28%) were obtained in the latter investigation, probably due to an improved method of isolation. Judging from the reported lignin contents that occasionally are unusually high, it is unlikely that any effort was made in either study to exclude compression wood from the samples used.

Although the yields of aldehydes were quite similar for the majority of samples, especially in the latter study (26.0 to 28.0%), significant structural differences in the lignin component may have caused the low yields obtained from Podocarpus macrophyllus v. Maki, Cephalodexus drupacea and Taxus baccata.

Syringaldehyde and p-hydroxybenzaldehyde represent only minor components of the aldehyde mixture (Table 3.9). Only in one case, Podocarpus

Table 3.9. Nitrobenzene Oxidation Products of Conifers, according to Leopold and Malmström (1)

Species	Klason lignin %	Total aldehydes % of K.L.	Total aldehydes % of MeO	p-Hydroxy benzaldehyde % of total aldehydes[a]	Syringaldehyde % of total aldehydes[a]
1. Araucaria araucana	27.6	27.2	31.1	*	-
2. Agathis alba	-	-	32.5	**	*
3. Dacrydium cupressinum	-	-	31.6	(*)	**
4. Podocarpus spicatus	-	-	31.1	**	-
5. P. totara	35.6	27.1	31.0	*	3.0
6. Phyllocladus romboidalis	32.3	26.6	31.1	**	*
7. Pseudotsuga taxifolia	26.0	26.9	31.9	**	**[b]
8. Tsuga heterophylla	-	-	31.2	(*)	-
9. Tsuga canadensis	31.2	26.6	32.0	3.5	**
10. Picea abies	27.5	28.0	31.6	***	*[b,c]
11. Larix sibirica	-	-	32.4	**	**
12. Cedrus deodara	31.1	26.4	30.5	***	**
13. Pinus sylvestris	26.9	27.3	31.7	***	*[b]
14. Sciadopitys verticillata	-	-	31.3	2.5	*
15. Sequoiadendron giganteum	35.4	27.3	32.3	3.5	(*)
16. Taxodium distichum	37.0	26.0	31.6	2.0	(*)
17. Cryptomeria japonica	-	-	32.6	2.5	*
18. Athrotaxis selaginoides	-	-	32.2	2.5	**
19. Widdringtonia juniperoides	-	-	31.0	(*)	-
20. Thuja plicata	33.2	27.4	30.3	*	(*)
21. Libocedrus chilensis	31.2	26.2	30.0	3.5	(*)
22. Cupressus macnabiana	-	-	32.0	3.0	(*)
23. Chemaecyparis nootkatensis	-	-	32.4	*	*
24. Juniperus communis	30.0	27.4	31.4	2.5	-
25. Cephalotaxus drupacea	36.1	21.4	25.0	*	-
26. Taxus baccata	32.5	23.6	27.4	3.5	-

a) - No spot on chromatogram, (*) identification uncertain, * 0.5% or less, ** 0.5-1%, *** 1.0-2.0%
b) Syringic acid detected
c) p-Hydroxybenzoic acid tentatively identified

totara, does the yield of isolated syringaldehyde indicate the presence of significant amounts of syringylpropane units in lignin. More importance is to be attached to the amounts of isolated p-hydroxybenzaldehyde which varies from 0.5 to 3.5%. p-Hydroxyphenylpropane units probably exist in lignin largely in condensed form and are therefore not extensively converted to p-hydroxybenzaldehyde. Consequently, the percentage of these units in lignin is likely to be much larger than the percentage of p-hydroxybenzaldehyde in the low-molecular nitrobenzene oxidation products. p-Hydroxybenzaldehyde yield has been found to be higher from lignin in 1-year old seedlings than mature wood lignin in Norway spruce (50).

There is an element of uncertainty involved in the interpretation of nitrobenzene oxidation results, due to the unknown effects of aging upon the yield of aldehydes. Results obtained by Kawamura and Higuchi (22) on Pinus thunbergi are quoted in Table 3.10 and suggest that the total aldehyde yield may be largest from lignin in newly formed wood, becoming progressively less in older growth rings within the tree, possibly as a consequence of condensation processes. On the other hand, it would be surprising that differences of this magnitude had escaped the attention of other workers, and more detailed studies are obviously required to clarify this point of potential importance.

Table 3.10. Effect of Age on the Yield of Aldehydes from Pinus thunbergii Wood (22)

Sample	Klason lignin % of wood	MeO, %	Total aldehyde in nitrob. oxidation % of lignin
1-year old growth ring	24.5	14.7	77.5
2-year old growth ring	25.4	14.9	31.5
Central part	28.3	15.0	12.0

All available evidence points out that ester groups are associated with conifer lignins in trace quantities only. Migita and coworkers (51) identified small amounts of vanillic and, occasionally, p-hydroxybenzoic acid among the alkaline saponification products of extracted woods from a number of conifer species, and trace quantities of the same two acids resulted from the alkaline hydrolysis of Brauns lignin from Douglas-fir (52).

5. Compression Wood Lignins. The exceptionally low methoxyl content of compression wood lignins of spruce (Picea excelsa) was first established by Hägglund and Ljungren (53) and confirmed by Bland (54) for the

Table 3.11. Properties of Normal and Compression Wood Lignins of Douglas-fir (Pseudotsuga menziesii) (55)

	Normal wood	Compression wood
Klason lignin, %	27.0	36.5 (Stem)
		36.3 (Branch)
MWL-yield, % of Klason lignin	58.0	55.3
C_9-formulae for MWL (corrected for carboh.)	C-9 H-8.7 O-2.72 OMe-0.90	C-9 H-9.4 O-2.96 OMe-0.71
Aldehydes from nitrobenzene oxidation (% of KL):		
Vanillin (Wood)	25.2	20.6
p-Hydroxybenzaldehyde (Wood)	None	Trace
Vanillin (MWL)	36.8	37.6
p-Hydroxybenzaldehyde (MWL)	None	14.3
H/V	—	0.47
(H)COCOMe/(G)COCOMe	0.10	0.36

Klason lignin isolated from Pinus radiata compression wood which contained 12.6% methoxyl instead of the 13.9 – 14.9% range observed for normal wood lignins. Bland also found that 21% of aldehydes obtained in the nitrobenzene oxidation of milled wood lignin from compression wood consisted of p-hydroxybenzaldehyde. Similar results were later obtained for compression wood lignins of Douglas-fir (55) (Table 3.11). The determination of Hibbert's ketones revealed that phenylpropane units of the p-hydroxyphenyl type were about three times as frequent in compression wood lignin as in normal wood lignin. The composition of milled wood lignin indicates the presence of approximately 29% of p-hydroxyphenylpropane units in compression wood lignin.

The differences in the UV-spectra between normal and compression wood lignins are, surprisingly, insignificant, with identical absorptivity values at 280 mµ (54, 56). The same is true for the infrared spectra, in which the only significant difference consists of slightly lower absorption at the 1720 cm^{-1} region in compression wood lignins. The measured carbonyl contents of both lignins are the same. Acidolysis and UV-evidence suggest that β-O-4 and β-5-linked structures may be less frequent and 5-5-linked structures more common in compression wood lignins (56).

6. Tissue Culture Lignins. Methods to grow cambial tissue cultures were developed by Jacquiot (57) and conifer cultures have been studied by Barnoud (58) and Barnoud and Higuchi (59). The cells grown in tissue

cultures generally remain undifferentiated, forming callus-like growth aggregates composed of round parenchymatous cells although in some cases, Sequoia for example, a fully developed plant may evolve. The lignin content is most conveniently estimated by conversion to lignin thioglycolic acids (59), because the high protein content (12 - 16%) makes Klason lignin determination meaningless. On the basis of the thioglycolic acid method, the lignin content of the tissue culture was found to be 12.5% for Pinus strobus and 9.7 to to 11.8% for Sequoia sempervirens, as compared with the contents 32.4 and 30.6% respectively for mature wood tissue.

The tissue culture lignin differed little from normal wood lignin yielding vanillin with a small amount of p-hydroxybenzaldehyde in nitrobenzene oxidation and the isolated Hibbert's ketones were exclusively of the guaiacyl propane type. In the infrared spectrum, however, the milled wood lignin isolated from the tissue culture lignin showed a more pronounced absorption at 1660 and 1720 cm^{-1} regions than the corresponding preparation from normal wood.

B. Cycad Lignins. Higuchi and Kawamura (5) have assigned among guaiacyl lignins a specific subclass (Type Nc lignins) to lignins present in cycads, a primitive gymnosperm family. The assignment was mainly based on a strong absorption maximum at 1740 cm^{-1} in the infrared spectrum of the milled wood lignin preparation. Although guaiacyl lignins generally display only a weak shoulder at this wavelength; pine lignins, as pointed out earlier in this chapter, do possess a strong maximum at 1735 cm^{-1} (49). The need for a specific subclass could therefore be questioned, especially since the lignin thioglycolic acid preparations, separately isolated from the leaf stems and vessel bundles of Cycas revoluta, possess essentially normal composition (Table 3.7). An exceptionally low vanillin yield (2.5% of Klason lignin) has been reported for the same species (2), but later work has failed to confirm this observation (5). Rather, nitrobenzene oxidation yielded vanillin, p-hydroxybenzaldehyde and syringaldehyde in approximately the same relative amounts as other guaiacyl lignins.

C. Lignins in Vascular Cryptogams. Among the lower vascular plants, many families are known that are heavily lignified whereas others, often closely related to the former species, show only faint signs of lignification. Clubmosses (Lycopodium), for example, give a strongly positive Wiesner reaction and display a sky-blue fluorescence in ultraviolet illumination, whereas in Isoetes both reactions are faint or absent (60). Common herbaceous ferns (Pteris) are heavily lignified, whereas horsetails (Equisetum) give only a faint Wiesner reaction around the vascular bundles (61).

Tropical tree ferns (Dicksonia, Alsophila, Cyathea) show a particularly interesting distribution of lignin: The woody trunks of these plants can reach the height of 30 feet and are enveloped by an extremely hard woody cortex resembling in outward appearance and mechanical properties heavily lignified tissues; yet, the cortex appears to be totally free of lignin. The

Wiesner reaction is negative and the ethanolysis products do not contain Hibbert's ketones (62). Polyphenolic materials other than lignin are present, as shown by relatively high contents of apparent Klason lignin—26% (62) and 50 to 55% (63). The isolated "Klason lignin" contains only 1% methoxyl (63). The polyphenolic materials may be isolated in the form of "milled wood lignin" which has an atypical IR spectrum devoid of the characteristic 1500 cm^{-1} maximum for example (62). These components yield neither alcohol-insoluble lignin thioglycolic acids nor dioxane acidolysis lignin, although small amounts of vanillin (3.5%) and a trace of p-hydroxybenzaldehyde can be isolated from nitrobenzene oxidation products (62). In contrast, Wiesner reaction is positive for localized regions of the soft inner tissues of trunks which also yield identifiable Hibbert's ketones (64). Milled wood lignins with IR spectra conforming with the typical guaiacyl-lignin pattern have been obtained from the soft tissues of Alsophila mertensina and Cyathea boninsimensis (5). These observations can perhaps be interpreted to mean that the presence of lignins in the primitive tree-ferns is strictly limited to conducting xylem tissues, while non-lignin polyphenols function as reinforcing components in the woody cortex tissues of these plants.

The Klason lignin contents of the leaf stalks of fourteen ferns, belonging to families Polypodiaceae and Osmundaceae, have been determined by Timell (65). The values obtained varied from 28 to 40%, with one exception (Matteuccia atruthiopteris) that had only 20.9% Klason lignin. Occasionally, values as high as 46% were obtained for leaves and leaf stalks, 52% for empty sporangia, and 60% for rhizomes. In comparison, Klason lignin contents of other vascular cryptogams were found to be 25% for Psilotum nudum, 11 to 14% for two Equisetum, and 20 to 29% for three species of Lycopodium (66). Some of the cited values for ferns appear suspiciously high for a true lignin component and suggest contribution by condensed proteins and unextractable polyphenolic constituents to the weight of Klason lignin. A substantial amount of non-lignin impurities must have been present in Klason lignins isolated by Higuchi and Kawamura (5), judging from the abnormally low methoxyl values obtained for their preparations (5 to 9%). On the other hand, the yields of lignin thioglycolic acids from ferns as well as clubmosses (Table 3.7) suggest the true lignin content to be of the order of 22 to 30%, a level still unusually high for herbaceous plants.

Milled wood lignin from the fern Polystichum munitum had the composition $C_9H_{9.4}O_{2.88}(OMe)_{0.83}$ and an infrared spectrum indistinguishable from conifer lignins (62). The same is true for the MWL spectrum of Lycopodium clavatum, a clubmoss, with the exception of a high maximum at 1740 cm^{-1} (5), possibly due to the presence of ester groups. The compositions of lignin thioglycolic acids obtained from ferns and clubmosses deviate little from the corresponding preparations from conifer woods, with the exception of the thioglycolic acid derivatives isolated from rhizones and pith which have low methoxyl and sulfur contents (see Table 3.7). In contrast, thioglycolic acid products from horsetails (Equisetum) are always too low in methoxyl (0.3 to 0.7%) and sulfur (1.2 to 4.6%) to be any real lignin derivatives (47).

All vascular cryptogams, including Psilotum and Equisetum (67) yield vanillin and small amounts of p-hydroxybenzaldehyde in nitrobenzene oxidation. The yield of vanillin from Lycopodium clavatum corresponds to 29 to 33% of Klason lignin (61). Syringaldehyde is generally absent, but a trace of it has been identified among the oxidation products of Isoetes (67). Members of the genus Selaginella form a notable exception from all other vascular cryptogams in that they yield substantial amounts of syringaldehyde and also display a clearly positive Mäule reaction (4, 68).

Possible ester groups associated with lignin have been studied in the case of Phylloglossum drummondi, a club moss (69). The acids obtained by saponification include protocatechuic, p-hydroxybenzoic, vanillic, ferulic and syringic. Only syringic acid was isolated from Lycopodium.

D. General Characteristics of Guaiacyl Lignins. On the basis of available information, guaiacyl lignins form a relatively uniform group of natural polymers in which the main differences are due to the variation in the content of p-hydroxyphenylpropane units. Some of these lignins may contain very small quantities of syringylpropane units and others, ester groups. Interspecies variation does not seem to exceed much the variation observed within a single plant (normal and compression wood lignins, for example), and consequently there is no clear chemical justification to divide guaiacyl lignins into definite subgroups. The first plants to contain guaiacyl lignin appeared during the Devonian period, approximately 300 million years ago, and it is of interest to note that the gradual evolution of conifers from ancient ferns has apparently not been accompanied by detectable changes in the lignin structure. In this respect guaiacyl lignins seem to share some of the "intrinsic immutability" of cellulose, which currently is believed to be chemically identical in all dry-land plants (70).

III. Guaiacyl-Syringyl Lignins.

A. Exceptional Conifers. The guaiacyl-syringyl nature of lignin in Ephedra procera was already apparent in the thioglycolic acid derivative (Table 3.12) isolated by Holmberg (47) as a part of his pioneering study. Later it was found by Schindler (71) that not only Ephedra species but members of all three genera of Gnetales gave a positive Mäule reaction. Creighton, Gibbs and Hibbert (2) subjected three species belonging to Gnetales, as well as three other conifers that have a positive Mäule test (Tetraclinis articulata, Podocarpus amarus and P. pedunculatus), to nitrobenzene oxidation, obtaining both vanillin and syringaldehyde in each case, as shown in Table 3.13.

It should be noted that some species of Podocarpus (e.g. P. spicatus, Table 3.9) yield no detectable syringaldehyde and consequently the vanillin-

Table 3.12. Analytical Compositions of Thioglycolic Acid Derivatives of Guaiacyl-Syringyl Lignins (47)

Species	Part of plant	Yield %	Anal. compositions C %	H %	MeO %	S %	MeO/C_9
Abnormal gymnosperms:							
Ephedra, Ephedra procera	Trunk wood	26	54.1	5.3	15.1	9.1	1.28
Woody angiosperms:							
Quaking aspen, Populus tremula	Trunk wood	17	55.3	5.6	16.7	8.9	1.38
Alder, Alnus glutinosa	Trunk wood	21	54.8	5.2	15.4	9.0	1.28
Ash, Fraxinus excelsior	Trunk wood	17	53.9	5.4	17.0	9.7	1.48
Birch, Betula verrucosa	Trunk wood	17	54.3	5.6	17.2	9.5	1.48
White beech, Carpinus betulus	Trunk wood	16	53.9	5.3	17.4	9.8	1.52
Red beech, Fagus Silvatica	Trunk wood	23	54.6	5.3	15.9	8.9	1.33
Elm, Ulmus montana	Trunk wood	17.5	54.3	5.3	14.8	9.4	1.24
Maple, Acer plantanoides	Trunk wood	19.5	55.2	5.7	17.2	9.4	1.44
Basswood, Tilia ulmifolia	Trunk wood	15.5	54.9	5.6	17.4	9.6	1.48
Oak, Quercus robur	Trunk wood	17	54.2	5.3	16.7	9.1	1.42
Magnolia, Magnolia soulangeana	Branch wood	24	54.3	5.3	16.4	10.0	1.41
Horse chestnut, Aesculus hippocastanum	Trunk wood	18.5	54.6	5.5	15.4	9.7	1.30
Sloe, Prunus spinosa	Trunk wood	16	54.2	5.2	16.2	9.4	1.38
Pear, Pyrus communis	Trunk wood	20	55.2	5.4	17.2	9.1	1.44
Hazelnut, Corylus avellana	Trunk wood	22.5	54.6	5.2	17.2	8.7	1.45
Corn rocket, Bunias orientalis	Stems	20	53.6	5.3	15.1	9.4	1.29
Laburnum, Cytisus laburnum	Branch wood	20.5	54.3	5.3	16.5	9.2	1.40
Walnut, Juglans regia	Branch wood	18.5	54.5	5.2	16.4	9.3	1.38
Saxaul tree, Haloxylon ammodendron	Branch wood	25.5	54.3	5.3	14.6	8.9	1.21
Wax palm, Copernicia australis	Trunk wood	23	55.7	5.4	15.3	7.1	1.20
Barberry, Berberis vulgaris	Trunk wood	16.5	54.0	5.4	16.4	9.8	1.42
Spiraea, Spiraea ulmaria	Stems	20	53.3	5.3	14.0	10.2	1.21
Heath, Calluna vulgaris	Trunk wood	21	54.4	5.2	13.7	9.5	1.13
Cocoa palm, Cocos nucifera	Nut shells	21	54.4	5.0	12.8	7.2	1.01
Hazelnut, Corylus avellana	Nut shells	33	54.9	5.1	10.9	9.7	0.88
Gooseberry, Ribes grossularia	Stems	20	54.1	5.3	14.3	10.1	1.21
Herbaceous angiosperms:							
Toadrush, Juncus bufonius	Stems,leaves	6.5	53.6	5.5	9.5	5.9	0.73
Asparagus, Asparagus officinalis	Stems	11	53.2	5.2	14.9	10.0	1.30
Orchid, Orchis mascula	Stems,leaves	10	52.6	5.2	9.7	6.7	0.77
Rhubarb, Rheum rhaponticum	Stems	11	52.9	4.9	16.7	9.5	1.49
Pea, Pisum sativum	Stems	17	53.2	5.2	11.4	9.0	0.94
Restharrow, Ononis repens	Stems	17	54.1	5.3	14.4	9.8	1.22
Acanthus, Heracleum sibiricum	Stems	17	52.0	5.1	13.9	8.7	1.21
Yellow birds nest, Monotropa hypopitys	Stems	10	54.7	5.0	7.1	7.1	0.62
Potato, Solanum tuberosum	Stems	12.5	50.7	5.0	9.7	6.1	0.80
Nightshade, Solanum dulcamara	Stems	19	54.9	5.4	13.1	9.6	1.07
Toothwort, Lathraea squamaria	Stems	7	52.9	5.0	2.0	4.0	0.14
Burdock, Arctium tomentosum	Stems	17.5	53.8	5.3	15.6	9.4	1.33
Artichoke, Helianthus tuberosus	Stems	11	53.4	5.3	13.8	9.9	1.18
Aquatic angiosperms:							
Pond-weed, Potamogeton natans	Stems	8.7	53.6	5.4	5.1	5.3	0.37
Sea-weed, Zostera marina	Stems,leaves	9	45.9	5.1	0.2	1.2	-
	Rhizomes	7	45.2	4.9	0.2	1.6	-
Grasses:							
Bamboo, Bambusa sp.	Fresh cane	16	55.7	5.1	13.1	7.1	1.00
	Aged cane	19	54.9	4.9	14.6	8.9	1.20
Reed, Phragmites communis	Cane	14	53.7	5.6	14.7	7.9	1.22
Reed, Calamus rotang	Cane	17	55.1	5.3	17.2	7.1	1.36
Rye, Secale cereale	Straw	14	56.2	5.0	14.6	7.6	1.14

Table 3.13. Yields of Aldehydes from the Nitrobenzene Oxidation of Exceptional Conifers (3)

	Klason Lignin %	Isolated aldehydes, % of Klason Lignin			Syringaldehyde-Vanillin ratio (S/V)	
		Total	Vanillin	Syringe-aldehyde		
1. Gnetales:						
Ephedra trifurca	23.3	39.6	9.3	27.7	3.0	
Gnetum indicum	24.7	26.8	6.8	16.5	2.5	
Welwitschia mirabilis	16.5	4.9	2.0	2.4	1.2	
2. Other Conifers:						
Tetraclinis articulata	29.5	29.2	15.5	12.0	0.8	
	31.8	36.8	–	–	~1	(2)
Podocarpus amarus	32.1	30.7	12.5	12.5	1.0	
Podocarpus pedunculatus	30.7	31.7	13.9	14.7	1.0	
P. totara D. Don	35.6	27.1	~24	3.0	0.13	(1)

syringaldehyde ratio within this genus varies from 1:0 to 1:1. Tetraclinis articulata, curiously enough, is the only known species in the family Cupressaceae that gives syringaldehyde in nitrobenzene oxidation. Many of the Gnetales contain both tracheids and vessels resembling morphologically angiosperm woods (4). The syringaldehyde-vanillin ratio 3:1 for certain species of Ephedra that corresponds to the average level of angiosperms is therefore not unanticipated.

The infrared spectrum reported for Ephedra gerardiana (5) exhibits all the characteristics of a typical guaiacyl-syringyl lignin and thus confirms the results from nitrobenzene oxidation. In addition, a strong maximum at 1740 cm^{-1} is apparent, giving some indication of associated acetyl groups.

B. Angiosperm Wood Lignins.

1. Distribution. Typical examples of the lignin contents of several hardwood species of the temperate zone are shown in Table 3.14. Generally, the lignin content of these hardwoods varies in the range from 16 to 24%. The same approximate range is indicated by the yields of lignin thioglycolic acids (Table 3.12). A somewhat higher range, 22.1±3.3% was reported by Kawamura (76) from a study of the lignin contents of 186 temperate-zone hardwoods. Since these estimates are based on Klason lignin contents determined without accounting for the soluble lignin part (~12% of insoluble Klason lignin) (77), the actual lignin contents may be of the order of 19 to 28%, a range still below that of conifer woods. The few available data on within-tree variation suggest the lignin content of sapwood to be equal to or slightly higher than that of heartwood, higher in early wood than late wood, and lower in tension wood than in normal wood, although the last-mentioned difference is quite small for some species (Table 3.14). In many species, lignification of wood tissue is nearly completed during the year of formation, whereas in other species, Robinia pseudoacacia for example, lignification during the first year of growth may reach only one-half of comple-

Table 3.14. Klason Lignin Contents of Angiosperm Woods

Species	Klason Lignin, % of extr. free wood			Ref.
	Sapwood	Heartwood	Tension wood	
White birch, Betula papyrifera	24.7	24.6	-	(24)
	18.9	-	-	(72)
Quaking aspen, Populus tremuloides	18	-	17	(73)
	21.4	-	-	(72)
Aspen, Populus canadensis	23	-	22	(74)
Elm, Ulmus americana	29	-	27	(73)
	24	-	-	(72)
Red alder, Alnus rubra	-	24.7^1	-	(15)
	-	23.0^2	-	(15)
American beech, Fagus grandifolia	22	-	-	(72)
White ash, Fraxinus americana	25.6^1	-	-	(15)
	23.5	-	-	(15)
Eucalyptus regnans	22	-	16	(75)
Eucalyptus nitens	21	-	15	(75)
Eucalyptus goniolyx	25	-	10	(75)

1) Early wood
2) Late wood

Table 3.15. Effect of Age on the Extent of Lignification in Hardwoods

Species	Age of growth ring	Klason lignin		Total aldehydes in nitrob. oxid. % of lignin
		% of wood	MeO, %	
Acacia mollissima (22)	1 yr	22.6	15.2	-
	2 yrs	24.0	15.6	-
	Mature wood	25.4	15.9	-
Robinia pseudoacacia (22)	1 yr	10.9	12.7	34.0
	2 yrs	17.1	17.7	48.5
	Mature wood	21.1	18.3	64.9
Eucalyptus botryoides (78)	Young wood	22.9	-	-
	Sapwood	29.7	-	-
	Heartwood	33.1	-	-
	Rootwood	30.2	-	-

tion (Table 3.15). The retarded increase in lignin content with age could be the consequence of very slow lignification of parenchyma tissues.

2. <u>Distribution in Single Cells.</u> Electron micrographs of sections from maple (Acer rubrum) after the removal of polysaccharides by 80% hydrofluoric acid suggest that a greater proportion of lignin is concentrated in the compound middle lamella, and the lignin network in the secondary wall appears less dense than in softwoods (34). The same conclusion has been drawn from UV micrographs of hardwood tissues (36). There are, however, uncertainties associated with both of these methods. Concentrated hydrofluoric acid, as other mineral acids (79), is likely to dissolve and remove a substantial portion of hardwood lignin. Likewise, the large difference between the absorptivities of guaiacyl- and syringylpropane units (see Chapter 7) makes the interpretation of UV micrographs uncertain.

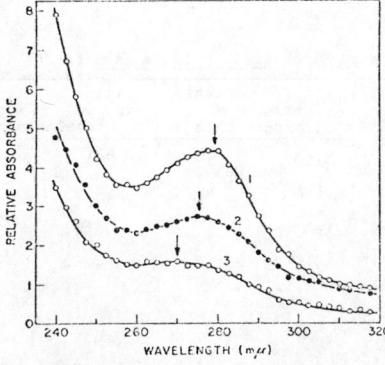

Figure 3.6. Birch UV spectral curves for the morphological regions. Vessel secondary wall (1), cell corner region of the compound middle lamella between fibers (2), and the fiber secondary wall (3) The arrows indicate the positions of the UV maxima (40).

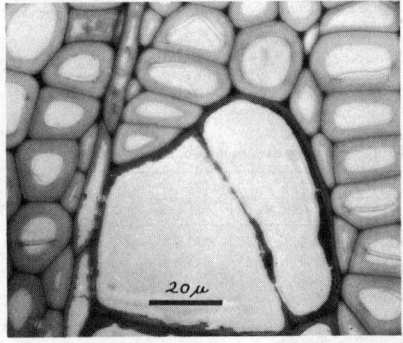

Figure 3.7. Cross section of paper birch early wood photographed in ultraviolet light of wavelength 240 mµ (40).

A very recent study by Fergus and Goring (40) on the lignin distribution in birch (Betula papyrifera Marsh) xylem has actually cast serious doubt on the earlier concepts. A detailed spectral analysis of the various types of cells by means of UV spectroscopy revealed that guaiacyl and syringyl lignins are, to a considerable extent, segregated in birch tissues (Figure 3.6). The cell walls and middle lamellae of vessels contain guaiacyl lignin, judging from the characteristic UV spectrum with a maximum at 279 mµ. The spectrum of lignin in the middle lamellae of fibers has a maximum at a lower wavelength (275 - 276 mµ), suggestive of guaiacyl-syringyl composition. The cell walls of fibers and ray cells exhibit a flat maximum at about 270 mµ characteristic of syringyl lignin. Starting from the somewhat arbitrary assumptions that lignin in the compound middle lamellae of fibers contains 50% (by weight) of of syringyl lignin, and that the cell walls of fibers and ray cells contain syringyl lignin only; Fergus and Goring were able to derive the probable lignin distribution from micrographs obtained using transmitted monochromatic illumination of wavelength 240 mµ. An example of such a micrograph is shown in Figure 3.7. The absorptivities of various cell elements and their volume

Table 3.16. Distribution of Lignin in Xylem Cells of Birch (Betula Papyrifera Marsh) (40)

Element	Morphological differentiation	Type of lignin	E_{280} cm^{-1}lg^{-1}	Relative absorbance	Lignin % of tissue	Tissue volume % of total	Lignin % of total
Fiber	Secondary wall	Syringyl	4.5	1.0	19	73	60
	Narrow sections of compound middle lamella	Syringyl-guaiacyl (1:1)	10.0	4.0	40	5	9
	Corner sections of middle lamella	Syringyl-guaiacyl (1:1)	10.0	10.0	85	2	9
Vessel	Secondary wall	Guaiacyl	15.6	4.8	27	8	9
	Middle lamella	Guaiacyl	15.6	7.6	42	1	2
Ray cells	Secondary wall	Syringyl	4.6	1.4	27	11	11

Table 3.17. Compositions of Milled Wood Lignins from Angiospern Woods

Wood species	Ref.	Analytical compositions				
		C, %	H, %	O, %	MeO, %	MeO/C$_9$
Birch, Betula verrucosa	(46)	58.8	6.5	34.0	21.5	1.58
Beech, Fagus silvatica	(46)	60.3	6.3	33.4	21.4	1.43
		63.47	5.75		21.56	1.36
Aspen, Populus tremula	(46)	60.4	6.2	33.0	21.4	1.43
Eastern cottonwood, Populus deltoides	(49)	59.28	5.70	35.02	18.05	1.20
Bigleaf maple, Acer macrophyllum	(49)	60.37	5.68	33.95	20.02	1.34
Cascara buckthorn, Rhamnus purshiana	(49)	60.22	6.05	33.73	21.43	1.44
Madrona, Arbutus menziesii	(49)	58.24	6.98	34.78	22.58	1.59

fractions were carefully measured, giving the lignin distribution shown in Table 3.16. It can be seen that the amount of lignin in compound middle lamellae is estimated to be only 19% of the total lignin. Although this figure is to be considered a tentative one only, there is hardly any doubt that the major part of birch lignin is localized in the cell wall.

3. <u>Relative Amounts of Guaiacyl- and Syringylpropane Units in Hardwood Lignins</u>. The methoxyl contents of hardwood lignins vary within wider limits than those of conifer lignins. A spread from 17.1 to 21.9%, with an average of 20.1%, was determined for the methoxyl contents of Klason lignins of seven Japanese hardwood species (46). From the analytical data on lignin thioglycolic acids (Table 3.12), it can be calculated that the average number of methoxyl groups per phenylpropane unit varies from 1.20 to 1.52. An almost identical range from 1.20 to 1.59 is valid for the few milled wood lignins so far analyzed (Table 3.17).

Only minute quantities of p-hydroxybenzaldehyde are detected among the nitrobenzene oxidation products of hardwoods (5); and, likewise, Hibbert's ketones of the p-hydroxyphenyl type are isolatable in trace quantities only (82). The frequency of p-hydroxyphenylpropane units in hardwood lignins appears consequently to be lower than in guaiacyl lignins; without great error, hardwood lignins may be treated as though they were copolymers of guaiacyl- and syringylpropane units only (49), and the relative amounts of these units obtained from the determined number of methoxyl groups per unit. For example, the MeO/C_9 value for madrona wood lignin (1.59, Table 3.17) indicates that 59% of the phenylpropane units are of the syringyl type; whereas this value is only 24% for the lignin thioglycolic acid from elm (Table 3.12).

Infrared spectra of milled wood lignins of hardwoods support the idea that these materials are, indeed, essentially guaiacylpropane-syringylpropane copolymers. Higuchi and Kawamura (5) brought into attention significant differences in the spectra of milled wood lignins of various hardwoods, and these differences were later shown by Sarkanen, Chang and Allan (49) to be due to absorption maxima associated with guaiacyl and syringyl nuclei. Interpolated absorptivities of maxima dominated by syringyl nuclei at 1130, 1235, 1335, 1430, 1470 and 1600 cm^{-1} all showed a linear ascending relationship with the MeO/C_9-values in a number of milled wood lignin preparations (including conifer lignins); whereas the maximum at 1275 cm^{-1}, typical of guaiacyl nuclei, showed a linear descending relationship (Figure 3.8). Infrared spectra can thus be used to determine the relative proportions of guaiacyl- and syringylpropane units in lignins. Also, the ultraviolet absorption at 280 mμ decreases in an approximately linear fashion with increasing number of syringylpropane units in borohydride-reduced preparations (see Figure 7.8 in Chapter 7).

The varying ratios of guaiacyl and syringyl moieties in hardwood lignins are reflected in the compositions of the nitrobenzene oxidation products of these lignins. Table 3.18 lists the results from the pioneering study by Creighton, Gibbs and Hibbert (2). The syringaldehyde-vanillin ratios (S/V) are given as molar ratios rather than weight ratios as in the original paper. It can be noted that the S/V values lie in the range 2.3 to 3.3, while the analytical compositions of milled wood lignins and lignin thioglycolic acids indicate molar ratios 0.3 to 1.5 of syringyl- to guaiacyl-propane units in hardwood lignins (see Tables 3.12 and 3.17). As a crude approximation, the S/V values may be roughly three times higher than the ratios of corresponding units in lignins, reflecting the more efficient conversion of syringylpropane units to syringaldehyde in nitrobenzene oxidation. This, in turn, is due to the absence of condensed structures among syringylpropane units, as pointed out by Freudenberg and Sidhu (80). While only 30% of guaiacyl-propane units in conifer lignins are recoverable as vanillin, the conversion of syringyl moieties to syringaldehyde may well be at the 90% level. The relationship between the S/V values and the composition of lignin certainly would deserve

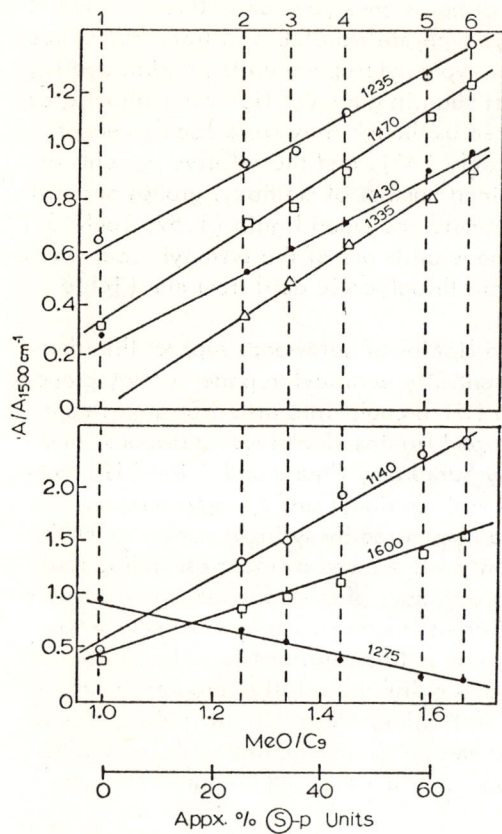

Figure 3.8 Relative absorptivities (A/A $_{1500\ cm^{-1}}$) of characteristic maxima in the IR spectra of borohydride-reduced milled wood lignins as a function of the molar percentage of syringylpropane units.

1. Western redcedar MWL
2. Eastern cottonwood MWL
3. Bigleaf maple MWL
4. Cascara MWL
5. Madrona MWL
6. Enzyme lignin from tension wood of Madrona (49).

to be studied more systematically. S/V values have generally been found to be close to the molar ratio of the corresponding Hibbert's ketones, usually expressed as syringoylacetyl-vanilloylacetyl ratio (SA/VA) (59).

There appears to exist an empirical correlation between the Klason lignin content of mature angiosperm wood and the composition of lignin. High-syringyl lignin is commonly associated with a low Klason lignin content and, conversely, extensively lignified hardwoods generally contain lignins predominantly of the guaiacyl type. This relationship is illustrated in Figure 3.9 in which the S/V values of several Eucalyptus species are plotted against their Klason lignin contents, based on results by Bland, Ho and Cohen (81). The total yield of aldehydes in nitrobenzene oxidation commonly increases with increasing percentage of syringyl units in lignin.

The determination of S/V values offers a convenient method for comparing lignins, and a large number of these determinations have been carried out

CLASSIFICATION AND DISTRIBUTION

Table 3.18. Nitrobenzene Oxidation Products of Angiosperm Woods (3)

Species	Klason lignin %	Total aldehydes % of K.L.	S/V (molar ratio)
Acer rubrum (Red maple) Whole wood	22.0	46.0	2.8
Heartwood	23.6	39.2	2.4
Sapwood	20.6	44.5	3.1
Acer saccharinum (Silver maple)	18.8	51.0	2.3
Populus tremuloides (Aspen)	17.4	44.7	2.8
	17.0	43.3	3.1
	20.6	41.5	3.1
Quercus borealis v. maxima (N. red oak)	18.4	47.2	2.8
Betula lutea (Yellow birch)	19.6	44.9	2.6
Fraxinus americana (White ash)	18.5	48.6	2.9
Ulmus fulva (Slippery elm)	23.0	41.7	2.7
Ulmus americana (American elm)	21.3	40.9	2.6
Juglans cinerea (Butternut)	19.0	45.2	2.3
Juglans nigra (Black walnut)	20.9	39.3	2.1
Tilia americana (Basswood)	23.8	35.2	2.3
Sorbus americana	18.5	49.0	3.3
Robinia pseudoacacia	21.2	39.8	2.4
Paulownia imperialis	18.9	44.6	2.6
Liriodendron tulipifera	20.9	49.2	2.7
Trachodendron sp.	24.0	40.4	2.6
Drimys winteri	24.3	35.3	2.3
Belliolum haplopus	36.9	29.6	0.9
Zygogynum vieillardi	35.5	32.8	1.0

for angiosperm woods. A whole spectrum of lignins with varying syringyl-guaiacyl ratios appears to exist in these plants. For example, Towers found that the S/V value was only 0.35 for a specimen of Zygogynum vieillardi Baill (67), while Bland et al. obtained the ratio 5.2 for Eucalyptus diversicolor F. Muell and E. maculata Hook (81).

Several botanical subgroups of angiosperm woods contain, as a rule, low-syringyl lignin. The last five species in Table 3.18 are representatives of so-called "primitive" dicotyledons that include the families Degeneriaceae, Magnoliaceae, Monimiaceae, Trochodendraceae, and Winteraceae and have the absence of vessels in common. Towers (67) has studied a number of species in this group, finding mostly low S/V values ranging from 0.35 to 1.7. However, three out of the five species in Table 3.18 form an exception to this rule. Higuchi and Kawamura (5) have proposed, as already mentioned, that lignins in tropical hardwoods should be classified as a distinct group of low-syringyl lignins. The Klason lignin contents of these woods are in the range 28.7±4.07% (76) as compared with the average lignin content of 22.1% for temperate-zone hardwoods. The methoxyl contents of Klason lignins from five tropical hardwood species varied from 14.7 to 18.4% (av. 16.9%) (82), all lower values than the average for the Klason lignins of temperate-zone hardwoods, 20.1%. The S/V values of these lignins are usually below 1.0, and the predominance of guaiacyl propane units is quite clear from the published IR spectra of milled wood lignins (5).

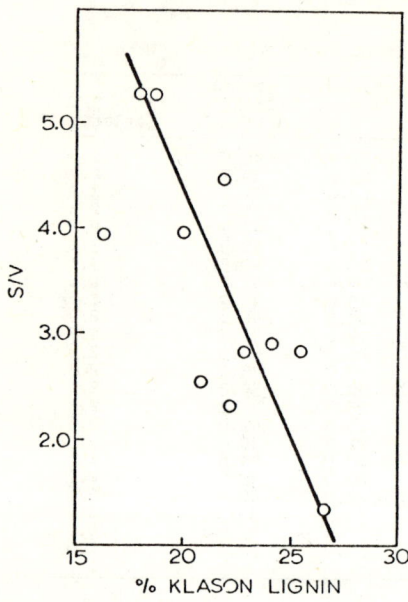

Figure 3.9. Relationship between the composition of nitrobenzene oxidation products and lignin content in various Eucalyptus species (81). S/V versus Klason lignin.

A comprehensive study has been carried out on the variation of S/V values within the maple family (Aceraceae) which includes approximately 150 species (4). The ratios were found to vary from 0.4 for Acer negundo and A. cissifolia to 3.3 for A. rubrum L.

High-syringyl lignins appear to be present in birches and in madrona, judging from the compositions of their milled wood lignins (Table 3.17) that suggest about 60 mole per cent of syringylpropane units. About equal amounts of syringyl- and guaiacyl propane units appear to be present in the lignin thioglycolic acids of ash, birch, white beech, and basswood (Table 3.12), while guaiacyl propane units are dominant in the investigated thioglycolic acid derivatives of other species. The S/V values 3.9 to 5.2 place lignins from five Eucalyptus species (E. regnans, E. gigantea, E. obliqua, E. maculata, and E. diversicolor) in the group of high-syringyl lignins, while this ratio is only 1.3 for E. marginata (81).

4. Within-Tree Variation in Hardwood Lignins. Bland (78) studied in detail the variation of S/V values in different parts of Eucalyptus botryoides, with results shown in Table 3.19. In mature wood, the total variation was not more than from 1.5 to 2.0. In Robinia pseudoacacia (22) lignins in the two outermost growth rings have a low methoxyl content (see Table 3.15), and the IR and UV spectra of isolated milled wood lignins demonstrate a lower content of syringylpropane units than in heartwood lignin (22). The

Table 3.19. Syringaldehyde-Vanillin Ratios from the Nitrobenzene Oxidation of Different Parts of Eucalyptus botryoides (78)

Part	Apparent Klason lignin %	Milled wood lignin yield % of K.L.	S/V
Foliage -Midrib	-	0.6	1.0
-Petiole	-	0.1	0.7
Young green bark	17.2	3.5	1.3
Young wood	22.9	3.0	1.9
Old bark	34.4	1.3	1.5
Sapwood	29.7	2.2	1.6
Heartwood	33.1	3.1	1.5
Root bark	23.3	2.4	1.9
Root wood	30.2	9.6	2.0

Table 3.20. Nitrobenzene Oxidation and Ethanolysis Products from Tissue Culture of Hardwoods (59)

Species	Lignin thioglycolic acid yield %	S/V	SA/VA
Paulownia tomentosa			
Tissue culture	9.5	0.01	0.02
Wood	22.6	1.5	1.4
Syringa vulgaris			
Tissue culture	11.4	0.32	0.38
Wood	16.0	2.1	1.7
Rosa wichuriana			
Tissue culture	8.7	0.04	0.09
Wood	16.4	1.6	2.1

S/V values are generally lower for young seedlings than for mature wood. This ratio increased from 0.86 to 1.70 during the first year of growth of an elm seedling (Ulmus campestris), while the ratio is 2.0 for mature elm wood (59). A one-year old vertical branch of a plum tree (Prunus) gave a 3.1 S/V ratio, to be compared with the value 4.7 for mature wood (83).

The so-called "tension wood" develops in hardwood species in the upper parts of branches and leaning stems. In some species with curved stems, tension wood formation can be so extensive that it is difficult to recover samples entirely free of tension wood. Tension wood lignins of birch and madrona contain more syringylpropane units than normal wood lignin (84).

Higuchi and Barnoud (59) characterized lignins in callus-like cambial tissue cultures of several angiosperms (Table 3.20) The S/V ratios, as well as the ratios of the corresponding Hibbert's ketones, were found to be exceptionally low, sometimes of the same order of magnitude as in conifer lignins. The low content of syringyl propane units is confirmed by infrared spectra that resemble the spectra of guaiacyl lignins, with the exception of high absorption at 1660 and 1740 cm^{-1} regions.

C. Lignins in Herbaceous Dicotyledons. Our current knowledge of the lignin component in herbaceous dicotyledons is appallingly scant. Mäule reaction was applied by Gibbs (4) and Towers (67) to a large number of species and was found to be positive in almost all cases. Those species that failed to give a positive reaction consisted of lightly lignified types, often more or less aquatic.

For all herbaceous plants, the Klason lignin determination is an unreliable means for the estimation of lignin contents unless proteins are removed prior to the determinations. The amount of protein material coprecipitated with the Klason lignin can be reduced, as shown by Kratzl (85), by reducing the sulfuric acid concentration from 72 to 57% in the lignin determination. Even then, the 2% nitrogen content found for the Klason lignin isolated in 15% yield from mature stems of potato indicates significant contamination by protein. An error in the negative direction may be caused by the soluble lignin lost in the filtrate.

Lignin thioglycolic acid preparations are perhaps best suited for the characterization of herbaceous lignins, and data provided by Holmberg are collected in Table 3.12. Judging from the analytical composition, most of the thioglycolic acid derivatives are not suspect of extensive contamination by non-lignin materials with the exception of toothwort and the three species of aquatic plants. The lignin content is indicated to vary generally between 10 and 17%, and the content of syringylpropane units is often low, as indicated by the MeO/C_9-values that are often below 1. It is probable, however, that many of the samples were not harvested at a mature state. The lignin contents of Jute (Corchorus) and Sassafras species may be as high as 22.3%, and the reported S/V ratios for these species are 2.5 and 2.2 - 2.6 respectively (3).

D. Lignins in Monocotyledons.

1. Lignin Contents of Mature Plants. The Klason lignin contents of a number of monocotyledonous plants are listed in Table 3.21, based on comparative studies by Higuchi (86). The values in Table 3.21 were not corrected for co-condensed protein; and some of them, particularly those pertaining to lignins with less than 10% methoxyl, are suspect for being artificially high. Corrected Klason lignin contents for mature cereal straws are often as low as 14% (87), and the yields of lignin thioglycolic acids vary from 14 to 19% (Table 3.12).

2. Chemical Composition. The analysis of lignin thioglycolic acids from several grasses (Table 3.12) indicated an average number of 1 to 1.36 methoxyl groups associated with phenylpropane units. No direct conclusions can be based on the magnitude of these figures, however. In contrast to angiosperm wood lignins, grass lignins contain significant amounts of constituents other than guaiacyl- and syringylpropane units. This was first shown by Creighton and Hibbert (3) who demonstrated that in addition to vanillin and syringaldehyde; corn, bamboo and rye straw yielded substantial amounts

Table 3.21. Nitrobenzene Oxidation of Monocotyledons (86)
Figures in parenthesis refer to plant
material extracted with cold 1 N NaOH.

Species	Klason lignin %	MeO in lignin %	Lignin oxid. products, % of K.L.			Molecular ratios		Acid from alkaline hydrolysis, % of lignin	
			Vanillin	Syring-aldehyde	p-hydroxy-benzaldehyde	S/V	H/V	p-Coumaric	Ferulic
1. Grasses:									
Miscanthus sinensis	23.5	6.4	7.7 (6.9)	4.2 (4.1)	3.0 (1.5)	0.46 (0.49)	0.49 (0.27)	4.0	2.0
M. sacchariflorus	21.4	13.4	11.5 (9.7)	8.1 (8.5)	4.9 (1.8)	0.59 (0.75)	0.53 (0.22)	4.9	1.9
M. condensatus 1	21.4	13.4	11.5 (9.7)	8.8 (8.5)	4.9 (1.8)	0.58 (0.75)	0.53 (0.22)	4.7	1.9
M. condensatus 2	17.7	11.8	10.9 (9.9)	7.7 (7.0)	6.1 (3.4)	0.58 (0.63)	0.69 (0.43)	5.7	2.4
M. tinctorius	18.3	11.7	6.0 (5.9)	5.0 (5.0)	4.0 (2.3)	0.66 (0.70)	0.75 (0.48)	6.1	2.5
Coix lachryma	22.7	8.3	6.0 (5.2)	2.8 (2.8)	2.4 (1.6)	0.37 (0.43)	0.51 (0.38)	3.9	2.9
Phragmites communis	20.6	10.3	9.5 (8.8)	9.5 (8.9)	6.7 (3.1)	0.84 (0.85)	0.95 (0.44)	6.2	1.1
Arundo donax	22.5	12.8	8.4 (7.1)	6.6 (7.2)	5.3 (2.6)	0.66 (0.85)	0.99 (0.46)	6.0	1.1
Imperata cylindrica	22.7	15.0	8.2 (7.5)	11.2 (12.7)	4.1 (2.9)	1.05 (1.41)	0.57 (0.48)	4.1	1.8
	22.0	6.9	4.4 (4.0)	1.9 (2.2)	1.9 (1.1)	0.34 (0.46)	0.54 (0.34)	2.7	1.3
2. Other monocotyledons:									
Dendrobium nobile	23.6	9.8	2.7 (3.4)	5.5 (6.8)	trace (1.0)	1.80 (1.68)	— (0.40)	trace	0.02
Asparagus officinalis	18.8	13.3	8.1 (13.0)	10.3 (10.3)	2.0 (1.9)	0.61 (0.66)	0.24 (0.18)	1.5	0.03
Trachycarpus excelsa	19.0	15.4	14.3 (13.5)	19.2 (17.9)	trace (1.3)	1.12 (1.11)	— (0.12)	0.03	0.03
Smilax china	24.9	13.5	15.9 (15.8)	11.6 (10.5)	1.8 (2.9)	1.63 (1.50)	0.37 (0.61)	—	—
Pandanus tectorius	19.8	19.3	13.6 (13.9)	33.5 (33.5)	2.7 (1.0)	2.05 (2.02)	0.25 (0.09)	0.13	0.02

of p-hydroxybenzaldehyde in nitrobenzene oxidation. They suggested that the p-hydroxybenzaldehyde was derived from p-hydroxyphenylpropane units in monocotyledon lignins. For a period of time, the high frequency of p-hydroxyphenylpropane units was considered to be a special characteristic of monocotyledon lignins, and this view received support from the isolation of Hibbert's ketones of the p-hydroxyphenyl type from the ethanolysis products of bamboo and rye by Kratzl and Claus (88).

These views have been recently revised by Higuchi, Kawamura and co-workers (86). They studied a number of mature monocotyledonous plants, finding that only members of the family Graminaceae yielded large amounts of p-hydroxybenzaldehyde as nitrobenzene oxidation products (Table 3.21). The yields for other monocotyledons were similar to those obtained for angiosperms in general.

Secondly, it was revealed that the yields of Hibbert's ketones of the p-hydroxyphenyl type were no larger from grasses than from conifers or angiosperm woods (89, 90). This finding focused the attention on ester groups as a potential source of p-hydroxybenzaldehyde. Earlier, Smith (91) had shown that Brauns lignin from bagasse contains as much as 11% p-coumaric acid, together with smaller amounts of ferulic, p-hydroxybenzoic, vanillic and syringic acids. These acids were found to be connected to the core lignin through an ester linkage. Later Kuc and Nelson (92) found 12.3% p-coumarate and 3.4% ferulate groups in mature corn stalk lignin. The p-coumaric ester groups consequently represented a major potential source for p-hydroxybenzaldehyde.

Higuchi, Kawamura and coworkers (86) were indeed able to prove that a major part of p-hydroxybenzaldehyde as well as a lesser amount of vanillin both stem from the associated ester groups. In the nine grass specimens studied, the p-coumarate groups varied from 2.7 to 6.2% and the ferulate groups from 1.3 to 2.9% of the weight of Klason lignin (Table 3.21). Removal of ester groups by saponification with cold alkali caused substantial reduction in p-hydroxybenzaldehyde yields, and the amount of vanillin also diminished. Even from the saponified specimens the yield of p-hydroxybenzaldehyde was larger than generally obtained from other angiosperms. The possible attachment of p-coumaric acid to the lignin core by unsaponifiable bonds, such as phenol-ether linkages, deserves consideration as an explanation.

3. Changes During Maturation. While in arborescent plants lignification becomes completed at a short distance from the cambium and then remains constant, lignification in the stems and leaves of annual plants is a gradual process that continues until maturity.

Phillips, Goss and Davis (93) characterized the course of lignification in young oat plants, and their results are shown in graphic form in Figure 3.10. The immature growth period lasts for about 50 days, and practically no lignification takes place during this period except in the root tissue. The protein content of the soft plant tissue is 35 to 45%. The immature stage is followed

CLASSIFICATION AND DISTRIBUTION

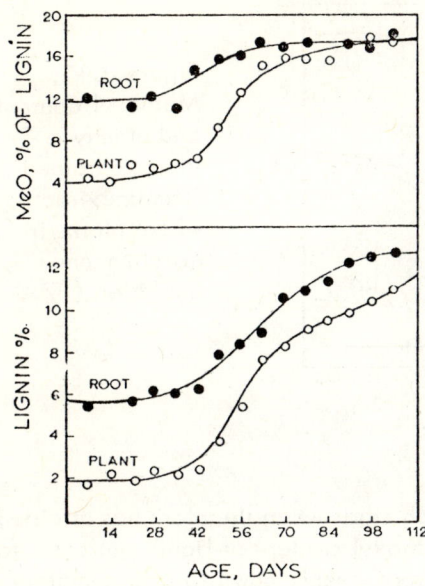

Figure 3.10. Lignin content and methoxyl in lignin as a function of age in oat plants (93). The plant material included stalks, sheaths and leaves. All values corrected for ash and protein.

Table 3.22. Nitrobenzene Oxidation Products from Bamboo and Yugao Fruit Shells at Various Stages of Maturation (94, 95)

Species	Plant material	Klason lignin %	MeO in Klason lignin %	Total yield, % of lignin	Aldehydes from oxidation Molar ratios	
					H/V	S/V
1. Bamboo:						
Parts of Phyllostachys heterocycla shoot, compared with mature stem	Upper part of shoot	2.3	5.2	11.3	1.25	0.37
	Medium part of shoot	6.2	7.8	19.6	0.67	0.40
	Basal part of shoot	7.8	8.3	20.3	0.50	0.55
	Mature plant	26.1	14.5	23.3	0.67	2.00
Parts of P. nigra shoot compared with mature plant	Upper part of shoot	2.1	4.3	9.6	1.00	0.30
	Medium part of shoot	6.9	5.9	18.3	0.55	0.61
	Basal part of shoot	7.1	7.5	19.5	0.45	0.64
	Mature plant	23.8	17.1	22.7	0.50	1.75
Parts of immature P. pubescens	2 meters from apex	-	-	-	1.40	0.51
	3 meters from apex	-	-	-	1.10	0.62
	4 meters from apex	-	-	-	0.95	0.67
	7 meters from apex	-	-	-	0.90	0.73
	8 meters from apex	-	-	-	0.91	0.77
2. Fruit shell of Yugao plant	Age: 12 days	1.4	-	-	-	-
	15 days	4.0	1.9	-	-	-
	18 days	11.8	7.3	-	-	-
	30 days	13.3	8.1	-	-	-
	50 days	23.1	11.6	-	-	-
	Mature shell	32.6	15.3	-	-	-

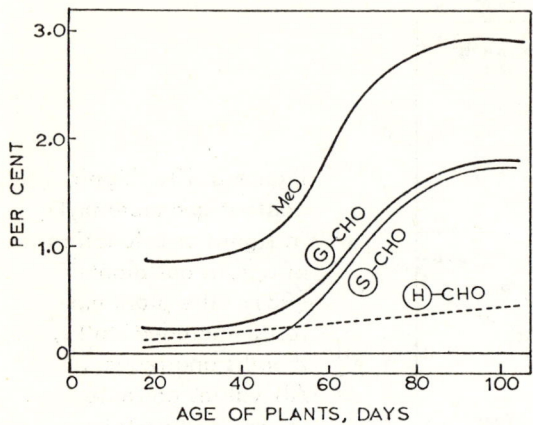

Figure 3.11. Methoxyl content and aldehyde yields obtained from unextracted wheat plants in nitrobenzene oxidation (87).

by a period of rapid lignification which starts when the plant has attained its final height. Simultaneously, the methoxyl content of lignin increases to its final level; and the protein content drops steeply down to approximately 2%. The pattern of lignification in young tissues of other plants appears to be quite similar, and analogous results have been obtained for wheat (87), for bamboo (94) and even for the lignification of the fruit shell of Yugao plant, a dicotyledon (95) (Table 3.22, page 77).

Stone (87) has determined the yields of aldehydes from young wheat plants as a function of age (Figure 3.11). At the immature stage, mainly vanillin and p-hydroxybenzaldehyde with very small amounts of syringaldehyde are obtained. The significance of p-hydroxybenzaldehyde is not quite clear because it may be formed not only from the p-hydroxyphenylpropane units or p-coumaric ester groups in lignin, but also from the tyrosine component of protein (87). Analogous results have been obtained by Higuchi for immature bamboos (Table 3.22). Histological studies have shown that the small amount of lignification that occurs during the immature stage is strictly limited to xylem vessels and some sclerenchyma fibers (96, 97, 98) and the lignin formed is guaiacyl lignin, in accordance with a positive Wiesner but negative Mäule reaction (98).

During the rapid lignification phase, the yields of vanillin and syringaldehyde increase drastically, but that of p-hydroxybenzaldehyde only slightly (87, 94). Lignification spreads from vessels to entire vascular bundles and fibers (96, 97). Parenchyma cells do not lignify, however, until lignification is completed elsewhere; and this process is not observed in bamboo until the plant has attained a height of 5 meters or more (97). In stems, lignification is most advanced at the base, diminishing progressively towards the growing apex (Table 3.22). Surprisingly, the opposite pattern has been

Table 3.23. Phenolic Acids Liberated on Alkaline Hydrolysis of Extracted Corn Plants at Successive Growth Stages (92)

Days after planting	Acids liberated, % of lignin			
	Normal corn plants		Mutant corn plants	
	p-Hydroxy-cinnamic acid	Ferulic acid	p-Hydroxy-cinnamic acid	Ferulic acid
56	5.56	3.21	3.41	3.95
66	8.02	3.09	3.77	3.03
80	6.73	3.43	5.91	4.07
96	5.75	2.49	2.99	2.23

often observed for leaves; the leaf tip being at a more advanced state of lignification than the leaf base (98).

Characteristic changes occur also in the contents of p-coumaric and ferulic ester groups, as shown by Kuc and Nelson (92) in their studies on the maturation of normal and mutant corn plants. Some of their results are quoted in Table 3.23. The amount of both ester groups in lignin increases initially, reaching a maximum during the early maturation period and finally declining. A similar, but more pronounced decline in the p-coumaric and ferulic ester contents of maturing bamboo lignins has been demonstrated by Higuchi et al. (90).

E. Guaiacyl-Syringyl Lignins in Cryptogams. It has been suggested that guaiacyl-syringyl lignins did not appear until in late evolution of plants and therefore are absent from lower plants. Such opinions have been expressed in obvious oversight of the important early findings by Siersch (99) showing that among the cryptogams, four species of Selaginella give a positive Mäule reaction. Later, Gibbs (4) established a positive reaction for 23 species of Selaginella, although it was often found for cortex only. Towers (67) studied the nitrobenzene oxidation of three Selaginella species and found substantial amounts of syringaldehyde in each case, in addition to vanillin and p-hydroxybenzaldehyde. Efforts to locate guaiacyl-syringyl lignins in cryptogams other than Selaginella have so far failed, and the phylogenetic significance of their seemingly premature appearance in the latter family remains to be elucidated.

IV. The Problem of First Appearance of Lignins in the Plant Kingdom. The existence of lignins as essential cell-wall constituents can be considered to be unambiguously proven for all vascular plants including the most primitive species, such as Psilotum. It seems clear, therefore, that in the phylogeny of plants, lignins appeared either concurrently with or prior to the development of conducting tissues. Which one of these alternatives should be accepted has not been solved in a satisfactory manner, and experimental evidence favoring one or the other of the opposing views has been brought forward by recent investigations.

Earlier investigators in this area were generally inclined to consider all non-vascular plants as lignin-free. Gjokic (100) and Linsbauer (101) applied Wiesner reaction to a number of plant species and found the reaction to be negative for all non-vascular plants studied including lichens, mosses, mushrooms and other fungi. Mäule reaction is likewise negative for these plants (4), and the cell walls of an alga (Fucus serratus) and of mosses (Sphagnum) were found to lack the characteristic blue fluorescence of lignified tissues (60). Holmberg (47) was unable to obtain lignin thioglycolic acids from algae, lichens, fungi and mosses; and Kondrat'ev did not succeed in isolating dioxane-acidolysis lignin from Sphagnum (102).

On the other hand, 5.5% of apparent Klason lignin is obtained from the alga Fucus serratus (103) and 13.5% from another alga, Durvillea antarctica, which also yields both vanillin and syringaldehyde in nitrobenzene oxidation (104). No aromatic aldehydes have been isolated from the oxidation of liverworts (Marchantia) or of Funaria (67), while Sphagnum invariably yields p-hydroxybenzaldehyde in amounts corresponding to 0.6 to 0.8% of the plant material (105, 106, 107). In addition, Lindberg and Theander identified minor amounts of vanillin (4% of aldehydes) and trace amounts of syringaldehyde (0.6%) and 5-formyl vanillin (0.1%) in the aldehyde mixture obtained from S. fuscum (105). Nilsson and Tottmar (108) applied cupric oxide oxidation to S. nemoreum obtaining p-hydroxyacetophenone as the main product, with smaller amounts of p-hydroxybenzaldehyde, acetoguaiacone, vanillin, p-hydroxybenzoic and vanillic acids. Farmer and Morrison (109) reported a yield of 4% p-anisic acid on permanganate oxidation of methylated S. cuspidatum. Klason lignin amounts to 5% in S. balticum, containing, however, only 1% methoxyl (105). All these observations are not incompatible with the idea that Sphagnum may contain a primitive lignin of some sort consisting mainly of units of the p-hydroxyphenyl type.

Towards the further clarification of this problem, Freudenberg and Harkin (110) isolated "milled wood lignins" from Sphagnum and Polytrichum commune, and removed associated carbohydrates from these preparations by conversion to their Klason lignins. The compositions of such Klason lignins corresponded to formulae $C_9H_{7.96}O_2[H_2O]_{0.90}[OMe]_{0.25}$ and $C_9H_{7.94}O_2[H_2O]_{0.78}[OMe]_{0.25}$, for Sphagnum and Polytrichum, respectively, and were taken to support the idea that these materials were indeed lignins. Most recently, Reznikov and Sorokina (111) studied a sample of "milled wood lignin" isolated in 0.43% yield from a mature specimen of S. medium. This preparation had a number average molecular weight of 25800, contained 2.33% methoxyl, 8.89% total and 1.3% phenolic hydroxyl groups. Nitrobenzene oxidation yielded p-hydroxybenzaldehyde (2.21%), vanillin (0.82%), p-hydroxybenzoic acid (0.75%) and vanillic acid

(0.5%). Most significantly, reduction with sodium in liquid ammonia gave C_6C_3- compounds, among which guaiacyl propane and guaiacylpropanol were identified.

However, all these data fall short of proving unequivocally the presence of "p-hydroxyphenyl-lignin" in Sphagnum, especially since some serious contradicting evidence has also been uncovered. Ethanolysis has so far been applied to only one moss species, Rhitadiadelphia loerious (62). In this case, the absence of Hibbert's ketones in the reaction products was clearly established. The IR and UV spectra of Sphagnum "lignins" (107, 108, 111) prove only that these preparations are quite different from guaiacyl lignins. Finally, Nilsson and Tottmar (108) have separated the "lignin" from S. nemoreum into several fractions by chromatography and found that while these fractions yielded p-hydroxybenzaldehyde and vanillin in oxidation, they also gave phloroglucinol, orcinol and resorcinol in alkali degradation. This observation conforms better with polyflavanoidal rather than lignin polymers. A polyflavanoid containing ferulic ester groups could, conceivably, yield the degradation products obtained by Reznikov and Sorokina from Sphagnum "milled wood lignin."

While it appears totally premature and probably inappropriate to call the polyphenolic materials isolated so far from Sphagnum and other non-vascular plants "lignins," it is not excluded that minor amounts of lignin polymers might be contained in these materials. Very recent studies by Siegel (111a) on the lignin-like constituents present in the gametophyte axes of the giant New Zealand mosses Dawsonia and Dendroligotrichum have shown that these materials give equally pronounced Wiesner reaction as Psilotum lignin, contain 5 to 8% methoxyl, and yield 14 to 18% of aromatic aldehydes in nitrobenzene oxidation. The exact interpretation of these findings is hardly permissible at the present time, and one can only speculate that perhaps both guaiacyl lignin as well as other polyphenolic constituents coexist as cell-wall constituents in giant mosses.

The exact nature of the non-lignin "structural polyphenols" remains to be evaluated. In Sphagnum, they appear to be chemically bonded to polysaccharides and possibly contribute to the mechanical reinforcement of the cell walls. There are several similarities between Sphagnum polyphenols and the polyphenolic constituents in the cortex of tree-ferns, which indeed appear to perform a reinforcing function similar to that of lignin in higher plants. Finally, there is some merit to considering actual phylogenetic relationship between these polyphenolic materials and the polyphenols isolated from tree barks which will be discussed in the section that follows.

V. Lignins and Phenolic Polymers in Tree Barks. The most recent review of the chemistry of bark (112) gives witness to the confusion on the bark lignin problem which exists in much of the literature. This is a result of the presence in extractive-free bark, especially that of conifers, of substantial quantities of phenolic polymers with a much lower methoxyl content than wood lignin. These materials are extractable with dilute alkaline solutions leaving a residue containing aromatic polymers with methoxyl contents approaching that of wood lignin. Both materials are determinable as Klason lignin, hence many investigators have referred indiscriminately to either or both of these materials as "bark lignin." Other investigators have termed the alkali-soluble polymers "phenolic acids" and suggested that they were structurally related to wood lignin except for the lower methoxyl content.

There appears to be increasing agreement, however, by prominent investigators in the field, i.e., Kurth (113), Higuchi (114), Soga (115) and Hata (116), with the proposal first made by Hergert (117-119): Extractive-free bark does, in fact, contain true lignin, in the sense of being derived from coniferyl, para-coumaryl and/or sinapyl precursors. The bark also contains phenolic polymers structurally related to the condensed tannins, biosynthesized from leucocyanidins, catechins and the like. The alkali-soluble "phenolic acid" fraction consists primarily of the latter along with minor amounts of low molecular weight lignin.

A. Histological Evidence for Lignin in Bark. Bark is anatomically more complex than wood. A number of different types of cells occur in bark. Sieve elements and parenchyma cells are the principal elements with lesser amounts of fibers, sclereids (stone cells) and periderm layers (cork). Relative size and amounts are highly species-dependent, as shown in the studies of Chang (120, 121). Staining studies for lignin in slash pine bark were reported by Lyness and Opdyke (122), while in a much more extensive study Srivastava (123) determined the distribution of lignin in bark and wood samples of 74 angiosperms and 22 gymnosperms by the Wiesner and Mäule reactions. Positive tests indicated presence of lignin in bark, mainly in the walls of primary and secondary phloem fibers, sclereids, some parenchyma cells in the phloem and cortex, and some cells in the periderm. Generally, the gymnosperm bark lignin was Mäule negative; while that in the angiosperm bark was positive, paralleling the behavior of wood lignin. Variability in staining reactions between different types of cells and even in wall layers of the same cell was particularly noticeable among angiosperms. This was interpreted as indicating variable ratios of guaiacyl and syringyl nuclei.

In most trees, outer parts of living bark are separated from inner parts by invasion of the periderm. These separated tissues die and, with the periderm, constitute the rhytidome. In some species, dead parenchyma cells give the lignin reaction which is not exhibited by living cells, suggesting a secondary deposition of lignin upon conversion to rhytidome. Douglas-fir bark, for example, shows a progressively stronger Wiesner reaction from the wood cambium to the periderm and an even stronger reaction in the rhytidome (124). In this connection, it is interesting to note that Holmes and Kurth (125) report 18.8, 31.7 and 79.3% Klason "lignin" for extractive-free, newly formed inner bark, mature inner bark, and outer bark of Douglas-fir, respec-

tively.
Srivastava (123) has concluded on the basis of his studies that the constitution of lignin in bark is comparable to that in wood. Furthermore, the variability of lignin reactions in different cells and considerations of the general structure and development of bark favor the view that lignin is synthesized from precursors generated within the differentiating cell rather than, as held by some lignin chemists, that precursors of lignin are present in cambium and move to differentiating wood and bark elements during their lignification. In support of this, it was pointed out that lignin often occurs in periderm cells of those species which clearly lacked all lignified elements in the living bark.

B. <u>Bark Phenolic Acids</u>. Characterization and degradation studies have been reported for phenolic acids from the bark of redwood (126), Japanese red pine (127 - 130), ponderosa pine (131), slash pine (117, 132), Douglas-fir (122 - 136), white fir (137, 138), Pacific silver fir (139), western hemlock (140), acacia (129, 130), oak (129 - 131), and chestnut (129, 130). These products were isolated, as previously noted, by mineral acid precipitation from a dilute alkaline extract (usually 1% sodium hydroxide at 100°C.) of bark previously extracted with neutral solvents to remove wax, tannins, etc. They are light brown in color and have solubility properties intermediate between phlobaphene and tannin extractives, e.g., partially soluble in water and soluble in polar solvents such as ethanol, aqueous dioxane, etc. In general, they are characterized by 1 - 4% aliphatic hydroxyl groups.

Degradation by nitrobenzene oxidation results in small amounts of typical lignin degradation products, e. g., vanillin, vanillic acid, 5-formyl vanillic acid and p-hydroxybenzaldehyde, and significant amounts of protocatechualdehyde. The latter, of course, is not found among the usual products of lignin degradation unless demethylating conditions are used. The presence of the catechol nucleus therefore has led to the belief that bark phenolic acids might have essentially the same structure as wood lignin with the exception that guaiacyl nuclei are substituted by catechol nuclei. If this were so, it should be possible to degrade bark phenolic acids by acidolysis or other methods to yield catecholic C_6C_3- degradation products. Recent experiments of this type (114 - 116) were completely unsuccessful, so this hypothesis has been rejected. Several investigators have also reported phloroglucinol as a minor but significant degradation product. This finding naturally invited comparison of the bark phenolic acids with the tannin or phlobaphene fraction.

The infrared spectra of bark phenolic acids have been of considerable utility in structural elucidation (117 - 119, 138 - 141). A typical comparison (139) with tannin is shown in Figure 3.12 for the phenolic acid from longleaf pine and derivatives in which the phenolic groups were methylated and all the hydroxyl groups acetylated. The only significant spectral difference is the presence of an absorption band at 1715 cm^{-1}, attributable to a carboxyl group, in the phenolic acids. Since treatment of catechin, synthetic polyleucocyanidin or conifer bark tannins with warm, dilute alkali results in

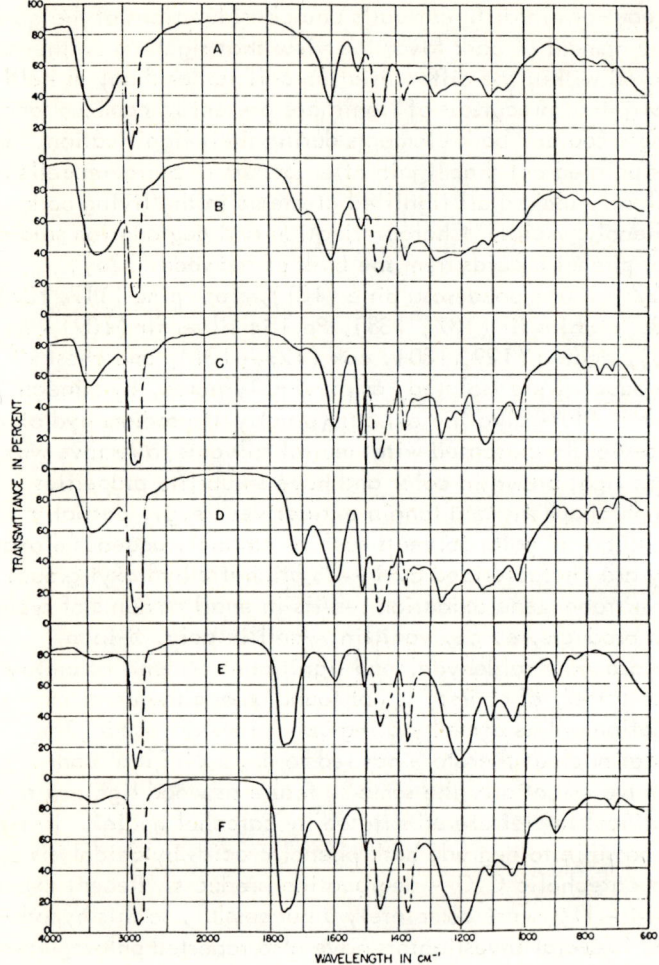

Figure 3.12. Infrared spectra (paraffin mulls) of A, longleaf pine tannin; B, longleaf pine phenolic acid; C, methylated tannin; D, methylated phenolic acid; E, acetylated tannin; and F, acetylated phenolic acid (124).

rearrangement and formation of carboxyl groups, it is very likely that the carboxyl group in bark phenolic acids is an artifact introduced during the isolation procedure. It is also possible that varying amounts of pectic acids are present in crude phenolic acid preparations. This suggests that degradative experiments might more profitably be carried out on extractive-free bark, especially inner bark, since it is less lignified than outer bark.

CLASSIFICATION AND DISTRIBUTION

Catechin has been detected as a product of a very mild acidolysis of extractive-free hemlock (142), pine, spruce and Amabilis fir (144) bark. Treatment with hot isopropanol-hydrochloric acid resulted in detection of cyanidin and delphinidin as major degradation products (117, 139, 140) from extractive-free inner bark. These same products are obtained from coniferous tannins through an oxidation-dehydration mechanism from flavan-3,4-diol nuclei (118). Similar treatment of coniferous bark after alkaline extraction did not yield any cyanidin or cyanidin-like products. This suggested that the product removed with alkali, bark phenolic acid, contained flavan-3,4-diol units prior to its removal. Hergert and coworkers (140) have concluded, therefore, that the phenolic acid fraction, as it exists in the bark, is similar in structure to the bark tannin and phlobaphene, but is not removed by solvent extraction due to molecular size or a three-dimensional polymeric network. Hemlock, pine, fir and, probably, most other coniferous tannins have the structure 1.

1

Coniferous bark phenolic acids in situ would be expected to have the same structure, possibly with an additional carbon-carbon linkage in the phloroglucinol ring to give a three-dimensional structure. Isolated phenolic acids will have a somewhat different structure resulting from the alkaline rearrangement. Further model compound work must be done before the exact nature of this rearrangement can be specified.

Since condensed and hydrolyzable tannins frequently co-occur in angiosperm bark, angiosperm phenolic acids could be expected to be more complex in structure. Acidolytic degradations on hardwood barks have not yet been reported, but it can be expected that gallic acid, ellagic acid, etc., types of degradation products might be expected in addition to anthocyanidins.

C. Bark Lignin. The most convenient method for isolation of lignin from bark is by extraction with dioxane-hydrochloric acid (115, 117, 133-135, 138-141, 143-146). Lignin has also been isolated from bark by

Table 3.24. Comparison of Dioxane-HCl Lignins from Amabilis Fir Wood and Bark

	% C	% H	% Methoxyl
Wood lignin	63.6	6.0	15.7
Bark lignin	64.3	6.0	16.3
Methylated wood lignin	66.1	6.4	24.3
Methylated bark lignin	66.6	5.5	24.8
Acetylated wood lignin	62.3	5.9	11.7
Acetylated bark lignin	63.5	5.7	12.2

Björkman's method (115, 129, 130). While the latter technique is less convenient, yields are lower, and phenolic acid contaminants more difficult to remove (124); bark lignins isolated by this method are undoubtedly less chemically altered than those isolated by the acidolytic procedure. Generally, the elemental composition and functional group content are very similar to lignin isolated in the same manner from wood of the corresponding species. A typical comparison is presented in Table 3.24 for dioxane-HCl lignins isolated from Amabilis fir wood and bark (139).

Bark lignin from some species has 1 − 2% lower methoxyl content than wood lignin. It is not certain whether this represents contamination from the co-occurring low methoxyl content phenolic acids or slightly higher ratios of p-hydroxyphenyl to guaiacyl nuclei in the coniferous species and guaiacyl to syringyl nuclei in hardwoods. The latter seems likely since some of the cells in angiosperm barks giving positive Wiesner tests give a negative Mäule reaction (123).

Treatment of wood or bark with dioxane-HCl mixtures at elevated temperatures not only results in extraction of polymeric lignin fractions but partial cleavage of the lignin into C_6C_3- monomers and dimers, usually in yields of 2-4% of the original lignin content (117). Acidolysis with dioxane-HCl is thus similar to the more widely known ethanolysis technique except that it does not suffer from etherification or acetal formation which takes place in the latter. Products obtained from treatment of longleaf pine, Amabilis fir, western hemlock and Sitka spruce bark were p-coumaraldehyde, coniferylaldehyde, vanilloyl methyl ketone, α-hydroxypropioguaiacone, guaiacyl acetone, vanillin, and acetoguaiacone, all of which are typical but distinctive wood lignin degradation products (117, 124, 139). Similarly, treatment of extractive-free black gum (Nyssa aquatica) bark yielded a mixture of these same products and the corresponding syringyl analogues (124). The formation of these products indicates the presence of the major wood lignin structural units, β-etherified guaiacyl or syringyl glycerol. Treatment of extractive-free longleaf pine and Amabilis fir bark with dioxane-water according to the technique

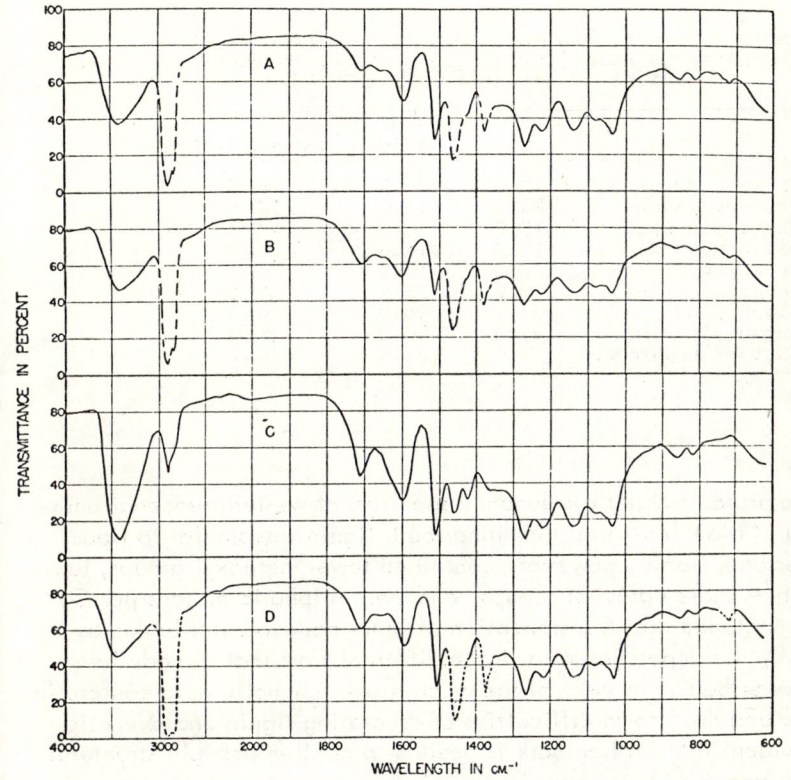

Figure 3.13. Infrared spectra dioxane acidolysis lignins isolated from A. Longleaf pine wood, B. longleaf pine bark, C. Amabilis fir wood and D. Amabilis fir bark. Spectras A, B and D measured as paraffin mulls; Spectrum C as a KBr wafer (124).

of Sakakibara and Nakayama (147) yielded coniferyl alcohol, para-coumaryl alcohol, β-hydroxyconiferyl alcohol and a substituted guaiacyl-glycerol derivative in small amounts (124). The same products were obtained from extractive-free wood. Fungal degradation of Björkman lignin from western white pine has been reported to yield the same products as obtained from wood (116, 148). Particularly significant was the detection of guaiacyl-glycerol and guaiacyl-glycerol-β-O-4-coniferyl alcohol. Degradation of bark lignins by nitrobenzene oxidation (114, 141, 143-145), copper chromite hydrogenolysis (115), ethanolysis (114), and permanganate oxidation (130) have also yielded the same products as the corresponding wood lignin.

Comparison of ultraviolet (neutral and alkaline) and infrared spectra of bark lignins isolated by dioxane-HCl (117, 128, 139, 141, 146) and the Björkman technique (129) with corresponding wood lignins has confirmed the essential identity of these materials. A comparison of the infrared spectra of bark and wood lignin from Amabilis fir and longleaf pine is shown in Figure 3.13.

Table 3.25. Lignin and Phenolic Acid Content of Coniferous Bark (152). (Percentages Based on Weight of Extractive-Free Bark)

	Lignin	Phenolic acid	Carbohydrate
Balsam fir	18.6	15.7	62.9
Western larch	24.5	13.0	62.8
Engelmann spruce	15.0	15.9	64.1
Black spruce	18.9	14.3	65.0
Jack pine	18.8	36.3	45.1
Lodgepole pine	9.9	17.2	72.2
Slash pine	31.7	28.9	38.3
Sugar pine	27.3	40.5	30.3
Eastern hemlock	27.6	21.6	52.2

A study of the proton magnetic resonance spectrum of western redcedar bark lignin by Swan (146) indicated that inner bark lignin was similar to wood lignin. Outer bark lignin, however, contained fewer methoxyl groups, fewer aromatic protons, more catechol groups, and more aliphatic protons per C_9 unit. Swan concluded that transformation of inner bark to outer bark was accompanied by the deposition of a lignin different from that already laid down in the inner bark. In view of the much higher phenolic acid content in the outer bark and the known difficulties of separating lignin and phenolic acids, it is evident that further work is needed to confirm possible structural differences in inner and outer bark lignin.

D. Proximate Analysis of Bark Lignin. Various procedures have been used in attempts to estimate the lignin content of bark (149-152) None of these is wholly satisfactory because of the current lack of satisfactory methods for isolating phenolic acids completely free of lignin impurities. Since extractive-free bark consists primarily of three components, polymeric carbohydrates (holocellulose), polymeric polyphenols (termed "phenolic acids" when isolated by alkaline extraction) and lignin, it is possible to estimate the relative amounts of the three fractions by a holocellulose determination and standard 72% insoluble "lignin" determinations before and after quantitative 1% sodium hydroxide extraction at 100°C. The 72% acid-insoluble residue after alkaline extraction is considered to represent the true lignin content of the bark. The phenolic acid content can be calculated by subtracting the percent acid-insoluble residue of the caustic-extracted bark (calculated back to a noncaustic-extracted basis) from the acid-insoluble residue of the original noncaustic-extracted bark (140). Using this procedure, the data of Chang and Mitchell (152) were recalculated to give the results for lignin content in a number of representative conifer barks as shown in Table 3.25.

Looking to the future, it is apparent that further experimental work will be required to completely resolve the bark lignin question. In particular, better methods for isolating phenolic acid fractions free of lignin are required. Finally, lignin isolation and characterization from discrete cell types, such as the study of sclereids by Haas and Kremers (150), will be a necessary adjunct.

REFERENCES

1. B. Leopold and I. L. Malmström, Acta Chem. Scand., 6, 49 (1952).
2. R. H. J. Creighton, R. D. Gibbs and H. Hibbert, J. Am. Chem. Soc., 66, 32 (1944).
3. R. H. J. Creighton and H. Hibbert, J. Am. Chem. Soc., 66, 37, (1944).
4. R. D. Gibbs, in The Physiology of Forest Trees (E. Thimann, Ed.) 1958, pp. 269-312.
5. L. Kawamura and T. Higuchi, Chim. Biochim. Lignine Cellulose Hemicelluloses, Grenoble, 1964, p. 469.
6. F. E. Brauns and D. A. Brauns, The Chemistry of Lignin Supplement Volume, Academic Press, 1960, pp. 157-167.
7. T. Arima, M. Sc. Thesis, Univ. of Washington, Seattle, 1967.
8. R. W. Kennedy and J. M. Javorsky, Tappi, 43, 25 (1960).
9. H. J. Kiefer and E. F. Kurth, Tappi, 36, 14 (1953).
10. G. A. Richter, Pulp and Paper Ind., 22 (4), 48 (1948).
11. R. Katzen, J. Pearlstein, R. E. Muller and D. F. Othmer, Tappi, 33, 67 (1950).
12. T. E. Timell, Tappi, 40, 30, 568 (1957).
13. D. Narayanamurti and N. R. Das, Holz Roh- u. Werkstoff, 13, 52 (1955).
14. K. Hata, J. Japan. For. Soc., 32, 257 (1950).
15. E. Anderson and W. W. Pigman, Science, 105, 601 (1947).
16. A. B. Wardrop, Australian J. Sci. Res. Ser. B., 4, 391 (1951).
17. B. J. Zobel and R. L. McElwee, Tappi, 41, 158 (1958).
18. W. A. Côté, Jr., and A. C. Day, in Cellular Ultrastructure of Woody Plants (W. A. Côté, Ed.), Syracuse Univ. Press, 1965, p.391.
19. W. A. Côté, Jr., N. P. Kutscha, B. W. Simson and T. E. Timell, Tappi, 50, 350 (1967).
20. T. E. Timell, in Cellular Ultrastructure of Woody Plants (W. A. Côté, Jr., Ed.) Syracuse Univ. Press, 1965, p. 127.
21. D. E. Bland, Holzforschung, 15, 102 (1961).
22. I. Kawamura and T. Higuchi, J. Japan.Wood Res. Soc., 8, 148 (1962).
23. B. Choulet, A. Robert, T. Higuchi and F. Barnoud, Chim. Biochim. Lignine Cellulose Hemicelluloses, Grenoble, 1964, p. 481.
24. G. J. Ritter and L. C. Fleck, Ind. Eng. Chem., 14, 1050 (1922);

15, 1055 (1923); 18, 608 (1926).
25. H. E. Dadswell and L. F. Hawley, Ind. Eng. Chem., 21, 973 (1929).
26. A. J. Bailey, Ind. Eng. Chem. Anal. Ed., 8, 52 (1936).
27. C. E. Curran, Pulp Paper Mag. Can., 37, 646 (1936).
28. N. Jansons, Latv. Psr. Zin. Akad. Izdev., 11, 127 (1950), quoted in ref. 31.
29. P. R. Larson, Forest Prod. J., 16, 37 (1966).
30. J. W. Wilson and R. W. Wellwood, in Cellular Ultrastructure of Woody Plants (W. A. Côté, Jr., Ed.) Syracuse Univ. Press, 1965, p. 551.
31. T. T. Wu and J. W. Wilson, Pulp Paper Mag. Can., 68, 159 (1967).
32. G. B. Squire, Ph.D. Thesis, Univ. of British Columbia, 1967.
33. R. W. Wellwood, G. Ifju and J. W. Wilson, Ref. 30, p. 539.
34. I. B. Sacks, I. T. Clark and J. C. Pew, J. Polym. Sci., Part C, No. 2, 213 (1963).
35. H. Meier, Holz Roh- u. Werkstoff, 13, 323 (1955).
36. P. W. Lange, Svensk Papperstidn., 57, 525, 533 (1954); Norsk Skogsind., 11, 425 (1957).
37. A. B. Wardrop and D. E. Bland, in Biochemistry of Wood (K. Kratzl and G. Billek, Eds.) Pergamon Press, 1959, p. 92.
38. A. R. Procter, W. Q. Yean and D. A. I. Goring, Pulp Paper Mag. Can., 68, T-445 (1967).
39. J. A. N. Scott, A. R. Procter, B. J. Fergus and D. A. I. Goring, Wood Science and Technology, in press.
40. B. J. Fergus, Ph.D. Thesis, McGill Univ., Montreal, Canada, 1968.
41. G. P. Berlyn and R. E. Mark, Forest Prod. J., 15, 140 (1965).
42. H. Harada and A. B. Wardrop, J. Japan.Wood. Res. Soc., 6, 34 (1960).
43. J. J. Balatinecz and R. W. Kennedy, For. Prod. J., 17, 57 (1967).
44. A. B. Wardrop, Tappi, 40, 225 (1957).
45. R. W. Kennedy and J. L. Farrar, in Cellular Ultrastructure of Woody Plants (W. A. Côté, Jr., Ed.) Syracuse Univ. Press, 1965, p. 419.
46. A. Björkman and B. Person, Svensk Papperstidn., 60, 158, 285, (1957).
47. B. Holmberg, Ing. Vetenskaps Akad. Handl., No. 103 (1930).
48. K. Kratzl and G. Billek, Holzforschung, 10, 161 (1957); K. Kratzl and W. Schweers, Monatsh. Chem., 85, 1046 (1954); Chem.Ber. 89, 186 (1956).
49. K. V. Sarkanen, H. M. Chang and G. G. Allan, Tappi, 50, 583, 587 (1967).
50. K. Freudenberg and B. Lehmann, Chem. Ber. 96, 1850 (1963).
51. H. Mikawa, K. Sato, C. Takasaki and K. Ebisawa, Bull. Chem. Soc. Japan, 29, 254 (1956).
52. D. C. C. Smith, J. Chem. Soc., 1955, 2347.
53. E. Hägglund and S. Ljungren, Papierfabrikant, 31, 35 (1953).
54. D. E. Bland, Holzforschung, 15, 102 (1961).
55. A. M. Latif, Ph.D. Thesis, Univ. of Washington, Seattle, 1968.

56. V. P. F. F. Lee, M. Sc. Thesis, Univ. of Washington, Seattle, 1968.
57. C. Jacquiot, Compt. rend. France, 236, 960 (1953).
58. F. Barnoud, Ph.D. Thesis, Univ. of Grenoble, 1962.
59. T. Higuchi and F. Barnoud, Chim. Biochim. Lignine Cellulose Hemicelluloses, Grenoble, 1964, p. 255; J. Japan Wood Res. Soc., 12, 36 (1966).
60. S. M. Manskaya and M. N. Koehneva, Doklady Akad. Nauk,S.S.S.R., 6, 505 (1948).
61. K. Kratzl and J. Eibl, Mitt. Österr. Ges. Holzforsch., 3, 77, (1951).
62. K. V. Sarkanen and A. M. Latif, unpubl. results.
63. I. M. Vener, Izvest. Akad. Nauk S.S.S.R., Otdel Tekh. Nauk.1947, 843; quoted in Ref. 6, p. 14.
64. T. Higuchi, Private comm.
65. T. E. Timell, Svensk Papperstidn., 65, 266 (1962).
66. T. E. Timell, Svensk Papperstidn., 67, 356 (1964).
67. G. H. N. Towers, Ph.D. Thesis, McGill Univ., 1951; quoted in Ref.4.
68. E. White and G. H. N. Towers, Phytochemistry, 6, 663 (1967).
69. E. White, A. Tse and G. H. N. Towers, Nature, 213, 285 (1967).
70. D. A. I. Goring and T. E. Timell, Tappi, 45, 454 (1962).
71. H. Schindler, Zeitschr. Wiss. Mikroskopie, 48, 289 (1931).
72. T. E. Timell, Wood Science and Tech., 1, 45 (1967).
73. L. P. Clermont and F. Bender, Pulp Paper Mag. Can., 59, No. 7, 139 (1958).
74. G. Jayme and M. Harders-Steinhauser, Papier, 4, 104 (1950).
75. A. B. Wardrop and H. E. Dadswell, Austr. Sci. J. Res., B 1, 3(1948).
76. K. Kawamura, Sci. Rept. Agr. For. School, Gifu Univ., No. 64, 25 (1948); No. 67, 1 (1949); quoted in Ref. 6, p. 140.
77. B. Swan, Svensk Papperstidn., 68, 791 (1965).
78. D. E. Bland, Holzforschung, 20, 12 (1966).
79. J. Uchida, J. Soc. Chem. Ind. Japan, 44, 1072 (1941).
80. K. Freudenberg and G. S. Sidhu, Holzforschung, 15, 33 (1961).
81. D. E. Bland, G. Ho and W. E. Cohen, Austral. J. Sci. Res. Ser.A, 3, 642 (1950).
82. I. Kawamura and T. Higuchi, J. Japan.Wood Res. Soc., 12, 178 (1966).
83. W. Tracey, K. V. Sarkanen and G. G. Allan, unpubl. results.
84. J. M. Casey and K. V. Sarkanen, unpubl. results
85. K. Kratzl, Experientia, 4, 100 (1948).
86. T. Higuchi, Y. Ito and I. Kawamura, Phytochem., 6, 875 (1967)
87. J. E. Stone, M. J. Blundell and K. G. Tanner, Can. J. Chem., 29, 734 (1951).
88. K. Kratzl and P. Claus, Monatsh. Chem., 93, 219 (1962).
89. T. Higuchi and I. Kawamura, Holzforschung, 20, 16 (1966).
90. T. Higuchi, N. Kimura and I. Kawamura, J. Japan Wood Res. Soc.,

12, 173 (1966).
91. D. C. C. Smith, Nature, 176, 267, 927 (1955).
92. J. Kuc and O. E. Nelson, Arch. Biochem. Biophys., 105, 103 (1964).
93. M. Phillips, M. J. Goss and B. L. Davis, J. Agr. Res., 59, 319 (1939).
94. T. Higuchi, Physiol. Plantarum, 10, 633 (1957).
95. I. Tachi, A. Hayashi and A. Sato, Bull. Agr. Chem. Soc. Japan, 21, 279 (1957).
96. W. J. Pigden, Can. J. Agr. Sci., 33, 364 (1953).
97. T. Higuchi, I. Kawamura and H. Ishikawa, J. Japan Forestry Soc., 35, 258 (1953).
98. A. B. Wardrop, Tappi, 40, 225 (1957).
99. E. Siersch, Mikrochemie, 4, 188 (1926).
100. G. Gjokic, Österr. Botan. Z., 45, 330 (1895).
101. K. Linsbauer, Österr. Botan. Z. 49, 317 (1899).
102. E. V. Kondrat'ev, Zh. Prikl. Khim., 22, 753 (1949).
103. S. M. Manskaya, Doklady Akad. Nauk S.S.S.R., 7, 611 (1947).
104. R. Ripa and L. M. Borquez, Scientie (Valparaiso, Chile), 34, 36 (1967); quoted in Chem. Abstr. 1969, 84125.
105. B. Lindberg and O. Theander, Acta Chem. Scand., 6, 311 (1952).
106. R. I. Morrison, J. Soil Sci., 9, 130 (1958).
107. V. C. Farmer and R. I. Morrison, Geochim. Cosmochim. Acta, 28, 1537 (1964).
108. E. Nilsson and O. Tottmar, Acta Chem. Scand. , 21, 1558 (1967).
109. V. C. Farmer and R. I. Morrison, Chem. Ind. (London), 1955, 231.
110. K. Freudenberg and J. M. Harkin, Holzforschung, 18, 166 (1964).
111. V. M. Reznikov and N. F. Sorokina, Zh. Prikl. Chim., 41, 176 (1968); Vestsi Akad. Navuk Belarus. S.S.R., Ser. Khim. Navuk, 1968, 104, quoted in Chem. Abstr. 1969, 20541.
111a. S. M. Siegel, Amer. J. Bot., 56, 175 (1969).
112. W. Jensen, K. E. Fremer, P. Sierila and V. Wartiovaara, "The Chemistry of Bark," in The Chemistry of Wood (B. L. Browning, Ed.), Interscience, New York-London, 1963, pp. 589-620.
113. E. F. Kurth, K. Aida and M. Fujii, Abstracts of Papers, 155 National Meeting Am. Chem. Soc., 22D (1968).
114. T. Higuchi, Y. Ito, M. Shimada and I. Kawamura, Cellulose Chem. Tech., 1, 585 (1967).
115. M. Sogo, T. Ishihara and K. Hata, J. Japan. Wood Res. Soc., 12, 96 (1966).
116. K. Hata, W. J. Schubert and F. F. Nord, Arch. Biochem. Biophys., 113, 250 (1966).
117. H. L. Hergert, Abstracts of Papers, 133 Meeting of the American Chemical Society, 7E (1958).
118. H. L. Hergert, For. Prod. J., 10, 610 (1960).

119. H. L. Hergert, "Economic Importance of Flavonoid Compounds: Wood and Bark," in The Chemistry of Flavonoid Compounds (T. A. Geissman, Ed.) Pergamon, Oxford, 1962, p. 573.
120. Y. Chang, Anatomy of Common North American Pulpwood Barks, Tappi Monograph Series No. 14, 1954.
121. Y. Chang, Bark Structure of North American Conifers, Tech. Bull. 1095, U. S. Dept. Agriculture, 1955.
122. W. I. Lyness and D. L. Opdyke, Abstracts of Papers, 133 Meeting of the Am. Chem. Soc., 6E (1958).
123. L. M. Srivastava, Tappi, 49, 173 (1966).
124. H. L. Hergert, unpublished work.
125. G. W. Holmes and E. F. Kurth, Tappi, 44, 893 (1961).
126. H. F. Lewis, F. E. Brauns, M. A. Buchanan and E. F. Kurth, Ind. Eng. Chem., 36, 759 (1944).
127. K. Hata and M. Sogo, J. Japan.Forestry Soc., 38, 473 (1956).
128. M. Sogo and K. Hata, J. Japan.Wood Research Soc., 8, 112 (1962).
129. M. Sogo, J. Japan.Wood Research Soc., 12, 293 (1966).
130. M. Sogo and K. Hata, J. Japan.Wood Research Soc., 13, 300 (1967).
131. E. F. Kurth, J. K. Hubbard, and J. D. Humphrey, For. Prod. Res. Soc., 3, 276 (1949).
132. W. F. Erman and W. I. Lyness, Tappi, 48, 249 (1965).
133. H. L. Hergert and E. F. Kurth, Tappi, 35, 59 (1952).
134. H. J. Kiefer and E. F. Kurth, Tappi, 36, 14 (1953).
135. E. F. Kurth and J. E. Smith, Pulp Paper Mag. Can., 55 (12), 125 (1954).
136. M. Fujii and E. F. Kurth, Tappi, 49, 92 (1966).
137. M. D. Fahey and E. F. Kurth, Tappi, 40, 586 (1957).
138. H. L. Hergert, For. Prod. J., 8, 335 (1958).
139. H. L. Hergert, paper presented at the Northwest Regional Meeting of the Am. Chem. Soc., 18 June 1959.
140. H. L. Hergert, L. E. Van Blaricom, J. C. Steinberg and K. R. Gray, For. Prod. J., 15, 485 (1965).
141. K. Hata and M. Sogo, J. Japan.Wood Research Soc., 6, 71 (1960).
142. H. L. Hergert, Abstracts of Papers, 155 National Meeting of the Am. Chem. Soc., 21D (1968).
143. K. Hata and M. Sogo, J. Japan.Wood Research Soc., 4, 5 (1958).
144. K. Hata and M. Sogo, J. Japan.Wood Research Soc., 4, 85 (1958).
145. M. Sogo, Tech. Bull. Fac. Agr. Kagawa Univ., 10 (1), 46 (1959).
146. E. P. Swan, Pulp Paper Mag. Can., 67 (10), T456 (1966).
147. A. Sakakibara and N. Nakayama, J. Japan.Wood Research Soc., 8, 157 (1962).
148. F. F. Nord and K. Hata, Current Aspects Biochem. Eng., 315 (1966).
149. B. L. Browning and L. O. Sell, Tappi, 40, 362 (1957).
150. B. R. Haas and R. E. Kremers, Tappi, 44, 747 (1961).

151. E. F. Kurth, Tappi, <u>32</u>, 175 (1949).
152. Y. Chang and R. L. Mitchell, Tappi, <u>38</u>, 315 (1955).

4

PRECURSORS AND THEIR POLYMERIZATION

K. V. Sarkanen

1. Biogenesis of Lignin Precursors. The biogenetic pathway from glucose to lignin precursors and the role of participating enzymes have received extensive clarification during recent times. The developments in this area have been reviewed more frequently than other areas of lignin chemistry, and more than fifteen outstanding treatises have appeared during the past ten years (1). The following presentation will attempt only to summarize the highlights of the most important aspects of the subject.

A. Application of Tracer Methods. The use of ^{14}C-labelled compounds provides the most important tool for the clarification of the biosynthetic pathway in lignin formation. For the determination of the overall effectiveness of incorporation, randomly labelled compounds generally perform equally well as those labelled at specific positions. Lignins adapt themselves particularly well to tracer studies because the final in vivo polymerization process is essentially irreversible. For the determination of the radioactivity of lignin, Klason lignin preparations have been found unsuitable (2) because they often contain condensed protein impurities. The three aldehydes, vanillin, syringaldehyde and p-hydroxybenzaldehyde, formed in the nitrobenzene oxidation, are well suited for the assessment of radioactivity and have been widely utilized for this purpose. Furthermore, they allow for separate determinations of incorporation in the three aromatic moieties in lignins. On the other hand, the C_6C_1-aldehydes contain only seven out of the nine skeletal carbons, and consequently the activity of the β- and γ- carbons in lignin remains unknown. In Hibbert's ketones all original carbon atoms are preserved, and they are thus the most appropriate products for the determination of the overall radioactivity.

Specifically labelled compounds are useful in establishing the correspondence of individual carbons in the lignin units on one hand and in the precursor molecules on the other. The absence of carbon shuffling can be ascertained in this fashion. Special methods have been developed to determine the distribution of radioactivity in C_6C_1-aldehydes and Hibbert's ketones. The procedure developed by Eberhard and Schubert (3) for vanillin is illustrated in Figure 4.1. The carbonyl-carbon is detached by means of Dakin's reaction in the form of sodium formiate which can be counted after conversion to barium carbonate. Nitro groups can be selectively introduced to positions 5, 6 or 2, by nitration of vanillin, veratraldehyde 1 or vanillin acetate 2, respectively, and the ring carbons carrying nitro groups are converted to bromopicrine 3 by treatment with bromine and calcium hydroxide. Finally, the radioactivity of methoxyl carbon atoms is assessed by counting tetramethyl ammonium iodide, obtained from the methyl iodide formed in the Zeisel determination of methoxyl groups.

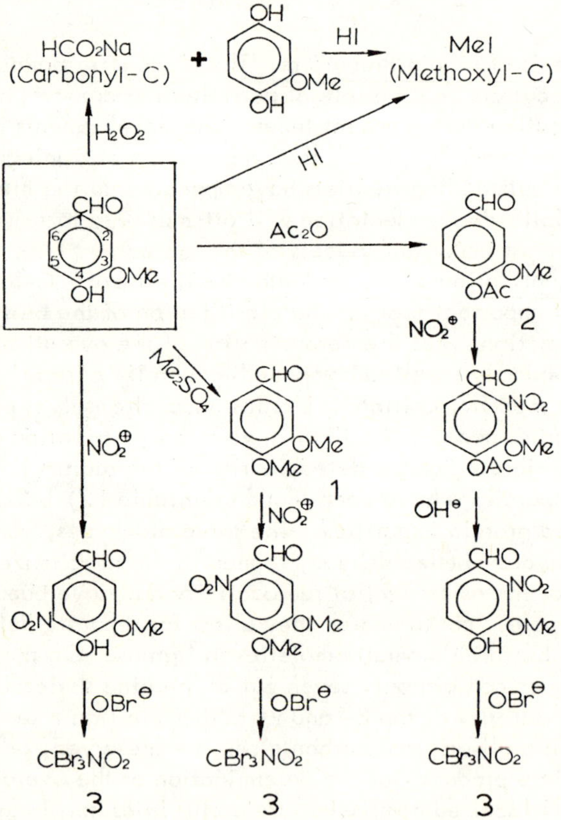

Figure 4.1. Chemical conversions utilized in the assessment of radioactivities of specific locations in vanillin, according to Eberhardt and Schubert (3,4).

PRECURSORS AND THEIR POLYMERIZATION

A method to determine the distribution of radioactivity in the side-chain carbons of Hibbert's ketones has been developed by Kratzl (5) and is shown in Figure 4.2. The carbon in the γ-position may be isolated as iodoform and counted. Both β- and γ-carbons can be obtained in the form of acetic acid by subjecting the diketone to periodate oxidation and separately assessed by successive conversions to acetone and iodoform. The α-carbon remains attached to the aromatic nucleus in the vanillic acid formed in the periodate oxidation.

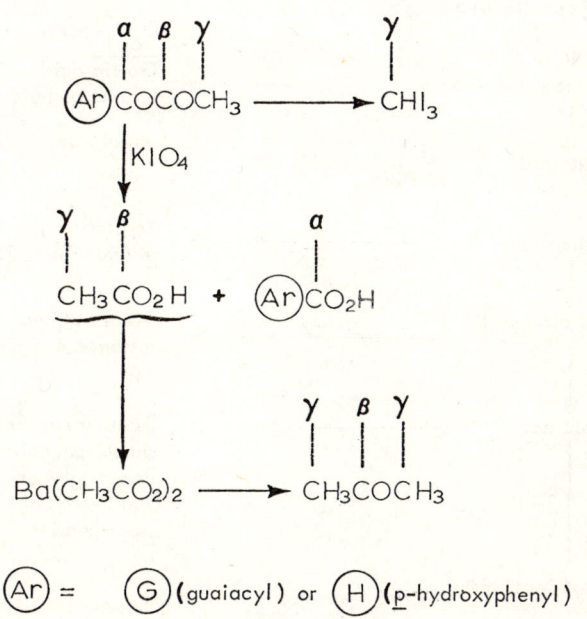

Figure 4.2. Assessment of radioactivity in the side-chain carbons of Hibbert's ketones, according to Kratzl et al (5).

B. **Shikimic Acid Pathway in Higher Plants.** Extensive tracer studies on the biogenesis of plant constituents have demonstrated that the majority of aromatic constituents in higher plants are generated through the shikimic acid pathway. These include many C_6C_1-compounds, tryptophan and C_6C_3-components, such as phenylalanine, tyrosine, cinnamic acids, coumarins and lignans, alkaloids and other aromatic nitrogen compounds, etc.

(6). The general sequence of transformations in the shikimic acid pathway is illustrated in Figure 4.3. Another group of aromatic constituents, the majority of phenolic compounds in fungi (7) for example, is generated through the acetate pathway. The latter are often recognizable from their 1,3-dihydroxylation pattern, while 1,2-dihydroxylation is more common in aromatics generated by the shikimic acid route.

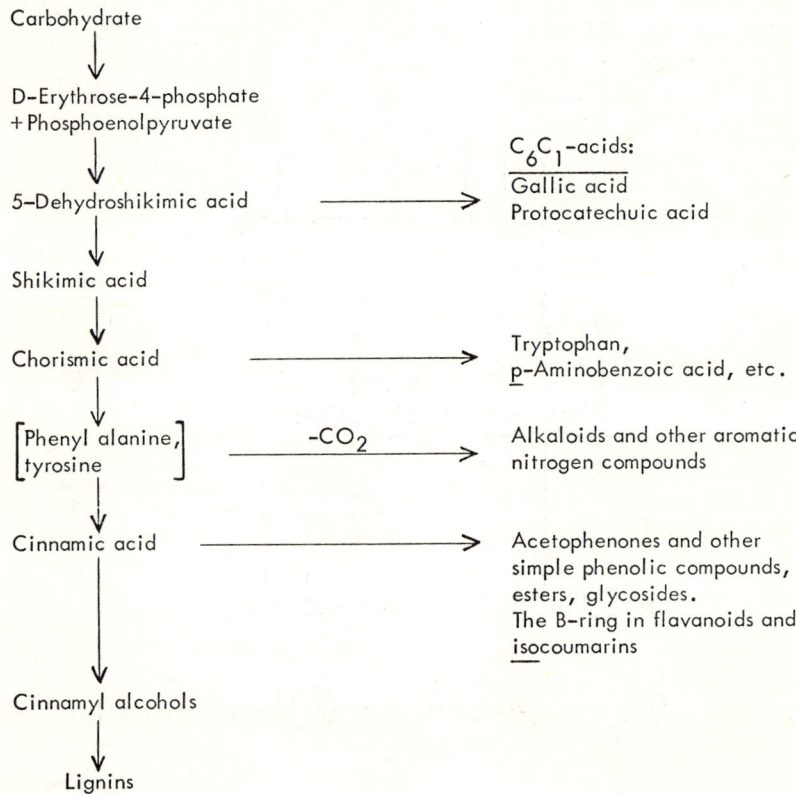

Figure 4.3. Generation of aromatic constituents in higher plants through the shikimic acid pathway.

Several outstanding reviews on the shikimic acid pathway have been published (8, 12) and only a short summary will be given here. Actually, the basic shikimic acid-phenylalanine and shikimic acid-tyrosine routes have not been studied in great detail in higher plants, and most of the details of intermediates and contributing enzymes have been obtained from studies on bacteria, notably Escherichia coli and its mutants (9) and Neurospora (10). The complete sequence of intermediates along the pathway is shown in Figures 4.4 and 4.5. The exclusiveness of the pathway in Escherichia has been convincingly demonstrated by Davis and coworkers (11). For example, a deficient mutant in which the formation of shikimic acid was blocked showed a quintuple nutritional requirement consisting of phenylalanine, tyrosine, tryptophan, p-aminobenzoic acid and p-hydroxybenzoic acid. All five of these compounds had to be added simultaneously to get good growth. On the other hand, the addition of shikimic acid alone was sufficient to replace the need for all five supplements. This observation demonstrated that shikimic acid played the role of an obligate intermediate in the synthesis of the five aromatic compounds.

Similar deficient mutants of higher plants are not available and cannot be readily obtained. This circumstance has prevented obtaining a rigorous proof for the exclusiveness of the shikimic acid pathway for higher plants, as has been possible in the case of bacteria. On the other hand, all results of both enzymatic and tracer studies are in full conformity with the idea that the synthesis of aromatics occurs in higher plants along a pathway similar to, if not identical with that observed for bacteria. 5-Dehydroquinase and dehydroshikimic acid reductase are examples of specific enzymes isolated and characterised from a number of plant sources (12). These enzymes appear to be clearly associated with the shikimic acid pathway catalyzing reactions e and f, respectively, Figure 4.4. The conversion of glucose-6-phosphate, or a mixture of phosphoenol pyruvate and erythrose-4-phosphate, to dehydroshikimic acid has been accomplished using cell-free extracts of mung beans (13). Tracer studies have shown that [^{14}C] shikimic acid can be converted to both phenylalanine and tyrosine in plants belonging to three different families (14).

The wide-spread occurrence of shikimic acid in plant tissues is in accordance with the importance of this intermediate in plant metabolism (15). In about two-thirds of 300 plant species, shikimic acid has been positively identified as a common constituent of leaves and young stems (15). Higuchi (16) has shown by means of histochemical methods that cross sections of woody stems show a strong positive reaction for shikimic acid in phloem and cambium, a weaker reaction in newly formed wood and no reaction in fully lignified wood.

C. Shikimic Acid Pathway in Lignin Formation. The first evidence for the operation of the shikimic acid pathway in lignin formation was provided by Brown and Neish (17) who demonstrated that randomly labelled shikimic acid and phenyl alanine were outstandingly efficient precursors for lignins in wheat (Triticum aestivum) and maple (Acer negundo). The evidence obtained from studies utilizing specifically labelled model compounds has been even more convincing. It should be noted here that shikimic acid formed in an organism from a radioactive glucose labelled at carbons 1 or 6

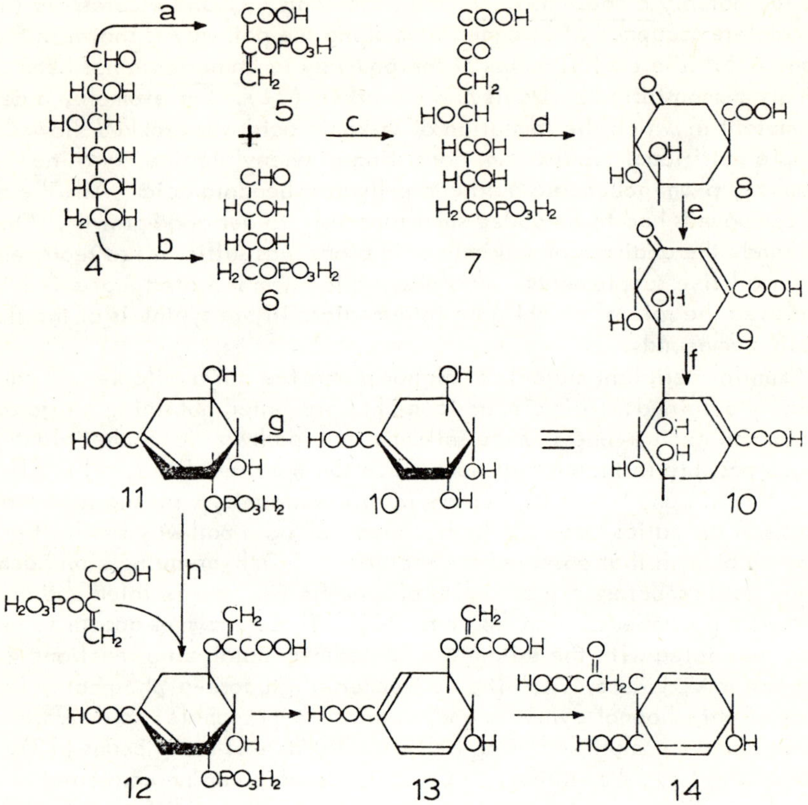

4 : D-Glucose
5 : Phosphoenolpyruvic acid
6 : D-Erythrose-4-phosphate
7 : 7-Phospho-3-deoxy-D-arabino-heptulosonic acid (DAHP)
8 : 5-Dehydroquinic acid
9 : 5-Dehydroshikimic acid
10: Shikimic acid
11: 5-Phosphoshikimic acid
12: 3-O-(α-Carboxyvinyl)-5-phosphoshikimic acid
13: Chorismic acid
14: Prephenic acid

a. Glycolytic pathway
b. Pentose phosphate pathway
c. DAHP- synthetase
d. 5-Dehydroquinate synthetase ($+Co^{++}$, NAD)
e. 5-Dehydroquinase
f. 5-Dehydroshikimic reductase (+NADP, NADPH)
g. Kinase (?) (+ATP)
h. 3-Enolpyruvylshikimate synthetase

Figure 4.4. Shikimic acid pathway from glucose to prephenic acid.

(or both) with ^{14}C, carried the label at positions 2 and 6. This is due to the fact that the carboxyl carbon in phosphoenolpyruvate $\underline{5}$ derives equally from positions 3 and 4 in glucose, the middle carbon from positions 2 and 5 and the methylene carbon from positions 1 and 6 (18) in accordance with the glycolysis mechanism. On the other hand, carbon atoms 2, 3 and 4 in erythrose-4-phosphate (6) are derived from hexose carbons 3 and 4, 2 and 5, and 1 and 6, respectively, except that in this case incorporation of carbons 4, 5 and 6 is somewhat greater than that of carbons 1, 2 and 3. Consequently, carbons 1 and 6 in glucose occupy positions 3 and 7 in 3-deoxy-D-arabino-heptulosonic acid 7-phosphate (DAHP) $\underline{7}$ and positions 2 and 6 in shikimic acid $\underline{10}$, equation 1.

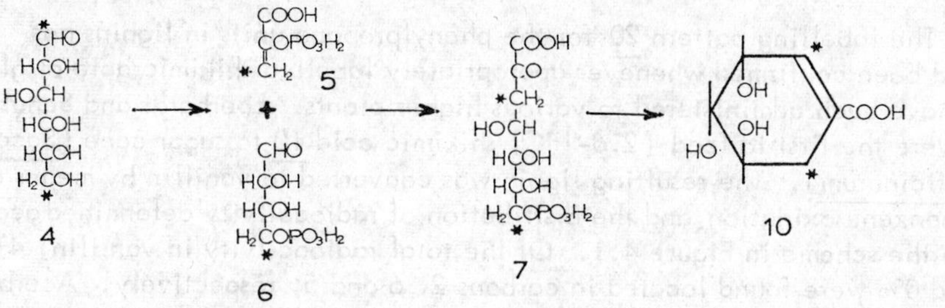

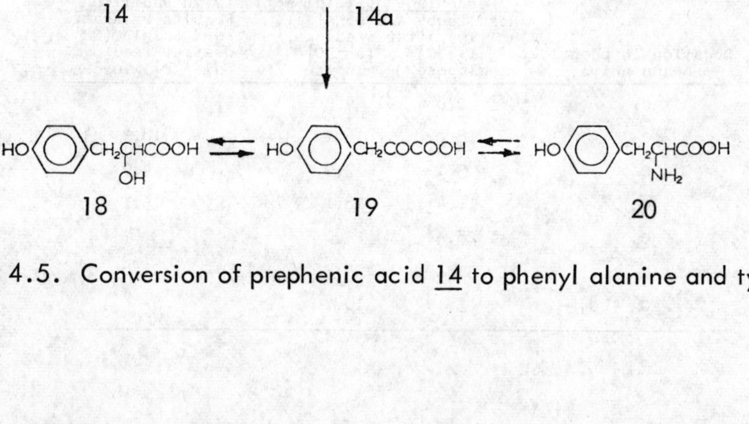

Figure 4.5. Conversion of prephenic acid $\underline{14}$ to phenyl alanine and tyrosine pools.

Shikimic acid produced from [1,6-^{14}C] glucose by Escherichia coli has indeed been found to contain 95% of total radioactivity at positions 2 and 6. Since the side-chain carbons in phenylpropanoids formed from shikimic acid acquire the three side-chain carbons from phosphoenolpyruvate 5, further conversion of active shikimic acid 10 may be expected to produce phenyl propanoids labelled at positions 2, 6 and α, Equation 2.

$$\underset{10}{\text{shikimic acid}} \longrightarrow \underset{20}{\text{phenylpropanoid}} \tag{2}$$

The labelling pattern 20 for the phenylpropane units in lignins has indeed been confirmed whenever appropriately labelled shikimic acid or glucose have been administered to various higher plants. Eberhardt and Schubert (3) were the first to feed [2,6-^{14}C] shikimic acid 10 to sugar cane (Saccharum officinarum). The resulting lignin was converted to vanillin by means of nitrobenzene oxidation and the distribution of radioactivity determined according to the scheme in Figure 4.1. Of the total radioactivity in vanillin, 41, 44 and 0% were found located in carbons 2, 6 and 5, respectively. Acerbo, Schubert and Nord (19) and Kratzl and Faigle (20), in independent studies, administered [6-^{14}C] glucose and [1-^{14}C] glucose to spruce (Picea abies) and determined the radioactivity distribution by the vanillin method. The results of both studies, Table 4.1, are in excellent accordance with each other and with the predictions of the shikimic acid route. The radioactivity associated with the methoxyl carbon is due to the fact that this carbon also stems from the one-carbon pool of glucose, through the S-adenosyl methionine route (21). (See Figure 4.6.)

Table 4.1. Distribution of Radioactivity in Lignins in Spruce, Fed with [1-^{14}C] Glucose and [6-^{14}C] Glucose

Location in phenyl propane units	Percentage of total activity in vanillin		
	Results by Acerbo et al. (19) after feeding [1-^{14}C] glucose	[6-^{14}C] glucose	Results by Kratzl and Faigle (20) after feeding [1-^{14}C] glucose
C_1	2.6	4.5	-
C_2	18.2	16.1	18.3
C_5	3.1	3.9	4.0
C_6	11.4	18.1	15.1
C_α	28.5	22.0	37.7
C_{MeO}	31.1	24.7	21.6
$C_3 + C_4$	5.1	10.7	-

PRECURSORS AND THEIR POLYMERIZATION

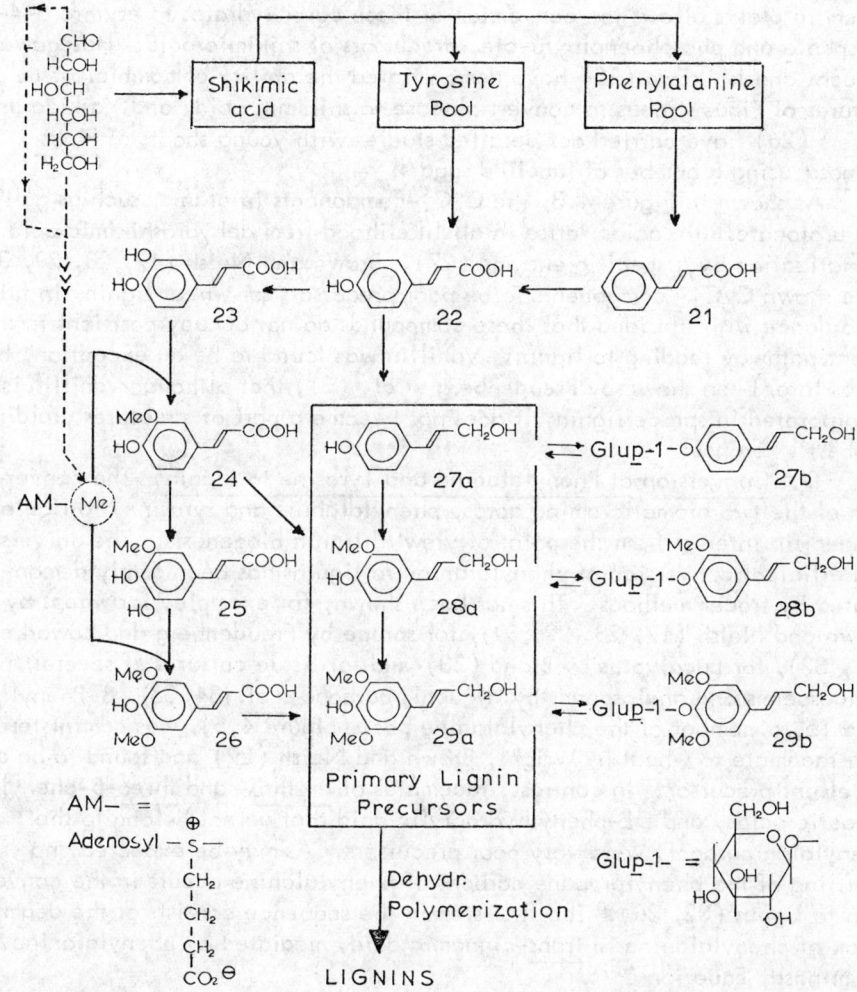

Figure 4.6. Pathway from amino-acid pools to primary lignin precursors.

Tracer studies with species other than Picea abies are in full accordance with the shikimic acid pathway. Hasegawa and Higuchi (22) demonstrated the conversion of labelled glucose into the lignin of Eucalyptus nitens. Carbohydrates other than glucose can also act as lignin precursors. Kratzl has shown the conversion of labelled erythritol to lignin in spruce (23), and Sergeeva and Kreitsberg have claimed the incorporation of radioactive pentoses into poplar lignin (24). It seems reasonable to assume that the metabolic system in plants allows the conversion of these carbohydrates to erythrose-4-phosphate and phosphoenolpyruvate, precursors of shikimic acid. Hasegawa, Higuchi and Ishikawa (25) have demonstrated the ability of cambial tissue cultures of Pinus strobus to convert glucose to shikimic acid, and Yashida and Towers (26) have carried out detailed studies with young shoots of Pinus resinosa using a number of labelled sugars.

As shown in Figure 4.3, the C_6C_1-components in plants, such as gallic and protocatechuic acids, arise in all likelihood from dehydroshikimic acid by aromatization by a specific enzyme (27). Brown and Neish (17, 28, 29, 30) have shown C_6C_1- components to be poor precursors for wheat lignin, in full accordance with the idea that these compounds do not occupy positions in the direct pathway leading to lignin. Vanillin was found to be an exception, but it has later been shown by Freudenberg et al. (31) that although vanillin is incorporated in spruce lignin, it does not become a part of structures yielding Hibbert's ketones.

D. Conversion of Phenylalanine and Tyrosine to Lignin. The conversion of the two aromatic amino acids, phenylalanine and tyrosine, to lignin is of specific interest from the point of view of lignin biogenesis. The universal and efficient conversion of phenylalanine to lignins has been amply demonstrated by tracer methods. This has been shown, for example, for wheat by Brown and Neish (17, 28, 29, 30), for spruce by Freudenberg and coworkers (31, 32), for Eucalyptus by Bland (33) and for tissue cultures of several gymnosperms and angiosperms by Higuchi, Barnoud et al. (34, 35). β-Phenyllacti acid 15, a member of the phenylalanine pool (Figure 4.5), was administered as a racemate to wheat by Wright, Brown and Neish (29) and found to be an excellent precursor. In contrast, racemates of erythro- and threo-β-phenylglyceric acids, and DL-phenylhydracrylic acid that do not belong to the phenylalanine pool, were very poor precursors. As may be expected, no shuffling of the phenylpropane carbons of phenylalanine occurs in the conversion to lignin (32, 36). The first step in the sequence consists of the deamination of phenylalanine to trans-cinnamic acid, mediated by phenylalanine deaminase, Equation 3.

$$\bigcirc\text{-}CH_2CHCOOH \quad \xrightarrow{\text{Phenylalanine deaminase}} \quad \bigcirc\text{-}CH=CHCOOH \qquad (3)$$
$$\quad\;\; NH_2$$

Phenylalanine deaminase was first isolated from acetone powders of barley by Koukol and Conn (37) and its presence in approximately forty plant species, representing gymnosperms, dicotyledons and monocotyledons, as well as in tissue cultures of redwood, lilac, rose and carrot has been demonstrated by Higuchi and Barnoud (38). The enzyme has a pH optimum in the region from 8.3 to 8.8. It does not catalyze the deamination of tyrosine to p-coumaric acid. On the contrary, the presence of tyrosine and coumaric acid exerts an inhibiting action on the activity of the enzyme.

Tyrosine is an efficient lignin precursor in the grass family only. Brown and Neish (28) showed that phenylalanine and tyrosine were about equally efficiently utilized for the synthesis of guaiacylpropane and syringylpropane units in wheat, but tyrosine lacked precursor properties outside of the grass family. β-(p-Hydroxyphenyl) lactic acid 18, a member of the tyrosine pool, was efficiently utilized for lignin formation in wheat (29), but not utilized in dicotyledons Fagopyrum tartaricum and Salvia splendens (39). Another member of the tyrosine pool, p-hydroxyphenyl pyruvic acid 19, was found to be an efficient precursor of sugar cane lignin by Acerbo, Schubert and Nord (4). Kratzl and Billek (40) found, however, that the same compound is a very poor precursor of lignin in spruce.

The explanation for all these observations is the circumstance that only grass species contain tyrase (tyrosine ammonia lyase), an enzyme capable of converting tyrosine to trans-p-coumaric acid, Equation 4.

$$HO-C_6H_4-CH_2CHCOOH \longrightarrow HO-C_6H_4-CH=CHCOOH \quad (4)$$
$$\quad\quad\quad\quad NH_2$$

In the absence of this enzyme, no pathway seems to be available from the tyrosine pool to cinnamic acids. Neish found tyrase in all members of Graminaceae that were studied such as sorghum, wheat, corn, barley, oats, rice and sugar cane. The enzyme could not be detected in peas, lupine, or sweet clover (41), or in tissue cultures of conifers and woody angiosperms (35, 38).

Grass lignins differ from lignins in other higher plants by virtue of their high content of p-coumaric ester groups (see Chapter 3) and there exists a probable connection between these groups and the presence of tyrase. It is possible, as proposed by Higuchi (42), that a large portion of p-coumaric acid generated through the deamination of tyrosine, becomes channeled to these ester groups.

E. Conversion of Cinnamic Acid Intermediates to Lignins. The pathway through cinnamic acid derivatives and cinnamyl alcohols to lignins is shown in Figure 4.6. The sequence along the pathway is suggested by numerous tracer experiments. Brown and Neish (43) tested labelled cinnamic 21, p-coumaric 22 and caffeic 23 acids as precursors for lignin in maple (Acer negundo) and wheat (Triticum aestivum) and found them to be comparable to phenylalanine in efficiency. Sinapic acid 26, an equally efficient precursor, was preferentially incorporated in syringylpropane units. On the other hand, m-methoxycinnamic acid was poorly incorporated, supporting the idea that hydroxylation at position 4 precedes that at position 3. The same authors also demonstrated that cinnamic acid was incorporated in wheat lignin without shuffling the side-chain carbon adjacent to the aromatic ring (28). Ferulic acid 24 showed efficient incorporation in guaiacyl lignin only in heading wheat plants. 5-Hydroxyferulic acid 25, on the other hand, was found by Higuchi and Brown (44) to be an efficient precursor of syringyl lignin in wheat. The conversion of labelled cinnamic, ferulic and p-coumaric acids to conifer lignins have been demonstrated by Smith and Neish (45), and Higuchi (34).

As shown in Figure 4.6, the cinnamic acid pathway consists of successive hydroxylation and O-methylation steps. Many of these steps appear to be partially reversible. Higuchi (34) has shown that sinapic acid 26, not an obvious precursor for conifer lignins, is partially incorporated in the guaiacyl lignin of tissue cultures of Pinus strobus. The partial reversibility is particularly well demonstrated by a study of McCalla and Neish (14) on cinnamic esters in Salvia splendens. In these species, p-coumaric, caffeic, ferulic and sinapic acids occur as esters of unknown composition. By feeding these acids in labelled form to the growing plant, it was established that they were not only converted to later intermediates in the sequence, but also slightly to the earlier ones.

The participation of the 5-hydroxyferulic acid 25 in the sequence of intermediates has been subject to some doubt because 5-hydroxyferulic has so far not been isolated from the plant kingdom. These doubts have been largely dissipated by a study by Higuchi and Brown (46) who demonstrated its precursor properties and also showed that when non-radioactive 5-hydroxyferulic acid was administered to wheat plants together with labelled sinapic acid, the recovered sinapic acid had an appreciably lower specific activity than when the labelled sinapic acid was administered alone. In a similar experiment of longer duration, radioactive 5-hydroxyferulic acid was recovered, indicating a partial reversibility of the O-methylation and hydroxylation reactions.

Tracer studies with labelled primary lignin precursors have been carried out using the phenyl-β-glucopyranoside of coniferyl alcohol, coniferin 28b (5, 31, 32, 40, 47). The incorporation of the three cinnamyl alcohol precursors as such has not been subjected to systematic studies as a consequence of their limited solubility in aqueous solutions. The absence of shuffling of the side-chain carbons in the conversion of coniferin to spruce lignin has been comprehensively demonstrated by Kratzl and coworkers (5, 40, 47, 48). Kratzl has also shown that syringin 29b and, surprisingly, syringaldehyde give rise to vanillin-yielding components in spruce (47, 48). Consequently, the partially reversible hydroxylation and O-methylation reactions may not be restricted to the cinnamic acids alone, and the possibility of cinnamyl alcohol precursors undergoing similar reactions as well should not be excluded.

PRECURSORS AND THEIR POLYMERIZATION

While the sequence of hydroxylation and methylation of the intermediate cinnamic acids is relatively well established, some recent evidence points out that certain esters of the cinnamic acids rather than the free acids might be the actual intermediates, at least in the members of the grass family. The conversion of labelled p-coumaric, ferulic and sinapic acids to their glucose esters by wheat plants was demonstrated by Higuchi and Brown (46) and similar conversions in young radish leaves were studied by Harborne and Corner (49). El-Basyouni, Neish and Towers (50) established that p-coumaric, ferulic and sinapic acids exist in wheat shoots in a bound form, probably as esters, part of which are extractable with 80% ethanol ("soluble esters"), whereas another part ("insoluble esters") remain in the extracted plant residue. The cinnamic acids may be isolated from the esters using alkaline hydrolysis. It was found that ^{14}C-labelled carbon dioxide, phenylalanine and tyrosine were incorporated more readily into the insoluble esters of the phenolic cinnamic acids, whereas the reverse was true for the incorporation of labelled cinnamic, p-coumaric, caffeic, ferulic and sinapic acids. There appeared to be a slow exchange between "soluble" and "insoluble esters," and both acted as intermediates in lignification. The authors concluded that the route involving the "insoluble esters" is probably more important in lignification because it was favored by all precursors preceding cinnamic acids, and proposed the tentative scheme in Figure 4.7 to rationalize their observations.

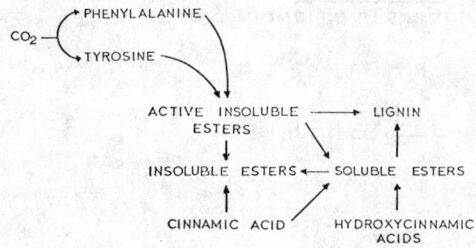

Figure 4.7. Proposed role of esters in lignin formation (51).

The nature of the "active insoluble esters" remains speculative. Neish (51) has brought into consideration sulfhydryl esters with co-enzyme A which could be insoluble in 80% ethanol, pointing out the requirement for activation of the carboxyl groups in the reduction of the p-hydroxylated cinnamic acids to the analogous cinnamyl alcohols. The "insoluble esters" in the scheme represent, in all likelihood, the p-coumarate and ferulate groups associated with grass lignins (See Chapter 3, Section III D).

Goldschmid and Hergert (52) have identified in the cambial sap of western hemlock (Tsuga heterophylla) a number of cinnamic acid esters which merit at least some consideration as potential lignification intermediates.

These esters were caffeoylshikimic acid, feruloylshikimic, feruloylquinic, p-coumaroylshikimic and, possibly, p-coumaroylquinic acids. The failure to isolate similar esters from the cambial saps of Abies and Pinus should perhaps not be interpreted as definitely indicative of the absence of these esters in the latter two species because of the often transitory existence of intermediates of this type.

Substantial gaps still remain to be filled in the knowledge of the enzymes mediating the conversion of cinnamic acids to lignin. The earlier failures in the isolation of the enzyme mediating the hydroxylation of cinnamic acid to p-coumaric acid 22 have found an explanation in the extreme instability of this enzyme. Nair and Vining (53), who finally isolated a cinnamic acid hydroxylase from spinach leaves, found that the enzyme lost its activity in one hour at 0 to 5°. The isolated cinnamic acid hydroxylase is distinct from phenylalanine hydroxylase, also present in spinach leaves (54).

The enzyme mediating the hydroxylation of p-coumaric acid to caffeic acid has not been isolated in a pure form, but is known to be present in the plant phenolase complex (55).

Catechol-O-methyltransferases have been isolated from cambial scrapings of apple tree and the woody shrub Pittosporum crassifolia (56), from pampas grass shoot tissue (Cortaderia selloana) (57), from etiolated wheat seedlings and young petunia leaves (58), and from immature bamboos (Phyllostaccus pubescens and P. reticulata) (59). These enzymes have been shown to mediate the methylations in equations 6 and 7.

$$HO\text{-}C_6H_4\text{-}CH\text{=}CHCOOH \longrightarrow \text{No methylation} \quad (5)$$
$$22$$

$$\underset{23}{HO\text{-}C_6H_3(OH)\text{-}CH\text{=}CHCOOH} \longrightarrow \underset{24}{HO\text{-}C_6H_3(OMe)\text{-}CH\text{=}CHCOOH} \quad (6)$$

$$\underset{}{HO\text{-}C_6H_2(OH)_2\text{-}CH\text{=}CHCOOH} \longrightarrow \underset{25}{HO\text{-}C_6H_2(OMe)(OH)\text{-}CH\text{=}CHCOOH} \longrightarrow \underset{26}{HO\text{-}C_6H_2(OMe)_2\text{-}CH\text{=}CHCOOH} \quad (7)$$

The enzyme is consequently not in position to methylate p-hydroxyl groups and differs, in this respect, from O-methyltransferases present in animal organisms (60) and in the fungus Lentinus lepideus (61). The specificity for the m-hydroxyl groups suggest strongly that the enzyme participates in the synthesis of lignin precursors, and also provides additional support for the pathway outlined in Figure 4.6.

The catechol-O-methyltransferases appear to be quite unspecific for substrate, methylating with equal ease vicinal polyphenols, catecholic acids and flavonoids (57). Hydroxyls meta- to a side-chain are generally preferentially methylated. Interestingly, in pyrogallol that lacks a carbon side-chain, the middle hydroxyl becomes methylated.

As in O-methylations in general, 5-adenosyl methionine, Figure 4.6, acts as the donor of the methyl groups. Byerrum and coworkers (21) have used methionine labelled by both ^{14}C and deuterium at the methyl groups to show that this methyl group is transferred to the lignins in tobacco and barley intact. Bland has demonstrated methionine to be the source of methoxyl groups also in Eucalyptus lignin (62). The methyl carbon in methionine in turn stems from the one-carbon pool of the plant, and carbons 1 and 6 in glucose (19, 20) as well as carbon -3 in serine, formaldehyde (63) and carbon -2 in glycine (64) have been found to form sources of the methoxyl carbons in lignin.

The reduction of hydroxycinnamic acids to primary lignin precursors is only partially understood. The presently available information conforms with the idea of a two-step reduction of p-coumaric, ferulic and sinapic acids to the corresponding cinnamyl alcohols, equation 8.

$$\underset{(OMe)}{\underset{(OMe)}{HO-\bigcirc-CH=CHCOOH}} \longrightarrow \underset{(OMe)}{\underset{(OMe)}{HO-\bigcirc-CH=CHCHO}} \longrightarrow \underset{(OMe)}{\underset{(OMe)}{HO-\bigcirc-CH=CHCH_2OH}} \qquad (8)$$

A reduction of this type has been demonstrated for ferulic acid only by Higuchi and Brown (46). After administering labelled ferulic acid to heading wheat plants, radioactive coniferyl aldehyde and coniferyl alcohol were recovered from water and ethanol extracts. The specific activity of the recovered coniferyl aldehyde was appreciably higher than that of the coniferyl alcohol. The same results were obtained with cambial tissue cultures of white pine (Pinus strobus). Acetone powders of wheat plants and from the pine tissue cultures effected reduction of ferulic acid to coniferyl aldehyde with no apparent formation of coniferyl alcohol. Consequently, two separate enzymes are probably involved in the reduction of the carboxyl group, and their isolation and characterization remain to be done.

F. The Role of Precursor Glucosides. The participation of the phenolic β-glucosides of the cinnamyl alcohol precursors (27a, 28a, and 29a) in the biogenesis of lignin has been debated over a number of years, and has already been discussed in Chapter 2. The idea of these glucosides acting as direct intermediates in lignification has been mainly promoted by Freudenberg and his group (31, 65). The glycosides were envisaged to be translocated from the cambial zone into the lignifying tissues (66). These views have been strongly criticized on the valid grounds that although the presence of the three glucosides 27b, 28b and 29b in the cambial sap of spruce (Picea excelsa) had been clearly demonstrated (67), and coniferin 28b has been identified in 15 conifer species belonging to six families (68, 69), these glucosides seem to be largely absent from angiosperms. Outside of the olive family, in which syringin 29b but no coniferin was encountered in five genera, syringin has been found only in black locust (68, 69) and all three glycosides seem to be absent in monocotyledons. Indeed, it is hardly necessary any more to assume separate sites for the synthesis of the cinnamyl alcohol glucosides and for their deposition as lignin after Hasegawa, Higuchi and Ishikawa (25) have demonstrated the ability of cambial tissue cultures of Pinus strobus to convert labelled glucose to lignin. Consequently, single lignifying cells appear to be in possession of all the necessary enzymes to transform carbohydrates to lignin.

If this view is accepted, it is logical to assign the cinnamyl alcohol glycosides in the sap of conifers the role of acting as a reservoir to augment the precursor supply of lignifying cells. The recent experiments of Freudenberg and Torres-Serres (70) are in full accord with this interpretation. It was found that when labelled L-phenylalanine was administered to spruce saplings, the three cinnamyl alcohol glucosides in the sap became radioactive a few days later. Subsequently, their radioactivity declined and the total administered activity became fixed with the newly deposited lignin. Manskaya (71) has observed inverse variations in the coniferin content of spruce sap and the peroxidase activity of lignifying cells.

Conifers utilize β-glucosidase to release cinnamyl alcohol precursors from their glycosides (31). The presence of β-glucosidase in association with the cell walls of lignifying cells of conifers has been demonstrated in histological studies carried out by Freudenberg and coworkers using the indican color reaction (72). The cells in the cambium in which the lignification has not yet started and fully lignified cells are devoid of β-glucosidase. The controlling action of the enzyme in lignification is demonstrated by the failure of labelled L-coniferin to become incorporated in spruce lignin (73). The presence of β-glucosidase in the lignifying cells of many hardwood species and one monocotyledon (Cordyline congesta) has been demonstrated, while others (Prunus sp.) do not give a clearly positive indican reaction (74). β-Glucosidase has also been shown to be present in the lignifying tissues of bamboos by Higuchi (75).

D-coniferin has been repeatedly claimed to stimulate lignification in tissue cultures of various plants which could be taken as indicative of β-glucosidase activity in lignifying cells. Wacek et al. (76) observed such a stimulation in carrot tissue cultures using staining reactions to compare the extent of lignification, and Barnoud et al. (77) found that lignification in tissue cultures of Syringa vulgaris and Rosa wichuraiana was stimulated both by coniferin and syringin.

In summary, the glucosides of cinnamyl alcohols act clearly as intermediates in lignification, whenever xylem cells of conifers utilize the reservoir of precursors present in the cambial sap. When lignin is formed from precursors generated within the same cell, the intermediacy of glycosides is less certain. The apparent lack of β-glycosidase in certain hardwood species gives some indication of the possibility of cinnamyl alcohols alone, or their derivatives other than glycosides acting as lignification intermediates in specific cases.

G. Enzymatic Aspects of Polymerization to Lignin. The dehydrogenative enzyme system causing the polymerization of the p-hydroxycinnamyl alcohols to the polymeric lignin molecule forms one of the focal points in lignin biosynthesis. Of the three potential enzyme systems, the polyphenoloxidases (tyrosinases), the laccases and the peroxidases, the first group of enzymes was eliminated from consideration by studies of Mason and Cronyn (78). These workers showed that a purified mushroom polyphenoloxidase did not effect dehydrogenative polymerization of coniferyl alcohol. It is more difficult to decide between laccases and peroxidases as polymerizing enzyme systems, and a brief discussion on the specific properties of these enzymes is therefore in order, on the basis of a recently published critical review by Brown (79).

Laccases mediate the oxidation of a large number of phenols to phenoxy radicals by air oxygen. The oxidative process appears to be strictly a one-electron transfer from the substrate phenol, and hydroxylations, common in tyrosinase-catalyzed oxidations, are therefore not observed. The molecular weights of laccases vary in the range from 64,000 to 120,000, and each molecule contains four coordinated cupric ions. The spectrum shows two characteristic absorption maxima, at 280 and at approximately 615 mµ. Laccases differ from tyrosinases by the lack of hydroxylation reactions, by not being inhibited by carbon monoxide and by converting hydroquinone to p-benzoquinone and guaiacol to colored products, both of which reactions tyrosinase fails to accomplish. On the other hand, laccase fails to act on tyrosine, which is readily converted to colored products by tyrosinase and produces colorless products from p-cresol, while colored products are obtained with tyrosinase as a catalyst. The differentiation between laccases and tyrosinases is easy on the basis of such chemical conversions. Unfortunately, only few specific differences are known in the oxidative actions of laccases and peroxidases.

Peroxidases mediate the oxidation of phenols to phenoxy radicals by hydrogen peroxide. The coenzyme part consists of an iron-porphyrin compound. Several enzymes of this type have been isolated in the crystalline state, the well-known horseradish peroxidase, for example. Peroxidase oxidizes tyrosine to dark colored melanins as well as to dityrosine, while laccase has no oxidative effect. Otherwise similar, if not identical products are obtained in oxidation with laccase.

In histological studies, laccase is identified by applying 0.05 M solutions of oxidizable compounds such as phenylenediamine, hydroquinone and catechol to the tissue material which thereby become oxidized to characteristic colored compounds by the action of air oxygen. To demonstrate the presence of peroxidase, solutions of benzidine, pyrogallol or 1-naphthol are applied in conjunction with 1% hydrogen peroxide. In lignifying tissues, both enzymes can be usually shown to be present, although the peroxidase activity appears to be more pronounced in bamboo shoots (75). Freudenberg and coworkers (80) have isolated both water-soluble oxidases and acetone powders from the cambial scrapings of a pine species (Araucaria excelsa) (72) and tested the effect of both enzyme mixtures in the dehydrogenative polymerization of coniferyl alcohol. In both cases, aeration caused a slow, and hydrogen peroxide, a rapid polymerization, which may be taken to indicate a more effective action by peroxidase than by laccase. On the other hand, the activity of peroxidase in enzymes isolated from spruce cambium was much weaker than that of laccase (80) and a combined action of both enzymes is currently visualized by Freudenberg (81). The solution of this problem must probably await a more comprehensive characterization of the plant phenolase complex.

H. Lignification in Maturing Plants. Growing bamboo plants form an excellent plant material to study the changes in the occurrence of lignification intermediates and in the enzymatic system accompanying the successive stages of lignification in the maturing plant tissue. Extensive studies in this area have been carried out by Higuchi and coworkers (42, 82, 83).

The growing apex of bamboos shows only faint signs of incipient lignification in the spiral vessels, and the uppermost 50 to 70 cm of the plant is essentially lignin-free (42, 82). Other types of vessels and sieve tubes become lignified in the next lower sections, and the tissue becomes consequently quite hard and woody at about 2 to 3 meters from the top. The parenchyma cells are the last to lignify, and their lignification does not generally start until the plant has attained a length of about five meters.

Several enzymes associated with the shikimic acid pathway were demonstrated to be present in the maturing tissues of bamboos. These included glucose-6-phosphate dehydrogenase and 6-phosphogluconate dehydrogenase, studied by Higuchi and Shimada (83). The activities of both enzymes were found to increase immediately below the apex and then remained constant, even in aged tissues. The C_6/C_1-ratio and oxygen uptake declined towards the basal part of the plant, parallel to an increase in the rate of lignification. These results may be taken as suggestive of the domination of the pentose phosphate pathway in the respiratory processes of lignifying tissues. The presence of 5-dehydroquinate hydrolyase (84) and that of 5-dehydroshikimic acid reductase (85) were likewise demonstrated. Both enzymes exhibited maximum activity at the locus immediately below the apex and a quite gradual decline in activity towards the base. The shikimic acid content of the bamboo tissue had a pronounced maximum below the apex as shown in Figure 4.8 (82). It appears that young tissues synthesize shikimic acid vigorously as a precursor for phenylalanine and tyrosine that are incorporated into protein. Later, the emphasis shifts to lignin synthesis for which a lower shikimic acid pool is sufficient.

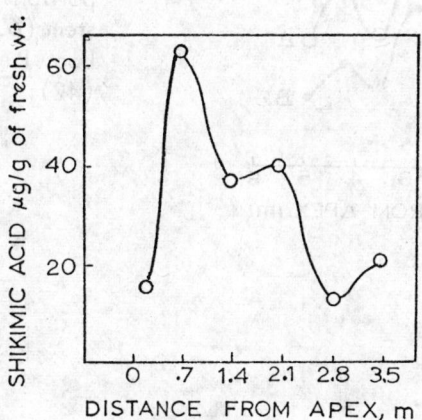

Figure 4.8. Variation of shikimic acid content in different parts of bamboo (82).

The activities of phenylalanine deaminase and tyrase can be anticipated to parellel closely the process of lignification. This expectation was first confirmed for buckwheat by Yoshida and Shimokoriyama (86). In a detailed study on bamboos by Higuchi (42) it was shown that in the immature section of the stem (~ 70 cm from the apex) the activities of both enzymes are almost nil and increase then rapidly reaching a maximum at about 2 to 3 meters from the top, as shown for a specimen of Phyllostachus reticulata in Figure 4.9. The activities of the two enzymes follow a parallel course along the stem of the plant, the activity of tyrase amounting to one-third to one-half of that of phenylalanine deaminase. Figure 4.10 illustrates a somewhat different pattern of enzyme activities in a specimen of Phyllostachus pubescens (42).

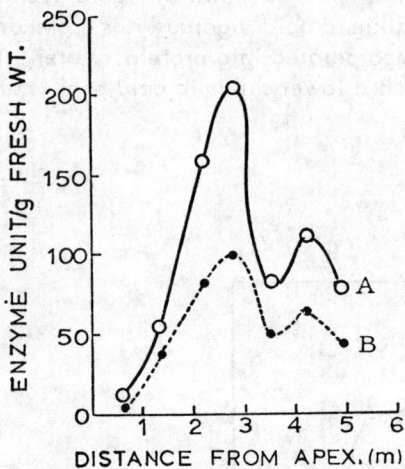

Figure 4.9. Activity of phenyl alanine deaminase (A) and tyrase (B) of different portions of immature stem (6.6m) of bamboo, P. reticulata (42).

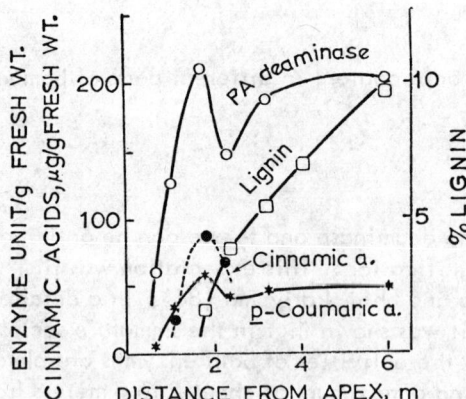

Figure 4.10. Activity of enzymes and amount of cinnamic acids and lignin of different portions of immature stem (8.8m) of bamboo, P. pubescens. (42).

It can be noted that the concentrations of trans-cinnamic and trans-p-coumaric acids in the plant tissue increase rapidly as a consequence of enhanced deaminase activity, and the first maximum in deaminase activity appears to trigger the rapid phase in the lignification process. A similar relationship between the rate of lignification and phenylalanine deaminase was established for Sequoia by Rubery and Northcote (87). It was found that the activity of this enzyme was highest in the young xylem ($E_{290}/h/g$ dry wt = 34.5), low in cambium (3.1) and practically absent in callus tissue. All these observations conform remarkably well with the biogenetic sequence derived from tracer studies. Higuchi and coworkers (59) also assessed the activity of O-methyl-transferase in bamboo stems obtaining results which for some reason do not correlate with lignification as well as deaminase activities, as shown in Figure 4.11.

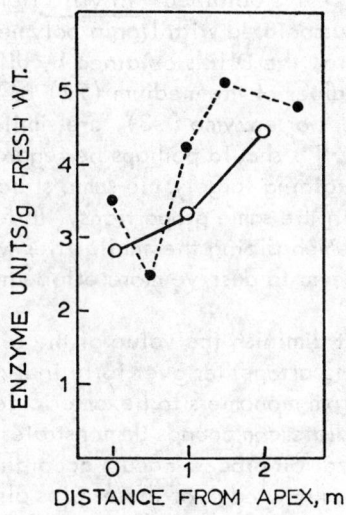

Figure 4.11. Changes in activity of O-methyl-transferase in bamboo shoots. P. reticulata was used as an enzyme source (59).

Length of shoots:
------ 280 cm.
——— 220 cm.

II. Dehydrogenative Polymerization to Lignin.

A. Introduction.
The dehydrogenative polymerization of primary lignin precursors to lignin is a complex process which is almost impossible to study in vivo. Our current knowledge of this polymerization process is almost entirely based on the extensive model dehydrogenations of coniferyl, and to some extent, p-coumaryl and sinapyl alcohols, carried out by K. Freudenberg and his group during the past twenty years (88). As a consequence of these studies, most of the known structural details of lignins may now be described as arising from various modes of coupling and quinone methide additions associated with the dehydrogenative polymerization.

Freudenberg's extensive studies were triggered by the observation (89, 90) that the dehydrogenation of coniferyl alcohol by air oxygen in an aqueous solution containing press juice from a mushroom, Psalliota campestris, afforded an insoluble polymer resembling, to an extraordinary degree, spruce lignin in terms of composition, reaction products and spectral properties. These polymers were called "Dehydrogenation Polymers" (DHP) by Freudenberg, and they were later also prepared using horseradish peroxidase and hydrogen peroxide (91).

It is perhaps presumptive to call the DHP's obtained "in vitro lignins," as long as the plant dehydrogenase system associated with lignin polymerization is not well-characterized. Furthermore, the DHP's obtained by different dehydrogenative agents, at different pH-values of the medium (92) or by varying the mode of addition of the monomer or enzyme (93), are similar but definitely not identical. Consequently, DHP's should perhaps be regarded as "polymer models" for natural lignins, containing largely the same structural features as the latter, but not necessarily in the same proportions. In fact, the elucidation of differences rather than emphasizing the similarities within the dehydrogenation polymers would now seem to deserve more effort than it has received in the past.

These reservations do not in the least diminish the value of the results by Freudenberg and coworkers in identifying altogether over forty individual intermediates of DHP formation, ranging from monomers to hexameric compounds (88). The nature of these intermediate compounds demonstrates clearly that the dehydrogenative polymerization process occurs according to certain well-defined rules, and our current knowledge of these rules already allows us to reconstruct much of the complex reactions occurring in the lignifying cell wall.

The recognition that dehydrogenative coupling processes are involved in the biosynthesis of a wealth of phenolic natural products other than lignin is of a more recent origin, and this area is the subject of a recent outstanding monograph (94). Studies carried out on stable phenoxy radicals (95) have provided new insight in the phenoxy coupling processes, and the development of new dehydrogenative polymerization processes to synthesize linear polyethers from 2,6-dimethyl phenol with outstanding mechanical and thermal properties (96) has added renewed interest to this area.

Many observations made in the coupling of simple phenoxyradicals have pertinence in the lignin field and deserve a brief discussion.

B. <u>Oxidative Coupling of Simple Phenols</u>. The generation of phenoxy radicals can be formalized either as hydrogen abstraction from the parent phenol or a one-electron transfer from the phenolate ion, equation 9.

$$\text{Aryl-O-H} \xrightarrow{-H \cdot} \text{Aryl-O} \cdot \quad (9)$$
$$\text{Aryl-O}^{\ominus} \xrightarrow{-e}$$

The exact mechanism of hydrogen abstraction has remained a matter of debate. Instead of a transfer of a hydrogen atom to the acceptor molecule, originally proposed by Haber and Willstätter (97), Kenner (98) favors electron transfer as the initiating step, followed by the release of proton. A third, more complex mechanism, based on ionic hydrogen abstraction, has been discussed by Dimroth (99).

A second electron transfer from the phenoxy radical gives a positively charged phenoxonium ion that can also be formed directly from the phenolate anion by a two-electron transfer process, equation 10.

$$\text{Aryl-O}^{\ominus} \xrightarrow{-e} \text{Aryl-O} \cdot \xrightarrow{-e} \text{Aryl-O}^{\oplus} \quad (10)$$

Therefore, the participation of phenoxonium ions in phenoxy coupling reactions merits serious consideration. As a matter of fact, reactions between phenolate and phenoxonium ions are expected to produce the same dimeric products as are formed in coupling reactions between phenoxyradicals. This is illustrated for the coupling of phenol in Figure 4.12. The phenoxy radical <u>30</u> has three different resonance structures, and the unpaired electron is therefore distributed between the phenol oxygen, two <u>ortho</u> positions and the <u>para-</u>position. One of the six conceivable coupling modes, that between two phenoxy oxygens to give a diarylperoxide, has never been found to occur in dehydrogenative coupling (100). This leaves five coupling modes, namely those between the phenoxy oxygen and the <u>ortho-</u>and <u>para-</u>carbons, and direct couplings between the two <u>ortho-carbons</u>, between the two <u>para-carbons</u> and between the o- and p-carbons, respectively. Two of the five coupling modes are illustrated in Figure 4.12 resulting in the formation of dimeric unsaturated ketones <u>31</u> and <u>32</u>. In protic solvents, these ketones readily tautomerize to p-phenoxy phenol <u>33</u> and 4,4'-dihydroxybiphenyl <u>34</u>, respectively. The dimeric ketones may, on the other hand, equally well be envisaged to be formed from the phenoxonium ion <u>35</u>, reacting in one of its resonance forms with the corresponding phenolate anion. The phenoxonium ion intermediate differs, however, from the phenoxyradical inasmuch as it can be readily hydroxylated to catechol 38 or hydroquinone 39, both of which may be further oxidized to the corresponding quinones. The absence of hydroxylated products and quinones can therefore be taken as a reasonably reliable indication of the absence of phenoxonium ions in the reacting system, while their actual formation suggests at least some contribution by phenoxonium ions.

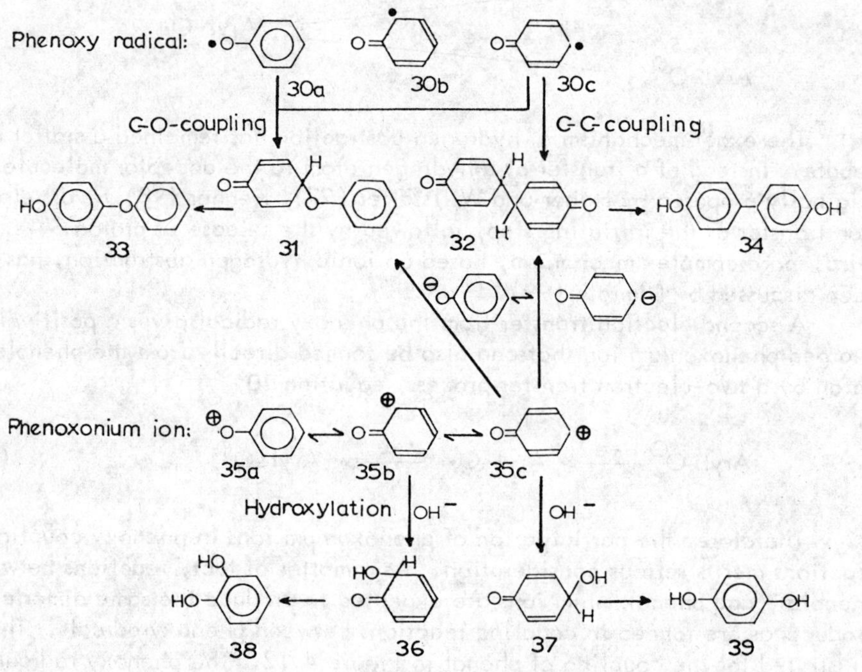

Figure 4.12. Formation of dimers from phenoxy radicals and phenoxonium ions. Competing hydroxylation of phenoxonium ions to catechols and hydroquinones.

Musso (101) has pointed out that reactions between phenoxy radicals and phenolate ions cannot be ruled out, particularly when the concentration of the former species is low. This aspect needs clarification from kinetic studies.

A large number of oxidants may be used to convert phenols to phenoxyradicals (102). In aqueous media, alkaline potassium ferricyanide has been found to be a versatile system. Often, a two-phase system with benzene, chloroform or ether is utilized to remove the coupling products in the organic phase immediately after their formation, to protect them against further oxidation. In oxidations in neutral and acidic media, ferric chloride has been extensively used. Anodic oxidation, particularly in alkaline media, has been successfully applied in a number of cases (102). In enzymatic oxidations, peroxidase-hydrogen peroxide is preferable to laccase-oxygen because the phenoxy radicals often react with oxygen forming peroxidic products.

In organic solvent systems (benzene, acetone, etc.), suspensions of freshly prepared PbO_2, Ag_2O, HgO and MnO_2 have been used extensively (102). Solutions of the stable radical 2,4,6-tri-$\underline{t}$-butylphenoxy may be applied for the conversion of other phenols to phenoxy radicals by means of radical exchange. Cuprous chloride complexes with pyridine in nitrobenzene in combination with air oxygen have been found to convert 2,6-dimethylphenol to a linear C-O-linked dehydrogenation polymer (103).

Many other dehydrogenation systems have been used, but are not entirely satisfactory. A too-high oxidation potential may cause the partial formation of phenoxonium ions, with undesirable overoxidation effects such as hydroxylation or quinone formation. Phenoxonium ions are also generated by two-electron transfer agents. Occasionally, radical species generated from the dehydrogenation agent may couple with the phenoxy radicals. For example, o-hydroxylation is common in cupric-ion catalyzed oxidations of phenols by air oxygen (104). Hydroxylations and usually ill-defined reaction products, are also common when hydroxyl radicals generated by Fenton's reagent ($FeSO_4 + H_2O_2$) are used for dehydrogenation. Air oxidation of phenolate solutions often generates hydroperoxides $\overline{40}$ and quinols $\overline{41}$, equation 11 (105).

$K_2S_2O_8$, H_2SO_5, peracids, lead tetra-acetate and Fremy's salt (106) may be considered largely as two-electron transfer agents, explaining, for example, the reactions shown in equations 12 and 13 (106, 107).

C. **Dehydrogenation of 2,4,6-Trisubstituted Phenols.** Phenoxy radicals substituted in positions 2, 4 and 6 possess an unusual degree of stability. A well-known example is the 2,4,6-tri-t-butyl phenoxy radical 42, discovered independently by Cook (108) and Müller and coworkers (109), which retains its radical properties both in solution and in the solid state, Figure 4.13. Galvinoxyl 43 likewise shows no tendency towards dimerization (110). 2,4,6-Triphenyl phenoxy radical 44, on the other hand, exists in a reversible equilibrium with its diamagnetic dimer in solution, and is totally dimerized in the solid state (111). The methoxyl analogue of galvinoxyl, syrinoxyl 45 shows more tendency towards dimerization than the latter (112).

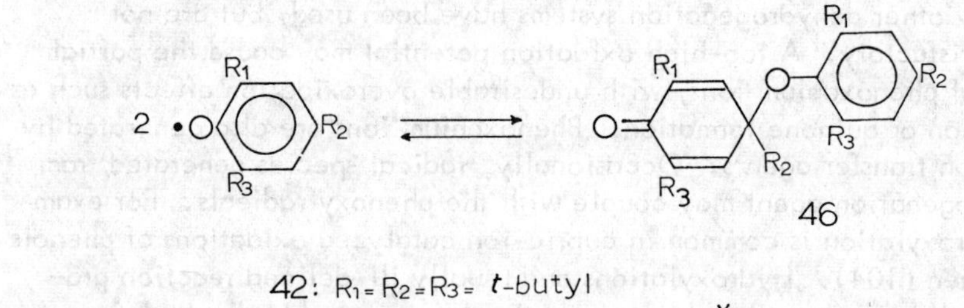

42: $R_1 = R_2 = R_3 = t$-butyl

43: $R_1 = R_3 = t$-butyl, $R_2 =$

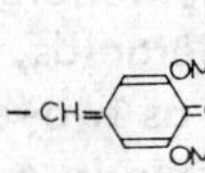

44: $R_1 = R_2 = R_3 =$ phenyl

45: $R_1 = R_3 =$ OMe, $R_2 =$

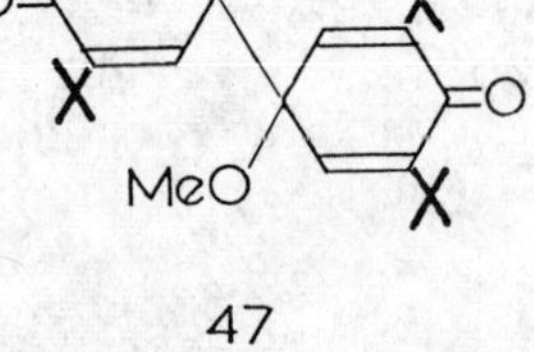

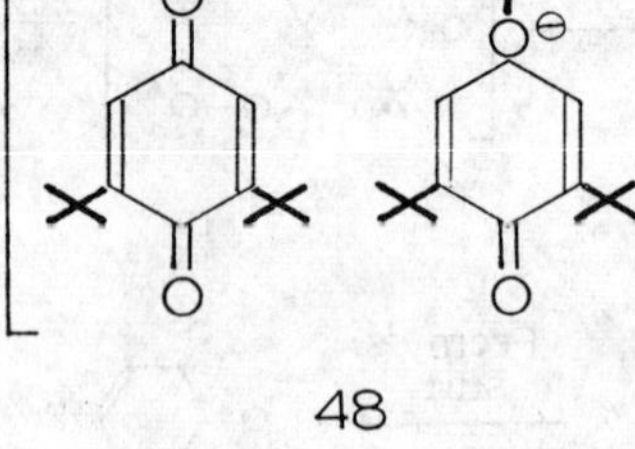

Figure 4.13. Postulated structures for reversibly formed diamagnetic dimers of 2,4,6-trisubstituted phenoxy radicals: a, quinolether 46; b, p-p' coupled dimer 47 (113); and c, charge transfer complex 48 (114).

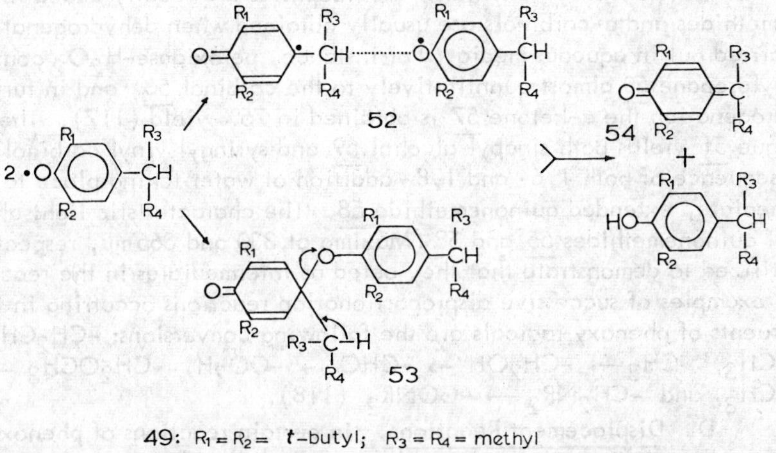

PRECURSORS AND THEIR POLYMERIZATION

49: $R_1 = R_2 =$ t-butyl; $R_3 = R_4 =$ methyl
50: $R_1 = R_2 =$ OMe; $R_3 = H$, $R_4 =$ ethyl
51: $R_1 = R_2 =$ OMe; $R_3 = H$, $R_4 =$ vinyl

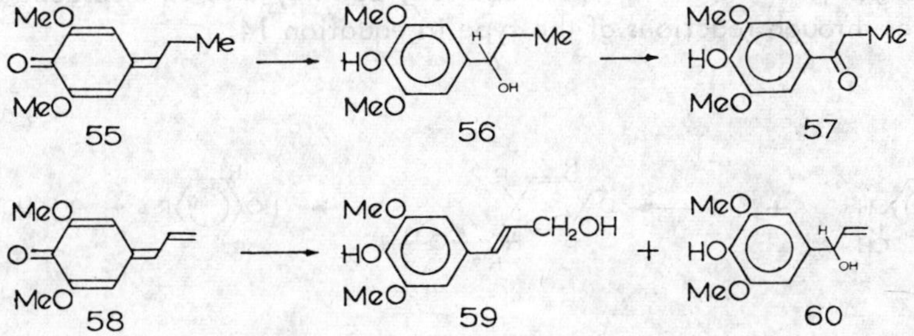

Figure 4.14. Disproportionation of 2,4,6-trisubstituted phenoxy radicals to quinone methides and their conversion products.

The diamagnetic dimers are usually formulated as quinol ethers 46, although this assignment is often not totally certain. Occasionally, other dimeric structures have been contemplated; for example, 47 for the dimer of 2,6-di-t-butyl-4-methoxy phenoxy radical (113), and the charge-transfer complex 48 for the dimer of 2,6-di-t-butyl-4-t-butoxyphenoxy radical (114).

When α-hydrogens are present on side-chain in phenoxy radicals, disproportionation to quinone methides 54 and the parent phenols occurs readily, Figure 4.14. These reactions have been reviewed (115). In non-polar media, quinone methides may be isolated in 70 to 90% yield. A head-to-tail transition state 52 has been proposed for the disproportionation reaction (116).

In protic solvents, electron shifts in the quinol ether dimer 53 deserve consideration as a possible mechanism. Nucleophiles are readily added to quinone methides and α-carbinols are usually obtained when dehydrogenations are carried out in aqueous media. For instance, peroxidase-H_2O_2 converts syringylpropane 50 almost quantitatively to the carbinol 56, and in further dehydrogenation the α-ketone 57 is obtained in 75% yield (117). The vinyl analogue 51 yields both sinapyl alcohol 59 and syringyl vinyl carbinol 60, as a consequence of both 1,6- and 1,8- addition of water taking place to the intermediate, extended quinonemethide 58. The characteristic light absorption of quinonemethides 55 and 58 (Maxima at 330 and 365 mμ, respectively) was utilized to demonstrate that they acted as intermediates in the reaction. Other examples of successive disproportionation reactions occurring in the 4-substituents of phenoxy radicals are the following conversions: $-CH_2CH_3 \longrightarrow -CO-CH_3$, $-CH_3 \longrightarrow -CH_2OH \longrightarrow -CHO \longrightarrow -CO_2H$, $-CH_2OCH_3 \longrightarrow -CO_2CH_3$, and $-CH_2NR_2 \longrightarrow -CONR_2$ (118).

D. <u>Displacement Reactions</u>. In certain reactions of phenoxy radicals o- and p-substituent groups may become displaced by an attacking radical. Even solutions of the stable 2,4,6-tri-t-butyl phenoxy radical generate isobutene $(CH_3)_2C=CH_2$ upon heating as a consequence of reactions of this type (118a). Carbinol groups in an α-position may not only be oxidized to carbonyl groups through disproportionation, but may also be displaced as aldehydes through reactions of the type in equation 14.

$$\cdot O\!\!-\!\!\underset{R''\ \ OH}{\overset{R'}{\bigcirc}}\!\!-\!\!CHR''' + \cdot R \longrightarrow O=\underset{R''\ \ HC-O-H}{\overset{R'\ \ \ R}{\bigcirc}}\!\!\longrightarrow HO\!\!-\!\!\underset{R''}{\overset{R'}{\bigcirc}}\!\!-\!\!R + R'''CHO \quad (14)$$

Displacements of side-chains with α-hydroxyl groups as aldehydes have been demonstrated by Pew and Connors (119) and by Schilling and Gratzl (120). For example, the reaction of the hydroxymethyl phenol 61 with ferricyanide yields diaryl ether 62 and biphenyl compound 63 in approximate 2:1 yield ratio (120). In the former case, the displacing entity is an aryloxy, in the latter, an aryl radical. The hydroxymethyl groups were displaced as formaldehyde, Equation 15.

Syringyl alcohol 64a and syringic acid 64b are both converted by a sequence of dehydrogenation reactions to C-O-coupled polymer 65 in dehydrogenation by ferricyanide (Equation 16) while syringaldehyde fails to react. In neither case was C-C-coupling with a double displacement observed, although it occurs as side-reaction in the dehydrogenation of 3,5-dimethyl-4-hydroxybenzyl alcohol (120). Laccase-O_2 differs from ferricyanide in converting syringic acid to 2,6-dimethoxy-p-quinone 66 in 45%

PRECURSORS AND THEIR POLYMERIZATION

yield (121) and also oxidizes vanillic acid to a similar product (122).

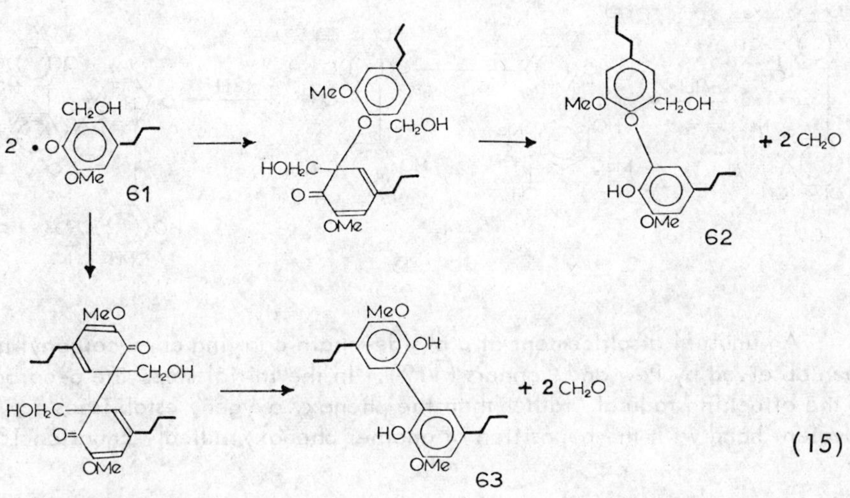

(15)

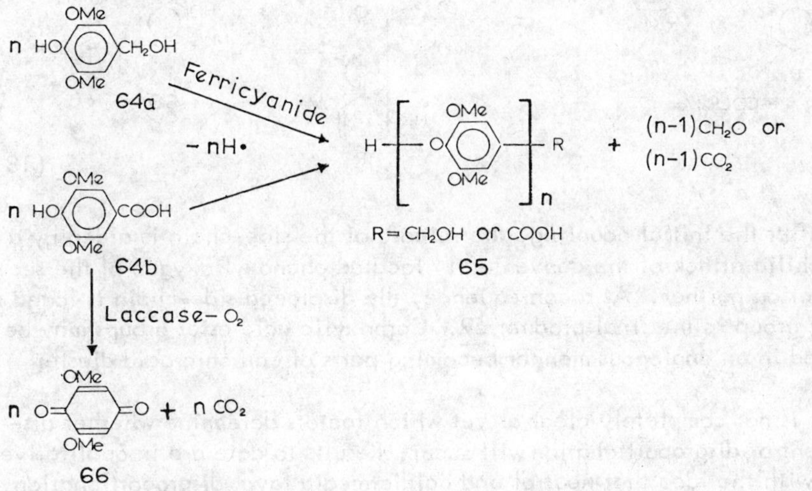

(16)

Pew and Connors (119) have shown that peroxidase–H_2O_2 causes side-chain displacement in dehydrodi-c-ethylvanillyl alcohol 67 generating propionaldehyde. The initial C–C-coupled product is not recovered, however, because it converts rapidly to dioxepin 68 by further dehydrogenation, Equation 17.

$$ (17) $$

An unusual displacement of a p-side-chain carrying an α-carbonyl has been observed by Pew and Connors (119). In the initial step, the o-carbon of the attacking radical, rather than the phenoxy oxygen, establishes a covalent bond with the p-position of another phenoxy radical, Equation 18.

$$ (18) $$

After the initial coupling, the release of the side-chain is aided by a nucleophilic attack of the conveniently located phenoxyl oxygen of the second reaction partner. As a consequence, the displaced side-chain is found as an ester group in the final product 69. Carboxylic acid ester groups may be displaced in an analogous manner becoming parts of carbonic acid diester groups.

It is not completely clear as yet which factors determine whether displacement or disproportionation will occur. Results to date are in qualitative accord with the idea that neutral and acidic media favor disproportionation, whereas displacement reactions are more frequent in basic solution. Structural effects have also been demonstrated by Schilling and Gratzl (120). When 3,5-disubstituted 4-hydroxybenzyl alcohols are oxidized with ferricyanide, those with a critical oxidation potential larger than 390 mV undergo disproportionation, forming aldehydes, whereas those with lower potentials are

PRECURSORS AND THEIR POLYMERIZATION

converted to polyethers with formaldehyde displacement.

Gratzl and Viehäuser (123) have found that p-hydroxybenzyl groups are also displaceable entities. The oxidation of di-(3,5-di-t-butyl-4-hydroxyphenyl)-methane 70 was shown to lead to 3,5-di-t-butyl-4-hydroxybenzyl alcohol and a diphenoquinone. A probable reaction mechanism is shown in equation 19.

(19)

By an interesting intramolecular rearrangement, illustrated in Figure 4.15, aryloxy radicals may displace another aryloxy group from the p-position (124). Dehydrogenation of 4-phenoxyphenol 71 using an amine complex of cuprous chloride and oxygen forms an intermediate quinol ether that usually dissociates to a trimeric and monomeric phenoxy radical. At lower temperatures, however, the dominant reaction is an intramolecular ionic rearrangement, illustrated by formula 73, whereby a linear tetramer 74 is formed. Repetition of the dehydrogenative coupling between a dimer and tetramer forms a linear hexamer, which again may be converted to an octamer by further coupling. Ultimately, very high molecular weights are reached, particularly when the 2- and 6- positions are occupied by methyl groups (125).

E. Coupling Reactions at Unsubstituted Sites of Phenoxyradicals. Whenever unsubstituted o- and p-positions exist in phenoxy radicals, coupling reactions usually take place preferentially at these sites, probably for steric reasons. This rule is, however, not without exceptions, and cases are known when coupling occurs preferentially at a site already carrying a substituent, although unsubstituted sites are available (e.g., the formation of "Pummerer's ketones," to be discussed shortly). After the initial coupling at an unsubstituted site is completed, the cyclohexadienone intermediate formed tautomerizes to the corresponding phenol (See Figure 4.12). This process takes place almost instantaneously in protic media and more slowly in non-polar solvents.

The coupling reactions of 2,6-dimethyl phenol have been extensively investigated. C-C- and C-O-couplings at the 4-position are competitive processes in this case, Equation 20. The former coupling reaction yields 4,4'-dihydroxybiphenyl 77, which is easily oxidized further to diphenoquinone 78. C-O-coupling gives the 4-phenoxyphenol derivative 79 which can be converted to a linear polymer in further dehydrogenation, in accordance with the mechanism outlined in Figure 4.15.

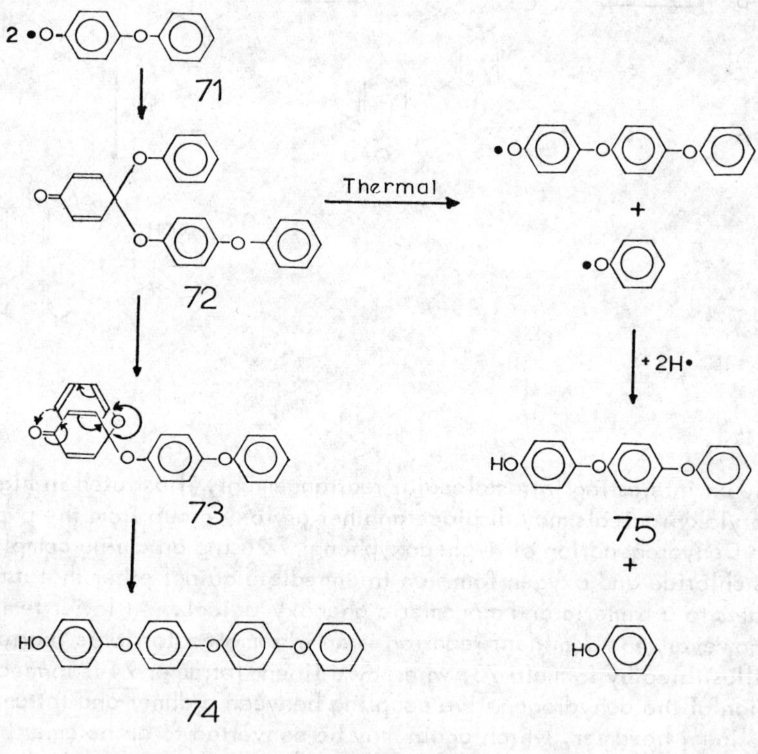

Figure 4.15. Quinol ether rearrangement in the coupling 4-phenoxyphenoxy radicals (124).

The diphenoquinone 78 is the main reaction product in dehydrogenations with air-oxygen and cupric ion in a neutral solution, together with small amounts of 2,6-dimethyl-p-benzoquinone (126). On the other hand, in basic

PRECURSORS AND THEIR POLYMERIZATION

dehydrogenation systems, such as silver oxide in benzene, ferricyanide (126) and cuprous chloride-amine complexes (127), the C-O-coupling dominates and the diphenoquinone is only a minor product of the reaction. By carrying out the dehydrogenation at low temperatures using specific copper-amine complexes, such as N,N-dimethylstearyl amine and copper (2:1), the formation of the diphenoquinone can be suppressed and the linear polyether recovered in excellent yields (127).

Polyethers are likewise obtained, if one of the o-methyls is replaced by ethyl, methoxyl, isobutyl or phenyl groups. However, C-O-coupling is totally hindered, probably for steric reasons, and the diphenoquinone is the principal product when one (or both) o-substituents are t-butyl groups, or, when both o-positions are attached to isobutyl, methoxyl or phenyl groups.

C-O-coupled products have rarely been obtained in enzymatic dehydrogenations of o-disubstituted phenols. Both laccase-oxygen (128) and peroxidase-hydrogen peroxide (129) oxidations of phenols substituted with methyl and methoxyl groups at the two o-positions lead to diphenoquinones, and 2-methyl-1-naphthol 81 gives an analogous quinone 82 in laccase-catalyzed oxidation, equation 21 (130).

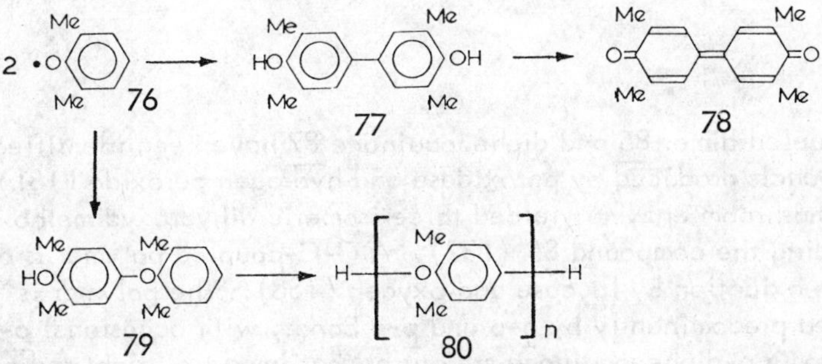

(20)

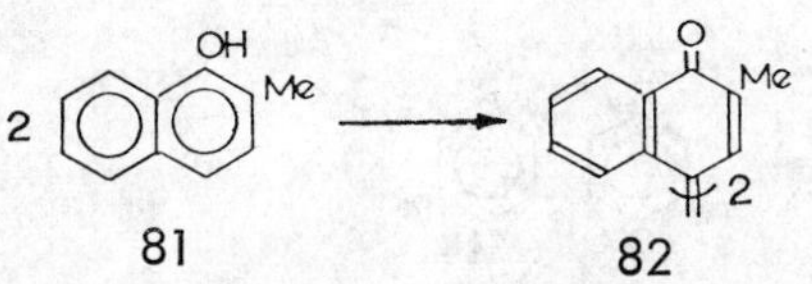

(21)

Guaiacol 83 has two sites available for coupling and, consequently, three separate coupling modes, p-p, o-o and o-p couplings are all possible, Equation 22.

$$(22)$$

o-o-Coupled dimer 85 and diphenoquinone 87 have been identified among the products produced by peroxidase and hydrogen peroxide (131) and oxidation by mushroom enzyme yielded three isomeric dihydroxydimethoxydiphenyls including the compound 85 (132). A C-C-coupled polymer is obtained from an oxidation by laccase and oxygen (133). The polymer is probably linked predominantly by o-o and p-p bonds, with occasional o-p linkages and some p-diphenoquinone structures that impart a bright red color to the polymer. In contrast, silver oxide and ferricyanide produce only minor amounts of C-C-linked products yielding mainly a C-O-coupled polymer 89.

o-o-Coupled dimers 90 have been commonly isolated from the oxidations of 4-substituted and 2,4-disubstituted phenols, Equation 23.

$$(23)$$

PRECURSORS AND THEIR POLYMERIZATION

$$
\begin{array}{lll}
\underline{91} & R' = CHO, & R'' = OMe \\
\underline{92} & R' = CH_3, & R'' = OMe \\
\underline{93} & R' = C_2H_5, & R'' = OMe \\
\underline{94} & R' = n\text{-}C_3H_7, & R'' = OMe \\
\underline{95} & R' = CO_2H, & R'' = OMe \\
\underline{96} & R' = CH_2\text{-}CH(NH_2)CO_2H, & R'' = H \\
\underline{97} & R' = CH_2CH_2NH_2, & R'' = H \\
\end{array}
\quad (23)
$$

For example, oxidation of vanillin 91 catalyzed by laccase from Russula species (134) or from Polyporus versicolor (122) or by peroxidase (135) yields dehydrodivanillin along with pale yellow dehydrogenation polymers. o-o-Coupled dimers have also been isolated from the oxidations of guaiacyl compounds 92, 93 and 94 with press juice from Psalliota campestris mushroom (136), laccase catalyzed oxidation of vanillic acid 95 (122) and from peroxidase-catalyzed oxidation of tyrosine 96 (137) and tyramine 97 (138). Interestingly, tyrosine 96 is not oxidized in the presence of laccase (139, 140).

Further examples of o-o-coupling are the dimerization of totarol 98 to podototarin 99 by laccase (141) (Equation 24) and that of pinoresinol 100 to dehydrodipinoresinol 101 by peroxidase-hydrogen peroxide (142) (Equation 25). A 25% yield of the dimerized product 103 has been reported for peroxidase-catalyzed dehydrogenation of "dihydrodehydrodi-isoeugenol" 102, Equation 26 (117).

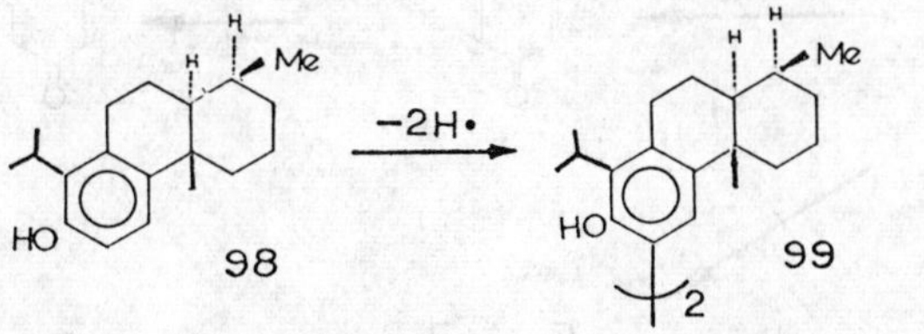

(24)

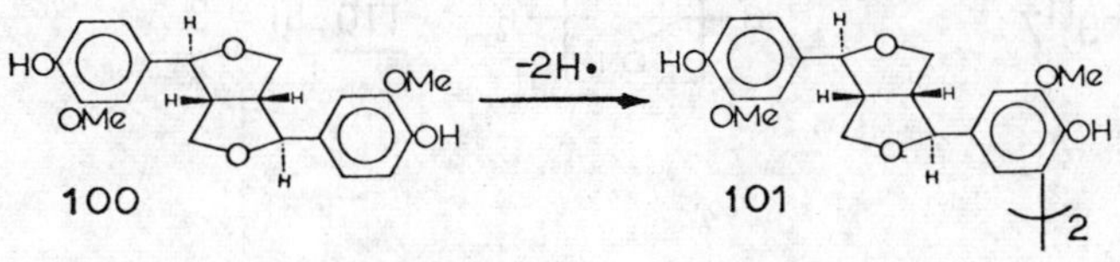

(25)

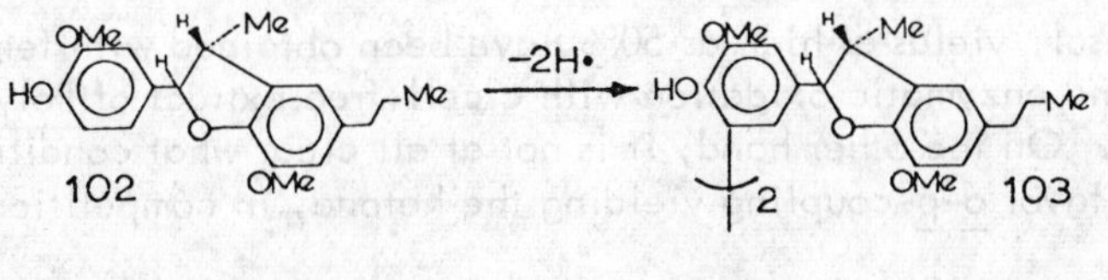

(26)

It would be a mistake, however, to assume that o-o-coupled dimers are always the major reaction products. Rather, they are easily isolated even from complex mixtures by virtue of their ready crystallization and insolubility. Furthermore, unlike p-p-coupled dimers, they are not oxidized further to diphenoquinones, but may form instead cyclic quinol ethers of the structural type 104 (143).

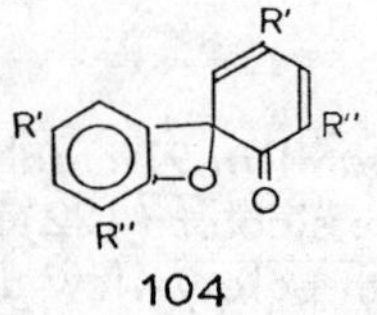

104

A major product in the oxidation of p-cresol (144, 145) 105, p-ethyl- 106 and p-n-propyl phenol (145) 107 is the "Pummerer's ketone" 108, formed through o-p-coupling followed by nucleophilic addition of phenolate to the cyclohexadianone system, Equation 27.

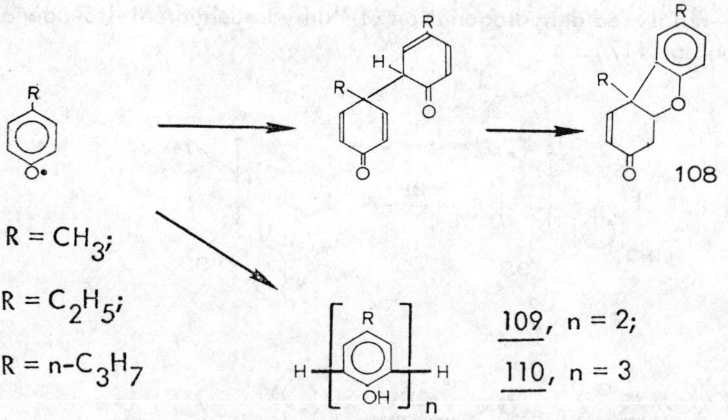

(27)

105, R = CH$_3$;
106, R = C$_2$H$_5$;
107, R = n-C$_3$H$_7$

109, n = 2;
110, n = 3

From p-cresol, yields as high as 50% have been obtained with ferricyanide (145) and enzymatic oxidation with a cell-free extract of Polyporus versicolor (140). On the other hand, it is not at all clear what conditions of dehydrogenation favor o-p-coupling yielding the ketone, in competition with the o-o-coupling yielding the dimer 109 and trimer 110. The latter appear to be the dominant products in oxidations with ferric chloride (146) and cupric-o-toluate (147) with only 1 to 3% of the ketone formed. Fenton's reagent yields 18% (148), but peroxidase-hydrogen peroxide only 13.4% (149) of the ketone. "Pummerer's ketone" has been obtained from 3,4-dimethylphenol (145) and in 2% yield from 2,4-dimethylphenol (150) with ferricyanide, but attempts to isolate it from the oxidation of 4-n-propyl-2-methoxyphenol with peroxidase-H$_2$O$_2$ totally failed (125). The conversion of methyl-phloroaceto-phenone 111 to usnic acid hydrate 112 by ferricyanide (151) serves as a further example of conversion to a coupling product of the Pummerer's ketone type, Equation 28.

$$\text{111} \xrightarrow{-2H\cdot} \text{112} \tag{28}$$

Pew, Connors and Kunishi have carried out detailed studies (117) on the dehydrogenation of 4-n-propyl-2-methoxyphenol (guaiacylpropane) 94 with peroxidase-hydrogen peroxide. In an aqueous-alcoholic medium, the o-o-coupled dimer 113 was obtained as a crystalline precipitate in 65% yield. The product composition changed radically, however, when the reaction was carried out in a purely aqueous medium (with added diatomaceous earth to maintain dispersion of insoluble reaction products), as shown in Figure 4.16. In this case, both C-O-coupling and disproportionation to α-hydroxylated products competed efficiently with o-o-coupling. The total yield of the o-o-coupled products (113 + 114) was only 21%, almost the same order of magnitude as that of C-O-coupled products (116 + 117 + 118), 17%. Interestingly, no products of the Pummerer's ketone type were isolated.

Pew's observations stress again the importance of the realization that the nature of the dehydrogenative system—hydrogen acceptor plus medium—exerts a profound influence on the course of various coupling, disproportionation and displacement processes. Other examples, cited earlier, include the effect of pH and dehydrogenation catalyst on the relative extent of C-C- and C-O-coupling and the bafflingly variable yields of Pummerer's ketone from the dehydrogenation of p-cresol. These observations are not readily compatible with the simplified assumption that the coupling processes taking place in the dehydrogenation of phenols consist entirely of simple recombinations of phenoxy radicals, and important gaps in the mechanistic knowledge of these reactions remain to be filled.

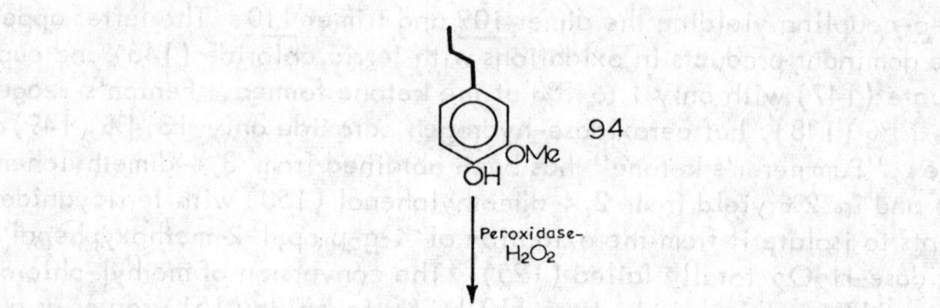

Figure 4.16. Dehydrogenation of 4-n-propyl-2-methoxyphenol by peroxidase-hydrogen peroxide in an aqueous medium.

F. **Dehydrogenation of p-Propenylphenol Derivatives.** In general, p-propenylphenoxy radicals possess more stability than phenoxy radicals without unsaturated groups at the o- and p-positions by virtue of one additional resonance structure R_β, Equation 29. Furthermore, the β-carbon provides an additional site for coupling reactions. The enzymatic dehydrogenation of trans-isoeugenol to "dehydrodiisoeugenol" 120 represents a classic case of reactions of this type, Equation 30.

The reaction was discovered by Cousin and Herissey (152) who used a crude enzyme mixture from the mushroom Russula delica. The structure of the crystalline reaction product 120 was determined by Erdtman (153) and its stereochemistry by Aulin-Erdtman (154), using nmr and oxidative degradation methods. Aulin-Erdtman also showed that dehydrodiisoeugenol is exclusively formed from the trans-isomer.

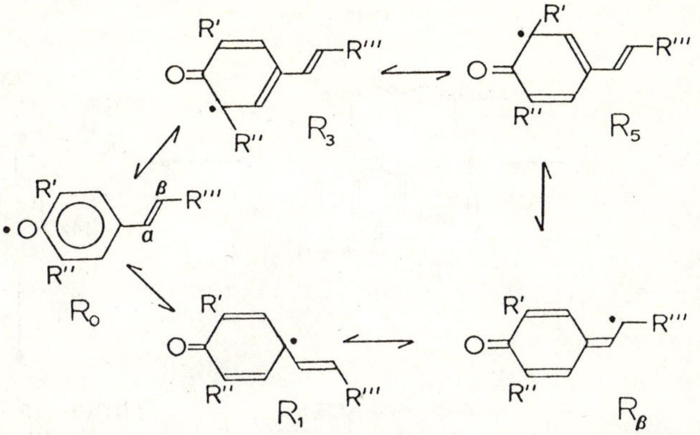

PRECURSORS AND THEIR POLYMERIZATION

(29)

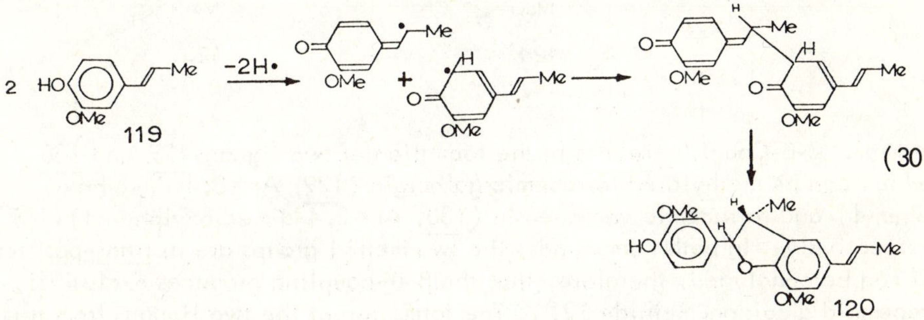

(30)

Lindberg (155) has provided evidence for the occurrence of β-β-coupling by isolating dehydroguaiaretic acid 123 from the dehydrogenation of isoeugenol with ferric chloride, Equation 31. The initially formed di-quinonemethide 121, after addition of one mole of water, undergoes acid-catalyzed condensation forming a tetrahydronaphthalene derivative. Removal of water and two additional dehydrogenation steps aromatize this intermediate to dehydroguaiaretic acid 123. Isoeugenol itself is quite sensitive to acids dimerizing to diisoeugenol 124 (156), which is commonly found as a side-reaction product in ferric chloride oxidations.

Neither dehydroguaiaretic acid nor diisoeugenol are formed with peroxidase-hydrogen peroxide in acetone-water mixtures, as demonstrated recently by Wallis and Sarkanen (157). By using 0.5 mole of hydrogen peroxide, both trans- and cis- isoeugenols may be converted quantitatively to dimers. The quantitative yields of each component are outlined in Table 4.2; the structures, in Figure 4.17. In accordance with earlier studies, the β-5-coupled dehydrodiisoeugenol 120 is the dominant product (65%) from the trans-isomer. β-β-Coupling results in the formation of two lignans 129 and 130 which can be methylated to racemic galbelgin (129, Ar = 3,4-dimethoxyphenyl) and to racemic veraguensin (130, Ar = 3,4-dimethoxylphenyl) (157), respectively. In both compounds, the two methyl groups are in trans-position. It can be concluded, therefore, that the β-β-coupling produces exclusively racemoid diquinonemethide 121. The formation of the two lignans from this intermediate occurs by water addition to one of the quinonemethide groups, after which the benzyl alcohol thus formed adds to the second quinonemethide, closing the tetrahydrofuran ring.

Table 4.2. Percentage Yields of Dimers Formed from Trans- and Cis-Isoeugenol by Peroxidase-Hydrogen Peroxide

	Per cent yield from	
	Trans-isoeugenol	Cis-isoeugenol
Dehydrodi-isoeugenol 120	65	None
Dehydrodi-isoeugenol 125	None	22
Threo-β-O-4 dilignol 127a	17	None
Erythro-β-O-4 dilignol 128a	5	None
Threo-β-O-4 dilignol 127b	None	40
Erythro-β-O-4 dilignol 128b	None	13
Lignan 129 (galbelgin type)	9	17
Lignan 130 (veraguensin type)	4	8

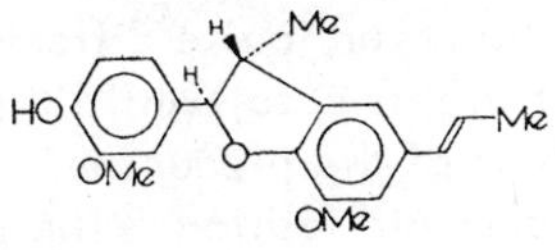

120

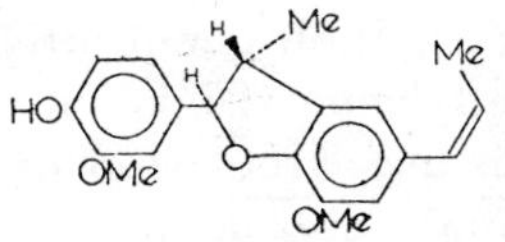

125

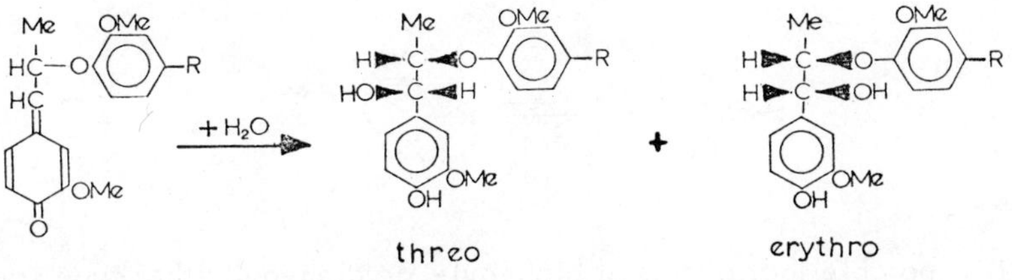

126a,b 127a,b 128a,b

a: R = trans-propenyl

b: R = cis-propenyl

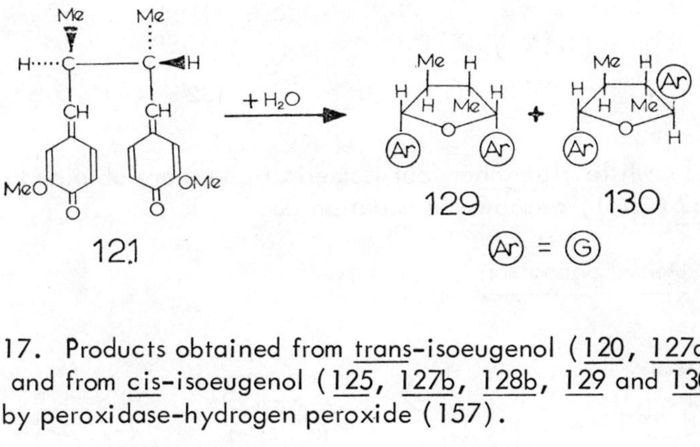

121 129 130

Ar = G

Figure 4.17. Products obtained from trans-isoeugenol (120, 127a, 128a, 129 and 130) and from cis-isoeugenol (125, 127b, 128b, 129 and 130) in dehydrogenation by peroxidase-hydrogen peroxide (157).

Coupling between position β and the phenoxy-oxygen produces the quinone methide 126a that is converted through water addition to a mixture of threo- (127a) and erythro- (128a) β-ethers. The former isomer predominates because its formation is favored for steric reasons.

The products from cis-isoeugenol are otherwise identical except that the propenyl side-chain in products 125, 127b and 128b retains its original cis-trans-position. Their quantitative amounts are, however, quite different. The β-5-coupled dimer 125 is only a minor product in this case, while the β-ethers 127b and 128b account for about one-half of the product mixture. β-β-coupling occurs again exclusively in the racemoid fashion. The possible conversion of the cis-radical to the trans-radical (or vice versa) must be a very slow process in comparison with the coupling reactions, Equation 32.

$$\tag{32}$$

It is notable that dimers of biphenyl- or diphenyl ether type were totally absent from the product mixtures. This suggests that the β-radical mesomer (R_β) is outstandingly reactive and is therefore a partner to practically all couplings that take place.

The trans-syringyl analogue of isoeugenol, 131, is dehydrogenated by peroxidase-hydrogen peroxide to two lignans of the types 129 and 130

(Ar = syringyl) while altogether four isomeric lignans are obtained from the cis-isomer 132 (157), as shown in equation 33.

132   Dehydrogenation
$$\longrightarrow \tag{33}$$

Consequently, racemoid β-β-coupling mode appears to be exclusive for the trans-isomer, while both racemoid and mesoid β-β-couplings occur in case of the cis-radical. Neither syringylpropene appears to yield products other than the β-β-coupled lignans with peroxidase, while ferric chloride oxidation of the trans-isomer leads mainly to β-ethers analogous to 127 and 128 (157).

PRECURSORS AND THEIR POLYMERIZATION

Müller and coworkers (158, 159) have studied the dehydrogenation of compounds 133 a-e with ferricyanide.

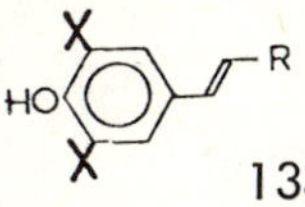

a. R = CO_2Me
b. R = CO_2Et
c. R = CO_2i-Pr
d. R = CN
e. R = CHO

These compounds undergo exclusively β-β-coupling, and the diquinone-methide 134 is isolated in excellent yields from 133 a, b and c, probably because the t-butyl groups hinder the addition of water to the quinonemethide groups. Dimer 134a has been separated to equal amounts of the racemoid and mesoid isomers (157), suggesting that the two coupling modes occur to approximately the same extent, Equation 34.

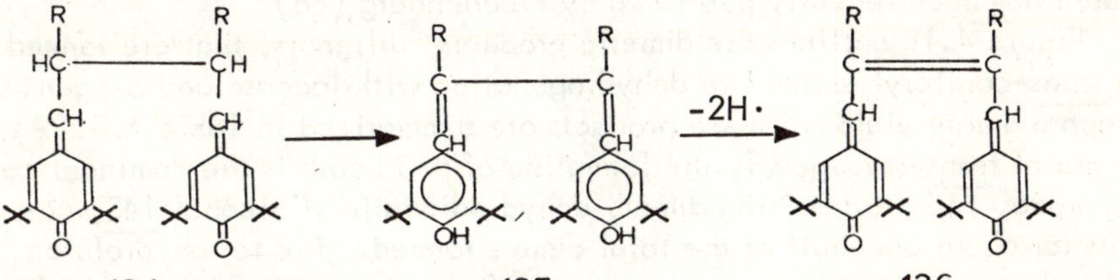

$$ (34) $$

134 R = CO_2Me, CO_2Et, CO_2i-Pr   135 R = CN, CHO   136

Diphenol 135 (160) instead of the diquinonemethide is obtained from 133 d and e, probably because of the activating effect of cyano and formyl groups on the hydrogens attached to the middle carbons of the side-chains, facilitating rearrangement to the more stable aromatic system. The diphenol, in turn, is readily oxidized further to the extended di-quinonemethide 136.

Trans-ferulic acid 24 is converted by ferric chloride partially to dimeric dilactone 138 through a racemoid β-β-coupling process (161). The carboxylic groups of the diquinonemethide intermediate 137 add internally to the quinone methide groups, Equation 35. Although the small yields obtained do not allow definite conclusions to be made on the exclusiveness of the racemoid coupling mode, this is probably the case because sinapic acid 26 is dehydrogenated to an analogous dilactone in 73% yield (162).

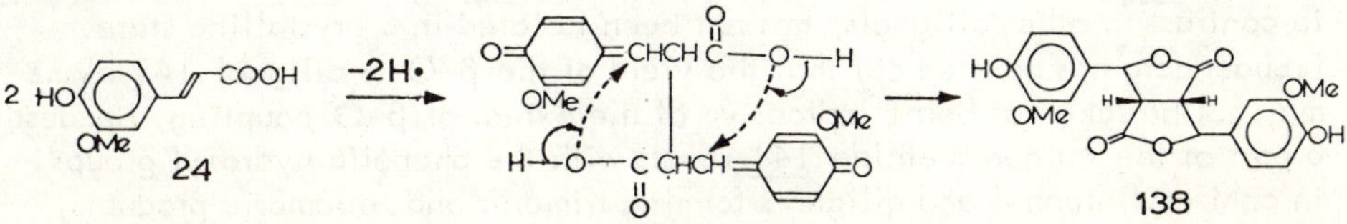

$$ (35) $$

Ferulic acid ethyl ester yields in enzymatic dehydrogenation as the main product crystalline dimer 139 that has the same stereochemistry as dehydrodi-isoeugenol, (163).

[Structure 139]

G. Dehydrogenations of Coniferyl and Sinapyl Alcohols. For the understanding of lignin polymerization in vivo, dehydrogenation experiments with primary lignin precursors provide the most convincing guidelines. This line of research has been pursued almost exclusively by Freudenberg and his school during the past twenty years, and an outstanding summary of this research has been recently published by Freudenberg (88).

Figure 4.18 outlines the dimeric products, dilignols, that are formed from trans-coniferyl alcohol in dehydrogenation with laccase and oxygen. The approximate yields of these products are summarized in Table 4.3. As in the case of trans-isoeugenol, the formation of β-5 bonds is the dominant coupling mode, and the resulting dimer, dehydrodiconiferyl alcohol 140, accounts for about one-half of the total dimers formed. The tetrahydrofuran ring has been shown to possess the same configuration as that in dehydrodiisoeugenol 120 (92) with the two hydrogens occupying trans-position. β-β-Coupling results in the formation of the diquinone-methide intermediate 141. Intramolecular additions of the γ-carbinol groups to the quinone methides lead to racemic pinoresinol 142 and, to a slight extent, to its isomer, racemic epipinoresinol 143. The structures of these two compounds demonstrate a racemoid β-β-coupling mode. Mesoid coupling would allow for the closure of only one of the five-membered rings in the central furofuran system, and no indication for the presence of products of this nature has been obtained. Coupling between the β-carbon and phenoxy oxygen yields quinonemethide 144 which is reasonably stable in a neutral aqueous medium. As other quinone methides, it is yellow in color with a characteristic light-absorption maximum at 312 mμ. Spectral measurements indicate half-lives as long as one hour for quinone-methide 144 in dilute solutions in neutral partially aqueous media (169). Addition of water converts it to β-O-4 dilignol 145, which is probably a mixture of threo- and erythro forms and, in contrast to other dilignols, has not been isolated in a crystalline state. Freudenberg has pointed out that the yield of the β-O-4 dilignol 145 alone may not be taken as being indicative of the extent of β-O coupling, because a part of the quinonemethide 144 reacts with the phenolic hydroxyl groups in coniferyl alcohol and dilignols forming trimeric and tetrameric products, to be discussed shortly.

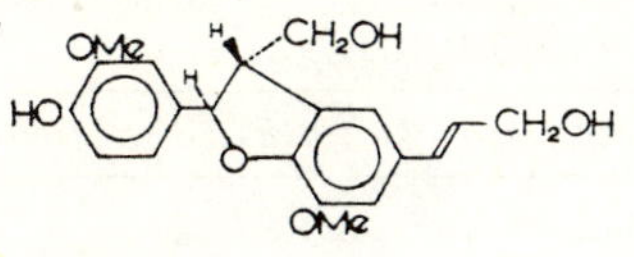

PRECURSORS AND THEIR POLYMERIZATION

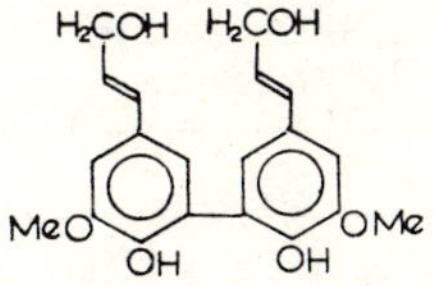

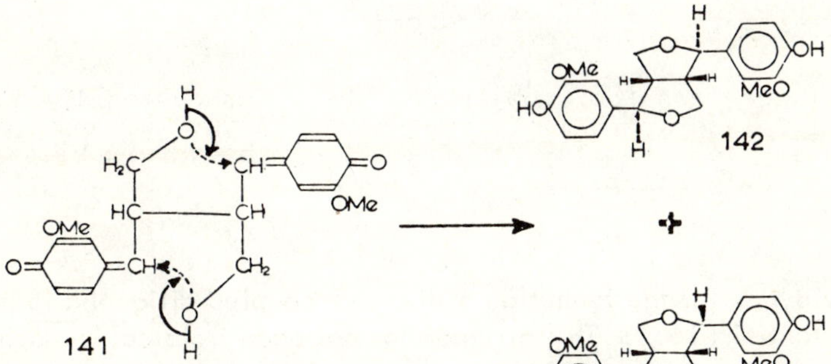

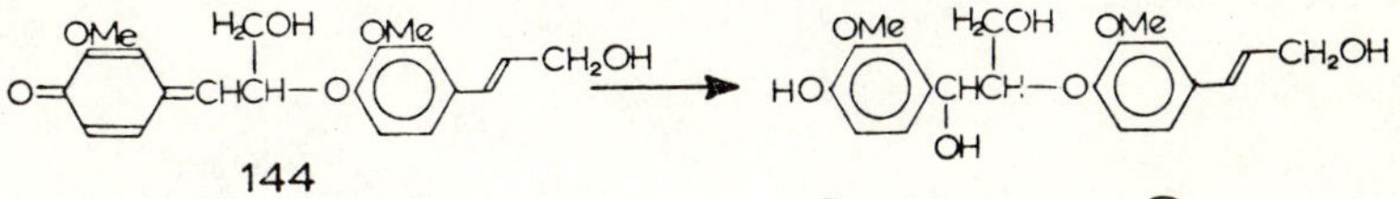

Figure 4.18. Products formed through direct coupling and disproportionation of coniferyl alcohol radicals in dehydrogenation by laccase and oxygen (164).

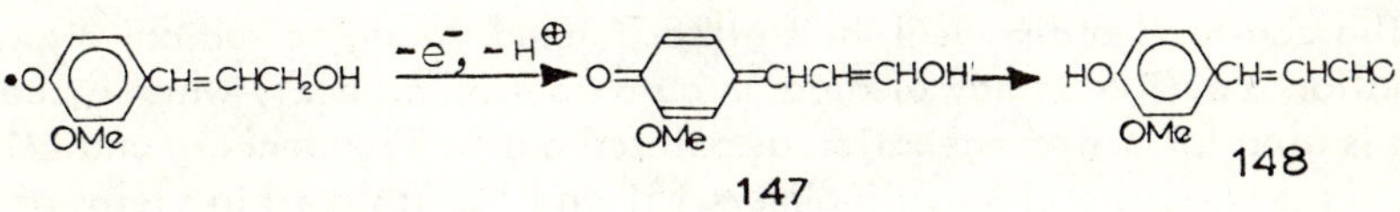

Table 4.3. Relative Amounts of Direct Coupling and Disproportionation Products in the Dehydrogenation of Coniferyl Alcohol by Laccase and Oxygen *)

Product	% of total	% of dimers	Ref.
Unchanged coniferyl alcohol	3	-	(164)
Polymer (DHP)	45	-	(164)
β-5 Dilignol 140	26 [1])	54	(164)
Rac. Pinoresinol 142	13	27	(164)
Rac. Epipinoresinol 143	trace	-	(165)
β-O-4 Dilignol 145	9 [2])	19	(164)
5-5 Dilignol 146	trace (?)	-	(166)
Coniferyl aldehyde 148	0.2	-	(164)

[1]) An approximate yield of 40% also reported (167)
[2]) Incomplete recovery

*) In his reviews (88, 168) Freudenberg has quoted 15% yield for 140, and approximately equal yield for 142 and the highest yield for 145, contradicting the data published in the original papers.

The evidence for the formation of the 5-5-coupled dilignol 146 is somewhat circumstantial (166). This dilignol has not been isolated, as such, from the dehydrogenation products of coniferyl alcohol, and neither could it be synthesized in a pure form. Paper chromatography suggested the presence of the tetrahydroderivative 149 in a hydrogenated sample of dehydrogenation products from coniferyl alcohol, but the published data do not reveal whether or not 149 or its dimethyl ether 150 were actually isolated. It should be also noted that dihydroconiferyl alcohol is a tenaceous impurity of coniferyl-alcohol preparations.

149, R = H

150, R = MeO

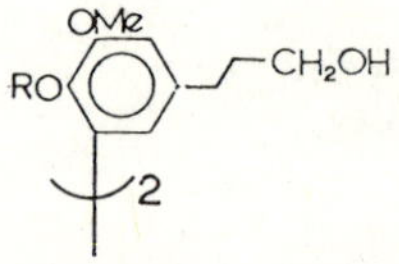

The coniferyl aldehyde 148 is either formed through a radical disproportionation reaction or may also be an autoxidation product, when laccase-oxygen is used for dehydrogenation as pointed out by Freudenberg and Lehman (165). Likewise, the aldehydic dimers 151 and 152 isolated in yields of 1.5 and 0.9% respectively, from laccase-oxygen dehydrogenations of coniferyl alcohol (165) may represent either cross-coupling products between coniferyl alcohol and -aldehyde radicals, or, alternatively, autoxidation products of β-5 (140) and β-O-4 (145) dilignols.

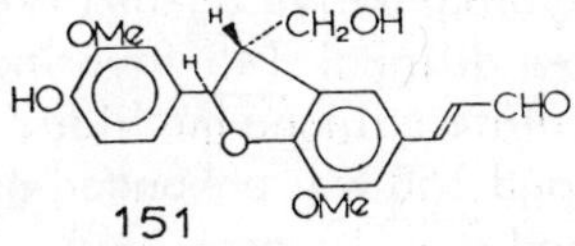

PRECURSORS AND THEIR POLYMERIZATION

151, 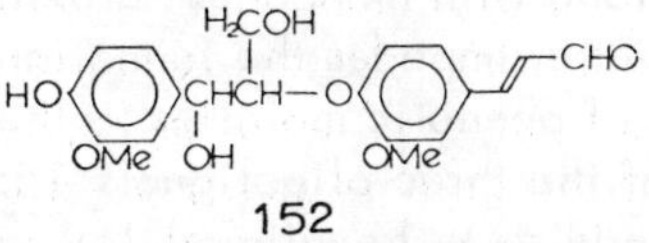 **152**

Coniferyl aldehyde itself is subject to further oxidation, yielding cis- and trans-ferulic acids, 24, 153, (171), traces of vanillin, 91, and vanillic acid, 95, (171), Equation 36. Ferulic acid radicals cross-couple with coniferyl alcohol radicals forming dimers, of which the β-β dilignols "pinoresinolide" 154 and "lignonolide" 155 have actually been isolated (171, 172). The formation of the exocyclic double bond in the latter dimer probably occurs through a similar mechanism as in the formation of dimers of the type 135 (equation 34).

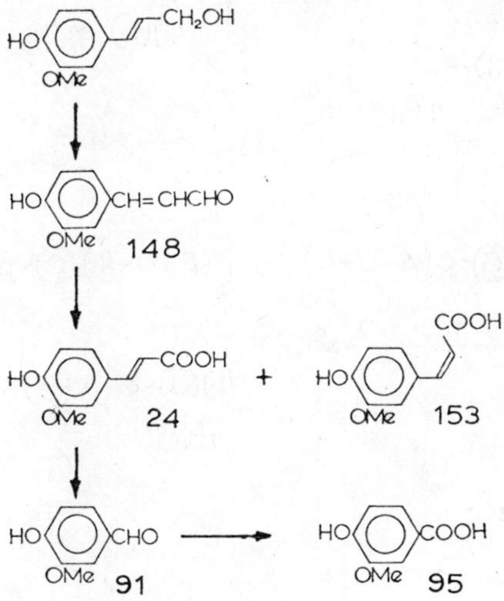

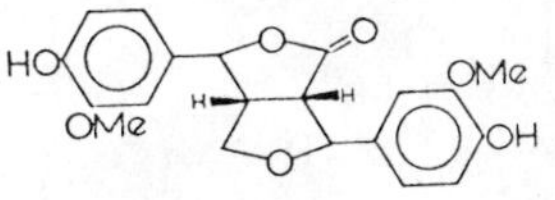

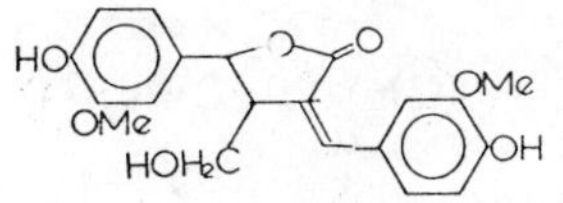

(36)

Freudenberg and coworkers (173, 174) have found strong indications for the presence of small amounts of trimer 156, tetramer 157 and pentamer 158 among the dehydrogenation products of coniferyl alcohol, Figure 4.19. Of these, 156 represents an addition product of coniferyl-alcohol to the β-O-4 linked quinone methide 144 and 157 that of the β-5 dilignol 140 to the same quinomethide. The formation of pentalignol 158 probably occurs through β-O-coupling of 157 with coniferyl alcohol.

In order to prepare the three oligolignols in larger quantities for characterization, dehydrogenations of coniferyl alcohol were carried out in water-free acetone with manganese dioxide as dehydrogenative agent (174). The lack of water impedes the formation of β-O-4 dilignol 145 favoring the addition of phenolic moieties to the intermediate quinone methide. The formation of the three oligolignols 156, 157, and 158 was enhanced as expected, and, in addition, hexalignol 159 was isolated and characterized. The formation of hexalignol 159 in aqueous systems is not very probable, as two successive phenolic quinonemethide additions are involved in its formation. Consequently it is not surprising that the hexalignol has not been positively identified among the dehydrogenation products in aqueous systems.

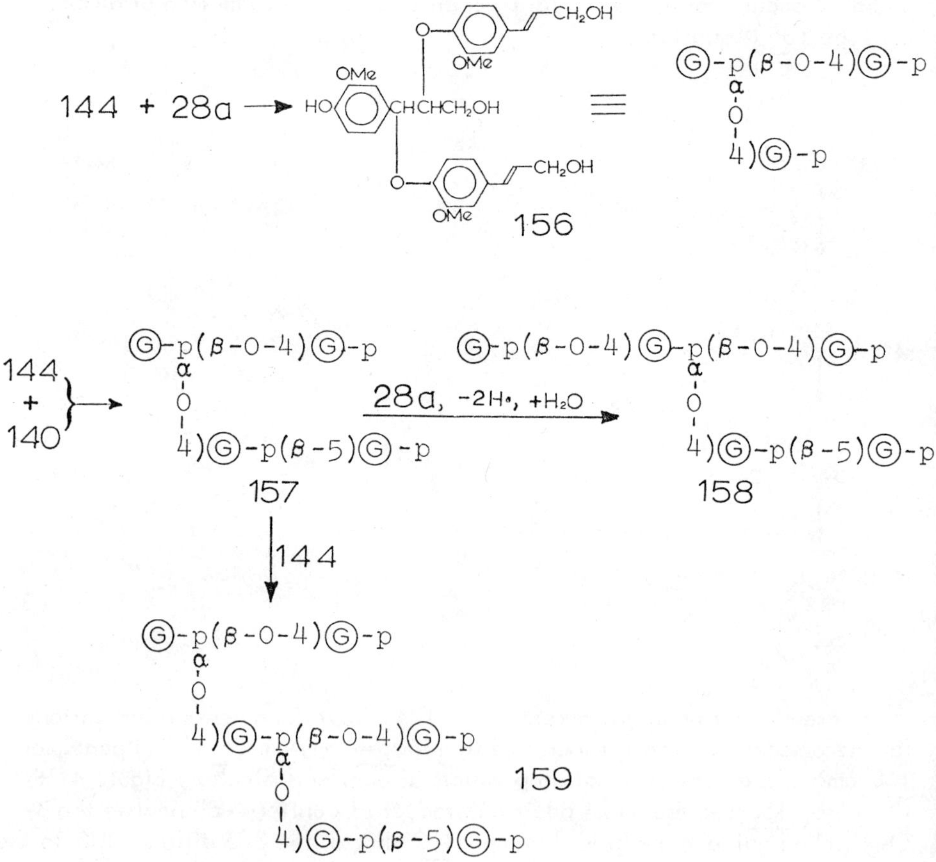

Figure 4.19. Products formed through the addition of mono- and dilignols to quinonemethide 144 (173, 174).

PRECURSORS AND THEIR POLYMERIZATION

Due to the extreme sensitivity of coniferyl alcohol towards acids, it is not practical to carry out dehydrogenations below pH 4 (92). Instead of dimers of the indane type that are formed in the acid-catalyzed dimerization of isoeugenol (156) and also from trans-ferulic acid ethyl ester (175), a polymeric material is obtained from coniferyl alcohol at pH 3 and higher acidities. The polymerization process appears to be of the ethylenic type, judging from the proposed structures of an isolated dimer 160 and trimer 161 (176).

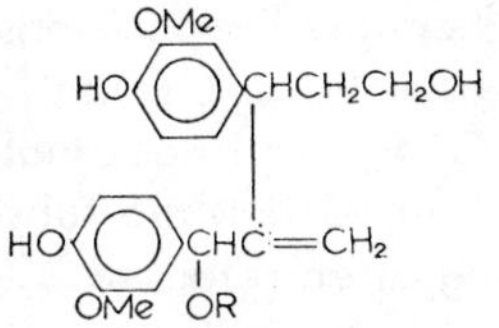

160

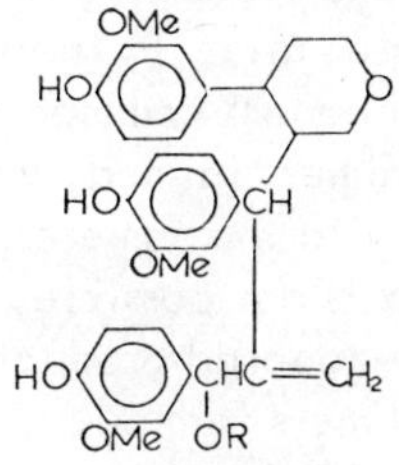

161

The dehydrogenation of sinapyl alcohol 29 leads only to a limited number of products because the 5-position is not available for coupling. Furthermore, the dual o-substitution forms a retarding barrier against β-O-coupling. As a consequence, racemoid β-β-coupling is the almost exclusive dehydrogenation reaction leading to racemic syringaresinol 162 in yields ranging from 80 to 88% in dehydrogenation by copper sulphate and air (80). Further dehydrogenation in the latter system produces 2,6-dimethoxy-p-benzoquinone 66 in 60% yield, Equation 37.

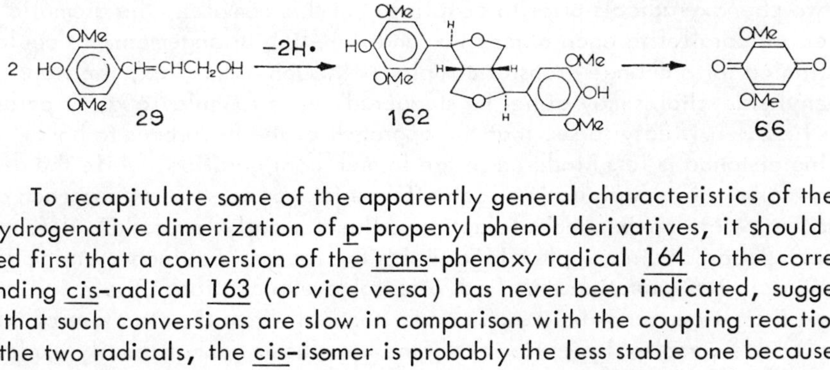

$$2 \; 29 \xrightarrow{-2H\cdot} 162 \longrightarrow 66 \qquad (37)$$

To recapitulate some of the apparently general characteristics of the dehydrogenative dimerization of p-propenyl phenol derivatives, it should be noted first that a conversion of the trans-phenoxy radical 164 to the corresponding cis-radical 163 (or vice versa) has never been indicated, suggesting that such conversions are slow in comparison with the coupling reactions. Of the two radicals, the cis-isomer is probably the less stable one because of the non-bonded interaction of the substituent groups attached to the terminal carbon of the side-chain with the aromatic ring. When both isomers are present, the initial coupling products are formed from the trans-phenol only (154), as a consequence of the equilibrium in equation 38 lying to the right.

$$\text{163} + \text{HO-Ar} \rightleftharpoons \text{HO-Ar} + \text{164} \quad (38)$$

Of the possible coupling modes, those producing a covalent bond at the β-carbon of at least one of the participating radicals are the dominant ones. Coupling modes producing diphenyl ether or biphenyl bonds are insignificant, if they occur at all. In trans-phenols with unoccupied o-positions, the β-5-coupling is the most common one, accounting usually for one-half of slightly more of the dimers formed, while β-β- and β-O-4- couplings usually occur approximately to the same extent. When both o-positions are substituted, β-5-coupling is not possible, and β-O-coupling often retarded. As a consequence, β-β-coupled products account commonly, but not always, for about 90% of the dimers formed.

The coupling processes favor the formation of certain stereoisomers over the others. β-5-couplings of both trans- and cis-propenylphenols are commonly followed by a ring closure to a coumaran system at which the γ-carbon and the aromatic ring occupy trans-positions. In case of the cis-phenols, a 180° rotation around the α–β bond is therefore necessary prior to the ring closure. In additions of water (and possibly phenols) to dimeric quinone methides of the β-O-4 type, the formation of threo-isomers appears to be favored over the erythro-forms. The often observed exclusive formation of racemoid dimers in β-β-coupling is not readily understandable in terms of a direct coupling mechanism, as pointed out by Wallis and Sarkanen (157), who have proposed an alternative mechanism illustrated in Figure 4.20.

According to this proposal, an intermediate complex is formed between the two phenoxyradicals prior to coupling. In this complex, the aromatic nuclei are parallel to each other in a sandwich-like arrangement, as could be anticipated for a change-transfer complex. In such a complex, the trans-propenyl side-chains may either be staggered, as in formula 165a, or parallel as in 165b. It is easy to see that the approach of the β-carbons to bond-forming distance is less hindered in the former configuration, while the α- and γ-carbons in configuration 166b offer resistance to such an approach as a consequence of non-bonded interaction. The sterically favored bond-forming between β-carbons in complex 165a yields the racemoid quinonemethide 166a, whose exclusive formation thus finds a feasable explanation. The racemoid coupling of cis-isoeugenol is likewise understandable from the structure of the intermediate complex 167. However, the hypothesis cannot readily explain why the syringyl analogue of cis-isoeugenol undergoes both racemoid and mesoid coupling while its trans-isomer undergoes only racemoid coupling. It should be also noted that the quinol ether 168 is a possible, although not very probable, alternative structure for the intermediate in β-β-coupling.

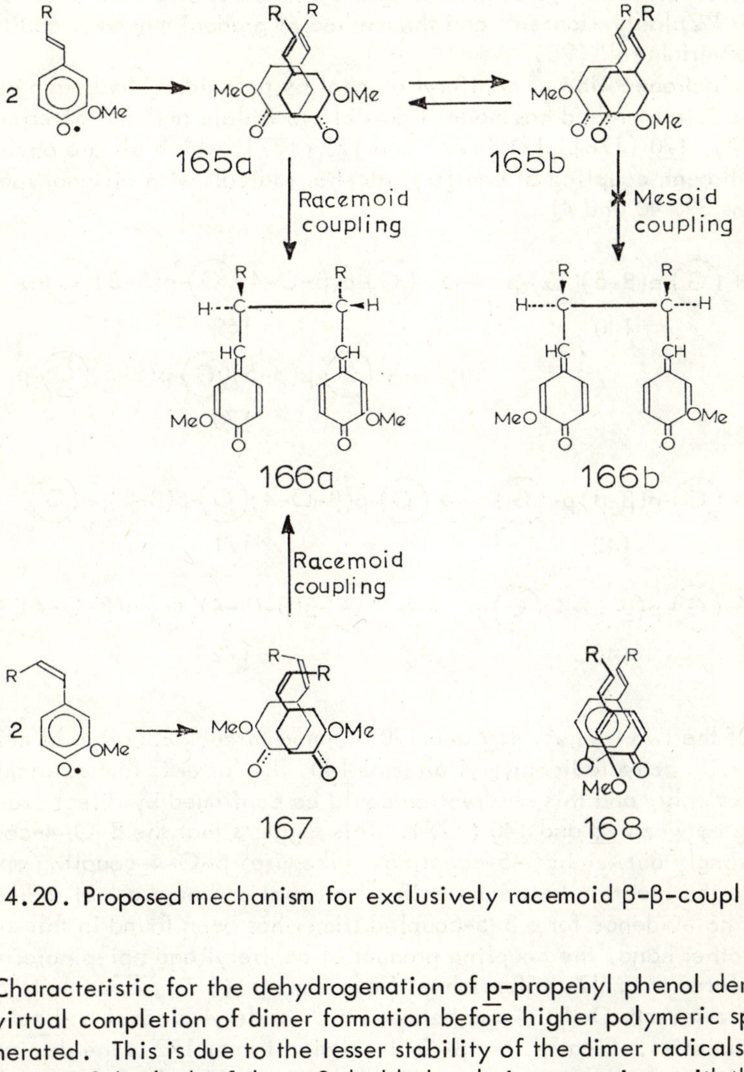

PRECURSORS AND THEIR POLYMERIZATION

Figure 4.20. Proposed mechanism for exclusively racemoid β-β-coupling (157).

Characteristic for the dehydrogenation of p-propenyl phenol derivatives is the virtual completion of dimer formation before higher polymeric species are generated. This is due to the lesser stability of the dimer radicals as a consequence of the lack of the α,β double bond, in comparison with the monomeric phenoxy radicals.

H. Cross-Couplings Between Mono- and Dilignols. In dehydrogenation systems, to which no additional monomeric phenol is added during the polymerization, coupling reactions between dimeric and monomeric radicals are uncommon due to the large difference in the stability of these two radical species. The probability of such couplings can be enhanced, however, by adding gradually fresh monomer to a dehydrogenation system that has reached the stage of dimerization. Freudenberg and coworkers have studied in detail both dehydrogenation systems, terming the method of one-time addition of monomer "Zulaufverfahren" and the method of gradual monomer addition "Zutropfverfahren" (93).

Dehydrogenation of coniferyl alcohol by peroxidase-hydrogen peroxide using the latter method has made it possible to isolate and characterize trimers 169 (177), 170 (178), 171 (179) and 172 (177), which all are obviously formed through coupling of coniferyl alcohol radicals with dilignol radicals, Equations 39, 40 and 41.

$$\text{(G)-p} + \text{(G)-p}(\beta\text{-}5)\text{(G)-p} \longrightarrow \text{(G)-p}(\beta\text{-O-4})\text{(G)-p}(\beta\text{-}5)\text{(G)-p}$$
$$ 28 140 169$$
$$+ \text{(G)-p}(\beta\text{-}5)\text{(G)-p}(\beta\text{-}5)\text{(G)-p} \quad (39)$$
$$ 170$$

$$\text{(G)-p} + \text{(G)-p}(\beta\text{-}\beta)\text{p-(G)} \longrightarrow \text{(G)-p}(\beta\text{-O-4})\text{(G)-p}(\beta\text{-}\beta)\text{p-(G)} \quad (40)$$
$$ 28 142 171$$

$$\text{(G)-p} + \text{(G)-p}(\beta\text{-O-4})\text{(G)-p} \longrightarrow \text{(G)-p}(\beta\text{-O-4})\text{(G)-p}(\beta\text{-O-4})\text{(G)-p} \quad (41)$$
$$ 28 145 172$$

Of the two trimers, 169 and 170, formed through coupling of coniferyl alcohol with dehydrodiconiferyl alcohol 140, the latter is formed in minor quantities only, and this observation could be confirmed by direct cross-coupling between 28 and 140 (177). This suggests that the β-O-4-coupling mode strongly outweighs β-5-coupling. Likewise, β-O-4-coupling appears to dominate the reaction between coniferyl alcohol and pinoresinol radicals because no evidence for a β-5-coupled trimer has been found in this case. On the other hand, the coupling product of coniferyl and epi-pinoresinol has been isolated from "Zutrofverfahren" experiments (179), and this trimer, too, possesses a β-O-4 linkage between the coniferyl alcohol and epi-pinoresinol moieties. Finally, the β-O-4-coupled trimer 172 is the dominant product from coupling between coniferyl-alcohol and β-O-4 dilignol 145.

In view of the ample possibility for stereo-isomerism, it appears natural that trimers 169 to 172 were isolated as amorphous preparations. Altogether eight racemates are probable for 172, four for both 169 and 171, while 170 ought to consist of two racemates only, assuming trans-configuration for the two phenyl coumaran systems. Actually, 170 could be induced to partial crystallization (178).

The finding of structural types corresponding to the unusual β-1 dilignol in conifer lignin by Nimz (180) and by Lundquist (181), and the later

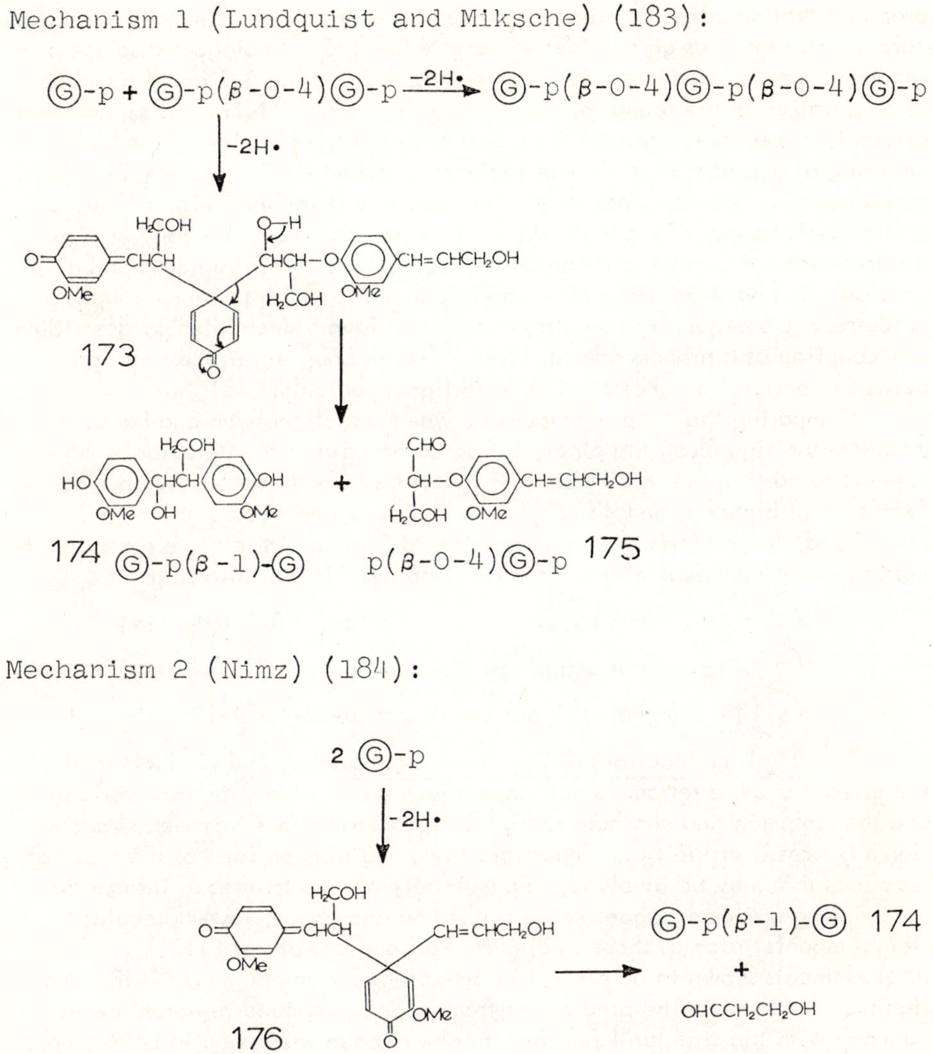

PRECURSORS AND THEIR POLYMERIZATION

Mechanism 1 (Lundquist and Miksche) (183):

$$\text{G-p} + \text{G-p}(\beta\text{-O-4})\text{G-p} \xrightarrow{-2H\cdot} \text{G-p}(\beta\text{-O-4})\text{G-p}(\beta\text{-O-4})\text{G-p}$$

$$\downarrow -2H\cdot$$

173

↓

174 G-p(β-1)-G   +   p(β-O-4)G-p **175**

Mechanism 2 (Nimz) (184):

$$2\ \text{G-p}$$

$$\downarrow -2H\cdot$$

176 → G-p(β-1)-G **174** + OHCCH$_2$CH$_2$OH

Figure 4.21. Two proposed mechanisms for the formation of β-1 dilignols (183, 184).

demonstration of the presence of dilignol <u>174</u> among the dehydrogenation products of coniferyl alcohol (182) have brought into consideration a new type of coupling reaction between coniferyl alcohol and dilignol <u>145</u> radicals, the β-1 coupling, mechanism 1, Figure 4.21. Lundquist and Miksche (183) have proposed that coupling at this position results in the displacement of the three-carbon side-chain as glyceraldehyde aryl ether 175. Analogous displacement reactions were discussed earlier in this chapter, and the β-1 coupling mode is also operative in the formation of Pummerer's ketones. Nimz has suggested an alternative mechanism for the formation of β-1 dilignols (184) based on β-1 coupling of two coniferyl alcohol radicals, mechanism 2, Figure 4.21. This proposition is, however, not very likely because it involves either a nucleophilic displacement (by a hydroxyl ion) at an unsaturated carbon, or else, the formation of a vinyl carbonium ion, both of which are rather improbable processes and have not been observed in connection with phenoxy coupling processes. Consequently, available evidence favors mechanism 1, describing β-1 coupling as a process competitive to β-O-4 coupling in the reaction between coniferyl alcohol and β-O-4 dilignol radicals.

Comparing the coupling modes between monolignols on one hand, and mono- and dilignols on the other, it can be noted that the β-radical form appears to participate in all coupling reactions in both instances, and the formation of biphenyl and diphenyl ether bonds appears to be absent. On the other hand, there seems to be an important difference in relative extent of the various coupling modes which, in all probability, is the following:

Ⓖ-Monolignol couplings: $\beta\text{-}5 > \beta\text{-}\beta \cong \beta\text{-}O\text{-}4$

Ⓢ-Monolignol couplings: $\beta\text{-}\beta >> \beta\text{-}O\text{-}4$

Ⓖ-Monolignol-dilignol couplings: $\beta\text{-}O\text{-}4 > \beta\text{-}1 > \beta\text{-}5$

1. Coupling Reactions of Di- and Oligolignols. The products from these reactions are tetramers or higher oligomers of wide structural variability, and the isolation and characterization of such compounds from dehydrogenation mixtures is virtually an impossible task. To form an idea of the types of reactions that may be involved, one must rely on two sources of information: i, Coupling processes suggested by studies on analogous, low-molecular weight models, such as those studied by Pew and coworkers (117); ii, Structural elements shown to be present in dehydrogenation polymers (DHP) and lignins. In general, the predictions from model compound research are in harmony with the structural features that have been identified in DHP-preparations and lignins, and receive a detailed discussion in Chapter 5.

Figure 4.22 summarizes the anticipated coupling reactions. Formula 177 represents a generalized dilignol molecule in which R stands for a side-chain (without an α-β double bond) in combination with another monolignol unit. C-C coupling between the 5-positions would lead to a tetramer with a biphenyl linkage 178. Dehydrodipinoresinol 103 is an example of a tetramer of this type and has been shown to be a component of dehydrogenation products from coniferyl alcohol by Freudenberg and Sakakibara (142). The alternative C-O coupling produces tetramer 179. Although tetrameric lignols of this type have not been isolated, 4-O-5 linkages are present in DHP's (166)

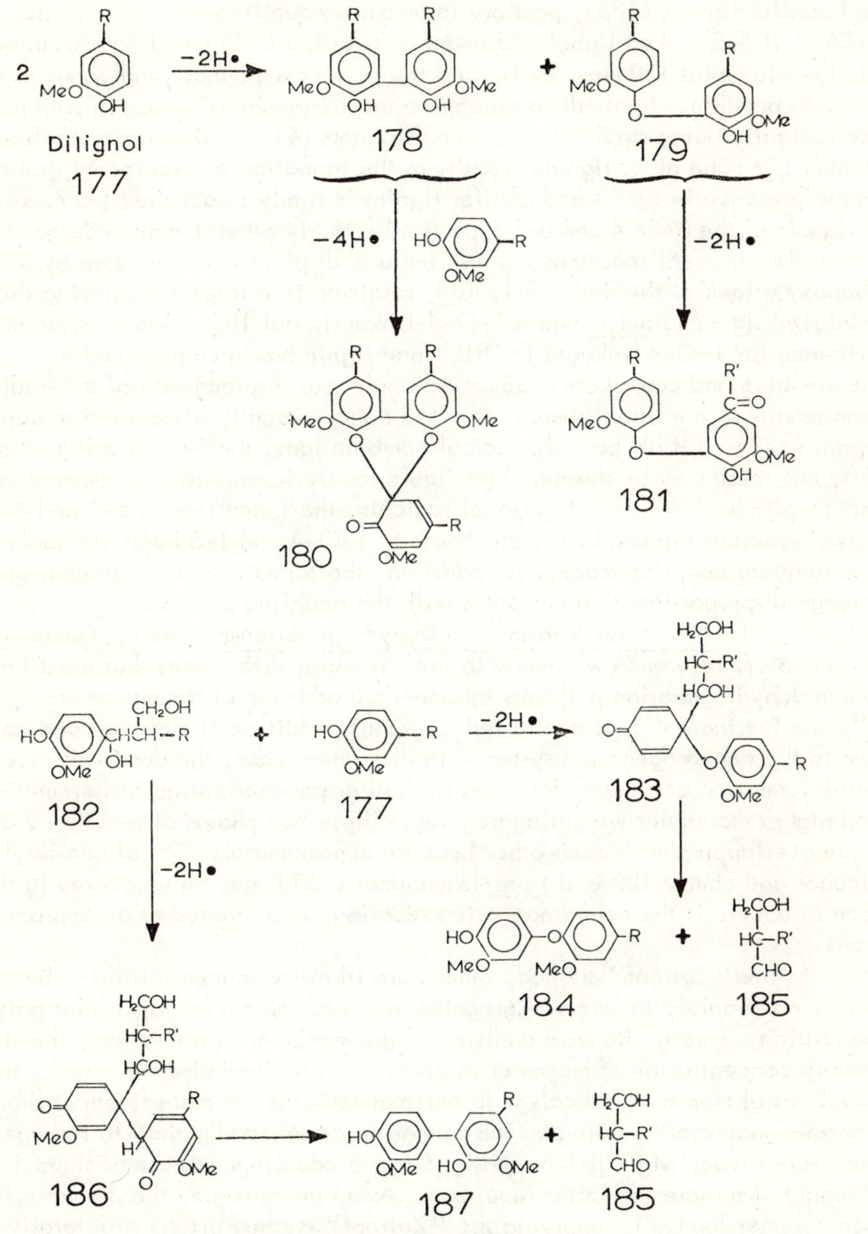

Figure 4.22. Coupling modes between dilignol radicals represented by 5-5- (178), 4-O-5- (179, 180), (184 and 185) and 5-1 (187 and 185) couplings. Competing disproportionation to α-carbonyl structures illustrated by tetramer 181. R and R': unspecified monolignol units.

and conifer lignins (185), possibly in amounts equalling the 5-5 linkages (186). If 5-5-linked lignol 178 undergoes further C-O coupling reactions, O, O- dioxepins 180 may result; and the p, p- compounds, analogous to 68, are also possible. Methods to determine the frequency of dioxepin structures are currently being studied by Pew and Connors (119). Disproportionation of lignols 179 (and other lignols) results in the formation of α-carbonyl groups whose presence in DHP's and conifer lignins is firmly established (117, 187). Dilignols of the β-O-4 and β-1 type (and possibly others) may undergo side-chain displacement reactions. These include displacement initiated by a phenoxy attack at the position 1, 183, resulting in a fragment carrying the displaced side-chain 185 and a 1-O-4-linked lignol 184. Again, structural evidence for 1-O-4 linkages in DHP's and lignin has been provided by Freudenberg and coworkers (188). An analogous displacement of a 5-radical form results in the establishment of a 1-5 bond, recently discovered in conifer lignin (189). Of all possible radical combinations, the 1-1 couplings remain left, but these may be deemed improbable for steric reasons. To summarize, in the couplings of di- and oligolignol radicals, the formations of 5-5 and 4-O-5 bonds represent the dominant, and those of 1-O-4 and 1-5 bonds probably the less frequent coupling modes. In addition, the formation of α-carbonyl groups through disproportionation competes with the coupling processes.

J. Structural Variance in Dehydrogenation Polymers. Earlier in this chapter, reference was made to the structural differences that exist between dehydrogenation polymers by one-time addition of the monomer ("Zulaufverfahren") as contrasted to gradual addition of monomer and catalyst to the dehydrogenation system. In the former case, the dominant reaction is the formation of dimers, followed by further polymerization to tetrameric and higher molecular weight aggregates. These two phases of reaction show little overlapping with each other because approximately 80% of coniferyl alcohol and almost 100% of trans-isoeugenol (157) may be recovered in the form of dimers if the dehydrogenation reaction is terminated at an appropriate stage.

In the "Zutropf" method, dimers are likewise formed initially, but they polymerize rapidly to larger aggregates which act as nuclei for further polymerization. Due to the slow addition of the monomer to the system, the stationary concentration of monomer radicals is maintained at a low level, and couplings of monomer radicals with polymer radicals are more frequent than monomer-monomer couplings. The former processes lead mainly to linkages of the β-O-4 type, while β-5 couplings (or β-β couplings in case of sinapyl alcohol) dominate the latter reactions. As a consequence, the dehydrogenation polymer formed by applying the "Zutropf" method differs structurally from that obtained in the "Zulauf" polymerization.

We will now undertake to derive the probable structure of a "purebred" dehydrogenation polymer formed in a "Zulauf" process from coniferyl alcohol, making use of the currently available rules for dehydrogenative polymerization

PRECURSORS AND THEIR POLYMERIZATION   151

as they are suggested from studies on simple phenol couplings. The polymer
will be termed the "bulk polymer" for the lack of a better expression, and its
specific characteristics will be compared with an "end-wise polymer" envisaged to be generated by an infinitely slow "Zutropf" polymerization.
Finally, the relationship of two polymers to natural lignins and the implications concerning the in vivo polymerization process will be elucidated.

The schematic formation of a "bulk polymer" from twenty initial coniferyl alcohol molecules is outlined in Figure 4.23. It is assumed that in the initial dimerization, β-5, β-β and β-O-4 dilignols are formed in the approximate ratios corresponding to the dehydrogenation of coniferyl alcohol and trans-isoeugenol (Tables 4.2 and 4.3). (For the sake of simplicity, none of the potential displacement reactions is included in the scheme, such as that between the β-O-4 and β-5 dilignols leading to the 1-O-4-linked trimer 184 and to the lignol 187). The final polymerization to an aggregate that will be a part of a continuous polymeric structure can, of course, occur in a number of ways, one of which is shown by formula 188. Other structures are possible for the "bulk polymers," but they have several characteristics in common, and it is of interest to compare these with the structural properties of natural lignins. The "bulk polymers" will all possess a large number of units with unattached side-chains that appear as coniferyl alcohol groups in polymer 188 and account for 35% of phenylpropane units, while in conifer lignins the total number of such units (coniferyl alcohol and -aldehyde groups) is estimated to be not more than 8% (190). On the other hand, the amount of units carrying a free phenolic hydroxyl group is only roughly 20% as contrasted with 30-33% for milled wood lignin preparations of conifers (191). β-5-linked units account for 25% of the monomer make-up of polymer 188, while 10-12% has been estimated for spruce MWL (192). The potential yield of Hibbert's ketones is negligible and that of vanillin very small from polymer 188, but these characteristics derive from the arbitrary manner in which it has been put together. An alternative bulk polymer, in which all available β-O-4 dilignol units would be attached to the rest of the polymer by α-O-4 bonds, would possess a potential for a maximum yield of 15% of Hibbert's ketones—more actually than the 8% yields obtained from conifer lignins (193).

The case of the end-wise polymerization is considerably simpler (Figure 4.24). Here it is assumed that the coniferyl alcohol molecules are released gradually to the medium, are dehydrogenated to radicals, and combine, one-by-one, with the radicals present on the surface of polymer particles. If the dominant β-O-4 and the much less frequent β-5 couplings would be the only reactions, totally linear polymer chains would be produced. However, a β-1 coupling process, whenever it occurs, disrupts the chain growth by blocking the end phenol group by a detached three-carbon side-chain and releases the dilignol 174. The latter is likely to be captured by intermediate quinone-methide groups in the terminal positions of other growing chains, forming a new growing branch grafted to another growing chain by an α-O-4 linkage. Thus, the termination of one growing chain may be anticipated to lead to the initiation of another, and the number of growing chain ends consequently remains constant. Coniferyl alcohol groups with an unattached allyl alcohol sidechain will not be present in an end-wise polymer unless the initial monomer occasionally escapes dehydrogenation, adding, as such, to the intermediate quinonemethide groups. Likewise, the number of units with free phenolic hydroxyl groups is very low in an end-wise polymer, because such units exist only at the terminal positions of growing chains, and, consequently, the fraction of such units becomes progressively less as the polymer grows. β-β-Linked pinoresinol units, 5-5 linkages, and 4-O-5 linkages are not present in end-wise polymers; while β-1 linkages are characteristic of this type of polymer alone and are not found in bulk polymers. End-wise polymers will produce high yields of Hibbert's ketones upon ethanolysis and large amounts of vanillin in nitrobenzene oxidation.

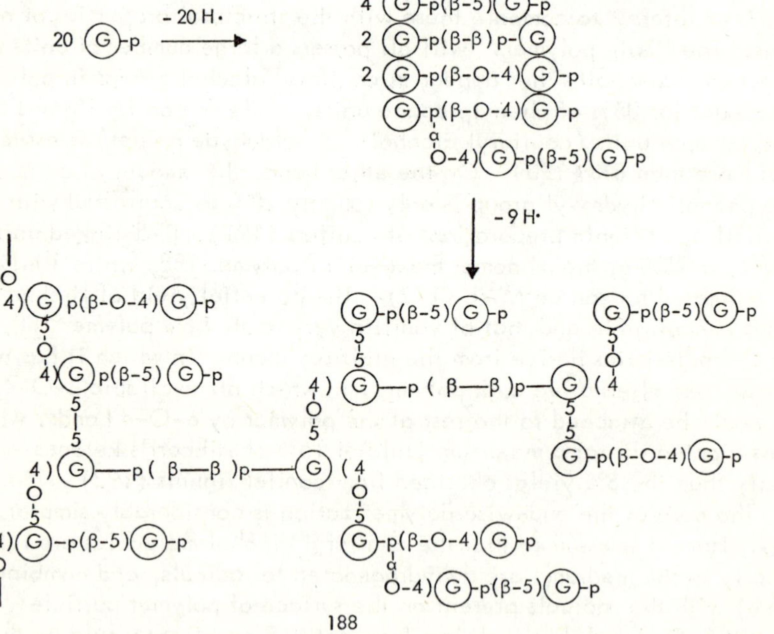

Figure 4.23. Projected bulk ("Zulauf") polymerization of coniferyl alcohol, illustrated for 20 molecules of the starting compound. The coniferyl molecules are converted simultaneously to radicals that couple forming dimers. These, in turn, are dehydrogenatively polymerized to tetramers (and some trimers), and finally, to a large macromolecule.

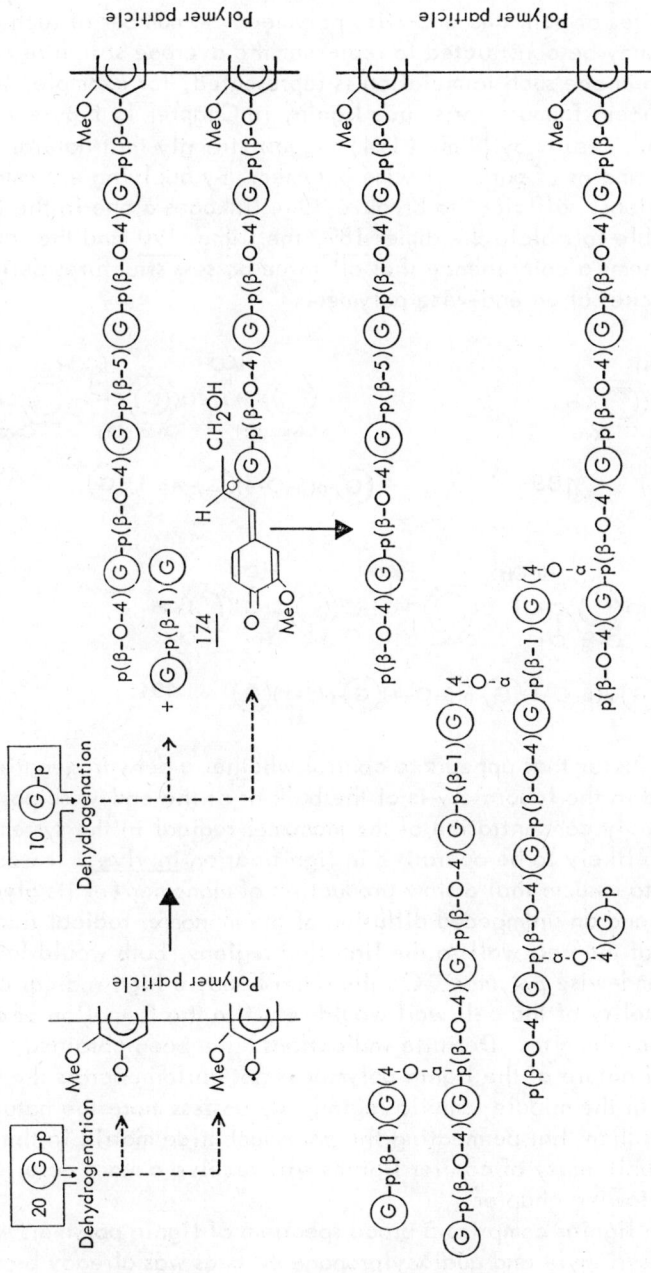

Figure 4.24. Projected end-wise ("Zutropf") polymerization of coniferyl alcohol. Coniferyl alcohol is added gradually to a system consisting of polymer particles and dehydrogenating enzyme. Coniferyl alcohol radicals couple with phenoxyradicals present on the surface of the polymer particles.

As will be discussed in greater detail in Chapter 5, the composition and characteristics of lignins known to date conform with the idea that they may represent composites of bulk and end-wise polymers. A number of such composite structures may be constructed to represent the average structure of natural lignins, and one such formulation is represented, for example, by the tentative Freudenberg formula for spruce lignin, in Chapter 1, Figure 1.1. On the other hand, results by Nimz (194) suggest strongly that natural lignins contain limited domains of pure end-wise polymer. By applying extremely mild hydrolytic conditions, sufficient to break α-O-4 linkages alone in the lignin matrix, he was able to isolate the dimer 189, the trimer 190 and the tetramer 191. It hardly seems a coincidence that all three possess structures derived for the growing branches of an end-wise polymer.

[Structure 189: dimer with (G)-p(β-1)-(G)]

[Structure 190: trimer with (G)-p(β-O-4)-(G)-p(β-1)-(G)]

[Structure 191: tetramer with (G)-p(β-O-4)-(G)-p(β-O-4)-(G)-p(β-1)-(G)]

The single factor that appears to control whether a dehydrogenation polymer produced in the laboratory is of the bulk or of the end-wise polymer type is the stationary concentration of the monomer radical in the system. This factor is also likely to be operative in lignification *in vivo*. It would seem reasonable to assume that a slow production of monomer (or its glycoside) in the cytoplasm and an unimpeded diffusion of the monomer radical from the peroxidase zone of the cell wall to the lignified regions, both would favor the formation of an end-wise polymer. On the other hand, a high radical concentration at any locality of the cell wall would result in the formation of a bulk polymer particle at the site. Definite indications have been obtained, indeed, that the chemical nature of the lignin polymer is not uniform across the cell wall, and lignin in the middle lamella region may possess more the nature of an end-wise polymer than that permeating the polysaccharide matrix in the S_2-layer. The non-uniformity of conifer lignins will receive a more detailed discussion in the following chapter.

Angiosperm lignins comprise a broad spectrum of lignin polymers with varying ratios of syringyl- and guaiacylpropane units as was already brought up in the preceding chapter. It may also be recalled that this ratio varies often within single cells, being higher in the cell wall than in the middle lamella regions. Syringyl propane units with the 5-position occupied by a methoxyl group have obviously a simpler bonding pattern than other units as linkages such as 5-5, β-5, 4-O-5 and 1-5 bonds are ruled out.

Freudenberg (196) has expressed doubts about the existence of lignins composed entirely of syringyl propane units. It was found (80, 92, 195) that sinapyl alcohol does not form a polymer in dehydrogenation by laccase-oxygen. In the initial stage, conversion to syringaresinol is almost quantitative (~ 85%) (80). In the further dehydrogenation disproportionation dominates, resulting in the formation of 2,6-dimethoxy-p-benzoquinone and other degradation products (80). On the other hand, "Zutropf"-polymerization of sinapyl alcohol, not as yet investigated, may well result in the formation of a polymeric material. Also, the existence in the fiber cell walls of angiosperms of what appears to be syringyl lignin, supports such a possibility. In lignins of this nature, structures derivable from the coupling properties of sinapyl alcohol could be of the type shown in Figure 4.25, for example. In general, angiosperm lignins yield larger amounts of Hibbert's ketones and aromatic aldehydes than conifer lignins (cf. Chapter 3, Section III B) which suggests a largely end-wise polymer type.

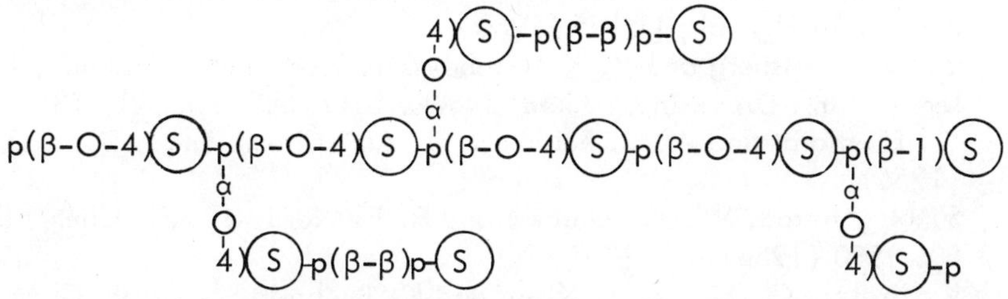

Figure 4.25. Illustration of probable structural types in syringyl lignins.

The speculative ideas developed here about the nature of end-wise and bulk polymers need, of course, to be tested on dehydrogenation polymers produced in the laboratory. There is hardly any doubt that an understanding of the dehydrogenative polymerization process as a whole, once developed, will contribute substantially to the elucidation of the macromolecular structure of natural lignins.

REFERENCES

1. K. Freudenberg and A. C. Neish, Constitution and Biosynthesis of Lignin. Springer-Verlag, Berlin-Heidelberg, New York, (1968); I. A. Pearl, The Chemistry of Lignin, Marcel Dekker, Inc., New York (1967) pp 136-148; J. M. Harkin, in Oxidative Coupling of Phenols,

(W. I. Taylor and A. R. Battersby, Eds.) Marcel Dekker, Inc. New York (1967), pp 243-321; S. A. Brown, Ann. Rev. Physiol., 17, 223 (1966); W. J. Schubert, Lignin Biochemistry, Academic Press, New York (1965); K. Freudenberg, Science, 148, 595 (1965); K. Kratzl, in Cellular Ultrastructure of Woody Plants, (W. A. Côté, ed.) Syracuse University Press, New York (1965) pp 157-180; The Formation of Wood in Forest Trees (M. H. Zimmerman, Ed.) Academic Press, New York (1964); S. A. Brown, in Biochemistry of Phenolic Compounds (J. B. Harborne, Ed.) Academic Press, New York (1964), pp 361-387; T. Higuchi, F. Barnoud and A. Robert, Assoc. Tech. Ind. Papetiere Bull. 18, 92 (1964); F. F. Nord and W. J. Schubert, in Biogenesis in Natural Compounds (P. Bernfeld, Ed.), Pergamon Press (1963), pp 695-725; 2nd (revised) Ed. (1967), pp 903-946; K. Freudenberg, in Fortschr. d. Chemie Organ. Naturstoffe (Zechmemeister, Ed.), 20, 41 (1962); F. F. Nord and W. J. Schubert, in Comparative Biochemistry, Vol. 4 (M. Florkin and H. S. Mason, Eds.), Academic Press, New York (1962), pp 65-106; K. Freudenberg, Pure and Appl. Chem., 5, 9 (1962); S. A. Brown, Science, 134, 305 (1961); F. E. Brauns and D. A. Brauns, The Chemistry of Lignin, Supplement Volume, Academic Press, New York (1960) pp 659-736; A. C. Neish, Ann. Rev. Plant Physiol., 11, 55 (1960).

2. Z. N. Kreitsberg and Y. K. Grabovskii, Trudy Inst. Lesokhoz, Problem i Khim. Drevesing; Akad. Nauk. Latv. S.S.S.R., 21, 131
3. G. Eberhardt and W. J. Schubert, J. Am. Chem. Soc., 78, 2835 (1956).
4. S. N. Acerbo, W. J. Schubert and F. F. Nord, J. Am. Chem. Soc., 80, 1990 (1958).
5. K. Kratzl, G. Billek, E. Klein and K. Buchtela, Monatsh. Chem., 88, 721 (1957).
6. K. Freudenberg and A. C. Neish, Constitution and Biosynthesis of Lignin. Springer-Verlag. Berlin-Heidelberg-New York 1968, p.9.
7. A. C. Neish, in Biochemistry of Phenolic Compounds (J. B. Harborne, Ed.) Academic Press, New York, p. 354 (1964).
8. J. M. Harkin, in Oxidative Coupling of Phenols (W. I. Taylor and A. R. Battersby, Eds.) Marcel Dekker Inc. New York, pp 256-263 (1968); A. C. Neish, in Biochemistry of Phenolic Compounds (J. B. Harborne, Ed.) Academic Press, New York, pp 313-340 (1964).
9. B. D. Davis, Experientia, 6, 41 (1950).
10. E. L. Tatum, S. R. Gross, G. Ehrensvärd and L. Garnjobst, Proc. Natl. Acad. Sci. U.S., 40, 271 (1954).
11. B. D. Davis, Advances in Enzymol., 16, 247 (1958); Arch. Biochem. Biophys., 78, 497 (1958).
12. E. C. Conn, in Biochemistry of Phenolic Compounds (J. B. Harborne, Ed.) Academic Press, New York, pp 402-403 (1964).

PRECURSORS AND THEIR POLYMERIZATION 157

13. M. Nandy and N. C. Ganguli, Biochem. Biophys. Acta, 48, 608 (1961).
14. D. R. McCalla and A. C. Neish, Can. J. Biochem. and Physiol.,37, 531, 537 (1959); D. L. Gamborg and A. C. Neish, Can. J. Biochem. and Physiol., 37, 1277 (1959).
15. M. Hasegawa, in Wood Extractives, (W. E. Hillis, Ed.) Academic Press, New York, pp 263-276 (1962).
16. T. Higuchi, Proc. Intern. Congr. Biochem. 4th Congr. Vienna, Vol.II, 161 (1959).
17. S. A. Brown and A. C. Neish, Nature, 175, 688 (1955).
18. D. C. Sprinson, Adv. Carbohydrate Chem., 15, 285 (1960).
19. S. N. Acerbo, W. J. Schubert and F. F. Nord, J. Am. Chem. Soc., 82, 735 (1960).
20. K. Kratzl and H. Faigle, Z. Naturf., 15b, 4 (1960).
21. R. V. Byerrum, J. H. Flokstra, L. J. Dewey and C. D. Ball, J. Biol. Chem., 210, 633 (1954).
22. M. Hasegawa and T. Higuchi, J. Japan.Forest. Soc., 42, 305 (1960).
23. K. Kratzl and J. Zauner, Holzforschung, 14, 108 (1963).
24. V. N. Sergeeva and Z. N. Kreitsberg, Trudy. Inst. Lesokhoz Problem in Khim. Drevesing; Akad. Nauk. Latv. S.S.S.R., 12, 245 (1959); through C. A. 52, 21070 (1958).
25. M. Hasegawa, T. Higuchi and H. Ishikawa, Plant and Cell Physiol., 1, 173 (1960).
26. S. Yoshida and G. H. N. Towers, Can. J. Biochem. Physiol., 41, 579 (1963).
27. S. R. Gross, J. Biol. Chem., 233, 1146 (1958) M. N. Zaprometov, Biokhimid, 27, 366 (1962).
28. S. A. Brown and A. C. Neish, Can. J. Biochem. and Physiol., 33, 948 (1955).
29. D. Wright, S. A. Brown and A. C. Neish, Can. J. Biochem. and Physiol., 36, 1037 (1958).
30. S. A. Brown and A. C. Neish, J. Am. Chem. Soc., 81, 2419 (1959).
31. K. Freudenberg, H. Reznik, W. Fuchs and M. Reichert, Naturwissenschaften, 42, 29 (1955).
32. K. Freudenberg and F. Niedercorn, Chem. Ber., 91, 591 (1958); K. Freudenberg and B. Lehman, Chem. Ber., 96, 1850 (1963).
33. D. E. Bland, Biochem. J., 88, 523 (1963).
34. T. Higuchi, Can. J. Biochem. and Physiol., 40, 31 (1962).
35. F. Barnoud, T. Higuchi, J. P. Joseleau and A. Mollard, Compt. rend., 259, 4339 (1960).
36. K. Freudenberg and F. B. Toribio, Chem. Ber., 102, 1312 (1969).
37. J. Koukol and E. E. Conn, J. Biol. Chem., 236, 2692 (1961).
38. T. Higuchi and F. Barnoud, Chim. Biochim., Lignine Cellulose Hemicellulose, Grenoble, p. 255 (1964); J. Japan.Wood Res. Soc., 12,

36 (1966).
39. S. A. Brown, D. Wright and A. C. Neish, Can. J. Biochem. and Physiol., 37, 25 (1959).
40. K. Kratzl and G. Billek, Monatsh. Chem., 90, 536 (1959).
41. A. C. Neish, Phytochemistry, 1, 1 (1961).
42. T. Higuchi, Agr. Biol. Chem., 30, 667 (1966).
43. S. A. Brown and A. C. Neish, Can. J. Biochem. and Physiol., 34, 769 (1956).
44. T. Higuchi and S. A. Brown, J. Biochem. and Physiol., 41, 613 (1963).
45. D. G. Smith and A. C. Neish, Phytochem., 3, 609 (1964).
46. T. Higuchi and S. A. Brown, Can. J. Biochem. and Physiol., 41, 621 (1963).
47. K. Kratzl and G. Hofbauer, Monatsh. Chem., 89, 96 (1958).
48. K. Kratzl, in Cellular Ultrastructure in Woody Plants (W. A. Côté, Ed.) Syracuse Univ. Press, p. 157 (1965).
49. J. B. Harborne and J. J. Corner, Biochem. J., 81, 242 (1961).
50. S. Z. El-Basyouni, A. C. Neish and G. H. N. Towers, Phytochemistry, 3, 627 (1964).
51. A. C. Neish, in The Formation of Wood in Forest Trees (M. H. Zimmermann, Ed.) Academic Press, New York, p. 562 (1964).
52. O. Goldschmid and H. C. Hergert, Tappi, 44, 858 (1961).
53. P. M. Nair and L. C. Vining, Phytochemistry, 4, 161 (1965).
54. P. M. Nair and L. C. Vining, Phytochemistry, 4, 401 (1965).
55. E. C. Conn, in Biochemistry of Phenolic Compounds (J. B. Harborne, Ed.) Academic Press, New York, pp 415-418.
56. B. J. Finkle and R. F. Nelson, Biochem. Biophys. Acta, 78, 747 (1963).
57. B. J. Finkle and M. S. Masri, Biochem. Biophys. Acta, 85, 167 (1964).
58. D. Hess, Z. Naturforsch., 19b, 447 (1964).
59. T. Higuchi, M. Shimada and H. Okashi, Agr. Biol. Chem., 31, 1459 (1967).
60. J. Pellerin and A. D. Iorio, Can. J. Biochem. and Physiol., 36, 491 (1958).
61. H. Shimazono and F. F. Nord, Arch. Biochem. Biophys., 78, 263 (1958).
62. D. E. Bland, M. Menshun and M. Szilagyi, Proc. Austr. Pulp Paper Res. Conf., 18, 22 (1961).
63. R. L. Hamill, R. V. Byerrum and C. D. Ball, J. Biol. Chem., 224, 713 (1957).
64. Z. N. Kreitsberg and Y. K. Brabovskii, Chem. Abstr., 59, 10478g (1963).
65. K. Freudenberg, in Biochemistry of Wood (K. Kratzl and G. Billek, Eds.) Pergamon Press, p. 121 (1959).

66. K. Freudenberg, in The Formation of Wood in Forest Trees (M. H. Zimmermann, Ed.) Academic Press, p. 203 (1964).
67. K. Freudenberg and J. M. Harkin, Phytochemistry, 2, 189 (1963).
68. R. E. Kremers, Tappi, 40, 262 (1957).
69. R. E. Kremers, Chim. Biochim. Lignine, Cellulose, Hemicellulose, Grenoble, p. 469 (1966).
70. K. Freudenberg and J. Torres-Serres, Liebigs Ann. Chem., 703, 225 (1967).
71. S. M. Manskaya, Doklady Akad. Nauk. S. S. S. R., 62, 369 (1948).
72. K. Beijeriuck, Kou Akad. Wetensch. Wis. en Naturk. Aja, 8, 572 (1900).
73. K. Freudenberg and F. Bittner, Chem. Ber., 86, 155 (1953).
74. K. Freudenberg, in Moderne Methoden der Pflanzenalalyse (K. Paech and M. V. Tracey, Eds.) Springer-Verlag p. 499 (1955).
75. T. Higuchi, Physiol. Plantarum, 10, 356 (1957).
76. A. V. Wacek, O. Härtel and S. Meralla, Holzforschung, 7, 58 (1953).
77. F. Barnoud, Compt. Rend. Acad. Sci. Paris, 243, 1545 (1956); P. Traynard and F. Barnoud, Holzforschung, 11, 85 (1957).
78. H. S. Mason and M. Cronyn, J. Am. Chem. Soc,, 77, 491 (1955).
79. B. A. Brown, in Oxidative Coupling of Phenols (W. I. Taylor and A. R. Battersby, Eds.) Marcel Dekker, Inc. New York, p. 167 (1968).
80. K. Freudenberg, J. M. Harkin, M. Rechert and T. Fukuzumi, Chem. Ber., 91, 581 (1958).
81. K. Freudenberg, Nature, 183, 1152 (1959).
82. T. Higuchi, I. Kawamura and H. Ishikawa, J. Japan.Forestry Soc., 35, 258 (1953).
83. T. Higuchi and M. Shimada, Plant and Cell Physiol., 8, 71 (1967).
84. T. Higuchi and M. Shimada, Agr. Biol. Chem., 31, 1179 (1969).
85. T. Higuchi and M. Shimada, Plant and Cell Physiol., 8, 61 (1967).
86. S. Yoshida and M. Shimokoriyama, Bot. Mag. (Tokyo), 78, 14 (1965); (from ref. 42).
87. P. H. Rubery and D. H. Northcote, Nature, 219, 1230 (1968).
88. K. Freudenberg and A. C. Neish, Constitution and Biosynthesis of Lignin, Springer-Verlag, Berlin-Heidelberg-New York, pp 82-97 (1968).
89. K. Freudenberg and H. Richtzenhain, Chem. Ber., 76, 997 (1943); Holzforschung, 1, 90 (1947); K. Freudenberg, Angew. Chem., 61, 228 (1949).
90. K. Freudenberg and W. Heimberger, Chem. Ber., 83, 519 (1950); K. Freudenberg and W. Siebert, W. Heimberger and R. Kraft, Chem. Ber., 83, 533 (1950).
91. J. M. Harkin, Unpubl. work, cf. Ref. 88, p. 84.
92. K. Freudenberg and H. Hübner, Chem. Ber., 85, 1181 (1952).
93. K. Freudenberg, Angew. Chem., 68, 508 (1956).

94. A. R. Battersby and W. I. Taylor, Oxidative Coupling of Phenols, Marcel Dekker, Inc. New York (1967).
95. E. R. Altwicker, Chem. Rev., 67, 475 (1967).
96. G. D. Cooper, A. R. Gilbert and H. Finkbeiner, Polym. Prep., 7, 166 (1966); D. A. Bolou, ibid., 7, 173 (1966); D. M. White, ibid., 7, 178 (1966).
97. F. Haber and R. Willstätter, Chem. Ber., 64, 2844 (1931).
98. J. Kenner, Tetrahedron, 3, 78 (1958); 8, 350 (1960).
99. K. Dimroth, Angew. Chem., 72, 714 (1960).
100. Ref. 95, p. 514.
101. H. Musso, in Oxidative Coupling of Phenols (W. I. Taylor and A. R. Battersby, Eds.) Marcel Dekker, Inc., New York, p.54 (1967).
102. Ref. 95, pp 477–481.
103. A. S. Hay, H. S. Blanchard, G. F. Endres and J. W. Eustance, J. Am. Chem. Soc., 81, 6335 (1959).
104. W. Brachman and E. Havinga, Rec. Trav. Chim., 74, 937, 1070, 1100, 1107 (1955).
105. K. Ley, Angew. Chem., 70, 74 (1958).
106. E. Adler and R. Magnusson, Acta Chem. Scand., 13, 505 (1959).
107. F. Wessely, J. Roy. Inst. Chem. 424 et Seq. (1959).
108. C. D. Cook, J. Org. Chem., 18, 261 (1953).
109. E. Müller and K. Ley, Chem. Ber., 87, 922 (1954).
110. M. S. Kharasch and B. S. Joshi, J. Org. Chem., 22, 1435 (1957); G. M. Coppinger, J. Am. Chem. Soc., 79, 501 (1957).
111. K. Dimroth and G. Neubauer, Angew. Chem., 69, 95 (1957); K. Dimroth, F. Kalk and G. Neubauer, Chem. Ber., 90, 2058 (1957) K. Dimroth and A. Berndt, Angew. Chem., 76, 434 (1964); Angew. Chem. Intern. Ed., 3, 385 (1964).
112. C. Steelink and R. E. Hanson, Tetrahedron Letters, 105 (1966).
113. E. Müller and K. Ley, Chem. Ber., 88, 601 (1955).
114. E. Müller and W. Schmidhuber, Chem. Ber., 89, 1738 (1956).
115. Ref. 95, p. 519.
116. C. D. Cook and B. E. Norcross, J. Am. Chem. Soc., 81, 1176 (1959).
117. J. C. Pew, W. J. Connors and A. Kunishi, Chim. Biochim. Lignine, Cellulose, Hemicellulose, Grenoble, p.229 (1964).
118. J. K. Becconsall, S. Claugh and G. Scott, Trans. Faraday Soc., 56, 459 (1960).
118a. C. D. Cook, N. G. Nash and H. K. Flanagan, J. Chem. Soc., 78, 1783 (1955); G. Müller, K. Ley and W. Kirdaisch, Chem. Ber., 87, 1605 (1954).
119. J. C. Pew and W. J. Connors, Nature, 215, 623 (1967).
120. P. Schilling, Ph.D. Thesis, University of Vienna, Austria (1968).
121. Ref. 94, p. 195.

122. W. Flaig and K. Haider, Planta Medica, 9, 123 (1961).
123. M. V. G. Viehäuser, Ph.D. Thesis, University of Vienna, Austria (1969).
124. G. D. Cooper, H. S. Blanchard, G. F. Endres and H. Finkbeiner, J. Am. Chem. Soc., 87, 3996 (1965).
125. A. S. Hay, G. F. Endres and J. E. Eustance, J. Polymer Sci., 58, 581 (1962); 58, 593 (1962); J. Org. Chem., 28, 1300 (1963).
126. B. O. Lindgren, Acta Chem. Scand., 14, 1203 (1960).
127. W. Krevelen, Allg. Prakt. Chem., 17, 231 (1966).
128. S. M. Bocks, B. R. Brown and A. R. Todd, Proc. Chem. Soc., 117 (1962).
129. B. P. Stark, Ph.D. Dissertation, Cambridge University, Cambridge (1958); quoted in Peroxidase (B. C. Saunders, A. G. Holmes-Siedle and B. P. Stark, Editors), Butterworth, London, p. 209 (1964).
130. B. R. Brown and A. R. Todd, J. Chem. Soc., 5564 (1963).
131. H. Booth and B. C. Saunders, J. Chem. Soc., 940 (1956).
132. Ref. 94, p. 187.
133. B. O. Lindgren, Acta Chem. Scand., 14, 2089 (1960).
134. R. Lerot, J. Pharm. Chem., 19, 10 (1904).
135. E. Bourquelot and M. Marchadier, Compt. Rend. France, 138, 1432 (1904).
136. H. Richtzenhain, Chem. Ber., 82, 447 (1949).
137. I. W. Sizer, Advan. Enzymol., 14, 129 (1953).
138. A. J. Gross and I. W. Sizer, Proc. Intern. Congr. Biochem. 3rd, Brussels, p. 29 (1955); J. Biol. Chem., 234, 1611 (1959).
139. B. G. Malmström, R. Mosbach and T. Vänngård, Nature, 183, 121 (1959).
140. B. R. Brown and S. M. Bocks, in Enzyme Chemistry of Phenolic Compounds (J. B. Pridham, Ed.) Pergamon Press, New York, p.129 (1963).
141. S. M. Bocks and R. C. Cambie, Proc. Chem. Soc., 143 (1963); S. M. Bocks, R. C. Cambie and T. Takahashi, Tetrahedron, 19, 1109 (1963).
142. K. Freudenberg and A. Sakakibara, Ann. Chem., 623, 129 (1959).
143. B. R. Brown, in Oxidative Coupling of Phenols (W. I. Taylor and A. R. Battersby, Eds.) p. 188 (1967).
144. R. Pummerer, D. Melamed and H. Puttfaicker, Chem. Ber., 55, 316 (1922).
145. C. G. Haynes, A. H. Furner and W. A. Waters, J. Chem. Soc., 2823 (1956).
146. K. Bowden and C. H. Reece, J. Chem. Soc., 2249 (1950).
147. W. W. Kaeding, J. Org. Chem., 28, 1063 (1963).
148. S. L. Cosgrove and W. A. Waters, J. Chem. Soc., 1726 (1951).
149. W. W. Westerfield and C. Lowe, J. Biol. Chem., 145, 463 (1942).
150. A. I. Scott, Quart. Rev., 19, 1 (1965).

151. D. H. R. Barton, A. M. Deflorin and O. E. Edwards, J. Chem. Soc., 530 (1956).
152. H. Cousin and H. Herissey, Compt. Rend., 147, 247 (1908); Bull. Soc. Chim. France, Ser. 4, 3, 1070 (1908); J. Pharmac. Chim., 28, 93 (1908).
153. H. Erdtman, Biochem. Z., 258, 172, 288 (1933); Ann. Chem., 503, 203 (1933); Proc. Roy. Soc., 143A, 177, 191, 223, 228 (1933); Svensk Kem. Tidskr., 40, 243 (1935); Svensk Papperstidn., 44, 243 (1941).
154. G. Aulin-Erdtman, Y. Tomita and S. Forsen, Acta Chem. Scand., 17, 535 (1963).
155. B. Lindberg, Svensk Papperstidn., 56, 6 (1953).
156. A. Müller, J. Org. Chem., 16, 481 (1951); 17, 787 (1952).
157. A. F. Wallis and K. V. Sarkanen, unpublished results.
158. E. Müller, R. Mayer, H. D. Spanagel and K. Scheffer, Ann. Chem. 645, 53 (1961).
159. E. Müller, H. D. Spanagel and A. Rieker, Ann. Chem., 681, 141 (1965).
160. I. Hagedorn, V. Eholzer and A. Luttringhaus, Chem. Ber., 93, 1584 (1960).
161. H. Erdtman, Ann. Chem., 503, 283 (1933); N. J. Cartwright and R. D. Haworth, J. Chem. Soc., 535 (1944); K. Freudenberg and H. Dietrich, Chem. Ber., 86, 1157 (1953).
162. K. Freudenberg and H. Schraube, Chem. Ber., 88, 16 (1955).
163. H. Nimz, K. Naya and K. Freudenberg, ibid, 96, 2086 (1963).
164. K. Freudenberg and H. Schluter, ibid., 88, 617 (1955).
165. K. Freudenberg and B. Lehman, ibid., 93, 1354 (1960).
166. K. Freudenberg and K. C. Renner, ibid, 98, 1879 (1965).
167. K. Freudenberg and D. Rasenack, ibid., 86, 755 (1953).
168. K. Freudenberg, Adv. in Chem. Ser., 59, 1 (1966).
169. K. Freudenberg, G. Grion and J. M. Harkin, Angew. Chem., 70, 743 (1958).
170. H. Nimz, Chem. Ber., 96, 478 (1963).
171. K. Freudenberg and H. Geiger, Chem. Ber., 96, 1265 (1963).
172. H. Nimz, Angew. Chem., 76, 597 (1964).
173. K. Freudenberg and M. Friedmann, Chem. Ber., 93, 2138 (1960).
174. K. Freudenberg and H. Tausend, ibid., 97, 3418 (1964).
175. K. Freudenberg and G. Shuhmacher, ibid., 87, 1882 (1954).
176. K. Freudenberg, Angew. Che., 76, 349 (1964).
177. K. Freudenberg and H. Tausend, Che. Ber., 96, 2081 (1963).
178. H. Nimz, Chem. Ber., 96, 478 (1963).
179. K. Freudenberg and H. Nimz, ibid., 95, 2057 (1962).
180. H. Nimz, Holzforschung, 20, 105 (1966).
181. K. Lundquist, Acta Chem. Scand., 18, 1316 (1964).

182. K. Freudenberg and H. Nimz, Chem. Commun., 132 (1966).
183. K. Lundquist and G. E. Miksche, Tetrahedron Letters, 2131 (1965).
184. H. Nimz, Chem. Ber., 99, 469 (1966).
185. K. Freudenberg, C. L. Chen and G. Cardinale, Chem. Ber., 95, 2814 (1962).
186. S. L. Larsson and G. E. Miksche, Acta Chem. Scand., 21, 197 (1967).
187. J. M. Harkin, in Oxidative Coupling of Phenols (W. I. Taylor and A. R. Battersby, Eds.) Marcel Dekker, pp 300-312 (1967).
188. K. Freudenberg, C. L. Chen, J. M. Harkin, H. Nimz and H. Renner, Chem. Commun. No. 11, 224 (1965).
189. E. Adler, Svensk Kemisk Tidskrift, 80, 279 (1968).
190. E. Adler and J. Marton, Acta Chem. Scand., 13, 75, 357, 370 (1961).
191. K. V. Sarkanen, in The Chemistry of Wood (B. L. Browning, Ed.) p. 273 (1963).
192. E. Adler, S. Delin and K. Lundquist, Acta Chem. Scand., 13, 2149 (1959); E. Adler and K. Lundquist, ibid., 17, 13 (1963).
193. L. Bückman, J. J. Pyle, J. C. McCarthy and H. Hibbert, J. Am. Chem. Soc. 61, 686 (1939).
194. H. Nimz, Chem. Ber., 99, 2638 (1966).
195. K. Freudenberg, K. Kraft and W. Heimberger, Chem. Ber., 84, 872 (1951).
196. K. Freudenberg and A. C. Neish, Constitution and Biosynthesis of Lignin, Springer-Verlag Berlin-Heidelberg-New York (1968), p.111.

5

ISOLATION AND STRUCTURAL STUDIES
Y. Z. Lai and K. V. Sarkanen

<u>I. Introduction</u>. The lack of adequate lignin isolation methods has been a continuing problem in attempts to determine the structure of this natural product. Initially, the acid lignin preparations were exclusively used in the chemical characterization, and it was soon determined that these differed from the original lignins in a number of ways, e.g., they were not converted to soluble lignin sulfonic acids in sulfonation (1) and aromatic aldehydes were obtained in nitrobenzene oxidation in yields that were much lower than those obtained from the original wood (2). Condensation reactions that always occur during acidic hydrolysis of lignin are responsible for changes of this nature and their effect may be reduced, but not eliminated, by applying milder conditions such as are used in the isolation of cuoxam lignins (3). Brauns lignins (4) that are obtained by direct extraction are not changed during their isolation, and studies on these preparations provided much useful information on the structure of lignins. Two basic errors were made, however, in interpreting results obtained on Brauns lignins. First, it was assumed that these preparations were free of non-lignin components, which is not always the case (5,6). The second error stems from the speculative proposition of Brauns envisaging lignins as homogenous polymers consisting of essentially identical pentameric units (s.c. "lignin building units")(7). Should this proposition be correct, it would be permissible indeed, as Brauns did, to generalize the results on the small isolated part of lignins to the over 90% of lignin that remains in the extracted plant material. As the idea of "lignin building units" has failed to gain general acceptance, Brauns lignins can be only expected to reflect the composition of the total lignin in an approximate manner. The relationship between these preparations and the total lignin will be elaborated in some detail in a later part of this chapter.

Remarkably, some of the exaggerated generalizations made in the interpretation of data on Brauns lignins are now being repeated, in a somewhat modified form, in studies on the milled wood lignin (MWL) preparations. It is true that the milling process that precedes the isolation of these preparations does not cause profound chemical modification, but it is equally true that some modification does occur. It has again been tacitly (although not explicitly) assumed that the 30-50% of lignin isolated in MWL preparations represents a structural average of the total lignin. In the section that follows, specific attention has therefore been devoted to the important question to what extent milled wood lignins—undoubtedly the most versatile lignin samples currently available—are representative of total lignin components <u>in situ</u>.

II. Milled Wood Lignins (MWL).

A. <u>Method of Preparation.</u> The first observations on the effect of milling on the solubility of wood were made by Staudinger and coworkers (8). They found that 12 to 40 hours grinding of dry spruce wood in a vibratory ball mill (German "Schwingmühle") converted 30 to 50% of the wood into a form soluble in cuprammonium solution. Hess and Heumann (9) studied vibratory milling in the presence of hydrazine hydrate, isolating 12 to 14% of lignin in spruce and beech woods, and 19% in flax, in the form of lignin-hydrazine compound. Grohn (10) demonstrated that dry vibratory milling for 48 hours makes a large part of the originally unextractable lignin soluble in organic solvents. An isolation procedure analogous to the preparation of Brauns lignins was applied to spruce wood, yielding 19.6% of a sample with a 15.8% methoxyl and a 10.7% total hydroxyl content. The yield from poplar wood was 26.4%, with 19.4% methoxyl and 9.6% hydroxyl. The analytical compositions of these preparations are quite close to those of milled wood lignin preparations, the isolation procedure of which was developed by Björkman in 1957 (5, 11). According to the original method, wood meal is first given a pregrinding treatment in a Lampen mill, and the final milling (48 h.) is done using 1 to 5g samples suspended in a non-swelling medium, such as toluene, in a vibratory mill developed by Forziati <u>et al</u> (12). Later work has shown, however, that neither the non-swelling medium nor the use of a vibratory ball mill is indispensable. Pew (13) has claimed vibratory ball milling at 2°C to be more efficient than milling in toluene suspension, while Pearl and Busche did not observe significant differences in millings at -78° and at ambient temperature (14). Brownell has demonstrated that ordinary ball milling in porcelain jars filled with flint pebbles produces adequate milling effect if the time of milling is extended to four weeks (15). Regardless of the time requirement, ordinary ball milling is often more convenient for the preparation of milled wood lignins because up to 40g of wood can be milled in a single one-gallon jar, and the acquisition of a specific vibratory mill is unnecessary. The gradual release of extractable milled wood lignin from conifer wood in ball milling is illustrated in Figure 5.1, based on results by Latif and Sarkanen (16). It can be seen that the amount of extractable milled wood lignin as a function of milling time follows an S-shaped curve levelling off at a point corresponding to slightly over one-half of the total lignin. The initial release of lignin is very slow during the first week of milling, which explains the earlier failure by Brauns (17) to increase the amount of extractable lignin by

ISOLATION AND STRUCTURAL STUDIES

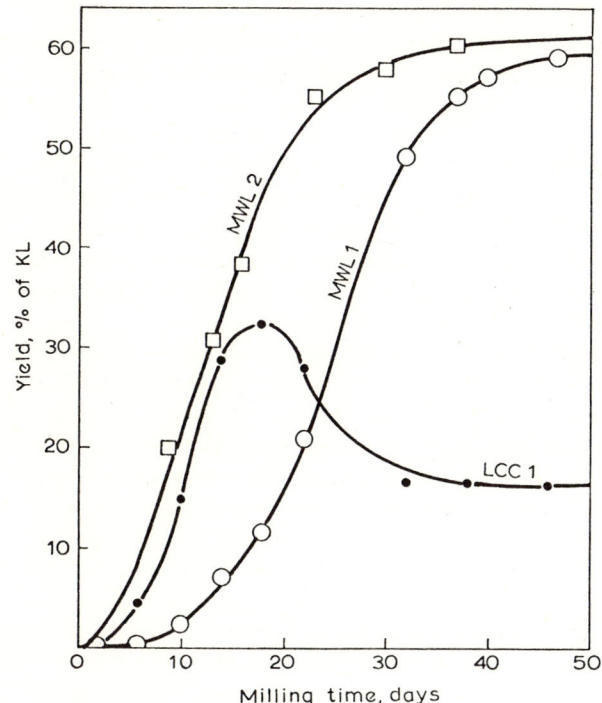

Figure 5.1. Yields of milled wood lignin (MWL) and lignin-carbohydrate complex (LCC) in the ball milling of normal Douglas-fir wood (38).
MWL 1; LCC 1; yields using glass balls.
MWL 2: yields using steel balls.

ball milling.

The maximum yields of milled wood lignin from conifer woods are usually approximately 50% (5), and the same yields are obtained from normal and compression woods (16). Higher yields have been obtained from hardwoods (18, 19); 87% from madrona wood, for example (19). A mild alkaline pretreatment of Eucalyptus woods increases the rate of release and yield of milled wood lignin, possibly as a consequence of saponification of ester bonds between lignin and hemicelluloses (18). On the other hand, Bland (20) was unable to obtain milled wood lignin from tension wood of Eucalyptus, while wood opposite to tension wood ("opposite wood") gave better yields than normal Eucalyptus wood. Bland has proposed that milled wood lignin comes from the middle lamella and not from the cell wall lignin as a probable explanation for his failure to isolate MWL from tension wood, and has supported this proposition by the observation that MWL yields are low from the midribs and petioles of Eucalyptus botryoides, in conformity with the absence of middle lamella lignin in these tissues (21). Interestingly, the highest yields were obtained from the rootwood of the same species. The virus disease causing "rubbery" wood condition in apple trees has been found to affect MWL yields (22). The same milling that releases 17% of normal apple tree lignin leads to the isolation of 31% of the lignin in rubbery wood. Milled wood lignins have been frequently isolated from herbaceous plants, but without reporting yields (23).

Milled wood lignins are usually isolated by the original method of Björkman (11), with minor modifications. Milled wood or plant material is stirred for two hours with dioxane-water (9:1 by volume) that dissolves crude milled wood lignin. After removal of solvent from the filtrate, the crude preparation is purified by precipitation of a solution in 90% acetic acid into water. The product is further purified by dissolving into 1,2 dichloroethane-ethanol mixture (2:1 by volume) and precipitation into ethyl ether. The purified milled wood lignin requires extensive drying under vacuum to bring it to a water-free condition. It is free of ash, but may contain from 2 to 8% of associated carbohydrates. The carbohydrate content, which is difficult to estimate accurately, is an undesirable feature from an analytical point of view, and a somewhat modified purification procedure has therefore been proposed by Harkin (24). According to his method, low molecular weight products are first removed from milled wood by cold ethyl acetate, and the crude milled wood lignin then extracted into 9:1 dioxane-water. Carbohydrate-containing materials are precipitated from the solution by gradual addition of 0.08 volumes of benzene together with a small amount of neutral alumina to facilitate filtration. The filtrate is concentrated under vacuum and freeze-dried. The yield of the purified product from spruce is reported to be 25% of Klason lignin, and the preparation is claimed to contain less than 0.2% of carbohydrates. Furthermore, the danger of introducing ethoxyl groups in lignin that may occur in dissolving the preparation in ethanol-dichloroethane in the original Björkman procedure is eliminated.

Finally, Brownell (15) has developed an additional isolation procedure. According to his method, milled wood is completely dissolved into 50% aqueous sodium thiocyanate solution, from which the lignin component may be transferred to an organic phase consisting of benzyl alcohol in admixture with an adjusted amount of dimethyl formamide. The recovery of milled wood lignin from the organic phase is somewhat cumbersome, requiring a lengthy dialysis step.

Of the lignin that remains in milled wood after dioxane-water extraction, a considerable part may be dissolved as lignin-carbohydrate complexes in dimethylformamide, dimethyl sulfoxide or, most efficiently, in aqueous acetic acid (1:1 by volume) (11, 25). Extractable lignin-carbohydrate complexes amount to approximately 30% of Klason lignin in conifers during the early phases of milling, levelling off to about 15% at the end (Figure 5.1). The Klason lignin content of the crude complexes is approximately 30%, varying from 20 to 30% in various isolated fractions (25). After isolation, the preparations are largely dispersible in water.

ISOLATION AND STRUCTURAL STUDIES

Some lignin-carbohydrate complexes may be formed when wood is disintegrated in an aqueous suspension. Brauns and Seiler (26) treated extracted spruce wood meal in water suspension in a Valley pulp beater for 8 to 10 hours, observing partial dissolution. The dissolved part ("colloidal wood") was probably a lignin-carbohydrate complex, judging from its analytical composition (5.3% MeO, 27.5% Klason lignin).

The isolation of lignin from milled wood by solvent extraction methods has obvious disadvantages. Not more than one-fourth of the lignin can be recovered as essentially carbohydrate-free milled wood lignin, and approximately one-fourth of it remains totally unextractable. To obtain preparations of the total lignin component from milled wood, Pew (13) has carried out studies to remove carbohydrates enzymatically using purified glycosidase preparations, such as Rohm and Haas No. 19 enzyme, and purified Trichoderma viride enzyme. While wood is only slowly attacked by these enzymes, 5 to 8 hours dry milling in a vibratory ball mill is sufficient to expose cell wall polysaccharides to their hydrolytic action, and 95% of carbohydrates can be removed in a period of 2 to 3 days (Figure 5.2). The residual 5% of carbohydrates is unfortunately inaccessible to enzyme action and the remaining insoluble lignin contains still 12 to 14% of associated carbohydrates. The carbohydrate content of the "Enzyme lignin" can be reduced to 6% by additional vibratory milling followed by hot water extraction. Some lignin is lost, however, in the extraction step.

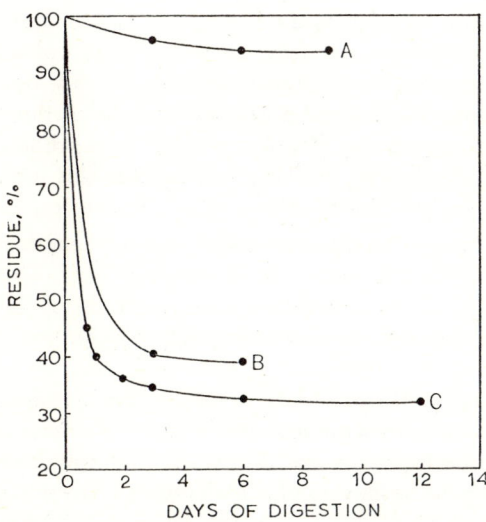

Figure 5.2. Digestion of spruce wood in Rohm and Haas No. 19 enzyme, 40°, pH 4.6. A, meal; B, milled wood, 1-hr grind; C, milled wood, 5-hr grind (13).

"Enzyme lignins" have been isolated from black spruce and poplar (13), and no indications of structural modification of lignin in the enzyme treatment have been obtained. More extensive characterization of enzyme lignins is needed, however, for convincing demonstration of this point; and if the results will be found confirmatory, there is good reason to believe that enzyme lignins will be utilized extensively in future investigations.

The previous discussion leaves several questions unanswered: How extensive is lignin modification during milling? Are milled wood lignin preparations representative of the total lignin component? Are the isolated lignin-carbohydrated complexes present in the original plant or are they artifacts formed during the milling? Although answers to these questions cannot be provided in unambiguous terms, consideration of the processes occurring during milling allows at least some tentative conclusions to be made.

B. Physical and Chemical Changes during the Milling Process. Pew has observed and described in detail some striking changes in the solubility properties of wood during vibratory ball milling (13). In 5 hours milling, wood becomes totally soluble in numbers of solvents in which it is originally insoluble. These include 2N sodium hydroxide, 50% aqueous sodium thiocyanate, 60% aqueous lithium bromide, and formic acid. Cold fuming hydrochloric acid, which usually dissolves only carbohydrates, produces total dissolution of milled wood. In contrast, the Klason lignin content of wood meal remains essentially unchanged during the milling. Brownell (15) found that 9 weeks' milling in a ball mill reduced the Klason lignin content of black spruce from 27.5 to 26.5% with absolutely no effect on the methoxyl content of wood.

It would be of considerable interest to determine whether or not the increased solubility may be caused by subdivision to particles of colloidal dimensions that may directly dissolve into suitable solvents. It has not been possible to determine the size of milled wood particles because of the tendency to aggromerate to larger aggregates ("caking") (27). Carbon black, on the other hand, is reduced in vibratory milling to particles less than 2000 Å in diameter (28), possibly as low as 300 to 600 Å according to another estimate (29). The diameter range corresponds roughly to the length of a cellulose chain composed of 80 glucose units, or to the length of a linear lignin segment 60 to 70 units long. Direct mechanical degradation of solid wood to particles of macromolecular dimensions is consequently a hypothesis meriting consideration.

Vibratory milling of cotton cellulose (12) results in rapid reduction in the degree of polymerization, with simultaneous conversion to amorphous cellulose. No change in the infrared spectra was observed, and only a small increase in carboxyl content could be detected. Milling of isolated lignin preparations increases their solubility, and the formation of certain functional groups have also been observed (13). Vibratory milling of hydrochloric acid lignin converts 65% of the preparation to a form soluble in aqueous dioxane.

ISOLATION AND STRUCTURAL STUDIES

The soluble part has a smaller molecular weight than milled wood lignin (27). Periodate lignin becomes soluble in 1% sodium hydroxide in 5 hours dry vibratory milling in air (13), and an increase in phenolic hydroxyl groups, particularly p-carbonyl phenols, is suggested by the UV spectra (Figure 5.3). An increase in the number of coniferyl aldehyde groups is also apparent. Vibratory milling of MWL preparations themselves produces oligomeric products, among which coniferyl alcohol has been identified (30). When isolated lignin-carbohydrate complexes are subjected to milling, a milled wood lignin preparation may be isolated from the product in an amount corresponding to 40% of the original bonded lignin (31). The isolated preparation has approximately the same methoxyl content (14.3%) and molecular weight (11,000) as MWL isolated directly from milled wood.

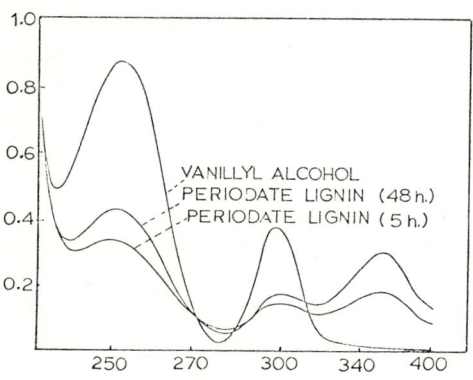

Figure 5.3. Difference spectra of spruce periodate lignins ground for 5 and 48 hours, respectively (13).

The last observation suggests that a part of isolated lignin-carbohydrate complexes are intermediates in the formation of milled wood lignin. The suggestion receives additional support from data shown in Figure 5.1 which demonstrate that the amount of lignin-carbohydrate complexes reaches a maximum during an early stage of milling and declines when the release of milled wood lignin becomes rapid. On the other hand, if lignin-carbohydrate complexes are intermediates of MWL formation as suggested, it is very unlikely that they would be artifacts rather than native elements of wood. Indeed, attempts to produce "artifact" complexes by milling MWL in admixture with hemicelluloses (31) or dioxane acidolysis lignin in admixture with holocellulose (15) totally failed. The behavior of lignin and polysaccharides in milling appears to be different from that of certain synthetic polymers which are converted to mixed graft- and block copolymers when milled together in an oxygen-free atmosphere (32).

It may now be fitting to try to piece together the available information on milling of wood to a coherent, although doubtlessly speculative, picture. The simple idea of viewing lignin in wood as being surrounded by polysaccharide matrix from which it can escape after this matrix has been sufficiently destroyed by milling obviously falls short of explaining the observed data. A more adequate concept may be developed by viewing the milling process as continuous mechanical subdivision of the material to progressively smaller particles. The accompanying generation of new surface area requires homolytic cleavage of both lignin and polysaccharide bonds. We will further assume that wood consists of alternating domains of pure lignin and lignin-free polysaccharides and that these domains are connected by lignin-carbohydrate bonds transgressing the domain boundaries. This situation is illustrated schematically in Figure 5.4, depicting a small isolated lignin domain (II) and a larger continuous region of lignin (I). The solid lines mark intermediate subdivision to particles of colloidal size. At this stage, most of the lignin fragments are contained in particles composed of both lignin and polysaccharide components. Such particles will be isolated as lignin-carbohydrate complexes if soluble. A small amount of particles essentially consisting of lignin (BC' and BD') are formed already at this stage and will provide a minor amount of milled wood lignin. As the subdivision approaches its final stage in continued milling, most of the original LCC particles are degraded, in part, to particles consisting of lignin alone. As a consequence, the amount of LCC decreases while the MWL fraction increases, in qualitative agreement with empirical observations in Figure 5.1.

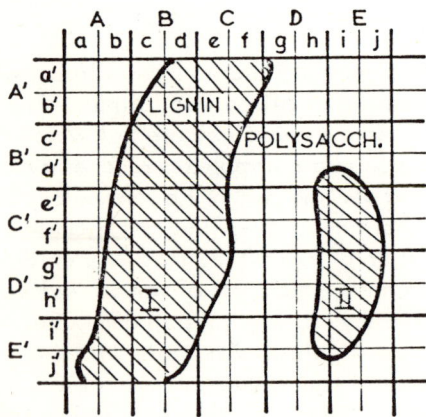

Figure 5.4. Schematic presentation of the effect of milling on the formation of MWL and LCC. Thick lines: Early subdivision of particles. Thin lines: Later subdivision.

ISOLATION AND STRUCTURAL STUDIES

The outlined picture of the milling process, approximate as it is, conforms rather well with the findings to date, and it is consequently of interest to derive from it some further tentative corollaries. On the basis of this concept, the initially released MWL may be expected to stem mainly from relatively large domains of lignin in the cell wall, notably from the corner sections of the middle lamellae, and this idea confirms well with the proposition made by Bland (21). The final yield of MWL (~ 50%) from conifers exceeds that of total middle lamella lignin (~ 25%) (33, 34) and consequently, these preparations must contain generally some cell wall lignin. The part of lignin not convertable to MWL may be surmised to be located in those parts of the secondary wall where the minimum diameter of pure lignin domains is of the same order of magnitude as (or less than) the average diameter of milled wood particles.

The residual carbohydrate composition of milled wood lignins supports the idea of middle lamella origin (Table 5.1). While the average carbohydrate composition of lignin-carbohydrate complexes is very close to the composition of total hemicelluloses, the residual carbohydrates of MWL contain more galactose and arabinose and less mannose (31). On the other hand, Meyer (35) has demonstrated that hemicelluloses containing galactose are located exclusively, and those containing arabinose, almost exclusively, in the middle lamella and S_1-layer of Norway spruce; while glucuronoarabinoxylans are relatively evenly distributed, and glucomannans reach highest concentrations in the S_2- and S_3- layers.

Table 5.1. Carbohydrate Composition in Milled Wood Lignins and Lignin Carbohydrate Complexes of Norway Spruce (31)

	MWL (average)	LCC (average)	Total hemicellulose
Galactose	17	9	8
Glucose	17	16	18
Mannose	31	48	46
Arabinose	10	4	4
Xylose	25	23	24

In chemical behavior, milled wood lignins display some definite differences from that of original lignins, casting doubt on the supposition that they represent the structural average of the total lignin. Far too little attention has been paid to to these differences in the past. It is true that spruce milled wood lignin is sulfonated to a soluble lignin sulfonic acid with approximately the same sulfur content (0.54 S/MeO) as lignin sulfonic acids prepared from wood (36); however, Björkman himself (37) has warned against making far-reaching conclusions on the basis of this observation. Lignin thioglycolic acids prepared from spruce milled wood lignin contain 15.4 to 17.1% sulfur and 8.9 to 9.7% methoxyl (36), in contrast with 10.1% sulfur and 11.3% methoxyl in lignin thioglycolic acids prepared from spruce wood. Likewise, milled wood lignin prepared from normal Douglas-fir wood yields 30% more vanillin than total wood lignin and 2.5 times more vanillin than lignin remaining associated with milled wood in extraction with 90% dioxane (Table 5.2). Similar observations were made for compression wood lignin with less striking differences, however. In water-dioxane hydrolysis at 175°, conifer milled wood lignin yields coniferyl alcohol and -aldehyde, and that from hardwoods also sinapyl alcohol and -aldehyde, while none of these compounds are obtained from the remainder lignins (39). One possible explanation to account for the differences between milled wood and "remainder" lignin could be based on the assumption that the cell wall regions yielding milled wood lignin, notably the middle lamella, contain lignin of the predominantly "endwise polymer" type (cf. Chapter 4); while lignins of the "bulk polymer" type are only incompletely converted to soluble form in milling. Conclusions of this nature must remain purely tentative, however, until further confirmatory (or contradictory) evidence has been obtained.

Table 5.2. Vanillin Yields from Douglas-Fir MWL and from Unextractable Lignin in Milled Wood, in Nitrobenzene Oxidation (38)

	Vanillin yield, % of lignin			
	Milled wood lignin	Unextractable lignin	Weighted average for milled wood	Wood
Normal wood	38.1	15.1	28.4	26.9
Compression wood	24.6	15.8	21.8	20.8

It is rather difficult to form a clear picture of the exact extent of chemical modification of lignin during the milling process. It seems reasonably certain that the degradation of the solid material occurs through homolytic cleavage of covalent bonds, and probably cleavages of both C-C and C-O bonds are involved. The rupture of covalent bonds produces radicals, whose presence in milled wood has been demonstrated by Kleinert and Marton using e.s.r. spectroscopy (40). As mentioned before, there is hardly any evidence for radical combination reactions (forming artifact lignin-carbohydrate complexes, for example), and the lack of such reactions may be attributed to the rigidity of the solid matrix. The radicals produced must become stabilized in one way or another, and hydrogen abstraction from the surrounding matrix or reaction with air-oxygen to form peroxy radicals are likely possibilities for such stabilization. One may speculate that all bond cleavages occur along the newly formed surfaces, and units located inside of the particles remain intact. An average molecular weight of 11,000 has been estimated for spruce MWL by means of ultracentrifuge (36) corresponding to a weight average degree of polymerization of approximately 60. If it could be ascertained, for example, that on the average five bond cleavages are necessary to release a molecule of this size from its association with the solid cell wall matrix, it could be concluded that approximately 8% of the units in milled wood lignin had suffered modification in milling. However, this is an entirely fruitless line of speculation in the absence of experimental guidelines, and merely points out that low-molecular weight fractions of milled wood lignins are probably more modified than the high-molecular weight part. The removal of oligomeric fractions from milled wood lignin preparations prior to chemical characterization, as recommended by Freudenberg (24), is consequently a desirable procedure.

Some approximate comparisons between MWL preparations and total wood lignins have been attempted by spectral means. Kolboe and Ellefsen (41) attempted subtracting the holocellulose contribution from the IR spectrum of spruce wood in a KBr pellet. The method could not be applied to the 1000 to 1200 cm^{-1} region, where carbohydrates absorb strongly, but elsewhere the obtained spectrum conformed well with the spectrum of milled wood lignin. Although the IR spectra do not reflect many of the structural details in lignins (cf. Chapter 7), the equal magnitude of conjugated carboxyl absorption at 1665 cm^{-1} in both spectra is noteworthy. On the other hand, the UV spectra of black spruce middle lamella lignin, obtained by Fergus and Goring using ultraviolet microscopy (33), exhibit less absorption in the 300 to 330 mμ region than milled wood lignin spectra. This suggests that conjugated side-chain structures such as α-carbonyl and α,β-ethylenic groups, might be more common in MWL preparations than in wood lignin.

In summary, MWL lignins are indeed "useful preparations for lignin chemists," quoting Björkman's expression (5), but certain restraint appears desirable in the interpretation of results on these preparations insofar that they pertain, in all likelihood, mainly to lignins located in the middle lamella and separate studies are required for the total characterization of the cell wall lignin. Associated carbohydrates are almost unavoidable contaminants of these preparations, requiring proper attention. In addition, many tree species contain unextractable phenolic heartwood constituents, such as ellagic acid in Eucalyptus species (42) and various polyphenolic materials in redwood, larches, Douglas-fir, and cedars that may contaminate MWL samples unless segregated sapwood material is used for their preparation (43).

III. Brauns Lignins and "Enzymicly Liberated Lignins."

A. Characteristics of Brauns Lignins. The extractives of many woods and plants contain a lignin component, often isolatable in reasonable purity by solvent fractionation. Originally somewhat misleadingly designated as "native lignins" (4), these preparations have more recently been called either "soluble lignins" or "Brauns lignins." The original procedure by Brauns (4) is based on thorough extraction of finely ground plant material (100 – 150 mesh) with 95% ethanol and precipitation of the crude Brauns lignin from the concentrated ethanol extract by water. The final purification was carried out by precipitating a solution of the crude product in dioxane into ethyl ether. Many modifications of the original procedure have been suggested. Freudenberg, for example, recommends 9:1 acetone-water mixture for the extraction, and cold ethyl acetate for the removal of other extractive components and oligolignols from the acetone extract (44).

Not many investigators have been able to duplicate in other species the yields of Brauns lignins originally reported for black spruce (8 to 10% of Klason lignin)(4). The isolated amounts vary rather erratically from one species to another, as shown by the following examples, reported as percentages of the total plant material: 1.6 (45) and 0.2 to 0.4% (46) for Norway spruce wood; 3.2% for Scots pine (47); 0.7% for aspenwood (48); 3.0% for Eucalyptus sieberiana (49); negligible yields for Eucalyptus regnans (20); 0.24% for the heartwood and 0.15% for the sapwood of black wattle (50); and 0.04% for wheat (51). Hergert (52) has shown by systematic studies that the yields of Brauns lignin are highest from the heartwood parts of gymnosperms, and often negligible for the sapwood (Table 5.3).

Table 5.3. Distribution of Brauns Lignin in Several Representative Conifers* (52)

Species	Type**	Yield, % of wood	
		Sapwood	Heartwood
Western hemlock	1	0.10	1.02
	2	0.21	1.38
Sitka spruce	1	0.02	1.18
Amabilis fir	1	0.06	0.29
Loblolly pine	2	0.12	0.14
Slash pine	2	0.12	0.14
Longleaf pine	2	0.06	0.07
Western red cedar	2	0.42	3.27
Baldcypress	2	0.62	1.74

* Procedure modified by precipitation into chloroform instead of ether in order to remove lignans.
** Type 1 is old, slow-growth tree (100 yrs.); Type 2 is young, second growth (25-50 yrs.).

Associated carbohydrates have been shown to be present in many Brauns lignins. Sugars were positively identified in the hydrolyzate of the Brauns lignins isolated from Norway spruce (5) but seemed to be absent in the corresponding hydrolyzate of beech (53). The xylan content of Brauns lignin from Eucalyptus sieberiana was estimated to be 2.1% (49). In addition, preparations from gymnosperms often contain extractive components as impurities, such as hydroxymatairesinol and other lignans in Norway spruce (54), leucoanthocyanins in Douglas-fir and western hemlock, and taxifolin in Douglas-fir (55). The amounts of lignan impurities can be substantially reduced by appropriate solvent treatments (54, 55).

Disregarding the impurities present, the major part of Brauns lignins from conifer woods display properties that leave little doubt about its lignin nature. There are no clearly significant differences in the elemental compositions of these preparations and milled wood lignins, as shown in Table 5.4. Brauns lignin from black spruce yields the same amount of Hibbert's ketones (10%) (56) and vanillin (24%) (57) as the wood lignin, and lignin thioglycolic acids from both sources have the same overall composition (58). The infrared spectrum of Brauns lignin from Cryptomeria is indistinguishable from the spectra of milled wood lignins of conifers, and that of red pine displays only minor differences (59). On the other hand, the phenolic hydroxyl content of Brauns lignins from conifers, 0.46/MeO (60, 61), is about 50% higher than that of milled wood lignins (0.29/MeO), and this difference is also clearly reflected in nmr spectra of the acetates of these preparations (62). The number average molecular weights of Brauns lignins are in the range from 850 to 1000 (63), sometimes even 4200 (64), while the Z-average molecular weight 11,000 has been obtained for milled wood lignin from spruce (31). Since the latter value corresponds to a number average molecular weight of 6,000 or less (cf. Chapter 17), the available data suggest a definite, but not very large difference in molecular weights.

Brauns lignins from angiosperms have not been characterized as extensively as conifer preparations. The methoxyl contents of these lignins are generally lower than those of milled wood lignins (Table 5.4). In the case of birch, BL and MWL preparations have quite divergent compositions, and the low methoxyl content of the former as well as that of the corresponding oak lignin are suggestive of guaiacyl lignin. This impression is supported by low syringaldehyde yields in nitrobenzene oxidation (1.7% for birch and 4.0% for oak lignin). On the other hand, the published IR spectra (65) of both preparations are quite abnormal, and methylation with diazomethane produces exceptionally high methoxyl contents (66). Consequently, massive contamination by some polyphenolic constituents must be suspected in these two cases. In contrast, black wattle preparations possess high methoxyl contents (50), and the infrared spectra of Brauns lignins from bamboo and beech (59) are of the same type as hardwood MWL preparations.

Brauns lignins frequently contain much larger amounts of ester groups than milled wood lignins. Smith has shown that a Brauns lignin preparation from aspen with 21.8% MeO contained more than 7% of p-hydroxybenzoic ester groups (67). The same amount of p-hydroxybenzoate groups is present

Table 5.4. Elemental Compositions of Brauns Lignins

Species	Ref.	Preparation	Carbon %	Hydrogen %	MeO %
Conifers:					
Black spruce	(4)	BL	63.6	6.2	14.8
(Picea mariana)	(36)	MWL	63.7	6.3	15.4
Norway spruce	(64)	BL (Total)	65.3	5.9	15.9
(Picea abies)	(64)	BL (Fractions)	61.8-67.3	5.5-5.9	14.9-15.7
	(5)	MWL	63.8	6.0	15.8
Scots pine	(66)	BL	64.0	6.3	14.5
(Pinus sylvestris)	(36)	MWL	64.0	6.1	15.7
Angiosperms:					
Maple	(66)	BL	61.0	5.6	17.4
	(43)	MWL	60.4	5.7	20.0
Aspen	(48)	BL	63.5	6.0	19.5
	(36)	MWL	60.4	6.2	21.4
Birch	(66)	BL	61.4	5.5	14.9
	(36)	MWL	58.8	6.5	21.5
Oak	(66)	BL	58.6	5.3	14.8
Kiri	(66)	BL	60.1	6.2	16.6
(Paulownia tomentosa)					
Eucalyptus sieberiana	(49)	BL	58.2	5.7	17.6
Black wattle	(50)	BL	61.1	6.2	21.0
			60.6	6.1	20.5
Bagasse	(66)	BL	61.5	5.7	15.3

in Brauns lignin of Populus maximowiczii (68). The presence of these groups influences the UV spectra which in neutral solution are devoid of the commonly observed maximum at the 280 mµ region (67, 48), and in alkaline solution display a pronounced maximum at 305 mµ due to the phenolate ion of p-hydrooxybenzoic ester groups (67). A strong maximum at 1694 cm^{-1} in the infrared spectrum is due to the carbonyl absorption of the same groups (67). In addition to p-hydroxybenzoic acid, vanillic, syringic, and ferulic acids have been identified among the saponification products of Brauns lignin from aspen (67), and the total ester groups therefore account for approximately 10% of the weight of these preparations. Only a fraction of this amount has been found in milled wood lignins (14).

Brauns lignins from bagasse exhibit an unusual UV absorption maximum at 315 mµ that shifts to 355 mµ in alkali (69). The uncommon spectral properties are imparted by p-hydroxycinnamic acid ester groups that account for more than 11 weight % of the preparation (70). In addition, esters of ferulic, vanillic, syringic, and p-hydroxybenzoic were present. p-Hydroxycinnamic and ferulic acids form the dominant ester groups in Brauns lignin from wheat, with traces of vanillic and syringic acids. The ester content of bagasse Brauns lignin clearly exceeds the range 5 to 9% of ester groups indicated for the total lignins in the grass family (see Table 3.21) In contrast, Brauns lignins from horsechestnut and Douglas-fir contain only small amounts of ester groups consisting of vanillic, p-hydroxycinnamic and ferulic acid ester groups in the former case, and the latter preparation contains vanillic and p-hydroxybenzoic ester groups only. The latter two acids, or vanillic acid alone, are also the only acids that could be isolated from the alkaline hydrolysis products of a number of extracted conifer woods (71).

B. "Enzymicly Liberated Lignins." Nord and coworkers have shown (72) that the action of pure cultures of certain brown-rot fungi on plant tissues increases the amount of lignin material that can be extracted by means of 95% ethyl alcohol. This material was isolated by the procedure developed for the isolation of Brauns lignins, and was found to possess the same chemical properties as the latter. This led Nord to propose that total lignin may actually be an identical material with Brauns lignin, but the majority of it was not extractable because of its association with the polysaccharide matrix. The action of brown-rot fungi upon the plant material was assumed to be limited to the enzymatic hydrolysis of polysaccharides only, and it was speculated that sufficiently long exposure to the decaying organisms would ultimately yield virtually all lignin as "enzymicly liberated lignin." Nord's proposition was proven to be incorrect by Pew (13), who demonstrated that the release of alcohol-soluble lignin from spruce stops at the level of approximately 13% of the total lignin, even when the decay by brown-rot fungi was continued to the point of 6% of residual carbohydrates. While both Brauns lignins and "enzymicly liberated lignins" thus only have some broad implications in terms of the bulk of the lignin component, the fact remains that the two preparations are outstandingly similar, as proven by extensive studies by Nord and coworkers (65, 73). The compositions of these preparations themselves, their phenylhydrazones, acetates and diazomethane methylated products were demonstrated to be the same within the limits of analytical error. Further evidence in support of essential identity was obtained from nitrobenzene oxidation studies, and from comparisons of UV (69) and IR (74) spectra and electrophoretic mobilities (75). These comparisons not only establish a close relationship between the two lignin preparations, but also suggest the absence of major lignin modification during the decay period.

The largest yield of enzymicly liberated lignin, 22.7% of total lignin (Table 5.5), has been obtained from Scots pine, while angiosperms yield from 7 to 10% of this material.

C. The Possible Nature of Brauns and Enzymicly Liberated Lignins. The excessive claims for significance originally placed on both Brauns lignins and enzymicly liberated lignins has prevented objective deliberation on the question of their role in lignification. It should be noted first that the amount of Brauns lignin increases drastically from sapwood to heartwood in conifers (Table 5.3), and that an analogous increase has been observed for black wattle (50). This could be explained by assuming that the parenchyma cells located at the heartwood-sapwood boundary exude Brauns lignin (or lignin

Table 5.5. Yields of "Enzymicly Liberated Lignins" from Various Plant Materials (66, 72, 74)

Plant species	Decay treatment		Lignin content, % of decayed material	Yield of isolated lignin, % of total lignin
	Organism	Time of decay, months		
White fir	Lentinus lepideus	7	39.3	12.5
Scots pine	Lentinus lepideus	7	44.9	15.2
Scots pine	Lenzites saepiaria	13	50.1	18.3
Scots pine	Poria vaillanti	15	52.5	22.7
Oak	Daedalea quercina	10	32.5	9.9
Birch	Daedalea quercina	10	35.0	8.6
Maple	Daedalea quercina	10	33.7	8.3
Bagasse	Poria vaillanti	8	50.4	6.5

precursors) to that surrounding tissue together with heartwood extractives. A far simpler hypothesis, however, is to regard Brauns lignins as a part of the originally insoluble lignin that is rendered soluble through hydrolytic processes associated with the aging of the tissue. The reduced amount of acetyl groups in the heartwood of Eucalyptus species (76) and the diminishing molecular weights of arabinogalactans in the centripetal direction of birch tree trunks (77) have been regarded to be caused by hydrolytic processes of this kind. Nimz (78) has recently shown warm neutral water solutions alone are in position to hydrolyze certain sensitive lignin bonds, possibly linkages of the α-O-4 type, releasing oligolignols, many of which have been identified. It would not be surprising, therefore, that exposure to the mildly acidic xylem sap for several decades could be the cause of a gradual formation of Brauns lignins. The additional release of similar lignin material by the action of brown-rot fungi could be envisaged simply as somewhat accelerated continuation of the hydrolysis of acid-sensitive bonds. It follows from a hypothesis of this kind that Brauns lignins would represent solely an arbitrary cut of the total lignin hydrolysis products which would also include smaller molecular weight oligolignols. The presence of oligolignols in the acetone extract of spruce has actually been reported by Freudenberg (44).

Brauns lignins are not determined as Klason lignin because benzene-alcohol extraction removes them from wood (49). Conversion of originally insoluble lignin to Brauns lignin would therefore reduce the apparent lignin content of aged wood tissue. In certain pine species, the determined lignin content of heartwood has been found to be lower than that of sapwood (cf. Chapter 3), and the formation of Brauns lignin merits consideration as possible explanation.

IV. Periodate Lignins. Of the known methods to isolate the total lignin from plant materials, the periodate method (79) avoids the condensation processes that cause changes in the acid lignins, but certain oxidative modification occurs instead. The method is based on the gradual conversion of monosaccharide units in polysaccharides to dialdehydes which are susceptible to hydrolysis by boiling water (79) as illustrated for a glucose unit in cellulose in Equation 1.

$$\text{Cell.}-\text{O}-[\text{glucose-OH}]-\text{O-Cell.} \longrightarrow \text{Cell.}-\text{O}-[\text{dialdehyde}]-\text{O-Cell.} \longrightarrow \text{Cell.}-\text{OCH}(\text{CH}_2\text{OH})-\text{CH}(\text{OH})-\text{CHO} + \text{OHC-CH}_2\text{OH-CHO} + \text{HO-Cell.} \quad (1)$$

The method is a tedious one requiring at least six successive treatments with periodate (4.5% aq. $Na_3H_2IO_6$, 24h at 20°), each followed by a 3 hour treatment with boiling water. Finally, periodate lignin is obtained as a light-brown powder, in which the morphological structure of wood is retained. Lignins become partially oxidized, notably the units with free phenolic hydroxyl groups. Adler has demonstrated that periodate converts such units first to o-quinones, then to muconic acid structures, as shown in Equation 2 (80).

$$\text{ArOH(OMe)} \longrightarrow \text{o-quinone} \longrightarrow \text{muconic acid (COOH, COOH)} + \text{MeOH} \quad (2)$$

Other oxidative action upon lignin has not been sufficiently clarified. Only a minute amount of formaldehyde (appr. 0.01 moles/phenylpropane unit) has been isolated from the periodate oxidation products of Brauns lignins from aspen (68), and the yields are negligible from various milled wood lignins (81). In contrast, the sum of formic acid and formaldehyde has been claimed to be 0.15 mole in the oxidation of Sugi (a conifer) lignin preparations (82).

The reaction in equation 2 releases methanol from methoxyl groups, and consequently the reported compositions of periodate lignins, shown in Table 5.6, allow estimates to be made on the extent of this reaction. The average number of methoxyl groups in a C_9-unit can be calculated to be 0.75 in spruce periodate lignin, and this suggests oxidative modification of about 20% of the original aromatic rings. Periodate lignins themselves consume periodate when subjected to additional treatments. Consequently, their composition is much dependent on at what exact point the periodate treatments are terminated in their isolation. Judging from the methoxyl content, 13.4%, reported for periodate lignins from two Japanese conifers (82), it may be possible to exert

Table 5.6. Compositions of Periodate Lignins (79)

Species	Yield %	C %	H %	MeO %	Ash %	Klason lignin %
Spruce	29.8	61.4	6.0	12.2	1.97	93.7
Maple	22.3	58.4	5.4	20.4	1.14	74.9
Birch	21.8	57.6	5.3	21.4	1.08	78.5
Beech	24.4	54.6	6.1	16.7	2.17	72.2

some control on the oxidative modification of lignin.

Periodate lignins are obviously of little value in the analytical characterization of lignins because of oxidative modification. On the other hand, approximately 80% or more of the lignin units remain apparently intact. Conifer periodate lignins have been shown to yield vanillin (25%) in nitrobenzene oxidation (79), propylcyclohexanol derivatives in hydrogenation with copper-chromite (79) and Hibbert's ketones (11.5%) in ethanolysis (56), and the reported yields are comparable to those from wood lignins and superior to those obtained from acid lignin preparations. Periodate lignins are also readily converted to water-soluble lignin sulfonates in acidic sulfonation (79). They are remarkably insoluble in organic solvents and aqueous sodium hydroxide. This property may well be a consequence of some cross linking by the reactive o-quinone structures formed (see Equation 2).

V. Isolation Methods Based on Acidic Hydrolysis of Polysaccharides.

A. Condensation Reactions Caused by Mineral Acids. It is well known that lignins are quite sensitive to even mild treatments with mineral acids. The changes that occur in the structure of lignin under the influence of acids have been broadly characterized as "condensation reactions" and their occurrence is manifested, among other things, in reduced yields of aldehydes obtainable in nitrobenzene oxidation (2) and reduced conversion to soluble lignin sulfonic acids in sulfonation (1). It is relatively easy to see the reason for the occurrence of reactions of this type (Figure 5.5). Lignin contains a number of structures convertible to resonance-stabilized carbonium ions. These include cinnamyl alcohol groups 1 that readily convert to allyl carbonium ion 2, cinnamaldehyde groups 3 that give rise to similar carbonium ions 4, and α-carbinol and -ether groups 5 giving benzyl carbonium ions 6. The role of cinnamyl carbonium ions has not been clarified experimentally so far. It is probably similar to that of carbonium ions formed from coniferylaldehyde which Pew (83) has shown to condense in the position 6 of another unit in fuming hydrochloric acid solution.

When Brauns lignin from spruce or thin sections of spruce wood are treated with cold fuming hydrochloric acid, a yellow color due to coniferyl

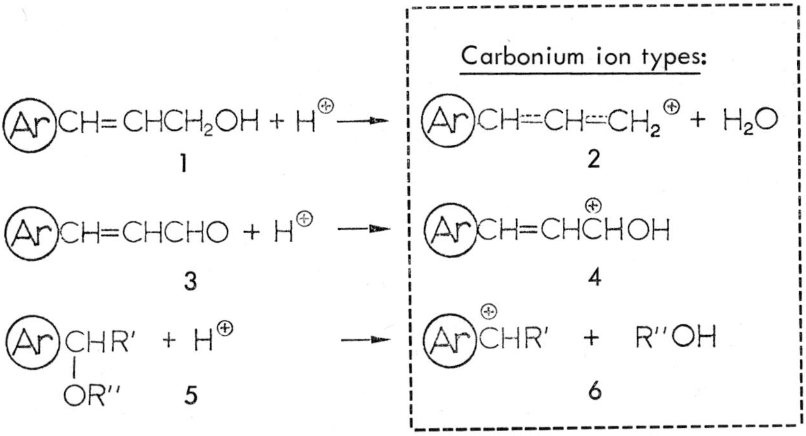

Carbonium ion types:

$(Ar)CH=CHCH_2OH + H^{\oplus} \longrightarrow (Ar)CH\!=\!\!=\!CH\!=\!\!=\!CH_2^{\oplus} + H_2O$
               1   2

$(Ar)CH=CHCHO + H^{\oplus} \longrightarrow (Ar)CH=CH\overset{\oplus}{C}HOH$
              3   4

$(Ar)\underset{OR''}{CHR'} + H^{\oplus} \longrightarrow (Ar)\overset{\oplus}{C}HR' + R''OH$
            5   6

a. Condensation reactions:

$R^{\oplus} + \underset{OR}{\underset{|}{\bigcirc}}_{OMe}^{C_3} \longrightarrow R\underset{OR}{\underset{|}{\bigcirc}}_{OMe}^{H\;C_3} \xrightarrow{-H^{\oplus}} R\underset{OR}{\underset{|}{\bigcirc}}_{OMe}^{C_3}$

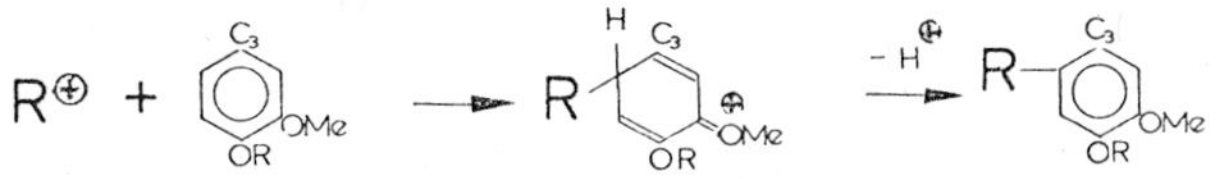

b. Competing reactions:

$R^{\oplus} + A^{\ominus} \longrightarrow RA$ (3)

$RO\overset{\oplus}{H}_2 + HA \longrightarrow RA + H^{\oplus} + H_2O$ (4)

$R^{\oplus} + HA \longrightarrow RA + H^{\oplus}$ (5)

$A^{\ominus} = OSO_3H^{\ominus}, Cl^{\ominus}, F^{\ominus}, OPO_3H_2^{\ominus}, SO_3H^{\ominus}$, etc.

$HA = H_2O$, $HSCH_2CO_2H$, MeOH, etc.

c. Acid-catalyzed rearrangements (see Chapter 9):

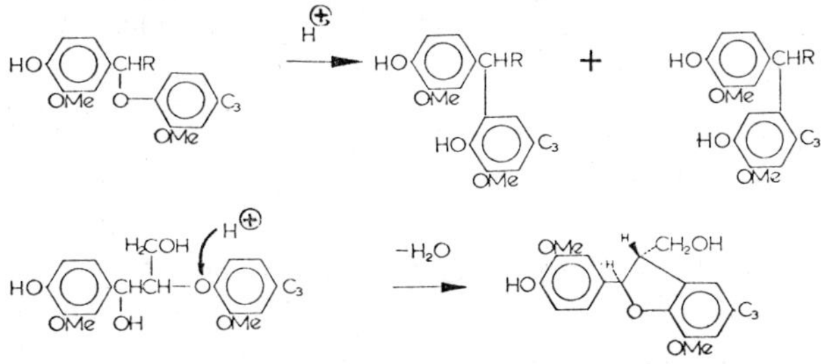

Figure 5.5. Formation of carbonium ions and the resulting condensation reactions in treatment of lignin with mineral acid.

aldehyde carbonium ion 7 develops immediately (Equation 6). The carbonium ion condenses with a phenylpropane unit in lignin, forming 8 that converts to a blue carbonium ion 9. As a consequence, the color shifts gradually through emerald green to dark blue, and such color shifts are always observed when wood or lignins are contacted with concentrated mineral acids. When associated with solid wood, chromophores 7 and 9 exhibit absorption maxima at slightly higher wavelengths than in solution. Spectrophotometric studies suggest that carbonium ion 7 does not condense at the 5-position.

$$(6)$$

Condensation reactions involving benzyl carbonium ions are illustrated by model reactions shown in Equations 7 (84) and 8 (85).

$$(7)$$

$$(8)$$

Again, condensation at position 6 predominates, although some occurs also at the 5-position. This is important to note because in much of the earlier lignin literature, position 5 was incorrectly assumed to be the sole site of condensation reactions. Condensation at the 6-position is also supported by the isolation of 4,5-dimethoxyphthalic acid (10) from the permanganate oxidation of acid-treated lignins, methylated prior to permanganate oxidation (86).

ISOLATION AND STRUCTURAL STUDIES

Additions of carbonium ions to the anions present in the solution compete with the condensation reactions. As a consequence, sulfuric acid half-ester groups are introduced in the lignin molecule by concentrated sulfuric acid treatment, and chlorine is a part of hydrochloric acid lignins. The amount of sulfur in sulfuric acid lignins has been reported to be in the range 1.8 to 3.2% (87), reduced to 0.3 to 0.8% by hydrolysis using refluxing in 3% H_2SO_4 or in 0.5% HCl. The chlorine content of hydrochloric acid lignin after hydrolysis has been found to be approximately 0.75% (88).

Nucleophilic species, such as thioglycolic acid, bisulfites, etc. reduce the amount of condensation reactions in dilute acid solutions, probably by virtue of converting the parent structures to stable groups through $S_N 2$ displacements shown in Equation 5.

The structural modifications caused by condensation reactions are often reflected in the IR spectra of lignins (cf. Chapter 7). The reduction in vanillin yield in nitrobenzene oxidation is a more sensitive measure of condensation reactions.

B. <u>Lignin Isolated by Means of Mineral Acids</u>. Comparing the properties of lignins prepared using various mineral acids, there are definite indications that the extent of condensation depends on the acid used. Sulfuric acid lignins are prepared generally by using 72% H_2SO_4 (89), although lower concentrations, 65% for example (87), are sometimes recommended. Extracted wood or plant material is treated with the acid (10 cc/g of wood) at ambient or lower temperature (15°). The color changes referred to above are observed; the polysaccharides first swell and then dissolve, forming, for a short period a very viscous solution which rapidly becomes fluid as the hydrolysis of the dissolved cellulose chains proceeds. After 2-hours reaction time, two volumes of water are added; and the precipitated lignin refluxed for 4 hours in 3% sulfuric acid to remove sulfuric acid ester groups and co-precipitated dextrans. Hydrochloric acid lignin is prepared in an analogous manner using concentrated HCl with specific gravity 1.22 (determined at 0°C) and reaction temperature 1 to 5°.

There exists some indication that the nature of the mineral acid enters as a factor determining the degree of condensation. Sulfuric acid lignins display infrared spectra that differ substantially from the spectra of milled wood lignins and an even greater deviation has been observed for lignins isolated by means of 65% perchloric acid (90). Sulfuric acid lignins are generally not convertable to soluble lignin sulfonic acids (1); and the yield of vanillin in nitrobenzene oxidation is drastically reduced, sometimes to as low values as 1.5% (2). As a consequence, sulfuric acid lignins are quite inadequate for structural and reactivity studies.

In contrast, hydrochloric acid lignins, for reasons that are presently not entirely clear, appear to be less condensated than sulfuric acid lignins; especially if they are prepared by methods requiring minimum exposure of lignin to acid. For example, Hägglund (91) has shown that treating extracted spruce wood with 41% hydrochloric acid for thirty minutes dissolves most of the polysaccharides leaving 24.4% of wood (<90% of total lignin) as insoluble hydrochloric acid lignin, while at least 10% of lignin dissolves. The hydrochloric acid lignin cannot be extensively condensed because as much as 57% of it can be gradually made to dissolve by successive treatments with boiling 5% hydrochloric acid, alternating with occasional treatments with cold 42% HCl (92). Treatment of spruce wood with fuming hydrochloric acid at 15° dissolves 26% of the total lignin (93) and milled spruce- and aspen woods are totally soluble in cold concentrated hydrochloric acid (13). In contrast, 72% sulfuric acid dissolves less than 1% of lignin in conifer woods, although 25 to 60% of lignins in hardwood species may dissolve in this medium (94).

Hydrochloric acid lignin from spruce has been reported to yield 19.3% vanillin in nitrobenzene oxidation as compared with 20.8 to 24.3% yields obtained from wood (95). The reported yield of Hibbert's ketones (11%) is more than obtained from wood and Brauns lignins (10%) (56). Acidic sulfonation converts appropriately prepared hydrochloric acid lignins quantitatively to soluble lignin sulfonic acids (96). The infrared spectra of spruce hydrochloric acid lignins are almost indistinguishable from those of Brauns and milled wood lignins (97, 98), while extensive condensation is reflected in the spectra of sulfuric acid lignins (cf. Chapter 7). Another indication for smaller amounts of condensed structures in hydrochloric acid lignins is provided by the 1.0% yield of benzene polycarboxylic acid obtained in direct permanganate oxidation, while 2.4% of these acids are obtained from sulfuric acid lignin (99).

The above examples were cited mainly to illustrate the varying properties of acid lignins which are commonly described as condensed materials, "insoluble in any solvent without further change (100)." The effects of condensation are not pronounced in the behavior of carefully prepared hydrochloric acid lignins in many reactions, and the above statement about the insolubility of acid lignins is approximately true for Klason lignins of conifers, but it is incorrect and misleading in reference to Klason lignins of angiosperm woods which are acetone-soluble to the extent of 27 to 49% (101). The lack of reported solubility data on the less-condensed hydrochloric acid lignins leads one to suspect that this property may not have been the subject of systematic studies. Generalizations within the class of hydrochloric acid lignins are, however, not warranted in view of the variance caused by the numerous methods of preparation. For example, hydrochloric acid lignin carefully prepared according to the original recipe of Willstätter and Kalb (103) refuses to become converted to soluble lignin sulfonic acids (102), while another preparation made by the milder procedure of Hägglund and Richzenhain does it willingly (96). There may still be room for improvements in the control of condensation reactions, and acid lignins should therefore not be written off as outmoded and useless preparations.

C. <u>Cuoxam Ligins</u> were developed by Freudenberg <u>et al</u>. to reduce the degree of condensation in acid lignins (104). The method of preparation is based on using cuprammonium hydroxide ("Schweizer's reagent") to remove the polysaccharidic components of wood. This is not possible without partial hydrolysis of polysaccharides, and therefore, alternating treatments with boiling 1% sulfuric acid (1-2h.) and extractions with copperammonia (12h.) are necessary. Four to five hydrolysis-extraction treatments are necessary to obtain reasonably carbohydrate-free preparations. Copperammonia dissolves a part of lignin, reducing the yield to about 80% in case of spruce (104) and to only 55% in case of beech (105). Cuoxam lignins are lighter in color than acid lignins in general, and the yield of vanillin from a spruce preparation in oxidation by m-nitrobenzoate is practically the same as from wood lignin (21%) (106). Because of the tedious procedure of preparation, incomplete yields, and repeated exposures to acidic and alkaline reagents, cuoxam lignins are hardly to be considered attractive preparations for the characterization of lignins.

D. Dioxane Acidolysis Lignins. The isolation of lignin preparations soluble in organic media requires a combination of mild hydrolytic conditions with a good lignin solvent. A 9:1 mixture of dioxane and dilute hydrochloric acid would appear ideal for such a purpose, and systematic studies for the isolation of dioxane acidolysis lignins have been carried out using both ambient (107, 63) and elevated (108, 109) temperatures. The reported results are not, however, very encouraging. Preparations isolated in treatments at ambient temperatures are obtained in appreciable yields (85% from beech wood) (63) containing, however, approximately 2% of chlorine. At refluxing temperatures, rapid condensation reactions take place resulting in substantially reduced yields of aldehydes in nitrobenzene oxidation (Figure 5.6) (109). Pepper and coworkers have tried to select conditions that avoid excessive condensation (110, 111), and their method yields 10 to 13% acidolysis lignin from conifer woods, 22 to 35% from angiosperm woods and 44 to 52% from grass species (112). Chlorine contents have generally not been determined for these preparations. The carbohydrate contents may be substantial at least in some preparations, judging from the reported xylose contents (1.6 to 7.5%) for dioxane acidolysis lignins isolated from a variety of woody and herbaceous species (101). At the moment, dioxane acidolysis lignins reported in the literature represent a heterogeneous group, and the amount of useful information in terms of characterizing the nature of the original lignin is meager. The same is true for acetic acid lignins that can be isolated in relatively large yields (95% from spruce) by treating wood with 90% acetic acid and magnesium chloride (115). These preparations contain acetyl groups, and 10.8% yield of vanillin has been recorded for a preparation from spruce (106).

Sakakibara and Nakayama (113,114) have found that heating wood in neutral 1:1 dioxane-water at elevated temperatures (175°, 2h.) dissolves 40% of lignin in spruce and 60% of that in beech. Dioxane lignins of this

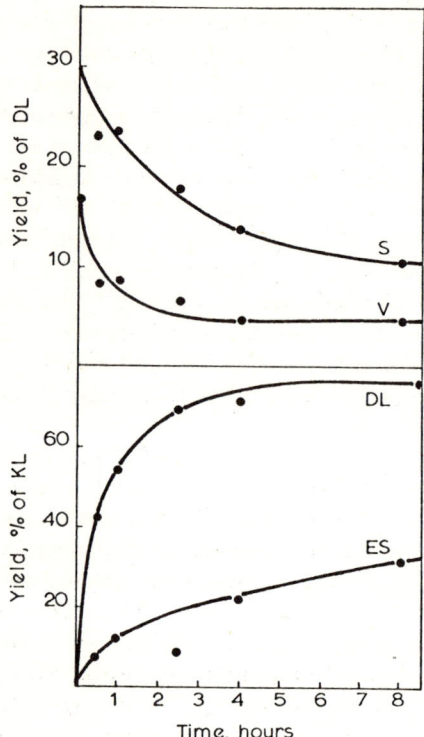

Figure 5.6. Yield of aldehydes in nitrobenzene oxidation of aspen dioxane acidolysis lignin. S, syringaldehyde; V, vanillin; DL, dioxane lignin; ES, ether-soluble dioxane lignin (109).

type have not been characterized, but undoubtedly deserve a closer look because condensation reactions under these conditions appear to be minor.

VI. Alkali Lignins. Wood lignins require rather drastic hydrolytic treatments (5% of NaOH, 130 to 170°) to become soluble in aqueous solutions of alkali hydroxides, and consequently the "alkali lignins" obtained through the acidification of such solutions are not particularly useful lignin preparations. In contrast, lignins in grass species can be obtained in substantial yields by mild alkaline treatments (116), even at ambient temperatures, as shown in Table 5.7.

Surprisingly, alkali lignins have only rarely been utilized for the characterization of grass lignins. The mild alkali treatment is not prone to cause much chemical modification beyond the saponification of p-coumaric and ferulic ester groups (23). Alkali lignins, in their crude form, may contain as impurities carbohydrates (117), silica (116) and protein material (118), especially if they are isolated from not totally mature plant tissue. These impurities, however, may be reduced by appropriate solvent treatments, and

Table 5.7. Yields of Alkali Lignins from Cereal Straws Calculated from Data by Beckmann et al. (116)

Species	Hydrochloric acid lignin, corrected for silica		Percent of lignin methoxyl recovered in alkali lignin, extracted with 1.5% NaOH		
	Yield %	MeO %	48 h at amb. temp.	Additional 6 h at 100°	Total
Rye	16.5	14.5	50	28	78
Oat	12.9	18.6	47	35	82
Barley	12.3	19.2	50	29	79
Rice	12.1	14.0	21	28	49

by removing the major part of proteins enzymatically (119) prior to the alkali treatment.

VII. Lignin Thioglycolic Acids. Of all lignin derivatives, lignin thioglycolic acids are best suited for the characterization of the total lignin (120). They are prepared by heating extracted plant material (4 g) with thioglycolic acid (3 g) in 2 N hydrochloric acid (40 ml) for 7 hours (121). Lignin is converted to lignin thioglycolic acid which remains insoluble and can be extracted by 2% sodium hydroxide from which it can be precipitated by acid. Lignin thioglycolic acids are rarely contaminated with polyphenol and protein impurities, and their IR spectra indicate absence of condensation reactions.

VIII. Limitations of Lignin Preparations. Milled wood lignins can generally be successfully isolated from woody plants, albeit in varying yields, and from mature specimens of grasses, ferns, etc. Immature herbaceous plants yield only small amounts of milled wood lignins with high carbohydrate contents. Hydrochloric acid lignins, prepared according to Hägglund and Johnson (122), deserve more attention in characterization and reactivity studies than they have received in the past. It should be noted, however, that lignins condense readily under acidic conditions both with proteins and polyphenols, such as tannins and polyflavanoids, and the hydrochloric acid method is not well applicable to specimens containing these constituents. The advantages gained from the isolation of cuoxam and periodate lignin hardly equal the toil and time expenditure required in their preparation. Because of the convenience of preparation, dioxane acidolysis lignins are excellent for orienting studies. Dioxane acidolysis lignins have also been utilized successfully for the characterization of lignins in immature tissues (123) and in conifer barks (124). Lignin thioglycolic acids and alkali lignins are particularly useful in studies on herbaceous plants.

IX. Quantitative Determination of Lignin.
A. Applicability of Klason Lignin Method to Wood Samples.

A critical and thorough review of lignin determinations has been published recently by Browning (125). Here, the discussion will be limited to the evaluation of the Klason lignin method and its application to different plant materials. The most commonly used procedure of Klason lignin determination is the Tappi Standard Method T-13 (125).

In brief, this method calls for an extraction of air-dry wood (1g, 40 mesh) with 95% ethanol (4h.) and with ethanol-benzene (1:2, 4h.), followed by digestion with hot water (400ml, 100°, 3h.). After drying in air, the wood is treated with 72% sulfuric acid (15ml, 18-20°, 2h.), diluted with water (560 ml.), and refluxed for 4h. The insoluble Klason lignin is filtered, dried to a constant weight at 105° and weighed. The method prescribes expressing the lignin content as percentage of the oven-dry unextracted wood, but lignin contents on an extractive-free basis are more commonly reported in the literature. Corrections for ash are usually unnecessary for woody materials with the exception of certain tropical species; while herbaceous plants, grasses in particular, often contain significant amounts of silica.

The extraction procedure in this method could stand some simplification. Freudenberg has pointed out (24) that extraction with acetone-water (9:1) is usually equally effective as successive extractions with ethanol and ethanol-benzene. In either case, the extractives removed contain the soluble Brauns lignin, and the question arises as to whether this lignin fraction should be regarded to be a part of the total lignin. It was pointed out earlier that Brauns lignin probably represents a low molecular weight part of the originally deposited lignin that becomes released from its association with the cell wall matrix through hydrolytic processes associated with aging. If this interpretation is correct, the amounts of Klason and Brauns lignins together could represent the closest figure to the lignin originally deposited. The matter is not trivial because Brauns lignins may amount to 10% of Klason lignins in heartwoods of certain conifer species. Quantitative methods for the estimation of Brauns lignins in total extractives ought to be relatively easy to develop on the basis of differential solubility.

It has been repeatedly shown that hot water extraction prior to Klason lignin determination reduces the lignin yield without affecting its methoxyl content (126, 127, 128). This is due to the dissolution of a small part of lignin in the hot water. Kratzl (129) has shown that a 2-hour hot water extraction dissolves 4 to 5% of extracted spruce wood and the dissolved material has a Klason lignin content of 30%. Lyness and Schenker (130) have demonstrated by analytical and spectral means that hot water extracts indeed contain a lignin component. Consequently, hot water extraction merely causes a negative error in the lignin determination corresponding to a reduction by 5% in case of spruce and by 10% (from 23.2 to 21.5%) in the case of maple (128). It can only be hoped that hot water extraction will be banned from future standard methods as a useless and harmful operation. The confusion which may arise from the application of hot water extraction is best illustrated by the following example: Suppose that a sample of spruce, after removal of 3% extractives by benzene-alcohol, gives 27.0% Klason lignin. Its lignin content may then be expressed as either 27.0%, on an extractive free-, or 26.2%, unextracted, basis. The inclusion of the hot water treatment in the extraction procedure, however, will change these numbers to 27.8 and 24.7%, respectively, representing a spread of 12%.

The 72% concentration of sulfuric acid has been selected on the basis that cellulose is not completely hydrolyzed at concentrations below 65% (87), and concentrations of 80% and above give rise to the formation of insoluble products from polysaccharides. The condensation reactions occurring in the sulfuric acid treatment result in a weight loss in the form of removed water. The extent of this weight loss is not known exactly; but it is probably not serious and is largely compensated by the weight gain imparted by the sulfuric acid ester groups, a part of which remains unhydrolyzed in refluxing with dilute mineral acid, as demonstrated by the small sulfur content (0.3 to 1.0%) (87) of Klason lignins.

Klason lignin determinations probably give the most accurate values for conifer woods because the recovery of lignin as Klason lignin is essentially quantitative (125). In some species, redwood for example, unextractable heartwood polyphenols may cause a positive error by condensing with lignin. Angiosperm woods behave differently. Treatment with 72% sulfuric acid brings 25 to 60% of the total lignin into solution (94), the dissolved lignin consisting predominantly of syringylpropane units, as shown in Table 5.8 (131).

Table 5.8. Lignin Dissolvable in 72% H_2SO_4 in Various Wood Species

Species	Lignin soluble in 72% H_2SO_4		Lignin insoluble in 72% H_2SO_4	
	% of total	MeO %*	% of total	MeO %
Spruce	2	–	98	16.0
White fir	3	–	97	16.1
Basswood	15	25.8 ?	85	18.9
Beech	55	23.5	45	18.9
White birch	59	22.5	41	19.7
Elm	68	23.3	32	17.1

* Calculated from the methoxyl contents of total and insoluble lignin.

Dilution with water and hydrolysis in boiling 3% sulfuric acid precipitates most, but not all, of the dissolved lignin; and part of it passes into the Klason lignin filtrate as "acid soluble lignin." Campbell and McDonald (132) isolated an acid-soluble lignin from maple by adsorbing it from the filtrate on a column of Zeo-Karb 215 resin. The low methoxyl (16%) and carbon contents of the material suggest the possibility of a lignin carbohydrate complex. Acid soluble lignins have also been isolated from Eucalyptus regnans (133) and birch (134). The quantitative determination of the amounts of acid-soluble lignin in Klason lignin filtrates has been attempted by UV spectroscopy, but this method involves several possible sources of error. The filtrates contain furfural and hydroxymethylfurfural (125) absorbing at 280 mµ. On the other hand, it has been claimed that these compounds largely evaporate if the hydrolysis is carried out by boiling rather than refluxing (134), and the lignin determination may also be based on absorption at 208 mµ (14, 136), where the absorption by furan aldehydes has less effect. Another difficulty consists of the estimation of a realistic absorptivity value which is dependent on the ratio of syringyl to guaiacyl nuclei (see Chapter 6). Finally, ester groups are partially hydrolyzed, forming the free acids; and p-hydroxybenzoic, vanillic and syringyl acids have been identified in the Klason lignin filtrates of aspen (14).

Acid soluble lignin has been estimated to amount to 16% of Klason lignin in Eucalyptus regnans (133) and to approximately the same amount in birch (134). Very small amounts of soluble lignin were claimed for beech and aspen woods by Richzenhain and Dryselius (135), while Pearl and Busche (14) demonstrated that the amounts were definitely significant for the aspen wood (Populus tremuloides) studied by them. It would not be surprising if the amounts of acid soluble lignin will be found variable for angiosperm woods. Until more detailed information becomes available, it should only be recognized that the determined Klason lignin values for angiosperm woods are, as a rule, lower than their true lignin contents.

Application of the standard Klason lignin method to certain Eucalyptus species gives unreasonably high lignin contents (28 to 55%); and the methoxyl contents of the isolated Klason lignins are abnormally low (13 to 15%), as shown in Table 5.9 (137). This circumstance is caused by the presence of an essentially methoxyl-free polyphenolic material, "kino," that permeates the vessels and parenchyma cells and partially fills the lumina of wood fibers (138). It is formed as a consequence of cambial injury and consists largely of polymerized leuco-anthocyanins (139). This polymeric material is not removed by the standard extraction procedure, and the sulfuric acid treatment condenses it with lignin. Cohen (137) has shown that the amount of the true lignin component in Eucalyptus Klason lignins may be estimated from the methoxyl content. To do this, the methoxyl content of "pure" Klason lignin needs to be determined by isolating a sample free of "kino" in the following way: A normally extracted wood sample is subjected to further extraction either by 0.125 N sodium hydroxide (100°, 1h.) or by 3% sodium sulfite (98°, 2h.). Both treatments remove kino quantitatively, dissolving also a part of hemicelluloses and lignins. However, the extraction probably has little effect on the methoxyl content of Klason lignin, as suggested by independent experiments on pine and maple (Table 5.9). Consequently, the

Table 5.9. Klason Lignin Determinations of Eucalyptus Species (137)

Species	Extraction loss, %		Klason lignin				Calc. lignin content %
	Benzene-alcohol & hot water	Additional Extr. with 0.125N NaOH	Extracted with benzenealcohol and hot water		Extracted with benzenealcohol, hot water & hot 0.125N NaOH		
			Yield %	MeO %	Yield %	MeO %	
Eucalyptus marginata	6.4	32.4	41.0	12.6	21.6	20.3	25.5
Eucalyptus regnans	10.5	18.0	17.7	21.3	15.2	23.6	16.0
Eucalyptus polyanthemos	25.2	40.3	29.5	15.2	18.6	20.8	21.6
Pinus radiata	3.5	10.6	26.2	15.5	24.7	15.9	25.3
Rock maple	10.2	20.5	20.1	21.0	18.5	20.9	20.1

methoxyl content of the Klason lignin from the alkali or sulfite extracted Eucalyptus sample is probably applicable for the estimation of the lignin component in the Klason lignin obtained according to the standard method.

B. Lignins in Bark Tissues. The barks of both conifers and angiosperm trees contain large amounts of polyphenols related to anthocyanins, flavonoids and phlobatannins ("bark phenolic acids," cf. Chapter 3, Section V) which resemble the kino of Eucalyptus trees in not being extractable by benzene-alcohol and in forming insoluble condensates with lignin. For example, cork tissue from Douglas-fir, freed from bast fibers by sieving and extracted with benzene-alcohol, shows an apparent Klason lignin content of 74.8% (140). Only a minor part of the insoluble material consists of lignin because 70% of the cork tissue is soluble in 1% sodium hydroxide and fails to give Hibbert's ketones upon ethanolysis. The alkali-soluble bark phenolic acids from conifer barks usually show a methoxyl content of 2 to 4.3% (141, 142). Klason lignin isolated from alkali-extracted bark with 10 to 11% methoxyl (142) currently represents the closest figure to the true lignin content of bark tissues.

C. Herbaceous Plants Including Grasses. The standard Klason lignin procedure needs to be modified extensively for the determination of lignin in herbaceous plants, and the same modifications are required in applications to immature tissues or foliage of trees. Rather surprisingly, the necessary modifications, although clarified thirty years ago by Phillips and Goss (143, 144), have escaped the attention of a number of investigators; and numerous examples of meaningless Klason lignin contents could be quoted from the recent literature omitting vital corrections for the content of ash or of co-condensed protein, or both. The main point to observe is the fact that protein, whenever present in plant tissue, condenses extensively with lignin in the treatment with sulfuric acid, imparting extra weight to Klason lignin. The amount of protein varies in plants according to their age. For example, the protein content of oat plants diminishes from the level of 38 to 45% in immature plants to 2.5% at the mature stage (143). The magnitude of the error caused by protein in Klason lignin determinations is quite appreciable. Hydrochloric acid lignins of 7 to 50 day-old barley (144) and oat plants (143) contain 30 to 50% of co-condensed protein on an ash-free basis, and the same degree of contamination is probable in Klason lignins. The amount of protein impurities in lignins isolated from mature plants may still be in the range of 8 to 12%.

The amounts of co-condensed protein in acid lignins may be estimated approximately by multiplying the nitrogen content by the factor 6.25 (145). This is probably an adequate procedure in correcting the ash-free Klason lignin contents of fully mature plants, but not recommendable for the analysis of immature plants because of the uncertainties associated with the use of the factor 6.25. This factor is derived from the composition of the original protein, which may not be the same as the composition of protein fragments condensed with lignin (146). Consequently, it is advantageous to remove a large part of proteins from extracted immature plant material by treatment with proteolytic enzymes. For example, the lignin determination method of the Association of Official Agricultural Chemists (147, 148) prescribes treatment with pepsin in 0.1 N hydrochloric acid (40°, ~ 12h.) for the removal of proteins, and treatments with trypsin (0.25% sodium carbonate solution, 35°, 18h.) (149) have also been attempted. The removal of proteins may remain incomplete, however; and Klason lignins from trypsin-treated vegetable materials have been found to contain 2.5 to 15% protein (149), requiring correction based on nitrogen analysis. As an alternative or complementary method of protein removal, refluxing either with 5% sulfuric (147) or 5% hydrochloric acid (149) for one hour has been employed, but the extent of possible removal of lignin hydrolysis products in such treatments has not been sufficiently clarified.

There are also other uncertainties associated with the determination of lignin in herbaceous plants that deserve continued studies. Herbaceous plants, as angiosperm woods, may contain some lignin that is dissolved in the lignin determination. Some errors may stem from the partial hydrolysis of p-coumaric and ferulic ester groups (cf. Chapter 3). Finally, it is by no means excluded that certain herbaceous species may contain polyphenolic constituents unextractable with organic solvents, analogous to those present in Eucalyptus wood and tree barks, that may cause a positive error by condensing with lignin. Alternative methods of estimating lignin in immature tissue, such as that based on the isolation of lignin thioglycolic acids, applied by Higuchi and Barnoud to the study of tissue culture lignins (150), deserve attention.

D. <u>Lignin in Pulps</u>. Various modifications of the Klason lignin

ISOLATION AND STRUCTURAL STUDIES 195

determination provide reliable estimates for kraft pulps of conifer woods, and the amount of soluble lignin is negligible, while 20 to 50% of residual sulfonated lignin in sulfite pulps becomes acid-soluble (151). Consequently, the soluble lignin in the Klason lignin filtrate of sulfite pulps requires a separate determination by means of UV absorption (cf. Chapter 7). The Klason lignin method is not suitable for the estimation of lignin in plant materials of pulps treated with such agents as chlorine, hypochlorites, chlorites or chlorine dioxide. In industrial practice, the lignin content of pulps is generally estimated indirectly, usually by determining the consumption of permanganate in an acidic solution (permanganate or K-number) (152) or that of gaseous chlorine (Roe-number) (153). These methods have been discussed in two recent reviews (154, 155).

X. Structure of Lignins. Conifer lignins have been the subject of more intensive structural studies than lignins in other species, and in Europe most of these studies have been concentrated around a single species, Norway spruce (Picea excelsa). Until very recently, hardwood lignins had been only cursorily studied, but currently an encouraging trend away from the somewhat one-sided preoccupation with conifer lignins seems to continue.

The concept of lignins representing dehydrogenation polymers of three basic cinnamyl alcohol precursors forms a solid framework in which the still-scattered pieces of structural information can be appropriately fitted; and frequent reference will be made, in the discussion that follows, to the tentative coupling rules that were developed in the preceding chapter. Of course, modifications to the above concept should be flexibly considered whenever there is sufficient reason to do so. For example, there is already now some evidence to suggest that the cinnamaldehydes and cinnamic acids that precede the cinnamyl alcohol precursors of lignin in the biogenetic sequence may, to a minor degree, participate in the dehydrogenative coupling reactions in the polymerization process leading to lignin.

A. End Groups with Unattached Side-Chains. The existence of cinnamyl alcohol (12) and cinnamaldehyde (11) end-groups in the lignin macromolecule is a direct consequence of the dehydrogenative polymerization mode (Figure 5.7). The genesis of the cinnamaldehyde groups is not completely clear. They can no doubt be formed through disproportionation of cinnamyl alcohol precursors, as demonstrated by the model dehydrogenation experiments of Freudenberg et al. (156). This does not rule out the possibility, however, that some cinnamaldehydes representing the immediately preceding stage in the biogenetic sequence leading to cinnamyl alcohols could directly participate in the polymerization. A third possibility, autoxidation of existing cinnamyl alcohol end-groups is suggested by model experiments (157).

Both cinnamyl alcohols and cinnamaldehydes have been repeatedly identified as hydrolysis products of lignins. Graefe (158) was the first to identify coniferyl aldehyde 11a among the hydrolysis products of spruce wood ("hadromal") obtained by treatment with a zinc chloride solution at ambient temperature, and his finding was later confirmed by Adler (159). Lundquist (160) has also demonstrated 0.5% yield of coniferyl aldehyde in the acidolysis of spruce milled wood lignin. Borohydride reduction prior to acidolysis reduces the yield to 0.03%, supporting the formation of coniferyl aldehyde from end-groups in lignin. Small amounts of coniferyl aldehyde have been repeatedly isolated from the steam hydrolysis products of various woods (161, 162, 163, 164) as well as from acetolysis products (165) and from neutral hydrolysis with dioxane and water at 170° (114). However, some coniferyl aldehyde may actually be formed from β-O-4-linked lignol structures under the conditions of high temperature hydrolysis (166). Coniferaldehyde together with smaller amounts of p-coumaraldehyde 11c is also formed in the fermentation of conifer wood with microorganisms (167), and it is likewise obtained from Brauns lignin and milled wood lignin in a similar treatment (168, 169). p-Coumaraldehyde has been identified among the steam hydrolysis products of western hemlock wood (162), and it is not impossible that a part of cinnamaldehyde end-groups might be of the p-hydroxyphenyl type. The existence of sinapaldehyde 11b end groups in hardwood lignins has not been established firmly as yet, although water hydrolysis of quaking aspen and beech woods (163, 164) has been reported to yield sinapaldehyde.

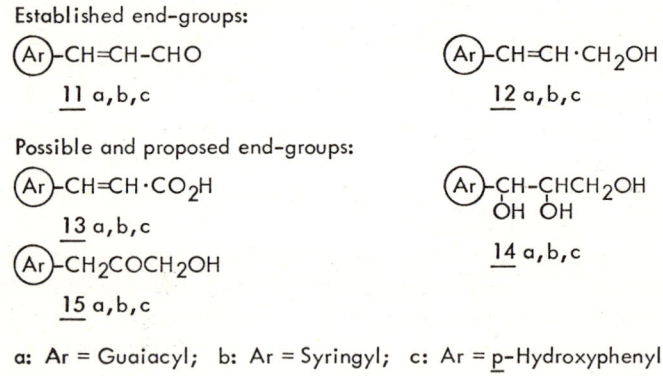

Figure 5.7. Lignin end-groups with an unattached side-chain.

The detection of cinnamyl alcohols among the acidolysis products of conifer lignins has so far failed (171), possibly due to the sensitivity of these compounds towards acidic reagents. High temperature hydrolysis with dioxane water (39, 113), water alone (161 - 164), or with 0.2 M Na_2S at 100° (170) yields coniferyl alcohol 12a and p-coumaryl alcohol 12c from conifer woods,

ISOLATION AND STRUCTURAL STUDIES

and coniferyl alcohol and sinapyl alcohol 12b from hardwoods (39). Coniferyl alcohol sulfonate 16a has been isolated in 0.2% yield from lignin sulfonates of western hemlock together with the isomeric sulfonate 17a, formed from the former through allylic rearrangement (172). The syringyl analogue of the latter 17b (173), sulfonate 18b, disulfonates 19b and 20b have been obtained from hardwood lignin sulfonates representing probable reaction products of sinapyl alcohol end groups (174).

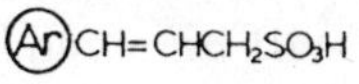

$$(Ar)CH=CHCH_2SO_3H$$
16a

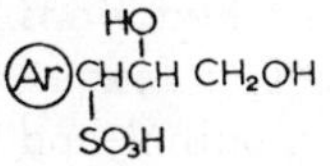

$$(Ar)\underset{SO_3H}{\overset{HO}{CHCH}}CH_2OH$$
18b

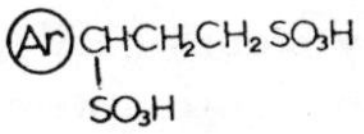

$$(Ar)\underset{SO_3H}{CHCH_2CH_2SO_3H}$$
19b.

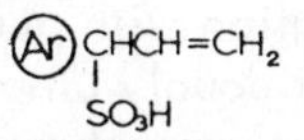

$$(Ar)\underset{SO_3H}{CHCH=CH_2}$$
17a,b

a: Ar = Guaiacyl
b: Ar = Syringyl

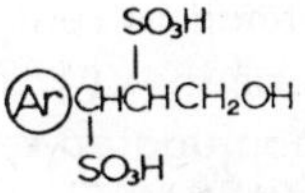

$$(Ar)\underset{SO_3H}{\overset{SO_3H}{CHCHCH_2OH}}$$
20b

The monomeric hydrolysis products formed from end-groups are probably released from their association with the lignin macromolecule through the hydrolysis of β-O-4- and possibly α-O-4-linkages. The former linkage has been demonstrated through the isolation of the dimeric hydrolysis products 21 and 22 from spruce wood by Nimz (175, 176, 177). In addition, unhydrolyzable β-5 linkages are also present, as shown by the isolated dimer 23 (176) which has also been detected in the cambial sap of spruce (179) and amongst the dehydrogenation intermediates of coniferyl alcohol (178), together with the corresponding aldehyde 24 (157).

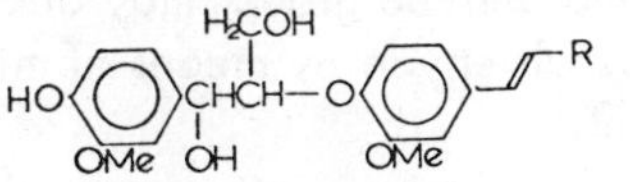

21, R = CH$_2$OH
22, R = CHO

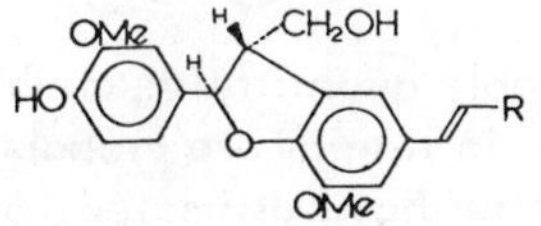

23, R = CH$_2$OH
24, R = CHO

The quantitative amounts of cinnamyl alcohol and -aldehyde groups in lignins are of vital interest for the evaluation of the total structure. It was pointed out in the previous chapter that end-groups of this type would account for as many as one-third of units present in a lignin molecule of the extreme "bulk polymer" type, while their amount ought to be small in an "end-wise polymer." The fraction of end-groups of the total number of units can thus serve as a sensitive index for the overall character of the lignin molecule. Currently available data are not sufficient for anything more than a crude evaluation of end groups (with unattached side chains) in one lignin, that of Norway spruce. For such an evaluation, it is necessary to separately determine the amounts of cinnamaldehyde and cinnamyl alcohol groups. The former groups can readily and often quantitatively be converted to chromophoric groups with strong absorption in the visible range, offering the opportunity for spectrophotometric determination. Reactions of this type have been reviewed in detail by Brauns (180), and only a few examples will be given here. The well-known reaction with phloroglucinol and hydrochloric acid leads to the purple chromophore 25 which has been used by Adler and coworkers (159, 181) for the determination of coniferyl aldehyde groups. Analogous chromophoric groups are formed in reactions with a large number of phenols, phenol ethers and aromatic amines (182). Base-catalyzed condensation with acetaldehyde extends the conjugated system, giving as main product doubly unsaturated aldehyde 27 (183). The latter forms a violet-colored chromophore in reaction with phloroglucinol hydrochloric acid. Both base- and acid-catalyzed condensations with indoxyl afford the chromophore 26 (184), and analogous condensation reactions may be performed with barbituric acid derivatives (185) and indole (186). The chromophoric properties of cinnamaldehyde groups themselves may be used to obtain an upper limit for these groups from light absorption at 350 mµ (see Chapter 6).

The determination of cinnamaldehyde groups may also be based on reversed aldol condensation to acetaldehyde by means of mild alkaline hydrolysis (187), as shown in Equation 9.

$$-(Ar)-CH=CH-CHO + H_2O \longrightarrow -(Ar)-CHO + CH_3CHO \quad (9)$$

In view of the variety of available quantitative methods, more precise estimations of cinnamaldehyde groups in lignins are probably possible than those reported to date. Adler and Elmer have estimated, on the basis of the Wiesner reaction, that coniferyl aldehyde groups account for 2 to 2.5% of units in spruce wood lignin (159) and for 3 to 4% of units in Brauns lignin from western hemlock (162). From UV absorption data, Adler and Marton (188) suggest 3 to 4% cinnamaldehyde groups for spruce milled wood lignin. The amount of cinnamaldehyde groups is only 1 to 2% in birch milled wood lignin according to Klemola (189), and Nord and De Baun found the amount

ISOLATION AND STRUCTURAL STUDIES

of these groups in Brauns lignins from hardwoods to be approximately one-half of that present in corresponding softwood preparations (190).

A colorimetric method for the determination of cinnamyl alcohol groups in lignin has been developed in the pioneering work of Lindgren and Mikawa (191). It was shown that 3,4-dimethoxy cinnamyl alcohol 28 in treatment with pyridine and tosyl chloride converts to pyridinium ion $\overline{29}$, affording the purple chromophore 30 (λ_{max} = 475 mμ) in further treatment with potassium cyanide and p-nitrosodimethyl aniline (Equation 10).

$$\text{MeO-C}_6\text{H}_3(\text{OMe})\text{-CH=CHCH}_2\text{OH} \longrightarrow \text{MeO-C}_6\text{H}_3(\text{OMe})\text{-CH=CHCH}_2\text{N}^+\text{C}_5\text{H}_5 \;\; \text{Cl}^-$$

28   29

$$\downarrow$$

$$\text{MeO-C}_6\text{H}_3(\text{OMe})\text{-CH=CHC(CN)=N-C}_6\text{H}_4\text{-NH}_2$$

30

(10)

Using this specific color reaction, it was possible to demonstrate the instability of coniferyl alcohol end groups toward air oxidation. While freshly cut and thoroughly extracted sections of spruce wood showed a strong reaction for coniferyl alcohol groups, the reaction was negative for sections stored in air for six months. When the coniferyl aldehyde groups in fresh sections were converted to coniferyl alcohol groups by reduction with borohydride, the intensity of the color reaction was increased by approximately 50%. This observation would indicate the amount of coniferyl alcohol groups to be twice that of coniferyl aldehyde groups in unoxidized lignins. As the coniferyl alcohol groups became destroyed by air oxidation, the intensity of the color reaction caused by reduced coniferyl aldehyde groups increased by 70%. This finding is suggestive of partial autoxidative conversion of coniferyl alcohol groups to coniferyl aldehyde units. Application of the reaction to a stored sample of Brauns lignin from spruce gave a negative reaction.

The estimation of end groups may also be based on the elimination of α, β-ethylenic double bonds upon hydrogenation, the extent of which can be estimated by means of UV spectroscopy. On the basis of such experiments, Adler and Marton (188) suggested that coniferyl-alcohol and -aldehyde groups accounted for 6% of units in spruce milled wood lignin.

The possibility of additional types of end-groups in lignins cannot be excluded. Cinnamic acid 13 end-groups may be present only in minute amounts in conifer lignins (192), but their occurrence in grass lignins could be more frequent (193). Glycerol side-chains exist in a number of hydrolysis products from conifer and angiosperm wood lignins, as shown by Nimz (194,

195). Guaiacyl glycerol 14a together with dilignol 31 were isolated by percolating spruce wood with water (194). In a similar treatment of beech wood syringyl glycerol 14b was obtained in yields of 2.5% based on lignin (164). Both arylglycerols represent optically inactive mixtures of threo- and erythro isomers, the former isomer predominating.

$$\text{HO-}\underset{\underset{\text{OMe}}{|}}{\bigcirc}\text{-CHCH-O-}\underset{\underset{\text{OMe}}{|}}{\bigcirc}\text{-CHCHCH}_2\text{OH}$$
$$\text{H}_2\text{COH} \quad \text{OH}$$
$$\text{OH} \quad \text{OH}$$

31

For many reasons, however, the existence of arylglycerol end-groups in lignins appears doubtful. Firstly, the isolated arylglycerol derivatives may well be intermediate products of hydrolytic cleavage of β-O-4 bonds (see Chapter 9) which at higher acidities are further converted to ketol structures 15 (170). Secondly, arylglycerol side-chains ought to be oxidized by periodate to formaldehyde and formic acid, but only minute quantities of these compounds have been isolated from oxidations of softwood and hardwood milled wood lignins (195).

Ketol end-groups of the type 15 have been included in the lignin formula proposed by Freudenberg (196), to account for the presence of non-conjugated carbonyl groups in approximately ten percent of arylpropane units in spruce milled wood lignin. This estimation was made by Adler and Marton, who assumed that the non-conjugated carbonyl groups were present in the form of β-carbonyl groups (187). With the discovery of side-chain displacement reactions (cf. Chapter 4), it now seems more likely that the aldehyde groups associated with detached side-chain structures are better candidates for the non-conjugated carbonyl groups. Furthermore, the minute yields of formaldehyde obtained in the periodate oxidation of milled wood lignins do not support the existence of ketol side-chains in significant quantities (196).

C_6-C_1-aldehydes have been encountered as such or as parts of dilignols amongst the products obtained in mild hydrolysis of lignins. These include vanillin, syringaldehyde and dilignol 32 (176). Reversed aldol condensation of original cinnamaldehyde groups probably accounts for the formation of such structures, but it is not clear whether this process takes place during the hydrolysis of the wood material or occurs gradually in the living tree as a consequence of aging. The possibility remains, therefore, for the existence of a limited number of end groups with a one-carbon formyl side-chain in lignins.

32

ISOLATION AND STRUCTURAL STUDIES

B. Units with Free Phenolic Hydroxyl Groups. The determination of the contents of phenolic hydroxyl groups in lignins has been attempted by means of methylation with diazomethane (36) ionization difference spectra (197), potentiometric titrations in basic solvents (198, 199) and conductometric titration (60, 200). A method applicable to guaiacyl lignins only (201), is based on the oxidation of phenolic groups with periodate to o-quinones and methanol, both of which are analytically determinable (Eq. 11).

None of these methods is suitable for the estimation of the average phenolic hydroxyl content of the total lignin component. The methods based on ionization difference spectra give results that are often ambiguous (60), due to the variance in the light absorption characteristics of phenolate groups. On the other hand, titration methods and the periodate oxidation method generally give concordant results in the absence of carboxylic acid groups (80). The determined percentages of phenolic units in spruce milled wood lignins are in the range from 30 to 35%, 32% representing the most probable value (202). However, there is hardly sufficient justification to project this value as a representative average of the total spruce wood lignin. Considerably higher contents of phenolic units, 46 to 48%, have been obtained for Brauns lignins of conifers by methylation with diazomethane (36), titration methods (200) and periodate oxidation (201). An appalling lack of data on the phenolic units in lignins other than conifer lignins remains to be filled, and the discussion will therefore be limited to guaiacyl lignins only.

The phenolic units in conifer lignins may be divided to those possessing an unsubstituted 5-carbon 33 often called "uncondensed" phenolic units, and others, in which position 5 carries an aryl 34 aryloxy 35 or alkyl 36 substituent, collectively called "condensed" phenolic units.

$$\underset{\text{OH}}{\overset{R}{\underset{R'}{\bigcirc}}}\text{OMe} \xrightarrow{IO_4^-} \underset{R'}{\overset{R}{\bigcirc}}=O + \text{MeOH} \qquad (11)$$

33: HO–⟨⟩–R, OMe

34: $C_5(C_1)$ — HO–⟨⟩–R, OMe

35: $O-C_4$ — HO–⟨⟩–R, OMe

36: $C_\beta(C_\alpha)$ — HO–⟨⟩–R, OMe

Adler and Lundquist (203) have developed a method to estimate the uncondensed phenolic units, based on the oxidation of these units by $ONO(SO_3K)_2$ (Fremy's salt) to o-quinoidal structures 37 that can be estimated spectrophotometrically. The method requires the elimination of units possessing a carbonyl or hydroxyl group at the α-position, and the lignin preparation is consequently reduced with borohydride and treated with methanolic hydrochloric acid prior to oxidation in order to convert these two groups to α-methoxyl groups.

$$\text{HO-C}_6\text{H}_3(\text{OMe})\text{-R} \xrightarrow{\text{ONO(SO}_3\text{K})_2} \text{O=C}_6\text{H}_3(\text{OMe})\text{-R (37)} \quad (12)$$

According to the results obtained, spruce milled wood lignin contains 15 to 18% of phenolic units of the uncondensed type, representing approximately one-half of the total phenolic units. Of the uncondensed phenolic units, those possessing a hydroxyl group in the α-position ("p-hydroxybenzyl alcohol groups") 38 can be assessed by means of a specific reaction, developed by Gierer (204). In all likelihood, phenolic groups of this kind belong to β-O-4 and β-1-linked structures. In mildly alkaline solution they react with quinoneimide chloride 39 that displaces the side-chain forming the colored indophenol 40 (Equation 13). The latter can be estimated spectrophotometrically.

$$\underset{38}{\ominus\text{O-C}_6\text{H}_3(\text{OMe})\text{-CHR-OH}} + \underset{39}{\text{Cl-N=C}_6\text{H}_4\text{=O}} \longrightarrow \text{O=C}_6\text{H}_3(\text{OMe})(\text{-N=C}_6\text{H}_4\text{=O})(\text{CHOH-R})} \longrightarrow \underset{40}{\ominus\text{O-C}_6\text{H}_3(\text{OMe})\text{-N=C}_6\text{H}_4\text{=O}} + \text{RCHO} \quad (13)$$

The measurements indicate 5 to 7% of units of the type 38 to be present in spruce milled wood lignin and 12 to 14% of such units for spruce Brauns lignin (205). The reaction is directly applicable to wood materials, and the yield of indophenol obtained is equivalent to 3% of p-hydroxybenzyl alcohol structures 38 in spruce wood lignin. The deviation from the larger content of these groups in milled wood lignin may not be real necessarily because lignin in wood can only partially be accessible to the reagent. Reaction of quinoneimide chloride with aspen wood has been demonstrated to release both indophenol 40 and its syringyl analogue. An aryl substituent in position 5 inhibits the formation of the indophenol, as shown by Pew (206).

Another displacement reaction of uncondensed phenolic units of the guaiacyl type was discovered by Gustafsson and Anderson (207) and is also directly applicable to wood materials. It is based on nitration in ethyl ether medium (Equation 14), and 4,6-dinitroguaiacol 41 can be isolated as the reaction product by means of paper chromatography. Unlike the indophenol reaction, conversion to dinitroguaiacol does not require the presence of an α-hydroxyl group, and gives reproducible results. The yields from spruce and pine woods were 1.66 and 1.52%, expressed as percentages of isolated dinitroguaiacol of Klason lignin, and were interpreted to indicate the presence of approximately 7% of uncondensed phenolic units in conifer lignins. The yields of dinitroguaiacol were 1.04 and 1.05% for lignins in birch and alder woods, respectively, in accordance with the lower guaiacyl content in these lignins. The conversion to dinitroguaiacol deserves undoubtedly further exploration as a diagnostic test.

$$
\begin{array}{c}
\text{HO–C}_6\text{H}_3(\text{OMe})(\text{CHR-OR}') \xrightarrow{NO_2^{\oplus}} \text{HO–C}_6\text{H}_2(\text{OMe})(\text{NO}_2)_2 \ (\mathbf{41}) \ \pm \ \text{RCHO} \ \pm \ \text{R'OH} \\
\downarrow \text{HBr in CHCl}_3 \\
\text{HO–C}_6\text{H}_3(\text{OMe})(\text{CHR-Br}) \ (\mathbf{42}) \xrightarrow{HCO_3^{\ominus}} \text{O=C}_6\text{H}_2(\text{OMe})(=\text{CHR}) \ (\mathbf{43})
\end{array}
$$

(14)

An alternative to the dinitroguaiacol test is hydrobromination in a chloroform solution, followed by the conversion of the α-bromo compound 42 to the chromophoric quinonemethide 43 that remains attached to the rest of the lignin molecule. The reaction, investigated by Adler and Stenemur (208), is limited in application to chloroform-soluble Brauns lignins.

Comparative studies on relative amounts of uncondensed and condensed phenolic units can be based on oxidative degradation with permanganate after protecting the phenolic hydroxyl group by diazomethane methylation. This approach was initiated by Richtzenhain (209), who oxidized methylated spruce wood with permanganate at pH 7, isolating veratric acid 44a in an amount corresponding to 4.9% of Klason lignin. This acid obviously stems from uncondensed phenolic units, while the 5,6-dimethoxyisophthalic acid ("isohemipinic acid") 45a, obtained simultaneously in 0.9% yield, must be produced from condensed phenolic units. Larsson and Miksche (210) improved Richtzenhain's procedure by carrying out the permanganate oxidation at pH 14 instead of pH 7, obtaining an improved recovery of the aromatic acids, Figure 5.8. The yields were determined by gas-liquid chromatography after conversion to methylester, and in addition to the esters of veratric and isohemipinic acids, those of 4,5-dimethoxyphthalic ("metahemipinic") acid 46 and of dimeric acids 47 and 48 were identified and estimated quantitatively. The yields of esters obtained from methylated spruce MWL are indicated by numbers in parentheses below the corresponding structural formulae.

Veratric methyl ester 44b accounts for somewhat more than one-half of the total esters, in qualitative agreement with earlier discussed results of oxidation with Fremy's salt. Dimeric acid 47a obviously stems from 5-5-linked dilignol structures in which both phenolic hydroxyls originally have been free.

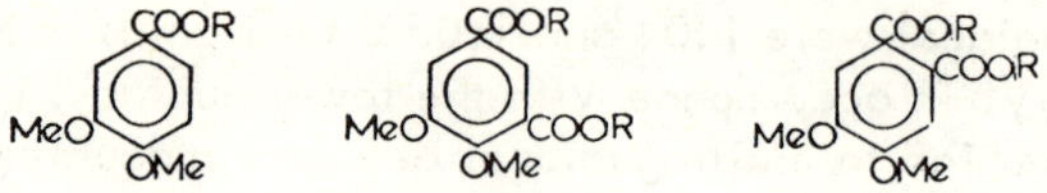

[Structures: 44a,b (7.7%); 45a,b (1.6%); 46a,b (0.8%); 47a,b (1.8%); 48a,b (1.2%)]

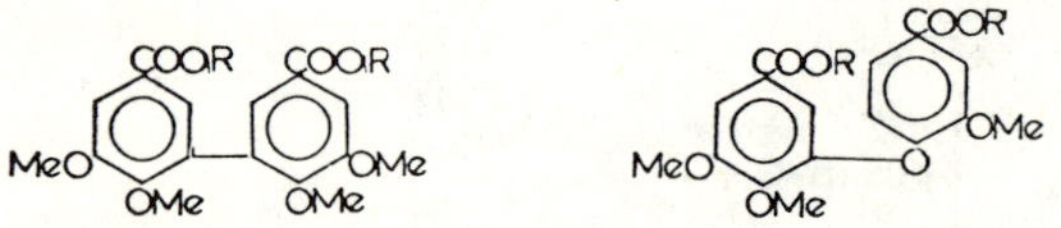

a: R=H
b: R=Me

Figure 5.8. Yields of carboxylic derivatives obtained from permanganate oxidation of methylated spruce MWL (210).

As its ester is obtained in a larger yield than dimeric ester 48b, o-arylphenol groups are probably more frequent than o-aryloxyphenol groups in the original lignin. The source of the 5- and 6-carboxyl groups is not clear in iso- 45a and metahimipinic 46a acids, respectively. The 6-carboxyl groups in the latter acid stems from a structure not directly derivable from the current ideas on dehydrogenative polymerization. On the other hand, the 5-carboxyl of isohemipinic acid could derive from a β-carbon of an "open phenylcoumarane structure" 49 which according to a claim by Adler and coworkers (211) may be present in 2 to 3% of units in spruce MWL.

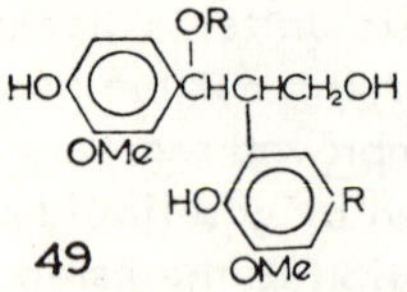

[Structure 49]

Alternative structures giving rise to the formation of isohemipinic acid include α-5 linked structures which can be formed through the rearrangment of original α-O-4 linkages (85), and 5-5 linked structures with a single free phenolic hydroxyl group. The isolation of isohemipinic acid gives, therefore, no direct support for the existence of open phenyl-coumaran structures; and the arguments in favor of such are, as a matter of fact, less convincing now than they were at the time of their presentation, as will be discussed later.

ISOLATION AND STRUCTURAL STUDIES

C. β-O-4 Lignol (Arylglycerol-β-Ether) Structures. The structures in lignins in which the side-chain is connected with the next unit through a β-O-4 linkage have been commonly called "arylglycerol-β-ether" structures. The presence of such structures was first postulated on the basis of ethanolysis experiments (see Chapter 9). Later, the formation of β-O-4 linkages in dehydrogenative polymerization was demonstrated by Freudenberg and coworkers (212), and finally Nimz succeeded in the isolation of a number of oligomeric hydrolysis products in which the original β-O-4 linkage had been preserved (174, 176, 194, 213). These products are summarized in Table 5.10.

Table 5.10. β-O-4 Lignols Isolated as Hydrolysis Products of Spruce Lignin by Nimz

Compound	Source	Appr. yield % of lignin	Ref. No.
(G)-p(β-O-4)(G)-CH=CH·CH$_2$OH	Spruce wood	0.1-0.2	175
(G)-p(β-O-4)(G)-CH=CH·CHO	Spruce MWL	0.035	177
(G)-p(β-O-4)(G)-CH-CH·CH$_2$OH 　　　　　　　　OH OH	Spruce wood	0.05	194
(G)-p(β-O-4)(G)-CHO	Spruce MWL	0.025	177
(G)-p(β-O-4)(G)-p(β-1)(G)	Spruce wood	0.3	213
(G)-p(β-O-4)(G)-p(β-O-4)(G)-p(β-1)(G)	Spruce wood	0.25	213

The β-O-4 linkages together with the α-O-4 bonds are amenable to hydrolysis reactions catalyzed by acids and bases, while most other bonds in lignin are largely unhydrolyzable. Consequently, the reactions of β-O-4 linked structures in acid- and base-catalyzed reactions, in sulfite and kraft processes and in hydrogenation are of utmost importance and are discussed in detail in the respective chapters. In this context, the extent of occurrence of β-O-4 linkages in lignins of various types, is of primary interest.

The β-O-4 linkages represent undoubtedly a major structural feature in all lignins, yet their exact determination is quite a difficult task. Most attempts in this direction have made use of either ethanolysis or acidolysis reactions (214) (see Chapter 9). These reactions convert a large part of β-O-4-linked structures __52__ to monomeric "Hibbert's ketones" __53__ (Equation 15).

$$-\text{(Ar)}\underset{\underset{\text{OR'}}{|}}{\overset{\overset{\text{H}_2\text{COH}}{|}}{\text{CHCH}}}-\text{O-(Ar)}-\quad\quad \text{Ar} = \text{guaiacyl, syringyl} \quad\quad R = C_2H_5 \text{ or } H$$
$$\mathbf{52} \quad\quad\quad \text{or } \underline{p}\text{-hydroxyphenyl}$$

$$\downarrow \quad\quad\quad\quad\quad\quad\quad\quad\quad\quad\quad\quad\quad\quad\quad\quad\quad\quad (15)$$

$$\underset{a}{\text{(Ar)COCHMe}} + \underset{b}{\text{(Ar)CHCOMe}} + \underset{c}{\text{(Ar)CH}_2\text{COMe}} + \underset{d}{\text{(Ar)COCOMe}} + \underset{\mathbf{54}}{\text{HO-(Ar)}-}$$
$$\underset{\text{OR}}{|}\quad\quad\quad \underset{\text{OR}}{|}$$
$$\mathbf{53}$$

Hibbert and coworkers made attempts to determine the total amounts of monomeric products ("distillable oils") obtained in the ethanolysis of various plant materials. Preliminary results (215) revealed that the yields were lowest (8 to 10% of Klason lignins) from conifer lignins, and higher but variable (15 to 43%) for angiosperm wood and grass lignins, and that Hibbert's ketones of the syringyl type were the predominant products. The yields even for a single species depend largely on the conditions of the reaction. For example, reported yields of "distillable oils" from maple wood lignin range from 19.2% (216) to 35.0% (215). When analogous procedures were applied to spruce and maple woods, the yields of monomeric oils were 8.3 and 25.4%, respectively, of corresponding Klason lignins, indicating a three-fold difference between the two wood lignins. More precise yield measurements have been carried out by Gardner (56) who obtained 10% ethanolysis monomers from spruce wood lignin. Adler and Lundquist (160) have applied acidolysis in aqueous dioxane to milled wood lignin from spruce, obtaining a 12% yield of identifiable monomers, isolated by means of gel chromatography. One-half of the monomers consisted of ω-hydroxyguaiacyl-acetone $\underline{55}$, the immediate hydrolysis product of a β-O-4-linked guaiacylpropane unit.

$$\text{HO}-\underset{\text{OMe}}{\bigcirc}-\text{CH}_2\text{COCH}_2\text{OH}$$
$$\mathbf{55}$$

During the ethanolysis or acidolysis procedure, some losses of Hibbert's ketones occur as a consequence of condensation reactions (216). Even accounting for these losses, the monomeric ketones can only be formed from such arylglycerol β-ether structures in which the Ar group is either an uncondensed unit with a free phenolic hydroxyl group or is linked through hydrolyzable α-O-4 or β-O-4 bonds with the rest of the lignin molecule. In all likelihood, however, the majority of arylglycerol-β-ether structures are linked through unhydrolyzable β-5, 5-5, 5-O-4 or 4-O-5 bonds. As a consequence, the majority of these structures are not converted to Hibbert's ketones; and it can be merely concluded, as Adler does (170), that the amount of β-O-4-

linked units in spruce milled wood lignin is probably about 25% or somewhat more. It should be noted that β-aryl ether structures of the syringyl type are more effectively converted to Hibbert's ketones, and renewed studies on angiosperm lignins appear desirable.

It has been shown by model compound experiments that all glycerol-β-ether side-chains are converted to ketone structures 53 in acidolysis with the exception of mono- and diketones 53c and d which are not formed from units with an ether group at position 4 (217). Equation 15 shows that one mole of terminal methyl groups frees one mole of phenolic hydroxyl groups and increases the amount of conjugated carbonyl groups. Determination of the changes in these functional groups can be used to estimate the total amount of β-O-4 linked structures. Adler, Pepper and Eriksoo (218) found that the increase in terminal methyl groups during the acidolysis of spruce milled wood lignin was equivalent to 25 to 33% of arylglycerol-β-ether groups. Schuerch and Sarkanen (219) studied the changes in the phenolic hydroxyl content of a spruce ethanolysis lignin subjected to further ethanolysis and found an increase from an initial value of 0.36/MeO to 0.62, corresponding to 26% of β-O-4-linked structures in the initial preparation.

A method to determine the relative amounts of guaiacyl, syringyl and p-hydroxyphenyl components of Hibbert's ketones has been developed by Kratzl (220). It involves the oxidation of the ketone mixtures to α-diketones 53d which can be separated from other components in the form of the nickel salts of their glyoximes. The regenerated three diketones can be analyzed by means of paper- (220) or gas-liquid (221) chromatography. The method is important in the characterization of plant lignins and several examples of the ratios of Hibbert's ketones have been quoted in Chapter 3.

Gierer (222) has shown that β-O-4 linked units with an ether group at position 4 (56) are hydrolyzed by alkali at elevated temperatures (2 N NaOH, 170°) to aryl glycerols 57. The amount of the latter can be estimated by means of periodate oxidation, as shown in Equation 16. Alkali-hydrolyzed spruce milled wood lignin was shown to contain 13% of released aryl glycerol units (222, 223, 224).

$$R'O{-}\phi(OMe){-}CH(OH){-}CH{-}O{-}\phi(OMe){-}R \xrightarrow{OH^{\ominus}} R'O{-}\phi(OMe){-}CH(OH){-}CHOH{-}CH_2OH \xrightarrow{IO_4^-} HCHO + HCOOH + R'O{-}\phi(OMe){-}CHO \quad (16)$$

$$\underset{56}{} \qquad \underset{57}{}$$

Phenolic arylglycerol-β-ether units 58 do not provide aryl glycerols in high-temperature alkaline hydrolysis but are converted to other products such as the alkali resistant vinyl ether structures 59 as shown by Gierer (225). On the other hand, Ashorn and Enkvist (226) and Ishiza et al (227) have demonstrated by means of model compounds that weakly alkaline hydrosulfide solution at 100° affords acetoguaiacone 60 (12 to 16%) propioguaiacone 61 (2 to 3%) and some coniferaldehyde 62 (226) (Equation 17). Applied to spruce wood (228, 229), this reaction produced acetoguaiacone in a yield corresponding to 6.3% of lignin, while propioguaiacone was obtained in 1.3% yield (226, 228). The isolated acetoguaiacone stems, in all likelihood, from such uncondensed arylglycerol-β-ether units in lignin that contain either a phenolic hydroxyl group or an easily hydrolyzable ether group at position 4 and include few, if any, such units that afford arylglycerol in high-temperature alkaline hydrolysis. Consequently, the sum of the yields of aryl glycerols and of aceto- and propio- guaiacones, 21%, points to approximately the same level of original aryl glycerol structures as estimated from ethanolysis and acidolysis reactions, if allowance is made for side-reactions and for some condensed structures that are not recovered as products of either reaction.

(17)

Nakano and coworkers (230) have compared yields of acetoguaiacone, applying the hydrosulfide reaction to various fir lignin preparations. The following yields were established: wood lignin 7.4%, Brauns lignin 1.5%, hydrochloric acid lignin 3.2% and sulfuric acid lignin 0.1%. Beech wood lignin afforded 1.1% acetoguaiacone and 1.8% acetosyringone.

D. β-1 Lignol Structures. Hydrolysis products containing β-1 lignol structures have been isolated by Nimz (213, 231, 232) from spruce and beech woods, and the currently available data are collected in Table 5.11. The results suggest that β-1-linked structures may be more common in angiosperm wood lignins and in these lignins, syringyl moieties are much more frequently associated with these structures than guaiacyl groups. A plausible explanation for this circumstance is obtained by considering the endwise polymerization (see Chapter 4) of sinapyl alcohol. The steric hindrance of the o-methoxyl groups probably retards the β-O-4 coupling mode; and β-1 coupling, with ensuing side-chain displacement, will therefore compete more efficiently with the former coupling mode than in the endwise polymerization of coniferyl alcohol.

ISOLATION AND STRUCTURAL STUDIES

Table 5.11. β-1 Lignols Isolated from Water Hydrolysis of Beech and Spruce Woods

Compound	Isolated from	Appr. yield % of lignin	Ref. No.
(G)-p(β-1)(G)	Beech wood	0.1	232
(G)-p(β-1)(S)	Beech wood	0.12	232
(S)-p(β-1)(S)	Beech wood	1.5	164
(G)-p(β-1)(G)-p(β-1)(G)	Spruce wood	0.3	213
(G)-p(β-O-4)(G)-p(β-O-4)(G)-p(β-1)(G)	Spruce wood	0.25	213

The presence of β-1 lignol structures in lignins is manifested in a number of dimeric products formed in various degradation reactions. The ready conversion of dilignol 63 to stilbene 64 has been demonstrated (170, 233) (Figure 5.9), and this compound has been obtained in alkaline hydrolysis of spruce lignin sulfonates (234) and from the reaction of red pine (235) and beech wood (236) with aqueous sodium hydroxide at 180°. For the quantitative determination of stilbene structures of this type, oxidation to stilbene quinone 65 (λ_{max} = 478 mμ) by cupric ion and hydrogen peroxide can be employed (237b). Divanilloyl 66, obtained by Pearl (237a) in the cupric oxide oxidation of lignin sulfonates, is probably a product formed from stilbene 64. Diguaiacyl-ethane 67 has been observed by Schuerch and coworkers to be one of the main components of the dimeric fraction obtained in the alkaline hydrogenation of spruce (238), while the corresponding dimers from maple contain also the syringyl analogue 68. Adler, Lundquist and Miksche (170, 233) have identified stilbene 64 together with dimeric ketones 69 and 70 as main components of the dimers formed in the acidolysis of spruce milled wood lignin, and traced their genesis to β-1 lignol structures (see Chapter 9).

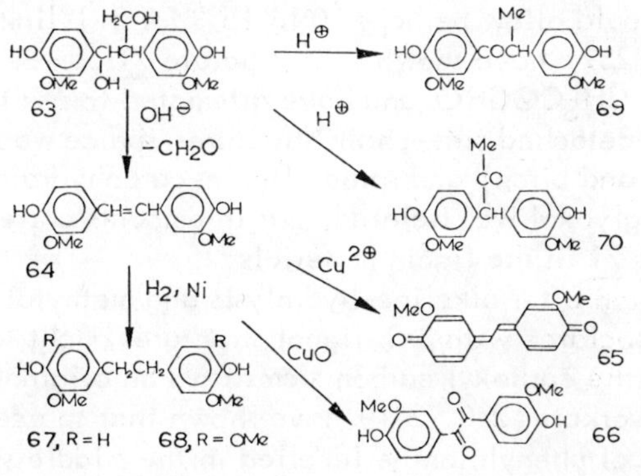

Figure 5.9. Dimeric products formed from β-1-lignol structures in acidolysis, alkaline oxidation and hydrogenation reactions.

The fact that β-1 lignol derivatives appear to be the principal components of acidolysis dimers derived from conifer lignins as well as the dominant dimers obtained from both conifer and hardwood lignins by alkaline hydrogenation, suggests that a quite significant number of monomeric units in guaiacyl lignins, and possibly even a larger number in guaiacyl-syringyl lignins, are associated with structures of this type. It may be recalled (Chapter 4) that the formation of β-1 linked structures in dehydrogenative polymerization probably occurs via side-chain displacement mechanism. As a consequence, an upper limit estimate for units associated with β-1 linked structures could be made on the basis of detached side-chain structures such as 71. The formation of these structures in dehydrogenative polymerization is recapitulated in Equation 18.

$$(18)$$

It is not excluded, however, that β-5- and β-β-dilignol structures could undergo analogous displacement reactions forming displaced side-chain structures 72a and 72b, respectively, characterized, again, by the presence of an aldehyde group. The unconjugated carbonyl groups which were estimated to be present in approximately 10% of the units in spruce milled wood lignin by Adler and Marton (187) could be in actuality the aldehyde groups in such structures 71, 72a and 72b. On this basis, the absolute maximum of all aromatic groups associated with structures formed through side-chain displacement would be 20%, which would allow perhaps 10 to 15% for β-1 linked structures. Lundquist et al. (239) have shown that structure 71 can be hydrolyzed to methyl glyoxal CH_3COCHO and have attempted to use this reaction for the estimation of detached side-chain structures spruce wood milled wood lignins of spruce and birch, and spruce lignin-carbohydrate complexes. In all cases, methyl glyoxal was isolated, but in amounts corresponding only to 0.3% of structures 71 in the lignin materials.

In permanganate oxidation after alkaline-hydrolysis and methylation, one of the aromatic groups associated with β-1 lignol structures ought to give rise to veratric acid in which the carboxyl carbon stems from an original β-carbon. Freudenberg and coworkers (240, 241) have shown that spruce lignin formed *in vivo* after injection of phenylalanine labelled in the middle side-

ISOLATION AND STRUCTURAL STUDIES

chain carbon with ^{14}C yields veratric acid which possesses only 1% of the radioactivity of the total lignin. This result again discourages the idea of extensive β-1 linking in lignins.

In addition to β-1 lignol groups, other structures are known in lignins, the formation of which is best envisaged to occur—at least at the current stage of knowledge—in coupling processes involving side-chain displacements of β-O-4 lignol structures. The nature of these groups is discussed below.

E. 4-O-1 and 5-1 Linked Structures. Linkages of the 4-O-1 and 5-1 types have not, as yet, been encountered among the identified hydrolysis products of lignins. This may not be surprising because they have probably been formed rather late in the in vivo polymerization process and may be located in such sites of the lignin molecule which are not easily degradable to simple products. The presence of 4-O-1-linked structures was demonstrated by Freudenberg and coworkers (175, 242), who isolated acids 73 and 74 together with 1,2,4-trimethoxybenzene 75 by the permanganate degradation method of spruce lignin. The formation of these products is illustrated in Equation 19.

Adler and coworkers (243) have recently identified acid 76 amongst the permanganate oxidation products of methylated spruce milled wood lignin. The postulated formation of this acid from 5-1 linked structures is shown in Equation 20.

F. β-5 Lignol Structures. Only two hydrolysis products containing β-5 structures have been identified. Dehydrodiconiferyl alcohol 23 was isolated by mild hydrolysis of spruce milled wood lignin by Freudenberg and coworkers (174, 175), and Lundquist (160) has obtained dimer 77 as an acidolysis product from the same material. The ketol side-chain of the latter suggests trimeric structure 78 as the probable source.

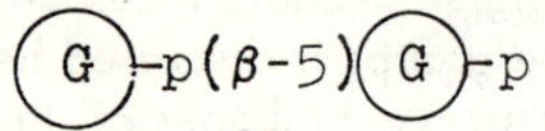

$$\boxed{G}-p(\beta-5)\boxed{G}-p$$
23

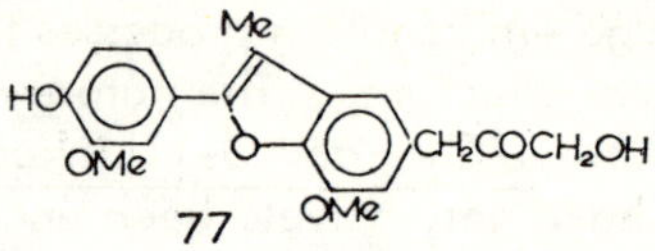

[Structure 77: HO-phenyl(OMe)-O-furan(Me)-phenyl(OMe)-CH$_2$COCH$_2$OH]

$$\boxed{G}-p(\beta-5)\boxed{G}-p(\beta-O-4)\boxed{G}-p$$
78

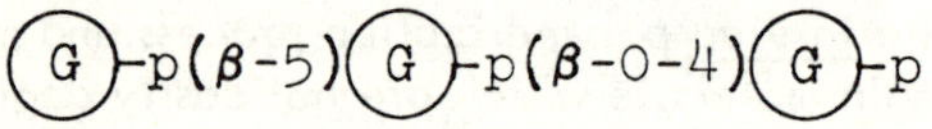

The presence of β-5 lignol structures (phenyl coumaran structures) is manifested in a number of isolated degradation products (Figure 5.10).

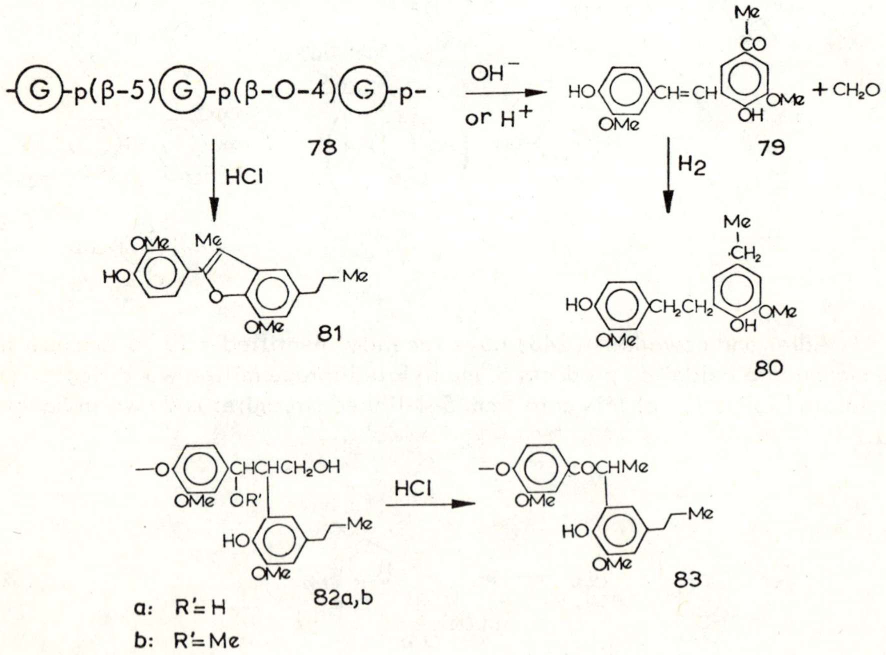

$$-\boxed{G}-p(\beta-5)\boxed{G}-p(\beta-O-4)\boxed{G}-p- \quad \xrightarrow[\text{or H}^+]{\text{OH}^-} \quad \text{HO-phenyl(OMe)-CH=CH-phenyl(OMe)(OH)-CO-Me} + CH_2O$$

78 → 79

↓ HCl ↓ H$_2$

81: HO-phenyl(OMe)(OMe)-furan(Me)-phenyl(OMe)-Me

80: HO-phenyl(OMe)-CH$_2$CH$_2$-phenyl(OMe)(OH)-CH$_2$-Me

82a,b: -O-phenyl(OMe)(OR')-CHCHCH$_2$OH ... HO-phenyl(OMe)-Me   →HCl→   83: -O-phenyl(OMe)-COCHMe ... HO-phenyl(OMe)-Me

a: R'= H
b: R'= Me

Figure 5.10. Dimeric products formed from β-5 lignol structures in acidolysis, alkaline hydrolysis and hydrogenation reaction.

ISOLATION AND STRUCTURAL STUDIES   213

A structure such as 78 in the original lignin can be expected to convert to stilbene 79 in high-temperature alkaline hydrolysis. p-, o- Stilbene structures of the type 79 have been identified in spruce kraft lignin on the basis of their light-absorption properties in alkaline solution (ε_{max} at 378 mμ) by Falkehag, Marton and Adler (244). The amount of units participating in p-, o'- stilbene structures was estimated to be approximately 7%. Schuerch and coworkers (238) have isolated from alkaline hydrogenation products of spruce the diarylethane 80, an obvious hydrogenation product of stilbene 79. Adler and coworkers (211, 245) have shown that dehydrodiconiferyl alcohol and its methyl ether are converted in acidolysis to phenyl coumaranes of the type 81 in approximately 75% yield, and, to some extent, to p-, o'- dihydroxystilbenes. The UV spectrum of the former structure is applicable for quantitative measurements, and such estimations of phenylcoumarane groups were carried out using acidolyzed spruce milled wood lignin. On the basis of these data, 11% of the units in the original lignin were estimated to possess β-5-linked side-chains, or, in other words, 22% of the units were participants of phenylcoumarane systems. Diazomethane methylation prior to acidolysis lowered the estimate from 11 to 8%. To explain this reduction, Adler proposed the existence of hypothetical "open" phenylcoumarane systems 82a which, after methylation, would convert to ketones 83 instead of phenylcoumaranes. However, this is an uncertain interpretation in the absence of independent confirmation of the presence of open phenylcoumaran structures. Furthermore, p-, p'- dihydroxystilbenes formed from β-1 lignol structures probably cause a small positive error in the determination of the total number of β-5-linked structures spectrophotometrically.

Permanganate oxidation of hydrolyzed and methylated spruce lignin yields isohemipinic acid 45a which has been assumed to derive almost exclusively from β-5 lignol units. Freudenberg and coworkers (175) have recently demonstrated that not more than one-half of isohemipinic acid stems from such structures. When this acid is isolated from lignin made radioactive by the administration of β-labelled coniferin, the measured radioactivity is only one-half of the radioactivity of the total lignin (175). On this basis, the yield of isohemipinic acid formed from phenylcoumaran structures in spruce MWL amounts only to 3% of the lignin. This again suggests only a minor occurrence of phenylcoumarane structures in milled wood lignin, but it should not be forgotten that the non-MWL part of lignin has not been explored to date and may possess structural characteristics different from milled wood lignin.

G. β-β Lignol Structures. D, L-Pinoresinol 84 and DL-Syringaresinol 85 are the only hydrolysis products obtained so far that contain β-β linkages. The former was obtained from mild hydrolysis of milled wood lignin (175), and the latter from the hydrolysis of beechwood by 2% acetic acid (175) or water (164), in a maximum yield of 1% of lignin (246). The isolation of pinoresinol has recently been reported from hydrolysates of spruce wood and milled wood lignin at pH 12, or in 1 N sodium hydroxide (247).

Freudenberg (248) has proposed that two other β-β lignol structures might be present in spruce lignin; namely, lignonolide 86 and pinoresinolide 87. So far, these compounds have been identified as trace intermediates in the dehydrogenative polymerization of coniferyl alcohol (156, 249), but not in lignins.

β-β-lignol elements in lignins are sensitive to acid treatments, condensing to derivatives of the type 88 (250, 251, 252). These condensation processes are revealed by the direct oxidation of acid-treated lignins with permanganate to benzenepentacarboxylic acid 89 in 0.6 to 2.0% yield (99), while only trace amounts (0.14%) are obtained from wood lignin. Brunow (250, 251) has demonstrated this reaction for pinoresinol 84 (R-H) which after treatment with 72% sulfuric acid yields 3.4% benzenepentacarboxylic acid and 0.5% of benzenehexacarboxylic acid (250). Pinoresinol has also been converted to aromatized dicarboxylic acid (94) which yields 52 to 73% of benzenepentacarboxylic acid upon oxidation (251). In contrast, conidendrin 95 gives only traces of this acid in permanganate oxidation, Figure 5.11. This observation does not support an earlier idea (54) that hydroxymatairesinol impurities in lignin, that are converted to conidendrin in acidic treatments, might form the source of benzenepentacarboxylic acid. Rather, this function must be ascribed to pinoresinol elements and possible other β-β-linked structures in lignin.

The yield of m-hemipinic acid 90, from hydrolyzed and methylated spruce milled wood lignin is approximately 1.1% in permanganate oxidation, and increases to 1.8% if an acidolysis treatment precedes the alkaline hydrolysis (239). The increase in m-hemipinic acid yields in exposure to acidic reagents was first observed by Richtzenhain (253) and appears to be general. Again, the condensation products of β-β lignol elements form a probable source of this acid. The syringyl analogue 91 has also been isolated (254).

Erdtman and Gripenberg discovered in 1947 (255) that nitration of pinoresinol leads to the formation of dilactone 92, through dual side-chain displacement by nitronium ion followed by oxidation. This reaction appeared to be ideal for the specific identification of pinoresinol units in lignins. The initial attempts to identify dilactone 92 among the nitration products of lignins

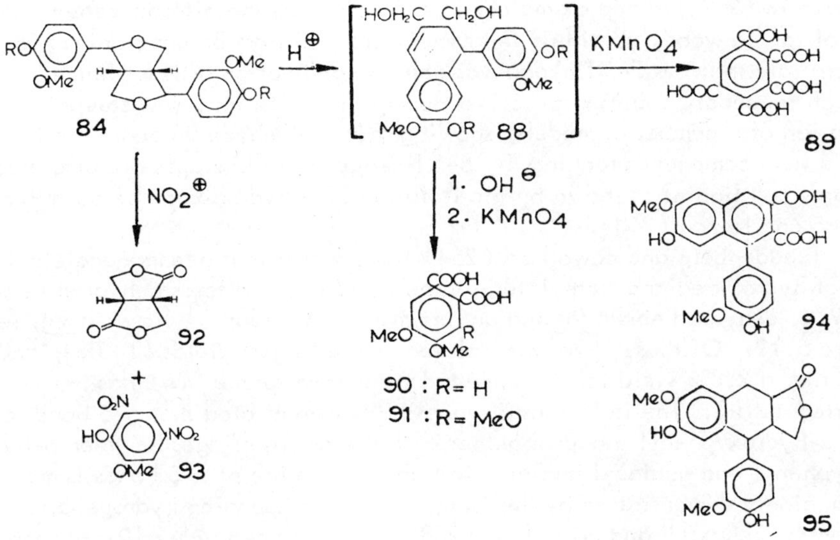

ISOLATION AND STRUCTURAL STUDIES 215

Figure 5.11. Products formed from pinoresinol in the oxidation and condensation reactions.

were not successful, however, and it was not until recently that Ogiyama and Kondo (256) were able to isolate crystalline 92, together with 4,6-dinitroguaiacol, from the nitrations of cryptomeria, red pine and white birch milled wood lignins, as well as of several other lignin preparations. Although the yields obtained (approximately 0.05%) were of low order of magnitude, they definitely established the presence of pinoresinol and syringaresinol units in these lignins.

Presently, no methods are known for the quantitative estimation of pino- and syringaresinol elements. Indirectly, Freudenberg (257, 258) has concluded from the estimated amounts of dialkyl ether linkages an approximate level of 10 to 12% of pinoresinol elements in spruce milled wood lignins. This estimate appears quite reasonable because the dehydrogenative dimerization of coniferyl alcohol produces twice as much dehydrodiconiferyl alcohol as pinoresinol, and the units associated with structural elements of the former type make up approximately 22% of spruce milled wood lignin. By the same token, relatively larger amounts of syringaresinol units may be anticipated in hardwood lignins on the basis of the 80% conversion of sinapyl alcohol to syringaresinol in dehydrogenative dimerization (cf. Chapter 4).

H. **5-5 and 5-O-4 Lignol Structures.** The presence of 5-5 linkages in conifer lignin was first proposed by Pew (57), who isolated dehydrodivanillin 96 (259) in approximately 2.2% yield from the nitrobenzene oxidation of spruce wood, obtaining the same yield also from Brauns lignin. To demonstrate that the 5-5 linkage was genuine and not an artifact formed through secondary coupling processes during the oxidation, Pew studied the oxidation of a number of models, showing that dialdehyde 96 could not be obtained from compounds lacking the 5-5 linkage. 5-5 Linkages are also present in hardwood lignins as shown by the isolation of dehydrodivanillic acid 97a by Tanaka and Kondo (260).

Freudenberg and coworkers (254) have shown that permanganate oxidation of hydrolyzed and methylated spruce wood lignin affords dehydrodiveratric 97a, dehydrodianisic 98 and an unsymmetrical dicarboxylic acid 99, see Figure 5.12. Of these, 97b was isolated in the largest yields (1.3%); and recently, a 2.8% yield for the dimethyl ester from spruce MWL has been reported (210). The isolations of 98 and 99 demonstrated that 5-5 bonds connect p-hydroxyphenyl groups together and also act as bridges between p-hydroxyphenyl and guaiacyl nuclei. The genuine nature of 5-5 bonds is most convincingly demonstrated by the isolation of corresponding hydrogenation products. Schuerch and coworkers (238) have obtained dimer 101, a probable product of tetrameric lignin structure 100, from the alkaline hydrogenation of spruce. An analogous phenylpropane dimer 102 was obtained by Nahum (261) from a hydrogenation utilizing dicobalt octacarbonyl as catalyst.

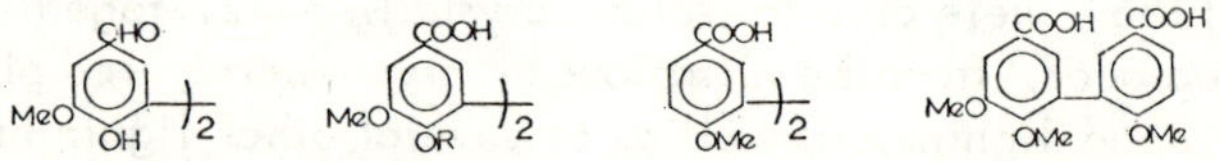

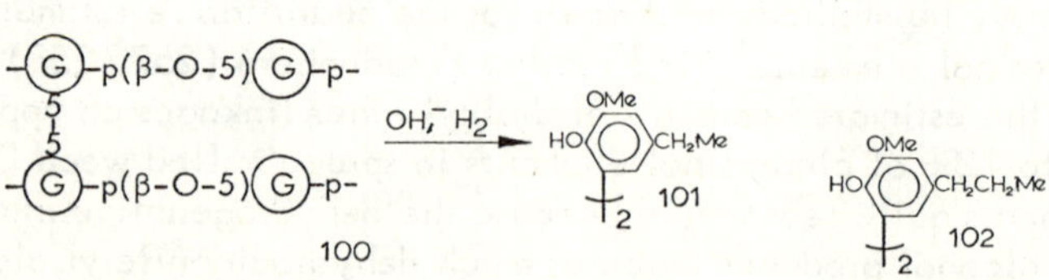

Figure 5.12. Dimeric products formed from 5-5-lignol structures in the oxidation of wood.

ISOLATION AND STRUCTURAL STUDIES

Although 5-5 bonds must be relatively common in conifer lignins, the estimation of their quantity is a difficult task. Aulin-Erdtman has found evidence for their existence in various soluble lignin preparations from spectral studies in the UV region (262), and Pew (263) has attempted to estimate the quantitative amount of 5-5 linked structures from the absorption maximum at 258 mµ, where structures of other types display a minimum. An approximate value of 25% for units involved in 5-5 linked structures was arrived at for spruce milled wood lignin, but must be considered to have tentative significance only, in view of the uncertainties involved in the calculation.

The evidence for the presence of linkages of the 4-O-5 type in lignin has been obtained exclusively from permanganate oxidation products. The dicarboxylic acid 103 was isolated from spruce milled wood lignin in approximately 1% yield by Freudenberg and Chen (242); and by an improved oxidation process, Larsson and Miksche were able to increase the yield of its dimethyl ester to 2.0% (210). Linkages of this type may therefore be almost as frequent as 5-5 bonds in conifer lignins. This conclusion is supported by experience with dehydrogenative polymerization of 4-propylphenols by Pew, discussed in detail in the previous chapter. It should be noted that the 3,4,5-trimethoxybenzoic acid 104 isolated from spruce lignins is not necessarily formed from syringyl moieties alone, but may also be formed from 5-O-4 linked guaiacyl units (Equation 21). Permanganate oxidation of hydrolyzed and methylated birch lignin affords the guaiacyl-syringyl dicarboxylic acid 105.

$$-p(G)-(5-O-4)(G)-p \xrightarrow{\text{Permang. method}} 103 + 104 \qquad (21)$$

105

I. **Minor Linkage Types in Conifer Lignins.** Freudenberg (254) has shown that permanganate oxidation products of spruce lignin contain minor amounts of acids 106, 107 and 108, which are all products of p-hydroxyphenylpropane units. The nature of these acids suggests a similar bonding pattern for methoxyl-free units as for the dominant guaiacyl propane elements, with the exception that two sites at the aromatic nucleus may occasionally be carbon-bonded, as indicated by the structure of the tricarboxylic acid 108.

The structure of 2,3,4-trimethoxybenzoic acid 109 suggests that the propyl side-chain has been located ortho to the carboxylic acid group, but has become eliminated in the oxidation. Otherwise, it is not clear whether this acid originated from syringyl or guaiacyl units. A puzzling 5-6 bond is present in the dicarboxylic acid 110 which has been obtained in trace quantities only (242). The most probable explanation for the generation of such abnormal linkages is provided by a dienone-phenol rearrangement (264, 265) of a 5-1 adduct during the formative polymerization of lignin, as illustrated in Equation 22.

$$(22)$$

Evidence for the presence of abnormal β-6 linkages has been obtained from tracer studies conducted by Freudenberg and coworkers (266). It was established that in vivo administration of β-labelled coniferin produced spruce lignin from which a partially radioactive (12% of the radioactivity level of lignin) m-hemipinic acid was isolated. Again, the formation of β-6 linkages is best interpreted in terms of a dienone-phenol rearrangement of an initial β-1 adduct.

J. Possible Covalent Bonds Between Lignin and Polysaccharides. It has not been possible, so far, to isolate such degradation products of original plant lignins in which chemical bonds between lignin and polysaccharides have been preserved in a clearly identifiable form. For this reason, the existence or non-existence of covalent bonds between these cell wall constituents has remained a matter of perennial debate; and there are still those who believe that the evidence for such bonds is not rigorous enough for the acceptance of their existence as a proven fact. On the other hand, circumstantial as the pieces of evidence for the occurrence of such bonds in the cell wall may seem as taken alone, the total weight of data certainly makes it very difficult to provide a reasonable picture of all pertinent observations without resorting to the hypothesis of occasional chemical bridges linking polysaccharides, notably amorphous hemicelluloses, with the interpenetrating lignin component

ISOLATION AND STRUCTURAL STUDIES

in the native cell-wall structure.

The feasibility of formation of lignin carbohydrate bonds in the dehydrogenation of coniferyl alcohol has been demonstrated by Freudenberg, Grion and Harkin (267, 268), who carried out the reaction in a concentrated aqueous sucrose solution. It was possible to isolate a product consisting of a β-O-4 dilignol in combination with one mole of sucrose, with a probable structure 112. This product could be formed through the addition of an undetermined hydroxyl group in sucrose to a quinone methide intermediate, as shown in Equation 23. Another product consisting of one coniferyl alcohol in combination with two sucrose molecules was also isolated, and a radical coupling between coniferyl alcohol and sucrose must be invoked to account for its formation. This observation would suggest that radical exchange reactions may occasionally generate carbohydrate radicals that could couple with lignin. The tentative formulation 111 depicts oxygen coupling at position β, but carbon-to-carbon coupling would also appear to be a feasible alternative. Obviously, there should be no limitation for the carbohydrate radicals to couple at certain other positions in the lignin aggregates, at position 5, for example, if they are indeed formed.

(23)

The experiment by Freudenberg and Harkin brings into consideration three potential types of lignin-polysaccharide linkage:

   i.   α-Ether linkages of the type 113 which could be formed by the addition of hydroxyl groups in various monosaccharide units in hemicelluloses to quinone methide moieties. Such linkages ought to be hydrolyzable by acids, but more resistant towards basic hydrolysis.

   ii. Ester linkages to glucuronic acid residues 114 may be anticipated to exist particularly in angiosperm woods rich in glucuronoxylans, since Freudenberg (269) has shown that carboxylic acids show a greater tendency towards addition to quinone methides than neutral carbohydrates. These linkages ought to be easily hydrolyzed by mild alkaline treatments.

   iii. Linkages formed through radical coupling processes may be both carbon-to-carbon and carbon-to-oxygen type, and would probably resist hydrolysis more effectively than the two previous types.

A fourth possibility, that of phenyl glycosidic bonds, also deserves consideration. The participation of arylglycosides in the in vivo lignification is is perhaps not very likely, but glycosidation of deposited lignin is not excluded in view of the presence of appropriate enzyme systems in the cell wall (270, 271).

K. <u>Isolated Lignin-Carbohydrate Complexes</u>. As pointed out earlier in this chapter, many lignin preparations such as milled wood lignins, enzyme lignins, Brauns lignins and others generally contain some carbohydrate constituents not removable by solvent fractionation, suggesting the presence of complexes of high lignin to carbohydrate ratio. It was also shown that at a certain stage of milling as much as thirty per cent of lignin in milled conifer wood may be obtained as lignin-carbohydrate complexes by extraction with dimethyl formamide. Björkman (31) has used also other solvents, such as dimethyl sulfoxide and 50% acetic acid, for their isolation; and Pew has shown (13) that the methyl cellosolve extract of conifer milled wood may contain as much as 31% of carbohydrates. Grohn and Schierbaum (272) found that approximately one-third of milled wood may dissolve in cold water and the determined 10% lignin content for the extract is suggestive of water-dispersible lignin-carbohydrate complexes. Similar water-soluble complexes were apparently isolated by Brauns and Seiler (26) in extenuated beating of spruce wood in a Valley pulp beater. Some lignin-carbohydrate complexes may be obtained from grass species without prior milling by extraction with 50% acetic acid for several days at ambient temperature, as shown by Hayashi and Tachi (273).

In most cases the extracts obtained are mixtures of lignin-carbohydrate complexes with either polysaccharidic or lignin components. Lindgren (274) has shown that crude lignin-carbohydrate complexes extracted from milled spruce wood could be divided into two fractions by means of electrophoresis. The fraction with slower mobility consisted of carbohydrates alone, while the faster moving fraction contained lignin and carbohydrates in about equal amounts, and mannose, glucose and xylose could be identified amongst the hydrolysis products of the latter. Consequently, the lignin contents of crude complexes obtained by Björkman (31) by extractions of spruce milled wood with dimethylformamide, dimethyl sulfoxide and 50% acetic acid, which range from 20 to 33%, must be lower than those of the actual lignin-carbohydrate complexes contained in these extracts. Brownell's partition experiments (15) suggest that a whole spectrum of complexes with varying lignin-to-carbohydrate ratios are present in these products.

The electrophoretic mobility of lignin-carbohydrate complexes is less than that of milled wood lignin, suggesting a higher molecular size for the former (274). This conclusion is supported by the liberation of milled wood lignin with normal electrophoretic mobility in the milling of lignin-carbohydrate complexes.

Lignin-carbohydrate complexes isolated from milled wood have not been the subject of sufficient systematic studies to clarify the nature of chemical bonding between the lignin and carbohydrate components. Brownell has observed that while aqueous sodium- and ammonium hydroxide solutions at ambient temperature do not hydrolyze these bonds, 1% sodium hydroxide at 100° releases the lignin component, as shown by the partitioning of the product between the phases in an aqueous sodium thiocyanatebenzyl alcohol-dimethylformamide system (15). On the other hand, Pew (13) has obtained evidence supporting the presence of highly resistant lignin-carbohydrate bonds in milled spruce and aspen woods. He has shown that enzymatic removal of polysaccharides from milled woods by means of hydrolysis with purified glycosidases stops totally when 95% of the carbohydrates have been removed. The remaining 5% appears to be firmly anchored to the remaining lignin, because solvent fractionations fail to reduce the carbohydrate content, and sugars are released only slowly by hydrolysis with boiling 4% sulfuric acid. The resistant carbohydrates of spruce enzyme lignin are composed of glucose (42%), mannose (23%), xylose (19%), galactose (13%) and arabinose (3%).

Various acid-catalyzed hydrolysis reactions have been repeatedly observed to yield products from which sugars can be released by further acidic hydrolysis. Traynard (275) has obtained from the water-hydrolysis of poplar wood a lignin-carbohydrate complex soluble in 96% acetone which was decomposed to lignin and reducing sugars on hydrolysis. Kawamura and Higuchi (276) have demonstrated the presence of xylose in the dioxane-acidolysis lignin of beech. The xylose moiety appears to be bound glycosidically to lignin, because it can be released enzymatically using diastase and hydrolysis after methylation yields 2,3,4-tri-O-methylxylose as the sole carbohydrate product. Aryl-glycosidic bonds have also been claimed to be present in lignin-carbohydrate complexes isolated from wheat straw by Hayashi (277). These preparations were found to release simultaneously new phenolic hydroxyl groups and reducing sugars upon hydrolysis with β-glucosidase. In contrast, both xylose and xylobiose were obtained by Kawamura and Higuchi in the hydrolysis of sugar-free acetolysis products of beech wood (278), and from the hydrolysis of similar products from wheat by Hayashi and Tachi (279). In the former case, hydrolysis after methylation yielded both 2,3,4-tri-O-methylxylose and 2,3-di-O-methylxylose.

These observations support the idea of a chemical combination between lignin and 4-O-methylglucuroxylans in angiosperms. Further support to this idea has been provided by Stewart and McPherson (280), who claimed the isolation of a component containing both lignin and uronic acids from the products of methanol extraction of Eucalyptus regnans at 150°; and by Aaltio and Roschier (281), who carried out successive extractions of aspen wood at 158° with aqueous butanol buffered to pH 7.0 and observed that the ratio of lignin to pentosan dissolved remained approximately constant ($\sim$ 0.85) for several successive extractions. Kawamura and Higuchi have also isolated apparent lignin-xylan complexes from acetylated kraft pulps of beech (282) and from the acetylated holocellulose preparation of the same species (283).

The presence of combined carbohydrates in lignin sulfonates has also been well documented. Richtzenhain (284) subjected spruce wood to neutral sulfonation at pH 4 to 6 at 135°, which leads only to a partial dissolution of the lignin sulfonates formed (see Chapter 15). Polysaccharide components were extracted from the sulfonated wood with cuprammonium solution in which 37% turned out to be insoluble. The residue was probably a lignin-carbohydrated complex because acid hydrolysis converted it to soluble lignin sulfonates and sugars, among which glucose and mannose were identified. Lignin sulfonates that dissolved in cuprammonium were also associated with hydrolyzable sugars. Pearl and Beyer (285) carried out elaborate fractionations of ammonium lignin sulfonates of aspen by means of ion exchange resin columns and countercurrent distribution in a butanol-ethanol-water system and found that all isolated fractions yielded sugars upon acid hydrolysis. Extension of these studies to other lignin sulfonates from softwood and hardwood species gave the same result (286).

Further information on hemicellulose lignin complexes has been obtained from studies on materials in which the lignin component has been degraded by chlorination or by treatment with chlorite. Traynard (275) has shown that acetone- and ethylacetate-soluble chlorolignins from poplar wood yielded identifiable amounts of xylose, arabinose and galactose. Kringstad and Ellefsen (287) have studied the water-soluble polysaccharides obtained in the chlorite treatment of spruce wood which consist mainly of galactoglucomannans. It was shown by sephadex fractionation that lignin and polysaccharide components were contained in the same molecular weight fractions and were therefore quite probably chemically combined. Kringstad has extended these studies to spruce wood delignified by treatments with chlorine and ethanolamine (288). Soluble complexes of glucomannan with 12% lignin and 4-O-methyl-glucuroxylan with 4% lignin were isolated. In these complexes the lignin component is present as branches (a in Figure 5.13) rather than

ISOLATION AND STRUCTURAL STUDIES 223

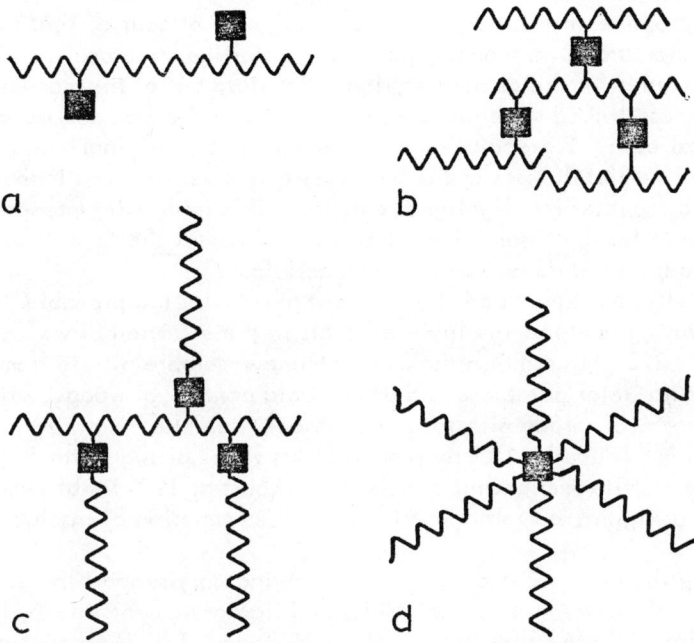

Figure 5.13. Possible structures for lignin-hemicellulose complex isolated from wood delignified with chlorite.

cross-links because further delignification caused only a minor shift in the molecular weight distribution curve. Linnell and Swenson (289) and Linnell, Thompson and Swenson (290) have recently carried out extensive studies on glucomannan-lignin complexes, obtained by alkali extraction of incompletely delignified chlorite holocellulose from black spruce. These preparations were acetylated and divided by fractional precipitation into fractions with $\overline{DP}_w$ values ranging from 82 to 557. It was shown by viscometric and sedimentation equilibrium experiments that the hydrodynamic properties of these fractions conformed with branched structures. For the quantitative evaluation of results, the treatment developed by Kilb (291) for a tetrafunctionally cross-linked model (b in Figure 5.13) was selected. On this basis, the number of cross-links per molecule was found to increase from 0.1 for the lowest molecular weight fraction ($\overline{DP}_w = 82$) to 7.1 for the highest one ($\overline{DP}_w = 557$). Consequently, only the lowest molecular weight fraction was free of cross-links, corresponding to structure a, while the other fractions conformed with cross-linked structures of the type b. The calculated values for the primary glucomannan chains were remarkably constant, with an average DP of 77, while the lignin branches and cross-links appeared to contain approximately two phenylpropane units only. It could be argued, of course, that the application of other structural models, notably the starlike structure d, would give higher DP-values for the associated lignin entities; but at the moment, at least, the cross-linked structure b appears to be in the best accord with the experimental data. Extrapolation of this model to the original wood structure suggests that hemicelluloses in the S_2-layer may exist in a continuous network cross-linked together by oligolignol entities. This interesting proposition would account for the insolubility of hemicelluloses in situ as well as for the often contradictory data on isolated hemicelluloses.

Cross-linked lignin-carbohydrate complexes are the probable components generating acetone-insoluble products in the nitration of wood. Timell and Snyder (292) have shown that the acetone-insoluble nitrate from balsam fir contains the total galactose and uronic acid content of wood. Abadie has shown that 59% of spruce nitrates are acetone-insoluble. However, their amount can be reduced to 4% by prehydrolysis with boiling water for 50 days. Mild neutral sulfite semichemical pulping (6 hours at 120°) that removed only one-fourth of lignin, completely eliminated the formation of insoluble nitrates (293, 294).

While the accumulated data rather convincingly support the existence of chemical bonds between polysaccharidic and lignin elements in woody tissues, the character of these bonds is not well understood. This is perhaps not surprising taking into account that in many isolated complexes, bonds between lignin and carbohydrates may not exceed 1 to 2% of the total bonds between monomers. A large part of these bonds appears to be hydrolyzable by acids or bases or both. Benzyl ether, benzyl ester and glycosidic bonds belong to this category, and the existence of the last-mentioned bonds is quite probable in both angiosperm woods and grasses. In addition, acetal bonds have been brought into consideration repeatedly (295, 296), but the arguments in favor of such bonds have so far fallen short of being very convincing. It would seem that the clarification of this problem, which is no doubt of great importance for the development of correct ideas of the nature of lignin in situ, is not at all impossible by systematic application of modern tools of molecular fractionation, enzymology and reaction kinetics, and it is to be hoped that new information will develop in this important area in the future.

L. Structural Implications of Compositions and Functional Group Analysis of Milled Wood Lignins. To relate the composition of lignins to the dehydrogenative processes involved in their formation, Freudenberg (297) has developed a system for the estimation of the average degree of dehydrogenation per aryl propane unit and for that of the number of added water molecules.

To carry out such estimations, it is convenient to replace the methoxyl groups in the C_9-formulae by hydrogen atoms as shown in Table 5.12. In this fashion, a C_9-formula of an equivalent p-hydroxyphenyl lignin is obtained.

ISOLATION AND STRUCTURAL STUDIES

The comparison of such a formula with that of p-coumaryl alcohol suggests first that any oxygen content beyond 2.0 atoms has been introduced during the dehydrogenative polymerization in the form of water added to quinonemethide intermediates. Subtracting this water from the C_9-formula leaves a hydrogen content that is lower than that of p-coumaryl alcohol, the difference representing an estimate for the degree of dehydrogenation. The latter is due to hydrogen loss in both coupling and oxidation processes. For example, dehydrogenative dimerization would give a degree of dehydrogenation of 1.0 and further coupling to infinitely long linear polymers, 2.0. On the other hand, "oxidized" structures such as coniferyl aldehyde or units with α-carbonyl groups possess a degree of dehydrogenation (D.D.) that is equivalent to 2.0, and thus make a separate contribution to the average D.D. value.

The average degree of dehydrogenation could be used as an important characteristic parameter of lignins, were it possible to determine it in a reliable fashion. Unfortunately, it must be computed as a difference between two large numbers, and inaccuracies in hydrogen analysis as well as the presence of associated carbohydrates are bound to cause large errors in these values. The carefully determined D.D.-value of Freudenberg for spruce milled wood lignin is 2.0, which is of reasonable magnitude, but computation based on the analytical values of Björkman (11) for a similar preparation gives the clearly too-low value of 0.95. Similar low D.D.-values have also been obtained for milled wood lignins of beech and birch. Freudenberg has attributed the error to associated carbohydrates, and has removed these by converting the MWL preparations of several species to their Klason lignins, obtaining D.D.-values ranging from 1.8 to 2.1. In theory, at least, this is a legitimate approach because condensations occurring in the preparation of Klason lignins ought to be without effect upon the D.D.-values. Still, the level of accuracy is probably not sufficient for a realistic comparison of the degree of dehydrogenation in various lignins.

The estimates for the amount of added water appear, in contrast, to be more reliable; and the values for angiosperm wood lignins are clearly higher than those of gymnosperm preparations. Probably β-O-4 and β-1 coupling modes that result in the formation of quinone methides capable of water addition are more important in the formation of hardwood lignins. Absorbed water could cause some positive error in these values, but the preparations for which the analysis is reported in Table 5.12 were meticulously dried. An independent estimate of added water may be made by summing up structural units containing α-hydroxyl groups, α-carbonyl groups and unconjugated carbonyl groups, which in all likelihood represent α-aldehyde groups in displaced sidechain structures. The summation below conforms indeed well with the idea of approximately 0.40 added water molecules.

Fractions of units containing:

α-Hydroxyl groups, according to Adler (298)   0.20

α-Carbonyl groups, according to Adler and Marton (188)   0.07

Unconjugated carbonyls, according to Adler and Marton (188)   0.10 to 0.13

Total: 0.37 to 0.40

Table 5.12. Analytic Comparisons of Milled Wood Lignins

Species	C_9-formula	Methoxyl-free formula	Degree of dehydrogenation	Moles of added H_2O	Ref. No.
p-hydroxycinnamyl alcohol	$C_9H_{10}O_2$	$C_9H_{10}O_2$	0	0	
Spruce	$C_9H_{7.92}O_{2.40}(OMe)_{0.92}$	$C_9H_{8.04}O_2(H_2O)_{0.40}$	1.96	0.40	301
	$C_9H_{8.83}O_{2.37}(OMe)_{0.96}$	$C_9H_{9.05}O_2(H_2O)_{0.37}$	0.95	0.37	11
Beech	$C_9H_{8.50}O_{2.86}(OMe)_{1.43}$	$C_9H_{8.21}O_2(H_2O)_{0.86}$	0.93	0.86	11
	$C_9H_{7.59}O_{2.53}(OMe)_{1.39}$	$C_9H_{7.92}O_2(H_2O)_{0.53}$	2.18	0.53	301
Birch	$C_9H_{9.03}O_{2.77}(OMe)_{1.58}$	$C_9H_{9.07}O_2(H_2O)_{0.77}$	0.93	0.77	11

Freudenberg, in several recent publications (299, 300), has attempted to combine the currently available analytical information on spruce milled wood lignin into an internally consistent structural concept. Since the data on many structural elements are quite approximate and on others may be totally missing, an endeavor of this type involves a number of more or less arbitrary assumptions. Freudenberg has pointed out, however, that the structural model he arrives at is in reasonable conformity with a number of known reactions of conifer lignins.

The structural description shown in Table 5.13 is a modification of the Freudenberg scheme (197). The most important deviations from the original scheme are the following: The unconjugated carbonyl groups are assumed to consist of aldehyde groups in displaced side-chain structures—a view that has been recently accepted by Adler (243). Furthermore, it is assumed that β-O-4, β-5 and β-β lignol structures have been equally susceptible to side-chain displacement processes. On an analogous basis, a random distribution of α-carbonyl groups between different structural types has been assumed. The amount of coniferyl alcohol groups has been assumed to be twice that of coniferyl aldehyde groups. For the lack of adequate quantitative information, many minor groups have been omitted from the scheme, such as lactone groups,

ISOLATION AND STRUCTURAL STUDIES

Table 5.13. Tentative Assignment of Structural Units for Milled Wood Lignin of Norway Spruce

1. Side-chain structures:

	"Normal" units	Mole Fractions of Units		
		Corresp. units[a] with α-carbonyls	Corresp. displaced[a] side-chains	
(Ar)-p(β-O-4)-	0.35[b]	0.04	p(β-O-4)-	0.06
(Ar)-p(β-5)-	0.10[c]	0.01	p(β-5)-	0.03
(Ar)-p(β-β)-	0.13[d]	0.01	p(β-β)-	0.03
(Ar)-p(β-1)-	0.13[e]	0.01	p(β-1)-	0.01
(Ar)-CH=CHCH$_2$OH	0.06[f]	----		
(Ar)-CH=CH,CHO	0.03[f]	----		
Totals:	0.80	0.07		0.13
			Grand total:	1.00

2. Aromatic groups:

"Uncondensed"	0.56[g]
(5-5)(Ar)-p-	0.25[h]
-(4-O-5)(Ar)-p-	
-(1-O-4)(Ar)-p-	0.05[i]
-(β-5)(Ar)-p-	0.14
Total:	1.00

3. Phenolic elements:

Phenolic OH	0.32[j]
β-O-4	0.45
α-O-4, open	0.08[k]
α-O-4, cyclic	0.10
5-O-4 / 1-O-4	0.05
Total:	1.00

a: Each of the structural types of side-chains (with the exception of coniferyl aldehyde and -alcohol) is assumed to contribute equally to α-carbonyl and displaced side-chain structures. The former amount 0.07 and the latter probably to 0.10 to 0.13, on the basis of carbonyl determinations (188). It should be noted that β-β and β-5 lignol units with α-carbonyl groups (or displaced side-chains) are non-cyclic.

b: Arbitrarily selected value above the minimum of 0.25, suggested by Adler (171) on the basis of acidolysis experiments.

c: Estimation by Adler (211, 245) is 0.09.

d: No direct estimations available. The indicated amount conforms approximately with the pinoresinol-dehydrodiconiferyl alcohol ratio obtained in the dehydrogenative dimerization of coniferyl alcohol (see Table 4.3).

e: Assumed to be equivalent to non-conjugated carbonyl groups (α-aldehydes), formed in side-chain displacement. The value is higher than the conservative estimate 0.10 of Adler and Marton (188).

f: The amount of coniferyl alcohol groups is assumed to be twice (191) that of coniferyl aldehyde groups (159, 188).

g: Estimations by nmr suggest approximately one-half of uncondensed nuclei (62, 197).

h: A very approximate value based on the tentative estimation by Pew (263). Aromatic groups with carbon-bonds to positions 6 and 2 also belong to this category (254).

i: These values have been obtained by difference and can be grossly inaccurate.

j: An average of several independent determinations ranging from 0.30 to 0.34 (202).

k: Based on the increase of phenolic hydroxyl in cold methanolysis (298, 295).

C_γ
- γ-OH, Coniferyl alcohol ... 0.06
- γ-OH, Other primary hydroxyl groups ... 0.78
- γ-O, Coniferyl aldehyde ... 0.03
- γ-O-α, in pinoresinol units ... 0.13
- Total ... 1.00

C_β
- β-O-4, Total ... 0.45
- β-5, Total ... 0.14
- β-β, Total ... 0.17
- β-1, Total ... 0.15
- β-C_α, Coniferyl alcohol and -aldehyde ... 0.09
- Total ... 1.00

C_α
- α-OH, β-O-4 and β-1 units ... 0.20
- α=O, α-keto groups ... 0.07
- α=O, aldehydes in displaced side-chains ... 0.13
- α-O-γ, in pinoresinol units ... 0.13
- α-O-4, in β-5 units ... 0.10
- α=C_β, in coniferyl alcohol and aldehyde ... 0.09
- α-O-4, in open ethers ... 0.08
- Various carbon-to-carbon linkages (?) ... 0.20
- Total ... 1.00

Figure 5.14. Rydholm-diagram illustrating the possible distribution of functional groups.

and structures linked to the 6- and 2- positions of the aromatic groups. Figure 5.14 illustrates the distribution of functional groups into phenylpropane units.

M. *Relationship Between Lignins and Extractives.* The relationship between lignins and a specific class of extractives, lignans, has intrigued natural-products chemists for a period of time. Lignans occur as common extractives in practically all woody plants (302) and represent various derivatives of 2,3-di-(p-hydroxybenzyl)-butane 115 in which the configuration around the middle side-chain carbons is almost invariably racemoid (303). "Bisepoxylignans" (116-119) resemble closely β-β-lignol structures in lignins. Common biogenetic formation is excluded, however, because all isolated lignans have been found to be optically active. Furthermore, lignans with piperonyl (118) or 3,4-dimethoxyphenyl groups 119 (302) lack counterpart structures in lignins. It should also be noted that while diequatorial conformations are predominant, if not exclusive, in β-β-lignol structures, examples of

ISOLATION AND STRUCTURAL STUDIES 229

all three possible conformations have been encountered among lignans of the piperonyl type, isolated from sesame oil (304), and both equatorial-equatorial and equatorial-axial isomers have been identified among guaiacyl- and syringyl lignans (305, 306). All these observations point out that while the formation of lignans in nature may well take place through dehydrogenative β-β-coupling of p-propenylphenol precursors, this process must be under stricter enzymatic control than that operating in the formation of lignins.

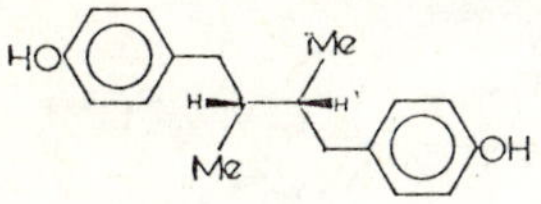

115

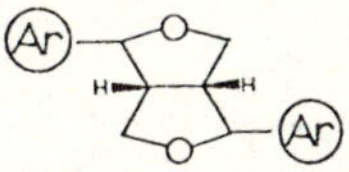

116: Ar = Guaiacyl   a: Aryls equatorial
117: Ar = Syringyl   b: Aryls equatorial-axial
118: Ar = Piperonyl   c: Aryls axial
119: Ar = 3,4-Dimethoxyphenyl

Other extractive components that possess at least superficial similarity with lignin units have been isolated. These consist of two optically active isomers of guaiacyl glycerol 120 from Pinus resinosa (307) and of the β-O-4-linked dimer 121 from western hemlock wood (308), for which no optical activity could be established.

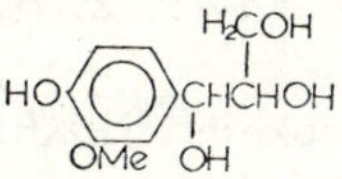

120

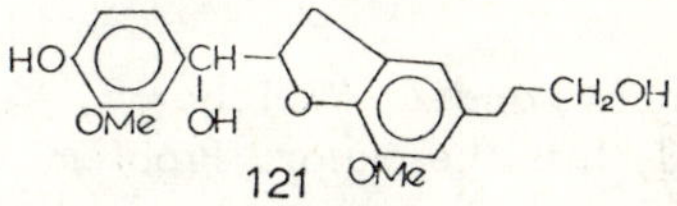

121

The optical activity of 120 and the n-propanol side-chain and unusual phenylcoumaran structure of 121 make a direct connection with lignin formation rather unlikely. Guaiacyl glycerol 120 deserves attention as the only known extractive capable of yielding Hibbert's ketones upon ethanolysis.

Paper chromatographic studies by Goldschmid and Hergert (309) have revealed several components in the sapwood extractives of hemlock that may represent either lignification intermediates or lignin hydrolysis products. These include ferulic acid, coniferaldehyde and notably, dehydrodiconiferyl alcohol, also reported to be isolated from the cambial zone of spruce (310). In comparison with Brauns lignins, the amount of such components appears to be minor.

It has not been clarified whether phenolic extractive components may or may not participate in the build-up of lignin macromolecules. Freudenberg (54) has considered the possibility that hydroxymatairesinol 122, a major lignan component of spruce and hemlock species, might become, in part, associated with lignin; but no experimental verification of this proposition is currently available. Adler and Lundquist (171) have isolated from the acidolysis products of spruce milled wood lignin divanillyltetrahydrofuran 123, which could well represent a copolymerized lignan component. The puzzling fact is, however, that lignan 123 has not been isolated from spruce extractives; although the related (-)olivil 124 has been obtained both from spruce and larch extractives, and the analogous piperolignan (-)cubecon 125 is well-known (311).

REFERENCES

1. E. Hägglund, Acta Acad. Aboensis, Math. et Phys., 2, No. 4, 10 (1922).
2. P. Odincovs and Z. Kreicberga, Voprosy Leosokhim. i. Khim. Drevesing Trudy Inst. Lesokhoz. Problem, Akad. Nauk Latv. S.S.S. R. 6, 51, 63 (1953).
3. K. Freudenberg and W. Durr, Chem. Ber., 62, 1814 (1929).
4. F. E. Brauns, J. Am. Chem. Soc., 61, 2120 (1939); Paper Trade J., 111, No. 14, 35 (1940).
5. A. Björkman, Svensk Papperstidn., 59, 477 (1956).
6. K. Freudenberg, Angew. Chem., 68, 84 (1956).
7. F. E. Brauns, The Chemistry of Lignin, Academic Press, New York, 1952, p. 4.
8. H. Staudinger, E. Dreher and A. af. Ekenstam, Chem. Ber., 69B, 1099 (1936).
9. K. Hess and K. E. Heumann, Chem. Ber., 75, 1802 (1942).
10. H. Grohn, Chem. Tech. (Berlin), 3, 240, 299 (1951).
11. A. Björkman and B. Person, Svensk Papperstidn., 60, 158 (1957).
12. F. H. Forziati, W. K. Stone, J. W. Rower and W. D. Appel, J. Res. Nat. Bur. Standards, 45, 109 (1950).
13. J. C. Pew, Tappi, 40, 553 (1957).
14. I. A. Pearl and L. R. Busche, ibid., 43, 961, 970 (1960).

15. H. H. Brownell, ibid., 48, 513 (1965); H. H. Brownell and K. L. West, Pulp Paper Mag. Can., 62, T374 (1961).
16. M. A. Latif and K. V. Sarkanen, unpublished results.
17. F. E. Brauns, The Chemistry of Lignins, Academic Press, New York, 1952, p. 688.
18. D. E. Bland and M. Menshun, Appita, 21, No. 1 (1967)
19. H. M. Chang and K. V. Sarkanen, unpublished results.
20. D. E. Bland, Proc., Royal Austr. Chem. Inst., 29, 116 (1962).
21. D. E. Bland, Holzforschung, 20, 12 (1966).
22. G. Scurfield and D. E. Bland, J. Horticultural Sci., 38, 297 (1963).
23. T. Higuchi, Y. Ito, M. Shimada and I. Kawamura, Phytochem., 6, 1551 (1967).
24. K. Freudenberg and A. C. Neish, The Constitution and Biosynthesis of Lignin, Springer-Verlag, Berlin-Heidelberg, New York, 1968, p. 64.
25. A. Björkman, Ind. Eng. Chem., 49, 1395 (1957).
26. F. E. Brauns and H. Seiler, Tappi, 35, 67 (1952).
27. A. Björkman, Svensk. Papperstidn., 60, 329 (1957).
28. J. G. M. Bremner and J. H. Colpitts, Trans. IRI, 24, 35 (1948).
29. R. A. Mott, J. Soc. Chem. Ind., 69, 346 (1950).
30. K. Freudenberg, J. Praktische Chemie, 4, 220 (1960).
31. A. Björkman, Svensk Papperstidn., 60, 243 (1957).
32. E. H. Immergut and H. Mark, Makromol. Chem., 18/19, 322 (1956).
33. L. Fergus, Ph.D. Thesis, McGill University, Montreal, Canada (1968).
34. G. B. Berlyn and R. E. Mark, Forest Prod. J., 16, 140 (1965).
35. H. Meyer, J. Polymer Sci., 51, 11 (1961).
36. A. Björkman and B. Person, Svensk Papperstidn., 60, 285 (1957).
37. A. Björkman, Chim. Biochim. Lignine, Cellulose, Hemicellulose, Grenoble, France, 1964, p. 317.
38. M. A. Latif, Ph.D. Thesis, University of Washington, Seattle, (1968).
39. A. Sakakibara and N. Nakayama, J. Japan.Wood Res. Soc., 8, 157 (1962).
40. T. N. Kleinert and J. R. Marton, Nature, 196, 334 (1962).
41. S. Kolboe and O. Ellefsen, Tappi, 45, 163 (1962).
42. D. E. Bland and M. Menshun, Holzforschung, 19, 33 (1965).
43. K. V. Sarkanen, H. M. Chang and G. G. Allan, Tappi, 50, 583 (1967).
44. K. Freudenberg and A. C. Neish, The Constitution and Biosynthesis of Lignin, Springer-Verlag, Berlin-Heidelberg, New York, 1968, p. 51.
45. K. Freudenberg and G. S. Sidhu, Holzforschung, 15, 33 (1961).
46. H. Nimz, Chem. Ber., 96, 478 (1963).
47. F. F. Nord and W. Schubert, J. Am. Chem. Soc., 72, 977 (1950).
48. M. A. Buchanan, F. E. Brauns and R. L. Leaf, J. Am. Chem. Soc., 71, 1297 (1949).
49. J. W. T. Merewether, Holzforschung, 8, 68 (1954).

50. P. R. Enslin, J. Sci. Food. Agr., 4, 328 (1953).
51. J. G. Stone and K. G. Tanner, Can. J. Chem., 30, 166 (1952).
52. H. L. Hergert, private communication.
53. I. Kawamura and T. Higuchi, Res. Bull. Fac. Agr. Gifu Univ., No. 2, 67 (1952).
54. K. Freudenberg and L. Knof, Chem. Ber., 90, 2857 (1957).
55. H. L. Hergert, J. Org. Chem., 25, 405 (1960).
56. J. A. F. Gardner, Can. J. Chem., 32, 532 (1954).
57. J. C. Pew, J. Am Chem. Soc., 77, 2831 (1955).
58. F. E. Brauns, The Chemistry of Lignin, Academic Press, New York, 1952, p. 492.
59. K. Sofue and S. Fukuhara, J. Chem. Soc. Japan, Ind. Eng. Sect., 61, 1070 (1958).
60. K. Freudenberg and K. Dall, Naturwissenschaften, 42, 606 (1955).
61. S. Hernestam, Svensk Kem. Tidskr., 67, 37 (1955).
62. C. H. Ludwig, B. J. Nist and J. L. McCarthy, J. Am. Chem. Soc., 86, 1186 (1964).
63. W. Stumpf and K. Freudenberg, Angew. Chem., 62, 537 (1950).
64. C. L. Hess, Tappi, 35, 312 (1952).
65. G. de Stevens and F. F. Nord, Fortschr. Chem. Forsch., 3, 70 (1954).
66. G. de Stevens and F. F. Nord, Proc. Natl. Acad. Sci., U. S., 39, 80 (1953); J. Am. Chem. Soc., 74, 3447 (1952).
67. D. C. C. Smith, J. Chem. Soc., 1955, 2347.
68. J. Nakano, A. Ishizu and N. Migita, Abstr. 3rd Lignin Symposium, Fukunoka (1958). Ref: F. E. Brauns and D. E. Brauns, The Chemistry of Lignin, Supplement Vol., p. 266.
69. G. de Stevens and F. F. Nord, J. Am. Chem. Soc., 73, 4622 (1951).
70. D. C. C. Smith, Nature, 176, 267 (1955).
71. J. Nakano, A. Ishizu and N. Migita, J. Japan.Wood Res. Soc., 4, 1 (1958).
72. F. F. Nord and W. J. Schubert, Holzforschung, 5, 1 (1951); W. J. Schubert and F. F. Nord, J. Am. Chem. Soc., 72, 977, 3835 (1950).
73. F. F. Nord and W. J. Schubert, Tappi, 40, 285 (1957).
74. S. F. Kudzin and F. F. Nord, J. Am. Chem. Soc., 73, 690, 4619 (1951).
75. F. F. Nord and G. de Stevens, Naturwissenschaften, 39, 479 (1952).
76. C. M. Stewart, J. F. Kottek, H. E. Dadswell, and A. J. Watson, Tappi, 44, 798 (1961).
77. T. E. Timell, private communication.
78. H. Nimz, Holzforschung, 20, 105 (1966).
79. P. F. Ritchie and C. B. Purves, Pulp Paper Mag. Can., 48, No. 12, 74 (1947).
80. E. Adler and S. Hernestam, Acta Chem. Scand., 9, 319 (1955).
81. E. Adler, Private communication.

82. H. Ishikawa and T. Nakajuma, J. Japan.Forestry Soc., 36, 104 (1954).
83. J. C. Pew, J. Am. Chem. Soc., 74, 2850 (1952).
84. B. O. Lindgren, Acta Chem. Scand., 17, 2199 (1963).
85. J. M. Harkin, Adv. Chem. Ser., 59, 65 (1966).
86. H. Richtzenhain, Acta Chem. Scand., 4, 589 (1950).
87. K. Freudenberg and Th. Ploetz, Chem. Ber., 73, 754 (1940).
88. L. Kalb and T. Lieser, Chem. Ber., 61, 1007 (1928).
89. E. Hägglund, Chemistry of Wood, Academic Press, New York, 1951 p. 326.
90. M. A. Latif, H. M. Chang and K. V. Sarkanen, unpublished results.
91. E. Hägglund, Cellulosechemie, 4, 84 (1923).
92. E. Hägglund and C. B. Björkman, Biochem. Z., 147, 74 (1924).
93. P. N. Odiutsov, Leschim. Prom., 5, 5 (1936).
94. N. Migita and M. Kawamura, J. Agr. Chem. Soc. Japan, 20, 348, 507 (1944).
95. K. Freudenberg, W. Lautsch and K. Engler, Chem. Ber., 73, 167 (1940).
96. E. Hagglund and H. Richtzenhain, Tappi, 35, 281 (1952).
97. K. Kratzl and H. Tschamler, Monatsh. Chem., 83, 786 (1950).
98. N. N. Shorygina, L. L. Sergeeva and B. V. Lopatin, Izvestiva Akademii Nauk SSR, Ser. Khim., No. 2, 372 (1967).
99. D. E. Read and C. B. Purves, J. Am. Chem. Soc., 74, 116, 120 (1952).
100. F. E. Brauns, The Chemistry of Lignin, Academic Press, New York, 1952, p. 54.
101. I. Kawamura and T. Higuchi, J. Japan.Wood Res. Soc., 12, 178 (1966).
102. F. E. Brauns and D. E. Brauns, The Chemistry of Lignin, Suppl. Vol. p. 329.
103. R. Willstätter and L. Kalb, Chem. Ber., 55, 2637 (1922).
104. K. Freudenberg, H. Zocher and W. Dürr, ibid., 62, 1814 (1929).
105. E. Wedekind and O. Müller, ibid., 69, 1517 (1936).
106. K. Freudenberg and E. Plankenhorn, ibid, 75, 857 (1942).
107. K. Freudenberg and L. Zechmeister, Progress in the Chemistry of Organic Natural Products, 11, 43 (1954).
108. D. F. Arsenau and J. M. Pepper, Pulp Paper Mag. Can., 66, No.8, T415 (1965).
109. J. M. Pepper, P. E. T. Baylis and E. Adler, Can. J. Chem., 37, 1241 (1959).
110. J. M. Pepper and M. Siddiqueullah, ibid., 39, 390 (1961).
111. J. M. Pepper and M. Siddiqueullah, ibid., 39, 1454 (1961).
112. J. M. Pepper and D. D. S. Wood, ibid., 40, 1026 (1962).
113. A. Sakakibara and N. Nakayama, J. Japan.Wood Res. Soc., 7, 13

(1961).
114. A. Sakakibara and N. Nakayama, ibid., 8, 153 (1962).
115. F. Schutz and W. Knackstedt, Cellulosechemie, 20, 15 (1941).
116. E. Beckmann, O Liesche and F. Lehmann, Biochem.Z., 139, 491 (1923).
117. H. Pringsheim and W. Fuchs, Chem. Ber., 56, 2095 (1923).
118. A. Okuda and S. Hori, Trans. Intern. Congr. Soil Sci. 5th Congr., Leopoldville, Vol. II, P. 255 (1954); Mem. Research Inst. Food Sci. Kyoto Univ. No. 7, 1 (1954).
119. F. E. Brauns and D. E. Brauns, The Chemistry of Lignin, Supplement Vol., Academic Press, New York, 1962, p. 128.
120. B. Holmberg, Ing. Vetenskaps Akad. Handl. No. 103 (1930).
121. E. B. Brookband and F. E. Brauns, Paper Trade J., 110, No. 5, 33 (1940).
122. E. Hägglund and T. Johnson, Biochem. Z., 202, 440 (1928).
123. T. Higuchi, Physiol. Plantarum, 10, 633 (1957).
124. K. Hata and M. Sogo, J. Japan.Forestry Soc., 38, 473 (1956); 39, 102 (1957); J. Japan Wood Research Soc., 4, 5 (1958).
125. B. L. Browning, Methods of Wood Chemistry, Vol. II, Chapter 34, Interscience, 1967.
126. A. G. Norman and S. H. Jenkins, Biochem. J., 28, 2147 (1934).
127. M. J. Goss and M. Phillips, J. Assoc. Offic. Agr. Chemists, 19, 341 (1936).
128. W. E. Cohen and E. E. Harris, Ind. Eng. Chem. Anal. Ed., 9, 234 (1937).
129. K. Kratzl, Mitt. öster. Ges. Holzforsch., 2, 135 (1950).
130. W. I. Lyness and C. Schenker, Tappi, 46, 79 (1957).
131. J. Uchida, J. Soc. Chem. Ind. Japan, 44, 1072 (1941).
132. W. G. Campbell and I. R. C. McDonald, J. Chem. Soc., 2644, 3180 (1952).
133. C. M. Stewart, D. H. Foster, W. E. Cohen, R. T. Leslie and A. J. Watson, Appita, 5, 267 (1951).
134. B. Swan, Svensk Papperstidn., 68, 791 (1965).
135. H. Richtzenhain and E. Dryselius, Svensk Papperstidn., 56, 324 (1953).
136. T. Kleinert and W. Wincor, Tappi, 38, 183A (1953).
137. W. E. Cohen, Austr. Counc. Sci. Ind. Res. Div. For. Prod. Tech. Paper, No. 14 (1934). Quoted in F. E. Brauns, The Chemistry of Lignin, p. 144.
138. W. E. Cohen and H. E. Dadswell, ibid., No. 3 (1931).
139. W. E. Hillis and A. Carle, Biochem. J., 74, 607 (1960).
140. M. Fujii and E. K. Kurth, Tappi, 49, 92 (1966).
141. E. K. Kurth, J. K. Hubbard and J. D. Humphrey, Proc. For. Prod. Res. Soc., 3, 276 (1949).

142. H. J. Kiefer and E. F. Kurth, Tappi, 36, 14 (1953).
143. M. Phillips, M. J. Goss and B. L. Davis, J. Agr. Res., 59, 319 (1939).
144. M. Phillips and M. J. Goss, ibid., 51, 301 (1935).
145. L. Paloheimo, Biochem. Z., 165, 463 (1925); 214, 160 (1929).
146. M. Phillips, J. Assoc. Offic. Agr. Chemists, 22, 422 (1939).
147. Association of Official Agr. Chemists. Official Methods of Analysis, 9th ed., Washington, D. C., p. 91.
148. B. L. Browning, Methods of Wood Chemistry, Vol. II, Interscience, 1967, p. 792.
149. E. R. Armitage, R. B. Ashworth and W. S. Ferguson, J. Soc. Chem. Ind., 67, 241 (1948).
150. T. Higuchi and F. Barnoud, Chim. Biochim. Lignine, Cellulose, Hemicellulose, Grenoble, 1965, p. 255.
151. V. Lorås and F. Löschbrandt, Norsk Skogind., 15, 302 (1961).
152. Tappi Standard Method T214.
153. Tappi Standard Method T202.
154. B. L. Browning, Methods of Wood Chemistry, Vol. II, Interscience, 1967, pp. 802-812.
155. I. A. Pearl, The Chemistry of Lignin, Marcel Dekker, Inc., New York, 1967, pp. 47-52.
156. K. Freudenberg and H. Schluter, Chem. Ber., 88, 617 (1955); Naturwiss., 41, 567 (1954); K. Freudenberg and H. Geiger, Chem. Ber., 96, 1265 (1963).
157. K. Freudenberg and B. Lehmann, Chem. Ber., 93, 1354 (1960).
158. V. Grafe, Monatsh. Chem., 25, 987 (1904).
159. E. Adler and L. R. Ellmer, Acta Chem. Scand., 2, 839 (1948).
160. K. Lundquist, Acta Chem. Scand., 18, 1316 (1964).
161. A. Sohn and P. Lenel, Papier, 3 (718), 109 (1949).
162. O. Goldschmid, Tappi, 38, 728 (1955).
163. D. A. Stanek, Tappi, 41, 601 (1958).
164. H. Nimz, Chem. Ber., 98, 3153 (1965).
165. T. Fukuzumi and M. Terasawa, J. Japan. Wood Res. Soc., 8, 77 (1962); T. Fukuzumi, S. Sakuma, H. Takahoshi, K. Tonita, F. Fujahara, Y. Isome and T. Shibamota, Holzforschung, 20, 51 (1966).
166. K. Kratzl, W. Kisser, J. Gratzl and H. Silbernagel, Monatsh. Chem., 90, 771 (1951).
167. T. Higuchi, I. Kawamura and H. Kawamura, J. Japan. Forestry Soc., 37, 298 (1955); F. F. Nord, Tappi, 47, 624 (1964).
168. H. Ishikawa, W. J. Schubert and F. F. Nord, Arch. Biochem. Biophys., 100, 131 (1963).
169. H. Katsumi, Holzforschung, 20 (5), 142 (1966).
170. S. Raisanen, Suomen Kemistilehti 40B, No. 1, 35 (1967); through Bull. Inst. Paper Chem. 38 (5), 3537.

171. E. Adler, K. Lundquist and G. E. Miksche, Adv. Chem. Ser., 59, 22 (1966).
172. V. F. Felicetta, D. Glennie and J. L. McCarthy, Tappi, 50, 170 (1967).
173. D. W. Glennie and J. S. Mothershead, Tappi, 47, 356, 519 (1964).
174. J. R. Parrish, Tetrahedron Letters, 1964, 555; J. Chem. Soc., 1967, 1145.
175. H. Nimz, Chem. Ber., 98, 533, 588 (1965).
176. K. Freudenberg, C. L. Chen, J. M. Harkin, H. Nimz and H. Renner, Chem. Commun., No. 11, 224 (1965).
177. H. Nimz, Chem. Ber., 100, 2633 (1967).
178. K. Freudenberg and H. H. Hübner, Chem. Ber., 85, 1181 (1952).
179. J. M. Harkin, Oxidative Coupling of Phenols, (W. I. Taylor and A. R. Battersby, eds.) Marcel Dekker, Inc., New York, 1967, p 272.
180. F. E. Brauns and D. E. Brauns, The Chemistry of Lignin, Supplement Vol., Academic Press, New York, 1960, pp. 33-59.
181. E. Adler, J. J. Björkquist and S. Häggroth, Acta Chem. Scand., 2, 93 (1948).
182. G. Ya. Vanag, Zhur. Anal. Khim., 5, 110 (1950).
183. T. Nakamura and S. Kitaura, Ind. Eng. Chem., 49, 1388 (1957).
184. T. Harada and Z. Nikuni, Mem. Inst. Sci. and Ind. Research Osaka Univ., 7, 163 (1950), quoted in ref. 180.
185. Y. Hachihama and S. Jyodai, Chemistry of Lignin, Nippon Hyoronsha Tokyo, 1946, p. 42, quoted in ref. 180.
186. T. Harada and Z. Nikuni, J. Agr. Chem. Soc. Japan, 23, 415 (1950).
187. K. Kratzl, G. Billek, E. Klein and K. Buchtela, Monatsh. Chem., 88, 721 (1957).
188. E. Adler and J. Marton, Acta Chem. Scand., 13, 75, 357 (1961); 15, 370 (1961); E. Adler, J. Marton and K. I. Person, Acta Chem. Scand., 15, 384 (1961).
189. A. Klemola, Suomen Kemistilehti, A41, 106 (1968).
190. R. M. DeBaun and F. F. Nord, Tappi, 34, 71 (1951).
191. B. O. Lindgren and H. Mikawa, Acta Chem. Scand., 11, 826 (1957).
192. J. Marton and E. Adler, Tappi, 46, 92 (1963).
193. H. A. Stafford, Plant Physiol., 37, 643 (1962); 39, 350 (1964).
194. H. Nimz, Chem. Ber., 100, 181 (1967).
195. E. Adler, private communication.
196. K. Freudenberg and A. C. Neish, The Constitution and Biosynthesis of Lignin, Springer-Verlag, Berlin-Heidelberg, New York, 1968, p 103.
197. G. Aulin-Erdtman, Svensk Papperstidn., 55, 745 (1952); 57, 745 (1954).
198. T. Enkvist, B. Alm and B. Holm, Paperi Puu, 38, 1 (1956); T. Enkvist, B. Alfredsson and E. Hägglund, Svensk Papperstidn., 55, 588
199. G. Gram and B. Althin, Acta Chem. Scand., 4, 967 (1950);

ISOLATION AND STRUCTURAL STUDIES 237

       K. Freudenberg, Angew.Chem., 68, 84 (1956); K. Freudenberg, J. H. Harkin and H. K. Wecker, Chem. Ber., 97, 909 (1964).
200. K. Sarkanen and C. Schuerch, Anal. Chem., 27, 1245 (1955).
201. E. Adler, S. Hernestam and I. Wallden, Svensk Papperstidn., 61, 641 (1958).
202. K. Freudenberg and A. C. Neish, Constitution and Biosynthesis of Lignin, Springer-Verlag, Berlin-Heidelberg, New York, 1968, p. 71.
203. E. Adler and K. Lundquist, Acta Chem. Scand., 55, 223 (1961).
204. J. Gierer, ibid., 8, 1319 (1954).
205. J. Gierer, Chem. Ber., 89, 257 (1956).
206. J. C. Pew, W. B. Connors and A. Kuniski, Chim., Biochim. Lignine, Cellulose, Hemicellulose, Grenoble, 1964, p. 229.
207. C. Gustafsson and L. Anderson, Paperi Puu, 37, 1 (1955).
208. E. Adler and B. Stenemur, Chem. Ber., 89, 291 (1956).
209. H. Richtzenhain, Svensk Papperstidn., 53, 644 (1950); Acta Chem. Scand., 4, 206, 508 (1950).
210. S. L. Larsson and G. E. Miksche, Acta Chem. Scand., 21, 1970 (1967).
211. E. Adler and K. Lundquist, ibid., 17, 13 (1963).
212. K. Freudenberg and H. Schlüter, Chem. Ber., 88, 617 (1955).
213. H. Nimz, Chem. Ber., 99, 2638 (1966).
214. E. Adler, Chim. Biochim. Lignine, Cellulose, Hemicellulose, Grenoble, 1964, p. 73.
215. L. Bückman, J. J. Pyle, J. L. McCarthy and H. Hibbert, J. Am. Chem. Soc., 61, 686 (1939).
216. W. B. Hewson, J. L. McCarthy and H. Hibbert, ibid., 63, 3045 (1941).
217. H. E. Fisher and H. Hibbert, ibid., 69, 1208 (1947).
218. E. Adler, J. M. Pepper and E. E. Eriksoo, Ind. Eng. Chem., 49, 1391 (1957).
219. K. V. Sarkanen and C. Schuerch, J. Am. Chem. Soc., 79, 4203 (1957).
220. K. Kratzl, G. Billek, E. Klein and K. Buchtela, Monatsh. Chem. 88, 721 (1957).
221. T. Higuchi and I. Kawamura, Holzforschung, 20, 16 (1966).
222. J. Gierer and I. Noren, Acta Chem. Scand., 16, 1713 (1962).
223. J. Gierer and I. Noren, ibid., 16, 1976 (1962).
224. J. Gierer and B. Lenz, Svensk Papperstidn., 68, 334 (1965).
225. J. Gierer and I. Noren, Paperi Puu, 43, 654 (1961).
226. T. Enkvist, T. Ashorn and K. Hästbacka, ibid., 44, 395 (1962).
227. A. Ishiza, J. Nakano and N. Migita, J. Japan.Wood Res. Sci., 8, 139 (1962).
228. T. Ashorn, Soc. Sci. Fennica Commentations Phy-Math., 25, No. 8 (1961).

229. A. Ishiza, C. Tokatsuka and N. Migita, J. Japan.Wood Res. Soc., 7, 121 (1961).
230. J. Nakano, S. Ka, K. Kato and N. Migita, ibid., 13, 108 (1967).
231. H. Nimz, Chem. Ber., 98, 3160 (1965).
232. H. Nimz, ibid., 99, 469 (1966).
233. K. Lundquist and T. E. Miksche, Tetrahedron Letters, 25, 2131 (1965).
234. H. Richtzenhein and C. Hofe, Chem. Ber., 72, 1890 (1939).
235. T. Ishihara and T. Kondo, Bull. Agr. Chem. Soc., Japan, 21, 250 (1957).
236. J. Tanaka and T. Kondo, J. Japan Tech. Assoc. Pulp & Paper, Ind., 11, 111 (1957).
237a. I. A. Pearl and D. L. Beyer, J. Am. Chem. Soc., 76, 6106 (1954);
237b. E. Adler and S. Häggroth, Svensk Papperstidn., 53, 287, 321 (1950).
238. P. E. Parker, R. L. Coalson and C. Schuerch, Adv. in Chem.Ser., 59, 249 (1966).
239. K. Lundquist, G. E. Miksche, L. Ericsson and L. Berndtson, Tetrahedron Letters, 46, 4587 (1967).
240. K. Freudenberg and B. Lehmann, Chem. Ber., 96, 1850 (1963).
241. K. Freudenberg and A. C. Neish, The Constitution and Biosynthesis of Lignin, Springer-Verlag, Berlin-Heidelberg, New York, 1968, p 100.
242. K. Freudenberg and C-L. Chen, Chem. Ber., 100, 3683 (1967).
243. E. Adler, Svensk Kemisk Tidskrift, 80, 279 (1968).
244. S. I. Falkehag, Paperi Puu, 43, 655 (1961); S. I. Falkehag, J. Marton and E. Adler, Adv. in Chem. Ser., 59, 75 (1966).
245. E. Adler, S. Delin and K. Lundquist, Acta Chem. Scand., 13, 2149 (1959).
246. H. Nimz and H. Gaber, Chem. Ber., 98, 538 (1965).
247. A. Von Wacek and H. Griengl, Holz Roh-Werkst., 25, 167, 225 (1967).
248. K. Freudenberg and A. C. Neish, The Constitution and Biosynthesis of Lignin, Springer-Verlag, Berlin-Heidelberg, New York, 1968, p.99.
249. H. Nimz, Angew. Chem., 76, 597 (1964).
250. G. Brunow, Finska Kemistsamfundets Medd., 74, 20 (1965).
251. G. Brunow, Soc. Sc. Fenn. Comment Phy-Math., 33, 51 (1967).
252. H. Nimz, Liebigs Ann., 691, 126 (1966).
253. H. Richtzenhain, Chem. Ber., 83, 488 (1950).
254. K. Freudenberg, C. C. Chen, Ct. Cardinale, Chem. Ber., 95, 2814 (1962).
255. H. Erdtman and J. Gripenberg, Acta Chem. Scand., 1, 171 (1947).
256. K. Ogiyama and T. Kondo, J. Japan.Wood Res. Soc., 11, 206 (1965); 14, 31 (1968); Tetrahedron Letters, 19, 2083 (1966).
257. K. Freudenberg, Holzforschung, 18, 3 (1964).
258. K. Freudenberg and J. M. Harkin, ibid., 18, 166 (1964).
259. B. Leopold, Acta Chem. Scand., 6, 38 (1952).

260. J. Tanaka and T. Kondo, J. Japan.Wood Res. Soc., 4, 34 (1958).
261. L. S. Nahum, Ind. Eng. Chem. Prod. Res. Develop., 4, 71 (1965).
262. G. Aulin-Erdtman, Svensk Papperstidn., 57, 745 (1954); 59, 363 (1956).
263. J. C. Pew, Nature, 193, 250 (1962); J. Org. Chem., 28, 1048 (1963).
264. W. Metlesics, F. Wessely and H. Budzikiewicz, Tetrahedon, 6, 345 (1959).
265. S. M. Bloom, J. Am. Chem. Soc., 80, 6280 (1958).
266. K. Freudenberg and A. C. Neish, The Constitution and Biosynthesis of Lignin, Springer-Verlag, Berlin-Heidelberg, New York, 1968, p.101.
267. K. Freudenberg and G. Grion, Chem. Ber., 92, 1355 (1959).
268. K. Freudenberg and J. M. Harkin, ibid., 93, 2814 (1960).
269. K. Freudenberg and A. C. Neish, The Constitution and Biosynthesis of Lignin, Springer-Verlag, Berlin-Heidelberg, New York, 1968, p.94.
270. K. Freudenberg, H. Reznik, H. Boesenberg and D. Rasenack, Chem. Ber., 86, 641 (1952).
271. K. Freudenberg, H. Reznik, W. Fuchs, Angew. Chem., 66, 109 (1954); Naturwissenschaften, 42, 29 (1955).
272. H. Grohn and F. Schierbaum, Holzforschung, 12, 65 (1958).
273. A. Hayashi and I. Tachi, paper presented at the 3rd Lignin Symposium of the Japanese Pulp and Paper Assoc., Fukuoka (Nov. 4, 1958).
274. B. O. Lindgren, Acta Chem. Scand., 12, 447 (1958).
275. P. Traynard, A. M. Ayroud and A. Eymery, Tech. Ind. Papetiere Bull., 45 (1953).
276. I. Kawamura and T. Higuchi, J. Soc. Textile and Cellulose Ind. Japan, 8, 335 (1952).
277. A. Hayashi, J. Agr. Chem. Soc., Japan, 35, 80, 83 (1961).
278. I. Kawamura and T. Higuchi, J. Soc. Textile and Cellulose Ind., Japan, 9, 9 (1953); 8, 442 (1952).
279. A. Hayashi and I. Tachi, J. Agr. Chem. Soc., Japan, 30, 442, 445, 448, 791 (1956).
280. C. M. Stewart and J. A. McPherson, Holzforschung, 9, 140 (1954).
281. E. Aaltio and R. H. Roschier, Paperi Puu, 36, 157 (1954).
282. I. Kawamura and T. Higuchi, J. Soc. Textile and Cellulose Ind., Japan, 9, 454 (1953).
283. I. Kawamura and T. Higuchi, ibid., 9, 157 (1953).
284. H. Richtzenhain, B. Abrahamsson and E. Dryselius, Svensk Papperstidn., 57, 473 (1954).
285. I. A. Pearl and D. C. Beyer, Tappi, 47, 458 (1964).
286. I. A. Pearl and D. C. Beyer, ibid., 47, 779 (1964).
287. K. Kringstad and O. Ellefsen, Papier, 18, 583 (1964).
288. K. Kringstad, Acta Chem. Scand., 19, 1493 (1965).
289. W. S. Linnell and H. A. Swenson, Tappi, 49, 444, 494 (1966).

290. W. S. Linnell, N. S. Thompson and H. A. Swenson, Tappi, 49, 491 (1966).
291. R. W. Kilb, J. Polymer Sci., 38, 403 (1959).
292. J. L. Snyder and T. E. Timell, Svensk Papperstidn., 58, 851 (1955).
293. F. A. Abadie-Maumert, Papeterie, 77, 255 (1955).
294. F. A. Abadie, Norsk Skogind., 3, 290 (1949).
295. B. Holmberg and S. Runius, Svensk Kem. Tidskr., 37, 189 (1925).
296. H. I. Bolker and N. Terashima, Adv. Chem. Ser., 59, 110 (1966); H. I. Bolker, Nature, 197, 489 (1963).
297. K. Freudenberg, Beitr. Biochem. Physiol. Naturstoffen, Festschrift Kurt Mothes, Jena, 1965, p. 165.
298. E. Adler, M. D. Becker, T. Ishihara and A. Stamvick, Holzforschung, 20, 3 (1966).
299. K. Freudenberg, J. M. Harkin and H. K. Werner, Chem. Ber., 97, 909 (1964).
300. K. Freudenberg, J. Polymer Sci., 48, 371 (1960); Pure Appl. Chem., 5, 9 (1962); Science, 148, 595 (1965).
301. K. Freudenberg and A. C. Neish, Constitution and Biosynthesis of Lignin, Springer-Verlag, Berlin-Heidelberg, New York, 1968, p. 113.
302. W. M. Hearon and W. S. MacGregor, Chem. Rev., 55, 957 (1955).
303. K. Weinges, in Oxidative Coupling of Phenols (W. T. Taylor and A. R. Battersby, eds.), Marcel Dekker, Inc., New York, 1967, pp. 323-353.
304. Ibid., p. 345.
305. K. Freudenberg and H. Dietrich, Chem. Ber., 86, 1157 (1953).
306. E. E. Dickey, J. Org. Chem., 23, 179 (1958).
307. E. V. Rudloff, Chem. Ind., 1965, 180.
308. E. Barton, private communication.
309. H. L. Hergert and O. Goldschmid, Tappi, 44, 858 (1961).
310. J. M. Harkin, in Oxidative Coupling of Phenols (W. I. Taylor and A. R. Battersby, eds.), Marcel Dekker, Inc., New York, 1967, p.272.
311. K. Freudenberg and K. Weinges, Tetrahedron Letters, 1959, 19.

6

ULTRAVIOLET SPECTRA

Otto Goldschmid

I. Introduction. In contrast to the polysaccharides of the cell wall, which are transparent in visible and near-ultraviolet light, lignin, owing to its aromatic nature, absorbs strongly in the ultraviolet range of the spectrum. Since 1927, when Herzog and Hillmer (1) discovered the characteristic ultraviolet absorption of lignin solutions, spectra of a large number of lignin preparations and lignin model compounds have been determined by these (2) and many other investigators (3 - 6). Aulin-Erdtman (7) and Jones (8) reviewed the results of this work in 1949. Although much was learned about the spectra of aromatic model compounds of known structure, attempts to establish the structure of lignin derivatives on the basis of their spectra were not very successful. Jones (8) concluded that the characteristic bands in the lignin spectrum are attributable to oxygen-substituted benzene rings, and that the relatively high absorption in the 300-450 mµ region suggests the presence of carbonyl groups or double bonds conjugated with the benzene ring. Aulin-Erdtman (7) suggested that the perhaps most important result of this work was that the ultraviolet absorption spectrum definitely proved the aromatic nature of lignin. Lange (9) extended this proof to lignin in wood when he obtained the characteristic ultraviolet absorption band of lignin by applying a microspectrophotometric technique (10) to thin sections of spruce wood. Lange's photomicrographs of such sections (11) taken in 280 mµ light clearly demonstrated both the aromatic nature of lignin and its location in the cell wall. Ultraviolet light microscopy has become an important tool for studying the topochemistry of lignification (see Chapter 2, Section V) and delignification (12, 13).

Work on the ultraviolet spectrum of lignin has been summarized by Brauns in 1952 (14) and by Brauns and Brauns in 1960 (15).

II. Origin of Ultraviolet Absorption Spectra of Aromatic Compounds. Simple benzenoid compounds, that is, those which contain only one ring and do not contain additional unconjugated chromophores, usually show three absorption bands in the ultraviolet region. In a systematic study of the spectra of substituted benzenes, Doub and Vandenbelt (16, 17) found that these bands form a regular progression of the absorption bands of benzene, considerably displaced towards the visible depending upon the number, nature, and position of the substituent groups. They designated the bands corresponding to the weak 254 mµ ($\varepsilon \sim 200$), the medium 203 mµ ($\varepsilon \sim 7,400$), and the intense 180 mµ ($\varepsilon \sim 50,000$) bands of benzene as the secondary, first primary, and second primary bands, respectively. These designations avoid both the confusion caused by some of the earlier band designations and the need to assign bands to certain transitions when such assignments are still in doubt.

The absorption bands of benzene arise from the transition of an electron from the highest occupied to the lowest vacant orbital of benzene $\pi \rightarrow \pi^*$ transition, (see (18) for a representation of different types of transition). Detailed molecular orbital calculations including configuration interaction of the electrons lead to the correct excitation energies of the three observed absorption bands (19). In the highly symmetrical benzene molecule the transition corresponding to the 254 mµ absorption is symmetry-"forbidden." The observed low intensity of this band is made possible by small vibrational distortions of the ring (20). Substitution on the ring further distorts the hexagonal electron symmetry of benzene and thus causes an even greater increase in the intensity of the secondary band.

Substitution causes changes in the transition energies as well as in the intensities. Doub and Vandenbelt (16) showed that when the substituent groups are divided into ortho-para directing (electron-donating) and meta-directing (electron-withdrawing) groups, the displacement of the first primary band of benzene increases in the order shown below for some of the groups:

o-p directing: $CH_3 < Cl < OH < OCH_3 < NH_2 < O^-$

m directing: $NH_3^+ < COO^- < COOH < COCH_3 < CHO$

The two series parallel the increasing electron flow between the substituent group and the ring. Irrespective of the direction of the flow, both types of substituents lower the transition energies by stabilizing polar excited states and thus displace the primary absorption band to longer wavelengths (21). Among these groups, the ionized hydroxyl group and the aldehyde group respectively, cause the greatest band displacement. Disubstitution of benzene with groups of opposite type causes greater band displacement than that with groups of the same type; the greatest displacement of the primary band occurs in the case of para-disubstitution with groups of opposite type (16). Table 6.1 illustrates some of these effects. The table includes data (22) for p-acetoxy benzaldehyde. In this compound the acetyl group prevents the electrons on the phenolic oxygen from interacting with the ring. As a result, the spectrum of p-acetoxy benzaldehyde is shifted back to a position close to that of benzaldehyde.

The absorption bands of certain trisubstituted benzene derivatives (23) often correspond to the most displaced bands in the spectra of the disubstituted derivatives that contain the same substituent groups. For instance, 3,4-

Table 6.1. Ultraviolet Absorption Bands of Some
Mono- and Para-Disubstituted Benzenes

R'—◯—R"

		First primary		Secondary		
R'	R"	λ_{max}	ϵ	λ_{max}	ϵ	Ref.
H	H	203.5	7,400	354	203	(16)
OH	H	210.5	6,200	270	1,450	(16)
O$^-$	H	235	9,400	287	2,600	(16)
H	CHO	249	12,000	279	1,400	(22)
OH	CHO	283.5	16,000	-	-	(16)
O$^-$	CHO	330	27,900	-	-	(16)
CH$_3$-CO-O	CHO	252	13,800	287	1,900	(22)

dihydroxy acetophenone absorbs at 274 and 305 mμ; the former wavelength corresponds to the 275 mμ absorption maximum of p-hydroxy acetophenone, the latter to the 308 mμ maximum of m-hydroxy acetophenone.

In addition to the empirical, semi-quantitative relationships developed by Doub and Vandenbelt, there exists a phenomenological theory capable of quantitatively predicting many of the intensity changes and band displacements in the spectra of homo-substituted benzenes. The theory has been developed by Sklar (24), Förster (25), Platt (26) and others and has recently been well summarized by Stevenson (27). In the absence of strongly perturbing substituents, this theory correctly predicts, 1: that para-disubstituted and 1,2,4,-trisubstituted benzene derivatives absorb more strongly than ortho- and meta-disubstituted benzenes; 2: that in the absence of steric effects, the spectra of ortho- and meta-disubstituted benzenes are nearly identical; 3: that in para-disubstituted derivatives, the transition moment corresponding to the secondary band is polarized in a direction perpendicular to the line joining the substituents, while that corresponding to the first primary band is polarized parallel to this line (28); and 4: that the symmetrical 1,2,3- and 1,3,5- trisubstituted benzenes have absorption bands of low intensity.

In a most systematic study, Aulin-Erdtman and Sandén (29) recently investigated the ultraviolet absorption of p-hydroxyphenyl, guaiacyl, and syringyl derivatives with identical side-chains in the p-position. Spectra of both the non-ionized and the ionized forms of these compounds were determined and presented in curves that permit a direct comparison between the three types of derivatives. The effects of methoxylation and ionization of the phenols on the position and intensities of their absorption bands were discussed and, where possible, band assignments to the originating electron transitions made. Table 6.2 summarizes the positions and intensities of the major absorption bands in the spectra of some of the p-hydroxyphenyl, guaiacyl, and syringyl derivatives compared in this work.

Table 6.2. Summary of Ultraviolet Absorption of Some p-Hydroxyphenyl, Guaiacyl, and Syringyl Derivatives Related to Lignins (29, 30)

R	H-R				G-R				S-R			
	Non-ionized		Ionized		Non-ionized		Ionized		Non-ionized		Ionized	
	λmax mμ	ε	λmax mμ	ε	λmax mμ	ε	λmax mμ	ε	λmax mμ	ε	λmax mμ	ε
H	269	1,580	234-5 286-7	9,100 2,500	274	2,400	239 289	7,800 3,640	267-8	940	245-6 283	7,900 3,600
$CH_2-CH_2-CH_3$	277	1,620	237 294-5	8,900 2,500	279-80	2,800	245 296-7	9,100 4,100	272-3	1,150	247-8 288-9	8,100 3,800
$CO-CH_2-CH_3$	276-8	14,800	327	24,600	—	—	—	—	299	10,500	357-8	20,000
$CO-CHOH-CH_3$	—	—	—	—	278-9 306	10,000 9,400	350	24,000	—	—	—	—
CHO	283-4	16,000	330	28,000	279 309	11,000 9,600	347-8	26,000	306-7	12,600	363-5	24,000
-$CH=CHCH_3$	259	20,900	284	22,400	260	15,100	286-7	17,400	271-2	14,000	301	15,500
-$CH=CHCH_2OH$	263	19,500	290	22,400	266	15,100	290 315	16,600 15,100	276	14,000	320	17,000
-CH=CHCOOH	308-9	20,400	333	24,000	320	17,400	346	24,000	326	19,500	356-7	24,000
-CH=CHCHO	323-4	27,600	383	36,400	337-8	21,800	414	35,500	341-2	21,800	421	30,900

III. Ultraviolet Spectra of Lignins.

Ultraviolet spectra of different lignin preparations are usually quite similar. The typical lignin spectrum (Figure 6.1) decreases from a maximum near 205 mμ to a shallow minimum near 260 mμ, with a pronounced shoulder near 230 mμ. The minimum is followed by a characteristic lower maximum near 280 mμ and a gradual decrease towards the visible range of the spectrum. This characteristic lignin spectrum is shown by such different preparations as milled wood lignin in methyl cellosolve (31), Brauns lignin in ethanol (7), spruce lignin sulfonic acid of different degrees of sulfonation in water (32), and sulfonated synthetic model lignin (Freudenberg's DHP) (33).

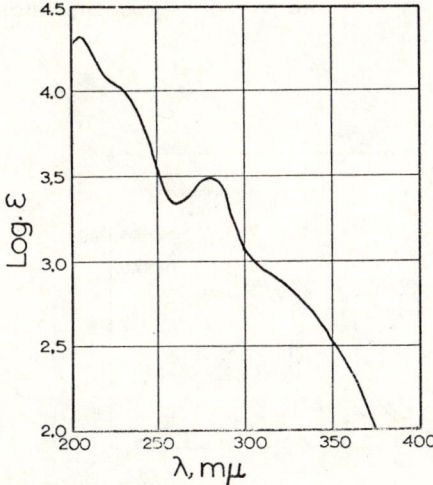

Figure 6.1. Ultraviolet absorption of lignin sulfonic acid (32).

Clearly, this spectrum is a composite of the absorption bands of the different phenylpropane units that constitute the lignin polymer. Hess (34) resolved the ultraviolet spectrum of spruce Brauns lignin into six symmetrical absorption bands; their maxima were at 228, 262, 282, 312, 331, and 351 mμ. Japanese authors (35) using a computer program (36) recently resolved the ultraviolet spectra of spruce milled wood lignin and of thiolignin into six bands with similar λ_{max} values. How can these bands be assigned to the structure in the lignin polymer that causes them?

One method that has greatly helped in the interpretation of the ultraviolet spectrum of lignin is the method of difference spectroscopy (37, 38). This method, which will be described in detail in the next section, was originally developed to study the effects on the lignin spectrum of the ionization of phenolic hydroxyl groups. However, the so-called $\Delta\varepsilon$-method has been applied successfully to spectral changes resulting from reactions other than ionization (39), such as ethanolysis (40), borohydride reduction (41), and hydrogenation (42). Other reactions that cause changes which may help in the interpretation of lignin spectra are: acetylation of free phenolic hydroxyls (Table 6.1) (22, 43), and acetal formation (44) as a means of detecting aldehyde groups.

A. Effect of Constituent Groups.

1. Phenolic Hydroxyl Groups.

Free and etherified hydroxyl groups contribute significantly to the characteristic absorption maximum of lignins near 280 mμ. As Pew (45) has shown, the absorption maxima of 14 unconjugated guaiacyl and 3,4-dimethoxy-phenyl model compounds fall within the narrow wavelength range from 277 to 282 mμ (Figure 6.2). In neutral solution, the spectra of free and etherified phenols are nearly identical; in alkaline solution, on the other hand, ionization of the hydroxyl group shifts the absorption bands of compounds with free phenolic groups towards longer wavelengths (46, 47).

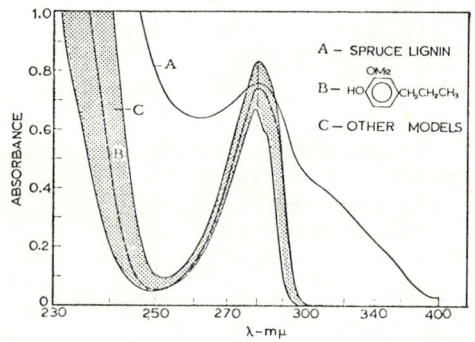

Figure 6.2. Ultraviolet absorption of spruce lignin (curve A) compared with 1-guaiacylpropane (curve B) and related, unconjugated models (curve C) (110)

Aulin-Erdtman noted a similar red-shift of the absorption of certain lignin preparations and lignin model compounds (7). Because the relatively few ionizable phenolic units contribute proportionately less to the total lignin spectrum than the large number of etherified, non-ionized units, the shift is small in the case of lignin. However, since the absorption of the non-ionizable units does not change when the solution is made alkaline, it can be eliminated by subtracting the spectrum of the neutral solution from that of the alkaline solution (37, 38). The resulting difference spectrum then represents essentially only the absorption of the ionizable phenolic units.

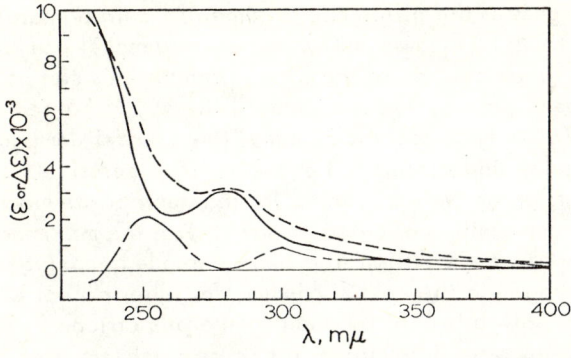

Figure 6.3. Ultraviolet absorption of spruce lignin sulfonic acid at pH 5 (——), pH 14 (---), and $\Delta \varepsilon_i$ curve (— - —) (111).

In the case of lignins (Figure 6.3), an ionization difference ($\Delta \varepsilon_i$) curve with maxima near 250 and 300 mµ is obtained. These maxima are similar in position and relative intensity to those of the $\Delta \varepsilon_i$-curves of many simple unconjugated lignin models (48); they evidently correspond to the first primary band and the secondary band, respectively, of compounds such as 1-guaiacylpropane (Table 6.2) shifted towards longer wavelengths as a result of ionization of the phenolic hydroxyl group in alkaline solution. $\Delta \varepsilon_i$-Curves of many lignin preparations show, however, in addition to these maxima a shoulder (Figure 6.3) or a maximum (Figure 6.4, curve 1) at longer wavelengths (300 to 400 mµ). Difference curve maxima in this region are similar to those in the $\Delta \varepsilon_i$-curves of model compounds (49) containing phenolic hydroxyls conjugated with carbonyl groups, carbon-carbon double bonds, or biphenyl groups. The maxima correspond to the shifted primary bands of compounds with α-carbonyl groups such as (G)-CO-CHOH-CH$_3$ (Table 6.2) with λ_{max} 350 mµ (ε 24,000) in alkaline solution. In neutral solution this compound absorbs in the 280 to 310 mµ region ($\varepsilon \sim 10,000$). As a result of their high molar absorptivity, the presence of even a small number of conjugated carbonyl groups will cause a perceptible shoulder in the difference curve of a lignin preparation, and will also contribute to the 280 to 310 mµ absorption in the neutral spectrum (38).

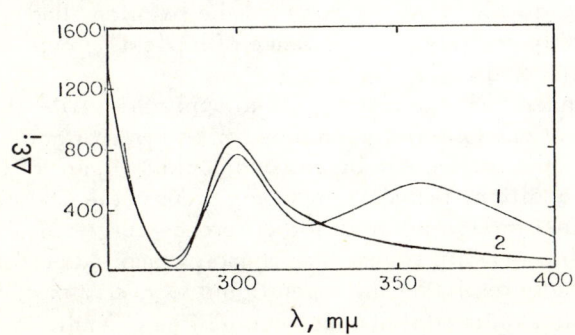

Figure 6.4. Ionization $\Delta \varepsilon$-curves. Curve 1: spruce milled wood lignin; Curve 2: NaBH$_4$-reduced spruce milled wood lignin (60).

By comparing the heights of the difference maxima of lignin preparations with those of the corresponding $\Delta \varepsilon_i$-maxima for appropriate model compounds, the free phenolic hydroxyl group content of the lignin preparations can be estimated (37). In a series of papers, Aulin-Erdtman (48, 49, 30) determined the neutral, alkaline, and difference spectra of many lignin model compounds representing both unconjugated and conjugated phenols. By successively subtracting the required proportion of the $\Delta \varepsilon$-curve for the appropriate model, the lignin difference spectrum can be corrected, starting with the maximum at the longest wavelength, for the contribution from more extended, conjugated chromophores and, proceeding with the next maximum, for other phenolic groups until the residual is negligible. In this manner the proportion of different types of phenolic groups were determined in a large number of lignin preparations including various derivatives of spruce lignin (50), Freudenberg's synthetic model lignin, DHP (51), and Brauns lignin from spruce and hemlock (52). The results of this work have been reviewed in detail by Brauns and Brauns (15).

In a simple method intended especially for the rapid determination of the relative phenolic hydroxyl content of technical lignin preparations, the difference curve was obtained directly by scanning the alkaline vs. the neutral lignin solutions, respectively placed in the sample and reference cells of a spectrophotometer (53). This technique is of course greatly facilitated by the use of a recording spectrophotometer (54). In this method the height of the difference curve maximum near 300 mµ was compared with an average value ($\Delta \varepsilon_i$ 4,100) for conidendrin and eugenol as the reference compounds. Conjugated α-carbonyl groups interfere with the determination because their presence lowers the 300 mµ difference curve maximum (55). The interference may be minimized either by using the 250 mµ maximum of the difference curve for the comparison (54) or by eliminating the carbonyl groups by borohydride reduction of the sample prior to the determination (56). Figure 6.4 illustrates the effect of borohydride reduction on the 300 mµ difference curve maximum.

Freudenberg and Dahl (57) compared the phenolic hydroxyl content of various lignin preparations obtained by the $\Delta \varepsilon$-method and by potentiometric titration in ethylene diamine. From the results, they concluded that the optical method generally gives 25 to 30% lower results than the titration, and that the discrepancy is probably caused by the presence of biphenyl groups and other "hindered" phenolic hydroxyls.

Aulin-Erdtman and Sandén (58) have attempted to apply the $\Delta \varepsilon$-method to the determination of phenolic groups in aspen milled wood lignin and aspen lignin sulfonates. The presence in aspen (and poplar) lignins of syringyl, guaiacyl, and, in addition, p-hydroxybenzoate groups (see Chapter 3), however, renders the exact interpretation of the difference curve somewhat doubtful because the chromophoric system is much more complicated than in softwood lignins. Preliminary results for the aspen lignin suggested that 25 to 40% of the phenylpropane units carried unconjugated free phenolic

ULTRAVIOLET SPECTRA

hydroxyl groups. These appear to consist of approximately equal numbers of guaiacyl- and syringylpropane units. In addition, the extended chromophores were interpreted to include unetherified p-coumaraldehyde groups and phenolic groups conjugated with α-keto groups or possibly with biphenyl or ethylenic groups.

2. Carbonyl Groups.

a. Conjugated carbonyl groups. It had long been thought that the high absorption in the 300 to 400 mμ region of the lignin spectrum is due to carbonyl groups or double bonds conjugated with the aromatic ring (56). Further evidence for phenolic conjugated α-keto groups was provided by the ΔΕ-method (37, 38, 49). Ionization difference curves for certain lignin preparations show a shoulder or a maximum near 350 mμ which agrees well with the maxima in the spectra of alkaline solutions of models such as vanillin (λ_{max} 353 mμ) and acetoguaiacone (λ_{max} 348 mμ) (47). The presence of carbonyl groups in lignin was confirmed by Smith (59) who found that borohydride reduction diminished the absorbance in alkaline solution near 350 mμ of four different Brauns lignin preparations. Lignin units of type 1 and type 2 absorb near 350 mμ; the former do so only in alkaline solution when the hydroxyl group is ionized, whereas the latter absorb near 350 mμ at all pH's.

R = H or OCH$_3$

By a study of appropriate model compounds, Smith found that α-keto groups and cinnamaldehyde groups conjugated with free phenolic hydroxyl groups are very slowly reduced by borohydride compared with their etherified analogues. Thus, by following the rate of change of the absorbance at 350 mμ during the reduction, he was able to estimate the relative amounts of type 1 and type 2 structures in the Brauns lignin preparations.

In a quantitative study of lignin ethanolysis, Sarkanen and Schuerch (40) determined the spectral changes occurring during ethanolysis by subtracting spectra of the original ethanolysis lignin samples from those of the same products after re-ethanolysis for various periods of time. The resulting difference curves with a shoulder near 270 mμ and a maximum at 350 mμ clearly indicated the formation of an α-carbonyl group during ethanolysis. Moreover, a single linear correlation between absorptivity change at 305 mμ and change in phenolic content for different re-ethanolysis samples strongly suggested that a phenolic hydroxyl group and an α-carbonyl group were produced in the same reaction.

Spectral changes occurring on borohydride reduction were used by Adler and Marton (41) in their thorough study of carbonyl groups in lignins. By subtracting spectra of spruce milled wood lignin samples reduced with alkaline borohydride for different lengths of time from those of the alkaline solution of the original samples, they obtained borohydride reduction difference ($\Delta\varepsilon_r$) spectra which they compared with similar $\Delta\varepsilon_r$-curves for a large number of model compounds. In this way the amounts of four different types of conjugated carbonyl groups (Table 6.3) could be estimated. The two structures with $\Delta\lambda_{max}$ values near 340 to 350 mμ can be distinguished by their greatly differing rates of reduction.

Table 6.3. Estimation of Various Conjugated Carbonyl Groups in Spruce Milled Wood Lignin by Alkaline NaBH$_4$ Reduction (41)

Structure	λ_{max} values for $\Delta\varepsilon_r$-curves mμ	Percentage of total C_6-C_3-units
(G)-CH=CH-CHO	400	<1
R-O-4-(G)-CH=CH-CHO	340	3
(G)-CO-R	350	<1
R-O-4-(G)-CO-R	310	5-6

In contrast to milled wood lignins which contain very few phenolic α-keto groups, kraft lignin studied by the same method (60) showed about 5% phenolic α-keto groups and no etherified aryl-α-keto groups or coniferyl aldehyde units.

b. Nonconjugated carbonyl groups. In the course of their study, Adler and Marton (41) also investigated the spectral characteristics of non-conjugated β-ketones. In neutral solution the ultraviolet spectra of the model compounds (G)CH$_2$CO-Me, (S)CH$_2$CO-Me, (G)CH$_2$CO-CH$_2$OH were similar to those of simple guaiacylpropane compounds. Upon alkaline borohydride reduction, the $\Delta\varepsilon_r$-curves of these compounds showed maxima near 270 to 310 mμ. However, their intensity was lower than that of the maxima of the phenolic conjugated α-ketones by a factor of 10. A possible explanation of this spectral effect is the formation of a few conjugated carbon-carbon double bonds by partial enolization in alkaline solution. In fact, the $\Delta\varepsilon_r$-curves of the β-ketones resemble alkaline-neutral ionization difference curves of isoeugenol (49) and coniferyl alcohol (30). β-Carbonyl groups, if present in lignins, thus have no effect on the neutral spectra, though they may have a slight effect on the alkaline spectra and thus on the 310 mμ region of the $\Delta\varepsilon_r$-curve where etherified aryl-α-carbonyl groups absorb.

3. α-β Double Bonds. In a very detailed study of Brauns lignin prepared from various coniferous woods, Aulin-Erdtman and Hegbom (52) investigated the spectral effects of a double bond conjugated with the aromatic ring. Difference spectra between coniferyl alcohol or isoeugenol and 1-guaiacylpropane, representing the elimination of such double bonds by hydrogenation ($\Delta\varepsilon_h$-curve) showed maxima near 260 and 300 mμ. Comparison of $\Delta\varepsilon_h$-curves for the Brauns lignins with these model difference spectra indicated, after correction for conjugated carbonyl structures, the presence of isoeugenol-type double bonds in 3 to 6% of the phenylpropane units in these preparations.

Marton and Adler (42) studied the effect of a mild catalytic hydrogenation on the ultraviolet absorption spectrum of spruce milled wood lignin. Hydrogenation markedly lowered the absorption in the 300 to 400 mμ region, and alkaline-neutral ionization difference spectra indicated that reduction of phenolic conjugated α-keto groups had occurred (Figure 6.5). Since alkaline borohydride reduction of the hydrogenated milled wood lignin produced no further decrease of absorption in the 300 to 400 mμ region, the non-phenolic α-keto and coniferyl aldehyde groups had also been reduced. Hydrogenation and borohydride reduction difference curves differed in the region below 300 mμ. By comparison with the $\Delta\varepsilon_h$-curves for appropriate model compounds, the number of ethylenic double bonds and other chromophores (Table 6.4) in spruce milled wood lignin could be estimated.

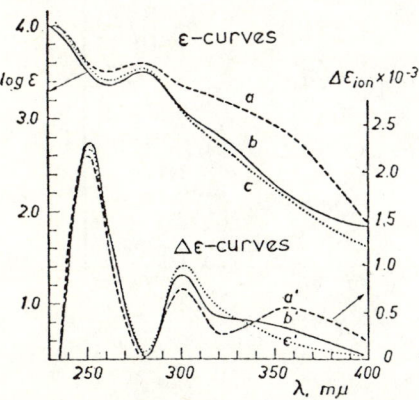

Figure 6.5. Ultraviolet absorption in neutral solution and $\Delta\varepsilon_i$-curves of spruce milled wood lignin (curves a and a') hydrogenated spruce milled wood lignin (curves b and b'), NaBH$_4$-reduced and subsequently hydrogenated spruce milled wood lignin (curves c and c'). Data by Marton and Adler (42).

Table 6.4. Estimation of Ethylenic Double Bonds and Other Chromophores in Spruce Milled Wood Lignin by Catalytic Hydrogenation (42)

Structure	λ_{max} values for $\Delta\epsilon_h$-curves		Percentage of total C_6C_3-units
	mµ	mµ	
R-O-4-(G)-CH=CH-CH$_2$OH	262-9	295-8	3
R-O-4-(G)-CH=CH-CHO	250	343	3-4
R-O-4-(G)-CO-R	270-80	305-8	5

The presence of ring-conjugated double bonds in milled wood lignin can be most clearly illustrated by first reducing the sample with borohydride and then subjecting the carbonyl-free product to catalytic hydrogenation (61). The spectral change on hydrogenation agrees well with the $\Delta\epsilon_h$-curve for coniferyl alcohol with maxima near 260 and 300 mµ (Figure 6.6).

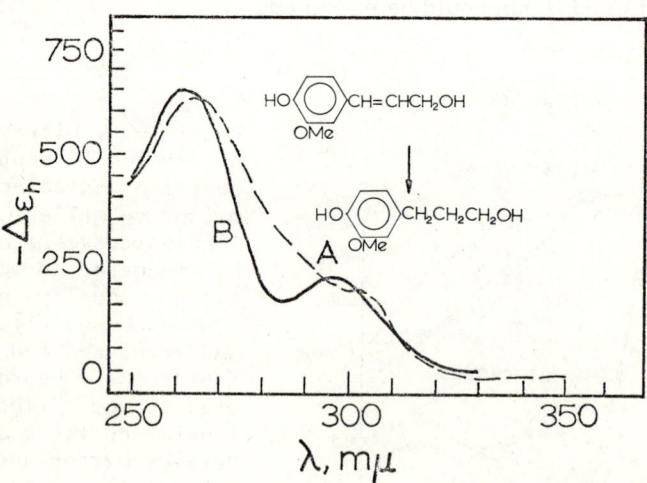

Figure 6.6. Hydrogenation $\Delta\epsilon$-curves. Curve a: NaBH$_4$-reduced spruce milled wood lignin; curve b: isopropylconiferyl alcohol (R = i-Pr) (61).

ULTRAVIOLET SPECTRA

Ring-conjugated double bonds are also responsible for the characteristic ultraviolet absorption spectra of the phenylcoumarone and stilbene structures that are formed by acid and alkaline treatment of lignins respectively. Upon heating the dimeric model 3 (λ_{max} 281 mμ, ε 5,900) with dilute HCl in methanol or dioxane-water (62), it is converted into the phenylcoumarone derivative 4 with λ_{max} 310 mμ (ε 26,400) in neutral, and λ_{max} 330 mμ (ε 31,800) in alkaline solution.

The product of a similar "acidolysis" (see Solvolysis, Chapter 9) of milled wood lignin, after $NaBH_4$-reduction to remove the interfering absorption of carbonyl groups, showed an "acidolysis" difference ($\Delta\varepsilon_a$) curve with a maximum at 310 mμ similar to that of the model coumarone 4. (The $\Delta\varepsilon_a$-curve was obtained by subtracting the spectrum of a similarly $NaBH_4$-reduced solution that had not been acid-treated from that of the acid-treated solution of the lignin sample.) Comparison of the $\Delta\varepsilon$-value at 310 mμ with the corresponding value for the model coumarone indicated that about 18% of the phenylpropane units in the milled wood lignin were present in dimeric elements of phenylcoumarane type 3.

By kraft cooking, the phenylcoumaran model 3 is converted into the trans-stilbene derivative 5 (63-65) which has a characteristic strong blue fluorescence and an ultraviolet absorption maximum in alkaline solution at 378 mμ. When spruce milled wood lignin was subjected to a similar kraft cook (63, 65), and subsequently reduced with $LiAlH_4$ in order to remove interfering carbonyl groups, the ionization $\Delta\varepsilon$-curve of the product had a similar maximum at 378 mμ (Figure 6.7). Comparison of the $\Delta\varepsilon_i$-value with that for model compound 5 indicated the presence of about 7 stilbene structures of this type per 100 phenylpropane units in the kraft lignin. Figure 6.7 also shows the $\Delta\varepsilon_i$-curve for the p,p'-dihydroxystilbene 6. This compound has been shown to be present in very small amounts in kraft lignin (63), where it is believed to originate from structures such as 7 which have recently been found in spruce milled wood lignin (66, 67).

A maximum at 377 mμ in the ionization difference curve of a borohydride-reduced pine kraft lignin (60) indicated the presence of stilbene units in the sample which had been obtained by acidification of an industrial kraft pulping liquor.

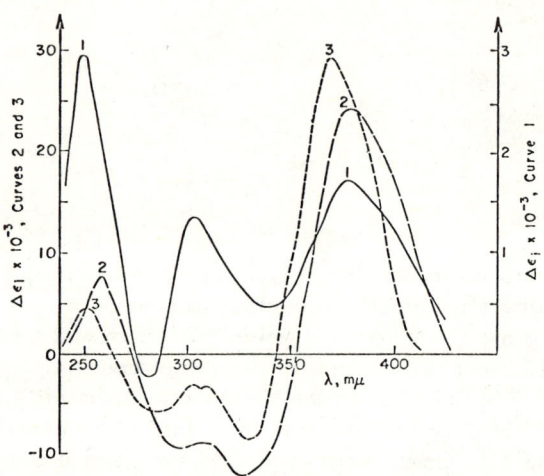

Figure 6.7. Ionization Δε-curves. Curve 1: LiAl$_4$-reduced kraft-cooked spruce milled wood lignin; curve 2: model compound 5; curve 3: model compound 6. Data by Falkehag, Marton and Adler (65).

4. Biphenyl Groups. By dehydrogenation coupling of simple guaiacyl models, Pew (45) prepared biphenyl derivatives such as 8 as models for various possible biphenyl structures in lignin.

Compared with the simple models, the maxima of the biphenyl derivatives were shifted towards longer wavelengths and the absorption at both 250

ULTRAVIOLET SPECTRA

and 300 mµ was high (Figure 6.8). From his investigation of the spectra of a large number of model compounds Pew concluded that the spectra of the biphenyl derivatives are more lignin-like than those of the simple models. He also suggested that by raising the absorption in the 250 and 300 mµ regions, a substantial number of biphenyl structures, if present in lignin, would explain much of the difference between the spectrum of lignin and that of the simple models seen in Figure 6.2. The presence of a small number of phenolic biphenyl units in western hemlock Brauns lignin and a larger number in Freudenberg's synthetic model lignin (DHP) had previously been established by the $\Delta \varepsilon$-method (51, 52). Aulin-Erdtman and Hegbom made these determinations by comparing the $\Delta \varepsilon_i$-curves of the lignin preparations in the 320 to 340 mµ region with that for model compound 8.

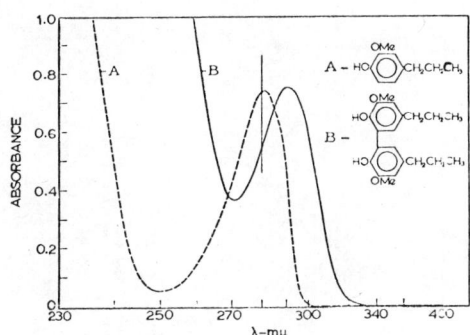

Figure 6.8. Ultraviolet absorption of 1-guaiacylpropane (curve A) and model compound 8 (curve B) (110).

Aulin-Erdtman and Sandén (68, 29) have investigated the effects of planarity and hydrogen bonding in various solvents on the ultraviolet absorption of biphenyl derivatives.

5. Summary. The ultraviolet absorption maxima of various guaiacylpropane structures containing the chromophoric groups discussed in the foregoing sections are listed in Table 6.5. The first column of values refers to structures with free phenolic groups in acidic or neutral solution, and to those with etherified phenolic groups in acid, neutral and alkaline solution. The second column refers to phenolic structures only in alkaline solution.

Table 6.5. Summary of Band Assignments for Guaiacyl Lignin Structures

Structure	Approximate λ_{max}-values, mμ	
	Non-ionized or etherified	Ionized
HO-⌬(OMe)-C-C-C	230 280	250 300
HO-⌬(OMe)-CO-C-C	240 280-310	250 350
HO-⌬(OMe)-C-C-C (OMe, 2)	220 250 290	230 265 310
HO-⌬(OMe)-CH=CH-C	260 300	290 320
HO-⌬(OMe, Me)-furan-⌬(OMe)-C	310	330
HO-⌬(OMe)-CH=CH-⌬(OMe, (HO))-OH (H) (C₃)	330	370-380
HO-⌬(OMe)-CH=CHCHO	250 350	260 400

B. Effect of Species.

Björkman and Person (31) determined ultraviolet absorption spectra in methylcellosolve solution of milled wood lignins from several softwoods and hardwoods. Except for small differences in absorbance in the 350 mμ region, the curves for two spruces (Picea abies and Picea mariana), scots pine (Pinus sylvestris), and western hemlock (Tsuga heterophylla) were very similar. Ionization difference spectra for the four preparations differed even less than the neutral curves. Similar spectra for aspen (Populus tremula), beech (Fagus sylvatica), and birch (Betula verrucosa) differed from the softwoods by having maxima at shorter wavelengths (near 275 instead of 280 mμ). The hardwood spectra differed from each other only in the 350 mμ region; their ionization difference curves, however, indicated somewhat larger differences in phenolic hydroxyl content.

The differences in the 350 mμ region probably indicate variations in the number of carbonyl groups introduced during the milling operation. The main difference between the softwood and hardwood lignins, the shift of the 280 mµ maximum to shorter wavelengths, results from the larger number of the syringyl groups in the hardwood lignins. In agreement with theoretical predictions (27) that symmetrical substitution displaces the absorption bands of benzene less than unsymmetrical substitution, the long wavelength bands of the unconjugated syringyl compounds generally appear at somewhat shorter wavelengths, and have appreciably lower absorptivities, than those of the corresponding guaiacyl derivatives. This is illustrated, for instance, by the spectra of 1-guaiacylpropane (29) λ_{max} 279-80 mµ, ϵ 2,800, and 1-syringylpropane (30) λ_{max} 272-3 mµ, ϵ 1,200. Ionization of the 4-hydroxy group (29) or conjugation with an α-carbonyl group destroys the similarity between the hydroxyl and methoxyl groups and, thus, the symmetry of substitution. Hence, ionized and conjugated syringyl derivatives tend to absorb at longer wavelengths than the corresponding guaiacyl derivatives.

In a recent series of papers on species variation in lignins, Sarkanen, Chang and Allan (69, 56) presented ultraviolet absorption data for various softwood and hardwood milled wood lignins. The methoxyl-to-phenylpropane group ratio for these samples varied from approximately 0.9 to 1.6. Data were obtained before and after borohydride reduction which eliminated the effects of conjugated carbonyl groups. The absorption at 280 mµ of the hardwood lignins was considerably lower than that of the softwood lignins. In fact, a plot of the 280 mµ maximum of the reduced softwood and hardwood lignins against their MeO/C_9 ratio resulted in a straight-line relationship (Figure 6.9). The single deviating point in this plot represented cottonwood (<u>Populus deltoides</u>) lignin in which the presence of <u>p</u>-hydroxybenzoate groups causes abnormally high absorption.

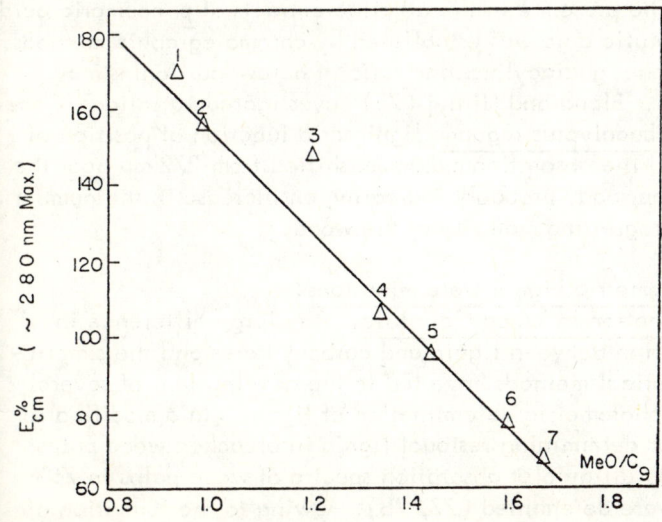

Figure 6.9. Absorptivities of the 280 mµ maxima of softwood and hardwood milled wood lignins as a function of the methoxyl : carbon ratio (1 and 2, conifer lignins; 2 through 7, hardwood lignins) (56).

Abnormal ultraviolet absorption spectra have previously been reported for Brauns lignins from aspen wood (Populus tremuloides) (70) and from sugar cane bagasse (71). Investigation of their spectra in alkaline solution proved, however, that the abnormal absorption was caused by the presence of ester linkages in the lignins of these species. Aspen Brauns lignin in alkaline solution (72) showed an abnormally high absorption band at 295 mµ which was stable in 0.01 N sodium hydroxide but which in 1.0 N sodium hydroxide was slowly replaced by a broad band at 280 mµ.

A similar change occurred on alkaline hydrolysis of methyl p-hydroxybenzoate; in alkaline solution, the ester absorbed at 295 mµ, the free acid, at 280 mµ. After alkaline hydrolysis of the aspen lignin, p-hydroxybenzoic acid was isolated in almost 10% yield, and the residue had a normal lignin spectrum. A Brauns lignin preparation from bagasse in alkaline solution (73) showed an absorption maximum at 355 mµ which shifted in 1.0 N sodium hydroxide to 330 mµ. After acidification of the solution, p-coumaric acid was isolated by ether extraction followed by recrystallization. The shift of the absorption band closely paralleled that observed during the alkaline hydrolysis of methyl p-coumarate suggesting that a similar ester linkage was present in the bagasse lignin. In both cases, the shift of the absorption maximum to shorter wavelengths which accompanies the transition from ester to acid salt results from the decreased electron-withdrawing effect of the carboxylate group.

The presence of similar ester linkages between p-coumaric acid and ferulic acid and lignin in milled wood lignin preparations from several other grasses has recently been confirmed (74) with the help of ultraviolet absorption. Neutral spectra of the grass lignins in methylcellosolve showed, in addition to the 280 mµ absorption, a maximum at 315 mµ which disappeared after saponification. The presence in the alkaline extracts of p-coumaric acid and small amounts of ferulic acid was established by chromatographic methods.

The syringylpropane:guaiacylpropane ratio of hardwood lignins may vary even within a tree. Bland and Hirnyj (75) investigated variations in the ultraviolet spectrum of Eucalyptus regnans lignin as a function of position of the sample in the tree. The absorption maximum shifted from 272 mµ near the pith to 262 mµ in the sapwood, probably indicating an increase in the number of syringyl units with progressing maturity of the wood.

IV. Spectrophotometric Lignin Determinations.

A. Determination in Wood Cellulose. The large difference in ultraviolet light absorption between lignin and carbohydrates and the simplicity and reliability of optical methods have led to the development of several methods for the spectrophotometric determination of lignin. In a modification of an optical method for determining residual lignin in bleached wood pulps using visible light (76), ultraviolet absorption spectra of wood pulps in 76% sulfuric acid solution were determined (77, 78). Owing to the formation of

5-(hydroxylmethyl)-furfural from hexoses and furfural from pentose, the absorbance of such solutions increases on standing. However, the increase can be minimized by carefully diluting the solution, after the sample is completely dissolved, to about 60% sulfuric acid. Changes in absorbance are most rapid near 280 mµ, close to the absorption maxima of both 5-(hydroxymethyl)-furfural (79, 80) and furfural (81), and least near 200 mµ, where lignin absorbs strongly. The absorbance at 220 mµ was found to correlate well with lignin content of the pulps. In another modification, wood pulps were dissolved in 80% phosphoric acid (82, 83). In this solvent, furfural formation is sufficiently slow for the 280 mµ maximum to be used as a measure of the lignin content of the sample. Sjöström and Enström (83a) determined residual lignin in wood pulps in amounts ranging from a few percent to about 0.05% of the pulps by measuring the absorbance at 280 or 300 mµ of their cadoxen solutions.

Ultraviolet absorption spectra played an important part in establishing the presence of soluble lignin in the acid filtrates from Klason lignin determinations (84-86). Examination of chlorite holocellulose preparations demonstrated (85, 86) that the presence of undetermined acid-soluble lignin in the holocellulose was the reason for the sum of the lignin in the wood and of the holocellulose corrected for acid-insoluble lignin to exceed 100%. Browning and Bublitz (86) showed that the error may be greatly diminished by estimating from ultraviolet absorbance the amount of soluble lignin present in the acid filtrates. They also suggested that the interference with this estimate by furan-type compounds, formed mainly during the refluxing step in the lignin determination, may be minimized by calculating the soluble lignin concentration from the absorbance of the filtrate at two wavelengths.

Using absorptivity values for lignin and for carbohydrate degradation products obtained, respectively, from spruce Brauns lignin and from a synthetic mixture of glucose, xylose, mannose, and glucuronolactone which had been subjected to the hydrolysis conditions of the lignin determination (86) the following equations may be written:

$$a_{280} = 0.68 c_D + 18 c_L$$

$$a_{215} = 0.15 c_D + 70 c_L$$

Where a_{280} and a_{215} are the absorbance values of the lignin filtrate, 0.68 and 0.15 the absorptivities of carbohydrate degradation products, 18 and 70 the absorptivities of lignin at 280 and 215 mµ, respectively, and c_D and c_L the concentrations in g/l of carbohydrate degradation products and of soluble lignin in the filtrate. By solving the simultaneous equations, the following expression for the soluble lignin concentration in the filtrate is obtained:

$$c_L = \frac{4.53\, a_{215} - a_{280}}{300}$$

The amount of acid-soluble lignin present in different types of samples varies appreciably. Coniferous wood generally contains only a few tenths of a percent of soluble lignin; chlorite holocellulose preparations, depending on the extent of delignification, may contain more than 10% soluble lignin. In such samples, the amount of soluble lignin greatly exceeds that of acid insoluble lignin and correcting the lignin content for soluble lignin is essential. Although the absorptivity values for lignin and carbohydrate degradation products may be somewhat uncertain, as stressed by Browning and Bublitz (86), this error is small compared with that resulting from not correcting for soluble lignin.

This correction is very important also in the case of hardwoods. Pearl and Busche (87) who studied the Klason lignin determination as applied to aspen wood (Populus tremuloides) found that approximately 15% of the total lignin content of this wood is acid soluble. The amount of soluble lignin was estimated from the absorbance at 208 mµ using an absorptivity value of 105 $1 g^{-1} cm^{-1}$. Swan (88) investigated the soluble lignin from Klason lignin determinations on birch (Betula verrucosa) and eucalyptus (E. globulus) wood. Using the absorbance of the filtrate at 205 mµ and an absorptivity of 110 $1 g^{-1} cm^{-1}$, Swan estimated the soluble lignin of both woods to correspond to 12-13% of the total lignin content.

Johnson, Moore and Zank (89) developed a spectrophotometric lignin determination in which a small sample of wood meal is treated with a mixture of acetyl bromide and acetic acid. After destruction and removal of excess reagent, the absorbance of the resulting solution is read at 280 mµ. With minor modifications, this method may be applied also to unbleached wood pulps (90). The method may be calibrated by dissolving a known amount of an appropriate lignin standard such as milled wood lignin, purified lignin sulfonate, or isolated kraft lignin in the acetyl bromide mixture and determining its absorptivity. Advantages of the method are the small amount of sample needed (5 to 35 mg) and the fact that the total lignin is determined spectrophotometrically by one method rather than as the sum of Klason and soluble lignin.

Ultraviolet absorption spectra of isolated lignins and of wood pulps in less than milligram amounts have been determined in pressed potassium chloride disks (91). Absorption at 220 and 280 mµ was found to be proportional to the lignin content of the pulps as measured by their kappa number, but particle size of the pulp remained a critical factor in obtaining satisfactory spectra.

B. Determination in Pulping Liquor. Ultraviolet spectrophotometric methods are especially useful for the determination of lignin or lignin derivatives present in solution. For instance, Patterson et al. (92) suggested using the absorbance at 280 mµ for the determination of lignin sulfonates in sulfite pulping liquor stating that the various other available methods were time consuming and poor in reproducibility. From a study of the short-wave length absorption of lignin, Kleinert and Joyce (93) concluded that, compared with the 280 mμ maximum, the maximum near 205 mμ is less influenced by carbohydrate degradation products and provides a more accurate means of following delignification during sulfite cooking. Experiments with a flow-microcuvette with a variable short light path and with a mixing device that automatically diluted digestor strength liquor 1:500 showed that the 205 mμ-absorbance could be measured continuously during technical sulfite cooks. The same method was found applicable also to kraft cooking (94). Schöning and Johansson (95), who reinvestigated the short-wave length absorption of sulfite cooking liquor, found that initially, when little lignin was present, the absorption maximum was due to SO_2 and occurred at 196 mμ. As the cook progressed and the SO_2 concentration diminished, the absorption maximum was displaced to longer wave lengths. At the end of the cook, the wave length became constant at about 204 mμ for spruce, and at about 206 mμ for birch sulfite cooks. Ivancic and Rydholm (96) investigated the effects of various reactions including acid condensation and chlorination on the ultraviolet spectrum and the visible color of a lignin sulfonate isolated from a mild sulfite cook. Acid condensation increased the ultraviolet absorption of the lignin sulfonate, especially in the 280 mμ region; chlorination affected the shape of the absorption spectrum but left the absorptivity at 280 mμ relatively unchanged at a chlorine consumption (5-6 Cl/OCH_3) that caused a large increase in absorption of visible light by the lignin sulfonate. A subsequent alkali treatment greatly diminished the absorption in the visible without an appreciable change of the ultraviolet spectrum, indicating that the alkali hydrolized the chlorine but did not degrade the aromatic rings.

The relative merits of using the maxima near 205 or 280 mμ for following the solubilization of lignin during sulfite cooking have more recently been discussed again (97, 98). Sjöström and Haglund (97) found that in both acid and neutral sulfite cooks, SO_2 contributed appreciably to the short-wave length absorption near 205 mμ, even toward the end of the cook. In acid-sulfite cooks, satisfactory results were obtained by measuring the absorption at 280 mμ, but not in bisulfite cooks at high temperature (160°C). In the latter cook, however, there was less interference from SO_2 near 205 mμ because the SO_2: lignin ratio was low at the end of the cook. Discussing the application of ultraviolet absorption measurements for determining the end point of a cook, Sjöström and Haglund concluded that for chemical pulps, the change in absorption at the end of the cook is much greater in the visible than in the ultraviolet range of the spectrum and that the change in ultraviolet absorption is probably too small for an accurate determination of the endpoint except perhaps in semi-chemical cooks.

An instrument for the continuous recording of the absorbance of cooking liquors has recently been designed (99) utilizing the strong emission line of zinc at 213.8 mμ as a light source. Laboratory and limited mill trials indicated the potential usefulness of this instrument for improving uniformity in cooking of eucalyptus kraft and P. radiata bisulfite pulps.

In spite of possible interferences, ultraviolet light absorption measurements are frequently used for the determination of lignin or lignin derivatives in solution. For instance, such measurements represent the most convenient means of estimating lignin sulfonate concentrations in spent sulfite liquor. Because of the presence of carbohydrate degradation products in these liquors, such measurements should not be made at 280 mµ. Since SO_2 may interfere with measurements at 205 mµ, the best choice of wavelength for this measurement perhaps is at the shoulder of the lignin spectrum near 230 mµ. The absorptivity to be used in such determinations depends on the type of wood being pulped and on the degree of sulfonation of the lignin sulfonate and must be determined using a representative sample of the particular lignin sulfonate carefully purified, i.e., by dialysis. For the estimation of purified, isolated lignin sulfonates, the 280 mµ maximum may safely be used. From spruce (P. abies) cooks of 2, 4, and 8 hours duration, Forss et al. (100) isolated lignin sulfonates of different molecular weights by gel permeation chromatography. They found that the absorptivity at 280 mµ, 3080 l (mole MeO)$^{-1}$ cm^{-1}, did not differ appreciably between fractions of less than 7,000 and more than 40,000 molecular weight, or for different cooking times. Although molecular weight does not affect the absorptivity of lignin sulfonates at 280 mµ, it may influence the shape of the absorption curve. Felicetta, Ahola and McCarthy (101) found that both the ratios of the absorbance at the 280 mµ maximum to that at the 260 mµ minimum, as well as that of the absorbance at 280 mµ to that at 310 mµ, decreased in an ordered way with increasing molecular weight of a series of lignin sulfonates. The second ratio showed a nearly linear decrease with the logarithm of molecular weight.

C. Determination in Receiving Waters. Increasing efforts towards maintenance of clean water have created the need for sensitive methods of estimating low concentrations of spent pulping liquors in receiving waters. The use of ultraviolet absorption measurements for this determination has been suggested (102). However, except in pure fresh water, high ultraviolet background absorbance, such as is found for instance in sea water, generally prevents the quantitative estimation of low lignin sulfonate concentrations by this method. In the Pacific Northwest of the United States, the Pearl-Benson method (103) is in common use (104). This method is based upon the reaction between acidified nitrite and phenolic units of the lignin which results in the formation of a nitrosophenol. Upon addition of alkali, a highly colored quinone monoxime is formed which has an absorption maximum at about 430 mµ The method has recently been standardized by Barnes et al. (105). Certain substances, such as most phenols, tannins, and some amines react with nitrous acid in a way similar to lignin sulfonates and therefore interfere with the determination (104, 106). In the absence of "background" absorbance by the water and of interfering substances, the presence of as little as 0.2 to 0.5 ppm of spent sulfite liquor solids can be determined by this method. In an interesting comparison between the nitroso lignin method and ultraviolet

ULTRAVIOLET SPECTRA

absorption for the estimation of spent sulfite liquor (107), it has been suggested that the two methods are best used in conjunction with one another.

Recent results obtained by fluorescence spectrophotometry (108, 109) indicate that the fluorometric method may permit the detection of lignin sulfonates in even lower concentrations than does either the nitroso lignin method or ultraviolet absorption spectroscopy. Excitation with light of 290 to 360 mμ results in a characteristic fluorescence of lignin sulfonates with a maximum intensity near 400 mμ. Although fluorescence is affected by pH and temperature, it is reported (108) that the fluorescence is insensitive to pH variations in the range of pH 4 to 7, and the temperature dependence is linear and relatively small between 5 and 30°C. It is claimed that fluorometry is at least ten times more sensitive than the nitroso lignin determination, and comparable in accuracy and precision. Moreover, use of selected wavelengths for excitation and fluorescence determination combined with the effect of pH changes is said to provide means for differentiating between kraft and sulfite lignins (109).

REFERENCES

1. R. O. Herzog and A. Hillmer, Chem. Ber., 60, 365 (1927); Z. Physiol. Chem., 168, 117 (1927).
2. R. O. Herzog and A. Hillmer, ibid., 62, 1600 (1929), 64, 1288 (1931); Papier-Fabr., 29, Festschrift, 40 (1931) 30, 205 (1932); A. Hillmer, Chem. Ber., 66, 1600 (1933); A. Hillmer and P. Schorning, Z. Physik. Chem., A167, 407 (1933), A168, 81 (1934).
3. E. Hägglund and F. W. Klingstedt, Svensk Kem. Tidskr. 41, 185 (1929); Z. Physik. Chem., 152, 295 (1931).
4. R. E. Glading, Paper Trade J., 3, No. 23, 32 (1940).
5. R. F. Patterson and H. Hibbert, J. Am.Chem. Soc., 65, 1862 (1943); ibid., 65, 1869 (1943).
6. G. Aulin-Erdtman, Svensk Papperstidn., 47, 91 (1944).
7. G. Aulin-Erdtman, Tappi, 32, 160 (1949).
8. E. J. Jones, Jr., Tappi, 32, 311 (1949).
9. P. W. Lange, Svensk Papperstidn., 47, 262 (1944).
10. T. Caspersson, J. Royal Microsc. Soc., 60, 8 (1940).
11. P. W. Lange, Svensk Papperstidn., 50, No. 11B, 130 (1947).
12. A. R. Procter, W. O. Yean, and D. A. I. Goring, Pulp and Paper Mag. Canada, 68, T-445 (1967).
13. G. Jayme and H. F. Torgersen, Holzforschung, 21, 110 (1967).
14. F. E. Brauns, The Chemistry of Lignin, Academic Press, New York, 1952, pp. 217-230.
15. F. E. Brauns and D. A. Brauns, The Chemistry of Lignin, Suppl. Vol., Academic Press, New York, 1960, pp. 199-219.
16. L. Doub and J. M. Vandenbelt, J. Am. Chem. Soc., 69, 2714 (1947).

17. L. Doub and J. M. Vandenbelt, ibid., 71, 2414 (1949).
18. A. I. Scott, Interpretation of the Ultraviolet Spectra of Natural Products, Macmillan, New York, 1964, p. 8.
19. H. H. Jaffé, D. L. Beveridge, and M. Orchin, J. Chem. Ed., 44, 383 (1967).
20. H. H. Jaffé and M. Orchin, Theory and Applications of Ultraviolet Spectroscopy, Wiley, New York, 1962, pp. 104, 129-30.
21. L. N. Ferguson, The Modern Structural Theory of Organic Chemistry, Prentice-Hall, New Jersey, 1963, p. 481.
22. H. L. Hergert, Personal Communication.
23. L. Doub and J. M. Vandenbelt, J. Am. Chem. Soc., 77, 4535 (1955).
24. A. L. Sklar, J. Chem. Phys., 10, 135 (1942).
25. Th. Förster, Z. Naturforsch., 2a, 149 (1947).
26. J. R. Platt, J. Chem. Phys., 17, 484 (1949).
27. P. E. Stevenson, J. Chem. Ed., 41, 234 (1964).
28. A. C. Albrecht and W. T. Simpson, J. Chem. Phys., 23, 1480 (1955).
29. G. Aulin-Erdtman and R. Sandén, Acta Chem. Scand., 22, 1187 (1968).
30. G. Aulin-Erdtman and L. Hegbom, Svensk Papperstidn., 60, 671 (1957).
31. A. Björkman and B. Person, Svensk Papperstidn., 60, 158 (1957).
32. G. Aulin-Erdtman, A. Björkman, H. Erdtman, and S. E. Häglund, Svensk Papperstidn., 50, No. 11B, 81 (1947).
33. K. Freudenberg and G. Schuhmacher, Sitzber. Heidelberg. Akad. Wiss. Math.-Naturw. Kl., 1953/1955, 127 (1956), Fig. 14.
34. C. L. Hess, Tappi, 35, 312 (1952).
35. K. Tiyama, J. Nakano, and N. Migita, J. Japan.Wood Res.Soc., 13, 125 (1967).
36. R. D. B. Fraser and E. Suzuki, Anal. Chem., 38, 1770 (1966).
37. G. Aulin-Erdtman, Svensk Papperstidn., 55, 745 (1952).
38. O. Goldschmid, J. Am. Chem. Soc., 75, 3780 (1953).
39. G. Aulin-Erdtman, Chem. & Ind., 1955, 581.
40. K. Sarkanen and C. Schuerch, J. Am. Chem. Soc., 79, 4203 (1957).
41. E. Adler and J. Marton, Acta Chem. Scand., 13, 75 (1959).
42. J. Marton and E. Adler, Acta Chem. Scand., 15, 370 (1961).
43. Ref. 18, p. 91.
44. E. P. Crowell, W. A. Powell, and C. J. Varsel, Anal. Chem., 35, 185 (1963).
45. J. C. Pew, J. Org. Chem., 28, 1048 (1963).
46. R. A. Morton and A. L. Stubbs, J. Chem. Soc., 1940, 1347.
47. H. W. Lemon, J. Am. Chem. Soc., 69, 2998 (1947).
48. G. Aulin-Erdtman, Svensk Papperstidn., 56, 91 (1953).
49. G. Aulin-Erdtman, ibid., 56, 287 (1953).
50. G. Aulin-Erdtman, ibid., 57, 745 (1954).
51. G. Aulin-Erdtman and L. Hegbom, ibid., 59, 363 (1956).
52. G. Aulin-Erdtman and L. Hegbom, ibid., 61, 187 (1958).

53. O. Goldschmid, Anal. Chem., 26, 1421 (1954).
54. A. S. Wexler, ibid., 36, 213 (1964).
55. O. Goldschmid and L. F. Maranville, Abstracts, 130th Meeting, Am. Chem. Soc., 12E (1956).
56. K. V. Sarkanen, H.-M. Chang, and G. G. Allan, Tappi, 50, 587 (1967).
57. K. Freudenberg and K. Dahl, Naturwiss., 42, 606 (1955).
58. G. Aulin-Erdtman and R. Sandén, Paper and Timber, Helsinki, 43, 671 (1961).
59. D. C. C. Smith, Nature, 176, 927 (1955).
60. J. Marton, Tappi, 47, 713 (1964).
61. E. Adler, Paper and Timber, Helsinki, 43, 634 (1961).
62. E. Adler, S. Delin, and K. Lundquist, Acta Chem. Scand., 13, 2149 (1959); E. Adler and K. Lundquist, ibid., 17, 13 (1963).
63. I. Falkenhag, Paper and Timber, Helsinki, 43, 655 (1961).
64. E. Adler, J. Marton, and I. Falkenhag, Acta Chem. Scand., 18, 1311 (1964).
65. S. I. Falkenhag, J. Marton, E. Adler in R. F. Gould, Ed., Adv. Chem. Ser., 59, Am. Chem. Soc., Washington, D. C., 1966, p. 75.
66. K. Freudenberg, C.-L. Chen, J. M. Harkin, H. Nimz, and E. Renner, Chem. Commun., 1965, 224.
67. K. Lundquist and G. E. Miksche, Tetrahedron Letters, 1965, 2131.
68. G. Aulin-Erdtman and R. Sandén, Acta Chem. Scand., 17, 1991 (1963).
69. K. V. Sarkanen, H.-M. Chang, and G. G. Allan, Tappi, 50, 583 (1967).
70. M. A. Buchanan, F. E. Brauns, and R. L. Leaf, J. Am. Chem. Soc., 71, 1297 (1949).
71. G. de Stevens and F. F. Nord, ibid., 73, 4622 (1951).
72. D. C. C. Smith, J. Chem. Soc., 1955, 2347.
73. D. C. C. Smith, Nature, 176, 267 (1955).
74. T. Higuchi, Y. Ito, M. Shimada, and I. Kawamura, Phytochem., 6, 1551 (1967).
75. D. E. Bland and B. Hirnyj, Nature, 179, 529 (1957).
76. H. W. Giertz, Svensk Papperstidn., 48, 485 (1945).
77. O. Goldschmid and D. W. Balkema, Abstracts, 114th Meeting, Am. Chem. Soc., 4D (1948).
78. F. Loschbrandt, Norsk Skogind, 4, 130 (1950).
79. M. L. Wolfrom, R. D. Schuetz, and L. F. Cavalieri, J. Am. Chem. Soc., 70, 514 (1948).
80. B. Singh, G. R. Dean, and S. M. Cantor, J. Am. Chem. Soc., 70, 517 (1948).
81. M. L. Wolfrom, R. D. Scheutz, and L. F. Cavalieri, J. Am. Chem. Soc., 71, 3518 (1949).
82. P. O. Bethge, G. Gran, and K. Ohlsson, Svensk Papperstidn., 55,

44 (1952).
83. D. E. Bland, Appita, 10, 287 (1956).
83a. E. Sjöström and B. Enström, Svensk Papperstidn., 69, 469 (1966).
84. A. v. Wacek and D. Schroth, Holz Roh-Werkstoff, 8, 7 (1951).
85. W. G. Campbell and I. R. C. McDonald, J. Chem. Soc., 1952, 2645, 3180.
86. B. L. Browning and L. O. Bublitz, Tappi, 36, 4521 (1953).
87. I. A. Pearl and L. R. Busche, ibid., 43, 961 (1960).
88. B. Swan, Svensk Papperstidn., 68, 791 (1965).
89. D. B. Johnson, W. E. Moore, and L. C. Zank., Tappi, 44, 793 (1961).
90. J. Marton, ibid., 50, 335 (1967).
91. H. I. Bolker and N. G. Somerville, ibid., 45, 826 (1962).
92. R. F. Patterson, J. L. Keays, J. S. Hart, R. K. Strapp, and P. Luner, Pulp Paper Mag. Can., 52, No. 12, 105 (1951).
93. T. N. Kleinert and C. S. Joyce, Tappi, 40, 813 (1957).
94. T. N. Kleinert and C. S. Joyce, ibid., 41, 372 (1958).
95. A. G. Schöning and G. Johannson, Svensk Papperstidn., 62, 646 (1959).
96. A. Ivancic and S. A. Rydholm, ibid., 62, 554 (1959).
97. E. Sjöström and P. Haglund, Tappi, 47, 286 (1964).
98. R. J. Orsler and D. F. Packman, Svensk Papperstidn., 67, 855 (1964).
99. D. J. Williams, Appita, 22, 46 (1968).
100. K. Forss, O. Schott, and B. Stenlund, Paper and Timber, Helsinki, 49, 525 (1967); K. Forss and K.-E. Fremer, Tappi, 47, 485 (1964).
101. V. F. Felicetta, A. Ahola, and J. L. McCarthy, J. Am. Chem. Soc., 78, 1899 (1956).
102. E. Treiber, T. Kleinert and W. Wincor, Holzforschung, 6, 101 (1952).
103. I. A. Pearl and H. K. Benson, Paper Trade J., 111, No. 18, 35 (1940).
104. V. F. Felicetta and J. L. McCarthy, Tappi, 46, 337 (1963).
105. C. A. Barnes, E. E. Collias, V. F. Felicetta, O. Goldschmid, B. F. Hrutfiord, A. Livingston, J. L. McCarthy, G. L. Toombs, M. Waldichuk, and R. Westley, Tappi, 46, 347 (1963).
106. O. Goldschmid and L. F. Maranville, Anal. Chem., 31, 370 (1959).
107. G. J. Jayme and E. Pohl, Das Papier, 21, 645 (1967).
108. R. F. Christman and R. A. Minear, The Trend, 19, 3 (1967).
109. W. C. Wilson, Proceedings of the 13th Pacific Northwest Industrial Waste Conference, Washington State U., Pullman, April 6-7, 1967.
110. J. C. Pew, W. J. Connors, A. Kunishi, Chim. Biochim. Lignine, Cellulose, Hemicellulose, Grenoble, 1964, p. 229.
111. G. Aulin-Erdtman, Svensk Kem. Tidskr., 70, 145 (1958).

7

INFRARED SPECTRA

Herbert L. Hergert

I. Introduction. To some people, infrared spectroscopy appears to be a fully developed technique and is overshadowed by some of the more recent analytical tools such as nuclear magnetic resonance and mass spectrometry. To others, as Whetsel notes (1), it is a rapidly growing and changing field marked by new developments in instrumentation, technique, data handling and application. In other words, infrared spectroscopy is a relatively old technique, but it is also a dynamic one that offers much promise for the future.

The first comprehensive study of lignin infrared spectra was made by Jones at the Institute of Paper Chemistry and was reported in 1948 (2, 3). This pioneering study was closely followed by Freudenberg's comparison of the infrared spectra of his synthetic lignin (DHP) with spruce wood lignin (4, 5), Kratzl's study of lignin in situ (6, 7), Schubert and Nord's comparison of Brauns lignin (BL) with enzymatically liberated lignin (8-11), Hess' investigation of the homogeneity of black spruce BL (12), and Hergert and Kurth's attempt to diagnose the nature of lignin carbonyl groups by measurement of the spectra of naturally occurring ketonic model compounds (13). Since that time more than 130 studies using infrared spectroscopy in whole or part for the investigation of lignin have appeared in the literature. While some of these studies have been reviewed in the previous lignin chemistry textbooks, the recent appearance of detailed assignment studies on lignin and lignin model compounds, such as that reported by Sarkanen and coworkers (14-16) and Sundholm (17, 18), invites a critical reappraisal of the literature which has appeared thus far.

Emphasis will be placed in this chapter on those facets of lignin chemistry in which infrared spectroscopy has uniquely served to provide answers to difficult problems. Foremost among these is the identity of a given isolated lignin preparation with lignin as it exists in wood. The only successful approach to this problem thus far has been the comparison of the spectrum of an isolated lignin with the difference spectrum between the original and a delignified wood section. Infrared spectroscopy has also been exceedingly useful in the comparison of lignin isolated by different techniques and in ascertaining possible differences in lignin isolated by the same procedure from different species. Discussion of this subject would not, of course, be complete without mention of those areas where exploration and future discoveries still remain to be made.

II. Methods of Measurement.
  A. Effect of State upon Spectral Measurement. The infrared spectrum is one of the most characteristic properties of a compound (19, 20). On the other hand, the spectrum in a gaseous or liquid state shows pronounced differences. In the gaseous state, the molecules are sufficiently distant from each other so that the spectrum consists mainly of rotation-vibration bands. The spectrum of a liquid usually contains no rotational fine structure, and irregular intermolecular interactions cause shifts and broadening of vibrational absorption bands. In the crystalline state, the interactions are the same for each molecule. This results in a sharpening of bands by going from the liquid to the solid state.

Spectral differences are usually noted in crystalline modifications of the same compound, hence solution spectra are normally preferred for infrared studies involving assignment of group frequencies. The criteria for a good solvent are few or no infrared absorption bands in the region of interest and minimal molecular interaction with the solute. Unfortunately, polymers usually either have low solubility in solvents of any kind or are only soluble in highly polar solvents which have a number of strong, interfering absorption bands. Consequently, the infrared spectra of polymers are normally obtained as films cast from solutions, pressed or melted films, microtomed samples, paraffin oil or flurocarbon mulls or potassium bromide wafers (21, 22).

Since lignin, depending upon the mode of isolation, is either insoluble in organic solvents or is only soluble in polar solvents which form strong hydrogen bonds with lignin, lignin infrared spectra are usually measured on finely ground solid samples dispersed in paraffin mulls and potassium bromide disks or as thin films obtained by evaporation of a solution on a salt plate. The only available solution studies are those of Jones (2, 3), Lindberg (23-25), Reznikov, Pilipchuk and Solovev (26), Ekman and Lindberg (27) and Smith (28, 29) in which the spectra of spruce and aspen lignins in dioxane were reported. Additionally, Michell (30) has recently made a comprehensive survey of the hydroxyl and carbonyl stretching frequencies of lignin and lignin model compounds in the solid state and in dichloromethane, carbon tetrachloride, dioxane, methyl cellosolve and pyridine solutions.

To minimize loss of incident radiation by scattering, the particle size of a solid sample must be minimized by grinding. Generally speaking, this is easier to accomplish with lignin by the paraffin mull technique than with potassium bromide wafers. As a consequence, paraffin mulls show somewhat better absorption band resolution, especially in the carbonyl stretching region. A second disadvantage of wafers compared with mulls is the pickup of

INFRARED SPECTRA

moisture which manifests itself as a particularly troublesome, interfering peak at about 1625 cm^{-1}. This problem can, however, be circumvented by deuterium oxide exchange according to the recently published procedure of Sarkanen, Chang and Allan (15). Wafer spectra of lignin, aside from the aforementioned defects, offer obvious advantages over mulls: Semiquantitative work can be carried out by controlling pressing conditions and the weight ratio of lignin and potassium bromide. Finally, potassium bromide possesses no interfering absorption bands in the 4,000-600 cm^{-1} region as do paraffin or fluorocarbon oil-mulling media.

B. <u>Near and Far Infrared Spectroscopy</u>. The spectral region between the visible and the rock salt area of the infrared has been designated as the near infrared (15,000-4,000 cm^{-1} or 0.8 - 2.5 μ). Spectra of solids in this region are usually measured in solution on quartz prism or grating-type instruments. Since the absorption bands observed in the near infrared largely originate as overtones of the fundamental stretching frequencies of C-H, O-H or N-H bonds together with various combination bands, any information to be gained concerning the molecular structure of lignin is likely to be limited. This probably explains why only one study of the near infrared spectrum of lignin has appeared in the literature thus far. Ziechman (31) has compared the near infrared spectrum of oak lignin with that of various humic substances and peat and has reported absorption bands at 1.42 and 1.92 μ (7,000 and 5,200 cm^{-1}) (see Figure 7.1) which were respectively assigned to an overtone of the O-H valence vibration and a combination of effects caused by hydroxyl and carbonyl vibrations. The anticipated absorption band for aromatic C-H at 7,700 cm^{-1} was obscured by the 7,050 cm^{-1} band. Shoulders at 5,900 cm^{-1}, ascribed to aromatic C-H, and at 4,750 cm^{-1}, assigned to phenolic hydroxyl groups, were also noted in the spectrum.

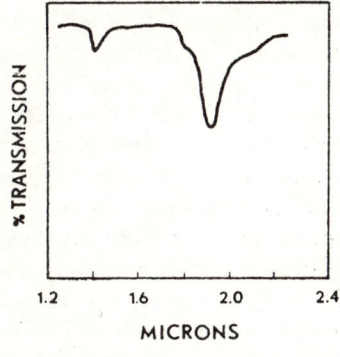

Figure 7.1. Near infrared spectrum of oak lignin (31).

The recent availability of improved filter-grating infrared spectrophometers has greatly facilitated measurements in the far infrared region (650 - 50 cm^{-1}). Various skeletal bendings are the most significant source of absorption from organic compounds in this region, but a high degree of uniqueness from compound to compound permits only a few broad generalizations concerning this part of the spectrum. Specifically, vibrations originating from carbon-bromine, carbon-iodine and carbon-chlorine bonds, out-of-plane carbon-hydrogen bonds, and inorganic and metal-oxygen or metal-nitrogen bonds in complexes are of importance in the 625-400 cm^{-1} region. A fundamental study of the infrared absorption of phenol (32) indicates in-plane monomeric C-OH bending at 400 cm^{-1}, out-of-plane monomeric OH bending at 320 cm^{-1}, and out-of-plane monomeric C-OH bendings at 272 and 235 cm^{-1}. These assignments should be of help in subsequent studies of the phenolic character of lignin.

As in the case of near infrared spectra, investigation of the far infrared region of lignin has been meager. Bolker and Somerville (33) report that softwood lignins generally have bands at 560 and 470 cm^{-1}, while hardwood lignins absorb at 535 cm^{-1}. Lignosulfonic acids have, in addition, bands at 655 and 540-520 cm^{-1} assignable to the sulfonic acid grouping. A very weak band at 630 cm^{-1} is found in thiolignins and is assigned to a C-S vibration. Eucalyptus lignin has been reported by Michell, Watson and Higgins (34) to have bands at 620, 530 and 470 cm^{-1}. The first of these was unassigned, while the 530 and 470 cm^{-1} bands were attributed to the C-O-C bending mode and deformation of the aromatic ring, respectively. Further study of lignin in the far infrared is obviously required to exploit the information which is potentially available.

III. <u>Interpretation of Spectra</u>. There has been a temptation when working with polymer spectra, especially in the field of natural products chemistry, to conclude that two or more products are identical because their spectra were "very similar." In actuality, even slight differences in spectra, i.e., relative heights of neighboring absorption bands, the presence of weak shoulders, etc., are of considerable significance. They may signal differences in the ratio of constituent monomeric units and the order in which they are linked. Depending upon the intractability of the material and the consequent ease in obtaining a good spectrum, up to 5% of a different monomer may be introduced into a polymer without significant changes in the spectrum. Nonetheless, study of the differences in spectra of various lignin products and their significance to structure is probably the most valuable area of study in the field of lignin spectra. To be fruitful, however, correct assignment of the absorption bands in the lignin spectrum is a necessary prerequisite.

A. <u>Lignin Absorption Band Assignment</u>. While there are significant differences in lignin infrared spectra which are a function of the method of isolation (see Figure 7.2), two major categories may be recognized, viz.,

INFRARED SPECTRA

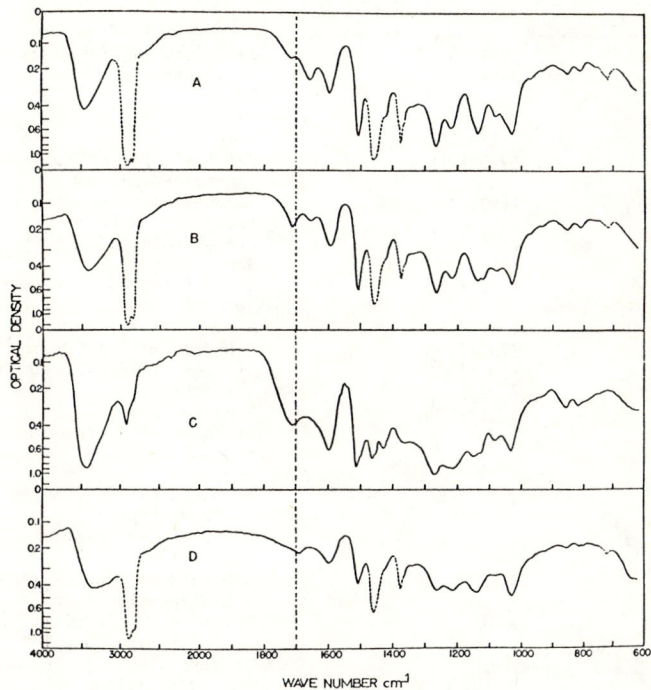

Figure 7.2. Infrared spectra of lignin isolated by different methods from western hemlock wood: (A) Milled wood lignin (mull), (B) Dioxane-HCl-lignin (mull), (C) Kraft lignin (KBr wafer), and (D) Lignosulfonic acid (mull); Dotted lines indicate bands due to paraffin in mulls (43).

guaiacyl and guaiacyl-syringyl lignin spectra. The major absorption band frequencies and the most probable assignment of each band in these two types of spectra are summarized in Table 7.1. Because of the amorphous nature of lignin, the complexity of individual monomeric units, and the random order in which they are linked, it is not possible to apply group theory to interpretation of the lignin spectrum. Band assignment is, therefore, empirical; but the variety of methods used by lignin investigators in assigning bands (14-18, 33-45) permits a considerable degree of confidence in all but two or three of the assignments tabulated in Table 7.1.

Table 7.1. Assignment of Infrared Absorption Bands in Mildly Prepared Wood Lignins

Position in cm.$^{-1}$		Band Origin
Guaiacyl L.	Guaiacyl-Syringyl L.	
3425-3400	3450-3400	OH stretching (H-bonded)
2920	2940	OH-stretch in methyl and methylene groups
2875-2850	2880	Same
2820	2845-2835	Same
1715	1715-1710	Carbonyl stretching - unconjugated ketone and carboxyl groups
1675-1660	1660	Carbonyl stretching - para substituted aryl ketone
1605	1595	Aromatic skeletal vibrations
1515-1510	1505	Same
1470	1470-1460	C-H deformations (asymmetric)
1460		Same
1430	1425	Aromatic skeletal vibrations
1370	1370-1365	C-H deformation (symmetric)
	1330-1325	Syringyl ring breathing with CO-stretching
	1235-1230	Same
1270	1275	Guaiacyl ring breathing with CO-stretching
1230		Same
1140	1145 shoulder	Aromatic C-H in-plane deformation, guaiacyl-type
	1130	Aromatic C-H in-plane deformation, syringyl-type
1085	1085	C-O deformation, secondary alcohol and aliphatic ether
1035	1030	Aromatic C-H in-plane deformation, guaiacyl-type, and C-O deformation, primary alcohol
970	970	=CH out-of-plane deformation (trans)
	915	Aromatic C-H out-of-plane deformations
855	860 shoulder	Same
815	835	Same
750-770 shoulder	750-770 shoulder	Same

B. Role of Model Compounds and Lignin Derivatives in Band Assignment. Absorption bands in the lignin spectrum are potentially assignable to a variety of groups. Since different groups may frequently produce absorption bands at the same frequency, structural assignments based on the lignin spectrum alone are subject to error. If the spectra of derivatives of lignin are also measured, the absorption bands of many groups are shifted or eliminated depending upon the type of derivative which has been prepared. In a given derivative, the magnitude of the shift varies with the type of functional group. Measurement of these shifts in model compounds and subsequent application of this information to lignin and lignin derivative spectra provides a method for determination of lignin functional groups (5, 13-14, 17-18, 23-25, 27-30, 35, 43, 46-56). Derivatization has included acetylation, methylation with diazomethane (for phenolic hydroxyl groups) or dimethyl sulfate (aliphatic and phenolic hydroxyl groups), deuteration, catalytic hydrogenation, reduction with sodium borohydride or lithium aluminum hydride, conversion to a salt (carboxyl groups) or phenolate, sulfidation, sulfonation, and treatment with alcoholic hydrogen chloride, i. e., alcoholysis. After assignments to pertinent functional groups in model compounds were made, comparison of lignin isolated and/or treated by the same methods has then been carried out.

1. <u>Hydroxyl Groups</u>. Hydroxyl groups give rise to absorption in two principal areas of the infrared spectrum, O-H stretching frequencies in the 3,700-3,000 cm^{-1} area, and O-H or C-O bending frequencies in the 1,200-900 cm^{-1} region. Generally, the O-H stretching frequencies of aliphatic hydroxyl groups occur at somewhat higher wave numbers than the phenolic hydroxyl group. On the basis of solvent effects, Lindberg (49-51) and Michell (30) have concluded that all hydroxyl groups appear to be involved in hydrogen bonds in the lignin macromolecule, and the phenolic groups are sterically hindered.

A partial determination of the types of hydroxyl groups present in a compound may be made by examination of the spectrum of the acetate derivative (30, 37, 43). The carbonyl stretching absorption of the acetoxy group occurs at different frequencies depending upon whether it is an acetate of an aromatic or aliphatic hydroxyl group. Usually, the stretching frequency of the carbonyl group in the acetate derivatives of phenols occurs at 1,765 to 1,760 cm^{-1} in aliphatic acetates except when the hydroxyl group is adjacent to a carbonyl group as in phenacyl alcohol or 3-hydroxy flavanones. In the latter case, the absorption occurs slightly higher, i. e., 1,750-1,745 cm^{-1}. The acetate derivatives of model compounds which have both aliphatic and phenolic hydroxyl groups usually show two absorption bands. For example, the acetate derivative of vanillyl alcohol shows two absorption bands, 1,735 and 1,760 cm^{-1}. The relative heights of these two bands in an acetate derivative of lignin provide a qualitative indication of the relative proportion of aliphatic and aromatic hydroxyl groups.

Deuteration of guaiacylic model compounds results in disappearance of two absorption bands in the 1,376-1,325 and 1,220-1,170 cm^{-1} regions, pointing to their hydroxylic character. Assignment to phenolic O-H deformation and C-O stretching is indicated. Lignin spectra show a weak band at 1,370 cm^{-1} which is diminished by deuteration (41) and a weak shoulder at 1,210-1,200 cm^{-1}. These two bands may be assignable to phenolic OH groups, but diazomethane methylated lignin, which contains no free phenolic hydroxyls, still retains the 1,370 cm^{-1} absorption band. The preferred assignment of this band to symmetric C-H deformation, as recommended by Sundholm (18), is therefore retained in Table 7.1, although it is recognized that a minor part of this band may arise from phenolic OH deformation.

C-O stretching bands in the 1,150 to 1,000 cm^{-1} region are frequently of value in distinguishing primary, secondary and tertiary hydroxyl groups in an unknown compound, or if the structure is known, equatorial and axial substitution may be distinguished. The usual frequency ranges are 1,070 to

1,010 cm^{-1} for primary alcohols, 1,120-1,075 cm^{-1} for secondary alcohols and 1,145-1,125 cm^{-1} for tertiary alcohols. These values are lowered by the presence of an adjacent phenyl group or by α, β unsaturation. Vanillyl alcohol shows an absorption at 1,000 cm^{-1} from the primary benzylic alcohol group, while the primary alcohol group in beta-hydroxy coniferyl alcohol absorbs at 1,038 cm^{-1} and at about the same frequency in coniferyl alcohol. Detection of primary and tertiary hydroxyl groups in lignin by the use of these bands is made difficult by the concurrent moderately strong bands at 1,140-1,130 cm^{-1} and 1,040-1,030 cm^{-1} originating from aromatic ring C-H in-plane deformations.

Lignin model compounds with secondary hydroxyl groups, i. e., α-hydroxypropiovanillone and apocynol (1-guaiacylethanol) show an absorption at 1,076 cm^{-1}. Diagnosis of secondary alcohols from this band is complicated by a strong band at 1,125-1,085 cm^{-1} and a weak band at 1,065-1,025 cm^{-1} originating from C-O-C deformation of aliphatic ethers. Thus, etherification of the secondary hydroxyl group in α-hydroxypropiovanillone causes an increase in the absorption at 1,125 cm^{-1} and a decrease in the relatively strong 1,075 cm^{-1} band to a weak doublet at 1,080 and 1,065 cm^{-1}. Etherification (methylation) of the secondary hydroxyl group of apocynol results in the loss of the 1,075 cm^{-1} band and the appearance of bands at 1,065, 1,090 and 1,115 cm^{-1}. One means of distinguishing between the aliphatic ether or secondary alcohol origination of a 1,090-1,075 cm^{-1} absorption in an unknown compound is to reexamine the spectrum after acetylation. If the band persists after acetylation, it originates from an ether linkage. If it disappears upon acetylation, it originated from an aliphatic hydroxyl group in the compound. Acetylation of 1-guaiacylethanol eliminates the 1,076 cm^{-1} absorption, but the 1,087 cm^{-1} aliphatic ether band is retained upon acetylation of 1-ethoxy-1-guaiacylethane. Acetylation of mildly prepared lignin diminishes but does not eliminate the 1,085 cm^{-1} absorption band. For this reason, this band is assigned (Table 7.1) to both secondary alcohol and aliphatic ether C-O deformation.

2. <u>Methyl and Methylene Groups.</u> Assignment of bands at 2,940-2,820, 1,470-1,460 and 1,370-1,365 cm^{-1} to aliphatic C-H stretching and bending is straightforward and based on a considerable body of experimental data. Unfortunately, it is not possible to readily distinguish C-methyl, methoxyl or methylene C-H in these bands of the lignin infrared spectrum (nmr spectra are of much greater utility in this area of measurement). Sarkanen, Chang and Ericsson (14) have found in their study of model compounds that a simple relationship does not exist between the number of methyl groups and the intensity of the absorption bands at 1,470-1,460 cm^{-1} which are usually ascribed to asymmetric bending of methyl groups.

It seems likely that the intensity of these bands are affected by geometry and electronic effects in the case of aromatic methoxyl groups.

3. <u>Carbonyl Groups.</u> The carbonyl group (and the region of

the infrared spectrum in which the stretching frequency is observed, 1,750-1,600 cm^{-1}) has undoubtedly received more attention than any other functional group presumed to be present in lignin. Reasons for this are several: A number of lignin investigators, notably Brauns, over the years have postulated a carbonyl group as a reaction site during sulfonation, ethanolysis, etc., but have been unable to unequivocally establish either the existence or precise type of carbonyl groups present in lignin in situ by chemical methods. Since the 1,750-1,600 cm^{-1} region of the spectrum is relatively free of absorption bands originating from other types of functional groups, and different types of carbonyl groups have characteristic frequencies; infrared spectroscopy is the method of choice for investigations involving the carbonyl group in lignin.

Saturated open-chain ketones have a characteristic carbonyl stretching frequency of 1,700-1,715 cm^{-1} in the solid state. The lignin model compounds, guaiacylacetone, β-hydroxyconiferyl alcohol and 1-ethoxy-1-guaiacyl-2-propanone, show absorption bands at 1,705, 1,709 and 1,710 cm^{-1}, respectively. The only other types of carbonyl groups likely to be mistaken for an unconjugated open-chain ketone group are aliphatic carboxyl (dimer or H-bonded) and aryl or α-β unsaturated esters. These can be readily distinguished from the ketone carbonyl by their non-reducibility in aqueous sodium borohydride solution, e. g., the carbonyl bands persist after attempted reduction and, in the case of the carboxyl group, by the shift of the carbonyl band to 1,590-1,560 cm^{-1} (carboxylate C=O stretching) upon conversion to a sodium salt. Unconjugated aldehydes usually absorb above and conjugated aldehydes somewhat below the 1,715-1,700 cm^{-1} region. Therefore, if a lignin product contains a carbonyl group which absorbs in the infrared at 1,715-1,700 cm^{-1}, and this absorption disappears from the spectrum after reduction of the product with aqueous sodium borohydride and subsequent acidification, it can be concluded to be an unconjugated ketone carbonyl group.

Care must be exercised in using alkaline treatment of lignin in conjunction with spectral studies. Some lignin products contain small amounts of esterified aromatic acids which may be hydrolyzed during alkaline treatment. The methyl esters of p-hydroxy benzoic acid, vanillic acid and p-hydroxycinnamic acid absorb at 1,694, 1,699 and 1,694 cm^{-1}, respectively (28, 29, 43). Some investigators have assigned an absorption peak at 1,720 cm^{-1} in hardwood lignins to aromatic ester groups (15, 16), but this value is clearly too high to have originated from this source. Assignment to aliphatic carboxyl groups is much preferable for the 1,720 cm^{-1} wavelength (27). The absorption band of the carboxyl group in aromatic acids varies with the ring substitution pattern. Thus, vanillic acid absorbs at 1,677 cm^{-1}, veratric acid (3,4-dimethoxybenzoic acid) at 1,672 cm^{-1} and acetylvanillic acid at 1,686 cm^{-1}, a value identical with that of benzoic acid.

A second problem with alkaline treatment is oxidative cleavage (43). Treatment of the model compounds, guaiacyl acetone and the keto form of β-hydroxyconiferyl alcohol, with mild alkali in the presence of air results in the loss of the 1,712 cm^{-1} ketone carbonyl and its subsequent replacement upon acidification with a carboxylic carbonyl group.

The structural character of conjugated ketone carbonyl groups can be rather accurately diagnosed provided spectra of appropriate derivatives are available. The carbonyl group in acetophenone, which is conjugated with a phenyl group, absorbs at 1,685 cm^{-1}. Addition of an OH to the para position of the acetophenone molecule causes a further shift to 1,645 cm^{-1} as a result of the electron-donating character of the para-phenolic hydroxyl group. Further addition of a methoxyl group to the nonconjugated meta position of the p-hydroxyacetophenone molecule, i.e., acetovanillone, or extension of the side-chain (p-hydroxypropiophenone) has only slight effect on the carbonyl frequency. Alteration of the para substituent, however, has a marked effect. This is illustrated by the series, sodium phenolate of acetoguaiacone, acetoguaiacone, acetoguaiacone methyl ether and acetoguaiacone acetate, which have carbonyl frequencies of 1,639, 1,653, 1,672 and 1,680 cm^{-1}, respectively. The value for acetoguaiacone acetate (1,680 cm^{-1}) is very near to that of unsubstituted acetophenone, so the acetoxy group (when in the para position) is a very weak electron-donor. The presence of a hydroxyl group or, to a lesser extent, an ether linkage (such as a methoxyl or ethoxyl group) on the carbon alpha to the carbonyl group increases the value by about 10 cm^{-1}, i.e., α-methoxyacetophenone absorbs at 1,697 cm^{-1} as compared with acetophenone (1,685 cm^{-1}); also, compare α-hydroxypropiovanillone with acetoguaiacone (1,664 vs. 1,653 cm^{-1}) and the methyl ethers of α-ethoxypropioguaiacone with acetoguaiacone (1,677 vs. 1,672 cm^{-1}).

In an unknown para-disubstituted aromatic compound, such as lignin, with a carbonyl absorption in the 1,650-1,660 cm^{-1} region, there may be difficulty in deciding whether the absorption originates from a ketone or an aryl aldehyde group (again, carboxylic groups would be eliminated by virtue of their nonreducibility in aqueous sodium borohydride). If the carbonyl group is a ketone conjugated with a p-hydroxyphenyl group, it should show the characteristic shifts upon acetylation or conversion to a sodium salt. If the carbonyl group is conjugated with a phenyl group etherified in the para position, the carbonyl frequency would not be lowered upon conversion to a sodium salt, and would not be appreciably increased by acetylation. A very slight increase may be expected in the latter case due to the removal of intermolecular hydrogen bonding (through acetylation of hydroxyl groups) which tends to lower carbonyl frequencies of compounds measured in the solid or liquid phase. If the carbonyl group is an aryl aldehyde or α,β-unsaturated aldehyde, it should show band splitting and characteristic shifts as described subsequently.

The aldehyde carbonyl in coniferyladehyde, a model for terminal groups in lignin, absorbs at 1,652 cm^{-1}. Methylation increases the value to 1,660 cm^{-1}, and acetylation nullifies the effect of the para hydroxyl group, increasing the frequency to 1,670 cm^{-1}. Difficulty may, therefore, be experienced in distinguishing between a para-etherified aryl ketone and a coniferylaldehyde residue. In this case, chemical color tests must be used as a supplement. Coniferylaldehyde and methylated coniferylaldehyde give a strong magenta coloration with phloroglucinol-hydrochloric acid reagent which may be measured spectrophotometrically. Ultraviolet measurements are also of help since the secondary band of the coniferylaldehyde group occurs at 340 mμ, while the corresponding band of a paraetherified aryl ketone absorbs at 303 mμ.

From the foregoing, it should be evident that determination of the nature of carbonyl groups in a given lignin product by infrared spectroscopy requires not only the spectrum of the original product but spectra of derivatives prepared by acetylation, neutralization with an alkaline buffer solution, methylation with diazomethane and reduction with aqueous sodium borohydride. Mild acid hydrolysis and subsequent chromatographic analyses for sugars is also useful since carbohydrate contaminants, especially in milled wood lignins, result in spurious absorptions at 1,740-1,735 cm^{-1} (acetyl and uronic ester groups) and 1,720-1,715 cm^{-1} (uronic acid carboxyl groups).

4. Ethylenic Double Bond. An α,β double bond conjugated with an aromatic ring is the only type of ethylenic double bond likely to be encountered in the lignin molecule. The C=C stretching vibration of aromatic conjugated trans-double bonds occurs in the range of 1,626-1,608 cm^{-1}, and the C-H out-of-plane deformation appears at 990-960 cm^{-1} (43). The C=C stretching band of coniferyl and sinapyl alcohol derivatives is not well defined and generally appears as a shoulder on the more intense aromatic stretching band at about 1,600 cm^{-1}. The only other functional group absorption interfering in the 990-960 cm^{-1} region is a weak band associated with a carbonyl group alpha to an aromatic ring such as is present in the model compounds, acetoveratrone, vanillic acid, p-hydroxypropiophenone, etc., which is not present in the corresponding alcohol. The series of spectra reported by Pearl (47) is particularly instructive in this regard. If the infrared spectrum of a lignin product shows a band in the 985-960 cm^{-1} region, and it still persists after reduction with sodium borohydride or lithium aluminum hydride, it may be concluded that the lignin contains a trans-double bond.

5. Aromatic Skeletal Bands. Guaiacyl and syringyl type model compounds have three absorption bands at 1,605-1,595, 1.515-1,505 and 1,450-1,420 cm^{-1} recognizable as aromatic skeletal vibration modes (14, 18). The first of these is a relatively pure ring-stretching frequency which may be strongly associated with the aromatic C-O stretching mode. Comparison of spectra of p-hydroxyphenyl-, guaiacyl-, and syringylpropane derivatives reported by Pearl (47) show amplification of the intensity of the 1,600 cm^{-1} band as the number of C-O bonds is increased. Conjugation with an α-carbonyl group, i.e., propioguaiacone, results in further amplification. In softwood lignins and unconjugated guaiacyl model compounds, the intensity

of the 1,600 cm^{-1} band is significantly lower than that of the 1,510 cm^{-1} band (see Figures 7.3 and 7.4). In hardwood lignins and unconjugated syringyl model compounds, the intensity of the two bands is approximately equal. The intensity ratio of these two bands may also be affected by the presence of p-hydroxyphenyl esters, carboxylate ions in carbohydrate impurities, moisture (absorption band at 1,625 cm^{-1}) or condensed tannin impurities, all of which tend to increase the relative intensity of the 1,600 cm^{-1} band with respect to the 1,510 cm^{-1} band.

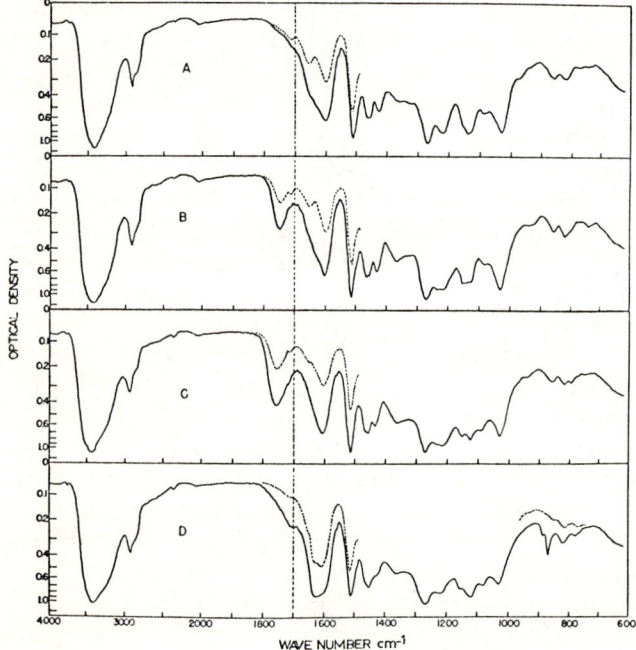

Figure 7.3. Infrared spectra of Brauns lignins: (A) longleaf pine, (B) Pacific silver fir, (C) western redcedar and (D) Douglas-fir. Spectra were measured as KBr wafers (solid lines) and mulls (dotted lines) (43).

On the basis of ring deuteration studies, the 1,510 and 1,430-1,425 cm^{-1} bands are considered to be ring-stretching modes strongly coupled by C-H in-plane deformations. The intensity of the latter band is sensitive to the nature of ring substituents. This is illustrated by a decrease in intensity proceeding from acetoguaiacone and vanillin to vanillyl alcohol to 4-methylguaiacol. The relative intensity of the 1,510 cm^{-1} ring-stretching band and the band at 1,460 cm^{-1} due to C-H bonds including methoxyl groups is reversed in softwood (guaiacyl) and hardwood (syringyl) lignins.

A considerable number of investigators in the past have assigned some of

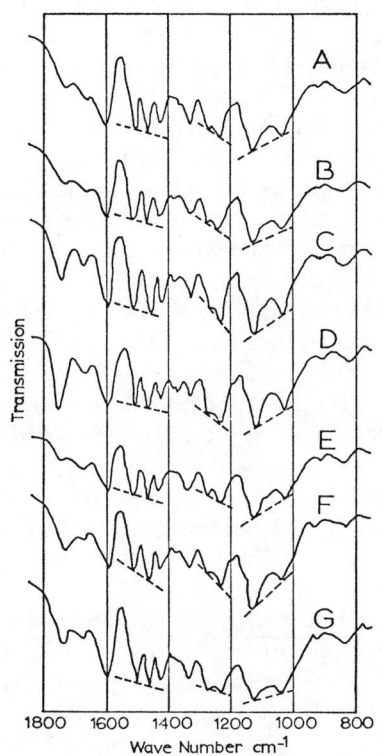

Figure 7.4. Infrared spectra of milled wood lignins of hardwoods (KBr wafers) from (A) Casuarina equisetifolia, (B) Populus euramericana, (C) Fagus crenata, (D) Trochodendron aralioides, (E) Robinia pseudoacacia (F) Baccharis halimifolia and Paulownia tomentosa (42).

the bands in the 1,400-1,000 cm^{-1} region of the spectrum to aromatic methoxyl groups. It is now believed that bands at 1,270, 1,230 and 1,130 cm^{-1} in guaiacyl compounds and at 1,335 and 1,235 cm^{-1} in syringyl derivatives are assignable to ring breathing with C-O stretching. Guaiacylic bands at 1,160 and 1,040 cm^{-1} and the syringyl bands at 1,130 are assigned to aromatic C-H in-plane deformation. Softwood lignins sometimes show two bands, at 1,160-1,150 and 1,135-1,130 cm^{-1}, assignable to both of the modes described above, or a single unresolved band. Hardwood lignins, containing both guaiacyl and syringyl nuclei, show a complex mixture of bands originating from both types of nuclei, but in general it is usually possible to distinguish them from softwood lignins on the basis of relative intensity of the bands. The 1,270 cm^{-1} band is more intense than the 1,230 cm^{-1} band, and the 1,035 cm^{-1} band is equal or greater in intensity than the 1,140 cm^{-1} band in softwoods. The reverse is true of hardwood lignins. The slopes of the maxima of these two sets of bands (with respect to the spectral baseline) can be semiquantitatively correlated with the ratio of guaiacyl and syringyl nuclei and, hence, the methoxyl content of the lignin (14-16, 42), see Chapter 3, Section III, Figure 3.7.

Unconjugated p-hydroxyphenyl, guaiacyl and syringyl model compounds show a series of two or three bands in the 900-750 cm^{-1} region attributable to aromatic C-H out-of-plane vibrations (18, 33, 35, 42, 43, 47). The 880-850 cm^{-1} and 855-800 cm^{-1} peaks in guaiacyl compounds are assigned to mono-hydrogen and two adjacent ring hydrogen deformations, respectively. Syringyl compounds have a band at 860-830 cm^{-1} assignable to the mono-hydrogen deformation and two weaker bands at 920-900 and 830-750 cm^{-1} which vary in wavelength depending upon the electronegativity of the side-chain. In their extensive study of milled wood lignins, Kawamura and Higuchi (42) note that softwood lignins generally show two absorption bands at 855 and 815 cm^{-1}, characteristic of the guaiacyl ring. It had been previously noted (43) that the intensity of these two bands varies in different types of lignin products suggestive of relative differences in the degree of substitution in the 5-position. Temperate hardwood lignins show a medium intensity band at 835 cm^{-1}, characteristic of the syringyl nucleus, or in the case of tropical hardwood lignins, a somewhat weaker 835 cm^{-1} band with shoulders at 855 and 815 cm^{-1}, indicating a higher guaiacyl to syringyl nuclei ratio than in the former.

C. <u>Problems and Pitfalls in the Use of Lignin Infrared Spectra</u>. Allusion has already been made in the first part of this section to some of the problems and pitfalls in the use of infrared spectra for reaching structural conclusions in various lignin studies. Specifically, the most common error is to conclude that two or more lignin products are "identical" because their infrared spectra are "similar." Rather, the opposite must be true, i. e., if the spectra are not <u>precisely</u> identical, it must be concluded that the products are not identical. A particularly glaring example of this, unfortunately, perpetuated in a recent textbook (57), are the assertions of Nord and coworkers (8-11, 58) that lignin products isolated by solvent extraction of wood are "identical in all respects" with enzymatically liberated, whole wood lignin. The infrared spectra reproduced in these studies clearly refute this claim, however, since significant differences in the spectra of the two types of products are readily apparent. The most significant of these are the increased absorption at 1,715 cm^{-1} and, especially, in the 1,600 cm^{-1} band, suggestive of a relatively high content of carboxyl groups, part of which are ionized, in the enzymatic lignins. Chemical studies by Grohn and Deters (59) have shown that enzymatic attack of wood results in chemical degradation of the lignin through methoxyl loss, etc., corroborating the spectral evidence. A further example of this problem is evident in comparisons of DHP and milled wood lignins by Freudenberg and coworkers. Although the artificial lignin is asserted to contain the same building units as spruce wood lignin, the infrared spectra (4, 60, 61) show important differences in carbonyl content (1,715 cm^{-1}), ethylenic double bonds (975 cm^{-1}), content of ether and hydroxyl groups (shape of 1,140 cm^{-1} band and intensity ratios of 1,140 and 1,035 cm^{-1} bands) and aromatic substitution patterns (peak ratios in 900-800 cm^{-1} region).

Another type of problem arises when attempts are made to establish the nature of chemical changes which have taken place during lignin isolation, particularly when difference spectra are used. This is best illustrated in Bolker's comparison (33, 62) of the difference spectrum between whole and delignified wood (holocellulose) and isolated lignins such as milled wood or dioxane lignins. It was concluded that there was no absorption band at 1,710 cm^{-1} in the difference spectrum and that the presence of this band in isolated lignins represented the liberation of a keto group present as an acetal linkage with carbohydrates in the original wood. Unfortunately, aliphatic acetyl groups absorb at 1,735 cm^{-1}, and since acetyl groups are lost during holocellulose preparation, the difference spectrum shows a substantial "absorption" at 1,735 cm^{-1} (44). Ascertaining the presence or absence of a weak 1,710 cm^{-1} band as a shoulder on an artifact 1,735 cm^{-1} band can be described as a precarious venture at best and hardly significant evidence for an acetal link between lignin and cellulose. Furthermore, the 1,710 cm^{-1} band in milled wood lignins may well be primarily carboxylic in nature, while the 1,710 cm^{-1} band in dioxane lignin most likely arises through acidic rearrangement of a guaiacylglycerol-β-aryl ether during isolation (43).

Another problem area is the assignment of structural features to certain absorption bands that are inconsistent with pertinent model compound spectra. This was common and understandable when infrared spectral studies were in their infancy, but the ample literature now available should eliminate this problem, at least in the 4,000-1,500 cm^{-1} region. Unfortunately, such is not the case. For example, in a recent study (63) of lignins isolated from decayed spruce wood, a progressively increasing absorption band at 1,727 cm^{-1} was attributed to the formation of para-etherified vanillic acid groups. The carbonyl group in 3,4-dimethoxybenzoic acid, an appropriate model for this structure, absorbs at 1,672 cm^{-1}. The structural assignment is obviously incorrect.

IV. Application of Spectra.

A. Polymers Related to Lignin. A number of years ago, Brauns discovered that extraction of wood with neutral, polar solvents resulted in the isolation of an amorphous polymer in yields up to 2% based on wood. The product was considered to be identical with whole wood lignin and was, therefore, named "native lignin." Subsequent investigators prefer the name "Brauns lignin" for such products. The infrared spectra of Brauns lignin from the following species have been illustrated in full or in part in the literature: western hemlock (43), pine (8, 64), spruce (2-4, 12, 46, 64-67), Douglas-fir (29), poplar (68, 69), aspen (3, 28, 70), chestnut (29), acacia (71), beech (64), oak (9), maple (9), sugar cane (11, 29), bamboo (64), wheat straw (29), and reed, Phragmites communis (72). Comparison of the published spectra, with due allowance for differences in sample preparation and instruments, shows that they may be placed in two broad categories, (a) conifers and (b) hardwoods, grasses, bamboos and reeds.

Within a given species, the Brauns lignin spectrum strongly resembles but is not identical with mildly prepared whole wood lignin (see Figure 7.2). The most notable differences are in the carbonyl stretching region, $1,775$–$1,625$ cm^{-1}, and in the area of the spectrum where the aromatic substitution pattern is of particular consequence, $1,250$–$1,100$ and 900–800 cm^{-1}. The differences may be interpreted as indicating a higher content of nonconjugated carbonyl groups, phenolic hydroxyl groups and trans-double bonds and a lower degree of substitution in the 5-position of the aromatic ring in Brauns lignins. Some of the hardwood Brauns lignins appear to contain a lower syringyl to guaiacyl ratio than the corresponding milled wood lignins. This raises the intriguing question of the physiological and structural relationship of Brauns lignin and residual whole wood lignin.

Some coniferous Brauns lignin spectra (see Figure 7.3), depending upon the genus from which the product is isolated, show an absorption band at $1,775$–$1,750$ cm^{-1} which is not present in milled wood lignin isolated from the sapwood of the same species. This band arises from a lactone carbonyl group and provides the clue for the relationship of Brauns lignin to whole wood lignin. Those conifers containing lignans with lactone groups also yield Brauns lignin with lactone carbonyl groups. When the extractives contain C_6-C_5-C_6 compounds such as sugiresinol, the Brauns lignin spectrum is suggestive of the inclusion of similar structures. A "hidden" carbonyl absorption at $1,620$ cm^{-1} is observed in Brauns lignins from Douglas-fir and larch, which also contain flavanonols with a carbonyl band at $1,620$ cm^{-1} among their extractives. This evidence, coupled with a study of the location of these products in wood (43), leads to the conviction that Brauns lignins are post mortem products generated within and upon the death of parenchyma cells (for an alternative interpretation see Chapter 5, Section III). Polymer fractionation studies on spruce Brauns lignin by Hess (12) and on poplar Brauns lignin by Desmet (68, 69) show that these products are heterogeneous. It must be concluded, therefore, that while structural investigations on Brauns lignin are of interest in understanding the physiology of wood as a whole, results cannot be directly translated to whole wood lignin.

Another type of product closely related to whole wood lignin but not identical with it, at least as evidenced by infrared spectral differences, is "lignin" isolated from callus tissue cultivated in vitro. Barnoud and coworkers (73-75) prepared milled wood lignins from pine, redwood, lilac, rose and and paulownia and compared them with similarly isolated milled wood "lignins" from cultured callus tissue. The products isolated from the latter primarily differed in much more intense carbonyl bands at 1,715 and 1,660 cm^{-1}. The spectra of hardwood callus tissue "lignin" much more closely resembled that from softwood than the corresponding hardwood lignin and thus indicated a lower syringyl to guaiacyl ratio in cultivated callus tissue than in the wood. The relationship between callus tissue and whole wood lignin is so reminiscent of that between Brauns lignin and whole wood lignin as to suggest that biosynthetic mechanisms operating in the genesis of "lignin" in parenchyma cells and cultured callus tissue are very similar.

Solvent extraction of wood after steam hydrolysis yields a lignin-like product in yields up to 10% of the whole wood lignin present prior to hydrolysis. Klemola (76) has studied this type of product from birchwood. While the overall spectrum generally resembles that of milled wood lignin from the same species, the 1,720 and 1,660 cm^{-1} carbonyl stretching bands are more intense. It was concluded that the product contained furan or related structures derived from carbohydrates. On the basis of infrared studies, Naveau (39) reached the conclusion that condensation of carbohydrate and lignin also takes place when wood is heated under conditions of forming particle boards.

As previously noted, infrared spectroscopy is invaluable in comparison of synthetic "lignins" with whole wood lignin. In 1947, A. Russell proposed a structure and synthesis of gymnosperm lignin based on intermolecular condensation polymerization of 2-hydroxy-3-methoxy-5-formylacetophenone to give a poly-8-methoxydihydrobenzopyrone presumably in equilibrium with its open chain tautomer. On the basis of physical and chemical properties, Russell claimed that his synthetic product was indistinguishable from wood lignin. The subsequent availability of the infrared technique, however, demonstrated the complete nonidentity of Russell's product with lignin (3).

The infrared spectra of the products prepared by the enzymatic "dehydrogenation-polymerization" of substituted cinnamyl alcohols by Freudenberg and coworkers, on the other hand, closely resemble those of natural wood lignin (4, 5, 48, 60-61, 63). These studies have also been of help in band assignments in the 1,300-1,200 cm^{-1} region of the spectrum. Nakano and Migita (48) prepared polymers from coumaryl, coniferyl and sinapyl alcohols. In the product derived from coniferyl alcohol, which contains only guaiacyl rings, the 1,270 cm^{-1} band is more intense than the 1,230 cm^{-1} band. In products prepared from mixtures of coniferyl and sinapyl or coniferyl or coniferyl, sinapyl and coumaryl alcohols, the intensity relationship was reversed. It is reasonable to conclude, therefore, that the greater intensity of the 1,230 cm^{-1} band compared with that at 1,270 cm^{-1} in hardwood lignins is a function of the syringyl and para-hydroxyphenyl ring content. The product prepared only from coniferyl alcohol also shows a weak band at 1,325 cm^{-1}. Hence, this band cannot be considered as diagnostic for hardwood lignins nor does it necessarily indicate the presence of small amounts of syringyl groups in softwood lignins as suggested in a recent study (15).

B. <u>Whole Wood Lignins</u>. The discoverer of the vibratory ball milling procedure for isolation of milled wood lignins (MWL), A. Björkman, reported that "the infrared spectrum of MWL has sharper bands and a more detailed fine structure than is normally obtained with lignin samples" (77), but did not illustrate nor further discuss their spectra. Since then, the infrared spectra of milled wood lignins have been extensively studied, the most comprehensive survey being that of Kawamura and Higuchi (42) which includes illustrated spectra of three species of pteridophytes, six species of gymnosperms, twelve dicotyledons and four monocotyledons. More recently, Sarkanen and coworkers (15, 16) have investigated the spectra of MWL from six conifers and four angiosperms. Other investigators have studied MWL of hemlock (43), spruce (40, 46, 60, 61, 63, 78-83), larch (81), sequoia (73-75), pine (81, 84), aspen (70), birch (76), eucalyptus (84, 85) and several grasses (86, 87). Typical temperate hardwood MWL spectra are shown in Figure 7.4.

At the present stage of lignin investigation, MWL undoubtedly represents the least degraded form of isolated lignin. Infrared spectra of these products or of difference spectra (described subsequently) usually form the base for studies of spectra of all other types of isolated lignin, e. g., any given infrared absorption band in the lignin being examined is more or less intense, more diffuse, shifted in frequency, etc. than the MWL of the corresponding species. The wavelength and band assignments in Table 7.1 are, therefore, based primarily on MWL as the best representational form of whole wood lignin. Several problems exist with MWL infrared spectra and must be properly understood if the maximum value is to be derived.

MWL as currently prepared and purified contains discrete amounts of hemicellulose, probably chemically bound, since no effective method has been discovered for its removal by physical methods. Pearl and Busche (70), for example, found that aspen MWL prepared by Björkman's standard procedure continued 4.3% xylose, 0.6% acetyl and significant, but not precisely determinable, amounts of uronic acid. Partially acetylated xylan polyuronide is the primary hemicellulose in hardwood and shows infrared absorption bands at 1,730, 1,460, 1,235, 1,055, 1,030, 990 and 895 cm^{-1} (44). The absorption bands at about 1,730 and 900 cm^{-1} in MWL spectra are best assigned to carbohydrate artifacts. The modestly strong 1,730 cm^{-1} band in MWL results in considerable difficulty in assigning the weak 1,715 cm^{-1} band, which is present as a shoulder and probably originates from both unconjugated ketone carbonyl and carboxyl groups. In some woody genera, phenolic acids such as p-hydroxybenzoic acid are esterified to lignin and result in amplification and shifting of this band to 1,700-1,690 cm^{-1}.

A second type of artifact which may contaminate MWL and alter the spectrum is commonly present in heartwood. Not only is Brauns lignin primarily a heartwood constituent, but additional quantities of polyphenolic polymers, nonextractable with neutral solvents, may be co-deposited. MWL isolated from extractive-free whole wood including the heartwood frequently shows spurious carbonyl stretching bands, e. g., at 1,750 cm^{-1} depending upon species, and changes in peak heights of the aromatic stretching bands at 1,600 and 1,515 cm^{-1}. The solution to this latter problem is to use only fresh sapwood as a raw material for MWL preparation and spectral investigation. Polymer fractionation studies on MWL, such as those carried out by Hess (12) and Desmet (68, 69) on Brauns lignin, followed by spectral comparison of individual fractions would be invaluable in establishing the homogeneity of MWL.

Infrared spectra of thin wood sections (6, 88-91) show typical aromatic absorption bands and were used as an unequivocal proof of the in situ aromaticity of lignin (6). When a thin wood section is placed in the sample beam of a double beam spectrometer and holocellulose (delignified wood) is placed in the reference beam, a differential infrared spectrum of lignin is obtained. Such spectra show almost total agreement with MWL from the corresponding species (33, 34, 41, 44, 62, 78). The difference spectrum of lignin usually shows a medium absorption band at 1,730 cm^{-1} and a weak band at 900 cm^{-1} which is best interpreted as resulting from the loss of some acetyl groups during the preparation of holocellulose and consequent imperfect compensation of these bands. Marchessault and coworkers (91) have shown that the removal of hemicellulose from whole wood results in a total loss of absorption in the 1,730-1,700 cm^{-1} region of the spectrum, while carefully delignified wood sections show that all of the 1,730 cm^{-1} absorption is essentially retained.

Treatment of wood with brown rot organisms and subsequent extraction with ethanol results in the isolation of "enzymically liberated lignin." Nord and Schubert (8-11, 57, 58) consider these products to be representative of whole wood lignin and therefore suitable materials for deductions concerning the structure of lignin. As previously noted, there are spectral differences between these products and the Brauns lignins isolated from the species. The published spectra are not identical with those of MWL or difference spectra primarily in the intensity ratio of aromatic stretching and C-H in-plane deformation bands. Furthermore, the peaks are not sharp and there is a variable and strong peak at 1,725 cm^{-1}, the strength of which appears to be dependent upon the duration of fungal treatment (63). These divergencies are indicative of extensive structural alteration during fungal attack.

One of the disadvantages of the use of MWL for structural studies, aside from carbohydrate contamination, is that preparation is exceedingly slow and results in relatively small amounts of material. These difficulties may be circumvented by the use of dioxane-hydrochloric acid as a lignin extraction medium. Dioxane-HCl lignin spectra of a variety of hardwoods, softwoods and grasses have been reported (26, 42, 43, 72, 76, 80, 81, 92). Spectra of four typical conifers are shown in Figure 7.5. While some investigators, notably Pepper (38, 92), consider lignin isolated from wood by the dioxane-HCl method to be relatively unchanged in structure, the spectra give evidence

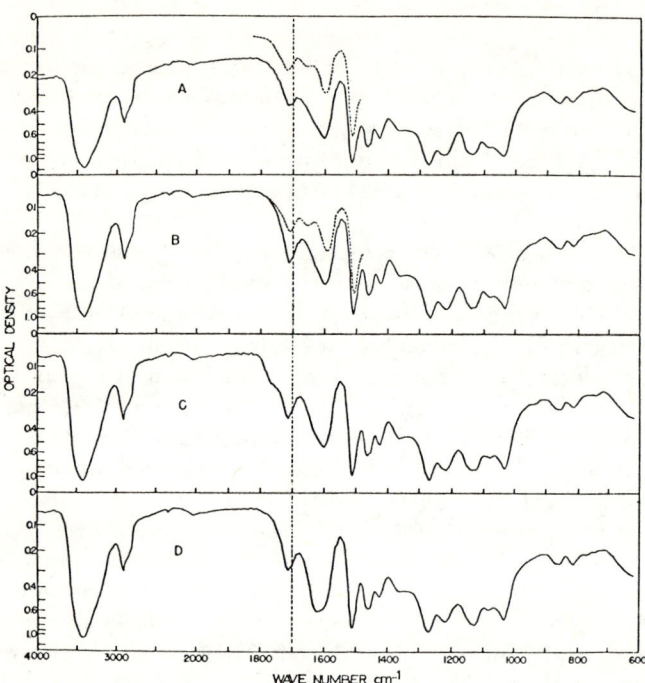

Figure 7.5. Infrared spectra of dioxane-HCl lignins: (A) longleaf pine, (B) Pacific silver fir, (C) western redcedar, and (D) Douglas-fir. Samples were measured as KBr wafers (solid lines) and mulls (dotted lines) (43).

of some structural rearrangement of the lignin during isolation. The 1,710 and 1,660 cm^{-1} bands are appreciably stronger and weaker, respectively, than the corresponding bands in MWL. Further differences in dioxane lignin and MWL are indicated by the higher phenolic hydroxyl content of dioxane lignin, as evidenced by a more intense 1,760 cm^{-1} phenolic acetate carbonyl stretching band in the acetate derivative, and differences in the 1,150-1,125 cm^{-1} bands which may be related to aromatic-aliphatic ether linkages. All of the spectral changes can be rationalized on the basis of acidic cleavage of β-aryl ethers and subsequent rearrangement of the guaiacylglycerol side-chain. The decrease in the 1,660 cm^{-1} absorption band can be accounted for by the cleavage and loss of coniferylaldehyde end units in the lignin, and the increase in the 1,710 cm^{-1} band by the generation of beta carbonyl groups. This is not only confirmed by appropriate model compound experiments, but dioxane-HCl treatment reintroduces absorption bands at 1,710 and 1,660 cm^{-1} in Brauns lignin or MWL previously reduced with sodium borohydride or lithium aluminum hydride to eliminate carbonyl groups (43).

Polymer fractionation studies (93) suggest chemical inhomogeneities in whole wood dioxane-HCl lignin with respect to ethylenic double bonds, carbon, hydrogen and methoxyl content, conjugated carbonyl groups, etc. It should be remembered, however, that the same problems incurred when MWL is prepared from whole wood can also occur in the preparation of dioxane-HCl lignin. The spectra in Figure 7.5 indicate the presence of lactone and flavonoid artifacts in western redcedar and Douglas-fir whole wood dioxane-HCl lignins, respectively. Comparison with the spectra of Brauns lignin (Figure 7.3), which as previously noted is primarily a heartwood "extractive," shows intensification of the artifact absorption bands in the latter. Improved chemical homogeneity can be anticipated in dioxane-HCl lignins that are prepared only from sapwood.

Another acidolytic technique for isolation of lignin, closely related to the dioxane-HCl method, is the use of anhydrous ethanol or methanol and hydrogen chloride. The spectra of "alcoholysis" lignins have been reported from spruce, pine, oak, beech, moss and reeds (94-98). Compared with MWL, a strong band appears at 2,960 cm^{-1}, the intensity of absorption at 2,890 and 2,850 cm^{-1} increases, and a series of changes in the 1,150-1,100 cm^{-1} is observed, all of which are consistent with alkylation of benzylic hydroxyl groups or hydroxyl groups alpha to an α-carbonyl group (as in α-ethoxypropioguaiacone). The absorption at 1,715-1,710 cm^{-1} is increased from that in MWL, but the intensity is somewhat weaker than that in the corresponding dioxane-HCl lignin. This band is attributable to a β-ketone carbonyl group (43, 66, 98) which is generated upon cleavage of β-aryl ether groups and subsequent acid-catalyzed rearrangement. Bolker and Marracini (94) note that the spectra of spruce ethanolysis lignin is very similar to ethoxylated dioxane lignin. In view of the fact that both acidolytic rearrangement and alkylation occur during ethanolysis, this type of lignin is even less suitable than dioxane-HCl lignin to draw conclusions concerning the structure of lignin.

Although dimethylsulfoxide is an excellent solvent for isolated lignin, treatment of wood at 100° C. with this solvent does not result in removal of lignin (99). Addition of as little as 0.2% mineral acid, however, results in effective delignifying action. Clermont (100) used dimethylsulfoxide and 5% sulfur dioxide to isolate a lignin product in about 80% yield of the original lignin in black spruce wood. The spectrum was very similar to that of dioxane-HCl lignin except for a generally higher intensity in the 1,900-1,650 cm^{-1} region and the presence of two weak peaks at 1,810 and 1,775 cm^{-1}. The absence of sulfur in the product rules out the origin of these two bands from dimethylsulfoxide impurities or reaction products. Further work will be required to evaluate the products isolated by this procedure for their suitability in lignin structural studies.

The spectra of lignins isolated with phenol and mineral acid catalyst show absorption bands at 860-850 and 770-735 cm^{-1} characteristic of ortho- and para-substituted phenols (101). The spectra are consistent with the postulated reaction of benzylic hydroxyl groups with phenol resulting in aromatic-aliphatic C-C links and the splitting out of water. Lignin may be isolated by solvolysis in acetals such as 2,2-dimethoxypropane in dioxane and hydrochloric acid (102). Reaction with the acetal is indicated both by analysis and the spectrum, so the technique appears to offer no advantage over the use of dioxane-HCl alone other than possible insight into the Bolker lignin-carbohydrate acetal linkage hypothesis (62).

The only practical methods for total isolation of wood lignin are treatment of wood with 72% sulfuric acid (Klason lignin), supersaturated hydrochloric acid (Halse lignin) or periodate. Infrared spectra have been reported for lignins isolated by these procedures from a wide variety of softwoods and hardwoods (37, 45, 94, 103-105). The spectra are generally more diffuse throughout the whole spectral range indicative of rearrangement and condensation during isolation. Pilipchuk and coworkers (37) consider Halse lignin to be less degraded than periodate lignin, and the latter less degraded than Klason lignin on the basis of spectral diffusivity. Kawamura and Higuchi (103) were still able to classify Klason lignins into gymnosperm and angiosperm categories, but somewhat less readily than when MWL was used as the basis of comparison (42). They found that somewhat sharper spectra could be obtained by using the acetone-soluble fraction of Klason lignin.

C. Commercial Lignins. Structural studies of the lignins isolated by acidification of soda or kraft pulping liquors have been conducted to gain insight into the alkaline cooking process and to aid in the marketing of isolated kraft lignins (23-25, 33-35, 43, 98, 106-118). Comparison of kraft softwood lignin with the corresponding MWL indicates intensification of absorption at 1,715, 1,600, 1,370 and 1,230 cm^{-1}, diminishment of absorption at 1,085 and 1,035 cm^{-1}, and splitting of the 1,140 cm^{-1} peak into absorption bands at 1,150 and 1,125 cm^{-1} in the latter. These differences may be interpreted as indicating that the kraft process causes an increase in phenolic hydroxyl content through ether cleavage and loss of methoxyl group and, possibly, some loss of aliphatic hydroxyl groups either through dehydration or sulfide formation. Bolker and Somerville (33) report a very weak band at 630 cm^{-1} in softwood thiolignin which they assign to C-S stretching. Michell, Watson and Higgins (34) note that both soda and kraft eucalypt lignins have weak bands at 640 and 620 cm^{-1}, neither of which are present in MWL. The 620 cm^{-1} band is slightly weaker in soda lignin indicating the possibility of a weak superimposed C-S stretching band in the kraft lignin. No absorption band was observed at 2,600 cm^{-1} that could be ascribed to S-H stretching of thiol groups. The question of the existence and nature of carbon-sulfur linkages in kraft lignin, at least as far as infrared spectra are concerned, must remain open until additional work is conducted.

The absorption peak at 1,720-1,715 cm^{-1} in kraft lignins predominantly originates from an unconjugated carboxyl group. As the temperature and pH of kraft digestion is increased, the intensity of this band is correspondingly intensified (114). Sulfidity of the cook might also be expected to affect the intensity of this group since lignins isolated from polysulfide cooking liquor have substantially stronger carbonyl peaks at 1,715 and 1,660 cm^{-1} than conventional kraft lignins (111). The mechanism for generation of carboxyl groups is not known with certainty, but treatment of Brauns lignin (43) or strong acid lignins (119) with base results in the loss of unconjugated ketone carbonyl groups and their replacement with carboxyl groups, probably by cleavage and oxidation. The model compound, 1-(4-hydroxy-3-methoxyphenyl)-2-propanone, behaves similarly. The intensity of the carboxyl peak in kraft lignins is significantly more intense than the unconjugated carbonyl peak in MWL, however, so at least part of these groups must be generated by oxidation of alcoholic hydroxyl groups. Gierer and Lenz (117) have used an infrared technique to demonstrate the formation of 1,2 glycol groups in MWL upon treatment with alkali at elevated temperatures. Upon periodate oxidation of borohydride reduced, diazomethane-methylated, alkali-treated MWL, carbonyl groups are formed as evidenced by a broad absorption band at 1,780-1,650 cm^{-1}.

Kraft lignins have been fractionated by progressive acidification of kraft pulping liquor (107) or an alkaline solution of a commercial kraft lignin (115). Spectra of the fractions show important differences, particularly in the carbonyl stretching region, and were taken to provide strong evidence of the heterogeneity of this form of lignin. Some of these differences may be a function of the extent to which free carboxyl groups are liberated. Kraft lignins precipitated by carbon dioxide (pH 7-8) showed a general absence of absorption at 1,720-1,660 cm^{-1} and enhanced absorption at 1,590-1,580 and 1,400 cm^{-1} typical of carboxylate ion (43). Subsequent acidification with sulfuric acid generated absorption bands at 1,720-1,715 and 1,660-1,650 cm^{-1} which were attributable to unconjugated carboxyl groups and a keto group conjugated through the aromatic ring to a para-phenolic hydroxyl rather than to a para-ether linkage as in the original wood lignin. To have greater significance, each fraction of kraft lignin fractionated by progressive acidification of an alkaline solution should be treated with mineral acid prior to spectral measurement. When this is done, better assessment of the extent of heterogeneity of kraft lignin will be possible. Due attention should also be paid in future work to the possibility of carbohydrate impurities as noted by Kleinert (109).

The infrared spectra of lignosulfonic acids have been reported by several investigators (33, 34, 43, 120-122). Barium or sodium lignosulfonates show little or no absorption in the carbonyl stretching region, thus indicating no free carbonyl groups. Free lignosulfonic acids prepared by ion exchange show a medium intensity absorption peak at 1,720 cm^{-1} (but somewhat less intense than that in kraft lignin) which originates from free unconjugated carboxyl groups (43,120). A strong, broad band at 1,210-1,190 cm^{-1} (evidenced by markedly increased absorption at this wavelength as compared with MWL), a medium intensity band at 655 cm^{-1} and a weak band at 540-520 cm^{-1} originate from and are characteristic of sulfonic acid or sulfonate groups in lignosulfonate spectra. Both the 1,200 and 1,040 cm^{-1} absorption bands occur at or very near the same frequency as absorption bands already present in both softwood and hardwood MWL, and the 540-520 cm^{-1} band occurs at the same frequency as a band present only in hardwood MWL, so these bands are not particularly valuable for diagnostic studies. The 655 cm^{-1} band does not occur in unsulfonated lignin and can be used for characterization.

The difference spectrum of lignin remaining in sulfite pulp shows the characteristic sulfonate absorption peaks, but they are weaker than those in isolated lignosulfonates (33). Dioxane-HCl lignin from a spruce sulfite pulp showed a much weaker absorption at 655 cm^{-1} than a spruce lignosulfonic acid. This indicates that the lignin remaining in pulp is much less sulfonated than the lignins dissolved during pulping. A similar situation prevails with neutral sulfite semichemical (NSSC) pulping. Lignins isolated from eucalyptus NSSC black liquor showed characteristic sulfonic acid bands (34). Difference spectra of lignin in NSSC pulp showed no absorption at 655 cm^{-1}, which led to the conclusion that sulfonated lignin was not present in the unbleached NSSC pulp in appreciable quantities.

The mechanism of pulp bleaching has been studied with the aid of infrared spectra of barium lignosulfonates treated with chlorine water (122). The chlorinated products show intense carboxylate ion peaks at 1,600 and 1,340 cm^{-1}. A peak at 1,700 cm^{-1} and a shoulder at 1,750 cm^{-1} are present but eliminated upon reduction with sodium borohydride, indicating either ketone or aldehyde groups. Aromatic absorption at 1,510 cm^{-1} is virtually eliminated. It may therefore be expected that aromatic rings are destroyed and carboxyl, carbonyl and/or o-benzo-quinone structures would be formed in the residual lignosulfonic acid in unbleached sulfite pulp when it is treated with chlorine water during the bleaching process. Chlorination of softwood hydrotropic or commercial acid hydrolysis lignins in a nonaqueous medium resulted in the appearance of a C-Cl absorption band at 750 cm^{-1}, intensification of carboxyl peaks at 1,720 cm^{-1} and elimination of aromatic guaiacyl peaks (105).

D. Bark Lignin and Humic Acids. Reference has been made in Chapter 3, Section V, to the use of infrared spectra in helping to establish the relationship between bark and wood lignin. Comparison is made difficult by the substantial content of polyleucocyanidins, pectin-like materials, and polyestolides (polymers of hydroxy fatty acids) in extractive-free bark, all of which may contaminate bark lignin to a varying extent. The spectra of dioxane-HCl lignins prepared from conifer bark do not differ from those of wood other than a slightly higher absorption at 1,715 cm^{-1} which very likely results from carboxylic impurities. Hardwood bark and wood lignin spectra show differences which can best be interpreted as resulting from a lower syringyl-guaiacyl ratio in lignin derived from bark as compared to those obtained from the corresponding species of wood.

Infrared spectroscopy has been used in a number of studies aimed at elucidating the role of lignin in the formation of soil humic acids, peat and coal (31, 45, 67, 95, 104, 123-130). Most investigators generally agree that humus is derived from lignin and proteins since the organic matter in soils and peats must arise either as modifications of the plant constituents or as synthetic products of the organisms, animal or microbial, which assisted in the decomposition of the plants. Of the plant constituents, lignin could be expected to be the most resistant (95). The spectra of humic materials vary, of course, depending upon the source, e.g., the type of soil or peat, but they usually show absorption bands at 3,400 (OH), 2,950-2,850 (CH), 1,720 (COOH), 1,610 (COO^-, C=C), 1,510 (aromatic, weak), 1,460 and 1,375 (CH), 1,250-1,150 and 850-800 cm^{-1}. Steelink (125) concludes that attempts to structurally characterize humic acid by infrared absorption spectrophotometry have been unsuccessful. However, the conversion of lignin to products having humic acid-like spectra and then brown coal spectra by progressive demethylation, dehydration, and condensation provide rather convincing evidence of the role of lignin in the process. The intensification of the 1,600 cm^{-1} absorption band and the decrease in the 1,510 cm^{-1} band is interpreted as a result of progressive substitution of the aromatic ring. Elofson (104) notes that the 1,510 cm^{-1} absorption is completely absent in hexamethylbenzene.

When peat or lignitic humic acids are heated at 300° C. for varying periods of time, the carbonyl absorption at 1,720 cm^{-1} gradually disappears and the OH absorption at 3,400 is greatly diminished. A series of three distinct bands appear at 860, 800 and 755 cm^{-1}, characteristic of bituminous coal. This is considered to be evidence for the in vitro conversion of humic acid to a material which has a structure which approximates very closely that of bituminous coal. It should be noted that high temperature charring of cellulose or synthetic hydroquinone humic acids also leads to the same type of spectra, so the establishment of the importance of lignin in formation of humic acid and coal certainly requires other types of evidence in addition to that provided by infrared spectra.

E. Chemotaxonomic Studies. In his essay on chemotaxonomic aspects of lignins, Brown (131) notes that infrared spectra are potentially one of the most important tools available for taxonomic studies based on lignins. As Pilipchuk and coworkers (81) aptly note, differences in infrared spectra of various species of softwood lignin are more dependent upon the mode of isolation than the species. This points out the importance of using the same isolation method when making taxonomic spectral comparisons. As previously noted, it is also important to avoid complications introduced from the use of heartwood in lignin preparation. Unfortunately, this precaution was not observed in any of the major investigations involving the comparison of spectra from various species (42, 43, 85, 103) except in the recent work of Sarkanen and coworkers (15, 16). Bland (84) has compared the spectra of MWL isolated from normal and compression or tension wood. Differences in the 1,750-1,650 and 1,200-1,000 cm^{-1} regions were evident and appear to be related to variable xylanpolyuronide content in the hardwood lignins and p-hydroxyphenyl content in the coniferous compression wood lignin. This indicates the importance of also excluding abnormal wood in samples used for taxonomic comparisons, a point which appears to have been overlooked in all the taxonomic studies cited above.

In spite of these problems, it is possible to classify lignin spectra into major categories (42), see Chapter 3, Section I. The spectral differences appear primarily to be a function of the ratio of guaiacyl, syringyl and p-hydroxyphenyl units in each type of lignin. Secondary esterification of the lignin also affects the spectra and appears to be a distinctive feature of certain genera. Structural differences in lignin isolated from different species of conifers are rather minor, at least as indicated by spectra, but are much more significant in dicotyledonous woods.

Kawamura and Bland (85) have made a very interesting study of lignin from five eucalyptus species growing at southern latitudes varying from 10° to 40°. While the spectra of lignin from a given species tended to fall into a temperate or tropical category, considerable variation within a species was noted depending upon the latitude of growth, altitude, mean annual temperature and rainfall. The spectral differences were primarily related to syringyl-guaiacyl ratios which increased as more temperate growing conditions were approached. This type of study could profitably be repeated with a conifer and several temperate dicotyledonous species tolerant of a wide variety of growing conditions but containing less nonlignin phenolic contaminants than eucalyptus species.

F. Quantitative Analyses in Wood and Pulp. Determination of the amount and nature of lignin remaining in unbleached pulp is of interest in both pulping and bleaching research. Determination by strong acid hydrolysis is rendered difficult by filtration problems and partial solubility of lignins in the filtrate. A number of investigators have accordingly proposed methods which involve infrared measurements of a unique absorption band of lignin in pulp (34, 132-137). The most convenient band is that occurring at 1,510-1,500 cm^{-1}. Pulp is pressed in potassium bromide wafers or measured on a sheet by multiple internal reflectance. Marton and Sparks (135) used the latter method and found that the intensity of the 1,510 cm^{-1} peak was proportional to bulk lignin rather than the surface lignin content of pulp fibers. They were able to obtain a precision of ± 4 relative percent for pulps with lignin contents equal to or greater than 4%. The 1,600 cm^{-1} lignin band is difficult to use because of water band interference at 1,625 cm^{-1} (134).

The ratio of the intensity of the 1,600 cm^{-1} band to that at 1,510 cm^{-1} is different in sulfite and kraft pulps. Bruun and Sjöholm (137) suggest that the 1,600 cm^{-1} band is relatively stronger in kraft than in sulfite pulp because of externally conjugated rings. In any event, an empirical correlation of peak area or intensity with a chemical measurement of lignin content is probably required with pulps prepared by different types of pulping processes.

REFERENCES

1. K. Whetsel, Chem. Eng. News, 82 (Feb. 5, 1968).
2. E. J. Jones, J. Am. Chem. Soc., 70, 1984 (1948).
3. E. J. Jones, Tappi, 32, 167 (1949).
4. K. Freudenberg, W. Siebert, W. Heirnberger, and R. Kraft, Chem. Ber., 83, 533 (1950).
5. K. Freudenberg, H. Dietrich and W. Siebert, ibid., 84, 961 (1951).
6. K. Kratzl and H. Tschamler, Monatsh. Chem., 83, 786 (1952).
7. K. Kratzl, ibid., 84, 406 (1953).
8. W. J. Schubert and F. F. Nord, J. Am. Chem. Soc., 72, 3835 (1950).
9. S. F. Kudzin, R. M. DeBaun and F. F. Nord, ibid., 73, 4615 (1951).
10. S. F. Kudzin and F. F. Nord, ibid., 73, 690, 4619 (1951).
11. G. DeStevens and F. F. Nord, ibid., 73, 4622 (1951).
12. C. L. Hess, Tappi, 35, 312 (1952).
13. H. L. Hergert and E. F. Kurth, J. Am. Chem. Soc., 75, 1622 (1953).
14. K. V. Sarkanen, H. M. Chang and B. Ericsson, Tappi, 50, 572 (1967).
15. K. V. Sarkanen, H. M. Chang and G. G. Allan, ibid., 50, 583 (1967).
16. K. V. Sarkanen, H. M. Chang and G. G. Allan, ibid., 50, 587 (1967).
17. F. Sundholm, Soc. Scient. Fenn. Comm. Phys.-Math., 32 (1965) 31 pp.
18. F. Sundholm, Acta Chem. Scand., 22, 854 (1968).
19. L. J. Bellamy, The Infrared Spectra of Complex Molecules, 2nd ed., Methuen, London, 1958, 425 pp.
20. K. Nakanishi, Infrared Absorption Spectroscopy, Holden-Day, San Francisco, 1962, 233 pp.
21. R. Zbinden, Infrared Spectroscopy of High Polymers, Academic Press, New York, 1964, 264 pp.
22. D. O. Hummel, Infrared Spectra of Polymers in the Medium and Long Wavelength Regions, Interscience, New York, 1966, 207 pp.
23. J. J. Lindberg, Papuri Puu, 37, 206 (1955).
24. J. J. Lindberg, Soc. Scient. Fenn. Comm. Phys.-Math., 20, 1 (1957).
25. J. J. Lindberg, Finska Kemist. Medd., 68 (1), 5 (1959).
26. V. M. Reznikov, Y. S. Pilipchuk and L. S. Solovev, Doklady Mezhvuz. Nauch. Konf. Spektroskopii i Spektr. Analizu, Tomsk. Univ.,

1960, 123; through Chem. Abst., 56, 118 (1962).
27. K. H. Ekman and J. J. Lindberg, Paperi Puu, 42, 21 (1960).
28. D. C. C. Smith, J. Chem. Soc., 1955, 2347.
29. D. C. C. Smith, Nature, 176, 267 (1955).
30. A. J. Michell, Aust. J. Chem., 19, 2285 (1966).
31. W. Ziechmann, Geochim. Cosmochim. Acta, 28, 1555 (1964).
32. S. Pinchas, D. Sadeh and D. Samuel, J. Phys. Chem., 69, 2259 (1965).
33. H. I. Bolker and N. G. Somerville, Pulp Paper Mag. Can., 64, T187 (1963).
34. A. J. Michell, A. J. Watson and H. B. Higgins, Tappi, 48, 520 (1965).
35. S. Wada, Chem. High Polym. Japan, 18, 617 (1961)
36. Y. S. Pilipchuk, R. Z. Pen and A. V. Finkelstein, Russ. J. Phys. Chem., 39, 938 (1965).
37. Y. S. Pilipchuk, R. Z. Pen and A. V. Finkelstein, Izv. Vyssh. Ucheb. Zaved., Les. Zh., 11, 131 (1968); through Chem Abs., 69, 378 (1968).
38. J. M. Pepper, Pulp Paper Mag. Can., 65 (2), T35 (1964).
39. H. P. Naveau, Can. J. Chem., 46, 1893 (1968).
40. J. Marton, E. Adler and K. I. Persson, Acta Chem. Scand., 15, 384 (1961).
41. S. Kolboe and O. Ellefsen, Tappi, 45, 163 (1962).
42. I. Kawamura and T. Higuchi, J. Japan.Wood Res. Soc., 10, 200 (1964).
43. H. L. Hergert, J. Org. Chem., 25, 405 (1960); unpublished work.
44. K. J. Harrington, H. G. Higgins and A. J. Michell, Holzforsch., 18, 108 (1964).
45. R. A. Durie, B. M. Lynch and S. Sternhell, Aust. J. Chem., 13, 156 (1960).
46. J. Polcin, B. Kosikova, P. Sipos, M. Dandarova-Vasatkova and J. Suchy, Chem. Zvesti, 17, 891 (1963).
47. I. A. Pearl, J. Org. Chem., 24, 736 (1959).
48. J. Nakano and N. Migita, J. Japan.Wood Res. Soc., 9, 62 (1963).
49. J. J. Lindberg and J. Kenttämaa, Suomen Kemistilehti, 32B, 193 (1959).
50. J. J. Lindberg, Paperi Puu, 43, 672 (1961).
51. J. J. Lindberg, Finska Kem. Medd., 69 (1), 11 (1960).
52. C. Juslen and T. Enkvist, Acta Chem. Scand., 12, 511 (1958).
53. T. Higuchi and I. Kawamura, Holzforsch., 20, 16 (1966).
54. K. Freudenberg and B. Lehmann, Chem. Ber., 93, 1354 (1960).
55. K. Freudenberg and H. Geiger, ibid., 96, 1265 (1963).
56. K. Freudenberg and W. Eisenhut, ibid., 88, 626 (1955).
57. W. J. Schubert, Lignin Biochemistry, Academic Press, New York,

1965, 131 pp.
58. F. F. Nord and W. J. Schubert, Tappi, 40, 285 (1957).
59. H. Grohn and W. Deters, Holzforsch., 13, 8 (1959).
60. K. Freudenberg, Croatica Chem. Acta, 29, 189 (1957).
61. K. Freudenberg, Bull. Soc. Chim., 1959, 1748.
62. H. I. Bolker, Nature, 197, 489 (1963).
63. H. Hata, Holzforsch., 20, 142 (1966).
64. H. Sobue and S. Fukuhara, J. Chem. Soc. Japan, Ind. Eng. Sect., 61, 1070 (1958).
65. T. Fukuzumi and T. Shibamoto, J. Japan.Wood Res. Soc., 4, 15 (1958).
66. E. Adler and J. Gierer, Acta Chem. Scand., 9, 84 (1955).
67. I. A. Breger, Fuel, 30, 204 (1951).
68. A. Robert and J. Desmet, Compt. rend., 251, 430 (1960).
69. J. Desmet, Bull. Assoc. Tech. Ind. Papetiere, 1961, No. 2, 88.
70. I. A. Pearl and L. R. Busche, Tappi, 43, 970 (1960).
71. P. R. Enslin, J. Sci. Food Agr., 4, 328 (1953).
72. C. Simionescu and J. Anton, Das Papier, 19 (4), 150 (1965).
73. F. Barnoud, T. Higuchi, J. P. Joseleau and A. Mollard, Compt. rend., 259, 3589 (1964).
74. T. Higuchi and F. Barnoud, in Chim. Biochim. Lignine, Cellulose, Hemicellulose, Grenoble, 1964, pp. 255-274.
75. B. Choulet, A. Robert, T. Higuchi and F. Barnoud, ibid., pp. 481-490.
76. A. Klemola, Suomen Kemistilehti, 41B, 152 (1968).
77. A. Björkman, Svensk papperstidn., 59, 477 (1956).
78. V. M. Nikitin, G. L. Burkov and V. M. Skachkov, Lesnoi Zh., 11, 121 (1968).
79. V. M. Reznikov, G. D. Ponurov and L. S. Solovev, Zh. Prik. Khim, 36, 1557 (1963).
80. V. M. Reznikov, L. G. Matusevich, I. V. Senko and T. V. Sukhaya, ibid., 40, 1397 (1967).
81. Y. S. Pilipchuk, R. Z. Pen and A. V. Finkelstein, Tr. Sib. Tekhnol. Inst., No. 38, 20 (1966) through Chem. Abst., 67, 9546 (1967).
82. E. Adler and J. Marton, Acta Chem. Scand., 13, 75 (1959).
83. Y. S. Pilipchuk, R. Z. Pen and A. V. Finkelstein, Izo. Vysshikh Uchebn. Zavedenii, Lesn. Zh., 8, 151 (1965) through Chem. Abst., 63, 11850 (1965).
84. D. E. Bland, Holzforsch., 15, 102 (1961).
85. I. Kawamura and D. E. Bland, Holzforsch., 21, 65 (1967).
86. M. Ujiie and T. Yoshikawa, J. Japan.Wood Res. Soc., 11, 59 (1965).
87. T. Higuchi, Y. Ito, M. Shimada and I. Kawamura, Phytochem., 6, 1551 (1967).
88. L. P. Kuhn, Anal. Chem., 22, 276 (1950).
89. F. E. Brauns and H. Seiler, Tappi, 35, 67 (1952).

90. D. E. Bland and G. Scurfield, Holzforsch., 18, 161 (1964).
91. C. Y. Liang, K. H. Bassett, E. A. McGinnes and R. H. Marchessault, Tappi, 43, 1017 (1960).
92. D. F. Arseneau and J. M. Pepper, Pulp Paper Mag. Can., 66, T415 (1965).
93. P. Karpovskaya, et al., Zh. Prikl. Khim., 37, 1318 (1964).
94. H. I. Bolker and L. M. Marraccine, First Canadian Wood Chemistry Symposium, Toronto, Sept. 1963, pp. 107-117.
95. V. C. Farmer and R. I. Morrison, Geochim. Cosmochim. Acta, 28, 1537 (1964).
96. I. M. Skurikhin, Khim. Prirod. Soed., 1967, No. 3, 208.
97. J. Polcin, B. Kosikova, J. Suchy and M. Vasatkova, Chem. svesti, 16, 562 (1962).
98. K. H. Ekman, Finska Kemists. Medd., 66, 115 (1957).
99. J. J. Lindberg, Paperi Puu, 1960, Sp. No. 4a, 193.
100. L. P. Clermont, Pulp Paper Mag. Can., 63, T402 (1962).
101. D. D. Maisaiya and G. P. Grigorev, Zh. Prikl. Khim., 41, 679 (1968).
102. H. I. Bolker and N. Terashima, in "Lignin Structure and Reactions," Adv. Chem. Ser., 59, 110 (1966).
103. I. Kawamura and T. Higuchi, J. Japan Wood Res. Soc., 12, 178 (1966).
104. R. M. Elofson, Can. J. Chem., 35, 926 (1957).
105. V. R. Jaunzems, V. N. Sergeeva, L. N. Mozheiko, Izv. Akad. Nauk Latv. SSR, Ser Khim., 1966, No. 6, 729.
106. S. Wada, Bull. Chem. Soc. Japan, 35, 707, 710 (1962).
107. S. Wada, T. Iwamida, R. Iizima and K. Yabe, Chem. High Polym. Japan, 19, 699 (1962).
108. C. Simionescu and J. Anton, Celluloza Hirtie, 14, 7, 348 (1965).
109. T. N. Kleinert, Pulp Paper Mag. Can., 67, T299 (1966).
110. P. F. Nelson, W. F. Forbes and J. A. MacLaren, Holzforsch., 17, 89 (1963).
111. J. Nakano, S. Miyao, C. Takatsuka and N. Migita, J. Japan Wood Res. Soc., 10, 141 (1964).
112. J. Marton, Tappi, 47, 713 (1964).
113. J. J. Lindberg, H. Tylli and C. Majani, Paperi Puu, 46, 521 (1964).
114. J. J. Lindberg, Finska Kemist. Medd., 64, 23 (1955).
115. L. Field, P. E. Drummond, P. H. Riggins and E. A. Jones, Tappi, 41, 721 (1958).
116. L. Field, P. E. Drummond and E. A. Jones, Tappi, 41, 727 (1958).
117. J. Gierer and B. Lenz, Svensk Papperstidn., 68, 334 (1965).
118. S. Fukuwatari and I. Nishikori, Bull. Shimane Agr. College, 15, 1 (1967) through Chem. Abst., 68, 7699 (1968).
119. S. Fukuwatari, Bull. Fac. Agr. Shimane Univ., 1967, No. 1, 124; Bull. Shimane Agr. College, 14, 113 (1965).

120. A. N. James and P. A. Tice, Tappi, 48, 239 (1965).
121. B. C. Fogelberg, K. Forss and S. Fugleberg, Paperi Puu, 49, 725 (1967).
122. K. Sato and H. Mikawa, J. Chem. Soc. Japan, Ind. Chem. Sec., 65, 477 (1962).
123. G. H. Wagner and F. J. Stevenson, Soil Sci. Am. Proc., 29 (1), 43 (1965).
124. T. Urbanski, W. Hofman, T. Ostrowski and M. Witanowski, Bull. Acad. Polon. Sci., 7, 861 (1959).
125. C. Steelink, J. Chem. Educ., 40, 379 (1963).
126. V. M. Reznikov and N. F. Sorokina, Vesti Akad. Navuk Belarus. SSR, Ser. Khim. Navuk, 104 (1968) through Chem. Abst., 69, 1945 (1968).
127. C. R. Kinney and E. I. Doucette, Nature, 182, 785 (1958).
128. W. Hofman, T. Ostrowski, T. Urbanski and M. Witanowski, Chem. and Ind., 1960, 95.
129. W. Flaig, in The Biochemistry of Wood, Vol. 2 (K. Kratzl and G. Billek, Eds.), Pergammon, London, 1959, pp. 227-244.
130. D. E. Bland, A. Logan, M. Menshren and S. Sternhill, Phytochem., 7, 1373 (1968).
131. S. A. Brown, Lloydia, 28, 332 (1965).
132. J. Vodnansky, M. Slabina and B. Schneider, Coll. Czech. Chem. Comm., 28, 3245 (1963).
133. O. Töppel, Schrift des Vereins der Zellstoff- und Papier-Chemiker und Ingenieure, 1. Eucepa-Symposium, 27, 249 (1958).
134. R. Sjöholm and H. H. Bruun, Suomen Kemistilehti, 39A, 121 (1966).
135. J. Marton and H. E. Sparks, Tappi, 50, 363 (1967).
136. G. Jayme and E. M. Rohmann, Das Papier, 19, 719 (1965).
137. H. H. Bruun and R. Sjöholm, Suomen Kemistilehti, 39A, 121 (1966).

8

MAGNETIC RESONANCE SPECTRA

C. H. Ludwig

I. Introduction. The application of magnetic resonance spectrophotometric techniques to investigations of problems in chemistry has shown phenomenal growth through the past decade. For example, this explosive expansion is reflected in the remarkable increase in the number of references cited in a series of biennial reviews of this subject. Only 83 references were needed for a thorough review of analytical applications of magnetic resonance up to 1958. Ten years later in the same series, 3,100 references were reviewed covering a two-year period and these did not include many of the routine applications of these techniques (1).

Nuclear magnetic resonance (nmr) and electron paramagnetic resonance (epr) spectrophotometric techniques have both been found to have utility in various studies of the chemistry of lignins. No review of these studies has appeared as yet in the literature, so an attempt has been made to offer a complete literature review through 1968 in this chapter.

II. Nuclear Magnetic Resonance (NMR).
    A. Introduction.
        1. Proton Magnetic Resonance (PMR) Spectroscopy. The usefulness of pmr in the study of organic molecules is primarily dependent on two phenomena: the "chemical shift," which allows an observer to detect the varying degrees with which bonding electrons interact with externally applied magnetic fields to "shield" the various protons of molecules, and "electron coupled spin-spin interactions between protons" which allow observations to be made regarding the interrelationships between protons within the molecule. A feature which makes these two phenomena especially valuable is the direct quantitative relationship between the intensity of each signal and the number of protons responsible for it.

Pople, Schneider and Bernstein (2) have prepared a text giving an excellent groundwork in the theory of nmr. Jackman (3) has discussed nmr in a way which is especially suited to organic chemists who wish to gain a working knowledge of the field. The theory of spin-spin splitting in nmr is presented in a simplified manner by J. D. Roberts (4), and Wiberg and Nist (5) have reported a group of theoretically calculated nmr spectra as an aid in the rapid analysis of spin-spin splitting patterns, and Bovey has published a concise summary of nmr theory and technique (6). Many examples of correlations between pmr spectra and structure such as those compiled by Varian Associates (8) continue to appear in the current literature.

Of the several ways currently used to express pmr chemical shifts for organic compounds, two systems are now the most popular. Most investigators prefer to report chemical shift values as the number of parts per million gauss (ppm) below (i.e., shifted toward the less shielded values) the signal given by tetramethylsilane (tms). These values may be referred to as delta (δ) values. The other method, which is just about as popular, follows the procedure suggested by Tiers (7) for expressing chemical shifts as tau (τ) values where $\tau = 10 - \delta$ ppm. Both of these systems have been used by lignin chemists.

2. <u>Special Problems with Polymers</u>. Polymers and other high molecular weight materials present special problems in pmr studies. Their solutions are often more viscous than those of simpler compounds. This decreases the mobility of the molecules and allows less chance for the signals to average out, which in turn results in broadening of the signals. Protons may also be shielded by other parts of the polymer in random ways causing additional broadening of signals. Despite these problems, polymeric substances other than lignins have been studied by pmr spectroscopy; for instance, asphalts (9), polymethylmethacrylate and polyethylene (10) and insulin (11). Bradbury and Crane-Robinson have reviewed the problems and potential of nmr studies of biopolymers, touching on several matters which are relevant to lignins (12). They emphasize the advantages of using spectrometers having high field strengths such as 100 MHz and 220 MHz.

3. <u>Selection of Solvents for Lignin Models and Lignins</u>. The experimental procedures applied to lignins are limited among other things by the solubility characteristics of the lignins or lignin derivatives. Consequently, the selection of solvents, which preferably should be aprotic for pmr studies, is of primary concern.

Early success in using deuterium oxide solutions of monomeric model sulfonates for pmr studies led to further attempts to use this solvent for higher molecular weight lignin sulfonates (13, 14, 15, 16). The spectra obtained, however, proved to be essentially featureless. It was suggested that in aqueous solutions of high molecular weight lignin sulfonates, coulombic forces and/or other factors hold the molecules in rather rigid states which do not allow for the rapid random molecular motions necessary for averaging out the pmr signals to the extent necessary for the production of resolved spectra.

Mildly extracted lignins such as Brauns lignins (BL), milled wood lignins (MWL) and dioxane acidolysis lignins (DAL), when acetylated, yield useful pmr spectra in chloroform-<u>d</u> solvent (13, 14, 15, 16, 17, 18). This

MAGNETIC RESONANCE SPECTRA   301

system has gained wide application in lignin studies. Most of the lignin model compounds have also been studied in chloroform-d, to relate the observed chemical shifts of signals to those found in spectra from lignins. The conditions which will be referred to as standard in this chapter include the use of about 15% solutions (8 to 20%) in chloroform-d with tetramethylsilane as an internal standard.

Less frequently used solvents include carbontetrachloride (14), dimethylsulfoxide-d_6 (18, 19, 19a), acetone-d_6 (20), trifluoroacetic acid and thionyl chloride (19). Care must be taken in interpreting results found in different solvents since some chemical shifts are notably dependent on solvent type as well as on concentration and temperature.

When deuteriated solvents are used, it is necessary to account for the signals from residual protons in the solvents. For the following solvents, the approximate location of such signals are: chloroform-d, 7.27δ; dimethylsulfoxide-d_6, 2.62δ; acetone-d_6, 2.17δ, and deuterium oxide, about 4.8δ. The latter are subject to much variation with changes in temperature, pH and sample type and concentration. Other commercially available deuteriated solvents of potential interest include acetic acid-d_4, acetonitrile-d_3, diethyl-d_{10}-ether, dimethylformamide-d_7, p-dioxane-d_8, ethyl alcohol-d_6, methyl alcohol-d_4, methylenechloride-d_2, pyridine-d_5, tetrahydrofuran-d_8 and tetramethylene sulfone-d_8.

B. Lignin Models and Lignin Degradation Products.

1. Spectra of Dimeric and Oligomeric Lignin Models. Chemical shifts found for protons contained in a number of lignin models are summarized in Table 8.1. These are further catalogued, Table 8.2, according to the kind of environment associated with various protons.

Table 8.1. Chemical Shifts of Protons in Lignin Models

I. Aromatic Protons. Total range of δ-values: 6.35 to 7.52 ppm

Structural type	R'	Ring position	δ-Value, ppm Range	δ-Value, ppm Average	Number of models surveyed	References
R'' =H,Me,Bz, Ac,C_β R''' =H,C_β,C_5	R-CO-	2 & 6 5	7.23-7.78 6.68-7.24	7.52 6.92	20	14,47
	C,H, -CH_2-	2,5&6	6.35-7.20	6.82	26	14,47
	Other	2,5&6	6.63-7.12	6.91	50	14,21,22, 44,47
R'' =H,Ac	R-CO-	2 & 6	7.24-7.27	7.26	2	47
	-CH_2-	2 & 6	6.35-6.50	6.42	6	47
	Other	2 & 6	6.50-6.61	6.56	7	14,44,47
	R-CO-	2 & 6 3 & 5	- -	7.83 7.02	1 1	14 14

Table 8.1. (Continued)

II. Vinylic Protons. Total range of δ-values: 5.40 to 7.79 ppm

R'	Config-uration	δ-Value, ppm Range	δ-Value, ppm Aver.	$J_{\alpha\beta}$, sec^{-1}	$J_{\beta\gamma}$, sec^{-1}	Number of models	Ref.
-COOH	trans α	-	7.79	16.0	-	1	47
	β	-	6.53	-	None		
-CH$_2$OH, -CH$_2$OAc	trans α	6.56-6.64	6.60	16.0	-	3	14,42
	β	6.12-6.21	6.16	-	6.0		
-CH$_2$OPh	trans α	6.69-6.74	6.72	16.0	-	2	14
	β	-	6.34	-	5.5		
-CH$_3$	trans α	6.27-6.42	6.36	15.6	-	12	14,42, 44,47
	β	5.75-6.18	5.96	-	4.9-5.3		
	cis α	6.27-6.42	6.36	11.8	-	4	14,44, 47
	β	5.40-5.92	5.72	-	7($\alpha\gamma$ 1.8)		
-	-	β 5.93-5.97	5.96	5.8	15.5 (trans)	3	14,47
		γ 5.02-5.09	5.06		8.3 (cis)		

III. Benzylic Protons. Total range of δ-values: 2.52-6.25 ppm

R'	R''	δ-Value, ppm Range	δ-Value, ppm Aver.	$J_{\alpha\beta}$, sec^{-1}	Number of models	Ref.
-OAc	-OC$_4$	6.00-6.25	6.12	6.1 threo, 4.9 erythro	9	14,21, 33, 47
-OAc	-C$_1$	5.70-6.25	6.04	-	6	24,33
-OAc	-OAc	5.67-5.96	5.82	-	2	22
-OH	-OC$_4$	4.93-5.04	4.99	7.9 (cis), 3.6 (trans)	8	14,23, 44,47
-OH	-Me	4.06-4.91	4.62	6.5	2	21
-OH	-H	4.44-4.58	4.53	None	3	14,47
-OH	-Et	-	3.82	?	1	14
-CH$_2$OAc	-CH(OAc)-C$_1$	3.1 -3.6	3.35	?	6	33,24
-H	-C$_1$	-	3.90	None	3	39
-H	-vinyl	3.28-3.34	3.31	5.8 & 6.5	2	14,47
-H	-H	3.19-3.35	3.26	None	3	39
-H	-CH$_2$OH	2.98-3.00	2.99	6.2 & 7.0	3	22,24
-H	-CH$_2$-C$_1$	2.84&2.88	2.86	None	2	39
-H	-Me	-	-	-	-	-
-H	-CH$_2$CH$_2$-	2.52-2.65	2.58	None	4	39
-CH$_2$OH	-	-	5.54?	7.2	1	14
-Me	-	4.91-5.12	5.05	8.8-9.5	6	14,47
-CH$_2$OAc	-	-	5.51	7.2	1	14
-Ar	-H	4.73-4.80	4.76	4.2	5	14,47
-H	-Ar	equatorial H:	4.85	5.0	1	14
		axial H:	4.44	6.8	1	14

Table 8.1. (Continued)

IV. Protons on Saturated β- and γ-Carbons.

Structure	R'	R''	δ-Value, ppm Range	δ-Value, ppm Aver.	$J_{\alpha\beta}$, sec^{-1}	Number of models	Ref.
H-C(R')(R'')-CO-(Ar)	-OAc	-H	5.23-5.29	5.29	-	4	14,47
	-OC$_4$						
	-OH	-Me	5.01-5.36	5.18	-	1	14
	-OC$_4$	-CH$_2$OH	4.46-4.62	4.54	-	1	14
	-H	-CH$_2$OAc	-	4.50	-	1	14
	-Me	-H	2.93-2.97	2.95	-	2	14,47
	-H	-H	2.54-2.57	2.56	-	4	14,47
H-C(R'')(OC$_A$)(CH-R')(Ar)	-OH	-H	3.89-4.40	4.06	7.9 & 3.6	5	14,47
	-OH	-CH$_2$OH	3.52-4.07	3.80	3.7?	2	14,47
	-OAc	-CH$_2$OAc	4.4 -4.8	4.65	5-6	4	14,21,47
	-OAc	-H	4.07-4.20	4.18	3.7	2	14,47
	-H	-H	3.8 -4.6	4.18	6.2	3	22,24
	-H	-CH$_2$OAc	-	1.82	?	1	24
OAc at C$_\beta$:	-OAc	-CH$_2$OAc	5.1 -5.5	5.3	-	3	22
H-C(R')(CH$_2$)(Ar)	-Me	-	-	1.60	?	1	39
	-H	-	1.18-1.25	1.22	?	2	39
(bicyclic Ar structure)	-	-	3.05-3.09	3.09	?	6	14,47
CH=CHR' / H-C-O(Ar)	-CH$_2$OH	-	-	4.19?	7.2	1	14
	-CH$_2$OAc	-	-	3.79	7.2	1	14
	-Me	-	3.33-3.77	3.42	8.8	8	14,47
R'-HC-H / CH / CH(Ar)	-OAc	-	4.67-4.68	4.68	$J_{\beta\gamma}$ 6.2	2	14,42
	-H	-	-	1.75	trans 5.2 cis 7.0	12	14,42,44,47
	-OCOPh	-	-	4.96	6.0 & 5.0	2	14
H-CHOAc / HC-R' / CHOAc (Ar)	-OC$_4$	-	3.8 -4.55	4.20	?	7	14,21,22,24
	-Cl	-	-	4.30	?	3	33
CH=CHR'' / -OAc	-OAc	-CH$_2$OAc	-	4.34	?	1	14
	-H	-CH$_3$	1.32-1.41	1.37	7.2	8	14,47
(Ar)H bicyclic -(G)=Ar	-(G)=Ar	-H	eq4.25-4.28 ax3.89-3.92	4.26 3.90	?	6	14,47
	-H	-(G)=Ar	2.72-4.08	3.40	?	1	14
	-(S)=Ar	-H	3.94-4.50	4.22	?	1	14
H-CH$_2$-CH$_2$-R'	-CO-Ar	-	1.12-1.22	1.19	7.0	4	47
	-CH$_2$Ar	-	0.90-0.95	0.93	6.8	7	14,39,47
	-CH$_2$Ar	-	0.65-1.05	0.85	?	1	14

Table 8.1. (Continued)

V. Methoxyl Protons.

	R'	R''	δ-Value, ppm Range	Aver.	Number of models	References
H-CH₂O / -O-Ph-R'	-CO-R	H, OMe, C_β, C_5	3.86-3.94	3.91	25	14,47
	-CH=CHCOOH	H	-	3.90	1	47
	-Me					
	-CH₂CH₂R	H or OR	3.65-3.71	3.67	5	47
	Other	H or OR	3.71-3.90	3.80	104	14,21,22, 39,44,47
Tri- and tetramers			3.55-3.9	-	3	22,25
Aliphatic methoxyl			3.33-3.47	3.42	4	14

VI. Acetoxyl Protons.

	δ-Value ppm Range	Aver.	Number of models	References
Aromatic acetoxyls, except those on 5-5 dilignols	2.18-2.34	2.26	40	14,21,22, 24,25,39, 44,47
Aromatic acetoxyls on 5-5 dilignols	2.07-2.10	2.09	6	14,47
Aliphatic acetoxyls	1.80-2.11	2.04	21	14,21,22, 24,25,47

VII. Miscellaneous Protons.

	δ-Value ppm Range	Aver.	Number of models	References
Carboxylic	-	11.50	1	14
Aldehydic	9.80-9.94	9.89	3	14,23
Aromatic hydroxyl	4.94-6.87	6.21	34	14,44,47
Aliphatic hydroxyl	2.53-4.05	3.53	16	14,47

The pmr spectra of several compounds in which two phenylpropane type units are joined by linkages presumably common in guaiacyl lignins are shown in Figure 8.1 (14). The signals in these spectra are labelled to identify the protons to which they are ascribed. Signals from protons attached to aromatic rings, methoxyl groups, aromatic acetoxyl and aliphatic acetoxyl groups are identified without difficulty.

The signal at about 6δ in the spectrum of the β-O-4 linked dilignol, 1, Figure 8.1A, has a chemical shift which is typical of protons attached to a benzylic type carbon which also bears an acetoxyl group, see Table 8.1-III. The unsymmetrical four-part signal consists of two doublets arising from the erythro and threo forms of this compound (14). Miksche has separated the pure erythro and threo forms of the related (G)-p(β-O-4)(G) H triacetate and observed that its benzylic proton in the threo form signals at 6.25δ with a $J_{\alpha\beta}$ of 6.1 Hz. while the erythro form signals at 6.11δ with a $J_{\alpha\beta}$ of 4.9 Hz. (21).

Wallis (21a) has developed pmr spectrophotometric procedures for measuring the relative amounts of erythro and threo β-O-4 type dimers formed by oxidative coupling of isoeugenol. He also finds that the C-α acetate protons of the threo isomer are slightly more shielded, 2.02 δ, than those of the erythro form, 2.10 δ. Schmid (21b) reports procedures for determining configurations of a number of phenylpropane type compounds. Lenz (19) and

MAGNETIC RESONANCE SPECTRA

Table 8.2. Ranges of Chemical Shifts in Acetylated Lignin Models

No.	Selected ranges δ-Values, ppm	Contributing proton types	δ, ppm	τ, ppm
1.	9.75 to 12.00 (Strongly deshielded protons)	a. Carboxylic b. Aldehydic	11.00-12.00 9.75-10.00	-2.00 to -1.00 0-0.25

2.	6.35 to 7.90 (Aromatic and α-vinylic protons)	a. Aromatic, ortho to carbonyl:
		  -Hydroxyphenyl type
		  -Guaiacyl type
		  -Syringyl type
		b. Other arom. protons
		  -Hydroxyphenyl type
		  -Guaiacyl type
		  -Syringyl type
		c. α-Vinylic
			7.20-7.90
			7.83
			7.23-7.78
			7.24-7.27
			6.35-7.24
			7.02
			6.35-7.24
			6.35-6.61
			6.36-6.72
				2.10-2.80
				2.17
				2.22-2.77
				2.73-2.76
				2.75-3.65
				2.98
				2.76-3.65
				3.39-3.65
				3.28-3.64

3.	5.75 to 6.35	a. Benzylic protons in β-O-4 and β-1 dilignol and arylglycerol acetates b. β-Vinylic	5.75-6.35 6.12-6.34	3.65-4.25 3.66-3.88
4.	5.30 to 5.70	Benzylic protons in β-5 dilignol acetates	5.30-5.70	4.30-4.70
5.	2.35 to 5.30	a. H-Cα in β-β dilignol and H-Cβ in β-O-4 dilignol acetates. H-Oγ in cinnamyl alcohol acetates b. H-Cγ in β-5-, β-1-, β-O-4- and β-β dilignol acetates c. Arom. methoxyl protons d. H-Cβ in β-1- and β-β dilignol acetates and aliph. methoxyl protons	4.40-5.30 3.95-4.40 3.55-3.95 2.35-3.55	4.70-5.60 5.60-6.05 6.05-6.45 6.45-7.65
6.	2.15 to 2.35 (Most arom. acetoxyl protons)	Arom. acetoxyl protons	2.15-2.35	7.65-7.85
7.	1.75 to 2.15 (Aliph. acetoxyl protons)	Aliph. acetoxyl and arom. acetoxyl in 5-5 dilignol acetates	1.75-2.15	7.85-8.25
8.	0.75 - 1.75 (Highly shielded protons)	Miscellaneous protons	0.75-1.75	8.25-9.25

Nimz (22, 23, 24, 25) have also reported pmr studies of β-O-4 linked lignin models.

In the spectrum of pinoresinol 2, Figure 8.1B, a nonacetylated β-β-linked dilignol, the typically broadened aromatic hydroxyl proton signal appears at 5.8 δ. Shifts for hydroxyl protons are, of course, dependent on factors which effect hydrogen bonding such as temperature, solvent type and concentration. Each of the other proton signals in this spectrum appears in about

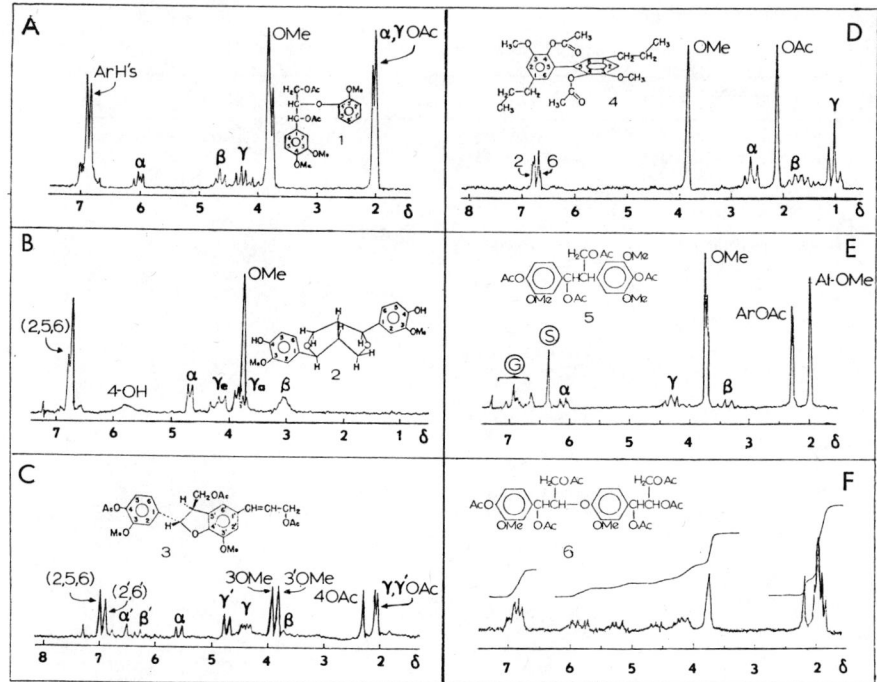

Figure 8.1. Pmr spectra of dimeric lignin model compounds.

the same location as found in the acetylated pinoresinol spectrum (14). The complex signal at about 3δ originates from the protons on the bridgehead carbons in the 3,7-dioxabicyclo-[3.3.0]-octane system. The signals from axial methylene protons overlap that from the methoxyl protons at about 3.8δ. General criteria for the assignment of stereochemistry of dimers having β-β linkages have been established by Birch (26), and pmr spectra of the related lignans, eudesmin, epieudesmin and diaeudesmin have been characterized by Atal, Dhar and Pelter (27). Ogiyama and Kondo (28) describe pmr evidence supporting the presence of stable intermolecular association via hydrogen bonding between pinoresinol molecules. Excoffier (29) has reported configurational studies of pinoresinol and related compounds, and Lenz (19) has tabulated chemical shifts for members of the same group in a variety of solvents.

The β-5-linked dilignol (dehydrodiconiferyl alcohol) triacetate, $\underline{3}$, yields the spectrum in Figure 8.1C. In it a signal appears at 5.6δ originating from the proton attached to the C-α of the furan ring. This is a uniquely located signal and thus has potential use in determining the amount of phenyl coumaran type linkage in lignins and oligolignol fractions. Signals from protons on alpha and beta vinylic carbons in this spectrum are found between 6 and 7δ. The quartet at 4.3δ indicates a probable restriction to the rotational freedom of the methylene group on C_γ. This was previously rationalized by suggesting a cis configuration between it and the vicinal guaiacyl substituent (14). Since then, the configuration of the closely related dehydrodiisoeugenol has been shown to be trans by degradation studies (30), and oxidative coupling studies of isoeugenol (21a) give added support to the idea that the acetylated dilignol, $\underline{3}$, actually has a trans configuration associated with the phenyl coumaran ring. The inhibition of rotation about the C_β-C_γ bond may be a result of any one or all of three factors. The π system of the aromatic ring may offer steric hindrance to the rotation; the acetoxyl group may be repulsed somewhat by coulombic forces from the ring oxygen and the π system; and/or the somewhat acidic hydrogens on the C_γ may hydrogen-bond to the ring oxygen, Figure 8.2.

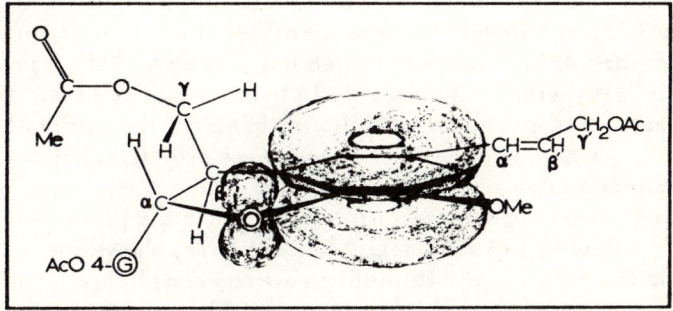

Figure 8.2. A representation of structural features which may inhibit rotation about the C_β-C_γ bond in molecules of acetylated dehydrodiconiferyl alcohol, $\underline{3}$.

Miller and Schuerch (30b) have used pmr to show the identity between synthetic 2,4'-dihydroxy-3,3'-dimethoxy-5-ethylbibenzyl and a pure crystalline dimer obtained by alkaline hydrogenation of wood from sugar maple (Acer saccharum). The latter is a product deriving from β-5 linked units in lignin.

The pmr spectra of related phenylcoumaran ring systems in hordantine glucoside derivatives have been analyzed by Stoessl (31), and Belorizky-Perret has made an extensive pmr study of benzofuran ring systems and their partly and fully hydrogenated derivatives (32).

The spectrum of the 5-5 linked model, 4, Figure 8.1D, contains a signal given by the aromatic acetoxyl protons which appears at about 2.1δ. This is in the range normally expected for aliphatic acetoxyl protons. A probable reason for this peculiarity is the strong shielding expected from the aromatic ring due to the conformation illustrated in the structural formula, 4, Figure 8.1D. The bulky acetoxy groups ortho to the 5-5 link no doubt force the rings out of the same plane causing the acetoxyl protons to be shielded by the opposite aromatic rings.

The aliphatic side-chain protons present in this model are not likely to occur in unmodified lignins. However, they do provide an example of the type of highly shielded protons which give rise to signals at low δ values.

The largely unused potential of pmr spectroscopy in investigating the structure of lignins is perhaps best illustrated by the remarkable studies made by Nimz (22, 23, 24, 25, 33) on acetylated, di-, tri- and tetra-lignols which he isolated from wood by very mild hydrolysis, see Chapter 9, Section II-A. The assignment of the correct structures, especially for the tri- and tetra-lignols, without pmr would have been an excessively difficult task.

Nimz (33) confirmed the presence of β-1 linkages in lignins when he reported pmr spectra for acetates of three beechwood hydrolysis products such as 5, Figure 8.1E, with variations only in the aromatic systems. The most unique signal in this spectrum is the quartet found for the tertiary hydrogen on C_β at 3.35δ. Subsequently, β-1 linkages were also detected among the dehydrogenation products of coniferyl alcohol (25).

The broadened appearance of the signals from the protons on side chain carbons of the β-O-4 linked dimer, 6, Figure 8.1F, shows how spin-spin splitting and the presence of diasterioisomers can complicate spectra of even relatively simple molecules related to lignins (22).

Nimz also used pmr to help identify two aldehydic β-O-4 linked dimers from hydrolyzed spruce wood (23).

The tetralignol octaacetate, 7, illustrated with its integrated pmr spectrum, the structure derived therefrom in Figure 8.3 (24, 25), represents the most elegant example of the application of pmr spectroscopy to purified lignols which has yet appeared in the literature. Here is a lignin fragment with unaltered structure and high enough molecular weight, 713 (prior to acetylation), to possess vital structural significance.

The high shielding, ~ 3.6 δ, of protons on one of the aromatic methoxyls in this compound and in the related trilignol hexaacetate (24, 25) would not be predicted from the values of about 3.8 δ usually found for dilignol acetates and related compounds, Table 8.1-V. This suggests that there is distant shielding of one of the methoxyl groups by one or more of the three other

MAGNETIC RESONANCE SPECTRA

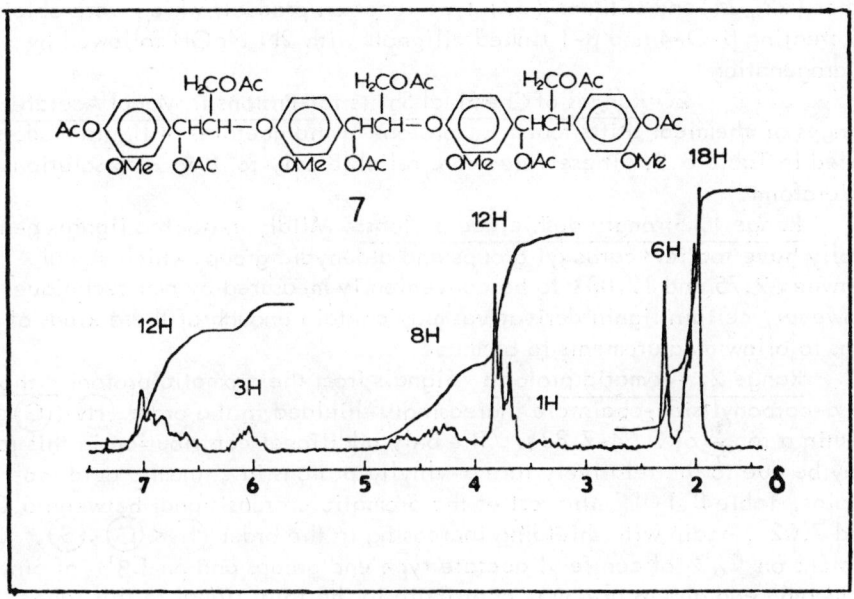

Figure 8.3. Pmr spectrum of a tetralignol octaacetate isolated from spruce wood hydrolyzate (25).

aromatic rings present in the tetralignol. An example of this type of distant shielding has been pointed out by Wallis (34) for the methoxyl protons on C-8 of the compound 8 which signal at 3.20 δ compared with 3.75-3.84 δ for the other methoxyls. This kind of distant shielding is one of the factors which makes the interpretation of the pmr spectra of lignin preparations subject to considerable uncertainty.

Nimz used pmr to characterize hydrogenated derivatives from vinyl ether (22, 24) and stilbene (33) type reaction products which were obtained by treating β-O-4 and β-1 linked dilignols with 2N NaOH followed by hydrogenation.

2. Ranges of Chemical Shifts for Protons in Model Acetates. Ranges of chemical shifts found for protons in the acetates of lignin models are listed in Table 8.2. These values are reliable only for 8 to 20% solutions in chloroform.

Range 1, Strongly deshielded protons: Mildly extracted lignins generally have too few carboxyl groups and aldehydic groups which signal between 9.75 and 12.00 δ to be conveniently measured by pmr techniques. However, certain lignin derivatives may contain enough of these kinds of protons to allow measurements to be made.

Range 2, Aromatic protons: Signals from the aromatic protons ortho to an α-carbonyl side-chain are increasingly shielded in the order (H)<(G)<(S) within a range of 7.24-7.83 δ. The only additional contribution in this range may be due to the relatively rare α-vinylic protons in cinnamic acid type side-chains, Table 8.1-II. The rest of the aromatic protons signal between 6.35 and 7.02, again with shielding increasing in the order (H)<(G)<(S). Vinyl protons on C_α's of coniferyl acetate type end groups and on C_β's of cinnamic acid type end groups also may contribute to the total signal in this range.

Range 3, Protons on acetylated C_α's: The major signals between 5.75 and 6.35 δ are those from protons on C_α's to which an acetoxyl group is also attached. Additional minor signals also may originate from vinylic protons on C_β's.

Range 4, Protons on C_α in α-O-4, β-5 type linkages: The protons on C_α in β-5-linked dilignols signal between 5.30 and 5.70 δ. There appear to be no serious interferring signals in this range.

Range 5a and b, Side-chain protons: The majority of protons on α, β and γ carbons in model acetates signal between 3.95 and 5.30 δ. Signals of major importance between 4.40 and 5.30 δ derive from protons on C_α's in β-β-linked dimers, on C_β's of β-O-4 linked dimers. $C\gamma$-protons of coniferyl acetate end groups also contribute to signals in this range. The range from 3.95 to 4.40 δ is governed by protons on $C\gamma$'s in β-5, β-1, β-O-4 and β-β-linked dimers.

Range 5c, Aromatic methoxyls: All aromatic methoxyl protons in the lignin models, Table 8.1, signal between 3.55 and 3.95 δ, the major signals being located in the range of 3.71 to 3.85 δ. An α-carbonyl or cinnamic acid type side-chain tends to deshield aromatic methoxyl at 3, 4 or 5 positions to about 3.90 δ. Condensation in β-5 or 5-5 linked dimers also tends to result in moderate deshielding of aromatic methoxyls. In Nimz's tri- and tetralignols, Figure 8.3, distant shielding occurs which causes some of the methoxyls to be shielded to values of 3.55 δ. Methyl, ethyl or ethylene (-CH_2CH_2-) groups at the 1 position also result in increased shielding of methoxyl groups to values of about 3.67. Otherwise, neither the position of methoxyl groups, nor the presence or absence of acetyl groups, appear to affect the proton signals of the former.

Protons on C_β's in acetylated β-5 dimers and axial protons on C_γ in β-β dimers also signal in this range and must be considered in any quantitative studies.

Range 5d, Protons on certain C_β's: The range including 2.35 to 3.55 δ covers signals from protons on C_β's in β-β and β-1 dimers.

In methanolysis lignins the protons on aliphatic methoxyls also register in this range. Lignins which have been subjected to alkaline hydrogenolysis yield signals in this range from protons in $-CH_2-CH_2-$ groups.

Range 6, Aromatic acetoxyl protons: The aromatic acetoxyl protons with the exception of those attached 5-5 linked dimers signal from 2.15-2.35 δ.

Range 7, Aliphatic acetoxyl protons: Aliphatic acetoxyl protons and aromatic acetoxy protons in 5-5 linked dimers signal from 1.75-2.15 δ.

There appear to be no significant interfering signals in ranges 6 and 7. Consequently, quantitative determinations of both aliphatic and aromatic acetoxyl groups are quite reliable.

Range 8, Highly shielded protons: Highly shielded protons in lignin models signal from 0.75-1.75 δ. All of these are attached to a carbon which has at least two carbons between it and any oxygen function or one carbon between it and a double bond or aromatic system. (See Chapter 9, Section II-B-3 for a possible source of highly shielded proton signals.)

The above relatively precise boundaries are applicable for acetylated model compounds, lignols and low molecular weight lignins. For higher molecular weight lignins it is generally wise to take into account the signal broadening effects and other uncertainties when such ranges are established. In Table 8.3 such ranges for lignins are given. These or similar values have been used in semi-quantitative pmr studies of such lignin derivatives as the acetates of MWL and BL preparations.

3. The Identification of Lignin Degradation Products. The first reported application of pmr spectroscopy to lignin chemistry confirmed the structures of two monomeric sulfonates isolated from western hemlock (Tsuga heterophylla) spent sulfite liquor. These were identified as (G)-CH($\overline{SO_3Na}$)-CH=CH_2 and (G)CH=CH-CH_2SO_3Na (13, 15, 18). Since then, pmr studies have been reported by Glennie and Mothershead (35) on related monomers recovered from red alder (Alnus rubra) spent sulfite liquor and by Parrish (36, 37) on similar products isolated from wattle wood (Acacia mearnsii). The relation of these structures to overall understanding of sulfonation of lignins is discussed in Chapter 15. Gellerstadt and Gierer have used pmr to examine products formed from dehydrodiisoeugenol by neutral sulfite pulping conditions (37a).

Parker, Coalson and Schuerch (38, 39) used pmr spectroscopy for the identification of dimeric products from the alkaline hydrogenation of maple

Table 8.3. Adjusted Ranges of Chemical Shifts for Protons in Acetylated Lignins (16)

Indicated ranges pertain to spectra in $CDCl_3$ with TMS-reference. Adjustments made to compensate for line broadening, signal overlap and other factors.

Number	Adjusted range δ, ppm	Proton types
1.	8.00 to 11.50	Carboxylic and aldehydic
2.	6.28 to 8.00	Aromatic and α-vinylic (as in structure 3)
3.	5.74 to 6.28	β-Vinylic (e.g. 3) and benzylic (e.g. 1 or 5)
4.	5.18 to 5.74	Benzylic, as in structure 3
5.	2.50 to 5.18	Methoxyl and most side-chain protons
6.	2.19 to 2.50	Most aromatic acetoxyl protons
7.	1.58 to 2.19	Aliphatic acetoxyl and aromatic acetoxyl ortho- to biphenyl, as in structure 4
8.	0.38 to 1.53	Highly shielded aliphatic (e.g. $H-CH_2CH_2O-$)

and spruce woods. They synthesized a group of dimeric models for comparison. Values for chemical shifts from these compounds are among those summarized in Table 8.1.

Hrutfiord and McCarthy (40) examined the ether-soluble product from the catalytic hydrogenolysis of red alder (Alnus rubra) in weakly acidic dioxane-water (1:1). The material represented about 26 to 39% of the lignin in the wood. The pmr spectrum of this fraction revealed the presence of dihydroconiferyl alcohol and dihydrosinapyl alcohol, and both of these compounds were, in fact, present in large amounts, see Chapter 12.

Pepper, Casselman and Karapalby (41) have used pmr to identify 3,3',4,4'-tetramethoxychalcone, a product from the cupric oxide oxidation of spruce wood meal.

4. <u>Reactivity Studies on Lignin Model Compounds</u>. Model compounds containing episulfide structures were synthesized by Gierer and Smedman (42) and their pmr spectra were analyzed. The models were treated with 2N sodium hydroxide to give dithiane ring systems which were also characterized by pmr spectroscopy and were suggested as possible structural features in kraft lignins, see Chapter 16, Section 2-C.

Pmr spectroscopy was found useful in determining the position and extent of deuteration of such lignin models as phenol, anisole, guaiacol, veratrole and 4-methylguaiacol by Ericsson, Nist and Sarkanen. They then used these models in rate studies to determine the reactivity towards electrophiles of the various positions on the aromatic rings (43). Chemical shifts determined for several chlorinated models for chloro-lignins have been reported by Braddon and Dence (43a). Procedures for determining quantities of o-, m- and p-cresols in their mixture have been developed using pmr spectroscopy by Wainai and Suzuki (43b). Procedures such as this would be helpful in analyzing products from certain lignin degradation procedures. Pmr has been used to help determine the structures of products from the mild alkaline hydrolysis of dehydrodiisoeugenol by Wacek and Voykowitsch (43c). Pew and Connors have subjected various lignin models to enzymic degradation reactions and used pmr studies to help in the determination of structures of products (43d).

C. Polymeric Lignin Derivatives.

1. Interpretation of Spectra. High resolution pmr studies are restricted to isolated lignins which are either soluble in suitable solvents or are capable of forming soluble derivatives. Such isolated lignins vary considerably in the degree to which they represent the total lignin in the wood.

In the following discussion the pmr spectra of relatively mildly extracted lignins will be treated first from a qualitative point of view. Early studies of such lignins have been reported by Bland and Sternhell (17, 18) and Ludwig, Nist and McCarthy (13, 16). The former authors studied lignins isolated from Eucalyptus regnans wood and Pinus radiata wood treated with methanol at 150°C for 4 hours and then purified using a liquid-liquid multistage extractor. The latter group studied lignins isolated from several gymnosperm woods by various mild treatments. Both groups acetylated the lignins and examined them as chloroform-d_6 solutions under standard conditions. Bland and Sternhell also report spectra of solutions of non-acetylated lignins both in chloroform-d and in dimethylsulfoxide-d_6.

Swan (20) reports that the phloem and bark dioxane acidolysis lignins from western redcedar (Thuja plicata Donn), even after acetylation are not soluble enough in chloroform-d. Instead, acetone-d_6 was used as a solvent to obtain pmr spectra.

Examples of pmr spectra of some of the isolated softwood and hardwood lignins are shown in Figures 8.4 and 8.5, respectively. The relatively broad envelopes noted in these spectra as compared with those of lower molecular weight compounds are probably caused in part by relaxation effects due to a tendency towards rigidity in lignin macromolecules resulting from cross linking and large ring structures (16). Spin-spin splitting and differences in the chemical environments of similarly located protons also contribute to signal broadening. Acetylated Brauns lignins, which are known to be of relatively low molecular weight, yield notably more sharply defined signals than those found in the higher molecular weight lignins, see Figure 8.4C.

Gagnaire (44) reports a pmr spectrum of Sequoia sempervirens milled wood lignin taken in dimethylsulfoxide-d_6 solution. His study of the hydroxyl content of this preparation will be discussed later.

A rather exotic pmr study of water insoluble lignins extracted over a

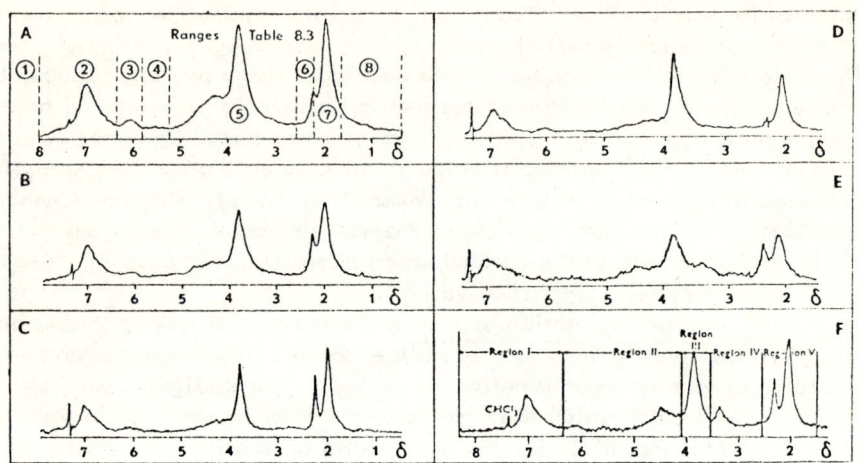

Figure 8.4. Pmr spectra of acetylated softwood lignins and their derivatives. A. Spruce milled wood lignin (16). B. Western hemlock dioxane acidolysis lignin (16). C. Spruce Brauns lignin (16). D. Diazomethane methylated spruce Brauns lignin (16). E. Spruce Brauns lignin which has been subjected to mild methanolysis (16). Pine methanolysis lignin (18).

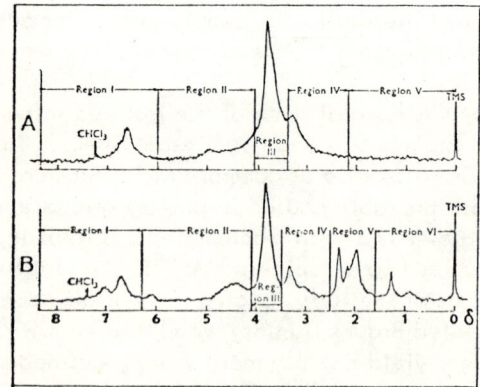

Figure 8.5. Pmr spectra of hardwood lignins. A. Eucalyptus methanolysis lignin (18). B. Acetylated eucalyptus methanolysis lignin (18).

period of three to nine years from oak barrels by cognac alcohol has been reported by Skurikhin (45). The spectra of acetylated derivatives suggest the presence of firmly bound carbohydrates in these lignins.

Klemola (46) has examined the spectra of acetylated milled wood lignin and dioxane acidolysis lignin from birch wood. These spectra, while similar to those in Figure 8.4A and B, show the expected variations due to the presence of a large proportion of syringyl nuclei in these hardwood lignins.

Lenz (19) has reported spectra of acetylated spruce milled wood lignin and acetylated pine and black gum dioxane acidolysis lignins. In addition, spruce milled wood lignin was examined in dimethysulfoxide-d_6 at the elevated temperature of 130°C to sharpen up the signals, Figure 8.6A. A time-averaged pmr spectrum of black gum lignin, Figure 8.6B was found to improve the signal to noise ratio. This work points the way to more accurate semi-quantitative results from lignin preparations in the future. Lenz (47) also has explored the potential of trimethylsilyl ethers of lignins for pmr spectroscopy. These derivatives are more easily soluble in many organic solvents than acetylated lignins and have the further advantage that the trimethysilyl groups give signals which do not interfere with those from other protons.

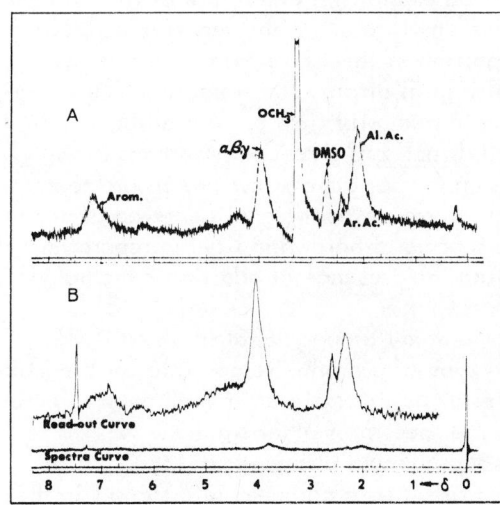

Figure 8.6. Special techniques in pmr spectroscopy. A. The pmr spectrum of acetylated spruce milled wood lignin in dimethyl-sulfoxide-d_6 solvent taken at high temperature (130° C.). Note especially the improved sharpening of side-chain and methoxyl proton signals (19).
B. A time-averaged spectrum of acetylated black gum dioxane acidolysis lignin.

Yokoyama et al. have compared spectra of milled wood lignins taken in dimethylsulfoxide-d_6 with those of the same lignins, acetylated and dissolved in chloroform-d (19a).

Prominent features generally common to spectra of acetylated lignins have been labelled in accordance with chemical shifts of signals in model spectra, see Tables 8.1 and 8.2. Signals from aromatic protons are centered at about 7.0δ in all of the spectra obtained from the acetylated gymnosperm lignins. The spectrum of a non-acetylated hardwood lignin from E. regnans, Figure 8.5A, on the other hand, shows a broad signal from aromatic protons centered at about 6.6 δ. Upon acetylation, the signal is divided, with one part retaining its original value and another part shifting to 6.95 δ. The less-shielded portion probably originates from protons associated with acetylated guaiacyl type rings which give signals centered at about 6.90 δ. The other portion originates from acetylated syringyl type rings signalling at about 6.56 δ, Table 8.1-I.

The spectra of acetylated lignins (except Brauns methanol acidolysis lignin), Figure 8.4E, have a small but definite maximum at 6.1 δ. With minor exceptions, signals in this range are caused by protons attached to acetylated α-carbons, Table 8.1-III. The spectrum of non-acetylated methanol lignin from E. regnans, Figure 8.5A, lacks the maximum at 6.1 δ but does have a small maximum at 5.0 δ where the counterpart of this type of proton in non-acetylated models is known to signal. The absence of a maximum at 6.1 δ in the acetylated Brauns methanol acidolysis lignin, Figure 8.4E, is consistent with the presence of benzylic methyl ether groups which block the reactive benzylic positions toward acetylation. The spectra of the acetylated methanolysis lignins prepared by Bland, in contrast, each have a maximum at 6.1 δ, see Figures 8.4F and 8.5B. This indicates that the extraction of lignins by neutral methanol at 150°C does not methylate all of the reactive benzylic positions, leaving a significant proportion of them free to be acetylated.

In general, the pmr spectra of lignins display low-intensity diffuse signals at about 3.5-6 δ. These originate primarily from protons other than those mentioned above which are attached directly to side-chain carbons in the lignins, Table 8.1-IV. The protons on aromatic methoxyl groups are responsible for all but a small portion of the very prominent signal centered at about 3.8 δ. This signal is quite strong in spectra of hardwood lignin preparations, Figures 8.5A & B and 8.6B, reflecting the presence of additional methoxyl groups in the syringyl type units.

Protons attached to aliphatic methoxyl groups signal at about 3.3 δ, Table 8.1-V. It is evident that this type of proton is responsible for the maxima appearing at this point in spectra of lignins isolated in methanol, Figures 8.4E & F and 8.5A & B, but not in the spectra of other lignins.

The spectra of models such as those shown in Figures 8.1B, C & E and 8.3 show that certain protons attached to side-chain carbons yield signals in the range from 2.5 to 4.0 δ. These are protons attached to C_β's in β-β linked pinoresinol type, the β-5 linked phenyl coumaran type and the β-1 linked diphenyl propane type linkages, Table 8.1-III & IV. These signals should be taken into account in any attempt to interpret the methoxyl signals in this range. It ought to be noted that when either acetone-d_6 or dimethylsulfoxide-d_6 are used as solvents, signals from their residual protons will appear at about 2.65-3.15 and about 2.45-2.80 δ respectively.

The spectrum of non-acetylated methanol lignin from E. regnans,

Figure 8.5A, shows very little apparent signal at less than 2.5δ. Acetylated lignins on the other hand yield spectra, Figures 8.4, 8.5B and 8.6, in which prominent signals due to acetoxyl protons appear within this range. The signals centered at 2.26δ are due to aromatic acetoxyl groups except for the 5-5 linked types and those centered at about 2.04δ are caused by aliphatic acetoxyl protons and, in addition, by aromatic acetoxyl protons in 5-5 linked units, Figure 8.1C and Table 8.1- VI. From the spectra of acetylated softwood lignins it is evident that these lignins contain comparable amounts of aromatic methoxyl and aliphatic acetoxyl groups indicating the parent lignins contain close to one aliphatic hydroxyl for each methoxyl group. The aromatic acetoxyl signals in these spectra are more variable in their intensity according to the type of lignin being examined. In the case of acetylated diazomethane methylated spruce native lignin, Figure 8.4D, the extremely small signal at about 2.25δ confirms the methylation of most of the phenolic hydroxyls in this derivative. The spectrum of acetylated lignin from E. regnans, Figure 8.5B, shows an aromatic methoxyl signal which as expected for syringyl type lignins is much larger than the aliphatic acetoxyl. The signals which appear in Region 6 of this spectrum appear neither in the spectra of acetylated gymnosperm lignins nor in the spectrum of non-acetylated E. regnans lignin. The origin of these signals is not readily apparent from the information at hand.

The spectra from Swan's acetylated phloem and bark lignins also have definite signals in the range of 0-1.5δ, giving evidence of the presence of highly shielded protons (20).

Several workers have conducted pmr studies of lignins which have been isolated by causing extensive structural changes. Interpretations of spectra from such products are subject to considerably greater uncertainty than the spectra from more mildly extracted lignins. Nevertheless, with the use of reasonable caution these spectra have yielded unique and valuable information.

Acetolysis lignins from spruce (Picea jezoensis Carr.) wood meal have been examined, using pmr and IR techniques, by Fukuzumi et al. (48). The crude preparation represented about 12% of the starting wood weight and was fractionated to yield ten individual fractions. Most of these fractions showed proportionately more aliphatic hydroxyls than acetylated spruce milled wood lignins, reflecting the probable formation of aliphatic acetoxyl groups during the acetolysis reaction. The lack of any noticeable signals at about 6.1δ shows that benzylic hydroxyl groups were lost (or at least not acetylated) in the process of acetolysis. This, the authors claim, can be explained by the formation of carbonyl functions, 2.5 times more frequent than in spruce MWL acetate. In some ether soluble fractions sharp signals were noted at about 1.25δ, and tentatively assigned to methyl groups having neighboring carbons to which no protons are attached.

Hrutfiord and McCarthy (40) subjected red alder (Alnus rubra) wood to catalytic hydrogenolysis both in 1:1 dioxane water mixtures (which become acidic during the reaction) and in 3% sodium hydroxide solutions, see Chapter 12, Section III. The pmr spectrum from one of their acetylated hydrogenolysis lignins is shown in Figure 12.4, Chapter 12. They also investigated model compounds especially suitable for comparison with these reduced lignins. Chemical shift data from these are summarized in Chapter 12, Table 12.4.

In another study of hydrogenolysis lignins, Parker (38) examined nondistillable, chloroform-soluble, hydrogenated lignin fractions obtained from sugar maple (Acer saccharum) and Norway spruce (Picea abies) woods. The pmr spectra of these lignins after acetylation showed unidentified signals between 1.16 and 1.50 δ.

Peracetic acid lignins from red alder and Douglas-fir have been analyzed by pmr (39). In studies as yet unpublished, Gupta claims to have obtained informative pmr spectra from lower molecular weight fractions of lignosulfonates dissolved in deuterium oxide and referenced to the water-soluble sodium 4,4-dimethyl-4-silapentane-1-sulfonate. This contrasts with the previously reported unsatisfactory results with higher molecular weight lignosulfonates (13, 14).

Klemola (46) hydrolyzed birch wood with steam at 180° C for two hours to obtain steam hydrolysis lignins (SHL) which were acetylated for pmr studies. It is readily evident from his spectra that steam hydrolysis increased the numbers of phenolic hydroxyls and markedly decreased the number of aliphatic hydroxyls.

Lenz (19) has investigated kraft and soda lignins from pine and black gum woods by pmr spectroscopy. The spectra he obtained from softwood and hardwood soda and kraft lignins exhibit certain features which are helpful in their characterization. The rather sharp signals obtained in these lignins reflect their low average molecular weight. The prominence of the aromatic acetoxyl peaks confirm the presence of increased numbers of phenolic hydroxyls in alkali lignins as compared with mildly reacted lignins shown in Figure 8.4. Yokoyama et al. have also used pmr in recent studies of thio lignins (19a).

2. Semi-Quantitative Functional Group Analyses. The direct relationship between signal intensity in pmr spectra and the number of protons responsible for the signal makes possible quantitative studies of pure, smaller molecular weight compounds and is the basis for a number of semi-quantitative studies of isolated lignins. Accurate quantitative pmr analysis of lignins is not possible because of the overlapping broadening of signals caused by relaxation effects, spin splitting, diastereoisomerism and uncertainty as to the kinds of environments for protons which are actually present in lignins.

Early semi-quantitative pmr studies of lignins extracted by various mild procedures and with one exception acetylated (13, 14, 16, 17, 26) have been followed by more recent ones of various lignin derivatives (19, 40, 44, 46).

The general procedure employed has been to use ranges of chemical shifts such as those shown in Table 8.3 and to determine the relative increase in signal intensity between the boundaries of each range of integration. The percent of total signal intensity which appears within each range can then be used, with the calculated total protons/C_9, to find the number of protons/C_9 signalling within each range. A typical integral curve is shown in Figure 12.3, Page 495.

This procedure is based on the assumption that lignins are entirely composed of phenylpropane units. However, as Pepper so aptly states, "... none of the known analytical data can be interpreted as proof that the entire lignin substance is of a phenyl propanoid character ..." (50). For instance, the β-1 linkage results in the detachment of some side-chains, and it is not known for certain whether all of the detached side-chains remain with the lignins as glyceraldehyde units, see Chapter 4, Section IV. Also, certain lignin derivatives, especially those subjected to alkaline treatments, are known to have lost one or more carbons from some of the side-chains.

Another factor which could affect the accuracy of quantitative interpretations is the possibility that relaxation effects might broaden some signals so that they no longer fall within the range of the spectrum and thus are not counted. Bland and Sternhell checked this possibility by comparing the signal strength of a known amount of sym-trinitrobenzene with that found for methanol lignins from E. regnans and from P. radiata in dimethylsulfoxide-d_6 solutions. They found by this means that the hydrogen contents of these lignins were 5.9% and 6.0%, respectively, compared with values of 6.10% and 5.79% obtained by conventional analyses and conclude that relaxation effects do not appreciably diminish the total amount of observed proton signal (18).

The data in Tables 8.1, 8.2 and 8.3 have generally been determined under the standard conditions described previously, Section II-A-3. Deviations from these conditions may have significant effects on the position of chemical shifts. For instance, apparent higher shieldings of methoxyl protons have been noted when certain models have been examined in carbon tetrachloride rather than in chloroform-d (14, 24). Lenz (19) used trifluoroacetic acid, dimethylsulfoxide-d_6 and thionyl chloride as solvents especially for non-acetylated lignins. He found noticeable differences in distribution of signals depending on the solvents used and suggests that when trifluoroacetic acid is used in place of chloroform-d_6 an adjustment of range boundaries to about 0.25 ppm downfield is appropriate. However, a close look at his model compound (47) data reveals that the variation in chemical shifts of particular protons caused by a change from chloroform-d solvent to trifluoroacetic acid ranges from -0.73 ppm to +0.78 ppm. The effect of trifluoroacetic acid solvent on chemical shifts of aromatic methoxyl protons of several compounds has been reported (50a).

Thionyl chloride apparently reacts with lignins, and dimethylsulfoxide-d_6 has the disadvantage of a rather strong residual proton signal at about 2.6 Thus, Swan's (20) use of acetone-d_6 adds an element of uncertainty to his results.

Hexamethyldisilazane has, on occasion, been used as an internal standard (14, 15, 18) because it has a higher boiling point than tetramethylsilane, and thus is easier to handle. The δ values must then be adjusted by compensation for the 0.052 shift of the hexamethyldisilazane.

In the semi-quantitative pmr studies of lignins reported to date, the total proton content has been determined by conventional analyses. Variations in these analyses even within the accepted limits are sufficient to cause problems in interpreting pmr data. Bland and Sternhell (17) attempted total hydrogen analyses by making comparisons between the signal intensity from a known quantity of sym-trinitrobenzene and the total proton signal from lignins. With recent improvements in instrumentation and by the use of time averaging techniques, very accurate elemental hydrogen analyses of lignins should become possible via pmr.

In Table 8.4, conventional analyses have been used to calculate the total hydrogen content excluding methoxyl and acetoxyl groups. Then semi-quantitative estimates of protons attached to aromatic and side-chain carbons have been derived from pmr data. Some general information about lignin structures can be gleaned from these determinations.

The range in which aromatic protons of lignin models signal, about 6.35 to 7.90δ, has been previously noted as being relatively free of interfering signals from other kinds of protons. This relative purity of the aromatic signal range allows estimations of protons/C_9 in range 2 to be made with more confidence than is true for much of the remaining portion of the spectrum. These values represent what to date is the only direct physical means for estimating the relative numbers of "condensed" and "non-condensed" units in gymnosperm lignin preparations. "Condensed" units contain only two aromatic protons as in the phenyl coumaran unit 3 or the biphenyl dimer 4, Figure 8.1, whereas "non-condensed" units have three aromatic protons/C_9.

In the case of typical hardwood lignins the situation is somewhat more complex because of the presence of a large proportion of syringyl units in the lignins. Furthermore, number of syringyl units relative to guaiacyl units in lignins varies widely among species and even between the lignins of the secondary wall and middle lamella of the same tree, see Chapter 3, Section III. As a consequence, while estimations of condensed units are available for both softwood and hardwood lignins, the former are undoubtedly more reliable. Such estimations are collected in Table 8.5.

The results show that whereas spruce milled wood lignins apparently contain only about 40-50% condensed units, lignins isolated under acidic conditions generally show about 70-80% condensed units.

The low contents of condensed units reported by Lenz (19) for kraft and soda lignins would seem to counterdict this. However, these were calculated on the basis of a C_6C_3 unit which may not be valid in alkali lignins. According to Marton, Chapter 16, Section V-D, it is probable that half or more of the terminal side-chain carbons are lost during kraft cooks. This would suggest an average unit of $C_6C_{2.5}$ or less for these lignins. In Table 8.4 the values for pine kraft lignins have been recalculated on such basis. With this adjustment, and even overlooking the interference by vinylic protons in stilbene units, there are, in the commercial kraft lignin, only 2.39 aromatic protons/$C_{8.5}$ unit. Thus at least 60% of the aromatic rings are probably "condensed."

Results found in calculating the percent condensed units in hardwood lignins indicate that in contrast to softwood lignins increase in condensation as a result of acid- and base-catalyzed reactions is not appreciable. Except for the black gum soda lignin spectrum (19), which may need adjustments for side-chain degradations such as those mentioned above for pine kraft lignins, most of the guaiacyl type units appear to be condensed to account for the low number of aromatic protons determined present in both mildly treated and modified lignins.

Hrutfiord and McCarthy (40) used a different approach to determine the relative numbers of syringyl type groups and condensed and uncondensed guaiacyl type groups in hydrogenolysis lignins from red alder (Alnus rubra). By setting ranges for aromatic protons of these three types, they determined the direct ratio of each to be present in the lignins, Table 8.5.

Total side-chain protons and highly shielded protons are also listed in Table 8.4. According to modern theory no protons such as the latter are proposed as being a part of the structure of unreacted lignins, yet almost all pmr spectra of lignins examined to date show significant signal intensity in this range. The spruce milled wood lignin examined by Lenz (19) seems to be an exception with only 0.06 "highly shielded" protons per phenylpropane unit, Table 8.4. Despite this exception which may or may not be more significant than the generally observed results, the fact remains that an appreciable signal has usually been observed in this range.

The total side-chain protons are indicated to be on the order of 4.0-4.8 H's/C_9 in softwood lignins. These figures are slightly higher than what Freudenberg's representation for spruce lignins would predict (3.8 side-chain H's/C_9). It may be noted that unusually large values are indicated for the dioxane acidolysis and Holmberg lignins from black gum wood. This does not seem to be generally true for hardwood lignins and may indicate the presence of unusual features in black gum lignins.

The milled wood lignins generally have low amounts of highly shielded protons, 0.06 to 0.46 protons/C_9. Spectra from Brauns lignins average somewhat larger amounts of these protons reflecting what may be a tendency on the part of these more soluble lignins to be associated with more non-phenylpropanoid material than is true for milled wood lignins. Compared with milled wood lignins, the dioxane acidolysis wood lignins also show relatively more (0.44 to 0.71) protons/C_9 in this highly shielded range.

The very high values reported by Swan for phloem and bark lignins of 2.69 and 3.67 protons/C_9 signalling in range 8 show that these products are significantly different from normal wood lignins. The acetylated methanolysis

Table 8.4. Hydrogen Balance in Acetylated Lignins Based on Conventional Analysis and FMR

	Ref.	Empirical C9-formulae					Aromatic (Range 2)	Hydrogens/C9-unit		
									Side-chain hydrogens	
		C	H	O	OMe	OAc		Total	Highly shielded (Range 8)	Benzylic H-CαOAc-type
1. Milled wood lignins										
Spruce	16	9	7.07	2.45	0.96	1.28	2.56	4.51	0.46	0.33
Birch	19	9	6.85	2.50	0.85	1.11	2.57	4.28	0.06	0.32
Birch	46	9	6.94	3.10	1.43	1.29	2.11	4.83	0.28	0.45
2. Brauns lignins										
Spruce	16	9	6.95	2.61	0.88	1.36	2.38	4.57	0.82	0.35
Western hemlock	16	9	6.37	2.71	0.88	1.54	2.32	4.05	0.73	0.31
3. Dioxane acidolysis lignins										
Western hemlock	16	9	6.68	2.39	0.94	1.37	2.39	4.29	0.46	0.33
Birch #1	46	9	6.88	2.83	1.48	1.30	2.09	4.79	0.44	0.47
Birch #2	46	9	6.39	2.56	1.39	1.24	2.04	4.35	0.53	0.34
Pine	19	9	6.96	2.42	0.81	1.04	2.28	4.68	0.71	0.48
Black gum	19	9	7.53	2.82	1.30	1.12	2.10	5.43	0.65	
Red cedar phloem	20	9	7.85	2.63	0.91	1.15	2.33	5.52	2.69	0.61
4. Methanolysis lignins										
E. regnans, unacetylated	18	9	8.08	2.70	1.86		1.83	4.86	0.36	
E. regnans	18	9	6.52	2.16	1.85	1.39	1.77	4.75	1.06	
P. radiata	18	9	7.11	2.48	1.14	1.00	2.30	4.81	1.46	
5. Lignin thioglycolic acids										
Pine	19	9	7.19	2.75	1.29	0.64 (SCH$_2$COOH 0.52)	2.16	4.01[a]	0.96	0.35
Black gum	19	9	8.07	2.29	1.37	0.60 (SCH$_2$COOH 0.49)	2.04	4.93[a]	1.42	0.35

MAGNETIC RESONANCE SPECTRA

6. Hydrogenolysis lignins								
Red alder	40	9	7.90	2.50	1.30	1.30	2.36	5.54
7. Steam hydrolysis lignins								
Birch	46	9	6.26	2.21	1.10	0.94	1.63	4.63	1.02	0.22
8. Soda lignins (Lab. pulpings)								
Pine	19	9	7.22	1.77	0.84	1.33	2.82	4.40	0.58	0.58
Black gum	19	9	6.80	2.15	1.28	1.14	2.28	4.52	0.42	0.39
9. Kraft lignins[b] (Lab. pulpings)								
Pine	19	9	7.71	2.29	0.82	1.29	2.82	4.89	0.30	0.41
		8.5	7.30	2.16	0.77	1.22	2.66	4.64	0.28	0.39
Pine	19	9	7.05	2.15	0.83	1.42	2.74	4.31	0.26	0.41
		8.5	6.52	2.03	0.78	1.34	2.58	3.94	0.25	0.39
Pine	19	9	6.82	2.17	0.86	1.41	2.69	4.13	0.34	0.39
		8.5	6.50	2.03	0.81	1.33	2.54	3.96	0.32	0.37
Pine	19	9	6.38	1.18	0.84	1.42	2.76	3.62	0.86	0.46
		8.5	6.04	1.12	0.79	1.34	2.60	3.44	0.82	0.44
10. Kraft lignins[b] (Indust. pulpings)								
Pine	19	9	7.00	1.87	0.77	1.26	2.52	4.48	0.85	0.50
		8.5	6.61	1.77	0.73	1.19	2.39	4.22	0.81	0.47

a. Corrected for methylene groups in thioglycolic acid residues.
b. 0.05 to 0.07 sulfur per C_9 in Kraft lignins (19).

Table 8.5. Condensed Units in Lignins by PMR Estimations

	Appr. % condensed units	Ref.
1. SOFTWOOD		
a. Milled wood lignin		
Spruce	40 to 50	16
	43 to 45	19
b. Dioxane acidolysis lignin		
Western hemlock	70	16
Pine	72	19
Western red cedar phloem	70	20
c. Methanolysis lignin		
P. radiata	70	18
d. Thioglycolic acid lignin		
Pine	84	19
e. Soda lignin		
Pine	18	19
f. Kraft lignin, lab. pulping		
Pine, 3 samples	34 to 46*	19
Kraft lignin, ind. pulping		
Pine	61*	19

	Guaiacyl-C_3 units		Syringyl-C_3 units	
	Condensed	Not Condensed		
2. HARDWOOD				
a. Milled wood lignin				
Birch	46	11	43	46
b. Dioxane acidolysis lignin				
Birch	43	9	48	46
	57	4	39	46
	60	10	30	46
c. Thioglycolic acid lignin				
Black gum	69	4	27	19
d. Hydrogenolysis lignin				
Red alder, ether soluble 1	43	23	34	40
ether soluble 2	30	22	48	40
ether insoluble	34	51	15	40

* Estimated on the basis of the number of protons/$C_{8.5}$, see text.

lignins, lignin thioglycolic acids and steam hydrolysis lignin generally show increased numbers of protons signalling in range 8, indicating extensive changes probably occurred during their formation. Part of the high shielded protons may be attached to terminal methyl groups formed during acidic solvolysis reactions, Chapter 9, Section II-A. The industrial kraft pulpings also form more of the "highly shielded" protons, although the milder laboratory cooks do not seem to show this effect, Table 8.4, IX and X.

The last column in Table 8.4 lists values reported for the range 3 where signals from protons on carbons attached to both an aromatic ring and an acetoxyl group occur. Values of 0.30 to 0.45 protons/C_9 seem normal in this range for mildly treated lignins. In steam hydrolysis lignins this drops to a value 0.22/C_9. According to Klemola (46), this decrease reflects the replacement of benzylic hydroxyls by α-carbonyls. The residual signals in this range may be due in part to additional β-vinylic protons from groups formed during the steam hydrolysis. The signal strength found for soda and kraft lignins in this range also may be due primarily to β-vinyl type protons formed during the pulping.

Attempts to determine total hydroxyl contents of lignins by measuring the acetoxyl proton content in ranges 6 and 7, Table 8.3, suggest the presence of more such groups than conventional analyses show. The data in Table 8.2 show that there should be no significant number of other interfering protons signalling in the same range in mildly extracted lignins. The answer to this mystery is not yet clear. In some cases signals from acetyl protons of carbohydrate impurities in milled wood lignins may be the source. Otherwise there may be some distant shielding effects or unrecognized type of proton in the lignin preparations which cause this effect. Of course, uncertainties in the conventional acetoxyl determinations may also contribute to this apparent discrepancy.

In a unique and elaborate study, Gagnaire (44) has determined the total hydroxyl content in a sample of very carefully prepared milled wood lignin from Sequoia sempervirens by dissolving it in dry dimethylsulfoxide-d_6 and exchanging all its protons with trifluoroacetic acid-$\underline{d}$. The exchanged protons signal at about 9.0 to 9.5δ. He compares the intensity of this signal as found by integration with the intensity of the signal at about 7.5-8.0δ from a known amount of pyrazine and calculates that there are about 1.52 ± 0.17 hydroxyls/C_9 in this lignin. This is somewhat higher than results determined for other conifer lignins by direct integration techniques.

Values found by pmr for the number of aromatic acetoxyl protons/C_9 have been shown to be generally slightly lower than values found by determinations of phenolic hydroxyl groups in unacetylated lignins (15), Table 8.6. The lower values are very likely due to the failure of aromatic acetoxyl protons in 5-5 linked structures to signal in the same range as other aromatic acetoxyl protons.

    3.   Broad Line PMR Studies. Kushchenko, Karakozov and

Table 8.6. Aromatic Hydroxyl Estimations by PMR and by Potentiometric Titration (16)

	Phenolic Hydroxyl, OH/OMe	
	N.m.r.[a]	Titration
Dioxane acidolysis lignin	0.37[c]	0.41[b]
Milled wood lignin	0.29[b]	0.33[b]
Brauns lignin	0.45[b] 0.70[c]	0.55 to 0.70[b]

a Range 6, acetoxyl protons/MeO-protons, Table 8.4
b Spruce
c Western hemlock

Mishchenko (51) report that a study of relaxation times by broadline pmr examination during conventional kraft pulping shows promise as an aid for controlling the process. They report that the relaxation time decreases slowly during the first forty to sixty minutes of pulping at 100°-120°C. Following this, a rapid decline in relaxation time is observed over a period of about one hour, suggesting that a relation exists between relaxation time and some process taking place at this time. During the final period the rate of decrease in relaxation time again becomes very slow. This behavior, they believe, parallels the process of diffusion and dissolution of reaction products.

D. Future Potential in Lignin Research. The full potential of nmr spectroscopy in lignin research has not been realized to date, and this method may well become one of the most powerful tools yet applied to the study of lignins. Innovations and improvements in instrumentation and techniques have come so rapidly that some of the newer methods have not been applied to lignins as yet. For example, most of the studies reported so far have utilized 60 MHz spectrometers. The use of the more powerful 100 and especially the 220 MHz spectrometers will doubtlessly allow future workers to acquire not only additional but also more reliable information from their spectra. This has already proven to be true for other biopolymers (12). The determination of the presently unknown nature of the protons signalling in the highly shielded range 8 would alone widen the scope applications to structural studies in the lignin field. A variety of new applications are anticipated from the use of such isotopes as ^2H and ^{14}C in combination with nmr spectroscopy. The sulfur isotope ^{33}S has potential for studies on lignosulfonates and kraft lignins.

The characterization of dimeric and oligomeric degradation products of lignins by pmr will obviously continue to be one of the most fruitful applications of this technique.

III. Electron Paramagnetic Resonance (EPR).

A. Introduction. The theory and development of electron paramagnetic resonance spectroscopy (epr) often referred to as electron spin resonance spectroscopy, (esr) parallels in many ways that of nmr. In principle each is dependent on interactions between externally applied magnetic fields and the magnetic fields associated with sub-atomic particles. In epr spectroscopy unpaired electrons play a similar role to that played by nuclei in nmr. However, because of its relatively much smaller mass/charge ratio the electron has an associated magnetic moment which is on the order of a thousand times as great as that of a typical nucleus in nmr. As a consequence of this difference, assuming the use of magnetic fields of the same order of magnitude as those used in nmr, the signals found in epr fall in the microwave region rather than the radio frequency region common to nmr.

Signals received in epr are generally recorded as the first derivative of the absorption curve and may contain from one to hundreds of lines. Unpaired electrons to be studied by epr may occur in atoms, metals, compounds of transition metals, point defects in solids, odd electron molecules such as NO and O_2 as well as in organic free radicals. It is the latter, however, that have provided the epr signals which have been studied in the field of lignin chemistry. The theory and practice of epr spectroscopy have been described and reviewed elsewhere (52).

Alger (53) has provided one of the best sources of general information regarding practical procedures used in epr studies.

Because of limitations caused by dielectric losses in polar solvents, lignins and related polymers most often have been studied in the solid state. Recent improvements in spectrometer design however have increased the signal to noise ratio and decreased dielectric losses to the point where even aqueous solutions of lignins now may be examined.

One of the most remarkable features of epr spectroscopy is its high sensitivity for measuring numbers of unpaired electrons. For instance, the signal of a $10^{-7}M$ potassium permanganate solution can be detected. In another way of viewing it 10^{10} spins/g can be measured under optimum conditions. It is important to remain aware of actual spin concentrations which are being observed in any given experiment. The behavior of only one thousandth to one trillionth of a given sample may not be indicative of that of the sample as a whole.

B. <u>The Measurement of Free Radical Content of Lignins and Related Materials</u>. Most of the epr studies of lignins and related materials to date have strongly emphasized the determination of total spin content of samples, though several workers also have made careful studies of g values, (spectroscopic splitting factors). These g values generally confirm that the sources of spin content in lignins are free radicals. However, small differences in g values are difficult to determine with great accuracy and even when determined have not as yet yielded much useful information to lignin chemists. Therefore, the following discussion will emphasize the spin content found for various lignins and related materials.

Most of the lignins subjected to epr so far have been from wood and none

Table 8.7. Free Radical Content of Lignins and Related Materials by EPR

	Spins/g×10^{17}		Approximate no. of C_9 units/spin×10^{-3}		
	Neutral	Na-salt	Neutral	Na-salt	Ref.
1. Lignins and derivatives:					
DHP	0.22	-	140	-	67
Spruce BL	0.5	50	60	.6	60
Spruce MWL	.23 to 1.0	15	30 to 60	2.0	60,65,67
Spruce DAL	.83	-	36	-	65
Spruce lignosulfonates	.84 to 3.0	-	10 to 36	-	60,65
Softwood kraft lignins	3.0 to 4.0	70 to 300	7 to 10	.10 to .43	60,65
Hardwood kraft lignins	8.0	550 to 880	3.8	.03 to .06	56
"Indulin AT"[a]	3.0 to 5.1	72	6 to 10	.42	60,64
Decayed spruce lignin[b]	1.4	-	22	-	65
Softwood Klason lignins	0.3 to 0.9	-	33 to 100	-	60
Softwood HCl-lignins:					
- Direct measurement	1.5 - 1.7	-	18 - 20	-	65,68
- After CH_2N_2-methylation	0.14	-	214	-	68
Spruce hydrolysis lignins:					
- Direct measurement	2.2	-	14	-	65
- After alkali activation	19.5	-	1.5	-	65
Aspen hydrotropic lignin	3.2	-	9.3	-	65
Cotton hull lignin	11.0	-	2.7	-	65
2. Woods:					
Spruce wood meal:					
- Direct measurement	0.04	-			56
- After vibratory	0.80	-			56
Saguro wood meal	0.5	-			62
Decayed western hemlock	0.9	-			60
3. Other related materials:					
Fulvic acid	3.0	7			62
Humic acids	3 to 20	14-800			62
	0.5 to 2.2	-			64

[a] Soda lignin, Westvaco Co.
[b] Isolated from spruce degraded by Fomes pinicola.

from grasses or barks. Spin contents for lignins and related materials are tabulated in Table 8.7.

1. **Lignocellulose in Wood.** Lignins as they are present in woods and other plant tissues are intimately associated and probably chemi-bonded to carbohydrates. Thus, it is of some interest to note whether or not wood itself is paramagnetic, and if so, whether the lignin portion of the wood contributes to this feature.

The first to attempt to measure spin content in wood by epr was Rex (54). He reported in 1960 that no signal was detected by epr spectroscopy of air-dried, undamaged fresh wood shavings from a number of species. About two years later Kleinert and Morton (55), perhaps by use of a more sensitive spectrometer, detected a small signal in undamaged black spruce (Picea mariana) wood, Figure 8.7a. They also found that the wood after passing through a Wiley mill gave a more intense signal, which was increased further by subsequent ball milling, Figure 8.7b & c. High purity bleached pulp, on

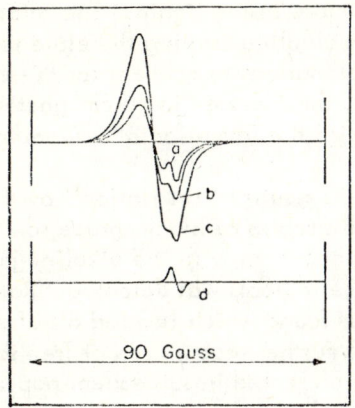

Figure 8.7. Epr signals from black spruce (Picea mariana) wood in progressive stages of subdivision. a. initial undamaged wood; b. wood ground through mesh 40 of a Wiley mill; c. wood ground according to (b) and then ball milled; d. high purity bleached pulp ball milled as in (c) (55).

the other hand, gave only a very small signal even after ball milling, Figure 8.7d. These authors suggested that either the radicals are formed through the effect of grinding upon the rigid lignin network in wood or more radicals are trapped in the ground lignin than in ground cellulose. Heating the ground wood or ground cellulose to 180° C decreased the epr signals to much lower values. The effect of milling on paramagnetism was confirmed by Steelink, (56) who reported that 0.04×10^{17} spins/g in spruce wood meal increased about 20 fold after a period of vibratory milling, Table 8.7-2.

Kleinert and Morton (55) also reported studies, which were continued subsequently in greater detail by Kleinert (57, 58, 58a), on alkali impregnated and freeze dried wood, alpha cellulose and holocellulose. When the samples were rapidly heated to 180° to 190°C, very strong epr signals were detected (55). The presence of oxygen, sulfur or a free radical scavenger such as 2,6-di-tert.-butyl-p-cresol suppressed strongly the development of the epr signals. From these results Kleinert suggests that sulfur functions as a free radical scavenger and may play a similar role in the kraft pulping process inhibiting the combination of macro-radicals formed in the process. Both the free radical formation upon heating alkalized wood (after correction for radicals from holocellulose) and the "bulk delignification" of the same wood in alkaline liquid phase were found to follow similar first order kinetics. From these observations Kleinert develops a rather unique hypothesis postulating that alkaline delignification proceeds first with a homolysis of the lignin, followed by a second stage "residual delignification" consisting of the more difficult dissolution of hemicelluloses which have been grafted onto the lignin by free radical processes during the earlier stages of the "bulk delignification." This view departs from commonly accepted reaction mechanisms for alkaline hydrolysis of wood and lignins, see Chapters 9 and 16. It is also well to remember that the spin measurement discloses no more than what is happening to only one C_6C_3 unit per thousand or more in the lignin. The relation of the esr signal to the major reaction path in alkaline pulping therefore remains speculative. Furthermore, alternative explanations to account for the esr signal should be considered. For instance, the increase in paramagnetism during alkaline pulping could be associated with the formation of semi-quinone ion radicals in lignin (61).

Kleinert also carried out a study of the neutral "sulfitation" by the rapid heating of sodium sulfite impregnated samples of black spruce sawdust, holocellulose and alpha cellulose (59). In contrast with the alkaline pulping experiments, no immediate formation of free radicals was detected. Rather a slow increase in relative spin content was found which leveled off after 25 minutes in the case of the spruce sawdust. Further elaboration of free radical mechanisms for delignification processes is presented in subsequent papers (59a, 59b).

2. <u>Isolated Lignins</u>. In 1960, Rex (54) described the first successful detection and measurement of paramagnetism in dioxane acidolysis lignins by epr spectroscopy. The lignins isolated from oak, pine, eucalyptus, redwood and spruce were found to contain on the order of 10^{18} spins/gram at g values of about 2.00.

Since this report, several other groups of workers have become intrigued with this phenomenon and have attempted both to measure and explain it. The number of spins/gram in isolated lignins and related materials are tabulated in Table 8.7-1. In order to illustrate the actual free radical content in the lignins, the approximate numbers of C_6C_3 units per spin have been included in the same table.

Except for the data reported by Nagar <u>et al</u>. (64) on humic acids, the values in Table 8.7 show that the spin content increases approximately in the order: mildly extracted lignins and DHP $\simeq$ Klason lignins < lignosulfonates $\simeq$ lignin from decayed wood $\simeq$ hydrolysis lignins < kraft and alkaline softwood lignins $\simeq$ hydrotropic aspen lignins < kraft hardwood lignins < cotton hull lignins < alkaline activated hydrolysis lignins < Na salts of lignins. Since spruce wood meal shows only 0.04×10^{17} spins/gram, it is evident that mechanical, biological and chemical attacks on lignins are all in position to create additional radical centers.

One free radical per 30,000 to 60,000 C_6C_3 units is the range found for values determined for mildly extracted lignins by Steelink, Reid and Tollin (60, 61) and Chudakov <u>et al</u>. (65, 66, 66a). The lower spin content found by Freudenberg and Harkin may be explained by a milder milling procedure. The latter value for the milled wood spruce lignin agrees very well with that found for synthetic DHP lignin from coniferyl alcohol and was cited as additional evidence for the similarity of the two (67).

In semi-quantitative studies Kleinert (69) measured relative areas under epr spectral curves from spruce wood lignins isolated by various procedures. He found that the spin content becomes larger in the order milled wood lignin = dioxane acidolysis lignin < conventional kraft lignin < oxidized kraft lignin < amine salt purified lignin sulfonic acid. He also detected rather weak signals from products obtained from coniferyl alcohol both by dehydrogenation using peroxidase and by the action of elemental oxygen in base at 30°C. These results are in general agreement with those in Table 8.7. It is interesting to note, however, that whereas the calcium lignosulfonates, Table 8.7, show the same or fewer free radicals compared with kraft lignins, the opposite is true for the purified free lignin sulfonic acid.

Raskin, Mal'tsev and Chudakov (69a) have studied the effect of subjecting dioxane acidolysis lignin to treatment with 1% hydrochloric acid for one hour at temperatures from 100°C to 350°C. Their results, Table 8.8, show that as the temperature of treatment was raised, the free radical content of the products increased from 3.4×10^{17} to 1.2×10^{20} spins/gram. The generation of free radicals was accompanied by an increase in electrical conductivity. The original dioxane acidolysis lignin contained $<10^{15}$ spins/gram. Sodium borohydride reduction prior to condensation treatment increased the extent of free radical formation, suggesting that the hydroxyl groups added by this means play a role in these reactions. The importance of phenolic hydroxyls in free radical formation was shown by the inertness of the diazomethane and dimethyl sulfate methylated lignins. Methylation in the benzylic positions on the other hand, merely decreased but did not completely suppress the free radical formation. Treatment of the dioxane acidolysis lignin in water at 180°C for one hour caused little effect when carried out under nitrogen. However, in the presence of air, appreciable free radical formation took place.

Table 8.8. Effect of Condensation Treatments on the Spin Content of Dioxane Acidolysis Lignin

	Temperature of condensation for one hour, °C	Spins/g$\times 10^{17}$
Starting material	–	<.01
Directly condensed preparations		
	100	3.4
	150	8
	180	12
	210	38
	250	400
	350	1200
Condensed in N_2-atm.	180	<.01
Condensed after		
- Hydrosulfite reduction	180	12
- Borohydride reduction	180	84
- Methylation with MeOH-HCl	180	3
- Methylation with CH_2N_2	180	<.01
- Methylation with Me_2SO_4	180	<.01

C. **Structural and Reactivity Studies of Lignins and Models.** In his pioneering study, Rex (54) speculated that the radicals he detected in lignins could be of a semi-quinoid type. Steelink and his coworkers have supported similar views in later studies (61). They noted an increase in free radical content of lignins after being converted to their salt form (61). While sodium borohydride reduction of Brauns and kraft lignins has little direct effect on the free radical content, the sodium salts of the reduced lignins have a much smaller spin content than the salts of the parent lignins, Table 8.9. These results are interpreted to indicate the presence of a semi-quinone type radical entity co-existing in lignins with a diamagnetic quinhydrone moiety. Infrared absorption spectra for lignins and humic acids as well, both in the free acid and salt forms, were found to be consistent with the presence of hydroxyquinone structures (63).

Table 8.9. Effect of Borohydride Reduction on Free Radical Content (61)

	Spins/g×10^{17}	
	Before reduction	After reduction
Brauns lignin, Spruce: Neutral	0.5	1.1
Na-salt	50	8.4
Kraft lignin, Yellow pine: Neutral	3.0	1.3
Na-salt	100 to 300	22.0

In Steelink's view, both biological and chemical oxidation of lignins would increase the number of quinone type moieties. Increased radical formation may also arise from enzymic or alkaline demethylation to form substituted catechols which could be oxidized to o-benzoquinones and then participate in quinhydrone formation. Even a small number of benzoquinone or quinonemethide species in typical lignins could explain the increase in free radical content upon salt formation.

As for the residual spin content in lignins themselves, Steelink believes that its most probable source is from free radicals entrapped and stabilized by the polymeric matrix. In order to gain insight into the possible structural features of such stabilized radicals, he and his coworkers have investigated a number of model compounds of the syringyl type. The first group of models has side-chains containing an alpha carbonyl. These models, when oxidized in benzene solution with one-electron oxidants (lead peroxide preferred), form radicals having half lives of five hours at room temperature. To test the possibility that such radicals form under biological conditions, Steelink treated aqueous solutions of syringaldehyde and aceto- and propiosyringones with hydrogen peroxide and peroxidase. Each of these solutions formed transient green species turning red within a few minutes. Freezing these solutions in liquid nitrogen a few seconds after mixing yielded a blue solid in each case. Upon epr examination the solids were found to contain free radicals, apparently identical to those found in benzene solution and having half lives of 30 to 60 seconds.

When similar experiments were attempted with guaiacyl analogues of these models, no epr signal was detected and only brown amorphous solids were recovered from the reaction mixtures. Apparently, the radicals formed undergo immediate coupling at their free 5-positions forming secondary products.

Caldwell, Fitzpatrick and Steelink (71) have recently reported studies on the peroxidase catalyzed oxidation of the syringyl type alcohols, 9, 10, 11 and 12 and their corresponding carbonyl analogues, 13, 14, 15 and 16. These oxidations proceed via the phenoxy radical intermediates, 9a through 16a. The epr spectra of the radicals 9a, 10a, 14a and 16a are shown in Figure 8.8 along with theoretical spectra calculated from the hyperfine splitting constants which were determined. Major splittings of 9.87 and 5.45 gauss by the alpha carbon protons cause the triple envelope for 9a and the double one for 10a, respectively. The fine splittings are the result of coupling with other protons in the structures. All of the models exhibited splitting constants of 1.35 to 1.45 gauss for the coupling with methoxyl and ring protons and from 0.67 down to less than 0.2 gauss for protons on beta and gamma carbons. These unique spectral properties made it possible to monitor the concentration of single radical species in solution on a continuous basis.

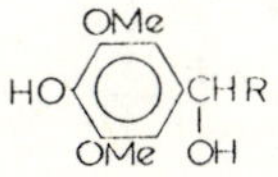

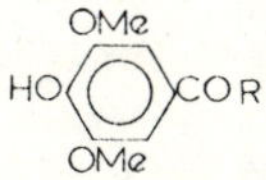

9 R=H
10 R=Me
11 R=CH$_2$Me
12 R=CH$_2$-O-Guaiacyl

13 R=H
14 R=Me
15 R=CH$_2$Me
16 R=CH$_2$-O-Guaiacyl

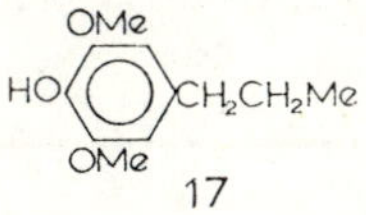

17

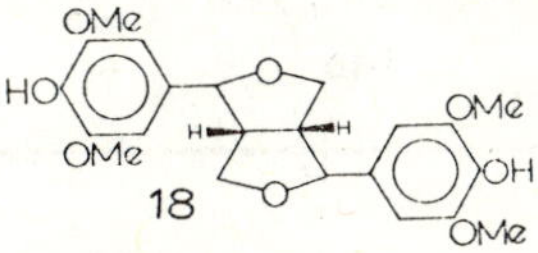

18

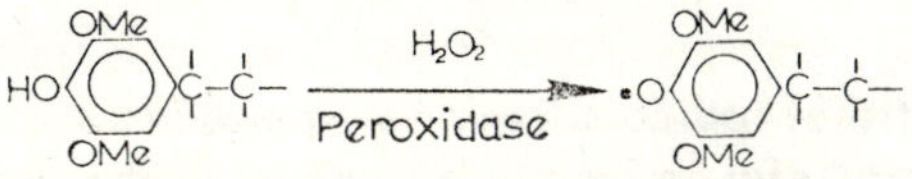

9-18 $\xrightarrow{H_2O_2, \text{Peroxidase}}$ 9a-18a

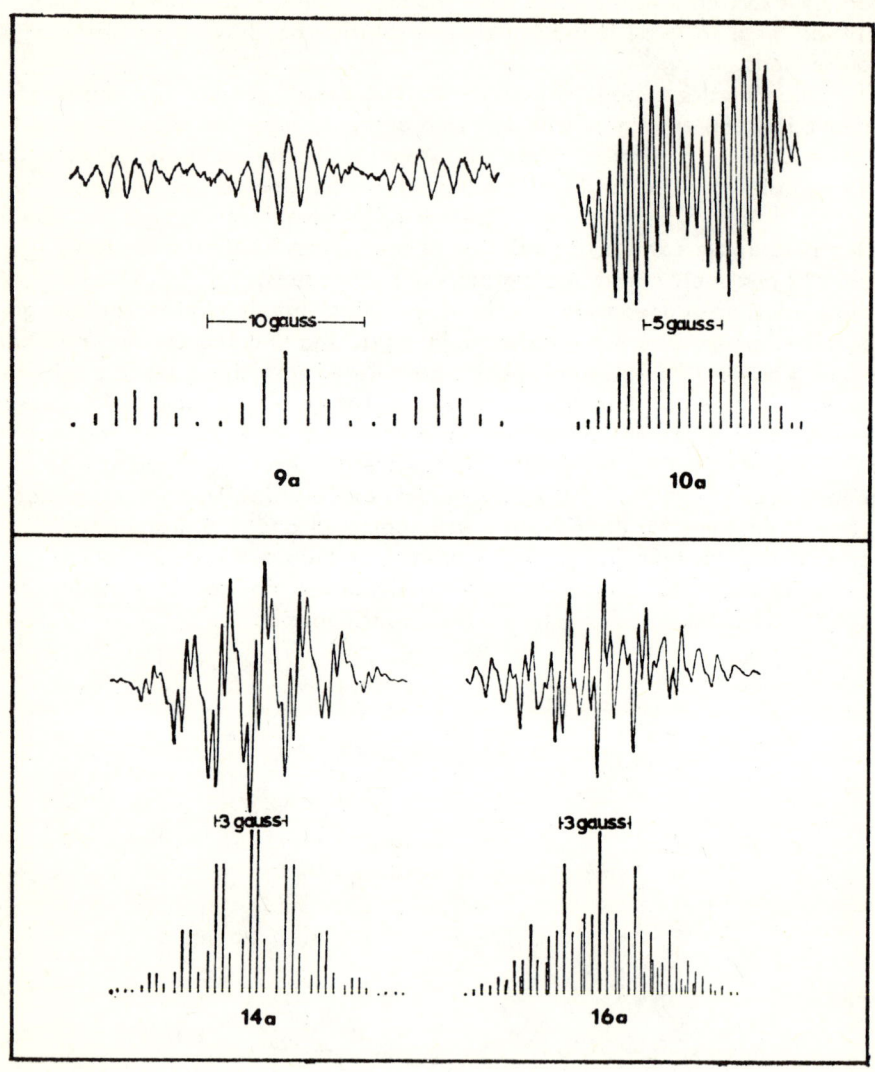

Figure 8.8. Epr spectra of radicals formed by peroxidase oxidation of model compounds. The numbers refer to structure numbers in the text (71).

MAGNETIC RESONANCE SPECTRA

By using epr in conjunction with optical spectroscopy and thin-layer chromatography, a rational sequence of events was worked out. When compound 9 was treated with H_2O_2 and peroxidase, consecutive reactions were observed as recorded in Figure 8.9.

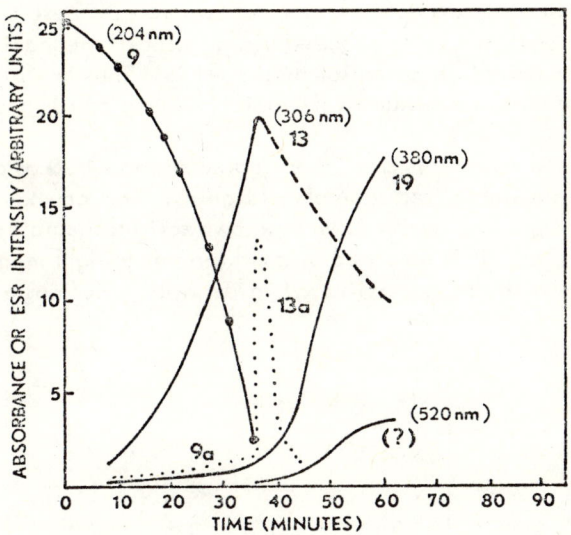

Figure 8.9. Time-concentration plots for oxidation of syringaldehyde.

As 9 decreased in concentration, a corresponding increase in 13 was noted until all of 9 had been converted to 13. At this point the nature of the radical signal changed abruptly from the triple envelope of radical 9a to the single envelope of 13a, Figure 8.8. The latter signal decayed by first order kinetics with the simultaneous appearance of 2,6-dimethoxyquinone 19 and small amounts of an unknown red substance absorbing at 520 nm.

The apparent overall reaction sequence is shown in equation 1. Compounds 10, 11 and 12 underwent the same steps in enzymatic oxidation.

$$9 \xrightarrow{slow} 9a \xrightarrow{fast} 13 \xrightarrow{intermediate} 13a \xrightarrow{slow} 19 \text{ and red compound} \quad (1)$$

Compound 17, similarly, was converted via 17a to 11 and then continued on in the sequence in equation 1. Syringaresinol, 18, likewise followed the same kind of course except at a much slower rate. These peroxidase catalyzed reactions apparently proceed via a "clock" mechanism. The system 9-9a has a lower oxidation potential than 13-13a. Thus any of compound 9 remaining in the reaction mixture will rapidly reduce any 13a which might be formed.

Steelink and coworkers also investigated syrinoxyl 20 as a speculative model for the more stable free radicals in lignins. This compound is methoxylated analogue of galvinoxyl 21, a well-known solid standard for epr measurements (72, 73, 74). It is obtained by oxidizing disyringyl methane in dichloromethane with PbO_2. Syrinoxyl, in its violet solutions, exists 66% in its radical form.

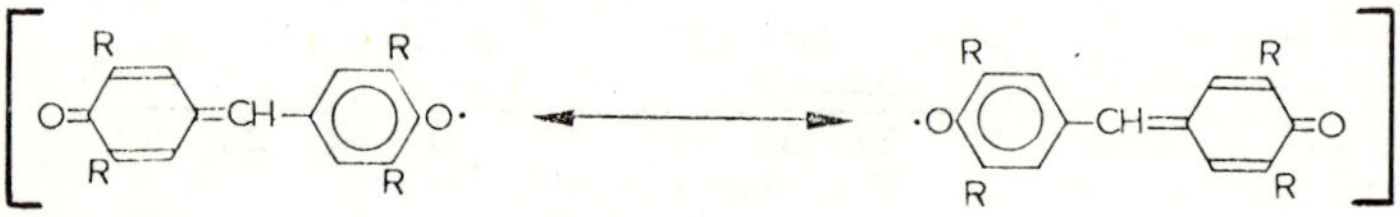

20, R = OMe

21, R = t-butyl

As a solid under nitrogen, syrinoxyl has a half life of one week, and even when exposed to oxygen its half life is two days. In methylene chloride solution, the epr spectrum of syrinoxyl has 130 lines compared with only nine for galvinoxyl, Figure 8.10. Of significance in the interpretation of epr spectra of solid lignin samples is the fact that single line epr spectra are obtained from solid samples of both 20 and 21, so the lack of fine structure in epr spectra of lignins does not rule out specific structures as their source (55, 69). The complex fine structure in the spectrum of syrinoxyl in solution was first rationalized incorrectly (75), but the interpretation was rectified later (76) on the basis of electron nuclear double resonance (endor) studies.

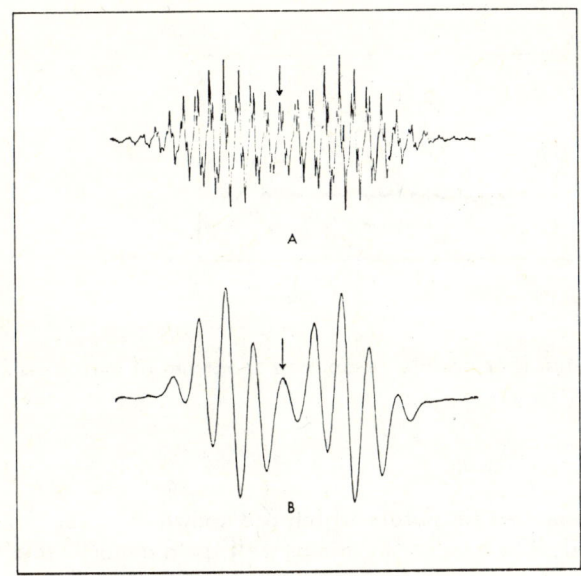

Figure 8.10.
First derivative epr spectra of (A) syrinoxyl in methylene chloride, and (B) galvinoxyl in methylene chloride (56).

Endor is, in a way, a hybrid of epr and nmr. It was first successfully applied to the investigation of free radicals by Hyde and Maki (77), who used it to study galvinoxyl. The technique measures the change in the height of a hyperfine line in an epr spectrum as a nuclear radio frequency signal is swept through the precision frequencies of the protons on the radical. Endor lines result whenever the radio frequency applied corresponds to the nuclear resonance frequency of protons which are coupled to unpaired electrons. Two lines equidistant from the center of the endor spectrum result from each group of chemically equivalent protons.

The endor spectrum of syrinoxyl is shown in Figure 8.11. By this method authors have shown that hyperfine coupling constants for syrinoxyl are 15.968 MHz for the methide proton, 3.450 MHz for the ring protons and 2.227 MHz for the methoxyl protons.

Efforts to determine the nature of free radicals in lignins have been continued by Chudakov and coworkers (65, 66, 78). The authors determined oxidation equivalents (indicated by the formation of iodine from potassium iodide in acetic acid), finding poor correlation with the radical content in lignins. Furthermore, the sodium dithionate reduction of lignins causes a complete disappearance of their spin content. The authors express the opinion that the paramagnetism of lignins may not be related to the presence of active radicals of the aryloxy or semi-quinone types but rather results from inter-

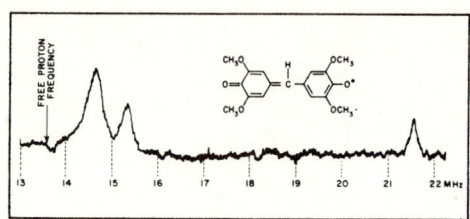

Figure 8.11. The electron-nuclear double resonance spectrum of syrinoxyl (76).

actions of intra- and inter-molecular nature which are known to occur in polymers with conjugated double bonds. In this as well as in a more recent paper (79), an attempt is made to find correlations between infrared spectra and spin contents of lignins. One problem in studies such as these is that the measurements of oxidation equivalents as well as infrared procedures are generally incapable of detecting the very low concentrations of functional groups which may be the source of even relatively strong epr signals. In another attempt to determine the nature of the free radicals in lignins, Koshijima et al. (68) found that the spin content of hydrochloric acid pine lignin is diminished by a factor of ten by the methylation of its phenolic hydroxyls with diazomethane. This result, they suggest, in contrast to the view expressed by Chudakov et al., indicates the presence of a large number of phenoxyl radicals in this lignin. They also established evidence of semi-conductive properties in these lignins as well as products formed from them by graft polymerization.

Reactions of lignin models with peracetic acid have been followed using epr spectroscopy by Hatakeyama, et al. (80). Of three models, vanillyl alcohol and -sulfonate, and vanillyl ethyl sulfide, detectable radicals were formed only in the first material.

Freudenberg, et al. (81) succeeded in detecting free radicals having half lives of 45 seconds during the dehydrogenation of coniferyl alcohol with peroxidase-hydrogen peroxide in 50% aqueous dioxane at 20°C. They failed, however, to detect any free radicals in aqueous solutions of coniferyl alcohol such as those used for biosynthetic lignins produced in vitro. Apparently the half lives of these radicals in aqueous media are too short to be detected by epr.

MAGNETIC RESONANCE SPECTRA

Kratzl (82) has discussed the role played by free radicals in the formation and chemical and biological degradation of lignins and how epr spectroscopy has been applied to this subject. He also described epr studies of a new plastic polyphenylene-oxide which is produced by dehydrogenation of phenols such as 2,6-dimethylphenol.

In recent work on alkaline solutions of commercial hardwood alkali lignins, Fitzpatrick and Steelink have detected and characterized discrete benzoquinone radicals (83). The spectrum, Figure 8.12a, was found upon examining 5% solutions of the lignins at several pH's, under nitrogen. The rate at which the signal assumes its full strength increases with increasing pH. At pH 13 the signal appears immediately. An identical spectrum is found when 2,6-dimethoxybenzoquinone, 19, is treated in the same manner. This signal then is replaced by that shown in Figure 8.12b, taking 17 hours for the conversion at pH 12, 2.5 hours at pH 13 and being instantaneous at pH 14. When 2-hydroxy-6-methoxy-p-benzoquinone was treated similarly, it too gave the spectrum in Figure 8.12b. The respective semiquinone radicals are 22 and 23.

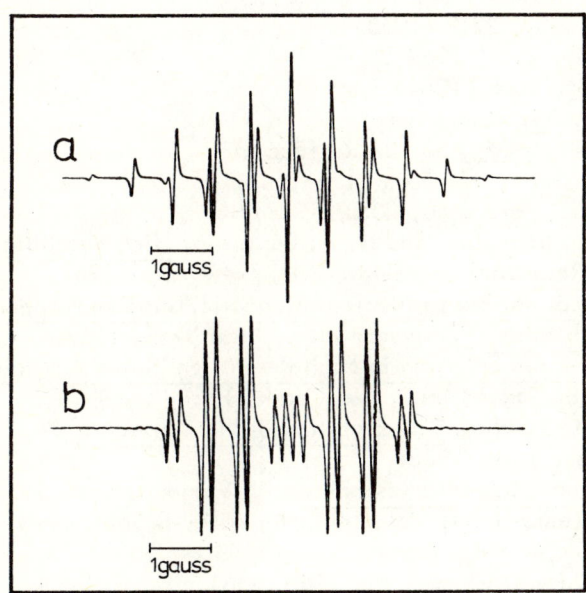

Figure 8.12. Epr spectra of semiquinones; a. structure 22 and b. structure 23.

Martensson and Karlsson (84) have recently applied quantum mechanical procedures to predict spin densities for the radicals p-coumaryl, coniferyl and sinaply alcohols. This line of research is of course of utmost importance for the understanding of the biogenesis of lignins, and it is only to be hoped that with improved instrumentation and experimental techniques, such as flow systems, the characterization of these radicals will become possible.

As a matter of fact, the application of esr spectroscopy to the study of lignin polymers has just fairly begun, and many additional avenues of research remain to be explored. For example, no attempts have been reported so far to study electrolytic reactions of lignins by epr though equipment is available for use in such studies (85). Other new avenues of research may include investigations of the nature of free radicals formed by radiation of lignins and the study of radical formation in lignins of varying botanical origin.

REFERENCES

1. C. A. Reilly, Anal. Chem., 30, 839 (1958).
   C. A. Reilly, ibid., 32, 221R (1960).
   H. Foster, ibid., 34, 255R (1962).
   H. Foster, ibid., 36, 266R (1964).
   Ernest Lustig, ibid., 38, 331R (1966).
   Dolan H. Eargle, Jr., ibid., 38, 371R (1966).
   Jerry P. Heeschen, ibid., 40, 560R (1968).
   Dolan H. Eargle, Jr., ibid., 40, 303R (1968).
2. J. A. Pople, W. G. Schneider and H. J. Bernstein, High Resolution Nuclear Magnetic Resonance, McGraw-Hill, New York, 1959.
3. L. M. Jackman, Applications of Nuclear Magnetic Resonance Spectroscopy in Organic Chemistry, Pergamon Press, New York, 1959.
4. J. D. Roberts, Spin-Spin Splitting in High Resolution Nuclear Magnetic Resonance, W. A. Benjamin, Inc., New York, 1961.
5. K. B. Wiberg and B. J. Nist, Interpretation of NMR Spectra, W. A. Benjamin, Inc., New York, 1962.
6. F. A. Bovey, Nuclear Magnetic Resonance, A. Standen, Ed., Encylopedia of Chemical Technology, Vol. 14, John Wiley & Sons, New York, 1965, pp. 40-74.
7. G. V. D. Tiers, J. Phys. Chem., 62, 115 (1958).
8. N. S. Bhacca, L. F. Johnson and J. N. Shoolery, NMR Spectra Catalog, Varian Associates, Palo Alto, Calif., 1962, Vol. 2, 1964.
9. J. W. Ramsey, F. R. McDonald and J. C. Peterson, Ind. Eng. Chem., Prod. Res. Develop., 6, 231 (1967).
10. Kang-Jen Liu, J. Polym. Sci., Part A-2, 5, 1199 and 1209 (1967).
11. B. Bak, C. Dambmann, F. Nicolaisen, E. J. Pedersen and N. S. Bhacca, J. Mol. Spectrosc., 26, 78-97 (1968).

12. E. M. Bradbury and C. Crane-Robinson, Nature, 220, 1079 (1968).
13. C. H. Ludwig, Ph.D. Thesis, University of Washington, Seattle, 1961.
14. C. H. Ludwig, B. J. Nist and J. L. McCarthy, J. Amer. Chem. Soc., 86, 1186 (1964).
15. S. W. Schubert, M. C. Andrus, C. H. Ludwig, D. Glennie and J. L. McCarthy, Tappi, 50, 186 (1967).
16. C. H. Ludwig, B. J. Nist and J. L. McCarthy, J. Amer. Chem. Soc., 86, 1196 (1964).
17. D. E. Bland and S. Sternhell, Nature, 196, 985 (1962).
18. D. E. Bland and S. Sternhell, Aust. J. Chem., 18, 401 (1965).
19. B. L. Lenz, Tappi, 51, 511 (1968).
19a. S. Yokoyama, T. Ishii, G. Takeya and A. Sakakibara, Kogyo Kagaku Zasshi, 72, 346 and 353 (1969).
20. E. P. Swan, Pulp Paper Mag. Can., 67, T456 (1966).
21. G. E. Miksche, Chalmers Tekniska Högskola, Göteborg, Sweden, personal communication, 1968.
21a. A. F. A. Wallis, Univ. Washington, Seattle, pers. comm.
21b. G. H. Schmid, Can. J. Chem., 46, 3415 (1967).
22. H. Nimz, Chem. Ber., 100, 181 (1967).
23. H. Nimz, ibid., 100, 2633 (1967).
24. H. Nimz, ibid., 99, 2636 (1966).
25. H. Nimz, Holzforschung, 20, 105 (1966).
26. A. J. Birch, P. L. Macdonald and A. Pelter, J. Chem. Soc. C, 1967, 1968; through Chem. Abstr., 67, 1085 72 (1967).
27. C. K. Atal, K. L. Dhar and A. Pelter, J. Chem. Soc. C, 1967, 2228.
28. K. Ogiyama and T. Kondo, J. Japan. Wood Res. Soc., 13, 345 (1967).
29. G. Excoffier, Ph.D. Thesis, Universite de Grenoble, Grenoble, France, 1966.
30. G. Aulin-Erdtman, Y. Tomita and S. Forsen, Acta. Chem. Scand., 17, 535 (1963).
30a. J. G. Miller and C. Schuerch, Tappi, 51, 273 (1968).
30b. G. Gellerstadt and J. Gierer, Acta. Chem. Scand., 22, 2029 (1968).
31. A. Stoessl, Tetrahedron Letters, 1966, 2287.
32. N. Belorizky-Perret, Ph. D. Thesis Universite de Grenoble, Grenoble, France, 1965).
33. H. Nimz, Chem. Ber., 99, 469 (1966).
34. A. F. A. Wallis, Tetrahedron Letters, 1968, 3281 and 5287.
35. D. W. Glennie and J. S. Motherhead, Tappi, 47, 358 (1964).
36. J. R. Parrish, Tetrahedron Letters, 1964, 555.
37. J. R. Parrish, J. Chem. Soc. C., 1967, 1145.
38. P. E. Parker, Ph. D. Thesis, State University College of Forestry, Syracuse, New York, 1967.
39. P. E. Parker, R. L. Coalson and C. Schuerch, The Structure of Dimers from the Alkaline Hydrogenation of Lignin, Chapter 17, in Advances

in Chemistry Series (Lignin, Structure and Reactions) (J. Marton, Ed.) No. 59, Amer. Chem. Soc., Washington, D. C., 1966.
40. B. F. Hrutfiord and J. L. McCarthy, Hydrogenolysis of Lignin, Chapter 15 in ibid.
41. J. M. Pepper, B. W. Casselman and J. C. Karapally, Can. J. Chem., 45, 3009 (1967).
42. J. Gierer and L. Smedman, Acta. Chem. Scand., 20, 1769 (1966).
43. B. Ericsson, B. Nist and K. V. Sarkanen, Chim., Biochim., Lignine, Cellulose, Hemicellulose, 1964.
43a. S. A. Braddon and C. W. Dence, Tappi, 51, 249 (1968).
43b. T. Wainai and U. Suzuki, Bunseki Kagaku, 17, 315 (1968).
43c. Von A. v. Wacek and A. Voykowitsch, Holzforschung, 22, 77 (1968).
43d. J. C. Pew and W. J. Connors, J. Org. Chem., 34, 580, 585 (1969).
44. D. Gagnaire and D. Robert, Bull. Soc. Chim. Fr., 1968, 281.
45. I. M. Skurikhin, Prikl. Biokhim, Microbiol., 4, 113 (1968).
46. A. Klemola, Suomen Kemistilehti, B., 41, 99 (1968).
47. B. L. Lenz, Owens-Illinois Technical Center, Toledo, Ohio, personal communication, 1968.
48. T. Fukuzumi, S. Sakuma, H. Takahashi, K. Tomita, K. Fujihara, Y. Isome and T. Shibamoto, Holzforschung, 20, 51 (1966).
49. P. R. Gupta, University of Washington, Seattle, personal communication, 1968.
50. J. M. Pepper, Pulp and Paper Mag., Can., 65, 35 (1964).
50a. R. G. Wilson and D. H. Williams, J. Chem. Soc. C., 1968, 2475.
51. V. V. Kushchenko, N. A. Karakozov and K. P. Mishchenko, Zh. Prikl. Khim., 40, 1159 (1967).
51a. T. Higuchi, Wood Research, 48, (Wood Res. Institute, Kyoto, Japan), 1 (1969).
52. J. M. Shoolery and H. E. Weaver, Ann. Rev. Phys. Chem., 6, 433 (1955); Varian EPR at Work, Varian Associates, Palo Alto, Calif., 1964.
M. C. R. Symons, Identification of Organic Free Radicals by Electron Spin Resonance, Advances in Physical Organic Chemistry, Vol. 1 (V. Gold, Ed.) Academic Press, New York, 1963, pp. 284-363
M. Bersohn and J. C. Baird, Introduction to Electron Paramagnetic Resonance, W. A. Benjamin, New York, 1966.
J. E. Wertz, Electron Spin Resonance, Encyclopedia of Chemical Technology, Vol. 7 (A. Standen, Ed.), John Wiley & Sons, New York, 1965, pp. 874-903.
G. A. Russell, Electron Spin Resonance in Organic Chemistry, Science, 161, 423 (1968).
H. Fischer, Fortschri Hochpolym. Forsch., 5, 463 (1968).
53. R. S. Alger, Electron Paramagnetic Resonance: Techniques and Applications, Wiley/Interscience, New York, 1968.

54. R. W. Rex, Nature, 188, 1185 (1960).
55. T. N. Kleinert and J. R. Morton, Nature, 196, 334 (1962).
56. C. Steelink, "Stable Free Radicals in Lignins and Lignin Products," Chapter 5 in Advances in Chemistry Series (Lignin, Structure and Reactions)(J. Marton, Ed.), No. 59, Amer. Chem. Soc., Washington, D. C., 1966.
57. T. N. Kleinert, Holzforschung, 18, 166 (1964).
58. T. N. Kleinert, Tappi, 49, 126 (1966).
58a. T. N. Kleinert, Pulp and Paper Mag. Can., 1966, T299.
59. T. N. Kleinert, Nature, 207, 631 (1965).
59a. T. N. Kleinert, Holzforsch. Holzverwert., 19, 60 (1967).
59b. T. N. Kleinert, Papier (Darmstadt), 21, 653 (1967).
60. C. Steelink and G. Tollin, Biochem. Biophys. Acta, 59, 25 (1962).
61. C. Steelink, T. Reid and G. Tollin, J. Amer. Chem. Soc., 85, 4048 (1963).
62. C. Steelink, Geochim. Cosmochim. Acta, 28, 1615 (1964).
63. G. Tollin and C. Steelink, Biochim. Biophys. Acta, 112, 377 (1966).
64. B. R. Nagar, N. P. Datta, M. R. Das and M. P. Khakhar, Indian J. Chem., 5, 587 (1967).
65. M. I. Chudakov, V. I. Mal'tsev, V. E. Bronovitskii, Gidrolizn. i. Lesokhim. Prom., 19, 13 (1966).
66. M. I. Chudakov, V. E. Bronovitskii, V. I. Mal'tsev and M. G. Okun, Sb. Tr. Gos. Nauch.-Issled. Inst. Gidroliz. Sul'fitno-Spirt. Prom., 15, 276 (1966).
66a. M. I. Chudakov, V. C. Bronovitskii and I. M. Krivoyez, Dokalady Akademi Nauk Uzbekskoy S.S.R., 25, 30 (1968).
67. K. Freudenberg and J. M. Harken, Holzforschung, 18, 166 (1964).
68. T. Koshijima, E. Muraki, K. Naito and K. Adachi, Nippon Mokuzai Gakkaishi, 14, 52 (1968); through Chem. Abstr. 68, 115, 808 (1968).
69. T. N. Kleinert, Tappi, 50, 120 (1967).
69a. M. N. Raskin, V. I. Mal'tsev and M. I. Chudakov, Gidrolizn. i. Lesokhim Prom., 21 (6), 8 (1968) and ibid., 21 (7), 16 (1968).
70. C. Steelink, J. Amer. Chem. Soc., 82, 2056 (1965).
71. E. Caldwell, C. Steelink, Biochem. Biophys. Acta, 184, 420 (1969).
72. G. M. Coppinger, J. Amer. Chem. Soc., 79, 501 (1961).
73. M. S. Kharasch and B. S. Joshi, J. Org. Chem., 22, 1435 (1957).
74. W. H. Thurston and J. J. Windle, J. Chem. Phys., 27, 1429 (1957).
75. C. Steelink and R. E. Hansen, Tetrahedron Letters, 1966 (1), 105.
76. C. Steelink, J. D. Fitzpatrick, L. D. Kispert and J. S. Hyde, J. Amer. Chem. Soc., 90, 4354 (1968).
77. J. S. Hyde and A. H. Maki, J. Chem. Phys., 40, 3117 (1964).
78. M. I. Chudakov and A. P. Samsonova, Sb. Tr., Gos. Nauch.-Issled. Inst. Gidroliz. Sul'fitno-Spirt. Prom., 15, 285 (1966).
79. V. E. Bronovitskii, M. I. Chudakov and L. L. Kalinskaya, Uzb. Khim.

zh., <u>11</u>, 63 (1967).
80. H. Hatakeyama, K. Suzuki, J. Nakano and N. Migita, Kogyo Kagaku Zasshi, <u>71</u>, 247 (1968); through Chem. Abstr., <u>68</u>, 115, 807 (1968).
81. K. Freudenberg, C. L. Chen, J. M. Harkin, H. Nimz and H. Renner, Chem. Commun., <u>1965</u>, 224.
82. K. Kratzl, Cellul. Chem. Technol. (Jassy), <u>1</u>, 379 (1967).
83. C. Steelink, Tetrahedron Letters, <u>1969</u>, No. 57 5041.
84. O. Martensson and G. Karlsson, Ark. Kemi, <u>31</u>, 5 (1969).
85. A. Zweig, W. G. Hodgson and W. H. Jura, J. Am. Chem. Soc., <u>86</u>, 4124 (1964).

9

SOLVOLYSIS BY ACIDS AND BASES
A. F. A. Wallis

I. Introduction. Acid and base-catalyzed solvolysis reactions of lignin are of paramount importance in many aspects of lignin chemistry. They are useful for structural determination, for preparation of lignin derivatives, and for degradation of the lignin polymer. The last-mentioned forms the basis of the commercial pulping processes, and degradation is achieved by a variety of displacement and cleavage reactions, often, however, accompanied by undesirable condensation reactions.

Holmberg (1) suggested that benzyl alcohols (1, R'' = H) and benzyl ethers (1) were the main reactive structures in lignin sensitive to solvolytic conditions. The reactions of these groups are illustrated for the guaiacyl and 3,4-dimethoxyphenyl derivatives in Figure 9.1. When the aryl moiety is guaiacyl, the p-quinonemethide (2) may be formed from 1 or the phenolate anion (3) in neutral or basic solutions. The quinonemethide (2) is very reactive towards nucleophilic attack (2) and gives rise readily to the α-substituted derivative (6). The formation in solvolytic reactions of quinonemethide intermediates provides an explanation for the reactivity at the benzylic carbon atom of units containing free para phenolic hydroxyl groups.

Under acidic conditions, nucleophilic replacement at the benzylic position of 1 may proceed through the intermediate 5 to give 6 by an S_N2 mechanism (3). A similar reaction path is operative for base-catalyzed replacement reactions of methylated guaiacyl units. When a strongly acidic solution is used, solvolysis reactions are to a large extent accompanied by condensation reactions of the carbonium ion (4) onto an aromatic nucleus.

II. Acid Solvolysis.
    A. Mild Acid Hydrolysis. Lignin solvolytic reactions in wood take place even by the action of water at elevated temperatures. This treatment liberates components from the wood which render the solution mildly acidic and catalyze the degradation process. For example, Sohn and Lenel (4) found that the aqueous extract from the reaction of pine wood with water at 150° for 1-2 hrs. had a pH of 3.6-4.0. After reaction of aspen wood with water at

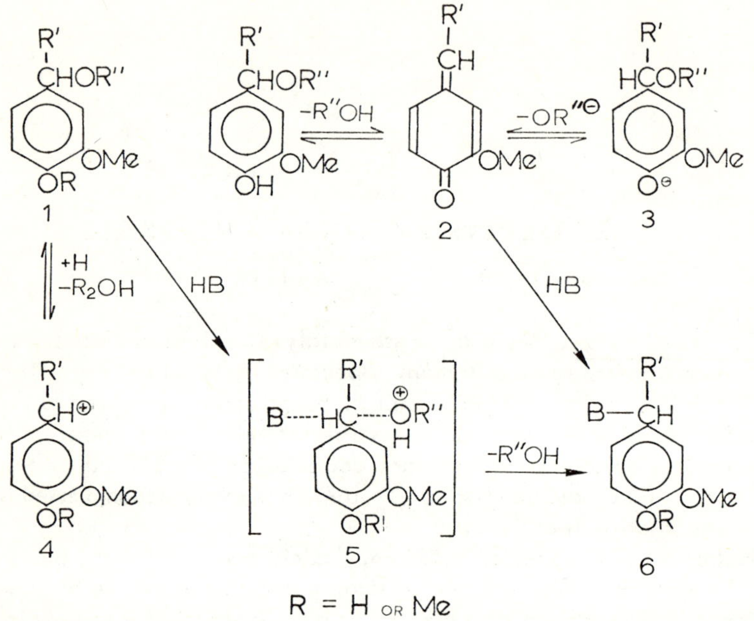

Figure 9.1. Solvolytic reactions of benzyl alcohol and benzyl ether groups.

185° for 4 hr., 45% of the residual lignin was rendered soluble by extraction with benzene-alcohol (5). Saizeva and Nikitin (6) extracted oak wood with water for 4 hr. at 100° 250 successive times, and observed that 75% of the lignin in the wood had been removed.

In the presence of aqueous dioxane, the amount of lignin removed from wood increases with increasing temperature, and hardwood lignins are removed to a greater extent than softwood lignins (7). Furthermore, as the temperature increases, condensation reactions leading to linkages between the benzylic carbon atoms and aromatic nuclei become more prominent. This is shown in Figure 9.2 by the decrease with increasing temperature in the amount of vanillin isolatable upon oxidative degradation of residual spruce wood lignin after treatment with water for 20 hr. (8). Concurrently, the yield of condensed polylignols increases with increasing temperature.

The high temperature reaction of wood with water under pressure has been used as a prehydrolysis stage of sulfate pulping for many years. This process is also employed prior to mechanical treatment of wood in the production of

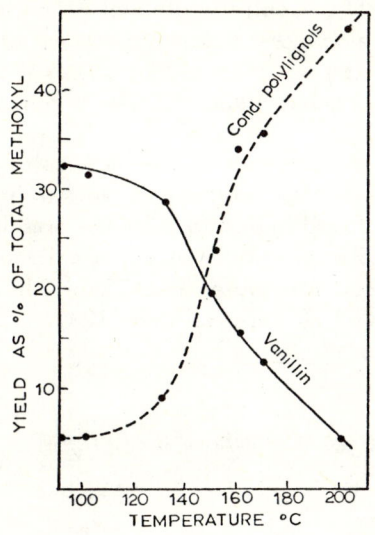

Figure 9.2. Yield of vanillin and condensed polyphenols in nitrobenzene oxidation of spruce wood pretreated with water for 20 hours at varying temperatures (8).

coarse fibers. The degradation reactions accompanying the high temperature aqueous treatment are considered to be partly responsible for the dissolution of lignin at 135° during sulfite cooking of wood, see Chapter 15.

The main reactions of lignin with water are the solvolytic splitting of α and β-ether linkages. Sakakibara et al. (9) have shown that the yield of guaiacol obtained on treatment of models 7 - 9 with water-dioxane (1:1) at 180° for 30 min. decreases from 88% in 7 to 3% in 9 in the order 7 >> 8a > 8b ≃ 9. Thus α-aryl ether linkages are split more rapidly than β-aryl ethers, and products from both these processes may be expected upon high temperature aqueous treatment of lignin.

Besides guaiacol, Kratzl and coworkers (10) identified vanillin (10b), coniferaldehyde (12b) and guaiacylacetone (14) by treatment of 8a with aqueous ethanol at pH 5 for 1 hr. at 175°. Coniferyl alcohol (13b) is also claimed to be one of the products of the aqueous dioxane hydrolysis of 8a (9). The presence of 10b, 12b and 14 among the reaction products is explicable as having arisen from 8a (see Section IIC-2), but the mechanism of formation of 13b remains obscure.

Numerous workers have reported the presence of low-molecular weight material in the hydrolysis liquors. After water treatment of western hemlock at 175° for 45 min., Goldschmid (11) found indications for the presence of the monomers 10b, 12a, 12b, 14 and 15 in the liquor by paper chromatographic analysis. After reaction of aspen wood with water at 170° for 2 hr., the monomers 10b, 10c, 11a-c, 12b, 12c and 15 were isolated (12).

$$\text{(Ar)-CHO} \quad \text{(Ar)-COOH} \quad \text{(Ar)-CH=CHCHO} \quad \text{(Ar)-CH=CHCH}_2\text{OH}$$
$$\quad 10 \qquad\qquad 11 \qquad\qquad\qquad 12 \qquad\qquad\qquad\qquad 13$$

$$\text{(a)} \quad \text{Ar=(H)} \quad \text{(b)} \quad \text{Ar=(G)} \quad \text{(c)} \quad \text{Ar=(S)}$$

$$\text{(G)-CH}_2\text{COCH}_3 \qquad\qquad \text{(G)-COCOCH}_3$$
$$\qquad 14 \qquad\qquad\qquad\qquad 15$$

Sakakibara and Nakayama (13, 14) found 10a, 10b, 11b, 12b, 13a and 13b from the reaction of spruce wood with water-dioxane (1:1) at 140-180°, and 10b, 10c, 11b, 11c, 12b, 12c, 13b and 13c from beech wood. They also noted (7) that softwoods characteristically gave 11b, 12b, 13a and 13b, whereas hardwoods yielded 11b, 11c, 12b, 13b and 13c by the same treatment.

In a detailed study of the hydrolysis products obtained by the reaction of birch wood with water for 2 hr. at 185°, Klemola (15) was able to identify two p-hydroxyphenyl, ten guaiacyl and sixteen syringyl monomers by gas chromatographic analysis (Table 9.1). Six catechol derivatives were also detected. This suggested some demethylation of guaiacyl nuclei had occurred during the course of the hydrolysis. From nmr spectra of the degraded lignin, it was concluded (16) that about 50% of the aromatic nuclei of birch lignin had condensed by formation of carbon-to-carbon bonds. Table 9.1 lists examples of the products detected as a result of acid-catalyzed cleavage, condensation and oxidation-reduction reactions which will be discussed in later sections.

By a series of investigations, Nimz has isolated and identified a number

SOLVOLYSIS BY ACIDS AND BASES 349

Table 9.1 Compounds detected after steam treatment of birch wood at 185° for 2 hr. (15).

(G)-	H CHO COOH CH_2COOH CH_2CH_2OH	$CH=CHCH_2OH$ CH=CHCHO $CH_2CHOHCH_3$ CH_2COCH_3 $COCOCH_3$
(S)-	H CHO COOH CH_3 CH_2CH_2OH $COCH_3$ CH_2COOH CH=CHCOOH	$CH=CHCH_2OH$ CH=CHCHO $CHOHCH_2CH_3$ $CH_2CHOHCH_3$ $CH_2CH_2CH_2OH$ CH_2COCH_3 $COCOCH_3$ CH_2CH_2COOH
(H)-	COOH CH_2COOH	
HO-(ring)-HO	H CH_3 CHO	COOH $COCH_3$ CH=CHCOOH

of low-molecular weight lignin degradation products by hydrolysis of finely-powdered wood with percolating water at 100° (17). The major reaction taking place under these conditions is the hydrolysis of the labile benzyl ether linkages.

Percolation of beech wood or spruce wood with water at 100° affected the removal of lignin in 40% and 10% yield respectively (18). From beech wood, compounds <u>12b</u>, <u>12c</u>, <u>13b</u>, <u>13c</u>, and <u>16-20</u> were obtained (19-24), while spruce wood has afforded <u>12b</u>, <u>13b</u>, <u>19</u>, and <u>21-25</u> (17, 19, 23-26). These lignols have been invaluable aids in conclusively demonstrating the presence of various linkages between lignin monomeric units.

(S)-p(β-β)p-(S)   (S)-CHOHCHOHCH₂OH   (S)-p(β-1)-(S)   (G)-p(β-1)-(G)
     16   17   18   19

(G)-p(β-)-(S)   (G)-p(β-O-4)(G)-p   (G)-CHOHCHOHCH₂OH
     20   21   22

(G)-p(β-O-4)-(G)R
     23: R = a, CHO; b, CHOHCHOHCH₂OH; c, CH=CH-CHO

(G)-p(β-O-4)(G)-p(β-1)-(G)   (G)-p(β-O-4)(G)-p(β-O-4)(G)-p(β-1)-(G)
         24   25

B. Hydroxyl Displacement Reactions.

1. Introduction of Alkoxy Groups. It has been known for a long time that when lignin is treated with alcohols under acidic conditions, new alkoxy groups are introduced (27). Adler and Gierer (28) found that lignin was readily methylated by methanolic hydrogen chloride at room temperature, and these conditions were considered mild enough to avoid lignin degradation and radical structural changes.

In order to determine the reactivity of various structural elements of lignin to methanolic hydrogen chloride at room temperature, a number of aryl propane model compounds were studied. Adler and coworkers (29) observed that vanillyl alcohols and ethers reacted at a higher rate than 3,4-dimethoxyphenyl models. Furthermore, the rates of hydroxyl and ether displacements of the aryl carbinol (26a), aryl carbinol aryl ether (26b), and aryl carbinol ether (26c) by methanol containing hydrogen chloride were in the order 26a > 26b > 26c (equation 1).

$$\text{MeO-C}_6\text{H}_2(\text{OMe})_2\text{-CH}_2\text{OR} \xrightarrow{\text{MeOH/HCl}} \text{MeO-C}_6\text{H}_2(\text{OMe})_2\text{-CH}_2\text{OMe} + \text{ROH} \quad (1)$$

26   a   R=H
       b   R=2-methoxyphenyl
       c   R=Et

Substituents in the β-position of the propyl side-chain decreased the rate of methylation. However, methylation of the β-ether (8b) was complete after 2-3 days (30). The cyclic ethers pinoresinol (27) and the phenylcoumaran (28) could not be methylated at room temperature.

The contribution of carbonyl-containing functions to the reaction was studied with the aid of model compounds (30). Ketols 29 and 30 reacted by etherification of their aliphatic hydroxyl groups, but only one of the ketols studied, ω-hydroxyguaiacyl acetone (31), reacted with the loss of carbonyl function to give a product containing more than one new methoxyl group. The α-carbonyl β-aryl ether (32) did not react with methanolic hydrogen chloride at room temperature.

(G)-p(β-β)p-(G)   (G)-p(β-5)(G)$CH_2CH_2CH_2OH$ (G)$CO-CH_2CH_2OH$
     27   28   29

(G)-COCHOHMe   (G)-CH_2COCH_2OH (G)-COCHCH_2OH
                                                                       O-4(G)-H
     30   31   32

(R-O-4)(G)-CH_2OH
33, a: R = H; b: R = Me

Spruce milled-wood lignin takes up 0.56 new methoxyl groups for every methoxyl group present (30). It was concluded that the majority of the methylatable structures in milled-wood lignin belong to the benzyl alcohol and non-cyclic benzyl ether types, which react via the intermediate 5. However, minor amounts of ketol, coniferyl alcohol and carboxyl groups may also contribute to the reaction.

At elevated temperatures, further acid-catalyzed reactions of lignin with alcohols take place by solvolytic degradation and replacement processes. These reactions will be discussed in Section IIC-2.

2. Mercaptolysis. As in the mild acid-catalyzed treatment of lignin with alcohols, the benzyl alcohol and benzyl ether functions are mainly responsible for the reaction of lignin with mercapto groups under acidic conditions. The strongly nucleophilic sulfur atom contributes to the ease of this reaction, and the α-hydroxyl and α-alkoxyl groups are readily displaced.

Holmberg (31) found that thioglycolic acid reacted readily with the α-alkoxy function in lignin to give alcohol-insoluble lignin thioglycolic acid derivatives (equation 2).

$$-O\text{-}Ar(\text{OMe})\text{-CHR-OR'} \xrightarrow[H^\oplus]{HSCH_2COOH} -O\text{-}Ar(\text{OMe})\text{-CHR-SCH}_2COOH + R'OH \quad (2)$$

This reaction has been widely used for the characterization of lignins, and is considered to follow a similar course, at least in the early stages, to the reaction of sodium sulfite and lignin, which is discussed in Chapter 15.

Kinetic studies have shown that the rate of reaction of vanillyl alcohol (33a) with thioglycolic acid (3), sodium sulfite (32), or with sodium thiosulfate (33) is first-order and passes through a minimum at pH 5. The rate-determining step is the formation of the quinonemethide (2) from 33a, followed by fast attack of a nucleophile to form 6. At lower pH values, the reaction is second-order, and thus proceeds through the path 1 ⟶ 5 ⟶ 6, Figure 9.1.

The reaction of the methyl ether (33b) with sodium sulfite or thioglycolic acid, where the formation of the intermediate quinonemethide is precluded, exhibits second-order kinetics regardless of the pH value. In neutral or alkaline solutions, the rate of reaction of vanillyl alcohol with sodium sulfite is far greater than the reaction of the methyl ether (33b) with sodium sulfite (34). This is a further indication that the quinonemethide intermediate enhances the rate of nucleophilic displacement reactions of 33a.

3. Water Elimination. During the acid-catalyzed hydrolysis of (β-5) or (β-1) dilignols, it has been shown that the γ-hydroxy function may be eliminated as water.

On heating with methanolic hydrogen chloride or under "acidolysis" conditions (heating with 0.2M hydrogen chloride in dioxane-water 9:1), the (β-5) dilignol (34a) was converted into the phenylcoumarone derivative

(39a) (35, 36) in 75% yield (37). Similarly the methyl ether (34b) was converted to 39b by the loss of water. The mechanism of the conversion of 34 to 39 is illustrated in Figure 9.3 (37). Initially, ring opening results in the formation of the benzyl carbonium ion (35), which loses a proton to form 36. Protonation of 36 followed by loss of water allows the formation of 37, which by reclosure of the hydrofuran ring gives 38. Finally, migration of the exocyclic double bond gives rise to 39.

The stilbene derivative (40) was formed as a minor product of the reaction by loss of formaldehyde from the intermediate 35 (broken lines) by a reversed Prins reaction (37).

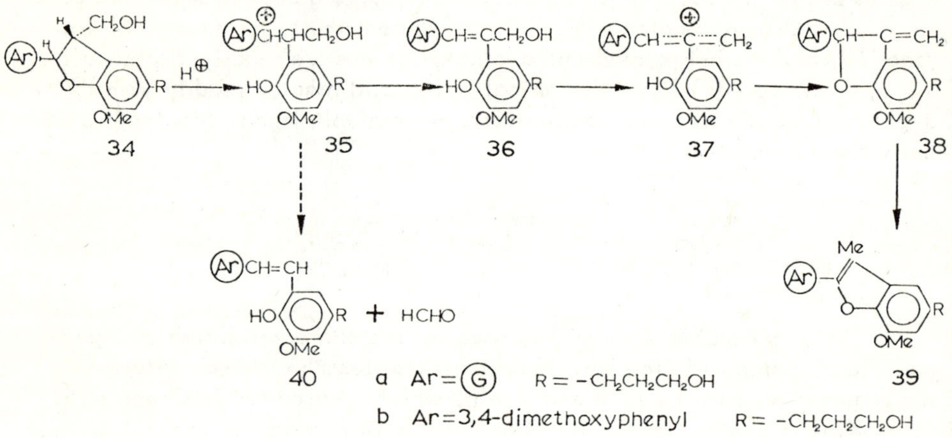

Figure 9.3. Acid-catalyzed hydrolysis of phenylcoumaran dimers.

Phenylcoumarone units have been detected in lignin treated either with methanolic hydrogen chloride or under acidolysis conditions by their characteristic ultraviolet and ionization-$\Delta\varepsilon$ spectra (38). This provides evidence in addition to that from the model compound study that the water elimination reaction occurs during acid-catalyzed solvolysis of lignin.

A second structural unit in lignin from which water may be eliminated during acidolysis is illustrated by the (β-1)-dilignol; 1,2-diguaiacyl-1,3-propandiol (41). Treatment of 41 with boiling aqueous dioxane containing hydrogen chloride gives rise to 1,2-diguaiacyl-1-propanone (46), 1,1-diguaiacyl-2-propanone (45) and the stilbene (43) (37, 39, 40).

The mechanism of formation of these compounds initially involves loss of water to give the benzyl carbonium ion 42 as in Figure 9.4. Loss of formaldehyde from 42 in a reversed Prins reaction yields the stilbene (43). Alterna-

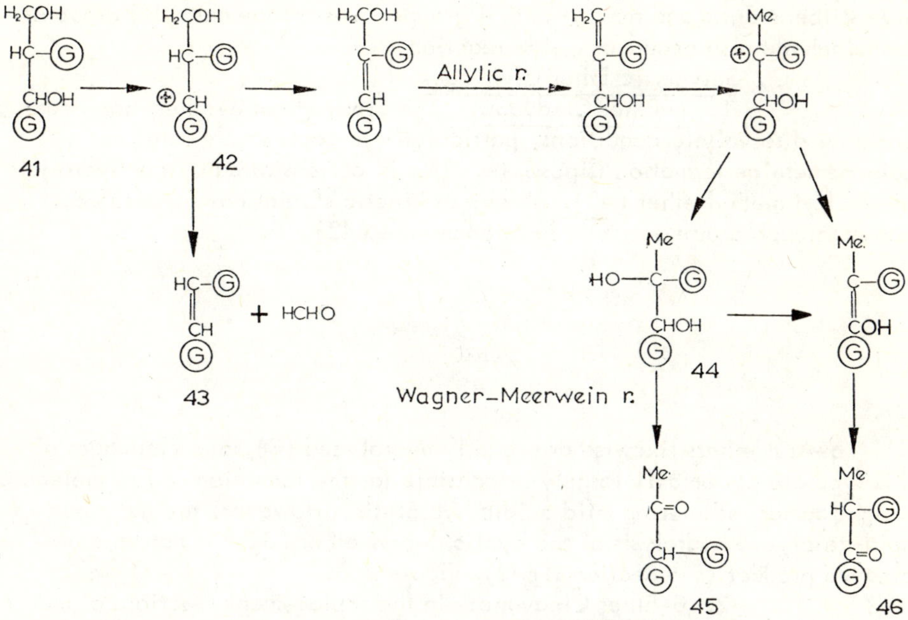

SOLVOLYSIS BY ACIDS AND BASES

Figure 9.4. Acid-catalyzed hydrolysis of (β-1) dilignols.

tively, a series of steps involving proton loss, allylic rearrangement and water addition leads to the diol intermediate 44. Elimination of the β-hydroxyl function as water gives rise to 46. The isomeric ketone 45 may also arise from 44 by elimination of the α-hydroxyl group accompanied by a Wagner-Meerwein shift of the β-guaiacyl substituent.

As further proof of the behavior of these structural features of lignin to acidolysis, Lundquist (37, 41) was able to isolate the stilbenes (43 and 47), the phenylcoumarone (48), 1,1-diguaiacyl-2-propanone (45) and 1,2-diguaiacyl-1-propanone (46) from acidolysis products of spruce milled-wood lignin.

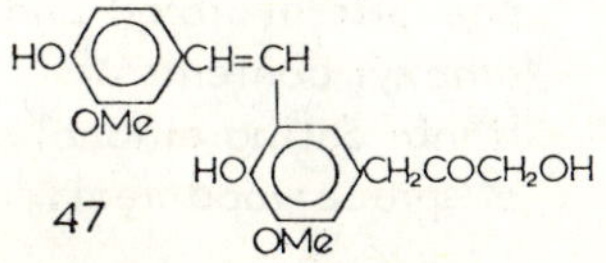

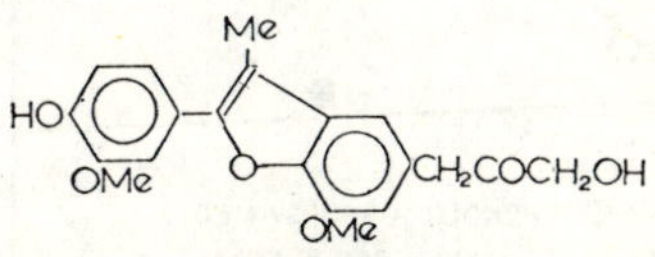

Thus it may be seen that propyl side-chains containing β-aryl linkages and a γ-methylol group react under acidic conditions with the loss of water to give stilbene units and new γ-methyl groups. Loss of the methylol group as formaldehyde also occurs as a side reaction.

C. Solvolytic Ether Cleavages.

1. α-Ether Cleavage. The cleavage of α-ethers proceeds under mild solvolytic conditions, particularly in cases where intermediate quinonemethide formation is possible. This is demonstrated by the hydrolysis of vanillyl methyl ether (49), which, as kinetic studies have indicated, passes through a quinonemethide intermediate (42).

HO–⟨ ⟩–CH$_2$OMe
OMe
49

α-Aryl ethers likewise are readily hydrolyzed (8), and cleavages of this type are presumably largely responsible for the formation of low molecular weight compounds during mild acidic hydrolysis. However, the net result of acid-catalyzed hydrolysis of the cyclic α-aryl ether (34a) is not an ether-cleaved product (see Section IIB-3), Figure 9.3.

2. β-Ether Cleavage. In the replacement reactions of methanolic hydrogen chloride on lignin at ambient temperature, the latter remains insoluble (Section IIB-1). In contrast, refluxing lignin in ethanol containing hydrogen chloride ("ethanolysis") results in dissolution of most of the lignin forming monomers (43) and soluble oligomers ("ethanolysis lignin"). This is due to a solvolytic degradation of the lignin molecule (44) which follows the rapid initial ethoxylation reaction.

It was shown that terminal methyl groups are produced during the course of ethanolysis (45), and that while the ethoxyl content decreases slightly, the phenolic hydroxyl content increases, Figure 9.5 (46). This indicates that phenyl ether linkages are cleaved during the reaction.

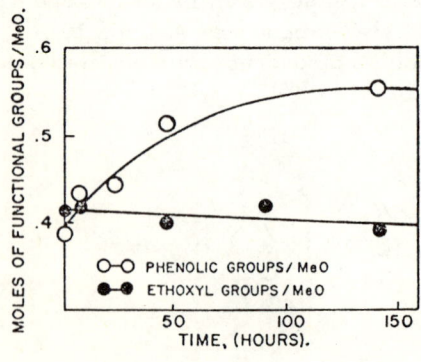

Figure 9.5. Change in phenolic hydroxyl and ethoxyl contents of lignin during ethanolysis of spruce woodmeal.

SOLVOLYSIS BY ACIDS AND BASES   355

Sarkanen and Schuerch (46) have further demonstrated that carbonyl groups are formed at the same rate as the new phenolic hydroxyl groups. They attributed the decrease in oxygen content of the lignin during ethanolysis to the loss of water during condensation reactions of C-α with C-5 and C-6.

As a result of a series of investigations concerning the water-soluble products obtained from the reaction of extractive-free sprucewood with boiling ethanol containing 2% hydrogen chloride, Hibbert and coworkers were able to isolate and identify the four guaiacyl ketones (57a, 58a, 59a and 60a) in amounts corresponding to 10% of the original lignin (43). Maple lignins gave, in addition to the guaiacyl derivates, the syringyl ketones (57b, 58b and 60b); and the total yield of monomeric products was higher (46). The monomers 55-60 are now commonly known as Hibbert's ketones, Figure 9.6.

Adler and coworkers (35) subjected milled-wood lignin to refluxing dioxane-water 9:1 containing 0.2M hydrogen chloride ("acidolysis" conditions) for 48 hr., and detected by paper chromatographic analysis the ketones 55a, 59a and 60a, analogous to Hibbert's ethanolysis monomers. In later experiments, the ketol 56a was also isolated from the acidolysis liquors (41, 48, 49).

Hibbert first suggested that the units in lignin giving rise to the monomeric ketones (55-60) under acid-catalyzed solvolytic conditions had the structure of β-hydroxy coniferyl alcohol 52a or its keto form 53a (50). Adler extended this assumption by demonstrating that the α-arylglycerol-β-ethers (50) yield the same ethanolysis and acidolysis monomers as lignin (35, 51). As the β-ether units (50) are recognized features of lignin structure, they are now considered to be the actual progenitors of the monomers.

Ethanolysis of the β-ether (50c) afforded 57c, 58c and some guaiacol (54) (51). From the guaiacyl derivative (50a), the ketones 58a, 59a and 60a were obtained (35). Treatment of 50a under acidolysis conditions (35) gave rise to guaiacol (54), ketones 55a, 59a and 60a, and in later studies the ketol 56a was identified (52).

The mechanism of formation of monomeric ketones from 50 is illustrated in Figure 9.6. Adler has suggested that the initial step is a slow acid-catalyzed dehydration to the enol ether (51) which is followed by a faster hydrolysis to β-hydroxy coniferyl alcohol (52) with the loss of guaiacol (54) (37, 53). The ketones 55 and 56 are formed from 52 by an allylic rearrangement (54, 55, 56), and several proton shifts. In the presence of ethanol, the ethoxyketones 57 and 58 are produced from 55 and 56. Finally, ketones 59a and 60a are formed by oxidation-reduction of the ketol 56a. The absence of 59c and 60c in the acidolysis products of non-phenolic ketols 55c and 56c (57) indicates that the quinonemethide derived from the phenolic benzyl alcohol 56a is involved in the reaction (53).

An alternative mechanism for the ether cleavage stage in the solvolysis of 50 involves direct nucleophilic displacement of guaiacol to form the α-arylglycerol (61) (10). Acid-catalyzed dehydration of 61 gives rise to β-hydroxy

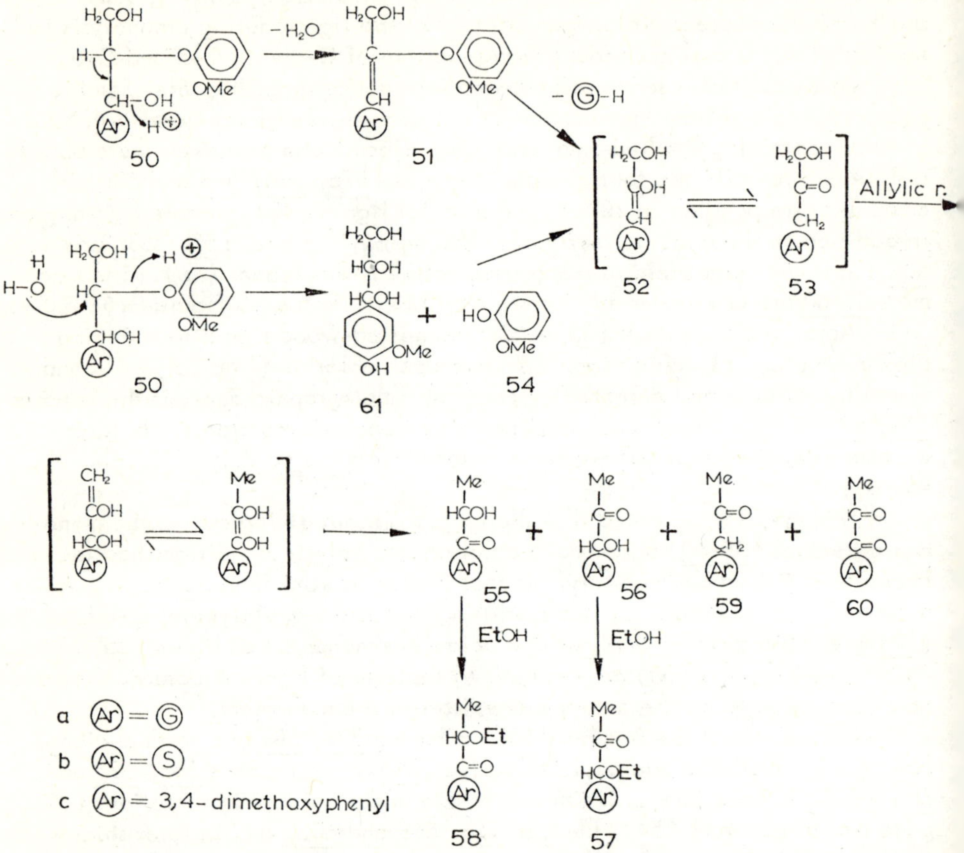

Figure 9.6. Acid-catalyzed solvolysis of α-arylglycerol-β-aryl ethers to form Hibbert's ketones, 55-60.

coniferyl alcohol (52), which may further react as discussed above.

In support of this mechanism, Nimz has isolated low molecular weight lignin degradation products with glyceryl side-chains as products of treatment of wood with percolating water at 100° (Section IIA). Thus the direct displacement mechanism seems operable under mildly acidic conditions. α-Arylglycerol structures do not survive at low pH values (57) and both mechanistic paths are possible for ether cleavage of 50 under acidolysis or ethanolysis conditions.

Convincing evidence in favor of the reaction scheme in Figure 9.6 was obtained by conversion of the 3,4-dimethoxyphenyl ketol (53c) to 57c and

58c on ethanolysis (54). Similar treatment of guaiacyl compounds resulted in the transformation of 56a (56), its diacetate (58), and the ketol (53a) (55) to 57a, 58a, 59a and 60a.

Ketols 53a and 53c have been isolated after brief acidolysis of 50a and 50c respectively (29, 59). Treatment of 56a under acidolysis conditions has given 55a, 60a and minor amounts of vanillin and vanillic acid (56). Traces of the latter two compounds were also detected in the acidolysis of 50a (37), and probably originate by oxidative rather than solvolytic processes.

In the acidolysis of the β-aryl ether models (50), the ether cleavage reaction is faster than the allylic rearrangement. Therefore, ketols 53a and 53c accumulate and were isolated in 53% and 34% yield after acidolysis of 50a and 50c for 4 hr. (37). In the same time, guaiacol was liberated in 74% and 42% yield, respectively.

The allylic rearrangement and formation of 59 and 60 have been shown to be irreversible, but equilibration of 55a and 56a to a 4:1 mixture proceeds in 30 min. under the conditions of acidolysis (60).

Kratzl et al. (10) have found coniferaldehyde as the major product of acid-catalyzed solvolysis of guaiacylglycerol-β-aryl ether (50a) at pH 5 and 175°. The mechanism of its formation may involve removal of a γ-proton from 50a and elimination of guaiacol, followed by elimination of the benzylic hydroxyl function as water (equation 3)

$$\text{50a} \longrightarrow \text{[intermediate]} \xrightarrow{-H_2O} \text{12b} \qquad (3)$$

Although coniferaldehyde may be obtained from spruce milled-wood lignin in a yield of 0.5% on acidolysis (37, 53), it is considered to originate from coniferaldehyde end-groups in lignin as the yield is only 0.03% from borohydride-reduced lignin. Thus under stronger acidic conditions, the above mechanism may be inoperative.

On further investigation of the acidolysis products of spruce milled wood lignin, Lundquist has been able to isolate products which lend support to the proposed acidolysis schemes being applicable to lignin. The ketol (53a) in 6% yield (48), and the dimers 47 and 48 with a primary ketol side-chain (37, 40) were among the products obtained after a reaction time of 4 hr. Products 47 and 48 obviously arise from units containing a phenylcoumaran structure as well as a glyceryl aryl ether side-chain (Section II B-3). Small amounts of p-hydroxyphenyl-ω-hydroxyacetone (62), coniferaldehyde,

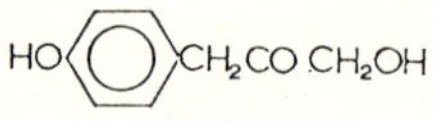

HO—⟨O⟩—CH₂CO CH₂OH
62

vanillin and vanillic acid were also isolated (37, 53).

During the acidolysis treatment, condensation reactions involving benzyl cations (4) also occur. These reactions are reflected in the time-dependent decrease in vanillin and syringaldehyde yield upon oxidative degradation of the isolated acidolysis lignin (61), as discussed in Chapter 5.

The reactions of the arylglycerol-β-aryl ether model compounds indicate that the main features of the acid-catalyzed solvolysis of these structural entities in lignin are the liberation of phenolic hydroxyl groups and the formation of γ-methyl groups and α- and β-carbonyl groups. Solvolysis in alcoholic media results in the formation of alkoxy groups. Condensation reactions forming new carbon-to-carbon bonds are also of importance.

The behavior of glyceraldehyde-2-aryl ether structures toward acidolysis was studied using the methyl acetal (66) as a model compound, whereby methylglyoxal (67) and p-creosol (68) were obtained (62). The ether (69) was found to be stable to acidolysis, and an attempt to prepare the glyceraldehyde derivative (63, Ar = 2-methoxy-4-methylphenyl) by reaction with formaldehyde resulted in the formation of a cyclic dimer, which, however, gave 67 and 68 on acidolysis (62) Figure 9.7.

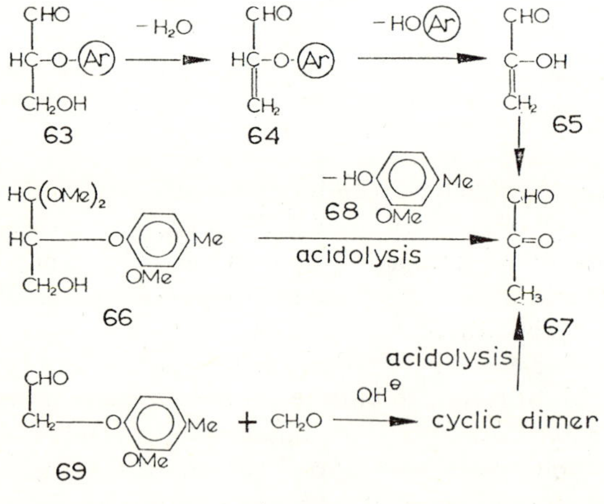

Figure 9.7. Acid-catalyzed hydrolysis of glyceraldehyde-2-aryl ether.

SOLVOLYSIS BY ACIDS AND BASES

The detection of methyglyoxal (67) among the acidolysis products of milled-wood lignin indicates that it may be formed from units of the type 63 in lignin. The mechanism of the reaction presumably involves loss of water from 63 to yield 64, which in turn is hydrolyzed by cleavage of the aryl ether to the enol of methylglyoxal (65).

In Figure 9.8, the reactions occurring during acidolysis of a portion of a guaiacyl lignin molecule are represented. The resulting oligomer (units 4 through 10) is a typical lower molecular weight part of dioxane lignin.

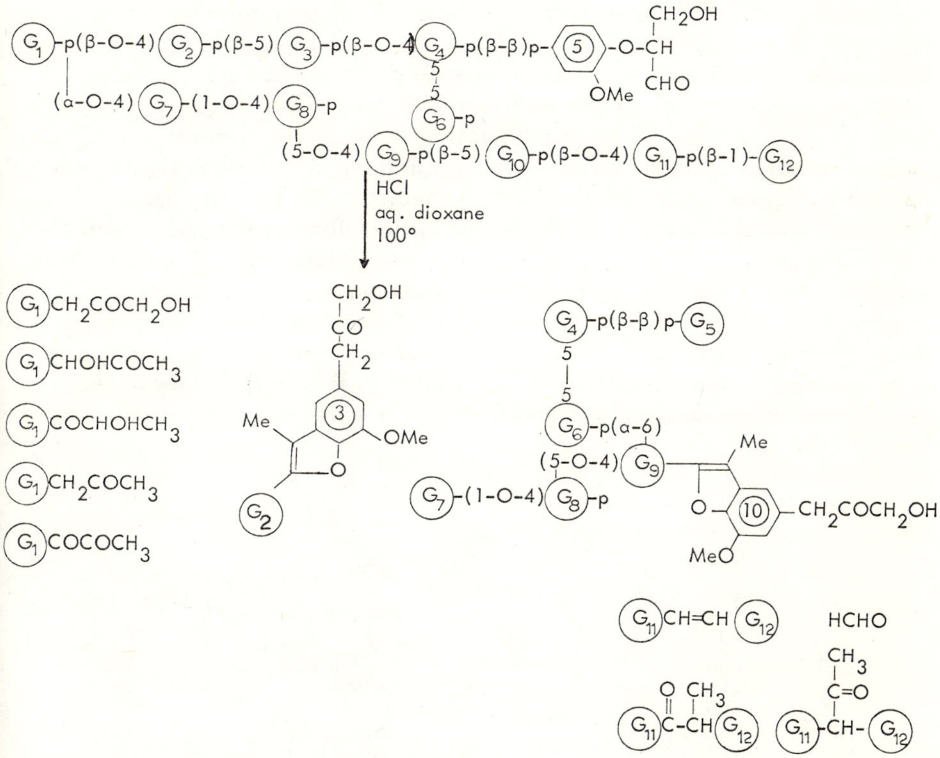

Figure 9.8. Acid-catalyzed hydrolysis of a guaiacyl lignin.

D. Solvolytic C-C Cleavages.

1. β-γ Cleavage. Hägglund (63) found that hydrochloric acid lignin, on heating with 12% aqueous hydrochloric acid, yields a volatile aldehyde, which Freudenberg and coworkers (64, 65) subsequently identified as formaldehyde. A number of workers have now shown that heating a variety of lignin preparations with 12% hydrochloric acid or 28% sulfuric acid produces formaldehyde in yields sometimes approaching 4% (66).

Freudenberg and Mueller (67) summarized the results of studies on the behavior of a series of model compounds to acid hydrolysis by stating that formaldehyde may be formed from phenylpropane derivatives containing a terminal methylol group, and substituted in the C-α and C-β positions with hydroxy or ether functions. The β-carbon atom alternatively may contain an aryl group.

The cleavage of the γ-methylol group in the phenylcoumaran (34) and 1,2-diguaiacylpropane-1,3-diol (41) structures as formaldehyde as minor reactions under acidolysis conditions has been discussed in Section II B-3. Adler and Yllner (68) have shown that from 1 mole of the β-ether model (50c), 0.19 moles of formaldehyde are formed on heating with 28% sulfuric acid. These reactions are examples of reversed condensations, and proceed via a benzyl carbonium ion, as in Figure 9.3, page 352.

    2. α-β Cleavage. Small amounts of vanillin and vanillic acid have been obtained in acidolysis reactions of both milled-wood lignin, and of the guaiacylglycerol-β-aryl ether (50a) (Section II C-2). The isolation of 23a by mild acid hydrolysis is also an example of formyl side-chain formation (24). The genesis of formyl groups has not been clarified, but it seems likely that they are the resultant of oxidative processes.

    E. Side-Chain Rearrangements. Nimz (69) found that treatment of the β-ether (70) with water at 100° for seven days gave the (β-5) dilignol (73) and dihydroconiferyl alcohol in yields of 32% and 41% respectively. Two higher molecular weight phenolic products were observed but not identified, Figure 9.9.

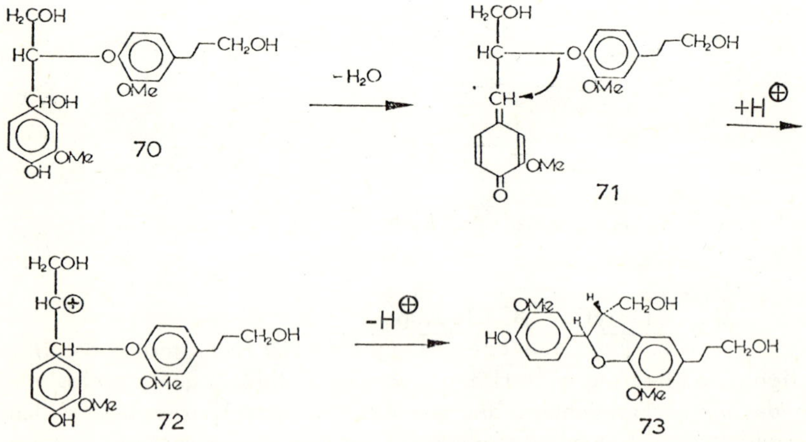

Figure 9.9. Rearrangement of an α-guaiacylglycerol-β-aryl ether in water at 100°.

This unusual rearrangement of a (β-O-4) to a (β-5) dilignol did not occur under acid-catalyzed solvolysis conditions. Furthermore, the methyl ether corresponding to 70 was found to be unreactive to boiling water, which suggests that a quinonemethide intermediate may be operative in the rearrangement reaction. The most probable reaction mechanism is formulated in Figure 9.9. The quinonemethide 71, formed by dehydration of 70, may undergo a 1,2-aryloxy shift to 72. Intramolecular condensation of the carbonium ion at C-β in 72 gives rise to the phenylcoumaran 73.

Another side-chain rearrangement that occurs in the acid-catalyzed solvolytic reactions of lignin is the allylic rearrangement of the γ-hydroxy group to the α-position. This rearrangement features in the mechanism of solvolysis of (β-O-4) and (β-1) dilignols, and is discussed in Sections II B-3 and II C-2.

III. Basic Solvolysis.

A. <u>Solvolytic Ether Cleavages.</u> Lignin undergoes a series of reactions leading to its dissolution under the influence of alkaline reagents at elevated temperatures. These reactions are mainly solvolytic cleavages of the ether linkages between phenylpropane units with the simultaneous formation of phenolic hydroxyl groups. They form the basis of the alkaline pulping processes discussed in Chapter 16. Alkaline solvolysis reactions have been utilized for the isolation of lignin as alkali lignin preparations.

Other processes also occur during alkali solvolysis of lignin, including cleavage of carbon-to-carbon bonds, and secondary condensation reactions.

1. <u>α-Ether Cleavage.</u> Studies with model compounds have shown that benzyl ether linkages are cleaved by alkaline reagents if there is a free phenolic hydroxyl group in the position <u>para</u> to the propyl side-chain (69). The intermediate formation of a quinonemethide species is believed to facilitate the reaction, and is supported by rate studies on the hydrolysis of the α-methyl ether of vanillyl alcohol (42).

Non-phenolic α-ethers are also cleaved provided there is a hydroxyl group in adjacent β-position (69). Cleavage of the ether linkage in this case probably receives assistance from the neighboring hydroxyl group with the simultaneous formation of an epoxide, as has been shown for non-phenolic β-ether cleavage (Section III A-2).

As a model for cyclic α-aryl ether structures, the phenolic phenylcoumaran (28) was subjected to both 2N sodium hydroxide and sodium hydroxide-sodium sulfide solutions at 170° (soda and kraft cooking respectively). In both cases, the hydrofuran ring opened and the γ-methylol group was split off as formaldehyde in a reversed aldol reaction of the quinonemethide intermediate (74), leaving the <u>trans</u>-stilbene (75) as the reaction product after acidification (70, 71) (equation 4).

The presence of stilbene structures in milled-wood lignin which has been subjected to kraft cooking conditions has been detected by ionization difference ultraviolet spectra (71, 72).

Methylated phenylcoumaran models, from which quinonemethide formation is precluded, were found to be unreactive to kraft or soda cooks (60, 70).

2. β-Ether Cleavage. Cleavage of the β-aryl ether linkage of non-phenolic (β-O-4) dilignol models during alkaline solvolysis occurs if there is any adjacent hydroxyl or carbonyl group in the α or γ-position (73). Reaction of the β-ether model 50c with 2N sodium hydroxide at 170° for 2 hr. afforded guaiacol (54) and the arylglycerol (77) (69) (equation 5).

The epoxide (76), formed by nucleophilic attack of the α-hydroxyl group on the β-carbon atom, was postulated as an intermediate in the reaction on the basis of the following evidence. Non-phenolic β-ethers devoid of adjacent free hydroxyl groups are not cleaved by alkali (69, 74). Hydrolysis of α-hydroxy β-ethers in the presence of piperidine enabled the reactive epoxide intermediates to be trapped as α-hydroxy piperidino compounds (69, 74). Finally, the reaction product has the same stereochemistry as the initial β-ether. This may be rationalized in terms of formation of the epoxide (79) from erythro-78, and further reaction of 79 by alkali to the erythro-glycol (80) with an overall two-fold inversion of configuration (75) (equation 6).

Confirmatory evidence that the mechanism of alkaline hydrolysis in lignin follows the same course as in model compounds was obtained by the lack of ether cleavage with the liberation of new phenolic hydroxyl groups in milled-wood lignin, in which both the aryl and alkyl hydroxyl groups were methylated, when treated under soda (73) or kraft (76) cooking conditions. In untreated milled-wood lignin, reaction with alkali afforded about 0.3 new phenolic

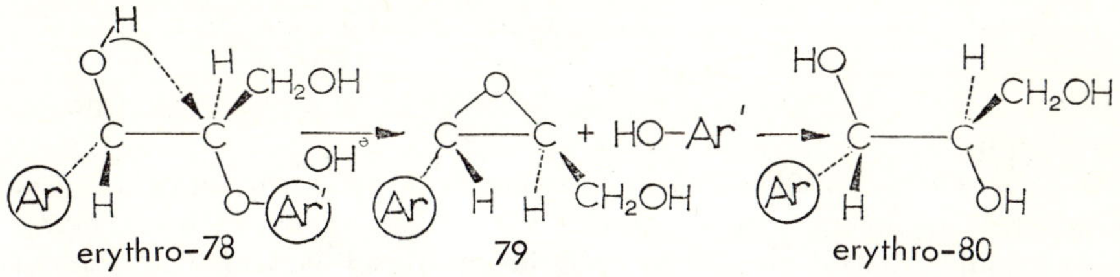

SOLVOLYSIS BY ACIDS AND BASES

[Structural diagram showing reaction (6):]

erythro-78 → 79 + HO-Ar' → erythro-80   (6)

Ar = 3,4-dimethoxyphenyl   Ar' = 2-methoxyphenyl

groups per methoxyl group.

The phenolic β-ether model 50a also undergoes ether cleavage when subjected to a soda cook. The yield of guaiacol resulting from this was only 20-30%, and the enol ether (82) was also formed (69). Treatment of 50a under kraft cooking conditions, however, led to the formation of guaiacol as the dominant reaction, and lesser amounts of the enol ether (82) (77).

The differences observed in the soda and kraft cooks of 50a may be rationalized as follows (53, 77, 78). In the soda cook, the epoxide mechanism may be invoked for the cleavage of guaiacol (54) from 50a (Figure 9.10).

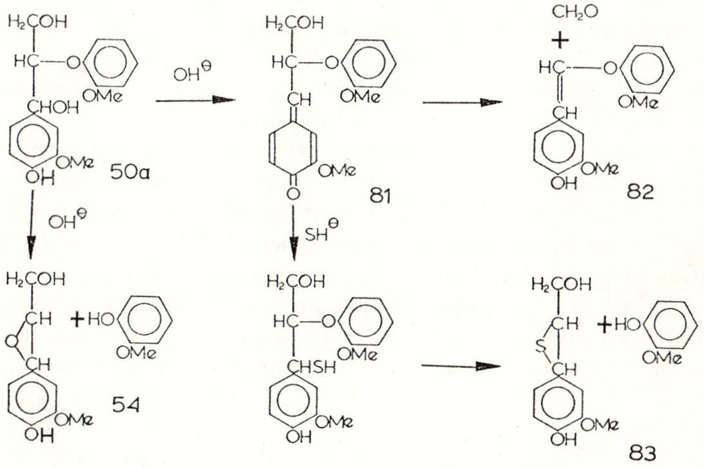

Figure 9.10. Alkaline degradation of α-guaiacylglycerol-β-aryl ethers.

Alternatively, dehydration leads to the quinonemethide (81), which loses formaldehyde in a reverse aldol reaction, giving the relatively alkali-stable enol ether (82). During the kraft cook, the strongly nucleophilic sulfide ions react with the quinonemethide intermediate (81) in competition to the reversed condensation, finally leading to the episulfide (83) (79) and guaiacol (54). Thus the base-catalyzed cleavage reaction of β-ether models is enhanced in the presence of sulfide ions, and may partly account for the favorable effect of sodium sulfide in the kraft delignification process (see Chapter 16, Section IIC).

A novel method of β-ether cleavage has been demonstrated by Jerkeman and Lindberg (80) for the p-tolylsulfone (84), which may be obtained from the corresponding benzyl alcohol by an acid-catalyzed nucleophilic displacement reaction with sodium p-tolylsulfinate.

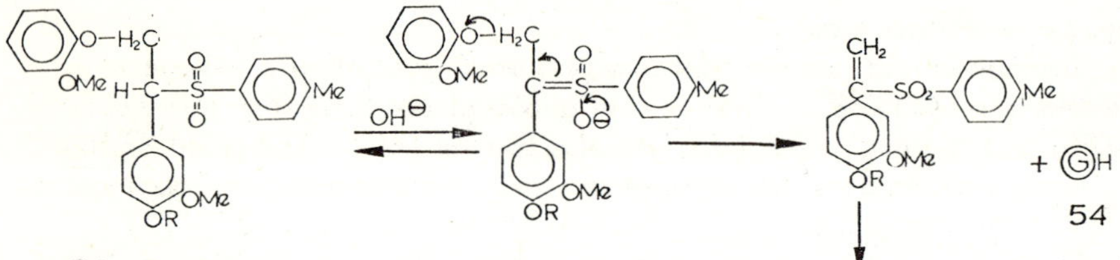

(7)

84a R = H
 b R = Me

products

By allowing the sulfone (84b) to react with 2N sodium hydroxide at 20° for 3 hr., guaiacol (54) was split off, presumably by a β-elimination mechanism as in equation 7. The guaiacyl compound (84a) also gave guaiacol on reaction with 2N sodium hydroxide at 100° for 2 hours. This method of degradation has not yet been utilized for the lignin polymer.

Gellerstedt and Gierer (81) have recently provided an example of β-ether cleavage under neutral solvolysis conditions. Treatment of 50a with aqueous sodium bisulfite at 180° for 3 hours has given among the reaction products trans-4-hydroxy-3-methoxystyrene-ω-sulfonic acid (85) and guaiacol. The formation of 85 implies that a direct displacement of the β-ether moiety as guaiacol by the nucleophilic bisulfite group and a reversed aldol reaction are operative. Further experimental work is required, however, before it is possible to elucidate the timing aspect of the mechanism of these reactions.

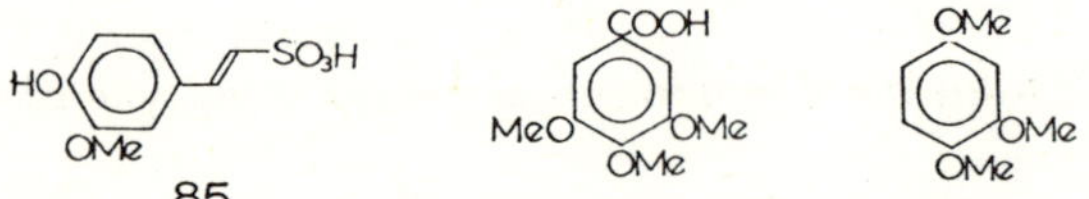

85   86   87

3. **Diaryl Ether Cleavage.** Small amounts of diaryl ether cleavage products have been obtained from the alkaline hydrolysis of lignin by

Freudenberg et al. Thus methylated artificial "D.H.P." lignin consisting exclusively of guaiacyl units, when reacted with 70% potassium hydroxide solution at 170° for 90 minutes, remethylated and oxidized with potassium permanganate afforded 3,4,5-trimethoxybenzoic acid (86)(82). The formation of 86 is explicable as having arisen from a (4-O-5)-coupled unit by rupture of the diaryl ether linkage.

By similar treatment, methylated spruce lignin has given rise to 1,2,4-trimethoxybenzene (87)(83), presumably by cleavage of a (4-O-1)linkage. It must be emphasized that these are only to be regarded as minor reactions in the alkaline solvolysis of lignin.

4. Methoxyl Group Cleavage. The methoxyl groups in lignin are generally resistant toward alkaline hydrolysis. However, they may be hydrolyzed to varying degrees at temperatures around 200°C., as shown by a large number of studies. Sarkanen and coworkers (84), in a kinetic study of the alkaline hydrolysis of lignin models, concluded that the methoxyl groups of strongly acidic phenols, e. g. vanillin, hydrolyze much more rapidly than those of weakly acidic phenols. They also found that the methoxyl hydrolysis in lignin consists of a rapid and slow phase. Hydrosulfide and methyl mercaptide ions, which are stronger nucleophiles than hydroxide ions, react more readily with methoxyl groups in lignin model compounds, converting them to methyl mercaptan and dimethylsulfide, respectively (85, 86). (See Chapter 19).

Croon and Swan (87) found that under slightly alkaline sulfonation conditions, some demethylation occurs. Reaction of 3,4-dimethoxybenzaldehyde with sodium sulfite at pH 8 and 170° for 20 min. gave vanillin in 4.5% yield. Vanillic acid also showed signs of demethylation under these conditions.

Demethylation reactions also occur during neutral sulfite cooking. Treatment of dimeric β-ether (81) and phenylcoumaran (88) model compounds containing aryl methoxyl groups with aqueous sodium bisulfite at 180° for 3 hrs. has given methane sulfonic acid and catechol derivatives, obviously by a methoxyl cleavage following bisulfite ion attack. Methane sulfonic acid has also been detected in the liquors from neutral sulfite cooking of isolated lignins (88).

B. Solvolytic C-C Cleavages.

1. β-γ Cleavage. Cleavage of the γ-methylol group as formaldehyde during the alkaline hydrolysis of the (β-5) and (β-O-4)-linked dimers (28 and 50a) has already been mentioned. The formaldehyde cleavage reactions take place only in structures containing free phenolic hydroxyl groups, where intermediate quinonemethide formation is possible.

The (β-1) dilignol (41) gives rise to formaldehyde by reversed condensation on hydrolysis with 2N sodium hydroxide at 100° (19). The reaction again proceeds via a quinonemethide (88), from which the stilbene (43) is formed by loss of formaldehyde (equation 8)

The stilbene (43) has been isolated among the alkaline hydrolysis

$$\text{41} \xrightarrow[-H_2O]{OH^{\ominus}} \text{88} \longrightarrow HCHO + \text{43} \qquad (8)$$

products of lignosulfonates (89) pine wood (90) and beech wood (91).

Further indication that β-γ cleavage takes place under alkaline hydrolysis conditions is obtained by examination of the products of alkaline hydrogenolysis of lignin. The major products of these reactions possess two carbon side-chains (see Chapter 12). An example of the type of product isolated from alkaline hydrogenolysis of lignin is provided in the bibenzyl (89) (92). Its formation obviously involved the loss of formaldehyde from a β-5 linkage, followed by hydrogenation.

89

The formaldehyde liberated from lignin under alkaline conditions may undergo electrophilic substitution onto an aromatic nucleus with the formation of a methylol derivative.

        2. α-β Cleavage. By the action of hot alkali on lignosulfonates under nitrogen, vanillin (93) and acetaldehyde (94) together with smaller amounts of formaldehyde (95) and acetoguaiacone (96) are formed. To demonstrate that vanillin and acetaldehyde result from an α-β cleavage, Kratzl and Hofbauer (97) subjected lignosulfonic acids obtained from lignins labelled in the α and β carbon atoms to alkaline hydrolysis and found among the products active vanillin and inactive acetaldehyde, and inactive vanillin and active acetaldehyde respectively.

        Adler and Häggroth (98) obtained vanillin and acetaldehyde by alkaline hydrolysis of spruce lignin alone. Their formation was assumed to be due to the coniferaldehyde end-groups in the lignin. The observation that lignosulfonates gave rise to vanillin in excess of that expected from the number of coniferaldehyde groups prompted investigations of the alkaline hydrolyses of model sulfonic acids.

        Kratzl and coworkers (99) found that the β-ether sulfonic acid (90b) afforded carbonyl compounds at the same rate as lignosulfonates. Both 90a and 90b react with alkali with the formation of vanillin and veratraldehyde respectively, guaiacol and acetaldehyde. In addition, acetoguaiacone and formaldehyde have been obtained from 90a (100, 101).

In order to demonstrate that the main reaction of 90a with alkali is an α-β cleavage, Kratzl and Spona labelled the γ-carbon atom and obtained inactive vanillin and active acetaldehyde on alkaline hydrolysis (102). Furthermore, the acetaldehyde had its activity on the carbonyl group.

The mechanism of formation of vanillin from 90a has been rationalized (100, 101) in terms of elimination of guaiacol to form the β-γ double bond, enol-keto tautomerism, substitution at C-α by a hydroxyl ion and reverse condensation, as shown in equation 9.

$$(9)$$

90 a R=H
   b R=OMe

The preferential removal of a proton from C-γ may be due to less steric hindrance to the approach of a hydroxyl ion compared with C-α proton removal. Also, the negatively charged sulfonic acid group hinders the approach of a negatively charged hydroxyl group at C-α (100, 101).

Another possible mechanism for the elimination of guaiacol from 50c is through the cyclic intermediate 91, whereby γ-proton removal is facilitated by the α-sulfonate anion.

The formation of acetoguaiacone and formaldehyde from 90a as minor alkaline hydrolysis products is more difficult to explain. It has been suggested (100, 101) that guaiacol elimination may give some 92 as a reaction intermediate from which cleavage may take place. However, the failure of 93 to react with alkali does not support this hypothesis.

A clue to the mode of formation of acetoguaiacone is obtained from the observation that formaldehyde is produced at a faster rate than acetaldehyde in the initial stages of the reaction and then more slowly. This suggests that the sample assumed to be 90a may contain some material resulting from sulfonation at C-γ as an impurity.

[Structures 91, 92, 93 shown]

From equation 10, it follows that acetoguaiacone and formaldehyde result from reversed condensation of 94 by a β-γ cleavage. Furthermore, as formaldehyde formation is faster than acetaldehyde, the overall rate of hydrolysis of the γ-sulfonic acid (equation 9) may be greater than the α-sulfonic acid (equation 10).

[Reaction scheme for compound 94]
$$\quad (10)$$

3. <u>α-1 Cleavage</u>. During alkaline hydrolysis of wood, some guaiacol is found in the cooking liquors. This is probably a reaction (equation 11) of a guaiacylpropane carrying an α-hydroxyl substituent, resulting in aldehyde formation.

$$\text{(G)}-\overset{\text{OH}}{\underset{}{\text{CH}}}-\text{R} \longrightarrow \text{(G)}-\text{H} + \overset{\text{O}}{\underset{}{\overset{\|}{\text{CH}}}}-\text{R} \quad (11)$$

4. <u>Cleavages at High Temperatures</u>. Enkvist and coworkers (103, 104) have carried out extensive studies on cleavage reactions at temperatures above 200° using NaOH-NaSH melts. The cleavage reactions appear to be predominantly homolytic in nature, resulting in a wide variety of products from lignins and model compounds. As an example, the identified compounds from phenylcoumaran 95 are summarized in Figure 9.11.

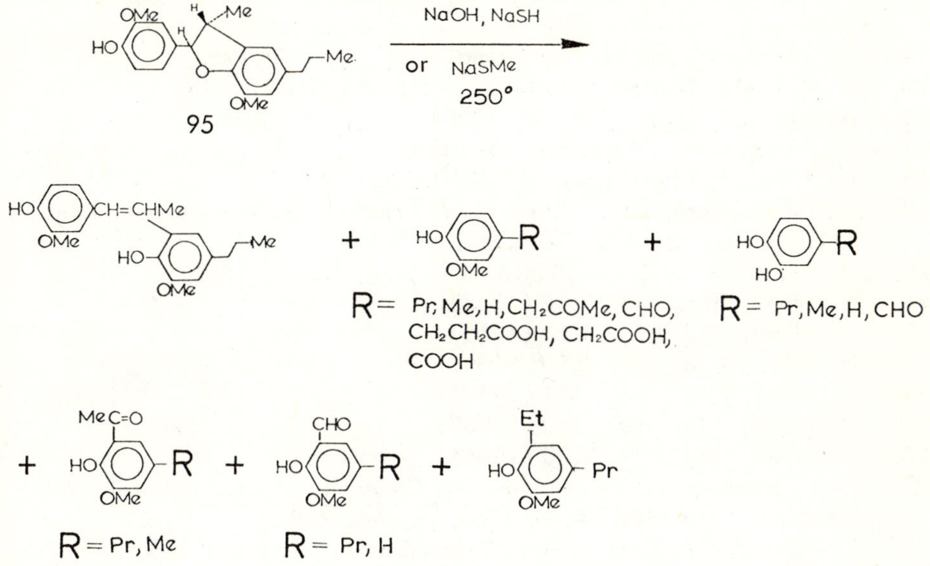

Figure 9.11. Reaction of phenylcoumaran 95 with NaOH, NaSH or NaSMe solutions at 250°.

REFERENCES

1. B. Holmberg, Svensk Papperstidn., 39, Special no. 113, 117 (1936).
2. L. J. Filar and S. Winstein, Tetrahedron Letters, 1960, No. 25, 9.
3. B. O. Lindgren, Acta Chem. Scand., 17, 2199 (1963).
4. A. W. Sohn and P. O. Lenel, Das Papier, 3, 109 (1949).
5. S. Aronovsky and R. Gortner, Ind. Eng. Chem., 22, 264 (1930).
6. A. F. Saizeva and N. I. Nikitin, J. Appl. Chem. U.S.S.R. (English translation), 24, 392, 427 (1951).
7. A. Sakakibara and N. Nakayama, J. Japan Wood Res. Soc., 8, 157 (1962).
8. K. Kratzl and H. Silbernagal, Monatsh. Chem., 83, 1022 (1952).
9. A. Sakakibara, H. Takeyama and N. Morohoshi, Holzforschung, 20, 45 (1966).
10. K. Kratzl, W. Kisser, J. Gratzl and H. Silbernagel, Monatsh. Chem., 90, 771 (1959).
11. O. Goldschmid, Tappi, 38, 728 (1955).

12. D. A. Stanek, Tappi, 41, 601 (1958).
13. A. Sakakibara and N. Nakayama, J. Japan Wood Res. Soc., 7, 13 (1961).
14. A. Sakakibara and N. Nakayama, ibid., 8, 153 (1962).
15. A. Klemola, Suomen Kemistilehti, 41B, 83 (1968).
16. A. Klemola, ibid., 41B, 99 (1968).
17. H. Nimz, Chem. Ber., 98, 533 (1965).
18. H. Nimz, Holzforschung, 20, 105 (1966).
19. K. Freudenberg, C.-L. Chen, J. M. Harkin, H. Nimz and H. Renner, Chem. Commun., 1965, 224.
20. H. Nimz and H. Gaber, Chem. Ber., 98, 538 (1965).
21. H. Nimz, ibid., 98, 3153 (1965).
22. H. Nimz, ibid., 98, 3160 (1965).
23. H. Nimz, ibid., 99, 469 (1966).
24. H. Nimz, ibid., 100, 181 (1967).
25. H. Nimz, ibid., 100, 2633 (1967).
26. H. Nimz, ibid., 99, 2638 (1966).
27. F. E. Brauns and D. A. Brauns, The Chemistry of Lignin, Suppl. Vol. Academic Press, New York, 1960, pp. 441-458.
28. E. Adler and J. Gierer, Acta Chem. Scand., 9, 84 (1955).
29. E. Adler, Paperi Puu, 11, 634 (1961).
30. J. Marton and E. Adler, Tappi, 46, 92 (1963).
31. B. Holmberg, Svensk Papperstidn., 33, 679 (1930).
32. L. Ivnäs and B. Lindberg, Acta Chem. Scand., 15, 1081 (1961).
33. M. Goliath and B. O. Lindgren, Svensk Papperstidn., 64, 469 (1961).
34. B. O. Lindgren, Acta Chem. Scand., 3, 1011 (1949).
35. E. Adler, J. M. Pepper and E. Eriksoo, Ind. Eng. Chem., 49, 1391 (1957).
36. E. Adler, S. Delin and K. Lundquist, Acta Chem. Scand., 13, 2149 (1959).
37. E. Adler, K. Lundquist and G. E. Miksche, Adv. Chem. Ser., 59, 22 (1966).
38. E. Adler and K. Lundquist, Acta Chem. Scand., 17, 13 (1963).
39. K. Lundquist and G. E. Miksche, Tetrahedron Letters, 1965, 2131.
40. E. Adler, Svensk Kem. Tidskr., 80, 279 (1968).
41. K. Lundquist, Acta Chem. Scand., 18, 1316 (1964).
42. S. Larsson and B. Lindberg, Acta Chem. Scand., 16, 1757 (1962).
43. E. West, A. S. MacInnes and H. Hibbert, J. Am. Chem. Soc., 65, 1187 (1943).
44. C. Schuerch, J. Am. Chem. Soc., 72, 3838 (1950).
45. W. S. MacGregor, T. H. Evans and H. Hibbert, J. Am. Chem. Soc., 66, 41 (1944).
46. K. Sarkanen and C. Schuerch, J. Am. Chem. Soc., 79, 4203 (1957).
47. M. Kluka, W. L. Hawkins and H. Hibbert, J. Am. Chem. Soc., 63,

2371 (1941).
48. K. Lundquist, Acta Chem. Scand., 16, 2303 (1962).
49. J. A. F. Gardner and H. MacLean, Can. J. Chem., 43, 2421 (1965).
50. H. Hibbert, Ann. Rev. Biochem., 11, 183 (1942).
51. E. Adler, B. O. Lindgren and U. Saeden, Svensk Papperstidn., 55, 245 (1952).
52. K. Lundquist, Svensk Kem. Tidskr., 75, 423 (1963).
53. E. Adler, Chim. Biochim. Lignine, Cellulose, Hemicelluloses, Grenoble, France, 1964, p. 73.
54. H. E. Fisher, M. Kulka and H. Hibbert, J. Am. Chem. Soc., 66, 598 (1964).
55. J. A. F. Gardner, Can. J. Chem., 32, 532 (1954).
56. J. A. F. Gardner, D. W. Henderson and H. MacLean, Can. J. Chem., 40, 1672 (1962).
57. E. Adler and S. Yllner, Svensk Papperstidn., 57, 78 (1954).
58. L. Mitchell and H. Hibbert, J. Am. Chem. Soc., 66, 602 (1944).
59. R. Lundgren, Paperi Puu, 43, 643 (1961).
60. K. Lundquist and K. Hedlund, Acta Chem. Scand., 21, 1750 (1967).
61. J. M. Pepper, P. E. T. Baylis and E. Adler, Can. J. Chem., 37, 1241 (1959).
62. K. Lundquist, G. E. Miksche, L. Ericsson and L. Berndtson, Tetrahedron Letters, 1967, 4587.
63. E. Hägglund and C. B. Björkman, Biochem. Z., 147, 74 (1924).
64. K. Freudenberg and M. Harder, Chem. Ber., 60, 581 (1927).
65. K. Freudenberg, M. Harder and L. Markert, ibid., 61, 1760 (1928).
66. F. E. Brauns, The Chemistry of Lignin, Academic Press, New York, 1952, p.440.
67. K. Freudenberg and H. G. Mueller, Ann. Chem., 584, 40 (1953).
68. E. Adler and S. Yllner, Svensk Papperstidn., 55, 238 (1952).
69. J. Gierer and I. Norén, Acta Chem. Scand., 16, 1713 (1962).
70. E. Adler, J. Marton and I. Falkehag, ibid., 18, 1311 (1964).
71. I. Falkehag, Paperi Puu, 43, 655 (1961).
72. S. I. Falkehag, J. Marton, and E. Adler, Adv. Chem. Ser., 59, 75 (1966).
73. J. Gierer, B. Lenz, I. Norén and S. Soderberg, Tappi, 47, 233 (1964).
74. J. Gierer and I. Kunze, Acta Chem. Scand., 15, 803 (1961).
75. J. Gierer and I. Norén, ibid., 16, 1976 (1962).
76. J. Gierer, B. Lenz, and N. H. Wallin, Tappi, 48, 402 (1965).
77. E. Adler, I. Falkehag, J. Marton and H. Halvarson, Acta Chem. Scand., 18, 1313 (1964).
78. J. Gierer, B. Lenz and N. H. Wallin, Acta Chem. Scand., 18, 1469 (1964).
79. T. Enkvist, T. Ashorn and K. Hästbacka, Paperi Puu, 44, 395 (1962).
80. P. Jerkeman and B. Lindberg, Acta Chem. Scand, 18, 1477 (1964).

81. G. Gellerstedt and J. Gierer, Acta Chem. Scand., *22*, 2510 (1968).
82. K. Freudenberg, C.-L. Chen and G. Cardinale, Chem. Ber., *95*, 2814 (1962).
83. K. Freudenberg and C.-L. Chen, ibid., *100*, 3683 (1967).
84. K. V. Sarkanen, G. Chirkin and B. F. Hrutfiord, Tappi, *46*, 375 (1963).
85. T. Enkvist, J. Turenen and T. Ashorn, Tappi, *45*, 128 (1962).
86. W. M. Hearon, W. S. MacGregor and D. W. Goheen, ibid., *45*, No. 1, 28A (1962).
87. I. Croon and B. Swan, Svensk Papperstidn., *67*, 177 (1964).
88. G. Gellerstedt and J. Gierer, Acta Chem. Scand., *22*, 2029 (1968).
89. H. Richtzenhain and C. von Hofe, Chem. Ber., *72*, 1890 (1939).
90. T. Ishihara and T. Kondo, Bull. Agr. Chem. Soc. Japan, *21*, 250 (1957).
91. J. Tanaka and T. Kondo, J. Japan Tech. Assoc. Pulp and Paper Ind., *11*, 111 (1957).
92. J. G. Miller and C. Schuerch, Tappi, *51*, 273 (1968).
93. V. Grafe, Monatsh.Chem., *25*, 987 (1904).
94. I. K. Buckland, G. H. Tomlinson and H. Hibbert, J. Amer. Chem. Soc., *59*, 597 (1937).
95. K. Kratzl, Monatsh.Chem., *80*, 314 (1949).
96. K. Kratzl, ibid., *78*, 173 (1948).
97. K. Kratzl and G. Hofbauer, ibid., *88*, 776 (1957).
98. E. Adler and S. Häggroth, Acta Chem. Scand., *3*, 86 (1949).
99. K. Kratzl, Paperi Puu, *43*, 1 (1961).
100. K. Kratzl and E. Risnyovszky, Chim. Biochim. Lignine, Cellulose, Hemicelluloses, Grenoble, France, 1964, p. 151.
101. K. Kratzl, E. Risnyovszky-Schäfer, P. Claus and E. Wittmann, Holzforschung, *20*, 21 (1966).
102. K. Kratzl and J. Spona, ibid., *20*, 27 (1966).
103. J. Turunen, Soc. Scient. Fenn. Comm. Phys. Mathem., *28*, No. 9 (1963).
104. T. Enkvist and J. Turunen, Chim. Biochim. Lignine, Cellulose, Hemicelluloses, Grenoble, France, 1964, p. 177.

10

HALOGENATION AND NITRATION*
C. W. Dence

1. Introduction. Nitration and, to an even greater extent, halogenation are among the most widely studied of all lignin reactions. The extent of the investigations in these areas is consistent with the fact that lignin is composed of phenolic nuclei and thus readily undergoes the halogenation and nitration reactions characteristic of such units.

An even greater incentive for studying the nitration and halogenation behavior of lignin is the fact that these reactions form the basis for important or potentially important technological and laboratory processes in the wood and wood-pulp industries. Chlorine is utilized commercially both as a pulping agent (Pomilio Process) and as a bleaching and/or delignifying agent in multi-stage pulp bleaching sequences (1). Chlorine also has been employed on a laboratory scale as a delignifying agent in the preparation of cellulose (2) and holocellulose (3) from wood and wood pulp. Although the nitration of lignin currently has found only limited industrial application, the practicability of using nitric acid as a pulping agent for wood continues to receive serious consideration (4, 5).

In the ensuing discussion, the main emphasis has been placed on the chlorination of lignin since it has been studied in far greater detail than other halogenation processes or nitration. As can be seen from the data in Table 10.1, halogenation and nitration treatments have been performed on a large variety of lignin substrates under widely diverse reaction conditions. As a consequence, there often is a lack of a common basis for comparison and interpretation of results. This applies not only to the analytical values in Table 10.1, but other properties of the halo- and nitrolignins which may reflect the heterogeneity of the imposed reaction conditions in an unpredictable and less obvious manner. In order, therefore, to establish a sound basis for evaluating

---

*The terms "halogenation" and "nitration" as used in the context of this chapter refer to the sum of all reactions occurring when a "halogenating" or "nitrating" agent is applied to lignin and not to an individual process such as substitution.

Table 10.1 Composition of Halo- and Nitrolignins Prepared Under Varying Conditions

Substrate	Halogenating or Nitrating Agent	Reaction Medium	Time of Treatment	Temperature of Treatment °C	C	H	Cl	Total N	NO_2	OCH_3	Reference
Spruce Wood	Chlorine	H_2O	Unspecified	Unspecified	42.5	4.12	25.0	-	-	4.7	6
Birch Wood	Chlorine	H_2O	10 hours	20	35.4	5.3	12.1	-	-	8.1	7
White Spruce Wood	Chlorine	CH_3OH	Undeterminable	Unspecified	43.1	3.4	35.4	-	-	16.95	8
Poplar Wood	Chlorine	CH_4	10 hours	20	40.4	2.6	28.0	-	-	9.65	9
Maple Wood	Chlorine	CH_3OH	2.5 hours	65	40.8	3.8	29.5	-	-	19.8	10
Spruce Hydrochloric Acid Lignin	Chlorine	H_2O	5 hours	Unspecified	-	-	23.8	-	-	-	11
Spruce Hydrochloric Acid Lignin	Chlorine	CH_3COOH	10 hours	Unspecified	-	-	31.1	-	-	-	11
Spruce Klason Lignin	Chlorine	CH_3OH	Unspecified	10	-	-	34.1	-	-	16.8	12
Maple Klason Lignin	Chlorine	CH_3OH	Unspecified	10	-	-	33.4	-	-	16.6	12
White Spruce Methanol Lignin	Chlorine	CCl_4	Unspecified	Unspecified	-	-	26.3	-	-	12.1	13
Softwood Lignosulfonic Acid	Chlorine	H_2O	9 hours	0	44.3	4.1	27.1	-	-	8.1	14
Red Pine Wood	Bromine	CCl_4	4 hours	2	-	-	-	-	-	-	15
Spruce Klason Lignin	Bromine	H_2O	Unspecified	Unspecified	-	-	-	-	-	6.4	16
Spruce Klason Lignin	Bromine	CCl_4	Unspecified	Unspecified	-	-	-	-	-	9.5	16
Spruce Ethanol Lignin	Bromine	CH_3COOH	1 hour	Unspecified	38.5	3.35	-	-	-	6.6	17
Spruce Wood	Nitric Acid	CH_3OH	2 hours	65	58.1	5.12	-	3.1	10.0	15.35	18
Spruce Wood	Nitric Acid	CH_3CH_2OH	2 hours	78.5	57.1	5.09	-	3.6	10.9	10.07	18
Poplar Wood	Nitric Acid	HNO_3	4 days	40	52.8	5.5	-	3.7	12.1	8.2	19
Beech Wood	Acetyl Nitrate	$CH_3COOH-(CH_3CO)_2O$	6 hours	0	-	-	-	5.9	8.2	5.0	20
Hydrochloric Acid Spruce Lignin	Nitric Acid	HNO_3	Unspecified	Unspecified	52.4	3.9	-	4.3	14.1*	9.6	21
Methylated Spruce Cuoxam Lignin	Nitrogen Dioxide	-	0.5 hour	Unspecified	58.4	5.7	-	4.0	13.1*	23.5	22
Pine Wood	Nitric Acid	$HNO_3-H_3PO_4$	4 hours	20	-	-	-	12.4	8.2	3.6	20

*Calculated assuming all nitrogen present in the form of nitro groups.

HALOGENATION AND NITRATION

and interpreting the halogenation behavior of lignin, a prior consideration of various halogenation and nitration systems and the halogenation and nitration behavior of lignin-like materials has been included.

II. Halogenation and Nitration Systems.

A. Halogenating Agents. Halogens most generally have been applied to lignin in their elemental (molecular) forms dissolved either in water (1, 8, 11, 23, 24, 25) or in a suitable organic solvent such as acetic acid, chloroform, carbon tetrachloride or methanol (8, 10, 24, 26, 27).

In aqueous solution, however, chlorine, bromine and iodine are hydrolyzed to varying degrees according to the general equation 1.

$$X_2 + 2H_2O \rightleftharpoons HOX + H_3O^+ + X^- \tag{1}$$

The hypohalous acid (HOX) thus formed is slightly dissociated in neutral solution, equation 2.

$$H_2O + HOX \rightleftharpoons H_3O^+ + OX^- \tag{2}$$

The concentrations of the various species in equilibria 1 and 2 are related as shown in expressions 3 and 4, respectively.

$$K_1 = \text{Hydrolysis Constant} = \frac{[HOX][H^+][X^-]}{[H_2O][X_2]} \tag{3}$$

$$K_2 = \text{Dissociation Constant} = \frac{[OX^-][H^+]}{[HOX]} \tag{4}$$

Typical values of K and K for equilibria involving chlorine, bromine and iodine are shown in Table 10.2.

Table 10.2. Typical Values for the Hydrolysis and Dissociation Constants of Chlorine, Bromine and Iodine (28)

	Hydrolysis Constant (K_1)	Dissociation Constant (K_2)
Cl	4.47×10^{-4} (25°C)	3.4×10^{-8} (25°C)
Br	5.8×10^{-9} (25°C)	2.1×10^{-9} (22°C)
I	4.6×10^{-13} (25°C)	4.5×10^{-13} (20°C)

The hydrolysis constants can be seen to decrease rapidly on proceeding from chlorine to iodine. As a consequence, the hydrolysis of bromine and iodine occurs only to such a limited extent that these halogens may be regarded as existing totally in their elemental form in neutral, aqueous solution. Under comparable conditions, chlorine is hydrolyzed to a much greater degree with the result that substantial amounts of hypochlorous acid are present in equilibrium with the elemental form. The interrelationships existing between pH, temperature, concentration and the composition of the chlorine-water system are shown in graphic form in Figure 10.1 (29). Since the various reactive species comprising the aqueous halogenation systems often respond differently toward a given substrate, pH adjustment serves as a convenient method of altering the halogenating system in such a way as to promote a desired reaction. As an example, this principle has been long recognized and exploited in commercial bleaching practice, where chlorine frequently is applied in neutral or alkaline aqueous solutions to utilize the oxidative (i.e., bleaching) capacities of the hypochlorous acid and hypochlorite ion, respectively.

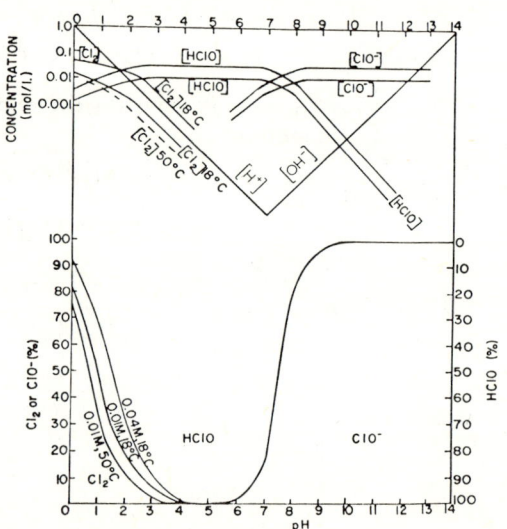

Figure 10.1. Concentration and existing forms of chlorine at varying pH and temperature (29).

In addition to the elemental halogens and the halogen-containing species resulting from their hydrolysis reactions, a few less common halogenating reagents have been reacted with lignin. These reagents function either by reacting directly with the substrate or by undergoing a prior *in situ* transformation to their elemental forms. Examples of halogenating systems which can be classified as being of either of these types are the mixtures of hydrochloric acid and potassium chlorate (2), hydrochloric and nitric acids (11) phosphorous pentachloride (30), acetyl bromide (31), thionyl chloride (32) sulfuryl chloride (33), dichlorourea (12), N-bromosuccinimide (34), and t-butyl hypochlorite (35).

The reactive species in halogenation systems may be classified as elemental halogen, X_2, other neutral halogenating agents XY, such as hypochlorous acid, and cationic agents carrying a positive charge. The reactivity of neutral halogenating species is related to the facility with which the heterolysis of the molecule to a polarized form X^+ --- Y^- occurs. It is well-known that Lewis acid catalysts, such as aluminum chloride frequently are employed to induce polarization of the halogen molecule and thereby increase its reactivity. However, more reactive aromatics, including the lignin polymers, are sufficiently reactive towards elemental chlorine and bromine without catalytic assistance. Recent evidence suggests that iodine, likewise, reacts in its molecular form (36) rather than as a positive entity as proposed earlier (37). In contrast, fluorine atoms instead of molecular fluorine are the species attacking aromatic compounds (38). A change from an ionic to a free-radical mechanism is of course possible with other halogens under proper conditions.

In protic solvents, elemental halogens may be transformed to other reactive forms, such as hypohalous acids, acetyl hypochlorite, acetyl hypobromite and chlorine and bromine monoxides. Acetyl hypobromite (39) and acetyl hypochlorite (40) are believed to arise when the corresponding hypohalous acids are dissolved in acetic acid containing trace amounts of water, equation 5.

$$HOAc + HOX \rightleftharpoons XOAc + H_2O \qquad (5)$$

Chlorine and bromine monoxide are present in equilibrium with the parent hypohalous acids, equation 6, and are reported to react with the ethylene groups of certain compounds forming halohydrins (41-44).

$$HOX + OX^- \rightleftharpoons X_2O + OH^- \qquad (6)$$

Such chlorinating agents as thionylchloride $SOCl_2$, sulfuryl chloride SO_2Cl_2 and phosphorous pentachloride function probably in a manner approximating that of the elemental halogens.

The halogenating agents are not equivalent in reactivity nor can they always be expected to function similarly with respect to a given substrate. Thus, depending on a number of factors, either oxidation or substitution may predominate. Preference for a particular mode of attack, however, is not necessarily an indication of a basic difference in the character of the attacking species, since, according to Levitt (45), the initial step for both oxidation and aromatic substitution processes involves the attack of an electrophilic species on the substrate.

The subject of cationoid halogenating agents has been reviewed recently by Berliner (46). Although direct evidence for these reactive species is limited (47), the reality of their existence has been supported by numerous kinetic studies (36, 48-51, 52-56). Thus far, however, no description of a deliberate attempt to react cationic halogenating agents with lignin has been reported.

The reactions of hypohalous acids in halogenide-free systems, acidified with sulfuric or perchloric acids, have been the subject of numerous kinetic studies (48, 50, 53, 55, 57). According to substrate reactivity, three types of kinetics have been established, shown in Equations 7 to 9.

$$\frac{-d[HOX]}{dt} = k(\text{substrate})[HOX][H^+] \tag{7}$$

$$\frac{-d[HOCl]}{dt} = k[HOCl] + k'[HOCl][H^+] \tag{8}$$

$$\frac{-d[HOCl]}{dt} = k[HOCl] + \{k' + k''(\text{substrate})\}[HOCl][H^+] \tag{9}$$

Equation 7 expresses the kinetics for moderately reactive aromatics and is consistent with a reaction involving hypohalous acidium ion $H_2\overset{+}{O}X$ and/or halogenium ion X^+, formed according to reaction 10.

$$HOX + H^+ \rightleftharpoons H_2\overset{+}{O}X \rightleftharpoons X^+ + H_2O \tag{10}$$

Kinetics in Equation 8 are obtained for reactive substances such as anisole. The expression is zero-order in substrate and suggests that chlorinium ion Cl^+ is formed through the heterolysis of either hypochlorous acid or its protonated form, reacting with the substrate as soon as it was formed. Still more reactive aromatics are capable to react directly with protonated hypochlorous acid and hence, the substrate concentration reappears in the corresponding rate, equation 9. Supporting evidence for these conclusions has been provided through determination of the isotope effect in deuterium oxide (57).

In an analogous manner, the addition of strong acids converts acyl hypohalites to their protonated forms, equation 11.

$$AcOX + H^+ \rightleftharpoons Ac\overset{+}{O}XH \rightleftharpoons X^+ + AcOH \tag{11}$$

Halogenonium ions solvated with acetic acid are more reactive than their

HALOGENATION AND NITRATION 379

unsolvated forms. In those instances where the reactive species are free halogenonium ions, heterolysis occurs prior to the attack on the substrate with the result that the energy requirement for the latter process is greatly reduced. When heterolysis occurs during or after the attack on the substrate, the reactivity of one halogenating species relative to another can be predicted on the basis of the extent to which the bond undergoing heterolysis is polarized in the resting state. As a general rule, the halogenating power of the electrophile, X, is directly proportional to the electronegativity of Y. On this basis, Ingold's arrangement (58) of the various reactive bromine species in the following order of decreasing activity is quite reasonable.

$$Br-OH_2^+ > Br-Br > Br-OAc > Br-OH$$

A similar order has been established for chlorine (53), and expanded by Sarkanen (59) to include other chlorinating agents. This series is analogous to a proton donor sequence:

Chlorinating Agents		Acids
Cl^+	↑	H^+
$Cl-Cl$		$H-Cl$
$AcO-Cl$		$H-OAc$
$ClO-Cl$	increasing chlorinating power	$H-OCl$
$HO-Cl$		$H-OH$
CH_3O-Cl		$H-OCH_3$
$RNH-Cl$		$H-NHR$
$O-Cl^-$		$H-O^-$
CH_3-Cl		$H-CH_3$

(with increasing acidity on the acid side)

The reactivity of any given halogenating agent, however, is also dependent on such factors as polarizability and strength of the bond undergoing fission with the result that exceptions to the general order are sometimes observed (60).

In the foregoing discussion, the principle that the species involved in the halogenation of lignin are neutral or ionic has been adopted as being consistent with the known facts concerning the halogenation behavior of aromatic compounds. At the same time, direct participation of halogen radicals in such reactions represents a reaction mode which cannot be overlooked entirely since several of the reagents applied to lignin are known to be sources of halogen atoms (61). Contributions of the latter species to the overall reaction in all likelihood are minor, since conditions favoring the formation of radicals, i.e., high temperatures, aprotic media, external radiation and/or other radical initiators, are seldom employed in the halogenation of lignin.

B. Nitrating Agents. Systems comprised of nitric acid diluted with an appropriate solvent or solvent mixture have been used almost exclusively for the nitration of lignin (21, 62-65). Depending on such experimental conditions as reaction temperature and composition of nitrating mixture, however, nitric acid may undergo partial reduction by the substrate yielding a series of nitrogen-containing intermediates which are nitrating or nitrosating reagents in their own rights. In some instances, urea (66) or urea nitrate (67) have been added to decompose nitrous acid, one of the aforementioned reduction products, thereby reducing the number of possible side reactions.

Among the solvents which, alone or as components of solvent mixtures, serve as diluents for the nitration process are water, sulfuric, phosphoric and acetic acids, acetic anhydride, ethanol, methanol, ethyl ether and carbon tetrachloride. Other reagents which have been used in the nitration of lignins are nitrogen dioxide (22, 68), nitrogen pentoxide (69, 70), and acetyl nitrate (71, 72). The last-named is presumed to be present in a mixture of concentrated nitric acid and acetic anhydride (73, 74). The action of the various nitrating agents on lignin does not necessarily produce equivalent results even under comparable or identical reaction conditions. This statement is perhaps best illustrated by the substitution reaction where the <u>ortho-para</u> ratio has been found to depend on the nature of the nitrating species (75).

In the overall reaction of nitrating species, which may include oxidation as well as substitution processes, a nitrogen-oxygen bond of the attacking reagent is heterolyzed forming a nitronium ion, NO_2^+, equation 12 (83).

$$HONO_2 + H^+ \rightleftharpoons H_2\overset{+}{O}NO_2 \rightleftharpoons H_2O + NO_2^+ \qquad (12)$$

The existence of nitronium ions is well supported by more direct evidence provided by crystallographic analysis of such solids as nitronium perchlorate NO_2ClO_4 (76) and dinitrogen pentoxide N_2O_5, (77) cryoscopic measurements (78, 79) and by spectroscopic measurements (80, 81).

Nitric acid is completely converted to nitronium ion in 90-100% sulfuric acid (82), while in pure nitric acid, nitronium ions are formed by self-dehydration to the extent of approximately 4% (83). In solvents such as nitromethane, acetic acid and chloroform only trace amounts of free nitronium ions are present. For the nitration of reactive substrates (e.g., lignin) the reaction rates are independent of the concentration of substrate, indicating that the rate determining step was the formation of the attacking electrophile. Since the self-protonization of nitric acid should be extremely rapid, the foregoing result indicated that neither molecular nitric acid nor its conjugate acid ($H_2O^+NO_2$) is particularly effective as a nitrating species.

During the nitration of reactive compounds such as phenol, anisole and aromatic amines, nitric acid is reduced to nitrous acid and oxides of nitrogen. The nitrous acid produced thereby has been found to catalyze the nitration (substitution reactions) of these reactive substrates (84, 85). The kinetic form of the catalyzed process supports the theory that the reactive substrate first undergoes nitrosation followed in turn by nitric acid oxidation of the nitroso to the nitro derivative (86, 87). The nitrosation is due both to the action of dinitrogen tetroxide N_2O_4 and the nitrosonium ion, NO^+ (88). The latter

HALOGENATION AND NITRATION

ion has been identified through Raman spectra (89). These compounds are formed by the action of nitric acid on nitrous acid according to the equilibria 13.

$$HNO_3 + HNO_2 \rightleftharpoons N_2O_4 + H_2O \rightleftharpoons NO^+ + NO_3^- + H_2O \quad (13)$$

The intervention of nitrous acid itself and nitrous acid acidium ion (H_2O^+NO) has been ruled out on the grounds that their presence is inconsistent with the form of the rate expression generally found to apply (88).

As an electrophilic agent, nitrosonium ion is far weaker than nitronium ion and, consequently, its contribution in the nitration process becomes evident only in those instances when reactive substrates are available. On the other hand, at low acidities the concentration of nitronium ion may be very low and nitrosation by nitrosonium ion can play an important, if not decisive role.

Other less commonly employed nitrating agents are dinitrogen pentoxide and acetyl nitrate. In aprotic solvents, such as carbon tetrachloride, covalent dinitrogen pentoxide appears to be the reactive form (90). Acetyl nitrate is formed in the reaction of nitric acid with excess acetic anhydride. Boardwell and Barbisch (74) obtained kinetic evidence suggesting that the protonated form of acetyl nitrate (Ac^+OHNO_2) is a reactive nitrating species. The protonated form was regarded as being considerably more effective as a nitrating agent than the unprotonated form.

The reactivity sequence for various reactive nitrating species has been observed (91) to be:

$$NO_2^+ > NO_2^+OH_2 > NO_2\text{-}NO_3 > NO_2\text{-}OAc > NO_2\text{-}OH.$$

III. Halogenation and Nitration Reactions of Lignin Model Compounds. In the foregoing discussion, the terms "halogenation" and "nitration" were broadly defined to include several distinctive reactions. Insofar as these individual reactions are considered applicable to lignin, they are outlined in Table 10.3, and discussed in detail below.

A. Reactions of the Aromatic Moiety.

1. Electrophilic Substitution. Nearly all aromatic nuclei in lignin have two or three positions available for substitution depending on whether or not the 5-position of the ring is initially occupied (see Figure 10.2). Theoretically, trisubstitution may take place on an uncondensed guaiacyl unit ($R_3 = H$), but only disubstitution may be expected in the case of syringyl ($R_3 = OCH_3$) and condensed (R_3 = carbon atom of adjacent monomer unit) nuclei.

The nitration of reactive substances such as phenols, phenol ethers and (presumably) lignin is complicated by the formation of nitrous acid which autocatalyzes the overall substitution process. This auto-catalysis is due to a

Table 10.3. Summary of Electrophilic Reactions Occurring During the Halogenation and Nitration of Lignin

Aromatic substitution:

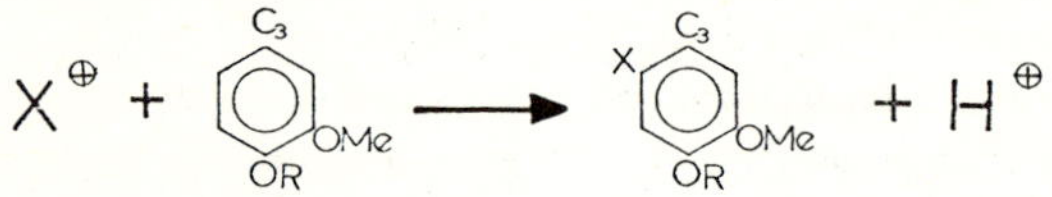

$X^\oplus$ + [aromatic ring with C₃, OMe, OR] ⟶ [aromatic ring with X, C₃, OMe, OR] + $H^\oplus$

Electrophilic displacement:

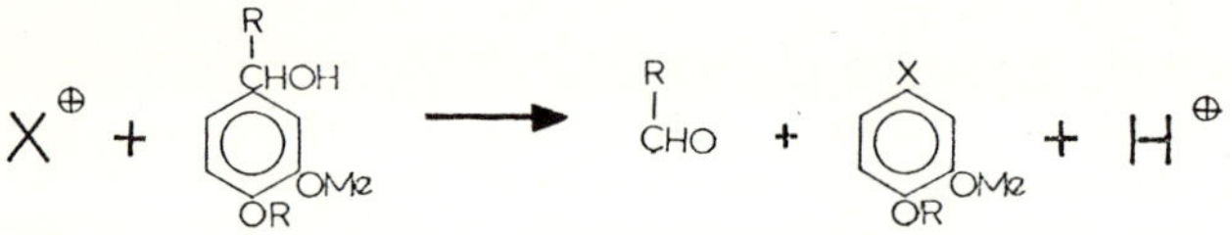

$X^\oplus$ + [aromatic ring with CHOH-R, OMe, OR] ⟶ [R-CHO] + [aromatic ring with X, OMe, OR] + $H^\oplus$

Aliphatic substitution:

$X^\oplus$ + $HC-CO-$ ⟶ $XC-CO-$ + $H^\oplus$

Addition:

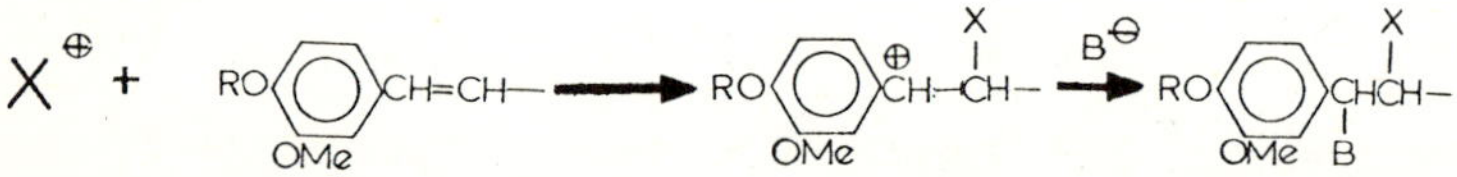

$X^\oplus$ + RO-[aromatic ring, OMe]-CH=CH- ⟶ RO-[aromatic ring, OMe]-CH⁺-CH(X)- $\xrightarrow{B^\ominus}$ RO-[aromatic ring, OMe]-CH(B)-CH(X)-

Oxidation:

$X^\oplus$ + $HOCH-$ $\xrightarrow{-H^\oplus}$ $XOCH-$ $\xrightarrow{-HX}$ $OC-$

$X^\oplus$ + $O=CH-$ ⟶ $XOCH-$ $\xrightarrow{OH^\ominus}$ $XOCH-(OH)$ $\xrightarrow{-HX}$ $OC-(OH)$

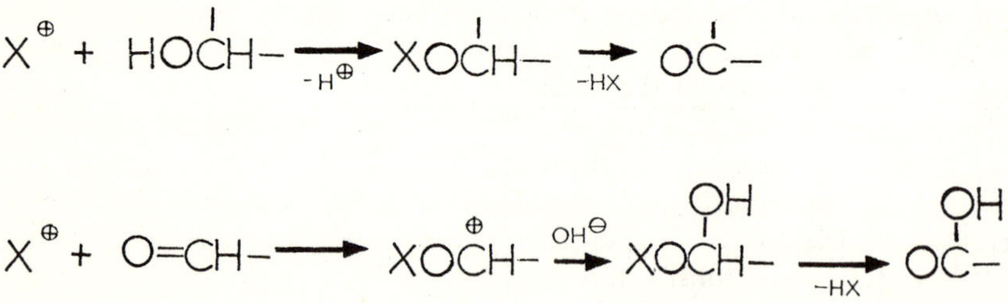

$X^\oplus$ denotes electrophile
R denotes alkyl substituent

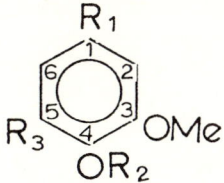

Figure 10.2. Generalized representation of lignin monomer unit.

rapid nitrosation reaction to a derivative which subsequently is oxidized by nitric acid to a nitro compound regenerating nitrous acid in the process (84, 86, 87, 92, 93). In addition, direct substitution of a nitro group via a nitronium ion may be superimposed on the nitrosation reaction. As pointed out earlier, nitrosation is more favored in dilute aqueous media. Since, from a qualitative viewpoint, nitrosation and nitration usually lead to similar results, no attempt has been made to distinguish between the two processes except in those instances where distinct and significant differences were detectable.

The generally accepted mechanism for electrophilic substitution in aromatic systems is depicted in equation 14.

$$\underset{}{\bigcirc} + \overset{\oplus}{X}(\text{or } \overset{\delta+}{X}\!-\!\overset{\delta-}{Y}) \longrightarrow \underset{(+)}{\overset{X\ \ H}{\bigcirc}} \xrightarrow{-H^{\oplus}} \underset{}{\overset{X}{\bigcirc}} \qquad (14)$$

It consists of an attack by an electrophilic species represented either as a formal cation (X^+) or as a neutral but polarized molecule on a carbon atom to form a cationic intermediate. Subsequent loss of a proton from the intermediate regenerates the aromatic ring and yields a halo- or nitro-aromatic compound. In kinetic studies involving the nitrosation of phenols (94, 95), a significant hydrogen isotope effect has been observed indicating proton loss as the rate-controlling process in the overall reaction.

Orientation of the entering substituents is determined primarily through the combined electronic (resonance or inductive) and steric influences of substituent groups on the ring. With few exceptions the substituent groups attached to the aromatic nuclei of protolignin are, in varying degree, <u>ortho-para</u> directing. Carbonyl groups situated alpha to the ring direct <u>meta</u> and constitute the single most important exception to the above generalization. Of the various substituents attached to the aromatic nucleus of lignin, phenolic hydroxyl and alkoxyl groups exert the greatest directive influence through the contributions of canonical forms, 1-4.

$$\underset{1}{\underset{OR}{\bigcirc}} \longleftrightarrow \underset{2}{\underset{\oplus OR}{\bigcirc^{\ominus}}} \longleftrightarrow \underset{3}{\underset{\oplus OR}{\bigcirc_{\ominus}}} \longleftrightarrow \underset{4}{\underset{\oplus OR}{\bigcirc_{\ominus}}}$$

An examination of similar contributing resonance forms for guaiacyl or syringyl nuclei leads to the conclusion that every unsubstituted position in these units is activated in varying degrees toward electrophilic attack and that there is considerable overlapping in the ortho-para directive influences of substituent groups.

The directive influence of phenolic hydroxyl and alkoxyl groups is marked on occasion by substitution via displacement of para-situated substituents attached to the ring by carbon-carbon bonds. Such reactions are more properly termed "electrophilic displacements" and are discussed as separate phenomena in the following section.

Derivatives of veratrole (Figure 10.2: $R_1 = R_3 = H$; $R_2 = CH_3$) frequently serve as models for the etherified guaiacyl units of softwood lignin. With such models, the positions para to the methoxyl groups are preferentially substituted by halogen and nitro groups (96-102). In their assessment of the influence of methoxyl groups on aromatic substitution, de la Mare and Vernon (103) found that, in the bromination of veratrole, the partial rate factor for para substitution was approximately 80 times that for ortho substitution. For a more realistic case where a substituent group (side-chain in lignin monomer unit) occupies the 1-position, the 6-position remains the single most reactive site in halogenation or nitration (104-111, 112, 113, 114, 115, 116) even when the substituent is a strongly meta-directing group such as -CHO (107, 116, 117-119).

The situation is somewhat altered in the case of guaiacol derivatives, since the phenolic hydroxyl group is more strongly ortho-para directing than the alkoxyl group. Consequently, it is not surprising to observe a greater preference for substitution in the 5-position when guaiacyl models are halogenated or nitrated (105, 108, 112, 113, 116, 120-124). As a pertinent example, Van Buren and Dence (112) determined that the sole primary electrophilic substitution product resulting from chlorination of 3,4-dimethoxyphenyl ethyl carbinol was the 6-chloro derivative whereas nearly equal amounts of the 5- and 6-chloro isomers were formed when the guaiacol analogue was chlorinated. When the strongly meta-orienting carbonyl adjoins the aromatic ring at the 1-position as in the case of vanillin (Figure 10.2: $R_1 = CHO$, $R_2 = R_3 = H$), substitution at the 5-position is enhanced even more (33, 105, 108, 122, 124-131).

Although halogen substitution occurs most readily when the halogen is in its elemental form, substitution of phenolic lignin model compounds also may occur under alkaline conditions where the prevailing species is the hypochlorite ion (132, 133).

The increased preference for substitution <u>ortho</u> to a phenolic hydroxyl group may be further heightened through the use of <u>t</u>-butyl hypochlorite (134, 135). Nevertheless, it appears probable that preferred substitution occurs at the 1- and 6-positions in lignins when nitric acid or the elemental form of the halogen (120, 136, 137, 138) is used. A comparison of the Hammett σ values (139) for <u>p</u>-hydroxyl and <u>p</u>-methoxyl groups (-0.37 and -0.268, respectively) indicates that the former substituent is more activating than the latter insofar as electrophilic substitution reactions are concerned. Thus, in the halogenation of guaiacol a preference for substitution at the position para to the phenolic hydroxyl group generally is observed (102, 120, 140, 141).

An alkyl group substituted in the 5-position could reasonably be expected to exert a directive influence comparable to that of the side-chain lignin monomer unit joined (condensed) at this position. Accordingly, such substituents would be expected to have a positive influence in directing entering electrophiles to the 2- and 6-positions. Similarly, substituents of the type generally present in the 1-position of guaiacyl or syringyl nuclei of lignin also would be expected to direct <u>ortho</u> and <u>para</u>. However, the directive influence of such substituents in all likelihood is weak in comparison with that of hydroxyl and alkoxyl groups.

The strong <u>para</u> directing influence of the additional methoxyl group is manifested in the halogenation behavior of syringyl derivatives. In this situation, the 2- and 6- positions are equally activated and substitution occurs at either site with equal facility (142-144). The 1- and 2,6-positions, however, are non-equivalent with respect to the other substituents, and the tendency is for the initial substitution to occur at the 2- and 6-positions (143, 145) as well as at the 1-position (102).

On the basis of the foregoing analysis, alkoxyl groups have been tentatively arranged in decreasing order with respect to their ability to activate the designated position toward electrophilic substitution.

$$\underline{p}\text{-OH} > \underline{p}\text{-OR} > \underline{o}\text{-OH} > \underline{o}\text{-OR} \quad (R = \text{alkyl})$$

The influence of methoxyl groups in determining the degree of substitution is considerable. Thus, using mild reaction conditions and a moderate excess of chlorine, the 6-position of a guaiacyl nucleus alone is readily substituted, whereas both the 2- and 6-positions of a syringyl unit are substituted under the same conditions (146). Complete substitution of available ring sites can be achieved through application of theoretical or greater amounts of halogen provided the reaction conditions are such that side reactions are minimized and the integrity of the aromatic nucleus is maintained (147-149). Although disubstitution of the aromatic nucleus occurs when lignin model compounds are nitrated (86, 108, 113, 114, 115, 116, 138), attainment of still higher levels of substitution is unlikely due to the deactivating influence of the nitro groups initially introduced. The effect of steric interaction between the entering group and those already present during the aromatic substitution reactions of lignin has not been studied systematically. Steric effects observed in the substitution reactions of relatively simple aromatic systems (150-153) suggest that the size of substituent groups attached to the aromatic nucleus of lignin is effective, if not decisive, in determining the side and extent of electrophilic attack taking place during halogenation and nitration.

The ring site occupied by the entering substituent also depends on the size of the attacking species. Thus the proportion of ortho substitution expressed, for example, by the partial rate factor is found to decrease as the size of the attacking species increases from that of the free halogenonium or nitronium ion to the dimensions of bulkier forms of the same reagent consisting of neutral or charged molecules (154). A decreased tendency for ortho substitution paralleling the increase in atomic dimensions also is observed, for example, when comparing the same reactive forms of different reagents. Thus, the ortho-para substitution ratios decrease in the order $Cl^+ > Br^+ > NO_2^+$ (154). Steric effects involving the entering group may be further complicated by the formation of elemental halogen-solvent complexes (155, 156).

2. Electrophilic Displacement of Ring Substituents.

In specific instances, substitution can occur by replacement of some group other than hydrogen. In such an event, the process is designated as an "electrophilic displacement" in order to distinguish it from the more general type of substitution.

A number of typical electrophilic aromatic substitution reactions involving reactive halogenation and nitration species are recorded in Table 10.4. As can be seen from this table, the reported displacements are restricted to chlorination, bromination and nitration. Electrophilic aromatic displacements are generally observed to occur when the replaceable group is situated ortho or para to a strongly activating (electron-donating) group such as -OH, -OCH$_3$ or -NH$_2$. Electrophilic displacement also is dependent on the nature of the displaced substituent and is facilitated when the latter possesses a high degree of resonance stabilization (169). On the basis of these observations and by analogy with aromatic substitution reactions, the mechanism for electrophilic aromatic displacement may be depicted as in equation 15.

$$\underset{R'}{\underset{|}{\overset{R}{\underset{|}{\bigcirc}}}} \xrightarrow{X^\oplus} \underset{\oplus R'}{\underset{|}{\overset{X\diagdown \;\; R}{\underset{|}{\bigcirc}}}} \longrightarrow \underset{R'}{\underset{|}{\overset{X}{\underset{|}{\bigcirc}}}} + R^\oplus \qquad (15)$$

(X^+ = reactive halogenation or nitration species)

Table 10.4. Selected Electrophilic Displacement Reactions with Chlorine, Bromine and Nitric and Nitrous Acids

R' = Me,	R'' = H	Yields: 30% (in H$_2$O) / 78% (in AcOH)	not det.	(157)
R' = Me,	R'' = C$_2$H$_5$	—	10-12%	(112)
R' = Me,	R'' = HOCH$_2$CH$_2$	Ca. 100%	not det.	(158)
R' = i-Pr,	R'' = H	15%	not det.	(146)

R = Me	Yields: 78%	not det.	(159)
R = i-Pr	14%	not det.	(112)

R = H	Yields: +	not det.	(157)
R = C$_2$H$_5$	−	4-6%	(112)

R = H	Yields: 66%	+	(158)
R = Me	10-15%	not det.	(158)

77%	+	(158)

Table 10.4. continued

[Reaction: 4-hydroxybenzaldehyde + X₂ → 3,5-dihalo-4-hydroxybenzene + CO]

	Yields:	
X = Cl	+	not det. (157)
X = Br	> 94%	+ (161)

[Reaction: 4-hydroxybenzoic acid (with R substituents) + X₂ → 3,5-dihalo-4-hydroxybenzene + CO₂]

		Yields:	
R = H	X = Cl	+	not det. (157)
R = Br	X = Br	55%	+ (162)

[Reaction: 4-hydroxybenzyl alcohol + Br₂ → 2,4,6-tribromophenol + CH₂O]

100%	not det. (160)

[Reaction: 3,4-dimethoxybenzoic acid + Br₂ → dibromo-dimethoxybenzene + CO_2 (163)]

[Reaction: guaiacyl-type compound → 2,6-dinitro product]

Reactant:	R:	Yield, %:	
HNO_3	$-CH_2OH$	30–36	(122, 164, 165)
HNO_2	$-CH_2OCH_3$	5	(108)
HNO_3	$-CHO$	12–14	(164, 165)
HNO_3	$-CO_2H$	58	(164, 165)
HNO_3	$-COCH_3$	20	(108)
HNO_3	$-CH(OH)CH_2CH_3$	+	(109)
HNO_2	$-CH(OH)CH_2CH_3$	14	(116)
HNO_3	$-CH(OMe)CH_2CH_3$	17	(113)

Table 10.4 continued

Reactant:	R:	Yield, %:	
HNO₂	-CH₂OH	9	(116)
HNO₃	-CHO	85	(167)
HNO₃	-CH(OH)CH₂CH₃	+*	(111)
HNO₂	-CH(OH)CH₂CH₃	11	(116)
HNO₃	-CH(OMe)CH₂CH₃	17	(113)

* Yield of 1,2-dimethoxy-4,5-dinitrobenzene 35% (113)

Reactant:	R:	Yield, %:	
HNO₂	H	60	(108)
HNO₃	Br	36	(168)

	Yield, %:	Yield, %:	
	22	4	(115)

R:	Yield, %:	Yield, %:	
-CH₂OH	11	11	(116)
-CHO	38	21	(116)
-CO₂H	52	22	(116)

In this generalized reaction sequence, R' represents a strongly activating group such as hydroxyl or alkoxyl which has been arbitrarily positioned para to a displaceable group, R. In nitrosation (114) and nitration (108, 113) reaction, the phenolic hydroxyl group appears to be more effective in promoting displacement than is the alkoxyl group. The opposite effect has been observed in chlorination (112). An explanation for this difference may be related to competing oxidation reactions (see section 4) which by converting phenolic to non-benzenoid units, cause a corresponding reduction of the displacement reaction. During halogenation in aqueous media, phenolic units are more susceptible to oxidation than etherified phenolic units and hence would be expected to undergo a correspondingly smaller degree of side-chain displacement.

The effect of structure on the displacement of substituents (R) during nitration has been studied by Gustafsson and Andersen (164) and Andersen (108). Some of their results are recorded in Table 10.5. The evidence available suggests that roughly the same stability trends prevail in halogenation systems as well. Etherification appears to stabilize both primary and secondary benzyl alcohol groups, but the primary to a greater extent (108). The extent of electrophilic displacement also appears to be directly or indirectly affected by such reaction conditions as temperature (161) and solvent (157).

Table 10.5 Effect of Side-Chain Structure on the Electrophilic Displacement During Nitration

Compound	Side-Chain Displacement* (% of Theoretical)
ⓖ-CO-Me	100
ⓖ-p(β-β)p-ⓖ	60
ⓖ-COOH	58
ⓖ-CH$_2$OH	30, 36
ⓖ-CHO	12, 14
ⓖ-CH$_2$OMe	5
ⓖ-CH = CHMe	5

*Based on the amount of the 4,6-dinitroguaiacol resulting from the displacement (108, 164).

As noted previously, these differences in displacement behavior can be traced to the varying degrees of resonance-induced stability exhibited by the departing cations, R^+. Thus, displacement of a primary carbinol group ($-CH_2OH$) occurs readily since the corresponding cation ($^+CH_2OH$) can be stabilized by decomposition to formaldehyde and a proton. Primary alkyl (170) and alkyl sulfonate (157) groups, on the other hand, remain attached to the aromatic ring since their corresponding carbonium ions are unstable and cannot be stabilized through the simple proton transfer mechanism cited above.

In protolignin, the side-chains consist mainly of benzyl alcohol and benzyl ether derivatives (Chapter 5) and can therefore be expected to undergo displacement during halogenation or nitration. Chemical treatment of lignin frequently produces structural modifications on the alpha carbon atom of the side-chain which, depending on the nature of change, may either facilitate or retard side-chain displacement in accordance with the principles outlined above.

Structures undergoing electrophilic displacement reactions are degraded, in effect, since primary (carbon-carbon) bonds are ruptured in the process. It therefore seems probable that the breakdown of lignin effected by various commercial and laboratory halogenation and nitration processes can, in part, be attributed to reactions of this type.

3. Cleavage of Alkyl Aryl Ether Linkages. Treatment with halogens or nitric acid has long been recognized to result in a decrease in the methoxyl content of lignin. On the basis of recent studies (146, 157, 158, 171), it appears that the reaction is not limited to methyl aryl ethers but applies as well to alkyl aryl ethers in general. Of the various halogens, only chlorine has been studied extensively as a reactant in the dealkylation process.

The splitting of alkyl aryl ether linkages is essentially a solvolytic reaction which results in the liberation of phenolic hydroxyl groups (14, 86, 106, 120, 123). When halogenation and nitration treatments of lignin are carried out in aqueous or partially aqueous solutions, the aliphatic product usually obtained is the alcohol, ROH, corresponding to the alkyl moiety, R. Depending on the composition of the reaction medium, however, the initially formed alcohol may undergo further reaction. Thus Sobolev (123), for example, isolated methyl nitrite rather than methanol from the reaction of 4-methylguaiacol with aqueous nitric acid.

When examined more closely, the cleavage process is found to possess certain characteristics which serve to delineate its reaction mechanism. First, the reaction appears to occur more readily in aqueous or partially aqueous solution than under anhydrous conditions. When chlorinations are carried out in glacial acetic acid or some other organic solvent, the alkoxyl groups remain largely intact (137, 147, 158). On the other hand, Shorygina et al. (172) have reported figures indicating extensive demethylation resulting from treatment of hydrochloric acid lignin with nitric acid in ether and carbon tetrachloride.

Secondly, because the products resulting initially from the dealkylation are alcohols or their derivatives (most often esters) rather than aldehydes, ketones, or carboxylic acids, an oxidation mechanism can be eliminated from consideration. Moreover, the possibility of acid-catalyzed hydrolysis of the phenolic ether bond as suggested earlier (173) is likewise untenable since, under conditions of comparable acidity, alkyl aryl ethers are stable in the absence of nitric and nitrous acids (123) and chlorine (157). In the absence of oxidative and substitution effects, it seems probable that the hydrolysis of alkyl aryl ether linkages is catalyzed by a cationoid species derived from the halogenating or nitrating agent.

Dealkylation in aqueous chlorination systems characteristically displays a sensitivity toward the pH of the reaction mixture. Cleavage of the ether linkage is more rapid and extensive in the low (1-3) pH range, where elemental chlorine predominates, than in higher pH regions where hypochlorous acid and hypochlorite ions prevail. This suggests that a cationic species derived from elemental chlorine is involved in the dealkylation process, since, in the latter form, chlorine is a more effective electrophile than either hypochlorous acid or hypochlorite ion (174, 175). Analogously, Bunton et al. (87) have ascribed the demethylation of p-chloroanisole during nitration to nitronium and nitrosonium ions. Additional evidence for the participation of nitrosonium ions in the dealkylation process has been provided by Sobolev (123) and by Schramm and Westheimer (86) who found the demethylation of 4-methylguaiacol and anisole, respectively, to be catalyzed by nitrous acid and inhibited by urea.

In accordance with the foregoing, the dealkylation mechanism for an aqueous halogenation system can be represented as shown in Figure 10.3 (123, 157). In this mechanism a concerted attack of the electrophile on the ether oxygen and of the solvent on the alkyl group produces a hypothetical intermediate. This intermediate subsequently is transformed to an aryl hypohalite which then undergoes rapid acidolysis regenerating the original electrophile. A similar mechanism also could apply for nitrosation and nitration.

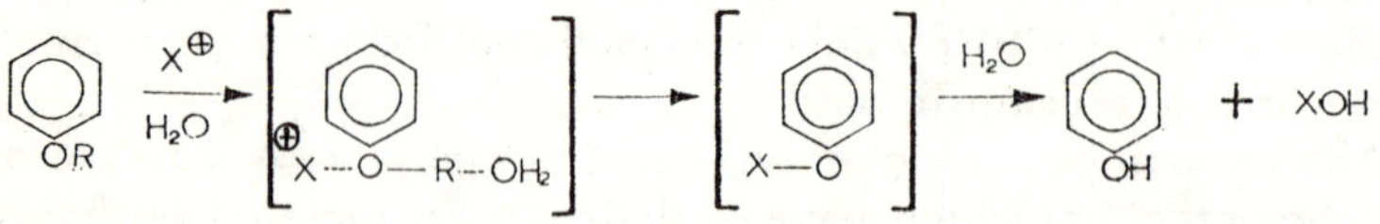

Figure 10.3. Proposed mechanism for the dealkylation of alkyl aryl ethers (123, 157).

The possibility of an alternative mechanism for chlorination was pointed out by Sarkanen and Dence (157), consisting of an attack of a chlorine molecule on the position para to the methoxyl group forming intermediates of type

HALOGENATION AND NITRATION

5 and 6 in aqueous solution.

R = H, Cl, Alkyl, etc.

Generation of methanol and regeneration of the chlorine molecule would then complete the process.

Indeed, the mechanism in Figure 10.3 has been considered to be the most probable one in the hydrolysis of methoxyl groups in chlorination (157, 171) as well as in nitration (87, 123), although its validity has not been demonstrated in unambiguous terms. Likewise, Bunton et al. (87) considered an analogous mechanism to be a reasonable alternative in demethylation by nitration and nitrosation. Bolker, Jung and Kee (102) have suggested a related, but more complex sequence of events to account for the conversion of 1,2,3-trimethoxybenzene to 2,6-dimethoxy-p-benzoquinone by nitrous acid.

The dealkylation behavior of representative lignin model compounds during chlorination in aqueous acidic media has been studied in considerable detail by Sarkanen and Strauss (159, 171) and by Dence and coworkers (146). Their investigations showed that, in the initial stage of chlorination, dealkylation took place very rapidly at room temperature, but subsequently leveled off or ceased completely even in the presence of a substantial excess of active chlorine. This behavior was typified by several of the model compounds shown in Figure 10.4. The extent to which dealkylation occurred was found to vary depending on the structure of the substrate, the amount of the substrate, the amount of applied chlorine and the composition of the solvent but, under optimum conditions, the yield of cleaved alkoxyl group (determined as the corresponding alcohol) often exceeded 60% of the theoretical value and frequently approached 100%.

In contrast to the extensive loss of alkoxyl groups resulting from halogenation in aqueous media, Kee (116) found no evidence for demethylation when veratryl-type model compounds were reacted with nitrous acid (pH 0.5, 100°C or pH 2.0, 70°C). Based on their response to hot, aqueous nitrous acid, therefore, ether linkages in veratryl compounds exhibited less susceptibility to rupture than those of etherified syringyl type nuclei as typified by 1,2,3-trimethoxybenzene. This observation was cited by Kee as a possible explanation for the more extensive degradation sustained by birch, as compared with spruce, dioxane lignin when subjected to the action of aqueous nitrous acid.

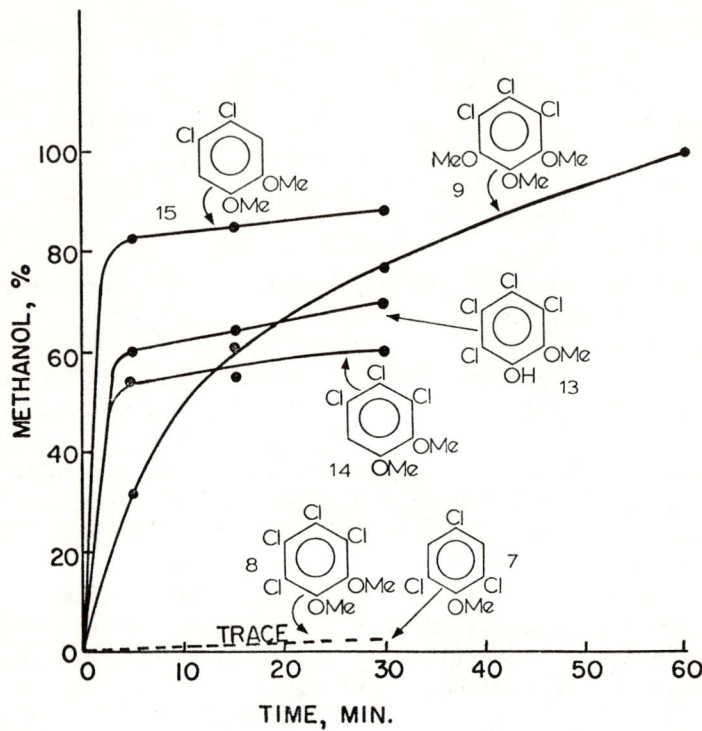

Figure 10.4. Formation of methanol in the chlorination of various model compounds in monochloroacetic acid–water medium (159).

Failure to achieve complete hydrolysis of alkyl aryl ether linkages has been ascribed in part to steric factors. In this connection, Dence et al. (146) showed that the extent of hydrolysis of various alkyl aryl ethers depended on the nature of the alkyl moiety and decreased as the size of the latter increased. Hydrolytic cleavage of the ether linkage was also inhibited when the alkoxyl group was flanked in the ortho positions by two chlorine atoms or by a chlorine atom and another alkoxyl group as shown for example by the behavior of 2,4,6-trichloroanisole 7 and tetrachloroveratrole 8, respectively, in Figure 10.4. By way of contrast, the etherified syringyl type model 9, Figure 10.4, was demethylated by aqueous chlorine, suggesting that the initial attack may have occurred at the middle methoxyl group.

Ortho substituents also were observed by Bunton and coworkers (87) to

HALOGENATION AND NITRATION

produce a similar retarding effect on the demethylation of anisole derivatives by nitric acid in glacial acetic acid medium. Thus p-chloroanisole 10 was found to yield substantial amounts of 2,6-dinitro-4-chlorophenol, while the sterically hindered derivatives, 2,6-dichloroanisole 11 and 2,6-dimethylanisole 12, were completely resistant toward demethylation by nitric acid.

10   11   12

The inhibition by ortho substituents was absent when one of these substituents was a phenolic hydroxyl group or a hydrogen atom. Thus, whereas tetrachloroveratrole 8 was completely stable toward chlorine, tetrachloroguaiacol 13 and 3,4,5-trichloroveratrole 14 were 50-60% demethylated within 5-10 minutes and 4,5-dichloroveratrole 15 more than 80% demethylated in the same time interval (Figure 10.4). A phenolic hydroxyl group situated ortho to an alkoxyl group could facilitate dealkylation by virtue of either its relatively small size or its favorable inductive influence. The readiness with which 2-alkoxyphenolic units undergo dealkylation may be interpreted as possibly indicating a susceptibility toward oxidative cleavage via a mechanistic sequence analogous to the one proposed by Adler (176) for periodate oxidation of guaiacol, equation 16.

(16)

In contrast to the hydrolytic mechanism described previously (Figure 10.3), this oxidative mechanism would lead to the formation of o-quinone rather than catechol derivatives.

As the foregoing discussion may already have suggested, the less than theoretical amount of alkoxyl cleavage frequently resulting from chlorination and nitration treatments is explicable on the basis of competing substitution and dealkylation processes. To the degree that substitution in the ortho position(s) precedes dealkylation, the latter process is completely or partially retarded. Ultimately then, the extent of dealkylation depends on the relative values of the rate constants for this process and for substitution in the position(s) ortho to the alkoxyl groups.

The phenolic hydroxyl groups formed through catalytic hydrolysis further activate the ortho and para portions on the ring, thereby promoting substitution reactions. In the example shown in equation 17, reported by Sato, Kobayashi and Mikawa (124), chlorine is substituted in the 2-position in structure 16, the site originally ortho to the methoxyl group.

$$\underset{OH}{\underset{|}{\underset{OMe}{C_6H_3(CH_2SO_3Ba/2)}}} \xrightarrow{Cl_2, H_2O} \underset{16}{\text{(CHO, Cl, OH)}} + \underset{17}{\text{(CHO, Cl, Cl, OH)}} + MeOH \quad (17)$$

Since the 2-position on the guaiacyl nucleus is not greatly activated toward substitution, it appears probable that substitution at this site occurred subsequent to hydrolysis of the adjacent methoxyl group. This reasoning is more tenuous, however, when applied to structure 17, since the methoxyl group effectively activates the 6-position to a degree that substitution at this site may be expected to precede demethylation.

The cleavage of alkyl aryl ether bonds by certain of the halogens and nitric and nitrous acids is particularly important in the case of lignin, for, as was emphasized in Chapter 5, a large fraction of lignin monomer units are joined through linkages of this type. Although definite proof is lacking, it appears certain that these inter-connecting ether linkages are cleaved in much the same manner as the corresponding groups in the model structures.

4. Oxidation and Other Reactions. Generally speaking, phenolic nuclei are quite susceptible to oxidative attack. Although the literature abounds with accounts of phenol oxidation reactions employing a wide variety of oxidants, references to the oxidation of di- and trihydric phenols (e.g. lignin-like structures) by halogens and nitric acid appear with comparative infrequence. This deficiency may be accounted for in part by the generally accepted fact that the reactions of these latter reagents with phenolic substrates are non-specific and often difficult to control, with the consequence that their application as oxidants in syntheses or theoretical investigations has been somewhat limited.

Although lacking thorough documentation in terms of product identification, it nevertheless seems certain that phenolic nuclei, such as are contained in lignin, are oxidized to non-aromatic structures by nitric acid and certain of the halogens through the general ionic mechanism proposed by Levitt (45). The intense color of reaction mixtures often resulting from the treatment of phenolic nuclei with these reagents has been interpreted as indicating the formation of quinonoid products. In all likelihood, the latter structures in part undergo further oxidation ultimately yielding dibasic acid fragments.

The oxidation of catechol (1,2-dihydroxybenzene) and pyrogallol (1,2,3-trihydroxybenzene) derivatives generally is regarded as proceeding through the formation of o- or p-benzoquinonoid intermediates.

The formation of o-benzoquinone intermediates from the treatment of lignin with aqueous chlorine has been proposed by several investigators (14, 106, 173, 177). In support of this theory, tetrachloro-o-quinone has been recovered from the oxidation of tetrachlorocatechol with aqueous chlorine (178) and 4,6-di-t-butyl-o-quinone from reaction of 4,6-di-t-butylguaiacol with bromine in acetic acid (179). Gess (180) succeeded in identifying 4-methyl-o-benzoquinone in the chloroform extract of the products obtained from the reaction of 4-methylguaiacol (creosol) with aqueous chlorine. The difficulties involved in the isolation of such intermediates can be ascribed to the instability of low-substituted halogenated o-benzoquinones in certain solvents, particularly water (181-183). This instability diminishes as the quinone ring becomes more highly substituted (184).

Bolker and coworkers (102, 114, 185) isolated 2,6-dimethoxy-p-benzoquinone from the separate reactions of nitrous acid with 1,2,3-trimethoxybenzene, syringyl alcohol, syringaldehyde or syringic acid. Based on the yield of quinone, the arrangement of side-chain substituents in order of ease of displacement and subsequent oxidation was -COOH > -CHO > -CH_2OH. Using Sobolev's nitration conditions (123), Kung (185) also obtained 2,6-dimethoxy-p-benzoquinone from the reaction of nitric acid with 1,2,3-trimethoxybenzene.

Sergeeva, Shorygina and Lopatin (186) isolated crystals of what appeared to be nitrated o-quinone 18 from the nitric acid oxidation of dihydroconiferyl alcohol 19 in ethyl ether, equation 18.

$$\text{19} \xrightarrow{HNO_3} \text{18} \qquad (18)$$

In this case, too, the o-quinone was observed to be unstable in the reaction mixture. Quinones of the above type are regarded by these authors as possible progenitors of carboxyl groups arising from the oxidation of lignin with nitric acid.

In his study on the nitric acid oxidation of 4-methylguaiacol, Sobolev (123) isolated 2-hydroxy-5-methyl-3-nitro-1,4-benzoquinone when the reactions were carried out in acetic acid. This quinone is tautomeric with the less stable ortho form, 21, equation 19.

$$\xrightarrow{HNO_3} [\quad] \longrightarrow \text{20} \longleftrightarrow \text{21} \qquad (19)$$

The formation of p-benzoquinone of the type isolated by Sobolev apparently involves a prior hydroxylation step. One of the several conceivable mechanisms for such a process has been discussed by Ley and Müller (187).

Benzoquinones, particularly the ortho forms, inherently are reactive compounds and therefore inclined to undergo secondary reactions in the halogenation or nitration system. Their susceptibility to further oxidation in these circumstances was commented on earlier. Benzoquinones also may be expected to participate in various 1,4-addition reactions typical of the conjugated dienone structure. Of primary interest in this connection is their disposition under certain conditions to form dimers or polymers whose monomer units are linked through carbon-carbon or carbon-oxygen-carbon bonds (188-193). Reactions of this type may account for the dimeric product 3,3'-dimethoxy-4,4'-dihydroxydiphenyl 22, isolated by Sato and Mikawa (120) from the product of the reaction of guaiacol with aqueous chlorine. It has been suggested (59) that the formation of a diphenyl compound can be rationalized on the basis of a phenoxonium ion intermediate as shown in equation 20.

$$(20)$$

Under appropriate conditions, benzoquinones also may be expected to function as oxidizing agents undergoing quinone-hydroquinone equilibria.

In aqueous media, lignin model compounds ultimately undergo partial conversion to dicarboxylic acids through the action of chlorine and nitric acid. Gess (180) succeeded in detecting methylmaleic, methylfumaric and oxalic acids among the products of the reaction of 4-methylguaiacol with chlorine water at room temperature. Oxalic acid has been recovered by Chudakov and Milovanov (194) and by Kung (185) after reacting both partially and completely etherified lignin models with hot, dilute nitric acid. In connection with the former study, the susceptibility of the compounds to oxidation was reported to depend on the nature of the side-chain and on the oxygen-containing substituent groups on the aromatic nucleus. Phenylcoumaran-type compounds were most resistant to oxidation while diphenyl rings of the dehydrodivanillin type were the most susceptible.

B. Reactions of the Side-Chain Moiety. As a consequence of the greater heterogeneity of the aliphatic as compared with the aromatic portion of lignin (Chapter 5), a description of the halogenation and nitration behavior of lignin side-chains in terms of accompanying structural changes involves, at best, a considerable degree of uncertainty. A number of lignin side-chains can be viewed as derivatives of glycerol for the purpose of outlining their

HALOGENATION AND NITRATION

reactions. Glycerol itself was reported by Hlasiwetz and Habermann (13) to undergo oxidation to D, L-glyceric acid when acted upon by aqueous chlorine at room temperature for an unspecified length of time. Gierer (158), on the other hand, found that glycerol was not attacked by chlorine in acetic acid at room temperature up to reaction periods of approximately 45 minutes. When substituted by a 3,4-dimethoxyphenyl group, the glycerol moiety underwent electrophilic displacement from the nucleus (Section IIIA-2) and was recovered as glyceraldehyde (158). The glyceraldehyde or glyceraldehyde derivatives formed in such reactions apparently exhibited little tendency to undergo further oxidation to the corresponding glyceric acid under the imposed reaction conditions.

Direct substitution of halogen for hydrogen on the side-chain via a free radical mechanism is rather improbable since the halogenation of lignin and lignin model compounds usually is carried out employing conditions unfavorable to the formation of radicals. More realistically, halogen substitution of the side-chain can be expected to occur at position(s) alpha to a carbonyl group through reaction of elemental halogen with the enol form of the latter group. Since reactions of this type are acid- as well as base-catalyzed, their occurrence should be favored because of the acidic conditions generally prevailing during the halogenation of lignin. In support of this proposed mode of substitution, Kratzl and Bleckmann (105) obtained α-bromo derivatives 23a and 23b when either propioguaicone 24a or its methyl ether 24b was brominated in boiling chloroform, equation 21.

$$RO\text{-C}_6H_3(OMe)\text{-COCH}_2CH_3 \xrightarrow{Br_2} RO\text{-C}_6H_2(Br^*)(OMe)\text{-COCHBrCH}_3 \quad (21)$$

24a R=H   23a R=H
24b R=Me   23b R=Me

*(Bromine substituted in the 5- or 6-position depending on whether R is H or CH_3).

Carbon-sulfur linkages in the side-chain moiety appear to be susceptible to attack by halogens. Kratzl and Bleckmann obtained the tribrominated derivative 25 by reacting the sodium salt of the sulfonate 26 with bromine in boiling chloroform, equation 22.

$$MeO\text{-C}_6H_3(OMe)\text{-CHCH}_2CH_3(SO_3Na) \xrightarrow{Br_2} MeO\text{-C}_6H_2(Br)(OMe)\text{-COCBr}_2CH_3 \quad (22)$$

26   25

A similar product was obtained when the corresponding guaiacyl compound was brominated. It is of interest to note that these reactions did not take place when the nucleus was unsubstituted. The aforementioned results can be interpreted as suggesting that the sulfonic acid group was first oxidized by the halogen to a keto group which then activated the adjacent methylene group toward halogen substitution as shown above. Thioacetate groups also appear to be susceptible to replacement (195). The reactions of sulfur compounds of the above type are of interest in connection with the bleaching of sulfite and kraft pulps.

Small amounts of double bonds are present in lignin (Chapter 5), and these may react with elemental halogen by addition as illustrated by the bromination of isoeugenol $\underline{27}$ (121), equation 23.

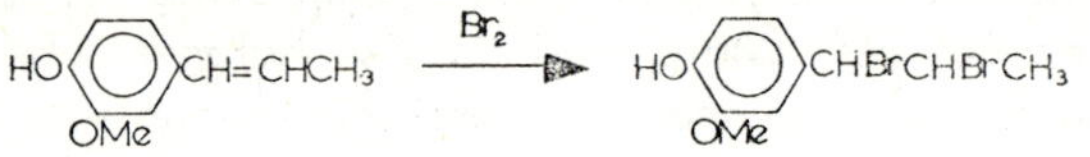

$$HO\text{-}C_6H_3(OMe)\text{-}CH=CHCH_3 \xrightarrow{Br_2} HO\text{-}C_6H_3(OMe)\text{-}CHBrCHBrCH_3 \quad (23)$$
$$\underline{27}$$

The nitration behavior of various guaiacyl propanol derivatives has been studied in detail by Shorygina and coworkers (109, 110, 111, 113–115). The nitrations were carried out under comparatively mild conditions (in carbon tetrachloride or ether at room temperature or lower) so as to minimize secondary degradation reactions. Under these circumstances, the type and amount of product isolated were found to depend on the solvent, on the molar ratio of nitric acid to substrate and on the concentration of the nitric acid.

The susceptibility of the propanol side-chain toward oxidation was found to be dependent on the position of the alcohol group on the side-chain (109). When substituted on the β-carbon atom, as in $\underline{28}$, the hydroxyl group was oxidized to a ketone, $\underline{29}$, in the presence of excess nitric acid, equation 24.

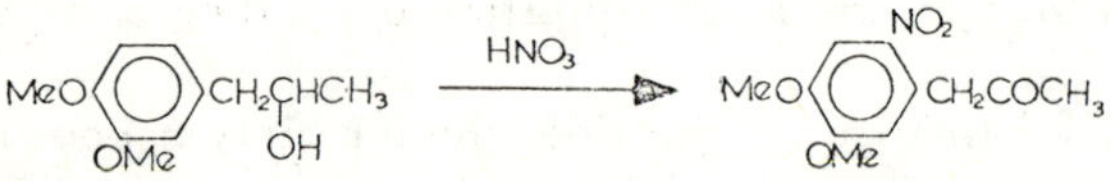

$$MeO\text{-}C_6H_3(OMe)\text{-}CH_2CH(OH)CH_3 \xrightarrow{HNO_3} (NO_2)MeO\text{-}C_6H_2(OMe)\text{-}CH_2COCH_3 \quad (24)$$
$$\underline{28} \qquad\qquad \underline{29}$$

When the hydroxyl group was substituted on the α-carbon atom as in $\underline{30b}$ (Figure 10.5, less than 1% conversion to the α-ketone compound $\underline{31b}$ resulted. Instead, a greater preference was shown for electrophilic displacement of the side-chain with the result that 4-nitroveratrole $\underline{32c}$ and 4,6-dinitroguaiacol $\underline{32d}$ were obtained in comparative high yields from $\underline{30a}$ and $\underline{30b}$, respectively. Under the same reaction conditions, a hydroxyl group substituted on the γ-carbon atom was not noticeably attacked by nitric acid.

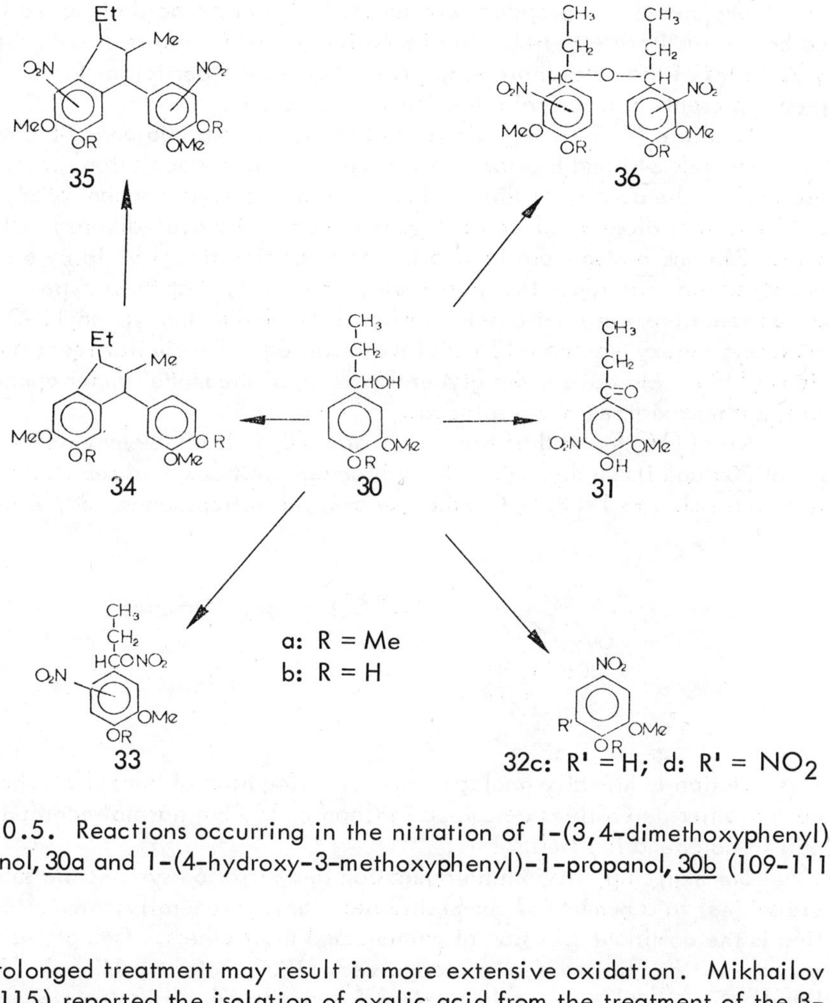

HALOGENATION AND NITRATION

Figure 10.5. Reactions occurring in the nitration of 1-(3,4-dimethoxyphenyl)-1-propanol, 30a and 1-(4-hydroxy-3-methoxyphenyl)-1-propanol, 30b (109-111).

Prolonged treatment may result in more extensive oxidation. Mikhailov et al. (115) reported the isolation of oxalic acid from the treatment of the β-guaiacyl ether of 3,4-dimethoxyphenylglycol with moderate to large amounts of nitric acid at 0°. According to the authors, the oxalic acid was derived in all probability from the side-chain moiety.

A more pronounced oxidation by nitrous acid was established by Kee (116) who found that treatment of 1-guaiacyl-1-propanol, 30b and its 4-methyl ether 30a with hot, aqueous nitrous acid resulted in partial conversion to corresponding α-ketones. Not unexpectedly, etherification of the benzyl alcohol group effectively reduced the amount of oxidation occurring at this site.

The conversion of 30a and 30b (Figure 10.5) to nitrate esters 33a and

33b was enhanced through the application of more concentrated solutions of nitric acid. Small amounts of these esters also were detected when 1-(3,4-dimethoxyphenyl)-3-propanol was reacted with nitric acid dissolved in cold carbon tetrachloride (114). The formation of nitrate esters at both the α and γ positions of the side-chain suggests that at least a portion of the nitrate ester groups present in nitrolignin are similarly situated.

In dilute (0.2-0.3N) nitric acid solutions both 30a and 30b dimerized to compounds 34a and b prior to nitration. These condensation reactions were ascribed to the acidity of the medium since dilute (and concentrated) sulfuric acid solutions alone produced the same dimers. The α,α'-dibenzyl ether derivatives 36a and b also were formed in small but significant yield by a competing condensation reaction. It is noteworthy, however, that these types of condensation reactions were not observed when 1-(3,4-dimethoxyphenyl)-2-(2-methoxyphenoxy)-propan-1,3-diol was nitrated under similar reaction conditions (110). Moreover, the glycerol portion of the model dimer apparently was neither oxidized nor displaced.

Kee (116) found that treatments of 1-(3,4-dimethoxyphenyl)-1-propanol 30a and its methyl ether 37 with aqueous nitrous acid resulted in a partial conversion to 1-(3,4-dimethoxyphenyl)-1-nitropropane, 38, equation 25.

$$\underset{\substack{30a \ R=H \\ 37 \ \ \ R=Me}}{MeO\text{-}C_6H_3(OMe)\text{-}CH(OR)CH_2CH_3} \xrightarrow[H_2O]{HNO_2} \underset{38}{MeO\text{-}C_6H_3(OMe)\text{-}CH(NO_2)CH_2CH_3} \quad (25)$$

This reaction is formally analogous to the sulfonation of benzyl alcohol and benzyl ether derivatives described in Chapter 15, but has no recorded counterpart in halogenation treatments.

Summarizing, in both halogenation and nitration systems the substrates are subject to a number of competing reactions. Generally, aromatic substitution is the dominant reaction of phenols and their ethers. Exceptions to this generalization, however, are recognized. Thus, Sato and Mikawa (120) isolated catechol in an unspecified yield from among the reaction products of aqueous chlorine with guaiacol, and Gess (180) detected 4-methyl-1,2-benzoquinone among the products of the reaction of aqueous chlorine with 4-methylguaiacol. These results are noteworthy since they demonstrate that demethylation is not necessarily preceded by substitution of the aromatic nucleus as was suggested earlier (177). In addition, Kee (116) has shown that in the reaction of certain veratryl-type model compounds with nitrous acid, oxidation and substitution of the side-chain occur in preference to aromatic substitution. The latter was observed, however, when the reaction was carried out at a lower pH and a higher temperature.

HALOGENATION AND NITRATION

In further competition with aromatic substitution and side-chain displacement, oxidative ring opening converts phenolic structures to non-aromatic ones. To the extent that such a transformation takes place, the occurrence of the above two processes is necessarily limited. Conversely, substitution and side-chain displacement are favored at the expense of oxidation in ethers of phenolic compounds (112). In addition to structural factors, the solvent medium also plays an important role. Thus, aromatic substitution generally is enhanced by a non-aqueous medium. Specific differences between chlorination (218) and nitrosation (116) are illustrated in Figures 10.6 and 10.7, respectively. It can be noted that guaiacyl units are substituted in both the 5- and 6-position during chlorination but only in the 5-position during nitration. Secondly, oxidation of the benzyl hydroxyl to a keto group was observed during nitrosation but was not detected as a result of chlorination.

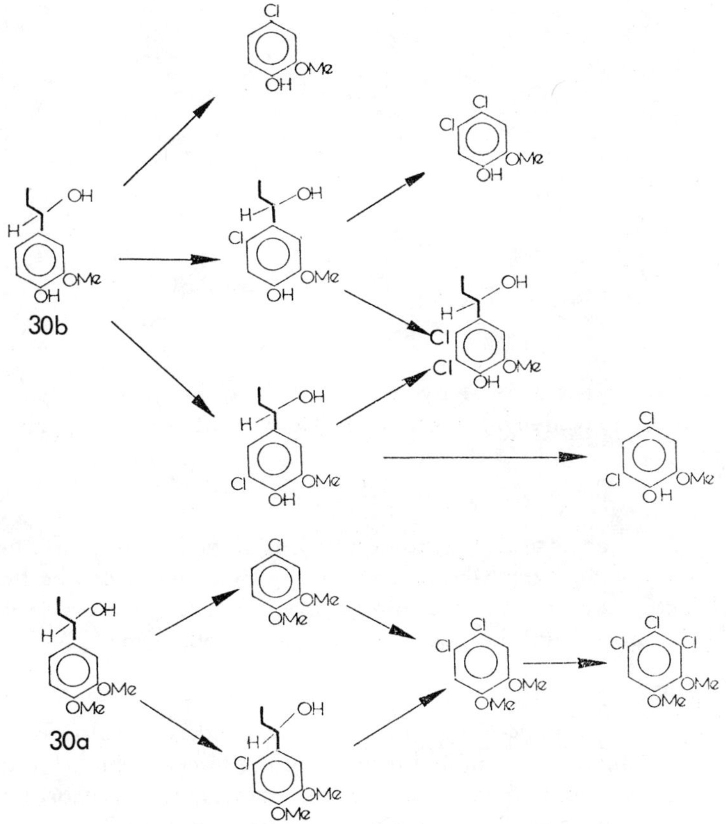

Figure 10.6. Reactions occurring in the chlorination of 1-(3,4-hydroxy-3-methoxyphenyl)-1-propanol, <u>30b</u> and 1-(3,4-dimethoxyphenyl)-1-propanol, <u>30a</u> with chlorine water (112).

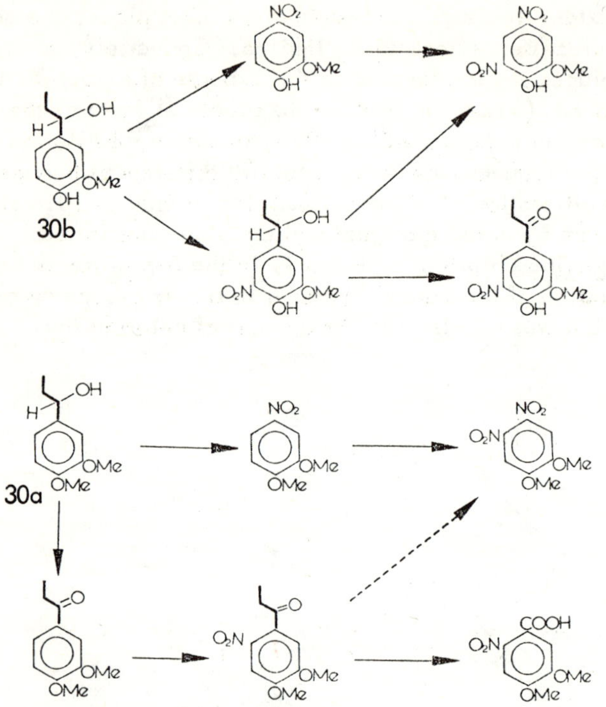

Figure 10.7. Reactions of 1-(4-hydroxy-3-methoxyphenyl)-1-propanol 30b and 1-(3,4-dimethoxyphenyl)-1-propanol 30a with aqueous nitrous acid (116).

Other divergencies relating to demethylation and side-chain substitution have also been noted. Whether these variations in behavior can be traced to superficial factors such as, for example, a specific reaction condition, or are inherent in the process has not been conclusively established.

IV. Halolignins.
A. Isolation and Purification. The procedures employed in the isolation of halolignins are determined largely by the form in which the starting material exists. In the case of wood meal the halogenated (primarily chlorinated) product is extracted with an organic solvent, usually 95% ethanol (7, 9, 23, 196). The extract is concentrated in vacuo and precipitated into water. To solubilize additional amounts of chlorolignin, the solvent-extracted wood is, in turn, extracted with a warm, dilute sodium hydroxide solution (7, 9,

197).The alkaline extraction may be accompanied by a loss of halogen atoms as described more fully in Section IV-D, Page 416.

Isolated lignins have been halogenated in suspension as well as in solution (11, 12, 16, 25, 198-202). The halolignins are then recovered directly by filtration or after precipitation by the addition of a non-solvent, usually water. Chlorinated lignosulfonic acid may be obtained directly by evaporation of ion-exchanged chlorination liquors (202) or by precipitation as a salt through addition of an organic base such as 1-(N-piperidinoacetylamino)-naphthalene (203) or in the form of a metal salt (usually calcium or barium) by precipitation from a concentrated aqueous solution with alcohol (173). Purification methods have been described utilizing precipitation from glacial acetic acid (204) or from aqueous sodium hydroxide (198) solution. The isolated halolignins, particularly bromolignin, tend to decompose in the dry state with the loss of halogen atoms (16, 31). At 135-140°C, jute chlorolignin is reported to decompose with the loss of hydrogen chloride (198).

B. Physical Properties. Isolated halolignins are amorphous substances (9, 17, 204) whose colors vary from different shades of yellow to orange and brown (6-9, 17, 23, 196, 204, 205) depending on the form of the starting material and the conditions of the treatment. Chlorolignins exhibit no distinct melting point but reportedly soften in the temperature range 140-180°C (7-9, 196), possibly with decomposition (6, 17).

The solubility characteristics of halolignin preparations are reasonably similar in spite of their varying origin (7-9, 196, 197, 199). They are commonly soluble in sodium hydroxide, sodium carbonate and ammonium hydroxide solutions, pyridine, acetic acid, ethyl acetate, alcohol, acetone, dioxane and dimethyl sulfoxide; but generally insoluble in water, chloroform, carbon tetrachloride, ether, benzene and petroleum ether. By virtue of the hydrophilic sulfonic acid group, chlorinated lignosulfonic acids are water-soluble (206). The solubilities of various pine chlorolignins in sodium, calcium and ammonium hydroxides, sodium silicate and sodium carbonate have been determined by Arnold, Simmonds and Curran (196).

The molecular weight range of various halolignins have been determined cryoscopically in phenol (198) nitrobenzene (9) and acetone (207), ebulliometrically in acetone (208) and by functional group analysis (7, 16) and found to lie in the range 1000 to 6000.

As illustrated in Figure 10.8, chlorination of softwood lignosulfonic acid causes a slight bathochromic shift and a flattening of the maxima at 200 and 280 mµ in the ultraviolet spectrum (173). If only moderate amounts (2-5 Cl/OMe) of chlorine are applied, the original minimum and maximum at 263 and 280 mµ, respectively, reappear at 270 and 285 mµ after reduction with sodium borohydride (203). The spectrum displays a general absorption in the visible region which decreases markedly with increasing application of chlorine (173). The ultraviolet spectrum of a chlorolignin prepared from a hardwood (poplar) exhibited an absorption maximum at 266 mµ, indicating a hypso-

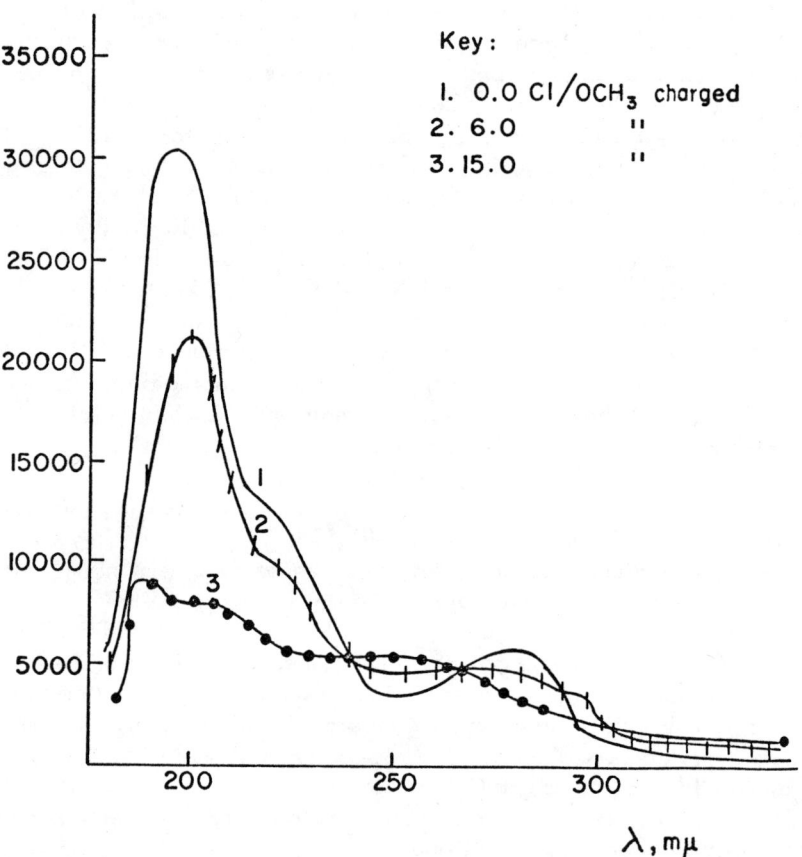

Figure 10.8. Effect of chlorination on the ultraviolet spectrum of lignosulfonic acid (173).

chromic shift as a result of chlorination.

C. **Chemical Constitution.** The lack of a systematic approach in the halogenation studies of lignins has resulted in the accumulation of a large amount of seemingly unrelated facts, the interpretation of which is not always obvious. In the succeeding sections, the reported results have been

HALOGENATION AND NITRATION

evaluated in terms of the established halogenation pattern of lignin model compounds.

The high halogen contents of isolated halolignins (6, 8-12, 16, 17, 23-25, 34, 62, 195, 198-201, 206, 208, 209-216) indicate the extensive occurrence of substitution reactions. These reactions are found to proceed most readily in an acidic environment, i.e., under conditions which favor the elemental form of the halogen (177, 217). The rate and extent of substitution is determined by the composition and structure of the lignin and by the nature of the solvent medium. The amount of iodine reacting with lignin by substitution is difficult to assess since in many instances the few reported values are based on indirect methods of estimation, the validity of which has been challenged (25). Using direct methods of measurement, Polcin (25) found that less than 1% iodine was bound to alkali sulfite lignin iodinated under neutral or acidic conditions and only 13% under alkaline conditions. On the other hand, through exhaustive halogenation treatments, lignins with chlorine and bromine contents of approximately 30 and 40%, respectively, frequently have been obtained. The corresponding degree of halogen substitution calculated on the basis of a C_9 unit generally is in the range 2.5 to 3.0. These values should be interpreted with caution, however, since calculations made on such a basis are inexact to the extent that the side-chain undergoes electrophilic displacement by halogen. Under the milder conditions utilized in technical pulp chlorinations, the uptake of chlorine by substitution has been estimated to be only 1.5 to 2 atoms per monomer unit for lignin in sulfite pulp (173). The corresponding range for pine kraft pulp (5% Klason lignin) has been estimated as 0.75 to 1.0 atom per monomer unit (218).

Halogen substitution may occur on the aromatic nucleus or on the side-chain of the lignin unit. Under mild halogenation conditions (low temperature, low halogen/substrate ratio) substitution is believed to occur primarily on the nucleus. The isolation of 6-chlorovanillin (201), 6-bromovanillin (219), 5-iodovanillin (219), 4,5-dibromoguaiacol (220), 5-bromovanillic acid (220) and a mixture of mono- and dibromovanillins (103) after halogenation and subsequent degradation of lignin supports this conclusion and, moreover, indicates that the 5- and 6-positions are reactive sites. This latter finding is compatible with the known substitution pattern of guaiacyl-derived lignin model compounds.

The aromatic substitution behavior of kraft and sulfite lignin during chlorination has been studied by Van Buren and Dence (218). In this investigation, kraft and spruce sulfite pulps containing approximately 5% Klason lignin were reacted with varying amounts of chlorine (1-7% based on the pulp) then degraded to monomeric fragments by digestion with alkaline cupric oxide. In addition to vanillin and acetoguaiacone, both 5- and 6-chlorovanillin were found among the degradation products. The amount of recovered 6-chlorovanillin exceeded by several fold that of 5-chlorovanillin, but the difference was considerably less in the case of the sulfite lignin. Disubstitution was indicated by the detection of significant amounts of 5,6-chlorovanillin after degradation of the chlorinated sulfite pulp. Failure to detect more than a trace of the same material after degradation of the chlorinated kraft pulp was viewed as a reflection of the unavailability of the 5-position for substitution due to prior condensation at the same site. Competing oxidation of aromatic nuclei prior to and/or following chlorine substitution appeared to limit effectively the extent of aromatic substitution as evidenced by the sharp decrease in the total recovery of aromatic products as the amount of applied chlorine was increased from 1 to 2%.

Reactions of halogens at aliphatic sites in the lignin unit are regarded as occurring to a significant extent mainly when the halogen is applied in excess of the amount readily consumed in aromatic substitution. The possible addition of halogen to the double bonds in coniferyl alcohol and coniferyl aldehyde units in protolignin (221, 222) and stilbene units in kraft lignin (223) can be cited. Moreover, when carbonyl groups are present on the lignin side-chain, substitution may occur at the alpha carbon as illustrated by equation 21.

The evidence for halogen substitution on the side-chain is indirect and based on the assumption that when subjected to the action of such chemicals as sodium carbonate, sodium hydroxide, sodium sulfite and alcoholic silver nitrate; the aliphatic-bound halogen atoms are selectively displaced leaving the aromatic-bound halogen atoms intact. This behavior of lignin-bound chlorine under alkaline conditions is dealt with in detail in Section IV-D.

Rather arbitrarily, several authors (7, 14, 224, 225) have designated the position alpha to the aromatic ring as a likely site for halogen substitution. However, because of the generally recognized ease with which benzyl halides undergo hydrolysis, even under neutral conditions, it appears unlikely that lignin units with α-halo substituents could survive isolation.

The results of kinetic studies conducted by Polcin (202) suggest substitution at positions other than the alpha carbon. Substitution at the beta position can be visualized as occurring by the addition of elemental halogen to the enol form of an α-keto unit and subsequent elimination of the hydrohalide (226) as was shown by Kratzl (105) in the bromination of propioguaiacone.

Shorygina et al. (224) have speculated that aliphatic chlorine may be present on the γ-carbon atom of the lignin side-chain. According to these authors, if the γ-carbon atom is also substituted with an aromatic ether group (which is not very likely), the ether linkage may be cleaved with boiling water or 5% aqueous sodium hydroxide solution liberating hydrogen chloride and producing free phenolic and aldehydic groups, equation 26.

$$-O\text{-Ar(OMe)}-CHCHCH(Cl)-O\text{-Ar(OMe)}-C_3 \xrightarrow[-HCl]{H_2O} -O\text{-Ar(OMe)}-CHCHCHO + HO\text{-Ar(OMe)}-C_3 \quad (26)$$

The occurrence of electrophilic displacement reactions during the halogenation of lignin has been demonstrated by the isolation of tetrachloroguaiacol after exhaustive chlorination of spruce wood meal in glacial acetic acid (106). The origin of the isolated compound was regarded to be an end group in the original lignin, equation 27.

$$\underset{R = H \text{ or Alkyl}}{\text{[Lig-CHOR-aromatic ring with OMe, OH]}} \xrightarrow{Cl_2} \text{Lig.—CHO} + \text{[tetrachloro-aromatic with OMe, OH]} \qquad (27)$$

Although the quantitative aspects of the displacement as it occurs in lignin have not been investigated, the results of lignin model compound studies lead to the prediction that its frequency of occurrence will vary over a wide range as determined by the nature of the functional groups attached to the alpha carbon atom of the side-chain.

The bromination and chlorination of lignin invariably is accompanied by a decrease in methoxyl content. By contrast, a similar observation has not been reported in the case of iodination treatments and, in view of the generally sluggish character of the reactions of iodine with aromatic substrates, demethylation is presumed to be minor or non-existent in the reaction of iodine with lignin. Since both aromatic substitution and demethylation are enhanced by the application of increasing amounts of chlorine, several authors (14, 177, 227, 228) have interpreted this as suggesting that the former process is a prerequisite for the latter. However, model compound studies support rather the idea of independent competing processes.

The demethylation behavior of protolignin and various lignin preparations in response to varying applications of aqueous chlorine has been studied in detail by Sarkanen and Strauss (159, 171). As shown in Figure 10.9, the demethylation patterns of various lignins closely resemble those observed for model compounds (Figure 10.4); i.e., the overall process consists of initial rapid loss of methoxyl followed by a second phase in which the rate of demethylation is drastically reduced. Although extensive in most instances, demethylation of lignin substrates is rarely quantitative, (see, however, reference 202). This fact was interpreted by Sarkanen and Strauss (171) as being indicative either of the presence of structures possessing methoxyl groups of exceptional stability or of steric effects originating from substituent groups ortho to the methoxyl group.

The demethylation of softwood and hardwood lignosulfonic acids as a function of the amount of applied chlorine has been studied by Sato and Mikawa (203) and a portion of their results is shown in Figure 10.10. The methoxyl loss was substantial in both cases and was directly proportional to the applied

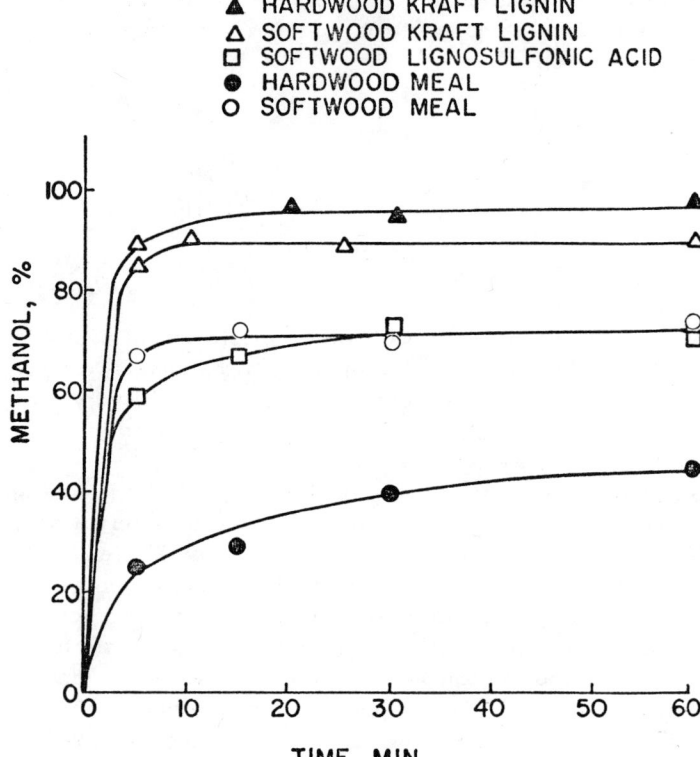

Figure 10.9. Formation of methanol in the chlorination of softwood and hardwood meals, softwood lignosulfonic acid, and softwood and hardwood kraft lignin in monochloroacetic acid-water medium (159).

chlorine up to a point where a maximum was reached. The decrease in methanol yield resulting from the application of additional chlorine was ascribed to the decomposition of methanol by the oxidizing action of excess chlorine.

The same factors which were critical in determining the rate and extent of demethylation for lignin model compounds also are operative in the demethylation of lignin by aqueous chlorine: loss of methoxyl is facilitated by the presence of water (10, 25, 106) and by acidic reaction conditions (100, 171) as shown in Figure 10.11. It should be pointed out, however, that a certain amount of demethylation is observed even when halogenation is conducted in an initially anhydrous medium (16, 33, 211) or at a pH in the vicinity of 10

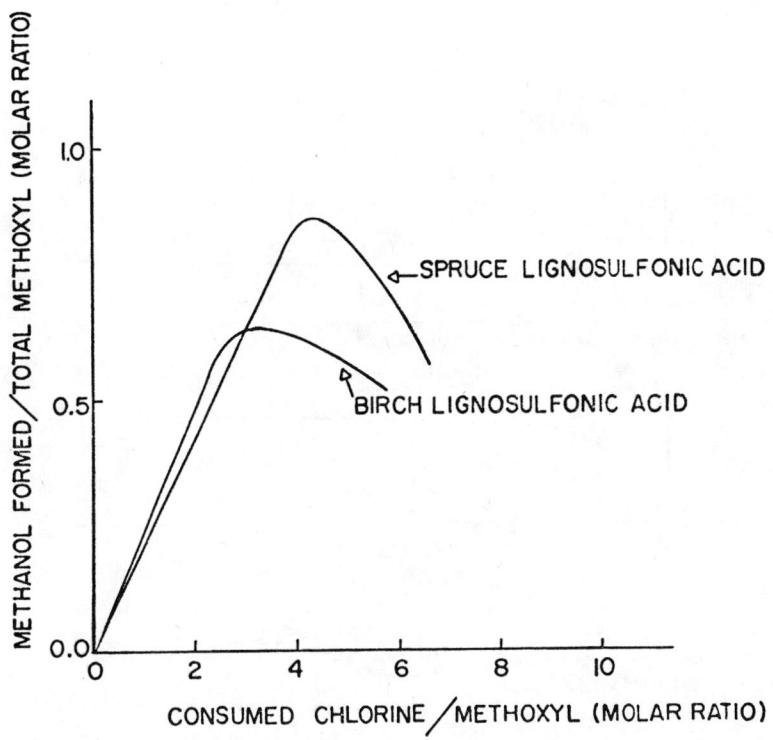

Figure 10.10. Formation of methanol from spruce and birch lignosulfonic acids by chlorination (203).

(25, 171, 229).

The cleavage of methoxyl groups in chlorination and bromination treatments also is affected by structural features of the lignin itself. As was shown in model compound studies (146, 159, 171), demethylation is promoted when a phenolic hydroxyl group is situated ortho to a methoxyl group. Thus softwood kraft lignin which has been shown to contain a high proportion of phenolic hydroxyl groups (230) undergoes demethylation more rapidly and extensively than a softwood lignosulfonic acid which contains a smaller amount of such units (Figure 10.9). As expected, diazomethane methylation of the phenolic hydroxyl groups in both softwood and hardwood kraft lignins causes a decrease in the initial rate of demethylation with aqueous chlorine. As suggested above,

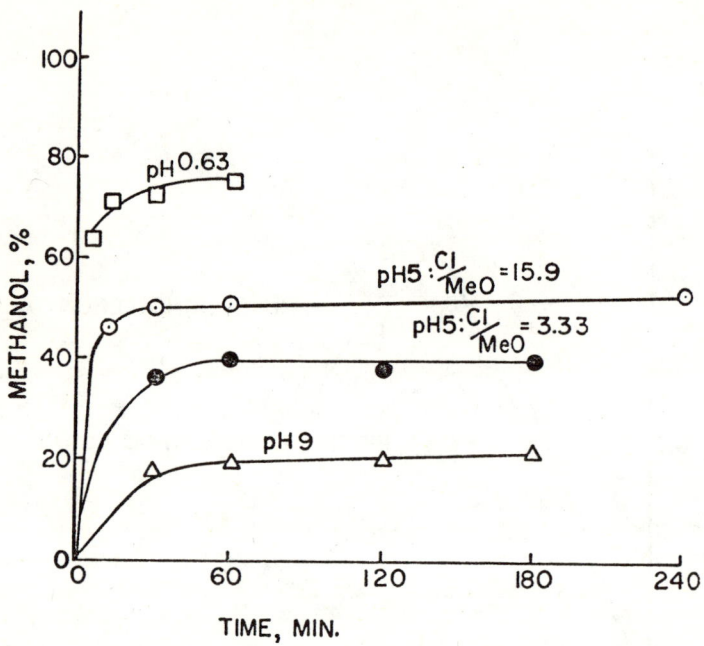

Figure 10.11. Chlorination of softwood lignosulfonic acid in aqueous solution at various pH levels (171).

steric factors also may be expected to have an effect in determining the extent of demethylation. In this connection, Strauss (159) noted that the demethylation pattern of birch wood meal (Figure 10.9) bore a striking resemblance to that of a syringyl type derivative, 1,2,3-trichloro-4,5,6-trimethoxybenzene (Figure 10.4), whose sluggish demethylation behavior was ascribed to steric hindrance.

Model compound studies (146) have also indicated, however, that aromatic ethyl and isopropyl ethers are cleaved by aqueous chlorine, suggesting that alkyl aryl ethers in general are susceptible to this reaction. If such is the case, then the extensive degradation of lignin by the action of aqueous chlorine and bromine becomes understandable.

According to the mechanism proposed for dealkylation (Figure 10.3), both aliphatic and phenolic hydroxyl groups are liberated as a consequence of the reaction. Isolated halolignins have been shown to contain both types of hydroxyl groups (6-9, 14, 199, 227, 231), although the amounts of such groups in the original untreated material are rarely available for comparison. However, as a result of chlorinating lignosulfonic acids in spent sulfite liquor, Newcombe and Marshall (14) were able to detect significant increases in phenolic and aliphatic hydroxyl contents. The increase of the former was considerably less than anticipated on the basis of the methanol recovered from the process. Probably, the vicinal phenolic hydroxyl groups produced by demethylation were subsequently oxidized to quinones by excess chlorine as suggested by other workers (106, 173, 177).

Shorygina et al. (208) chlorinated hydrolysis and hydrochloric acid lignin suspended in 10% hydrochloric acid with gaseous chlorine and subsequently compared the composition of the products with the original materials. The chlorinated hydrochloric acid lignin showed a 10 to 20-fold increase in carboxyl content and a loss of 2/3 of the original methoxyl groups. Slightly more than two atoms of chlorine/C_9 were introduced as a result of the treatment.

The presence of chlorocatechol units in chlorinated pine kraft and spruce sulfite pulps has been demonstrated by Van Buren and Dence (218). The chlorinated pulps were ethylated and subjected to degradation by hot alkaline cupric oxide, equation 28.

$$\text{HO-C}_6\text{H}_2(\text{Cl})\text{-OH} \xrightarrow{\text{Et}_2\text{SO}_4/\text{OH}^\ominus} \text{EtO-C}_6\text{H}_2(\text{Cl})\text{-OEt} \xrightarrow{\text{CuO}/\text{OH}^\ominus} \text{HO-C}_6\text{H}_2(\text{Cl})(\text{CHO})\text{-OEt} \quad (28)$$

Detection of 6-chloro-3-ethoxy-4-hydroxy-benzaldehyde among the decomposition products was interpreted as constituting evidence for chlorocatechol units in the chlorinated pulp. The estimated amount of ethoxylated product from each pulp was not large, however, nor was it significantly increased by reduction of the chlorinated pulp prior to ethylation. In view of the extensive loss of methoxyl groups sustained by the pulps during chlorination, it appears that both quinone and catechol units are relatively short-lived in the chlorination sequence.

Measurement of the hydrogen halide produced in the chlorination and bromination of wood and wood pulp has provided indirect evidence for the occurrence of extensive oxidation. The overall oxidation process proceeds with constant velocity throughout the treatment except in its initial stages where it is more rapid (217, 232, 233). By comparison, consumption of halogen in substitution reactions also is rapid in the initial phases of the treatment but, unlike oxidation, thereafter ceases for all practical purposes (217). In the case of chlorination, the proportion of chlorine consumed in oxidation reactions increases with increasing temperature and application of chemical. Estimates of the fraction of applied chlorine consumed in oxidation reactions cover a wide range, but a value in the vicinity of 50-60% can be considered typical. These values should be interpreted cautiously, however, in view of the uncertainties involved in the methods of measurement. The procedures employed in the determination of residual active chlorine and hydrogen chloride formed by substitution and oxidation reactions are open to question on the following grounds:

    1. In the iodometric procedure used to determine active chlorine, quinones also liberate iodine thus producing erroneous results.

    2. When used as a precipitant for chloride ion, silver nitrate may react with and displace organically bound chlorine.

    3. o-Chloroquinones are unstable in aqueous solutions and may decompose or condense with the liberation of hydrogen chloride (181).

Although there is little doubt but that oxidation plays an important role in the halogenation of lignin, information pertaining to the exact nature of the oxidation reaction still is very fragmentary. Chlorolignin preparations have been shown to contain both carbonyl (7, 9, 199, 203, 234, 235) and carboxyl (6, 7, 9, 203, 208, 231, 237) groups which are, in part, formed as a result of the halogenation treatment. The origin of these functional groups with respect to the aromatic and aliphatic moieties of the lignin monomer unit has only rarely been established.

    A proposed reaction sequence for the oxidation of the aromatic moiety of lignin by halogen is shown in Figure 10.12. Quinone formation has been proposed by several authors (14, 106, 173, 177) as a means of explaining the orange and red reaction mixtures resulting from halogenation treatments. Some direct evidence for the formation of such units has been provided by Cross and Bevan (2) who sublimed a quinone from chlorinated jute lignin which after reduction was identified as trichloropyrogallol. Also, it is noteworthy that behavior consistent with that of quinonoid systems such as susceptibility to reduction, instability in aqueous solutions, and absorption of visible light (173) also are characteristic of halogenation reaction mixtures.

    The formation of dicarboxylic acids through oxidative cleavage of o-benzoquinone rings has been established for such oxidizing agents as peracetic acid (236), sodium periodate (238) and alkaline hydrogen peroxide (239). However, the evidence for similar reactions occurring with elemental halogens is meager. Indeed, Sato and Mikawa (203) have shown that when spruce lignosulfonic acid is chlorinated with varying amounts of chlorine and then reduced with sodium borohydride, the absorbance values at 285 mµ remained essentially constant. This result suggests that although aromatic nuclei may have been transformed to quinone rings, the latter are not ruptured for the most part, during the chlorination treatment. Ivancic and Rydholm (173) found that although chlorination initially produced relatively small changes in the ultraviolet portion of the spectrum of lignosulfonic acid, the decrease in absorbance at 457 mµ was much more pronounced after 5 moles of chlorine per mole of methoxyl were consumed. The authors interpret this result as indicating the

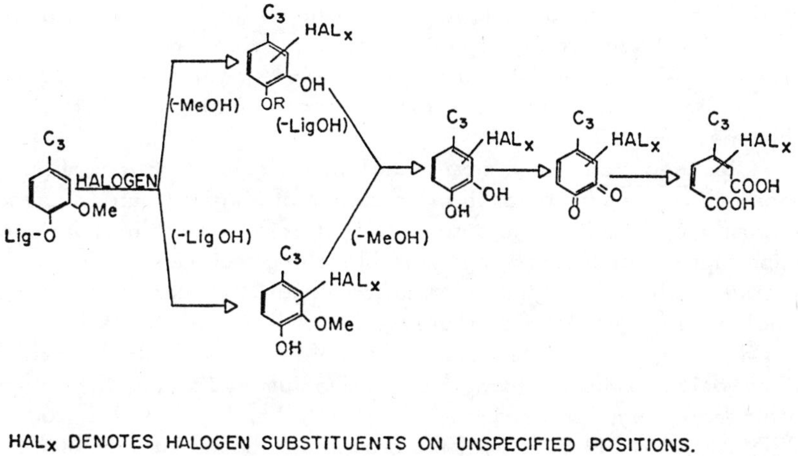

HALOGENATION AND NITRATION

HAL$_x$ DENOTES HALOGEN SUBSTITUENTS ON UNSPECIFIED POSITIONS.

Figure 10.12. Proposed reaction sequence for the oxidative cleavage of the aromatic ring of a softwood lignin unit by halogen.

the destruction of chromophoric (quinonoid) groups.

The oxidation of lignosulfonate by halogens is accompanied by a displacement of the sulfonate group which can be recovered as sulfuric acid (14, 25, 105, 202, 203, 227, 240). As suggested by the model compound studies of Kratzl and Bleckmann (103), carbonyl groups undoubtedly are introduced at the displacement sites. The displacement appears to be extensive and, in a typical reaction, Sato and Mikawa (203) found that 50% of the sulfonic acid groups were lost from spruce lignosulfonic acid even after mild chlorination. A substantial proportion of the sulfur also was split off from thiolignin during chlorination (240).

Other reactions proposed to account for the introduction of carbonyl groups into the lignin side-chain lack verification and must be regarded as speculative. These include the direct formation of carbonyl and carboxyl groups through oxidation of alcohol groups and electrophilic displacement of the side-chain with resultant formation of an aldehydic group (Section IIIA-2). These latter units might be anticipated to undergo further oxidation to carboxyl groups. Gierer (158), however, found no indication of such a reaction when the chlorination was conducted in glacial acetic acid. Shorygina and Kolotova (234) have ascribed the formation of carbonyl groups to the hydrolysis of α-chloroether units.

The importance of oxidation reactions in the degradation and solubilization of kraft lignin during the chlorination stage of a pulp bleaching sequence has been stressed by Giertz (241), Grangaard (232) and Rydholm (173). Apparently the condensation reactions occurring during the alkaline pulping process necessitates the added degrading effect provided by oxidation to effect solubilization. Grangaard (232) has further expressed the belief that oxidation also performs an important function in the degradation of sulfite lignin with chlorine.

D. <u>Reactions</u>. The most widely investigated reactions of halolignin are concerned with its behavior in the presence of alkaline materials, especially sodium hydroxide. In the case of chlorolignin, such studies are particularly important since, in multistage pulp bleaching sequences, a hot alkaline treatment generally follows the chlorination stage. By means of this treatment, additional amounts of chlorolignin are solubilized.

As a consequence of the treatment of halolignin with alkalis such as sodium hydroxide, sodium carbonate and ammonium hydroxide, the halogen content undergoes a marked reduction (6, 7, 9, 14, 15, 173, 198-200, 205, 207, 227, 240). A loss of 50% of the organically bound halogen atoms is not uncommon in such instances. Aliphatic-bound halogen atoms generally are regarded as being more alkali labile than those attached to the aromatic ring (9, 202, 215). Accordingly, a number of investigators (7, 14, 34, 242) have assumed that the decrease in halogen content resulting from alkaline treatment of halolignin is due exclusively to the hydrolysis of halogen atoms substituted on the side-chain. However, on the basis of their studies on the reactions of hot alkali or sulfurous acid with chlorinated lignin model compounds, Migita and coworkers (144) concluded that chlorine substituted on the side-chain was not always less stable than that substituted on the aromatic nucleus thus indicating that lability towards alkali is an unreliable test for determining the position of chlorine in chlorolignin. However, conversion of the phenolic hydroxyl to methoxyl groups stabilized the aromatic chlorine to a degree where it could be distinguished from aliphatic chlorine.

Based on the fact that one-half the chlorine on tetrachloro-o-benzoquinone is removed by a mild treatment with alkali (243), Rydholm (173) has suggested that chlorine similarly is lost from o-benzoquinone residues during caustic extraction of chlorinated sulfite pulp. The mechanism can be represented by equation 29.

$$\underset{O}{\overset{O}{\bigcirc}}\!\!\!\!\!\overset{Cl}{\underset{C_3}{\bigcirc}} \xrightarrow{OH^\ominus} \underset{O}{\overset{HO}{\ominus O}}\!\!\!\!\!\overset{Cl}{\underset{C_3}{\bigcirc}} \xrightarrow{-Cl^\ominus} \underset{O}{\overset{O}{\bigcirc}}\!\!\!\!\!\overset{OH}{\underset{C_3}{\bigcirc}} \qquad (29)$$

The alkaline hydrolysis of chlorine-substituted model compounds has been studied in detail by Braddon and Dence (244). As expected, chlorine substituents on side-chain moieties were rapidly and completely removed by treatment with 0.5M sodium hydroxide at 60°C. and extensively removed by aqueous sodium bicarbonate and water alone at the same temperature. However, chloro-o-benzoquinones and chloromuconic acids also sustained extensive losses of chlorine under the same conditions. Aromatic chlorine substituents were resistant to alkali with the exception of those associated with chlorocatechol units. The anomalous behavior of such compounds was ascribed to their prior conversion to the corresponding chloro-o-benzoquinone by trace amounts of oxygen dissolved in the alkali. Taken collectively these observations indicate that halogen atoms substituted on lignin exhibit a wide range of alkali lability. This predication has been borne out in practice where large variations in halogen loss are recorded (196, 227, 242, 245). Hydrolysis occurs even during relatively mild treatment with hot water (7, 202, 235) or sodium hydroxide at room temperature (198, 245), although to a lesser extent than observed in hot, dilute sodium hydroxide solutions.

The phenolic hydroxyl content of chlorinated lignosulfonic acid has been shown to increase as a result of treatment with aqueous sodium hydroxide (14). In addition, alkali treatment of chlorinated lignosulfonic acid has been found to cause an intensification in light absorbance, particularly in the near visible region of the spectrum, which was only partly reversible after acidification (173). Speculatively, the increase in absorbance may be attributed to contributions from phenolic ions overlapping from the ultraviolet region or to the formation of unsaturated groups conjugated with the aromatic nucleus as a result of dehydrohalogenation of the side-chain. The absorbance maximum at approximately 285 mμ undergoes a small bathochromic shift and is intensified as a result of the alkaline treatment (173). This result is consistent with the introduction of phenolic hydroxyl groups. Shorygina and Kolotova (234) and Hilpert (205) noted that alkaline treatment of chlorolignin and chlorinated spent sulfite liquor, respectively, was accompanied by an increase in total hydroxyl content but did not specify whether the hydroxyl groups were phenolic or aliphatic. Newcombe and Marshall (14) found the aliphatic hydroxyl content of chlorinated lignosulfonic acid to decrease on treatment with alkali and attributed the direction of change to oxidation reactions.

The removal of halogen atoms from halolignin also occurs during reduction with zinc and acetic acid and as a result of the action of sodium acetate on bromolignin (212). As a means of improving its utility, chlorolignin has been modified chemically by processes involving essentially the replacement of chlorine atoms. Thus, by heating chlorolignin with ammonia under pressure, Polcin (242) obtained a nitrogen-containing product with a greatly reduced chlorine content. Nitrogen-containing products resulting from the reaction of 1,6-hexamethylenediamine and thiourea with chloromethylated hydrotropic lignin also have been described (247). In other modification treatments, nitrile (246) and sulfonate (248) groups have been introduced into chlorolignin. The latter process in all likelihood involves the loss of aliphatic chlorine.

V. Nitrolignin. An evaluation of the results reported for the nitration of lignin is subject to extensive uncertainties due to the wide variety of nitrating agents and reaction conditions that have been applied. The effects of some parameters during nitration were noted briefly earlier. For a more comprehensive discussion of these and other theoretical aspects of nitration, the reader is referred to the excellent reviews authored by Ingold (249), de la Mare and Ridd (250) and Norman and Taylor (251).

A. Isolation. The nitration of wood meal most commonly is conducted in a non-aqueous medium such as glacial acetic acid. In these instances the nitrolignin is precipitated from the reaction mixture by dilution with water (18-20, 63, 69, 252-254). Nitrolignin also can be recovered from nitrated wood meal by extraction with dilute alkali (69, 255-260) or with an organic solvent such as acetone or methanol (18, 261).

Isolated lignin preparations are nitrated in the form of a suspension or after solution in a suitable medium. Soluble nitrolignins are precipitated by dilution of the reaction mixture with a non-solvent (62, 71, 72, 199, 262), while insoluble products are recovered simply by filtration (21, 67, 70, 102, 177, 262). Nitrolignins have also been precipitated by the addition of gelatin and alkaloids (261) and heavy metal salts (19, 21, 214, 262).

B. Physical Properties. Nitrolignins are amorphous solids (see, however, (19) ranging in color from canary yellow to dark red or brown depending on the previous history of the treatment (20, 21, 69, 72, 102, 172, 199, 214, 252, 255, 262). When prepared from isolated lignin, they are reported to be generally darker in appearance than those obtained by direct nitration of wood (261). Nitrolignin preparations undoubtedly contain molecules of a wide range of sizes (263), and molecular weight values in the 600-2,000 range have been reported (19, 21, 72, 263, 264) based largely on cryoscopic methods. Nitrolignins melt or soften with decomposition at widely varying temperatures (19, 252, 255) which generally exceed the range found for halolignins.

Nitrolignin preparations are soluble in alkaline solutions and in such common organic solvents as methanol, ethanol, acetone, acetic acid, ethyl acetate and dioxane, but are insoluble or only slightly soluble in ether, benzene, petroleum ether, chloroform, carbon tetrachloride, water and dilute mineral acids (19, 21, 66, 69, 252, 261, 262). These solubility characteristics are very similar to those observed for halolignin preparations.

The infrared spectra of nitrolignin preparations are characterized by bands at approximately 1,350 and 1,550 cm^{-1} which have been assigned to aromatic nitro groups (19, 69, 116, 172, 265). Under certain conditions, nitrolignin has been prepared with an absorption band at 1,042 cm^{-1} ascribed to nitrate ester groups (172). The presence of hydroxyl groups is clearly indicated by a maximum at 3,450 cm^{-1} (19, 116, 172, 265).

C. Chemical Constitution. Treatment of protolignin and isolated lignin with a variety of nitrating agents results in the formation of products containing organically bound nitrogen. The nitrogen content of nitrolignin generally lies in the range 2 to 4% (19-21, 63, 67, 109, 116, 199, 254, 262), reaching occasionally values as high as 6.7% (172).

The nitrogen in nitrolignin preparations is present largely in the form of aromatic nitro groups (18, 21, 63, 64, 69, 116). For example, in nitrolignin prepared by reacting birch and spruce dioxane lignins with aqueous nitrous acid, Kee (116) found that 76 and 80%, respectively, of the total nitrogen was present in nitro groups, while only 1-2% of the total nitrogen could be accounted for as nitroso groups. The form of the remaining nitrogen was not determined. Small amounts of organically bound nitrogen may be present as nitrate ester groups (63, 72, 172, 266). The formation of these latter groups is favored when excess nitric acid is available (266) and by acidic, anhydrous nitrating conditions as demonstrated by Lieser and Schaack (261) and Sergeeva and coworkers (267). The nitrogen content of wood meal nitrated under the latter conditions often approaches 12%. The increase in nitrogen content over that found in isolated lignin preparations can be ascribed, for the most part, to the formation of nitrate esters of cellulose and other carbohydrate material (20, 63, 69, 261).

The existence of aromatic nitro substituents in nitrolignin has been convincingly established by the isolation and identification of monomeric nitroaromatic compounds. Several investigators, for example, have recovered small but significant amounts of 4,6-dinitroguaiacol from among the products of the reaction of nitric and nitrous acids with lignin (102, 111, 116, 164, 172, 268). Kee (116) reacted spruce and birch dioxane nitrolignins with 1N sodium hydroxide at 80° and recovered 5-nitrovanillin and 5-nitrovanillic acid from both nitrolignin preparations and 6-nitrosyringic acid from the birch nitrolignin. Other nitro-aromatic products obtained include 3,5-dinitrobenzaldehyde from the nitric acid pulping of bagasse (269), 4-hydroxy-3-nitro-benzaldehyde and 4-hydroxy-3-nitro-benzoic acid from bagasse lignin (270) and 5-nitrobenzoic acid obtained following permanganate oxidation of rice straw nitrolignin (270).

These results indicate that the preferred sites for substitution of nitro groups are <u>ortho</u> and <u>para</u> to a free phenolic hydroxyl group. Although model compound studies previously cited have shown conclusively that substitution in the 6-position occurs when etherified phenolic units are nitrated, the expected 6-nitro-substituted products have not been detected in the degradation products of nitrolignin. Kee (116) has attributed the apparent anomaly to the inability of nitrous or nitric acids to release such products through splitting the interunitary ether linkages in the lignin network. Reference has been made to the limited demethylation (or dealkylation) observed during nitration or nitrosation later in this section.

In addition, significant amounts of degradation products containing no aromatic nitro substituents have been recovered from alkali-degraded nitrolignin (116). It is therefore not surprising that analytical data indicate an average degree of substitution of generally less than one aromatic nitro group per phenylpropane unit (19, 63, 70, 172, 214, 267, 268, 271).

The site(s) at which nitrate ester formation may be expected during the nitration of lignin has not been established. In this connection it is of interest to note that Sergeeva, Shorygina and Lopatin (111) nitrated 1-(2-nitro-4,5-dimethoxyphenyl)-1-propanol in carbon tetrachloride and isolated the corresponding nitrate in 67% yield.

Based on the results of Kee's study (116) of the nitrosation behavior of lignin model compounds, the alpha carbon atoms of lignin itself may become substituted with nitro groups when subjected to a similar treatment. These possibilities are outlined in equation 30.

$$\text{Lig}\cdots\text{O}\underset{\text{OMe}}{\bigcirc}\underset{\text{OH}}{\overset{\text{CH}-}{|}} \xrightarrow[\text{HNO}_2,\ \text{pH2},\ 70°]{\text{HNO}_3,\ \text{CCl}_4,\ 0°\ \text{or}} \text{Lig}\cdots\text{O}\underset{\text{OMe}}{\bigcirc}\underset{\text{NO}_2}{\overset{\text{CH}-}{|}} \qquad (30)$$

During the nitration of lignin, nitric acid frequently undergoes reduction forming a variety of products including nitrous acid (22, 133, 272). The latter in turn may be expected to introduce nitroso groups on the aromatic nucleus Failure to detect such groups in nitrolignin may be related to the general reactivity of such intermediates in nitrating mixtures (273).

The recovery of 4,6-dinitroguaiacol (102, 111, 116, 161, 172, 268) and 4-nitroguaiacol (102, 116) from the reaction mixtures of nitrated and nitrosated lignin, respectively, indicates that electrophilic displacement reactions contribute to the overall degradation. The reported yields of 4,6-dinitroguaiacol and 4-nitroguaiacol rarely, if ever, exceed 5%, based on the lignin. This observation suggests that side-chain displacement may be confined mainly to those aromatic nuclei in which the side-chain is situated para to a phenolic hydroxyl group.

Nitration of lignin is accompanied by a widely varying decrease in methoxyl content (19, 22, 64-67, 69, 72, 102, 116, 172, 267, 268, 271). Kee (116) found that the extent of methoxyl loss increased with increasing reaction time and temperature and decreasing pH of the reaction mixture. Shorygina et al. (172) noted that, irrespective of reaction conditions, the methoxyl loss from fir hydrochloric acid lignin did not exceed 1/3 of the original methoxyl content. Since approximately 1/3 of the aromatic units in lignin contain free phenolic hydroxyl groups, it is tempting to suggest that demethylation occurred primarily at those units with free phenolic hydroxyl groups ortho to the methoxyl groups (i.e., guaiacyl units). The methoxyl group is recovered either as methanol (22) or methyl nitrite (272). The latter is believed to originate from a secondary reaction of methanol with nitrous acid (123).

On the basis of its acetylation and methylation behavior, nitrolignin

contains a substantial number of hydroxyl groups (18, 19, 21, 199, 256, 268).
There is a reasonable certainty that of the total, a portion are phenolic in
character (19, 116, 256). The presence of aliphatic hydroxyl groups in nitrolignin has been claimed (18) but not thoroughly verified.

According to a comparison made by Traynard and Robert (19), the total
number of hydroxyl groups in poplar nitrolignin is only 40% of the amount in
spruce Brauns lignin.

Under severe reaction conditions, lignin is extensively degraded to
small, water-soluble fragments such as oxalic and acetic acids and carbon dioxide (21, 68, 260, 267, 274-276). Chudakov et al. (278) oxidized several lignin preparations with hot, 30% nitric acid and isolated maleic in
addition to oxalic acid. The authors speculated that both acids were derived
from quinones formed from the oxidation of non-condensed lignin units. Kee
(116) also isolated oxalic acid in 1-2% yields by reacting spruce and birch
dioxane lignins with nitrous acid at 70°C. Under somewhat milder conditions
the degradation accompanying oxidation is suppressed. Nitro-lignin preparations obtained under these conditions contain substantial amounts of carboxyl
groups (18, 19, 64, 172, 214, 267). Thus, for example, a 10 to 20-fold
increase in the carboxyl content of fir hydrochloric acid lignin was observed by
Shorygina (172) after treatment with nitric acid in carbon tetrachloride and
ether. At the present time it has not been established whether these groups
originate from the aliphatic or aromatic portion of the lignin unit or from both.
After treatment of dry hydrochloric acid beech lignin with nitrogen dioxide in
petroleum ether, Schaarschmidt (260) claimed to have isolated a dicarboxylic
acid with the empirical formula $C_{11}H_{10}O_{12}$ but the material was not characterized further. Traynard and Robert (19) have obtained results which pointed
to the presence of methyl ester groups, as well as other ester or lactone groups
in poplar nitrolignin. Methyl ester groups might of course arise through the
oxidative splitting of guaiacyl nuclei.

A number of investigators have cited evidence for the occurrence of
carbonyl groups in nitrolignin (18, 19, 214). These groups may arise through
oxidation of side-chain alcohol and ether groups or through conversion of aromatic nuclei to quinones. Analogous with the results obtained during halogenation of lignosulfonic acids, desulfonation also occurs during nitration treatments (214). This strongly suggests that some attack occurs on the side-chain
at the benzylic position. It is worth noting in this connection that nitrogen
tetroxide is reputed to be an excellent reagent for the oxidation of benzyl
alcohols to aldehydes (277).

D. Reactions of Nitrolignin. Recently, Chudakov et al. (278,
279) have isolated what they believe to be quinone nitropolycarboxylic acids
from the reaction of nitric acid (cl. 1.35) with hydrolysis lignin at temperatures ranging from 50 to 150°C. These materials, with contemplated structures
39 and 40, are reported to be effective plant growth stimulators. Additional
support for the formation of quinones was recently reported by Bolker (102)

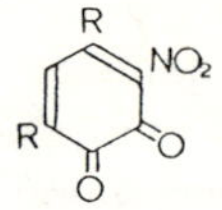

39 R= CH₂COOH

40 R= CH₂CH₂COOH

(structure: substituted nitro-quinone with R groups)

who isolated 2,6-dimethoxy-p-benzoquinone from the products of the reaction of birch dioxane lignin with aqueous nitrous acid. The analogous product of softwood lignin (methoxy-p-benzoquinone) was not detected, however.

Support for the existence of aromatic nitro groups in nitrolignin was obtained through reduction with Raney nickel (256), zinc and acetic acid (19), and sodium amalgam (72). The resulting aminolignins were subsequently diazotized and coupled with various phenols forming dyes. When pine aminolignin was titrated conductometrically with dilute acid, the profile of the resulting curve was consistent with that of a product containing both phenolic hydroxyl and amino groups (256). Partial conversion of nitro to amino groups through the action of sodium sulfite on nitrated Willstatter (257) and hydrolysis (258) lignins also has been reported.

Nitrolignins have been methylated (19, 22, 256), acetylated (21, 199) and brominated (22). On heating with aromatic amines, nitrolignins form high molecular weight azohydroxy compounds which oxidize an excess of the amine to the corresponding azohydrocarbon (280). Attempts also have been made to form methylol derivates of nitrolignin (281).

REFERENCES

1. C. W. Dence, "Chlorination" and J. M. McEwen, "Hypochlorite Bleaching," in The Bleaching of Pulp, Tappi Monograph No. 27, W. H. Rapson, Ed., Technical Association of the Pulp and Paper Industry, New York, 1963.
2. E. J. Bevan and C. F. Cross, J. Chem. Soc., 38, 666 (1880).
3. W. G. Van Beckum and G. J. Ritter, Paper Trade J., 105, No. 18, 127 (1937).
4. Anon., Chem. and Eng. News, 38, No. 11, 47 (1960).
5. Anon., Chem. Week, 93, No. 12, 60 (1963).
6. O. A. Müller, Paper-Fabr., 32, 329, 338, 341, 354 (1934).
7. A. Robert and Ch. de Choudens, Assoc. tech. ind. papetiere, Bull., 18, 107 (1964).
8. G. V. Jansen and J. W. Bain, Can. J. Research, 15B, 279 (1937).
9. A. M. Ayroud, Ph.D. Thesis, University of Grenoble, 1954.
10. R. Katzen, J. Pearlstein, R. E. Muller, and D. F. Othmer, Tappi, 33, 67 (1950).

11. B. Rassow and P. Zickmann, J. prakt. Chem., [2] 123, 214 (1929).
12. E. E. Harris and L. J. Lofdahl, J. Am. Chem. Soc., 63, 112 (1941).
13. H. Hlasiwetz and H. Habermann, Ann. Chem., 155, 120 (1870).
14. A. G. Newcombe and H. B. Marshall, Can. J. Tech., 33, 152 (1955).
15. H. Ishikawa, J. Japan Forestry Soc., 35, 363 (1953); Chem. Abstr., 48, 13216 (1954).
16. E. E. Harris, E. C. Sherrard and R. L. Mitchell, J. Am. Chem. Soc., 56, 889 (1934).
17. A. Friedrich and J. Diwald, Monatsh. Chem., 46, 31 (1925).
18. K. Kürschner and H. Peikert, Tech. Chem. Papier Zellstoff-Fabr., 31, 1, 17, 53, 69, 73, 85 (1934); Chem. Abstr., 29, 5441 (1935).
19. Ph. Traynard and A. Robert, Bull. soc. chim., France, 1952, 746.
20. H. Friese and H. Fürst, Chem. Ber., 70, 1463 (1937).
21. F. Fischer and H. Schrader, Brennstoff-Chem., 2, 2713 (1930).
22. K. Freudenberg and W. Dürr, Chem. Ber., 63, 2713 (1930).
23. E. Heuser and R. Sieber, Z. angew. Chem., 26, 801 (1913).
24. H. Ishikawa, J. Japan Forestry Soc., 35, 121 (1953); Chem. Abstr., 48, 6117 (1954).
25. J. Polcin, Chem. zvesti, 9, 254 (1955).
26. Ph. Traynard and A. M. Ayroud, Bull. soc. chim., France, 1952, 1001.
27. E. F. Kurth and A. A. Swelim, Tappi, 46, 591 (1963).
28. J. W. Mellor, A Comprehensive Treatise on Inorganic and Theoretical Chemistry, Supplement II, Part 1, Longmans, Green and Co., London, 1956, pp. 544, 547, 708, 751, 837, 870.
29. S. A. Rydholm, Pulping Processes, Interscience, New York, 1965, p. 921.
30. F. Paschke, Cellulosechem., 3, 19 (1922).
31. P. Karrer and F. Widmer, Helv. Chem. Acta, 6, 817 (1923).
32. A. Friedrich, Z. physiol. Chem., 176, 127 (1928).
33. G. M. Telysheva, A. Kalnins and V. N. Sergeeva, Latv. PSR Zinat. Akad. Vestis, Khim. Ser., 1966, 745; Chem. Abstr., 68, 14206m (1968).
34. T. L. Fletcher and D. M. Ritter, J. Am. Chem. Soc., 74, 3297 (1952).
35. J. Nakano and C. Schuerch, ibid., 82, 1677 (1960).
36. R. P. Bell and E. Gelles, J. Chem. Soc., 1951, 2734.
37. E. Grovenstein and N. S. Aprahamian, J. Am. Chem. Soc, 84, 212 (1962).
38. L. A. Bigelow, Chem. Rev., 40, 88, 89 (1947).
39. P. B. D. de la Mare and J. L. Maxwell, J. Chem. Soc., 1962, 4829.
40. P. B. D. de la Mare, I. C. Hilton and C. A. Vernon, ibid., 1960, 4039.
41. N. P. Kanyaev, Sbornik Statei Obshchei Khim., 2, 1172 (1953); Chem. Abstr., 49, 5087 (1955).
42. E. A. Shilov, G. V. Kupinskaya and A. A. Yasnikov, Doklady Akad.

Nauk S.S.S.R., 81, 435 (1951); Chem. Abstr., 46, 3376 (1952).
43. G. C. Israel, J. K. Martin and F. G. Soper, J. Chem. Soc, 1950, 1282.
44. K. D. Reeve and G. C. Israel, ibid., 1952, 2327.
45. L. S. Levitt, J. Org. Chem., 20, 1297 (1955).
46. E. Berliner, J. Chem. Ed., 43, 124 (1966).
47. K. Gonda-Hunwald, G. Graf and F. Korosy, Nature, 166, 68 (1950).
48. E. A. Shilov and N. P. Kanyaev, Compt. rend. acad. sci. U.R.S.S., 24, 890 (1939); Chem. Abstr., 34, 4062 (1940).
49. W. J. Wilson and F. G. Soper, J. Chem. Soc., 1949, 3376.
50. D. H. Derbyshire and W. A. Waters, ibid., 1951, 73.
51. D. H. Derbyshire and W. A. Waters, ibid., 1950, 564.
52. P. B. D. de la Mare, E. D. Hughes and C. A. Vernon, Research (London), 3, 192, 242 (1950).
53. P. B. D. de la Mare, A. D. Ketley and C. A. Vernon, J. Chem. Soc., 1954, 1290.
54. P. B. D. de la Mare and J. T. Harvey, ibid., 1956, 36.
55. S. J. Branch and B. Jones, ibid., 1954, 2317.
56. E. Berliner, J. Am. Chem. Soc., 72, 4003 (1950).
57. C. G. Swain and A. D. Ketley, J. Am. Chem. Soc., 77, 3410 (1955).
58. C. K. Ingold, Structure and Mechanism in Organic Chemistry, Cornell University Press, Ithaca, N. Y., 1953, p. 290.
59. K. V. Sarkanen, personal communication.
60. P. B. D. de la Mare, I. C. Hilton and S. Varma, J. Chem. Soc., 1960, 4044.
61. C. Walling, Free Radicals in Solution, Wiley, New York, Chapter 8, 1957.
62. E. Hägglund, Arkiv. Kemi, 7, 20 (1918).
63. L. Brissaud and S. Ronssin, Assoc. tech. ind. papetiere, Bull., 1953, 107.
64. V. I. Ivanov, A. A. Chuksanova and L. L. Sergeeva, Izvest. Akad. Nauk S.S.S.R., Otdel. Khim. Nauk, 1957, 503; Chem. Abstr., 51, 14261 (1957).
65. S. Poller, Faserforsch. Textiltech., 19, 124 (1968).
66. P. B. Sarkar, J. Indian Chem. Soc., 11, 407 (1934); Chem. Abstr., 29, 130 (1935).
67. H. Friese and W. Lüdecke, Chem. Ber., 74, 308 (1941).
68. A. Schaarschmidt, P. Nowak and W. Zetsche, Z. angew. Chem., 42, 618 (1929).
69. W. Elias and L. D. Hayward, Tappi, 41, 246 (1958).
70. K. Freudenberg, W. Lautsch and G. Piazolo, Cellulosechem., 21, 95 (1943).
71. M. Erfan Ali and M. H. Khundkar, J. Indian Chem. Soc., 31, 471 (1954); Chem. Abstr. 49, 4987 (1955).

72. H. Hibbert and L. Marion, Can. J. Research, 3, 130 (1930).
73. J. Chédin, S. Fénéant and R. Vandoni, Mém. services chim. état (Paris), 35, 53 (1950); Chem. Abstr., 46, 2887 (1952).
74. F. G. Bordwell and E. W. Garbisch, Jr., J. Am. Chem. Soc., 82, 3588 (1960).
75. P. B. D. de la Mare and J. H. Ridd, Aromatic Substitution: Nitration and Halogenation, Academic Press, New York, 1959, p. 49.
76. E. G. Cox, G. A. Jeffrey and M. R. Truter, Nature, 162, 259 (1948).
77. P. E. Grison, K. Ericks and J. L. deVries, Acta Cryst., 3, 290 (1950).
78. R. J. Gillespie, J. Graham, E. D. Hughes, C. K. Ingold and E. R. A. Peeling, Nature, 158, 480 (1946).
79. R. J. Gillespie, J. Chem. Soc., 1950, 2493.
80. D. J. Millen, ibid., 1950, 2600, 2606.
81. L. Médard, Compt. rend., 199, 1615 (1934).
82. Reference 75, p. 61.
83. E. D. Hughes, C. K. Ingold, and R. I. Reed, J. Chem. Soc., 1950, 2400.
84. H. Martinsen, Z. physik Chem., 50, 385 (1904).
85. F. M. Lang, Compt. rend. Acad. Sci., Paris, 226, 1381 (1948); Chem. Abstr., 42, 7263 (1948).
86. R. M. Schramm and F. H. Westheimer, J. Am. Chem. Soc., 70, 1782 (1948).
87. C. A. Bunton, E. D. Hughes, C. K. Ingold, D. I. H. Jacobs, M. H. Jones, G. J. Minkoff and R. I. Reed, J. Chem. Soc., 1950, 2628.
88. E. L. Blackall, E. D. Hughes, and C. K. Ingold, ibid., 1952, 28.
89. W. R. Angus and A. H. Leckie, Proc. Royal Soc., A150, 615 (1935).
90. V. Gold, E. D. Hughes, C. K. Ingold and G. H. Williams, J. Chem. Soc., 1950, 2452.
91. C. K. Ingold, Structure and Mechanism in Organic Chemistry, Bell and Sons, London, 1953, p. 282.
92. F. Arnall, J. Chem. Soc., 1923, 3111.
93. S. Veibel, Chem. Ber., 63B, 1577 (1930).
94. K. M. Ibne-Rasa, J. Am. Chem. Soc., 84, 4962 (1962).
95. B. C. Challis and A. J. Lawson, Chem. Comm., 1968, 818.
96. L. Jurd, Aust. J. Sci. Research, 2A, 111 (1949).
97. G. Castelfranchi and G. Borra, Ann. chim. (Rome), 43, 293 (1953); Chem. Abstr., 48, 10650 (1954).
98. R. A. B. Bannard and G. Latremouille, Can. J. Chem., 31, 469 (1953).
99. L. Weinberger and A. R. Day, J. Org. Chem., 24, 1451 (1959).
100. G. A. Holmberg, Acta Chem. Scand., 8, 728 (1954).
101. N. L. Drake, H. C. Harris and C. B. Jaeger, Jr., J. Am. Chem. Soc., 70, 168 (1948).
102. H. I. Bolker, F. L. Kung and M. L. Kee, Tappi, 50, 199 (1967).
103. P. B. D. de la Mare and C. A. Vernon, J. Chem. Soc., 1951, 1764.

104. L. Rosenkranz and M. Perez, Ciencia (Mex.), 6, 364 (1945); Chem. Abstr., 40, 7179 (1946).
105. K. Kratzl and Ch. Bleckmann, Monatsh. Chem., 76, 185 (1947).
106. C. W. Dence and K. V. Sarkanen, Tappi, 43, 87 (1960).
107. R. Pschorr, Ann. Chem., 391, 23 (1912).
108. L. Andersen, Finska Kemistamfundets Medd., 66, 1 (1957); Chem. Abstr., 52, 5329.
109. A. A. Chuksanova, L. L. Sergeeva and N. N. Shorygina, Izvest. Akad. Nauk S.S.S.R., Otdel. Khim. Nauk, 1959, 2219; Chem. Abstr., 54, 10935 (1960).
110. A. A. Chuksanova and N. N. Shorygina, Izvest. Akad. Nauk, S. S. S. R., Otdel. Khim. Nauk, 1960, 1511; Chem. Abstr., 55, 1499 (1961).
111. L. L. Sergeeva, N. N. Shorygina and B. V. Lopatin, Izvest. Akad. Nauk, S. S. S. R., Otdel. Khim. Nauk, 1962, 1295; Chem. Abstr., 58, 5552 (1963).
112. J. B. Van Buren and C. W. Dence, Tappi, 50, 553 (1967).
113. L. L. Sergeeva and N. N. Shorygina, Izvest. Akad. Nauk S. S. S. R. Ser. Khim., 1905, 1630; Chem. Abstr., 64, 1992 (1966).
114. L. L. Sergeeva, N. N. Shorygina and B. V. Lopatin, ibid., 1967, 2114; Chem. Abstr., 69, 3835m (1968).
115. N. P. Miklailov, N. N. Shorygina and B. V. Lopatin, Izvest. Akad. Nauk, S. S. S. R., Ser. Khim., 1967, 1709; Chem. Abstr., 68, 49232v (1968).
116. M. L. Kee, Ph.D. dissertation, McGill Univ., Montreal, Canada, 1968.
117. D. Ginsberg, J. Am. Chem. Soc., 73, 702 (1951).
118. G. Kubiczek, Monatsh. Chem., 76, 54 (1946).
119. C. A. Fletscher, Org. Syn., 33, 65 (1953).
120. K. Sato and H. Mikawa, Bull. Chem. Soc. Japan, 33, 1736 (1960).
121. Th. Zincke and O. Hahn, Ann. Chem., 329, 1 (1903).
122. Ph. Traynard and A. Robert, Bull. Chim. Soc. France, 1954, 1364.
123. I. Sobolev, J. Org. Chem., 26, 5080 (1961).
124. K. Sato, A. Kobayashi and H. Mikawa, Kami-pa Gikyoshi, 17, 431 (1963); Chem. Abstr., 61, 12188 (1964).
125. A. E. Menke and W. B. Bentley, J. Am. Chem. Soc., 20, 316 (1898).
126. R. M. Hann and G. C. Spencer, ibid., 49, 535 (1927).
127. R. A. McIvor and J. M. Pepper, Can. J. Chem., 31, 298 (1953).
128. A. W. Sohn, Holzforschung, 7, 1 (1953).
129. J. M. Pepper and J. A. MacDonald, Can. J. Chem., 31, 476 (1953).
130. L. C. Raiford and J. G. Lichty, J. Am. Chem. Soc., 52, 4576 (1930).
131. L. C. Raiford and D. E. Floyd, J. Org. Chem., 8, 358 (1943).
132. C. Y. Hopkins and M. J. Chisholm, Can. J. Research, 24B, 208

(1946).
133. G. D. Thorn and C. B. Purves, Can. J. Chem., 32, 373 (1954).
134. M. Anbar and D. Ginsberg, Chem. Rev., 54, 925 (1954).
135. D. Ginsberg, J. Am. Chem. Soc., 73, 2723 (1951).
136. L. C. Raiford and R. E. Silker, J. Org. Chem., 2, 246 (1937).
137. M. Matell, Acta Chem. Scand., 9, 1017 (1955).
138. J. Ehrlich and M. T. Bogert, J. Org. Chem., 12, 522 (1947).
139. D. H. McDaniel and H. C. Brown, ibid., 23, 420 (1958).
140. R. H. Rosenwald, J. Am. Chem. Soc., 74, 4602 (1952).
141. P. W. Robertson, J. Chem. Soc., 93, 788 (1908).
142. K. R. Kavanaugh and J. M. Pepper, Can. J. Chem., 32, 216 (1954).
143. E. C. Horning and J. A. Parker, J. Am. Chem. Soc., 74, 2107 (1952).
144. N. Migita, J. Nakano and A. Ishizu, J. Japan Wood Res. Soc., 1, 55 (1955); Chem. Abstr., 50, 7450 (1956).
145. A. A. Levine, J. Am. Chem. Soc., 48, 797, 2719 (1926).
146. C. W. Dence, J. A. Meyer, K. Unger and J. Sadowski, Tappi, 48, 148 (1965).
147. R. Fort, J. Sleziona and L. Denivelle, Bull. soc. chim. France, 1955, 810.
148. F. Brüggem nn, J. Prakt. Chem., 53, 250 (1896).
149. C. L. Jackson and H. A. Torrey, Am. Chem. J., 20, 395 (1898).
150. A. F. Holleman, Chem. Rev., 1, 187 (1924).
151. C. K. Ingold, Quart. Rev., 11, 1 (1957).
152. M. S. Newman, Steric Effects in Organic Chemistry, Wiley, New York, 1956, pp. 167-182.
153. Ref. 58, pp. 256-259.
154. Ref. 75, pp. 142, 143.
155. L. M. Stock and A. Himoe, Tetrahedron Letters, 1960, No. 13, 9.
156. A. E. Favorskii, J. Russ. Phy. Schem. Soc., 38, 741 (1906).
157. K. V. Sarkanen and C. W. Dence, J. Org. Chem., 25, 715 (1960).
158. J. Gierer and H. F. Huber, Acta Chem. Scand., 18, 1237 (1964).
159. R. W. Strauss, Ph.D. Thesis, State Univ. of New York College of Forestry, 1960.
160. I. W. Ruderman, Ind. Eng. Chem., Anal. Ed., 18, 753 (1946).
161. A. W. Francis and A. J. Hill, J. Am. Chem. Soc., 46, 2498 (1924).
162. E. Grovenstein and U. V. Henderson, Jr., J. Am. Chem. Soc., 78, 569 (1956).
163. K. U. Matsmoto, Chem. Ber., 11, 122 (1878).
164. C. Gustafsson and L. Andersen, Paperi Puu, 37, 1 (1955).
165. W. B. Bentley, Am. Chem. J., 24, 171 (1900).
166. I. A. Pearl, J. Am. Chem. Soc., 68, 1100 (1946).
167. A. H. Salway, J. Chem. Soc., 95, 1155 (1909).
168. H. Erdtman and J. Gripenberg, Acta Chem. Scand., 1, 71 (1947).

169. J. Hine, Physical Organic Chemistry, 1st ed., McGraw-Hill, New York, 1956, p. 344.
170. W. Qvist, Chem. Ber., 60, 1847 (1927).
171. K. V. Sarkanen and R. W. Strauss, Tappi, 44, 459 (1961).
172. N. N. Shorygina, L. L. Sergeeva and B. V. Lopatin, Izvest. Akad. Nauk S. S. S. R., Ser. Khim, 392; Chem. Abstr., 67, 3871d (1967).
173. A. Ivancic and S. A. Rydholm, Svensk Papperstidn., 62, 554 (1959).
174. Ref. 169, p. 340.
175. Ref. 58, p. 294.
176. E. Adler, I. Falkehag and B. Smith, Acta Chem. Scand., 16, 529 (1962).
177. E. V. White, J. N. Swartz, Q. P. Peniston, H. Schwartz, J. L. McCarthy and H. Hibbert, Tech. Assoc. Papers, 24, No. 1, 179 (1941).
178. C. W. Dence, unpublished results.
179. E. Müller, K. Ley and W. Kiedaisch, Chem. Ber., 87, 1605 (1954).
180. J. Gess, State Univ. of New York College of Forestry, unpublished results.
181. R. Willstätter and H. E. Müller, Chem. Ber., 44, 2182 (1911).
182. R. Willstätter and A. Pfannenstiel, ibid., 37, 4745 (1904).
183. C. L. Jackson and R. D. MacLaurin, Am. Chem. J., 38, 127 (1907).
184. J. Cason, in Organic Reactions, Vol. IV, R. Adams, Ed., Wiley, New York, 1948.
185. F.-L. Kung, Ph.D. Dissertation, McGill Univ., Montreal, Canada, 1967.
186. L. L. Sergeeva, N. N. Shorygina and B. V. Lopatin, Izvest. Akad. Nauk S. S. S. R., Otdel, Khim. Nauk, 1964, 1254; Chem. Abstr., 61, 11915 (1964).
187. K. Ley and E. Müller, Chem. Ber., 89, 1402 (1956).
188. W. Flaig, Sci. Proc. Roy. Dublin Soc., Ser. A1, No. 4, 149 (1960); Chem. Abstr., 55, 3896 (1961).
189. W. Flaig, Suomen Kemistilehti, 33A, 229 (1960).
190. H. Erdtman, Proc. Roy. Soc., (London), A143, 177, 191, 228 (1933).
191. H. Erdtman and M. Granath, Acta Chem. Scand., 8, 811, 1442 (1942).
192. H. Erdtman and N. E. Stjernstrom, ibid., 13, 653 (1959).
193. J. Eliasek and A. Jungwirt, Collection Czech. Chem. Commun., 28, 2163 (1963); Chem. Abstr., 60, 443 (1964).
194. M. I. Chudakov and A. V. Milovanov, Khim. Perarabotka Drevesing, Ref. Inform. No. 26, 9 (1965); Chem. Abstr., 65, 4093 (1966).
195. A. v. Wacek, K. Kratzl and A. v. Bézard, Chem. Ber., 75, 1348 (1942).
196. G. C. Arnold, F. G. Simmonds and C. E. Curran, Paper Trade J.,

107, No. 10, 32 (1938).
197. F. Loschbrandt and C. U. Wetlesen, Svensk Papperstidn., 61, No. 18B, 656 (1958).
198. P. B. Sarkar, J. Indian Chem. Soc., 11, 777 (1934).
199. W. J. Powell and H. Whittaker, J. Chem. Soc., 125, 357 (1924).
200. K. Freudenberg, W. Belz and C. Niemann, Chem. Ber., 62, 1554 (1929).
201. N. N. Shorygina and A. A. Chuksanova, Doklady Akad. Nauk S. S. S. R., 86, 1135 (1952); Chem. Abstr., 47, 12296 (1953).
202. J. Polcin, Chem. zvesti, 10, 450 (1956); Chem. Abstr., 51, 7004 (1957).
203. K. Sato and C. Mikawa, Bull. Chem. Soc., Japan, 35, 477 (1962).
204. F. C. Cross and E. J. Bevan through F. E. Brauns, Chemistry of Lignin, Academic Press, New York, 1952, p. 310.
205. S. Hilpert, Papier-Fabr., 24, 145 (1926).
206. J. H. Pedersen and H. K. Benson, Pacific Pulp and Paper Ind., 14, No. 8, 48 (1940).
207. P. Waentig and W. Gierisch, Z. angew. Chem., 32, 173 (1919).
208. N. N. Shorygina, O. P. Grushnikiv and V. D. Tychima, Izvest. Akad. Nauk S. S. S. R., Ser. Khim., 1967, 317; Chem. Abstr., 67, 3869 (1967).
209. F. Loschbrandt, Bleaching of Sulfate Pulp, Technical Assoc. of the Pulp and Paper Industry, New York, 1941, 11. 37-43.
210. P. Waentig and W. Gierisch, Z. physiol. Chem., 103, 87 (1918).
211. E. E. Harris, J. Am. Chem. Soc., 58, 894 (1936).
212. W. Fuchs, Brennstoff-Chem., 9, 348, 363 (1928).
213. A. Friedrich and E. Peliken, Biochem. Z., 239, 461 (1931).
214. C. Doree and L. Hall, J. Soc. Chem. Ind., 43, 257 (1924).
215. J. Polcin, Chem. zvesti, 12, 60 (1958); Chem. Abstr., 52, 11410 (1958).
216. N. N. Shorygina and L. I. Kolotova, Zhur. Obschei Khim., 27, 1641 (1957); Chem. Abstr., 52, 1605 (1958).
217. H. W. Giertz, Svensk Papperstidn., 46, 152 (1943).
218. J. B. Van Buren and C. W. Dence, unpublished results.
219. W. Lautsch and G. Piazolo, Chem. Ber., 73, 317 (1940).
220. T. Nakamura, J. Okuma, I. Sasaki and S. Kitaura, Kogyo Kagaku Zasshi, 61, 1073 (1958); Chem. Abstr., 55, 21578 (1961).
221. J. Marton and E. Adler, Acta Chem. Scand., 15, 370 (1961).
222. B. O. Lindgren and H. Mikawa, Acta Chem. Scand., 11, 826 (1957).
223. T. Enkvist, Svensk Kem. Tid., 72, 93 (1960).
224. N. N. Shorygina, T. V. Izumrudova, N. M. El'khones and K. M. Starostina, Gidroliz, i Lesokhim. Prom., 11, No. 6, 8 (1958); Chem. Abstr., 53, 709 (1959).
225. T. Nakamura and S. Kitaura, Abstr. 3rd Lignin Symposium, Fukuoka

(1958) through F. E. Brauns and D. A. Brauns, The Chemistry of Lignins, Supplement Volume, Academic Press, New York, 1960, pp. 295-296.
226. E. S. Gould, Mechanism and Structure in Organic Chemistry, H. Holt and Co., New York, 1959, p. 374.
227. L. L. Larson, Paper Trade J., 113, No. 21, 25 (1941).
228. N. N. Shorygina and L. I. Kolotova, Izvest. Akad. Nauk S. S. S. R., Otdel, Khim. Nauk, 1953, 526; Chem. Abstr., 47, 12806 (1953).
229. J. Polcin, Chem. zvesti, 8, 227 (1954); Chem. Abstr., 49, 7240 (1955).
230. J. Marton, Tappi, 47, 713 (1964).
231. K. Sato, A. Kobayoshi and H. Mikawa, Kami-pa Gikyoshi, 17, 647 (1963); Chem. Abstr., 61, 14886 (1964).
232. D. H. Grangaard, Tappi, 39, 270 (1956).
233. K. Nakajima, M. Okubo and S. Onoe, Kami-pa Gikyoshi, 18, 205 (1964); Chem. Abstr., 61, 16296 (1964).
234. N. N. Shorygina and L. I. Kolotova, Zhur. Obshchei Khim., 23, 2037 (1953); Chem. Abstr., 48, 7896 (1954).
235. G. J. Ritter, R. L. Mitchell and R. M. Seborg, Ind. Eng. Chem., 24, 1285 (1932).
236. J. Böeseken, Proc. Acad. Sci., Amsterdam, 35, 750 (1932).
237. A. M. Petrova, S. A. Berezkina, L. G. Kudryavtseva and Ya. N. Sukhushin, Tr. Tomskogo Gos Univ., Ser. Khim, 170, 73 (1964); Chem. Abstr., 63, 18446 (1965).
238. E. Adler and S. Hernestam, Acta Chem. Scand., 9, 319 (1955).
239. C. W. Bailey and C. W. Dence, Tappi, 52, 491 (1969).
240. K. Sato, K. Ebisawa and H. Mikawa, J. Chem. Soc. Japan, Ind. Chem., Sect., 61, 1090 (1958); Chem. Abstr., 55, 21579 (1961).
241. H. W. Giertz, Tappi, 34, 209 (1951).
242. J. Polcin, Chem. zvesti, 12, 108 (1958); Chem. Abstr., 52, 10569 (1958).
243. E. Erdmann, J. prakt. Chem., 22, 282 (1880).
244. S. A. Braddon and C. W. Dence, Tappi, 51, 249 (1968).
245. V. A. Vekhov, L. P. Garanzha, Z. D. Nekrasova and N. F. Kulish, Gidroliz, i. Lesokhim. Prom., 17, 16 (1961); Chem. Abstr., 61, 2046 (1964).
246. A. M. Petrova, Tr. Tomskogo Gos. Univ., Ser. Khim., 170, 58 (1964); Chem. Abstr., 63, 11851 (1965).
247. V. Jannzems, V. N. Sergeeva and L. N. Mozheiko, Latv. PSR Zinat. Akad. Vestis Khim. Ser., 1967, 630; Chem. Abstr., 68, abstr. no. 14209 r (1968).
248. J. Wolf, Sbornik Vyzhumnich praci z oboru celulosy a papiru, 1, 206 (1956); Chem. Abstr., 51, 13388 (1957).
249. Ref. 58, Chapter 6.

250. Ref. 75, Chapter 5.
251. R. O. C. Norman and R. Taylor, Electrophilic Substitution in Benzenoid Compounds, Elsevier, London, 1965, Chapter 3.
252. J. S. Carpenter and H. K. Benson, Pacific Pulp and Paper Ind., 14, No. 12, 17 (1940).
253. K. Kurschner, Cellulosechem., 12, 281 (1931); Chem. Abstr., 26, 3103 (1932).
254. Ph. Traynard and A. Robert, Assoc. tech. ind. papetiere Bull., 5, 401 (1951); Chem. Abstr., 46, 4223 (1952).
255. O. Routala and J. Sevón, Cellulosechem., 7, 113 (1926).
256. J. Okabe and Y. Hachihama, J. Chem. Soc. Japan, Ind. Chem. Sect., 58, 779 (1955); Chem. Abstr., 50, 10400 (1956).
257. T. V. Izumrudova, Zh. Prihl. Khim., 38, 2614 (1965); Chem. Abstr., 64, 6883 (1966).
258. T. M. Kalmykova, Gidroliz Lesokhim. Prom., 18, 16 (1965); Chem Abstr., 68, Abstr. no. 4118w (1968).
259. K. Shinra, J. Soc. Chem. Ind. Japan, 46, 1094 (1943); Chem Abstr., 43, 1973 (1949).
260. A. Schaarschmidt and P. Nowak, Cellulosechem., 13, 143 (1932).
261. Th. Lieser and W. Schaack, Chem. Ber., 83, 72 (1950).
262. F. König, Cellulosechem., 2, 93, 105, 117 (1921).
263. A. A. Chuksanova, O. P. Grusnikov and N. N. Shorygina, Izvest. Akad. Nauk S. S. S. R., Otdel Khim. Nauk, 1961, 1810; Chem. Abstr., 56, 7549 (1962).
264. K. Kürschner, Zellstoff-Faser, 32, 87 (1935); Chem. Abstr., 29, 6755 (1935).
265. A. Robert, Ph. D. Thesis, University of Grenoble, France, 1957; Abstr. Bull. Inst. of Paper Chem., 28, 1494 (1958).
266. N. N. Shorygina, N. P. Mikhailov and B. V. Lopatin, Khim. Prirod. Soedim., Akad. Nauk S. S. S. R., 2, 58 (1966); Chem. Abstr., 65, 909 (1966).
267. L. L. Sergeeva, A. A. Chuksanova and N. N. Shorygina, Izvest. Akad. Nauk, S. S. S. R., Otdel Khim. Nauk, 1957, 653; Chem. Abstr., 51, 14261 (1957).
268. A. A. Chuksanova, L. L. Sergeeva and N. N. Shorygina, Izvest. Akad. Nauk, S. S. S. R., Otdel Khim. Nauk, 1956, 250; Chem. Abstr., 50, 9737 (1956).
269. Y. Hachihama and M. Onishi, J. Soc. Chem. Ind., Japan, 39, Suppl. binding 362 (1936), Chem. Abstr., 31, 2423 (1937).
270. Y. Hachihama, S. Jodai and H. Sato, J. Chem. Soc. Japan, Ind. Chem. Sect., 52, 325 (1949); Chem. Abstr., 46, 7324 (1952).
271. K. Shinra, J. Soc. Chem. Ind. Japan, 47, 779 (1944); Chem. Abstr., 43, 1974 (1949).
272. F. J. Stevenson and M. A. Kirkman, Nature, 201, No. 4914, 107

(1964).
273. Ref. 75, p. 98.
274. E. Heuser, H. Rösch and L. Gunkel, Cellulosechem., 2, 13 (1921).
275. M. Phillips and M. J. Goss, J. Am. Chem. Soc., 55, 3466 (1933).
276. V. Stanik and J. Wolf, Sb. Vyskum. Prac. Odborn Celulosy Papiera, 10, 125 (1965); Chem. Abstr., 65, 13942 (1966).
277. B. B. Field and J. Grundy, J. Chem. Soc., 1955, 1110.
278. M. I. Chudakov, A. V. Antipova, A. B. Polyak and M. N. Raskin, Dokl. Akad. Nauk S. S. S. R., 164, 598 (1965); Chem. Abstr., 64, 908 (1966).
279. M. I. Chudakov and A. V. Antipova, U. S. S. R. Patent No. 181,083, April 15, 1966; Chem. Abstr., 65, 8827 (1966).
280. K. Kürschner, Cellulosechem., 18, 70 (1940); Chem. Abstr., 35, 6441 (1941).
281. H. Mikawa, E. Nokihara and K. Sato, J. Chem. Soc. Japan, 53, 94 (1950); Chem. Abstr., 46, 8852 (1952).

11

OXIDATION

H.-M. Chang and G. G. Allan

I. <u>Introduction</u>. Being phenolic in nature the lignin macromolecule is prone to oxidation by either homo- or heterolytic pathways (1) depending on the oxidant and the reaction conditions used. Furthermore, many of these oxidation reactions are not only of academic but are also of considerable technical importance. However, with few exceptions these have not been studied carefully from a mechanistic viewpoint, and in most cases relatively little detailed information is available. Nevertheless, it is felt that the best approach to a study of lignin oxidation reactions is to examine the mode of transfer of electrons from the substrate to the oxidant. The oxidations of lignin by various classes of oxidants have been treated accordingly throughout this chapter. It should be noted that although the mechanisms proposed are based on rather scanty experimental data and may require future modification, they do provide a better current understanding of the nature of these reactions.

Since the lignin macromolecule is susceptible to a wide variety of oxidants, the oxidation reactions have been arbitrarily classified into three categories according to the degree of lignin degradation achieved. These comprise oxidations, a. degrading lignin to aromatic carbonyl compounds and carboxylic acids, b. degrading aromatic rings, and c. limited to specific groups.

The first, which leads to the formation of aromatic carbonyl compounds and carboxylic acids is of special importance in the characterization of lignin as well as in the commercial production of aromatic compounds from pulping wastes. This category includes oxidations using nitrobenzene, molecular oxygen or metal oxides, all in alkaline medium, as well as permanganate degradation of methylated lignin. The second category is also of practical technical importance and covers the action of the strong oxidants including peracetic and nitric acids, chlorine, chlorine dioxide and the oxidizing anions of hypochlorous and chlorous acids. The final category garners the more exotic oxidants employed in structural elucidation and functional group analysis exemplified by periodic and nitrosodisulfonic acid salts, dichlorodicyanobenzoquinone and the alkali peroxides.

The particular importance of the chlorination and nitration reactions has merited a separate treatment in Chapter 10, and likewise the enzymatic degradation of lignin, which also involves an oxidation process, is more appropriately included in Chapter 18.

II. Oxidations Degrading Lignin to Aromatic Carbonyl-Containing Compounds.

A. Alkaline Nitrobenzene Oxidation.

Among all of the oxidations in this category, the most illuminating has been the alkaline two-electron transfer process of nitrobenzene degradation which over the years has become a mainstay of lignin investigators. This procedure was introduced (2) as a diagnostic device by the Heidelberg group around 1940, being an adaptation of a long known technique (3) for the conversion of isoeugenol 1 to vanillin 2 (Figure 11.1).

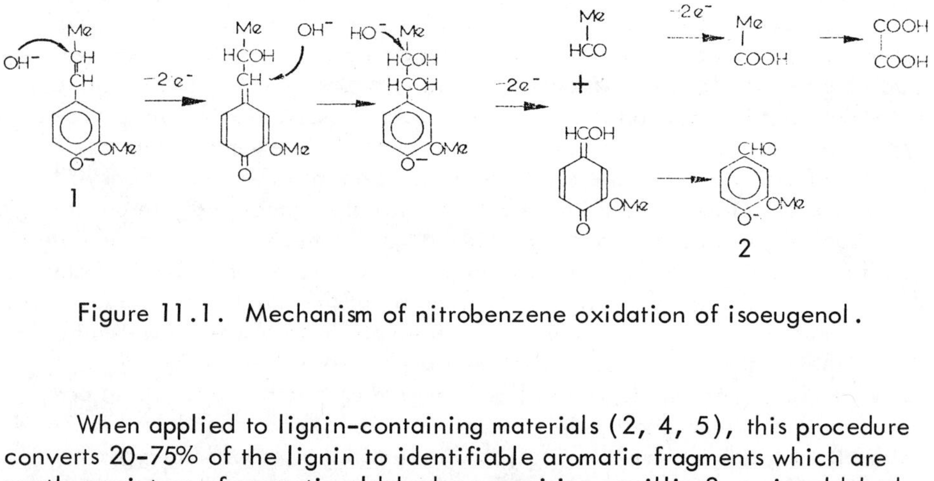

Figure 11.1. Mechanism of nitrobenzene oxidation of isoeugenol.

When applied to lignin-containing materials (2, 4, 5), this procedure converts 20-75% of the lignin to identifiable aromatic fragments which are mostly a mixture of aromatic aldehydes comprising vanillin 2, syringaldehyde 3 and p-hydroxybenzaldehyde 4, Table 11.1. The amounts of each aldehyde contained in this mixture are determined by the origin and structure of the lignin sample and their estimation is thus of crucial importance in structural studies. The importance of this mode of investigation has also been apparent in a botanical context where it has been adapted for taxonomical purposes (6-11). In addition, the commercial conversion of lignosulfonates from pulp wastes to vanillin has been allocated much attention and is the basis of current production of this flavoring agent (Chapter 19). More recently the reaction has been evaluated as a new approach to pulping (12).

The oxidation is normally carried out in the laboratory by heating the lignin-containing material with nitrobenzene and 2N sodium hydroxide in a stainless steel pressure vessel at elevated temperature. Optimum reaction

Table 11.1. Nitrobenzene Oxidation Products from Norway Spruce (16,17).

Oxidation Products	R'	R''	Yield, % of Klason Lignin	
			1[a]	2[b]
Vanillin 2	CHO	H	27.5	25.8
Syringaldehyde 3	CHO	OMe	0.06	-
p-Hydroxybenzaldehyde 4	-	-	0.25	-
5-Formylvanillin 5	CHO	CHO	0.23	-
Dehydrodivanillin 6	CHO	½	0.80	2.2
Vanillic acid 7	COOH	H	4.8	1.3
Syringic acid 8	COOH	OMe	0.02	-
5-Formylvanillic acid 9	COOH	CHO	0.1	-
5-Carboxyvanillin 10	CHO	COOH	1.2	0.6
Dehydrodivanillic acid 11	COOH	½	0.03	-
Guaiacol 12	H	H	-	-
Acetoguaiacone 13	COMe	H	0.05	-

a According to Leopold (17)
b According to Pew (16)

conditions have been developed as a result of model compound studies by Leopold (13), and these have been endorsed by investigations carried out by Kavanaugh and Pepper (14), using aspen wood as the lignin substrate. In both cases the best yields were secured at a reaction temperature of 170-180° after a 2-2.5 hour heating period.

Under these conditions the nitrobenzene apparently functions as a two-electron acceptor and undergoes conventional stepwise reduction to nitrosobenzene, phenylhydroxylamine and aniline as classically indicated in Equation 1.

$$PhNO_2 \longrightarrow PhNO \longrightarrow PhNHOH \longrightarrow PhNH_2 \quad (1)$$

Secondary condensations between these reduction intermediates are well-known and azoxybenzene, azobenzene and p-hydroxyazobenzene as also depicted in Equation 1 have all been isolated from lignin oxidation mixtures (15, 16).

These reductive convolutions of the nitrobenzene lead of course to the concomitant oxidation of the lignin which, if derived from gymnosperms, yields predominately vanillin. This behavior of the softwoods is to be contrasted with the hardwoods where a slightly more complex picture emerges. Thus, in the case of the dicotyledon angiosperms a mixture of vanillin and syringaldehyde is obtained while the monocotyledonous grass lignins also afford p-hydroxybenzaldehyde as a third component. Furthermore, both gymno- and angiosperms form many other aromatic moieties in relatively minor amount. Thus, a typical nitrobenzene oxidation of a softwood (16, 17) affords the twelve products 2-13 collated in Table 11.1.

In spite of these and other efforts in the study of model compounds (16-26) which have been reviewed by Leopold (27) the mechanism of this unusual reaction is not firmly established. It probably proceeds through two steps, of which the first is the alkaline hydrolysis of the alkylaryl ether linkages combined with side-chain modification. The second step is the oxidation of the side-chains with the generation of aromatic aldehydes of various types which as previously indicated (Chapter 3) will thus reflect structural features of the parent lignin. These two steps can be formulated as in the sequence following, Equation 2.

$$\text{[structure with } C_3, \text{OMe]} \xrightarrow{OH^-} \text{[structure with } C_2, \text{OMe, } O^-\text{]} \xrightarrow[OH^-]{PhNO_2} \text{[structure with CHO, OMe, } O^-\text{]} \qquad (2)$$

The significance of the presence of p-hydroxyl groups in the oxidative conversion of the propyl side-chain to the aldehyde function was clearly demonstrated by the results of oxidations carried out using the model compounds as enumerated in Table 11.2.

Thus it was found that those compounds without a free phenolic hydroxyl group in the position para to the side-chain gave a negligible amount of oxidation products. Exceptions to this generalization were provided by those models which carried side-chains containing an α-carbonyl group and also by those capable of undergoing a reverse aldol condensation (28). Furthermore, in both exceptions aromatic acids rather than aldehydes were the resulting oxidation products.

The formation of an aromatic acid as a consequence of the retroaldol cleavage can be rationalized by invoking oxidation of the aldehyde released by the scission. It is well recognized that aromatic aldehydes without a stabilizing free phenolic hydroxyl group are labile under drastic alkaline conditions. Thus, veratraldehyde 14 readily undergoes a classical Cannizaro

Table 11.2. The Effect of p-Hydroxyl Group on Alkaline Nitrobenzene Oxidation (18-21).

Side chains	Yield of aromatic aldehydes and acids with different nuclei					
	Phenyl		Guaiacyl		3,4-dimethoxyphenyl	
	Aldehydes %	Acids %	Aldehydes %	Acids %	Aldehydes %	Acids %
$-CH_2-CH_2-CH_3$	0	0	6	0	-	-
$-CH_2-CO-CH_3$	0	1	-	-	0	8
$-CH(SO_3H)-CO-CH_3$	0	Traces	45	Traces	0	Trace
$-CO-CH(SO_3H)-CH_3$	0	52	31	5	0	31
$-CO-CH_2-CH_2(SO_3H)$	0	13	-	-	0	46
$-CH=CH-CH_3$	0	2	89	0	0	0
$-CH=CH-CO_2H$	0	0	60	10	-	-
$-CH=CH-CHO$	0.5	86	86	4	-	-
$-CH(SO_3H)-CH_2CHO$	1	81	-	-	-	-
$-CH=CH-CO-CH_3$	Traces	40	-	-	-	-
$-CH(OH)-CH_2-CH_3$	0	44	-	-	-	-
$-CO-CH(OH)-CH_3$	-	-	26	19	-	-

reaction, Equation 3.

$$\underset{14}{\text{CHO-Ar(OMe)(OMe)}} \xrightarrow{OH^-} \underset{15}{\text{COOH-Ar(OMe)(OMe)}} + \underset{16}{\text{H}_2\text{COH-Ar(OMe)(OMe)}} \qquad (3)$$

In the presence of nitrobenzene it may be oxidized to veratric acid via analogous hydride transfer (29, 30). Aldehydes containing phenolic hydroxyl substituents, on the other hand, are stable under such reaction conditions (18-21) and the stability of vanillin, syringaldehyde and p-hydroxybenzaldehyde reflects the resonance stabilization of the carbonyl group by the phenolate anion which tends to protect it from the nucleophilic attack of hydroxide anion which is of course the initial step in the base-catalyzed Cannizzaro and related reactions.

These facts taken with the observation that vanillin and not vanillic acid is the major oxidation product from softwood lignin lead inevitably to the conclusion that the oxidation must be preceded by the hydrolysis necessary to generate the stabilizing phenolate anion. Further support for this conclusion is provided by the reduction in vanillin yield obtained by oxidation of a softwood lignin which had been previously methylated with diazomethane (25). The results of the model compound studies by Leopold (22-25) and Pew (16), which are summarized in Tables 11.3, 11.4, 11.5 and 11.6, have demonstrated that the structural features of the side-chain determine the yields and nature of the oxidation products.

Table 11.3. Nitrobenzene Oxidation Products of Lignin Model Compounds (16,22-25).

Compounds			Yield %	
Formula	R or X	n	Vanillin	Vanillic Acid
	$-CH_2OH$	-	82	3
	$-CH=CH_2$	-	80	6
	$-CO-CH_3$	-	81	4
	$-CO-CH_2OH$	-	66	10
	$-CH_2-CH_2-CH_3$	-	17	-
	$-CO-CH_2-CH_3$	-	30	-
	$-CO-CH(OH)-CH_3$	-	33	20
	$-CH_2-CH=CH_2$	-	88	-
	$-CH=CH-CH_3$	-	90	Trace
	$-CH=CH-CHO$	-	86	4
	$-OH'$	2	87	0
	$-OH$	3	74	4
	$-OH$	4	72	-
	$-SO_3 Ba/2$	2	83	0
	$-SO_3 Ba/2$	3	70	-
	$-SO_3 Ba/2$	4	71	-
	H	2	63	11
	H	3	60	12
	H	4	60	-
	CH_3	2	88	0
	CH_3	3	71	4
	CH_3	4	73	-

Table 11.4. Nitrobenzene Oxidation Products of Lignin Model Compounds (16,22-25).

Compounds			Yield %			
	R	R'	Vanillin	Vanillic Acid	5-Formyl-Vanillin	5-Formyl Vanillic Acid
	CH_2OH	CH_2OH	0.8	-	13	6
	$CH_2CH=CH_2$	CH_2OH	2.2	-	9	11
	CHO	$CH_2CH=CH_2$	1.6	-	20	6
	CHO	CHO	1.5	-	12	26
	CHO	CO_2H	2.0	-	-	-

Table 11.5. Nitrobenzene Oxidation Products of Lignin Model Compounds (16,22-25).

Compounds			Yield %				
	R	R'	Vanillin	Vanillic Acid	5-Formyl-vanillin	5-Carboxy-vanillin	Dehydro-divanillin
	CHO	CO_2H	1.9	0	1.9	85	0
	$CH_2CH=CH_2$	CH_2OH	2.6	0	10.0	25	0
	$CH_2CH=CH_2$	$CH(OH)CH_3$	3.3	0	7.2	24	0
	$CH_2CH=CH_2$	$CH(OH(CH_3)_2$	3.2	0	7.9	17	0
	$CH_2CH=CH_2$	H	73.1	2.0	0	0.3	0
6-allyl-4-hydroxy-3',4',8-trimethoxyflavan			14.6	-	0.6	14	0
6-allyl-4-hydroxy-8-methoxyflavan			Trace	0	0.6	21	0
Eugetic acid			9.7	-	0	65	0
Dehydrodieugenol			0	0	0	0.4	69
Dehydrodiisoeugenol			22.6	1.4	0	7.5	0
Dihydrodehydrodiisoeugenol			11.9	Trace	0	Trace	0

Table 11.6. Nitrobenzene Oxidation Products of Lignans (16,22-25).

Lignans	Vanillin	Vanillic acid
Olivil 17	83	3
Pinoresinol 18	31	9
Lariciresinol 19	63	5
Matairesinol 20	15	2
Isoolivil 21	3	-
Conidendrin 22	1	-

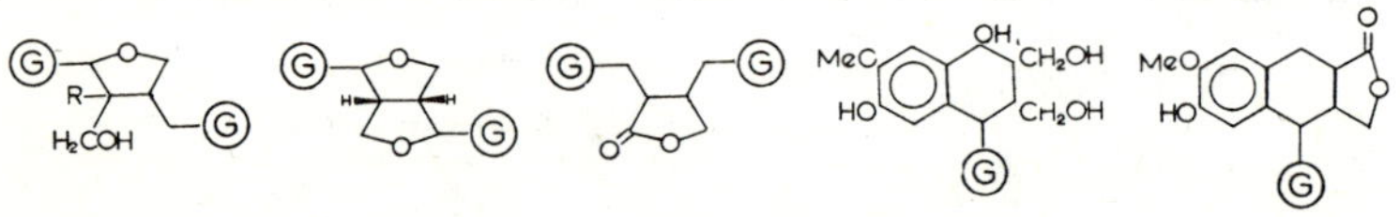

17, R=OH
19, R=H

The following conclusions can then be drawn:

1, Guaiacyl units having α-hydroxyl group or α,β-ethenoid structure in the side-chain should give high yields of vanillin (75-90%) on oxidation with alkaline nitrobenzene. The presence of α-carbonyl functions will lower the yield of vanillin and generate appreciable amounts of vanillic acid. A similar low yield of vanillin is obtained when the α-carbon carries an ether substituent; 2, Guaiacyl units with additional alkyl substitution ortho to the phenolic hydroxyl group are most resistant to degradation and yield only 25-35% of their weight of isolable and identifiable oxidation products. These comprise 5-formyl- and 5-carboxy-vanillin as well as both the corresponding vanillic acids. Both 5-formyl- and 5-carboxy-vanillin are quite stable under the reaction conditions and only negligible amounts can be converted to vanillin by decarboxylation; 3, The work of Leopold (23, 26) on the condensation products of resorcinol with guaiacyl compounds also demonstrated that guaiacyl units having an aromatic nucleus on the α-carbon of the side-chain, as exemplified by 23, are stable to oxidation with this alkaline nitrobenzene system. This explains the reduction in vanillin yield when lignins prepared under acidic conditions are subject to oxidation (31).

These results are probably best interpreted by the mechanism depicted in

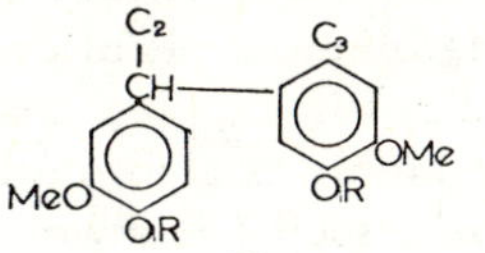

OXIDATION

[Structure 23: a diphenylmethane-type structure with C₂–CH–C₃ bridge, MeO, OR substituents]

23

Figure 11.1, using isoeugenol as an example. The reaction initially involves the interaction of the negative moiety in the phenolate anion with nitrobenzene. The actual oxidation probably proceeds from the phenolate-transitional complex by two-electron transfer through the quinone methide intermediate 24 which, by the assimilation of a hydroxide ion, affords 1-guaiacylpropane-1, 2-diol 25. Further oxidation of 25 through the same mechanism yields vanillin (see page 434).

The role of the phenolic hydroxyl group in the postulated mechanism is obvious since it allows the formation of a quinone methide intermediate. The significance of the presence of phenolic hydroxyl groups has been demonstrated by the results of model compound studies (cf. Table 11.2). For example, while isoeugenol 1 affords vanillin in excellent yield, neither its methyl ether 26 nor the analogous propenylbenzene 27 is oxidized under these reaction conditions. Oxidation of other model compounds containing no potential phenolate site also yielded no aldehydes (18-21).

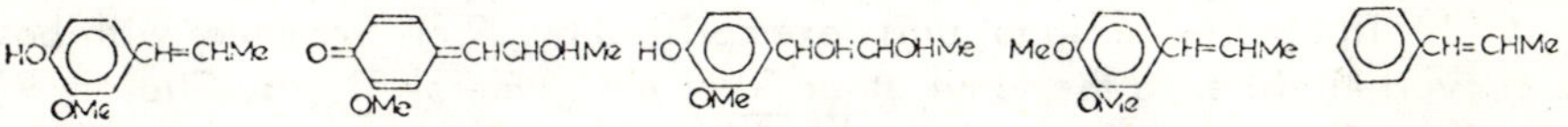

[Structures 1, 24, 25, 26, 27 shown with substituted aromatic rings]

1: HO-C₆H₃(OMe)-CH=CHMe
24: O=C₆H₃(OMe)=CHCHOHMe
25: HO-C₆H₃(OMe)-CHOHCHOHMe
26: MeO-C₆H₃(OMe)-CH=CHMe
27: C₆H₄(OMe)-CH=CHMe

The yields of vanillin and the oxidation behavior of the tetrahydrofuran derivatives 17 through 22 (cf. Table 11.6) can also be accommodated by this mechanism. Thus the cyclic ether linkages in pinoresinol 18 are relatively stable to alkaline hydrolysis while those in olivil 17 and lariciresinol 19 can be cleaved as a consequence of the presence of the primary alcohol function. Subsequent oxidation of the vinyl structure formed gives high yields of vanillin, Equation 4.

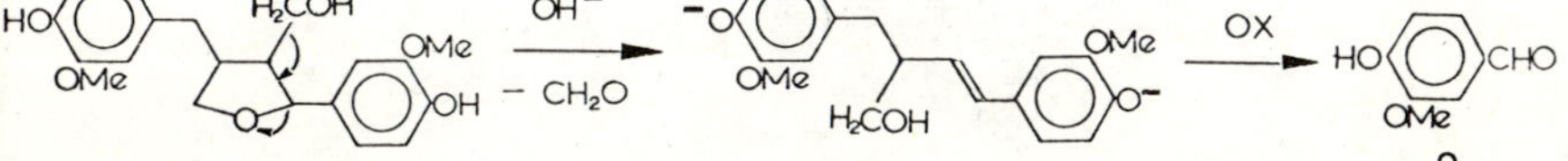

[Equation 4: Structure 19 (lariciresinol with H₂COH, OMe, OH) + OH⁻ → intermediate (with OMe, H₂COH, O⁻) − CH₂O → OX → vanillin 2 (HO-C₆H₃(OMe)-CHO)]   (4)

The oxidation of the α-carbonylic acetoguaiacone 13 and guaiacylpropanone 28 to vanillin 2 can also be rationalized by a closely related mechanism. The carbonyl group in these ketones is of course deactivated by the para phenolate anion which delocalizes its dipole, and the electron-withdrawing nature of this functional group prohibits electron transfer through a quinone methide intermediate. However, such a pathway can be envisaged if the ketones are in their enol forms as depicted in Figure 11.2 for guaiacylpropanone. A non-enolic degradation would be expected to afford vanillic acid exclusively.

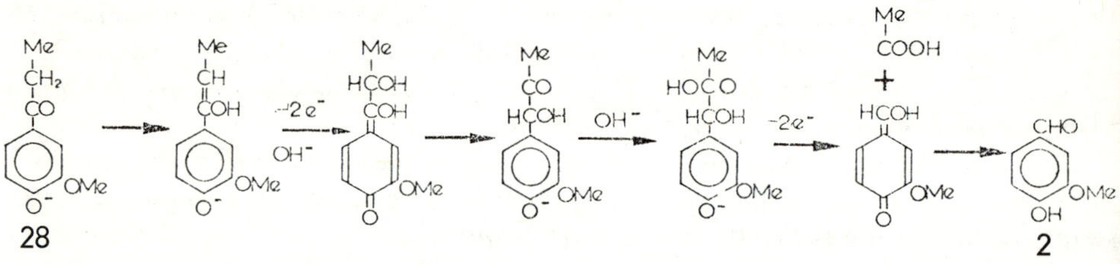

Figure 11.2. Mechanism of nitrobenzene oxidation of guaiacylpropanone.

When these model compound results are applied to structural units in lignin the following picture emerges:

(1) The β-O-4 dilignol units exemplified by 29 on treatment with base are converted either to the vinyl ether 30 or the glycerol 31 (cf. Chapter 9). Both of these intermediates are capable of yielding vanillin 2 on further oxidation, Equation 5.

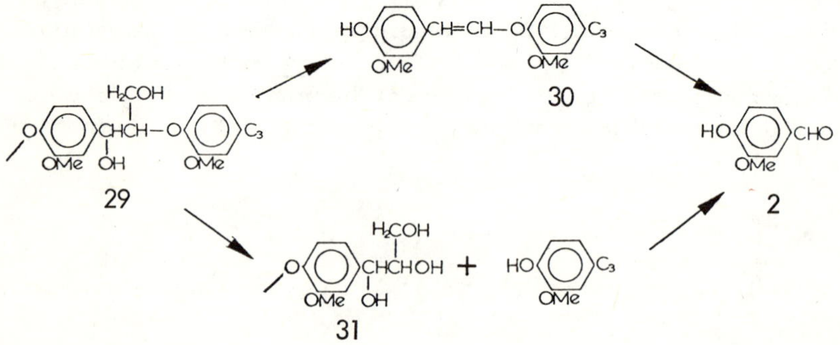

OXIDATION

(2) The loss of formaldehyde experienced by the β-O-4 dilignol is paralleled by the β-5 dilignol structures represented by 32. The vinyl group formed (Equation 6) spans the two aromatic rings and is readily cleaved to afford vanillin 2 and its 5-formyl relative 9.

$$\text{32} \longrightarrow \text{(diarylethylene)} \longrightarrow \text{9} + \text{2} \quad (6)$$

(3) The formation of the bridging vinyl linkage with the exclusion of formaldehyde also originates from the presence of β-1 dilignol 33 structures, Equation 7. In this case both segments of the diarylethylene 34 are capable of yielding vanillin and no 5-formyl vanillin is encountered.

$$\text{33} \longrightarrow \text{34} \longrightarrow 2\;\text{2} \quad (7)$$

(4) No vanillin is formed when the 5,5 dilignol moieties 35 are oxidized. Instead, the degradation proceeds in duplicate to afford dehydrodivanillin 6, noteworthy for its extreme insolubility and high melting point. It does however crystallize beautifully from pyridine (32).

(5) No facile oxidation occurs with the β-β dilignol unit classically represented by pinoresinol 18 itself. The stability of the cyclic oxygen heterocycles effectively tends to protect this entire unit from degradation and the vanillin yield is thereby considerably lowered.

(6) Other oxidation resistant units in lignin can be represented by the 5-O-4 36 and 1-O-4 37 dilignol type moieties. While no degradation

products have been identified from these precursors it can be assumed that such aldehydic structures as 38 will eventually be isolated.

B. Metal Oxide Oxidations.

1. Probable Mechanism of Metal Oxide Oxidation.

Like alkaline nitrobenzene, the alkali solutions of some heavy metals also oxidatively degrade lignin materials without destroying the aromatic nuclei. These oxidants include cupric, mercuric, silver and cobalt oxides. Aromatic aldehydes or aromatic carboxylic acids or mixtures of both are the major oxidation products, and their ratio depends on the particular oxidant used. Among the metallic oxides, silver has the highest oxidizing potential and hence gives acids as the major oxidation product. Cupric oxide, on the other hand, is a relatively weak oxidant and gives aldehydes instead while mercuric oxide occupying an intermediate position gives a mixture of both.

Compared to the alkaline nitrobenzene oxidation, metal oxide oxidations have not been studied at all extensively, and most of the work reported has been concentrated on the degradation of lignosulfonates. Little therefore is known concerning the precise nature of the reactions between the metal oxides and lignin macromolecule.

Unlike the alkaline nitrobenzene oxidation of lignin which is a two-electron transfer process, the metallic oxides employed are known to be one-electron transfer oxidants. However, it is pertinent to note that the oxidation products are rather similar to those of the nitrobenzene oxidation and show no indication of the presence of material resulting from oxidative coupling (33) which is of course characteristic of the oxidative treatment of phenolic compounds by one-electron transfer reagents (34). Reaction via three body collisions may be excluded, since occurrences of this sort are extremely infrequent. Kratzl, Gratzl and Claus (34) have shown recently that the primary step in the air oxidation of monohydric phenols such as 4-hydroxymethyl-guaiacol in aqueous alkali is the formation of resonance-stabilized free phenoxy radicals. It seems very likely that the oxidation of lignin and lignin models by metal oxides will proceed via the same initial step, that is, the formation of resonance-stabilized free phenoxy radicals 39 as depicted in Equation 8 for vanillyl alcohol. However, before a phenoxy radical can couple with another the intervention of a second electron transfer from the radical to another cupric oxide molecule may occur which initiates the formation of a quinone methide intermediate 40.

$$\underset{39}{\overset{\text{OMe}}{\underset{}{\cdot O\langle O\rangle CH_2OH}}}\overset{-OH^-}{\xrightarrow{-e^-}}\underset{40}{\overset{\text{OMe}}{\underset{}{O=\langle O\rangle =CHOH}}}\overset{-OH^-}{\longrightarrow}\underset{2}{\overset{\text{OMe}}{\underset{}{HO\langle O\rangle CHO}}} \qquad (8)$$

Vanillin is then formed by the rearrangement of the quinone methide intermediate 40 which is formally identical with the key intermediate of the alkali nitrobenzene oxidation pathway.

Indeed, it is very probable that the oxidation of phenolic compounds of this sort by metal oxides is a competitive reaction between the oxidative coupling sequence and the two consecutive one-electron transfer reactions. Both oxidation routes have a common initial step which is the formation of the phenoxy radical. Since the resulting phenoxy radical is far less stable than the corresponding phenolate anion, the electron transfer reaction in the initial step must be much slower than both the successive electron transfer (step II) and the coupling of the radicals (step III). In other words, the formation of the phenoxy radicals (step I) is the rate-determining step in the overall reaction mechanism. This also implies that the concentration of radicals must be very low in the reaction mixture. Therefore, if an excessive amount of metal oxide is used in the oxidation, the possibility of the coupling of the radicals is substantially limited since the rate of the coupling is proportional to the second power of radical concentration whereas the rate of the successive electron transfer processes is proportional to the product of the cupric oxide and radical concentrations as expressed in Equations 9 and 10.

$$\text{Rate of radical coupling} = k_c \cdot [\text{phenoxy radical}]^2 \qquad (9)$$

$$\text{Rate of successive electron transfer reactions} = k_t \cdot [CuO][\text{phenoxy radical}] \qquad (10)$$

Furthermore, it is well-known that while the coupling of radicals requires no activation energy, the electron transfer reactions must require some energy, although this may not be very high in this particular case. Therefore it appears obvious that reactions carried out at high temperatures will increase the rate of the electron transfer processes without affecting the rate of coupling of the phenoxy radicals. Thus, the amount of material proceeding along the coupling pathway can be minimized to a negligible amount by carrying out the reaction at high temperatures in the presence of a considerable excess of the metallic oxide. This postulate provides a satisfying explanation for the fact that the highest vanillin yields are obtained when 13.5 moles of cupric oxide are employed per lignin building unit at 170-190° (35, 36). Since it is well established that oxidative coupling of phenolic compounds such as coniferyl alcohol and isoeugenol can be effected by many one-electron transfer oxidants such as silver oxide, ferricyanide, manganese dioxide, persulfate, ferric chloride and oxygen at fairly low temperatures (34, 37), the reduced vanillin yield observed in oxidations with lesser amounts of cupric oxide at lower temperature (38) can presumably be ascribed to the occurrence of the coupling reaction to some degree.

2. <u>Cupric Oxide Oxidation</u>. Cupric oxide as an oxidant is milder than the other two commonly used metal oxides of mercury (II) and silver. Due probably to the competitive nature of the metal oxide oxidation, as outlined previously, the oxidation potential of the oxidant will be a very important factor. A weak oxidant will tend to favor the oxidative coupling mechanism while a stronger oxidant will further oxidize the intermediate oxidation products. Cupric oxide probably has an appropriate oxidation potential to strike a suitable balance between these two opposing factors, and it has therefore been investigated rather more extensively than the other two metallic oxides. The use of this oxidant for lignin materials was first reported by Pearl in 1942 (38) and his continued studies on the oxidation of various lignosulfonate materials constitute the principal effort in this special area (35, 36, 39-47). If carried out under optimum conditions, the cupric oxide method gives results comparable to those attained with alkaline nitrobenzene. The nature of the oxidation products is also similar except that usually the yields of acetoguaiacone and vanillic acid are somewhat higher and this is at the expense of the yield of vanillin in the case of the oxidation of softwood lignin (35, 48). Similar results were also obtained with the syringyl moieties originating from the oxidation of hardwood lignin (44-46, 48). In this connection it is to be noted that Pepper and coworkers (48) have demonstrated recently that both acetoguaiacone and acetosyringone are stable under the cupric oxide oxidation conditions but are oxidized to vanillin and syringaldehyde, respectively, under the conditions prevailing in alkaline nitrobenzene oxidations. It has been suggested that the oxidation of acetoguaiacone by alkaline nitrobenzene may be preceded by the enolization of the ketone. The reason why this material is stable under the cupric oxide oxidation conditions is not completely clear, and it has been ventured (48) that this stability may be due to the fact that cupric oxide is a milder oxidant than nitrobenzene. For comparison, the yields of oxidation products obtained from spruce and aspen lignins by both methods are collected in Table 11.7. It should be noted that the vanillin yield from the nitrobenzene oxidation achieved in this work are lower than those recorded by Leopold (17). Furthermore, no acidic products such as vanillic or syringic acid were isolated and the yield of the former acid from spruce lignin has been separately reported to be as high as 4% (17).

In addition to the oxidation products isolated from the alkaline nitrobenzene oxidation, cupric oxide degradation has yielded three new products, 4,4'-dihydroxy-3,3'-dimethoxychalcone <u>41</u> (0.08%), 4,4'-dihydroxy-3,3'-dimethoxybenzil (vanillil) <u>42</u> (0.1%) and 4,4'-dihydroxy-3,3'-dimethoxybenzophenone <u>44</u> (0.1%), Equation 11.

Table 11.7. Yields of Oxidation Products of Spruce and Aspen Lignins with Nitrobenzene and Cupric Oxide Oxidants

Oxidation Products	Yield, % of Klason Lignin	
	Nitrobenzene	Cupric Oxide
Spruce Lignin		
Guaiacol	-	1.40
Vanillin	19.1	15.9
Acetoguaiacone	-	3.91
Total	19.1	21.21
Aspen Lignin		
p-hydroxybenzaldehyde	-	1.75
Vanillin	12.4	7.82
Acetoguaiacone	-	1.96
Syringaldehyde	30.0	20.0
Acetosyringone	-	5.34
Total	42.4	36.87

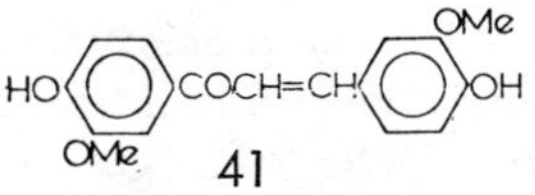

HO-⟨⟩-COCH=CH-⟨⟩-OH
    OMe   OMe
         41

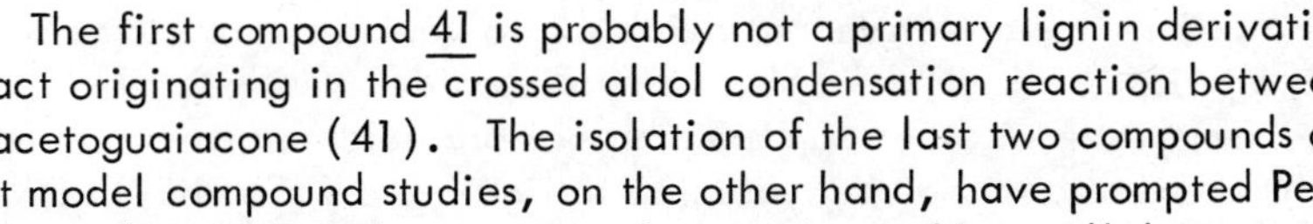

HO-⟨⟩-COCO-⟨⟩-OH ⟶ HO-⟨⟩-C(COOH)(OH)-⟨⟩-OH ⟶ HO-⟨⟩-CO-⟨⟩-OH   (11)
   OMe   OMe   OMe   OMe   OMe   OMe
       42   43   44

The first compound 41 is probably not a primary lignin derivative but an artifact originating in the crossed aldol condensation reaction between vanillin and acetoguaiacone (41). The isolation of the last two compounds and subsequent model compound studies, on the other hand, have prompted Pearl and his coworkers (39, 42, 43) to propose the presence of bivanillyl structures 45 in the lignin macromolecule.

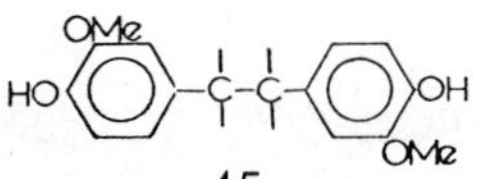

HO-⟨⟩-C-C-⟨⟩-OH   HO-⟨⟩-CHCH(CH₂OH)-⟨⟩-OH
   OMe   OMe   OMe OH   OMe
       45 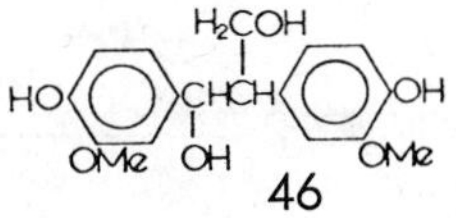   46

However, it should be emphasized that lignin building units linked together by a β-1 coupling mode (cf. Chapter 4) as in 46 may also yield vanillil 42 and the benzophenone derivative 44 on cupric oxide oxidation. Compound 44 is believed (39) to be formed from vanillil by a benzylic acid rearrangement. (Equation 11) followed by oxidative decarboxylation of the α-hydroxyacid 43.

Cupric oxide has also been recommended as preferable to nitrobenzene for lignin structural study because of its milder and more selective degrading action (48). Moreover, the isolation and identification of the oxidation products is somewhat simpler in the case of cupric oxide due to the absence of the interfering organic by-products generated from the nitrobenzene itself.

However, before this recommendation can be fully accepted, more model compound studies are needed to clarify the precise nature of the reaction. An additional motivation for further research in this area is provided by the fact that cupric oxide is a potential oxidant for the commercial production of vanillin from spent liquor since 90 to 95% of the cupric hydroxide can be regenerated and reused (49).

3. Other Metal Oxide Oxidations. The oxidation of lignosulfonate with silver oxide has also received some attention, again from Pearl and his coworkers (44-46, 50-52). With this oxidant, vanillic acid rather than vanillin is formed as a major oxidation product since the aldehyde is converted quantitatively to vanillic acid under the reaction conditions employed (53,54) In addition to vanillic acid itself, its 5-carboxy derivative is also a significant oxidation product from lignosulfonates.

Compared to cupric oxide, silver oxide is of course a stronger oxidant. The oxidation of the lignin is usually carried out in an alkaline suspension of silver oxide at refluxing temperatures (105°). Nonetheless, even at this temperature, vanillic and syringic acids have been shown to be unstable under these reaction conditions (50). Usually, silver oxide degradation yields less ether-soluble material than the corresponding cupric oxide oxidation (44-46, 50). However, the yield of vanillic and/or syringic acids from the silver-containing reaction may exceed the combined yields of vanillin and vanillic acid obtained using cupric oxide (44-45). This may be attributed to the fact that silver oxide is less selective and oxidizes, to vanillic acid, some of the intermediate products which are stable to cupric oxide.

Maximum yields of ether-soluble material are obtained when the molar ratios of silver oxide and sodium hydroxide to lignin are 3.75 and 21.0, respectively, and the reaction temperature is 105° (50).

Mechanistically the silver oxide conversion of vanillin to vanillic acid has been shown to proceed via a Cannizzaro reaction (54) as represented by Equations 12 and 13.

$$2\,RCHO + NaOH \xrightarrow{NaOH,\ Ag} RCH_2OH + RCOONa \qquad (12)$$

$$RCH_2OH + Ag_2O + NaOH \longrightarrow RCO_2Na + 2Ag + H_2 + H_2O \qquad (13)$$

OXIDATION

Overall these equations can be combined into Equation 14.

$$2\,RCHO + Ag_2O + 2\,NaOH \xrightarrow{NaOH} 2\,RCO_2Na + 2\,Ag + H_2 + H_2O \quad (14)$$

and it has been shown that the active metallic silver produced in the above reaction catalyzes the Cannizzaro reaction, converting vanillin to equivalent amounts of vanillic acid and vanillyl alcohol. In the presence of alkali alone, vanillin does not undergo the Cannizzaro reaction (55).

A related metal oxide oxidation utilizes mercuric oxide and gives a mixture of vanillin and vanillic acid as the principal oxidation products (51, 56). The combined yield of vanillin and vanillic acid was as high as 31.3% when a lignosulfonate (56) from softwood was used as the starting material. Prolonged treatment with this oxidant further converts the vanillin formed into vanillic acid (56, 57). It has been separately demonstrated that the reaction of vanillin with mercuric oxide yields 60 to 70% vanillic acid though the reaction is much slower than with silver oxide (57).

The effect of some other metal oxides and hydroxides have also been briefly studied (58, 59). Thus, vanadium, chromium and nickel derivatives have an oxidizing action which is dependent on the valency of the metal in the oxide. The lower valence oxides increase the yield of lignin acids while the higher valency metal derivatives augment the production of ether-soluble substances at the expense of these acids. The rapidity of the oxidation with the higher valency metal derivatives can be attributed to their ability to remove two or even three electrons from the substrate. However, although the high initial radical concentrations lead to rapid oxidation rates, termination is also facile. This reduces the extent of oxidation and the formation of carbonyl compounds predominates. The lower valency oxides on the other hand are comparable in their effect to the copper and silver oxides.

C. Molecular Oxygen and Alkali. For nearly half a century it has been known that lignin is very readily oxidized by oxygen in alkaline media. Even at room temperature, prolonged treatment of this sort will cause part of the lignin to dissolve with the loss of some of its methoxyl content (60). Under the more drastic conditions of elevated temperatures (circa 200°) and high pressures (55 atm) oxygen is continuously absorbed by hydrochloric acid lignins for example, and after 16 hr. all the lignin has passed into solution. Acetic, oxalic, formic and benzenepentacarboxylic acids are the major oxidation products (61, 62). However, such extensive degradation can be avoided, and Lautsch, Plankenhorn and Klink were able to show some twenty years later (63) that at moderate temperatures and pressures, pure vanillin could be obtained in about 10% yield by the air oxidation of either spruce wood or sulfite spent liquor. Processes based on these findings are now employed by several industrial chemical plants in North America for the commercial production of vanillin from lignosulfonates. The possibility of applying this process to alkali lignin has been also investigated (64).

The mechanism of vanillin formation by the oxidation of lignin with molecular oxygen in alkaline medium is probably quite similar to that of the metal oxide oxidations; that is, oxidation occurs by way of two consecutive one-electron transfers. However, subsequently this reaction must become far more complicated because both the diradical oxygen and the peroxy radicals formed during the first one-electron transfer are capable of reacting with the resonance-stabilized free phenoxy radicals in several different ways, as shown in Equation 15.

$$(15)$$

Thus an electron may be abstracted from a phenoxy radical, causing the formation of vanillin, as represented in Path I. Alternatively, the oxygenous radicals can couple with a phenoxy radical at the 3-position, forming a peroxide which on subsequent hydrolysis will afford the orthoquinone shown in Path II. This reaction is analogous to that proposed by Kratzl, Gratzl and Claus (65) for the degradation of biphenyl structures. Finally, coupling of the oxygen or peroxy radicals at the 1-position as illustrated in Path III and the subsequent cleavage of the peroxide so formed initiates the displacement of the side-chain and the formation of a p-quinone structure. This side-chain displacement has been recently demonstrated (66) in dilute alkali at 70° using as lignin models α-carbonyl and -carbinol guaiacyl compounds. All the o- and p-quinones formed are of course subject to further oxidation to various dibasic acids and carbon dioxide and the dissolution and degradation of hydrochloric acid lignin at high temperature (61) can probably be ascribed to this type of reaction. Reactions not involving the peroxy radical include the coupling of two phenoxy radicals. For guaiacyl derivatives with an uncondensed 5 position, 5-5 coupling is the predominant mode (65), see Figure 11.3. In contrast, for syringyl derivatives, phenoxy coupling is less likely since, of course all of the available ortho and para positions are blocked by substituents. Moreover, Steelink (67) has demonstrated that α-carbonyl syringyl derivatives are oxidized to remarkably stable radicals in solution while their labile

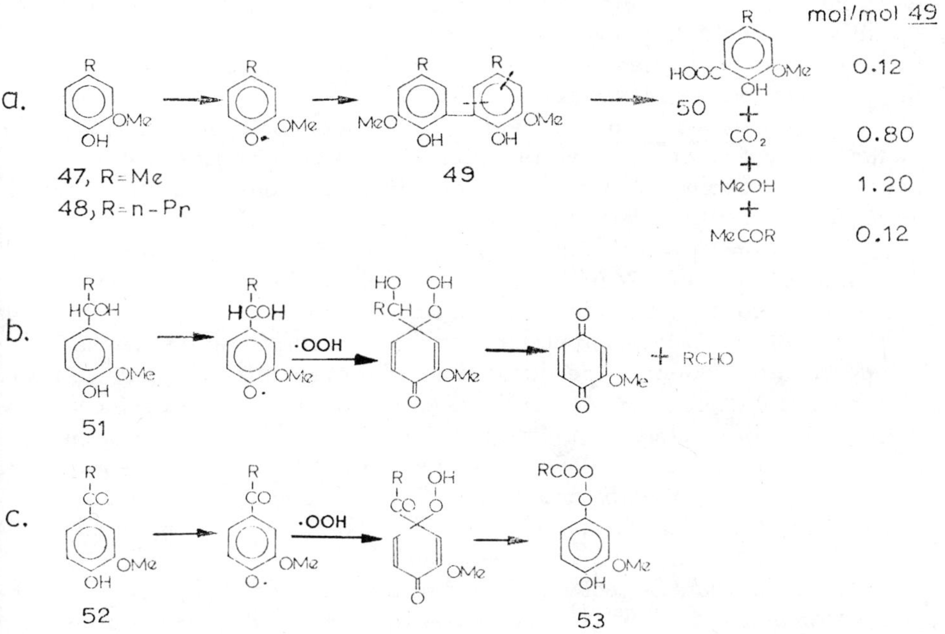

OXIDATION

Figure 11.3. Effect of side-chain on the air oxidation of guaiacyl models.

guaiacyl analogues yield coupling products (68).

The predominance of any one of the foregoing competitive reactions is probably governed mainly by the reaction temperature and side-chain structure. Higher temperatures favor electron transfer (69) and consequently the formation of vanillin. Both vanillin formation and side-chain replacement occur readily when benzylic alcohol groups are present. A saturated alkyl substitution in the side-chain, on the other hand, inhibits these two reactions. Thus model compound studies by Kratzl et al. (65, 66) have shown that the oxidation of methylguaiacol, 47, at 70° results in the dimerization through the coupling of phenoxy radicals to yield 49, whereas the oxidation of α-carbonyl 52 and α-carbinol 51 derivatives leads to the elimination of side-chains through Dakin-type reactions as shown in Figure 11.3.

Since the initial presence of a free phenolic hydroxyl group is necessary for all of the reaction pathways, the oxidation of lignin or its derivatives for the production of vanillin usually are carried out at 170° in 2N sodium hydroxide solution to simultaneously hydrolyze the aryl-ether linkages in lignin.

Under those conditions, the temperature probably is high enough to secure the hegemony of vanillin formation. Nonetheless, it is to be emphasized that at elevated temperatures (circa 200°) vanillin is unstable under these reaction conditions. Similar hydrolytic pretreatments have also been employed by Russian workers to augment the yields of oxalic and protocatechuic acids from the air oxidation of hydrolysis lignin (70).

D. Permanganate Oxidation. The structural information obtained by the degradation of lignin with alkaline nitrobenzene or metal oxides has been usefully supplemented by the use of neutral solutions of potassium permanganate as a mild oxidant.

This technique which was initially developed and later refined by Freudenberg et al. (71-74) first calls for a preliminary alkaline hydrolysis at 170-180° which is followed by a partial stabilization of the lignin by methylation. This of course serves to protect the aromatic rings with free phenolic groups against oxidation by permanganate. However, this stabilization is only successful to a degree and nuclear degradation can still occur as evidenced by the formation of succinic acid (75-78). Notwithstanding, the oxidation of methylated spruce wood at 50-95° under neutral conditions affords a mixture of veratric 54, isohemipinic 55 and dehydrodiveratric 56 acids in yields of 20, 6 and 4%, respectively. From a methylated hardwood (beech) on the other hand, the yields of aromatic acids were much lower and less than 10% of the lignin oxidized could be recovered as veratric (3%) trimethylgallic 57 (5%) and isohemipinic 55 (1.5%) acids. These low yields reflect the relative instability of the oxidation products and exemplify one of the drawbacks of the permanganate oxidation system.

Compensatory factors for these instabilities in neutral media have been evaluated by subjecting the pure aromatic acids to the same oxidizing reaction. These factors are collected in column 3 of Table 11.8 and their magnitude clearly prohibits the use of this oxidation method as a quantitative tool. Nevertheless, considerable structural information has been gleaned using this approach. Noteworthy examples include the isolation of isohemipinic 55 and dehydrodiveratric 56 acids by Freudenberg and his coworkers, which constitute the first direct evidence of the presence of the condensed guaiacyl units in spruce protolignin (71-73, 79-82, 85-89).

More recently Richtzenhain (75, 76, 83, 84) has extended the permanganate degradation to methylated spruce wood and various lignin preparations

Table 11.8. Stability of Aromatic Acids to Permanganate Oxidation

Aromatic Acid	Recovery after Oxidation	Compensatory Factor
Veratric	71.5%	1.3
Isohemipinic	11.1	9.0
Dehydrodiveratric	100	1.0
Trimethylgallic	57.2	1.75

Table 11.9. Permanganate Oxidation Products of Various Methylated Lignins, With and Without Preliminary Alkaline Hydrolysis (71-73,75,76,79-84)

Methylated Lignins	Oxidation Products %			
	Veratric Acid	Isohemipinic Acid	Metahemipinic acid	Dehydrodi-Veratric Acid
Protolignin				
with alkaline hydrolysis	20	6	-	4
without alkaline hydrolysis	4.9	0.9	-	-
Hydrochloric acid lignin				
with alkaline hydrolysis	7.8	3.4	0.9	1.5
without alkaline hydrolysis	2.9	1.9	1.3	0.2
Ethanol lignin				
without alkaline hydrolysis	9.2	0.95	1.3	0.5
Lignosulfonate	3.2	1.2	0.8	-
Thiolignin	10.2	6.65	-	3.8

which had not been subjected to the preliminary alkaline hydrolysis. The results secured in these studies are compared with those obtained by the Heidelberg school in Table 11.9.

It will be noted that the isolation of the somewhat elusive metahemipinic acid was actually first observed in oxidations of lignins which had been prepared using acidic treatments. Later, small amounts of metahemipinic acid were isolated from the oxidation products of a methylated fir wood (77), a milled wood lignin (78) and a DHP preparation (78).

In another application of the permanganate degradative technique, Kyogoku, Fujii and Hachihama studied the nature of the substitution on the p-hydroxyphenylpropane moieties of a beech lignin (90). Among the mixture of oxidation products, 6-methoxyisophthalic 58 and methoxytrimesic 59 acids were identified, whereas p-anisic acid 60 was conspicuous by its absence. It therefore follows that all of the p-hydroxyphenylpropane units in this hardwood lignin are highly condensed.

A similar high level of condensation in a p-coumaryl artificial lignin produced in vivo on potato parenchyma cells has also been demonstrated by the same oxidative technique (91).

In yet another elaboration of Freudenberg's basic approach, Hayashi and Namura (92) used a combination of ethylation, methylation and vapor phase chromatography to investigate the occurrence of o-dihydroxyphenyl groupings in a softwood lignosulfonate. Thus the lignin substrate was treated with diethyl sulfate before oxidation with the permanganate. The degradation products were then esterified using methanol and vapor phase chromatograms of these mixed esters disclosed the presence of methyl 3,4-diethoxybenzoate, 61, and methyl 3-methoxy-4-ethoxybenzoate, 62. The isolation of the diethoxy methyl ester 61 provides direct evidence of the presence of the catechol group which plays such an important role in the dichromate gelation of lignosulfonates (e.g. Chapters 13, 17).

In very recent investigations, Larsson and Miksche (93, 94) have found that considerably higher yields of aromatic carboxylic acids could be obtained if the oxidation was carried out at pH 12 instead of within the conventional range of pH 6-7. However, appreciable amounts of phenylglyoxylic acids are formed under these conditions though they can be further degraded to the corresponding aromatic acids by subsequent treatment with 5% H_2O_2 at pH 9-10. After methylation with diazomethane, the acids can be separated quantitatively by vapor phase chromatography. By applying this modified technique, these authors were able to obtain the methyl esters of veratric, isohemipinic, metahemipinic and 5-5'-dehydrodiveratric acids in yields of 23.4, 5.8, 1.1 and 2.8%, respectively, from prehydrolyzed and methylated Björkman lignin. In addition, using a combination of vapor phase chromatography and mass spectroscopy, Larsson and Miksche were also able to identify 2,5',6'-trimethoxy-4, 3'-dicarboxydiphenyl ether, 63, 3',4',5,6-tetramethoxybiphenyl-3-carboxylic acid, 64, and 2',3',5,6-tetramethoxybiphenyl-3-carboxylic acid, 65. The identification of 63 and 64 suggests the occurrence of 4-O-5 and 5-1 coupling of phenoxy radicals, respectively, during the lignin formation (cf. Chapter 4). The possibility of the presence of diphenyl ether linkages in lignin has previously been considered (95, 96) and indeed the diphenyl ether acid 63 has also been isolated in very small yield from the oxidative degradation products of methylated, prehydrolyzed and remethylated wood meal (78, 97). However, the unusually high yield (2.0%) of 63 obtained by Larsson and Miksche (94) led to the conclusion that diphenyl ethers, like biphenyl hydrocarbons, actually are structural units of appreciable importance in coniferous lignin.

In addition to 63, two other diphenyl ether acids, 4-carboxy-2,3',4'-trimethoxydiphenyl ether, 66, and 4,3'-dicarboxy-2,4',5'-trimethoxydiphenyl ether, 67, were isolated recently by Freudenberg and Chen (97) from the oxidation of methylated spruce lignin.

The formation of phenylglyoxylic acids 73 under the alkaline conditions probably results from the oxidation of the α-β-unsaturated units 68 formed in the alkaline prehydrolysis, such as vinyl ether and stilbene structures, see Equation 16. The analogous oxidation of styrene to phenylglyoxylic acid by alkaline permanganate has been reported (98). One can only speculate on the mechanistic details of this reaction because permanganate is an oxidant with manifold reaction paths and may function as either a one- or two-electron transfer oxidant (99, 100). Consequently, the sequence in Equation 16 is a tentative one.

[Structures 68, 69, 70, 71, 72, 73 with reaction scheme] (16)

There is hardly any doubt that the formation of a cyclic manganese ester 69 is the first step in the reaction leading, under mild reaction conditions, to the free glycol 70 (101). Cleavage will ensue from further reaction, with the ultimate production glyoxylic acid esters 71 and 72 representing possible intermediates.

Under acidic or neutral conditions the phenylglyoxylic acid formed undergoes oxidative decarboxylation to an aromatic aldehyde which is readily oxidized further to the corresponding acid. It is probable that the benzylic positions also are a major site for oxidative attack. Wiberg and Fox have suggested that the oxidation proceeds by abstraction of a hydrogen atom leaving a radical pair (100) which rearranges to an alkyl hypomanganate ester, Equation 17.

$$R_3CH + MnO_4^- \longrightarrow [R_3C \cdot MnO_4H^-] \longrightarrow R_3COMnO_3H^- \qquad (17)$$

This ester can decompose to the corresponding alcohol which in the case of lignin is often really a hemiacetal susceptible to further degradation, Equation 18.

[Reaction scheme of structures] (18)

OXIDATION

Finally, since ketones are also readily attacked by alkaline permanganate, their formation during oxidation does not stabilize the methylated lignin macromolecule or fragments thereof. Their oxidative degradation almost certainly proceeds via the corresponding enol (105)(Equation 18).

In unmethylated lignins, alcoholic hydroxyl groups provide a third side for the attack of permanganate (102, 103, 104). In neutral and mildly acid solution these groups are slowly attacked, but in basic media the rate of oxidation is rapid and is proportional to the hydroxyl ion concentration (102, 103), Equation 19.

$$\text{Thus,} \quad \frac{-d[\text{Lignin OH}]}{dt} = k[\text{Lignin OH}][\text{MnO}_4^-][\text{OH}^-] \quad (19)$$

III. *Oxidative Processes Degrading Aromatic Nuclei.* Being phenolic in nature, the aromatic nuclei of lignin are quite susceptible to the oxidative attack of a wide variety of reagents. Strong oxidants such as permanganate and dichromate in acidic solution not only oxidize lignin completely to carbon dioxide and dibasic acids but also cause extensive degradation of associated carbohydrate material. Some of the less drastic oxidants, however, may selectively attack lignin only and therefore are commonly used for holocellulose preparation and for the bleaching of commercial pulp. This class of substances includes chlorine, nitric acid, chlorine dioxide, sodium hypochlorite, peracetic acid, hydrogen peroxide and ozone. Unlike the mild oxidants discussed in Section II, which reacted mainly with the side-chains of lignin, this group of reagents attack primarily the aromatic nuclei of the lignin macromolecule. Of the reactants mentioned, chlorine and nitric acid have of course already been treated in detail in Chapter 10.

A. *Acid-Catalyzed Peracetic Acid Oxidation.* Peracetic acid has been recommended for the preparation of holocellulose (106-108) as a consequence of the pioneering research by Poljak (109-111). Although it has not yet found commercial acceptability, mainly because of cost, peracetic acid and its salts have been shown to be effective agents for the bleaching of pulps in general and for high yield pulps (112-119) in particular.

Although free radical reactions of peroxides are common, many reactions of peroxidic compounds often proceed through polarized intermediates (120) as a consequence of an inherent tendency for their weak peroxidic bonds to break heterolytically as depicted in Equation 20 for a protonated peracid.

$$\underset{\underset{H^+}{O}}{\overset{R}{\diagdown}}C\text{-}O\text{-}OH \rightleftharpoons \underset{O}{\overset{R}{\diagdown}}C\text{-}OH + OH^+ \quad (20)$$

It is more likely, however, that the polarized (or protonated) peracid rather than OH^+ (or AcO^+) (121) is the reactive species in two-electron transfer oxidations by peracetic acid (122). The heterolytic nature of the peracetic acid oxidation of lignin and lignin model compounds has been supported by recent kinetic studies (123, 124). Thus, Sakai and Kondo (123) have found that the peracetic acid oxidations of guaiacol, veratrole, 2,6-dimethoxyphenol and pyrogallol trimethylether follow second order kinetics being first order in peracetic acid and first order in substrate.

Endorsement of this view is provided by another model compound study in which Hatakeyama, Nakano and Migita (129) were unable to detect free radical formation in the peracetic acid oxidation of barium vanillylsulfonate and ethylvanillylsulfide by electron spin resonance spectroscopy. In apparent contradiction, using vanillyl alcohol as the substrate, free radicals could be detected which had an induction period of 50 min. However, genesis of the free radicals appeared to be independent of the oxidation by peracetic acid and was attributed to the polymerization of the vanillyl alcohol by an unspecified mechanism.

1. <u>Peracetic Acid Oxidation of Model Compounds</u>. The peracetic acid oxidation of lignin model compounds has been the subject of extensive studies in the past few years (118, 121, 123-135). In general, phenols and their ethers undergo rapid nuclear substitution reactions in available o- and p-positions. Oxidations by peracetic acid are commonly initiated by <u>ortho</u> and <u>para</u>-hydroxylation reactions. Catechols formed by the former reaction are converted first to <u>ortho</u> quinones which in turn are oxidized to muconic acids, Equation 21 (159, 135b). Hydroxylation at the <u>para</u> site gives hydroquinones subsequently converted to p-quinones and then to maleic and fumaric acid derivatives (135b). Substituent at the p-position of phenols often hinders oxidation of this site. For example, although p-cresol can afford a muconic acid and a corresponding lactone (Equation 22) via the o-hydroxylation initiation, it cannot be transformed to a p-quinone derivative (135b, 159).

(21)

(22)

OXIDATION

β-Naphthylmethyl ether reacts in a similar fashion, although the recovery of the original methoxyl group as the ester function in muconic acid monomethyl ester (Equation 23) clearly eliminates the necessity for an o-quinone intermediate and proves the occurrence of direct ring cleavage (135b). Similar reaction mechanisms also operate in the perbenzoic acid oxidation of 1,2-dimethoxybenzene (124) where cis, cis- and trans, trans-dimethyl muconates were isolated as degradation products (Equation 24).

$$\text{(23)}$$

$$\text{(24)}$$

In contrast, etherified phenols closer akin to major component units of lignin yield p-quinones through the intermediacy of hydroxylation at position 6 (Equation 25) (126).

$$\text{(25)}$$

Other loci for the initiation of the peroxidative degradation of phenol ethers may comprise the alkyl-aryl ether groups. This group includes, of course, the methoxyl and the etherified phenolic hydroxyl groups, and these can be split either oxidatively or hydrolytically. The existence of these cleavage modes is suggested by the oxidative formation of vanillyl alcohol from veratryl alcohol (125) and by the production of protocatechualdehyde from both vanillyl alcohol and apocynol (127) (Table 11.10).

In addition to the extensive effect of oxidation on the aromatic nuclei, it has been shown that when vanillyl alcohol, apocynol or guaiacylglycerol are treated with peracetic acid at low temperatures (25-40°), vanillic acid and vanillin are formed (125, 127, 128). These findings, as summarized in Table 11.10, indicate that side-chain oxidation may occur at those units containing a hydroxyl or carbonyl group in the α-position. Side-chain displacement, on the other hand, has not been convincingly confirmed in peracetic acid oxidation. Thus, the formation of 2-methoxyhydroquinone from apocynol and other lignin models (Table 11.10) may well be interpreted in terms of oxidation (131).

Table 11.10. Products Formed in the Peracetic Acid Oxidation of Model Compounds

Model Compounds	Oxidative Products					Miscellaneous
General Type: (R₂, OMe, OR)	CHO / OMe / OH	COOH / OMe / OH	CHO / OH	OH / OMe / OH	quinone (O=/OMe/=O)	
R_1 = OMe R_2 = CO_2H						Veratraldehyde, veratric acid and vanillyl alcohol (125)
R_1 = H R_2 = CH_2OH	+	+				β-carboxymuconic, maleic, fumaric, oxalic and acetylvanillic acids; vanillyl alcohol diacetate (125)
R_1 = H R_2 = CHO			+	+		
R_1 = H R_2 = COMe			+	+		Muconic, maleic, fumaric and oxalic acids (127)
R_1 = H R_2 = -CH(OH)-Me	+		+	+	+	
R_1 = H R_2 = CH(OH)CH(OH)CH$_2$OH	+		+			Products claimed (128): 2-hydroxy-1-guaiacylpropane-1-one 1-guaiacylpropane-1,2-diol 1-guaiacylpropane-1,3-diol

Syringyl lignin models have not been investigated in detail. Recent kinetic studies (123, 124) indicate that syringyl nuclei have a higher reactivity with peracids in terms of moles of peracid consumed per mole of compound. As a rule, of course, electrophilic substitution of aromatic rings is facilitated by the presence of electron-donating groups and impeded by electron-withdrawing groups. In conformity with these considerations, the ease of peracetic acid oxidation for lignin model compounds was found to follow the order 2,6-dimethoxyphenol > 1,2,3-trimethoxybenzene > guaiacol > 1,2-dimethoxybenzene (123), and for the oxidation by perbenzoic acid, 1,2,3-trimethoxybenzene > 1,2-dimethoxybenzene > anisole (124).

2. *Peracetic Acid Oxidation of Lignins and Wood.* The most important feature of peracetic acid oxidation of lignin-like compounds is the oxidative rupture of the aromatic nuclei to dibasic acids. The relevance of

OXIDATION

this degradation to lignin itself was first demonstrated by Sarkanen and Suzuki (134) in the oxidative degradation of Douglas-fir lignin to a water-soluble polyacid. Their conclusion was supported by the infrared characteristics and the lowered methoxyl content and ultraviolet absorption at 280nm of the water-soluble lignin formed. The peracetic acid lignins isolated from both soft- and hardwoods afford similar infrared and ultraviolet spectra and had unexpectedly similar compositions (121, 130), especially in methoxyl content (121). This observation was rationalized in terms of proposed mechanisms illustrated in Equations 26 and 27 (121).

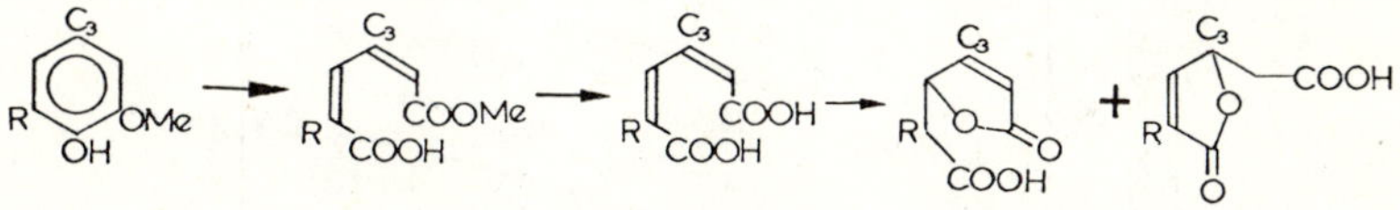

(26)

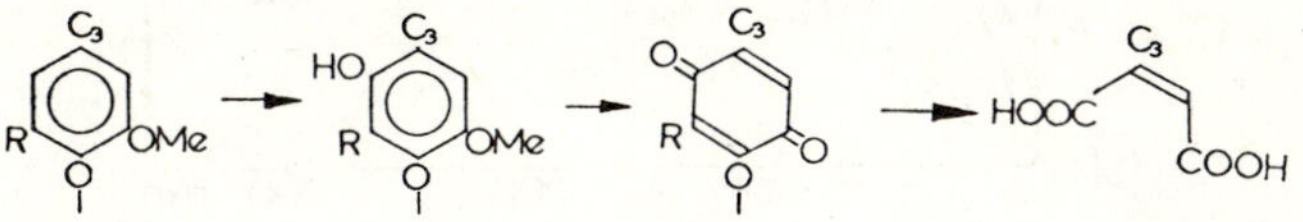

(27)

Equation 26 is probably the predominant oxidation route for guaiacyl units containing a free phenolic hydroxyl group, and the results (121, 134) indicate that the lactones rather than the acyclic form of the muconic acids are the main reaction products. It should be noted that some of the methoxyl groups in lignin become ester functions in the final oxidation products, and a conceivable pathway for this reaction is shown in Figure 11.4 (121, 134).

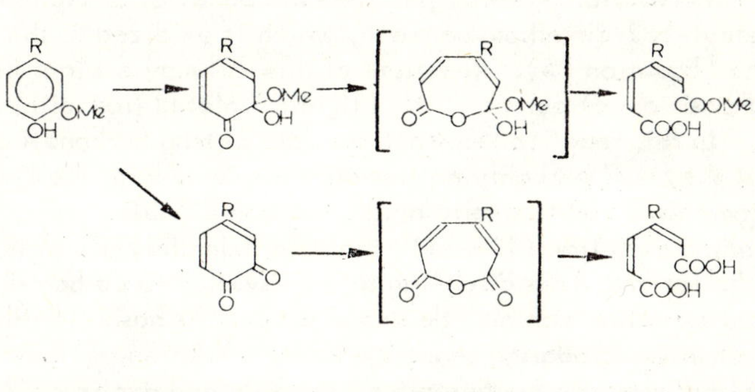

Figure 11.4. Degradation pathway for peracetic acid oxidation of guaiacyl derivatives (118, 129).

The reaction was further investigated by Sakai and Kondo (130), and their results obtained from the oxidation of pine dioxan lignin are illustrated in Figure 11.5. In this figure, curve I is the total MeOH recovered in saponification and curve II is the methanol formed by direct demethylation. The difference (curve III) represents the so-called loosely bound MeO which probably originates in the methyl esters of muconic acid.

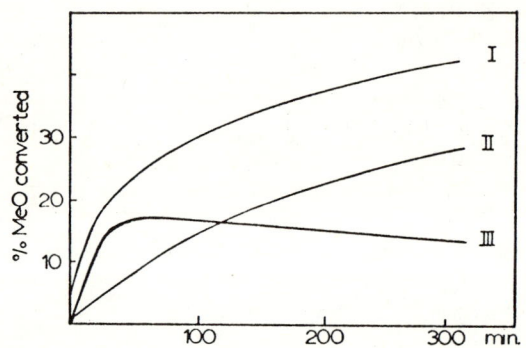

Figure 11.5. Conversion of methoxyl groups in pine dioxan lignin to methanol and methyl ester groups by the action of peracetic acid (130). Curve I: Percent MeO recovered as MeOH and ester-methoxyl; Curve II: Recovery as MeOH; Curve III: Conversion to ester-methoxyl.

Equation 27 is likely to predominate during the reactions of syringyl and etherified guaiacyl propane units which are initiated by hydroxylation at the 6-position. This reaction pathway parallels the behavior of the model compound, 4-methyl-1,2-dimethoxybenzene, which is oxidized to the corresponding p-quinone (Equation 25). Reactions of this sort may explain the nearly identical compositions of peracetic acid lignins isolated from either softwoods or hardwood. In this reaction sequence the fate of ring carbons 4 and 5 is unknown, but they will probably not remain associated with the lignin moiety and may become oxidized to oxalic acid or carbon dioxide.

The finding of Poljak (109-111) that approximately 4% of the carbon present in Scholler and Willstätter lignins is converted to carbon dioxide upon peracetic acid oxidation can also be explained on this basis. Furthermore, the involvement of this degradation sequence in the oxidation of lignin can be considered to be confirmed by the detection of maleic and oxalic acids among the peracetic acid oxidation (Table 11.10) although Hatakeyama, Nakano and Migita have suggested that these identified products are derived from the further oxidation of the muconic acid structures. The latter view, however, is doubtful because of the stability of such muconic acid derivatives under the reaction conditions (135a).

The reaction sequences summarized in Equations 26 and 27 serve as an interpretation of the substantial demethylation in the oxidation of lignin observed by Poljak (109-111). Thus, a methoxyl group is excised by hydrolysis of the muconic acid monomethyl ester and also during the formation of the 3,6-quinone intermediate. In addition, a direct demethylation occurs as evidenced by the detection of protocatechuic acid among the lignin degradation products (131) and the results obtained from the studies of model compounds (Table 11.10). Sakai and Kondo (130) also showed that the amount of lignin solubilized during peracetic acid oxidation is directly related to the amount of demethylation.

In accordance with kinetic behavior of model compounds in peracid solution (123, 124) Lai and Sarkanen (121) were able to show that syringyl groups react much faster than guaiacyl groups with peracetic acid during mild oxidation of hardwood periodate lignins.

The detection of vanillic acid (131) among the lignin degradation products suggests the occurrence of side-chain oxidation which may take place at those units containing a hydroxyl or carbonyl group in the α-position (Table 11.10). The extent of such side-chain oxidation in lignin may not be pronounced (121, 134).

B. <u>Sodium Hypochlorite, Chlorine Dioxide and Sodium Chlorite</u>. The chlorine derivatives, sodium hypochlorite, chlorine dioxide and sodium chlorite are of practical importance since they are commercially used as lignin removing pulp bleaching agents (136a). In the laboratory, they are also used for the preparation of holocellulose.

Hypochlorite bleaching is usually carried out at pH 11 or above. Under these conditions the hypochlorite anion is the predominant species, and very little (<0.1%) hypochlorous acid is present in the solution (<u>cf</u>. Figure 10.1). Hypochlorite oxidations of lignin preparations and lignin model compounds have shown, as illustrated in Figure 11.6, that only those nuclei having free phenolic hydroxyl groups are oxidized by hypochlorite ions. The etherified nuclei are not oxidized. Thus, guaiacol consumes 6 moles of hypochlorite, forming presumably methanol, oxalic acid and maleic acid, whereas veratrole is stable under the same reaction conditions, and no hypochlorite consumption was observed (137). Furthermore, the side-chains of both types of nuclei are oxidized if unsaturated groups are present or if oxygen-containing substituents are located on the α-position. Thus, while acetoguaiacone consumes 9 moles of hypochlorite and is completely oxidized, acetoveratrone takes up 3 or 4 moles of hypochlorite, forming presumably veratric acid which is stable under the reaction conditions (137).

Kinetic studies have shown that nucleic oxidation is a very rapid reaction (Figure 11.6), whereas the rapidity of side-chain oxidation varies somewhat with the constitution of these catenae. Thus, the side-chains of both

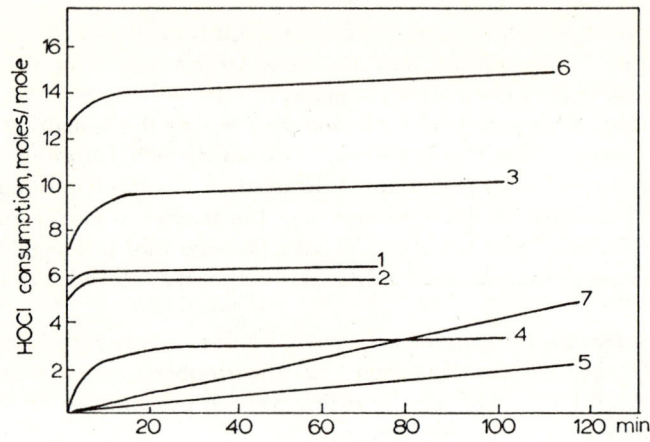

Figure 11.6. Consumption of hypochlorite by lignin model compounds (137).
1. Vanillic acid; 2. guaiacol; 3. isoeugenol; 4. isoeugenol methyl ether;
5. eugenol methyl ether; 6. guaiacylglycerol-β-(2-methoxyphenyl) ether;
7. 4-methyl ether of the γ-methyl analogue of the previous compound.

isoeugenol and its methyl ether are oxidized readily, but those of eugenol methyl ether and of some etherified dimers are oxidized only slowly (138, 139).

Model compound studies have also revealed that the nucleic oxidation is preceded by chlorination even at pH 11. Since there is almost no molecular chlorine present at this pH, it may be presumed that the reactive species in the chlorination reaction is hypochlorous acid or even perhaps the hypochlorite anion. Two experimental results support this presumption. First, chlorine is known to be able to chlorinate both phenols and phenol ethers (cf. Chapter 10). However, there is no hypochlorite consumption when veratrole and veratric acid are oxidized with hypochlorite (137). Second, chlorine is also known to catalyze the hydrolysis of phenol ethers (140). No such hydrolysis was observed in the hypochlorite oxidation of veratryl derivatives (137). Thus, it seems apparent that hypochlorous acid is the active chlorination species in the hypochlorite oxidations, and, since it is such a weak electrophile, it can only chlorinate phenolate anions at their ortho- and para-positions, as depicted in Equation 28.

Additional support for this conclusion is provided by the fact that in hypochlorite bleaching of pulp the maximum brightness is obtained at pH 9-9.5 which is a compromise range necessitated by the controlling pK values of hypochlorous acid (pK ~ 7.5) and the phenols (pK ~ 10.5) involved.

OXIDATION

$$\text{HO-Cl} + \underset{O^-}{\underset{|}{\bigcirc}}\text{-OMe} \xrightarrow{-OH^-} \underset{O}{\underset{\|}{\overset{H}{\underset{Cl}{\bigcirc}}}}\text{-OMe} \longrightarrow \underset{Cl\quad OH}{\bigcirc}\text{-OMe} \qquad (28)$$

Since the introduction of a chlorine atom on the aromatic nucleus deactivates the ring toward further substitution and chlorination is also competitive with a rapid nuclear oxidation, the latter usually takes place immediately after the chlorination. The oxidation of the nucleus is believed to proceed through an oxidative demethylation with the formation of an o-quinone followed by its ring opening to muconic acid derivatives and further oxidation to maleic and oxalic acids (cf. Chapter 10).

Hypochlorite treatment of various lignin preparations (137) shows an initial rapid consumption of the oxidant followed by a slower reaction (Figure 11.7).

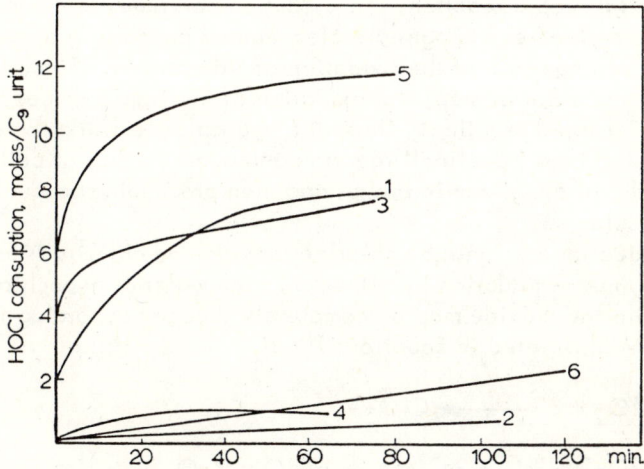

Figure 11.7. Consumption of hypochlorite by methylated and unmethylated lignins (137). 1, HCl lignin, unmethylated; 2, HCl lignin, methylated 29% MeO); 3, lignosulfonate, unmethylated; 4, lignosulfonate, methylated (23.6% MeO); 5, kraft lignin, unmethylated; 6, kraft lignin, methylated (26.4% MeO).

At 95°, a hydrochloric acid lignin or a lignosulfonate consumed 6-8 moles, whereas kraft lignin adsorbed 12 moles of hypochlorite per phenylpropane unit. Most of the consumption occurs during the first ten minutes. Methylation of the lignin preparations with diazomethane causes a remarkable reduction in the consumption of hypochlorite (Figure 11.7). In these methylated cases, the oxidations are probably limited to the side-chain and to those units whose phenolic ether linkages are split as a result of such side-chain oxidation.

At 20° a lignosulfonate consumed 2 moles of hypochlorite per lignin monomer unit during the first rapid reaction (137). At this stage it was found that only 0.2 mole of chlorine per lignin monomer was attached to the lignin, about one-third of the methoxyl groups having been cleaved. It is noteworthy that the amount of the methoxyl split off is about the same as the amount of the free phenolic hydroxyl group in the lignosulfonate. The lower chlorine content is probably due to the competition of the oxidation of the nucleus since chlorination is not feasible after quinone formation. It was also noted that at this stage the lignin was not degraded appreciably and the UV adsorption spectrum was almost unchanged. Further, slower oxidation caused a total consumption of 7 moles of oxidant per lignin monomer unit, and a deep-seated change in the UV adsorption spectrum. A satisfying explanation of the above studies (137-139) postulates that the hypochlorite oxidation of lignin begins at the monomers containing free phenolic hydroxyl groups and proceeds rapidly. After this initial, rapid reaction, the oxidation continues at the adjacent unit whenever the aryl-ether linkages are cleaved and phenolic groups are liberated, possibly as a consequence of the oxidation of side-chains. Since the oxidation of side-chains is a slower step, the oxidation of the lignin proceeds slowly after the initial rapid reaction. Thus, the hypochlorite oxidation of lignin can be considered to be a "peeling" reaction analogous to that of cellulose (136a) starting rapidly at the phenolic groups and then proceeding gradually throughout the molecule.

The oxidation of lignin by chlorine dioxide and chlorite is much more complicated than degradation by either chlorine water or hypochlorite. Theoretically, chlorine dioxide may be completely reduced according to the stepwise reactions formulated in Equations 29-31.

$$ClO_2 + e^- \longrightarrow ClO_2^- \qquad (29)$$

$$ClO_2^- + 3H^+ + 2e^- \longrightarrow HClO + H_2O \qquad (30)$$

$$HClO + H^+ + 2e^- \longrightarrow Cl^- + H_2O \qquad (31)$$

As implied by these equations, the last two steps are catalyzed by acid, whereas the first step is believed to dominate at pH values above 7. Therefore, in alkaline media, chlorine dioxide is reduced according to reaction 29 involving one oxidation equivalent. A similar reaction (Equation 32) also occurs in which two chlorine dioxide molecules disproportionate to form chlorate and chlorite ions in the presence of hydroxyl ions (141).

OXIDATION

$$2ClO_2 + 2OH^- \longrightarrow ClO_3^- + ClO_2^- + H_2O \qquad (32)$$

In acidic media chlorine dioxide is completely reduced and hence affords the five oxidation equivalents provided by the summation of the three stepwise reactions 29-31. Chlorite, on the other hand, is stable under neutral and alkaline conditions and hence has to be acidified to be an effective oxidizing agent. Obviously, chlorous acid rather than chlorite ion is the reactive species. In acidic media the chlorous acid formed is reduced according to two stepwise reactions, 30 and 31, and as a whole, therefore, provides four oxidation equivalents. The hypochlorous acid formed as an intermediate in the reduction of chlorous acid has a higher oxidation potential than chlorous acid and hence can oxidize the latter to chlorine dioxide via the reaction 33.

$$HClO + 2HClO_2 \longrightarrow 2ClO_2 + H_2O + H^+ + Cl^- \qquad (33)$$

Furthermore, hypochlorous acid, in mobile equilibrium with elemental chlorine, and chlorine dioxide may also be formed from chlorite in acidic media by the decomposition depicted in Equation 34.

$$8HClO_2 \longrightarrow 6ClO_2 + HClO + HCl + 3H_2O \qquad (34)$$

It is therefore evident that in acidic media not only both chlorine dioxide and chlorous acid but also hypochlorous acid and chlorine are all involved in the reaction no matter whether chlorine dioxide or chlorite is used as the oxidizing agent. The similarity of the two oxidants has long been recognized (136c, 141). However, the differences between the two reagents have also been noted (142, 143). It is to be emphasized that chlorite reacts slowly and gradually and obviously only with the phenylpropane units containing free phenolic hydroxyl groups (144). Chlorine dioxide, on the other hand, reacts rapidly with both free and etherified units at 60-70° but only sluggishly with etherified units at room temperature (140, 142, 143, 145, 146). Studies on model compounds showed that demethylation, ring cleavage, chlorination and o- and p-quinone formation all take place (140, 142, 146). Examples of the few products isolated from complex reaction mixtures are shown in Figure 11.8.

The oxidation products obtained from lignins are unstable and difficult to characterize (140, 143). Chlorine dioxide oxidation probably involves both one- and two-electron transfer processes, in contrast with the peracetic acid oxidation which involves only a two-electron transfer. The one-electron transfer process (Equation 29) in chlorine dioxide oxidation may be the dominant reaction when the pH of the reaction medium is higher than 6, at which pH the reduction of chlorous acid is negligible. However, it has been found that even at pH values as high as 7 a considerable amount of chloride is formed in the oxidation medium indicating the consumption of all five oxida-

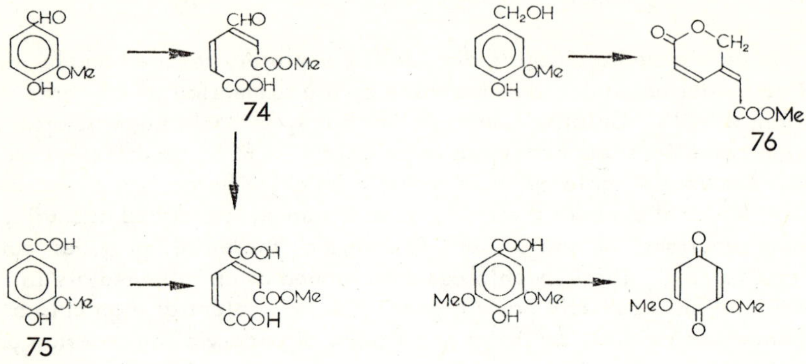

Figure 11.8. Reactions of chlorine dioxide with lignin models (140, 142, 143, 146).

tion equivalents (142, 144, 147). In acidic solutions, on the other hand, elemental chlorine is one of the reaction intermediates and will give rise to reactions characteristic of this species, such as ring substitution and side-chain displacement as well as hydrolysis of alkyl-aryl ether bonds (147-151) discussed in detail in Chapter 10.

C. Acid-Catalyzed Reaction of Hydrogen Peroxide. It has been shown that hydrogen peroxide may act either as a nucleophile or as an electrophile, depending on the pH of the reaction medium (152). The electrophilic species is represented by protonated hydrogen peroxide $H_3O_2^+$ (153), while peroxy anion HO_2^- may serve as a nucleophile, Equation 35.

$$H_3O_2^+ \underset{}{\overset{H^+}{\rightleftharpoons}} H_2O_2 \underset{}{\overset{OH^-}{\rightleftharpoons}} HO_2^- + H_2O \quad (35)$$

In the intermediate state represented by neutral solutions, there is evidence indicating that hydrogen peroxide is more acidic than water. The ions in aqueous hydrogen peroxide solution are, therefore, principally H_3O^+ and OOH^- (154). However, it has also been shown that although the oxidation reactions in which H_2O_2 acts as an electrophile are catalyzed by acids, hydrogen peroxide itself can also serve as an OH^+ donor for negatively charged substrates in neutral solution (155). Accordingly it is very likely (153) that oxidation kinetics in aqueous solution are of the form in Equation 36, in which the substrates are expressed as nucleophilic species. Rate constant k_3 is about two orders of magnitude larger than k_2.

$$\text{Rate of oxid.} = k_2[H_2O_2][\text{nucleophile}^-] + k_3[H_2O_2][\text{nucleophile}][H^+] \quad (36)$$

OXIDATION

Even in neutral solutions lignins may be oxidized by hydrogen peroxide, and although prolonged treatment at room temperature does not cause much degradation, methylene groups are converted into carbonyl groups, and these in turn are susceptible to conversion to carboxyl functions (156). However, when the reaction is carried out at elevated temperatures, the lignin macromolecule can be extensively degraded and dissolved. Oxalic, malonic, acetic and formic acids as well as other dibasic acids are isolated, indicating an extensive breakdown of the aromatic nucleus (157-159).

The acid-catalyzed oxidation of lignin by hydrogen peroxide can be expected to be largely analogous to that of peracetic acid oxidation. In support of this view, Ishikawa and Oki (160) found that oxidation of either vanillyl alcohol, vanillyl ethyl ether, acetoguaiacone or guaiacyl methyl carbinol with hydrogen peroxide at pH 2 affords in each case vanillin, vanillic acid and protocatechualdehyde as shown in Equation 37. These findings indicate that side-chain oxidation and demethylation have taken place.

$$\underset{\underset{OH}{\overset{R}{\bigcirc}}}{\bigcirc}\text{OMe} \longrightarrow \underset{\underset{OH}{\overset{CHO}{\bigcirc}}}{\bigcirc}\text{OMe} + \underset{\underset{OH}{\overset{COOH}{\bigcirc}}}{\bigcirc}\text{OMe} + \underset{\underset{OH}{\overset{CHO}{\bigcirc}}}{\bigcirc}\text{OH} \qquad (37)$$

R = CH$_2$OH, CH$_2$OEt, COMe, CH(OH)Me

The hydrogen peroxide oxidation of lignin in alkaline media proceeds by different mechanism, to be discussed in Section IV-D of this chapter.

D. **Other Oxidation Processes Degrading Aromatic Rings.** Strong oxidants such as potassium permanganate, dichromate and ozone, for example, oxidize lignin readily, degrading it to carbon dioxide and carboxylic acids.

As has been mentioned earlier, phenols are easily attacked by permanganate in neutral solution even at room temperature, but phenol ethers are oxidized only slowly. Thus, aromatic acids such as veratric, isohemipinic and metahemipinic acids can be obtained only if lignin is first hydrolyzed and then methylated before oxidation so as to protect lignin from too drastic an attack by the oxidant. Without the prehydrolysis procedure, only a small amount of aromatic acids can be derived from methylated lignin preparations (161). In a weakly alkaline medium a large part of the methylated lignin is oxidized to carbon dioxide even at 3-5°, and oxalic acid, in yields up to 26.5%, is the only oxidation product isolable (162, 163). Oxidation of the unmethylated lignin preparations with an alkaline permanganate solution, on the other hand, yields a mixture of benzenepolycarboxylic acids, 77-79, in addition to the aliphatic oxalic and acetic acids (164-168). Bone and coworkers (165) oxidized the hydrochloric acid lignin from pine, fir, beech, poplar and bagasse with an alkaline permanganate solution and reported 13-15% yields of a mix-

                77   78   79

ture of benzenepolycarboxylic acids. However, the yields secured by other investigators were much lower (164), and the values reported by Read and Purves (166, 167) are summarized in Table 11.11.

Table 11.11. Permanganate Oxidation Products of Various Lignins (166,167)

Lignin Preparation	Benzenepolycarboxylic Acids (%)			
	tetra- 77	penta- 78	hexa- 79	total yield
Spruce Wood	-	0.14	-	0.14
Periodate Lignin	0.17	0.62	-	0.8
Willstätter	1.21	0.76	-	1.0
Klason	0.37	2.04	-	2.4
Alkali Aspen	0.46	3.20	0.17	3.8

Not surprisingly, the total yield of benzenepolycarboxylic acids increases with increasing degree of condensation within the lignin macromolecule The conclusion is supported by the results of Chudakov and his coworkers (168). These investigators oxidized a coniferous hydrolysis lignin with an alkaline permanganate solution and obtained a 2.4% yield of benzenepolycarboxylic acids. When the lignin was prehydrolyzed with alkali at 180° and 270°, the yields were 8.0 and 25.2%, respectively. In accord with these findings, Brunow (169) has shown that an acid treatment of pinoresinol as a lignin model affords a condensation product which yields over 3% of these polycarboxylic acids. Presumably, the tetrahydrofuranyl ethers are cleaved to transient polycyclic intermediates capable of subsequent oxidation.

Although the origin of these aromatic polyacids is not completely understood, it has been postulated that aromatic rings in lignin containing oxygen atoms are destroyed, and the benzenepolycarboxylic acids are formed from the side-chains of the condensed units having conidendrin-like structures 22 as a result of dehydrogenation (170). It should be noted that the structure 80 , which is believed to be present in condensed lignins (cf. Chapter 4) and which may be formed through the acid treatment should give benzenepentacarboxylic acids on oxidation.

(structure 80)

In contrast, permanganate ion in acidic solution is a much stronger oxidant and hence degrades lignin completely. The amount of potassium permanganate consumed per gram of pulp in strongly acidic solution has been used as a standard method to determine the extent of delignification of pulp (171).

Ozone is a strong oxidant which is reactive enough to attack even benzene at room temperature. The aromatic monomer units of lignin are highly activated toward electrophilic attack by virtue of the presence of two more oxygenated substituents and are therefore readily susceptible to ozone attack. Thus, it is well-known that when wood or lignin preparations are oxidized with ozone in neutral or acidic suspension, carbon dioxide, oxalic, acetic and formic acids are formed (172-178). In all of these cases oxalic acid was the major monomeric oxidation product.

The observations that in the absence of water, ozone had very little effect on lignin (179) and that oxidation is much faster in acidic media than in water suspension (175) strongly support the idea of an electrophilic attack by O_3H^+. It is known that reactions of this sort are catalyzed by the presence of Lewis acids (180).

In their recent model compound studies on the oxidation of vanillyl and veratryl alcohols, Hatakeyama, Tonooka, Nakano and Migita (181) have found that the δ-lactone of β-hydroxymethyl muconic acid monomethyl ester 76 is the major product under acidic oxidation conditions. This compound had been identified earlier among the chlorine dioxide oxidation products of vanillyl alcohol, see Figure 11.8. It was thus concluded that rupture of the guaiacyl nucleus takes place between C_3 and C_4, and no demethylation at C_3 proceeds prior to the ring opening reaction. Some side-chain oxidation was also observed. From an alkaline oxidation medium, muconic acid lactone 81 rather than its methyl ester 76 was obtained. In addition, protocatechuic aldehyde and protocatechuic acid were also obtained in fairly high yield, indicating that demethylation occurs at both the 3- and 4-positions and that side-chain oxidation takes place more extensively than in acidic solution. In both cases, continued oxidation would probably yield oxalic acid as the end product.

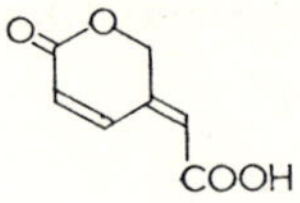

81

IV. Oxidation Processes Limited to Specific Groups.

A. Periodate Oxidation. The significance of the free phenolic hydroxyl group as a locus for oxidative attack on the lignin macromolecule was again demonstrated when Adler and Hernestam (182) found, using guaiacyl compounds as lignin models, that such compounds are quantitatively oxidized by sodium periodate to o-quinones with the concomitant liberation of the methoxyl group as methanol. Since the methanol formed could be determined quantitatively, this reaction was applied (183) to determine the free phenolic hydroxyl content of lignin preparations. The values obtained were in good agreement with those obtained by UV spectroscopy (184).

In a later study on the oxidation of guaiacol and hydroquinone monomethyl ether in H_2O^{18}; Adler, Falkehag and Smith (185) found that in both cases 90% of the o- and p-benzoquinone obtained, respectively, were labelled with O^{18}, and no $MeO^{18}H$ was formed in either case, indicating demethoxylation rather than demethylation is actually the prevalent mode of reaction. The authors therefore concluded that the periodate aryl ester 82 is converted directly or via its aroxy cation 83 by the electrophilic attack of a water molecule to the quinone half ketal 84 which then decomposes into the quinone as depicted in Equation 38.

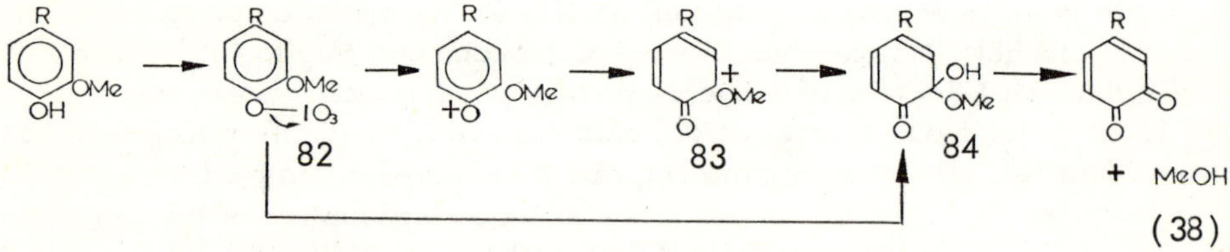

(38)

B. Fremy's Salt. A comparable formation of o-quinones without demethylation can be achieved using the inorganic radical ion, potassium nitrosodisulfonate, $ON(SO_3K)_2$, colloquially known as Fremy's salt. This unusual oxidant effectively converts guaiacyl compounds to o-quinones if the 5-position is not blocked. The overall oxidation is believed to consist of two consecutive one-electron transfer reactions (186) as depicted in Equation 39.

OXIDATION

$$(39)$$

This transformation has been used to determine the uncondensed guaiacyl units in lignin since the o-quinones formed can be determined spectrometrically (186).

C. Dichlorodicyano-p-quinone. Another somewhat exotic oxidizing agent, dichlorodicyano-p-quinone, is capable of converting benzyl alcohols to the corresponding aryl ketones (187, 188) as shown by Equation 40.

$$(40)$$

The aryl ketones so formed can be conveniently assayed by spectroscopic techniques, and the reaction has therefore been used to determine quantitatively the number of benzyl alcohol groups in the lignin macromolecule.

D. Hydrogen and Sodium Peroxides in Alkaline Media. Alkaline peroxide bleaching of pulp has been widely employed since its first commercial introduction in 1944 (189). This process provides substantial brightness gains for high yield pulps as well as moderate brightness increases with chemical pulps (190). It may be employed alone or in conjunction with other bleaching agents. Generally, the alkaline peroxide bleaching of pulp results in good brightness stability with small pulp yield losses. These economic and technical advantages make this process comparable to other bleaching techniques, especially for high yield pulps.

As bleaching chemicals, peroxides in alkaline media have been classified as lignin bleaches rather than lignin-removing reagents, since they selectively attack some specific chromophoric groups within the lignin under normal bleaching conditions (191). In reality, however, the reactions of chromophoric groups account only for a fraction of the total chemical processes in peroxide bleaching.

The oxidation of lignin with peroxides in alkaline media has been studied recently using model compounds (192-196) and the reactions observed fit into two separate mechanistic types. The first group of reactions is initiated by a hydroperoxide anion HO_2^- at electron deficient loci within the molecule.

The oxidative degradation of cinnamaldehyde-type compounds demonstrated by Reeves and Pearl (192) in Equation 41 affords a good example of transformations of this type. The initial attack of the nucleophilic hydroperoxide ion probably occurs at the carbon adjacent to the aromatic ring and results in the formation of benzaldehyde derivative, possibly via an oxirane intermediate. The reaction requires activation of the double bond by a carbonyl group because ethylenic compounds of other types, including cinnamic acids, are quite stable under oxidation conditions.

$$(Ar)-CH=CH-CHO + O_2H^- \longrightarrow (Ar)-CH-CH=CH-O^- \xrightarrow{-OH^-}$$
$$\underset{O-O-H}{|}$$
(41)

$$\left[(Ar)-CH-CH-CHO \right] \xrightarrow{OH^-} (Ar)-CHO + HOCH_2CHO$$
$$\underset{O}{\diagdown\diagup}$$
(?)

The degradations of quinoidal structures (194, 195), aryl acetones (193), α-keto acids and α,β-diketones (Equations 42-46) also fall into the category of reactions initiated by hydroperoxide ions, although the attack occurs in these cases often at the carbonyl carbon. The oxidation of methoxy-p-benzoquinone, Equation 42, to malonic, oxalacetic, hydroxysuccinic, maleic and oxalic acids can be inferred from studies both by Ishikawa and Oki (194) and by Bailey and Dence (195). The latter authors also found, while studying the oxidation of cresol, that 4-methyl-o-benzoquinone undergoes an oxidative ring-opening to muconic acids as well as to methylmaleic, methylfumaric, β-acetylacrylic and oxalic acids. It should be noted that the initially formed cis-cis-muconic acid readily isomerized to the cis-trans form which also contributes to the composition of the ultimate oxidation products.

The second group of reactions is initiated, in all likelihood by an attack of a phenolate anion upon unionized hydrogen peroxide, now playing the role of an electrophile. The most important reactions of this type were discovered in the model studies of Ishikawa and Oki (194) and are illustrated in Equation 47. Thus, it was shown that vanillyl alcohol, apocynol, and acetoguaiacone all yield methoxyhydroquinone 85 after treatment with hydrogen peroxide at pH 12. Excess peroxide yields methoxy-p-benzoquinone, which may be further converted to the dicarboxylic acids formulated in Equation 42. These results have been confirmed by later workers (192, 195).

$$\underset{OMe}{\text{[benzoquinone]}} \xrightarrow{OOH^-} \begin{array}{l} HOOCCH_2COOH \\ + HOOCCH_2COCOOH \\ + HOOCCH_2CHCOOH \\ \quad\quad\quad\quad\quad | \\ \quad\quad\quad\quad\quad OH \end{array} + \begin{array}{l} COOH \\ COOH \end{array} + \begin{array}{l} COOH \\ | \\ COOH \end{array}$$
(42)

OXIDATION

[Structural equation showing oxidation of methyl-substituted quinone to muconic acid derivatives and products including MeCOOH + CO₂] (43)

$$(Ar)CH_2COCH_3 \longrightarrow (Ar)CHO + CH_3COOH \quad (44)$$

$$RCOCOOH \longrightarrow RCOOH + CO_2 \quad (45)$$

$$RCOCOR' \longrightarrow RCOOH + R'COOH \quad (46)$$

[Structural equation showing conversion of substituted phenol to hydroquinone 85 to quinone] (47)

R = -CH₂OH, -CHOHCH₃, or -COCH₃

The conversion of acetoguaiacone to the hydroquinone 85 appears to be an example of a Dakin reaction in which side-chain carbons are recovered in the form of acetic acid, Equation 48.

[Structural equation showing acetoguaiacone + OOH⁻ → intermediate → hydroxyl product + CH₃COOH] (48)

Only minor amounts of acetoguaiacone and acetic acid were detected by Bailey and Dence (195) in the peroxide oxidation of apocynol. Rather, side-chain carbons were recovered as acetaldehyde, as shown in Equation 49. Consequently, the dominant reaction is the displacement of side-chain as acetaldehyde, and the dakin reaction plays a subordinate role only.

$$\underset{\underset{OH}{\overset{Me}{\overset{|}{HCOH}}}}{\bigcirc}_{OMe} \xrightarrow{OH^-} \underset{\underset{O}{\overset{Me}{\overset{|}{HCOH}}}}{\bigcirc}_{OMe}^{\ominus} \xrightarrow{HOOH} \underset{\underset{O}{\overset{Me}{\overset{|}{HCOH}}}}{\bigcirc}_{OMe}^{OH} \longrightarrow \underset{\underset{OH}{\overset{OH}{\bigcirc}}}{\bigcirc}_{OMe} + CH_3CHO \quad (49)$$

In addition, demethylation of apocynol competes with the oxidative side-chain removal. The extent of demethylation can be appreciated from the fact that at all stages of the oxidation reaction, the yield of methanol exceeds the combined yields of acetaldehyde and acetic acid (Figure 11.9). This tendency to dealkylation is particularly illustrated by cresol, whose methyl side-chain resists oxidation and which therefore exclusively follows the demethylation route.

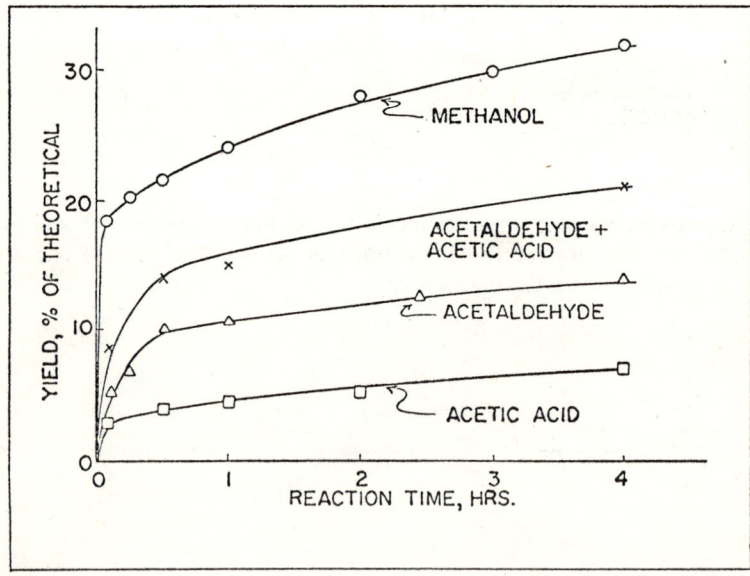

Figure 11.9. Rate of formation of acetaldehyde, acetic acid and methanol in the reaction of α-methylvanillyl alcohol with alkaline hydrogen peroxide (195)

OXIDATION

Application of the model compound studies to the behavior of the lignin macromolecule itself suggests that unetherified units generally resist the action of alkaline peroxide, with the exception of cinnamaldehyde- and certain carbonyl-containing structures. For this reason the degradation of lignin remains incomplete, and this explains the well-known fact that only minor amounts of lignin are removed in the peroxide bleaching of pulps. While the destruction of chromophoric groups has a notable visual bleaching effect, these reactions account for only a minor part of the peroxide consumption. The major amount of reactant is probably consumed in the degradation of aromatic units with free phenolic hydroxyl groups. Experiments by Bailey and Dence (195) on the oxidation of spruce groundwood and groundwood milled wood lignin support this idea. In both cases, 16 to 17% of the original methoxyl groups were cleaved to methanol suggesting that approximately one-half of the phenolic lignin units had been degraded. Oxalic, malonic, maleic, methoxysuccinic (or hydroxysuccinic) and oxalacetic acids were identified among the reaction products, and the carboxyl content of the lignin was shown to increase substantially during the oxidation.

Recently, growing interest has developed towards the utilization of peracetates as non-delignifying bleaching agents. The optimum pH range for pulp brightening is between 7 and 9 (115). Alternatively, the treatment may be initiated at lower pH values such as 3.5-4.0 where peracetic acid presumably acts as a partial delignifying agent. Subsequent elevation of the reaction mixture to pH 7-8 achieves final brightening (117). While the oxidative action of peracetates has not been clarified, reactions broadly analogous to those occurring in peroxide bleaching may be anticipated, and this area certainly deserves future studies.

V. Photoxidation of Lignin. The lignin macromolecule can also be oxidized by radiation as are other phenolic materials. Thus, when wood is exposed to light, color changes occur depending on the incident wavelength. Irradiation using wavelengths < 385 nm causes a darkening; while above 480 nm, the opposite effect is observed (197-200). Between these two wavelengths initial brightening is followed by darkening. A plot of the relative spectral sensitivity of production of color against the irradiating wavelengths affords a curve closely similar to the ultraviolet absorption spectrum of lignin (200, 201). This coincidence clearly demonstrates that lignin is intimately involved in the photochemical reactions of wood which also requires the presence of oxygen (198, 202-204).

This requirement implies that the phenolic hydroxyl groups are probably involved, and this conclusion finds support in the observation that esterification or etherification of the lignin increases the light stability of the wood (205-209).

It can therefore be anticipated that the lignin in wood will be subject to facile degradation by oxygen in the presence of light of an appropriate wavelength and a substantial decrease in the Klason lignin content of wood indeed occurs during irradiation (210). This can be attributed to gravimetric losses occasioned by the increased solubility of the irradiated lignin as a consequence of oxidative demethoxylation. A severe drop in the methoxyl content of irradiated lignin has been repeatedly noted (202, 204, 210, 211), and the extent of darkening has been correlated with methoxyl cleavage (199). The expected formation of methanol has also been confirmed (212).

The first step in this degradation will probably be the photo-induced production of free radicals as in Equation 50, since e.s.r. spectroscopy confirms that these are formed in several wood species on irradiation. Combination with oxygen affords the peroxy radical (Equation 51), which by a chain transfer process yields a hydroperoxide and regenerates a free radical, Equation 52.

$$\text{Lignin} \xrightarrow{h\nu} \text{Lignin} \cdot \tag{50}$$

$$\text{Lignin} \cdot + O_2 \longrightarrow \text{Lignin} - OO \cdot \tag{51}$$

$$\text{Lignin} - OO \cdot + \text{Lignin-H} \longrightarrow \text{Lignin} - OOH + \text{Lignin} \cdot \tag{52}$$

The hydroperoxide can then proceed to oxidatively attack the lignin macromolecule by the electrophilic routes previously described in Section III.

REFERENCES

1. W. A. Waters, Mechanisms of Oxidation of Organic Compounds, John Wiley and Sons, Inc., New York, 1964, pp. 1-5.
2. K. Freudenberg, W. Lautsch and K. Engler, Chem. Ber., 73, 167 (1940).
3. L. Schulz, German patent 693,350 (1940); U.S. Patent 2,187,366 (1940).
4. J. G. Bicho, E. Zavarin and D. L. Brink, Tappi, 49, 218 (1966).
5. R. Z. Pen, Zh. Anal. Khim., 20, 277 (1965).
6. R. H. J. Creighton, J. L. McCarthy and H. Hibbert, J. Am. Chem. Soc., 63, 312 (1941).
7. R. H. J. Creighton, J. L. McCarthy and H. Hibbert, ibid., 63, 3049 (1941).
8. R. H. J. Creighton, D. Gibbs and H. Hibbert, ibid., 66, 32 (1944).
9. D. E. Bland and W. E. Cohen, Australian J. Sci. Research 3A, 642 (1950).
10. B. Leopold and I. L. Malmström, Acta Chem. Scand., 6, 49 (1952).
11. T. Higuchi, I. Kawamura and I. Morimoto, Trans. 62nd Meeting Japan. Forestry Soc., J. Japan. Forestry Soc. Spec. Issue, 279 (1953).
12. D. W. Clayton, L. M. Marraccini and A. Sakai, Svensk Papperstidn., 71, 857 (1968).
13. B. Leopold, Acta Chem. Scand., 4, 1523 (1950).

14. K. R. Kavanagh and J. M. Pepper, Can. J. Chem., 33, 24 (1955).
15. Z. N. Kreitsberg, P. N. Odintsovs and C. A. Sobolevskii, Trudy Inst. Lesokhoz. Problem, Akad. Nauk Latv. S.S.R., Voprosy Lesokhim. i Khim. Drevesiny, 12, 213 (1957).
16. J. C. Pew, J. Am. Chem. Soc., 77, 2831 (1955).
17. B. Leopold, Acta Chem. Scand., 6, 38 (1952).
18. A. v. Wacek and K. Kratzl, Cellulosechem., 20, 108 (1942).
19. A. v. Wacek and K. Kratzl, Chem. Ber., 76, 891 (1943).
20. A. v. Wacek and K. Kratzl, ibid., 77, 516 (1944).
21. A. v. Wacek and K. Kratzl, Österr. Chem. Ztg., 48, 36 (1947).
22. B. Leopold and I. L. Malmström, Acta Chem. Scand., 5, 936 (1951).
23. B. Leopold, ibid., 5, 1393 (1951).
24. B. Leopold, ibid., 6, 55 (1952).
25. B. Leopold, Svensk Papperstidn., 55, 816 (1952).
26. B. Leopold, Acta Chem. Scand., 6, 1294 (1952).
27. B. Leopold, Svensk Kem. Tid., 64, 18 (1952).
28. A. v. Wacek and K. Kratzl, J. Polymer Sci., 3, 539 (1948).
29. J. B. Hlava and F. E. Brauns, Holzforschung, 7, 62 (1953).
30. W. J. Brickman and C. B. Purves, J. Am. Chem. Soc., 75, 4336 (1953).
31. P. N. Odintsovs and Z. N. Kreitsberg, Voprosy Lesokhim. i Khim. Drevesiny, Trudy Inst. Lesokhoz. Problem, Akad. Nauk Latv. S.S.R., 6, 63 (1953).
32. G. G. Allan, unpublished work.
33. G. G. Allan, P. Mauranen, A. N. Neogi and C. E. Peet, Chem. & Ind., 1969, 623.
34. K. Kratzl, L. J. Gratzl and P. Claus, Advances in Chem. Series, 59, 157 (1966).
35. I. A. Pearl and D. L. Beyer, Tappi, 33, 544 (1950).
36. I. A. Pearl, J. Am. Chem. Soc., 2309 (1950).
37. J. M. Harkin, Oxidative Coupling of Phenols, W. J. Taylor and A. R. Battersby, Eds., Marcel Dekker, Inc., New York, 1967.
38. I. A. Pearl, J. Am. Chem. Soc., 64, 1429 (1942).
39. I. A. Pearl and E. E. Dickey, ibid., 74, 614 (1952).
40. I. A. Pearl and D. L. Beyer, ibid., 76, 6106 (1954).
41. I. A. Pearl and D. L. Beyer, Tappi, 39, 171 (1956).
42. I. A. Pearl and D. L. Beyer, J. Am. Chem. Soc., 76, 2224 (1954).
43. I. A. Pearl, ibid., 78, 5672 (1956).
44. I. A. Pearl and D. L. Beyer, Tappi, 42, 800 (1959).
45. I. A. Pearl and D. L. Beyer, Forest Prod. J., 11, 442 (1961).
46. I. A. Pearl, D. L. Beyer and D. Whitney, Tappi, 44, 479 (1961).
47. I. A. Pearl and D. L. Beyer, Forest Prod. J., 14, 316 (1964).
48. J. M. Pepper, B. W. Casselman and J. C. Karapally, Can. J. Chem., 45, 3009 (1967).

49. A. A. Sokolova, N. A. Brannova and E. V. Nazareva, through ABIPC, 29, 961 (1959).
50. I. A. Pearl and D. L. Beyer, Tappi, 33, 508 (1950).
51. I. A. Pearl, J. Am. Chem. Soc., 71, 2196 (1949).
52. I. A. Pearl, ibid., 72, 1427 (1950).
53. I. A. Pearl, ibid., 67, 1628 (1945).
54. I. A. Pearl, ibid., 68, 429 (1946).
55. I. A. Pearl, ibid., 68, 1100 (1946).
56. R. F. Davis, E. T. Reaville, Q. P. Peniston and J. L. McCarthy, ibid., 77, 2495 (1955).
57. I. A. Pearl, ibid., 70, 2008 (1948).
58. L. A. Pershina and V. P. Vasileva, Izv. Tomsk. Polytekh. Inst., 136, 33 (1965).
59. L. A. Pershina, V. L. Kuksina and V. P. Vasileva, ibid., 136, 36 (1965).
60. H. Schrader, Ges. Abhandl. Kenntnis Kohle, 5, 276 (1920).
61. F. Fisher, H. Schrader and A. Friedrich, ibid., 6, 1 (1921).
62. F. Fisher, H. Schrader and W. Treibs, ibid., 5, 221 (1920).
63. W. Lautsch, E. Plankenhorn and F. Klink, Z. Angew. Chem., 53, 450 (1940).
64. I. A. Pearl and D. L. Beyer, Advances in Chem. Series, 59, 145 (1966).
65. K. Kratzl, J. Gratzl and P. Claus, ibid., 59, 157 (1966).
66. K. Kratzl, W. Schafer, P. Claus, J. Gratzl and P. Schilling, Monatsh. Chem., 98, 891 (1967).
67. C. Steelink, Advances in Chem. Series, 59, 51 (1966).
68. J. C. Pew, Nature, 193, 250 (1962).
69. G. A. Russell, Peroxide Reaction Mechanisms, J. O. Edwards, Ed., John Wiley and Sons, New York, 1962, p. 107.
70. A. N. Zavyalov and S. S. Frolov, Russian patent 168,672 (Mar. 1965).
71. K. Freudenberg, A. Janson, E. Knopf and A. Haag, Chem. Ber., 69, 1415 (1936).
72. K. Freudenberg, M. Meister and E. Flickinger, ibid., 70, 500 (1937).
73. K. Freudenberg, K. Engler, E. Flickinger, A. Sobek and F. Flink, ibid., 71, 1810 (1938).
74. K. Freudenberg and H. F. Müller, ibid., 71, 1821 (1938).
75. H. Richtzenhain, Acta Chem. Scand., 4, 206 (1950).
76. H. Richtzenhain, Svensk Papperstidn., 53, 644 (1950).
77. K. Freudenberg and C. Chen, Chem. Ber., 93, 2533 (1960).
78. K. Freudenberg, C. Chen and G. Cardinale, ibid., 95, 2814 (1962).
79. K. Freudenberg and F. Niedercorn, ibid., 89, 2168 (1956).
80. K. Freudenberg and F. Niedercorn, ibid., 91, 591 (1958).
81. K. Freudenberg, K. Jones and H. Renner, ibid., 96, 1844 (1963).
82. K. Freudenberg and B. Lehmann, ibid., 96, 1850 (1963).
83. H. Richtzenhain, Acta Chem. Scand., 4, 589 (1950).

84. H. Richtzenhain, Chem. Ber., 83, 488 (1950).
85. B. Ericsson, B. Nist and K. V. Sarkanen, Proc. Intern. Symp. Chem. & Biochem., Lignin, Cellulose and Hemicellulose, Grenoble, 1964, p.59.
86. K. Freudenberg, Brennstoff-Chem., 44, 328 (1963).
87. K. Freudenberg and H. K. Werner, Chem. Ber., 97, 579 (1964).
88. K. Freudenberg, J. M. Harkin and H. K. Werner, ibid., 97, 909 (1964).
89. K. Freudenberg, Science, 148, 595 (1965).
90. Y. Kyoguku, S. Fujii and Y. Hachihama, J. Chem. Soc. Japan, Ind. Chem. Sect., 64, 2023 (1961).
91. D. E. Bland and A. F. Logan, Biochem. J., 95, 515 (1965).
92. A. Hayashi and Y. Namura, Bull. Chem. Soc. Japan, 33, 512 (1965).
93. S. Larsson and G. E. Miksche, Acta Chem. Scand., 21, 1970 (1967).
94. G. E. Miksche and S. Larsson, ibid., 23, 917 (1969).
95. H. Erdtman and C. A. Wachtmeister, Festschrift Arthur Stoll, Birkhauser, Basel, 1957, p. 144.
96. K. Freudenberg, V. Jovanovic and F. Topfmeier, Chem. Ber., 94, 3327 (1961).
97. K. Freudenberg and C. L. Chen, Chem. Ber., 100, 3633 (1967).
98. C. D. Hurd, R. W. McNamee and F. O. Green, J. Am. Chem. Soc., 61, 2979 (1939).
99. R. Stewart, in Oxidation in Organic Chemistry, Part A, K. B. Wiberg, Ed., Academic Press, New York, 1965, pp. 1-68.
100. K. B. Wiberg and A. S. Fox, J. Am. Chem. Soc., 85, 3487 (1963).
101. W. L. Evans, ibid., 45, 171 (1923).
102. R. Stewart, ibid., 79, 3057 (1957).
103. R. Stewart and R. van der Linden, Trans. Faraday Soc., 1966, 211.
104. J. L. Kurz, J. Am. Chem. Soc., 86, 2229 (1964).
105. A. Y. Drummond and W. A. Waters, J. Chem. Soc., 1953, 435.
106. H. Haas, W. Schech and U. Strole, Papier, 9, 469 (1955).
107. B. Leopold, Tappi, 44, 230 (1961).
108. N. Thompson and O. Kaustinen, Tappi, 47, 157 (1964).
109. A. Poljak, Angew. Chem., 60, 45 (1948).
110. A. Poljak, Holzforschung, 50, 31 (1951).
111. A. Poljak, Angew. Chem., 66, 302 (1954).
112. R. L. McEwen and F. R. Sheldon, Canadian patent 461,242 (Nov. 22, 1949).
113. K. Prett and K. Thurnher, Canadian patent 645, 944 (July 31, 1962).
114. M. Wayman, C. B. Anderson and W. H. Rapson, Tappi, 48, 113 (1965).
115. C. W. Bailey and C. W. Dence, ibid., 49, 9 (1966).
116. C. B. Christiansen, G. L. Burroway and W. F. Parker, ibid., 49, 49 (1966).
117. W. P. Stevens and R. Marton, ibid., 49, 452 (1966).
118. H. Ishikawa, K. Okubo, T. Oki and S. Watanabe, J. Japan.Tappi, 19,

393 (1965).
119. K. Sakai and S. Kishimoto, J. Japan. Wood Res. Soc., 14, 411 (1968).
120. E. J. Behrman and J. O. Edwards, Progress in Physical Organic Chemistry, A. Streitwieser and R. W. Taft, Eds., Interscience Publishers, New York, 1967, p. 93.
121. Y. Z. Lai and K. V. Sarkanen, Tappi, 51, 449 (1968).
122. Ref. (1), pp. 41-42.
123. K. Sakai and T. Kondo, J. Japan. Wood Res. Soc., 12, 57 (1966).
124. S. L. Fries, A. H. Soloway, B. K. Morse and S. C. Ingersoll, J. Am. Chem. Soc., 74, 1305 (1952).
125. H. Hatakeyama, J. Nakano and N. Migita, J. Chem. Soc. Japan, Ind. Chem. Sect., 68, 972 (1965).
126. H. Davidage, A. Davis, J. Kenyon and R. F. Mason, J. Chem. Soc., 4569 (1958).
127. H. Ishikawa, T. Oki and K. Ohkubo, J. Japan. Tappi, 20, 435 (1966).
128. Y. Kinoshita, T. Oki and H. Ishikawa, J. Japan. Wood Res. Soc., 13, 319 (1967).
129. H. Hatakeyama, K. Suzuki, J. Nakano and N. Migita, J. Chem. Soc. Japan, Ind. Chem. Sect., 71, 247 (1968).
130. K. Sakai and T. Konda, J. Japan. Wood Res. Soc., 12, 310 (1966).
131. H. Hatakeyama, J. Nakano and N. Migita, J. Chem. Soc. Japan, Ind. Chem. Sect., 70, 957 (1967).
132. H. Hatakeyama, K. Suzuki, G. Meshizuka, J. Nakano and N. Migita, ibid., 70, 1399 (1967).
133. H. Ishikawa, Y. Kinoshita, T. Oki and K. Ohkubo, J. Japan. Tappi, 21, 945 (1967).
134. K. V. Sarkanen and J. Suzuki, Tappi, 48, 459 (1965).
135a. H. Ishikawa, T. Oki and K. Ohkubo, J. Japan. Tappi, 20 , 485 (1966).
135b. J. Boeseken and R. Engleberts, Proc. Acad. Sci. Amsterdam, 34, 1292 (1931); 35, 750 (1932).
135c. J. Boeseken, C. F. Metz and J. Plum, Rec. Trav. Chim., 54, 345 (1935).
136. S. A. Rydholm, Pulping Processes, Interscience Publ., New York, 1965, (a) p. 839, (b) p. 211, (c) p. 975.
137. H. Richtzenhain and B. Alfredson, Acta Chem. Scand., 7, 1177 (1953).
138. H. Richtzenhain and B. Alfredson, ibid., 8, 1519 (1954).
139. H. Richtzenhain and B. Alfredson, ibid., 10, 719 (1956).
140. C. W. Dence, M. K. Gupta and K. V. Sarkanen, Tappi, 45, 29 (1962).
141. W. H. Rapson, Tappi Monograph Series, No. 27, p. 131 (1963).
142. R. M. Husband, C. D. Logan and C. B. Purves, Can. J. Chem., 33,

68, 82 (1955).
143. G. Gionola and J. Meybeck, Assoc. Tech. Ind. Papetiere, Bull. No. 1: 25 (1960).
144. R. Soila, D. Lehtikoski and N. E. Virkola, Svensk Papperstidn., 65, 632 (1962).
145. R. A. Murphy, K. Kahehi and K. V. Sarkanen, Tappi, 44, 465 (1961).
146. K. V. Sarkanen, K. Kakehi, R. A. Murphy and H. White, ibid., 45, 24 (1962).
147. J. Paulson, ibid., 45, 933 (1962).
148. K. V. Sarkanen and C. W. Dence, J. Org. Chem., 25, 715 (1960).
149. C. A. Bunton, E. D. Hughes, C. K. Ingold, D. I. H. Jacobs, M. H. Jones, E. J. Minkoff and R. I. Reed, J. Chem. Soc., 2628 (1960).
150. K. V. Sarkanen and R. W. Strauss, Tappi, 44, 459 (1961).
151. H. Fernholz, Chem. Ber., 84, 110 (1951).
152. E. J. Behrman and J. O. Edwards, Progress in Physical Organic Chemistry, Vol. 4, A. Streitwieser, Jr., and R. W. Taft, Eds., Interscience Publishers, New York, 1967, p. 93.
153. J. O. Edwards, Peroxide Reaction Mechanisms, J. O. Edwards, Ed., Interscience Publishers, New York, 1962, p. 67.
154. E. S. Shanley, ibid., p. 129.
155. J. O. Edwards, J. Phys. Chem., 56, 279 (1952).
156. P. Klason, Chem. Ber., 55, 448 (1922).
157. O. Anderzén and B. Holmberg, ibid., 56, 2044 (1923).
158. H. Richtzenhain, ibid., 75, 269 (1942).
159. G. G. Henderson and R. Boyed, J. Chem. Soc., 97, 1659 (1910).
160. H. Ishikawa and T. Oki, J. Japan. Tappi, 18, 477 (1964).
161. K. Freudenberg, M. Meister and E. Flickinger, Chem. Ber., 70, 500 (1937).
162. E. Heuser and S. Samuelsen, Cellulosechem., 3, 78 (1922).
163. H. Urban, ibid., 7, 73 (1926).
164. F. Fisher, H. Schräder and A. Friedrich, Ges. Abhandl. Kenntnis Kohle, 6, 22 (1921).
165. W. A. Bone, L. G. B. Parsons, R. H. Sapiro and C. M. Gwocock, Proc. Roy. Soc., London, A. 148, 492 (1935).
166. D. E. Read and C. B. Purves, J. Am. Chem. Soc., 74, 116 (1952).
167. D. E. Read and C. B. Purves, ibid., 74, 120 (1952).
168. M. I. Chudakov, S. I. Sukhavovskii and M. P. Akimova, Zh. Priklad. Khim., 32, 608 (1959).
169. G. Brunow, Finska Kemists. Medd., 74, 20 (1965).
170. F. E. Brauns and D. A. Brauns, The Chemistry of Lignin, Supplement Volume, Academic Press, New York, 1960, p. 517.
171. Tappi Standard T236 M60, Published by Tappi.
172. C. Dorée and M. Cunningham, J. Chem. Soc., 103, 677 (1913).

173. F. Konig, Cellulosechem., 2, 93 (1921).
174. H. Richtzenhain, Chem. Ber., 75, 269 (1942).
175. K. Freudenberg, F. Sohns and A. Janson, Ann. Chem., 518, 62 (1935).
176. A. Bell, W. L. Hawkins, G. F. Wright and H. Hibbert, J. Am. Chem. Soc., 59, 598 (1937).
177. R. M. Dorland, W. L. Hawkins and H. Hibbert, ibid., 61, 2698 (1939).
178. M. Phillips and M. J. Goss, ibid., 55, 3466 (1933).
179. M. Cunningham and C. Dorée, J. Chem. Soc., 101, 497 (1912).
180. Ref. (1), p. 129.
181. H. Hatakeyama, T. Tonooka, J. Nakano and N. Migita, J. Chem. Soc. Japan, Ind. Chem. Sect., 70, 2348 (1967).
182. E. Adler and S. Hernestam, Acta Chem. Scand., 9, 319 (1955).
183. E. Adler, S. Hernestam and I. Wallden, Svensk Papperstidn., 61, 641 (1958).
184. G. Aulin-Erdtman, ibid., 55, 74 (1952).
185. E. Adler, I. Falkehag and B. Smith, Acta Chem. Scand., 16, 529 (1962).
186. E. Adler and K. Lundquist, ibid., 15, 223 (1961).
187. H. D. Becker and E. Adler, ibid., 15, 218 (1961).
188. E. Adler and T. Ishihara, Paperi Puu, 43, 662 (1961).
189. J. A. Lee, Chem. Met. Eng., 51 (8), 106 (1944).
190. P. C. Holladay and R. J. Salari, Tappi Monograph Series No. 27, p. 180 (1963).
191. Ref. (136), p. 885.
192. R. H. Reeves and I. A. Pearl, Tappi, 48, 121 (1965).
193. D. D. Jones and D. C. Johnson, J. Org. Chem., 32, 1402 (1967).
194. H. Ishikawa and T. Oki, J. Japan. Tappi, 18, 477 (1964).
195. C. W. Bailey and C. W. Dence, Tappi, 52, 491 (1969).
196. C. A. Bunton, Peroxide Reaction Mechanisms, J. O. Edwards, Ed., Interscience Publishers, New York, 1962, p. 11.
197. W. Sandermann, F. Schlumbom, Holz als Roh-und Werkstoff, 20, 245, 285 (1962).
198. P. Nolan, J. A. Van den Akker and W. A. Wink, Paper Tr. J., 121, 101 (1945).
199. G. F. Leary, Tappi, 50, 17 (1967).
200. S. Claeson, E. Olson and A. Wennerblom, Svensk Papperstidn., 71, 335 (1968).
201. J. A. Van den Akker, H. F. Lewis, G. W. Jones and M. A. Buchana Tappi, 32, 187 (1949).
202. H. F. Callow, Current Science, 16, 286 (1947); Nature, 159, 309 (1947).
203. R. M. Sinclair, T. A. Vincent, N.Z. J. Sci., 7, 196 (1964).

204. G. F. Leary, Tappi, 51, 257 (1968).
205. P. L. D. Peill, Nature, 158, 554 (1946).
206. D. F. Manchester, J. W. McKinney and A. A. Pataky, Svensk Papperstidn., 63, 699 (1960).
207. D. H. Andrews and P. Des Rosiers, Pulp Pap. Mag. Can., 67, T119 (1966).
208. R. P. Singh, Tappi, 49, 281 (1966).
209. H. J. Callow, Nature, 159, 309 (1947).
210. L. V. Forman, Paper Tr. J., 111, 34 (1940).
211. H. F. Lewis and D. Fronmüller, ibid., 121, 25 (1945).
212. G. O. Phillips and A. C. Jett, Jr., Textile Res. J., 64, 497, 572 (1964).

12

REDUCTION AND HYDROGENOLYSIS

B. F. Hrutfiord

I. Introduction.

　　A. The Impact of Hydrogenolysis Studies on the Elucidation of Structure. As used in this chapter the terms reduction and hydrogenolysis refer to the use of a number of reagents and catalysts applied to lignin to bring about structural changes under reducing conditions. Early studies in this area were aimed at converting wood into products of commercial importance and not at the determination of lignin structure. Lindblad (1), for example, was able to liquefy wood, cellulose and isolated lignins by hydrogenation using a variety of catalysts. Several of his results in fact pointed the way for later structural studies. It was established that cellulose forms a tar consisting of essentially neutral components, while lignins, when hydrogenated below 300°C, form tars soluble in alkali and phenolic in character.

　　Harris and Adkins (2) discovered in 1938 that an isolated lignin could be converted by hydrogenation to propylcyclohexane derivatives in unexpectedly high yields. The finding lent strong support to the yet speculative concept of viewing lignins as polymers composed of C_6C_3 units. The positions of oxygen substitution on the C_6C_3 monomers were also established to include the 3- and 4-positions on the C_6 ring relative to the propyl side-chain, as well as the γ position on the propyl side-chain. Later, Hibbert and coworkers (3) isolated identical cyclohexane derivatives by direct hydrogenation of wood. Lautsch and Piazolo (4) used an alkaline hydrogenation medium and isolated guaiacyl ethane from the hydrogenolysis product of spent sulfite liquor. This finding provided early evidence in support of the aromatic character of spruce lignin. Brewer, Cooke and Hibbert (5) initiated a productive phase in hydrogenation studies by the characterization of "hydrol lignins," formed by mild acidic or alkaline hydrogenation of wood. These studies, continued later by Pepper (6) and by Schuerch (7), led to the identification of numerous monomers of the guaiacyl and syringyl types with C_2 or C_3 side-chains. The isolation of guaiacyl propane derivatives with β as well as α-hydroxyl groups from products of the treatment of wood and lignin with sodium in liquid ammonia (8, 43) reinforces the impression of the presence of an oxygen function at all three side-chain carbons of the original unit in lignin. Thus, early hydrogenation studies suggested already that the monomeric products were derived from lignin structures of the type later specifically identified as arylglycerol-β-ether structures.

More recently dimeric molecules have been isolated from hydrol lignin (9) and structural details of the polymeric hydrol lignin are beginning to be elucidated by spectral means (10). Specific reductions of functional groups have also been used recently in connection with spectral methods to obtain useful structural data on lignin (11).

B. The Concurrence of Hydrolysis and Hydrogenation. Either acid- or base-catalyzed hydrolysis of lignin usually plays an important role in hydrogenations. Thus, in the conversion of wood lignin to "hydrol lignins," acidic (or alkaline) hydrolysis provides the essential primary breakdown of the lignin macromolecule, while hydrogenolysis and hydrogenation provide further breakdown and stabilization of the degradation products. Similarly, in most studies on isolated lignins, the conditions of isolation, i.e., methanolysis, sulfite pulping, kraft pulping, all involve hydrolytic conditions. Interpretation of hydrogenation results of necessity must take into account the structural changes which have already occurred prior to the hydrogenation treatment. Such hydrolysis reactions have been described in detail in Chapter 9. In studies using milled wood lignins and in hydrogenations performed in organic solvents, the contribution of hydrolytic processes may be minor.

II. Non-Degradative Reduction of Lignin. Treatment of various lignins by alkaline metal hydrides to change specific structural features of the molecule is the most widely used non-degradative reduction method. This usually results in the conversion of a carbonyl group to an alcohol with concurrent and easily detectable changes in ultraviolet and infrared spectra.

Adler and Marton (11) made an extensive study of carbonyl groups in milled wood lignin utilizing alkaline sodium borohydride reduction combined with differential ultraviolet spectroscopy. Approximately 0.2 carbonyl groups per methoxyl was established as the total content. Detailed examination of the $\Delta \varepsilon_{(H)}$ curves, using many model compounds to establish positions in the spectra where individual types of carbonyl groups may be determined, showed that MWL contained only traces of phenolic coniferaldehyde and phenolic aryl α-carbonyl groups, 0.03 non-phenolic coniferaldehyde groups and 0.06 non-phenolic aryl α-carbonyl groups per methoxyl. The difference, about 0.1 carbonyl/methoxyl, was attributed to non-conjugated carbonyl groups. Also, 0.03 ethylene group per methoxyl was established by a combination of borohydride and catalytic hydrogenation over a palladium-barium sulfate catalyst and $\Delta \varepsilon_{(H)}$ determinations. More recently, Chang, Allan and Sarkanen have used borohydride reduction to eliminate the contribution of carbonyl groups to infrared and ultraviolet spectra and to determine various types of carbonyl groups in softwood and hardwood lignins (12).

III. Mild Catalytic Hydrogenolysis of Lignin.

A. Hydrogenolysis. The conditions used by Brewer, Cooke and Hibbert (5) in the isolation of "hydrol lignin" consisted of hydrogenating maple woodmeal in a 1:1 ethanol-water solvent using a Raney nickel catalyst and a temperature of 160-170°C. The hydrogenated lignin was recovered from the reaction product as a chloroform solution containing about 75% of the original lignin as viscous oily product. The total product contained phenolic and methoxyl (24-25%) groups, and the three aromatic monomers 1, 2 and 3 could be isolated from it in yields of 8.8, 0.84 and 0.83%, respectively, based on the Klason lignin content of the wood. The isolation of both syringyl and guaiacyl aromatic units agrees well with results from earlier oxidation studies (13).

$(S)-CH_2CH_2CH_2OH$ $(G)-CH_2CH_2CH_2OH$ $(S)-C_3H_7$
        1   2   3

In the neutral hydrogenations in ethanol-water or dioxane-water, the final pH is 5.5 as a consequence of the release of acetic and other carboxylic acids during the hydrogenolysis. Further studies of the influence of pH by Pepper and Hibbert (14) established that alkaline conditions (3% NaOH) give the best yield of chloroform-soluble products. The new aromatic compounds 4, 5 and 6 were isolated from maple wood in yields of 2.16%, 15.38% and 6.2% of Klason lignin, respectively. Rehydrogenation of the residual hydrol lignin gave additional yields of the cyclohexane derivatives 7, 8 and 9 in yields of 1.87, 0.86 and 0.82%, respectively.

$(G)-CH_2CH_3$ $(S)-CH_2CH_3$ $(S)-CH_2CH_2OH$
       4   5   6

$HO-\langle\rangle-CH_2CH_3$ $HO-\langle\rangle-CH_2CH_2CH_3$ $\langle\rangle-CH_2CH_2CH_2OH$
         7   8   9

Thus, alkaline hydrogenation results in substantial loss of the terminal side-chain carbons forming phenyl ethane derivatives. The appreciable yield of 6 gives evidence for oxygen substitution in the β position of the side-chain of the precursor monomers.

A number of additional studies have been made to establish the optimum conditions in hydrogenation (6, 15-18). Most studies have utilized hardwood species since these provide better yields of hydrol lignin than softwoods. The optimum temperature range using aqueous solvents is 160-170°, while higher temperatures tend to reduce the methoxyl content. On the other hand, Pepper et al. have reported that temperatures of 190-195° are advantageous in obtaining maximum yields of identifiable monomers (18). Temperatures above 200°C result in increasing amounts of cyclohexane derivatives. Several catalysts may be used, but none of them appears superior to Raney nickel in obtaining stabilized products. Reaction times of four to five hours are near optimum.

The solvent has a profound influence on the nature of the reaction products. Several workers have shown that some water must be present for successful hydrogenolysis (14, 15, 19) because the use of organic solvents alone results in complete failure of delignification. Ethanol-water mixtures result in ethanolysis of the lignin as well as hydrogenolysis (20) and are therefore somewhat inferior to 1:1 dioxane-water. Water alone is effective under alkaline conditions. The yields of hydrol lignin are better using alkaline rather than neutral or acidic conditions, especially with softwoods.

B. <u>Characterization of Hydrogenolysis Lignin</u>. Hydrogenolysis lignin on nitrobenzene oxidation yields the usual aldehydes vanillin and syringaldehyde, but in only about one-half of the yield obtained from lignin as it exists in wood (15). No Hibbert's ketones are obtained on ethanolysis (20).

Fractionation of hydrogenolysis lignin by Schuerch demonstrated that hydrogenolysis lignin consists of a continuous spectrum of molecules ranging in size from monomeric units up to material of over 5000 in molecular weight (20). Portions of the material also differed in acidity and could be fractionated on this basis by partition using various buffers.

C. <u>Monomeric Degradation Products from Hydrogenolysis Lignins</u>. In addition to C_6C_3- and C_6C_2-compounds, C_6C_1- and C_6-compounds and related carboxylic acids have been obtained in later studies on hydrol lignins (5, 6, 10, 14, 21-23) as shown in Table 12.1. The phenylpropane (C_6C_3-) compounds are the major monomeric products in yields of up to 25% of Klason lignin. The main product from softwood lignins is dihydroconiferyl alcohol 2 accompanied by lesser quantities of propylguaiacol 11 and minor amounts of ethyl and methyl guaiacols 4 and 10. Hardwoods yield small quantities of guaiacyl compounds and give dihydrosinapyl alcohol 12 as the major product along with syringyl propane 3, -methane 15 and -ethane 5, in decreasing order. The C_6-compounds are represented by traces of guaiacol and 2,6-dimethoxyphenol.

Alkaline hydrogenolysis conditions yield identical monomers, although most of the products are of the C_6C_2 type. Results from alkaline hydrogenolysis of the hardwoods, aspen, red alder and maple are summarized in Table 12.2.

Bhattacharya et al. have reported traces of aromatic acids including p-hydroxybenzoic, vanillic and syringic acid together with their homoacids 13, 14, 16 (21).

The analysis of the monomer fraction of hydrogenolysis lignins has been simplified by gas chromatography, first applied to this material by Coscia (24). This kind of analysis reveals that several minor components remain to be identified. Gas chromatograms of hydrogenolysis products from red alder, formed under both neutral and alkaline conditions are shown in Figure 12.1. In the two chromatograms, cyclohexyl derivatives appear in the 3-6 minute period,

Table 12.1. Yields of Monomeric Lignin Hydrogenolysis Products

	Yield, % of Klason Lignin				
	Hardwoods[a]		Softwoods		
Compound	alkaline[b]	neutral[b]	alkaline[b]	neutral[b]	Reference
ⓖH	<1	<1	<1	<1	6
ⓖCH_3	<1	<1	<1	<1	6
ⓖCH_2CH_3	2-5	-	9.4	-	6,14,22
ⓖC_3H_7	1.3	-	-	1.2(5.1)*	6
ⓖCH_2CH_2OH	tr	-	-	-	6,10,23
ⓖ$CH_2CH_2CH_2OH$	-	6-11	-	13.5	6,10
ⓖCO_2H	tr	-	-	-	21
ⓖCH_2CO_2H	0.12	-	-	-	21
ⓖ$CH_2CH_2CO_2H$	0.14	-	-	-	21
ⓢH	-	<1	-	-	6
ⓢCH_3	1.5	1	-	-	6
ⓢC_2H_5	9-15	2	-	-	6,10,14
ⓢC_3H_7	<1	12*	-	-	6,10
ⓢCH_2CH_2OH	4-6	-	-	-	6,14
ⓢ$CH_2CH_2CH_2OH$	1.5	13	-	-	5,6,10
ⓢCO_2H	0.14	-	-	-	21
ⓢ$CH_2CH_2CO_2H$	0.17	-	-	-	21
ⓗCO_2H	tr	-	-	-	21

*0.1M HCl added.

a) Values from reference 6 and from reference 10 were estimated by using 1/4 of the reported % chromatographable value.

b) Conditions were generally 4-5 hours reaction time, 500 psig H_2 pressure, Raney nickel catalyst, 1:1 dioxane-water solvent for neutral conditions. In alkaline hydrogenations, 3-5% NaOH were added to the solutions.

ⓖ-CH_3 ⓖ-$(CH_2)_2CH_3$ ⓢ-$CH_2CH_2CH_2OH$ ⓗ-CH_2CO_2H
10   11   12   13

ⓖ-CH_2CO_2H ⓢ-CH_3 ⓢ-CH_2CO_2H
14   15   16

Table 12.2. Alkaline Hydrogenolysis Products from Hardwoods

	Yields, % Klason Lignin		
	Aspen (6)	Alder (10)	Maple (21,23)
G-H	0.5	0.1	-
G-CH_3	0.5	1.0	Trace
G-C_2H_5	4.9	4.3	.2
G-CH_2CH_2OH	Trace	0.5	Trace
G-C_3H_7	1.3	1.9	-
G-$CH_2CH_2CH_2OH$	0.4	0.9	Trace
S-H	Trace	0.1	-
S-CH_3	1.5	0.7	Trace
S-C_2H_5	9.1	10.7	15.4
S-CH_2CH_2OH	4.0	0.6	6.2
S-C_3H_7	0.6	1.4	-
S-$CH_2CH_2CH_2OH$	1.7	3.0	Trace

followed by guaiacyl and syringyl type hydrocarbons and then by guaiacyl and syringyl type alcohols. Examination of these curves shows a number of unknown maxima. The nature of some of the unidentified compounds may be anticipated on the basis of relative frequencies of aromatic nuclei and side-chain structures as they are generally obtained in hydrogenation (see Figure 12.2). For instance, one would expect a very small amount of p-methyl phenol, based on the occurrence of the p-hydroxyphenyl nucleus and a methyl "side-chain."

D. The Isolation of Dimers from Hydrogenolysis Lignin. Schuerch reported evidence for a dimer-rich fraction from the alkaline hydrogenolysis of maple woodmeal as early as 1952 (19), and further research led to the isolation of seven crystalline dimers 17 through 23 (9, 25).

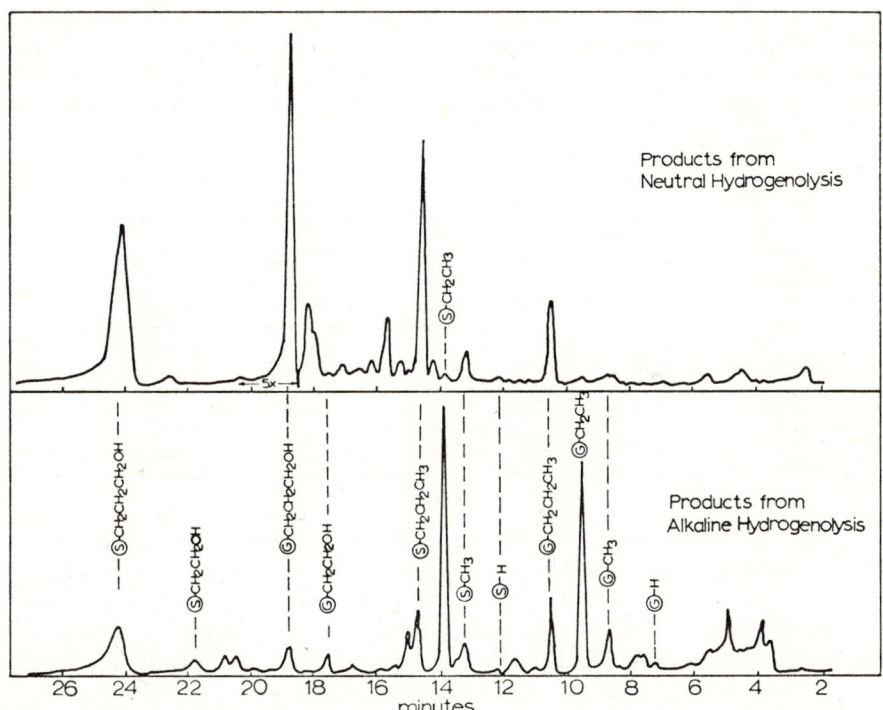

Figure 12.1. Products from neutral and alkaline hydrogenolysis of red alder as separated by gas chromatography (10).

(Ar)CH_2CH_3 ≷ (Ar)$CH_2CH_2CH_2OH$ >> (Ar)$CH_2CH_2CH_3$ > (Ar)CH_2CH_2OH
(Ar)CH_3 > (Ar)H > (Ar)CO_2H ⋚ (Ar)$CH_2CH_2CO_2H$ ≷ (Ar)CH_2CO_2H

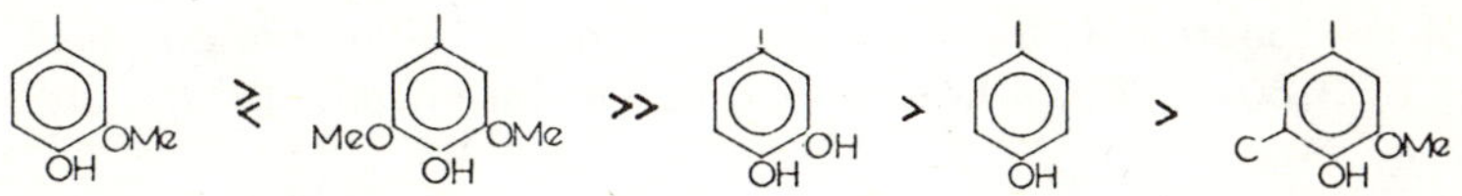

Figure 12.2. Relative frequencies of aromatic nuclei and side-chain structures from hydrogenolysis.

E. Polymeric Hydrogenolysis Lignins. At least half of the material in hydrogenolysis lignins is more highly polymerized than the dimer fraction. Schuerch (20) has separated the ether-insoluble material of maple hydrogenolysis lignin into a series of fractions by precipitation from neutral solvent mixtures and characterized the individual fractions. The results are summarized in Table 12.3.

Table 12.3. Polymeric Fractions from Maple Hydrogenolysis Lignin (20)

Fraction Number	% Alkoxyl	M.W.	% Alkali Soluble	% Alkoxyl	M.W.
1	16.5	(high)	-	-	-
2	25.8	(high)	38.6	20.3	-
3	25.2	2600-1500	-	-	-
4	25.2	1600-1450	49	19.8	1580-1400
5	24.7	870-700	-	-	-
6	23.9	720-640	55	21.3	950-940

Individual fractions were then partitioned between ethanolic chloroform and ethanolic aqueous alkali and the subfractions were again characterized. Generally, the alkali solubility increases as the molecular weight of the main fraction decreases. Of special interest here is the main fraction 1, which is high in molecular weight and low in alkoxyl content, 16.5%, which is the expected value for a hydrogenated guaiacyl lignin.

Pepper has determined the methoxyl content of a polymeric hydrogenolysis lignin from aspen to be 21.8%, and the infrared spectrum was similar to that of dioxane acidolysis lignin from the same species. Hrutfiord and McCarthy applied nmr spectroscopy to the study of hydrogenolysis lignin from red alder (10), isolated using neutral conditions in dioxane-water and consisting of ether-soluble (65%) and ether-insoluble (35%) fractions. The spectrum of the acetylated ether-insoluble hydrol lignin is shown in Figure 12.3. An unusual feature of the spectrum is the presence of highly shielded protons appearing in the range of 2.80-0.78δ. The assigned values for chemical shifts are listed in Table 12.4.

Of particular interest in the nmr spectra are the signals from aromatic protons associated with guaiacyl and syringyl nuclei. In the spectra of unacetylated guaiacyl compounds usually four lines are observed in the range of 6.53-6.96δ. The syringyl nucleus contains two identical protons which produce a singlet in the range of 6.3-6.5δ. A guaiacyl nucleus with an alkyl substituent attached to the 5-position (e.g., a condensed guaiacyl nucleus) usually produces a singlet in the range of 6.48-6.62δ. These δ values are somewhat variable according to the state of oxidation of the α-carbon atom on the side-chain (see Chapter 8 for details). The estimated distribution of variou types of protons in the polymeric hydrogenolysis lignin is compared with those

REDUCTION AND HYDROGENOLYSIS

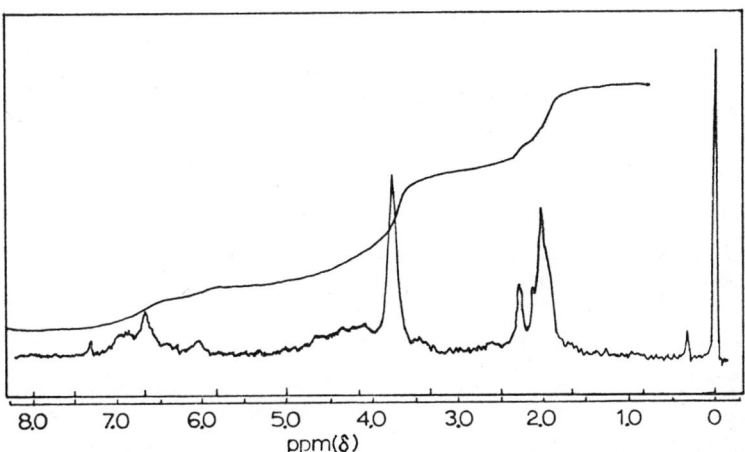

Figure 12.3. Nuclear magnetic resonance spectra of acetylated ether-insoluble hydrogenolysis lignin from red alder (10).

Table 12.4. Ranges of δ Values for Protons in Compounds Related to Hydrol Lignin (10).

Type of Proton	Ranges of Chemical Shifts
Aromatic	7.3-6.28
Guaiacyl	7.3-6.73
Condensed guaiacyl	6.73-6.48
Syringyl	6.48-6.28
Aliphatic-Substituted	6.28-3.50
Acetylated benzylic	6.28-5.74
α,β,γ	5.74-3.50
γ-CH_2OH	3.80-3.50
γ-CH_2OAc	4.28-3.97
Methoxyl	4.00-3.72
Acetoxyl	2.42-1.88
Aromatic	2.42-2.20
Aliphatic	2.20-1.88
Aliphatic hydrocarbon	2.80-0.78
α to Phenyl	2.80-2.34
β to Phenyl	2.12-1.03
β-Methylene to CH_2OH	2.12-1.66
β-Methylene	2.00-1.30
β-methyl	1.37-1.03
γ-Methyl	1.18-0.78
Hydroxyl	4.40-3.60(?)

of acetylated Björkman spruce and western hemlock dioxane-acidolysis lignins in Table 12.5. It can be noted that appreciable amounts of benzylic hydroxyl groups are present in this preparation. The retention of these easily reduced groups indicates that a significant part of them probably has not been accessible to the hydrogenation catalyst.

Table 12.5. Distribution of Types of Protons in Hydrogenolysis Lignin Based on NMR Data (10)

Type of Proton and Range	Spruce M.W.L.	Hemlock Acidolysis Lignin	Hemlock Hydrogenolysis Lignin	Ether Insoluble Alder Hydrogenolysis Lignin
Aromatic 7.3-6.28	18.6	17.2	15	12
Benzylic 6.28-5.74	4.8	5.5	-	3.6
Methoxyl 4.0-3.72	20.2	20.7	19	20
Acetoxyl 2.42-1.88	29.5	28.0	24	28
Aliphatic	26.9	28.6	42	37

The spectrum of the ether-insoluble fraction of alder hydrogenolysis lignin, Figure 12.3, is nearly identical with a spectrum obtained by Ludwig for a conifer acidolysis lignin (26). The most striking feature is the dominant condensed guaiacyl proton signal in the 6.5-6.75δ region. The relative abundance of aromatic units was estimated to be 15% syringyl, 51% condensed guaiacyl, 34% uncondensed guaiacyl units. In addition, there is a strong signal in the range 5.74-6.28 characteristic of benzylic acetyl protons. Signals appearing in the region 4.0-5.74δ are probably caused by protons associated with dimeric units of the β-β, β-5, and β-O-4 types (cf. Chapter 8). The ratio of aliphatic to aryl hydroxyl groups is estimated to be 4:1 as compared with 2:1 for the total hydrol lignin, suggesting the presence of relatively few phenolic hydroxyl groups in the insoluble fraction. Thus, the high molecular weight fraction displays the characteristics of largely undegraded and highly condensed guaiacyl type of lignin. It appears that the syringyl units are preferentially removed from the original lignin by hydrogenolysis, while guaiacyl and condensed guaiacyl units remain. Perhaps the isolated syringyl monomers originate in the predominantly syringyl lignins in the secondary walls, while the guaiacyl lignins in the middle lamellae and vessel elements of hardwoods (see Chapter 4, Section III) may provide the ether-insoluble part of hydrogenolysis lignins.

F. Mechanism of Formation of Hydrogenolysis Lignin. The formation of hydrogenolysis lignin from lignin in wood may best be explained according to the scheme shown in Equation 1.

$$\text{Wood} \xrightarrow[\text{hydrolysis}]{H^+ \text{ or } OH^-} \text{solubilized lignin fragments} \xrightarrow{\text{further hydrolysis and hydrogenolysis}} \text{stabilized lignin fragments or "Hydrogenolysis Lignin"} \quad (1)$$

The first part of the reaction involves partial breakdown of the lignin and cleavage of some lignin carbohydrate bonds producing rather large fragments that are solubilized and diffuse into the aqueous solution. Several points of evidence support this view. First, a certain amount of water must be present in the solvent system for successful hydrogenolysis. Secondly, alkaline hydrogenation pulping studies (17) showed that wood chips could be effectively delignified in a system where the catalyst was physically separated from the wood chips and these conditions produced a hydrogenolysis lignin typical of that usually found under alkaline conditions (16). Neutral conditions using water and dioxane result in lowering the pH to 5 as a consequence of the formation of acetic acid and other acids liberated by hydrolysis of ester linkages. These acids then promote an acidolysis of the lignin to soluble fragments which may in turn be degraded further.

The solubilized lignin fragments are in turn degraded to the monomeric, dimeric and polymeric material described earlier. This degradation probably involves a significant amount of hydrogenolysis as well as continuing hydrolysis as principal factors. The extensive body of knowledge on acidic and alkaline hydrolysis of lignin reviewed in Chapter 9 is in full accord with the idea that it is the lignin monomer units linked together by β-O-4 ether linkages which are responsible for a majority of the monomeric products found in hydrol lignins. Among the few model studies in this area, Pepper's investigations (6) on reactions of several reduced monomers in hydrogenolysis deserve attention and are summarized in Table 12.6. These results demonstrate that the conversion of syringyl to guaiacyl nuclei is possible through hydrogenative demethoxylation and point out that the hydrocarbon ethyl and propyl side-chains may be formed from a propyl alcohol type of side-chain. Of even more significance are the results on lignin model compounds 24 to 27 reported in the same paper. All four compounds contain highly oxygenated side-chains and were hydrogenated under neutral and alkaline conditions. Although no details were reported, Pepper states that "in each case the products were very complex mixtures, the recoveries varied widely and ring reduction occurred in some cases. All the monomeric products that have been obtained from lignin were found here, and, similar to lignin, the yields from alkaline reactions were always greater than from the neutral runs."

Although the actual precursors of hydrogenolysis monomers of lignin are

Table 12.6. Hydrogenolysis of Model Compounds (59).

Compound	Reaction Conditions	Structures	% of Product
(G)-$CH_2CH_2CH_2OH$	Neutral	(G)-$CH_2CH_2CH_2OH$	72
		(G)-$CH_2CH_2CH_3$	28
	Alkaline	(G)-$CH_2CH_2CH_2OH$	100
(S)-$CH_2CH_2CH_2OH$	Neutral	(G)-$CH_2CH_2CH_2OH$	85
		(G)-$CH_2CH_2CH_2OH$	8
		(S)-CH_2CH_3	4
		(S)-$CH_2CH_2CH_3$	3
	Acidic	(S)-$CH_2CH_2CH_2OH$	72
		(S)-$CH_2CH_2CH_3$	21
		(G)-$CH_2CH_2CH_2OH$	6
(S)-$CH_2CH_2CH_3$	Neutral	(S)-$CH_2CH_2CH_3$	98
		(G)-$CH_2CH_2CH_3$	2
(G)-CH_2CH_2OH	Alkaline	(G)-CH_2CH_2OH	34
		(G)-CH_2CH_3	66

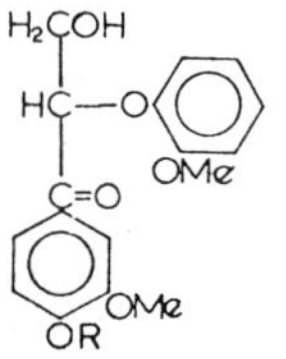

24, R = CH_2Ph
25, R = Me

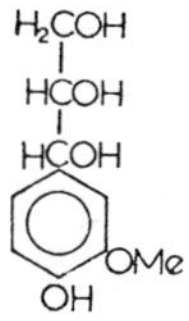

26

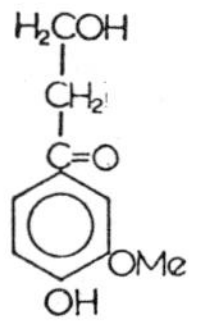

27

thus difficult to identify, tentative mechanisms of formation have nevertheless been formulated by Parker, Coalson and Schuerch (9). A mechanism perhaps more in accord with the present concepts of lignin hydrolysis (cf. Chapter 9) is shown in Equation 2. A major source of guaiacyl propanol 2 is etherified units 28 in lignin which in alkaline hydrolysis are converted to an epoxide through β-aryloxy elimination. Guaiacyl propane 11, on the other hand, derives from phenolic units 29, present as such or formed through the hydrolysis of phenol ether bonds. The probable intermediate is the extended quinone methide 31 in alkaline hydrogenation. A more probable route, however, is the loss of the γ-carbon as formaldehyde from quinone methide 30, yielding in reduction

C_6C_2-monomers 32 and 4. The C_6C_1-monomers such as methyl guaiacol 10 may arise from vanillin formed by reversed aldol condensation of coniferyl aldehyde groups. The latter may also be formed in the neutral hydrolysis of phenolic structures 29.

$$(2)$$

Most of the monomeric degradation products resulting from neutral or acidic hydrogenolysis are of the C_6C_3 type, and their formation from the β-O-4' linked unit is analogous to ethanolysis with stabilization of the fragments by reduction rather than ketone formation, Equation 3.

The dimers of the diphenyl methane type (17, 19, 21) are probably artifacts formed from vanillyl and syringyl alcohols, via quinone methides as shown in Equation 4. The two alcohols are probably derived from degradation of unsaturated aldehyde structures.

The other dimeric products reported by Parker et al. (9) most likely arise directly from structural elements present in the lignin molecule. The 5,5-biphenyl dimer 22 is analogous to the well-known dehydrodivanillin from oxidative degradation (27), and isolation of this type of dimer from alkaline as well as inert neutral hydrogenation is further evidence for the existence of 5-5 bonds in lignin. The bibenzyl dimers 18 and 20 probably arise from β-1 dilignol structures (28), Equation 5.

Finally, the unsymmetrical dimer, 23, may arise from a phenyl coumaran unit by loss of the γ carbon from both monomer units as formaldehyde followed by reduction (25), Equation 6.

G. **Hydrogenolysis of Milled Wood Lignin.** Most of the studies cited above involved hydrogenolysis under hydrolytic conditions. Coscia, Nord and Schubert (24) have subjected milled wood lignin (MWL) from oak and birch to catalytic hydrogenation in anhydrous dioxane in an attempt to degrade the lignin without structural alteration by condensation reactions. At a temperature of 220-250°C, the hydrogen uptake was at least 4 moles per phenyl propane unit, and product analysis showed no loss of aromatic rings or elimination of methoxyl groups. The products were vacuum distilled and analyzed by gas chromatography, and about 21% (based on milled wood lignin) of the lignin was recovered as identified guaiacyl or syringyl derivatives. The results are quoted in Table 12.7.

Table 12.7. Reaction Products from Non-Hydrolytic Hydrogenolysis of Milled Wood Lignins.

Compound	Yield % of MWL			
	Birch (24)	Oak (24)	Blue Spruce (29)	White Pine (29)
(G)-H	-	-	0.3	0.3
(G)-CH_3	1.1	1.0	3.5	2.9
(G)-CH_2CH_3	0.9	0.6	2.1	2.9
(G)-$CH_2CH_2CH_3$	2.3	2.6	5.9	5.3
(G)-$CH_2CH_2CH_2OH$	2.0	0.6	8.1	7.3
(S)-CH_3	2.0	3.1	-	-
(S)-CH_2CH_3	1.1	0.7	-	-
(S)-$CH_2CH_2CH_3$	3.9	7.3	-	-
(S)-$CH_2CH_2CH_2OH$	7.9	0.8	-	-
Totals	21.2	16.7	19.9	17.7

Olcay (29) applied the same hydrogenolysis conditions to blue spruce and white pine and obtained the expected guaiacyl derivatives only (Table 12.7). About two-thirds of the distillable hydrogenolysis product representing 20% of the original lignin was identified. It is interesting to note the close concordance in the identified reaction products from pine and spruce, indicating extensive structural similarity of the MWL preparations, as contrasted with the divergence in products from oak and birch. Formation of the monomeric products in this type of hydrogenolysis has been explored also by model compound studies. Olcay (30) studied hydrogenolysis of various models, and the results are summarized in Table 12.8. The good yield of 4-n-propyl guaiacol found in MWL hydrogenolysis products probably arises from the β-O-4' linked monomers discussed earlier. The high yield of 4-methyl guaiacol, 10, is somewhat puzzling. Coscia, et al. has proposed formation of 4-methyl guaiacol from phenyl coumaran units (24).

Olcay also investigated the higher molecular weight product from hydrogenolysis of softwood MWL. The amorphous powder isolated by solvent frac-

Table 12.8. Reaction Products from Non-Hydrolytic Hydrogenolysis of Model Compounds by Copper Chromite as 230-240°C

Model Compound	Products	Yields
(G)-CH=CHCH$_3$	(G)-CH$_2$-CH$_2$-CH$_3$	100 %
(G)-CH$_2$CH=CH	(G)-CH$_2$CH$_2$CH$_3$	100
(G)-COCH$_3$	(G)-CH$_2$CH$_3$	100
(G)-CH$_2$COCH	(G)-CH$_2$-CHOH CH$_3$	95.5
	(G)-CH$_2$CH$_3$	4.1
	(G)-CH$_2$CH$_2$CH$_3$	0.1
(G)-CH$_2$CH$_2$OH	(G)-CH$_2$CH$_2$OH	70
	(G)-CH$_2$CH$_3$	30
(G)-CH=CHCH$_2$OH	(G)-CH$_2$-CH$_2$-CH$_2$OH	40
	(G)-CH$_2$CH$_2$CH$_3$	12
	(G)-CH$_2$CH$_3$	0.4
	(G)-CH$_2$CH$_2$CH$_3$	53.1
	(G)-H	40.5

tionation represented 60% of the original MWL. Its infrared and ultraviolet spectra were similar to those of the unhydrogenated MWL except for loss of conjugated carbonyl groups. More significantly, practically no differences were found in the C, H and methoxyl analysis. Thus, a resistant "core" lignin is found in milled wood lignin, similar to that observed with the hydrogenolysis of wood lignin described earlier.

H. The Isolation of a C$_6$C$_3$ Biphenyl Dimer. Nahum (31a) has hydrogenated woodmeal in benzene using the hydroformylation system CO + H$_2$ with a cobalt octacarbonyl catalyst. The usual guaiacyl hydrocarbon products were identified, and in addition the diphenylic C$_6$C$_3$-dimer 32a was found. From structural point of view, the isolation of this compound provides the most convincing evidence for the presence of the biphenyl linkage in lignin.

32a

REDUCTION AND HYDROGENOLYSIS

I. Other Hydrogenolysis Studies Producing Aromatic Monomers. Hachihama (31b) in a series of papers has reported the results of his investigations of more vigorous hydrogenation in the 200°C range using a variety of nickel catalysts. Wood meal, hydrochloric acid and other lignins gave mixed products, i. e. both aromatic and cyclohexane derivatives. Aromatic products reported were guaiacylpropane, 11, protocatechuic acid, pyrocatechol, p-hydroxybenzoic acid, guaiacol and creosol, 10. The presence of alkali caused demethylation to occur rather extensively at the high temperatures used. Lautsch and Piazolo (4) likewise studied hydrogenolysis of lignin materials using nickel catalysts. In the 250° temperature range, spent sulfite liquor gave about 50% yields of phenolic products, 30% of which were monomeric phenols. Isolated aromatic compounds were guaiacol and guaiacylethane 4, catechol and p-cresol, 10. A comprehensive review of hydrogenation studies aimed at industrial utilization of lignin has been presented by Brauns and Brauns (31c).

Brown and Neish (32) have made an interesting application of hydrogenolysis to biogenetic studies on lignin. The method was to isolate phenyl propane type of lignin degradation products and determine their radioactivity in order to assess the incorporation of labeled precursors.

IV. Hydrogenation of Lignins to Cyclohexane Derivatives.
A. Studies on Isolated Lignins. Propyl cyclohexane derivatives were first obtained in good yields from isolated aspen lignin in 1938 by Harris and Adkins (2). The lignin was hydrogenated in dioxane over a copper-chromium oxide catalyst at 250-260° for eighteen hours using 200-350 atmospheres of hydrogen. These vigorous conditions resulted in the formation of methanol, 4-n-propylcyclohexanol-1 8, 3-(4-hydroxycyclohexyl)-propanol-1 33, and 4-n-propylcyclohexanediol-1,2 34 in yields of 26.5, 11, 32 and 24%, respectively, based on the lignin subjected to the hydrogenation. Comparison with earlier model work (33) makes it clear that the majority of original bonds broken are C-O bonds, and the cyclohexane compounds are probably formed from aryl propane units according to Equation 7.

The hydrogenation results of Harris and Adkins are often mistakenly cited as showing that 70% of lignin is convertible to monomers. This interpretation is contrary to results, and the authors have stated this clearly. In reality, the "methanolysis" lignin used for hydrogenation was isolated from aspen wood in a 60% yield. Since the yield of hydrogenolysis monomers was about 60% of methanolysis lignin, the yield of identified monomers is 36% of wood lignin, which is in line with the monomer yields of other degradation products, i.e., oxidation, isolated from hardwood lignins (33-36).

In addition, similar hydrogenations were carried out on aspen sulfuric acid and soda lignins, and on spruce methanolysis lignin (34, 35). Monomeric products were again isolated, but in much lower yields. All methoxyl groups were cleaved from the hardwood preparation, but about 10% methoxyl remained in the product from spruce, probably due to the greater degree of condensation in the latter and its diminished accessibility to the catalyst. Comparative studies using nickel catalysts were also done (36).

The influence of the hydrolytic pretreatment of the lignin was shown clearly in a study of hydrogenation of soda lignin from a hardwood mixture (35). The yield of monomer fraction was only about one-fifth of that obtained from methanol lignin, and whereas the methanol lignin yielded 25% of 3(4-hydroxycyclohexyl) propanol-1 33, none could be detected from soda lignin, nor were more than trace amounts of any C_6C_3 compound found. Cyclohexanol, 4-methyl cyclohexanol and 4-ethyl cyclohexanol were found in small quantities. Products of higher molecular weight than C_6C_3 dominated in the product mixture, most of which was distillable.

Hibbert and coworkers (37) showed that wood could be completely liquefied when treated under Adkin's conditions and that 4-n-propyl cyclohexanol 8 (19.5%) and 3(4-hydroxycyclohexyl) propanol-1 33 (5.8%) could be readily isolated, representing together 36% of the Klason lignin of the maple wood (37). It was later shown (38) that the material reported as 4-n-propylcyclohexanol was in fact a mixture of this compound with 3-cyclohexylpropanol-1 9. Thus, two-thirds of the cyclohexane monomers from lignin have a terminal CH_2OH group.

Saeman and Harris (39) solubilized methanol aspen lignin over Raney nickel recovering 24% yield of volatile products including 3(4-hydroxycyclohexyl) propanol-1, 4-ethyl-cycylohexanol, 4-n-propyl-cyclohexanol and 2-methoxy-4-ethyl-cyclohexanol. In later work, soda lignin was hydrogenated with copper chromite at 325°C. yielding 14% water, 8% methanol, 8% tar acids including catechol, 4-methyl catechol and 4-propyl catechol, phenol, guaiacol, p-cresol, 4-alkyl-2-methoxyl phenols and xylenols, 13.3% oxygen compounds, 24% hydrocarbons and 28% heavy oil (40).

V. Hydrogenolysis of Lignin by Potassium and Sodium in Liquid Ammonia.
  A. Isolation and Characterization of Reaction Products. Sodium or potassium metal dissolves readily in liquid ammonia forming blue solutions with

REDUCTION AND HYDROGENOLYSIS

a strong reducing power. The reagent, which is similar to lower alcohols in solvent capacity for organic materials, is effective in reduction, especially in ether cleavage reactions.

Shorygina and coworkers have carried out extensive studies of the effect of sodium in liquid ammonia on isolated lignins and on wood. In the initial study (41) cuoxam lignin was subjected to nine successive treatments with the reagent. About 90% of the lignin was converted into materials soluble in water or ether, and the methoxyl content of the insoluble lignin decreased from 15.31% to 10.07%, while the hydroxyl content remained fairly constant at about 16.5%. From spruce, aspen and pine, 4-n-propyl guaiacol 11 was isolated in good yields (8% of the cuoxam lignin from spruce). Additional products included 1-guaiacyl-propanol-2 35 and a dimeric compound with the tentative structure of 2,3-bis(hexahydrobenzyl)butane 36 (42). Cuoxam aspen lignin gave 4-n-propylguaiacol 11 and the syringyl analogue 1-(4-hydroxy-3,5-dimethoxyphenol)-propane 3. The ether-soluble fraction from aspen was obtained in 18% yield, about twice as much as from spruce. In more recent work on spruce and pine woodmeal total conversion to ether-soluble products was accomplished in only one treatment, and two additional compounds, 1-guaiacyl propanol-3 2 and 1-guaiacyl propanol-1 37, were identified (43).

Freudenberg and coworkers (44) have treated wood and isolated lignin with potassium and potassium amide in liquid ammonia. Spruce lignin showed loss of methoxyl groups from 15% to 6% in two treatments and some increase in hydroxyl content.

B. Mechanism. Of the reactions of organic compounds with alkali metals in liquid ammonia, ether cleavages and hydrogenation reactions are of interest in this content. Generally, aliphatic ethers are stable, but aryl alkyl ethers are readily cleaved. Birch has reported the ether cleavages shown in Table 12.9 using sodium in ammonia (45). Freudenberg has also treated several phenyl ethers with potassium in ammonia at 20°C under more vigorous conditions than those used by Birch (46).

The mechanism of the reaction involves cleavage by addition of electrons followed by stabilization of the hydrocarbon, Equation 8. These results indicate that partial demethylation of lignin should occur, but it is apparent that vigorous conditions are needed to cleave a methoxyl group ortho to a phenolic

Table 12.9. Ether Cleavages by Potassium in Liquid Ammonia.

Ether	Product	Yield	Reference
anisole	phenol	27	45
o-methyl anisole	o-cresol	17	45
m-methyl anisole	m-cresol	9	45
p-methyl anisole	p-cresol	4	45
veratrole	guaiacol	89	45
m-methoxy anisole	m-methoxyphenol	71	45
p-methoxy anisole	p-methoxyphenol	25	45
anisole	phenol	--	46
veratrole	guaiacol and catechol	--	46
propyl veratrole	guaiacyl propane and 4-methoxyl-3-hydroxyphenolpropane	--	46
dihydroeugenol	unchanged	--	46
vanillin	vanillyl alcohol + polymer	--	46
vanillic acid	unchanged	--	46

$$\underset{OMe}{\bigcirc} + 2e^- \longrightarrow \underset{O^-}{\bigcirc} + CH_3^- \xrightarrow{NH_3} HCH_3 + NH_2^- + \underset{O^-}{\bigcirc} \tag{8}$$

hydroxyl. This concept is in harmony with the results on lignin. The mild system of Shorygina showed a loss of about 1/3 of the methoxyl groups, while Freudenberg's more vigorous treatment resulted in the elimination of about 2/3 of methoxyl groups.

The monomeric degradation products undoubtedly arise from units held together in lignin by β-O-4 linkages. Again, the main monomeric product from softwood lignin is 4-n-propyl guaiacol, and the α, β and γ-hydroxyl derivatives of this compound all occur in small amounts (47, 48).

Shorygina has shown that coniferyl alcohol yields 4-n-propylguaiacol (86%) and some dihydroconiferyl alcohol on reduction with sodium in liquid ammonia. Formation of the benzyl alcohol, 1-guaiacyl-n-propanol, is unusual in reductive systems but is expected here since model studies show vanillyl alcohol as a product from vanillin (Table 12.9). The survival of the β-hydroxyl in 1-guaiacyl-n-propanol-2 is also unexpected because alkyl-aryl ethers in reductive cleavage by sodium and ammonia generally yield phenols and alkyl hydrocarbons. On the other hand, the isopropyl alkyl group is the most difficult to cleave, and the presence of an o-methoxyl unit has been shown to aid carbon-oxygen cleavage ortho to the methoxyl (49). Thus a side reaction in the reduction of β-O-4 lignol structures could well produce the β-alcohol together with a m-substituted anisole in addition to the main products 4-n-propyl guaiacol and another guaiacol derivative, as shown in Equation 9.

(9)

REFERENCES

1. A. Lindblad, Ing. Vetenskaps Akad. Handl., No. 107, 7 (1931).
2. E. E. Harris and H. Adkins, Paper Trade J., 107, 20, 38 (1938).
3. H. P. Godard, J. L. McCarthy and H. Hibbert, J. Am. Chem. Soc., 62, 988; 63, 3061 (1941).
4. W. Lautsch and G. Piazolo, Chem. Ber., 76, 486 (1943).
5. C. P. Brewer, L. M. Cooke and H. Hibbert, J. Am. Chem. Soc., 70, 57 (1948).
6. J. M. Pepper, W. F. Steck, R. Swoboda and J. C. Karapally, Adv. in Chem., 59, 238 (1966) and references cited therein.
7. J. G. Miller and C. Schuerch, Tappi, 51, 273 (1968) and references cited therein.
8. T. Y. Kefeli and N. N. Shorygina, J. Gen. Chem. U.S.S.R, 20, 1199 (1950).
9. P. E. Parker, R. L. Coalson and C. Schuerch, Adv. in Chem., 59, 249 (1966).
10. B. F. Hrutfiord and J. L. McCarthy, ibid., 59, 226 (1966).
11. E. Adler and J. Marton, Acta Chem. Scand., 13, 75 (1959); 15, 357, 370 (1961).

12. K. V. Sarkanen, H. M. Chang and B. Ericsson, Tappi, 50, 572; and G. G. Allan, 50, 583, 587 (1967).
13. R. H. J. Creighton, R. D. Gibbs and H. Hibbert, J. Am. Chem. Soc., 66, 32, 37 (1944).
14. J. M. Pepper and H. Hibbert, ibid., 70, 67 (1948).
15. J. M. Pepper and D. C. Hagerman, Can. J. Chem., 32, 614 (1954).
16. I. Sobolev, H. G. Arlt and C. Schuerch, Ind. and Eng. Chem., 49, 1399 (1957).
17. I. Sobolev and C. Schuerch, Tappi, 41, 545 (1958).
18. J. M. Pepper and W. Steck, Can. J. Chem., 41, 2867 (1963).
19. M. Granath and C. Schuerch, J. Am. Chem. Soc., 75, 707 (1953).
20. C. Schuerch, J. Am. Chem. Soc., 72, 3838 (1950).
21. A. Bhattacharya, E. Sondheimer and C. Schuerch, Tappi, 42, 446 (1959).
22. J. M. Pepper, C. J. Brownstein and D. A. Shearer, J. Am. Chem. Soc., 73, 3316 (1951).
23. H. G. Arlt, S. K. Gross and C. Schuerch, Tappi, 41, 64 (1958).
24. C. J. Coscia, W. J. Schubert and F. F. Nord, J. Org. Chem., 26, 5085 (1961).
25. J. G. Miller and C. Schuerch, Tappi, 51, 273 (1968).
26. C. H. Ludwig, B. J. Nist and J. L. McCarthy, J. Am. Chem. Soc., 86, 1196 (1964).
27. I. A. Pearl, ibid., 68, 429 (1946).
28. H. Nimz, Holzforschung, 20, 105 (1966).
29. A. Olcay, J. Org. Chem., 27, 1783 (1962).
30. A. Olcay, Holzforschung, 17, 105 (1963).
31a. L. S. Nahum, Ind. Eng. Chem., Prod. Res. Develop. 4(2), 71 (1965).
31b. Y. Hachihama, S. Jyodai et al., see F. E. Brauns and D. A. Brauns, The Chemistry of Lignin, Supplement Volume, Academic Press, New York, 1960, pp. 486-489.
31c. F. E. Brauns and D. A. Brauns, ibid., Chapter 18, pp. 486-497.
32. S. A. Brown and A. C. Neish, J. Am. Chem. Soc., 81, 2419 (1959).
33. H. Adkins, Reactions of Hydrogen, Wisconsin Press, 1937.
34. E. E. Harris, J. D. Ianni and H. Adkins, J. Am. Chem. Soc., 60, 1467 (1938).
35. H. Adkins, R. L. Frank and E. S. Bloom, ibid., 63, 549 (1941).
36. E. E. Harris, J. Saeman and E. C. Sherrard, Ind. and Eng. Chem., 32, 440 (1940).
37. L. M. Cooke, J. L. McCarthy and H. Hibbert, J. Am. Chem. Soc., 63, 3056, 3062 (1941).
38. J. R. Bower, L. M. Cooke and H. Hibbert, ibid., 65, 1192 (1943).
39. J. F. Saeman and E. E. Harris, ibid., 68, 2507 (1946).
40. E. E. Harris, J. F. Saeman and C. B. Bergstrom, Ind. and Eng. Chem., 41, 2063 (1949).

41. N. N. Shorygina and T. Ya. Kefeli, J. Gen. Chem. U.S.S.R., 17, 2058 (1947); 18, 528 (1948).
42. N. N. Shorygina, T. Ya. Kefeli and A. F. Semechkina, ibid., 19, 1558 (1949).
43. A. F. Semechkina and N. N. Shorygina, ibid., 28, 3265 (1958).
44. K. Freudenberg, E. Flickinger, A. Sobek and F. Klink, Chem. Ber., 72, 217 (1939).
45. A. J. Birch, J. Chem. Soc., 1947, 102.
46. K. Freudenberg, W. Lautsch and G. Piazolo, Chem. Ber., 74, 1879 (1941).
47. A. J. Birch, Quart. Rev., 12, 17 (1958).
48. C. D. Hurd and G. L. Oliver, J. Am. Chem. Soc., 81, 2795 (1959).
49. F. J. Sowa, P. A. Sartonetto, A. L. Kranzfelder and J. J. Verbanc, ibid., 59, 603, 148811 (1937); 60, 94 (1938).

13

MODIFICATION REACTIONS
G. Graham Allan

I. Introduction. Although lignin is one of the most abundant natural macromolecules and it is available at exceedingly low cost, it has received relatively little attention as a polymer per se. Indeed, standard polymer texts (1,2) often do not even acknowledge its existence as one of the important natural polymers. When lignin is allowed to briefly emerge from the shadows, it finds itself categorized as "not a real high polymer," (3) and it is therefore scarcely surprising that many chemists are only vaguely aware of this phantom macromolecule.

The amelioration of lignin's reputation as a mysterious and intractable material is thus long overdue, and it is an object of this chapter to contribute to this goal by reintroducing the various lignins to respectable chemical society. Lignins should be regarded as a family of three-dimensional polymers, spherical in solution, containing a variety of functional sites and capable of a surprising selection of modifying reactions. It is intended that this chapter primarily present the organic chemistry to complement the physico-chemical treatment of the lignin macromolecule in Chapter 17 and serve to excite polymer chemists at large with the essentially untapped potential of lignin, a renewable natural resource.

II. Reactive Sites Available in Lignins. The constitutional model of a softwood lignin (see Chapter 1) (4,5) gives the broad picture of the reactive groups available in the native lignin. These consist of ethers of various types, primary and secondary hydroxyl groups, carbonyl groups, carboxyl and ester functions, ethylenic linkages and the sulfur-containing groups such as thiols and sulfonic acids, introduced as a consequence of sulfide or bisulfite pulping. Since lignin is a phenylpropanoid polymer, there also exists a number of typical aromatic sites and activated aliphatic locations capable of involvement in modification reactions.

A. Ether Linkages. Of the total array of functional groups the ether linkage is certainly the most characteristic in all lignins and several different types have been recognized. These include the simple phenolic ethers exemplified by alkyl-aryl bonds between monomeric units, methoxyl groups associated with guaiacyl and syringyl groups, cyclic ethers of the tetrahydrofuran type (7) and diaryl ethers (6).

It is to be emphasized that many of these ether functions are located on the α-carbon of the propanoid side-chain. They are therefore benzylic ethers and as such are rather reactive. An approximate total of about 0.17 benzylic ether moieties per methoxyl group are found in conifer lignins (8) and the inclusion of the comparably reactive benzylic hydroxyl group (0.06 per methoxyl) is appropriate in modification considerations.

While the methoxyl groups are not generally considered a reactive function, they can be split off from lignin units 1 using the conventional hydriodic acid technique, albeit at somewhat elevated temperatures (9). This is, of course, the basis of the Zeisel methoxyl determination, Equation 1. Attempts to circumvent the employment of hydriodic acid have been made by substituting a mixture of phosphoric acid and iodide (10) which also affords methyl iodide as a by-product. Unfortunately, the polymeric residues 2 from the methoxyl determinations of lignin do not seem to have been studied. The apparent increase in functionality by the formation of the reactive phenolic hydroxyl groups will, however, be substantially offset by acid-induced condensation.

$$\underset{1}{\text{ArOMe}} \xrightarrow[\Delta]{HI} \underset{}{\text{Ar-O}^+\text{HMe}} \longrightarrow \underset{2}{\text{ArOH}} + \text{MeI} \qquad (1)$$

The methoxyl group is also susceptible to cleavage using chlorine (11) nitric acid (12) and hydroxide or hydrosulfide ions (13, 14). These modification reactions usually result in the elimination of the methoxyl group as methanol though other methane derivatives can also be formed. The cleavage of the methyl moiety of the methoxyl by hydrosulfide ion and the subsequent transformation to dimethylsulfide has been investigated at some length (15-17). The demethylated lignin residue has been proposed as a phenol substitute in resin manufacture (18).

Although under these cleavage conditions, the emergent methyl radical could not be captured as methylamine (15) this can be accomplished by other means. Thus, treatment of a kraft lignin in sulfolane solution with boron tribromide affords a complex 3 which can be decomposed with water, ammonia or piperidine to yield methanol, methylamine and N-methyl-piperidine, respectively (15, 19). The reaction can probably be formulated as in Equation 2. The nature of the cleavage residue 4 from this and other types (20) of ether

MODIFICATION REACTIONS

cleavages was not examined.

$$1 \xrightarrow{BBr_3} 3 \xrightarrow[2. H_2O]{1. base^*} 4 + MeOH \qquad *bases: H_2O, NH_3, piperidine \qquad (2)$$

The comparable reaction does not appear to be practical with aluminum trichloride which, however, is effective in cleaving anisole (21). This can probably be attributed to the difficulty of finding a solvent which can swell the lignin and also dissolve the aluminum trichloride. The usual solvents for this reaction, nitrobenzene and carbon disulfide, do not dissolve lignin and sulfolane (tetramethylenesulfone), a good lignin solvent, for example, reacts with aluminum trichloride (22). Likewise, pyridine hydrochloride, which cleaves anisole to phenol in 82% yield, does not seem to have been explored for lignin reactions though it has been used to study lignin hydrolysis products (23).

The known cleavage reactions of ethers by sodamide (24), sodium hydroxide (25) and by sodium in pyridine (26) apparently have not been systematically applied to lignin though model compounds exemplified by anisole, phenetole, phenylbenzyl ether and diphenylether are converted to phenol in yields above 90% by the latter procedure. While the above cleavage chemistry is fairly well established, its application to lignin results in simultaneous changes in the bulk of the lignin which are not quite so clear and deep-seated molecular condensation may further occur. This is particularly illustrated by the case of aniline hydrochloride. Grangaard has shown (27) that in addition to the normal demethylation of lignin a more fundamental reaction occurs which results in the formation of catechol in considerable yield. Subsequent experiments carried out on model compounds indicate that the formation of catechol could stem from structural elements of the dihydroxydiphenylmethane $\underline{5}$ type. The decomposition of these units is illustrated in Equation 3.

$$\underset{5}{HO-C_6H_4-CH_2-C_6H_4-OH} \xrightarrow[-PhOH]{HCl} HO-C_6H_4-CH_2-Cl \xrightarrow{PhNH_2} \qquad (3)$$

$$HO-C_6H_4-CH_2-C_6H_4-NH_2 \xrightarrow[-PhOH]{HCl} H_2N-C_6H_4-CH_2-Cl \xrightarrow{PhNH_2} H_2N-C_6H_4-CH_2-C_6H_4-NH_2$$

Electronic equivalents such as eugenol and compounds containing elements capable of homocondensation or reaction with aniline to form substituted diphenylmethane moieties also yield catechol **6**. The reaction with a lignin unit can therefore be formulated as in **1** to **6**, Equation 4.

$$\text{(Equation 4)}$$

In the alkaline hydrolysis of lignin ether linkages other than the methoxyls are more readily cleaved (28-30). The splitting of alkyl-aryl ether linkages in position 4 under these conditions is well established and has also been demonstrated with guaiacyl ether model compounds (cf. Chapter 9). The generation and isolation of carbonyl compounds such as vanillin, acetaldehyde, formaldehyde and acetoguaiacone by such cleavages (31-35, 37) opens up the possibility of modifying lignin by simultaneous aldol or Michael-type condensations which are catalyzed by reaction conditions similar to those used in the alkaline hydrolysis (36).

Cleavage of the tetrahydrofuranyl ether linkages of the pinoresinol and phenylcoumaran types probably also occurs during pulping by the bisulfite process (38). In such cleavages, sulfonic acid groups are introduced to the benzylic positions (39-41). The reactivity of the ether group has also been investigated by Shorygina and her colleagues using sodium and liquid ammonia (42-45). This is a particularly interesting reaction system because of the solvent parameters of liquid ammonia (46-48) and because the modified lignin products should be more amenable to modification reactions. The various ether linkages are theoretically cleaved by the sodium and liquid ammonia as shown in Equation 5.

$$ROR' \longrightarrow RONa + R'Na \longrightarrow ROH + R'H \qquad (5)$$

Using the cuoxam lignin obtained from spruce or aspen, Shorygina and Kefeli successively applied (9 times) the sodium and liquid ammonia reduction for eight to twelve days. After these nine treatments, almost 90% of the lignin had been degraded to low molecular weight compounds including monomeric 4-n-propylguaiacol **7**, its syringyl analogue **8** and 1-(4-hydroxy-3-methoxylphenyl)-2-propanol **9** (cf. Chapter 12).

$$\text{(G)}-CH_2CH_2CH_3 \qquad \text{(S)}-CH_2CH_2CH_3 \qquad \text{(G)}-CH_2\underset{OH}{CH}\ CH_3$$
$$\qquad 7 \qquad\qquad\qquad 8 \qquad\qquad\qquad 9$$

The modified lignin concomitantly formed by these cleavages has received little attention although Brauns and Brauns have suggested it would be interesting to test if it would still undergo classical diagnostic lignin reactions (49). It is therefore worthy of speculative comment. Before the sodium and liquid ammonia reduction, the cuoxam spruce lignin contains about 9.3% hydroxyl and 15.3% methoxyl groups. After the first reaction, the hydroxyl content has increased to 16.86%, and the methoxyl content has decreased. Although these changes can be attributed to general ether cleavages within the lignin molecule, the effect on the possible sites for modification can only be gauged after a consideration of the various ether cleavages possible. Normally, dialkyl ethers typified by diethyl ether, 1,2-dimethoxyethane, dioxan and tetrahydrofuran are inert toward solutions of alkaline metals in liquid ammonia (50, 51) even in the presence of alcohols (52) and indeed are often used as co-solvents in metal-ammonia reductions (53–57). However, carbon-oxygen fission can be achieved with alkyl, benzyl (58, 59), aryl or enol ethers. Lignin, of course, contains all of these particular ether types as exemplified by phenylcoumaran and pinoresinol structures.

That this increase in sites for modification involves more than a straight-forward demethylation reaction is endorsed by the fact that remethylation gives a higher methoxyl value (60). This was initially interpreted as indicating the presence of methylene dioxybenzene groups since catechol acetals have been shown (61, 62) to be hydrogenolyzed to the corresponding phenols. This view is not now accepted (63). Normally, aromatic methyl ethers undergo cleavage to the corresponding phenols, but in no case has the fate of the alkyl portion been determined. Presumably, it is liberated as methane (Equation 5a).

$$\text{ArOMe} \xrightarrow{2e} \text{ArO}^- + \text{Me}^- \xrightarrow{2H^+} \text{ArOH} + \text{MeH} \qquad (5a)$$

Birch has investigated (52) the ease of fission of a range of methyl aryl ethers by determining the yield of phenol (Table 13.1) and concordant results have been obtained by Freudenberg and his colleagues using potassium in liquid ammonia (61, 62). From these data, it will be realized that alkali metals in liquid ammonia will cause considerable but not complete demethylation. The fact that o-dimethoxybenzenes resist complete demethylation indicates that a phenolic hydroxyl protects an adjacent methoxyl against fission, probably because of salt formation. In agreement, it was found that dihydroeugenol, vanillin and vanillic acid also resist demethylation (62). Vanillin underwent partial reduction of the carbonyl group giving the previously known (63, 64) vanillin-vanillyl alcohol condensation product 10.

Table 13.1. Fission of Methyl Aryl Ethers to Monophenols (61,62)

Methyl Aryl Ether	Monophenol	Yield %	Related Function Present in Lignin
Methoxybenzene	Phenol	27	No
2-Methylmethoxybenzene	o-cresol	17	No
3-Methylmethoxybenzene	m-cresol	9	Yes
4-Methylmethoxybenzene	p-cresol	4	No
1,2-Dimethoxybenzene	2-methoxyphenol	89	Yes
1,3-Dimethoxybenzene	3-methoxyphenol	71	Yes
1,4-Dimethoxybenzene	4-methoxyphenol	25	No

The increase in hydroxyl content of the lignin in the initial stages is therefore most likely not to have originated exclusively from demethylation. The ether groups may suffer other fates than cleavage as exemplified by the conversion of catechol ether 11 to anisole 12, Equation 6, by loss of oxygen from the ring (52, 66).

$$\underset{11}{\text{C}_6\text{H}_4(\text{OMe})(\text{OCH}_2\text{OMe})} \xrightarrow[-\text{CH}_2\text{O},\ -\text{MeOH}]{\text{Na, NH}_3} \underset{12}{\text{C}_6\text{H}_5\text{OMe}} \qquad (6)$$

The functionality of the lignin can also be modified by the reductive elimination of methoxyl groups from aromatic systems which are particularly susceptible to nuclear reduction as illustrated by the transformation (65) of veratric acid 13 to 3-methoxycyclohex-2-en-1-carboxylic acid 14, Equation 7. Though these precise conditions have not yet been used on lignin, it is interesting to note that reduction of anisole with sodium and ethanol and liquid ammonia affords the conjugated cyclohexenone 15 via the sequences (6) in Equation 8.

$$\underset{13}{\text{veratric acid}} \xrightarrow{\text{Red.}} \underset{14}{\text{3-methoxycyclohex-2-en-1-carboxylic acid}} \qquad (7)$$

MODIFICATION REACTIONS

$$\text{12} \xrightarrow{} \text{(OMe)} \xrightarrow{} \text{(=O)} \xrightarrow{} \text{15} \quad (8)$$

The possibilities of condensation during the sodium-liquid ammonia reduction must not be overlooked. The conversion of benzylphenyl ether to phenol and 1,2-diphenylethane (67) represents a relevant example. The dimerized hydrocarbon is probably formed by the action of the benzyl anion upon benzylphenyl ether as expressed by Equation 9:

$$PhCH_2^- + PhOCH_2Ph \longrightarrow PhO^- + PhCH_2CH_2Ph \quad (9)$$

Since there is a number of benzylic ether functions in lignin (see Chapter 5), it seems reasonable to assume that liberated benzyl anions could attack intact benzylic ethers to form carbon-carbon bonds.

Overall, the most noteworthy aspect of the sodium and liquid ammonia reaction is the dramatic increase in the hydroxyl group content of the lignin which is almost doubled with essential maintenance of molecular weight. On the basis of analytical determinations of Björkman spruce lignin, it has been established that about 30% of the guaiacyl propane lignin units contain a free phenolic group (68). The remaining 70% of the units should accordingly be cleavable ethers. If all of these ether linkages were converted to hydroxyl groups, then the percentage hydroxyl increase would be 7% which corresponds excellently to the increase in initial reduction step.

The sodium and liquid ammonia reduction can also be expected to reduce the αβ double bond of any coniferaldehyde end groups even though this type of reduction usually employs lithium (69). The resulting aryl-substituted propionaldehyde (70) contains an α-methylene group capable of undergoing aldol-type condensation. The presence of such moieties no doubt accounts for the ability of the low-molecular weight reaction products to reduce Fehling's solution.

Demethylation can also be accomplished by oxidative rather than reductive cleavage, and Marton and Adler (71) have patented a process whereby the hydroxyl functionality of lignin is increased in order to improve its reactivity toward aldehydes useful in resin manufacture. The reaction proceeds via an o-quinone intermediate 16 which is rapidly reduced to the orthodiphenol 17 by the sequential addition of sulfur dioxide (Equation 10). This reduction avoids the dimerization of the orthoquinone to the polyquinone 18 or its isomers. Some o-diphenolic sites for modification may also be generated during the course of sulfite pulping (72) by demethylation reactions and Hayashi and Namura (73) have convincingly demonstrated their presence in coniferous lignosulfonates. A method for the quantitative determination of such catechol moieties based on their oxidation with ammoniacal silver nitrate to o-quinones has been proposed (74).

$$\text{R}\underset{\text{OH}}{\overset{\text{OMe}}{\bigcirc}} \xrightarrow[\text{NaIO}_4]{\text{H}^\oplus} \text{R}\underset{\text{O}}{\overset{\text{O}}{\bigcirc}} \xrightarrow{\text{SO}_2} \text{R}\underset{\text{OH}}{\overset{\text{OH}}{\bigcirc}}$$
16 → 18
17

(10)

Additional, although nonaromatic, vic-glycols may be available for modification from milled wood lignin as a result of alkali treatment (2N NaOH) at 170°. Their presence was demonstrated by Gierer and Lenz using the periodate method (75). It is pertinent to note that the aliphatic vic-glycols presumably originate from intermediate oxiranes, and in a kraft pulping these will be subject to facile cleavage by the strongly nucleophilic thiol species (76). β-Hydroxyalkyl sulfides may thus be formed instead of the vic-glycols.

Formation of quinones also occurs during oxidation with potassium nitrosodisulfonate, $ON(SO_3K)_2$ (Fremy's salt). In this case (77), demethylation is not involved, and lignin model compounds with an unblocked 5-position 19 are readily converted to stable 3-methoxy-1,2-benzoquinones 20, Equation 11, if the 5-position is not deactivated by a meta-carbonyl function. The presence of a meta-carbinol function also prevents o-quinone formation and guaiacyl ethyl carbinol 21 affords a p-quinone 22 and propionaldehyde, Equation 12. Similarly, dehydrodivanillyl alcohol 23 was oxidized to the di-p-quinone 24, Equation 13.

$$19 \xrightarrow{ON(SO_3K)_2} 20 \qquad (11)$$

$$21 \xrightarrow{ON(SO_3K)_2} 22 + \text{MeCH}_2\text{CHO} \qquad (12)$$

$$23 \xrightarrow{ON(SO_3K)_2} 24 \qquad (13)$$

A similar transformation may have occurred where lignin has been treated with nitrosodimethylaniline (78). The product which apparently contained nitrogen (11.74%) gave a red-violet color in water, and this is reminiscent of the colors produced by Fremy's salt oxidation of lignin model compounds.

B. Hydroxyl Groups. The second most abundant function available as modification sites in the lignin macromolecule is the hydroxyl group. Both phenolic and all aliphatic types are present. Although the total amount of such groups is difficult to determine accurately, useful working figures are available and are collected in Chapter 5.

The modification reactions which these groups can undergo can be classified into two groups, those in which all types of groups participate indiscriminately and those in which selectivity can be achieved. The former will be discussed first. It should be realized that many lignins including those derived from kraft and sulfite pulping will contain catechol type units (73, 79). Many alkylations, especially methylations, have been performed on the hydroxyl groups in a variety of lignins. Using dimethylsulfate (80-82), both the aliphatic and aromatic hydroxyls will be methylated, though the drastic conditions (concentrated alkali at 60°) probably cause skeletal condensation or the liberation of some phenolic groups. Some tertiary aliphatic hydroxyl groups may escape alkylation. Such products by virtue of their increased methoxyl contents have a new range of solubility due to the reduction in their polyionic character except in the case of the lignosulfonates where the sulfonic acid groups, which dominate the physical properties, are not esterified under these conditions. Methylation using diazomethane has also been extensively employed as a more moderate experimental approach to avoid secondary reactions, but of course aliphatic hydroxyl groups are not methylated using this reagent in nonpolar solvents. However, diazomethane also reacts with carbonyl groups in the milled wood lignin of spruce (83). Model compounds with α- and β-carbonyl groups afforded the corresponding 1.1-disubstituted oxiranes. The analogous formation of epoxides in the lignin was evidenced by the reduced carbonyl content and analysis using sodium iodide.

The nonalkylative reaction of the complete molecule of diazomethane with double bonds to form pyrazolines (84) must also be considered and will be discussed later. Higher alkyl homologues of the methyl derivatives of the various lignins have been prepared using alkyl halides and the lead or thallium salts of alkali lignins (85-90), but thus far no significant uses for these macromolecules have been developed. The hydroxyl groups are also capable of conventional tritylation (86, 88-90) and benzylation (87). The latter treatment afforded benzene-soluble macromolecules which were capable of reaction with formaldehyde. The difunctional analogue of benzyl chloride, p-xylylene dichloride gave similar products.

A number of other functional alkyl groups have, however, been added to replace or uniformize the hydroxyl content. Thus, Senzyu and Ishikawa (91) prepared a hydroxyethyl lignin by reaction of extractive-free wood treated with sodium hydroxide with ethylene oxide. Hydrolysis of the modified wood afforded in 29% yield the hydroxyethyl lignin which had a methoxyl content of 15.5%. The ethylene oxide probably reacts with the phenolic (or enolic) rather than the aliphatic hydroxyls because of the much greater nucleophilicity of the phenoxide anions. Nonetheless, it need not follow that all the phenolic hydroxyls are easily hydroxyethylated and because of the presence of water, it is unlikely that polyethyleneoxy chains emanating from a phenolic (or enolic) hydroxyl could be propagated (92). Thus, Ishikawa, Oki and Fujita (93) in model compound studies have observed that phenolic hydroxyl groups which are not conjugated through the benzene ring with carbonyl or unsaturated moieties in the side-chain react quantitatively with ethylene oxide. Phenolic hydroxyl groups which are so conjugated, on the other hand, are resistant to hydroxyethylation. This behavior presumably reflects the reduced nucleophilicity of the derived phenoxide anion originating in the electron withdrawing character of the para-substituent (94).

The structure of "ethylene oxide lignin" is therefore probably well represented by the unit 25 in Equation 14 derived from the hydroxyethylation of phenolic groups in the lignin. Methylation with diazomethane (95) increased the percent methoxyl from 15.6% to 18.3% which corresponds to only 1.5% free phenolic groups. The original phenolic hydroxyl content would be about 4% (96). More drastic methylation using dimethyl sulfate increases the methoxyl group content by a further 9.5%. This methylation covers the original aliphatic hydroxyl groups plus the hydroxyls of the introduced hydroxyethyl group. These views are in accord with earlier work on hydroxyethylation of a hydrochloric acid spruce lignin (97) where the added hydroxyethyl functions could be cleaved with hydriodic acid (ether cleavage) but not with 5% sulfuric acid (ester cleavage)(93-94). A supporting result was also forthcoming from Ishikawa's degradation of acetylated beech wood which had been hydroxyethylated. Hydrolysis with 5% H_2SO_4 afforded the hydroxyethylated lignin in almost quantitative yield (98). A considerable amount of Japanese work has also gone into a study of this hydroxyethyl lignin, and Ishikawa has determined by methylation and acetylation that one out of every seven hydroxyl groups is phenolic (95). Bromination of the hydroxyethylated lignin followed a somewhat different path than that of chlorination. Thus, hydrolysis of the hydroxyethyl lignin showed that no bromination had occurred at position 1 of the guaiacyl or syringyl units (99) although derivatives containing bromine at the 3, 5

MODIFICATION REACTIONS   521

and 6 positions were identified. The combined bromine content of 35% was not all attached to aromatic sites, and some 5% could be removed by refluxing with alcoholic potassium hydroxide (100). Bromine atoms susceptible to displacement were assumed to be located in the original or introduced side-chains (101). The most probable sites for this bromine are α to carbonyl groups preexisting in the lignin or formed as a result of oxidation of the hydroxyethyl function $\underline{25}$ to the corresponding aldehyde $\underline{26}$, Equation 14. In a further extension of these studies the hydroxyethyl lignin was reacted with phenol (102-104). before bromination.

$$\underset{25}{\text{Ar-C}_6\text{H}_3(\text{OMe})\text{OCH}_2\text{CH}_2\text{OH}} \xrightarrow{\text{Oxid.}} \underset{26}{\text{Ar-C}_6\text{H}_3(\text{OMe})\text{OCH}_2\text{CHO}} \qquad (14)$$

The conversion of the phenolic hydroxyl groups to aliphatic hydroxyl functions enables the lignin macromolecule to be used as the polyol component in polyurethane foams (105). Thus, reaction of kraft or sulfite lignin with propylene oxide affords oils with viscosities and hydroxyl numbers suitable for mixing and reaction with diisocyanates in typical rigid foam formulations (106). This is in contrast to the unmodified lignin which was shown not to perform satisfactorily after a detailed investigation by Kratzl <u>et al</u>. (107). Before lignin itself was used for condensations, the University of Vienna researchers carried out a series of experiments with model compounds (see Table 13.2) to determine the reactivities of the various functional groups which are present in lignin. Since the alcoholic and phenolic hydroxyls differ in reactivity, Kratzl and his group were able to prepare both a mono- and a di-urethane from guaiacylpropanol-2 using phenyl isocyanate. With hexamethylenediisocyanate, a slower reaction was observed and selectivity between alcoholic and phenolic hydroxyls could not be achieved and bifunctional hydroxyl model compounds yielded only polymeric oily tarry products. However, naphthalene diisocyanate, a commercially important reactant, yielded diurethanes in every case. When these studies were translated to technical hydrolysis (hydrochloric acid) lignins reactions with isocyanates were described as very poor and this was attributed to a paucity of reactive positions. Attempts were made to increase the number of reactive positions by α-condensation with polyalcohols exemplified by glycerol (108). These new hydroxyl functions did not apparently raise the reactivity to a commercially interesting level. However, lignin urethanes of apparent commercial utility as molding resins have been obtained by the reaction of a kraft pine lignin with excess phenyl isocyanate (109). The product claimed, characterized by its strong urethane absorption in the infrared and

and absence of isocyanate functions, contained 1.93% N. It still possessed a high percentage of the original hydroxyl content (5.22 versus 6.3 meq.) indicating a rather low level of reaction. More effective reaction between lignin and isocyanates seems to have been obtained by Nichols (110) who used a pine wood alkali lignin to cross-link a diisocyanate condensation polymer (MW 800-5000) thus achieving improvements in the elastomeric properties. Better physical properties were obtained using an oxidized lignin which probably reflects the higher functionality available.

Table 13.2. Reactions of Lignin Model Compounds with Isocyanates (107)

Model compound	Product obtained using		
	Phenyl isocyanate	1,6-Hexane-diisocyanate	Napthalene-1,5-diisocyanate
HO-C6H3(OMe)	Monourethane	Monourethane Diurethane	Diurethane
HO-C6H3(Me)(OMe)	Monourethane	Diurethane	Diurethane
HO-C6H3(CH2CHOHMe)(OMe)	Monourethane Diurethane	Oil	Diurethane
HO-C6H3(CH2COMe)(OMe)	Monourethane	–	Diurethane
HO-C6H3(CH2CH2Me)(OMe)	Monourethane	Diurethane	Diurethane
HO-C6H3(CHO)(OMe)	–	Monourethane Diurethane	Diurethane
HO-C6H3(CH2CH=CH2)(OMe)	–	Diurethane	Diurethane
HO-C6H3(CH2OH)(OMe)	–	Oil	Diurethane
HO-C6H3(COCHOHMe)(OMe)	–	Oil	Diurethane
HO-C6H3(COCH2Me)(OMe)	–	–	Diurethane
HO-C6H5	Monourethane	Resins	–
HO-C6H4(COCH2Me)	Monourethane	Monourethane	–

MODIFICATION REACTIONS

Most recently in this area, the modification of thiolignin with tolylene diisocyanate has been found (111) to give a polyurethane which is a useful aluminum adhesive when cross-linked with m-phenylenediamine or a polyethyleneglycol of molecular weight 400. This uncrosslinked isocyanate-terminated thiolignin polyurethane functions as a good wood adhesive without the addition of cross-linking agents. Presumably, the hydroxyl groups in the wood surface can act in this capacity. The facile reaction of the thiolignin with the diisocyanate was shown by IR spectroscopy to result in the formation of carbamate and allophanate ($NH_2CONHCOOH$) moieties.

Modification of the lignin macromolecule can also be achieved by ring opening of a small ring nitrogen heterocycle in place of the alkylene oxides. Thus, the achievement of the Dow Chemical Company in making ethyleneimine a low-cost commercial commodity (112-113) was followed by Doughty who briefly investigated and patented (114) the lignin reaction products obtained by modification using this extremely reactive aziridine. Aminoethylation of active hydrogens by ring opening of the aziridine is the principal reaction which will occur with lignin and which can be represented as 27 using phenol as the substrate, Equation 15.

$$\text{[structures]} \tag{15}$$

Unlike the ethylene oxide system where chain propagation is restricted in aqueous media, the aminoethyl function can polymerize more aziridine to give grafted and branched chains of polyethyleneimine. Ethyleneimine can also add to lignin by virtue of its ready reaction with carbonyls, olefins and quinones, Equations 16-18, all of which groupings are present in lignins.

$$\text{[structures]} \tag{16}$$

$$RCH=CH_2 \xrightarrow{HN\triangleleft} RCH_2CH_2N\triangleleft \tag{17}$$

$$\text{benzoquinone} + HN\triangleleft \xrightarrow[\text{Excess ethyleneimine}]{\text{Excess quinone}} \text{products} \quad (18)$$

Doughty found (114) that alkali lignins from alkaline pulping of hardwoods (Meadol) and an alkaline sulfide (kraft) cook of southern pine (Indulin A) reacted with ethyleneimine under both mild and drastic conditions. Under the former conditions in an inert organic solvent at room temperature, the modified kraft lignin was found to contain about 3% nitrogen. Under the more drastic conditions where the alkali and kraft lignins were dissolved in ethylenimine at room temperature, about 6.5 and 8.5% nitrogen was found in the products. The modified ligninimine had markedly changed solubility and it seems probable that under the mild conditions only the more strongly acidic functions or aliphatic carbonyl compounds react with ethylenimine. Under the more drastic conditions, other less reactive groups such as alcoholic hydroxyls may begin to participate.

Because of the amine constituent, the ethylenimine modified lignins have been proposed for use as anionic ion exchange materials, as antioxidants, as anticorrosive agents, and as surface active materials. The modification of the lignin macromolecule with ethylenimine need not be carried out in nonaqueous media as Allan and Halabisky (115, 116) have shown using aqueous solutions of lignosulfonates. The modified lignosulfonates have augmented coagulation characteristics.

Other alkylations of the phenolic hydroxyl groups to introduce new functional groups have employed the condensation of lignins with halogenoalkanoic acids under typical phenolic ether synthesis conditions (81, 82). By reacting a thiolignin from a commercial kraft black liquor with chloroacetic acid and its methyl, ethyl and propyl homologues, a series of α-lignin acetic acids was obtained. These acids which all melt about 200° have been somewhat confusingly designated (49) as β-lignopropionic, γ-lignobutyric and δ-lignovaleric acids, but they are more correctly represented structurally by the general formula 28, since they are derived from α-halogeno acids. Since thiolignin contains about 7% of sites capable of modification by this reaction (96) the carboxyl content of these products must be in the neighborhood of 14%. A number of uses of the metal salts have been disclosed in the patent literature covering these compounds (117). These include insecticidal and fungicidal activity (118, 119). The latter claim has been somewhat substantiated by Ali and coworkers (120), who have found that the acetic acid derivatives (R=H) show promising fungicidal properties in controlling jute-destroying fungi, particularly the Rhizoctonia species and Colletrichum corchori. However, it seems likely that this activity is due to some of the monomeric aryloxyacetic acids formed from monomeric phenolic compounds present in the thiolignin mass (121).

Lignin-O-CHRCOOH
28

This type of phenyl ether forming reaction has also been applied in a bifunctional sense by Imoto and Hachihama (122) who reacted an alkali lignin with methylene iodide to obtain a product which was no longer soluble in alkali or in organic solvents. While some cyclic methylenedioxide structures may be formed from suitably placed vicinal hydroxyl groups (123, 124) within a single lignin molecule, Brauns' suggestion (49) that this is a condensation reaction in which separate lignin molecules are combined by methylene bridges seems quite reasonable. This view is endorsed by the formation of an insoluble product by application of the same reaction to a lignosulfonate. Its suggested use (118) as an ion exchange resin would probably be frustrated technically by its low capacity—among other reasons. This type of condensative crosslinking to achieve insolubility of the lignosulfonate and the formation of the anionic resin has also been attempted by Hachihama (125) using alkyl dihalides but conceptually and practically this suffers from the same drawbacks as outlined above. Polyalkylation of the hydroxyl groups has also been carried out using epoxides. Thus, an alkali lignin (Indulin AT, a product of Westvaco) can be reacted with the diglycidyl ether of bisphenol-A 29, allyl glycidyl ether 30, diglycidyl glycerol 31 or higher homologues so that the acidic hydrogens in the lignin can open the oxirane ring (126, 127).

The epoxy group can thereby be regarded as a vehicle whereby the lignin molecules can be chemically interlinked with each other to produce a complex resinous molecule.

The compatibility of the lignin with epoxy resins can be improved by a preliminary partial alkylation of the lignin hydroxyls with benzyl chloride or diethyl sulfate (126). These modified lignin epoxies can still be cured using conventional epoxy amine hardeners. In spite of satisfactory physical properties, these efforts to use alkali lignin as an epoxy resin component do not seem to have met with significant commercial success as yet. Further reaction of the lignin-epoxy composition with unsaturated fatty acid mixtures such as tall or drying oils affords xylene-soluble mixtures useful as paints which can be cured using metal naphthenates (127). Esterification of the lignin hydroxyls with phthalic anhydride before reaction with the epoxies is also helpful (126). Blends of the lignin epoxy with polysulfide polymers obtained from the reaction of dichlorodiethylformal and an alkali polysulfide can also be prepared and cured to give tough, slightly flexible products. Rheinau lignin, which is obtained as a by-product of the production of wood sugars, could be employed to only a very limited extent because of its low solubility in the epoxy resin intermediates. Demethylated lignin, however, can be used. Lignosulfonates, by virtue of their predominately ionic character, are substantially insoluble and cannot be used.

Some efforts have also been expended (128, 129) to use kraft lignin as the backbone for epoxy resins derived from epichlorohydrin. Although glycidyl groups could be attached to the lignin macromolecule presumably at the phenolic hydroxyl sites, the lignin epoxy resins were not soluble in conventional organic solvents and the epoxy equivalents were rather low (0.14-0.19 equivalents/100g).

Attempts to improve the solubility characteristics by pre-reaction of the lignin with ketones to form lignin analogues of bisphenol-A were not particularly helpful although up to 0.6 mole of acetone per lignin methoxyl could be combined. This modification of lignin by condensation with ketones has been the subject of an extensive investigation by the Vienna group (130) which will be discussed later in this chapter.

Phenolated lignin gave somewhat higher epoxy equivalents (0.28 equivalents/100g) after conversion to its glycidyl derivative but its solubility was still unsatisfactory. These lignin epoxy resins, however, can be cured with phthalic anhydride (but not with diethylenetriamine) to give acceptable aluminum and wood adhesives.

Modification of the hydroxyl groups probably also occurs during the reaction of an alkali lignin with acetylene. Nikitin and coworkers (131) carried out this reaction in aqueous alkali for a prolonged period and obtained a resin in 190% yield which was soluble in a range of organic solvents including ether and which softened at 60-90°. The high yield is partly attributable to the polymerization of acetylene itself (132) under the reaction conditions which also cause ready vinylation of phenols (133). Guaiacol, for example, afforde its vinyl ether in over 90% yield (134). Simultaneously, the lignin will be continuously undergoing alkaline degradation and recondensation. This will be analogous to the type of degradation studied by Enkvist and his colleagues (13

This total reaction product can therefore be considered as a mixture of polymerized acetylene, vinylated low-molecular weight phenolic compounds such as catechol and vinylated demethylated recondensed alkali-degraded lignin.

A somewhat related reaction comprising simultaneous modification and degradation is the base-catalyzed alkylation of lignin by an activated alkene, such as acrylonitrile (135). The cyanoethyl products which are thermoplastic contain about 2.8% nitrogen and are not completely alkylated as evidenced by their solubility in base. Assuming that the alkylation of the phenolic hydroxyls occurs preferentially then, based on the phenolic hydroxyl content (equivalent wt., 252) of the alkali lignin used (Indulin A), the alkylation is probably not much more than 60% complete.

The hydroxyl groups of lignin also offer the opportunity of modification by acylation reactions, but as yet no useful counterparts to the cellulose esters have emerged from this second macromolecule of wood. However, a rather large number of acetylations for example have been carried out. The conventional acylation methods are applicable, such as the Schotten-Baumann acylation in aqueous alkali as well as other base- and acid-catalyzed acylations (136). The acetyl groups introduced correspond to the total hydroxyl group content and the solubility characteristics are considerably changed toward solvency in less polar solvents. Callow (137), in studies of jute, suggested that the lignin is more readily acetylated than the cellulose. He also pointed out that prolonged acetylations tended to afford products having much lower methoxyl contents (and higher acetyl values) which were interpreted as being due to methoxyl cleavages. The products isolated from wood using formic or acetic acid can probably be regarded as the formates and acetates of esterifiable hydroxyl groups since Freudenberg and Fuchs (138) have shown by using radioactive acids that the introduced groups can be hydrolyzed to afford lignins which were not radioactive. Recent NMR studies (139) on the structures in acetylated lignins have indicated that acetylated acetic acid lignin contains more aliphatic acetyl than acetylated Brauns native lignin.

The higher homologues of the formate and acetate esters have also received some attention, and Brauns and his coworkers (140, 141) have prepared a rather complete homologous series of straight chain esters of an alkali hardwood lignin. These esters are all soluble in acetone, chloroform, dioxan and ethyl acetate, but as might be surmised, their solubility decreases in polar solvents and increases in nonpolar solvents with ascent of the series. No uses appear to have been suggested for any of the members of this series. Other series of lignin esters have also been prepared by McNair and Jahn (142) including the acetate, butyrate, oleate, crotonate, cinnamate, succinate, benzoate and furoate esters using a lignin obtained from white fir by hydrolysis with 70% sulfuric acid. Since this lignin starting material must correspond closely to the highly condensed Klason lignin, it is not surprising that the esters obtained were insoluble in all solvents.

Acylation with alkanoyl halides has also been applied to lignosulfonates.

Dorée and Hall (143) attempted to acetylate a sodium lignosulfonate, but failed. Sulfur dioxide was evolved and this acetylation could simply be reversing the sulfonation reaction. Acetylation, of course, is difficult to achieve with the lignosulfonates because of the solubility characteristics of the starting material and the Schotten-Baumann technique is the most readily applicable in this case. p-Toluenesulfonyl chloride is reported to esterify almost all the hydroxyl groups (144). Substituted benzoyl or benzenesulfonyl halides such as p-bromobenzoyl, chlorobenzenesulfonyl, m-nitro-p-toluenesulfonyl, 3,5-dinitrobenzoyl, p-nitrobenzoyl have also been used in studies to determine the hydroxyl content of lignosulfonic acid (145a). The tosyl derivatives of lignin have also been used by Freudenberg in iodine exchange reactions designed to determine the amount of primary hydroxyls (144) present.

Lignin modification reactions peculiar to the secondary alcoholic groups have also been recorded. Thus, those in positions alpha to an unetherified guaiacyl unit can be determined quantitatively by condensation with quinoneimide chloride in alkaline solution (146). The product formed from these p-hydroxybenzyl alcohol units is an indophenol susceptible to convenient spectrophotometric measurement. p-Hydroxybenzyl alcohols are also converted to quinone methides by the successive action of hydrobromic acid and sodium bicarbonate (147).

Some other functional esters of lignin have also been described and acylation certainly offers a ready means of altering the functionality. Thus Nikitin and Rudneva (148) prepared a chloroacetyl lignin by reaction with chloroacetic acid itself at 100° for several hours. Acidolysis reactions may intervene here due to the drastic reaction conditions employed.

More recently Allan (149) has synthesized a number of esters of lignin using pesticidally active acids. The combinations typified by the 2,4-dichlorophenoxyacetate of lignin 32 show no initial biological activity. However, in use these polymers degrade by a number of mechanisms providing a controlled release of the active pesticide in minute amounts over prolonged periods of time. This sustained release concept is not confined to pesticide acids, and any pesticide containing a replaceable hydrogen atom can be chemically attached to the lignin macromolecule by suitable bridging entity. Thus, O,O-dimethyl (N-methylcarbamoyl)methylphosphorodithoate 33 can be, for example, combined with a kraft lignin by reacting with tolylene diisocyanate 34 or pyromellitic dianhydride 35.

Carboxyl functions can, of course, be introduced into the lignin molecule by using dibasic acid anhydrides (150). Sandermann (151) found that maleic and succinic anhydrides reacted readily with a variety of lignins with the weight increase reaching a maximum after about one hour. These products are probably half esters which on prolonged heating at 170° lose weight to give products formulated as diesters by Brauns (49). Since this would involve the thermal elimination of a mole of maleic or succinic acid, a more attractive explanation for the weight loss would be the conversion of the free carboxyl groups to give acid anhydrides by dehydration. Similar half-ester products have also been obtained by the reaction of phthalic, hexahydrophthalic, dodecenylsuccinic, methylnadic, tetrachlorophthalic and pyromellitic anhydrides (126).

The participation of the hydroxyl groups of lignin in xanthation reactions has also been studied. Glover and Bain (152) made the initial exploration in this area and found that an alkali spruce lignin gave two sulfur-containing products. The low sulfur content indicated that not more than one xanthate group was present for each two phenylpropane units, and this can be rationalized assuming that phenolic hydroxyl groups do not form xanthates. Any xanthates formed would then be attributable to the primary and secondary alcoholic hydroxyl content (153) or to reactions of the carbon disulfide with other reaction sites such as alpha hydrogen atoms exemplified (154, 155) by the reactions: $\underline{36}$ to $\underline{37}$, Equation 19, and $\underline{38}$ to $\underline{39}$, Equation 20.

$$\underset{36}{\begin{array}{c}\text{Ph}\\|\\\text{CH}_2\\|\\\text{CO}\\|\\\text{CH}_2\\|\\\text{Ph}\end{array}} \xrightarrow{\underset{\text{KOH}}{CS_2}} \underset{37}{\text{Ph}\diagup\overset{O}{\underset{S\diagdown S \diagup S}{C}}\diagdown\text{Ph}} \qquad (19)$$

$$\underset{38}{\begin{array}{c}\text{Me}\\|\\\text{CO}\\|\\\text{CH}_2\\|\\\text{Ph}\end{array}} \xrightarrow{CS_2} \underset{39}{\begin{array}{c}\text{Me}\quad\quad\text{Me}\\|\quad\quad\quad|\\\text{CO}\diagup S\diagdown \text{CO}\\C=C\diagdown\;\;\diagup C=C\\\text{Ph}\quad S\quad\text{Ph}\end{array}} \qquad (20)$$

In experiments with the less condensed cuoxam spruce lignin, Lieser and Schwind (156) were able to achieve sulfur contents (12%) corresponding to xanthation of 75% of the alcoholic hydroxyl groups. Oxidation of the xanthate with iodine yielded a product whose sulfur content (16%), origins and properties are consistent with the view that it is formed by conversion of the xanthate to a xanthide according to Equation 21.

$$\text{ROCSSNa} \longrightarrow \text{ROCSS-SSCOR} \qquad (21)$$

It is also worthy of note that in experiments under the same conditions, phloroglucinol was used as a model compound and allegedly afforded a mixture of a di- and a trixanthate. Since phenols do not generally form xanthates, this observation may emphasize the problems implicit in the choice of model compounds and phloroglucinol may not be functioning as a trihydric phenol. This type of peculiarity of phloroglucinol has also been noted in numerous unsuccessful attempts to prepare the triglycidyl ether of phloroglucinol by reaction with epichlorohydrin (157). Xanthates can be obtained by routes other than the alkaline reaction of carbon disulfide and Freudenberg and Dietrich (158) have prepared methyl xanthates of cuoxam spruce and beech lignins by acylation using methyl dithiochlorocarbonate, CH_3SCSCl, in pyridine.

Esterification of the hydroxyl groups can also be accomplished to obtain esters of inorganic acids. Attention has already been drawn to the boron and aluminum derivatives and several groups (159-162) have investigated the reaction of lignin with phosphorus acid derivatives. Using an alkali lignin obtained by a soda or kraft pulping process, Doughty found that a reaction could be achieved with phosphorus halides, oxyhalides, thiohalides, oxides and sulfides. Not unexpectedly, the reaction of a polyfunctional hydroxyl macromolecule with a trifunctional acid halide or equivalent tends to lead to insoluble three-dimensional esters and in fact Doughty points out that no solvent has been found for the lignin phosphorus esters except strong aqueous alkali solutions which give only limited solubility. Probably some concomitant hydrolysis occurs during this dissolution.

However, if the lignin macromolecule is treated with phosphorus halides in equimolecular proportions such that cross-linking is discouraged, the formation of insoluble three-dimensional networks can be avoided. Subsequent treatment with mono- and poly-hydroxy compounds affords soluble film-forming resins which are flame retardant by virtue of their phosphorus content (162). Phosphorus oxychloride under drastic reaction conditions can also cause demethylation, but the residual lignins are highly condensed (163). Other phosphorus-containing lignins have been prepared by the more exotic route employing esters and amides of trivalent phosphorus (164). The products and the reactions can probably be formulated as in Equations 22 and 23.

$$\text{Lignin-OH} + P(OR)_3 \longrightarrow \text{Lignin-OP}(OR)_2 + ROH \qquad (22)$$

$$\text{Lignin-OH} + P(NR_2)_3 \longrightarrow \text{Lignin-OP}(NR_2)_2 + R_2NH \qquad (23)$$

In earlier work, comparable condensations had been carried out using both phosphorus (165) and antimony pentachloride (166). However, the products were not regarded as esters of inorganic acids, and the antimony-containing product, for example, was hydrolyzed without detailed examination.

More recently, acid hydrolytic lignins and their chloro- and nitro-derivatives have been reacted with thiophosphoryl and diethylthiophosphoryl chlorides to afford thiophosphates insecticidal to the common housefly (167). These are probably derived from low-molecular, phenolic components in the lignin degradation mixtures. The corresponding phosphates were probably formed when Kashima and Oiwa (168) treated soda lignins with phosphoryl chloride before oxidation with mercuric or cupric oxide. The oximes of nitrolignin derivatives have also been reacted with diethylthiophosphoryl chlorides to obtain O-iminyl-thiophosphates (169).

Analogous esterifications with Group V oxyhalides probably occurred when lignins were treated with sulfuryl (165, 170) and thionyl (168, 171) chlorides. Paschke (165), for example, obtained a product having the empirical composition $C_{37}H_{42}O_{12}S_3Cl_3$ by treatment of an alkali straw lignin with sulfuryl chloride. These products, however, are generally too inadequately characterized to permit useful structural speculation. More recently the sulfuryl chloride reaction has been reinvestigated using a hydrolysis lignin as the substrate and vanillin as a lignin model. The latter was converted to 5-chlorovanillin in high yield while the lignin underwent chlorination, demethylation and degradation. Approximately half of the introduced chlorine was located in the aromatic moieties (172).

Similar esterification reactions were observed (159) with the phosphoryl and thiophosphoryl halides. The lignin used was also reported to react with phosphorus oxides and sulfides and both reactions are proposed to proceed by cleavage of the P-O-P or P-S-P bonds yielding primary and secondary esters of the lignin. All of the phosphorus esters of lignin reported by Doughty are brown powders containing about 1-8% phosphorus which do not melt but sinter around 250-300°. Their uses as flame retardants for cellulosic materials has been suggested. Again by the use of suitable reaction ratios phosphoryl, thiophosphoryl and chloromethylphosphonic chlorides afford reactive intermediates which can be capped by isopropanol, glycerol, polypropyleneglycol (MW 400), 2,3-dibromobut-2-en-1,4-diol and resorcinol. The modified phosphorus-containing lignin polymers were all organic-solvent soluble and are film-formers. They are also miscible with a wide range of commercial resins including epoxy, melamine, phenolic and furfuryl alcohol polymers (162).

C. Carbonyl Groups. Carbonyl groups are present in all types of lignins and are manifested, for example, in the infrared spectra. Quantitative determinations of the amounts of these groups are still clouded by some uncertainties. Probably values of the order of 0.2 carbonyl group per monomer unit encompass all types of lignins including milled wood lignin.

Attempts have been made to allocate the total carbonyl content to specific functional entities and evidence has been proposed (173) to suggest that 10 to 15% of the carbonyl functions are represented by coniferaldehyde end groups. Another 30% has been located in the α-position of the side-chains in milled wood spruce lignin by means of changes in light absorption caused by sodium borohydride reductions (174), and Komshilov and coworkers (175) have suggested that kraft pulping converts most of the lignin side-chains into benzyl-methyl ketones by oxidation-reduction reactions. If these carbonyl group contents are accepted, it follows that a substantial number of unconjugated carbonyl groups should exist in lignin. Oxidative deterioration of phenolic hydroxyls of course tends to increase the number of quinone carbonyls.

The typical carbonyl reagents such as hydroxylamine and phenylhydrazine convert lignin carbonyls to the expected oximes (176, 177) and phenylhydrazones (143, 178). A variety of other derivatives such as the azines (179), thiobenzhydrazones (180-182), nitrophenylhydrazones (88-90) and 2,4-dinitrophenylhydrazones (183-184) have also been prepared. A "phenylosazone" of an alkali eucalyptus lignin has also been described by Merewether (185), and taken as being indicative of the presence of an α-ketol grouping. The carbonyl groups of a variety of lignins have also been reacted with Grignard reagents, usually only as a means of characterization or masking (186–189) and in the extraction of lignin from spruce and birch with acetone acetals some transacetalization may occur (190). Some of these keto components in addition may very well activate adjacent hydrogens since condensation with benzaldehyde or m-nitrobenzaldehyde afforded arylidene derivatives (185, 191).

Very little effort has been made to utilize the lignin macromolecules by modification of the carbonyl function though such approaches have been successful in the starch field (192, 193).

Most of the attention given to the carbonyl reactivity has focused on the color reactions of lignin where a large number of characteristic chromogenic reagents (49, 194) have proliferated over the years. Most of the early mysteries have been dispelled, and a rational explanation of the hues produced from lignin by phenols (195), amines (196), concentrated hydrochloric acid (197) and dicarbonyl compounds (198) etc., can now be offered in terms of the condensative reactivity of coniferaldehyde 40 end groups. Earlier reports that compounds related to coniferaldehyde also give colors have been traced to the inadvertent presence of coniferaldehyde as an impurity (199).

Thus, the colored compound formed in the Wiesner reaction (200) from the action of phenols such as phloroglucinol probably has the structure 42 derived through the intermediacy of the carbonium ion 41, Equation 24. Analogous structures can be assigned to the condensation products obtained from resorcinol and orcinol (5-methylresorcinol).

(24)

MODIFICATION REACTIONS

This formulation finds endorsement in the detailed absorption studies of Pew comparing reduced chalcones and flavanones with the colored lignin products (197).

The observation that phloroglucinol trimethylether can also participate in these types of lignin modification reactions provides evidence that the presence of free hydroxyl group in the phenol is not required (201).

Additional support for assignment of coniferaldehyde end groups as the chromogenic functions is provided by Adler's demonstration of the presence of such groups in lignosulfonates. The dilemma that certain lignosulfonates did not appear to contain coniferaldehyde end groups and yet gave the color reactions was neatly resolved when a mild alkali pretreatment at room temperature caused the missing functions to appear (37). This appearance can be attributed to the cleavage of a loosely combined sulfurous acid which had been attached 42a to the double bond of the coniferaldehyde end groups 40, Equation 25a.

$$42a \xrightarrow{(a)\ base\ 20°C} 40 \xrightarrow{(b)\ CH_2N_2} 43 \qquad (25)$$

Further endorsing evidence for the condensative significance of coniferaldehyde end groups comes from the observation of Kitaura and coworkers (202, 203) that brominated wood does not react with phloroglucinol because of the addition of the halogen across the side-chain double bond. The possibility of aldehyde oxidation would appear to be excluded because treatment of the wood with hydriodic acid restored the color-forming properties. The coniferaldehyde color-forming ability can also be reversibly inhibited by reaction with acetic anhydride. The reaction can be formulated as in Equation 26. Hydrolysis reaffords the αβ-unsaturated aldehyde.

$$RCH=CHCHO \rightleftharpoons RCH=CHCH(OAc)_2 \qquad (26)$$

Irreversible inhibition of the color-forming reaction is illustrated by the failure of diazomethane-methylated lignins to react with phloroglucinol. This can be ascribed to the genesis of pyrazoline derivatives such as 43 by addition of the diazoalkane as discussed earlier to the ethylenic moiety of the coniferaldehyde unit, Equation 25b.

Other lignin modifications can be realized when extended conjugative functions are interposed between the coniferaldehyde end groups of lignin and the phenol coreactant. Thus, precondensation of wood with acetaldehyde under aldol-forming conditions, which convert coniferaldehyde methyl ether to 44, gives a modified wood which still reacts with phloroglucinol to produce a deep violet colored compound (202) which can probably be formulated as 45, Equation 27.

$$\text{44} \xrightarrow{\text{Wiesner rx.}} \text{45} \qquad (27)$$

Other types of carbon-carbon bonds can also be formed from the coniferaldehyde end groups during color forming reactions. For example, barbituric, thiobarbituric and diphenylthiobarbituric acid react (204, 205), and these products can probably be formulated as 46a, 46b and 47 in analogy to their known condensation with aldehydes (195). The reaction of lignin with indoxyl to give a deep red color also possibly occurs as proposed by Harada and Nikuni (206, 207) via nucleophilic attachment of the indoxyl to the positive component of the carbonyl dipole of the coniferaldehyde end group affording the final chromophore 48. The reaction of lignin with indole, pyrrole or furfuryl alcohol very likely follows a similar pattern where the products are considered to be 49, 50 and 51, respectively.

The violet colors obtained by Seifert (198) by the reaction of dicarbonyl compounds such as acetylacetone, benzoylacetone, ethyl acetoacetate and dibenzoyl methane can also be considered as falling in this active methylene group category and can be represented by the general formulation 52.

These reactions can also be used to increase the carbonyl functionality as is shown by the formation of an oxime containing 2.55% nitrogen from the lignin isolated from pre-extracted pinewood using acetylacetone. The nitrogen content corresponds to a carbonyl group content of over 0.3 per methoxyl.

The color reactions with amines (196), on the other hand, involve a condensation with the formation of carbon-nitrogen bonds. According to Harada and Nikuni (206, 207) and in conformity with the foregoing structures, the pigments formed by the reaction with amines can be represented as the Schiff base structure 53.

$$HO-\text{Ar(OMe)}-CH=CHCH=\overset{\oplus}{\underset{H}{N}}-\text{Ar} \quad 53$$

The aldehyde and carbonyl groups can of course be reduced to the corresponding alcohols by a variety of reagents thus increasing the total hydroxyl functionality or they can be completely removed by hydrogenolysis (see Chapter 12).

Lignins also give color reactions with a variety of inorganic reagents, but these in general do not involve the coniferaldehyde end groups and indeed are not associated with any one location within the lignin macromolecule. However, they do involve modification of the lignin structure, and it is therefore appropriate to briefly review them at this juncture. For example, Nakano, Ishizu, Sasaki and Migita (208) have shown that the colorations obtained with vanadium pentoxide in phosphoric acid is a consequence of the presence of phenolic hydroxyl groups. They are therefore probably due to vanadate ester formation while the chromogenic action of potassium chlorate in hydrochloric acid is a result of oxidation and quinone formation. The cobalt thiocyanate color reaction is in contrast explained as being due to π-complex generation by the Co^{++} ions and the aromatic rings, while the colors produced by the combinations of ferric sulfate, potassium ferrocyanide and mercuric acetate-ammonium sulfide are said to originate in the carbonyl groups of the hemicellulosic material and not in lignin itself (209). Of all the inorganic color reactions, the best known is the Mäule reaction (200) which occurs as a consequence of oxidative chlorination.

It has also been recently tentatively suggested that part of the carbonyl content of lignin may originate in o-quinone methides and a satisfying explanation (210) of their genesis has been given in terms of the basic transformation 54 to 56, Equation 28.

$$54 \xrightarrow{H^{\oplus}} 55 \xrightarrow{Oxid.} 56 \quad (28)$$

Dehydrogenation of certain compounds of this type can afford semiquinones which help to explain the origin of part of the free radical content of lignin (211-213). While no reactions directly attributable to the presence of this lignin function have so far been described, the reaction between phenols and formaldehyde in basic medium to yield diarylmethanes 59, Equation 29, is closely related (214) and relevant. Thus, it is established that a hydroxybenzyl alcohol 57 is initially formed which subsequently probably forms an intermediate quinone methide 58 which then captures a free phenol to give 59.

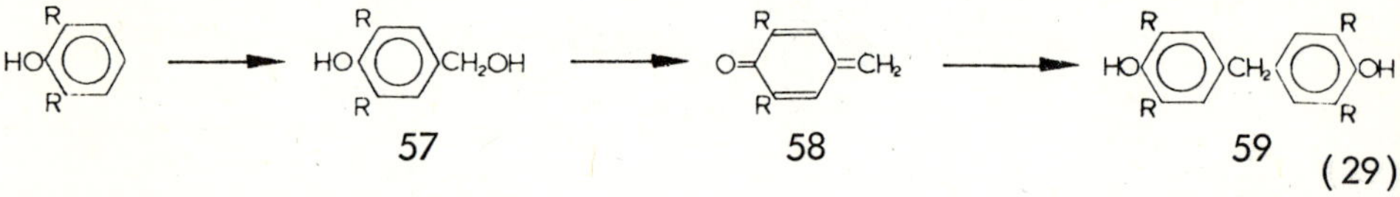

(29)

Although this mechanism deserves to be considered for the formation of condensed structures such as 55 in lignin, alternatively the formation of the diarylmethane may also occur by reaction between two molecules of the hydroxymethylphenol. An example of this is offered by the reaction of 1-methoxymethyl-2-naphthol 60 in basic solution (214). The isolation of the diarylmethane 61 as the only product indicates that 2-naphthol is not a required intermediate, and the reaction mechanism can be written as in Equation 30.

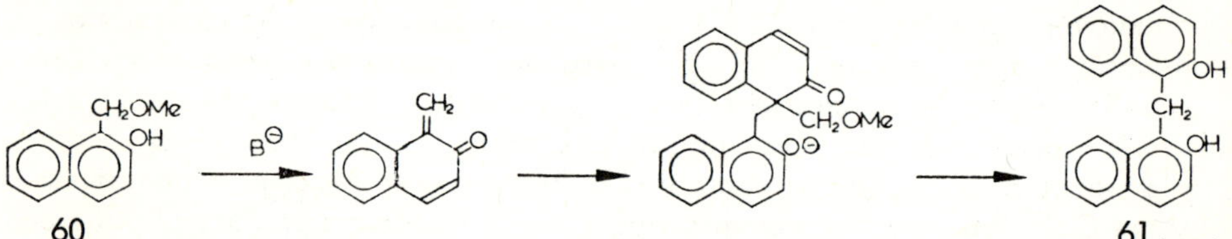

(30)

If applicable to lignin, this mechanism would lead to the formation of the p-analogue of 61 as a condensation unit generated by the sequence 62 to 63 via the quinonoid intermediate, Equation 31.

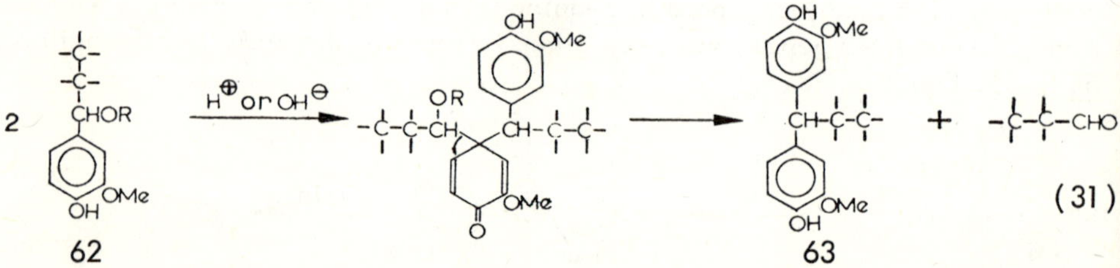

(31)

D. Carboxyl and Ester Functions. Neither carboxyl nor ester groupings are very abundant in most lignins (see Chapters 3 and 5). Recently, spectroscopic studies have shown the presence of small amounts of carboxyl groups in Björkman lignin (about 0.05 per methoxyl) (215), in lignosulfonates (~0.14-0.29) (216) and in a pine kraft lignin (0.14-0.17) (217). These generally aliphatic carboxyl groups (218-220) will have a pK of about 4-5 and naturally participate in reactions such as esterification. They do not however offer much scope for the modification of the lignin macromolecule and no specifics have been reported other than an unsuccessful attempt to characterize the carboxyl function by reaction with benzidine. Diazotization of the free amino group and coupling with β-naphthol did not give a characteristic color (221). Ester groups (222-225) are significant in grass lignins (222) and aspen woods (223) only and are of little interest for the purpose of structural modification.

E. Ethylenic Moieties. Early attempts to determine the amount of ethylenic linkages (226-229) in lignins were not successful, while more recent studies have shown the presence of about 0.06 ethylenic double bonds in spruce milled wood lignin (230-233) in association with coniferyl-alcohol and -aldehyde groups. The presence of stilbene linkages in kraft lignins has also been demonstrated (231, 232). Marton has also suggested (217) on spectrophotometric grounds (220) that kraft lignin contains conjugated double bonds of coniferyl alcohol and stilbene (7 per 100 C_6C_3- units) and has made efforts to quantitatively determine the latter. The presence of ethylenic linkages of the quinonoid type in kraft lignin in trace amounts has also been inferred from the electron paramagnetic resonance spectra (210, 234). The present position must therefore be that the ethylenic groups are rather few in number and do not offer a broad range of possibilities for modification reactions within the lignin macromolecule. Nonetheless, Kratzl and Wittman (235) have shown using cinnam- and 3,4-dimethoxycinnamaldehyde as model compounds that such linkages are capable of participating in addition reactions. Treatment with diazomethane afforded pyrazolines by addition of the diazoalkane (see section II-B) and a similar treatment of lignin afforded a nitrogenous lignin in which the nitrogen was assumed to be present in the heterocyclic form. The formation of a pyrazoline under diazomethane methylation conditions had also been reported by Spencer and Wright (236).

F. Sulfur-Containing Groups. Sulfur-containing lignins are artificial macromolecules resulting from kraft or sulfite pulping processes. The chemistry of delignification in both of these processes can be found in Chapters 15 and 16. Kraft lignin usually contains about 1-4% of sulfur, the nature of which has not been clearly established (237). Field and coworkers (238, 239) concluded that four types of sulfur bonds were present, one of which was especially stable. Infrared absorption spectra of the methylated thiolignins before and after oxidation suggested the presence of sulfoxide and probably sulfone sulfur. Only a negligible fraction appeared to be present in the form of thiol, thiocarbonyl, disulfide, polysulfide or dialkyl sulfide. Some of this sulfur may be contained in p-dithiane structures since Gierer and Smedman (240) found that model compounds containing thiarane rings dimerized to the six-membered heterocycle 64 on treatment with 2N sodium hydroxide. The p-dithianes were also obtainable by the same treatment of the related β-mercaptoalkyl aryl ethers. In spite of these interesting structures, however, the sulfur content of kraft lignin per se does not offer a broad opportunity for modification reactions.

64

A much more clear-cut situation prevails with the lignin macromolecules resulting from sulfite pulping. Here the bulk of the sulfur can be located in the sulfonic acid sites (241) and a substantial number of these groups (0.3-0.6 per methoxyl) are formed (242-245). The sulfonation process has also been shown (246, 247) to cleave the cyclic ether linkages. The nature of the unallocated, so-called "excess sulfur" which can be as much as 0.1 per methoxyl in a stepwise delignification is unknown (242).

The high polarity of the sulfonic acid groups make this macromolecule water-soluble, which is often disadvantageous in attempts to achieve modification reactions in non-aqueous media. This difficulty can frequently be circumvented by the use of salts of organic bases which can be chosen to afford a range of solubilities in non-polar solvents (145, 248-254). Indeed, alkaloids, chitin, dimethylaminodiphenylmethane, bis (2-methyl-4-amino-6-quinolyl) carbamide, 1-(N-piperidinoacetylamino)anthraquinone and diisobutylcresoxyethoxyethyldimethylbenzyl ammonium chloride can be used to isolate lignosulfonic acids from aqueous systems (252-258). Lignosulfonic acids can also be precipitated (259-262) using metal hydroxides or with complex metal salts such as hexamine cobalt chloride or nitrate. Anion exchange resins can also be used to extract lignosulfonates and macroporous beads give the best results (263). This can be attributed to the clogging of the smaller pores in conventional resin beads by the "jack-in-the-box" expansion of the hydrodynamic volume of the lignin macromolecule occasioned by pH changes (264). Other ion exchange variations have been claimed (265). A related approach (266) uses a precondensate of formaldehyde with an amide, thioamide or amidine to precipitate the lignosulfonate with which it may also react by virtue of its methylol group content.

Although the sulfonic acid group is available for modification reactions, its potential has hardly been evaluated. Indeed, the entire chemistry of sulfonic acids in general is sparsely studied in comparison with that recorded for carboxylic acids. It is noteworthy that heating of an ammonium lignosulfonate renders this material water-insoluble without significant loss of nitrogen or sulfur (267). Speculatively, this may be attributed to the formulation of sulfonamide linkages analogous to the formulation of carboxamides from ammonium carboxylates. Similar reactions, although poorly documented, have been reported in the older literature (268, 269). Application of this reaction to substituted amine salts of lignosulfonic acid should afford substituted lignosulfonamides. This appears to be the case since 1,2,4-triazyl-3-ammonium lignosulfonate yielded a water-insoluble derivative on heating at 150° for 20 hr. This product contained 7.8% of combined 3-amino-1,2,4-triazole and acted as a sustained release herbicide (149). Other amines can also be used and the dehydration can be effected in organic solvents so that the amine salts lose their water solubility and become soluble in polar organic solvents such as dimethylformamide. Sulfonamide groups are also capable of reaction with formaldehyde (270), and these modified lignins show increased formaldehyde reactivity (267).

If a solid lignosulfonic acid is subjected to ethanolysis, it passes into solution, and the product contains ethoxyl groups (271, 272). These are probably formed by the ethanolysis of benzyl ether linkages. However, Nikitin and Nikitin (273) have examined the reaction of 4,5-dihydroxy-2,7-naphthalenedisulfonic acid with a variety of aliphatic mono- and polyols and have claimed the formation of esters in all instances. Under the same conditions a lignosulfonic acid was 30% esterified using ethanol. Resulfonation causes the almost quantitative splitting off of the ethoxyl groups and their apparent replacement by sulfonic acid groups. The ethoxyl groups can also be removed by mineral acid hydrolysis (49). Sulfonation can also be applied (274) to water-insoluble lignins to convert them to soluble lignosulfonates.

III. Modifications at Structural Sites.

A. On Side-Chains. The benzylic carbons of the three-carbon side-chains exceed, by far, the reactivity of the β- and γ-carbons. The alcoholysis and thioalcoholysis are examples of typical reactions at benzylic sites. While the main basic background of these solvolytic reactions is detailed in Chapter 9, their modification aspects are appropriately covered in this chapter. These reactions are very facile and may even take place at room temperatures. Thus, spruce milled wood lignin treated with methanolic hydrogen chloride in dioxan increased its methoxyl content by more than half (31). Three reactions have been considered (176, 275) as possible causes of the observed introduction of alkoxyl groups: (1) acetalization of carbonyl groups, (2) etherification of enol groups and (3) etherification of alpha-hydroxy or alpha-ether structures—with the last being the most important. However, some as yet unidentified carbon-carbon bonds appear to be formed between the alcohol and the lignin since alkoxylation with radioactive methanol gave a product which was still radioactive after demethylation under the reaction conditions employed in the Zeisel methoxyl analysis (276). Until this point is clarified, the major alcoholysis reaction can be best represented by conversion to 65, Equation 32,

65, R = Et
66, $(CH_2CH_2O)_nH$
67, CH_2CH_2Cl
68, $CH_2CH_2NH_2$
69, CH_2CHCH_2Me with NO_2
70, $CH_2CH_2OCH_2CH_2OH$
71, C=CHCOOEt with Me

(32)

which is fully consistent (277) with the ethanolysis of lignosulfonic acid. In contrast, in the case of kraft lignin, the modification by methanolysis seems to be mainly due to esterification of its carboxyl groups and only a small degree to etherification or transetherification of benzyl alcohol, ether or ketol groups (278). Moreover, lignin in wood is not solubilized by cold alcoholysis: an elevated temperature and an appropriate solvent are necessary to dissolve most of the lignin. This dissolution results from the solvolytic degradation (277) which follows the rapid initial alkoxylation reaction. The increased content of phenolic hydroxyl groups indicates that phenol ether linkages are cleaved in this reaction (279). A part of the lignin is converted into monomers (280) discussed in Chapter 9. Obviously, the nature of the lignin isolated will be modified by the identity of the alcoholyzing component, and the majority of so-called "organosolv" lignin are probably often simply analogues of the methanolysis or ethanolysis lignins. This affords the opportunity to modify the functionality of the lignin macromolecule and the known organosolv lignins isolated using ethylene glycols (281, 282), glycol chlorohydrin (283), ethanolamine (284) and 2-nitrobutanol (285) may serve as examples. According to the scheme above, these products should contain free hydroxyl, chloro and amino groups as represented in structures 66 through 69, Equation 32. Certain of these introduced functions such as the amino group would of course be capable of further condensation reactions. Aminolignins may also be formed by reduction of nitrolignins.

Dioxan lignins formed by extraction of plants with dioxan containing hydrochloric acid (286-288) could theoretically very well contain units such as 70 formed by the hydrolysis of the dioxan to diethyleneglycol and its subsequent condensation. This view is supported by the lowered p-hydroxybenzyl alcohol group content in dioxan spruce lignin (146) and by the failure to extract lignin with anhydrous dioxan (145, 6), but it is contraindicated by experiments with radioactive dioxan which reportedly did not condense with the lignin (289). The degradation occurs during the dioxan-acid extraction with the formation of monomeric phenylpropane derivatives related to Hibberts ketones (290) (see also Chapter 9).

The broad class of "organosolv" lignins also includes products obtained using chloral and bromal hydrates (291). These, unlike other organosolv lignins, are insoluble in sodium hydroxide which may be due to etherification of the phenolic hydroxyl groups by halogen displacement in addition to the alcoholysis reaction.

A second case of noninvolvement of the solvent is exemplified by the isolation of a sulfur-free lignin by extraction using sulfur dioxide in dimethylsulfoxide (292, 293). A similar extraction using hydrochloric or sulfuric acid as the proton source gave a dioxan-soluble lignin containing 14-15% methoxyl which comprised 90.5% of the original Klason lignin and which did not react with sulfur dioxide or calcium bisulfite under the conditions of a sulfite cook (294).

The ethyl acetoacetate lignin isolated by Lemmel (295) and by Visasoro (296) can perhaps be regarded as analogues typified by 71 in which the enol form of the ethyl acetoacetate functions as the alcoholyzing component though of course aldol-type condensations with the doubly activated methylene will also occur (52). This same type of modification reaction can be achieved when a phenolic hydroxyl replaces the aliphatic hydroxyl of the samples cited above, and this aspect has been rather more definitively studied.

Thus, wood treated with phenol containing 0.2% of hydrogen chloride at 90-100° is appreciably delignified within one hour (297). The resulting phenol lignin contains a substantial amount of phenol in a condensed form, and the analytical composition agrees with the approximate formula $C_{42}H_{32}O_6(MeO)_5(OPh) \cdot 3PhOH$. Comparable products can also be obtained by the condensation of lignin from wood saccharification with phenol alone under pressure (298).

The condensation has been assumed to take place mainly between the lignin and the p-position of the phenol. However, at least a partial condensation to the ortho position is indicated by the isolation of salicylic acid from the nitrobenzene oxidation products (299) . This ortho condensation formulation is supported by (300) the model reaction in which benzoin 72 and phenol afford 2,3-diphenylbenzofuran 74 presumably via the intermediacy of 73, Equation 33.

$$\underset{72}{\text{PhC(O)CH(OH)Ph}} \xrightarrow{\text{PhOH}} \underset{73}{\text{intermediate}} \longrightarrow \underset{74}{\text{2,3-diphenylbenzofuran}} \quad (33)$$

However, ortho reaction is not invariable (301) and in the case of modification by the o- and p-phenolsulfonic acids the para condensation product predominated. Using dehydrodiisoeugenol as a model for the phenylcoumarane linkage in lignin, Chudakov and Milovanov (302) demonstrated that the reaction with phenol opened the heterocyclic ring with the formation of a carbonium ion which then undergoes nucleophilic phenol addition to yield 75. Condensation between wood lignin and phenols under the action of mineral acids occurs even in the cold; the condensation product, however, does not become soluble. As the condensation proceeds, yields of vanillin obtained by nitrobenzene oxidation gradually decrease from 26 to 10% of the weight of the lignin (303). This is in part due to demethylation which occurs during phenolation (304). As mentioned in Chapter 15, lignin condensed with phenols cannot be dissolved in the sulfite process because the groups reacting in sulfonation are probably identical with those condensing with phenols. The phenol condensation reaction products formed from lignin units can therefore be consistently formulated as 76 and 77, which is in concordance with the representations of the alcoholysis lignins. Functionally substituted phenols can of course replace the simple phenol usually employed; and p-chlorophenol, for example, affords the corresponding product which contains 10.6% chlorine (299). Phenols containing alkyl, nitro, bromo, amino, carboxyl and aldehyde functions have also been qualitatively examined (305) and Nakajima and Hachihama (306) have developed a procedure for the determination of lignin in coniferous woods based on the quantitative reaction of o-phenolsulfonic acid with Brauns' native lignin. Milder acids such as boric acid can also replace the hydrochloric acid (305). Efforts to develop useful lignin-phenol-formaldehyde resins based on lignin phenolysis have been reported (307, 308). Thus, acid condensation of a hydrolysis lignin with phenol followed by alkaline condensation with formaldehyde afforded liquid resins containing up to 35% lignin. Polyfunctional phenols have also been condensed, and the reaction of resorcinol has been quite extensively studied (308, 309), particularly with regard to the inhibition of pulping. Rozenberger (310) has shown that in the pulping of spruce the inhibition of lignin solubilization was directly related to the concentration of the resorcinol and the energy of activation of the condensation is higher than that of sulfonation (30 vs 18 kcal/mole). Following the scheme above, the modification product is probably adequately represented by the lignin units 78-79 since resorcinol has been found to be mostly active in the two and four positions under acid conditions (311). The products obtained from orcinol (5-methylresorcinol can be analogously formulated (206, 207).

	R_1	R_2	R_3
76,	H	OH	H
77,	OH	H	H
78,	OH	H	OH
79,	OH	OH	H

MODIFICATION REACTIONS 543

The presence of excess resorcinol probably largely suppresses further intra- and inter-molecular condensations and in pulping experiments the condensation of lignin has been reduced by the addition of phenol, catechol or resorcinol with the latter being the most effective (312). Nitrobenzene oxidation of this resorcinol lignin affords vanillin in yields decreased by as much as 65% relative to that of uncondensed lignin (308). In another investigation of phenolated lignin in situ the yield of vanillin was reduced in proportion to the resorcinol used, and this reduction was approximately 1.5 moles of vanillin per mole of resorcinol (313). These observations can be interpreted as additional support for the previous formulation of the phenolysis reaction since reaction at side-chain sites other than the alpha position would probably not affect the yield of vanillin. Also, consistent with this structural viewpoint is the fact that phenolysis is not inhibited by methylation of both the phenolic and aliphatic hydroxyl groups. Hydroxyl groups are therefore not necessarily involved in the phenol condensation (314).

The classic Wiesner color reaction of lignin with phloroglucinol (200, 315) will probably encompass a further extension of the modification given by phenol and by resorcinol and accordingly, the product can be represented as having the nonchromogenic units exemplified by 80 in addition to the coniferaldehyde color forming units described in Section IIC. A related product has also been prepared using 2,4-dimethylphloroglucinol (316).

Comparison of these two derivatives in pulping experiments demonstrated that the effect of the 2,4-dimethyl derivative was far less than its unsubstituted parent (303, 317). This may be due to the considerable reactivity of the attached phloroglucinyl moiety which may undergo further intermolecular condensation. This particular type of modification reaction caused by polyphenolic compounds probably is significant in the historic difficulties encountered in pulping certain heartwoods and is more explicitly covered in Chapter 15. Lignin also reacts with the trimethylether of phloroglucinol to give 81, indicating that the presence of a free hydroxyl apparently is not mandatory (318) for condensation.

	R_1	R_2	R_3
80,	OH	OH	OH
81,	OMe	OMe	OMe

Reactions very akin to phenolysis and alcoholysis can be carried out on a lignin unit using thioalcohols or thiophenols. Although the mechanism of this reaction is reviewed in Chapter 9, it is also appropriate to briefly consider its modification implications here. The vast majority of work in this area has concentrated on the reaction of thioglycolic acid, $HSCH_2COOH$, with lignin and a considerable effort has been expended in comparing this process with the sulfonation of lignin. Lignin thioglycolic acids do exhibit certain similarities to lignosulfonates, and they are probably formed mainly from the same reactive groups as lignosulfonic acids (319). These derivatives may also contain thiohydroxyacetate functions formed by esterification of some of the three types of hydroxyl groups in lignin. This representation is also rather well supported by 320, 321) the diversity of model compound reactions representing functions present within the lignin macromolecule, as in Equations 34–38.

$$MeO\text{-}Ar(OMe)\text{-}CH_2OH \longrightarrow MeO\text{-}Ar(OMe)\text{-}CH_2SCH_2COOH \longleftarrow MeO\text{-}Ar(OMe)\text{-}CH_2OCH_2\text{-}Ar(OMe) \tag{34}$$

$$HO\text{-}Ar(OMe)\text{-}CH_2CH=CH_2 \longrightarrow HO\text{-}Ar(OMe)\text{-}CH_2CH_2CH_2SCH_2COOH \tag{35}$$

$$HO\text{-}Ar(OMe)\text{-}CH=CHMe \longrightarrow \underset{82}{HO\text{-}Ar(OMe)\text{-}CH_2\underset{SCH_2COOH}{CHMe}} + \underset{83}{HO\text{-}Ar(OMe)\text{-}\underset{SCH_2COOH}{CHCH_2Me}} \tag{36}$$

$$\underset{84}{MeO\text{-}Ar(OMe)\text{-}\underset{OH}{\overset{H_2COH}{CHCH}}\text{-}O\text{-}Ar(OMe)} \longrightarrow \underset{85}{MeO\text{-}Ar(OMe)\text{-}\underset{SCH_2COOH}{\overset{H_2COH}{CHCH}}\text{-}O\text{-}Ar(OMe)} \tag{37}$$

$$86 \longrightarrow 87 \tag{38}$$

MODIFICATION REACTIONS   545

From these model transformations (322-324), it is apparent that the thioanion can replace a hydroxyl group (Equations 34, 37), cleave a cyclic (86) or a noncyclic (Eq. 34) benzylic ether and add to a terminal (Eq. 35) or nonterminal (Eq. 36) ethenoid linkage. In addition to these reactions, the formation of thioketals from the carbonyl groups in lignin will also occur. Although this general thio-modification can be theoretically carried out with other thio-compounds, few examples have been reported (325-327), and these have mostly been homologous mercaptoalkanoic or closely related dialkanoic acids comprising β-thiopropionic, thiolactic, α-mercaptoisobutyric, thiomalic and thiocitramalic acids though butyl and benzyl mercaptan have also been used (328). No other functional thiolignins seem to have been described. The reaction of thiophenol with lignin (329) is ostensibly one further such mercaptolytic example, but it is complicated by the fact that the thiophenol is also theoretically capable of reacting nuclearly analogously to phenol itself. At the present time, it does not appear to be possible to make an unequivocal decision between these possibilities since Dreyfus and Schnieder have shown (330) that thiophenol can condense with formaldehyde analogously to phenol. On the other hand, Douglas and Johnson have pointed out (331) that thiophenol, unlike phenol, is oxidized rather than nuclearly chlorinated in aqueous systems.

A more clear-cut case is the condensation where thiourea is the sulfurous component. Here Mikawa and Sato (332) studied the reaction of guaiacyl benzyl alcohols and the formation of the thiouronium product, Equation 39. When woodmeal was treated in the same way, about 0.3 isothiourea groups were introduced per methoxyl group.

$$RO\text{-}C_6H_3(OMe)\text{-}CHR\text{-}OH \xrightarrow{H_2N\overset{S}{C}NH_2} RO\text{-}C_6H_3(OMe)\text{-}CHR\text{-}S\overset{\oplus}{C}(=NH_2)NH_2 \qquad (39)$$

88

B. **Reactive Sites on Aromatic Rings.** The vacant sites of the aromatic rings in lignins comprise at most the 2, 5 and 6 positions, and the amount of reactivity available at a given position will be a function of its environment and contingent on the mode of preparation of that lignin. Of the reactions studied, the most extensively explored have been chlorination, nitration and the condensation with formaldehyde. Chlorination and nitration are detailed in Chapter 10, where the reactions of chloro- and nitro-lignin are approximately covered. It does, however, bear emphasizing in this chapter that these chloro and nitro functions offer two other degrees of freedom in the modification possibilities of the lignin macromolecule.

In the condensation with formaldehyde, a general repeated aim has been to develop methylolated lignins as potential substitutes for the widely used phenol-formaldehyde adhesive resins (333, 334). Unfortunately, to date the technicoeconomic success of this approach has been somewhat dubious. All lignins appear to react with alkaline formaldehyde to a greater or lesser degree, although it has not always been generally appreciated that in alkaline media only those lignin units having a free phenolic hydroxyl group are capable of nuclear reaction. Thus, the relative paucity of reactive sites in guaiacyl- (and of course particularly in syringyl-) lignin units should be recognized since both the para- and ortho-positions are blocked by the alkyl side-chain and the methoxyl groups, respectively. The distribution and location of hydroxymethyl groups has been studied and about 0.15 to 0.5 mole of these groups per methoxyl group can be introduced into spruce milled wood or kraft pine lignin, respectively (335). Based on the reaction of thiolignin and formaldehyde at various pH levels Mikawa has suggested that this condensation is a second order reaction (336). These reactions can be conventionally formulated as occurring at the five-position as shown by the modification of the lignin anion affording 89, Equation 40. It should also be realized that the molecular weight of lignin may increase through the formation of methylene 90 and dimethylene ether bridges. However, Allan and Halabisky (337) in the attempted synthesis of high molecular weight lignosulfonates were unable to raise the intrinsic viscosity by prolonged alkaline condensation with formaldehyde. In contrast, this particular modification goal could be readily attained by reaction in an acid medium.

In acid solution the protonated formaldehyde can react with the electron-rich six-position of lignin units, whether or not it contains a free phenolic hydroxyl group, to afford a methylolated lignin 91 by electrophilic mechanisms as in Equation 41 analogous to those of halogenation discussed in Chapter 10. Modified lignosulfonates with intrinsic viscosities of 0.5 dl/g were thus synthesized from commercial material of intrinsic viscosity 0.04 dl/g. The potential of the acid-catalyzed formaldehyde modification of lignin is emphasized by the observation that Klason lignin, a highly condensed insoluble material, may be completely solubilized if the reaction mixture contains formaldehyde (338, 339), Equation 41.

It must not, however, be overlooked that methylol groups can in certain cases be introduced into the side-chain if an activated hydrogen is available. Such an instance is illustrated by the lignin unit 92 where the position alpha t

MODIFICATION REACTIONS 547

a carbonyl is susceptible to methylolation to afford the keto-alcohol unit 93, Equation 42.

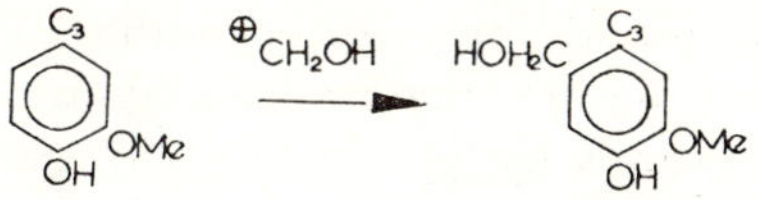

$$\text{(41)}$$

91

$$\text{(42)}$$

92   93

The extent of this type of Tollens reaction (340) can be determined, for example, in kraft lignin by first eliminating the activating carbonyl by reduction with sodium borohydride when methylolation can be attributed to phenolic activation.

Alternatively, the phenolic groups may be deactivated by methylation with diazomethane to permit methylolation to occur exclusively at the sidechain sites. The total of the separately determined methylol groups was in satisfactory concordance with the original methylolation achieved. The occurrence of these reactions can probably lead to condensative modification of kraft- and soda-lignins originating from alkaline hydrolysis of wood by virtue of the formaldehyde liberated during these processes. The methylolated lignins 89 can of course undergo further condensation to diphenylmethane derivatives 90 typical of phenolic-formaldehyde resins (341).

Efforts have also been made (342, 343) to combine the methylolation modification of lignin with concomitant condensation with phenols, but definitive studies have not been published. The kinetics of three-dimensional condensation polymerization (1) coupled with the higher reactivity of phenol towards formaldehyde suggests that the grafting of methylolated phenol units to lignin will not be extensive before the gel point. In support of this speculation Ivanenko and Nikitin (344) have presented evidence which indicates that isolated kraft lignin does not react with alkaline phenol-formaldehyde solutions. Attempts to use these modification reactions to convert lignins into ion exchange materials have been repeatedly made. Usually, a lignosulfonate has been employed and details of the commercial implications of these reactions are to be found in Chapter 19.

The scope of any one of these condensations involving the vacant five-position will of course be diminished in the case of hardwood lignins by the methoxyl located in position 5 of the syringyl unit.

Modification of the lignin macromolecule can also be carried out with reagents similar to formaldehyde using these same functional sites. Thus, Mikawa and coworkers (345) applied the Mannich reaction (343, 346) using formaldehyde and dimethylamine or piperidine. Using guaiacyl derivatives as models, they were able to show that the dimethylaminomethyl group was attached at the five-position as represented by 94. With model compounds having no vacant five-position, reaction did not occur.

94, R = NMe$_2$
95, R = SO$_3$H
96, R = P$^{\oplus}$(CH$_2$OH)$_3$ Cl$^{\ominus}$

When other guaiacyl compounds having a carbinol or carboxyl para to the phenolic hydroxyl were subjected to Mannich conditions using dimethylamine as the base, these substituents were displaced by the entering dimethylaminomethyl function. Etherification of the phenolic hydroxyl group prevented the Mannich reaction from occurring. Application of this reaction to a thiolignin and a phenolated thiolignin showed that the nitrogen content of the latter was more than twice that of the product from the original thiolignin (304, 343, 347). A similar result was obtained using formaldehyde alone. This type of material may be useful as a cationic surfactant for bitumen emulsions (348).

Sulfomethylation is another example of this type of condensation, and Gaslini (349) has used this procedure to improve the tanning properties of lignin. The structural modification achieved is exemplified by the lignin unit 95. In another similar reaction kraft lignin has been modified by tetrakishydroxymethyl phosphonium chloride. The first reaction product 96 must undergo further bridging since the phosphorus-containing lignin is increased in molecular weight (115). These modified lignins are effective coagulants for suspended material in water.

A related condensation involving formaldehyde has been reported by Borisek (350) who found that reaction of hexamethylenetetramine with a spent sulfite liquor afforded a lignin product containing about 4% nitrogen. Under more drastic reaction conditions, the sparingly soluble product was found to contain only 2.4% nitrogen indicating that further condensation had taken place. Other lignins subjected to similar but milder reaction conditions also combined with varying amounts of nitrogen.

Since the primary products formed by the action of hexamethylenetetramine on dicarbonyl compounds (351) and phenols (352) are molecular compounds, it can be expected that analogous structures will be formed by the carbonyl and the free phenolic sites in the lignin macromolecule. These phenol-hexamethylenetetramine adducts constitute a large clan of compounds which

MODIFICATION REACTIONS

may contain 1, 2 or 3 moles of the phenol per mole of hexamethylenetetramine (353). Hexamethylenetetramine triphenol is a well-characterized (354, 355) member of this group and on heating evolves ammonia with the concomitant formation of an insoluble, infusible resinous mass reminiscent of a "C" type phenol-formaldehyde resin (356). In general (357, 358), hexamethylenetetramine reacts with phenols like an ammonoformaldehyde differing from formaldehyde principally in that ammonia is evolved instead of water when methylene linkages are formed between phenol molecules.

Efforts to modify the lignin macromolecule have also been made using nonaldehydic carbonyl compounds usually with the aim of changing lignin's solubility in the more common organic solvents. The Japanese glycidylation work where lignin is reacted with acetone in the presence of sulfuric acid has already been mentioned. Gratzl and Lonsky (359) have investigated this reaction in considerable detail and have uncovered some unusual reactions in model compounds which are probably also generated within lignin itself.

Thus treatment of bicreosol 97 with a ketone in the presence of boron trifluoride etherate afforded a fluorene 98 as an apparently general reaction product (Equation 43). In the case of acetone only, the product was the benzopyran derivative 99, the first representative of a new heterocyclic ring system.

(43)

Additional reactions which occur at the five-position include mercuration and coupling with diazonium compounds. The former reaction was applied to methylated cuoxam (360) and ethyleneglycol spruce lignins (361) and to a lignosulfonate partially desulfonated with ammonium hydroxide (362). Mercurated lignin derivatives were obtained in each case, and the modified lignosulfonate has low fungitoxic characteristics. That the mercury is substantially located at the five-position follows from its quantitative conversion (363) to an iodolignin which yields 5-iodovanillin after oxidation with nitrobenzene or alkaline cobaltic hydroxide (364).

The coupling of the diazonium compounds can be located at the five-position as in <u>100</u>, Equation 44, only on purely theoretical grounds in many cases since in the early work the products obtained by treating lignins with such materials as diazotized anthranilic acid have been rather incompletely examined (365). More modern studies (366, 367) indicate that coupling occurs with dye formation, and this is accompanied by arylation of the lignin with concomitant evolution of nitrogen. The latter reaction is favored at high alkalinities. In addition, the reduced methoxyl content of the reaction products suggest demethylation as a third reaction (368). By the use of functional diazo compounds (369), a variety of modified lignins could be obtained. For example, anthranilic and sulfanilic acids (370) afforded lignins containing acidic groups, while the naphthylamines (371) yielded benzene-soluble lignin derivatives. Diazotized diamines caused extensive crosslinking and gave infusible products (372). Reductive cleavage of the azolignins produced aminolignins <u>101</u>, with the amino group located at the five-position ortho to the phenolic hydroxyl (373).

$$\underset{}{\text{HO-Ar(OMe)-}} \xrightarrow{RN_2Cl} \underset{100}{\text{HO-Ar(N_2R)(OMe)-}} \xrightarrow{H_2} \underset{101}{\text{HO-Ar(NH_2)(OMe)-}} \quad (44)$$

Stable carbon-carbon bonds can also be formed in the modification of lignin without the mediation of added molecules such as formaldehyde. Thus, wood lignins when heated in mildly acidic aqueous medium undergo not only a hydrolytic degradation but also condensation reactions which are analogous to the alcoholysis processes described in Chapter 9.

The monomeric products isolated from both reactions originate from a guaiacylglycerol-β-arylether structure in the precursor lignin. The analogous course of the alcoholysis and hydrolysis degradation is endorsed by the increase of the end-methyl group content in both cases which can be attributed to the side-chain rearrangement of guaiacylglycerol units (374). There is, nonetheless, a crucial difference between the two reactions. Whereas considerable yields of monomer are afforded by the ethanolysis degradation, only traces of such products are isolable from the acid hydrolysis. This can be explained in terms of the proclivity of the ketol structures to undergo condensation reactions under the acidic conditions used. These will be particularly liable to involve the 6- and 5-positions of the aromatic rings. These condensation phenomena are more effectively precluded in the alcoholysis by virtue of the etherification of both the labile α-hydroxyl and ketol functions.

These bond-forming reactions indeed appear to predominate over the acidolytic conditions and tend to overshadow the breakdown effects of hydrolysis. Especially sensitive are the p-hydroxybenzyl alcohol groups which can be destroyed by mild acid treatment, even at room temperature (375). The decreased oxygen content of Klason and Willstätter lignins (376) reflects their acidolytic genesis and the modifications resulting therefrom.

Heating pulverized wood in water in 100-200°, which is tantamount to a mildly acidic hydrolysis, causes condensation within the native lignin macromolecule. This is evidenced by the lowered yields of vanillin obtained on nitrobenzene oxidation (377), the loss of thermoplastic characteristics of the lignin, and its susceptibility to dissolution under sulfite pulping conditions (378). Condensations under the acid conditions of sulfite pulping do occur as detailed in Chapter 15 and are most rapid in the last period of pulping (379), but the reaction is not extensive because of the intervention of the sulfonic acid groups which block potential sites for condensation. Under abnormal pulping conditions where partial sulfonation is associated with high local acidities, delignification can be completely inhibited. Bockman (380) has suggested that this is partially due to reactions with carbohydrates and particularly with furfuraldehyde formed during the pulping process.

The condensing groups which interact with active sites in lignin under acid conditions probably comprise the coniferaldehyde and coniferyl alcohol moieties in addition to the α-hydroxyl groups and, of course, both intra- and inter-molecular condensations can occur. Up to 100° there is no significant change in the functional group content of a dioxan lignin exposed to 1% sulfuric acid. With increasing temperature a reduction in the total hydroxyl group occurs. The content of carbonyl, benzylic and γ-hydroxyls also decreased. The reduction was severe above 180°, and the temperature range of 130-150° marks the onset of extensive condensation (381). The intramolecular condensation structures can be exemplified by the isolation of metahemipinic acid from the permanganate oxidation products of methylated acid-treated lignins (382). The formation of this acid from both acid lignins and lignosulfonates in addition to the usual veratric and isohemipinic acids has been verified. Likewise, condensation increases the yields of benzene polycarboxylic acids in direct permanganate oxidation (383, 384).

Condensative treatment has been elaborated by Chudakov (385-389) into a patented process for producing benzene polycarboxylic acids. The 15% yield normally obtained by direct oxidation of pinewood lignin is reportedly doubled by autoclaving with sulfuric acid before oxidation. Adler and Lundquist (390) have also shown the importance of acidolytic modifications undergone by the phenylcoumaran units present in milled wood lignin which can rearrange to benzofuran units (see Chapter 9). The full extent of condensative modification of lignins is still far from clear (391-393).

Hardwoods with syringyl units in their lignin unlike softwoods can be pulped using alkali alone (394-398). These differences between the gymnosperm and angiosperm lignins implicate the 5-position and, in addition, the intrinsic differences in macromolecular structure should not be overlooked.

Adkins, Frank and Bloom (399) were also able to demonstrate this rather convincingly for hardwoods using aspen lignin. Here it was found that aspen soda lignin did not yield a monomeric fraction on hydrogenolysis, whereas the methanol lignin from the same wood is converted to hydrogenated monomers in 44% yield.

These condensations are largely averted in the kraft pulping process in which part of the hydroxide base has been replaced by sulfhydryl ion. Diminished condensation of the lignin during pulping is the beneficial consequence (400, 401). Free radical processes and their inhibition by sulfhydryl ions have also been speculatively considered (402-404).

Free radicals are more certainly involved in another type of condensation which leads to the formation of carbon-carbon bonds within the lignin molecule. Thus, it has been known for more than half a century that treatment of sulfite liquors with dichromate ions affords gel-like structures (405), and this reaction has been of some civil engineering interest as a means of increasing the cohesiveness of granular soils (406) and also promoting the effectiveness of lignosulfonates as tanning agents (407). Speculation as to the structure of these condensation products has led to the view that they are simply chromic complexes formed with (408) or without (409) concomitant elimination of sulfonate groups and their replacement by carboxyl functions. The time for gel formation can be accelerated by the addition of Al, Fe or Cu ions (410).

Cupric oxide oxidation of the gels (411) did not contribute significantly to the elucidation of their structure which has been recently studied by James and Tice (412). In a comprehensive investigation they showed that the hydrogel had a pH of 6.6 and was insoluble in organic solvents, dilute acid and aqueous alkali. Although the methoxyl and sulfonate contents were unchanged from the starting material, about 60% of the phenolic hydroxyl groups had disappeared and the infrared spectrum indicated the formation of about 7 mole percent of carboxyl groups. Concurrent with these changes, the average molecular weight had increased about twentyfold. The possibility that the molecular weight increase was attributable to aldehyde-phenol condensation reactions was discounted by the low reactivity of formaldehyde with guaiacol and sodium lignosulfonate under the gel-forming reaction conditions. The molecular weight increase can perhaps be explained by a phenoxy coupling mechanism between positions ortho to a phenolic hydroxyl group (413, 415-420). These views are supported by the recent work of Tanaka, Abe and Senju (414), who demonstrated that masking of the phenolic hydroxy groups by methylation or carboxymethylation lowered the gelation rate. When the phenolic group content was increased by chlorination, the gel rate was augmented.

Esterification does not seem to occur during molecular weight buildup because subsequent hydrolysis does not change the intrinsic viscosity of the product. This view is supported by the formation of a gel on treatment of sodium lignosulfonate with manganese dioxide, a reagent known to dehydrogenate phenols (418).

The gel was also disintegrated and eluted with various salt solutions when it was found that about 87% of the chromium was removed by treatment with N acetic or 2N hydrochloric acid solutions. The gel has then contracted to about one-half of its original volume, but expanded to about six times its original volume when washed with water. Readsorption of chromic ions caused recontraction, and this behavior is typical of an ion exchange resin with low cross-link density (421). The depleted gel still contained 1.3% (7.5 mole %) of firmly bound chromium which could only be removed by drastic treatment with ethylenediaminetetracetic acid which caused the gel to disintegrate. This residual chromium is therefore probably coordinated to the lignin macromolecule by carboxyl groups which are known to form more stable chromium chelates than sulfonic acid functions (412), and this satisfactorily accounts for the two different types of chromium within the gel (422).

James and Tice also assumed that the gels behaved as rubber-like bodies and applied Equation 45, derived from the kinetic theory of rubber elasticity to their data.

$$Z_c = \frac{cRT}{G_e M_o} \quad (45)$$

Where Z_c = average D. P. per network strand
   c = concentration of polymer, g/cm^3
   M_o = monomeric molecular weight
   G_e = pseudo-equilibrium shear modulus

The theory requires that the polymer chains be of high molecular weight and regularly cross-linked although real and randomly cross-linked polymers have also been found to fit the equation. A second requirement is that the cross-links must be comparable in type to the main chain bonds and weak bonds would not be taken into account by this expression.

The shear modulus data then indicated that the gel contained one cross-link for every 270 phenylpropane units, assuming that the monomeric molecular weight of the chromium-containing phenylpropanoid unit was 338. The picture of the gel which therefore emerges is that of a condensed lignosulfonate containing carboxyl groups and coordinated to hydroxylated chromium ion species by all types of its acid functions. The bulk of the chromium (87%) is held by bonds analogous to those in a simple ion exchange resin while the remainder (13%) acts as links holding the lignin macromolecules together to form a continuous network. It has been further suggested by James and Tice that the cross-links every 270 phenylpropanoid units may be due to binuclear chromium bridges, and from this it is inferred that only about 1% of the residual 13% chromium is necessary to maintain the macrogel: the majority of binuclear complexes being assumed to form intra- rather than inter-molecularly in analogy with the chemistry of chrome tanning where only a small portion of chromium actually cross-links the collagen fibers (423, 424). The cross-links can of course very well originate from oxidative coupling reactions described earlier, and Hayashi and Goring (425, 426) have pointed out that reactions between dichromate and catechol are similar to those of the dichromate-lignosulfonate system. Such catechol type units are present in coniferous lignosulfonates as demonstrated by the isolation of methyl 3,4-diethoxybenzoate after ethylation, permanganate oxidation and esterification with methanol (427).

Gels can, however, be obtained by the oxidation of lignins without the intervention of chromium compounds. Thus, high molecular weight lignins suitable as coagulants can be synthesized (116) by treatment with a variety of oxidants known to favor coupling (419).

Phenoxy coupling has been extended by Allan, Mauranen, Neogi and Peet (428) to provide a new method of grafting to lignocellulosic fibers. The reaction is fundamentally a lignin modification reaction in which lignin radicals are generated in the presence of various phenols. Coupling and polymerization ensue as formulated in Equation 46. This is in effect a synthetic counterpart of the lignification which occurs in the cell wall (Chapter 4). Since the pK for phenols is about 10, the observed grafting maximum at pH 10 implies the intermediacy of the phenolate anion rather than the phenol itself in the oxidation of the anion to a free phenoxy radical. These are then coupled rapidly and irreversibly to dimeric and polymeric products under kinetic control.

$$\text{(46)}$$

It is noteworthy that for many phenols no facile and complete termination of the grafting reaction can occur. Phenolic oxidative polymerization can therefore be regarded as producing a sort of "living" polymer. The amount of homopolymer formed therefore falls to zero with increasing time as required by kinetic considerations. However, the polymerization is a two-step process requiring a repetitive dehydrogenation for continual reactivation of the growing polymer chain which otherwise lapses into quiescence. The comatose macromolecule within the fiber can be repeatedly stimulated and induced to continue irreversible polymerization and grafting by the addition of more oxidant and phenol. Their designation as cataleptic polymers has therefore been proposed.

MODIFICATION REACTIONS 555

Lignin itself is of course a naturally occurring member of this class, and the recently introduced poly(phenylene)oxides (417, 420) are its synthetic counterparts.

If applied to a pulp waste lignin alone, then clearly lower molecular fractions will become grafted to the higher molecular material. The net result will be a decrease in the polydispersity and an increase in the molecular weight. Modified lignins of this type have been shown to function as coagulation agents in water purification (116).

Grafting to lignocellulosic fibers can also be achieved by another route based at least in part on a modification reaction of the lignin macromolecule. Thus, fiber reactive dyes exemplified by 102 react readily with the phenolic hydroxyl group of the lignin in ground wood or pulp fibers (429) to afford the modified lignin 103, Equation 47. The reaction need not be carried out within the fiber matrix (116). Lignin in solution readily reacts with cyanuric chloride 104, the parent of the fiber reactive dye, to give a wide range of modified lignins, Equation 48, and the steric requirements of this versatile modification procedure has been investigated by Allan, Allan, Mattila and Mauranen (430).

The more conventional grafting of vinyl monomers as a means of modifying lignin has also been studied, particularly by Japanese researchers. However, attempts to directly graft styrene, vinyl acetate or methyl methacrylate to lignin using γ-radiation have not met with very encouraging results although lignin-styrene graft copolymers have semi-conductive properties (431).

Nonetheless, using hydrochloric acid lignins from both softwood (Pinus densiflora) and hardwood (Betula tauschii) Koshijima and Muraki (432) were able to graft methyl methacrylate using mixtures of monomer to lignin of 3:1 or 10:1. Maximum grafting of 24.1% onto the softwood lignin was obtained. Under similar conditions maximum grafting for the hardwood lignin was only 9.5% which implies that grafting in softwood lignins involves the uncondensed five-position on guaiacyl units to a considerable extent. No grafting of styrene or vinyl acetate could be secured. The low or nonexistent grafting can be attributed to the extreme resistance of lignin to high energy radiation (49, 433, 434) and even gross dosages have no detectable effect (434). As might be expected, phenolic hydroxyl groups are not reactive sites for radical grafting because of the resonance stabilization of the radicals formed or because of the inhibiting action of the related quinone structures (1). Thus, for lignin and its derivatives the number of radicals produced per 100 eV of radiation, the G_R value, is 0.6-0.7 or about one-tenth the G_R value of methyl methacrylate (435). Therefore, grafting will be a predominantly surface phenomenon since reaction will preferentially occur at sites where methyl methacrylate is located (432).

These findings imply that masking of the phenolic functions will increase the rate of grafting (432). This conclusion is qualitatively substantiated by the increased grafting rates (38%) resulting from acetylation of the lignin. A more quantitative validation is provided by the corresponding increases in grafting rates with increasing methylation of the lignin summarized in Table 13.3.

Table 13.3. Variation of Radiation Grafting Rate with Degree of Methylation

Methylation procedure	Methoxyl content	Rate of grafting
CH_2N_2/ether	20.4	67
CH_2N_2/ether/dioxan	23.1	116-150
$(CH_3)_2SO_4$/alkali	33.3	44

In addition to the effects of the alkylation or acylation the nature of the solvent will contribute to the mode and rate of graft copolymerization (436). Thus, the use of benzene or hexane favors the formation of styrene homopolymer over grafting. On the other hand, for alcoholic solvents including methanol,

ethanol, isopropanol and n-butanol, the reverse was the case. These differences are probably related to the swelling characteristics of the solvents. The nonpolar hydrocarbons are poor solvents for lignins in comparison to the polar alcohols and therefore do not promote distension of the spherical macromolecule. Furthermore, styrene has a lower G_R value (0.3-0.7 depending on dose-rate) than lignin and in the case of homopolymerization, benzene and hexane, with G_R values of 0.75 and 7.6 (435), can act as chain transfer agents. With solvating polar irradiation media such as methanol the concentrations of solvent and monomer become of some importance. Thus, because of its high G_R value of 13.5, methanol functions both as a polymerization initiator and as a chain transfer agent in the grafting reaction. At the 2% concentration level, methanol permits the maximum grafting of styrene to lignin to be achieved (430%). Further augmentation of the methanol content causes a rapid decrease in the amount of grafting.

Grafting onto lignin of course can also be accomplished without γ-radiation by free radicals generated by typical polymerization initiators such as benzoyl peroxide, α-azobisisobutyronitrile or cumene hydroperoxide (437) and the products and their precursors have received some scrutiny as semi-conductors (431).

The structure of the grafted lignin have been investigated to some extent (438) and partly elucidated using the nitrobenzene oxidation technique (439). Thus, if grafting involves the guaiacyl units usually convertible to vanillin, then on oxidation a reduced yield of this aldehyde should be obtained. A collection of oxidation experiments on grafted lignin is summarized in Figure 13.1 from which the following conclusions can be drawn: 1, Initially and up to 50% grafting, the yield of vanillin is inversely proportional to the amount of grafting. 2, After the amount of grafting reaches 150%, the vanillin yield remains constant. Grafting is then occurring on sites which do not afford vanillin on oxidation, and these are most probably the already grafted polymer branches. 3, A part of the lignin corresponding to 5.6% vanillin apparently does not participate in the grafting reaction. Since the yield of vanillin from the ungrafted lignin is about 17.5%, it can be concluded that about 68% of the lignin macromolecule participates in the grafting reaction. Koshijima (439) has considered that grafting can occur at the following locales.

[1] .C-OH sites originating in the breakdown of alcohols
[2] .CH$_2$O sites stemming from methoxyl groups
[3] .C-Ō-C sites generated from alkyl aryl ethers
[4] active centers on the aromatic nuclei

In case [1] a secondary alcohol in the original lignin would be converted into a tertiary alcohol as a result of grafting. Since such an alcohol is resistant to acetylation, a drop in acetylatable hydroxyl content of the grafted polymer would be evidence for this type of grafting site. No such decrease is observed.

Analogously, if grafting originated from methoxyl groups as in case [2], the methoxyl content of the grafted lignin should be decreased. It remains unchanged. Supporting evidence is provided by the lower grafting rates obtained with hardwood lignins which contain additional methoxyl groups. Moreover, since this methyl ether grafting site is not formed, the equally probable case [3] of the alkyl aryl ether can also be discounted as a grafting site. This leaves the aromatic nucleus as the only likely grafting site and allows the styrene units per guaiacyl unit to be roughly estimated at between 5 and 10. When the amount of grafting is between 100 and 400%, the failure of 32% of the vanillin yielding lignin units to participate in the grafting must be ascribed to the presence thereon of unetherified phenolic hydroxyl groups which are converted to stabilizing phenoxy radicals.

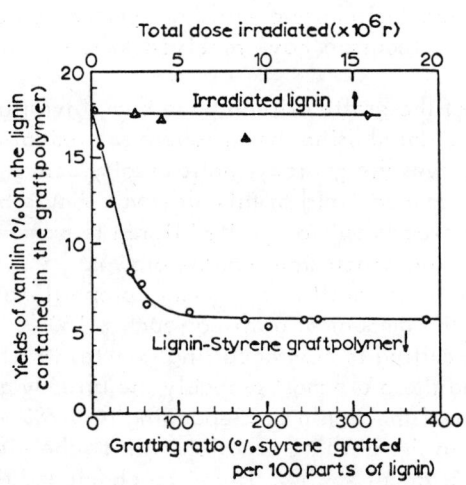

Figure 13.1. Yields of vanillin from lignin-styrene graftpolymers and irradiated lignin by nitrobenzene oxidation (439).

REFERENCES

1. P. J. Flory, Principles of Polymer Chemistry, Cornell University Press, Ithaca, New York, 1953.
2. C. Tanford, Physical Chemistry of Macromolecules, John Wiley & Sons, Inc., New York, 1961.
3. B. Jørgensons, Natural Organic Macromolecules, Pergamon Press, New York, 1962, p. 140.
4. K. Freudenberg, Science, 148, 595 (1965).
5. K. Freudenberg, C. L. Chen, J. M. Harkin, H. Nimz and H. Renner, Chem. Comm., 1965, 224.
6. K. Freudenberg and K. C. Renner, Chem. Ber., 98, 1879 (1965).
7. K. Ogiyama and T. Kondo, Nippon Mokuzai Gakkaishi, 11, 206 (1965).
8. E. Adler, H. D. Becker, T. Ishihara and A. Stamvik, Holzforschung, 20, 3 (1966).
9. K. Freudenberg, W. Belz and C. Niemann, Chem. Ber., 62, 1561 (1929).
10. B. D. Bogomolov, G. F. Prokshin and E. Gelfand, U.S.S.R. Patent 182,128 April 1, 1965.
11. C. Dence and K. Sarkanen, Tappi, 43, 87 (1960).
12. K. Freudenberg and W. Dürr, Chem. Ber., 63, 2713 (1930).
13. T. Enkvist, J. Turunen and T. Ashorn, Tappi, 45, 128 (1962).
14. T. Enkvist and T. Lindfors, Finska Kemists. Medd., 75, 1 (1966).
15. D. W. Goheen, Forest Products Journal, 12, 471 (1962).
16. W. M. Hearon, W. S. MacGregor and D. W. Goheen, Tappi, 45, 28A (1962).
17. G. F. Prokshin and B. D. Bogomolov, Lesnoi Zh., 9, 125 (1966).
18. Z. I. Chupka and A. V. Obolenskaya, Sb. Materialy Nauch — Tekh. Konfer. Leningrad. Lesotekh. Akad., (4), 203 (1966).
19. G. G. Allan, Tappi, 51, 224 (1968).
20. F. C. Benton and T. E. Dillon, J. Am. Chem. Soc., 64, 1128 (1942).
21. L. F. Fieser and M. Fieser, Advanced Organic Chemistry, Reinhold Publishing Corp., New York, 1961, p. 309.
22. Shell Chemical Company, Technical Information Bulletin "Sulfolane."
23. V. Prey, Chem. Ber., 74B, 1221 (1941); 75B, 350 (1942).
24. M. Mottier, Helv. Chim. Acta, 18, 840 (1935).
25. H. E. Ungnade and K. T. Zilch, J. Org. Chem., 15, 1109 (1950).
26. V. Prey, Chem. Ber., 76B, 156 (1943).
27. D. H. Grangaard, Tappi, 44, 433 (1961).
28. H. Adkins, R. L. Frank and E. S. Bloom, J. Am. Chem. Soc., 63, 549 (1941).
29. Q. P. Peniston and J. L. McCarthy, ibid., 70, 1329 (1948).
30. K. Freudenberg and H. F. Müller, Chem. Ber., 71, 1281 (1938).

31. E. Adler, Proceedings of the Fourth International Congress of Biochemistry, Vienna, 1958, Pergamon Press, 1959, p. 137.
32. K. Kratzl, Monatsh. Chem., $\underline{80}$, 314 (1949).
33. K. Freudenberg and E. Plankenhorn, Chem. Ber., $\underline{75}$, 857 (1942); ibid., $\underline{80}$, 149 (1947).
34. E. Adler, K. T. Björkqvist and S. Häggroth, Acta Chem. Scand., $\underline{2}$, 93 (1948).
35. A. v. Wacek and K. Kratzl, J. Polymer Sci., $\underline{3}$, 539 (1948).
36. R. B. Wagner and H. D. Zook, Synthetic Organic Chemistry, John Wiley and Sons, Inc., New York, 1953, pp. 49, 174.
37. E. Adler and S. Häggroth, Acta Chem. Scand., $\underline{3}$, 86 (1949).
38. B. O. Lindgren and V. Saedén, ibid., $\underline{6}$, 91 (1952).
39. B. O. Lindgren, ibid., $\underline{1}$, 779 (1947); $\underline{3}$, 1011 (1949).
40. N. Migata, H. Mikawa, J. Nakano and M. Ichino, J. Japan.Tech. Assoc. Pulp and Paper Ind., $\underline{8}$, 352, 482 (1954).
41. N. Migata, R. Mitsukawa, J. Nakano and M. Ichino, J. Japan.Tech. Assoc., Pulp and Paper Ind., $\underline{8}$, 2,566 (1954).
42. N. N. Shorygina and T. Ya Kefeli, J. Gen. Chem. U.S.S.R. (English translation), $\underline{17}$, 2058 (1947).
43. N. N. Shorygina and T. Ya Kefeli, Zhur. Obshchei Khim., $\underline{18}$, 528 (1948).
44. A. F. Semechkina and N. N. Shorygina, ibid., $\underline{28}$, 3265 (1958).
45. A. F. Semechkina and N. N. Shorygina, Izv. Akad. Nauk S.S.S.R., Ser. Khim., $\underline{5}$, 715 (1963).
46. C. Schuerch, J. Am. Chem. Soc., $\underline{74}$, 5061 (1952).
47. C. Schuerch, Forest Prod. J., $\underline{14}$, 377 (1964).
48. C. Schuerch, New Scientist, $\underline{20}$, 366, 484 (1963).
49. F. E. Brauns and D. A. Brauns, The Chemistry of Lignin, Supplement Vol., Academic Press, New York, 1960.
50. W. Hückel and H. Bretschneider, J. prakt. Chem., $\underline{151}$, 61 (1938).
51. C. A. Kraus and G. F. White, J. Am. Chem. Soc., $\underline{45}$, 768 (1923).
52. A. J. Birch, J. Chem. Soc., $\underline{1947}$, 102.
53. A. J. Birch and H. Smith, ibid., $\underline{1951}$, 1882.
54. P. M. Dean and G. Berchet, J. Am. Chem. Soc., $\underline{52}$, 2823 (1930).
55. T. H. Vaughn and J. A. Nieuwland, Ind. Eng. Chem. Anal. Ed., $\underline{3}$, 274 (1931).
56. A. L. Wilds and N. A. Nelson, J. Am. Chem. Soc., $\underline{75}$, 5360 (1953).
57. C. B. Wooster and J. F. Ryan, ibid., $\underline{56}$, 1134 (1934).
58. L. F. Audrieth and J. Kleinberg, Nonaqueous Solvents, John Wiley and Sons, Inc., New York, 1953, pp. 42, 53.
59. A. J. Birch, B. Milligan, E. Smith and R. N. Speaks, J. Chem. Soc., $\underline{1958}$, 4771.
60. K. Freudenberg, K. Engler, E. Flickinger, A. Sobek and F. Klink, Chem. Ber., $\underline{71}$, 1810 (1938).

61. K. Freudenberg, F. Klink, E. Flickinger and A. Sobek, ibid., 72, 217 (1939).
62. K. Freudenberg, W. Lautsch and G. Piazolo, ibid., 74, 1879 (1941).
63. J. S. Blair, J. Am. Chem. Soc., 48, 96 (1926).
64. A. St. Pfau, Helv. Chim. Acta, 22, 550 (1939).
65. A. J. Birch, J. Cymerman-Craig and M. Slaytor, Aust. J. Chem., 8, 512 (1955).
66. A. J. Birch, J. Chem. Soc., 1944, 430; 1946, 593.
67. P. Shorygin and S. A. Skoblinskaja, Compt. rend. acad. sci. U.S.S.R, 14 505 (1937); Chem. Abs., 31, 5777 (1937).
68. S. Hernestam and E. Adler, Svensk Kem. Tidskr., 67, 37 (1955).
69. F. Sondheimer, R. Yashin, G. Rosenkranz and C. Djerassi, J. Am. Chem Soc., 74, 2096 (1952).
70. D. H. R. Barton and C. H. Robinson, J. Chem. Soc., 1954, 3045.
71. J. Marton and E. Adler, U. S. Patent 3,071,570 (to West Virginia Pulp and Paper Co.) Jan. 1, 1963.
72. I. Croon and B. Swan, Svensk Papperstidn., 67, 177 (1964).
73. A. Hayashi and Y. Namura, Bull. Chem. Soc. Japan, 38, 512 (1965).
74. G. F. Prokshin, Lesnoi Zh., 7, 149 (1964).
75. J. Gierer and B. Lenz, Svensk Papperstidn., 68, 334 (1965).
76. J. Gierer and N.-H. Wallin, Acta Chem. Scand., 19, 1502 (1965).
77. E. Adler and K. Lundqvist, ibid., 15, 15 (1961).
78. F. Paschke, Cellulose Chem., 3, 19 (1922).
79. B. D. Bogomolov and E. D. Gelfand, Bum. Prom., 3, 3 (1966).
80. F. E. Brauns, J. Am. Chem. Soc., 61, 2120 (1939).
81. F. E. Brauns, Paper Trade J., 111, 35 (1940).
82. H. Urban, Cellulose Chemie, 7, 73 (1926).
83. J. Gierer and N.-H. Wallin, Acta Chem. Scand., 20, 2059 (1966).
84. T. L. Jacobs in Heterocyclic Compounds, Vol. 5 (R. C. Elderfield, Ed.) John Wiley and Sons, Inc., New York, 1957, p. 72.
85. F. E. Brauns, H. F. Lewis and E. B. Brookband, Ind. Eng. Chem., 37, 70 (1945).
86. G. Jones and F. E. Brauns, Paper Trade J., 119, 108 (1944).
87. T. Lindfors and T. Enkvist, Finska Kemists. Medd., 74, 29 (1965).
88. J. W. T. Merewether, Tappi, 37, 483 (1954).
89. J. W. T. Merewether, Holzforschung, 8, 116 (1954).
90. J. W. T. Merewether, Australian J. Chem., 7, 75 (1954).
91. R. Senzyu and H. Ishikawa, J. Agr. Chem. Soc. Japan, 22, 72 (1948).
92. J. Furukawa and T. Saegusa, Polymer Reviews, Vol. 3, "Polymerization of Aldehydes and Oxides," (H. F. Mark and E. H. Immergut, Eds.) Interscience Publishers, New York, 1963.
93. H. Ishikawa, T. Oki and F. Fujita, Nippon Mokuzai Gakkaishi, 7, 85 (1961).
94. H. Ishikawa, T. Oki and K. Yamashita, ibid., 6, 102 (1960).

95. H. Ishikawa, J. Japan. Forest. Soc., 33, 331 (1951).
96. K. Sarkanen and C. Schuerch, Anal. Chem., 27, 1245 (1955).
97. N. I. Nikitin and T. I. Rudneva, J. Applied Chem. (USSR), 8, 1176 (1935).
98. H. Ishikawa, J. Japan. Forest. Soc., 33, 329 (1951).
99. H. Ishikawa, ibid., 35, 121 (1953).
100. H. Ishikawa, ibid., 35, 363 (1953).
101. H. Ishikawa, Memoirs of the Ehime University, Sect. VI (Agr.), Vol. IV, No. 1, Published by the College of Agriculture, Ehime Univ., Matsuyama, Japan, Nov., 1958, pp. 21-23.
102. Reference 101, entire volume.
103. H. Ishikawa and T. Nakajima, Trans. 62nd Meeting Japan. Forestry Soc., J. Japan. Forest. Soc., Spec. Issue, 1953, p. 281.
104. H. Ishikawa and T. Nakajima, J. Japan. Forest. Soc., 36, 106 (1954).
105. G. G. Allan, U. S. Patent 3,476,795, Nov. 4, 1969.
106. J. H. Saunders and K. C. Frisch, Polyurethanes, Chemistry and Technology, Part I, "Chemistry," Interscience Publishers, New York, 1962.
107. K. Kratzl, K. Buchtela, J. Gratzl, J. Zauner and O. Ettinghausen, Tappi, 45, No. 2, 113 (1962).
108. E. Adler in Biochemistry of Wood, K. Kratzl and B. Billek, Eds., Pergamon Press, London, 1958, p. 137.
109. T. R. Santelli and R. T. Wallace, U. S. Patent 3,072,634 (to Owens-Illinois Glass Co.), Jan. 8, 1963.
110. R. F. Nichols, U. S. Patent 2,854,422 (to B. F. Goodrich Co.), Sept. 30, 1958.
111. S. Tai, T. Sawanobori, J. Nakano and N. Migita, Nippon Mokuzai Gakkaishi, 14, 46 (1968).
112. K. V. Sarkanen, in Wet Strength in Paper and Paperboard, Tappi Monograph Series No. 29, J. P. Weidner, Ed., 1965, p. 38.
113. Dow Chemical Co. Technical Bulletin, "Ethylenimine."
114. J. B. Doughty, Forest Products Journal, 13, 413 (1963).
115. G. G. Allan, U. S. Patent 3,470,148, Sept. 30, 1969.
116. G. G. Allan and D. D. Halabisky, unpublished work.
117. J. S. Pierce, U. S. Patent 2,503,297 (to Albemarle Paper Mfg. Co.), April 11, 1950.
118. V. M. Nikitin and T. M. Kroshilova, Lesnoi Zh., 7, 149 (1964).
119. V. R. Yaunzems, V. N. Sergeeva and L. N. Mozheiko, Izv. Akad. Nauk Latv. SSR, Ser. Khim. (6), 729 (1966).
120. M. E. Ali, Q. A. Ahmed and M. H. Khundkar, Pakistan J. Sci. Ind. Research, 1, 79 (1958).
121. B. F. Hrutfiord, Tappi, 48, 48 (1965).
122. M. Imoto and Y. Hachihama, J. Soc. Chem. Ind. Japan, 50, 132 (1947).
123. J. Gierer and B. Lenz, Svensk Papperstidn., 68, 334 (1965).

124. J. Gierer and N. H. Wallin, Acta Chem. Scand., 19, 1502 (1965).
125. Y. Hachihama, S. Jyodai, H. Sato, E. Fukui and T. Nakamura, Mem. Inst. Sci. and Ind. Research Osaka Univ., 7, 167 (1950).
126. F. J. Ball, W. K. Dougherty and H. H. Moorer, Jr., Canadian Patent 654,728 (to Westvaco) Dec. 25, 1962.
127. M. Mihailov and S. Gerdjikova, Compt. Rend. Acad. Bulgare Sci., 18, 829 (1965).
128. S. Tai, M. Nagata, J. Nakano and N. Migita, Nippon Mokuzai Gakkaishi, 11, 100 (1967).
129. S. Tai, J. Nakano and N. Migita, ibid., 13, 257 (1967).
130. J. Gratzl, private communication, 1969.
131. N. I. Nikitin, S. D. Antonov'skii and M. A. Miklailova, Zhur. Priklad. Khim., 30, 750 (1957); J. Applied Chem. USSR, 30, 792 (1957).
132. S. D. Antonov'skii, Trudy Leningrad Lesotekh. Akad. im S. M. Kirova, 91, 247 (1961); Chem. Abs., 55, 26504h (1961).
133. A. O. Zoss, W. E. Hanford and C. E. Schildknecht, Ind. Eng. Chem., 41, 73 (1949).
134. S. D. Antonov'skii and N. I. Nikitin, Trudy Leningrad Lesotekh. Akad. im S. M. Kirova, 75, 83 (1956); Chem. Abs., 10745i (1959).
135. H. M. Walker, U. S. Patent 2,816,100 (to Monsanto Chemical Co.) Dec. 10, 1957.
136. F. E. Brauns, The Chemistry of Lignin, Academic Press, New York, 1952, p. 427.
137. H. J. Callow, J. Textile Inst., 43, T423 (1952).
138. K. Freudenberg and W. Fuchs, Chem. Ber., 87, 1824 (1954).
139. T. Fukushimi, S. Sakuma, H. Takahashi, K. Tomita, K. Fujihara, Y. Isome and T. Shibamoto, Holzforschung, 20, 51 (1966).
140. J. C. Clark and F. E. Brauns, Paper Trade J., 119, No. 6, 35 (1944).
141. H. F. Lewis, F. E. Brauns, M. A. Buchanan and E. B. Brookbank, Ind. Eng. Chem., 35, 1113 (1943).
142. J. J. McNair and E. C. Jahn, Paper Trade J., 117, No. 8, 29 (1943).
143. C. Dorée and L. Hall, J. Soc. Chem. Ind., 43, 257T (1924).
144. K. Freudenberg in Progress in the Chemistry of Organic Natural Products (L. Zechmeister, Ed.), 11, 43 (1954).
145. Ref. 136, (a) p. 249, (b) p. 92. H. E. Hergert, private communication.
146. J. Gierer, Acta Chem. Scand., 8, 1317 (1954).
147. E. Adler and B. Stenemur, Chem. Ber., 89, 291 (1958).
148. N. J. Nikitin and T. I. Rudneva, Zhur. Priklad. Khim., 10, 1915 (1937).
149. G. G. Allan, Belgian Patent 706,509 (Dec., 1967).
150. H. Pauly, Chem. Ber., 81, 392 (1948).
151. W. Sandermann, Svensk Papperstidn., 52, 365 (1949).

152. R. L. Glover and J. W. Bain, Can. J. Research, 143, 65 (1936).
153. E. H. Rodd, Editor, Chemistry of Carbon Compounds, Vol. 1B, Elsevier Publishing Co., New York, 1952, p.897.
154. H. Apitzch, Chem. Ber., 37, 1599 (1904); 38, 2888 (1905).
155. P. Yates and D. R. Moore, J. Am. Chem. Soc., 80, 5577 (1958).
156. T. Lieser and V. Schwind, Ann., 532, 104 (1937).
157. T. Mika, Shell Chemical Co., personal communication.
158. K. Freudenberg and G. Dietrich, Ann., 563, 146 (1949).
159. J. B. Doughty, U. S. Patent 3,081,293 (to Westvaco) March 12, 1963.
160. A. I. Iosilevich, K. U. Usmanov and O. Ioannidis, Uzbek. Khim. Zh., 7, (5) 61 (1963).
161. G. Telysheva, V. Sergeeva and A. Kalninsh, Latv. PSR Zinat. Akad. Vestis, Kim. Ses., No. 6, 744 (1968).
162. G. G. Allan, unpublished work.
163. Z. N. Kreitsberg, V. N. Sergeeva, and Y. K. Grabovskii, Khim. Perarabotka i Zashchita Drevesiny, Riga, Akad. Nauk. Latv. SSR, 81 (1964).
164. E. E. Nifanter and I. V. Fursenko, U.S.S.R. Patent 181,108 (Jan. 13, 1965).
165. F. Paschke, Cellulosechem., 3, 19 (1922).
166. H. Tropsch, Gos. Abhandl. Kenntnis Kohle, 6, 301 (1923).
167. B. V. Tronov, L. A. Pershina, V. M. Morozova, A. V. Kovapenko and A. I. Galochkin, Gidroliz. i Lesokhim. Prom., 14, 5, 10 (1961).
168. K. Kashima and K. Oiwa, Yuki Gosei Kagaku Kyokai Shi, 17, 221 (1959); 17, 441 (1959).
169. B. V. Tronov, L. A. Pershina and A. T. Galochkin, U.S.S.R. Patent 164,276 (Dec. 23, 1963).
170. F. J. L. Aparicio and J. A. L. Sastre, An. Quim., 55, 191 (1969).
171. A. Friedrich, Z. physiol. Chem., 176, 127 (1928).
172. G. M. Telysheva, A. I. Kalninsh and V. N. Sergeeva, Izv. Akad. Nauk Latv. SSR Ser. Khim., 6, 745 (1966).
173. E. Adler and L. Ellmer, Acta Chem. Scand., 2, 839 (1948).
174. E. Adler and J. Marton, ibid., 13, 75 (1959).
175. N. F. Komshilov, N. G. Dzhurinskaya and M. N. Letonmyaki, Tr. Karelsk. Filiala Akad. Nauk SSSR, 38, 31 (1963).
176. E. Adler and J. Gierer, Acta Chem. Scand., 9, 84 (1955).
177. J. Gierer and S. Söderberg, ibid., 13, 27 (1959).
178. K. Freudenberg, H. Zocher and W. Dürr, Chem. Ber., 62, 1814 (1929).
179. K. Hess and K. Heumann, Chem. Ber., 75, 1802 (1942).
180. A. Björkman and B. Person, Svensk Papperstidn., 60, 285 (1957).
181. B. Holmberg, ibid., 50, No. 11B, 14 (1947).
182. B. Holmberg, Arkiv Kemi, 13, 211 (1958).
183. T. Enkvist and E. Hägglund, Svensk Papperstidn., 53, 85 (1950).

184. H. Ishikawa and T. Nakajima, J. Japan.Forest. Soc., 36, 104 (1954).
185. J. W. T. Merewether, Australian J. Sci. Research, 2A, 117; 600 (1949).
186. A. Bell and G. F. Wright, Can. J. Research, 27B, 505 (1949).
187. E. Hägglund and H. Richtzenhain, Tappi, 35, 281 (1952).
188. M. Lieff, G. F. Wright and H. Hibbert, J. Am. Chem. Soc., 61, 865 (1939).
189. G. F. Wright and H. Hibbert, ibid., 50, 125 (1937).
190. H. Bolker and N. Terashima, "Lignin Structure and Reactions," Advances in Chemistry Series, 59, American Chemical Society, Washington, D.C., 1966, p. 110.
191. N. L. Nikitin and I. M. Orlowa, J. Applied Chem. (USSR), 8, 1402 (1935).
192. C. L. Mehltretter, T. E. Yeates, G. E. Hamerstrand, B. T. Hofreiter and C. E. Rist, Tappi, 45, No. 9, 750 (1962).
193. G. E. Hamerstrand, B. T. Hofreiter, D. J. Kay and C. E. Rist, ibid., 46, 400 (1963).
194. I. Kawamura and T. Higuchi, J. Japan.Wood Res. Soc., 11, 19 (1965).
195. R. Kitamura and S. Suzuki, J. Pharm. Soc. Japan, 57, 659 (1937).
196. G. Ya Vanag, Zhur. Anal. Khim, 5, 110 (1950).
197. J. C. Pew, J. Am. Chem. Soc., 73, 1678 (1951); 74, 2850 (1952).
198. K. Seifert, Faserforsch.u.Textiltech., 4, 342 (1953).
199. K. Kratzl, Monatsh. Chem., 78, 173 (1948); 80, 437 (1949).
200. L. M. Srivastava, Tappi, 49, 173 (1966).
201. E. Hägglund and T. Johnson, Biochem. Z., 187, 98 (1927).
202. T. Nakamura and S. Kitaura, Ind. Eng. Chem., 49, 1388 (1957).
203. T. Nakamura, T. Kitaura, T. Nagaski and S. Kitaura, Repts. Research Inst. Sci. Ind. Kyushu Univ., No. 22, 43 (1957).
204. Y. Hachihama and S. Jyodai, Chemistry of Lignin, Nippon Hyoronsha, Tokyo, 1946, p. 42.
205. T. Pavoline, F. Gambarin and C. Zanini, Ann. Chim. (Rome), 41, 438 (1951).
206. T. Harada and Z. Nikuni, J. Agr. Chem. Soc. Japan, 23, 415 (1950).
207. T. Harada and Z. Nikuni, Mem. Inst. Sci. & Ind. Research Osaka Univ., 7, 163 (1950).
208. J. Nakano, A. Ishizu, A. Sasaki and N. Migita, Nippon Mokuzai Gakkaishi, 8, 144 (1962).
209. J. Nakano, K. Hirakawa, A. Ishizu, A. Kobayashi and N. Migita, ibid., 5, 50 (1959).
210. J. M. Harkin, "Lignin Structure and Reactions," Advances in Chemistry Series, 59, American Chemical Society, Washington, D. C., 1966, p.65.
211. C. Steelink, T. Reid and G. Tollin, J. Am. Chem. Soc., 85, 4048 (1963).

212. C. Steelink and R. E. Hansen, Tetrahedron Letters, 1966, 105.
213. G. Tollin and C. Steelink, Biochem. Biophys. Acta, 112, 377 (1966).
214. A. Merijan and P. D. Gardner, J. Org. Chem., 30, 3965 (1965).
215. K. H. Ekman and J. J. Lindberg, Paperi Puu, 10, 1 (1960).
216. H. N. James and P. A. Tice, Tappi, 48, No. 4, 239 (1964).
217. J. Marton, ibid., 46, 92 (1963); 47, 713 (1964).
218. T. M. Vasileva, G. P. Grigorev and K. P. Mishchenko, Zh. Prikl. Khim., 38, 2757 (1965).
219. E. D. Gelfand and B. D. Bogomolov, Lesnoi Zh., 9, 128 (1966).
220. S. O. Thompson and G. Chesters, Anal. Chem., 36, 655 (1964).
221. M. Krajcinovic, Arch. Kem., 20, 122 (1948).
222. D. C. C. Smith, Nature, 176, 267 (1955).
223. J. Nakano, A. Ishizu and N. Migita, Tappi, 44, 31 (1961).
224. J. Nakano, A. Ishizu and N. Migita, J. Chem. Soc. Japan, 1955, 2347.
225. J. Nakano, A. Ishizu and N. Migita, Nippon Mokuzai Gakkaishi, 4, 1 (1958).
226. E. Hägglund, Chemistry of Wood, Academic Press, N.Y., 1951, p.289.
227. H. Hibbert and C. A. Sankey, Can. J. Research, 4, 110 (1931).
228. K. R. Gray, F. E. Brauns and H. Hibbert, ibid., 13B, 48 (1935).
229. R. G. D. Moore and H. Hibbert, ibid., 14B, 404 (1936).
230. K. Kuerschner and G. Hostomsky, Holzforschung, 16, 180 (1962).
231. E. Adler and J. Marton, Acta Chem. Scand., 15, 357, 370, 384 (1961).
232. I. Falkehag, Paperi Puu, 43, 655 (1961); 46, 92 (1963).
233. G. Aulin-Erdtman, Svensk Papperstidn., 47, 91 (1944); 55, 745 (1952); 56, 91, 287 (1953); 57, 745 (1954); 59, 363 (1956).
234. C. Schuerch, Ind. Eng. Chem. Prod. R. & D., 4, 61 (1965).
235. K. Kratzl and E. Wittman, Monatsh. Chem., 85, 7 (1954).
236. E. Y. Spencer and G. F. Wright, J. Am. Chem. Soc., 63, 2017 (1941).
237. B. D. Bogomolov and O. F. Gorbunov, Lesnoi Zh., 9, 143 (1966).
238. L. Field, P. E. Drummond, P. H. Higgins and E. A. Jones, Tappi, 41, 721 (1958).
239. L. Field, P. E. Drummond and E. A. Jones, Tappi, 41, 727 (1958).
240. J. Gierer and L.-A. Smedman, Acta Chem. Scand., 20, 1769 (1966).
241. Y. V. Nikitin and V. M. Nikitin, Bum. Prom., 5, 12 (1965).
242. O. Samuelson, Svensk Kem. Tidskr., 60, 128 (1948).
243. H. Erdtman, B. O. Lindgren and T. Petterson, Acta Chem. Scand., 4, 228 (1950).
244. H. Erdtman, Svensk Papperstidn., 48, 75 (1945).
245. H. Mikawa, K. Sato, C. Takasaki and K. Ebisawa, Bull. Chem. Soc. Japan, 28, 649 (1955); 29, 209 (1956).
246. H. Erdtman, Svensk Papperstidn., 46, 226 (1943).

247. B. O. Lindgren and U. Saedén, Acta Chem. Scand., 6, 91 (1952).
248. E. W. Eisenbraun, Tappi, 46, 104 (1963).
249. Y. Kojima, A. Hayashi, K. Hirashitsuji and I. Tachi, Kami-pa Gikyoshi, 15, 607 (1961).
250. Y. Kojima, A. Hayashi and I. Tachi, ibid., 15, 652 (1961).
251. K. Hata and K. Nakamura, Kogyo Kagaku Zasshi, 67 (12), 2126 (1964).
252. F. E. Brauns, U. S. Patent 3,297,676 (Jan. 10, 1967).
253. D. H. Grangaard, U. S. Patent 3,251,820 (May 17, 1966).
254. G. R. Quimby and O. Goldschmid, Tappi, 49, 562 (1966).
255. B. Leopold, Svensk Kem. Tidskr., 63, 260 (1951).
256. H. Erdtman, ibid., 53, 201 (1941).
257. H. Erdtman, ibid., 45, 315 (1945).
258. A. Noll, Papierfabr. Wochbl. Papierfabr., 1944, 39.
259. R. Borisek and V. Stanik, Tappi, 41, 188A (1958).
260. L. Jantzen, German Patent 1,199,257 (March 24, 1966).
261. T. Morimoto, U. S. Patent 3,270,002 (Aug. 30, 1966).
262. M. N. Tsypkina and I. M. Balashova, Zhur. Priklad. Khim., 32, 166 (1959).
263. J. Seidl, Chem. Prumsyl., 16, 273 (1966).
264. K. Akagane, G. G. Allan, T. Mattila, A. N. Neogi and W. M. Reif, Nature, 225, 175 (1970).
265. W. J. Greiner, U. S. Patent 3,175,880 (Mar. 30, 1965).
266. R. Carlander and F. J. Elder, Swedish Patent 194,176 (Feb. 2, 1965).
267. G. G. Allan, unpublished research.
268. E. H. Nollau and L. C. Daniels, J. Am. Chem. Soc., 36, 1885 (1914).
269. H. Morren and R. Lehmann, J. Pharm. Belg., 1, 127 (1942).
270. W. H. Moss and B. B. White, Canadian Patent 319,729 (Feb. 9, 1932).
271. K. Kratzl and E. Klein, Monatsh. Chem., 86, 847 (1955).
272. K. Kratzl and W. Schweers, Monatsh. Chem., 85, 1046 (1954); 85, 1166 (1954); Chem. Ber., 89, 186 (1956).
273. V. M. Nikitin and Y. V. Nikitin, Lesnoi Zh., 8, 132 (1965).
274. S. W. Schubert, M. G. Andrus, C. Ludwig, D. Glennie and J. L. McCarthy, Tappi, 50, 186 (1967).
275. F. E. Brauns and H. Hibbert, Can. J. Research, 13B, 28 (1935).
276. D. E. Bland, G. Billek, K. Gruber and K. Kratzl, Holzforschung, 13, 6 (1959).
277. C. Schuerch, J. Am. Chem. Soc., 72, 3838 (1950).
278. J. Marton and E. Adler, Tappi, 46, 92 (1963).
279. K. Sarkanen and C. Schuerch, J. Am. Chem. Soc., 79, 4203 (1957).
280. E. West, A. S. McInnes and H. Hibbert, ibid., 65, 1187 (1943).
281. K. R. Gray, E. G. King, F. E. Brauns and H. Hibbert, Can. J. Research, 13B, 35 (1935).

282. A. A. Konkin and Z. A. Rogovin, Bum. Prom., 28, No. 9, 15 (1953).
283. F. Schutz, Cellulosechem., 19, 23 (1941).
284. L. E. Wise, F. C. Peterson and W. M. Harlow, Ind. Eng. Chem., Anal. Ed., 11, 18 (1939).
285. A. Bailey, Paper Ind. & Paper World, 30, 1196 (1948); 29, 1606 (1948).
286. E. Kondrat'ev, Zhur. Priklad. Khim., 22, 753, 882 (1949).
287. K. Freudenberg and W. Stumpf, Angew. Chem., 62, 537 (1951).
288. D. F. Arseneau and J. M. Pepper, Pulp Paper Mag. Can., 66, T415 (1965).
289. W. Stumpf, W. Weygand and O. A. Grosskinsky, Chem. Ber., 86, 1391 (1953).
290. K. Lundquist, Acta Chem. Scand., 18, 1316 (1964).
291. A. Ogait, Cellulosechem., 22, 15 (1944).
292. L. P. Clermont, Pulp Paper Mag. Can., 63, T402 (1962).
293. L. P. Clermont and F. Bender (to Canadian Patents and Development Ltd.) U. S. Patent 3,218,226 (Nov. 16, 1965).
294. B. D. Bogomolov and O. P. Alekseeva, Izv. Vyssikh Vchebn. Zavedenii Lesn. Zh., 5, 155 (1962); Chem. Abs., 58, 7024 (1963).
295. L. Lemmel, Anales Soc. expañ. fis. quim., 33, 389 (1935).
296. E. Visasoro, Anales Soc. cient argentina, 133, 191 (1942); Anales asoc. quim. argentina, 30, 54 (1952).
297. I. K. Buckland, F. E. Brauns and H. Hibbert, Can. J. Research, 13B, 61 (1935).
298. A. Rieche, Angew. Chem., 66, 132 (1954).
299. A. v. Wacek and H. Daubner-Rettenbacher, Monatsh. Chem., 81, 266 (1951).
300. A. v. Wacek, Intern. Holzmarkt., 20, 50 (1949).
301. F. E. Brauns and D. A. Brauns, The Chemistry of Lignin, Academic Press, New York, 1960, p. 480.
302. M. I. Chudakov and A. V. Milovanov, Gidroliz. i Lesokhim. Prom., 19, 17 (1966).
303. T. Nakamura, T. Kawvano, M. Kawasaki, K. Tominaga, T. Awa and S. Kitaura, Research Rept., Dept. Chem., Div. Agr., Kyushu Univ., No. 12, 43 (1945).
304. A. Kobayashi, T. Haga and K. Sato, Nippon Mokuzai Gakkaishi, 13, 306 (1967).
305. A. Hillmer, Cellulosechem., 6, 169 (1925).
306. K. Nakajima and Y. Hachihama, J. Chem. Soc. Japan, Ind. Chem. Sect., 67, 838 (1964).
307. M. G. Okun, M. I. Chudakov, I. V. Sokolova and P. I. Kantor, Sb. Tr. Vses Nauch.-Issled Inst. Gidroliz. i Sulfitno-Spirt. Prom., 13, 254 (1965).
308. A. Reiche and L. Redinger, East German Patent 32,605 (Nov. 15, 1964)

309. P. Lipsitz and R. M. Lollar, J. Am. Leather Chemists' Assoc., 46, 301 (1951).
310. N. A. Rozenberger, Bum. Prom., 2, 3 (1967).
311. H. Zollinger, Chem. & Ind., 1965, 891.
312. M. Ya Zarubin and D. V. Tishchenko, Zh. Prikl. Khim., 35, 2724 (1962).
313. R. Z. Pen and I. L. Shapiro, ibid., 39, 1668 (1966).
314. H. Erdtman, Svensk Papperstidn., 48, 217 (1945); Research (London), 3, 63 (1950).
315. J. Wiesner, Sitzber Akad. Wiss. Wien, Math.-naturw. Klasse, 17, 511 (1878).
316. T. Nakamura and S. Kitaura, Repts. Research Inst. Sci. & Ind., Kyushu Univ., No. 8, 1 (1952).
317. Y. Ujioke, T. Nakamura, T. Awa and S. Kitaura, ibid., No. 4, 12 (1951).
318. E. Hägglund and T. Johnson, Biochem. Z., 187, 98 (1927).
319. B. Holmberg, Chem. Ber., 69, 115 (1936).
320. E. Adler, B. O. Lindgren and U. Saedén, Svensk Papperstidn., 55, 245 (1952).
321. B. O. Lindgren and H. Mikawa, Acta Chem. Scand., 8, 954 (1954).
322. Y. Hachihama and S. Jyodai, J. Chem. Soc. Japan, Ind. Chem. Sect., 52, 386 (1949).
323. Y. Hachihama, S. Jyodai, H. Sato, E. Fukui and T. Nakamura, Mem. Inst. Sci. and Ind. Research Osaka Univ., 7, 167 (1950).
324. H. Hachihama and Y. Kyogoku, J. Chem. Soc. Japan, Ind. Chem. Sect., 56, 208 (1953).
325. B. Holmberg, Chem. Ber., 69, 115 (1936); 75, 1760 (1942).
326. B. Holmberg, Öster. Chem.-Ztg., 43, 152 (1940).
327. B. Holmberg, Ark. Kemi, Mineral. Geol., 21, No. 10 (1945); 24, No. 29 (1947).
328. F. E. Brauns and M. A. Buchanan, Paper Trade J., 122, 49 (1946).
329. F. E. Brauns and W. H. Lane, ibid., 122, 38 (1946).
330. C. Dreyfus and G. Schnieder, Can. Patent 325,064 (to Camille Dreyfus) (Aug. 9, 1932).
331. I. D. Douglas and T. B. Johnson, J. Am. Chem. Soc., 60, 486 (1936).
332. H. Mikawa and K. Sato, J. Chem. Soc. Japan, Ind. Eng. Sect., 61, 1078 (1958).
333. J. H. Carroll and H. C. Wallin, Can. Patent 707,382 (Apr. 6, 1965).
334. F. J. Ball, J. B. Doughty and W. G. Vardell, U. S. Patent 3,185,654 (May 25, 1965).
335. J. Marton, E. Adler, T. Marton and S. I. Falkehag, "Lignin Structure and Reactions," Advances in Chemistry Series, 59, 125 (1966).
336. H. Mikawa, E. Nokihara and K. Sato, J. Chem. Soc. Japan, Ind. Chem. Sect., 53, 94 (1950); 53, 134 (1950).

337. G. G. Allan and D. D. Halabisky, Pulp & Paper Mag.Can., 71, T50 (1970).
338. J. H. Ross and A. C. Hill, ibid., 27, 541 (1929).
339. A. A. Sokolova, L. A. Semakova and E. V. Nazareva, Izv.Vysshikh Uchebn. Zavedenii, Lesn. Zh., 8, 153 (1965).
340. H. Krauch and W. Kunz, Organic Name Reactions, John Wiley & Sons, Inc., New York, 1964, p. 455.
341. R. W. Martin, The Chemistry of Phenolic Resins, John Wiley & Sons, Inc., New York, 1956.
342. Y. A. Gugnin, Lesnoi Zh., 9, 127 (1966).
343. A. Kobayashi, T. Haga and K. Sato, Nippon Mokuzai Gakkaishi, 13, 312 (1967).
344. A. D. Ivanenko and V. M. Nikitin, Lesnoi Zh., 9, 125 (1966).
345. H. Mikawa, K. Sato, C. Takasaki and K. Ebisawa, Bull. Chem. Soc. Japan, 29, 259 (1956).
346. C. Mannich and W. Krösche, Arch. Pharm., 350, 647 (1912).
347. A. Ishizu, J. Nakano, T. Ogino and N. Migita, Nippon Mokuzai Gakkaishi, 11, 53 (1965).
348. M. J. Borgfeldt (to California Research Corp.) U.S. Patents 3,123,569 (Mar. 3, 1964) and 3,126,350 (Mar. 24, 1964).
349. F. Gaslini, Tappi, 41, 162A (1958).
350. R. Borisek, Chem. Zvesti, 5, 338 (1951).
351. M. V. Ionescu and V. N. Georgescu, Bull. Soc. Chim. France (4), 41, 692 (1927).
352. J. Altpeter, Das Hexamethylenetetramine und Seine Verwendung, Halle, Verlag von Wilhelm Knapp, 1931, pp.80-83.
353. A. A. Sokolova and R. S. Zhdanova, U.S.S.R. Patent No. 175,657 (1965).
354. H. Moschatos and B. Tollens, Ann., 272, 250 (1893).
355. M. T. Harvey and L. H. Baekeland, Ind. Eng. Chem., 13, 135 (1921).
356. H. Lebach, Angew. Chem., 22, 1600 (1909).
357. J. F. Walker, Formaldehyde, Reinhold Publishing Corp., New York, 1964, p. 542.
358. C. Ellis, The Chemistry of Synthetic Resins, Reinhold Publishing Corp., New York, 1935, pp. 307-310.
359. J. Gratzl and W. Lonsky, Private communication.
360. K. Freudenberg, F. Sohns, W. Durr and C. Niemann, Cellulosechem., 12, 263 (1931).
361. K. R. Gray, F. E. Brauns, A. Hibbert, Can. J. Research, 13B, 48 (1935).
362. J. Polcin, Holzforschung, 20, 121 (1966).
363. K. Freudenberg and H. F. Müller, Chem. Ber., 71, 1821 (1938).
364. W. Lautsch and G. Piazolo, ibid., 73, 317 (1940).
365. W. Küster and R. Daur, Cellulosechem., 11, 4 (1930).

MODIFICATION REACTIONS

366. K. Schwabe and E. Preu, ibid., 21, 1 (1943).
367. J. Polcin, Sbornek Vyzkumnych Praci Z.Oburv Celulosy a Papiru, 3, 198 (1958); Chem. Abs., 54, 13650 (1960).
368. T. M. Kroshilova and V. M. Nikitin, Isv Vysshikh Uchebn. Zavedenii, Lesn. Zh., 7, 147 (1964); 8, 124 (1965).
369. A. M. Petrova, Tr. Tomskogo Gos. Univ., Ser.Khim., 170, 58 (1964).
370. V. M. Nikitin, Dokl. Akad. Nauk SSR, 160, 359 (1965).
371. V. M. Nikitin, U.S.S.R. Patents 161,775 and 161,776 (Aug.31,1962).
372. T. M. Kroshilova and V. M. Nikitin, Bum. Prom., 2, 3 (1965).
373. N. N. Shorygina, I. V. Izumrudova, A. P. Khovanskaya, V. D. Tychina and N. P. Mikkailov, Nauch Trudy Leningrad. Lesotekh. Akad., No. 91, 2, 211 (1961).
374. E. Adler, J. M. Pepper and E. Erickson, Ind. Eng. Chem., 49, 1391 (1957).
375. J. Gierer, Chem. Ber., 89, 257 (1956).
376. R. Willstätter and L. Zechmeister, ibid., 46, 2401 (1913).
377. K. Kratzl and H. Silbernagel, Monatsh. Chem., 83, 1022 (1952).
378. A. J. Corey and O. Maass, Can. J. Research, 13B, 149 (1935).
379. Y. Kyogoku, I. Suzuki and Y. Hachihama, Koygo Kagaku Zasshi, 65, 96 (1962).
380. O. C. Bockman, Norsk. Skogind., 16, 320 (1962).
381. M. Lacan and D. Matasovic, Kem. Ind. (Zagreb), 14, 933 (1965).
382. H. Richtzenhain, Svensk Papperstidn., 53, 644 (1950).
383. D. E. Read and C. B. Purves, J. Am. Chem. Soc., 74, 120 (1952).
384. G. Brunow, Finska Kemists. Medd., 74, 20 (1965).
385. M. I. Chudakov, M. G. Okuń, A. P. Samsonova, M. N. Raskin and Y. S. Pilipchuk, Sb. Tr.,Vses. Nauch -Issled. Inst. Gidroliz. i Sulfitno-Spirt. Prom., 14, 238 (1965).
386. M. I. Chudakov, G. D. Georgievskaya, A. V. Antipova and I. V. Sokolova, ibid., 13, 227 (1965).
387. M. I. Chudakov, Doklady Akad. Nauk SSR, 137, 1389 (1961).
388. M. I. Chudakov, U. S. S. R. Patent 134,084 (Jan. 10, 1961).
389. M. I. Chudakov, Ref. Zh. Khim., 16, 521 (1963).
390. E. Adler and K. Lundquist, Acta Chem. Scand., 17, 13 (1963).
391. K. Freudenberg and F. Bittner, Chem. Ber., 86, 155 (1953).
392. K. Freudenberg and F. Niederkorn, ibid., 89, 2168 (1956).
393. K. H. Ekman, Tappi, 48, 398 (1965).
394. C. Schuerch, Forest Products Journal, 8, 150 (1958).
395. H. G. Arlt, Jr., S. K. Gross and C. Schuerch, Tappi, 41, 64 (1958).
396. I. Sobolev and C. Schuerch, ibid., 41, 447, 545 (1958).
397. A. Bhattacharya, E. Sondheimer and C. Schuerch, ibid., 42, 446 (1959).
398. A. Bhattacharya and C. Schuerch, ibid., 43, 360 (1960).
399. H. Adkins, R. L. Frank and E. S. Bloom, J. Am. Chem. Soc., 63,

549 (1941).
400. T. Enkvist, Tappi, 37, 350 (1954).
401. T. Enkvist, M. Moilanen and B. Alfredsson, Svensk Papperstidn., 52, 517 (1949).
402. T. N. Kleinert, Tappi, 48, 447 (1965).
403. T. N. Kleinert, Papier, 23, 135 (1969).
404. T. N. Kleinert, Holzforsch. Holzverwert., 19, 60 (1967).
405. W. Haage, German Patent 228,721 (1908).
406. B. K. Hough and J. C. Smith, Chem. Eng. News, 30, 74 (1952).
407. R. G. Banner and J. H. Pierce, U. S. Patents 2,995,415 and 3,039,841 (to Diamond Alkali Co.) (May 4, 1960 and June 19, 1962).
408. E. Aaltio and R. H. Roschier, Paperi Puu, 29, 2 (1957).
409. R. H. Twining, Can. Patent 492,647 (1953); U.S. Patent 2,874,545 (to L. H. Lincoln & Son, Inc.) (Feb. 24, 1959).
410. M. Akahane and H. Yano, Kogyo Kagaku Zasshi, 66, (10), 1533 (1963).
411. D. L. Beyer and I. A. Pearl, Tappi, 42, 800 (1959).
412. A. N. James and P. A. Tice, ibid., 47, 43 (1964).
413. K. Elbs and K. Lerch, J. prakt. Chem., 93, 1 (1916).
414. H. Tanaka, H. Abe and R. Senju, J. Chem. Soc. Japan, Ind. Chem. Sect., 69, 1968 (1966).
415. D. H. R. Barton, A. M. Deflorin and O. E. Edwards, J. Chem. Soc., 1956, 530.
416. D. H. R. Barton and G. W. Kirby, J. Chem. Soc., 1962, 806.
417. H. Finkbeiner, A. S. Hay, H. S. Blanchard and G. F. Endres, J. Org. Chem., 31, 549 (1966).
418. T. A. Davidson and A. I. Scott, J. Chem. Soc., 1961, 4075.
419. W. I. Taylor and A. R. Battersby, Oxidative Coupling of Phenols, Marcel Dekker, Inc., New York, 1967.
420. H. Lee, D. Stoffey and K. Neville, New Linear Polymers, McGraw-Hill, Inc., New York, 1967, p. 65.
421. R. Kunin, Ion Exchange Resins, J. Wiley & Sons, Inc., New York, 1958, Second Edition, p. 23.
422. J. D. Ferry, Viscoelastic Properties of Polymers, John Wiley & Sons, Inc., London, 1961, p. 392.
423. K. H. Gustavson, Chemistry of Tanning Processes, Academic Press, New York, 1956.
424. K. H. Gustavson, J. Soc. Leather Tr. Chem., 34, 259 (1950).
425. A. Hayashi and D. A. I. Goring, Pulp Paper Mag., Can., 66, (3) T154 (1965).
426. A. Hayashi, Nippon Mokuzai Gakkaishi, 11, 158; 212; 218 (1965).
427. A. Hayashi and Y. Namura, Bull. Chem. Soc. Japan, 38, 572 (1965).
428. G. G. Allan, P. Mauranen, A. N. Neogi and C. Peet, Chem. & Ind., 1969, 623.

429. G. G. Allan, P. Mauranen, M. D. Desert and W. M. Reif, Paperi Puu, 50, 539 (1968).
430. F. J. Allan, G. G. Allan, T. Mattila and P. Mauranen, Acta Chem. Scand., 23, 1903 (1969).
431. T. Koshijima, E. Muraki, K. Naito and K. Adachi, Nippon Mokuzai Gakkaishi, 14, 52 (1968).
432. T. Koshijima and E. Muraki, ibid., 10, 110 (1964).
433. E. J. Lawton, W. D. Bellamy, R. E. Hungate, M. P. Bryant and E. Hall, Science, 113, 380 (1951).
434. Y. Hachihama and S. Jyodai, Mem. Inst. Sci. and Ind. Research Osaka Univ., 6, 74 (1948).
435. A. Chapiro, Radiation Chemistry on Polymeric Systems, High Polymers Vol. XV, Interscience Publishers, John Wiley & Sons, New York, p. 82, 93, 173.
436. T. Koshijima and E. Muraki, Nippon Mokuzai Gakkaishi, 12, 139 (1966).
437. T. Koshijima and E. Muraki, ibid., 13, 355 (1967).
438. T. Koshijima and E. Muraki, ibid., 10, 116 (1964).
439. T. Koshijima, ibid., 12, 144 (1966).

14

HIGH ENERGY DEGRADATION
G. Graham Allan and Tapio Mattila

I. Introduction. This chapter reviews the degradative decomposition of lignin under conditions in which the energy input is high in comparison with conventional organic reactions. Such energetic reactions are characteristically solvent-free systems and include pyrolysis and fusion in high melting media. High temperature degradations in which water is an added component are also appropriately included in this classification. The effect of γ-radiation, however, is discussed in Chapter 13.

II. Thermal Decomposition of Lignin. The original pyrolysis of wood including lignin as a component is lost in the mists of antiquity although the ancient Egyptians seem to have been among the first applied wood chemists in their use of the pyrolysis products to embalm the dead. Of course, the thermal decomposition of wood to produce charcoal has been extensively practiced in more recent eras on an industrial scale and has been fairly intensively studied (1, 2). In its response to thermolysis (3,4), wood behaves approximately as if it were a mixture of cellulose, hemicellulose and lignin, and the products obtained from purely cellulosic materials have been the subject of a number of modern investigations (5-7), particularly with reference to fire control (8) and radiation damage (9).

This situation has not prevailed in the case of lignin where the bulk of the reported research was carried out at least a quarter of a century ago (10). During that period, considerable efforts were expended to in effect thermally depolymerize lignin, presumably to obtain useful materials. Although these experiments were carried out on softwood, hardwood and grass lignins isolated by a variety of techniques (11-24), the products are remarkably similar.

A. The Nature of the Pyrolysis Residue. The principal product of pyrolysis is coke, which is obtained in about 55% yield. Chemical characterization of this carbonaceous residue has generally not been attempted. An exception is provided by the work of Gillet and Urlings (25), where a Brauns lignin and an ethanolysis lignin were pyrolyzed under nitrogen for one hour at 550°. The residues obtained had analytical compositions analogous to those found for a Rhenish lignite and contained carbon, hydrogen and oxygen in amounts consistent with the formula $(C_{40}H_{38}O_{11})_2$. The authors expressed their ideas of a possible chemical composition in the form of the speculative structure 1.

However, Brauns and Brauns pointed out (26a) that such a lignite would not be expected to have a methoxyl content corresponding to the 18% required by 1. Moreover, the vicinal diol moieties, with their constituent benzylic hydroxyl functions, would scarcely be expected to survive a pyrolysis at 550°. Nonetheless, these carbonaceous residues may very well have highly condensed structures of this general type, with the oxygen as an integral part rather than as a functional group.

B. The Composition of the Aqueous Distillate. The second major component of the pyrolysis of lignin is the aqueous distillate, which corresponds to the classical pyroligneous acid fraction of whole wood distillation. The yield of this aqueous distillate varies rather more widely than the coke yield but approaches 20% of the lignin charged in most cases. The main constituents, in addition to water, are methanol, acetone and acetic acid. The methanol, which in the case of softwood lignins amounts to about 1% of the lignin charged, presumably originates from the methoxyl groups in lignin (27). In support of this view, hardwood lignins (21) that contain dimethoxylated syringyl moieties give about twice the yield of methanol afforded by softwood lignins (15, 19). Also, the yields of acetic acid from hardwood lignins (21) are significantly higher than those isolated from softwood lignins (19). Assuming that the acetic acid is derived by disruption of the propanoid side-chains, then two factors which may contribute to the difference in the amount of isolable acid are: (a) the lower molecular weight of the hardwood lignin which should facilitate thermolysis in preference to internal condensation and (b) presence of the additional methoxyl groups at position 5 which should be capable of functioning as blocking substituents, thus further reducing internal condensation. Acetone amounts to less than one-tenth of the combined quantity of methanol and acetic acid, both in the case of angiosperm and gymnosperm lignins. The acetone may not only originate in the propanoid side-chains, but could also be formed by the thermal decomposition of the acetic acid since good yields of this ketone can be obtained by passing the vapor of the acid over heated surfaces such as barium carbonate or pumice stone (28). Formic acid has been only occasionally isolated (29,30) from aqueous distillates. Possibly it is formed from the decomposition of oxalic acid originating from carbohydrates associated with the lignin preparation. The material used by Fletcher and Harris was probably in this category since it was obtained from Douglas-fir using the Madison wood sugar process (29, 30).

In addition to the aliphatic compounds regularly found in the aqueous distillate, aromatic compounds such as catechol have been occasionally recognized (11, 29, 31), but this is likely a consequence of a steam distillation effect and such aromatics are more properly considered as part of the tar fraction.

C. <u>The Constituents of the Thermolysis Tar</u>. This particular tarry thermolysis product is potentially the most valuable and is usually obtained in about 15% yield based on the lignin charged. Somewhat higher yields (22%) can be secured by continuous reactor systems (32, 33). It contains a phalanx of phenolic compounds, none of which occurs in a predominating quantity. The preponderance of these compounds are closely related to phenol, guaiacol and 2,6-dimethoxyphenol and carry substituents which are usually located in the position para to the hydroxyl group and which may be methyl (22, 29, 31, 34, 35), ethyl (22, 29, 34, 35), propyl (22, 29, 34, 35), isopropyl (22), vinyl (22, 31, 35), allyl (14, 22, 35-37), propenyl (22, 35, 38) carboxyl (22, 39, 40), carboxymethyl (31, 38) or carboxyaldehyde (20, 39, 40).

It is noteworthy that the side-chains consist of three carbon atoms or less and are almost invariably in the <u>para</u> position to the phenolic hydroxyl site. Thus, in more specific terms, Fletcher and Harris (29, 30) isolated from a softwood tar the compounds 2 through 8, while Phillips and Goss (31) were also able to isolate 2, 3 and 5 as well as the less degraded phenols exemplified by 9 and 10 from a destructive distillation of an alkali lignin from corncobs.

This type of experiment has been updated (35, 41, 42) using modern analytical instrumentation. With a platinum spiral pyrolysator coupled to a vapor phase chromatographic unit lignin samples can be conveniently examined, and indeed Kratzl, Czepel and Gratzl have shown that the technique can be made quantitative enough for repetitive analytical purposes (35).

Application of this procedure to a variety of lignins afforded guaiacol and methylguaiacol as the major products although the previously isolated phenols were again observed in minor amount. Carboxylic acids, however, were not detected, and this is attributable to the nonoxidizing pyrolysis atmosphere of nitrogen and to the decarboxylation of carboxylic acids originally present in the lignin substrate. Moreover, since under these pyrolysis conditions only hardwood-derived lignins afforded pyrogallol derivatives, the method may be used to diagnose the botanical origin of a specimen.

It is to be emphasized that many of the thermolysis tars were obtained from impure or prehydrolyzed lignins, and their composition will therefore differ in quantity and quality from the components obtainable from the tars of MWL, for example.

It may be concluded, therefore, that a distinctive difference exists between the energy barriers of demethoxylation and rupture of carbon-carbon bonds in the side-chain. Thus, destructive distillation of lignin at 400-450° primarily causes intrachain cleavage, and rapid pyrolysis at 900° also yields methoxyl containing phenols (35). That is, pyrolysis of guaiacyl and syringyl lignin units, unlike hydrolysis, does not furnish catechol or pyrogallol derivatives. Where catechol has been isolated, its formation is a result of lignin hydrolysis which has occurred during an acidic pretreatment such as in wood saccharification (30) or pulping (43). These conclusions are validated by model compound studies in which anisole, veratrole, 4-methylguaiacol and 2,6-dimethoxyphenol underwent mainly demethoxylation on pyrolysis. Demethylation occurred to a minor extent only (44) and is probably a consequence of hydrolytic reactions since the addition of water vapor substantially increased the amount of demethylated material. However, complete demethylation does not readily occur, and syringyl units are either not converted into pyrogallol or this trihydric phenol is too unstable to survive as an isolable species. The mechanism of lignin pyrolysis can therefore be primarily regarded as homolytic cleavage with a very minor hydrolytic cleavage contribution occasioned by adventitious water, perhaps formed by dehydration during the earlier stages of heating. The homolytic fissions will, of course, generate free radicals, although a lack of bond energy data does not permit an assignment of their location with increasing energy inputs. Nonetheless, it is clear that phenyl radicals will be relatively easily formed, and these will be capable of participation in chain transfer processes or combination with methyl radicals to afford the products $\underline{2}$, $\underline{3}$, $\underline{4}$ and $\underline{5}$, for example.

The various separations of the components of the thermolysis tar are not all generally well described, but the quantities of various constituents are probably rather akin to the figures recorded in the idealized Table 14.1, which is derived from data obtained for the vacuum distillation of lignin (20).

Table 14.1 Composition of Tar from the Vacuum Distillation of Lignin (20)

Component	% of Tar
Alkali insoluble oil	8
Bisulfite soluble compounds	2
Ether insoluble phenols	13
Ether soluble phenols	25
Ether insoluble carboxylic acids	26
Ether soluble carboxylic acids	26

D. Gaseous Pyrolysis Products. The fourth and final component of the pyrolysis of lignin consists of gases which amount to about 12% of the lignin charged. These comprise, in descending order of abundance, carbon monoxide, methane, carbon dioxide and ethane. The yields of these individual bases from various lignin sources have not often been determined, but they presumably parallel the approximate yields obtained from spruce HCl lignin (15), which are collected in Table 14.2. The high content of methane is noteworthy (32, 34, 45), since it indicates that demethanation may be a significant reaction. If sulfidic sulfur is present in the lignin, methyl mercaptan may also be formed (16-18).

Table 14.2 Composition of Gas from Dry Distillation of Spruce HCl Lignin (15)

Constituent	% of Gas
Carbon monoxide	50
Methane	38
Carbon dioxide	10
Ethane	2

E. Effect of Additives on the Pyrolysis. In attempts to improve the destructive distillation process, a number of additives have been admixed with the lignins. These have comprised either acidic and basic salts or oxides or elemental metals such as silver (46), nickel (38), zinc (47) or palladium (48) intended presumably to function as dehydrogenation catalysts. The effect of the metals cannot be claimed to have been especially exciting although Karrer and Bodding-Wiger (47) were able to obtain a slightly increased yield of tar (17%) by distillation of a hydrochloric acid spruce lignin which had been admixed with zinc dust. A similar slight improvement in yield, using the same metal catalyst, was obtained by Phillips in his studies on alkali lignin from corncobs (49, 50), but the products were apparently not different from those obtained without the use of the metal. The use of silver (46) to promote degradation of the lignin resulted in the isolation of hydroxymethyfurfural and melene. The former must be a degradation product of contaminating hexosans, but the latter is more of an enigma. Melene, which has the formula $C_{30}H_{60}$, is one of the highest alkenes known (51, 52), and it has also been obtained by the pyrolysis of beeswax and by the vacuum distillation of lignite, Montrambert coal and Galician petroleum.

An unusual pyrolysis product, pentamethylphenol 11, is also obtained when the so-called Rauma lignin is treated with methanol and calcium oxide at elevated temperatures and under high pressure (53, 54). Rauma lignin, itself named geographically after the town of Rauma in Finland, is prepared from a lignosulfonate or spent sulfite liquor by desulfonation with lime. The raw yield is 19% of the lignin charged, which falls to 7% as the phenol is purified. Oxidation of pentamethylphenol 11 yields duroquinone 12, a potentially valuable difunctional molecule.

F. Development of Technical Thermolytic Processes. Attention to thermolytic degradation of lignin has been recently renewed in the U.S.S.R., where the utilization of the lignin by-product of wood saccharification industry is the subject of significant research efforts.

The lignin starting material used in these investigations is usually a hydrolysis lignin which may be derived either from wood or cotton hulls. The experimental procedure employs anthracene oil as a swelling and heat transfer medium in which the lignin is suspended. Petroleum oils are not so effective, and this can probably be attributed to the lack of nitrogenous polycyclic bases which are undoubtedly present in the anthracene oil and which are known to be excellent lignin solvents.

Heating of the lignin suspension under reduced pressure (50 mm Hg) to 440° for about two hours affords a 25% yield of liquid products based on the weight of lignin charged. This material is reported to consist of about 11% acids, 40% phenols, 42% neutral material and 7% alkaline substances. The alkaline substances must obviously be derived from the pyridine-type components of the anthracene oil. The composition of the phenolic mixture closely parallels those obtained in the much earlier German and American thermolyses with guaiacol as the major product (36%). These modern studies have apparently reached a miniature pilot plant stage where briquets of a lignin-anthracene oil mixture are heated for 90 minutes at 480° under a reduced pressure of 260 mm. These drastic and prolonged conditions afford an activated carbon and a yield of phenols of only 7%.

Other Russian researchers (55-58) using superheated steam at 500° have degraded lignin from wood saccharification to afford 10% of its weight of a complex tar of which less than half of the components are phenols. The figures given indicate that in a continuous plant one ton of lignin yields about 80 pounds of phenols, 40 pounds of "terpinaceous light oils," 500 pounds of activated carbon and 10,000 cubic feet of gas usable as fuel.

G. Thermogravimetric Analysis. Some efforts to study the pyrolysis of lignin on a kinetic basis have been included in programs primarily concerned with the flammability of wood (59, 60). Thus, for example, Tang and Eickner (61, 62) have reported dynamic thermogravimetric and differential thermal analyses which show that a lignin derived from spruce by treatment with sulfuric acid pyrolyzed more slowly than wood (Figures 14.1 and 14.2). The existence of a dual segment mechanism is assumed (Figure 14.3), and kinetic parameters are calculated for a variety of pyrolytic conditions.

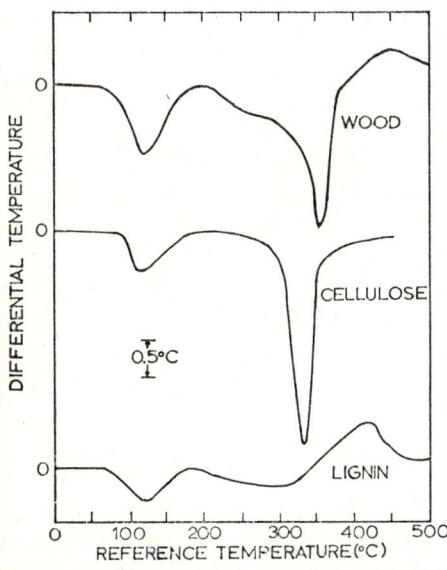

Figure 14.1. DTA thermograms of pyrolysis of untreated wood, cellulose and lignin corrected for baseline drift. Weight and form of samples in stainless steel capsules: Wood, 100 milligrams, 40-mesh ponderosa pine particles; Cellulose, 100 milligrams powdered Whatman cellulose; Lignin, 100 milligrams powdered (sulfuric acid processed spruce); Heating rate, 12°C per minute by conduction; Atmosphere, helium flow 5 milliliters per minute at 27° C atmospheric pressure (62).

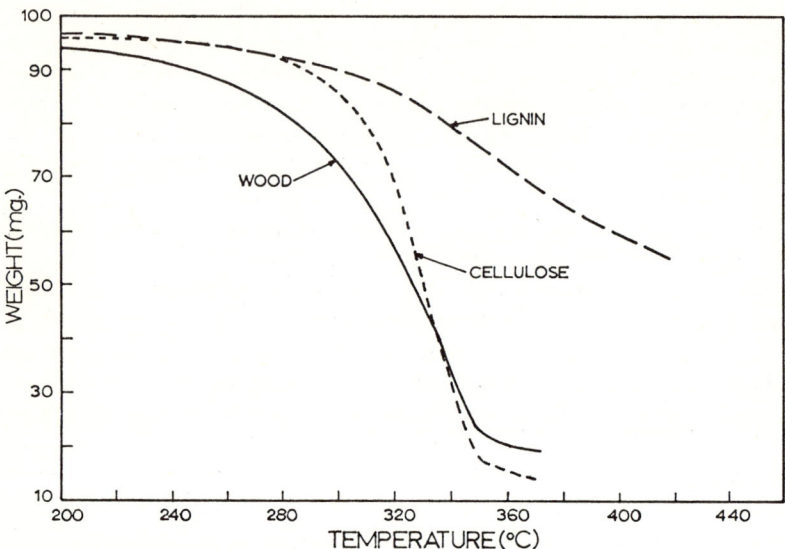

Figure 14.2. Relationship of increased temperatures to weight of untreated wood, cellulose and lignin (62).

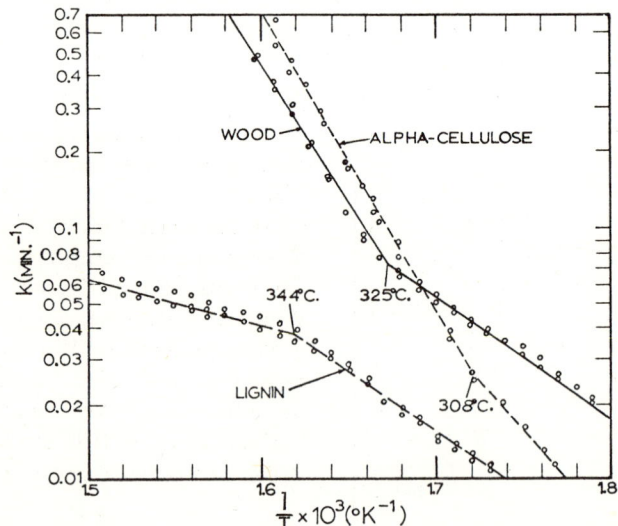

Figure 14.3. Arrhenius plot for pseudo first-order kinetics of pyrolysis of wood, alphacellulose and lignin with no treatment, from data of dynamic thermogravimetric analyses and rate of weight loss (62).

Sanderman and Augustin have previously suggested that during the heating of wood the lignin undergoes condensation to a thermally more stable form. The isolated lignin used by Tang and Eickner was, of course, highly condensed in comparison with the lignin in the wood itself, and this is reflected in the general stability of the isolated lignin which was only 27% volatilized at 360°. Under the same conditions the wood was 70% volatilized. The reduced amount of volatiles obtained from the lignin is attributed to the relative intricacy of the aromatic polymer in contrast to the simpler linear structure of the cellulose. Because of this unfortunate choice of a highly artificial condensed lignin, it is questionable if the pyrolysis data established are applicable to the less condensed pulping waste lignins normally of interest in lignin utilization research. However, the conclusion that the heat of pyrolysis is initially endothermic and later becomes exothermic is probably valid although more highly oxygenated lignins will require a greater initial heat input to sustain the preliminary endothermic dehydrations. This view is consistent with the findings that the effects of addition of various fire retardants on the pyrolytic behavior of the sulfuric acid lignin were rather minimal since such salts primarily function by dehydrative mechanisms, and the sulfuric acid lignin is already substantially dehydrated. Moreover, the pyrolytic temperatures employed did not exceed 500°, and as a consequence of the equipment used, these were attained only slowly. These are not conditions conducive to the development of a facile thermal depolymerization of lignin.

In a recent, more extensive comparison (63) the thermograms of milled wood-, dioxan- and Klason lignins were determined in a nitrogen atmosphere (Figure 14.4). All showed an endothermic minimum at 100-200° and a broad exothermic maximum at 390-420°. However, in air the minimum disappeared and maxima at 270-290 and 400-420° were recorded (59, 63, 64). There is also the possibility that lignin may react with alumina diluent often used in thermogravimetric work (65). Lignin appears to be rather sensitive to oxidation within these regions, while in the temperature range of 310-420° the greatest yields of gas and distillate occur (64). Considering the data in both atmospheres, the maximum around 400° seems to indicate the cleavage of carbon to carbon bonds (66). Unfortunately, thermodynamic data on the scission of carbon-oxygen bonds in demethoxylation is scarce and obscure, and possible cleavage temperatures ranging from 225 to 600° (32, 34, 67, 68) can equally well be rationally selected.

H. Fast Cracking. At the other end of the experimental scale, Goheen and Henderson (69) have studied the pyrolysis of a variety of lignins at very high temperatures for short periods of time. This research was apparently motivated by a concern over the neglect of our replenishible resources as typified by cellulose and lignin in relation to the world's finite supply of crude petroleum. Taking an estimate of wood waste of about 300 million tons per year, these investigators have justified efforts to convert wood and lignin to acetylene and have achieved a considerable measure of technical success (70).

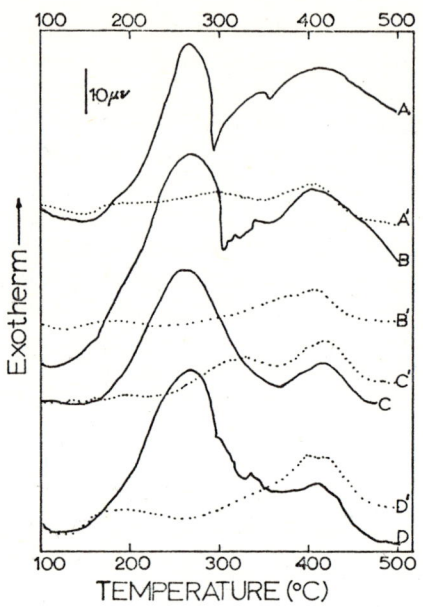

Figure 14.4. Thermograms of various isolated lignins. A (—): MWL of Buna (Fagus sp.) in air, A' (····): in nitrogen; B (—): Dioxan lignin of Shinanok (Tilia sp.) in air, B' (····): in nitrogen; C (—): MWL of Sugi (Cryptomeria sp.) in air, C' (····) in nitrogen; D (—): Dioxan lignin of Hinoki (Chamaecyparis sp.) in air, D' (····): in nitrogen (63).

In the first qualitative experiments, simply blowing a suspension of dry powdered lignin in nitrogen through an electric arc afforded a mixture of gases comprising carbon monoxide and dioxide, hydrogen cyanide and acetylene. The formation of the hydrogen cyanide was attributed to the endothermic reaction of carbon and hydrogen with the nitrogen sweep gas and was not further explored. Quantitative experiments using helium as the sweep gas gave acetylene in 14% yield based on the lignin charged. Other types of lignin and also wood and cellulose were also tried in subsequent experiments, and the results secured are collected in Table 14.3.

These experiments indicated that lignin, wood and cellulose gave unexpectedly large yields of acetylene when heated to high temperature. It had been previously thought that the presence of oxygen already combined with the carbon and hydrogen in the lignocellulosic material would prevent the formation of acetylene in any significant amount. By replacing the simple arc with a tungsten coil at 2,000-2,500° to achieve better temperature control, the

Table 14.3 Yields of Acetylene from Electric Arc Treatments (70)

Material Pyrolyzed	Wt. of Charge, grams	Volume of Product Gas, ml	% C$_2$H$_2$ in the Gaseous Product	Yield of C$_2$H$_2$ as % of Charge
Precipitated kraft lignin	0.4	1900	2.5	14
Desugared calcium base spent liquor solids	0.6	1600*	0.3	0.9
Desulfonated calcium base spent liquor solids	0.4	1750	2.5	13
Lignocarbon from heating calcium base spent liquor with acid	0.4	1600*	0.8	4
Powdered Douglas-fir wood	0.4	1700	2.5	13
Powdered cellulose	0.5	1800	3.5	15

*Gases contained a considerable amount of hydrogen sulfide.

yield of acetylene was raised to 23%. In all these experiments acetylene was the major organic product, and only small amounts of methane and ethylene were co-produced. A variety of further attempts to improve the process, including the use of carbon monoxide as the sweep gas and temperatures of up to 3,000° obtained using R.F. heating, did not raise the yield of acetylene but instead led to the formation of 15-20% methane and 2-7% ethylene. These gases were shown to be formed as a consequence of the increasing instability of acetylene as its concentration in the gas phase rose to above 2.5%. Since theoretical calculations indicated that the yield of acetylene could be as much as 40% of lignocellulosic material charged, a small pilot plant was built in the expectation that some of the problems associated with laboratory equipment would vanish. After a series of designs, yields of the acetylene actually formed from the sawdust charged were indeed about 40%, but concurrent decomposition within the reactor reduced the concentration of the acetylene in the product stream to 12% in association with 2% ethylene.

III. Alkali Fusion of Lignin.
A. Formation of Protocatechuic Acid, Catechol and Oxalic Acid.
Fusion with alkali, a classical technique of organic chemistry, was first applied to lignin in wood over a hundred years ago (71). These early studies inspired many investigators to study the separate roles of cellulose and lignin, and Erdmann (72) in 1866 was apparently the first to use an isolated lignin. The degradation products isolated were protocatechuic acid 13 and catechol 14

and in spite of subsequent, more sophisticated studies (73-75), this modest list of aromatic fusion products has not been extended. Thus, Hägglund (76, 77) and later Heuser, Rösch and Gunkel (78) were able to confirm protocatechuic acid as a degradation product of a carbohydrate-free hydrochloric acid spruce lignin. In more detailed investigation of the yields obtainable, Heuser and Winsvold (79-81) demonstrated that a secondary oxidation to oxalic acid occurred. This diminution of the yield of aromatic components is significantly affected by the oxygen available in the fusion reactor. With the exclusion of oxygen, the secondary formation of oxalic acid, amounting to about 20% is almost completely suppressed. Protocatechuic acid and catechol were then obtained in yields of 19 and 9%, respectively. The oxalic acid apparently originates almost entirely from the catechol since on admission of oxygen its yield becomes trivial while that of the protocatechuic acid is substantially unchanged. This difference in oxidation susceptibility probably is expressed by the classical deactivation of an aromatic ring by an electron-withdrawing group reinforced by the increased anionic character due to the carboxylate function.

This early work also disclosed the importance of the nature of the material of construction of the fusion reactor and the fortuitous use of iron crucibles nearly tripled the yield of catechol. No oxalic or protocatechuic acid could be isolated under these conditions. The 23% catechol actually isolated originates in part from the protocatechuic acid which is decarboxylated in this particular reaction media. Catechol is probably also at least partly decomposed during the fusion period, but it may be stabilized to some extent by the formation of cyclic iron chelates which would tend to prevent aromatic ring cleavage between the vicinal hydroxyl groups. Actual additions (82) of iron powder to the alkali fusions in nonferrous reactors increased the yields of both catechol and protocatechuic acid and favored the decarboxylation of the latter.

Attempts to reverse the conversion of 13 to 14 by addition of ammonium carbonate and the use of a carbon dioxide atmosphere were not definitively successful.

In a more modern setting, the Russian researchers Chudakov and Sukhanovskii have admixed hydrolysis lignin with alkali and water to obtain an extrudable mass (83). Pyrolysis of filaments containing less than 5% moisture in petrolatum at 200° converted about one-third of the lignin charged to oxalic acid. At lower temperatures protocatechuic acid was obtained in unspecified yield.

B. Formation of "Lignin Acids." In addition to the discrete isolation of protocatechuic acid, catechol and oxalic acid, about one-third of the lignin can be converted to "lignin acids." These materials, which are formed under less drastic fusion conditions, are completely uncharacterized and are presumably the result of partial degradation (84). Their further fusion at more elevated temperatures affords small yields of protocatechuic and oxalic acids. The formation of acetic and butyric acid was also noted. The genesis of the acetic acid can be readily explained since alkaline hydrolysis of lignin releases acetaldehyde by a retroaldol condensation (see Chapter 4). Also capable of oxidation to acetic acid are the C-methyl groups (85) present in structures exemplified by acetovanillone 15

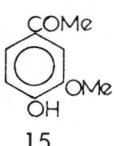

15

which are also generated by the alkaline hydrolysis of various lignins. The construction of a satisfying explanation for the formation of butyric acid poses more of a problem. If Rassow and Zickman (84) did indeed isolate butyric acid, and their identification is not satisfactory by modern standards, then the four carbon acid probably originates from a breakdown of the catechol by oxidation-reduction processes. The plausibility of this view is enhanced by the isolation of adipic acid from fusion of a lignosulfonic acid (86) with potassium hydroxide. The butyric acid could conceivably also originate via 1-hydroxycyclopentane carboxylic acid which is formed from catechol by the sequence outlined (87, 88) in Figure 14.5. Butyric acid has also been reportedly isolated by alkali fusion of a hydrochloric acid jute lignin in rather considerable amounts, and this finding has been interpreted as indicative of the presence of a side-chain of at least four carbon atoms in the lignin molecule (26b). However, reference to the original literature (89) shows that the identification of the butyric acid was by smell alone, and until more definitive evidence is collected, this claim must be regarded as extremely provisional.

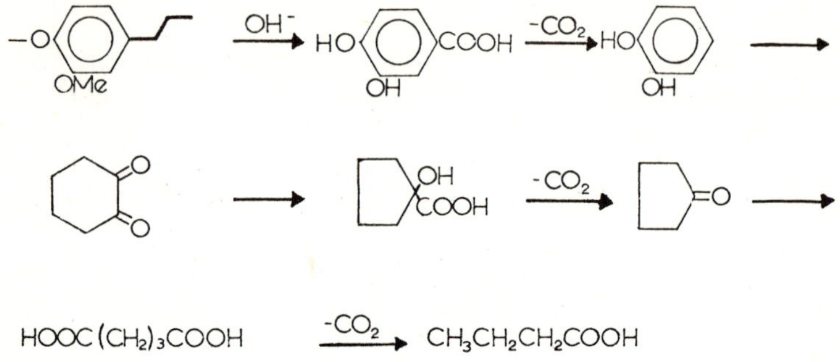

Figure 14.5. A probable reaction sequence for the production of butyric acid by alkaline fusion of lignins (87, 88).

C. Fusion of Various Lignin Preparations. These early results on alkali fusions seem to have a wide application to a variety of lignins. Thus, the grass lignins occurring in oats and wheat, which supposedly show more structural variance with species (see Chapter 4), give similar amounts of the same products (90-93). Lignins isolated from both commercial sulfite spent (80) and soda black (94) liquor also give the same products, though in some cases rather outstanding yields of protocatechuic acid and catechol have been claimed (80).

This matter of yields has been a subject of some conflict in the literature, and fusion reactions are notoriously irreproducible and often rather inexactly described.

In an effort to clarify these yield figures, Freudenberg, Harder and Markert (95) attempted some well-controlled fusions. The actual yield of protocatechuic acid, isolated as veratric acid, was less than 5%. However, because protocatechuic acid itself was about 50% destroyed when subjected to identical fusion reaction conditions these researchers concluded that the actual yield was closer to 10%.

Further work by Freudenberg and Müller (96) on hardwood lignins gave about equal amounts of protocatechuic and acetic acid; a result consistent with the guaiacyl-syringyl structure of the basic hardwood lignin polymer.

A preliminary investigation of alkali fusion of a spruce chlorolignin containing about 25% chlorine has also been carried out (97). In addition to hydrogen and carbon dioxide, the reaction products identified were methanol, acetic acid and oxalic acid. These were obtained in yields of 6, 8 and 23%, respectively, based on chlorine lignin. Unlike previous fusions, admixture of iron had no effect on the product distribution, but oxalic acid could not be

HIGH ENERGY DEGRADATION

isolated when aluminum had been added to the fusion reaction. The aromatic components, as expected, contained no protocatechuic acid or catechol, and none was firmly identified though the presence of terephthalic acid was suggested.

Such claims and the numerous other indications (26b) of the presence of uncharacterized aromatic products on alkaline degradation suggest that this entire area is ripe for re-investigation and clarification using modern analytical techniques.

The experimental difficulties of fusion reactions can be somewhat reduced by the addition of small amounts of water and of course hydrolytic reactions must surely intervene. Thus, Pearl and Beyer (98, 99) have fused lignin preparations with aqueous potassium hydroxide melts at 180-190° C. Anhydrous potassium hydroxide has a melting point of 360° C. At these lower temperatures vanillic acid was isolated in addition to the more usual protocatechuic acid and only 30-50% of the lignin was degraded. The addition of sodium sulfide to the aqueous potassium hydroxide melt at 200° C was not advantageous and the extent of lignin degradation dropped to 10%. The hegemony of hydrolysis over the fusion in the presence of water is further emphasized by the oxidative fusions carried out (99) on a softwood lignin using an aqueous smelt of potassium hydroxide at 225-230° containing cupric oxide. The aromatic products comprised protocatechuic 13, p-hydroxybenzoic 16 and vanillic acids 17. The same treatment of a hardwood lignin at 190° caused little breakdown although syringic acid 18 in addition to 16 and 17 could be isolated.

Undoubtedly the most persistent and vigorous investigations in such lignin degradations have been made by Enkvist and a series of collaborators at the University of Helsinki. In the early stages of their research, degradative conditions were essentially a continuation of kraft pulping after the removal of the cellulosic fibers (85, 100-106). Small amounts of sodium sulfide and hydroxide were added so that the alkali charge was about 30% of the organic matter in the solution. Pressure heating for 10-20 min. at 290° then converted about half of the lignin into ether-soluble degradation products. The reaction conditions are quite critical and the residue is a demethylated lignin (DL). Simultaneously, methyl sulfide is generated from the methoxyl groups in lignin, and its conversion to dimethylsulfoxide via dimethylsulfide is an important industrial reaction which is described in Chapter 19.

More recently the chemical engineering of the process has been further developed so that waste liquor is first continuously spray-dried with additional sodium hydroxide in a nitrogen atmosphere (87, 107-111). A few minutes subsequent heating at 250-300° at atmospheric or reduced pressure affords the volatile products listed in Table 14.4 and the sodium salts of the lignin materials and the aliphatic acids which are completely water-soluble. The aliphatic acids originate mainly but not exclusively from the carbohydrate material, while the enigmatic cyclopentanones stem from catechols via cyclohex-4-ene-1,2-dione, cyclohexane-1,2-dione and 1-hydroxycyclopentane carboxylic acid (88). After acidification of the crude reaction mixture, the products formed by the degradation of the lignin can be separated into three fractions; viz.,

1, A fraction (EWS) containing compounds of low molecular weight, soluble both in water and in ether. Among these are found pyrocatechol and its homologues, 4-methyl- and 4-ethyl-catechol amounting to about 25 pounds per ton of pulp. Protocatechualdehyde and 4-acetocatechol, as well as a mixture of homoprotocatechuic acid and other demethylated phenol carboxylic acids, are also found in this water-soluble fraction.

2, The largest fraction of the degraded lignin (ES) is soluble in ether, but not in water. This material consists of species with average molecular weights ranging between 250 and 1200 which can be further degraded by rapid cracking at 600-650° (108, 111). Elementary analysis, with calculation of the empirical formulae on a C-9 basis, shows a variation in composition from about $C_9H_9O_{2.7}$ in the species of higher molecular weight to $C_9H_{11}O_2$ in those of low molecular weight. The NMR-spectrum of the acetylated product reveals the absence of methoxyls and the presence of a high content of phenolic groups, a small proportion of alcoholic groups, and some terminal methyl groups. This fraction is soluble in neutral sodium molybdate solution as a result of complex formation (112) and also has a high capacity for reducing Fehling's solution. It therefore must contain a high percentage of o-diphenolic compounds. The infrared spectrum is rather peculiar, showing, inter alia, the presence of carbonyl groups in the phenolic fractions.

3, A smaller fraction (DL) of the degraded lignin is insoluble in ether and water, but is soluble in butanone and to a large extent also in sodium molybdate solution.

The formation of these lignin products seems to be mainly a consequence of conventional hydrolytic cleavage of both carbon-oxygen and carbon-carbon bonds, as detailed in Chapter 9, followed by oxidation or reduction. These reactions have been studied in detail by Enkvist and his coworkers using lignin model compounds (113-116).

The complexity of the product mixture makes its fractionation and commercial exploitation unlikely. However, catechol and its derivatives react swiftly with formaldehyde and should in theory be well suited for the manufacture of rapid-curing wood adhesives (108) and this in fact has been recently confirmed (117). Nonetheless, if these materials have to be solvent-separated

Table 14.4 Yields of Enkvist Degradation Products in Kg per Metric Ton of 90% Cellulose[1] (110).

		In Original Black Liquor	After Treatment
I.	Lignin Products		
	Ether- and water-soluble (EWS)	51	186
	Ether-soluble, water-insoluble (ES)	30	330
	Ether- and water-insoluble (DL)	422	55
II.	Volatile Products[2]		
	Dimethyl sulfide	2.3	15
	Methyl mercaptan	--	4
	Ethanol, acetone	1.3	2.3
	Methanol	12	2.9
	Cyclopentanone and homologues	--	0.9
	Guaiacol and other volatile phenols	?	3.4
III.	Aliphatic Acids		
	Acetic	46	55
	Formic	40	85
	Propionic	--	10
	Oxalic	--	87
	Succinic	--	18

[1] Experiments on amounts of dry material ranging up to 10 Kg. The figures given are, with a few exceptions, the means of results obtained from a considerable number of replicate runs.

[2] Distilled off during alkali heating; analyses mainly by gas chromatography.

from the demethylated lignin their cost would probably be excessive. This extraction step may be avoided since upon acidification the monomeric phenols are tenaciously included within the sludge-like demethylated lignin precipitate. The total aromatic material can thus be easily separated from the aqueous phase by centrifugation (118). The phenolic material in Douglas-fir bark behaves similarly to lignin in the Enkvist degradation giving a phenolic fraction containing guaiacol, vanillin catechol, p-methylcatechol and p-ethylcatechol as well as six other unidentified minor components (118).

More typical fusion reactions of lignin have recently been described in which base impregnated lignins are pyrolyzed to obtain low molecular weight materials (119-123). The demethylation of a dry basic lime treated lignosulfonate by heating in a rotary furnace at 330-340° is illustrative (119). The calcium content of the lignosulfonate influences the number of phenolic groups generated, and as much as 30-70% of ethanol-soluble material could be formed depending upon the reaction temperature employed. The degraded lignin consisted of carboxylic acids (50%), phenols (40%) and neutral material (10%).

The promise of these new efforts in fusion degradation of the lignin macromolecule is, however, clouded by the inherent difficulties of utilizing or cheaply separating a mixture of degradation products. Further research from both ends of the problem is clearly needed.

REFERENCES

1. A. W. Goos, Wood Chemistry, L. E. Wise and E. C. Jahn, Eds., Reinhold Publishing Corp., New York, 1952, p. 826.
2. B. L. Browning, The Chemistry of Wood, B. L. Browning, Ed., Interscience Publishers, New York, 1963, p. 85.
3. A. J. Stamm, Ind. Eng. Chem., 48, 413 (1956).
4. H. W. Eickner, Forest Prod. J., 2, 194 (1962).
5. S. L. Madorsky, V. E. Hart and S. Straus, J. of Research, National Bureau of Standards, 56, 343 (1956).
6. S. L. Madorsky, ibid., 60, 343 (1958).
7. E. J. Murphy, Trans. Electrochemical Soc., 82, 161 (1943).
8. W. K. Tang and W. K. Neill, J. Poly. Sci., 6, 65 (1964).
9. S. Martin, R. and D. Report, U.S.N.R.D.L., D1-TR-102-NS081-001 (1956).
10. F. E. Brauns, The Chemistry of Lignin, Academic Press, Inc., New York 1952, p. 581.
11. E. Erdmann, Z. Angew. Chem., 34, 309 (1921).
12. E. Heuser and A. Brötz, Papier-Fabrik., Fest und Ausland Heft, 23, 69 (1925).
13. F. Fischer and H. Schrader, Ges. Abhandl., Kenntnis Kohle, 5, 106 (1922).
14. A. Pictet and M. Gaulis, Helv. Chim. Acta, 6, 627 (1923).
15. E. Heuser and C. Skiöldebrand, Z. Angew. Chem., 32, 41 (1919).
16. J. D. Wethern, Can. Patent 776,171, Jan. 16, 1968.
17. D. W. Goheen, U. S. Patent, 3,326,980, June 20, 1967.
18. V. G. Vedernikov, V. F. Maksimov and V. I. Roshchin, U.S.S.R. Patent 202,124, June 22, 1966.
19. E. Hägglund, Arkiv Kemi, Mineral. Geol., 7, No. 8 (1918).
20. H. Tropsch, Ges. Abhandl. Kenntnis Kohle, 6, 293 (1923).
21. G. Szelenyi and A. Gömöry, Brennstoff-Chem., 9, 73 (1928).

22. G. L. Bridger, Ind. Eng. Chem., 30, 1174 (1938).
23. F. Yoshimura and M. Maruoka, J. Chem. Soc. Japan, Ind. Chem. Sect., 58, 571 (1955).
24. G. E. Domburg and V. N. Sergeeva, Isv. Akad. Nauk Latv. SSR, Ser. Khim., 62, 509, 624 (1967).
25. A. Gillet and J. Urlings, Chim. & Ind. (Paris), 68, 55 (1952).
26. F. E. Brauns and D. E. Brauns, The Chemistry of Lignin, Supplement Vol., Academic Press, Inc. New York, 1960, a: p.574, b: p.551.
27. E. Beglinger, Hardwood-Distillation Industry, USDA Forest Service Report 738, Feb. 1956.
28. P. Karrer, Organic Chemistry, Elsevier Publishing Co., Inc., New York, 1947, p. 168.
29. T. L. Fletcher and E. E. Harris, J. Am. Chem. Soc., 69, 3144 (1947).
30. T. L. Fletcher and E. E. Harris, Tappi, 35, 536 (1952).
31. M. Phillips and M. J. Goss, Ind. Eng. Chem., 24, 1436 (1932).
32. S. I. Sukhanovskii, E. I. Akhmina, T. A. Podgornava and Z. I. Lisina, Sb. Tr. Vses., Nauch-Issled. Inst. Gidroliz. i Sulfitno-Spirt. Prom., 13, 274 (1965).
33. D. W. Goheen and J. B. Martin, U. S. Patent 3,375,283, Mar.26, 1968.
34. S. I. Sukhanovskii, E. I. Akhmina, T. A. Podgornaya, E. S. Bezmosgin, A. G. Nemchenko and Y. D. Yudkevich, Gidroliz. i. Lesokhim. Prom., 17, 17 (1964).
35. K. Kratzl, H. Czepel and J. Gratzl, Holz als Roh-und Werkstoff, 23, 237 (1965).
36. B. Rasson and P. Zickmann, J. Prakt. Chem., (2), 123, 214 (1929).
37. M. Phillips, J. Am. Chem. Soc., 51, 2420 (1929).
38. K. Freudenberg and K. Adam, Chem. Ber., 74, 387 (1941).
39. K. Kürschner, Brennstoff-Chem., 6, 117, 177, 188 (1925).
40. K. Kürschner, Mikrochemie, 3, 1 (1925).
41. H. Watanabe and K. Kitao, Wood Res. (Kyoto), 38, 40 (1966).
42. I. Z. Kirsbaum, G. E. Domburg and V. N. Sergeeva, Latv. PSR Zinat. Akad. Vestis, Ser. Khim., 680 (1969).
43. Y. K. Shaposhnikov and L. V. Kosyukova, Gidroliz. i Lesokhim. Prom., 20, 16 (1967).
44. Y. K. Shaposhnikov and L. V. Kosyukova, Khim. Pererabotka Drevesiny, Ref. Inform., 3, 6 (1965).
45. J. M. Derfer, The Chemistry of Petroleum Hydrocarbons, B. T. Brooks, S. S. Kurtz, Jr. C. E. Boord and L. Schmerling, Eds., Reinhold Publishing Corp., New York, 1954, p.592.
46. W. Fuchs, Chem. Ber., 60, 957, 1131 (1927).
47. P. Karrer and B. Bodding-Wiger, Helv. Chim. Acta, 6, 817 (1923).
48. W. Lautsch and G. Piazolo, Chem. Ber., 76, 486 (1943).
49. M. Phillips, J. Am. Chem. Soc., 53, 768 (1931).

50. M. Phillips, Science, 73, 568 (1931).
51. J. Marcusson and F. Bottger, Chem. Ber., 57, 633 (1924).
52. A. Pictet and M. Bouvier, ibid., 48, 930 (1915).
53. R. Monnberg, U.S. Patent 2,608,561, August 26, 1952.
54. R. Monnberg, Svensk Papperstidn., 56, 46 (1953).
55. E. I. Akhmina, E. S. Bezmozgin, V. N. Lapin, R. M. Levit, S. I. Sukhanovskii and Y. Z. Sorokin, Sb. Tr., Vses. Nauch-Issled.Inst. Gidroliz. i Sulfitno-Spirt. Prom., 14, 258 (1965).
56. E. F. Morozov, Tr. Sibirsk, Tekhnol. Inst. Sb., 38, 137 (1966).
57. S. I. Sukhanovskii, E. I. Akhmina, E. B. Evstifeeva, T. A. Kalninsh, E. M. Abele, I. A. Alsup, A. I. Kulkevich and O. V. Kiselis, Khim. Pererabotka i Zashchita Drevesiny, Riga, Akad. Nauk Latv. SSR, 1964: 87.
58. V. G. Kashirskii and N. B. Lobacheva, Gidroliz. i Lesokhim. Prom., 14, 8 (1961).
59. R. Domansky and F. Rendos, Holz als Roh-und Werkstoff, 20, 473 (1962).
60. W. Sandermann and H. Augustin, ibid., 22, 377 (1964).
61. W. K. Tang, Effect of Inorganic Salts on Pyrolysis of Wood, α-Cellulose and Lignin, Determined by Dynamic Thermogravimetry, USDA Forest Service Research Paper FPL 71, Jan. 1967.
62. W. K. Tang and H. W. Eickner, Effect of Inorganic Salts on Pyrolysis of Wood Cellulose and Lignin Determined by Differential Thermal Analysis, USDA Forest Service Research Paper FPL 82, Jan. 1967.
63. F. Abe, J. Japan.Wood Res. Soc., 14, 98 (1968).
64. A. Kuriyama, ibid., 4, 30 (1958).
65. G. E. Domburg and V. N. Sergeeva, Izv. Akad. Nauk Latv. SSR, Ser. Khim., 377 (1967).
66. W. Pryor, Free Radicals, McGraw-Hill Co., New York, 1966, p. 58.
67. V. N. Sergeeva and Yaunzems, Tr. Inst. Lesokhoz. Problem Khim. Drevesiny, Akad. Nauk Latv. SSR, 24, 79 (1962).
68. G. E. Domburg and V. N. Sergeeva, Izv. Akad. Nauk Latv. SSR, Ser. Khim., 62, 744 (1967).
69. D. W. Goheen and J. T. Henderson, to be published.
70. W. M. Hearon, J. T. Henderson and D. W. Goheen, U. S. Patent 3,148,227, Sept. 8, 1964.
71. L. Possoz, Dingler Polytech. J., 150, 127, 382 (1858).
72. J. Erdmann, Ann., 138, 1 (1866).
73. F. Hoppe-Seyler, Chem. Ber., 4, 15 (1871).
74. F. Hoppe-Seyler, Z. Physiol. Chem., 13, 66 (1889).
75. G. Lange, ibid., 14, 215 (1890).
76. E. Hägglund, Ark. Kemi, Mineral. Geol., 7, No. 8, 1 (1919).
77. E. Hägglund and C. J. Malm, Acta Acad. Åboensis, Math. Phys., 2, No. 4 (1922)
78. E. Heuser, H. Rösch and L. Gunken, Cellulosechem., 2, 13 (1921).

79. E. Heuser and A. Winsvold, ibid., 2, 113 (1921).
80. E. Heuser and A. Winsvold, ibid., 4, 49, 62 (1923).
81. E. Heuser and A. Winsvold, Chem. Ber., 56, 902 (1923).
82. E. Heuser and F. Hermann, Cellulosechem., 5, 1 (1924).
83. M. I. Chudakov and S. I. Sukhanovskii, ABIPC, 28, 1116 (1958).
84. B. Rassow and P. Zickmann, J. Prakt. Chem., 123, 189 (1929).
85. Th. Ashorn and T. Enkvist, Acta Chem. Scand., 16, 548 (1962).
86. K. H. A. Melander, Svensk Papperstidn., 24, 440, 461 (1921).
87. T. Enkvist and J. Turunen, Chemie et Biochimie de la Lignine, de la Cellulose et des Hemicelluloses, Actes de Symposium International de Grenoble, Universite de Grenoble, 1964, p. 177.
88. O. Wahlroos and T. Enkvist, Acta Chem. Scand., 22, 3203 (1968).
89. P. B. Sarkar, J. Indian Chem. Soc., 10, 263 (1933).
90. E. Beckman, O. Liessche and F. Lehmann, Biochem. Z., 139, 491 (1923).
91. R. Willstätter and W. Mieg, Ann., 408, 134 (1915).
92. M. Phillips and M. J. Goss, J. Biol. Chem., 114, 557 (1936).
93. M. Phillips and M. J. Goss, ibid., 125, 241 (1938).
94. B. Holmberg and T. Wintzell, Chem. Ber., 54, 2417 (1921).
95. I. Freudenberg, H. Harder and L. Markert, Chem. Ber., 61, 1760 (1928).
96. K. Freudenberg and H. F. Müller, ibid., 71, 1821 (1938).
97. O. A. Müller, Papier-Fabr., 32, 347 (1934).
98. I. A. Pearl and D. L. Beyer, Lignin Structure and Reactions, R. F. Gould, Ed., Advances in Chemistry Series, American Chemical Society, Washington, D. C., 1966, p. 145.
99. I. A. Pearl and D. L. Beyer, J. Org. Chem., 27, 2287 (1962).
100. T. Enkvist, Finska Kemistsamfundets Medd., 64, 28 (1955).
101. T. Enkvist, Svensk Kem. Tidskr., 72, 93 (1960).
102. T. Enkvist, Teknillisen Kemian Aikakausilehti, 17, 127 (1960).
103. T. Enkvist, B. Holm, J. Turunen and K. Turunen, Finska Kemistsamfundets Medd., 70, 5 (1961).
104. T. Enkvist, Paperi Puu, 43, 657 (1961).
105. T. Enkvist, Th. Ashorn and K. Hästbacka, Paperi Puu, 44, 395 (1942).
106. T. Enkvist, J. Turunen and Th. Ashorn, Tappi, 45, 128 (1962).
107. T. Enkvist, Suomen Kemistilehti, 36A, 57 (1963).
108. T. Enkvist and T. Lindfors, Paperi Puu, 48, 639 (1966).
109. T. Enkvist and T. Lindfors, Finska Kemistsamfundets Medd., 75, 1 (1966).
110. T. Enkvist, private communication.
111. T. Enkvist, Kemian Teollisuus, 26, 615 (1969).
112. J. Halmekoski, Suomen Kemistilehti, 32B, 170 (1959); 33B, 74 (1960); 34B, 85 (1961); 36B, 24, 58 (1963).
113. J. Turunen, Paperi Puu, 43, 663 (1961).
114. Th. Ashorn, Soc. Scient. Fennica Comm. Phys. Mathem., 25, 8 (1961).

115. K. Hästbacka, ibid., 26, 4 (1963).
116. J. Turunen, ibid., 28, 9 (1963).
117. H. G. Freeman, G. F. Baxter and G. G. Allan, U. S. Patent 3,518,159, June 30, 1970.
118. G. G. Allan, unpublished work.
119. V. Schmidt, Zellstoff Papier, 14, 73 (1965).
120. L. V. Panasyuk, V. G. Panasyuk, M. Y. Zarubin and D. V. Tishchenko, Lesnoi Zh., 5, 132 (1962).
121. A. P. Runtso and T. V. Murashkevich, U.S.S.R. Patent 166,039, Oct. 30, 1961.
122. V. G. Panosyuk, S. I. Sukhanovskii, D. V. Tishchenko, L. V. Panasyuk, V. P. Repka and M. Y. Zarubin, U.S.S.R. Patent 158,284 (1963).
123. A. Hayashi and T. Uekita, Mokuzai Gakkaishi, 14, 387 (1968).

15

REACTIONS IN SULFITE PULPING
D. W. Glennie

I. Introduction. Lignin sulfonation has received more attention than any other reaction in lignin chemistry. This is largely a result of the extensive research directed to the improvement of sulfite pulping and to utilization of spent sulfite liquor. At the same time, considerable attention has focused on the sulfonation reaction as a means of deducing certain structural aspects of lignin.

In lignin sulfonation three major reactions usually proceed simultaneously: sulfonation, hydrolysis and condensation. Minor accompanying reactions of lignin may include oxidation, reduction, rearrangement, dehydration, thiosulfation and sulfidation. In addition, hydrolysis and condensation reactions can occur between lignin and associated wood components.

II. Early Research on Lignin Sulfonation. The sulfite pulping process was discovered first by Tilghman in 1866 (1). However, it appears that research on lignin sulfonation did not begin until 1890, when Pedersen (2) realized that appreciable sulfur in sulfite spent liquor was organically combined (3). In 1892, Lindsey and Tollens (4) concluded that sulfur was attached to lignin as a sulfonic acid group.

In 1893, Klason (5) proposed that ethylenic groups in lignin might react with sulfurous acid to give sulfonic acid groups. Later, Freudenberg and Sohns (6) considered the possibility that ethylene groups are formed as intermediates by dehydration of lignin during sulfonation. However, from the beginning there was some doubt about this because methylation studies (6, 7) indicated the total hydroxyl group content of lignin to be unchanged after sulfonation.

It was not until 1935 that Holmberg (8, 9) proposed that sulfonation, as well as thioglycolysis and ethanolysis, proceeds by reaction involving phenylcarbinol groups in lignin, Equation 1. Concurrent studies on methylated lignin pointed to a relationship between ease of sulfonation and presence of both phenylcarbinol groups and phenolic elements in lignin (10).

In the early days of research on lignin sulfonates, it was difficult to separate lignin sulfonates from other components in spent sulfite liquor. Com-

$$\underset{\substack{|\\ \text{CHOH}\\ }}{\overset{R}{}}\!\!\!\begin{array}{c}\text{OMe}\\ \text{O-}\end{array} \xrightarrow{H_2SO_3} \underset{\substack{|\\ \text{CHSO}_3\text{H}\\ }}{\overset{R}{}}\!\!\!\begin{array}{c}\text{OMe}\\ \text{O-}\end{array} \qquad (1)$$

monly, lignin sulfonates were precipitated by addition of salts such as calcium or sodium chloride. Klason referred to the precipitated and unprecipitated lignin sulfonates as α- and β-lignin sulfonates, respectively (11). For the former, a molecular weight of 6000 was obtained by a cryoscopic method (13). When Proctor and Hirst (12) discovered in 1909 that aromatic amines caused partial precipitation of lignin sulfonates, the two classes again were designated α- and β-lignin sulfonates.

The concept of lignin sulfonates as a polydisperse system of guaiacylpropane polymers began to emerge about 1930, especially through studies by Hägglund and coworkers (14-17). It was found that in bisulfite pulping of wood at pH 4 to 6, lignin was sulfonated in solid phase and the subsequent rate of solubilization increased with increasing acidity of sulfite solution. It was noted that lignin sulfonates produced under different sulfonation conditions differed in degree of dialyzability and precipitatability which seemed to be related to their relative molecular size. Gradually it became apparent that hydrolysis and condensation reactions accompanied lignin sulfonation.

As early as 1904, Grafe (18) found that heating lignin sulfonates with lime not only caused desulfonation, in agreement with results of Klason (5), but also produced vanillin in small amount. A little later, Melander demonstrated that lignin sulfonates behaved in the same way as wood lignin in forming protocatechuic acid on alkaline fusion (19). However, determination of structural aspects of lignin sulfonates through alkaline degradation reactions did not get well under way until about 1936, beginning with studies by Hibbert and coworkers (20). They found that alkaline hydrolysis of β-lignin sulfonates gave higher yields of vanillin than α-lignin sulfonates. In addition, hot caustic treatment of methylated lignin sulfonates produced veratraldehyde. This was regarded by Hibbert as convincing evidence for presence of free phenolic elements in lignin sulfonates.

This presents a few landmarks in research on lignin sulfonation spanning about 70 years from 1866 to 1936. Other worthy contributions during this era have been reviewed (3, 10, 21-25).

III. Methods of Sulfonating Lignin.

A. Technical Processes. The "acid sulfite" process produces an easily bleachable pulp in about 50% yield. Sulfite pulping liquors are made by passing SO_2 over a packed tower of limestone ($CaCO_3$) or magnesium oxide

(MgO) or into a solution of ammonia or caustic and are respectively referred to as Ca-, Mg-, NH_3- or Na-base acid-sulfite pulping liquors. The total SO_2 concentration by weight in solution is about 4-7%, which includes about 1% "combined SO_2," calculated as the amount combined with the base as sulfite. Thus, a calcium-base acid-sulfite liquor containing 6% total and 1% combined SO_2 contains 0.875% CaO, which is equivalent to 3.1% $Ca(HSO_3)_2$. Total SO_2 is measured iodometrically, and the combined SO_2 is the difference between the total SO_2 and the "free SO_2" (5% in the above example). Free SO_2 is measured by titration with a standard base to the phenolphthalein endpoint.

The liquor to wood ratio in sulfite pulping is commonly 4:1. Therefore, at 6% total and 1% combined SO_2 there is 24% total and 4% combined SO_2 based on dry wood. In practice, the digester is relieved after reaching the pulping temperature, SO_2 is recovered and the total SO_2 decreased to about 4% concentration in the liquor. Pulping temperatures are in the range of 120-140°C for a total reaction time of 5-20 hr.

In a study to determine the effect of the type of base in acid-sulfite pulping, it was found that: The pulp yield was independent of the base, sodium and ammonium bases gave significantly greater delignification at a given pulp yield, the greatest removal of pentosan occurred with Mg base, and the highest pulp viscosity was obtained with sodium base (26).

In the so-called "bisulfite" process, total SO_2 may be about 4-5%, and combined SO_2 one-half the total to give a pulping liquor of sodium, magnesium or ammonium bisulfite having a pH of about 4.5. The relationship between the amount of combined SO_2, pH, and type of pulping system is summarized in Table 15.1.

Table 15.1. Classification of Sulfite Pulping Processes

Combined SO_2, % of total	pH	System
0	1-1.6	sulfurous acid
15 to 25	1.8-3.1	acid sulfite
~50	4.3-5.0	bisulfite
~75	7.0	bisulfite-sulfite
100	9.5	sulfite

Some of the changes in acid-sulfite pine pulp composition toward the end of pulping are reported by Makkonen as shown in Figure 15.1. The results are contrasted with those for pine kraft pulp (27).

Also, two-stage sulfite processes have been developed to produce full chemical pulps. In the Stora process, for example, the first stage is carried out at about pH 4-8 and then SO_2 injected to complete pulping under acid sulfite conditions.

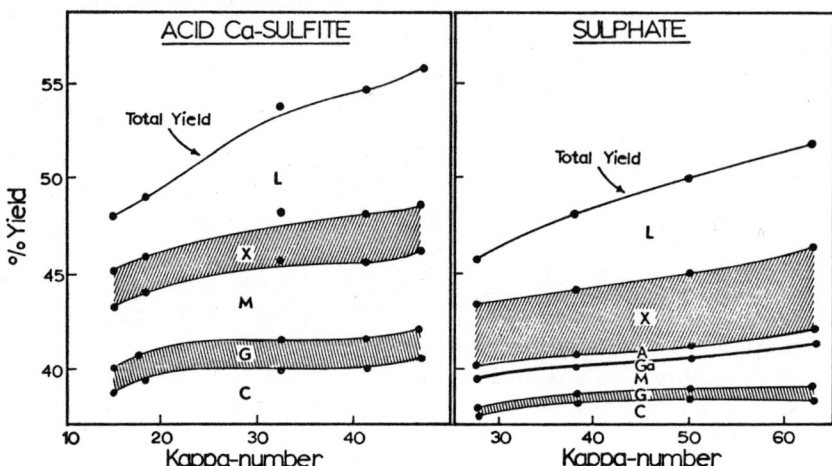

Figure 15.1. Changes in composition of pine pulp toward the end of the pulping reaction; yield as a percentage of dry wood at a given degree of delignification indicated by the Kappa number: L: lignin, X: xylan, A: arabinan, Ga: galactan, M: mannan, G: glucan and C: cellulose (27).

Conversely, in the Sivola process the first stage of pulping may be carried out with $NaHSO_3$-SO_2 at about pH 3-5 and then soda ash or caustic injected in the second stage, giving a bisulfite-sulfite system at about pH 6-9. Some advantages of two-stage sulfite processes are discussed by Makkonen (27) and by Rydholm (25).

All varieties of the sulfite process may be used to produce pulp in about 60-75%, or even up to 90%, yield. Such "high yield pulps" are only partially delignified and require refining for papermaking; therefore, they are referred to as semichemical pulps.

Recently, there has been interest in production of very high yield pulps by acid sulfite, bisulfite or neutral sulfite systems. For example, groundwood may be presulfonated or postsulfonated to improve pulp strength with only minor reduction in pulp yield (28, 29). An outstanding discussion of all aspects of sulfite pulping processes has been presented by Rydholm (25).

Continuous pulping methods for the conventional sulfite process have been developed recently which may increase its competitive potential (30).

B. Experimental Processes. Sulfonation of wood with bisulfite at about pH 5 and 135° leaves about 70% of the lignin as an insoluble "low-sulfonated lignin" or "Kullgren lignin" in the pulp (31, 32). Subsequent heating with mineral acid causes hydrolysis to yield a soluble "kullgren acid" with a low degree of sulfonation. The solubilization also can be effected in various solvents such as methanol-HCl (33).

Similarly, lignin sulfonation has been achieved by pulping wood with SO_2 in various alcohols as well as in acetone (34). In contrast, however, although SO_2 in dimethyl sulfoxide delignifies wood, the lignin is scarcely sulfonated (35, 36). From this it is apparent that the system SO_2-DMSO is a solvolysis pulping reaction and not a sulfonation, at least not until the water content is increased to above 20% in DMSO (37).

Sulfonates also can be prepared from isolated lignins such as milled wood lignin (MWL) (38, 39). Convenient procedures for the sulfonation of wood in the laboratory have been developed (40, 41). In order to obtain products with increased sulfonate content, sulfomethylation with alkaline sulfite and formaldehyde has been applied to certain lignins (42, 43).

IV. Methods of Isolating Lignin Sulfonates. The commercial "lignin sulfonates" usually contain substantial amounts of extraneous substances such as sugars, sugar acids, sulfonated sugars, alcohols, aldehydes, terpenes, lignans, and sulfite and sulfate salts (44).

Dialysis is an effective means of isolating high molecular weight lignin sulfonates from spent sulfite liquor (41, 45). However, if polysaccharides are present, isolation of pure lignin sulfonates may be improved by use of an electrophoresis-convection process (41). Electrodialysis has even been investigated as a possible commercial separation method (46). Ion exclusion gives partial separation as well as fractionation of lignin sulfonates (47, 48). Precipitation of lignin sulfonate salts in alcohol also gives partial separation and fractionation (45, 49). High molecular weight lignin sulfonates can be precipitated with amines such as 4,4'-bis(p-dimethylaminodiphenyl)methane or "bis" reagent (50). Extraction of lignin sulfonic acids with a long-chain aliphatic amine in amyl alcohol gives pure lignin sulfonates (51). Also, it has been shown that the dicyclohexylamine salt of lignin sulfonic acids can be fractionated in methanol-ethyl acetate (52). Sugars are separated from lignin sulfonates by passage of spent liquor over a column of diethylaminoethyl cellulose, but sugar acids may be incompletely separated. Separation of lignin sulfonates and molecular size fractionation may be obtained by ion exclusion and gel filtration (53). Behavior of lignin sulfonates in gel filtration on cross-linked dextran (Sephadex) was found to be sensitive to electrolyte concentration (54).

Also, it has been found that about 80% of lignin sulfonates can be precipitated from spent sulfite liquors with various long-chain substituted quaternary ammonium salts (55, 56, 57). Precipitated quaternary ammonium lignin sulfonates are soluble in methanol and can be fractionated by careful stepwise addition of water (57). The method is attractive for isolation of large amounts of pure lignin sulfonates and as a preliminary step for investigations of low molecular weight lignin sulfonates which remain dissolved. For example, the long-chain quaternary ammonium salt of sulfonated vanillyl alcohol is water-soluble (58).

Lignin sulfonates, especially α-lignin sulfonates, are adsorbed to some extent by various proteins. Thus, they are adsorbed by hide powder and act as tanning agents (23). The use of leather scrap to separate lignin sulfonates from spent sulfite liquor forms the basis for a commercial process in Sweden (59). Purification by adsorption of lignin sulfonates on crab chitin also has been proposed (60).

Other methods for isolating lignin sulfonates have been described (22, 23, 25), and further improvements in separation and fractionation techniques may be expected.

V. Mechanism of Sulfonation and Delignification.
A. Rate Studies and Reactive Groups in Lignin.
1. Wood Lignins.
In normal sulfite pulping of softwoods, commonly performed at pH 1-2 and at 135°, about one-half of the monomeric units assume a sulfonic acid group and approximately 90% of lignin becomes water-soluble. Extended sulfonation of the soluble lignin sulfonates can boost the level of sulfonation up to one group per monomer unit.

An entirely different behavior is observed if the pH of sulfonation is raised to the nearly neutral pH 5-6, as first shown by Hägglund and Kullgren (10). Only a part (~20%) of lignin becomes soluble even in extended sulfonation treatment while the bulk of lignin remains associated with wood and its degree of sulfonation levels off at 0.3 SO_3H/OMe. It was found, however, that this "solid lignin sulfonic acid" (or "Kullgren lignin") could be made water-soluble by means of mild acidic hydrolysis of lignin-carbohydrate and/or lignin-lignin bonds, yielding a "low-sulfonated lignin sulfonic acid" (or Kullgren acid") with approximately 0.3 sulfonic groups/C_9. In contrast with the "solid lignin sulfonic acid," "low-sulfonated lignin sulfonic acids" can be further sulfonated even at neutral pH. This suggests that the hydrolysis process used for solubilization produces new sites for sulfonation.

Erdtman (61, 62) rationalized these findings by postulating the existence of two broad classes of sulfonatable groups. The first class, the so-called "A-groups," are sulfonatable in the pH range 4-9 and are present to the extent of about 0.3/OMe. A-groups are also sulfonatable in an acidic medium, but preferentially condense with reactive polyphenols such as resorcinol, pinosylvin or phoroglucinol if such compounds are present. This type of condensation provides an explanation for the failure to pulp pinosylvin-rich pine heartwood by acidic sulfonation.

The groups that require acidic conditions for sulfonation were termed "B-groups" and their amount may be as high as 0.7/OMe. Acidic hydrolysis presumably converts B-groups to B'-groups capable of sulfonation throughout the pH range 1.5 to 9.

Subsequently, Lindgren, Adler and Mikawa (63-68) have undertaken the clarification of structure of A- and B-groups. It can be seen from Figure 15.2 that the sulfonation of A-groups under neutral and slightly basic conditions involves a rapid and a slow phase. Accordingly, the corresponding reactive groups were termed the "X"- and "Z"-groups, respectively, both present in an approximate amount of 0.15/OMe.

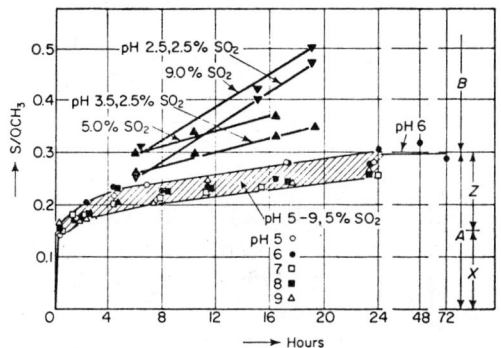

Figure 15.2. Sulfur uptake (S/OMe) by spruce woodmeal in sulfite solutions of varying pH at 135°C. Values of S/OMe for sulfonation at pH 2.5 and 3.5 were determined in solution for "low-sulfonated lignin sulfonic acids" (64, 68).

Model compound studies, initially limited to guaiacyl derivatives with an α-carbinol or -ether group, revealed that models with a free phenolic hydroxyl group exhibited the reactivity of the rapidly reacting X-groups. Of course, in all likelihood softwood lignins contain such "p-hydroxybenzyl alcohol" and -ether units in amounts larger than 0.15/OMe. The situation has become complicated, however, by the discovery of other rapidly reacting models. The etherified α-keto model 1, Equation 2, sulfonates in the γ-position to give 2. Furthermore, at pH 6-8 the rate is of the same order of magnitude as that of X-groups in lignin, as shown in Figure 15.3 (64). As things stand now, there are simply too many candidates for the X-groups to allow for reliable structural identification.

$$\text{1} \xrightarrow{HSO_3^-} \text{2} \qquad (2)$$

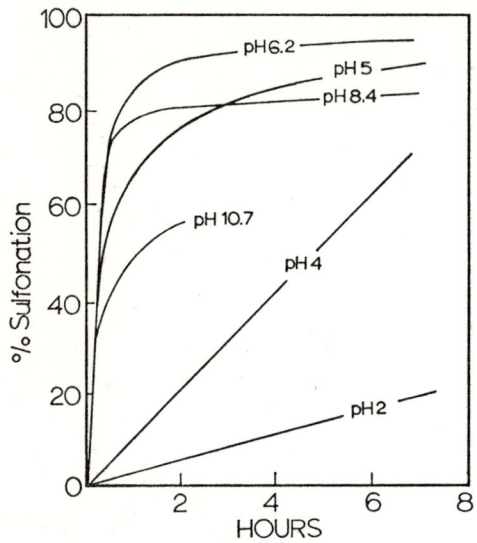

Figure 15.3. Rate of sulfonation of the ketol β-ether 1 at 135°C as a function of time and pH (64).

Suitable models to describe the reactivity of the more slowly reacting Z-groups were found among etherified guaiacyl derivatives containing an α-hydroxyl group. Therefore, in lignin, Z-groups can possibly be represented by certain etherified β-O-4- and β-1-lignol structures.

An outstanding model for B-groups was found in the etherified β-ether model 3 which incorporates an ether group at the α-position. Like the B-groups, this compound does not become sulfonated in the pH range 5-9 and is readily hydrolyzed by acid to compound 4 which has reactivity characteristics common with the B'-groups.

3, R = Et
4, R = H

Kinetic comparison between the B'-group model 4 and the actual B'-groups in low-sulfonated lignin sulfonic acids is illustrated in Figure 15.4 (65). The same figure also demonstrates the steric hindrance effect caused by the β-aryloxy group. It should be noted that some hydrolysis of the β-ether bond occurs as a minor side reaction in the acidic sulfonation of 4, resulting in the formation of Hibbert's ketones.

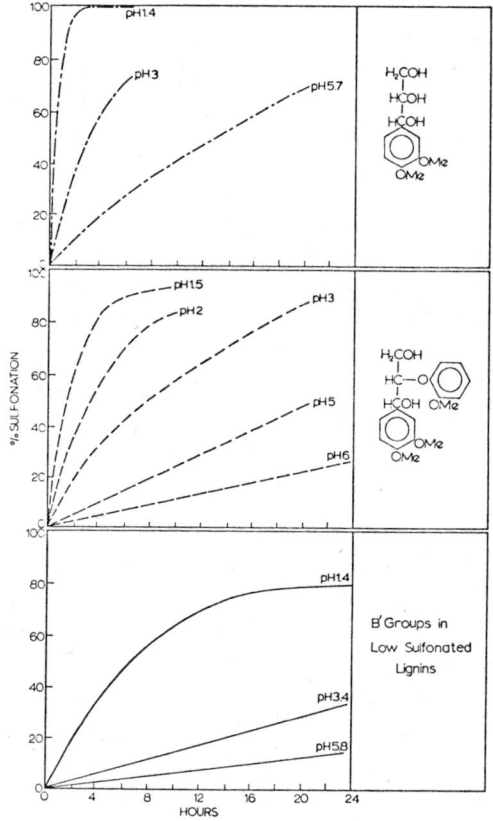

Figure 15.4. Rate of sulfonation at 135°C, with varying pH for (a) 3,4-dimethoxyphenyl glycerol, (b) its β-2-methoxyphenyl ether, and (c) B'-groups in low-sulfonated lignin (65). The rate of sulfonation of B'-groups is taken from the work of Lindgren (64) wherein low-sulfonated lignin (S/OMe = 0.36) was assumed to contain 1−(S/OMe) = 0.64 B'-groups (65).

Thus, there is relatively firm experimental evidence for the mechanism in Equation 3 to represent the hydrolysis prior to sulfonation of B-groups in lignin. On the other hand, additional structures may also belong to the B-group category, such as the etherified phenylcoumaran and pinoresinol structures in lignin.

$$5 \xrightarrow{H^\oplus} 6 \xrightarrow{H_2SO_3} \quad (3)$$

There are still several important gaps in our knowledge concerning the structure of lignin sulfonates, and these gaps become quite apparent if one attempts to construct structural formulations for these macromolecules. In the representation of structure for softwood lignin, Figure 15.5, there are several elements whose sulfonation behavior has not been clarified by rate studies. These include, for example, coniferaldehyde and coniferyl alcohol groups as well as detached side-chain structures, represented by units 1, 15 and 3', respectively. It could be surmised, rightly or wrongly, that in neutral sulfonation coniferaldehyde groups form disulfonates, coniferyl alcohol groups remain unreacted, and detached side-chain structures react analogously to α-carbonyl units. If such assumptions are accepted, the formula in Figure 15.6 can be considered to represent the "solid lignin sulfonic acid." On equally speculative grounds, the formula in Figure 15.7 may be considered for a product obtained in extensive acidic sulfonation. The uncertainties in both formulations are obvious and call for continued structural investigations in this area.

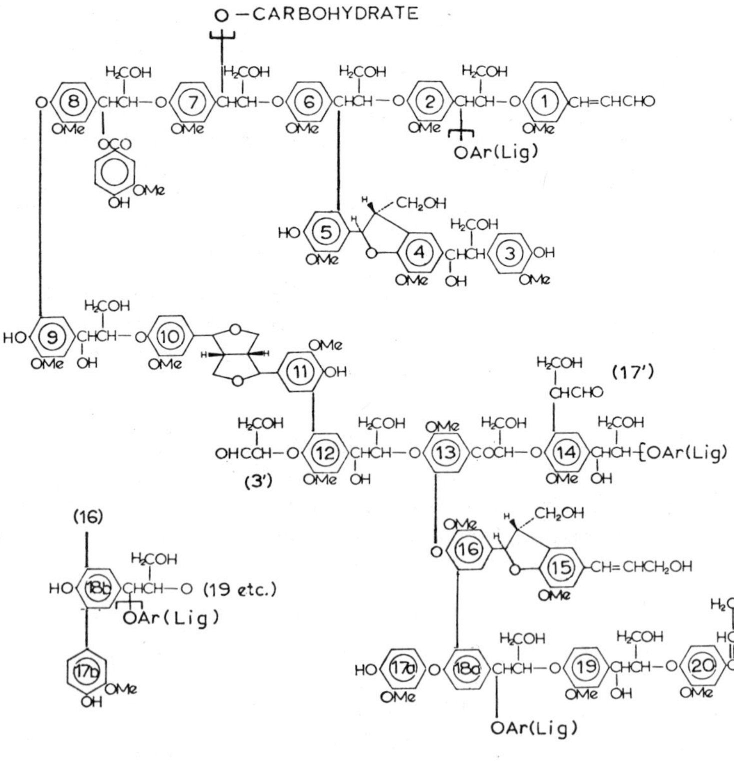

Figure 15.5. Schematic structure for a segment of spruce lignin.

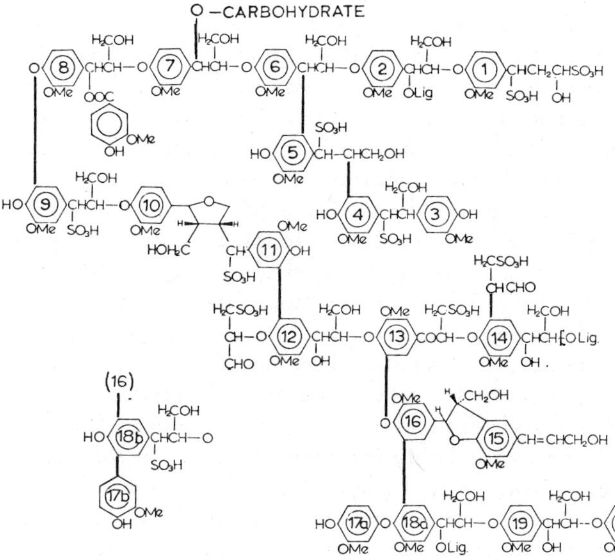

Figure 15.6. Structural sketch for insoluble lignin sulfonate, formed by sulfonation at pH 6 and 135°.

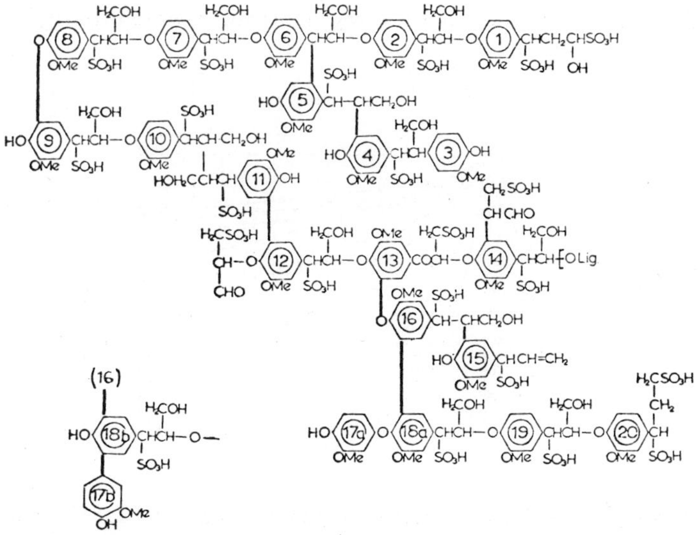

Figure 15.7. Structural sketch for extensively sulfonated lignin sulfonic acid.

Rather unexpectedly, the sulfonation pattern of milled wood lignins differs drastically from that of wood lignin. Thus, spruce milled wood lignin rapidly attains a value of about 0.51-0.57 S/OMe over a wide pH range during sulfonation (69), and about one-third of lignin is almost immediately dissolved (70). It is not easy to see why milled wood lignin should contain 3 to 4 times the amount of X-groups present in the total lignin in wood. It would seem more reasonable to assume that accessibility factors, together with chemical reactivity, determine the sulfonation pattern of wood.

Schuerch has offered an alternative explanation to account for the low level of wood lignin sulfonation. The sulfonated lignin in wood is present as a polyelectrolyte gel which is likely to repel bisulfite or sulfite ion in a Donnan membrane equilibrium (71). An apt example cited by Schuerch is a comparison of rates of saponification of polyvinyl acetate and polymethylmethacrylate. In the former case, the product is not ionized and the reaction goes to completion. In the latter case, apparently carboxylate anions attached to the polymeric backbone repel hydroxyl ions and prevent complete saponification.

Acidic sulfonation probably increases the level of sulfonation and extent of delignification in two ways: by hydrolysis of B-groups 5 to more reactive B'-groups 6, Equation 3, and by partial cleavage of the lignin polymer to permit swelling of the gel and increase the accessibility to bisulfite ion.

Rate of delignification by sulfonation of spruce wood chips at 135° is shown in Figure 15.8 (64, 68, 25). It may be noted that a minimum rate is obtained at about pH 7, which is slightly higher than the pH for the minimum rate for the sulfonation of vanillyl alcohol. Even at pH 7, about 10% of the lignin is dissolved in 3 hours and about 20% after 21 hours. Lignin sulfonates dissolved under these conditions are low-molecular-weight β-lignin sulfonates.

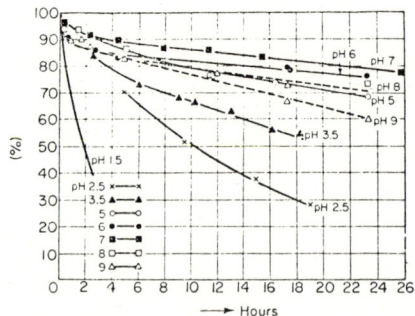

Figure 15.8. Sulfite delignification of spruce wood at 135°C as functions of pH and reaction time expressed as loss of methoxyl content in wood as % of original methoxyl content (25, 64, 68).

Rate of delignification of spruce and hemlock wood under acid sulfite pulping conditions is illustrated in Figure 15.9 (72). It is of interest to see that with 8% free SO_2, about 90% of the lignin was dissolved in about 2 hours; whereas with 4% free SO_2, the same degree of delignification required about 6 hours. Also, at about 90% lignin removal, the average molecular weight for dissolved lignin sulfonates is close to the maximum, as measured by diffusion (72), and decreases if sulfonation is continued.

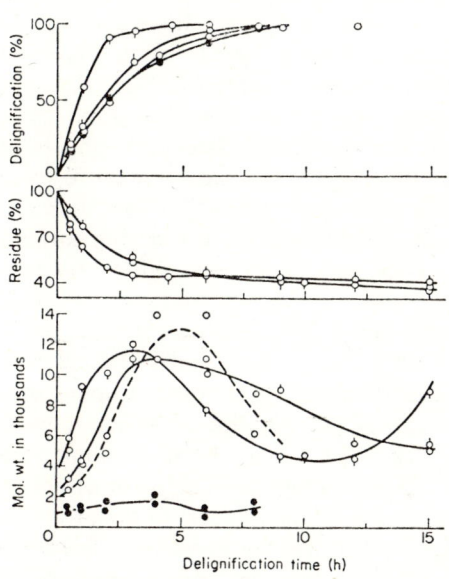

Figure 15.9. Percentage of delignification and of wood residue, and average molecular weights of dissolved lignin sulfonates, with time, at 135°C, by sodium-base acid-sulfite (72).

○ spruce, 3.1% free SO_2
● maple, 3.1% free SO_2
⊙ hemlock, 4.0% free SO_2
⊖ hemlock, 8.0% free SO_2

This suggests that hydrolyzable lignin-lignin (and possibly lignin-carbohydrate bonds are of at least two types, one relatively easily hydrolyzable in acidic aqueous solution and the other not. Lower molecular weight lignins dissolve first during delignification because the distribution of hydrolyzed linkages may be approximately random, and the smaller molecular weight lignins diffuse out from the wood residue into solution more rapidly than the larger fragments with the result that the average molecular weight of the dissolved lignin sulfonates increases as delignification proceeds. Acidic hydrolysis of the lignin sulfonates continues even after they have become dissolved so that although the average molecular weight reaches a maximum at about the time that the delignification is completed, thereafter the molecular weight decreases as hydrolysis proceeds. The average molecular weight reaches a minimum when presumably all of the hydrolyzable linkages have been hydrolyzed. An increase may then become evident in the average molecular weights of the lignins as a result of progress of irreversible condensation reactions which have presumably continued throughout the delignification process and can finally bring about condensation of the lignin sulfonates to a water-insoluble state (73).

Surprisingly, it was found that rate of delignification for maple wood was similar to that for spruce or hemlock, despite the much lower molecular weight of maple lignin sulfonates (72). This would seem to rule out diffusion of lignin sulfonates into solution as a rate-controlling step in sulfite delignification. Also, it was noted that delignification rate during sulfonation was about the same for wood chips as for wood meal reduced to less than 150-mesh particles (74).

Using 5.3% total and 1.3% combined SO_2 at pH 2, and observing sulfur uptake with time at 120°, it was found that softwoods reached a value for S/OMe of 0.3 in the undissolved lignin after 1 hour, increasing to 0.5 S/OMe in 12 hours reaction time (75). On the other hand, hardwoods (beech and birch) gave 0.1 and 0.2 S/OMe after 1 and 12 hours treatment, respectively (75). Allowing for the higher methoxyl content of hardwood lignin, S/OMe values of 0.1 and 0.2 probably correspond to about 0.13 and 0.26 SO_3H groups for each monomer unit, on the average. Therefore, it seems likely that hardwood lignins contain a lower proportion of A-groups than softwood lignins.

By comparing the rate of delignification in acid sulfite pulping with the additional delignification obtained by acid hydrolysis, it was demonstrated that both hydrolysis and sulfonation control the delignification rate (76). At low acidity, hydrolysis is rate-determining; and at high acidity, sulfonation may be rate-determining (25). Overall, the rate of delignification shows a good linear correlation with the partial pressure of sulfur dioxide regardless of base concentration (76a). Rydholm has pointed out, however, that there has been some disagreement on whether delignification is proportional to excess SO_2 as in Equation 4 or whether it is proportional to the free SO_2 as in Equation 5. The excess and free SO_2 are related to the sulfurous acid concentration and SO_2 vapor pressure, governed by the relationship in Equation 6.

$$\text{excess } SO_2 = [\text{total } SO_2] - 2[\text{combined } SO_2] \qquad (4)$$

$$\text{free } SO_2 = [\text{total } SO_2] - [\text{combined } SO_2] \qquad (5)$$

$$[SO_2]\text{ gas} = k[H_2SO_3]\text{ liq.} = k'[H^+][HSO_3^-] \qquad (6)$$

In Equation 6, H_2SO_3 stands for the sum of sulfurous acid with other hydrated SO_2 species (77). While Equation 7 is in accordance with available data, it is not clear whether or not undissociated SO_2 also may effect hydrolysis of lignin as well as participate in sulfonation (sulfitolysis) (25).

$$\text{Rate of delignification} = k''[H^+][HSO_3^-] \qquad (7)$$

In the case of technical sulfite pulping, reaction kinetics are complicated further by partial decomposition of pulping reagents and by side reactions with associated wood components, as disucssed in some detail by Rydholm (25) and by Gardner and Hillis (78).

    2. Kinetics of Sulfonation of Model Compounds. The kinetics of sulfonation of lignin model compounds have only been studied in detail in the case of vanillyl alcohol (79, 80). The rates obtained at different pH levels conform with the mechanism shown in Equation 8.

$$\underset{7}{\text{vanillyl alcohol}} \rightleftarrows \underset{8}{\text{quinone methide}} \xrightarrow{HSO_3^-} \underset{9}{\text{sulfonate}} \qquad (8)$$

In the pH range 3.5-7.0, the establishment of the equilibrium between vanillyl alcohol 7 and the corresponding quinone methide 8 is a slow process and governs the rate of the overall reaction. Both below pH 3.5 and above pH 7, sulfonation to 9 becomes rate-determining and consequently a gradual shift from first to second order kinetics has been observed (80). The reported activation energies range from 16.5 (79) to 24 kcal/mol (80). The sulfonation of vanillyl alcohol is roughly thirty times faster than that of X-groups in lignin (79). Similar kinetics may hold for formation of the two isomeric monosulfonate derivatives, 12a and 12b of coniferyl alcohol 10 via the extended quinone methide intermediate 11, as shown in Equation 9. In this case, the reaction is complicated by the formation of disulfonates and possibly dimeric or polymeric sulfonates, as discussed in Section VI-D.

$$\underset{10}{\text{coniferyl alcohol}} \rightarrow \underset{11}{\text{extended quinone methide}} \rightarrow \underset{12a}{\text{12a}} + \underset{12b}{\text{12b}} \qquad (9)$$

Lindgren made the significant observation that vanillyl alcohol is stable in toluene in a nitrogen atmosphere up to 130° but subject to polymerization on heating in water (81). This may help explain the resistance of wood to sulfonation after steam pretreatment (82, 83, 25). Furthermore, since the sulfonate, 9, is much more resistant to condensation reactions than vanillyl alcohol, it follows that delignification will be facilitated by sulfonation of condensation-prone p-hydroxybenzyl alcohol groups in lignin. To prevent condensation, it is important to achieve good penetration of wood with sulfite cooking liquor before raising the temperature. Some lignin units are hydrolyzed during sulfite pulping to release reactive p-hydroxybenzyl alcohol units which are then subject to condensation reactions, especially at low bisulfite concentration. Finally, despite good initial penetration of pulping liquor and use of high bisulfite concentration, portions of lignin not accessible to bisulfite may become condensed or cross-linked.

B. Topochemistry of Sulfonation and Delignification. It has been shown that lignin is a polydisperse polymer in wood (84, 85), but the disposition of low- and high-molecular-weight lignin is unknown. The fact that low-molecular-weight lignin is obtained first during sulfonation of wood has been taken to imply presence of two types of lignin in wood (25, 50, 86, 87).

Sulfite liquor penetrates wood about 50-100 times faster in the fiber direction than the transverse direction (88, 89). However, reduction of wood particle size had little effect on rate of delignification (74). Presumably, the coalescence of lignin (90), the repulsion of bisulfite and sulfite anions by insoluble sulfonated lignin (71), and pore size (91) all limit the rate of delignification. As sulfonation and hydrolysis of lignin proceed, soluble and insoluble lignin sulfonates may become less cross-linked and the carbohydrate matrix may expand to increase pore size and enable the larger lignin sulfonates to diffuse into solution (92). It has been suggested that the swelling effect may contribute to actual disruption of cell-wall structure. This effect may explain why sulfite pulp fibers possess inferior strength relative to kraft pulp fibers (91).

It is believed that initial flow of pulping liquor occurs in ray cells, followed by diffusion through pit membranes in cell lumen to contact sequentially the secondary wall, middle lamella and cell corners. Studies by various investigators have indicated that this topochemical sequence is followed in delignification (25, 93, 94). For the first 50% of lignin removed in acid sulfite and kraft pulping, it was estimated that lignin concentration was reduced by about 50% and 30% for the secondary wall (SW) and middle lamella (ML), respectively (94). If the original concentration in wood was about 12% and 70% in the SW and ML, respectively, it follows that at 50% delignification the residual lignin had decreased to about 6% in the SW and to 49% concentration in the ML region. It was suggested that lignin gels in the SW and ML differ in their resistance to dissolution by pulping liquor and that this difference may account for the observed topochemical pattern (94).

Following 50% delignification, there was a rapid increase in rate of removal of ML lignin which appeared to coincide with splitting of the ML and separation of fibers to open up the matrix to delignification (94). After about 80% delignification, residual lignin was predominantly in the cell corners and secondary wall (94).

In practice, the sulfite process is usually terminated at about 90-95% delignification to avoid loss of pulp yield and strength (Figure 15.1). Some information on the nature of insoluble sulfonated lignin remaining in an unbleached sulfite pulp is given in Table 15.2 (95). About 40% of the residual lignin sulfonates were removed from sulfite pulp by beating and 20% by extraction with dimethyl sulfoxide. It was concluded, as supported by the results in Table 15.2, that residual lignin sulfonates are probably most resistant to removal by virtue of their high molecular weights.

Table 15.2. Properties of Dissolved and Residual Lignin Sulfonates (96)[1]

	S/OMe	Phenolic Hydroxyl		Mol. Wt.
		Total	Condensed	
Dissolved LSA	0.4 -0.55	0.17-0.37	0.0 -0.03	4,000-40,000
Residual LSA	0.25-0.40	0.15-0.18	0.12-0.15	60,000-110,000

1. Ranges of values for fractionated lignin sulfonates in sulfite cooking liquor and residual insoluble LSA obtained by milling and solvent extraction.

The sulfite delignification of hardwoods is somewhat more complex. Goring and coworkers have shown that vessels contain guaiacyl lignin, secondary walls contain syringyl lignin and the rays and middle lamellae contain a mixture or copolymer of guaiacyl and syringyl lignins (95a). The sequential dissolution of these lignins in sulfonation may be followed by means of nitrobenzene oxidation. For example, lignin sulfonates dissolved in the first stage of neutral sulfite treatment of aspen wood gave only vanillin after oxidation with nitrobenzene and alkali (96). After about 40% delignification, lignin sulfonates yielded almost twice as much syringaldehyde as vanillin. It may be concluded that guaiacyl lignin in aspen (Populus tremuloides) is more reactive to sulfonation and delignification than guaiacyl-syringyl lignin (96). In the light of Goring's results (95a), the rapidly dissolved lignin sulfonates may derive from the vessel elements. Similar results were found from oxidation of lignin sulfonates obtained by successive treatment of aspen with bisulfite (97).

In the case of red alder wood, about 87% of the lignin was removed in three stages with bisulfite (98). Ultraviolet spectra for lignin sulfonates soluble in 80% aqueous ethanol (about one-half the total LSA) indicated a ratio of about 3:2 syringyl/guaiacyl units in lignin sulfonates and an indication of a slight trend to a higher syringyl/guaiacyl ratio with increasing degree of delignification (98).

It is interesting that after 60% delignification of aspen wood with aqueous butanol at 158°, extracted lignin contained about 20.5% methoxyl, decreasing in succeeding extractions to about 15% (99). These results indicate a reversed sequence of extraction of guaiacyl- and guaiacyl-syringyl lignins in comparison with sulfonation.

C. Use of Tracers in Sulfonation and Delignification Studies. It has been pointed out that ^{35}S is a relatively long-lived (half life: 87 days), low energy (0.166 Mev) beta emitter which is readily available and not particularly hazardous (100). Other radioactive isotopes of sulfur such as ^{31}S and ^{37}S have short half-lives and are less suitable for tracer work (101).

An interesting application of tracers was demonstrated by treating a low-sulfonated lignin (0.49 S/OMe) with $^{35}SO_2$, giving a radioactive lignin sulfonate with 0.68 S/OMe (102). The tagged sulfonic acid groups were preferentially released on heating with 0.2N NaOH at 100° (102). Thus, the last sulfonate groups to enter lignin were the first to leave. As pointed out by Brauns (23), however, there is question as to whether bound sulfonic acid groups can exchange with bisulfite ions in solution (101, 102).

In recent studies on lignin sulfonation, $^{35}SO_2$ has been used to monitor lignin sulfonates fractionated by gel filtration and to relate sulfur content to sulfonic acid group content (53, 103-109).

The use of tracer elements in pulping studies is still in its infancy and offers new opportunities for investigation. For example, the combination of ultraviolet microscopy and radiography would seem to be a valuable technique for obtaining more information on the topochemistry of sulfite treatments of wood using $^{35}SO_2$-bisulfite systems.

VI. Functional Groups in Lignin Sulfonates.

A. Relationship Between Functional Groups and Molecular Weight. In theory, lignin sulfonates represent an ideal type of lignin derivative for the clarification of the macromolecular structure of the parent lignins. Sulfonation is a reaction in which only few of the original lignin-lignin bonds are hydrolyzed and the extent of secondary bond-forming reactions is low. Furthermore, high-molecular-weight lignin sulfonates are readily soluble in water and thus can be characterized by such methods as light-scattering and viscometry (Chapter 17). For example, average molecular weights ranging from 10,000 to 40,000 are common among isolated lignin sulfonates, and values over 100,000 have been obtained for isolated high-molecular-weight fractions. An approximate degree of polymerization can be assigned by taking into account that lignin sulfonates usually contain one -SO_3Na group for two monomer units, and the average unit weight is consequently about 223 in lignin sulfonates deriving from softwoods. Thus, a DP-value of 50 is equivalent to a molecular weight of 11,200.

In the past, the main part of the characterization of lignin sulfonates has been carried out by McCarthy and his group at the University of Washington

(73, 110, 111) and by Forss and coworkers in Finland (53, 103-108). It should perhaps be noted that the latter group interprets their results in the framework of some exceptional ideas and terminology. For example, softwood lignin sulfonates are classified into the discreet groups of low-molecular-weight "hemilignin" (sulfonates) and "true" lignin sulfonates. The former correspond to about 20% of the original Klason lignin and are roughly equivalent to the classic β-lignin sulfonates. The "true" lignin sulfonates are presumed to derive exclusively from a single "repeating unit" consisting of 18 phenylpropane units. Hence, they can only have DP's that are multiples of 18. The idea of both compositionally and structurally identical "repeating units" is reminiscent of the hypothetical tetrameric "lignin building units" of F. E. Brauns. While the division of lignin sulfonates into two groups is supported by experimental evidence, the "repeating unit" hypothesis is on a shaky foundation and has received much criticism. Still, apart from the unusual interpretations and some obviously impossible structural formulations, the careful and detailed experimental results by Forss and coworkers deserve serious evaluation from a more conservative standpoint.

End-group analysis represents a useful tool in the characterization of natural and synthetic polymers. The application of this method to lignin sulfonates is excessively difficult, however. An example of lignin end-group is the coniferaldehyde group 13 which in lignin sulfonates exists as the sulfonated derivative 14 or sulfonated bisulfite addition product 15, Equation 10. While the quantitative determination of these groups is experimentally feasible, such determinations probably have little significance in the characterization of the polymer properties of lignin sulfonates for the following reasons: First, some lignin sulfonate molecules are devoid of coniferaldehyde end-groups, while others may contain more than one of these groups, as terminating branch ends for example; second, it is not excluded that some coniferaldehyde groups may exist in a bifunctionally linked form such as 16.

No known methods are currently available for the reliable determination of phenolic end-groups, i.e., monofunctionally linked units with a free phenolic hydroxyl group. On the other hand, the total number of phenolic hydroxyl groups can be determined, and from these values the number of phenolic ether bonds in various molecular weight fractions can be estimated. In this fashion it is possible to study whether or not interunit connections involving phenolic ether bonds (such as phenylcoumaran, β-O-4, α-O-4 and 5-O-4 linkages) represent the same fraction of interunit bonds, regardless of molecular size. The fraction of phenolic ether bonds may be calculated for various molecular weight fractions from Equation 11, valid for linear and branched lignin polymers alike. Of course, interunit unit bonds outside of the phenol ether category are represented by such carbon-carbon bonds as the β-β and 5-5 linkages.

$$\text{Phenolic ether bonds as a fraction of the total interunit connections} = \frac{(1 - \text{PhOH})\overline{DP_n}}{\overline{DP_n} - 1} \qquad (11)$$

where PhOH is the fraction of monomer units with a free phenolic hydroxyl group and $\overline{DP_n}$ is the number average degree of polymerization

Felicetta, Ahola and McCarthy (110) have studied a number of lignin sulfonates obtained by fractional dissolution in ethanol-water mixtures obtained from a purified acid sulfite liquor. Molecular weights were determined by diffusion, and phenolic hydroxyl content estimated by ultraviolet difference spectra (111). The results are summarized in Table 15.3, which also includes calculated values for the fraction of phenolic ether bonds. The latter values show considerable variance between low- and high-molecular-weight fractions.

Table 15.3. Phenolic Hydroxyl Content in Fractionated Lignin Sulfonates (110)[1]

Fraction No.	Cum.% Yield	Mol. Wt.	Degree of Polymerization[2]	Ph-OH/MeO[2]	Fr. of Phenol Ether Bonds
4	7.6	500	2.27	0.52	0.86
7	14.5	940	4.27	0.48	0.68
11	27.9	2500	11.37	0.30	0.77
14	53.6	10000	45.5	0.22	0.80
17	63.5	13000	59.0	0.16	0.85
21	87.8	31000	141.0	0.14	0.86

1. From acid sulfite treatment of western conifers and fractional precipitation in ethanol.
2. Assuming a molecular weight of 220/unit of methoxyl.

However, the results in Table 15.3 are in contradiction with those obtained by Forss and Fremer (104) on fractions obtained by gel filtration of spruce lignin sulfonates, see Figure 15.10. In this study, molecular weights of individual fractions were not determined, although that of the highest fraction was presumed to be over 40,000. Phenolic hydroxyl contents were in this case determined by the periodate method. The results by Forss and Fremer are suggestive of a remarkably constant value of 0.71 for the fraction of phenolic ether bonds in the majority of lignin sulfonate fractions. This value is also in good accord with the phenolic hydroxyl content of milled wood lignins.

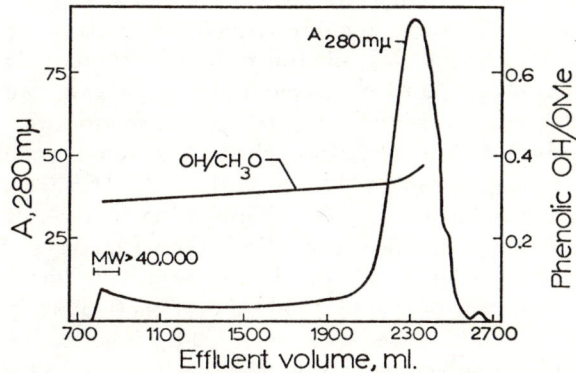

Figure 15.10. Plot of ultraviolet absorbance at 280 mµ and phenolic hydroxyl content of lignin sulfonates with elution volume by gel filtration (104).

In view of the inherent uncertainties of the UV method in the determination of phenolic hydroxyl contents, more reliance should be put on the results of Forss and Fremer. The UV method fails to detect weak phenolic hydroxyl groups, such as are present in compounds of the type 17 and 18, unless solutions are sufficiently alkaline (about pH 14) to cause complete ionization (112). Since the proportion of condensed units increases with increasing molecular weight of lignin sulfonates (23, 25, 63, 109), the phenolic hydroxyl contents of high-molecular-weight sulfonates measured by ultraviolet difference spectra are probably low. Even the periodate method may not fully account for the phenolic hydroxyl content of lignin sulfonates because hydroxyl groups in catechol units 19 remain undetermined.

B. Phenolic Hydroxyl Groups.

1. Content in Lignin Sulfonates from Bisulfite and Acid Sulfite Treatments. Mikawa and Ebisswa have used the difference in acidity to determine phenolic hydroxyl content of non-condensed units (Type I), which are conductometrically titratable, and phenolic hydroxyl content of condensed units represented by units 4 and 15 in Figure 15.7 (Type II), which are not titratable. It was found that α-lignin sulfonates from acid sulfite treatment of softwood gave a value for II/I of about 7/3 (113). A similar ratio was reported for purified lignin sulfonates from various commercial spent-sulfite liquors (114). Since low-sulfonated lignin obtained by sulfonation at pH 5-9 (A-groups only) gave a value for II/I of about 1:1, similar to that for untreated wood lignin, it was concluded that neutral sulfite treatment did not release free phenolic hydroxyl groups. Increase in the ratio for α-lignin sulfonates was attributed to acid sulfite hydrolysis of B-groups of the phenyl coumaran type as demonstrated by model compound experiments (115), Equation 12. (Compare also units 4 and 15 in Figures 15.6 and 15.7). Acid sulfonation of low-sulfonated lignin was estimated to release about 0.1 phenolic hydroxyl groups per methoxyl in this manner (116). This result is in good agreement with data for increase in phenolic hydroxyl content of milled wood lignin from 0.34 to 0.44/MeO after sulfonation with SO_2 in dioxane, as determined by cobaltamine titration (38). However, Freudenberg and coworkers ascribe the increase to hydrolysis of p-hydroxy and p-alkoxybenzyl ether bonds and not to phenylcoumaran units, since phenolic hydroxyl was not released from 20 under these sulfonation conditions (38). Instead, hydrolysis or "acidolysis" of the phenylcoumaran, 20, to phenylcoumarone, 21, Equation 13, is a likely process (117).

2. Phenolic Hydroxyl in Lignin Sulfonates from Neutral Sulfite Pulping.

Gierer and Gellerstedt (117a) have recently studied the behavior of model compounds during treatment with neutral sulfite at 180° and shown that unlike bisulfite or acid-sulfite sulfonation at 135°, appreciable cleavage of β-arylether linkages can occur (117a), Equation 14. Thus treatment of compounds 22 and 23 with neutral sulfite (pH 7) at 180° for 3 hours gave about 80% cleavage of the β-arylether bond with release of styrene-ω-sulfonic acid 24 and guaiacol, with minor amounts of catechol and methane sulfonic acid (117a). In contrast, these cleavage reactions are absent in acidic sulfonation at 135° where the α-sulfonates are the main reaction products. Methylation of the free phenolic hydroxyl in 22 and 23 reduces the cleavage to about 13%, which establishes the activating effect of a free phenolic hydroxyl group on the side-chain cleavage (118).

As pointed out by the authors (117a), the mechanism formulated for lignin in Equation 15 suggests that neutral sulfite pulping of softwoods should result in extensive delignification. Since this is not the case, it is proposed that the polymerization of styrene-ω-sulfonic acid structures interferes with delignification. The formation of catechol groups contributes to the dark color typical of high yield neutral sulfite pulps.

C. The Sulfonate Group.
 1. Relationship of Sulfonate Group Content to Sulfonation Conditions and Polydispersity of Lignin Sulfonates. Incremental delignification has been applied to western hemlock wood to study the detailed characteristics of dissolved fractions with the results shown in Table 15.4 (118). The liquor was separated after each of five successive stages of delignification, pulp yield and content of S and MeO in pulp measured, and molecular weights of dissolved lignin sulfonates determined. Molecular weights also were determined for each stage after hydrolysis with dilute hydrochloric acid.

Table 15.4. Incremental Delignification of Hemlock Wood (118)[1]

	Increment No.				
	I	II	III	IV	V
Wood Residue:					
Weight, %	-	82.1	72.9	61.1	51.5
S, %	1.08	1.15	1.09	0.67	0.25
OCH_3, %	3.76	3.85	2.81	1.40	0.40
S/OCH_3	0.29	0.30	0.39	0.48	0.62
Delignification, %	26.7	36.3	55.6	81.3	95.5
Diss. Lignin Sulfonates:					
Mol. wt.	3,400	6,600	14,500	30,000	20,500
M.W. after hydrolysis	1,900	4,800	7,200	10,500	13,000

1. Stepwise acid sulfite treatment to give five stages of wood residue and dissolved lignin sulfonates.

It may be seen that the sulfonate content increased from 0.29 to 0.62 S/OMe for undissolved lignin sulfonates, and the molecular weight of dissolved lignin sulfonates increased from 3400 to a maximum of 30,000 in stage IV, decreasing to 20,500 in stage V. However, after hydrolysis with HCl, molecular weights show essentially a linear increase from 1900 to 13,000 molecular weight with degree of delignification as shown in Figure 15.11. Further acid sulfonation of a non-dialyzable lignin sulfonate (average mol. wt. 44,000) gave the results in Table 15.5 (118). Here, it may be seen that extended acid sulfonation of higher molecular weight lignin sulfonates resulted in hydrolysis (including formation of dialyzable lignin sulfonates below about 5,000 M.W.) and uptake of sulfonate groups so that the final S/OMe approached 0.8 after 48 hrs

It appears that the molecular weights of the dialyzable lignin sulfonates fromed in each sulfonation treatment were quite low, in order to account for the large difference in molecular weights of lignin sulfonates before and after dialysis and the relatively small amount of lignin sulfonates rendered dialyzable. Similar results were reported for incremental delignification of spruce wood by continuous percolation of acid sulfite liquor and re-sulfonation of lignin sulfonates (87).

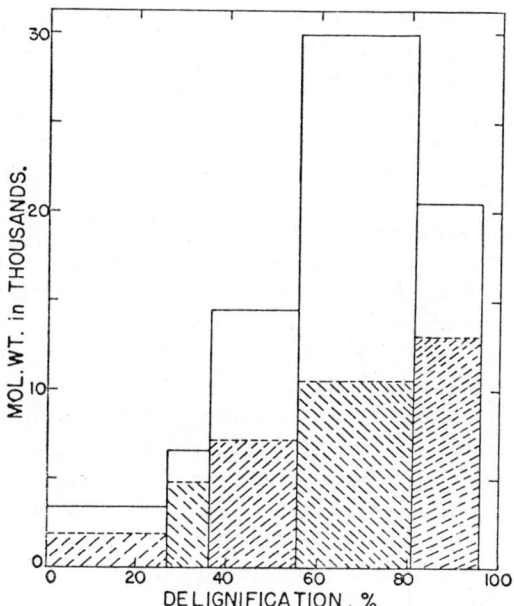

Figure 15.11. Change in molecular weight of lignin sulfonates obtained by incremental delignification with acid sulfite treatment of hemlock wood (118).

  ⊓ before hydrolysis

  ▨ after hydrolysis
    with 0.01N HCl

Table 15.5. Acid Sulfonation of Western Hemlock Lignin Sulfonates (118)[1]

Property	Sulfonation time in hours						
	0	1	2	4	8	24	48
Total LSA							
Av. Mol. wt.	44,000	25,000	19,500	17,000	16,000	12,000	14,000
Non-dialyzable LSA							
Wt. loss in dialysis,%	0	14	9	15	10	18	15
S/OMe	0.52	0.52	0.54	0.56	0.60	0.70	0.80
Av. Mol. wt.	44,000	--	31,000	30,000	28,000	23,000	16,000

1. Treatment of a non-dialyzable lignin sulfonate (M.W. 44,000) with 0.75% combined and 4.7% free SO_2 at 130° for 1-48 hr; each treatment starting with the same LSA (M.W. 44,000) followed by dialysis.

Similarly, Forss and coworkers recently obtained evidence indicating that re-sulfonation of high-molecular-weight lignin sulfonates with acid sulfite yields dialyzable lignin sulfonates containing a high proportion of very low-molecular-weight lignin sulfonates, possibly of monomeric and dimeric size (108). On the other hand, acid hydrolysis of the same lignin sulfonates with dilute HCl likewise yielded low-molecular-weight products, but the amount of sulfonates was much reduced. One may conjecture that in acid-sulfite cooking liquor, reactive groups such as p-hydroxybenzyl alcohol groups are stabilized through sulfonation; whereas mineral acid hydrolysis permits some condensation reactions to occur. By extending the acid sulfonation reaction, S/OMe can increase from about 0.5 to 1.0 and hydrolysis and condensation reactions increase the proportion of both low- and high-molecular-weight lignin sulfonates. Thus, the polydispersity of lignin sulfonates increases in extended sulfonation.

2. Presence of Non-Sulfonic Acid Sulfur in Lignin Sulfonates. It was noted by Samuelson and Westlin that lignin sulfonates contain more sulfur than can be accounted for in sulfonic acid groups (119). The amount of "neutral" sulfur, according to Mikawa and coworkers (120), can be as high as 0.11/OMe. The formation of "neutral" sulfur is not connected directly with sulfonation but is rather a consequence of a reaction of lignin with thiosulfate. The latter forms gradually in sulfonation liquors through the reduction of sulfite by sugars or, sometimes, by phenolic extractives (25).

Goliath and Lindgren (121) have studied the formation of "neutral" sulfur in the sulfonation of vanillyl and veratryl alcohols in the presence of thiosulfate. Due to the pronounced nucleophilicity of the latter species, thiosulfation competes effectively with sulfonation as shown in Equation 16. The thiosulfonate 25 formed reacts further forming divanillyl (or diveratryl) sulfide 26. However, only the sulfide formed from veratryl alcohol is stable, while divanillyl sulfide 27 reacts with excess sulfite generating the normal sulfonate, Equation 17.

$$Ar-CH_2OH \xrightarrow{H_2S_2O_3} \underset{25}{Ar-CH_2S_2O_3H} \xrightarrow{Ar-CH_2OH} \underset{26}{(ArCH_2)_2S} \quad (16)$$

$$\underset{27}{(\text{G}-CH_2)_2S} \xrightarrow{2H_2SO_3} 2\,\text{G}-CH_2SO_3H + H_2S \quad (17)$$

The cleavage of the sulfide, Equation 17, is followed by formation of polythionate, which accompanies the oxidation of H_2S. The fact that bisulfite sulfonation of MWL (135°, 3 hr.) in the presence of added thiosulfate results in some loss of thiosulfate but no increase in polythionate was interpreted to

REACTIONS IN SULFITE PULPING 623

mean that only stable sulfide linkages are formed; that is, only those of the veratryl sulfide type. Thus, it may be concluded that the Z- and B'-groups may be linked to some extent by sulfide bonds, depending on the extent of thiosulfate formation. As pointed out by Rydholm, thiosulfate formation is more likely to be a factor in technical sulfite liquors than in laboratory liquors (25).

Additional model compound work in this area appears desirable. For example, the sulfide 28 is less stable to strong acid hydrolysis than is veratryl sulfide (122). From this it seems likely that it would also be less stable to sulfonative cleavage or sulfitolysis.

$$(MeO-C_6H_3(OMe)-COCH_2-)_2S$$
28

An equivalence of cation (for example, Ba^{2+}) and total sulfur content for a given lignin sulfonate does not necessarily imply equivalent contents of cation and sulfonic acid groups since neutral sulfur may be counterbalanced by content of cations associated with carboxyl groups, see Section E.

3. Location of the Sulfonate Groups. On the basis of model compound work, sulfonate groups are generally believed to occupy α- or γ-positions in lignin units. Gierer, Alfredsson and Soderberg have attempted to clarify the actual location of these groups using acidic hydrolysis (123). They demonstrated that about one-third to one-half of the sulfonate groups in lignin sulfonates are quite resistant to acid cleavage using HI, HCl or H_3PO_2 at 140°. Reactions of various model compounds indicated substantial differences in desulfonation rates. While α-sulfonates deriving from phenylcoumaran and pinoresinol models, 29 and 30, respectively, desulfonated very rapidly; the rate for the β-arylether α-sulfonate 31 was considerably slower, Figure 15.12. A β-carbonyl group likewise retarded the desulfonation of α-sulfonic acid groups. α-Keto-γ-sulfonate 2 desulfonated extremely slowly, and the same was true for a number of β-sulfonate models.

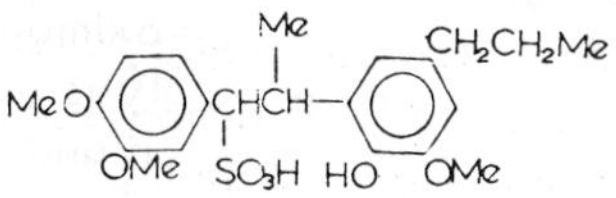

29 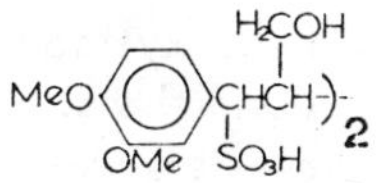   30 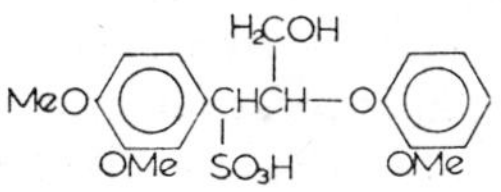   31

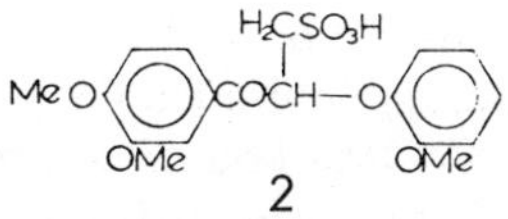

2

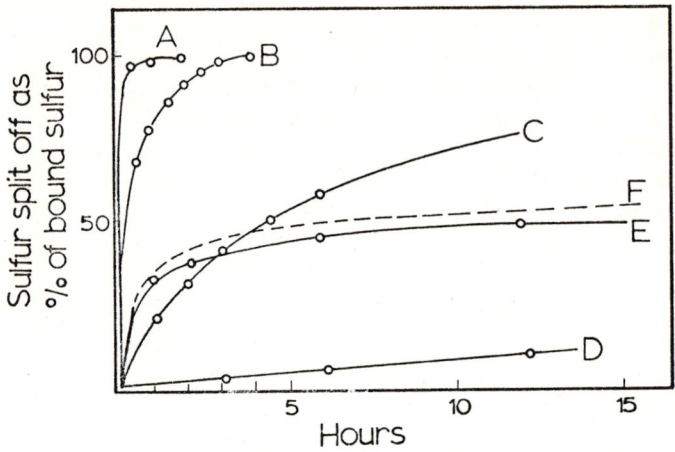

Figure 15.12. Rate of sulfur removal by acid cleavage using HI, HCl or H_3PO_2 at 140°: Compound 27 (Curve A), 28 (Curve B), 29 (Curve C), 11 (Curve D), a 1:1:1:3 mixture of 27, 28, 29 and 11, respectively (Curve E), and of "low-sulfonated lignin sulfonic acids" (Curve F) (123).

The rate of sulfur removal from a low-sulfonated lignin ("Kullgren acid," curve F, Figure 15.12) and that calculated from desulfonation rates for a 1:1:1:3-mixture of compounds 29, 30, 31, and 2 (curve E) conformed very well with each other. Of course, no firm conclusions can be drawn from this observation, but it is not impossible that one-half of the sulfonate groups in "Kullgren acid" are represented by α-keto-γ-sulfonate structures.

4. Alkaline Hydrolysis of Lignin Sulfonates. The free phenolic hydroxyl content of unsulfonated lignin is about 0.33/OMe and is not appreciably increased by mild alkaline hydrolysis. However, there is approximately a two-fold increase in phenolic hydroxyl content after alkaline hydrolysis of lignin sulfonates under moderate conditions. Data obtained by Peniston and McCarthy (124) for alkaline hydrolysis of non-dialyzable lignin sulfonates (>5000 M.W.) with 5% NaOH at 100° for 5 hours are summarized in Table 15. About 19.1% of the total non-dialyzable lignin sulfonates were rendered water-insoluble (Fraction A); and about one-half of the soluble lignin sulfonates became dialyzable (B_d), with the other half remaining non-dialyzable (B_{nd}). It was estimated that before hydrolysis the lignin sulfonates contained a phenolic hydroxyl content of about 0.2/OMe and an S/OMe ratio of about 0.5, changing upon hydrolysis to 0.66 and 0.41, respectively.

Table 15.6. Alkaline Hydrolysis of Lignin Sulfonates (124)[1]

Property	Fraction		
	A	B_{nd}	B_d
% of total LS	19.1	32.4	32.3
S/OMe	0.22	0.43	0.54
Ph-OH/Meo	0.41	0.52	0.95

1. Treatment of a non-dialyzable lignin sulfonate with 5% NaOH at 100° for 5 hr; A: insoluble LS, B_{nd}: soluble non-dialyzable lignin sulfonates; B_d: soluble dialyzable lignin sulfonates.

What is at least a partial explanation for these results is revealed through studies on alkaline hydrolysis of sulfonated lignin models by Kratzl and co-workers (125-127). Specifically, it was found (127) that while β-O-4 lignol structures are resistant to alkaline cleavage, under the same conditions the corresponding α-sulfonates, illustrated by the model, 32, are hydrolyzed to vanillin and acetaldehyde, Equation 18.

$$-O\langle\bigcirc\rangle\underset{\underset{SO_3H}{OMe}}{\overset{\overset{H_2COH}{|}}{CHCH}}-O\langle\bigcirc\rangle C_3 \xrightarrow{OH^-} HO\langle\bigcirc\rangle CHO + CH_3CHO + HO\langle\bigcirc\rangle C_3 \quad (18)$$
$$\underset{32}{} \qquad\qquad OMe \qquad\qquad OMe \qquad\qquad\qquad OMe$$

After alkaline hydrolysis, the three lignin sulfonate fractions in Table 15.6 have a composite value of about 0.41 S/OMe, indicating a loss of about 0.1 S/OMe during alkaline hydrolysis. This is much less than the increase of phenolic hydroxyl groups, about 0.4/OMe, and contradicts the equivalence of both losses predicted by Equation 18. The low degree of desulfonation in the alkaline hydrolysis of lignin sulfonates has been confirmed by Turunen (128) in connection with his studies on the Rauma process (acid sulfite pulping followed by an alkaline hydrolysis stage). In general, the S/OMe-ratios of lignin sulfonates (0.5 to 0.6/OMe) were found to remain essentially unchanged in the alkaline hydrolysis, regardless of increasing in phenolic and aliphatic hydroxyl groups, as well as in carbonyl and carboxyl groups. Turunen has proposed that the desulfonation reaction, which could well be of the type shown in Equation 18, could be counterbalanced by a resulfonation process; that is, by a recapture of the liberated sulfite by other reactive groups in the lignin sulfonate molecule. While this proposition merits consideration, further detailed studies are needed for its verification.

D. The Carbonyl Group.

1. Behavior and Formation of Carbonyl Groups During Sulfonation. Carbonyl groups in protolignin are represented by γ-carbonyls in coniferyl aldehyde groups, α-carbonyl groups and, presumably, aldehyde groups in detached side-chains, representing the "non-conjugated" category of carbonyl groups. Examples of these groups are shown in units 1 and 13, and detached side-chains 3' and 17' in Figure 15.6.

Coniferaldehyde units in lignin and in lignin sulfonates have been studied extensively (129-139). Hydrolysis of hemlock wood yielded coniferaldehyde (136) and, in an analogous manner, sulfonates of coniferaldehyde and sinapaldehyde were shown to be present in lignin sulfonates from red alder (137). Under mild alkaline conditions, sulfonated coniferaldehyde units are readily desulfonated and this provides a means of determining the amount present (129-132). It has been estimated that about 0.01 coniferaldehyde units contain a free phenolic hydroxyl group and about 0.03 units are etherified (129-132). Coniferaldehyde units presumably act as A-groups during sulfonation, but reaction kinetics have not been examined in detail.

As discussed in Section V-A, model experiments suggest that units containing α-carbonyl groups become sulfonated at the γ-position (Equation 2), and this reaction is especially rapid in the neutral pH region. In absence of model experiments, the sulfonation behavior of detached side-chain structures is uncertain. By analogy with the α-carbonyl structures, sulfonation of the primary carbinol groups in these side-chains is the probable reaction mode. Tentatively, at least, both α-carbonyl and detached side-chain structure appear to belong to the category of the so-called A-groups.

In acidic sulfonation, acidolysis of β-O-4 lignol structures may augment the number of carbonyl groups through the formation of side-chain structures of the Hibbert's ketone type (142) such as the acyloin side-chain in structure 33.

HO–⟨○⟩–CO-CH-CH$_3$
OMe   OH
33

Model ketol 33 has actually been found to be resistant to sulfonation (26, 140). Consequently, secondarily formed carbonyl groups are unlikely to enhance sulfonation. Combined acidolysis and sulfonation could also produce some guaiacylglycerol-α-sulfonic acid structures, but the significant presence of the latter groups is overruled by the proven absence of vicinal hydroxyl groups in lignin sulfonates (0.05/MeO) (141). On the other hand, threo- and erythro-syringylglycerol-α-sulfonates have been obtained in the sulfonation of black wattle wood (138, 139).

Attempts have been made to determine reducing groups in fractionated lignin sulfonates (103, 104, 143). The results indicated a decrease with increasing D.P. and decreasing sulfur content. The interpretation of these

REACTIONS IN SULFITE PULPING 627

results, however, remains obscure because the possible contribution by catechol groups as well as by associated reducing sugar moieties has not been clarified.

Additional reactions forming carbonyl groups in sulfonation are not ruled out. It is well-known that sulfites may act as both reducing agents and oxidizing agents, depending on the substrate. For example, the inhibiting action of dihydroquercetin on sulfite cooking of Douglas-fir has been attributed to oxidation of dihydroquercetin to quercetin and reduction of bisulfite to thiosulfate (144).

2. *Effect of Reduction on Sulfonatability*. It was noted by Gierer that Brauns lignin and lignin in wood meal after reduction with sodium borohydride ($NaBH_4$) was more resistant to sulfonation than the untreated lignins (145). Similarly, Sanyer, Itoh and Keller reported that delignification of Douglas-fir wood proceeded at a slower rate following borohydride reduction. They used a pulping liquor of sulfite or a combination of sulfite-soda ash, sulfite-sulfide and sulfide-caustic (146).

The decrease in sulfonation as a consequence of borohydride reduction is not easily interpreted in chemical terms. Reduction of α-keto groups would, of course, cancel the sulfonatability of the γ-position, but would at the same time introduce a new sulfonatable group in the α-position. Likewise, the effect of reducing coniferaldehyde to coniferyl alcohol groups should not be major because the latter groups can assume at least one, often two, sulfonic acid groups. However, reduction of detached side-chain structures 34 and 36 would result in unsulfonatable groups 35 and 37, as shown in Equations 19 and 20. It is not clear at this point whether or not reductions of the last type can adequately explain the lowered reactivity of reduced lignin.

$$\underset{34}{\underset{\text{CHO}}{\underset{|}{\overset{H_2COH}{\overset{|}{\underset{}{CH-O}}}}}\text{-}\!\!\bigcirc\!\!\text{-}C_3\,\text{OMe}} \quad \xrightarrow{NaBH_4} \quad \underset{35}{\underset{H_2COH}{\underset{|}{\overset{H_2COH}{\overset{|}{\underset{}{CH-O}}}}}\text{-}\!\!\bigcirc\!\!\text{-}C_3\,\text{OMe}} \qquad (19)$$

$$\underset{36}{\overset{HOH_2C\ CH\ CHO}{\underset{}{-O\!\!\bigcirc\!\!-C_3\,OMe}}} \quad \xrightarrow{NaBH_4} \quad \underset{37}{\overset{HOH_2C\ CH\ CH_2OH}{\underset{}{-O\!\!\bigcirc\!\!-C_3\,OMe}}} \qquad (20)$$

It has been estimated that coniferyl alcohol units are present in softwood lignin to the extent of less than 0.05/OMe (147, 148). These units are believed to be converted to monosulfonates 38 and 39 identified as monomers among lignin sulfonates from extractive-free western hemlock, Equation 21 (149-151).

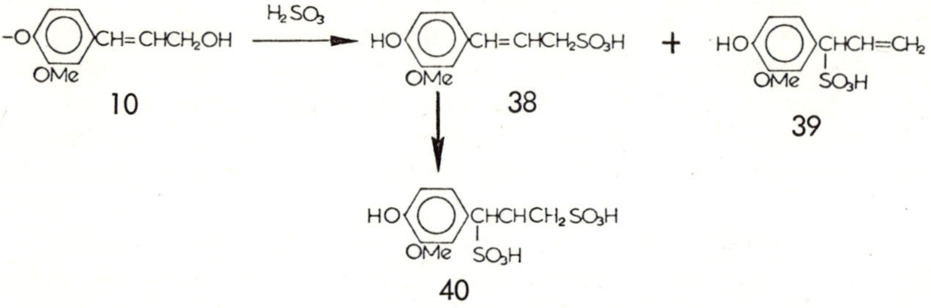

$$\text{-O-C}_6\text{H}_3(\text{OMe})\text{-CH=CHCH}_2\text{OH} \xrightarrow{\text{H}_2\text{SO}_3} \text{HO-C}_6\text{H}_3(\text{OMe})\text{-CH=CHCH}_2\text{SO}_3\text{H} + \text{HO-C}_6\text{H}_3(\text{OMe})\text{-CHCH=CH}_2\;(\text{SO}_3\text{H})$$

10 → 38 + 39

↓

40: HO-C$_6$H$_3$(OMe)-CHCHCH$_2$SO$_3$H (SO$_3$H)

(21)

In an analogous manner, monosulfonate 42, identified first among lignin sulfonates of wattle wood (138, 139), and later in lignin sulfonates of red alder (137), may be formed by sulfonation of etherified sinapyl alcohol units, Equation 22. The fact that 42 was obtained by fairly moderate sulfite treatment of red alder (pH 4.5, 132°, 4.5 hr) suggests that certain sinapyl alcohol units may be attached through the phenolic hydroxyl group by a benzylic ether bond or some other bond which is relatively easily hydrolyzed. It has been shown that sinapyl alcohol also forms the 1,3-disulfonate 43, which is found among the monomeric lignin sulfonates from wattle wood (139). This finding supports evidence reported for the formation of the desulfonate 40 by sulfonation of coniferyl alcohol and monosulfonates 38 and 39 (152).

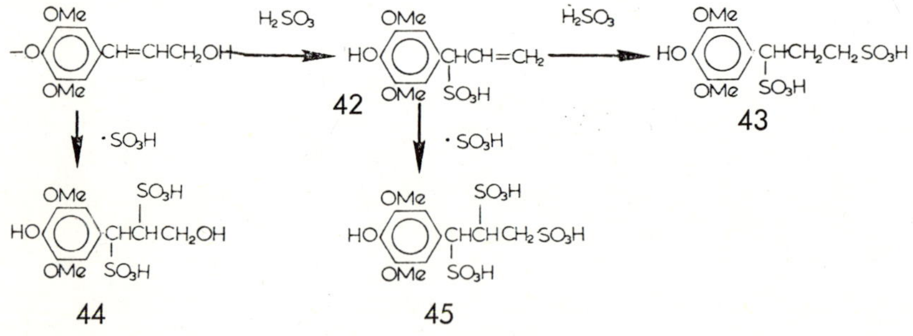

Equation (22): sinapyl alcohol derivative 10 → 42 → 43, with side products 44 and 45.

No evidence was found for presence of dimeric or polymeric sulfonates in products from sulfonation of sinapyl alcohol (138). However, there is some evidence to suggest that coniferyl alcohol may yield dimeric and possibly higher molecular weight sulfonates (152) formed in all likelihood through acid-catalyzed polymerization prior to sulfonation. In this connection, it is interesting to note the elementary composition of lignin sulfonates obtained from spruce wood, an enzymatic dehydrogenated polymer of coniferyl alcohol (DHP) and coniferin, the glucoside of coniferyl alcohol, Table 15.7 (153). The dimeric

Table 15.7. Elementary Composition of Various Lignin Sulfonates (153)[1]

Source	C	H	N	S	OMe
Spruce wood	53.20	5.38	3.38	7.16	13.08
DHP	54.60	5.58	3.65	6.62	13.65
Coniferin	53.40	5.82	4.18	8.59	13.26

1. Lignin sulfonates prepared by heating the source at 105° for 3 days and precipitating LSA with dihydrobenzacridine hydrochloride and subsequently converting LSA to ammonium salts.

sulfonate formed from coniferyl alcohol has about the same composition found for the sulfonated product of coniferin.

It was shown that 39 is converted to the conjugated isomer 38 by treatment with sodium sulfite, but the corresponding conversion of the syringyl analogue 42 probably gives only the disulfonate 43 (138, 152). From this it may be concluded that neutral sulfite pulping of softwoods should give predominantly 38, whereas in bisulfite or acid sulfite pulping the isomer 39 should predominate (148, 152). Extended sulfonation of hardwoods and/or softwoods yields the disulfonates 40 and 43.

Free radical sulfonation of sinapyl alcohol has been studied by Parrish (138) using addition of nitrite to the sulfonation system. The disulfonate 44 with two sulfonate groups adding to the double bond was obtained as the major product. Also, a minor product was isolated which may be the trisulfonate 45 formed by addition of two sulfonate groups to 42, Equation 22.

E. Carboxyl Groups in Lignin Sulfonates. Mikawa and coworkers have attempted to determine carboxylic acid groups in lignin sulfonates by conductometric titration and ultraviolet difference spectra before and after methylation (120). It was assumed that the "excess" barium found in their preparation was required by presence of 0.075 CO_2H/OMe; that is, about one carboxyl group in 13 guaiacylpropane units (120). Later analysis of various technical lignin sulfonates by James and Tice (114) has given higher values for carboxyl content, ranging from 0.14 to 0.29/OMe. That titrated weak acids are chiefly carboxyl groups and not strong phenolic hydroxyl groups was demonstrated from infrared spectra of lignin sulfonic acids. The carboxyl peak at 1710 cm^{-1} was prominent in lignin sulfonic acids at pH 2.2 but was shifted under the aromatic absorption peak at high pH due to formation of the carboxylate anion. Semiquantitative estimates of carboxylic acid groups can be based on this shift. Although lignin sulfonic acids were shown to be free of sugars (114), sulfonated sugar acids may have been present in appreciable amount (154, 155, 156), which would add to the value for carboxyl content.

It has been suggested that carboxyl groups may be attached at the 1- or

5-position of guaiacyl nuclei in lignin sulfonates (114, 154), but the evidence for this is not very compelling. It has been shown, however, that sulfonation of veratraldehyde (pH 4, 155°, 2 hr.) produced a 6.3% yield of vanillic acid (157) and α-keto groups in lignin may give rise to carboxyl groups (158).

Another possible source of carboxyl groups in lignin sulfonates may be the original carboxylic acid and lactone groups in lignin as proposed by Freudenberg (159, 160). It is also possible that sulfonated lignan lactones deriving from matairesinol (158) might contribute to the carboxyl content of lignin sulfonates from such species as western hemlock.

F. Catechol Groups Formed by Demethylation. Presence of catechol nuclei in lignin sulfonates has been suspected for some time (157). In studies on gelling of lignin sulfonates, conductometric titration with ferric chloride indicated the presence of catechol groups in an amount of 0.1/OMe (161). This has been confirmed by ethylating lignin sulfonates with diethyl sulfate, oxidizing with permanganate and determining the yields of carboxylic acids formed by gas chromatographic analysis of their methyl esters (162). The yield of methyl 3,4-diethoxybenzoate (from catechol units) was about 10% of that for methyl 4-ethoxy-3-methoxy benzoate (from guaiacyl units) (162). Acid sulfite treatment of vanillyl alcohol yielded methanol in an amount corresponding to about 4% demethylation (162).

Sulfonation has also been found to effect minor degrees of demethylation in other model phenols and phenol ethers (163, 164). Unexpectedly, an exceptionally high degree of demethylation (13.5%) was established for eugenol. In all the experiments cited above, the extent of demethylation was estimated from the formation of methanol alone. Gierer (117a) has subsequently demonstrated that methane sulfonic acid is generated from methoxyl groups during neutral sulfite treatments of model compounds, which constitutes another pathway for catechol formation.

To explain the gelling reaction of lignin sulfonates with dichromate, various mechanisms involving catechol units have been proposed (165-167). Hayashi and Minari have suggested that dichromate itself may also be capable of demethylating guaiacyl units and that this reaction is accelerated by addition of ferric chloride and cupric chloride (168).

These same authors also report that ferric ion associated with lignin sulfonates contributes to a darker color (168). About one-half of this ferric ion could be removed by passage over a strong acid ion exchange resin as is true for a ferric complex with "tiron" (catechol disulfonic acid). However, the other half is more strongly bound and may be attached to another type of chelating group in lignin sulfonates.

β-1-Lignol units, 46, in lignin are the precursors of stilbenes, 48, often isolated as alkaline and acidic hydrolysis products (169), Equation 23. Moderate sulfonation probably converts these units to α-sulfonates, 47, while acidic sulfonation conditions generate stilbene, 48, which actually has been isolated from sulfite spent liquors (171). Furthermore, it was found that the amount of stilbene was highest in spent liquors with a high sulfur content, which in turn may be an indication of higher acidity.

The presence of α-sulfonates, 47, can be inferred from the formation of stilbene 48 in the alkaline hydrolysis of lignin sulfonates (170). Stilbene structures, in turn, are probably converted to the α-diketone, 50, which has been obtained by oxidation of lignin sulfonates (172). This reaction probably proceeds via the intermediate, 49.

VII. Concluding Remarks. It would be sheer exaggeration to describe our present knowledge of lignin sulfonates as satisfactory; on the contrary, gaps in our knowledge more than equal the accumulated solid information. This is a natural consequence of superimposing the manifold reaction patterns of highly reactive sulfites upon an intrinsically complex natural macromolecule. The progress in this field, however, does not need to remain slow, and important new information may be anticipated to emerge from the characterization of dimeric and oligomeric lignin sulfonates, from kinetic studies on model compounds and from more extensive application of tracer methods. Lignin sulfonates remain important in the characterization of the polymeric nature of lignins, and technological benefits in the form of improved pulping processes will undoubtedly accrue from the total progress in this area.

REFERENCES

1. B. C. Tilghman, British patent 2924 (1866).
2. N. Pedersen, Papier-Ztg., 15, 422 (1890).
3. L. F. Hawley and L. E. Wise, The Chemistry of Wood, A.C.S. Monograph Series, The Chemical Catalog Co., Inc., New York, 1926.

4. J. B. Lindsey and B. Tollens, Ann., 267, 341 (1892).
5. P. Klason, Tekn. Tidskr. Avd. Kemi, 23, 49, 53 (1893).
6. K. Freudenberg and F. Sohns, Chem. Ber., 66, 262 (1933).
7. E. Hägglund and G. E. Carlsson, Biochem. Z., 257, 467 (1933).
8. B. Holmberg, Svensk Kem. Tid., 57, 257 (1935).
9. S. Heden and B. Holmberg, ibid., 48, 207 (1936).
10. E. Hägglund, Holzchemie, Academic Publishers, Leipzig, 1939.
11. P. Klason, Ark. Kem. Mineral. Geol., 3, No. 5, 9 (1908).
12. H. R. Procter and S. Hirst, J. Soc. Chem. Ind. London, 28, 293 (1909).
13. P. Klason, Beitrage z. Kenntnis d. chem. Zusammuensetzung des Fichtenholzes, Berlin, 1911, p. 18.
14. E. Hägglund, Svensk Kem. Tid., 37, 116 (1925).
15. E. Hägglund and T. Johnson, Biochem.Z., 202, 439 (1928).
16. E. Hägglund, Svensk Kem. Tid., 42, 159 (1930).
17. E. Hägglund, T. Johnson and H. Busch, Finnish Paper Timber J., 16, 282 (1934).
18. V. Grafe, Monatsh. Chem., 25, 1001 (1904).
19. K. H. A. Melander, Tekn. Tid. Avd. Kemi, 48, 147 (1918).
20. G. H. Tomlinson and H. Hibbert, J. Am. Chem. Soc., 58, 345, 348 (1936).
21. E. Hägglund, Chemistry of Wood, Academic Press, New York, 1951.
22. F. E. Brauns, The Chemistry of Lignin, Academic Press, New York, 1952.
23. F. E. Brauns and D. A. Brauns, The Chemistry of Lignin, Supplement Vol., Academic Press, New York, 1960.
24. C. E. Libby, Editor, Pulp and Paper Science and Technology, Vol. 1, McGraw-Hill Book Co., New York, 1962.
25. S. Rydholm, Pulping Processes, Interscience, J. Wiley and Sons, New York, 1965.
26. R. K. Strapp, W. D. Kerr and K. E. Vroom, Pulp Pap. Mag. Can., 58, 277 (1957).
27. H. Makkonen, Paperi Puu, 49 (7), 437 (1967).
28. P. K. Christensen, 4th Intl. Pulp Bleaching Conf., Tappi, May, 1967.
29. H. J. Kvisgaard, Norsk Skog., 19 (4), 155 (1965).
30. S. Rydholm, Continuous Pulping, Tappi Seminar, University of Washington, Seattle, September, 1968.
31. C. Kullgren, Svensk Kem. Tidskn., 42, (79)(1930); 44, 15 (1932).
32. H. Erdtman, Svensk Papperstdn., 48, 75 (1945).
33. C. Kullgren, ibid., 55, 1 (1952).
34. P. Schorning, Faserforsch. Textil tech., 8, 487 (1957).
35. H. W. Giertz and J. McPherson, Norsk Skog., 10, 348 (1956).
36. L. P. Clermont and F. Bender, Pulp Pap. Mag. Can., 62, T-28 (1961).
37. B. Philipp, F. Melms, H. Alsleben, W. Anders, R. Hedler and V. Jacopian, Zell. Papier, 17, 99 (1968).
38. K. Freudenberg, J. M. Harkin and H.-K. Werner, Chem.Ber., 97,

909 (1964).
39. N. Hartler, P. Ronstrom and L. Stockman, Svensk Papperstidn., 64, 694 (1964).
40. S. Yllner, K. Ostberg and L. Stockman, ibid., 60, 795 (1957).
41. W. Q. Yean and D. A. I. Goring, Tappi, 47 (1) 16 (1964).
42. Z. Kin, Przeglad Papier, 16 (5) 131 (1960).
43. C. M. Suter, R. K. Bair and F. G. Bordwell, J. Org. Chem., 10, 470 (1945).
44. B. F. Hrutfiord and J. L. McCarthy, Tappi, 47, 381 (1964).
45. A. E. Markham, Q. P. Peniston and J. L. McCarthy, J. Am. Chem. Soc., 71, 3599 (1949).
46. G. A. Dubey, T. R. McElhinney and A. J. Wiley, Tappi, 48, 95 (1965).
47. V. F. Felicetta and J. L. McCarthy, Tappi, 40, 851 (1957).
48. J. Benko, ibid., 44 (11) 771 (1961).
49. J. L. Gardon and S. G. Mason, Can. J. Chem., 33, 1477 (1955).
50. B. Leopold, Acta Chem. Scand., 6, 64 (1952).
51. E. E. Harris and D. Hogan, Ind. Eng. Chem., 49, 1393 (1957).
52. Y. Kojima, A. Hayashi, K. Higashitsuji and I. Tachi, Japan. Tappi, 15, 607 (1961).
53. W. Jensen, K. E. Fremer and K. Forss, Tappi, 45, 122 (1962).
54. P. R. Gupta and J. L. McCarthy, Macromolecules, 1, 236 (1968).
55. L. Sato, Science (Japan), 13, 403 (1943).
56. I. Croon and B. Swan, Svensk Papperstidn., 66, 812 (1963).
57. G. R. Quimby and O. Goldschmid, Tappi, 49, 562 (1966).
58. D. W. Glennie, Unpublished data.
59. L. Jantzen, U. S. patent 2,838,483, April 1, 1955.
60. F. E. Brauns, U. S. patent 3,297,676, January 10, 1967.
61. H. Erdtman, Tappi, 32, 303, 346 (1949).
62. H. Erdtman, Research (London), 3, 63 (1950).
63. K. V. Sarkanen, "Wood Lignins," in B. L. Browning, The Chemistry of Wood, Interscience, J. Wiley and Sons, New York, 1963, Chapter V.
64. B. O. Lindgren, Svensk Papperstidn., 55, 78 (1952).
65. E. Adler, B. O. Lindgren and U. Saeden, ibid., 55, 245 (1952).
66. H. Mikawa, Bull. Chem. Soc. Japan, 27, 53 (1954).
67. E. Adler, Svensk Kem. Tid., 80, (9) 279 (1968).
68. B. O. Lindgren, Acta Chem. Scand., 5, 603 (1951).
69. A. Björkman and B. Person, Svensk Papperstidn., 60, 285 (1957).
70. A. Björkman, Chem., Biochem., Lignin, Cellulose et Hemicelluloses, Proceedings of the International Symposium, Grenoble, France, July, 1964, pp. 317-26.
71. C. Schuerch, Ind. Eng. Chem., 4, 61 (1965).
72. E. Nokihara, M. J. Tuttle, V. F. Felicetta and J. L. McCarthy, J. Am. Chem. Soc., 79, 4495 (1957).
73. D. A. Pearson, E. O. Ericsson and J. L. McCarthy, "Sulfite Pulping,"

in **Pulp and Paper Science and Technology**, Vol. 1, C. E. Libby, ed., McGraw-Hill Book Co., New York, 1962, Chapter 9.
74. D. A. I. Goring and A. Rezanowich, Can. J. Chem., 36, 1653 (1959).
75. K. Kitao, Mokuzai Kenkyu, 8, 5, 9 (1952).
76. S. A. Rydholm and S. Lagergren, Svensk Papperstidn., 62, 103 (1959).
76a. J. M. Calhoun, F. H. Yorston and O. Maass, Can. J. Res., 15, Sec. B, 457 (1937).
77. O. Ingruber, Svensk Papperstidn., 65, 448 (1962).
78. J. A. F. Gardner and W. E. Hillis, **Wood Extractives**, W. E. Hillis, Ed., Academic Press, New York, 1962, Chapter 11.
79. N. Migita, R. Mitsukawa, J. Nakano and M. Ichino, J. Japan. Tappi, 8, 2 (1954).
80. L. Ivnas and B. Lindberg, Acta Chem. Scand., 15, 1081 (1961).
81. B. O. Lindgren, ibid., 1, 779 (1947); 3, 1011 (1949).
82. A. J. Corey, et al., Can. J. Res., B15, 168 (1937).
83. A. J. Corey and O. Maass, ibid., B 13, 149 (1935).
84. A. Björkman, Svensk Papperstidn., 60, 158, 243 (1957).
85. K. H. Ekman and J. J. Lindberg, Paperi Puu, 48, 241 (1966).
86. B. Abrahamson, B. O. Lindgren and E. Hägglund, Svensk Papperstidn., 51, 471 (1948).
87. W. Q. Yean and D. A. I. Goring, Pulp Pap. Mag. Can., 64, T-127 (1964).
88. H. W. Johnston and O. Maass, Can. J. Res., 3, 140 (1930).
89. J. H. Sutherland, H. W. Johnston and O. Maass, ibid., 10, 36 (1934).
90. T. N. Kleinert, Holzforschung, 18, (5) 139 (1964).
91. J. E. Stone and A. M. Scallan, Pulp Pap. Mag. Can., 69 (12) 69 (1968).
92. J. G. McNaughton, W. Q. Yean and D. A. I. Goring, Tappi, 50, 548 (1967).
93. A. B. Wardrop, Svensk Papperstidn., 66, 231 (1963).
94. A. R. Proctor, W. Q. Yean and D. A. I. Goring, Pulp Pap. Mag. Can., 68, T-445 (1967).
95. K. Sato, A. Kobayashi and H. Mikawa, J. Japan. Tappi, 16, 42, 108 (1962); Bull. Chem. Soc. Japan, 35 (4) 662 (1962).
95a. D. A. I. Goring, "Microscopic Patterns of Lignin Removal During Chemical Pulping," Tappi Paper Physics Seminar, 1969.
96. J. E. Stone, Tappi, 38, 610 (1955).
97. D. E. Marth, ibid., 42, 301 (1959).
98. D. W. Glennie and J. S. Mothershead, ibid., 47, 356 (1964).
99. E. Aaltio and R. H. Roschier, Paperi Puu, 36, 157 (1954).
100. A. Rezanowich, G. A. Allen and S. G. Mason, Pulp Pap. Mag. Can., 58, 153 (1957).
101. W. N. Tuller, Ed., **The Sulphur Data Book**, McGraw-Hill Co., New York, 1954.

102. A. Edhborg, H. Erdtman and B. Leopold, Acta Chem. Scand., 6, 450 (1952).
103. K. Forss and K. E. Fremer, Tappi, 47, 485 (1964).
104. K. Forss and K. E. Fremer, Paperi Puu, 47, 443 (1965).
105. K. Forss, K. E. Fremer and B. Stenlund, ibid., 48, 565, 669 (1966).
106. W. Jensen, B. C. Fogelberg, K. Forss, K. E. Fremer and M. Johanson, Holzforschung, 20, 48 (1966).
107. K. Forss, O. Schott and B. Stenlund, Paperi Puu, 49, 525 (1967).
108. B. C. Fogelberg, K. Forss and S. Fugleberg, ibid., 49, 725 (1967).
109. I. A. Pearl, The Chemistry of Lignin, M. Dekker, Inc., New York, 1967.
110. V. F. Felicetta, A. Ahola and J. L. McCarthy, J. Am. Chem. Soc., 78, 1899 (1956).
111. O. Goldschmid, Anal. Chem., 26, 1421 (1954).
112. G. Aulin-Erdtman, Svensk Papperstidn., 56, 287 (1953).
113. H. Mikawa and K. Ebisawa, Bull. Chem. Soc. Japan, 29, 209 (1956).
114. A. N. James and P. A. Tice, Tappi, 48, 239 (1965).
115. K. Freudenberg, M. Meister and E. Flickinger, Chem. Ber., 70, 500 (1937).
116. H. Mikawa, K. Sato, C. Takasaki and K. Ebisawa, Bull. Chem. Soc. Japan, 29, 245 (1956).
117. E. Adler, S. Delin and K. Lundquist, Acta Chem. Scand., 13, 2149 (1959).
117a. G. Gellerstedt and J. Gierer, ibid., 22, 2510 (1968).
118. V. F. Felicetta and J. L. McCarthy, J. Am. Chem. Soc., 79, 4499 (1957).
119. O. Samuelson and A. Westlin, Svensk Papperstidn., 651, 179 (1948).
120. H. Mikawa, K. Sato, C. Takasaki and K. Ebisawa, Bull. Chem. Soc. Japan, 28, 649, 653 (1955).
121. M. Goliath and B. O. Lindgren, Svensk Papperstidn., 64, 109 (1961).
122. J. Gierer and B. Alfredsson, ibid., 62, 434 (1959).
123. J. Gierer, B. Alfredsson and S. Soderberg, ibid., 63, 201 (1960).
124. Q. P. Peniston and J. L. McCarthy, J. Am. Chem. Soc., 70, 1329 (1948).
125. K. Kratzl, Paperi Puu, 43, 643 (1961).
126. K. Kratzl, E. R. Schafer, P. Claus and E. Wittman, Holzforschung, 20, 21 (1966).
127. K. Kratzl and J. Spona, ibid., 20, 27 (1966).
128. J. Turunen, Paperi Puu, 49, 151 (1967).
129. E. Adler, J. Bjorkvist and S. Häggroth, Acta Chem. Scand., 2, 93 (1948)
130. E. Adler and L. Ellmer, ibid., 2, 839 (1948).
131. K. Kratzl, Monatsh. Chem., 78, 173 (1948).
132. G. Aulin-Erdtman, Svensk. Papperstidn., 56, 91 (1953).

133. G. Aulin-Erdtman and L. Hegbom, ibid., 61, 187 (1958).
134. G. Aulin-Erdtman and R. Sanden, Paperi Puu, 43, 671 (1961).
135. E. Adler, ibid., 43, 634 (1961).
136. O. Goldschmid, Tappi, 38, 728 (1955).
137. J. S. Mothershead and D. W. Glennie, ibid., 47, 519 (1964).
138. J. R. Parrish, J. Chem. Soc. (C), 1967, 1145.
139. J. R. Parrish, Tetrahedron Letters, No. 11, 555 (1964).
140. A. Wacek, K. Kratzl and A. Bezard, Chem. Ber., 75, 1348 (1942).
141. B. O. Lindgren and U. Saeden, Acta Chem. Scand., 6, 963 (1952).
142. W. B. Hewson and H. Hibbert, J. Am. Chem. Soc., 65, 2371 (1941).
143. W. Jensen, B. C. Fogelberg, K. Forss, K. E. Fremer and M. Johanson, Tappi, 48, 174 (1965).
144. W. H. Hoge, ibid., 37, 369 (1954).
145. J. Gierer, Svensk Papperstidn., 61, 648 (1958).
146. N. Sanyer, T. Itoh and E. L. Keller, Tappi, 47, 323 (1964).
147. B. O. Lindgren and H. Mikawa, Acta Chem. Scand., 11, 826 (1957).
148. E. Adler, Paperi Puu, 43, 634 (1961).
149. V. F. Felicetta, D. W. Glennie and J. L. McCarthy, Tappi, 50, 170 (1967).
150. S. W. Schubert, M. G. Andrus, C. Ludwig, D. W. Glennie and J. L. McCarthy, ibid., 50, 186 (1967).
151. S. W. Schubert and J. L. McCarthy, ibid., 50, 202 (1967).
152. D. W. Glennie, ibid., 49, 237 (1966).
153. K. Freudenberg, G. Maercker and H. Nimz, Chem. Ber., 97, 903 (1964).
154. H. Hardell and O. Theander, Svensk Papperstidn., 68, 482 (1964).
155. B. Lindberg, J. Tanaka and O. Theander, Acta Chem. Scand., 18, 1164 (1964).
156. S. Yllner, ibid., 10, 1251 (1956).
157. I. Croon and B. Swan, Svensk Papperstidn., 67, 177 (1964).
158. H. L. Hergert, J. Org. Chem., 25, 405 (1960).
159. K. Freudenberg, Science, 148, 595 (1965).
160. J. M. Harkin, in Oxidative Coupling of Phenols, W. I. Taylor and A. R. Ballersby, Eds., Marcel Dekker, Inc., New York, 1967.
161. A. Hayashi and D. A. I. Goring, Pulp Pap. Mag. Can., 66, No. C, T-154 (1965).
162. A. Hayashi and Y. Namura, Bull. Chem. Soc. Japan, 38, 512 (1965).
163. A. Hayashi and Y. Namura, Nippon Mokuzai Gakkaishi, 12, 44 (1966).
164. A. Hayashi and Y. Namura, ibid., 13, 24 (1967).
165. A. Hayashi, ibid., 11, 212, 253 (1965).
166. A. Hayashi and Y. Namura, ibid., 12, 300 (1966).
167. A. Hayashi, Y. Namura and T. Uekita, ibid., 13, 194 (1967).
168. A. Hayashi and H. Minari, ibid., 13, 198 (1967).

169. H. Nimz, Chem. Ber., 98, 533 (1965).
170. H. Richtzenhain and C. Hofe, ibid., 72B, 1890 (1939).
171. E. A. Kvasnicka and R. R. McLaughlin, Can. J. Chem., 33, 637 (1955).
172. I. A. Pearl and D. L. Beyer, J. Am. Chem. Soc., 76, 2224 (1954).

16

REACTIONS IN ALKALINE PULPING
Joseph Marton

I. Morphological and Topochemical Considerations.

    A. Introduction. In 1967, the world pulp production was over 98 million short tons (1). The United States alone produced 36 million short tons of pulp; 85 percent of this was made by chemical or semichemical pulping. The great majority of the chemical pulps was obtained by alkaline pulping: 75 percent by sulfate pulping, and less than one percent by soda pulping.

    The first patent for the production of pulp by an alkaline process was issued to Watt and Burgess (2) in 1854, followed by the development of sulfate pulping by Dahl, a German chemist, in 1889. He replaced the sodium carbonate in the recovery cycle with sodium sulfate. The sodium sulfate was reduced by the burning of the organic matter to sodium sulfide and was used in this form to prepare the pulping liquor. There have been a number of significant innovations such as continuous pulping and the application of sulfate pulping to hardwoods, but the essence of the process has remained substantially unchanged. The main object of any delignification process is to facilitate the disintegration of wood into its fibrous components. To achieve this, the insoluble cross-linked protolignin must be degraded by breaking bonds within the lignin macromolecule; the fragmented lignin has to be solubilized and removed from the individual fibers. The alkaline pulping process is quite complex and affected by many elementary physical and chemical factors such as the morphological characteristics of the wood, the location of lignin in the woody tissue, the penetration pathway of the chemicals, the swelling or thermal coalescence of the lignin-containing interface, the cleavage of lignin-carbohydrate bonds, the chemical degradation of the lignin macromolecule, the tendency of the fragments to enter secondary condensation reactions, the stabilization of the fragmented molecules through solvation and by means of electrostatic repulsive forces, the removal of reaction products from the reaction zone by diffusion-transport, and the simultaneously occurring undesirable carbohydrate degradation reactions.

    In the soda process, as it is still applied to pulping of hardwoods, sodium hydroxide is the major pulping chemical. In the sulfate process, a mixture of sodium hydroxide and sodium sulfide is used for delignification of both softwoods and hardwoods. The sulfate pulps produced vary from high yield coarse pulps to low yield bleachable pulps. With regard to the outstanding strength of pulps produced in this process, it is commonly referred to as the "kraft" (the German word for strength) process. The terms kraft and sulfate pulping are interchangeable.

B. Pulping Chemicals. Kraft pulping liquor (white liquor) contains sodium hydroxide and sodium sulfide as its main components, with sodium carbonate, sodium thiosulfate, sodium polysulfide, etc. as minor constituents. See Table 16.1 for a typical composition (3).

Table 16.1. Initial Composition of Cooking Liquor in a Normal Kraft Cook (3).

Compound	Formal composition			Actual composition
	g/l as Na_2O	g/l as NaOH	g/l of compound	g/l of compound
NaOH	27	35	35	38
NaSH	-	-	-	4
Na_2S	12	15	15	9
Na_2CO_3	6	8	11	11
Na_2SO_4	0.1	0.1	0.2	0.2
Na_2SO_3	0.1	0.1	0.2	0.2
$Na_2S_2O_3$	0.2	0.2	0.4	0.4

Sulfidity 30%, effective alkali 42.5 g/l NaOH, active alkali 50 g/l NaOH.

In American literature, the total alkali designates the sum of sodium hydroxide, sodium sulfide, sodium carbonate and sodium sulfite, expressed as grams Na_2O per liter. Active alkali gives the sum of sodium hydroxide and sodium sulfide in grams Na_2O per liter; the term effective alkali designates the sum of sodium hydroxide + 1/2 sodium sulfide concentrations calculated as grams Na_2O per liter. In Scandinavian countries, the concentrations are expressed as grams NaOH instead of Na_2O. The sulfidity is defined as $100[Na_2S]/([NaOH] + [Na_2S])$, where the concentrations are expressed in equivalents of Na_2O or NaOH, respectively.

The fundamental and technical aspects of alkaline pulping are well presented in the magnificent book of Rydholm on pulping processes (3). The inorganic reactions of the pulping liquor have recently been reviewed by Hartler (4) and by Wenzl (5).

Sodium sulfide undergoes hydrolysis in aqueous solutions that are governed by pH-dependent equilibria shown in Equations 1 and 2.

$$S^= + H_2O \rightleftharpoons SH^- + OH^- \tag{1}$$

$$SH^- + H_2O \rightleftharpoons H_2S + OH^- \qquad (2)$$

Since the pK_a value of hydrogen sulfide is about 7 at room temperature, the amount of the un-ionized species is insignificant at the high pH of the kraft process. A pK_a value of 13.5 has frequently been assigned for hydrosulfide ions (4), with scanty experimental foundation, however. Actually, a considerably higher pK_a value for hydrosulfide ions is not excluded. Whatever the case, no outstanding effects deriving from equilibrium 1 have been observed in kraft systems, as illustrated by the following example (6). When spruce wood was treated with sodium hydrogen sulfide solutions at various pH values, the yield of alcohol-soluble thiolignin followed the increase of SH^- concentration between pH 5.5 and 8.5, see Figure 16.1 (6). From pH 8.5 upwards, the yield remained constant; and if $S^=$ concentration did increase, its effect was certainly no different from that of HS^- ions.

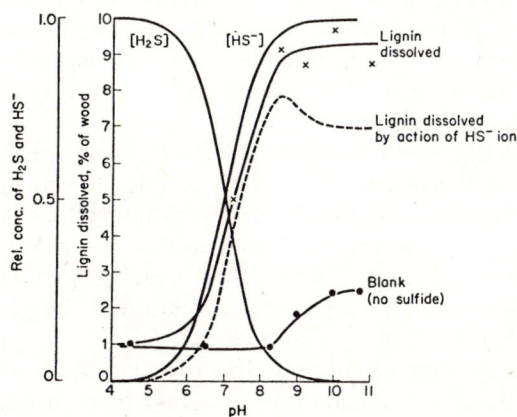

Figure 16.1. Delignification of softwood by hydrosulfide at varying pH, according to Enkvist et al. (6).

The lignin component alone reacts with sulfide in kraft process, forming thiolignin. In addition, methyl mercaptan, CH_3SH, and methyl sulfide, CH_3SCH_3, are formed from lignin methoxyl groups. Still, 65-80 % of the sulfide is not consumed, while only a fraction of active alkali remains after completed pulping.

Alkali is consumed: 1, in reactions with lignin; 2, in dissolution of carbohydrates; 3, in neutralizing various organic acids, both those present in the original wood and those produced during pulping; 4, by the resins of the wood;

and 5, to some extent, by adsorption on the fibers.

Most of the alkali required for pulping is consumed in the neutralization of the saccharinic acids formed in the degradation of hemicelluloses. These degradation reactions start at around 100°C and are largely completed (3, 7) when the temperature reaches 150°C, where delignification becomes the main reaction. In a normal softwood kraft cook, giving a pulp yield of 47%, of which 3% is lignin, the dissolved material consists of roughly 3% resin, 24% lignin, 24% degraded carbohydrates, and 2% acetyl groups. With a sulfidity of 30%, the consumption of active alkali is around 14% Na_2O, or 11.6% Na_2O effective alkali based on o.d. wood. The dissolved lignin requires about 3.1% Na_2O (on wood) for neutralization. The remaining 8.5% Na_2O (on wood) is utilized for hydrolysis of acetyl groups (1.15%), and neutralization of carbohydrate degradation products (7.35%). The latter figure roughly corresponds to 1.6 moles of acids per hexose unit (3). Lignin is thus responsible for only about a quarter of the alkali consumption of the kraft cook. A minimum residual alkali (pH$\geqslant$9) is required to keep the dissolved lignin in solution.

In a short cycle (20 min.) vapor-phase cooking of pressure impregnated spruce chips, the total consumption of effective alkali was reduced to 9.5% Na_2O at 50% pulp yield (8), and the sulfide consumption equaled about 2.5% Na_2O. In conventional liquid-phase cooking, the appreciably higher (ca. 12.5% Na_2O) alkali consumption is presumably caused by more extensive degradation of hemicelluloses.

In the white liquors of the alkaline cooking processes, about 15 to 25% Na_2O is applied on the weight of oven dry wood. The sulfidity varies from about 5% in the "soda" mills to about 20-40% in kraft mills. Recently, higher sulfidities have been suggested for the polysulfide and alkafide pulping processes. High yield kraft pulps (50-55%) are usually made by shortening the reaction time and/or by reducing the alkali charge. The yield increase is primarily due to higher residual lignin content.

C. <u>Penetration of Pulping Liquor.</u> The alkaline pulping systems are heterogenous; and unless the conditions are very favorable (effective preimpregnation, thin chips, and sufficiently high alkalinity throughout the cook), chemical transport phenomena (liquid penetration and diffusion) may control the rate of delignification (9, 10). Alkali sorption in the outer layers of the wood can be a major cause of non-uniformity in the alkali distribution within the chips, particularly at low alkali concentrations. Sulfur usually is not absorbed from the kraft liquors except from dilute sodium sulfide solutions (11). The non-uniformity of alkali penetration with liquors of low alkalinity limits the attempts to increase pulp yield by decreasing the alkali charge. In hardwood chips, the longitudinal penetration of white liquor (12-15) takes place through the vessels via the pits to adjacent fibers and rays. Laterally, the liquor proceeds through the rays via the pits to the fibers. In the morphologically less sophisticated softwoods, the initial penetration starts from the surface of the chips and proceeds through the open ends into the lumina of several tracheids, reaching adjacent cells through the pores of the bordered pit membranes. Laterally, the liquor moves through the rays to yet further tracheids. In each tracheid, penetration continues into the cell wall by diffusion from the lumen. Contact with the middle lamella is established at an early stage of the pulping process (14) as a consequence of longitudinal diffusion from the pit membrane.

Penetration is less of a problem in the alkaline processes than in the acidic sulfite pulping. Alkaline pulping liquor can diffuse into the wood at an approximately equal rate in all structural directions. The degree of swelling of the fibers also is considerably higher in alkali, producing a several hundred-fold increase in the potentially reactive surface area which increases further as the delignification proceeds.

Frey-Wyssling (16) claims that the fiber cell walls are not porous, but pores may develop during drying. Data of Stone and Scallan (17) indicate that significant porosity develops in the cell wall as the pulping proceeds. The average pore size diameter was 30 Å at 25% delignification level and reached 42 Å at 83% lignin removal from chips of black spruce. The size of the pores may limit the size of the dissolved lignin fragments that can escape from their location in the cell wall. Goring and his associates found (19) that the weight average molecular weight of kraft lignin increased from 1800 for the first fraction of a continuous cook of black spruce sawdust to 51,000 for the final fraction, roughly corresponding to the pore diameters found by Stone and Scallan (17). As the greater part of the lignin is dissolved, fibers begin to separate along the middle lamellae; and the developing fissures further enhance the removal of reaction products.

D. Distribution of Lignin in the Wood and Pulp Fibers. Besides the penetration pattern, the actual distribution of the cell-wall components exerts a major effect on the practical rate of delignification. The distribution of lignin in the different layers of the cell wall has been discussed in Chapters 2 and 3. The critical review of the problem by Berlyn and Mark (18), as well as the improvements introduced by Goring et al. (20-23) to the ultraviolet microspectrophotometric method of Lange (24-26), has resulted in a reliable concept of lignin distribution in softwood and hardwood cells. In general, the results agree with the electron micrographs of lignin "skeletons" of softwood tracheids (27, 28).

Studies on the organization of microfibrillar structure of cell wall also suggest that lignins in the cell wall and in the middle lamellae have different morphology (14). Yean and Goring (21) have proposed the existence of two types of protolignin in their study of the lignin fractions obtained by sulphonating spruce wood meal in a continuous-flow digester. Application of Flory's theory of trifunctional polymerization in reverse to the problem of the degradation of lignin polymer network during pulping (22) gave indication that the middle lamella and secondary wall lignins have a different rate pattern of degradation in chemical pulping. The initial investigation (20) with black spruce chips revealed that in both kraft and acid sulfite pulping, lignin was preferably removed from the secondary wall first, Figure 16.2. At about 50% delignification, the middle lamella lignin and lignin in cell corner areas dissolved rapidly, leaving the residual lignin in the secondary wall at the end of the cook. This finding contradicts an earlier observation of Bixler (29), who subjected very thin (20μ thick) spruce wood sections to the action of pulping liquors at 150-175° C. His microscopic observations indicated that in sulfite pulping, the delignification proceeded with the same rate both in the middle lamella and in the secondary wall; while in alkaline pulping, lignin was completely removed from the middle layer before any delignification could be observed in the cell wall.

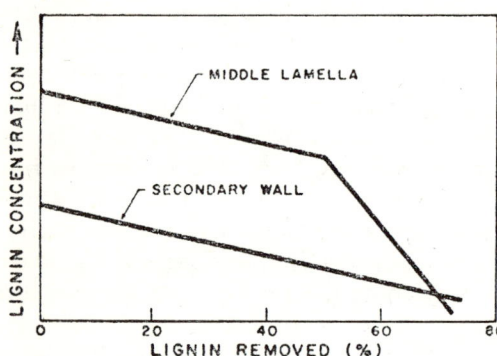

Figure 16.2. Decrease of concentration of lignin in middle lamella and secondary wall of spruce earlywood during delignification of kraft process, according to Goring et al. (20).

The final distribution of the residual lignin in the fibers determines many important papermaking properties such as bleachability, stiffness, fiber bonding, etc. Jayme and von Koppen (30) compared by light microscopy the lignin distribution in sulfite and sulfate spruce fibers of pulps from mill-size chips by removing the polysaccharides by hydrolysis with a mixture of concentrated sulfuric and hydrochloric acids. In sulfate fibers, the residual lignin (3.6%) appeared to be uniformly distributed in the entire cell wall; while in sulfite fibers, most of the residual lignin was located in the outer part of the wall $(P + S_1)$, the lignin content significantly decreasing in the middle part of the secondary wall (S_2) and increasing somewhat in the S_3-layer at the lumen side. Kallmes (31) found a similar lignin distribution in unbleached spruce sulfite fibers. Sulfite fibers swell in water more easily than sulfate fibers in spite of the higher surface lignin content of the former. This is connected—according to Jayme (30)— with the lower DP and higher amount of hemicellulose on the surface of sulfite fiber as opposed to the sulfate fiber.

Luce (32) determined the radial distribution of polysaccharides throughout the cell wall with an ingenious chemical peeling technique. He found that the DP of his kraft pulp was quite uniform throughout the fiber cell wall, while in sulfite pulps the fibers were more degraded on the outside than in the inside of the fiber. The surface of the kraft fibers contained considerably more xylan and also more mannan than the inner layers, while the hemicelluloses were uniformly distributed in the sulfite fiber. These results indicate that the molecular weight and distribution of polysaccharides as well as of lignin influence the swelling and other important papermaking properties of the fibers (33).

E. <u>Changes at Fiber-Liquor Interface</u>. The reactions of delignification take place at the solid-liquid interface of lignin and pulping liquor. One would expect that a large interface would accelerate the dissolution of lignin. Two properties are to be considered here which cause opposite interfacial changes: swelling of lignin and its thermal coalescence.

Isolated lignins swell extensively in water. Odintsov (34, 35) found that swollen acid or kraft lignins (after solvent exchange) had surface areas of several hundred m^2/g, only slightly less than the area of never dried swollen cellulose, while the surface area of dry lignin powders was only a few m^2/g. In contrast, lignin in wood exhibits very low swelling. The inhibiting effect of lignin on fiber swelling has been indicated by Jayme and Mohrberg (36), who showed that as delignification proceeds the swellability of spruce fibers in water increases. The removal of lignin between the coaxial lamellae in the cell wall may be expected to loosen the association between adjacent lamellae so that the flexibility of the fiber can increase (37).

It is obvious that some kind of swelling will precede the dissolution of lignin from the polysaccharide matrix during pulping; and at this stage, the number of reaction sites available for further degradation will markedly increase unless softening and thermal coalescence cancel this favorable effect. Goring (38) measured the softening points of isolated wood components, see Chapter 17, and suggested that hemicelluloses start to soften in the pulping liquor at around 50°C, the glassy transition of lignin starts from 90°, and cellulose softens at considerably higher temperatures. Since about half of the woody substance becomes a soft gel-like matrix, the interface available for reaction should decrease. It could be demonstrated, however, that the rate of chemical reaction was not controlled by this structural change (23). Kleinert and Maraccini (39) subjected pulps to sulfur dioxide adsorption in order to assess the surface available during different pulping conditions. They concluded that coalescence of lignin is far more important a factor in acid sulfite than alkaline pulping.

F. <u>Kinetics of Delignification, Reaction Mechanisms</u>. The favorable effect of sulfide on the rate of delignification is illustrated by Figure 16.3, based on experiments carried out by Hägglund (83). Spruce wood was delignified at 160° with 19% active alkali with varying sulfidities. It can be seen that the overall delignification process consists of a rapid phase, usually termed

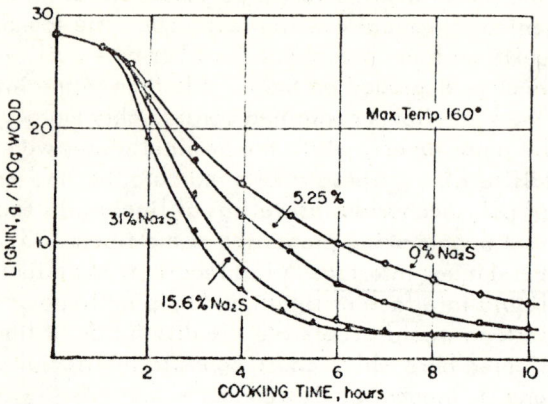

Figure 16.3. Delignification of spruce wood in soda and kraft cooks varying the sulfidity. Data of Hägglund (83).

"bulk delignification," removing most of the lignin from wood, and of a slow "residual delignification" phase. The last 10% of lignin can be removed only very slowly.

Kleinert (45, 46) has shown that if the pulping of thin wafers is performed isothermally, the bulk- and residual-delignification phases resolve to two intercepting straight lines in a logarithmic plot, illustrated in Figure 16.4. Thus, both reaction phases exhibit an apparent first-order rate behavior, the bulk delignification being more rapid for hard- than for softwoods. It may be noted that in the high-temperature (185°) pulpings depicted in Figure 16.4, bulk delignification is completed within ten minutes reaction time.

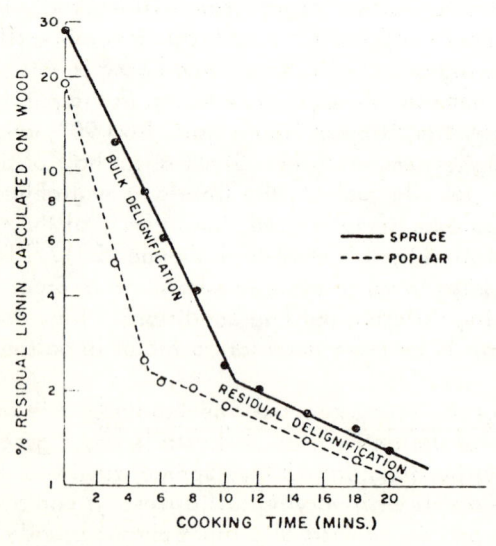

Figure 16.4. Kraft delignification of spruce and poplar wafers at 185°, according to Kleinert (46).

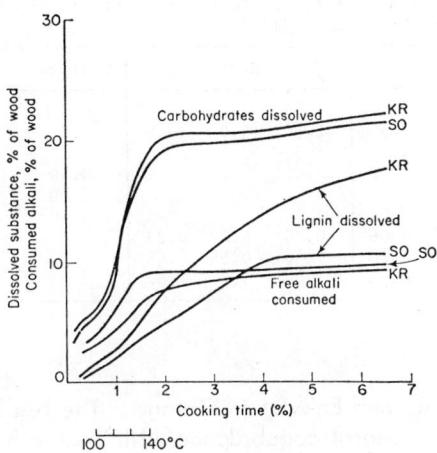

Figure 16.5. Alkali consumption and dissolution of wood substance in low-temperature (140°C) kraft (KR) and soda (SO) cooks of spruce. Data by Enkvist (93).

In comparison, delignification is much slower in low-temperature (140°) cooks studied by Enkvist (111), Figure 16.5. At this temperature, soda delignification becomes quite sluggish before one-half of the lignin has gone into solution, while delignification at 30% sulfidity level proceeds relatively unhindered. The amount and rate of carbohydrate dissolution is exactly the same in both types of cooks.

In view of the complexity of the physical and chemical steps involved in delignification, a kinetic description of the alkaline pulping process can only be very approximative (40-43). The determined values for the energy of activation vary from 24 to 38 kcal/mole for alkaline pulping; 29 and 32 kcal/mole representing perhaps the most reliable values for kraft and soda delignifications, respectively. In sulfite pulping, the energy of activation increased from 16 to 22 kcal/mole at the end of the cook. As a practical rule, reaction velocity increases almost threefold for an increase in maximum temperature of 10°C in a soda cook, 2.5 times in kraft pulping, and doubles in acid sulfite pulping.

Table 16.2 gives a very rough comparison (3) of relative reaction velocities of the three pulping processes at different temperatures and degrees of delignification. The approximate velocity constants were calculated on the assumption that delignification is a first-order reaction on lignin. Sulfite delignification requires the lowest pulping temperature (usually around 135°C), while the regular kraft process at around 170°C is about twice as fast as the soda process. Hardwoods require lower pulping temperatures than softwoods.

Much of our present knowledge about the mechanism of alkaline pulping reactions is based on extensive research carried out by research groups centered

Table 16.2. Reaction Velocities (in h^{-1}) for Different Pulping Processes, Assuming First Order Reactions (3).

Pulping Process	Soda			Kraft			Sulfite
Residual lignin, % of wood	2.5	5.0	7.5	2.5	5.0	7.5	2.5
Temperature, °C							
110							0.10
120							0.20
130							0.40
150			0.06	0.20	0.21	0.23	
160		0.20	0.20	0.46	0.55	0.65	
170	0.41	0.53	0.54	1.0	1.2	1.6	

around Adler and Gierer in Sweden, and Enkvist in Finland. The results obtained by these three groups are in general concordance with each other and suggest that the delignification process, in the chemical sense, is a consequence of certain ionic fragmentation reactions. However, quite divergent views have been expressed by Kleinert (45, 46), advancing the importance of homolytic processes in pulping, and by Chirkin and Tishchenko (107), who emphasize the possible reductive cleavage by hydrosulfide of secondarily formed diphenyl methane bonds.

According to Kleinert, the bulk delignification is caused by rapid fragmentation of the lignin macromolecule through rupture of bonds by thermal homolysis. The free radicals formed may also undergo secondary reactions such as condensation, grafting, termination and radical transfer reactions. The slow residual delignification is understood in terms of a slow alkaline degradation of cellulose portions onto which lignin was grafted during pulping. The role of sulfide (and elemental sulfur formed therefrom) is visualized to be that of a free radical scavenger. These sulfur species may form persulfide radicals (46) that combine with free radical centers formed in the lignin, preventing them from condensation and grafting. Experimental evidence in support of these ideas is provided by electron spin resonance measurements that demonstrate the formation of free radical species in alkaline pulping.

It is perhaps somewhat difficult to accept the view of homolytic cleavages as major reactions in alkaline pulping since such cleavages have not been observed in model compound studies performed at normal pulping temperatures. On the other hand, some free radical species are clearly present in alkaline pulping. They may possibly derive from intermediate quinone methides, and their reactions may exert a significant effect on the course of the pulping process. As it is, Kleinert's hypothesis needs further elaboration and more direct proof.

Chirkin and Tishchenko (107) assume that one of the main reactions in

REACTIONS IN ALKALINE PULPING

alkaline pulping is lignin condensation due to formation of hydroxydiphenyl methane structures. Such structures arise through the secondary formation of α-5 and, possibly, α-1 linkages in condensations of p-hydroxybenzyl alcohol or ether structures with phenolic elements. Also, some α-6-linked diphenyl methane structures might be present in the original lignin. It is visualized that the main role of sulfide in kraft pulping is the reductive cleavage of these diphenyl methane linkages, as shown in Equation 3.

$$Ar-CH_2-Ar + H_2S \longrightarrow Ar-CH_3 + HAr + S \qquad (3)$$

To test the validity of this hypothesis, Gierer et al. (108) subjected six dihydroxydiphenyl methane model compounds to soda and sulfate cooks. In each case, the models were recovered unchanged after the cook. Thus, the hypothesis of reductive cleavage was deemed unfounded.

Chirkin and Tishchenko (107) also claimed that addition of certain reducing agents such as hydrazine, sodium stannite, hydroxylamine, hypophosphite, dithionite and sodium sulfite to the alkaline cooking liquor is as effective as the usual sodium sulfide addition. However, earlier efforts by Enkvist et al. (109) to add reducing agents (pyrocatechol, pyrocatechol, pyrogallol, zinc powder, stannous chloride plus sodium hydroxide, ascorbic acid, arsenites, formates and hydrazine) did not show any favorable effect. From these findings Enkvist concluded that the role played by sulfur hardly can be explained by virtue of its reducing effect.

Dihydroxydiphenyl methane structures, while not playing the major role assigned to them by Chirkin and Tishchenko, may influence kraft process in a different way, as pointed out by Harkin (110). For example, 5-vanillyl-creosol 1 is readily oxidized to form reactive quinone methide 2, Equation 4. Analogous chromophoric groups may contribute to the color of kraft lignin.

$$(4)$$

In fact, heating of 1 to about 170° causes spontaneous conversion to the quinone methide with loss of hydrogen. Thus, o,p'-dihydroxydiphenyl methane structures may be dehydrogenated in alkaline pulping merely by the effect of heat alone, especially if traces of heavy metals are present. The hydrogen released may cause reduction reactions.

Quinone methide 2 adds water in strong alkali, is instantaneously reduced by borohydride, and may be oxidized further to the corresponding radical 3. Speculatively, radicals of this type could well contribute to the esr signal observed by Kleinert (45, 46).

II. Base Catalyzed Fragmentation of Lignin.

A. Reactive Structures. The essential step of delignification is the fragmentation of the lignin macromolecule by the action of alkali and other nucleophiles. With some exception, it is generally true that carbon-to-carbon bonds are stable in alkali, but carbon-to-oxygen linkages can be cleaved under pulping conditions. The most significant reaction of this type is the liberation of new phenolic hydroxyl groups by the cleavage of aryl alkyl ether bonds.

Model experiments have been applied extensively in delignification studies. Initially, Holmberg's hypothesis on benzyl alcohols and ethers (44) as main reactive structures of lignin stimulated much model compound work. We have to realize, nonetheless, that a simplification of conditions by a selected model system also contains an arbitrary restriction of conditions because in the real pulping system reactive groups may and do react with other types of structural elements of lignin (or polysaccharides) necessarily absent from the model systems. Model studies aimed at elucidating redox type of reactions suffer most from this limitation. Indeed, the most "appropriate" model of protolignin is, in many instances, isolated wood lignin itself.

Secondly, the initial complexity of lignin structure (47) makes it difficult to represent all significant structures by model compounds. While arylglycerol-β-aryl ether elements are undoubtedly the most important reactive structures, the number of functional groups associated with the α-carbon in this and other elements is quite extensive. The α-carbon may carry a hydroxyl, an aryl ether, or an aliphatic ether group. The aryl ether groups may be non-cyclic (52, 53) or, as in phenylcoumaran structures (54), cyclic ethers. A cyclic aliphatic ether group is present in pinoresinol elements, and non-cyclic aliphatic ether groups may likewise be found in lignin (48-51, 55). Finally, aliphatic ether bonds may connect the α-carbon to carbohydrates (56). The reactivity of all α-substituents depends in a decisive manner on whether the unit itself is phenolic or non-phenolic. A non-phenolic unit may become a phenolic one during the pulping if the aromatic ether linkage is hydrolyzable, a β-O-4 linkage, for example. This is not possible, however, if the unit is linked by a 5-O-4 diaryl ether bond (47, 52, 53).

The mechanism of various alkaline cleavage reactions can be viewed from a common starting point by assuming that in phenolic units the hydrolysis proceeds through the intermediate formation of reactive p-quinone methide structures. Formation of o-quinone methides also is possible; e.g., from certain (α-5) linked structures, see Equation 4. Quinone methides are formed only from units containing a free phenolic hydroxyl group. Thus, phenolic units constitute the most reactive structures in alkaline pulping. The non-phenolic structures are, in general, resistant to alkaline attack but may under specific circumstances (57, 58) undergo cleavages of alkyl aryl ether bonds.

B. Hydrolysis of α-Ethers. The alkaline hydrolysis of phenolic α-aryl ethers proceeds very easily (50) because formation of a quinone methide intermediate is feasible. Non-phenolic α-aryl bonds are stable (50) unless a free aliphatic hydroxyl group is present in the neighboring β-position of the side-chain, in which case the cleavage can proceed (57, 58) through α,β-epoxide formation in a manner analogous to a β-ether cleavage, to be discussed in Section C.

As a model for cyclic benzyl aryl ether (phenylcoumaran) structure in lignin, dihydro-dehydrodiconiferyl alcohol 4 was subjected to soda and kraft cooks at 170°C (59, 60). In both cases, the hydrofuran ring opened and the hydroxymethyl substituent was lost as formaldehyde, yielding the o,p-dihydroxy stilbene 6 in good yield. The quinone methide intermediate 5 (57) loses the methylol group presumably through a reverse (vinylogous) aldol reaction (59), Equation 5. In contrast, the aromatic monomethylether of 4 remained practically unaffected when heated with kraft white liquor (59).

$$(5)$$

The crystalline stilbene 6 shows a very characteristic ionization difference ($\Delta \varepsilon_i$) maximum at 378 mµ (61). Kraft lignin prepared from spruce milled wood lignin exhibited similar absorption spectrum after removal of carbonyl groups by reduction with lithium aluminum hydride (Figure 16.6). The intensity of the 378 mµ maximum in kraft MWL indicated the presence of .08 stilbene groups/MeO (61, 62). In the original milled wood lignin about the same number of phenylcoumaran units were found by acid-catalyzed conversion of phenylcoumarans to phenylcoumarones (54).

Gierer et al. heated the γ-methyl analogue of 4, dihydrodehydrodiisoeugenol with sodium hydroxide solution (57, 58) as well as kraft liquor (69). In both cases, the hydrofuran ring was cleaved and the corresponding methyl-substituted stilbene 7 formed. Removal of the γ-carbon as formaldehyde is thus

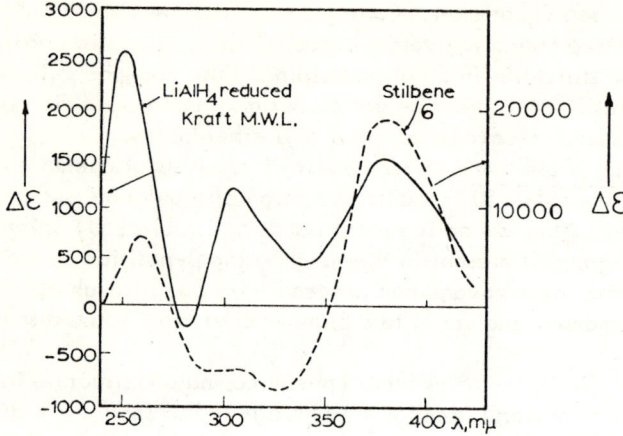

Figure 16.6.
Ionization-Δε
curves: Stilbene,
6, and LiAlH₄-
reduced kraft
milled wood
lignin (MWL).
Data by Falkehag,
Marton and Adler
(61,62)

not a necessary prerequisite for stilbene formation. Methylation of phenolic hydroxyl in dihydrodehydrodiisoeugenol again stabilized the phenylcoumaran system towards alkali.

The major part of the phenylcoumaran units probably are non-phenolic in wood lignin. As more and more units become phenolic in the course of pulping, almost all of these structures are converted to stilbene without the scission of C-C connecting bonds between units. The main contribution of phenyl coumaran structures to delignification is the formation of new phenolic hydroxyl groups, making the lignin more soluble in the pulping medium.

Reactions of non-cyclic benzyl alkyl ethers also have been investigated to some extent. It was shown that the alkaline hydrolysis of vanillyl alcohol methyl ether to vanillyl alcohol proceeds through intermediate formation of a quinone methide (71). Other phenolic models containing benzyl alkyl ether groups also were cleaved by alkali (57, 58). The cyclic benzyl alkyl ether dimer, pinoresinol, was degraded to a mixture of seven phenolic compounds, while pinoresinol dimethyl ether was completely alkali-stable in conformity with other non-phenolic benzylalkyl ethers (57).

C. Reactions of β-O-4 Alkyl Aryl Ethers. The cleavage of β-O-4 ether linkages probably is one of the most important single reactions of alkaline delignification. The non-phenolic β-ether model 8 is hydrolyzable by alkali if the alcoholic hydroxyl groups in the neighboring α and/or γ positions are free or if the α-position contains a carbonyl group (65).

Gierer and coworkers (63, 64) presented convincing evidence that the mechanism of cleavage involved a nucleophilic attack of the neighboring hydroxyl groups on the β-carbon atom, resulting in the formation of an epoxide 9 and simultaneous removal of the aryl ether substituent as phenoxide ion. The epoxide is then further hydrolyzed to the arylglycerol 10, Equation 6. Glycerol ether 11 undergoes hydrolysis by the same mechanism. The presence of benzyl thiol in alkaline hydrolysis of this compound produces sulfide 12 from the epoxide intermediate (106).

Methylation of the adjacent alcoholic hydroxyl groups in models 8 and 11 with dimethyl sulfate prevented the epoxide formation and rendered the β-aryl ether bonds in both model compounds, as well as in methylated milled wood lignin, stable towards hydrolysis with alkali (65) or kraft white liquor (66). Normally, soda or kraft cook of milled wood lignin results in the formation of about 0.3 new phenolic hydroxyl group. Periodate oxidation of soda-cooked milled wood lignin indicated that this hydrolyzed lignin contained new 1,2-glycol groups (about $0.1/OCH_3$) in probable association with non-phenolic arylglycerol groups 13 formed during the alkaline treatment (67). This finding constitutes strong evidence for the operation of the epoxide mechanism in the hydrolysis of lignins at 170°.

The reactions of phenolic β-O-4 linked units are more complex. Soda cook of the β-O-4 linked dimer 14 gave a moderate yield (20-30%) of guaiacol. With a mixture of alkali and sodium sulfide, the guaiacol cleavage was more extensive. In both cases a comparatively alkali-stable enol aryl ether 16 was also formed by competitive reaction dominant in the soda cook (68, 69). The reactions of compound 14 in alkaline cook are interpreted as shown in Equation 7 (60, 68, 69).

(7)

The phenolic β-aryl ether model 14 first converts to quinone methide 15 which may lose its terminal methylol group as formaldehyde yielding the comparatively stable enol aryl ether 16. β-Aryl ether cleavage takes place only as a side reaction (in about 25% yield), possibly with simultaneous formation of the epoxide 17. However, if sulfide is present, the strongly nucleophilic hydrosulfide ions add rapidly to the quinone methide 15, with the formation of benzyl mercaptan 18. This reaction competes with the alkali-catalyzed removal of the terminal methylol group from 15 and formation of enol ether 16. The strongly nucleophilic mercaptide anion then displaces the aryl ether substituent from the neighboring β-carbon atom with the formation of an episulfide structure 19 (70) and a new phenolic hydroxyl group. Under conditions of kraft cooking, the latter reaction pathway (14→15→18→19) is favored. The β-aryl ether cleavage is more extensive (over 60%), though a smaller amount of enol ether 16 also is formed.

The episulfide-containing side-chain in 19 undergoes further changes which are quite complex and not fully understood. Enkvist et al. (70) have shown that treatment of the β-ether 14 in sodium hydrogen sulfide solution at 100° for eighteen days produces propioguaiacone 20, coniferyl aldehyde 21

REACTIONS IN ALKALINE PULPING

and acetoguaiacone 22. Spruce wood in a similar treatment gave the same three products in 1.3, 2.0 and 6.5% yields, respectively. While it is quite legitimate to point out that the applied treatment was different from that of kraft pulping, it is pertinent to note that propio- and acetoguaiacones have also been isolated from kraft pulping liquors of spruce wood (70). Homovanillic acid 23, homovanillyl alcohol 24, dihydroferulic acid 25 and dihydroconiferyl alcohol 26 were likewise isolated from the same source, and all four compounds qualify as plausible candidates for the final reaction products of original guaiacylglycerol-β-aryl ether structures. Whether or not these compounds are actually formed in the kraft cook of β-ether model 14 cannot be unequivocally decided at this point for the lack of sufficiently detailed studies. To date, only homovanillic acid 23 has been identified, but thin-layer chromatography (51, 68) indicates the presence of a variety of low- and high-molecular-weight phenolic substances in the products. Condensations with the released formaldehyde undoubtedly contribute to the complexity of the product mixture. Speculatively at least, 23 and 24 could be dismutation products of homovanillin 27. Ketol 28 also deserves consideration as a potential short-lived intermediate. Finally, the occurrence of redox processes is suggested, and such processes have indeed been observed in later model studies.

[Structures 20–28: guaiacyl compounds with various side chains]

20: HO-C₆H₃(OMe)-COCH₂CH₃
21: HO-C₆H₃(OMe)-CH=CHCHO
22: HO-C₆H₃(OMe)-COCH₃
23: HO-C₆H₃(OMe)-CH₂COOH
24: HO-C₆H₃(OMe)-CH₂CH₂OH
25: HO-C₆H₃(OMe)-CH₂CH₂COOH
26: HO-C₆H₃(OMe)-CH₂CH₂CH₂OH
27: HO-C₆H₃(OMe)-CH₂CHO
28: HO-C₆H₃(OMe)-CH₂COCH₂OH

The reactions are less puzzling in the alkaline cook (140°, 24 hrs., .5N NaOH) of the erythro form of the β-ether 29 (51) which has a less reactive side-chain than β-ether 14. The isolated products include the threo form of the starting material 29, guaiacol, enol ether 30 and guaiacyl acetone 31, in full accordance with the sequential reactions in Equation 8. Guaiacyl acetone 31 is quite reactive and undergoes condensation reactions to further products to be discussed in Section 3.

Gierer et al. (69, 90-92) have studied the behavior of the guaiacyl glycol β-ether model 32 in great detail. In this case, a kraft cook causes an almost complete release of guaiacol as shown in Figure 16.7.

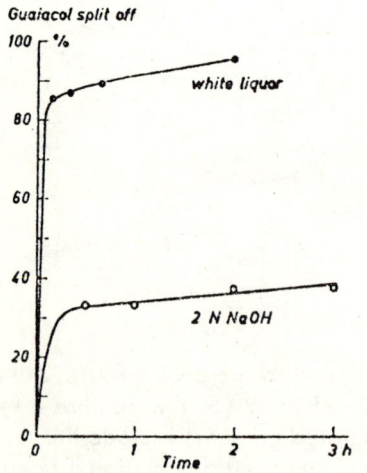

Figure 16.7. Cleavage of the β-O-4 ether bond in the guaiacyl glycol β-ether model, 32, by 2N NaOH and by white liquor at 170°, according to Gierer, Lenz and Wallin (69).

Equation 9 summarizes several model compound reactions under a variety of conditions. The intermediate quinone methide 33 was successfully synthesized from the corresponding p-hydroxybenzyl bromide; and 33, in turn, was converted to the benzyl thiol 34, via a thioacetate derivative. While the original ether, 32, required temperatures from 150 to 170° for successful hydrolysis, the quinone methide, 33, was cleaved by kraft liquor at 75° within half an hour to the extent of 55%. The benzyl thiol, 34, was even more reactive. Treatment with 2N sodium hydroxide at 60° for 15 minutes afforded a quantitative yield of guaiacol. On the other hand, alkali alone caused only minor hydrolysis of 32 and 33, converting both compounds mainly to enol ether, 16. A facile cleavage analogous to that of thiol 34 was also observed for the simplified model, 38.

Treatment of benzyl thiol, 34, in the form of its thioacetate, with alkali at 100° produces an amorphous product in 77% yield, presumably the benzyl alcohol expected by adding a mole of water to quinone methide, 36 (90). Acidification converts this product to stable dimeric dithiane, 37, also obtained directly by alkali treatment of episulfide 35 at ambient temperature (92). Kraft cook converts the dithiane to a mixture of products reducing the sulfur content from the original 17.6% down to 5.4%. The original β-ether, 32, yielded in kraft cook several unknown products with the same overall content of sulfur.

Analogous reactions have been demonstrated for the γ-methyl analogue 29 (50, 92a). In addition, the characterization of products from the kraft cook was successful. Compound 29 was heated with white liquor at 140° for 3 hours, and guaiacol as well as trans-isoeugenol, 41, were obtained in 99 and 64% yields, respectively. The latter compound was also obtained in the soda cook of dithiane, 39. The quinone methide, 40, appears to be a probable intermediate, Equation 10. It should be noted, however, that isoeugenol was also obtained as a side product (20% yield) from the soda cook of compound 29.

In analogy, the formation of coniferyl alcohol, 43, in the kraft cooks of both guaiacylglycerol-β-ether 14 and lignin appears likely; coniferyl alcohol, however, is not stable under kraft cooking conditions, giving rise through the reactive intermediate 44 to a number of further reaction products. Nimz (92b) has recently observed the formation of coniferyl alcohol in the alkaline treatment of thiolignin with Raney-nickel at 115°. The reaction may proceed via quinone methide 42, Equation 11, bearing similarity to the reaction pathways discussed above.

III. Condensation Reactions.

A. The Extent of Condensation in Delignification. Lignin condensation is too frequently used as a vague term in order to explain sudden decrease of reactivity, diminishing solubility, or any change which can be associated with an increase of molecular weight. Commonly, the term condensation has been applied to reactions in which a free 5-position of a lignin unit becomes substituted by a carbon atom of another unit. Another type of condensation has been suggested by recent studies (51), which indicate that condensation reactions may also link side-chain positions together forming α-α carbon-carbon bonds.

Condensation invariably accompanies lignin fragmentation as a side reaction to delignification. However, the exact extent of condensation reactions is extremely difficult to estimate. Thus, the observation of Goring and his associates that the molecular weight of kraft softwood lignin gradually increased in successive fractions of continuous cook is probably the consequence of three independent factors: first, the topochemical nature of the pulping reaction (18); second, the gradual increase of porosity, i. e., permeability of the cell wall (17) during the cook; and third, a continuing condensation of reactive lignin fragments before and after dissolution. Some indication of the extent of condensation may be inferred from the effect of stabilizing the reactive intermediates by hydrogenation. In a study of alkaline pulping of maple wood in the presence of hydrogen and a nickel catalyst, Schuerch and coworkers (72, 73) found that almost half of the lignin was converted to low-molecular-weight substances. The high yield of monomeric and dimeric phenols suggests that catalytic hydrogenation is indeed an effective tool for stabilizing the reactive molecular fragments. A similar study with softwood lignins would be even more informative, by using possibly a soluble catalyst if precautions could be taken to prevent hydrogenolytic side reactions.

B. Condensation Reactions Involving Ring-Position 5. From the standpoint of aromatic reactivity (electrophilic substitution) of lignin units in alkaline medium, the most important ring position is the position 5, ortho to the phenolic hydroxyl. Leopold (74), by alkaline nitrobenzene oxidation, and Freudenberg (75), by deuterium exchange, established that almost half of the 5-positions in spruce wood lignin are initially occupied by substituent groups. Adler and Lundquist (76) showed by oxidizing the phenolic part of spruce milled wood lignin with Fremy's salt that the free 5-positions are evenly distributed among the etherified and phenolic units, the latter constituting about one-third of the aryl propane units in softwood lignins.

The original suggestion that sulfate lignin consists mostly of units with condensed 5-position was based on an investigation of Richtzenhain (77). He found that diazomethane-methylated wood, when subjected to permanganate oxidation, gave isohemipinic acid, 45, in a yield of 0.9%, based on the amount of lignin; while methylated sulfate lignin yielded as much as 6.65%.

$$\underset{45}{\text{HOOC}\overset{\text{COOH}}{\underset{\text{OMe}}{\bigcirc}}\text{OMe}}$$

It was also found that if Willstätter lignin was treated with strong hot alkali prior to methylation, the yield of isohemipinic acid on oxidation increased from an initial 1.9% to 3.4%. The hot alkali treatment was assumed to bring about an opening of phenyl coumaran structures (see Equation 5), thereby increasing the structural elements yielding isohemipinic acid. The higher yield obtained from sulfate lignin seemed to indicate appreciable condensation during the delignification process with the formation of new carbon bonds to the 5-position. It would certainly be of general interest to repeat this basic experiment with improved oxidation techniques (78, 47), using modern chromatographic separation methods (79).

While Richtzenhain speculated that sulfate lignin was mainly condensed with carbohydrates, Enkvist (80) suggested that the high yield of isohemipinic acid was the result of an alkaline resol type condensation between the benzyl alcoholic group of phenyl propane units 46 with the free 5-positions of another unit 47, creating new α-5 linkages, 48, Equation 12.

$$\underset{46}{\text{CHOH-Ar(OMe)(OH)}} + \underset{47}{\text{C}_3\text{-Ar(OMe)(OH)}} \xrightarrow{\text{OH}^-} \underset{48}{\text{HC(Ar(OMe)(OH))-Ar(C}_3)(OMe)(OH)} \quad (12)$$

Model experiments for the secondary formation of α-5-bonds have been described by Migita and coworkers (81). Vanillyl alcohol (R=H) and its methyl- or guaiacyl ether 49 readily condensed with acceptor phenols unsubstituted in 5-position such as creosol 50 but did not react with etherified (nonphenolic) derivatives. The etherified model veratryl alcohol did not react with phenols. Obviously, two prerequisites for condensation are that a quinone methide intermediate can form from the reacting p-hydroxy benzyl alcohol and that the acceptor be a phenol unsubstituted in position 5, Equation 13.

$$\underset{49}{\text{HO-Ar(OMe)-CH}_2\text{OR}} + \underset{50}{\text{Me-Ar(OMe)(OH)}} \longrightarrow \underset{}{\text{HO-Ar(OMe)-CH}_2\text{-Ar(Me)(OMe)(OH)}} \quad (13)$$

Hästbacka (82, 70) investigated the kinetics of self-condensation of vanillyl alcohol 51 at 75° to 96° C in alkali with and without sodium sulfide present. In the presence of Na$_2$S, the main product was vanillyl monosulfide 55 in 87% yield; while in the absence of sodium sulfide, diguaiacyl methane 53 formed in 20% yield together with polycondensation products, Equation 14 and figure 16.8.

The consumption of vanillyl alcohol proceeded according to first order kinetics. The rate of consumption of vanillyl alcohol, Figure 16.9, was twice

REACTIONS IN ALKALINE PULPING

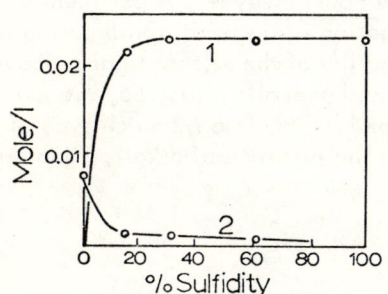

(14)

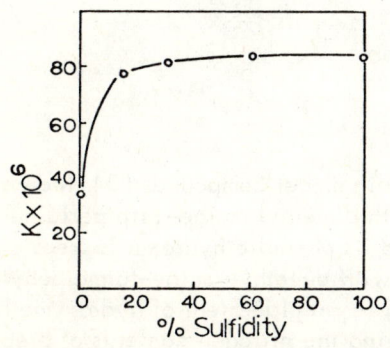

Figure 16.8. The yield of vanillyl monosulfide, 1, and diguaiacyl methane, 2, as a function of the sulfidity after a reaction time of 8 hrs. at 95.6°C. Initial concentration of vanillyl alcohol: 0.0648 mole/l. Active alkali content: 13.4g Na_2O/l (82).

Figure 16.9. The effect of sulfidity on the rate of consumption of vanillyl alcohol at 95.6°C. Concentrations, see Figure 16.8 (82).

as fast with sulfide present as with sodium hydroxide alone. It was assumed that the formation of quinone methide 52 is the rate-determining step (k_1). The net reaction rate depends on the relation of k_1 to k_2. If SH^- ion, which is a powerful nucleophile, is present, the hydrosulfide ion can rapidly react with 52 to form vanillyl mercaptan 54. This explains why the sulfidation dominates even at low sulfidities. The next step is probably the rapid formation of vanillyl sulfide 55 by the addition of the vanillyl mercaptan to the quinone methide. The formation of diguaiacyl methane 53 is suppressed because 54 is a much stronger nucleophile than vanillyl alcohol 51.

These experiments provide support for the idea of the blocking action of sulfur by sulfidation of the reactive benzyl alcoholic groups of lignin. This concept was originally proposed by Hägglund (83) to explain the differences between soda and sulfate pulping. However, Gierer (69) has expressed some reservation concerning the effectiveness of the blocking action by sulfide. He observed that p-hydroxybenzyl alcohol and p-hydroxybenzyl sulfide yielded chromatographically similar condensation products when treated under the conditions of kraft pulping.

C. Reactions of Kraft Lignin with Formaldehyde. A pertinent answer concerning the extent of lignin condensation in ring-position 5 during the sulfate pulping is obtained by direct investigation of the sulfate lignin recovered from the black liquor. The non-condensed phenolic units, 56, are expected to react with dimethyl amine and formaldehyde in a Mannich type of reaction to form 57, or with formaldehyde in the presence of alkali, Equation 15.

A study of Mannich reaction with lignin model compounds (84) indicated that a guaiacyl compound readily reacts with dimethyl amine- or piperidine-formaldehyde reagent in position 5 provided its phenolic hydroxyl is free. Vanillyl alcohol and vanillic acid reacted with dimethyl amine-formaldehyde both in position 5 and position 1 of the ring (by replacement of hydroxymethyl carboxyl groups, respectively). By comparing the nitrogen contents of preparations from pine sulfate lignin and diazomethane-methylated sulfate lignin,

Mikawa (84) concluded that 25-40% of the phenolic (guaiacyl) nuclei had no substituent in 5 position. His calculation included the assumption that the COOH groups of sulfate lignin are of the vanillic acid type. More recent data suggests that the carboxyl groups of lignin are aliphatic (85); thus, they would be non-reactive in the Mannich reaction. This correction would increase somewhat the number of free 5 positions in Mikawa's calculation.

Kraft lignin reacts with formaldehyde in the presence of alkali. Hydroxymethyl groups $\underline{58}$ are formed in position 5 which give a characteristic blue color reaction (absorption maximum at 590 mµ) with $FeCl_3$. Spectrophotometric investigation indicates that about every third phenolic unit has free 5 position available for phenol alcohol formation (86).

About 0.15 mole CH_2OH/OCH_3 was introduced into spruce milled wood lignin and about 0.5 mole into kraft pine lignin by the alkaline formaldehyde reaction (86). Unlike in the Mannich reaction, methylol groups were also introduced into certain side-chain positions, activated by neighboring carbonyl groups, as in $\underline{60}$, to form $\underline{61}$ (Tollens reaction), Equation 16.

$$60 + CH_2O \xrightarrow{OH^-} 61 \tag{16}$$

Kraft lignin was treated with $NaBH_4$ in order to remove the activating carbonyl groups from the side-chains, therewith eliminating the possibility of Tollens reaction; 0.4 CH_2OH/OCH_3 was introduced into the reduced lignin. In another experiment, the formaldehyde-reactive phenolic units were converted to non-reactive etherified ones by diazomethane, allowing only the carbonyl-containing side-chains to react in subsequent formaldehyde treatment; 0.1 CH_2OH/OCH_3 (orig.) was introduced into the methylated lignin. The sum of introduced hydroxymethyl groups in these separate experiments was in agreement with the amount found in reaction of untreated lignins, and good indication was obtained as to the location of the methylol groups in formaldehyde-treated kraft lignin. When correction is applied to account for formaldehyde reacted with catechol groups, it is found in accordance with the spectrophotometric study that about one-third of the phenolic nuclei are unsubstituted in position 5 of the kraft pine lignin against one-half of the units (phenolic and non-phenolic) in milled softwood lignin. These observations tend to indicate that an appreciable, but not too extensive condensation had occurred in the 5-positions during the kraft cook.

Besides the resol condensation, a further reaction possibility has to be considered. During alkaline hydrolysis of wood, formaldehyde is formed from phenylcoumaran (59) or guaiacyl glycerol aryl ether structures of lignin (68, 69). Ekman (87) has suggested that formaldehyde, which is continuously released during delignification, may react with the phenolic units forming methylol lignin. Under alkaline conditions the formation of diphenyl methane bridges 59 may occur more readily between two methylol groups with loss of formaldehyde and water than between a phenol alcohol and free nuclear positions (88), Equation 16.

Either one of the suggested condensation reactions involving free 5-positions would be less extensive for hardwood than for softwood lignin by reason of the presence of a methoxyl substituent in position 5 of the syringyl units. Accordingly, the molecular weight of hardwood kraft lignin is less than that of pine kraft lignin; $\overline{M}_n$ was found (89) to be about 1000 for hardwood kraft lignin and about 1600 for pine kraft lignin.

D. Condensations Between Side-Chain Positions. In the alkaline cleavage of the guaiacylglycol-β-ether model 29 (see Equation 8), the formation of an interesting condensation product, 63, was observed (51). The formation of this compound is best understood in terms of an addition of the enolate of guaiacyl acetone, 31, to the quinone methide of 29, 62, producing the diphenyl ethane derivative, 63, Equation 17. The second condensation product isolated, 64, represents simply a normal condensation of the product guaiacol with the starting material.

$$(17)$$

Condensed structures of the type 63 are possible in soda and kraft lignins in which they could have been formed in reactions between reactive β-carbonyl derivatives, such as homovanillin 27 or ketol 28, and quinone methide intermediates. While little is known of condensations of this type at the moment, their potential significance in alkaline pulping should not be underestimated.

REACTIONS IN ALKALINE PULPING

IV. The Role of Sulfide in Kraft Pulping.

A. Differences Between Soda and Kraft Pulping. The complexity of reactions in kraft and soda pulpings make the interpretation of direct observations an excessively difficult task. Consequently, it is more useful to try a different approach by attempting to apply results available from model compound research to the lignin macromolecule. In this way, a tentative and undoubtedly incomplete picture of the two delignification processes can be set forward and this picture may then be compared with observations made in direct pulping experiments. To do this, it is convenient to consider separately the behavior of phenolic and non-phenolic elements, representing roughly 30 and 70% of the monomeric units in softwood lignin, respectively.

In soda pulping, phenolic elements with benzylic hydroxyl groups are converted to the corresponding quinone methides in a reversible reaction resulting in an equilibrium. Aliphatic or aromatic ether groups associated with the α-carbon are rapidly hydrolyzed in a generally irreversible reaction with the formation of the quinone methides. (However, quinone methides formed from phenolic phenyl coumaran structures have the opportunity of reverting back to the cyclic ether). Further reactions of the quinone methide intermediates include the following: 1, The γ-carbons of quinone methides deriving from phenolic arylglycerol-β-ether-, phenylcoumaran- and β-1-linked structures are largely released as formaldehyde as a consequence of a reversed condensation process. The formaldehyde may be captured by phenolic units elsewhere in the lignin molecule, introducing a hydroxymethyl group in the 5 position of these units. The resulting o-hydroxybenzyl alcohol structures are quite reactive by virtue of their potential to form o-quinone methides and may generate diaryl methane cross-linkages at a later stage of the process. Alternatively, formaldehyde may be captured by side-chain carbons adjacent to a carbonyl, introducing hydroxymethyl groups to these positions also; 2, Concurrently with the release of formaldehyde (or in an independent reaction), quinone methide intermediates rearrange to α,β-unsaturated structures. Thus, enol ethers are formed from initial aryl glycerol β-ethers, o,p'-dihydroxystilbenes from phenyl coumarans and p,p'-dihydroxystilbene derivatives from β-1-linked elements. In a competitive hydrolytic process, original aryl glycerol-β-ethers may release also homovanillin, ketol 28 or derivatives of both, all of which react rapidly further. This hydrolysis reaction, as well as the formation o,p'-dihydroxystilbenes, generates new phenolic units subject to the same reactions as the original phenolic elements; and 3, In competition with the two processes mentioned, phenolic elements may participate in condensation reactions. Secondary α-5 linkages can be formed in reactions with other phenolic elements. Also, side-chain carbons adjacent to a carbonyl may establish a bond to the methylene carbon of the quinone methide.

The presence of sulfide in the kraft pulping modifies the reactions of the soda process probably in the following fashion: The quinone methide intermediates are largely converted benzyl thiols reducing the opportunity for condensation reactions. There is nothing to indicate that the reactions of phenylcoumaran and β-1-linked elements would be otherwise altered. The reaction path of arylglycerol-β-ethers, however, is profoundly changed. The release of formaldehyde from these units is reduced and the formation of stable enol ethers largely replaced by the hydrolytic cleavage of the β-ether bond. Thus, more β-ether linkages are cleaved and more new reactive phenolic units are generated in the kraft process than in soda pulping.

Of the non-phenolic units in lignin, only the arylglycerol-β-ethers are directly reactive. They are converted to a free glycerol side-chain and a new phenolic units in a reaction which requires relatively high temperatures (160-170°) and receives virtually no assistance from hydrosulfide ions. While hydrosulfide ions remain passive in the primary cleavage reaction, they again exert their beneficial influence upon the course of reactions experienced by the newly generated phenolic unit.

Thus, the function of sulfide in kraft pulping appears to be two-fold. On one hand, it promotes and accelerates the cleavage of ether linkages in phenolic units, aiding the fragmentation of the lignin macromolecule to soluble entities. On the other hand, it also reduces the extent of undesirable condensation reactions. Remarkably, the effect of reduced condensation was already postulated in the early papers of Hägglund (83), and Enkvist (104) formulated the two functions of sulfide in an essentially correct way at a time when little supporting evidence was available from model studies. The overall picture of soda and kraft processes fits also well together with kinetic experiments carried out by Gierer et al. (66) on milled wood lignin and its diazomethane methylated derivative, shown in Figure 16.10. The release of new phenolic groups was determined as function of time in soda and kraft cooks. It may be noted that soda cook released less phenolic hydroxyl groups in a slower reaction from milled wood lignin in comparison with its methylated derivative. This is probably caused by the exclusive presence of non-phenolic units in the methylated lignin reducing the chance for the formation of alkali-stable enol ethers.

B. Intermediate Thiolignins and Their Conversion to Kraft Lignin. The broad chemical features of alkaline delignification appear reasonably firm, but much of the finer detail remains obscure. For example, the possible influence of free radical reactions (45, 46) in the pulping process needs to be firmly established.

Attempts have been made to characterize the intermediate stages of the kraft process by studies on the so-called "thiolignins." These are formed, as first shown by Enkvist (93), upon treating wood with sodium hydrogen sulfide solution at pH 8-9, at 100°. The lignin combines with approximately 8% sulfur and only a small part of it dissolves, Figure 16.1. When the treatment is done at somewhat higher temperature, the thiolignin melts and flows out into the lumina of tracheid cells. This thiolignin, when heated with sodium hydroxide at 160°, loses the greater part of its sulfur, forming a product very similar to kraft lignins, with a sulfur content of 2-3%. The reported sulfur contents of

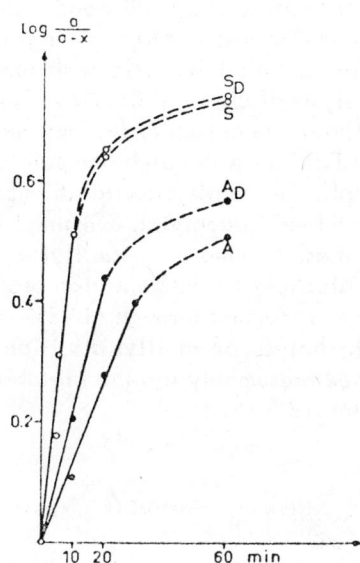

Figure 16.10. First-order rate plots for the splitting of aryl ether linkages in milled wood lignin and in diazomethane-methylated milled wood lignin by 2N sodium hydroxide and by white liquor at 170°C (66). S_D, sulfate cook of diazomethane-methylated milled wood lignin; S, sulfate cook of milled wood lignin; A_D, alkali cook of diazomethane-methylated milled wood lignin; A, alkali cook of milled wood lignin. Data by Gierer, Lenz and Wallin (66).

kraft lignins vary generally from 1.5-3%. Consequently, it appears probable that in the formation of thiolignins, kraft delignification is arrested at its early stage. Thiolignins are therefore well suited for studies on the initial sulfidation of lignin and on the later decomposition of sulfur-containing species as a separate step, using the reactions of pertinent model compounds as reference.

This approach has led to some ambiguous results, however. Gierer and Alfredsson (102) observed that milled wood lignin could be made to combine readily with 3% of sulfur, but the same reaction conditions failed to introduce any sulfur to the β-ether model 14, dihydrodehydrodiisoeugenol and pinoresinol. Rather, the first model underwent dehydration, and the two others were reclaimed unchanged. The lacking reactivity of the models was explained in terms of the steric hindrance offered by the bulky side-chain substituents present in the three models, but this left the sulfidation reaction of the milled wood on an obscure basis. It was observed (102, 103) that an appropriate mineral acid treatment removes the benzylic sulfur from a number of models, and this reaction was applied to the thio derivative of milled wood lignin. Two-thirds of the combined sulfur was removed in this fashion, giving some indication, at least, that the majority of sulfur occupied benzylic positions.

The alkaline hydrolysis of dithians 37 and 39 and related intermediates (see Equations 9 and 10) provide currently the best model experiments for the interpretation of desulfurization in the high-temperature stage of kraft pulping. Earlier model studies employing vanillyl sulfide 55 (Equation 14) and -disulfide 65 illustrate the risks involved in applying oversimplified models to the study of complex reactions. Of these two models, vanillyl sulfide reacts more slowly (94), yielding vanillyl alcohol, its condensation products, and vanillin (97, 94). Vanillin in this case is formed through air-oxidation (82); while in the more rapidly occurring hydrolysis of vanillyl disulfide 65 at 100°, it is a direct hydrolysis product, formed presumably via the unstable sulfenic acid intermediate 66 (95, 96), Equation 18.

$$2\ HO\text{-}C_6H_3(OMe)\text{-}CH_2SH \xrightarrow{OX.} HO\text{-}C_6H_3(OMe)\text{-}CH_2S\text{-}]_{/2} \xrightarrow{OH^-} HO\text{-}C_6H_3(OMe)\text{-}CH_2SH + HO\text{-}C_6H_3(OMe)\text{-}CH_2SOH$$
54 · · · · · · · · · · · · · 65 · 66

$$\downarrow OH^-$$

$$HO\text{-}C_6H_3(OMe)\text{-}CHO + HS^-$$

(18)

On the basis of these and other (98, 99) model studies, Mikawa has speculated that intermediate benzyl sulfide and disulfide groups may decompose in the kraft process to yield α-carbonyl, α-carboxyl groups and α,β-double bonds in the side-chains of sulfate lignin. However, recent studies (100, 101) have failed to detect significant numbers of α-carbonyl groups in kraft lignin, and the carboxyl groups present are not directly bonded to aromatic rings.

During the kraft process, the overall consumption of sulfide has been claimed to be one-fifth to one-half of the charged sulfide, depending on the sulfidity of the white liquor (111). The lost sulfur is mainly found as residual sulfur in the sulfate lignin. The location of this sulfur is still a matter of controversy. At one end, Zhigalow and Tishchenko (114) have questioned whether sulfate lignin contains organically bound sulfur at all. They claimed that sulfur is present only in elemental form and can be removed completely by extraction with potassium cyanate. The existence of lignin-sulfur bonds in kraft lignins has been eloquently defended by Enkvist (115), and the level of organically bound sulfur is generally placed at about 1.5%. At least five different types of linkages are indicated in pine kraft lignin, among them sulfones and sulfoxides, but thiol-, thiocarbonyl-, disulfide-, polysulfide- or dialkyl sulfide

REACTIONS IN ALKALINE PULPING

groups seem to be absent (112). Nelson (113) also confirmed the absence of thiol- and disulfide groups in eucalypt kraft lignin.

C. <u>Cleavage of Methoxyl Groups</u>. In soda pulping, the hydrolysis of methoxyl groups to methanol and a free phenolic hydroxyl is insignificant (117); in kraft pulping, however, the strongly nucleophilic hydrosulfide ions cleave a part of methoxyl groups to methyl mercaptan 67, generating catechol groups 68 in the kraft lignin, Equation 19. The anion of methyl mercaptan is an even stronger nucleophile than hydrosulfide (Cf. Chapter 19), and reacts also with methoxyl groups, generating methyl sulfide 69 and additional catechol groups, Equation 20 (119).

$$SH^- + \text{MeO-Ar(C}_3\text{)-OR} \longrightarrow \text{MeSH} + {}^-\text{O-Ar(C}_3\text{)-OH} \quad (19)$$
$$\qquad\qquad\qquad\qquad 67 \qquad\qquad 68$$

$$MeS^- + \text{MeO-Ar(C}_3\text{)-OR} \longrightarrow \text{MeSMe} + {}^-\text{O-Ar(C}_3\text{)-OH} \quad (20)$$
$$\qquad\qquad\qquad\qquad 69 \qquad\qquad 68$$

Kinetically, both reactions are straightforward S_N2-displacements, being first order in nucleophile and methoxyl concentrations (118). More methyl mercaptan and methyl sulfide are generated from hard- than from softwoods due to a rapid initial surge in the formation of these compounds from hardwoods, the nature of which has not been clarified. Methyl mercaptan and -sulfide are the main components of the characteristic unpleasant odor of kraft mills. The extent of their formation amounts to only 5-6% of original methoxyl groups in a normal kraft cook. The activation energies of the formation processes are lower than that of delignification, and this explains their reduced formation in high (185°) temperature vapor phase continuous pulping process (45, 46).

Protocatechualdehyde 70 and protocatechuic acid 71 (70) are examples of low-molecular-weight catechol derivatives formed in kraft pulping. Lindberg (122) has also detected several sulfur-containing catechol derivatives among the water-soluble degradation products of thiolignin. The catechol groups present in kraft lignin may be the source of important side reactions (condensation, o-quinone formation). The demethylation of guaiacyl and syringyl compounds becomes extensive at temperatures above 200°, and kraft lignin can be effectively demethylated further in sodium sulfide solutions under pressure (120) or simply by heating highly concentrated black liquor in a melt (121).

HO-C₆H₃(OH)-CHO   HO-C₆H₃(OH)-COOH

     70   71

Another interaction between sulfide and phenolic elements deserves to be mentioned. Lignin in black liquor has been found to catalyze the air oxidation of sulfides to polysulfide (123, 124) and the further oxidation of polysulfides to thiosulfate. Phenoxy- and semiquinone radicals as well as quinoidal intermediates may contribute to the electron-transfer process involved in the oxidation. The reaction has industrial importance in the black liquor oxidation stage often associated with the recovery of chemicals in kraft pulping and may find further application in connection with polysulfide pulping processes.

V. Structure and Properties of Kraft Lignin.

A. Isolation. In an average kraft cook, about 70-90% of the original lignin content of the wood is dissolved. The dissolved lignin (in the form of its sodium salt) constitutes about half of the organic matter in the black liquor. Additional components include other organic substances dissolved from the wood, the inorganic chemicals charged (mainly in the form of salts with organic acids), inorganic components of the wood substance (most importantly, calcium in hardwoods), and eventually chemical additives (defoamers, drainage aids) used for facilitating the pulp washing. The total solids content may vary from 15-23%. The approximate composition of a normal black liquor from a pine kraft cook is shown in Table 16.3 (3). The liquor is concentrated in multiple-effect evaporators up to a solids content of 50-70% in order to achieve proper burning of the organic substance in the recovery furnace.

Table 16.3. Typical Composition of Pine Kraft Black Liquor (3). (Total Solids about 20%, pH around 11)

	% of total solids
Alkali lignin (including soluble phenols)	41
Extractives (fatty and rosin acids)	3
Hydroxy acids and lactones	28
Acetic acid	5
Formic acid	3
Methanol	1
Sulfur	3
Sodium	16
TOTAL	100%

A smaller part of the wood lignin is retained in the pulp fiber and can only be removed by oxidative bleaching. Bolker (125) did not find any appreciable difference between the infrared spectra of the original lignin in wood and of the residual lignin in kraft fibers. Internal reflectance infrared spectra of lignin in kraft fibers are also similar to that of the dissolved kraft lignin fraction (126). Infrared spectroscopy is not a sufficiently sensitive tool, however, to recognize fine structural differences.

The lignin content of black liquor can be estimated by measuring its UV-light absorption at 280 mµ. An approximative absorptivity value (127) for calculating the total lignin content is 25 cm^{-1} x l x g^{-1}, after diluting the sample with 1N NaOH.

The greater part of the aromatics (80-90%) precipitate when the pH is lowered below 3, but the precipitate is difficult to filter. A two-step acidification procedure is therefore preferred. The sodium lignate is precipitated first with sulfuric acid or carbon dioxide at around pH 9-10, the precipitate coagulated below the boiling point and filtered. The sodium lignate is then resuspended in water and acidified with sulfuric acid to pH 2-3. The acid-insoluble lignin fraction is easily filtered and washed, and contains two-thirds of the aromatic components of black liquor.

Merewether (128-130) investigated the conditions for precipitation of eucalyptus kraft lignin, and Nikitin and associates (131) described the precipitation of lignin by using carbon dioxide. Ball and Vardell (132-134) have patented a number of processes related to the commercial isolation of lignin from kraft black liquors. The non-precipitable aromatic material is composed of lower lignin oligomers, phenolic dimers, and monomers. Part of the acid-soluble guaiacyl and catechol derivatives have been identified by paper chromatography (70). Guaiacol, vanillin, vanillic acid and acetoguaiacone are the main components, each in amounts of 2-5 pounds per ton of pulp.

The literature on different sulfate lignins is voluminous (135, 136), but most of the lignin preparations are poorly characterized. Enkvist (137) studied commercial kraft and soda softwood lignins by fractionation with organic solvents and by acidification. The small lignin fraction obtained at pH 2, after precipitation of the main part of lignin at pH 7, showed unusually high carboxyl content and a low content of methoxyl (9.6%). Wada (138) acidified stepwise a spruce kraft black liquor and characterized the fractions obtained, Table 16.4.

B. Analytical Composition. Wood source, pulping conditions, and the method of isolation influence the chemical properties of isolated kraft lignins; hardwood lignins especially are different in many important structural details. It would be impossible, at present, to judge just what significance is to be given to the diverging data published for different preparations. The discussion of structural properties of kraft lignin will therefore be based mainly on data (101) gathered in the author's laboratory unless otherwise indicated. The respective kraft lignin samples were isolated from commercial batch pulpings (23% sulfidity) of southern pine wood (mostly loblolly pine) and a mixture of hardwoods (oak and gum). The analytical compositions of two commercial (101) and one laboratory-prepared kraft lignin (139) samples are given in Table 16.5.

Table 16.4. Fractional Precipitation of Kraft Black Liquor from Spruce Wood (138)

Precipitation pH	% of Total Prec.	OCH_3 %	Phenolic OH* per OCH_3	COOH*	S%
10.75	27	13.5	0.64	0.05	1.16
10.5	32	14.2	0.67	0.07	2.07
10.2	10	13.1	0.70	0.08	
10.0	7	12.9			1.44
9.9	4	12.8			
9.5	4	11.6	0.74	0.37	
9.4	3				1.50
8.2	2	11.2			3.87
7.8	1				
1.5	10	10.3	0.90	0.37	6.46

*Determined from OCH_3 change by methylation and subsequent soponification.

Table 16.5. Representative Composition of Kraft Lignins (101,139).

Percent	Pine	Hardwood	Kraft M.W.L. Spruce
C	65.87	65.43	65.81
H	5.82	5.77	6.20
O	25.90	25.31	
S	1.56	1.50	1.44
OCH_3	14.04	19.24	15.7
ash	0.96	0.50	1.4

Table 16.6. Calculated C_6-C_3 Units in Lignins

Lignin	Calculated C_6-C_3 Units	Unit Weight	Reference
M.W.L. spruce[a]	$C_9H_{8.8}O_{2.37}(OCH_3)_{0.96}$	185	(140)
M.W.L. birch[b]	$C_9H_{8.24}O_{2.95}(OCH_3)_{1.38}$	206	(139)
Kraft pine[c]	$C_9H_{7.9}O_{2.1}S_{0.1}(OCH_3)_{0.82}$	178	(101)
Kraft hardwood	$C_9H_{7.2}O_{1.8}S_{0.1}(OCH_3)_{1.15}$	183	(139)
Alkali spruce	$C_9H_{7.3}O_{2.2}(OCH_3)_{0.9}$	179	(141)
Kraft M.W.L. spruce	$C_9H_{7.3}O_{2.2}S_{0.1}(OCH_3)_{0.9}$	179	(139)

a) corrected for carbohydrates;
b) not corrected for carbohydrates; the OCH_3 content probably is higher;
c) the average unit in Figure 16.22 has a composition of $C_{8.35}H_{6.4}O_{2.1}(OCH_3)_{0.78}$, with a unit weight of 168.

REACTIONS IN ALKALINE PULPING 673

The average C_6C_3-formulae of milled wood and kraft lignins are presented in Table 16.6. The comparison is not exact because in reality the composition of an average unit in kraft lignins probably is closer to a C_6-C_2 than to a C_6-C_3 unit as a consequence of the release of formaldehyde from side-chains during pulping (59, 68, 87).

The most apparent changes in the analytical composition of lignin due to kraft pulping are losses of hydrogen, oxygen and some methoxyl groups, and the introduction of sulfur into about every tenth unit.

C. <u>Functional Groups</u>. The functional groups found in pine kraft lignin (101) are compared with the corresponding groups of spruce milled wood lignin in Tables 16.7 and 16.8. The comparison is based on 100 C_6-C_3 units.

Table 16.7. Functional Groups of Lignin (101).

Groups in 100 C_6-C_3 units	M.W.L. Spruce	Kraft pine	Method Used	Method Ref.
Total OH	120	120	Acetylation	(86,142,143)
Guaiacyl OH	30	60	KIO_4 oxidation	(144)
2 X Catechol	-	12	Fe^{2+} colorimetry	(62)
Aliphatic OH[a]	90	48		
COOH	5	16	Methylation	(85)
Total CO	20	15	NH_2OH, HCl	(145)
			volum, KBH_4	(146)
Coniferyl aldehyde	3	-	Red. $\Delta\epsilon$	(147)
α-CO	7	5	Red. $\Delta\epsilon$	(147)
β and/or other -CO[a]	10	10		

a) by difference

Table 16.8. Some Reactive Structures in Lignin (101)

Groups in 100 C_6-C_3 units	M.W.L. Spruce	Kraft pine	Method Used	Method Ref.
Benzyl alcohol and ether				
Noncyclic	42	<6	Methylation	(85)
Cyclic				
Ph. coumaran	11	3	H^+, Spectrophoto-	(54)
Pinoresinol	ca. 10	ca. 5	Cat. acetylation	(155)
C=C double bond				
Coniferyl type	7	+	H_2-Red.	(156)
Stilbene	-	7	Spectrophotom.	(61,62)
Other	-	+		

1. Hydroxyl Groups. The amount of total hydroxyl groups is about the same in both lignins. A considerable loss of aliphatic hydroxyl groups is compensated in kraft lignin by an increase in phenolic hydroxyls. The hydrolytic cleavage of alkyl aryl ether linkages approximately doubles the number of phenolic guaiacyl nuclei. Ferrous sulfate color reaction (62) indicates that six aromatic rings per hundred contain o-dihydroxy groups due to demethylation during the pulping. The phenolic groups of kraft lignin have an average pK_a 10.5 to 11.0 as a consequence of some weakly acidic sterically hindered phenols (149).

The wide spectrum of ionization constants makes it difficult to differentiate the acidic groups by means of conductometric or potentiometric titration. The total number of acidic groups, determined potentiometrically by titration with potassium methoxide in dimethyl formamide (150), was for pine kraft lignin 4.65 mequiv./g (83 phenolic and carboxyl groups/100 units) and for hardwood soda lignin 3.5 mequiv./g. Conductometric titrations with lithium hydroxide (151) and lithium metaborate (152) indicated the presence of 4 and 4.8 mequiv. acidic hydroxyls per g in pine kraft lignin, respectively; while the latter method gave 5.85 mequiv per g for a hardwood soda lignin preparation (152). The cation exchange capacity of pine kraft lignin was determined by exchange with calcium and barium ions (105). The obtained value of 15 mequiv. per 100 g lignin at pH 7 is very low probably as a consequence of the poor swelling ability of lignin in the calcium and barium salt solutions. NMR analysis of alkali lignins confirmed (148) the average distribution of aliphatic and phenolic hydroxyls determined by conventional analytical methods, Tables 16.9 and 16.10.

Table 16.9. Protons per C_9 Structural Unit in Acetylated Lignins as Determined by the Integration on NMR Spectra (148)

Range	Values	Protons known to give signals within range	MWL Spruce	Lab Soda Blackgum	Lab Soda Pine	Lab. Kraft Pine	Mill Kraft Pine
1 and 2	8.03-6.28	Aromatic and α-vinyl	2.57	2.28	2.82	2.76	2.52
3	6.28-5.74	Total β-vinyl and $α_1$-acetylated benzylic	0.32	0.58	0.39	0.46	0.50
4	5.74-5.18	$α_3$-acetylated benzylic from coumaran system	0.34	0.43	0.38	0.38	0.36
5	5.18-2.50	Total protons	5.81	6.57	5.44	5.59	5.16
5a		Methoxyl, calculated (5,5a)	2.55	3.79	2.52	2.54	2.32
5b		All other α, β and γ protons found in models	3.26	2.78	2.92	3.05	2.84
6	2.50-2.19	Acetoxyl, aromatic	0.83	1.42	1.33	1.28	1.34
7	2.19-1.58	Acetoxyl, aliphatic	2.78	2.14	2.68	2.73	2.36
8	1.58-0.38	Aliphatic, highly shielded	0.06	0.58	0.42	0.86	0.85
		Total protons per C_9 structural unit	12.77	14.01	13.48	14.08	13.09

Table 16.10. Estimation of Aromatic Units and Hydroxyl Groups in Various Lignins by NMR Spectroscopy (148)

Lignin Prep.	Wood	Aromatic protons		Type of Aromatic Unit			Phenolic Hydroxyls		Aliphatic hydroxyls by NMR (per C_9)
		Theoretical (per C_9)	Actually found (per C_9)	Free guaiacyl	Condensed guaiacyl per 100 C_9	Syringyl	by NMR (per C_9)	by IO_4^- oxid. (per C_9)	
Acetyl M.W.L.	Spruce	3	2.57	57	43	--	0.27	0.26	0.92
Mill kraft	pine	3	2.52	52	48	--	0.44	0.43	0.79
Lab kraft	pine	3	2.76	76	24	--	0.43	0.47	0.89
Lab soda	pine	3	2.82	82	18	--	0.44	--	0.89
Lab soda	black gum	2.7[a]	2.28	28	44	28	0.47	--	0.71

a) based on 70% guaiacyl and 30% syringyl aromatic nuclei

2. Carboxylic Acid, Benzyl Alcohol and -Ether Groups.

Methylation by diazomethane converts phenolic elements in kraft lignin to their methyl ethers and esterifies the carboxylic acid groups. The methyl ester groups, in turn, are saponifiable by alkali and reducible to primary carbinol groups by lithium aluminum hydride. The methoxyl loss in such reactions can be used to determine the actual number of original carboxylic acid groups, which have been found to amount to about 16/100 units (85, 153).

It should be noted that if the carboxylic groups were aromatic in kraft lignin as proposed by Mikawa (99), the above reduction would generate new benzylic alcohol groups, methylatable by methanolic hydrogen chloride. Since the new carbinol groups were found not methylatable (85, 157), the carboxylic acid groups must be aliphatic rather than aromatic; that is, be derived from the β- and/or γ-carbons of protolignin. The aliphatic nature of the carboxylic acid groups is also supported by their infrared absorption (154) and methylation behavior (85). An estimate of benzylic hydroxyl groups can be based on the circumstance that these groups are unreactive towards diazomethane, but methylatable by methanolic hydrogen chloride. It has been found (85) that six new methoxyl groups (per hundred units) may be introduced to diazomethane-methylated kraft lignin by the latter treatment. However, the six methylatable groups represent only a ceiling estimate for benzylic hydroxyls because a host of other groups, including benzyl ethers, ketol hydroxyls and coniferyl alcohol residues, share the same behavior towards methylation. In milled wood lignin, groups of this type amount to 50/100 units, and their drastic reduction to 6 reflects a profound structural modification in kraft pulping. If a major part of the quoted value is indeed represented by benzylic hydroxyl groups as appears probable, they are likely to reside in non-phenolic guaiacyl glycerol units 13.

Direct methylation of kraft lignin with methanolic hydrogen chloride introduces 22 new methoxyl groups/100 units (85). Subtracting the contribution of benzylic hydroxyl and related methylatable groups leaves sixteen methoxyls as carboxylic acid esters in excellent agreement with the estimate based on reduction by lithium aluminum hydride.

Only a small number of residual phenylcoumaran structures were found in kraft lignin by using the coumaran-coumarone conversion method of Adler and Lundquist (54). Spectrophotometric evidence suggests, Figure 16.6, that the majority of phenylcoumarans have been converted to o,p-dihydroxystilbenes 6 during the cook (61, 62). By using Freudenberg's $HClO_4$-catalyzed acetylation technique (155), a conjecture has been obtained supporting the reduction of original pinoresinol units by one-half during the kraft pulping (139).

3. Unsaturated Structures.

Redinger (141) determined ten C=C double bonds in every hundred C_6-C_3 units in an alkali lignin, reacting them with peroxyphtalic acid. The presence of about 7-8 stilbene type double bonds has been demonstrated in kraft lignin by UV-spectroscopy (61, 62). Additional double bonds must be present in alkali-stable vinyl ether groups of the type 16 (68, 69). By electron spin resonance (ESR) technique, 10^{-4}

radical was found (158, 159) per aryl propane unit in kraft lignin. In the sodium salt of kraft lignin, the radical content was, by a factor of one hundred larger, 10^{-2} per aryl propane unit, which would indicate the presence of one semiquinone type of structure in every hundred aryl propane units of the kraft lignin. On the other hand, the presence of roughly 25 to 30 quinonoid groups was claimed by Chirkin and Tishchenko (160), who reacted alkali lignins with o-phenylendiamine. However, this estimate appears unreasonably high and should be viewed with some scepticism. Harkin (110) has recently proposed o-quinone methide structures of the type 2 as additional candidates for minor unsaturated structures in kraft and soda lignins.

4. Carbonyl Groups. Milled wood lignin contains about 20 carbonyl groups per 100 aryl propane units (147). Of these, coniferyl aldehyde groups (4/100) are destroyed during the alkaline pulping. Infrared absorption of kraft lignin derivatives suggests the presence of both conjugated and non-conjugated carbonyl groups. The total number of carbonyl groups in kraft lignin is about 15/100 units as determined (101) by oximation with hydroxylamine hydrochloride and by the volumetric potassium borohydride method. Migita and coworkers (100) measured 7-13 carbonyls in a sulfate spruce lignin and 10-16 in a soda lignin by the oximation method.

Much information can be obtained about the conjugated carbonyl groups in kraft lignin by ultraviolet difference spectroscopy (161, 162). The ionization difference spectrum in Figure 16.11a indicates a substantial amount of phenolic α-carbonyl units absorbing at 350 mμ in kraft lignin, but their amount cannot be estimated because of the strong overlap in absorption by phenolic stilbenes with a maximum at 370 mμ. Instead, a spectrophotometric reduction $\Delta\epsilon$—method, based on the use of borohydride, was employed (147) to differentiate between various types of conjugated carbonyls, Figure 16.11b and Table 16.7. It is important to note that etherified aryl-α-carbonyl groups ($\Delta\epsilon_r$ max. at 340 mμ) are reduced very rapidly, while the reduction of phenolic aryl-α-carbonyl groups may take several days ($\Delta\epsilon_r$ max. around 355 mμ).

Kraft lignin obviously does not contain any easily reducible carbonyls but mainly phenolic aryl-α-carbonyl groups with $\Delta\epsilon_r$ at 355 mμ, corresponding to about 5 α-carbonyl groups per 100 aryl propane units (101). By using the same (147) $\Delta\epsilon_r$ method, Migita (100) and coworkers found 5 and 6 phenolic α-carbonyl groups/100 units in sulfate and soda lignins, respectively. They noticed (100) that the $\Delta\epsilon_r$ curves of alkali lignins exhibit a weak second maximum ($\Delta\epsilon_r \simeq 300$) at around 420 mμ, indicative of the reduction of a carbonyl group with long conjugation. They also oxidized sulfate lignin with perbenzoic acid (Baeyer-Villiger method), and, from the formation of acetic acid, it was suggested that all the α-carbonyl groups should belong to ortho-acetoguaiacone type of structures, the methyl ketone group being in the 5-position (163).

Unlike in milled wood lignin (156), practically all α-carbonyl groups in kraft lignin are located in phenolic units. In addition, less than ten non-conjugated carbonyl groups/100 units are indicated to be present in kraft and soda

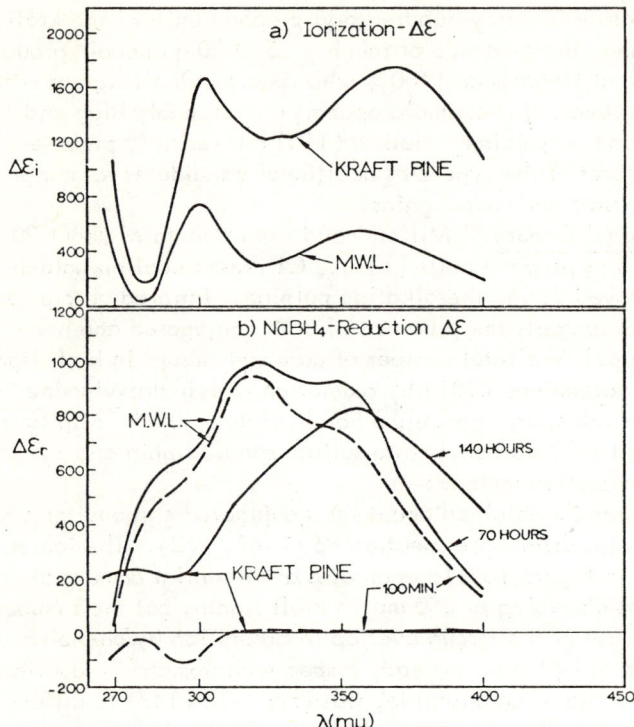

Figure 16.11. Ionization and NaBH$_4$-reduction difference curves of MWL and kraft pine lignin (101).

lignins on the basis of the total carbonyl determination. In all probability, the latter are β-ketone groups.

5. <u>The Aromatic Ring</u>. Determination (101) of o-methoxyphenols by periodate oxidation (144) indicated that about two-thirds of guaiacyl units in pine kraft lignin are phenolic. In accordance with Mannich reaction (84) and alkaline formaldehyde condensation (86), one-third of the phenolic units possess free 5-positions, while two-thirds are substituted, presumably by a carbon atom of another unit.

From the color reaction of lignin with ferrous sulfate, it was concluded (62) that kraft lignins contain about .06 catechol units/MeO. Judging from the more extensive demethylation of hardwood lignins in kraft pulping, indicated by data in Table 16.6, the number of catechol groups in these lignins might be higher.

Some information (164) on the 2,6-positions of kraft pine lignin has been obtained in the following way: Kraft lignin was carefully oxidized with sodium

REACTIONS IN ALKALINE PULPING 679

metaperiodate, converting about half of the phenolic guaiacyl units 72 to o-quinones. The latter groups were reduced to catechol units 73 by sulfur dioxide. This oxidized-reduced lignin preparation was more reactive in alkaline-formaldehyde condensation than untreated kraft lignin. Analysis indicated that about two hydroxymethyl groups were introduced in every catechol group formed. This suggests that the great majority of the 2,6-positions, at least in the phenolic units, are unsubstituted, Equation 21. This conclusion is in agreement with the results obtained from the permanganate oxidation of methylated kraft lignins (77).

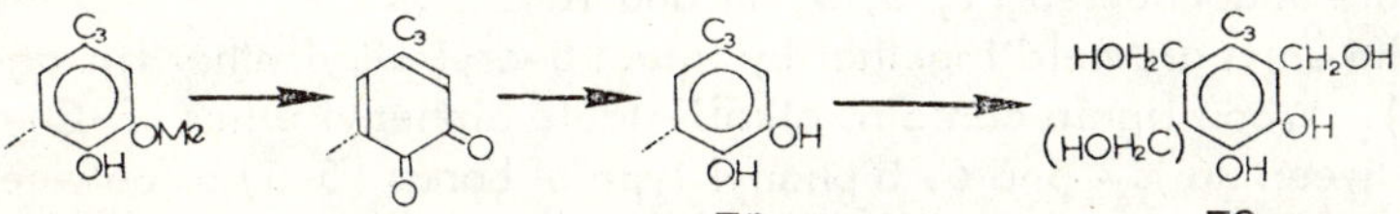

$$\underset{}{\text{Ar-OMe/OH}} \longrightarrow \underset{72}{\text{o-quinone}} \longrightarrow \underset{}{\text{catechol-OH}} \longrightarrow \underset{73}{(HOH_2C)\text{-Ar}(CH_2OH)_2\text{-OH}} \qquad (21)$$

D. Formulation for Kraft Lignin. The structural information on softwood kraft lignin has been summarized in the form of an idealized structural scheme by Marton (101). Of course, the present knowledge is quite insufficient to allow for constructing a reliable formulation. Thus, Figure 16.12 is being proposed with the understanding that it is not to be considered as a structural formula of softwood kraft lignin. The suggested formulation is simply a scheme compatible with our present stage of knowledge which needs to be modified and refined as soon as new information becomes available (105).

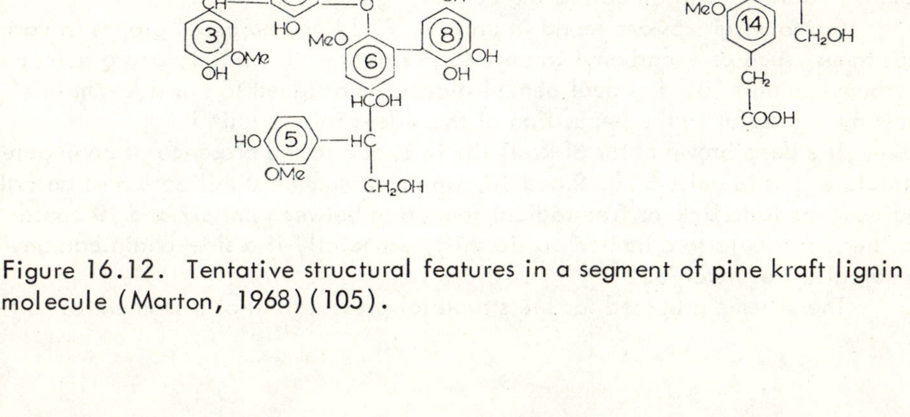

Figure 16.12. Tentative structural features in a segment of pine kraft lignin molecule (Marton, 1968) (105).

The average molecular weight of kraft lignin probably is much lower than that of wood lignin and also substantially influenced by the conditions of the cook. Figure 16.12 depicts a segment of the average pine kraft lignin molecule consisting of fourteen units, about half of them having intact C_6-C_3 skeleton. Part of the units lost side-chain carbon atoms (units 2, 8, 9, 10, 13 and 14) lowering the average composition of an aryl propane unit to about C_6-$C_{2.4}$. The kraft lignin formulation contains non-phenolic units (6, 7, 11 and 14) which may make up about 30% of all units according to the analytical data (101). Twenty percent of the aromatic rings do not contain methoxyl groups (4, 8 and 11). The majority of the nuclei are substituted in the 5-position, and about 35% are uncondensed (1, 3, 5, 11 and 13).

Some of the units are held together by intact β-aryl alkyl ether linkages (12, 13 and 14). Wood lignin contains alkali-stable biphenyl ether (5-O-4) links such as between units 4 and 6, biphenyl type of bonds (5-5) as between units 6 and 8, and diphenyl ethane type (β-1) linkages as between units 5 and 6. All these linkages withstand the attack of pulping chemicals. Phenyl coumaran (β-5) structures of wood lignin have been converted to stilbene structures, 8 and 9, which also are quite stable in alkali though sensitive to oxidative or reductive agents, expecially in the presence of light. Some of the phenyl coumaran structures may have reacted further by condensation (units 3 and 4). Part of the phenolic aryl glycerol β-aryl ether structures have been converted to alkali-stable vinyl ether linkages such as between units 10 and 11. Part of the non-phenolic pinoresinol type of structures of wood lignin may also survive and still connect neighboring units with β-β bonds in the side-chain and with cyclic benzyl alkyl ether (α-O-γ) links such as between units 4 and 7.

Condensation reactions may result in such new linkages as α-5 links in units 3-4 and 9-10 as well as 5-CH_2-5 links between units 11 and 12 deriving from condensation with formaldehyde. Side-chain ketone condensation may also lead, in units 12-13, to diphenyl-methane links of the α-1 type. Diphenyl ethane type (α-α) links connect the side-chains of units 1 and 2. Formaldehyde split off from the side-chains has reacted with the free 5-position in unit 12 and in position ortho to catechol hydroxyls in unit 11, forming reactive phenol alcohols which subsequently condensed. Formaldehyde has also entered the side-chain in unit 11 with an activating carbonyl group.

Carboxyl groups are found in units 2, 7, 14 and carbonyl groups in various forms, such as α-carbonyl in unit 4, β-carbonyl in unit 11, and a quinone carbonyl in unit 10. Residual benzyl alcohol is assigned to the side-chain of unit 6, and sulfur to the β-position of the side-chain in unit 1.

The deep brown color of kraft lignin is due to the presence of conjugated structures, as in units 6, 8, 9 and 10, where resonance stabilization of an ortho quinone methide link or free radical formation between units 9 and 10 could further contribute to a bathochromic shift, especially if a side-chain carbonyl group also was involved.

The scheme proposed for the structural presentation of a segment of softwood kraft lignin molecule reflects the results of analytical studies concerning the amount and distribution of various functional groups in kraft lignin. The selected sequence of units is more or less arbitrary, though not in a completely random way. The analytical composition of an average unit in the formulation would correspond to $C_{8.35}H_{6.4}O_{2.1}(OCH_3)_{0.78}$. The hydrogen content of this unit is somewhat lower than that ($H_{7.3}$) from the analysis of kraft lignin (see Tables 16.9 and 16.10).

E. Spectral Properties. The main features of the infrared spectra of both milled wood lignin and kraft pine lignin are similar in many respects. The most apparent differences are in the C=O region (5.7-6.0μ). Milled wood lignin has a medium-weak shoulder at 5.8μ and a strong composite band at 6μ, corresponding to its non-conjugated carbonyl and α-carbonyl plus coniferyl aldehyde groups, respectively. Kraft lignin exhibits a strong composite band at 5.8μ, and the coniferyl aldehyde absorption at 6μ is missing, Figure 16.13. Sodium borohydride reduction removes the great majority of C=O groups and lithium aluminum hydride removes them completely. In the reduced lignin, the residual C=O stretching band is indicative of carboxyl groups. The latter can be removed by reduction of their esters, leaving a spectrum free of carbonyl frequencies (85). Tentative interpretations of the absorption bands outside the carbonyl region have been attempted, see also Chapter 7 (165, 167).

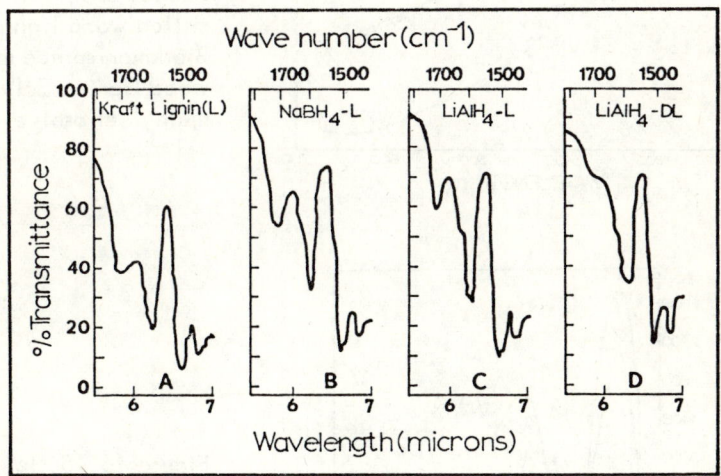

Figure 16.13. Infrared absorption spectra of various kraft lignin preparations in the C=O region (85).

Multiple internal reflectance (MIR) spectra of unbleached kraft pulps (126) contain the characteristic absorption bands of both pine kraft lignin and bleached kraft pulp. The intensity of lignin absorption at 1510 cm^{-1} has been utilized for quantitative determination of lignin in unbleached pulps (126).

The absorption curves of milled wood lignin and two kraft lignins in the ultraviolet and visible region are compared (62) in Figure 16.14. The high absorptivity of kraft lignins above 300 mµ is due to such chromophores as vinyl ethers, α-carbonyls, stilbenes, quinonoid structures and their conjugated combinations. Residual kraft lignin in kraft pulp can be determined by dissolving the fiber in acetyl bromide-acetic acid mixture and relating the absorbance of the solutions at 280 mµ to the absorptivity of reference lignin preparations (127).

Differential ionization spectra of borohydride reduced milled wood and kraft lignins are compared in Figure 16.15. The higher maximum of kraft lignin at 300 mµ reflects the larger number of phenolic elements, and that at 372 mµ, the presence of hydroxystilbene structures.

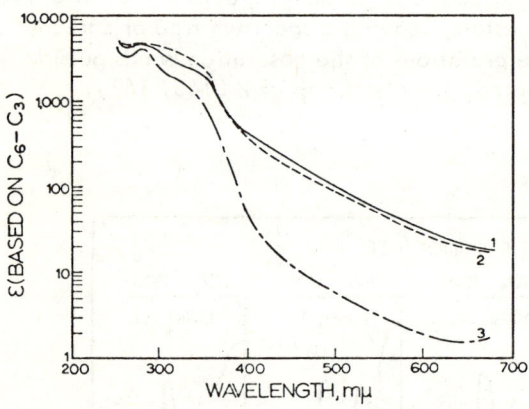

Figure 16.14. Absorption curves for softwood lignins (62). (1) kraft lignin (Indulin AT), (2) kraft-cooked Björkman spruce milled wood lignin, (3) Björkman spruce milled wood lignin. Solvent: methylcellosolve-ethanol, 1:1.

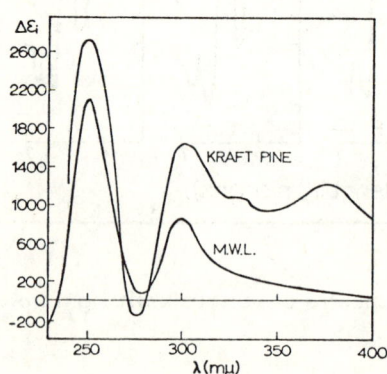

Figure 16.15. Ionization-Δε curves of NaBH$_4$-reduced lignin preparations (101).

REACTIONS IN ALKALINE PULPING

Nmr spectroscopy has been successfully applied to confirm several structural features of kraft lignin detected earlier by other analytical means. Table 16.9 summarizes the assignments of protons to various structures in pine and black gum kraft lignins, and Table 16.10 gives an estimate of the aromatic units and of hydroxyl groups in the two lignin preparations (148).

The color of powdered lignin can be characterized by its reflectance curve. The reflectance of a powder (R_∞) is related to its specific absorption (k, m^2/g) and scattering (s, m^2/g) coefficients through the Kubelka-Munk equation, 22 (167, 168).

$$F(R_\infty) \equiv \frac{(1-R_\infty)^2}{2R} = \frac{k}{S} \qquad (22)$$

Figure 16.16 presents the $F(R_\infty)$ reflectance functions (62) of kraft lignin and milled wood lignin. These curves characterize the typical colors of the lignin preparations as they appear. Some useful optical data on kraft lignin and its solutions are presented in Table 16.11.

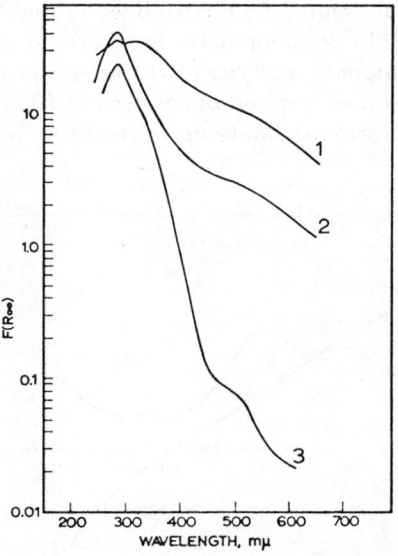

Figure 16.16. Kubelka-Munk reflectance functions. (1) Softwood kraft lignin, (2) $NaBH_4$-reduced softwood kraft lignin, (3) Björkman spruce milled wood lignin (62).

Table 16.11. Some Optical Data for Alkali Lignins

Data		Pine Kraft	Hardwood Kraft	Spruce Sulfate	Spruce Soda
Dielectric constant	(169)	5.15	4.55		
Refractive index, n for dry powder	(89)	1.635	1.63-1.64		
Refractive index increment, dn/dc, in Tetrahydrofuran (THF)	(89)	0.204	0.207		
Partial spec. vol., in THF	(89)	0.704	0.693		
dn/dc, in Pyridine	(170)	0.132			
dn/dc, in 0.1 N NaOH	(170)	0.249			
dn/dc, in Dioxane	(171)			0.188	0.191
dn/dc, in DMSO	(171)			0.116	

F. <u>Macromolecular Properties</u>. Besides the chemical structure, the physico-chemical and colloidal properties of kraft lignin are equally important factors which affect its behavior. It is interesting to note that in spite of their probably cross-linked structure, both milled wood and kraft lignins are thermoplastic. Pine kraft lignin does not melt, but starts to shrink at around 195-200°C. Differential thermal analysis (101) indicates that some tenaciously adsorbed or structural water is given off at around 105-110°C, after which the pine kraft lignin preparation is stable up to its shrink point, Figure 16.17.

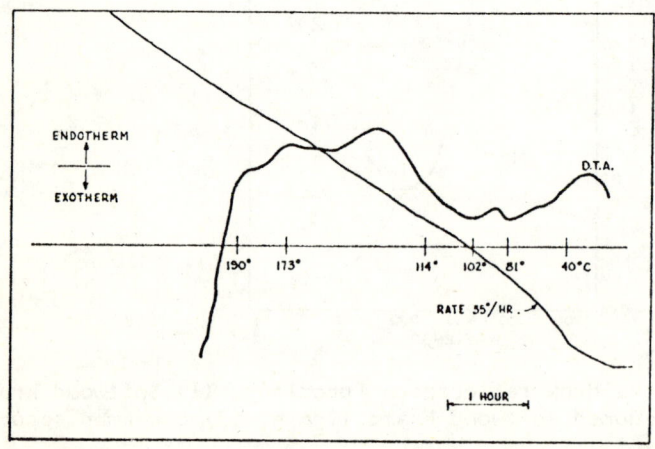

Figure 16.17. DTA curve of kraft pine lignin (101).

Like other polyphenolic materials, kraft lignin is very susceptible to air oxidation in alkaline solution, which brings about a considerable increase in its molecular weight (105). This might serve as a partial explanation for the rather divergent data published by different investigators, Table 16.12.

Table 16.12. Molecular Weight of Alkali Lignins

Lignin Preparation	Acetone Solubility,%	$\overline{M}_n$	$\overline{M}_w$	$\overline{M}_w/\overline{M}_n$	Gel. Filtr.	Reference
Kraft pine	65-75	1600	3500	2.2		(89)
Kraft pine; acetone sol. fract.	100	1000				(105)
Kraft pine; mildly oxidized in alkaline solution	25-15	3000-4300				(105)
Kraft hardwood	90	1050	2900	2.8		(89)
Sulfate, spruce	46		5600		10,000	(171,172)
Soda, spruce	51		82,000		20,000	(171,172)
Sulfate, spruce (Continuous flow process)	(Consecutive fractions)		1800-51,000			(19)

The number average molecular weights ($\overline{M}_n$) of pine and hardwood kraft lignins were determined by thermoelectric method (89), and the weight average molecular weights ($\overline{M}_w$) by equilibrium sedimentation in ultracentrifuge (89, 19). These methods indicate that an average pine kraft lignin molecule is composed of more than 20 units and the average molecule of a hardwood kraft lignin contains more than 15 units. The ratio $\overline{M}_w/\overline{M}_n = 2.2$ is close to what would be expected from a polymer degraded through a random process.

Molecular weight distribution by gel filtration (172) and by light scattering technique (171) furnished higher values (Table 16.12), especially noticeable with the acetone-soluble fraction. Weight average molecular weights ($\overline{M}_w$) calculated from light scattering data are unproportionally affected by the high molecular weight fraction present even in a relatively low amount. Number average molecular weights, ($\overline{M}_n$), on the other hand, are strongly influenced by the presence of low molecular weight substances. Measurements of Goring and associates (19) indicated that the molecular weight of kraft lignin increased from 1800 for the first fraction to 51,000 for the final fraction as consecutive fractions were taken during a continuous kraft cook of spruce sawdust in a flow reactor.

The viscosities of dilute solutions of pine kraft lignin are only a few centipoise in good solvents, but increase (Figure 16.18) steeply after a certain concentration has been reached where the lignin gels (173). It is interesting

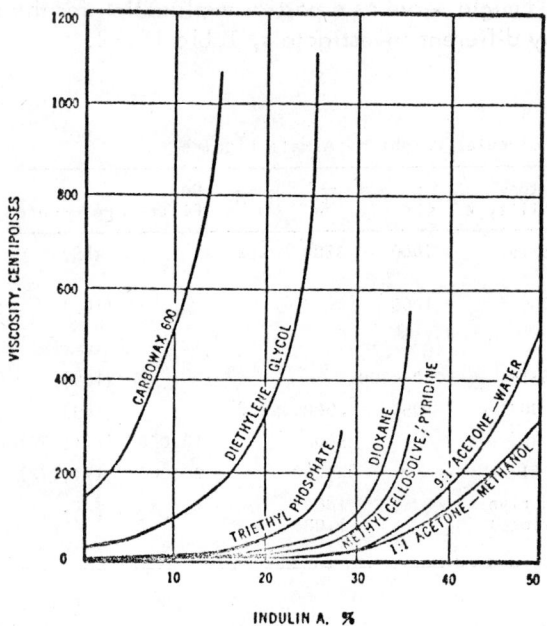

Figure 16.18. Viscosities of some organic solutions of Indulin A at room temperature (173).

to note that both the dilute and rather concentrated solutions behave like true Newtonian liquids. The intrinsic viscosities of softwood alkali lignins are 0.06 and 0.09 dl.g^{-1} in dioxane (171) and pyridine (174), respectively.

The intrinsic viscosity of polymers is related to the molecular weight by Equation 23, in which the magnitude of exponent a depends on the shape of the molecule.

$$[\eta] = KM^a \tag{23}$$

According to Lindberg (171), $K = 1.68 \times 10^{-2}$ and $a = 0.12$ for spruce sulfate lignin, which means that sulfate lignin behaves in dilute solution as a compact spherical particle consisting of cross-linked microgel swollen by the solvent (175), see Chapter 17. The low value of exponent a renders the viscosity of dilute kraft lignin solutions rather insensitive to variations in molecular weight. The situation changes conpletely in concentrated solutions due to the increasing association of the dissolved lignin molecules. Studies on the melt viscosity of polymers have shown (176) that a can reach values as high as 3.5; similarly, small variations in molecular weight of lignin may produce considerable changes in the viscosity of concentrated solutions.

Alkali lignins are soluble in dilute alkali, in most basic solvents, and in some neutral polar solvents, Table 16.13. The solvent property most important for lignin solubility is hydrogen bonding capacity (178). Secondly, the solvent should have (177) a cohesive energy density (Hilderbrand's solubility parameter) around 11. Cohesive energy densities either much lower (ether, C. E. D. 7.3) or much higher (glycerol, C. E. D. 14; water, C. E. D. 23) can result in poor lignin solubility even though the hydrogen bonding capacity of the solvent is high. The hydrogen bonding capacity can be characterized (177), for instance, with the wave length shift ($\Delta\mu$) of the O-D absorption band in infrared when CH_3OD is mixed with the solvent. Solvents with $\Delta\mu > 0.14$ and $\delta \approx 11$ are generally good lignin solvents (177, 178).

Table 16.13. Solubility of Alkali Lignins in Simple Solvents (178)

Solvent	C.E.D. (cal/c.c.)	$\Delta\mu$	Kraft Pine Lignin	Hardwood Soda Lignin
Diethylene glycol	9.1	H	Sol	Sol⁻
Dioxane	10.0	0.14	Sol	Sol
Acetone	10.0	0.14	P	P
Pyridine	10.7	0.27	Sol	Sol
Methyl Cellosolve	10.8	H	Sol	Sol
Acetonitrile	11.9	0.09	Sol⁻	P
Ethanol	12.7	H	Slight	Slight
Ethylene glycol	14.2	0.31	Sol	Sol
Methanol	14.3	0.28	P	P
Water	23.4	H	I	I

Abbrev.: H=High, Sol=Soluble, I=Insoluble, P=Partially Soluble

Alcohols are poorer solvents than expected; but when mixed with other polar solvents of low hydrogen bonding strength, the solvent power of the mixture is much greater than that of either component, e. g., chloroform-ethanol mixture (178). Likewise, addition of a small amount of water increases considerably the solubility of lignins in acetone, dioxane or acetic acid. The presence of a small amount of phenol or of the low-molecular-weight phenolic fraction of lignin can increase the solubility of the higher molecular-weight-fractions in dioxane and tetrahydrofuran in which they are otherwise insoluble (89). It seems that these additives, which may facilitate the initial swelling of the particles, also increase the accessibility of the molecules for the solvent.

Ekman and Lindberg (179) studied the correlation between solubility and molecular weight of lignin, and the effect of the dielectric constant of binary organic solvent mixtures on solubility.

The dissolving power of the solvents also can be characterized by simply measuring the amount of water needed (Table 16.14) to precipitate the lignin from the solution (173).

Table 16.14. Quantity of Water Required to Precipitate INDULIN A from 25% Solution (173)

Solvent	Parts H_2O per 100 parts of solvent
Ethylene glycol	5
Formamide	27
1:1, Acetone-methanol	40
9:1, Acetone-water	40
Methyl cellosolve	53
Dioxane	53
Triethyl phosphate	80
Carbowax 600	80
Pyridine	106
Triethanolamine	>1000

In aqueous solutions, lignin behaves like a hydrocolloid. Light scattering studies (101) show that lignin is soluble in water at low concentrations in the absence of dissolved salts, Figure 16.19. Nuclei start to form and grow as the pH is lowered, and an aggregate precipitates at around pH 3 where the decreasing negative charge on the particles can no longer stabilize them. In the presence of an electrolyte, the aggregation is enhanced; and lignin precipitates around pH 6.

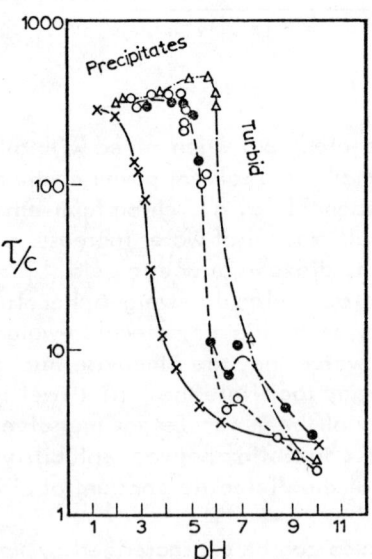

Figure 16.19. Specific turbidity of pine kraft solutions in water. X, no salt; O, 0.02M Na_2SO_4; ●, 0.04 NaCl; △, 0.10M Na_2SO_4. Lignin concentration: 0.1 g/l (101).

Overall, the general polymer properties of kraft lignins, including their behavior as polyelectrolytes, are known only superficially. There is great hope that increasing knowledge in this area will open new avenues for the utilization of these abundant by-products of the pulping industry.

REFERENCES

1. Anonymous, Pulp and Paper, 42, 18th Annual World Review, 1968.
2. C. Watt and H. Burgess, U. S. Patent 1,448 and 1,449 (1854).
3. S. A. Rydholm, Pulping Processes, Interscience Publ., New York, 1965.
4. N. Hartler, Svensk Papperstidn., 69, 191 (1966).
5. H. F. J. Wenzl, Kraft Pulping Theory and Practice, Lockwood Publ. Co., New York, 1967.
6. T. Enkvist, M. Moilanen and B. Alfredsson, Svensk Papperstidn., 52, 517 (1949).
7. S. Yllner, K. Östberg and L. Stockman, ibid., 60, 795 (1957).
8. T. N. Kleinert and L. M. Marraccini, Tappi, 48, 170 (1965).
9. N. Hartler, Paperi Puu, 44, 365 (1962).
10. T. N. Kleinert and L. M. Marraccini, Tappi, 48, 165 (1965).
11. T. N. Kleinert, L. M. Marraccini and E. J. Dostal, ibid., 43, 201 (1960).
12. A. B. Wardrop and G. W. Davies, Holzforschung, 15, 129 (1961).
13. A. B. Wardrop, Svensk Papperstidn, 66, 231 (1963).
14. A. B. Wardrop, Paper given at ESPRA meeting in Syracuse, New York, October 16, 1968.
15. F. Ruch and H. Hentgartner, Beih. Z. Schweiz, Forstver, 30, 75 (1960).
16. A. Frey-Wyssling, Wood Science and Tech., 2, 73 (1968).
17. J. E. Stone and A. M. Scallan, J. Polymer Sci., C11, 13 (1965).
18. G. P. Berlyn and R. E. Mark, For. Prod. J., 15, 140 (1965).
19. J. G. McNaughton, W. Q. Yean and D. A. I. Goring, Tappi, 50, 548 (1967).
20. A. R. Procter, W. Q. Yean and D. A. I. Goring, Pulp Paper Mag. Can., 68, T445 (1967).
21. W. Q. Yean and D. A. I. Goring, ibid., 65, T127 (1964).
22. A. Szabo and D. A. I. Goring, Tappi, 51, 440 (1968).
23. W. Q. Yean, T. Higgs, G. Suranyi and D. A. I. Goring, PPRIC. Tech. Report No. 458 (1966).
24. P. W. Lange, Svensk Papperstidn., 57, 501 (1954).
25. P. W. Lange and A. Kjaer, Norsk Skogsind., 11, 425 (1957).
26. P. W. Lange, in E. Treiber, Die Chemie der Pflanzenzellwand, Springer Verlag, Berlin, 1957, p. 259.
27. I. B. Sachs, I. T. Clark and J. C. Pew, J. Polymer Sci., C2, 203 (1963).

28. W. A. Côté, T. E. Timell and R. A. Zabel, Holz, Roh und Werkstoff, 24, 432 (1966).
29. A. L. M. Bixler, Paper Trade J. (Tappi), 107, 171 (Oct., 1938).
30. G. Jayme and A. von Köppen, Papier, 4, 455 (1950).
31. O. Kallmes, Tappi, 43, 143 (1960).
32. J. E. Luce, Pulp Paper Mag. Can., 65, T419 (1964).
33. S. Yllner and B. Enström, Svensk Papperstidn., 60, 549 (1957).
34. P. Odintsov and P. P. Erin'sh, Izv. Akad. Nauk. Latv. SSR, 2, 103 (1961).
35. P. N. Odintsov and P. P. Erin'sh, Tr. Inst. Lesokhoz. Prolb.I.Khim. Drevesiny, Akad. Nauk. Latv. SSR, 24, 3 (1962).
36. G. Jayme and W. Mohrberg, Papier, 3, 153 (1945).
37. H. W. Emerson, Fundamentals of the Beating Process, B.P.B.I.R.A., Kenley, 1957, p. 123.
38. D. A. I. Goring, Pulp Paper Mag. Can., 64, T517 (1963).
39. T. N. Kleinert and L. M. Marraccini, Tappi, 47, 605 (1964).
40. C. W. Carroll, ibid., 43, 573 (1960).
41. H. D. Wilder and S. T. Han, ibid., 45, 1 (1962).
42. H. D. Wilder and E. J. Daleski, Jr., ibid., 47, 270 (1964).
43. H. D. Wilder and E. J. Daleski, Jr., ibid., 48, 293 (1965).
44. B. Holmberg, Svensk Papperstidn., 39, 113 (1963) Spec. Nr.
45. T. N. Kleinert, Tappi, 48, 447 (1965).
46. T. N. Kleinert, Pulp and Paper Mag. Can., 67, T299 (1966).
47. E. Adler, Svensk Kem. Tidskr., 80, 279 (1968).
48. E. Adler, Ind. Eng. Chem., 49, 1377 (1957).
49. E. Adler, Paperi Puu, 43, 634 (1961).
50. E. Adler, Private communication.
51. G. E. Miksche and B. Johansson, to be published.
52. K. Freudenberg, Holzforschung, 18, 3 (1964).
53. K. Freudenberg, Advances in Chem. Series, 59, 1 (1966).
54. E. Adler and K. Lundquist, Acta Chem. Scand., 17, 13 (1963).
55. T. Ishihara, Paperi Puu, 43, 662 (1961).
56. K. Freudenberg and J. M. Harkin, Chem. Ber., 93, 2814 (1960).
57. J. Gierer and I. Norén, Acta Chem. Scand., 16, 1713 (1962)
58. J. Gierer, Paperi Puu, 43, 654 (1961).
59. E. Adler, J. Marton and I. Falkehag, Acta Chem. Scand., 18, 1311 (1964).
60. E. Adler, Chim., Biochim., Lignine, Cellulose et Hemicelluloses, Grenoble, 1964, p. 73.
61. S. I. Falkehag, Paperi Puu, 43, 655 (1961).
62. S. I. Falkehag, J. Marton and E. Adler, Advance in Chem. Series, 59, 75 (1966).
63. J. Gierer and I. Kunze, Acta Chem. Scand., 15, 803 (1961).
64. J. Gierer and I. Norén, ibid., 16, 1976 (1962).

65. J. Gierer, B. Lenz, I. Norén, and S. Soderberg, Tappi, 47, 233 (1964).
66. J. Gierer, B. Lenz and N.-H. Wallin, ibid., 48, 402 (1965).
67. J. Gierer and B. Lenz, Svensk Papperstidn., 68, 334 (1965).
68. E. Adler, I. Falkehag, J. Marton and H. Halvarson, Acta Chem. Scand., 18, 1313 (1964).
69. J. Gierer, B. Lenz and N.-H. Wallin, ibid., 18, 1469 (1964).
70. T. Enkvist, T. Ashorn and K. Hästbacka, Paperi Puu, 44, 395 (1962)
71. S. Larsson and B. Lindberg, Acta Chem. Scand., 16, 1757 (1962).
72. I. Sobolev and C. Schuerch, Tappi, 41, 545 (1958).
73. P. E. Parker, R. L. Coalson and C. Schuerch, Advances in Chem. Series, 59, 249 (1966).
74. B. Leopold, Svensk Kem. Tidskr., 64, 18 (1952).
75. K. Freudenberg, V. Jovanovic and F. Topfmeier, Chem. Ber., 34, 3227 (1961).
76. E. Adler and K. Lundquist, Acta Chem. Scand., 15, 223 (1961).
77. V. H. Richtzenhain, Svensk Papperstidn., 53, 644 (1950).
78. K. Freudenberg, C. L. Chen and G. Cardinale, Chem. Ber., 95, 2814 (1962)
79. S. Larsson and G. E. Miksche, Acta Chem. Scand., 21, 1970 (1967).
80. T. Enkvist and B. Alfredsson, Tappi, 36, 211 (1953).
81. A. Ishizu, J. Nakano, H. Oya and N. Migita, J. Japan.Wood Res. Soc., 4, 176 (1958).
82. K. Hästbacka, Thesis, Univ. of Helsingfors, Finland, 1961; Soc. Sci. Fenn. Comm. Phys.-Math, 26, No. 4; Paperi Puu, 43, 665 (1961).
83. E. Hägglund, Tappi, 32, 241 (1949).
84. H. Mikawa, K. Sato, C. Takasaki and K. Ebisawa, Bull. Chem. Soc. Japan, 29, 259 (1956).
85. J. Marton and E. Adler, Tappi, 46, 92 (1963).
86. J. Marton, E. Adler, T. Marton and S. I. Falkehag, Advances in Chem. Series, 59, 125 (1966).
87. K. H. Ekman, Tappi, 48, 398 (1965).
88. J. H. Freeman and C. W. Lewis, J. Am. Chem. Soc., 76, 2080 (1954).
89. J. Marton and T. Marton, Tappi, 47, 471 (1964).
90. J. Gierer and L. A. Smedman, Acta Chem. Scand., 19, 1103 (1965).
91. J. Gierer and N.-H. Wallin, ibid., 19, 1502 (1965).
92. J. Gierer and L. A. Smedman, ibid., 20, 1769 (1966).
92a. G. Brunow and G. E. Miksche, ibid., in press.
92b. H. Nimz, Chem. Ber., 102, 799 (1969).
93. T. Enkvist, Tappi, 37, 350 (1954).
94. H. Mikawa, Bull. Chem. Soc. Japan, 27, 50 (1954).
95. T. Enkvist and M. Moilanen, Svensk Papperstidn., 52, 183 (1949).
96. T. Enkvist and M. Moilanen, ibid., 55, 668 (1952).
97. J. J. Lindberg and T. Enkvist, Soc. Sci. Fenn. Comm. Phys. Math., 17, 4 (1953).

98. T. J. Wallace and A. Schriesheim, Tetrahedron, 21, 2271 (1965).
99. H. Mikawa, K. Sato, C. Takasaki and K. Ebisawa, Bull. Chem. Soc. Japan, 29, 265 (1956).
100. A. Ishizu, A. Sento, J. Nakano, T. Kagino and N. Migita, J. Japan. Wood Res. Soc., 9, 189 (1963).
101. J. Marton, Tappi, 47, 713 (1964).
102. J. Gierer and B. Alfredsson, Acta Chem. Scand., 11, 1516 (1957).
103. J. Gierer and B. Alfredsson, Chem. Ber., 90, 1240 (1957).
104. T. Enkvist and E. Hägglund, Svensk Papperstidn., 53, 85 (1950).
105. J. Marton and T. Marton, unpublished.
106. J. Gierer and L. A. Smedman, Acta Chem. Scand., 18, 1244 (1964).
107. G. Chirkin and D. Tishchenko, Zhur. Priklad. Khim., 35, 153 (1962).
108. J. Gierer, S. Söderberg and S. Thorén, Svensk Papperstidn., 66, 990 (1963).
109. T. Enkvist, B. Holm, A. Kourula and J. E. Marttelin, Paperi Puu, 39, 297 (1957).
110. J. M. Harkin, Advances in Chem. Series, 59, 65 (1966).
111. T. Enkvist, Svensk Papperstidn., 60, 616 (1957).
112. L. Field and P. E. Drummond and E. A. Jones, Tappi, 41, 727 (1958).
113. P. F. Nelson, W. F. Forbes and J. A. Maclaren, Holzforschung, 17, 89 (1963).
114. Y. V. Zhigalov and D. V. Tishchenko, Zhur. Priklad. Khim., 35, 147 (1962).
115. T. Enkvist and T. E. Rinaman, Paperi Puu, 45, 649 (1963).
116. A. Ishizu, J. Nakano and N. Migita, J. Japan.Wood Res. Soc., 8, 139 (1962).
117. K. V. Sarkanen, G. Chirkin and B. F. Hrutfiord, Tappi, 46, 375 (1963).
118. W. T. McKean, B. F. Hrutfiord and K. V. Sarkanen, ibid., 48, 699 (1965).
119. J. Turunen, Paperi Puu, 43, 663 (1961).
120. E. K. M. Hägglund and T. U. Enkvist, Swedish Patent 136,268 (1952).
121. F. J. Ball and R. Puschel, U. S. Patent 2,976,273 (1961).
122. J. J. Lindberg, Suomen Kemistilehti, 31, 35 (1958).
123. E. Bilberg and P. A. Landmark, Norsk Skogsind., 13, 375 (1959).
124. P. A. Landmark, P. J. Kleppe and K. Johansen, Tappi, 48, 5:56A (1965).
125. H. I. Bolker and N. G. Somerville, Pulp Paper Mag. Can., 64, 187 (1963).
126. J. Marton, Tappi, 50, 363 (1967).
127. J. Marton, ibid., 50, 335 (1967).
128. J. W. T. Merewether, Holzforschung, 15, 168 (1961).
129. J. W. T. Merewether, ibid., 16, 26 (1962).
130. J. W. T. Merewether, Tappi, 45, 159 (1962).

131. U. M. Nikitin and A. V. Obolenskaya, Tr. Leningr. Lesotekhn. Akad., 85, 13 (1960).
132. F. J. Ball, Canadian Patent 609,565 (1960); U. S. Patent 3,017,404 (1962).
133. F. J. Ball and W. G. Vardell, U. S. Patents 2,997,466 (1961) and 3,048,576 (1962).
134. F. J. Ball and W. G. Vardell, Can. Patent 666,818 (1963).
135. I. A. Pearl, The Chemistry of Lignin, Chapters 2 and 5, Marcell Dekker, Inc., New York, 1967.
136. F. E. Brauns and D. A. Brauns, The Chemistry of Lignin, Suppl. Vol., Chapters 8 and 14, Academic Press, New York, 1960.
137. T. Enkvist, C. Majani, H. Tylli and B. Widlund, Acta Polytechn. Scand., Ch. 53 (1966).
138. S. Wada, T. Iwamida, R. Iijima and K. Yabe, Chem. High Polymers (Kobunshi Kagaku), 19, 699 (1962).
139. E. Adler, S. I. Falkehag and J. Marton, unpublished.
140. A. Björkman and B. Person, Svensk Papperstidn., 60, 158 (1957).
141. L. Redinger, Monatsber. Deut. Akad. Wiss. (Berlin), 3, 571 (1961).
142. A. Verley and F. Bölsing, Ber. dtsch. Chem. Ges., 34, 3354 (1901).
143. K. Freudenberg and H. Schluter, ibid., 88, 618 (1955).
144. E. Adler, S. Hernestam and I. Waldén, Svensk Papperstidn., 61, 641 (1958).
145. J. Marton, E. Adler and K.-I. Persson, Acta Chem. Scand., 15, 384 (1961).
146. J. Gierer and S. Söderberg, ibid., 13, 127 (1959).
147. E. Adler and J. Marton, ibid., 13, 75 (1959).
148. B. L. Lenz, Tappi, 51, 511 (1968).
149. J. J. Lindberg, Finska Kemistsamfundets Medd., 68, 5 (1959).
150. J. P. Butler and T. P. Czepiel, Anal. Chem., 28, 1468 (1956).
151. K. V. Sarkanen and C. Schuerch, ibid., 27, 1245 (1955).
152. F. Gaslini and L. Z. Nahum, Svensk Papperstidn., 62, 520 (1959).
153. K. H. Ekman, Soc. Sci. Fenn. Comm. Phys. Math., 23, No. 1 (1958).
154. K. H. Ekman and J. J. Lindberg, Paperi Puu, 42, No. 1 (1960).
155. K. Freudenberg, H. Wilk, H.-U. Leuck, L. Knof and T. H. Fung, Liebigs Ann. Chem., 630, 1 (1960).
156. J. Marton and E. Adler, Acta Chem. Scand., 15, 370 (1961).
157. J. Marton, Paperi Puu, 43, 665 (1961).
158. C. Steelink, T. Reid and G. Tollin, J. Am. Chem. Soc., 85, 4048 (1963).
159. C. Steelink, Advances in Chem. Series, 59, 51 (1966).
160. G. Chirkin and D. Tishchenko, Bumazh. Prom., 39, 3 (1964).
161. G. Aulin-Erdtman, Svensk Kem. Tidskr., 70, 145 (1958), summarizing paper.
162. O. Goldschmid, J. Am. Chem. Soc., 75, 3780 (1953).

163. T. Kagino, J. Nakano, A. Ishizu, T. Ogino and N. Migita, J. Japan. Wood Res. Soc., 9, 85 (1963).
164. J. Marton and E. Adler, U. S. Patent 3,071,570 (1963).
165. H. L. Hergert, J. Org. Chem., 25, 405 (1960).
166. S. Wada, Chem. High Polymers (Japan), 18, 617 (1961).
167. P. Kubelka and F. Munk, Z. techn. Physik, 12, 593 (1931).
168. P. Kubelka, J. Optical Soc. Am., 38, 448 (1948).
169. W. H. Leith, M. S. Thesis, Univ. of South Carolina, 1951.
170. E. Matijevic, Private communication.
171. J. J. Lindberg, H. Tylli and C. Majani, Paperi Puu, 46, 521 (1964).
172. J. J. Lindberg, K. Penttinen and C. Majani, Suomen Kemistilehti, B38, 95 (1965).
173. Anonymous, "Indulin," Tech. Bull., No. 100 of Polychemicals Dept., West Virginia Pulp and Paper Co. (Westvaco), 1957, p. 19.
174. L. R. Lawson and J. B. Doughty, Chem. Eng. Data Ser., 3, 128 (1958).
175. D. A. I. Goring, Pure Appl. Chem., 5, 233 (1962).
176. F. Buche, Physical Properties of Polymers, Chapter 10, Interscience, New York, 1962.
177. C. Schuerch, J. Am. Chem. Soc., 74, 5061 (1952).
178. C. Schuerch, Ind. Eng. Chem., Product R & D, 4, 61 (1965).
179. K. H. Ekman and J. J. Lindberg, Suomen Kemistilehti B, 39, 89 (1966).

17

POLYMER PROPERTIES OF LIGNIN AND LIGNIN DERIVATIVES

D. A. I. Goring

1. Introduction. Wood is comprised mainly of three macromolecular species—cellulose, hemicellulose and lignin. In numerous ways, each of these major components has been isolated from wood and studied as a reasonably pure polymeric material. However, right at the start we should recognize an important limitation of wood polymer science. There is a considerable literature describing the behaviour of individual wood constituents as polymers or colloids. In contrast, very little is known of the detailed molecular architecture of wood itself. Properties such as molecular weight, molecular shape, crystallinity, orientation, glass temperatures, density, etc., have been investigated almost exclusively with isolated products and the characteristics of the wood polymers in situ can be deduced only by inference.

Despite the dearth of exact knowledge, it is useful to have a speculative picture of the molecular morphology of the three polymeric species as they occur in the cellular structure of wood. For cellulose, the picture is fairly well established. The undegraded cellulose molecule is found to be a linear chain of about 10,000 anhydroglucose units (1). Steric hindrance at the β-glycosidic linkage results in a rather extended conformation (2, 3). The isotactic regularity of the chain promotes crystallinity. In wood, the basic morphological unit of cellulose is the elementary microfibril, 35 Å in diameter and of indefinite length (4). Most of the cellulose microfibrils are laid down in the familiar pattern of spirally arranged strands which make up the secondary wall of the cell as shown in Figure 17.1 (5). The microfibrils in the primary wall, however, tend to be randomly arranged. Recent work by Manley (6, 7) has suggested that the microfibril is completely crystalline. The most important unanswered question is the exact arrangement of the cellulose molecules in the elementary microfibril.

Hemicelluloses are much lower in molecular weight than cellulose, usually having a degree of polymerization between 100 and 200 (8). Their polysaccharide chains often contain two or more types of sugar unit and can be linear or branched. They are usually isolated as amorphous materials which

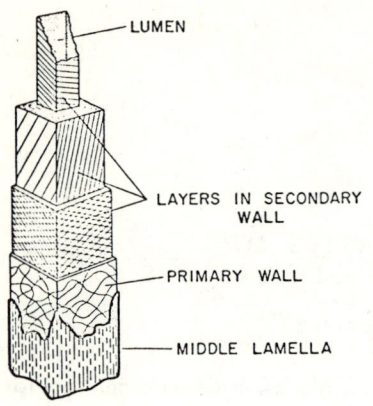

Figure 17.1. Diagrammatic representation of a single fibre of wood. The cellulose microfibrils are arranged in different patterns in the various layers of the cell. The intercellular substance in the middle lamella is largely lignin (5).

show strong swelling or complete solubility in water. Hardwoods contain considerable quantities of glucuronoxylan. Softwoods have glucuronoxylan and glucomannan. In addition, some woods have galactans and arabinogalactans. The location of hemicellulose molecules in the cellular structure of wood has recently been studied by Meier (9) who showed that xylan and glucomannan are found in all the cell wall layers while galactan and arabinan seem to be concentrated more in the middle lamella and primary wall. Thus the xylan and glucomannan hemicelluloses may be considered part of the amorphous matrix of the cell wall in which the crystalline microfibrils are embedded. Marchessault (10) has suggested, however, that the essentially linear xylan molecules, though not crystalline, are arranged in lateral order between the cellulose microfibrils as shown in Figure 17.2.

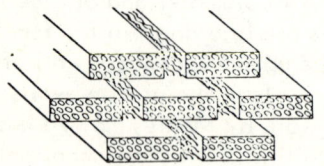

Figure 17.2. Crystalline cellulose microfibrils (rectangular) with oriented xylan molecules in the interstitial spaces, according to Marchessault (10).

Lignin is laid down in wood when surface growth of the cell wall has ceased. Wardrop and Bland (11) have shown that initial deposition of lignin takes place in the primary wall at the cell corners, see Chapter 2, Section IV.

Lignification then extends to the intercellular spaces and also into the secondary wall where the lignin is deposited around the cellulose microfibrils. When lignification is complete, the middle lamella is rich in lignin (12). However, electron microscopy on wood sections from which the carbohydrate has been removed has revealed that the lignin in both softwoods and hardwoods in distributed throughout the secondary wall (13, 14). This distribution is illustrated by the electron micrograph of red spruce shown in Figure 17.3 published by Côté, Timell and Zabel (14).

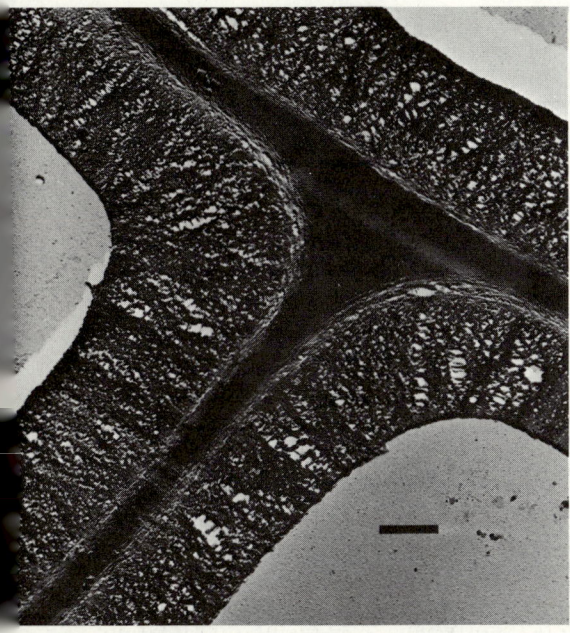

Figure 17.3. An electron micrograph of the cross-section of a tracheid corner of red spruce (14). The carbohydrate was removed with 80% hydrofluoric acid. ▬▬ = 1 μ. Photograph by Côté, Timell and Zabel (14).

In a recent note, Berlyn and Mark (15) have pointed out that the volume fraction of the compound middle lamella in coniferous wood is only about 12%. Since the lignin content of the wood is between 25 and 30%, the middle lamella could at most contain less than half of the total lignin.

The prediction made by Berlyn and Mark (15) has been proved correct by a quantitative determination of the distribution of lignin in black spruce by ultraviolet microscopy (16). The results in Table 17.1 show that although the concentration of lignin in the middle lamella (ML) is high, the secondary wall (S) of the tracheid contains about 70% of the total lignin. Surprisingly, a not very different pattern of lignification was found for the fibres in white birch (17), see Table 3.16, Chapter 3. If allowance was made for the low ultraviolet absorption of the syringyl lignin located in the fibre secondary wall (18), the concentration of lignin was 0.17g/g in this dominant morphological feature

Table 17.1. The Distribution of Lignin in Black Spruce Earlywood (16)

Tissue Type	Tissue Volume (%)	Lignin (% total)	Lignin Conc. (g/g)
S	87	72	0.22
ML	9	16	0.50
ML_{cc}	4	12	0.85

and it contained about 60% of the lignin in the wood. Thus it seems likely that most of the xylem lignin in both softwoods and hardwoods is contained in the secondary walls of the fibres.

It has not yet proved possible to remove the lignin from wood by treatment with a neutral solvent. Most methods for dissolving lignin change the chemical nature of the molecule. The mildest method known is that of Björkman (19) in which the wood is ball milled for 48 hours in dry toluene and then extracted with dioxane. Even here, however, it is not unlikely that some degradation takes place due to the severe mechanical disruption of the wood.

Chemically, softwood lignin is known to consist largely of guaiacylpropane units linked together in a bewildering variety of ways (20, 21). This is illustrated in Figures 1.1 and 1.2, Chapter 1, Section I, in which some of the structures and groupings believed to be present in spruce lignin are given. The monomers are joined predominantly by glycerol-aryl ether linkages but there are several types of C-C bonds which may serve as cross-links between relatively short, linear chains made up of guaiacyl propane units.

From the above, there emerges a speculative concept of protolignin as a three-dimensional network polymer, deposited between the cells and throughout the cell wall. The protolignin is reactive because of the considerable occurrence of aromatic, hydroxyl, ether, carbonyl and vinyl groups in the lignin structure. It is insoluble because the network is essentially infinite.

In the rest of this chapter, the polymeric properties of isolated lignins will be interpreted with the network concept of protolignin as a background. Topics covered will include molecular weight and shape, polyelectrolyte and colloidal behaviour, thermal transitions in solid lignin and solubility, accessibility and sorption. The chapter ends with a discussion of the state of the lignin in wood. Certain recent results are presented which indicate that the network theory, even though approximately correct, is a considerable oversimplification of the molecular morphology of the protolignin.

II. Molecular Weight.
  A. Molecular Weight and Molecular Weight Distributions of Lignin Derivatives. The molecular weights of soluble lignin derivatives cover an im-

mense range. There are many reports of molecular weights below 1000 (22) and values greater than 10^6 have been published for both lignosulphonates (23) and alkali lignins (24). Apparently, lignin can be dissolved as an entity small enough to be a pure chemical compound or as a particle large enough to show the behaviour of a high polymer or a colloid (25-28).

Usually, any single sample of a soluble lignin derivative is polydisperse with respect to molecular weight. This has been established most convincingly for softwood lignosulphonates. When spent sulphite liquor from a softwood cook is dialysed through cellophane, somewhat less than half the lignosulphonate diffuses through the membrane (29-31). The molecules which pass through the cellophane probably have a molecular weight of 5000 or less (29, 30, 32). Those retained by the membrane are found to have molecular weights from 5000 to 100,000 or even higher (30, 32, 33). Gardon and Mason fractionated a spent sulphite liquor from a mixed cook of balsam and fir and obtained eight fractions with molecular weights from 3700 to 58,000 (30). McCarthy and co-workers used column elution and non-solvent precipitation to fractionate hemlock and fir spent liquors and reported molecular weights ranging from 400 to 150,000 (34). Recently, a spruce sulphite liquor was purified of low molecular weight material by dialysis through cellophane and then fractionated by a series of prolonged dialyses against Millipore filters of graded pore size (32). Nine fractions were obtained with weight average molecular weights between 37,000 and 1.3×10^6.

Such wide polydispersity leads to a certain degree of indeterminacy in ascribing a single value of the molecular weight to a given sample of lignin. A number average molecular weight, M_n may be an order of magnitude smaller than the weight average molecular weight, M_w. A useful procedure is to measure a number and weight average by appropriate methods and then to regard the ratio M_w/M_n as a measure of the polydispersity of the sample. For a monodisperse system $M_w/M_n = 1$. When the molecular weight of a linear polymer is reduced by random sission, M_w/M_n for the degraded product approaches a value of 2 (35). Ideally, a distribution of molecular weights should be determined. If this is done by fractionation, reasonably correct values of the two averages may be obtained from Equations 1a and 1b in which $w_1, w_2, \ldots w_i$ and $M_1, M_2, \ldots M_i$ are the weights and molecular weights of the individual fractions.

$$M_n = \frac{\sum_i w_i}{\sum_i w_i/M_i} \quad \text{and} \quad M_w = \frac{\sum_i w_i M_i}{\sum_i w_i} \quad (1)$$

(a) (b)

Some examples of the ratio M_w/M_n for various lignin fractions and preparations are given in Table 17.2.

Table 17.2. Molecular Weight Polydispersity of Some Soluble Lignins

Wood	Method of Isolation	Method of Fractionation	Range of Molecular Weight	M_w/M_n	Ref.
Spruce	Sulphonation	Dialysis	324-20500	6.5	36
Balsam and Fir	Sulphonation	Dialysis and Ultrafiltration	3700-58000	2.2	30
Hemlock and Fir	Sulphonation	Elution from Column	400-150000	7.0	34
Hemlock	Sulphonation	Precipitation with Non-solvent	440-58000	6.4	34
Spruce	Sulphonation	Successive Cooking	5300-131000	3.1	32
Spruce	Alkali Cook of Periodate Lignin	Precipitation with Ba^{2+}	$35000-17 \times 10^6$	5.7	24
Spruce	Dioxane + HCl	Successive Extraction	4300-85000	3.1	37
Pine	Kraft	--	$M_w = 3500$	2.2	38
Hardwood	Kraft	--	$M_w = 2900$	2.8	38
Spruce	Kraft	Continuous Extraction	1800-51000	>3	39

Polydispersity has been reported in alkali and kraft lignins although their pattern of molecular weight distribution is not yet as well-known as that for the lignosulphonates. A water-soluble alkali lignin which was made by a mild caustic cook of spruce periodate lignin contained material of molecular weight as high as 50×10^6 and showed a wide polydispersity (24). Much lower molecular weights were found by Marton and Marton (38) for a pine kraft lignin although, as shown in Table 17.2, the sample also exhibited polydispersity. Lindberg, Tylli and Majani (40) fractionated sulphate and soda lignins from spruce and obtained a considerable spread in M_w as measured by light scattering. However, when the same fractions were subjected to gel filtration, the molecular weights estimated from the elution patterns were somewhat lower (41).

Lignins extracted with organic solvents are also polydisperse. Hess (42) reported molecular weights of 2800 to 6700 for various fractions of native black spruce lignin. Marton and Marton (38) obtained a number average molecular weight of 2100 for spruce Björkman lignin compared with reported weight average values of 7100 (37) or 11,000 (19). Spruce dioxane lignins prepared by successive acidolysis by the method of Pepper, Baylis and Adler (43) gave molecular weights of from 4300 to 85,000 (37).

The polydispersity in molecular weight of the softwood lignins is accompanied by rather small changes in other physical or chemical properties. Methoxyl content, UV absorption, partial specific volume, refractive index or refractive index increments have been found to vary only slightly, if at all, with molecular weight (24, 37, 39, 44-47). Thus, soluble softwood lignins, whatever their method of extraction from the wood, are composed of molecules of similar chemical constitution but differing widely in size.

Most of the molecular weight measurements on soluble lignins have been made on fractions prepared from softwoods. The few results for hardwood lignins have indicated that these are of lower molecular weight than softwood lignins but still polydisperse. Marton and Marton (38) reported an M_w/M_n value of 3.8 for a kraft lignin from a hardwood black liquor. Sjöstrom, Haglund and Janson (31) found that 30% of a birch lignosulphonate was retained on dialysis. This showed that the sample was polydisperse although of lower molecular weight than a pine lignosulphonate of which 64% did not pass through the membrane. Kyogoku, Ida and Hachihama (48) have reported that birch lignosulphonates are heterogeneous with respect to molecular weight in the same manner as lignosulphonates from softwoods. Similar results have been reported for lignosulphonates from beechwood by Hata and Nakamura (49).

Hardwood lignins contain both guaiacyl and syringyl residues. Pepper and Siddiqueullah (50) have obtained fractions of aspen dioxane lignin which have differed markedly in the ratio of syringaldehyde to vanillin produced by nitrobenzene oxidation. A similar effect has been reported by Kyogoku et al. (48) for beech lignosulphonate. Here, the low molecular weight fractions yielded a larger proportion of syringaldehyde which would be expected because of the greater likelihood that the guaiacyl residue will be biopolymerized into high molecular weight structures.

B. Reasons for the Polydispersity of Soluble Lignins. How does the polydispersity of soluble lignins arise? First of all it should be noted that naturally occurring macromolecules are not always polydisperse. The molecular homogeneity of the globular proteins is well-known (51). Recently, xylan from birch was found to have a rather narrow distribution of molecular weights as indicated by an M_w/M_n value of 1.2 (52). There has never been any convincing evidence of substantial polydispersity in undegraded celluloses. There is, however, a class of macromolecule which occurs in nature in a highly polydisperse state. These are branched polysaccharides such as amylopectin and glycogen. Here the glucose unit is trifunctional and the polymers behave as if they are formed by random condensation (53). Not only are these materials polydisperse in their native state but their polydispersity is retained on degradation by acid (53) or by enzyme systems (54). In some cases the ratio of M_w/M_n can exceed 100 (53).

A wide polydispersity of molecular weight is characteristic of nonlinear polymers (55). Under certain conditions a polyfunctional condensation leads to the formation of a gel fraction which consists of an infinite network. Flory has shown that M_n is finite but M_w/M_n tends to infinity at the point of incipient gelation (55). It is therefore a reasonable proposition that the random scission of a network into soluble fragments will produce a wide range of molecular sizes unless degradation is taken to the stage at which the product consists of monomer and very low molecular weight oligomers.

Thus the most straightforward explanation of the polydispersity of soluble lignins is the concept that lignin exists in the wood as a network made up of short linear chains, cross-linked in a variety of ways to give an infinite, three-dimensional structure. To make the lignin soluble, the network must be broken. This is usually effected by a chemical agent which breaks covalent bonds and reduces the network to fragments of various sizes which then become soluble. During processes in which the lignin becomes water-soluble, hydrophilic groups such as $-SO_3^-$ or $-COO^-$ are generated on the network and on the fragments as they dissolve.

The dissolution of the lignin in wood is a relatively slow process and several studies have been made of the molecular weight of the lignin made soluble at various stages of the delignification (37, 39, 46, 47, 56-60). The results are most readily interpreted if the lignin is removed from the site of the reaction soon after it becomes soluble. For softwoods, the low molecular weight material is first extracted with successively higher molecular weights being dissolved as the cook proceeds. This pattern has been found in successive fractions of spruce dioxane lignin (37). Similar behaviour was also noted during the acid sulphite treatment of spruce periodate lignin (47) and spruce and hemlock wood (56).

In a recent series of investigations (39, 46, 59), spruce sawdust was delignified in a stream of pulping liquor. The liquor was cooled shortly after it left the reaction vessel and was collected as a series of successive fractions. The molecular weights of the nondialysable lignins are shown in Figure 17.4.

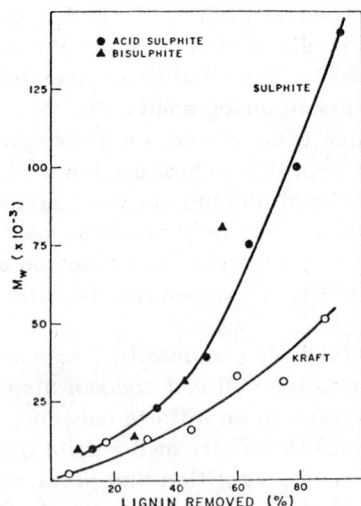

Figure 17.4. Increase of molecular weight of nondialysable fractions of spruce lignin with degree of delignification by acid sulphite, bisulphite and kraft pulping (59)

The fractions were all rather similar with respect to methoxyl group content and ultraviolet absorption, but showed a regular increasing trend in weight average molecular weight with removal of lignin. Interestingly, this trend is not limited to chemical delignification. Brown, Falkehag and Cowling (60) have shown that progressive enzymatic degradation of wood releases lignins of increasing molecular size.

Random sission of a network and subsequent solubilization of the fragments might be expected to produce the trends in molecular weight shown in Figure 17.4. Small molecules would be dissolved more or less as soon as they were free of the network whereas the larger fragments would need to be further degraded and made more hydrophilic before solution was possible.

These results are analysed in a more quantitative fashion in the last section of this chapter.

III. Molecular Shape.

A. Conformity of Viscosity, Sedimentation and Diffusion with Behaviour Expected for Spherical Microgel. The predominant physicochemical characteristic of a lignin solution is its small viscosity. This is illustrated in Table 17.3, in which the intrinsic viscosities of several lignins are compared with values for polysaccharides and synthetic polymers at equal molecular weight. The intrinsic viscosities of the lignins are about 1/40 of those of the polysaccharides and about 1/4 those of the synthetic polymers.

Table 17.3. Intrinsic Viscosities Corresponding to a Molecular Weight of 50,000 for Soluble Lignins and Other Macromolecules

Macromolecule	Solvent	$[\eta](dl.g.^{-1})$	Ref.
Dioxane-HCl lignin	Pyridine	0.08	37
Kraft lignin	Dioxane	0.06	40
Alkali lignin	0.1M Buffer	0.04	24
Lignosulphonate	0.1M NaCl	0.05	32
Polymethylmethacrylate	Benzene	0.23	61
Polymethylstyrene	Toluene	0.24	62
Xylan	CED	2.16*	63
Cellulose	CED	1.81	64

*From Eq. 7 in ref.63
CED ≡ Cupriethylenediamine

The intrinsic viscosity $[\eta]$ is simply an index of the effective hydrodynamic specific volume of the macromolecule when the solution is subjected to shear. For an incompressible sphere, $[\eta]$ in $dl.g.^{-1}$ is given by the Einstein equation in the form, 2,

$$[\eta] = 0.025 \bar{v} \qquad (2)$$

in which $\bar{v}$ is the specific volume of the material of the sphere in $ml.g.^{-1}$.

If a spherical macromolecule has solvent molecules attached to its surface or if it is swollen by solvent, then Equation 2 may still be considered to apply provided $\bar{v}$ is taken to be the effective hydrodynamic specific volume of the swollen or solvated macromolecule. In this sense, $\bar{v}$ and therefore $[\eta]$ will increase with solvation or swelling. A flexible linear polymer will immobilize a considerable volume of solvent within its randomly coiling chains. Although the overall shape of the molecular domain in solution is approximately spherical, its effective hydrodynamic specific volume is much greater than the specific volume of the bulk polymer. Thus, the intrinsic viscosities of synthetic linear polymers are relatively high as shown for polymethylmethacrylate or polymethylstyrene in Table 17.3. The intrinsic viscosity also increases when the macromolecule is nonspherical in shape, for example when it is a rod. The high values of $[\eta]$ for the polysaccharides in Table 17.3 probably reflect the asymmetric character of these molecules at molecular weights of about 50,000 (3, 63).

The low values of $[\eta]$ shown for the lignins in Table 17.3 suggest that these molecules behave in solution like Einstein spheres. The specific volume of lignin in bulk is about 0.7 ml.g.^{-1} which according to Equation 2 would result in an $[\eta]$ of 0.018 dl.g.^{-1} for spherical particles of solid lignin. The experimental values of $0.04 \longrightarrow 0.08 \text{ dl.g.}^{-1}$ indicate that at a molecular weight of 50,000, the effective hydrodynamic radius of the macromolecule in solution is only 30-60% greater than the radius of the dry, void-free macromolecule. Compared with the polymeric systems, this solvent-induced increment in dimensions is small and indicates that lignins exist in solution as compact non-asymmetric particles.

The conformation of the macromolecule in solution can also be deduced from the variation of $[\eta]$ and other hydrodynamic properties with molecular weight. The intrinsic viscosity can be expressed by the well-known Mark-Houwink equation, 3,

$$[\eta] = K_\eta M^a \tag{3}$$

in which K_η and a are constants characteristic of the polymer species under study. The intrinsic viscosity is independent of the molecular weight for an Einstein sphere for which, therefore, $a = 0$. For a non-freedraining random coil in a "poor" or θ-solvent, $a = 0.5$ and tends to increase to $0.7 \longrightarrow 0.8$ when the coil expands in a good solvent (68). An exponent of unity is expected for a completely freedraining coil (69). For rods, $a \approx 1.8$ (71).

Relationships similar to Equation 3 can be written for the diffusion constant, D, and the sedimentation coefficient, s, as in Equations 4 and 5,

$$D = K_D M^{-b} \tag{4}$$

$$s = K_s M^c \tag{5}$$

where K_D, K_s and b,c are constants analogous to K_η and a in Equation 3. The values of b and c characteristic of various model shapes are given in Table 17.4.

Table 17.4. Values of Exponents a, b and c for Soluble Lignins and Other Macromolecular Systems.

Macromolecule	Solvent	a.	b	c	Ref.
Einstein Sphere	-	0*	0.33†	0.67†	-
Hydrotropic lignin	Dioxane	0.12	-	-	65
Dioxane-HCl lignin	Pyridine	0.15	0.36	-	37
Alkali lignin	Dioxane	0.12	-	-	40
Alkali periodate lignin	0.1M buffer	0.32	-	0.52	24
Lignosulphonate amine	Methanol	0.13	-	-	66
Lignosulphonate	Methanol	0.54	-	-	66
Lignosulphonate	2M NaCl	0.47	-	-	67
Lignosulphonate	0.5M NaCl	-	0.56	-	67
Lignosulphonate	0.02M NaCl	-	0.57	-	33
Lignosulphonate	0.1M NaCl	0.32	0.33	-	32
Non-freedraining coil	-	0.5-0.8	0.5-0.6	0.4-0.5	68
Polymethylmethacrylate	Benzene	0.73	0.57	0.43	61
Polymethylstyrene	Toluene	0.71	-	-	62
Cellulose	Cadoxen	0.75	0.61	0.40	69
Cellulose nitrate	Ethylacetate	0.99	0.86	0.17	64
Xylan	DMSO	0.94	0.79	-	63
Freedraining coil	-	1	1	0	68
Xylan	CED	1.15	-	-	63
Carboxymethylcellulose	0.001M NaCl	1.40	-	0.11	70
Rods	-	1.8	-	-	71

* Equation 3
† Stokes law

CED = Cupriethylenediamine
DMSO = Dimethylsulphoxide

The classical approach is to fractionate a polydisperse macromolecular system and measure M, [η], D and s on the fractions. From the appropriate double logarithmic plots, the constants of Equations 3, 4 and 5 may be derived. Provided a given conformational model is applicable to all the fractions and that the polydispersities of individual fractions do not differ too greatly, it is possible to make certain deductions concerning the shape of the macromolecule from the values of the exponents a, b and c. This experimental procedure has been carried out on several lignin systems. Examples of the log/log plots corresponding to Equations 3 and 4 for lignosulphonates are given in Figure 17.5. Results for the exponents are included in Table 17.4.

From Table 17.4, it can be seen that the exponents for the polysaccharides are near to those expected theoretically for a freedraining coil. For xylan in cupriethylenediamine or carboxymethylcellulose in 0.001M NaCl, the

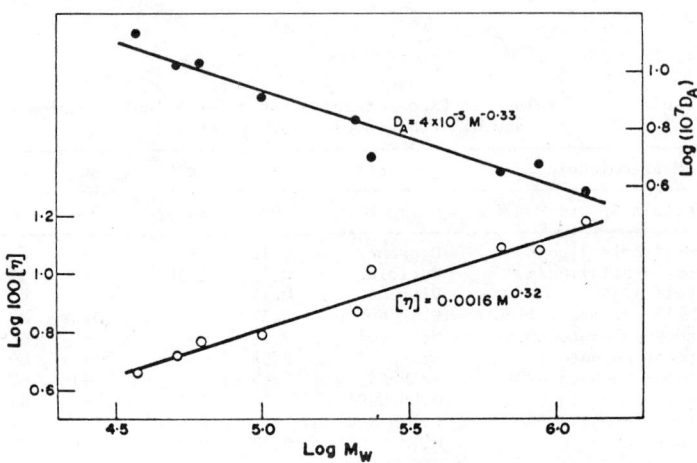

Figure 17.5. Logarithmic plot of the intrinsic viscosity and the diffusion coefficient vs. the molecular weight for spruce lignosulphonates (32).

value of a is greater than unity, which indicates an asymmetric conformation of the macromolecule. In contrast, a, b and c for lignins lie between the values anticipated for an Einstein sphere and a non-freedraining random coil in a θ-solvent. This is pretty well what would be expected if soluble lignin macromolecules result from the fragmentation of a cross-linked network. The protolignin would be peptized to give irregularly shaped pieces of microgel which, hydrodynamically, would approximate solvent-swollen spheres. The solvent molecules within the domain of the microgel would be immobilized. However, the freely flexible chains on the surface and within the microgel would impart a partially Gaussian distribution of segments which would produce the molecular weight dependence of $[\eta]$, D and s shown in Table 17.4.

Further examination of the values of the exponents in Table 17.4 reveals that alkali lignins and the organosolve lignins behave more like Einstein spheres than do the lignosulphonates. This suggests that the restriction on the freedom of movement of individual segments of the lignin chain is greater in alkali or organosolve lignins than in lignosulphonates. Thus sulphonation appears to produce fragments with a rather open network whereas cooking in alkali or solubilization in organic solvents results in a more tightly bonded structure.

B. Shape of Lignin Macromolecules as Determined by Electron Microscopy. If soluble lignins are spherical in shape, then an individual

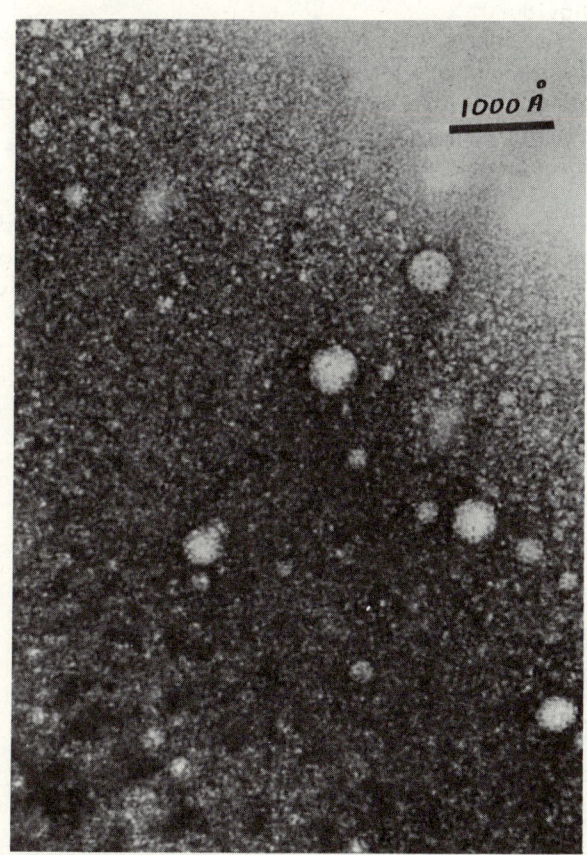

Figure 17.6. Electron micrograph of high molecular weight fraction of sodium lignosulphonate. Bright areas correspond to lignosulphonate macromolecules (72).

macromolecule of a high molecular weight sample should be well within the resolution afforded by modern electron microscopy (72). An electron micrograph of a high molecular weight fraction of sodium lignosulphonate is shown in Figure 17.6. The negative-staining technique was used by means of which molecules of the specimen are embedded in an electron-opaque film of phosphotungstate. Thus the bright areas in the picture correspond to the organic substrate, which is transparent to electrons. It is evident that the lignosulphonate is in the form of spherical particles of a wide range of sizes. The larger particles seem to be made up of aggregates of the smaller ones. Somewhat similar results were obtained for alkali lignins and a dioxane lignin (72). Of course, a spherical shape in the anhydrous environment of the electron microscope does not necessarily confirm a spherical shape in solution. However, the correspondence of the evidence in Figure 17.6 with the conclusions drawn from the hydrodynamic behaviour of soluble lignins does support the compact, roughly spherical shape of the macromolecule.

IV. Polyelectrolyte Behaviour.

A. Relationship of Viscometric Behaviour to Ionic Strength. Gardon and Mason (67) were the first to show that lignosulphonates behave as expanding polyelectrolytes. In neutral solution the sulphonate groups are ionized and the molecule carries a strong negative charge. This charge is balanced by a positively charged cloud of cations dispersed in the aqueous medium in the classical diffuse double layer. A pictorial representation of the lignosulphonate macromolecule is shown in Figure 17.7. If the lignosulphonate is dissolved in a neutral aqueous solution containing a high concentration of small ion (e.g. 1M NaCl), the potential of the double layer is reduced and the macromolecule shrinks. On the other hand, if the polyelectrolyte is dissolved in distilled water, the width of the double layer increases, the potential increases, and the resulting mutual repulsion between the charged sections of the chain causes the molecule to expand.

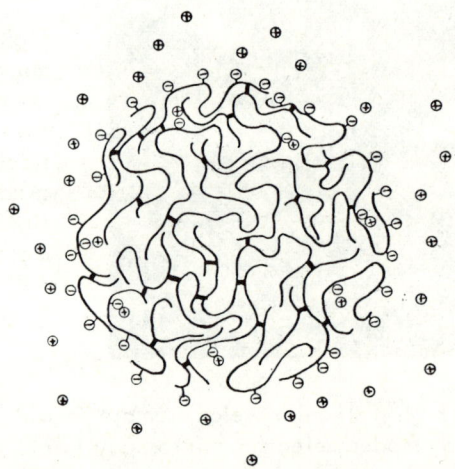

Figure 17.7. Diagrammatic representation of the lignosulphonate macromolecule (23)

The viscosity-increasing capacity of a macromolecular solute is a measure of its volume in solution, as shown in Equation 2. Thus it would be expected that the intrinsic viscosity of a given fraction of lignosulphonate would be a function of the ionic strength of the medium in which it is dissolved. That this the case is shown in Figure 17.8, in which the Huggins plots for a sodium lignosulphonate at a series of ionic strengths are given. The intrinsic viscosity, as derived from the intercept, increases regularly with decrease in the ionic strength. In this particular investigation (23), the size of the macromolecule

POLYMER PROPERTIES

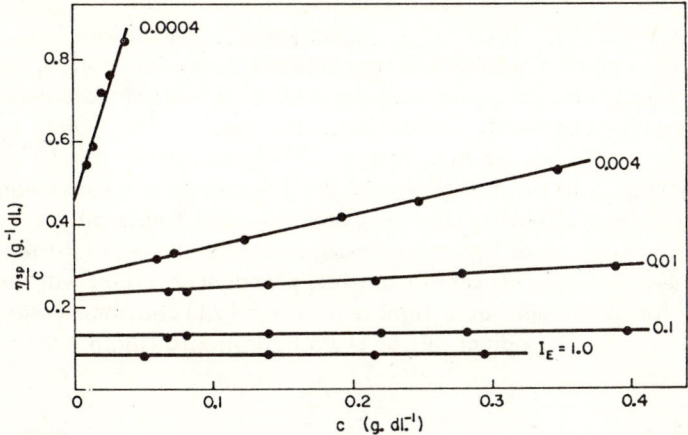

Figure 17.8. Huggins viscosity plots for a lignosulphonate fraction by isoionic dilution if various concentrations of sodium chloride. The ionic strength, I_E, is included by each line (23).

was measured by the light scattering technique and a corresponding increase in the radius of gyration was confirmed.

Several theoretical treatments of polyelectrolyte expansion are available but it was found that the behaviour of the lignin sulphonates fitted none of them. This is probably because the molecule is a microgel rather than the linear chain model on which the theories are based. At a relatively low degree of swelling, portions of the network will be stretched to such an extent that the chains can no longer obey random flight statistics. It is also likely that the effective charge is on the surface of the microgel and that the sulphonate groups buried deep within the molecular domain are neutralized by pairing with adjacent cations. Considerations such as the foregoing led to the proposal of the model in Figure 17.7 for a high molecular weight lignosulphonate molecule. Linear chains of about 20 phenyl propane units are cross-linked to give a netted, approximately spherical structure. The ionizing sulphonate groups are attached mostly to the surface of the microgel. The sulphonate groups within the microgel are neutralized by ion pairing. At low molecular weights this model is no longer strictly applicable because there may not be sufficient chains or cross-links to produce a three-dimensional structure. Kojima, Nakai and Tacki (73) have also studied the viscosity behaviour of lignosulphonates as polyelectrolytes and have come to the conclusion that the degree of branching is greater in the higher than in the lower molecular weight fractions.

In order to obtain linear plots of η_{sp}/c vs. c of the type shown in Figure

17.8, care must be taken to dilute the solution in such a manner that the ionic strength remains constant. If lignosulphonate is dissolved in and diluted with solvent in the usual way, then at low ionic strength the cations from the polyelectrolyte itself make a substantial contribution to the ionic strength. Thus the counter ion concentration decreases with the lignosulphonate concentration and the macromolecule swells on dilution. The result is the typical polyelectrolyte upsweep of η_{sp}/c shown in Figure 17.9 for dilution with 10^{-5} sodium chloride. At higher ionic strengths, the small ion concentration swamps the cation supplied by the lignosulphonate, and the usual linear plots of η_{sp}/c vs. c are obtained. This type of behaviour in lignosulphonates was first reported by Gardon and Mason (67). A similar but less pronounced effect was reported subsequently for alkali periodate lignins in water (74) and, more recently, for aqueous solutions of the sodium salt of birch hydrotropic lignin (75).

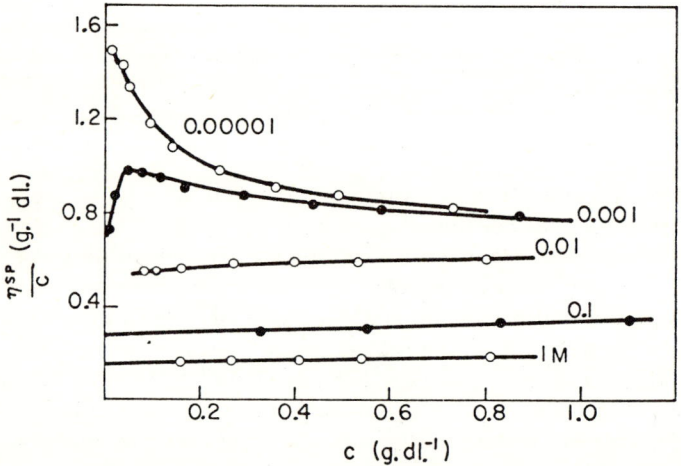

Figure 17.9. Graphs of η_{sp}/c vs. c for solvent dilution of lignosulphonate with various molarities of sodium chloride (23).

B. Effect of Polyelectrolyte Swelling on Gel Filtration of Lignosulphonates. An elegant demonstration of the polyelectrolyte swelling of lignosulphonate macromolecules was reported recently by Gupta and McCarthy (76), who fractionated lignosulphonates using Sephadex gel columns eluted with distilled water or with 10^{-1} to 10^{-4}M NaCl. Molecular weights were measured in the ultracentrifuge by the sedimentation equilibrium method. Gupta and McCarthy found that the effective hydrodynamic radius of the macromolecule approximately doubled as the electrolyte environment is changed from 0.1M NaCl to pure water. Over a wide range of ionic strength and molecular weight, the fraction of solvent imbibed by the Sephadex gel and available to macromolecules of a given size could be correlated well, Figure 17.10, with the hydrodynamic radius determined viscometrically.

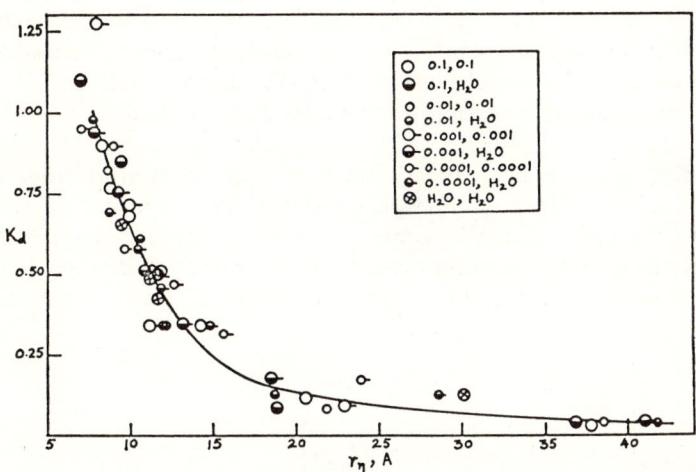

Figure 17.10. Plot of r_η (hydrodynamic radius determined viscometrically) vs. K_d (fraction of imbibed solvent available to macromolecules of a given size) for lignosulphonates. The molecular weights ranged from <1000 to 70,000, and the ionic strengths of the eluting liquid from <10^{-4} to 0.1M NaCl. Data by Gupta and McCarthy (76).

C. <u>Electrophoresis</u>. The polyelectrolyte nature of water-soluble lignins invites characterization by electrophoretic methods. Surprisingly little work has been reported along these lines, probably because electrophoresis is not sensitive to differences in molecular weight (77) in which property resides the major diversity of soluble lignins. Free boundary electrophoresis of fourteen lignosulphonates successively extracted from periodate lignin revealed in all fractions a single peak of equal mobility (78). Degradation of the sample produced slightly more complex patterns, and considerable heterogeneity was evident in low molecular weight lignosulphonates isolated from a softwood spent liquor (78). Similar results have been reported by Felicetta, Ahola and McCarthy for the electrophoresis of low molecular weight lignosulphonates in agar gel (34). McCarthy and coworkers have also used the method to identify compounds such as vanillin and acetovanillone produced by alkaline cleavage of lignosulphonates (79). Schubert et al. (80) found native and enzyme-liberated lignins from oak, maple and cork homogeneous while lignin from white Scots pine showed a small secondary component. Lindgren (81) has used glass fiber sheet electrophoresis to establish the heterogeneity of the lignin-carbohydrate complex (LCC) fraction prepared by Björkman (82) from milled spruce wood. Lindgren was able to separate the LCC fraction into two components, one of which was carbohydrate and the other a lignin-carbohydrate complex. The existence of the latter was taken to evidence of a chemical linkage between lignin and carbohydrate. Bolker and Wang (83) have used a similar technique to show that the LCC fraction from birch wood is split into its separate lignin and xylan moieties only after a mild acid hydrolysis. Loschbrandt and Wetlesen (84) have used paper electrophoresis to establish the polydispersity of sulphate pulp chlorolignin.

The principle of electrodialysis has been applied at pilot plant scale for the recovery of lignosulphonates, sugars and cooking chemicals for spent sulphite liquor (85, 86). In the laboratory, separation of lignosulphonates from water-soluble hemicelluloses in dialysed spent liquor is readily achieved by electrophoresis-convection (87). The principle of the method is shown in Figure 17.11.

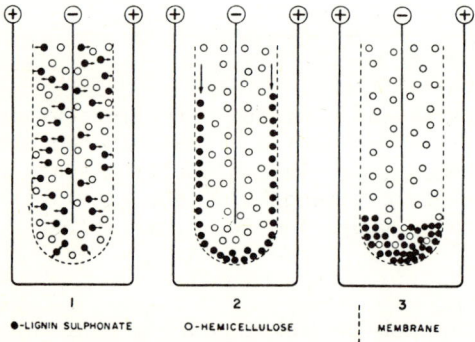

Figure 17.11. Principle of the separation of lignosulphonates from hemicellulose by electrophoresis-convection (87).

The solution of lignosulphonate and hemicellulose is contained in a bag membrane with a cathode and an anode inside and outside the bag, respectively. The negatively charged lignosulphonate macromolecules migrate anodically to the inner surface of the membrane. The increased solution density causes the lignosulphonate-rich solution to fall to the bottom of the bag where it is concentrated into a dark-brown layer about 1/2" in depth. The uncharged hemicelluloses do not migrate in the field and can be separated by removal of the water-clear solution in the top of the bag. The method has been tested for mixtures of methyl cellulose and lignosulphonate (87). It was possible to recover between 80-90% of each component, with a purity better than 99%.

The charge-bearing properties of the lignosulphonates lend these materials to the manufacture of inexpensive ion exchange resins. Lignosulphonates have been rendered insoluble by heating spent liquor with formaldehyde (88, 89). Kraft lignins have been condensed with furfural and then sulphonated (90). One novel application by Enkvist and coworkers (91) was to treat several lignin preparations with ammonia at 200 ⟶ 250° C. A considerable amount of nitrogen was introduced and the lignin developed a small anion exchange capacity. However, ion-exchanges prepared from lignin derivatives have rather low capacities, usually in the range of 1-2 m. equiv./g., and therefore this application has never become of great importance.

V. Colloidal Behaviour.

A. The Dispersing Properties of Lignosulphonates and Alkali Lignins; Theory.

Soluble lignin derivatives are increasingly sold as cheap industrial colloids. They find use as dispersants, adhesives, extenders and gelling agents (92-97). What are the properties of the macromolecule which are important in these applications? Some answers to this question have come from recent studies of the dispersive (98) and gelling (99) properties of the lignosulphonates.

Dispersion was measured by the decrease in the viscosity of an aqueous suspension of rutile titanium dioxide when a known weight of lignosulphonate was dissolved in the medium. A standard dispersion number, SDN, was determined using Equation 6,

$$\text{SDN} = \frac{(\tau) \text{ control}}{(\tau) \text{ dispersed}} \qquad (6)$$

in which the shear stress, τ, was measured in a rotating cylinder viscometer at a fixed shear rate.

The standard dispersion number was found to be remarkably dependent on the intrinsic viscosity and the molecular weight of the sample as shown in Figures 17.12 and 17.13. Good dispersion was favoured by lignins of low molecular weight. At higher molecular weight the dispersing power dropped off until for very large molecules, SDN was less than unity, which meant that addition of the soluble lignin caused the suspension to become more viscous. A few fractions of alkali lignin were tested and were found to follow the same trend as for the lignosulphonates.

At molecular weights below 10,000, the standard dispersion number decreased, which indicated that maximum dispersion was produced by lignosulphonates of weight average molecular weight between 10,000 and 40,000. The TiO_2 was found to adsorb from 90-100% of most of the fractions studied (98). Exceptions were the very low molecular weight materials which were taken up to only 60%. Thus the maximum in Figure 17.3 was probably due in part to the inferior uptake of some of the low molecular weight fractions. Rather similar effects have been reported by Ishikawa et al. (100) in the dispersion of

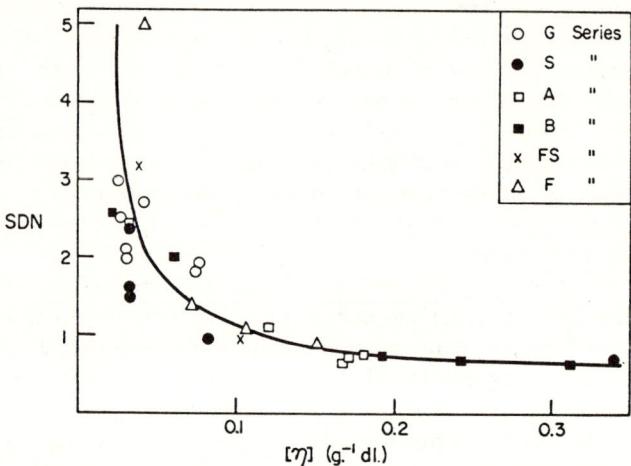

Figure 17.12. Standard dispersion number vs. intrinsic viscosity for aqueous TiO_2 dispersed by various soluble lignins (98).

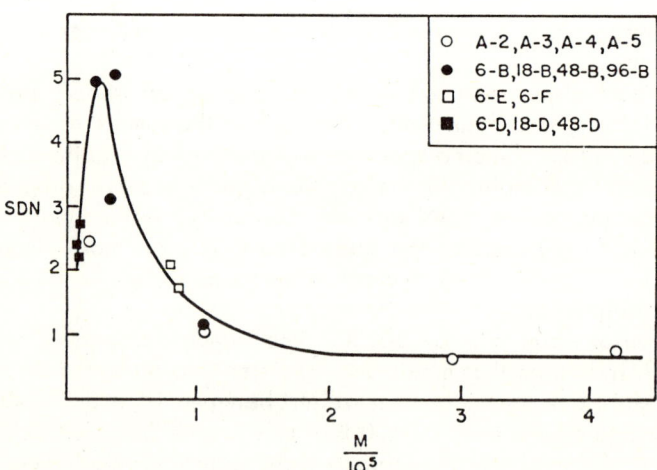

Figure 17.13. Standard dispersion number vs. molecular weight for aqueous TiO_2 dispersed by various soluble lignins (98).

ZnO slurries by lignosulphonates. Optimum dispersion corresponded to molecular weights of between 8,000 and 20,000, and the amount of lignin sorbed increased with increasing molecular weight. Kobayashi et al. have also found an optimum molecular weight for the dispersion of both $\overline{BaCO_3}$ (101) and cement slurries (102) by lignosulphonates. Low adsorption on TiO_2 of low molecular weight lignosulphonates in sodium acetate solution at pH 7.3 has been reported by Beeckmans, who also found the adsorption markedly pH-dependent (103).

The dispersion mechanism shown in Figure 17.14 was proposed to explain these results (98). Lignosulphonate molecules are adsorbed by the TiO_2 particles. The ionized sulphonate groups on the adsorbed lignosulphonate molecules are strongly electronegative. Interparticle repulsion is produced, and the structure-forming tendency of the TiO_2 is decreased. The suspension then sustains a lower stress at a given rate of shear; i.e., its apparent viscosity decreases. Low molecular weight lignin gives a good coverage and therefore acts as a dispersant at low concentrations. High molecular weight lignin gives less effective coverage and can cause interparticle adsorption at low concentration. This effect produces an increase in the apparent viscosity of the solution.

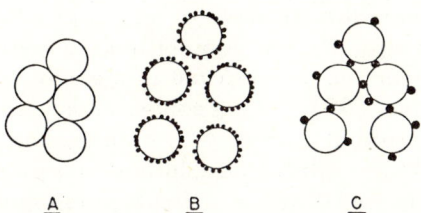

Figure 17.14. Pictorial illustration of the dispersion of TiO_2 by a lignosulphonate microgel (98). The large empty circles are the TiO_2 particles. The black circles are the lignosulphonate macromolecules. A: An aggregate in the undispersed suspension; B: dispersion on adsorption of a low molecular weight lignin; C: failure of a high molecular weight lignin to disperse the aggregate.

A logical consequence of the above concept is that at a high enough concentration the large molecules will cover the entire particle and produce effective dispersion. That this was so is shown in Figure 17.15, in which the dispersion number for a good, medium and poor dispersant is plotted against the concentration of lignosulphonate added. It is seen that the good dispersant lowers the viscosity of the suspension at all concentrations. The poor dispersant gives a slight decrease in the dispersion number at low concentration and becomes an effective dispersant at high concentration.

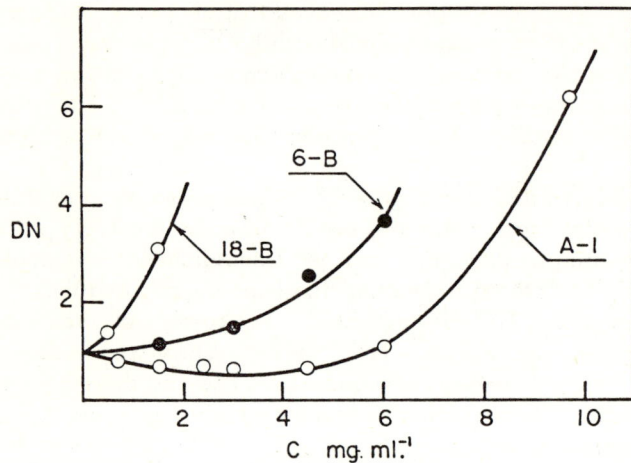

Figure 17.15. The effect of lignin concentration on the dispersion number for a good (18-B), medium (6-B), and poor (A-1) dispersant (98).

B. The Gelling Reaction of Lignosulphonate with Dichromate. We may therefore conclude that the molecular weight of a lignin derivative is an important factor in its effectiveness as a colloidal dispersant. The molecular weight has also been found to be important in the gelling reaction of lignosulphonates with dichromate (99). When potassium dichromate is added to a 12% solution of lignosulphonate at a pH of 4-5, an immediate dense greenish-brown colour is produced. After a while the solution thickens and then begins to gel. Finally, a firm gel is formed which is difficult to redisperse and from which it is not possible to remove all the chromium by prolonged dialysis or electrodialysis. Polymerization by phenolate coupling and electrostatic cross-linking by the trivalent chromic cation have been suggested by James and Tice as an explanation for this phenomenon (104).

A rheological study indicated that the reaction occurred in three distinct phases (99). This is shown in the top section of Figure 17.16. There is first a rather small increase in the viscosity of the solution. During this period, fairly active chemical processes take place as shown by the consumption of dichromate and the variation of pH and conductivity of the system (lower sections in Figure 17.16). During stage 2, the viscosity increases more quickly, the solution becomes non-Newtonian and a yield stress, τ_f, appears. In the final stage, a rapid increase in η and τ_f occurs and the solution gels. The pH and the conductivity are virtually constant in the final stage of the reaction.

The reaction was, undoubtedly, some type of interlinking of the lignosulphonate macromolecules to form a gel-like network which was extremely hydrophilic because of the sulphonate groups residual on the lignin. It was

POLYMER PROPERTIES   717

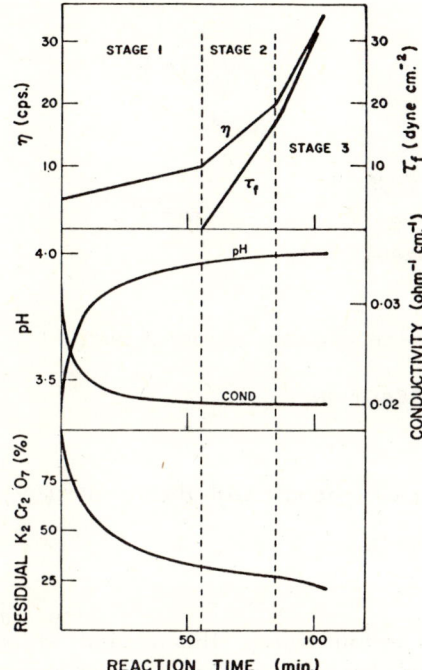

Figure 17.16. Graphs of viscosity, yield stress, pH, conductivity and dichromate consumption vs. reaction time for gelling solution of lignosulphonate, potassium dichromate and acetic acid (99).

shown that the active sites involved about 1/3 of the phenolic hydroxyl groups. Once these were used up in a gelling reaction, the remaining 2/3 were not effective in promoting gelation with dichromate. From this and other evidence, it was proposed that pyrocatechol groups were formed during sulphonation by demethylation of the pendent guaiacyl residues (99, 106, 107). In the gelling reaction the bichromate ion complexed with the catechol groups to form cyclic esters having a spiran type of structure analogous to that proposed for the boron-catechol complex (105). The chrome-complexed active sites could then dimerize or polymerize to give intermolecular cross-links which would in the final stages of the reaction produce gelation. The reaction scheme is shown diagrammatically in Figure 17.17. The gelling reaction is accompanied by a more general oxidation of the lignin substrate which produces chromic ion.

About one in ten of the C_6-C_3 units in the lignin were found to be active sites. This corresponded to an equivalent weight per catechol group of 2500. For a fairly small, compact molecule such as lignosulphonate, more than two cross-links per molecule would be necessary to induce gelation. Thus it would be expected that it would not be possible to produce a gel with a lignosulphon-

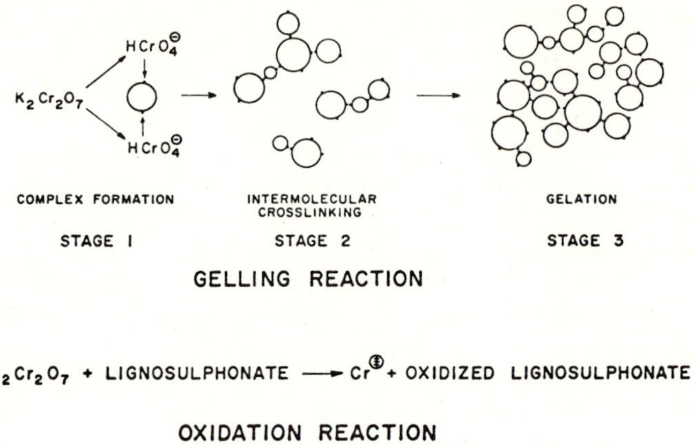

$K_2Cr_2O_7$ → $HCrO_4^\ominus$ / $HCrO_4^\ominus$

COMPLEX FORMATION	INTERMOLECULAR CROSSLINKING	GELATION
STAGE 1	STAGE 2	STAGE 3

GELLING REACTION

$K_2Cr_2O_7$ + LIGNOSULPHONATE ⟶ $Cr^\oplus$ + OXIDIZED LIGNOSULPHONATE

OXIDATION REACTION

Figure 17.17. Reaction scheme of lignosulphonate with dichromate (99).

ate having a molecular weight below a certain value. This was indeed found to be true. As shown in Table 17.5, gels became weaker for molecular weights below 10,000, and no gel at all could be produced with a lignosulphonate having a molecular weight of 2900.

Table 17.5. Variation of Catechol Content and Gelling Tendency with Molecular Weight for Fractions of Lignosulphonate (99).

Molecular Weight	Catechol Equivalent	Catechol Gps. per Molecule	Gel Test
60,000	2300	26	+
34,000	2900	12	+
9,500	2900	3.3	Weak Gel
5,400	2100	2.6	Weak Gel
2,900	1740	1.7	0

An alternative to the formation of pyrocatechol groups during sulphonation is the demethylation of the guaiacyl unit during the dichromate reaction itself. Hayashi, Namura and Uekita (108) have shown that model compounds such as vanillin and guaiacol in addition to lignosulphonate are demethylated with dichromate. These authors propose that the pyrocatechol groups thus generated can make cyclic chromate esters and contribute to the gelling reaction by intermolecular cross-linking as described above.

VI. Molecular Weight Determination.

A. <u>Number Average Molecular Weights; Methods</u>. We have seen that soluble lignins can exist in a wide range of molecular weights. Apparently, the molecular weight is also an important factor in the utilization of lignins as industrial colloids. The question then arises as to the best methods of measuring the molecular weight of a particular sample of lignin.

The determination of number average molecular weights of organosolve lignins presents little difficulty. Gross, Sarkanen and Schuerch (109) obtained reproducible results with no indication of association when number average molecular weights of ethanol, hydrol and kraft lignins were measured by cryoscopy in ethylene carbonate. In a similar molecular weight range, Marton and Marton, using the vapour pressure osmometer, obtained highly consistent data for kraft lignins in various organic solvents (38). There is no reason why conventional membrane osmometry should not work for organosolve lignins of molecular weights above 10,000. Some problems may arise due to diffusion of the low molecular weight species through the membrane, but there are experimental procedures developed to compensate for such errors (110).

Number average measurements are considerably more difficult with water-soluble lignins, particularly lignosulphonates. Accompanying the macroion in solution are a variable number of small counter ions, each of which contributes to any observed colligative effect. Similarly in osmometry, Donnan effects, even though swamped by a high concentration of supporting electrolyte, tend to cause large error in the measurement of molecular weight. One way to avoid this difficulty is to convert the water-soluble lignin to an organosolve lignin and then to measure M_n in an organic solvent. Quimby and Goldschmid (111) precipitated lignosulphonates with Hyamine 10-X to give the quaternary ammonium salts and then determined their number average molecular weights in methanol by vapour pressure osmometry. An alternative procedure is to fractionate the water-soluble lignin by some suitable method and then to calculate M_n from the molecular weight of each of the fractions by means of Equation 1.

B. <u>The Light Scattering Method; Limitations</u>. The light scattering method has been used to measure weight average molecular weights in several investigations of water-soluble lignin derivatives (23, 24, 32, 33). Both the intense colour and the fluorescence of the solutions present some difficulties, but these can be overcome by fairly straightforward experimental procedures (24, 33, 112). The major disadvantage of the light scattering technique is that the measurement is extremely sensitive to traces of colloidal debris in the solution (52). This is a general problem and must be considered in the application of light scattering to any macromolecular substance derived from a biological source. Rather small proportions of cellular debris can cause the light scattering method to yield unrepresentative results. Removal of such impurities is difficult and sometimes the results obtained depend strongly on minor differences in the purification procedures. Thus light scattering cannot be recommended as a method of characterizing soluble lignins of unknown molecular weight. On the other hand, the method can be used to follow changes in samples already characterized by other methods, e. g., changes in molecular weight due to association-dissociation reactions (32) or changes in molecular dimensions due to polyelectrolyte expansion (23).

C. Ultracentrifuge Methods. Of all the methods of measuring molecular weights of polymers, the ultracentrifuge is, perhaps, the most reliable. It has been used for the observation of both the sedimentation velocity (82, 113-115) and sedimentation equilbrium (32, 37-39, 46, 59, 76) of soluble lignins. As with light scattering, the colour of the solution presents a problem because the concentration gradients developed in the ultracentrifuge cell are usually detected optically by changes in the refractive index of the solution. Unless the colour is extremely intense, this difficulty can be overcome by matching the lignin solution with a solution of a low molecular weight dye which shows negligible sedimentation (37, 114). This then permits simultaneous photographic recording of the schlieren patterns for the solution and the solvent and thereby facilitates an accurate analysis of the sedimentation process.

The ultracentrifuge has been used to demonstrate the wide polydispersity of certain alkali periodate lignins (114). Some examples of the distribution of sedimentation coefficients are given in Figure 17.18.

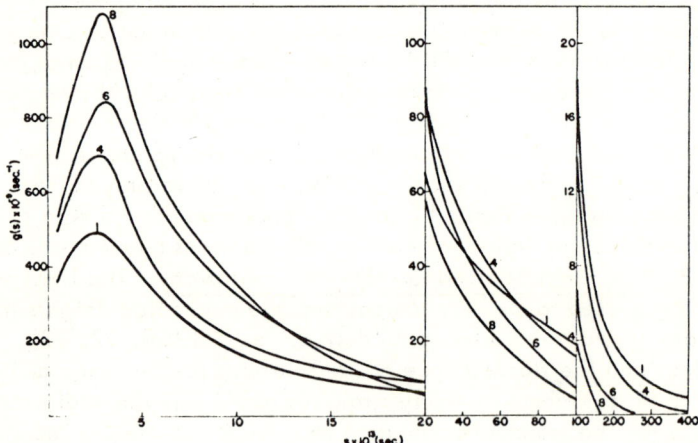

Figure 17.18. Distribution of sedimentation coefficients for fractions of alkali periodate lignin (114).

It can be seen that the value of the sedimentation constant, s, covers a range of from less than unity to over 400. Since s is proportional to the one-half power of the molecular weight, these results indicate a very large spread in molecular weight. Strikingly similar results have been reported by Bryce, Cowie, Greenwood and Jones for molecular weight distributions of degraded glycogens (54).

Due to the wide polydispersity, a special effect can occur in the measurement of the sedimentation velocity of soluble lignins (116). The rate of movement of the boundary (as indicated by the maximum of the refractive index gradient) is markedly dependent on the speed at which the centrifuge is run. Thus the maximum-ordinate sedimentation constant is field dependent. If s_m1 and s_m2 are maximum ordinate sedimentation constants measured respectively at fields of 12,600 g and 52,600 g, then the ratio s_m1/s_m2 can be as high as 9. Also, s_m1/s_m2 decreases with a decrease in the square root, σ, of the second moment of the distribution as shown in Figure 17.19 (116). At low fields the high molecular weight material contributes to the boundary, thus causing s_m1 to be high. With increasing centrifuge speed, the heavy material is spun down to the bottom of the cell where it forms a layer of gel. The boundary layer then contains only low molecular weight materials which gives a small value of s_m2. The extent of the discrepancy will depend on the degree of polydispersity as shown in Figure 17.19.

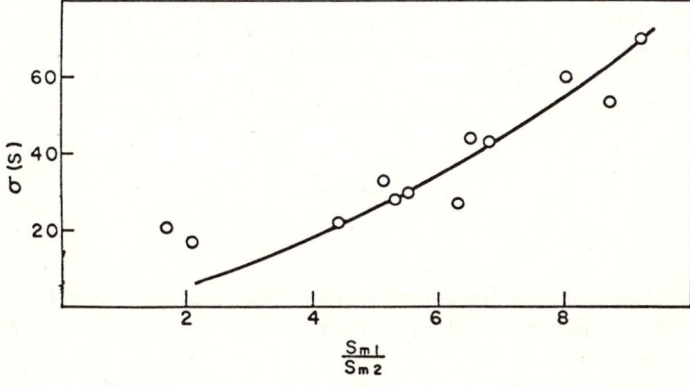

Figure 17.19. Increase of the ratio s_m1/s_m2 with increased polydispersity of the sample (116).

The dependence of the sedimentation constant on the centrifugal field is negligibly small for more monodisperse macromolecular systems (116). The substantial anomaly with lignin means that it is not possible to calculate molecular weights from the sedimentation constant determined at one speed unless the sample is known to possess a reasonably narrow distribution of molecular weights.

Even if a correctly representative value of the sedimentation constant is obtained, it alone is not sufficient for a determination of the molecular weight, M_{ww}. It is necessary to know the diffusion constant and then to substitute s and D in the well-known Svedberg equation, 7 (117).

$$M_{ww} = \frac{sNkT}{D(1 - \overline{V}\rho)} \qquad (7)$$

in which N is Avagadro's number, k the gas constant, T the absolute temperature, V the partial specific volume of the solute and ρ the density of the solution.

When s and D are weight averages, the value of M given by Equation 7 is a weight-weight average which is usually between a number and a weight average.

The method of sedimentation equilibrium has the advantage that it consists of a single series of sedimentation experiments from which a weight average molecular weight is obtained. Higher averages such as the Z-average can be derived by a more complete analysis of the distribution of concentration at equilibrium. A crippling disadvantage of the method used to be the long time required for equilibrium to be established in the centrifuge cell. However, since Van Holde and Baldwin introduced the short column technique, the time to equilibrium has been reduced from weeks to hours (118). In the short column method, the height of the solution in the ultracentrifuge cell is usually less than 1 mm. Examples of the schlieren pattern of refractive index gradient are given in Figure 17.20 (32). The molecular weight is calculated from Equation 8 (118),

$$M_w = \frac{NkT}{\overline{r}c_o \omega^2 (1-\overline{V})} \left(\frac{dc}{dr}\right)_{r=\overline{r}} \qquad (8)$$

in which r is the distance from the center of the rotor and c is the concentration of solute; c_o is the initial concentration and $\overline{r}$ is the value of r at the midpoint of the column.

Even with the equilibrium method, the polydispersity of the lignin created an unusual anomaly. It was found that the molecular weight obtained was a function of the speed of the centrifuge (32). At high fields, M_w was less than at low fields. At the same time, it was observed that a gel layer was always present at the base of the cell and that this was greater at high fields than at low fields. This effect is shown by the ultracentrifuge diagrams taken at different speeds shown in Figure 17.20. Evidently some of the high molecular weight material was immobilized as a layer of gel thereby causing a reduction of both the concentration and the average molecular weight of the fraction which remained in solution and which produced the schlieren pattern. At increased fields, the gel layer was thicker and the observed value of M decreased

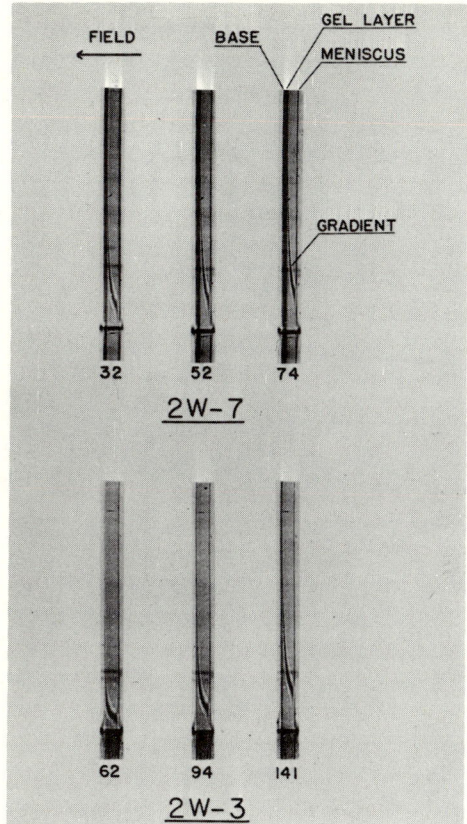

Figure 17.20. Schleiren photographs of short column sedimentation of two fractions of lignosulphonate. The numbers beneath each photograph represent the centrifuge speed in 100 revs. per min. Note the marked gel layer which increases in depth with increase in centrifugal field for the high molecular weight fraction, 2W-7 (32).

This difficulty was overcome by the empirical method of extrapolating to zero speed. A series of molecular weights were determined at several speeds and concentrations. The customary extrapolation of $1/M_w$ vs. c was then made for each centrifuge speed as shown in Figure 17.21. The intercepts $(1/M_w)_{c=0}$ were then plotted against g (Figure 17.22) and by extrapolation $(1/M_w)_{c=0, g=0}$ was obtained. Molecular weights determined by this method were found to agree pretty well with light scattering molecular weights on a series of carefully prepared fractions of lignosulphonate (32).

D. <u>Combination of Viscosity and Diffusion Methods</u>. The methods so far discussed are absolute in the sense that one obtains from them a molecular weight which does not depend on a separate calibration of the parameters concerned. Often, it is more convenient to measure a single parameter and then to derive a molecular weight from previously established relationships such as those in Equations 3, 4 and 5. The intrinsic viscosity, $[\eta]$, in particular, is

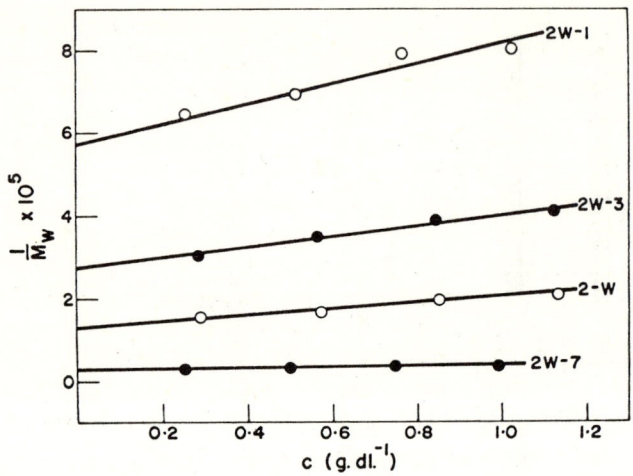

Figure 17.21. $1/M_w$ vs. c for fractions of lignosulphonate by sedimentation equilibrium (32).

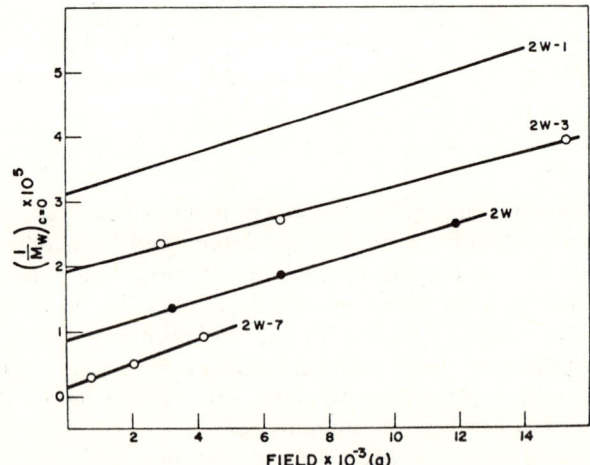

Figure 17.22. $(1/M_w)_{c=0}$ vs. centrifugal field for fractions of lignosulphonate by sedimentation equilibrium. The three points for 2W-1 were colinear but off the graph because of the higher values of field used with this low molecular weight fraction (32).

widely used in this manner to characterize synthetic polymers and has been applied to the determination of the molecular weight of lignins (119). It should always be remembered, however, that $[\eta]$ is essentially an effective specific volume and it is only for a chain conformation such as a linear random coil that an interpretable and reproducible relationship can be expected between the intrinsic viscosity and the molecular weight. For the lignin molecule with its gel-like structure, the intrinsic viscosity will depend not only on the molecular weight but also on the degree of cross-linking, and in the case of water-soluble lignins, the ionic strength and the surface charge.

A further disadvantage of the use of the intrinsic viscosity as a measure of the molecular weight is the small value sometimes found for the exponent, a, in Equation 3. For alkali and organosolve lignins, a is between 0.12 and 0.32 as shown in Table 17.4. Thus large changes in M produce only small changes in $[\eta]$ (120). This difficulty is somewhat less acute for the lignosulphonates for which values of a between 0.32 and 0.54 have been reported, Table 17.4. Thus it may be a useful expedient to determine the molecular weights of lignosulphonates by substitution of suitable constants in Equation 3. Caution should be exercised in applying the method to samples which have been modified in such a way that major changes in their charge or conformation would be expected.

The diffusion constant, like the intrinsic viscosity, can be used as a relative measure of the molecular weight. McCarthy and coworkers (121, 122) have developed an elegant method of measuring the distribution of D by diffusion of lignosulphonate from a solution to an agar gel. The method was calibrated by light scattering (33). The reliability of the method was tested by measurement of the distribution of D in a series of fractions, and comparison of their summation with the distribution measured on the unfractionated material (123). Good agreement was found as shown in Figure 17.23. Gardon and Mason measured the diffusion of various fractions of lignosulphonate through a porous disk (30). This method has also been adopted by Benko who used it to determine the relative molecular weights of various types of lignin (124, 125). Diffusion at a free boundary has been applied to lignosulphonates (32) and to organosolve lignins (37).

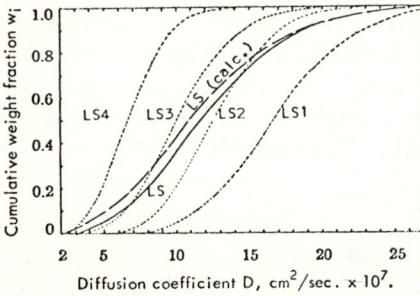

Figure 17.23. Distribution of diffusion coefficients for fractions (LS1-LS4) of lignosulphonate and parent unfractionated material (LS). Note agreement between LS measured and LS computed by summation of the distribution of the fractions. Data by McCarthy et al. (123).

The diffusion constant is inversely proportional to the frictional constant, f, which is the force required to produce unit translational velocity of a macromolecule in a liquid, Equation 9.

$$D = \frac{kT}{f} \tag{9}$$

For a sphere, f is related to the hydrodynamic radius, r, by Stokes law, Equation 10,

$$f = 6\pi\eta_o r \tag{10}$$

where η_o is the viscosity of the medium.

From Equations 9 and 10 it is quite clear that the diffusion constant, like the intrinsic viscosity, is a measure of the effective hydrodynamic volume of the macromolecule. For any particular type of lignin the constants in Equation 4 will depend on the degree of swelling, cross-linking or polyelectrolyte expansion shown by the macromolecule. The diffusion constant can therefore be used as a relative measure of molecular weight in a series of lignin fractions whose conformational properties are all alike. But the method is not absolute and should not be used to compare the molecular weight of soluble lignins from widely diverse sources or prepared by different methods.

A combination of Equations 2, 9 and 10 leads to the relationship, 11,

$$M = \left(\frac{kT}{\eta_o}\right)^3 \frac{N}{6480\pi^2 [\eta] D^3} \tag{11}$$

which for a spherical molecule can be used to calculate an absolute value of the molecular weight. The correctness of Equation 11 depends on the validity of the Einstein and the Stokes relationship. Both have been verified for macroscopic spherical particles (126, 127) and Polson (128, 129) and Edward (130, 131) have indicated that the approximate validity of Equation 11 extends to proteins and even smaller molecules.

Equation 11 has been applied with reasonable success to the calculation of the molecular weight of dioxane lignins (37). Similar relationships have been used to measure M for lignosulphonates (132, 133). We have seen that the conformation of the lignosulphonates approaches that of a non-freedraining coil in a θ-solvent. Thus a more precise relationship for lignosulphonates may be the empirical equation derived for flexible synthetic polymers by Mandelkern and Flory (134) in the form of Equation 11. This becomes Equation 12.

$$M = \left(\frac{kT}{\eta_o}\right)^3 \frac{N}{38600 [\eta] D^3} \tag{12}$$

E. Gel Filtration. Currently, the most rapidly developing method of measuring the size of macromolecules is based on molecular sieving in a chromatographic column. For synthetic polymers in organic solvents the technique is known as "gel permeation chromatography" (135), and when applied to water-soluble biopolymers it is called "gel filtration" (136).

The procedure is to pass a solution of macromolecules through a column of solvent-filled gel. Ideally, there is no specific interaction between the solute and the gel. Depending on their size, the macromolecules can diffuse into varying proportions of the porous volume of the column. Thus the elution volume of any particular fraction is a function of the dimensions of the macromolecules and the size of the pores in the gel.

According to Laurent and Killander (137), a gel filtration column can be characterized in terms of a parameter, K_{av}, which is the volume fraction of the gel phase available for a macromolecule of a given size (137). K_{av} is related to the elution characteristics of the column by Equation 13,

$$K_{av} = \frac{V_e - V_o}{V_t - V_o} \qquad (13)$$

where V_e is the eluate volume, V_o is the volume of the liquid phase in the column and V_t is the total volume of the gel bed.

The parameter K_{av} is also related to the hydrodynamic radius of the macromolecule, r, by Equation 14,

$$K_{av} = \exp[-\pi L(r+r_r)^2] \qquad (14)$$

in which L and r_r are characteristic dimensions of the chain molecules which constitute the skeleton of the swollen gel (137). Thus if V_e, V_t and V_o are measured and L and r_r are known, the hydrodynamic radius of a particular fraction can be determined.

Using the above treatment, Laurent and Killander were able to establish relationships between K_{av} and r for a wide range of solutes in various Sephadex gels of cross-linked dextran.

Gel filtration has been used by many investigators to fractionate soluble lignins. In some cases a relatively smooth distribution is noted while in others rather definite peaks indicate the presence of discrete fractions. An example of the latter taken from the work of Gupta and McCarthy (76) is shown in Figure 17.24.

Useful though gel filtration is for fractionating lignins, there is no reason why the method cannot be applied quantitatively for the determination of size and size distribution. Lindberg et al. (41) have used the method to estimate the molecular weights of fractionated kraft and soda lignins. McNaughton et al. (39) used gel filtration on Sephadex to obtain the distribution of radii in

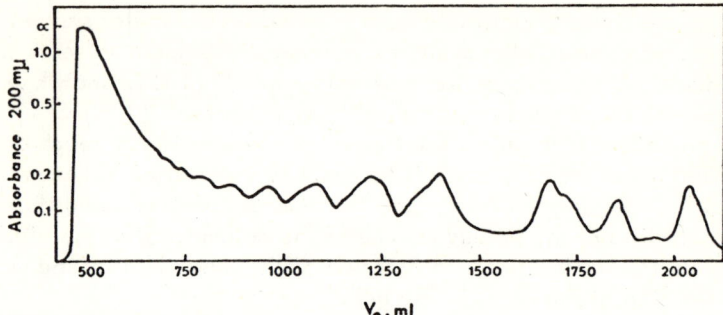

Figure 17.24. Gel filtration of low molecular weight lignosulphonate on Sephadex G-50, according to Gupta and McCarthy (76).

fractions of kraft lignin made soluble in a continuous flow process. As shown in Figure 17.25, the spread found in molecular size was rather wide. As mentioned in an earlier section, Gupta and McCarthy (76) have used gel filtration to demonstrate the increase in molecular size due to polyelectrolyte swelling of ligno-

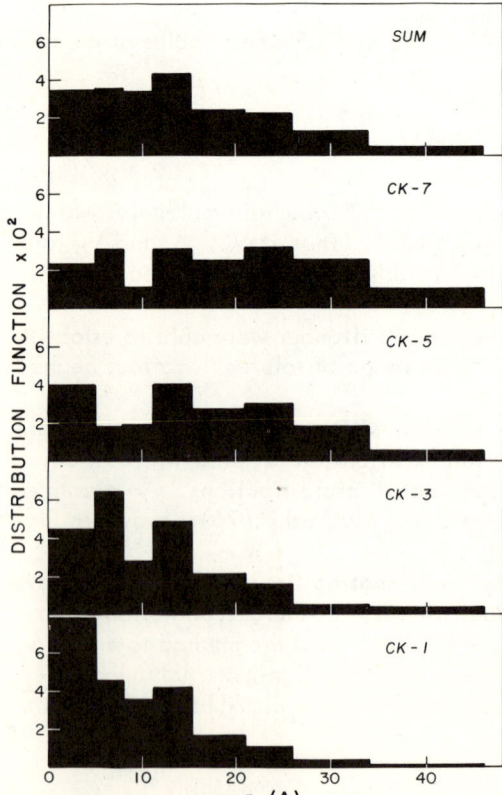

Figure 17.25. Normalized distribution of radii determined by gel filtration of four fractions of kraft lignin. The distribution at the top is the sum for all the fractions obtained by a continuous extraction of spruce wood (39).

sulphonates at low ionic strength. Recently, James, Pickard and Shotton (138) combined gel filtration with thin-layer chromatography to develop an elegant and rapid technique for measuring the molecular size distribution of lignosulphonates.

At this stage, it should be emphasized that gel filtration does not measure molecular weight as such but gives the effective hydrodynamic radius of the macromolecules. To convert this into a molecular weight, the relationship between size and weight of the macromolecular solute should be known. Grubisic, Rempp and Benoit (139) have pointed out that the hydrodynamic specific volume corresponding to the intrinsic viscosity is the molecular characteristic which determines retention in a gel filtration column. Little practical use can be made of this, however, since the intrinsic viscosity is itself usually a function of the molecular weight, the degree of cross-linking and (for aqueous solutions) the ionic strength.

Perhaps the simplest procedure is to run monodisperse fractions of known molecular weight through the column and thus establish the relationship between M and elution volume. Unknown samples can then be run and from the variation of concentration with elution volume, the molecular weight distribution can be determined. This approach has recently been taken by Soundararajan and Wayman (140), who used gel filtration on Sephadex to measure the molecular weight distribution in several milled wood lignins.

VII. Glass Transitions.

A. Softening Temperatures, Effects of Molecular Weight and Water Content.

It is useful to remember that much more lignin is sold as a constituent of newsprint and board than in any other manner. This trend is growing with the increasing use of high yield and refiner pulps in place of the more costly low yield fiber. In newsprint and board, lignin is a solid; and therefore knowledge of the bulk properties of lignin, such as density, thermal expansion, refractive index or thermal softening, becomes pertinent to the efficient preparation and processing of these materials.

Perhaps the most important bulk property of an amorphous polymer is its ability to undergo a glass transition. Below -70° C, rubber changes from a soft elastic material to a hard brittle solid. Above 70° C, polymethylmethacrylate will flow like a highly viscous liquid. This well-known behaviour arises out of the freezing in of the motion of the polymer chains below a certain temperature. As the temperature is raised, the solid polymer absorbs energy and the chains develop more violent motion until a temperature is reached at which intermolecular bonds are broken and the macromolecules become capable of large-scale displacement with respect to each other. The mechanical properties of the polymer change rapidly in this temperature region and the solid undergoes what is known as a glass transition. At temperatures below the transition, it is a glassy solid. A change to a rubbery or plastic material takes place above the transition.

The thermoplastic behaviour of lignin has been discussed by several authors. Kleinert and Tayenthal (141) reported the softening and melting of alcohol lignin in hot water. Lagergren, Rydholm and Stockman (142) have considered the effect of lignin softening in the defibreing of wood by different methods. The softening and steam plasticizing of lignin has been recognized for some time in the various processes for making board and dimensionally stabilized wood (143, 144). Higgins (145) observed a rapid decline in the yield value of plywood between 120 and 160°C which he ascribed to the thermal softening of the lignin.

The significance of lignin softening in chemical pulping has been pointed out by Björkman and Person (146) and by Kleinert (147); and in mechanical pulping, by Luhde (148) and by Atack and Pye (149). Recently, Stone and Scallan have demonstrated a large decrease in the surface area of a dioxane lignin at its softening temperature (150).

The glass transition behaviour of isolated lignins has been studied by observation of the heat-induced collapse of a column of powder under a constant gravitational load (151). The device used for these measurements in shown in Figure 17.26. Powdered lignin is compressed in a glass capillary by a plunger which supports a tungsten weight. The apparatus is sealed, placed in an oil bath and the temperature raised at a constant rate. As the glass transition is approached, the lignin softens and the plunger moves downward. The type of result obtained is shown in Figure 17.27 (152). The sharp maximum in the plunger velocity curve corresponds to the softening temperature, T_s^o of the lignin.

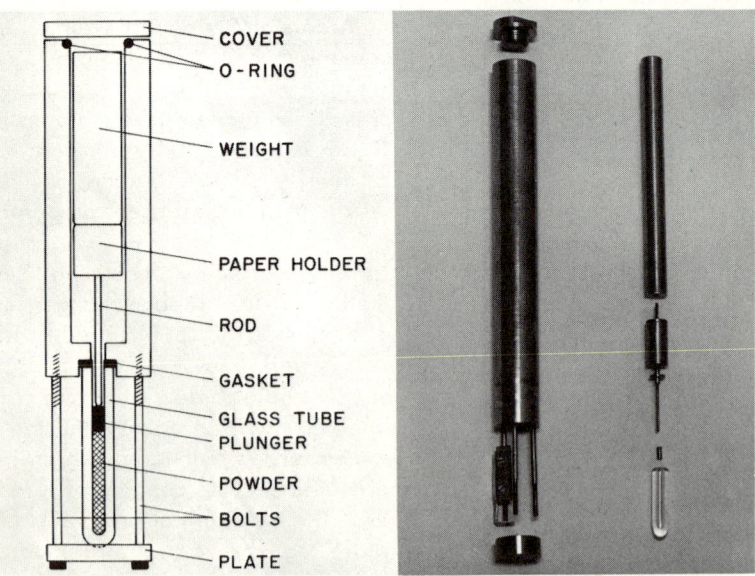

Figure 17.26. Drawing of the softening point apparatus and photograph of its parts (151).

POLYMER PROPERTIES

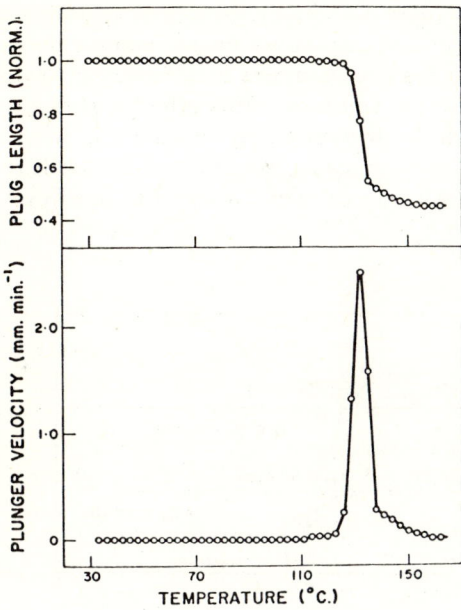

Figure 17.27. Length of powder column and plunger velocity vs. temperature for dry aspen dioxane acidolysis lignin (152).

Softening temperatures for five samples of lignin are given in Table 17.6. The lowest is 127°C for a low molecular weight spruce dioxane lignin I, and the highest is 193°C for spruce periodate lignin. This indicates that the glass transition temperature of isolated lignins vary considerably, depending on their source and method of isolation. One cause of variation is the molecular weight of the sample (151). Values of T_s^o for dioxane lignins were found to decrease from 176°C for $M_w = 85,000$ to 127°C for $M_w = 4300$. This trend is in agreement with the variation of the glass transition with molecular weight for synthetic polymers (153).

Table 17.6. Softening and Bonding Temperatures for Dry and Moist Lignins (151)

Sample	Dry (°C) T_s^o	Dry (°C) T_b^o	Water Uptake (g./100 g.)	Moist (°C) T_s	Moist (°C) T_b^*
Spruce Periodate Lignin	193	190	12.8	116	70
Birch Periodate Lignin	179	180	12.2	128	70
Spruce Dioxane Lignin I	127	110	7.7	72	50
Spruce Dioxane Lignin III	146	150	7.8	92	90
Aspen Dioxane Lignin	134	120	7.2	78	-

*T_b measured on specimens prepared in presence of excess water and then dried.

Water was found to have a pronounced effect on the softening behaviour of the lignin. This is shown, in Figure 17.28, by the decrease with water content of the maximum in the plunger velocity curve for periodate lignin. Evidently the water was plasticizing the lignin chains and thereby allowing the glass transition to occur at a lower temperature. This effect was found for all lignins. A comparison of T_s^o with T_s the softening temperature measured on the moist sample after equilibration over saturated sodium chloride is shown in Table 17.6. The lignins which took up most water seemed to show the greatest decrease in softening temperature.

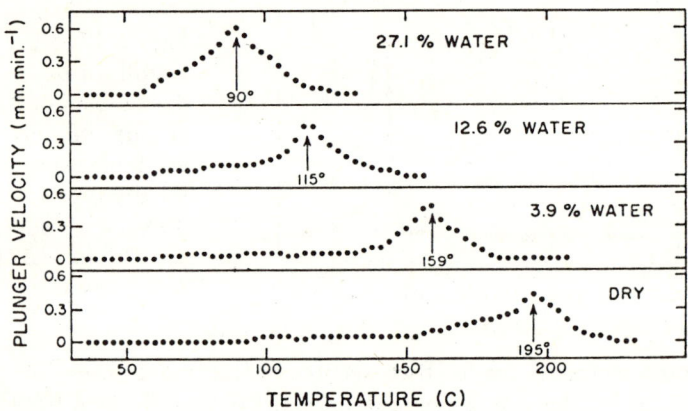

Figure 17.28. Plunger velocity vs. temperature for spruce periodate lignin containing various amounts of sorbed water (151).

The lowering of the glass transition of a polymer by the absorption of a low molecular diluent has been treated by Fujita and Kishimoto (154) who derive the relationship shown in Equation 15,

$$T_g^o - T_g = (\beta/\alpha_2)C \qquad (15)$$

in which T_g^o = glass transition temperature of dry polymer, T_g = glass transition temperature of polymer containing diluent, C = concentration of diluent in g/g, α_2 = change in thermal expansion coefficient below and above T_g (α_2 = 4.8 x 10^{-4} per deg. C), and β = parameter representing the contribution of the diluent to the increase in free volume.

In Figure 17.29, the water uptake is plotted against the lowering of the softening temperature for the lignins shown in Table 17.6 and some others. The

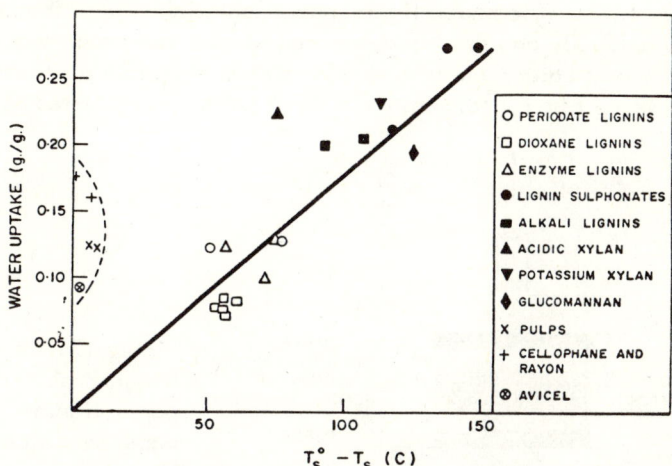

Figure 17.29. The decrease in softening temperature with water sorption over saturated sodium chloride for various lignins. Results for hemicelluloses and celluloses are included as well (151).

data can be represented, very approximately, by a straight line through the origin. From its slope, an average value of β is calculated to be 0.27, which is not very different from the values of 0.37 and 0.30 derived by Fujita and Kishimoto (154) from the effect of sorbed water on the glass transition temperature of polyvinylacetate and polymethylmethacrylate, respectively.

Figure 17.29 contains results for hemicelluloses and celluloses as well as for lignins. In passing, it is interesting to note that hemicellulose behaves rather like lignin in that its thermal softening temperature is markedly lowered by the presence of sorbed water. Cellulose was found to soften at temperatures of over 230° C, and, because of its crystalline nature, its softening temperature was scarcely affected at all by the presence of moisture. This behaviour is illustrated in Figure 17.29, where the points for cellulose are well off the straight line representing the results for the lignins and the hemicelluloses. Rather similar behaviour has been reported by Takamura for wood constituents isolated from a pine Asplund pulp (155).

B. Glass Transitions and Adhesion Properties. When a polymeric material softens, it often becomes tacky and exhibits autoadhesion. This is due both to the increased area of contact made possible by the conformability of the surface and also by the interdiffusion of the polymer chains caused by the increased molecular motion above the glass transition temperature (156). It would not be unexpected that the adhesive behaviour of lignin was a function of the temperature. This was tested by pressing lignin powder between two disks of paper (152, 157). The paper-lignin-paper sandwiches were then mounted as shown in Figure 17.30, and the force required to pull them apart was measured. Specimens were pressed dry and also soaking wet at temperatures from 25°C to 230°C. The wet-pressed specimens were dried before being tested to rupture.

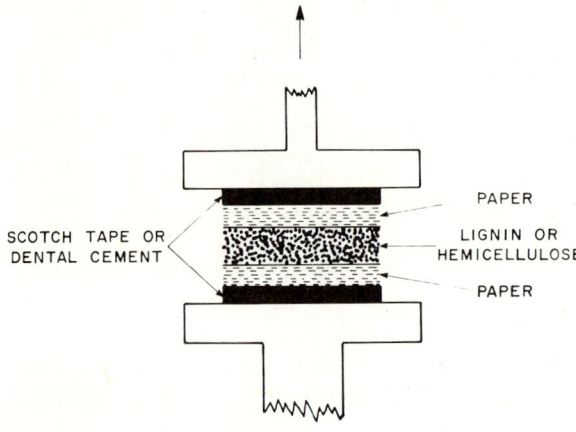

Figure 17.30. Diagram of a paper-lignin-paper sandwich. Hemicelluloses were tested as well (152).

Some results for dioxane lignins are shown in Figure 17.31. Under dry pressing, no bonding occurred until a certain temperature was reached after which the bond strength increased with temperature. The analogy between bonding and softening is quite clear when T_s^o is compared in Table 17.6 with the temperature, T_b^o, at which bonding developed. As was found for thermal softening, T_b^o increases with increase in the molecular weight of the dioxane lignins. Water had a marked effect on the adhesive behaviour as shown by the considerably lower values of the bonding temperature when the samples were pressed wet. It is interesting to note that birch xylan and pine glucomannan were found to show adhesive behaviour similar to that of the lignins (152).

The agreement between T_s^o and T_b^o in Table 17.6 indicates that both the thermal softening and the adhesive behaviour of isolated bulk lignins are manifestations of glass transition behaviour. The parallel decrease in both the softening and bonding temperatures with water sorption proves that water acts as a low molecular diluent in lowering the temperature of the transition.

C. The Significance of Glass Transitions in Paper and Board Manufacture. The significance of the above results in papermaking is fairly obvious. Lignin and hemicellulose together may comprise up to 50% of a finished sheet of newsprint. It is quite possible that in the dryers and calendar stacks of a paper machine, temperatures and moisture contents are reached which correspond to the glass transitions of the amorphous components of the sheet. If the paper is subjected to pressure when its lignin component is undergoing a glass transition, it is likely that the density, opacity and strength of the finished sheet will be affected.

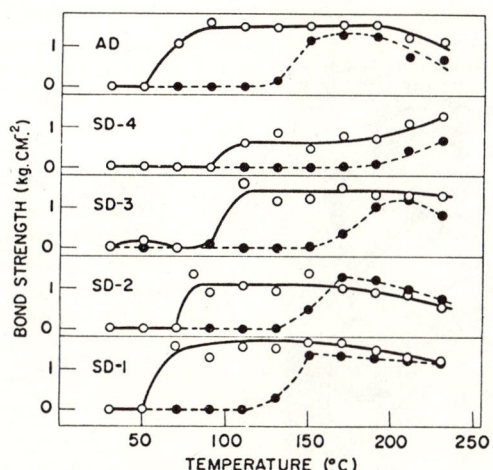

Figure 17.31. The effect of temperature of pressing on the bond strength for dioxane lignins. Dotted lines and filled circles are for dry pressing, while full lines and empty circles are for pressing wet. The molecular weight increases from spruce dioxane lignin 1 (SD-1) to spruce dioxane lignin 4 (SD-4); AD is an aspen dioxane lignin (157).

An interesting principle emerges from a comparison of the temperature dependence of chemical reaction rates and of physical relaxation. The rate of organochemical reactions are approximately doubled for every 10°C rise in temperature, whereas relaxation times decrease a thousandfold for a similar temperature increment near the glass transition. Thus, if it were possible to heat a lignocellulose sheet very rapidly, manipulate it at high temperature and then cool it down rapidly, it might be possible to take advantage of the thermoplastic properties of the lignin without incurring the penalty of excessive degradation. An interesting demonstration of this principle was recently given by Johanson and Back (158), who rapidly heated Asplund hardboard and hand formed the strips into the shapes shown in Figure 17.32. The material regained its stiffness and most of its strength properties on cooling.

Thermal softening of lignin might also be expected to influence the rate of delignification in chemical pulping (146, 147, 151). When wood is pulped, a mixture of chips, water and chemicals is heated to temperatures as high as 180°C. It is quite apparent that the physical nature of the lignin will change profoundly during the heating process. Above 100°C, the lignin and hemicellulose will consist of a soft gel-like matrix in which the cellulose portion is

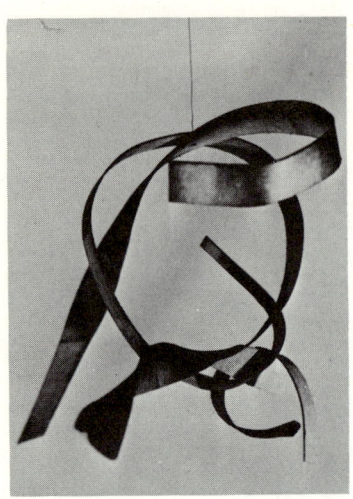

Figure 17.32. Asplund hardboard strips of 5 cm width, bent and twisted after external heating to 400° for 5 sec. Photograph by Johanson and Back (158).

essentially unaffected. Under such conditions it is likely that the intricate morphological detail of the wood structure will simplify owing to coalescence of the amorphous constituents. Loss of surface could then lead to a deactivation of the lignin to any chemical attack which depended on a surface reaction. In an attempt to demonstrate the coalescence effect, Yean et al. (159) pressed wet wood at elevated temperatures and then removed the lignin by acid sulphite pulping. Somewhat surprisingly, it was found that the rate of delignification was virtually unchanged by pretreatment of the wood at pressures up to 500,000 p.s.i. and temperatures up to 200° C. This held for both spruce and birch sawdust and also for birch from which the xylan had been removed by alkaline extraction. If coalescence of the lignin occurs when normal wood is pulped, one might expect an increase in the effect of coalescence on pulping pressed wood. The absence of any change suggests that either coalescence of lignin does not occur or, more likely, that if it does, it has little effect on the rate of delignification during chemical pulping.

It is in the production of the various types of fibreboard that the thermal softening of lignin would be expected to be particularly important. Pulps for fibreboard are usually made by a thermomechanical process (160, 161). The wood in the form of chips is first steamed and shredded in a disc refiner to give a coarse pulp, the individual strands of which are made up of fibre bundles. The fibreboard itself is manufactured by pressing the shredded pulp between heated platens. The temperatures and pressures applied are high enough to compress the pulp into a lignocellulose continuum with a density of up to $1.0 g ml^{-1}$. Clearly, the material properties of the finished board will be governed to a large extent by the thermoplastic and adhesive properties of the pulp from which the board is formed.

POLYMER PROPERTIES

Pulps made from steamed wood usually yield stronger board than can be produced with pulps from raw wood (161). Steaming also reduces the power required for refining, particularly in the case of hardwood use (161). An interesting question is whether or not the steaming affects the three-dimensional polymer structure of the wood in such a manner that the thermoplastic and adhesive properties of the fibreboard pulp are changed. In an attempt to answer this question, thermal softening of steamed and unsteamed wood was observed in the powder collapse apparatus (162). The results for poplar shown in Figure 17.33 indicate that steaming produced a marked change in the softening behaviour of the wet wood. Pulp from untreated poplar gave a single transition near that for cellulose at about 230°C. The corresponding plunger velocity curve for moist pulp from steam-cooked poplar was flat but showed two well-defined transitions at 108° and 208°C, respectively.

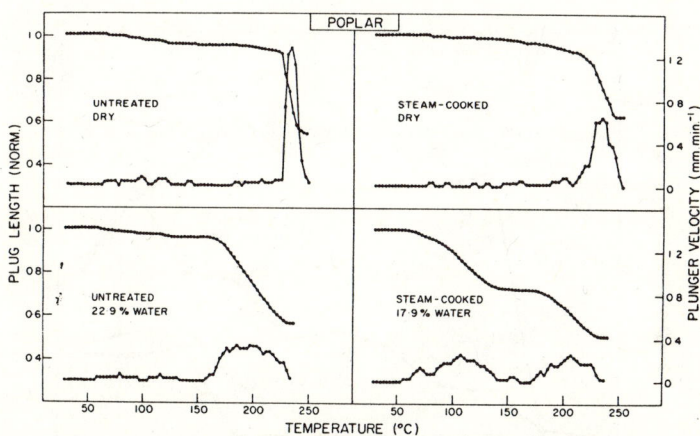

Figure 17.33. Thermal softening of hardboard pulp from untreated and steam-cooked poplar chips. Data are given for dry and moist samples. In each graph, the upper curve is the plug length normalized to unity at room temperature. The lower curve is the velocity of the plunger at various temperatures (162).

The appearance of two transitions in the moist steamed poplar can be interpreted by means of Björkman's (163) concept of occasional covalent bonding between the macromolecular constituents of wood. In such a spot-welded composite of wood polymers, one would expect thermal softening as observed by the powder collapse method to be dominated by the cellulose component. Even when sorbed moisture lowers the transition of the amorphous constituents, the powder would retain its rigidity up to high temperatures because of the overriding effect of the cellulose framework which is not plasticized by water molecules. During steaming, the structure may be loosened by the severance of the bonds between the three polymeric species. The moist amorphous matrix of lignin and hemicellulose can then flow if heated above its transition temperature, and thus can contribute independently of the cellulose to the thermal softening of the wood.

Interestingly, birch wood gave the same results as poplar, but the effect was much smaller in the case of spruce which appeared to be more resistant than the hardwoods to this type of structural loosening (162).

Flow of the amorphous constituents of structurally loosened wood would be expected to alter considerably the adhesive behaviour of the shredded wood. The results in Figure 17.34 show that this indeed is the case. The bonding properties of poplar show a dramatic increase in adhesion when wet pressed at temperatures above 100°C. Similar results were found for birch, but, as expected, spruce showed smaller effects (162).

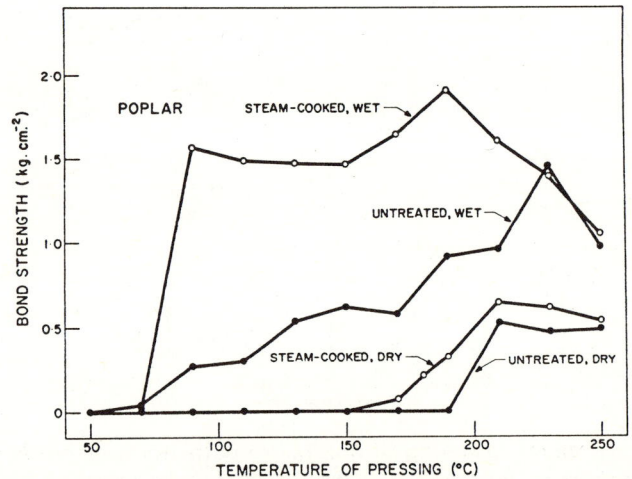

Figure 17.34. The development of bond strength with temperature of pressing for hardboard pulp from untreated and steam-cooked poplar chips. Data are included for laminates pressed both dry and wet (162).

Some support for the concept that steaming loosened the interpolymer spot-welds in wood was supplied by the results obtained for a mechanical mixture of cellulose, hemicellulose and lignin. As shown in Figures 17.35 and 17.36, respectively, the behaviour observed paralleled that of the steam-cooked hardwoods both in thermal softening and adhesive behaviour.

POLYMER PROPERTIES

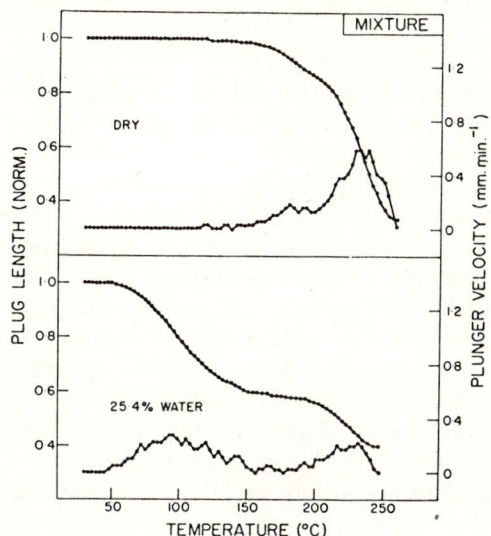

Figure 17.35. Thermal softening of a 50:25:25 mixture of powdered cellulose, birch xylan and periodate lignin. Data are given for the dry and moist mixture (162).

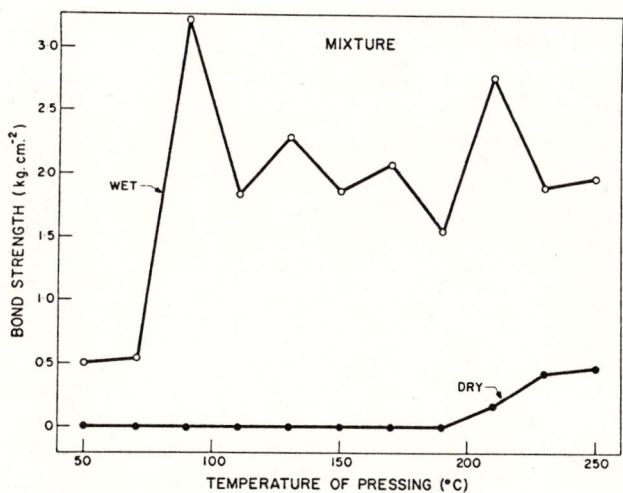

Figure 17.36. The development of bond strength with temperature of pressing for a 50:25:25 mixture of powdered cellulose, birch xylan and periodate lignin. Data are included for laminates pressed both dry and wet (162).

D. Thermal Expansion and Second Order Transition Phenomena.

Glass transitions are accompanied by second order transitions in certain thermodynamic properties of the solid polymer. A plot of volume vs. temperature shows a change of slope in the region of the glass transition (164). Similar effects are found in the temperature dependence of the refractive index (164) or the specific heat (165). A second order transition is often used as a means of detecting the glass transition of a synthetic polymer. It is difficult to measure a second order transition at the glass transition temperature of lignin or other wood polymers. Pyrolysis of wood polymers occurs at temperatures well below the values of $T_s^°$ listed in Table 17.6 (166). Chemical changes and the evolution of small quantities of gas confuse the interpretation of calorimetric or dilatometric data.

Additional second order transitions often occur at temperatures well below the glass transition temperature (167). The onset of restricted molecular motion such as the vibration of short segments of the chain or the rotation of groups pendant to the backbone can cause a low temperature second order transition. Recently, attempts were made to detect such transitions in lignin by dilatometry (167). The lignin was pressed into pellets, thoroughly degassed and sealed into the dilatometer. Mercury was used as the confining liquid. A volume temperature curve for periodate lignin is shown in Figure 17.37. Two transitions were detected, T_1 at 25-33°C and T_2 at 101-108°C. Dioxane lignin gave a similar result except that the second transition occurred at lower temperature. Transition temperatures and thermal expansion coefficients above and below the transitions are given in Table 17.7.

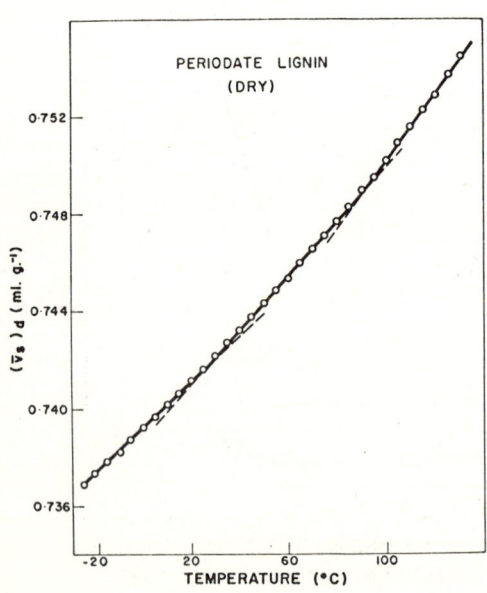

Figure 17.37. Volume vs. temperature for dry periodate lignin (167).

Table 17.7. Thermal Expansion and Transition Temperatures of Lignins both Dry and in the Water-Swollen (Wet) State (167).

Sample	Transition Temperatures (°C)		Thermal Expansion x 10^5 (ml.g.$^{-1}$°C^{-1})	
	T_1	T_2	Below T_1	Above T_1
Dry Periodate Lignin	25-33	101-108	9.3	11.4
Dry Dioxane Lignin	27-32	80-85	10.4	13.4
Wet Periodate Lignin	15-25	-	39.5	33.7
Wet Dioxane Lignin	25-35	-	22.9	29.9

Dilatometric measurements were also made with water as the confining liquid, i.e., with the sample soaking wet. The transition at T_1 was still evident, but the thermal expansion decreased above the transition for periodate lignin as shown in Table 17.7. Significantly, the transition remained positive for dioxane lignin, but in both samples the apparent thermal expansion coefficients in the wet state were considerably higher than those for the dry lignins.

The effects shown in Figure 17.37 and in Table 17.7 were not peculiar to lignin but were found for cellulose, hemicellulose and particularly for simple sugars such as glucose and cellobiose (167, 168). The 25°C transition for cellulose has been confirmed by ultrasonic measurements on paper by Back et al. (169), and Kubat and Pattyranie have reported a further cellulose transition at -30°C (170). The T_1 transition in the dry state is probably due in some way to bond breaking and motion of the hydroxyl group which is common to all the samples which showed the transition. The difference in the thermal expansion coefficients in the dry and the wet states was interpreted in terms of the breaking down of the water structure by the -OH groups on the wood polymers. Of course, any attempt to explain these phenomena must be regarded as tentative, particularly in view of the recently revived interest in the properties of water as a structural liquid (171, 172). However, it is clear that statements such as "Lignin is hydrophobic" or "Water plasticizes lignin" are gross oversimplifications of a rather complex interaction at the molecular level.

VIII. Solubility, Accessibility and Sorption.

A. Hildebrand's Solubility Parameter and Hydrogen Bonding Capacity. The lignin polymer in wood behaves as an insoluble, three-dimensional network. Chemical reactions of the protolignin are usually heterogeneous. Nakano and Schuerch (173) have formulated an important question dealing with the reactivity of lignin. Is the lignin network completely accessible to a reagent or does the reaction take place by removal of successive layers from the surface of microscopic pockets of solid protolignin? A conclusive answer cannot be given because of the dearth of information on the molecular morphology of the lignin and its submicroscopic distribution in wood tissue. However, investigation of the solubility of isolated lignins and the accessibility of lignin in wood have elucidated this problem to some extent.

Isolated lignins are known to be soluble in some solvents but not in others. Organosolve lignins are soluble in dioxane or pyridine but not in water. Schuerch (174) investigated the solubility of lignin in a number of solvents and found that solvent power for lignin was a function of both the cohesive energy density, c.e.d. and hydrogen-bonding capacity of the solvent. In terms of Hildebrand's solubility parameter, $\delta(\delta = \sqrt{c.e.d.})$, Schuerch noted that maximum solvent effects occurred for values of δ around 11 (cal./cc.)$^{1/2}$. Solvents with δ-values far removed from 11 (cal./cc.)$^{1/2}$ did not dissolve the lignin. This effect is illustrated in Table 17.8. Schuerch (174) also discovered that if a series of solvents were chosen all having δ-values between 10 and 11 (cal./cc.)$^{1/2}$ then their solvent power for lignin increased with increase in their capacity to form hydrogen bonds. This trend is shown in Table 17.9 in which the measure of hydrogen-bonding capacity is the infrared shift, $\Delta\mu$, in the O-D band when the solvent is mixed with CH_3OD. It can be seen that the best solvents are those which show the highest value of $\Delta\mu$. It was also found that the higher molecular weight fractions required better solvents to effect solution (174).

Table 17.8. The Solubility of Isolated Lignins in Single Solvents Selected from Schuerch's data (174).

Solvent	δ (cal./cc.)$^{\frac{1}{2}}$	$\Delta\mu$ (microns)	"Indulin" Kraft Pine	Native Aspen
Ether	7.5	0.19	Ins	Ins
Ethyl acetate	9.1	0.12	Sli	-
Dioxane	10.0	0.14	Sol	Sol
Methyl cellosolve	10.8	High	Sol	Sol
Methanol	14.3	0.28	Par	-
Water	23.4	High	Ins	Ins

Ins = Insoluble
Sli = Slightly soluble
Par = Partially soluble
Sol = Soluble

Table 17.9. The Effect of Hydrogen Bonding Capacity on Lignin Solubility from Schuerch (174). Solvents are listed in Order of Increasing Solvent Power for Lignins.

Solvent	δ (cal./cc.)$^{\frac{1}{2}}$	$\Delta\mu$ (micron)
1. Carbon disulphide	10.0	Low
2. Nitrobenzene	10.0	0.04
3. Acrylonitrile	10.5	ca. 0.08
4. Methyl formate	10.16	ca. 0.12
5. Acetone	10.0	0.14
6. Dioxane	10.0	0.14
7. Pyridine	10.7	0.27

Schuerch interprets these results to mean that there is nothing special or particular about the problem of dissolving low molecular weight isolated lignins. These are soluble in solvent systems which would be expected to dissolve an aromatic polymer containing many -OH groups. This, according to Schuerch, is evidence that the insolubility of lignin in the wood must be due to the fact that it is a network polymer and therefore requires fragmentation for its solution.

In the light of Schuerch's findings, Lindberg (175) has investigated the solubility of various lignins as a function of the intermolecular hydrogen-bonding capacity of the lignin itself. The excess of unbonded oxygen was calculated from the difference between the total oxygen content and the total hydroxyl content and was taken to be a measure of the unused hydrogen-bonding capacity of the molecule. The solubility was measured by the ease of precipitation of the lignin from dioxane solution by addition of chlorobenzene or carbon tetrachloride. Lindberg found, as shown in Figure 17.38, that the tendency to be precipitated increased regularly with the excess hydrogen-bonding capacity of the molecule. He points out that this effect may be the reason why small quantities of carbohydrate impurities cause lignins to be particularly insoluble.

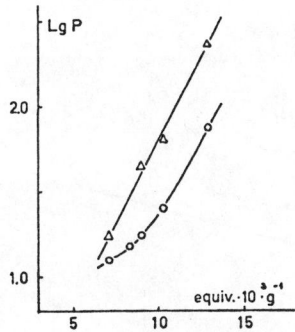

Figure 17.38. Logarithm of precipitability, P vs. meq./g. excess of ether and carbonyl oxygen for samples of soda, sulphate and ethanol lignins (175).
Δ, Precipitation with chlorobenzene;
O, Precipitation with carbon tetrachloride.

The interpretation and prediction of lignin solubility has been extended to binary solvent mixtures by Ekman and Lindberg (176), who measured the solubility of milled wood, dioxane and soda lignins in several solvent mixtures over a wide range of compositions. The dielectric constant, D, of the solvent mixture was also measured and D was converted to the solubility parameter, δ, by an empirical equation. The work confirmed that, usually, the most suitable solvent mixtures for lignin possessed δ-values in the range 10.5–12.5 (cal./cc.)$^{1/2}$. Marked solubility maxima at approximately 50/50 composition were found for the 1,2-dichloroethane ether and 1,2-dichloroethane-ethanol mixtures. The authors note that this behaviour may be of considerable practical interest. Recently Lindberg

(177) has applied the Flory-Huggins theory to the problem of lignin solubility and found that if the solubility parameter is split into its non-polar and polar components, the solubilities in pure solvents can be predicted. However, from vapour pressure osmometry on kraft lignin in various solvents at different temperatures, Brown (178) concluded that changes in the solvent structure rather than the nature of the lignin-solvent interaction governed the magnitude of the thermodynamic parameters.

B. Relation Between Accessibility and Solubility. The solubility of isolated lignins leads directly to the concept of the accessibility of the insoluble lignin in the wood. It is surprising that so much work has been reported on the accessibility of cellulose but very little on the accessibility of lignin. Yet the industrially important delignification reactions such as kraft or sulphite pulping must depend on the accessibility of the lignin to the chemical agent used. Nakano and Schuerch (173) have made a study of the way in which the accessibility of lignin in wood depended on the solvent system used. Aryl chlorination with $\underline{t}$-butyl hypochlorite was chosen as the accessibility reaction, and the authors were able to demonstrate by experiments with model compounds that the solvent effect in completely accessible systems was rather small. With wood, however, chlorination was strongly influenced by the choice of solvent as shown in Figure 17.39.

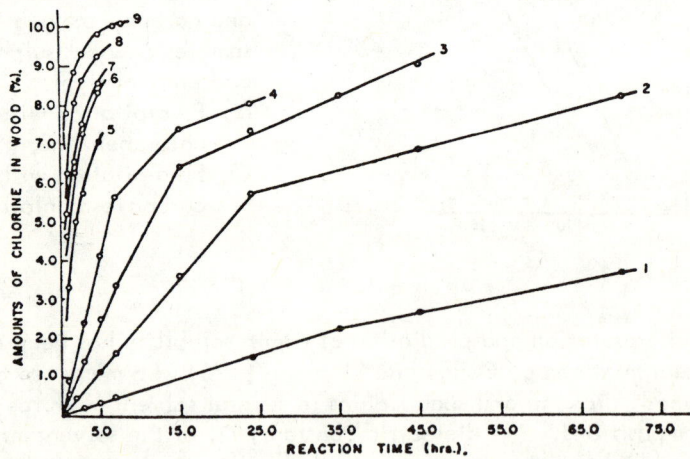

Figure 17.39. Rates of chlorination of woodmeal by $\underline{t}$-BuOCl in various solvents (173). 1 ≡ 1-nitropropane, 2 ≡ ethyl acetate, 3 ≡ nitroethane, 4 ≡ methyl ethyl ketone, 5 ≡ ethylene glycol dimethyl ether, 6 ≡ acrylonitrile, 7 ≡ acetone, 8 ≡ acetonitrile, and 9 ≡ nitromethane.

POLYMER PROPERTIES

Nakano and Schuerch (173) were then able to classify the solvents listed in Figure 17.39 according to whether the solvent promoted a fast, slow or intermediate reaction of the t-BuOCl with the lignin in the wood. This is shown in Figure 17.40 which also includes the solubility parameter and hydrogen-bonding capacity of each solvent. Comparison of the data in Figure 17.40 with the previous tests of the solubility of lignin showed that all solvents which promoted a rapid reaction dissolved lignin to some extent whereas those in which the reaction was negligibly slow were nonsolvents for lignin. The authors then draw attention to the fact that water is a poor lignin solvent and that reactions in aqueous media may be restricted by inadequate accessibility.

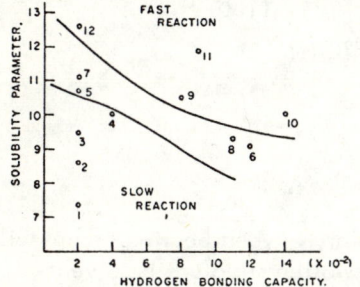

Figure 17.40. Relation between solvent properties and chlorination rate. Numbers correspond to solvents as listed in Figure 17.39 (173).

C. <u>Sorption of Vapours</u>. Several investigators have studied the sorption of vapours by lignin. Some lignins will absorb about as much water as a cellulose even though lignin contains a substantially smaller proportion of -OH groups in its chemical make-up (151, 179). The isotherm is reproducible but shows hysteresis as illustrated by the results of McKnight and Mason (180) in Figure 17.41. Similar isotherms were found for methanol and ethanol. Benzene was adsorbed to a much smaller extent by periodate lignin air-dried from water. However, if the water-swollen lignin was solvent exchanged to benzene and then freeze-dried, the sorption of benzene increased by a factor of 50. Thus, the open structure produced by the solvent exchange and freeze-drying process permitted a far greater uptake of a non-swelling sorbate. Significantly, McKnight and Mason (180) also found that solvent-exchanged lignin air-dried from benzene sorbed at saturation about 6 times as much benzene as lignin air-dried from water. A possible explanation of the result is that when the lignin polymer dries, the groups exposed depend on the liquid being evaporated. When lignin is swollen by a polar liquid like water, the chains in the network might be expected to undergo the conformational adjustments necessary to allow hydrophobic bonding between the nonpolar sites of the molecule (171,

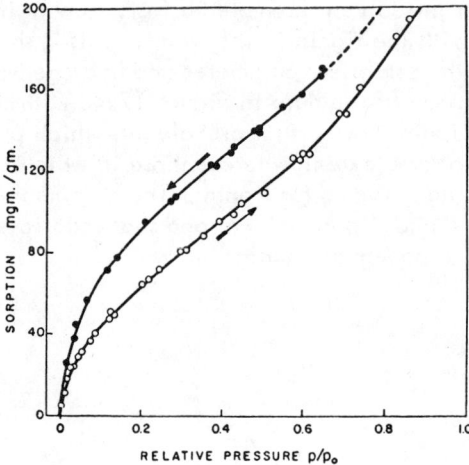

Figure 17.41. Water sorption isotherm on lignin. Points shown are for first, third and fifth cycles and illustrate the reproducibility. Data by McKnight and Mason (180).

181). On drying from water, mainly the polar, oxygen bearing groups will be exposed. However, when the open, water-swollen structure is solvent-exchanged to benzene, the polar groups on the chains now attract each other and drying from benzene, therefore, leaves a predominantly nonpolar internal surface. Such a surface would be expected to show a much greater sorptive capacity for nonpolar solvents than the surface produced by removal of water.

McKnight and Mason also reported marked sorption of SO_2 by periodate lignin and found that a small part of the gas was permanently absorbed probably due to chemisorption (180). Kleinert and Marraccini (182) used the sorption of SO_2 from solution to measure the internal surface of lignin. These authors found the sorption to be completely reversible and discovered a marked loss of surface due to coalescence of the lignin when the wood was heated in water at 180-185° C. McKnight and Mason (180) also noted a considerable decrease in the water sorption of periodate lignin when this material was given a 150° C treatment in water prior to measurement of sorption.

D. Sorption of Ions. The sorption by periodate lignin of alkali metal ions from solution has been studied by Screaton and Mason (183). The sorption increased rapidly with concentration at low concentrations of alkali as shown in Figure 17.42. Sorption levelled off at a plateau corresponding to 1.5 meg./g. of lignin. At higher concentrations of NaOH, both sorption of alkali and swelling of the lignin were found to decrease. This trend was probably due to a decrease in the uptake of water by the lignin because of the low activity of the water in the concentrated solution of NaOh. The primary heat of sorption show

POLYMER PROPERTIES

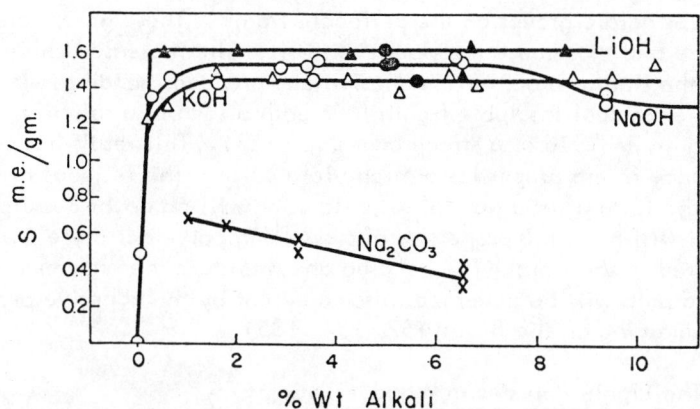

Figure 17.42. Apparent sorption of alkali metal ions on periodate lignin. Data by Screaton and Mason (183).

approximately the same trend as the sorption, rising rapidly to a fairly constant value and then at higher concentrations of NaOH decreasing again.

Sorption of minute quantities of sodium ion has been shown to increase the adhesive properties of periodate lignin. This effect is shown in Figure 17.43,

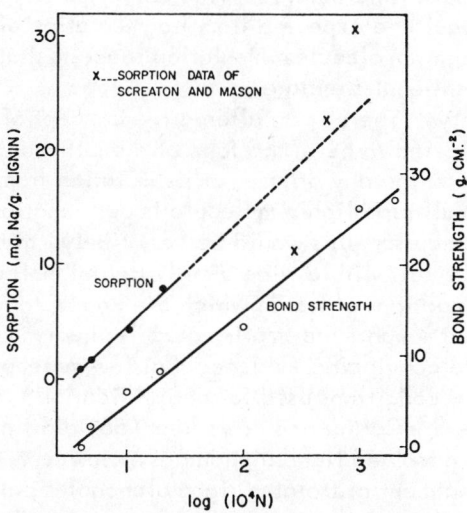

Figure 17.43. Relationship between NaOH concentration and bond strength (lower line). The upper line is the sorption of Na^+ on the periodate lignin (157). The points denoted by crosses are from the sorption data of Screaton and Mason (183).

in which the bond strength developed by powdered periodate lignin is increased by soaking the lignin in dilute solutions of NaOH (157). Adsorption of sodium ion neutralizes acidic groups on the periodate lignin. These groups are then able to ionize and thus cause the lignin to swell. The increased chain mobility then causes the lignin to act as an adhesive and produces bonding. It was also found that the residual insoluble lignin from both a soda or a sulphite cook of periodate lignin behaved as a strong adhesive (157). This result has considerable significance in the properties of high yield pulps. The residual lignin remaining on the fibre after a neutral sulphite cook will no doubt be sulphonated. This material will have the properties of a swelling polyelectrolyte though it remains anchored to the fibre. The bonding and therefore the papermaking qualities of such a pulp will be governed to some extent by the adhesive properties of the lignin remaining on the fibre (157, 184, 185).

IX. The Lignin Polymer in Wood.

A. Network Theory. Thus far, the present chapter has been written with the network theory of protolignin as its underlying concept. Lignin in wood is visualized as a three-dimensional structure, consisting of short linear chains cross-linked by a variety of interchain, covalent bonds. What can be said concerning the validity of this theory from the properties of wood and of isolated lignins?

The strongest argument in favour of the network theory is that lignin in wood is insoluble in all neutral solvents. Schuerch (174) has shown that there is nothing unusual about the solubility of isolated lignins. Thus the protolignin must be anchored by strong bonds to the macroscopic continuum of wood. The network theory claims that these bonds are covalent and mostly lignin-lignin. A second plausible argument arises out of the observed wide polydispersity of isolated lignins. This distribution would be expected from fragmentation of a network. Moreover, the shape of lignin molecules in solution suggests that they are "hunks" cut out of a three-dimensional structure.

B. Small-Molecule Theory. There is an alternative concept of lignin in which the protolignin is considered to be in the form of a small reactive molecule which tends to polymerize during any attempt at its isolation from the tissue of wood. If the supposedly small protolignin molecule is polyfunctional (and from what we know of lignin precursors this would be very likely) then a network type of high polymer would result with a wide distribution of molecular weight. The various condensation reactions of lignin which are known to make delignification of wood more difficult support the small-molecule theory.

However, there has been little convincing evidence that low molecular weight lignins are polymerized under conditions used for delignification. At the end of a sulphite cook of spruce, McCarthy and coworkers (56) did find an increase in the molecular weight of dissolved lignosulphonates. However, prolonged sulphonation or mild acid treatment of isolated lignosulphonates produce only degradation (186). In a more recent investigation (46), an initially extracted low molecular weight lignosulphonate from spruce was resulphonated. Only a reduction in molecular weight was noted even when the reaction was carried out at a concentration of 20% lignosulphonate to correspond to the swollen gel-like lignin substance in the wood during the reaction. This result indicates that even relatively low molecular weight lignins show little tendency to polymerize under the conditions used for their extraction from wood.

C. Lignin-Carbohydrate Bonds and Possible "Snake Cage" Effects.
It is difficult to explain the universal insolubility of protolignin if it consists only of small molecules. If each molecule of lignin is connected covalently to the carbohydrate part of the wood then the protolignin would tend to be insoluble. However, lignin-carbohydrate bonds in sufficient abundance to make a low molecular weight protolignin insoluble should have been detected with certainty by now. It seems more likely that if there exist covalent bonds between lignin and carbohydrate, they are comparatively rare and, as pictured by Björkman (163), may be likened to occasional spot welds which enhance the mechanical permanence and stability of three interlocking polymeric structures in wood. It should be noted here that the difficulty of separation of lignin impurities from a carbohydrate preparation or carbohydrate impurities from a lignin preparation need not be due to a covalent bond between the lignin and the carbohydrate. In the lignin preparation, linear carbohydrate molecules may be held in the gel-like lignin matrix by molecular entanglement similar to the "snake cage" structures proposed for mixtures of polymers (187, 188, 189). This is shown pictorially in Figure 17.44A. Similarly an isolated hemicellulose may retain on its chain small beads of the lignin network which would be extremely difficult to remove except by fairly severe degradative procedures, Figure 17.44B.

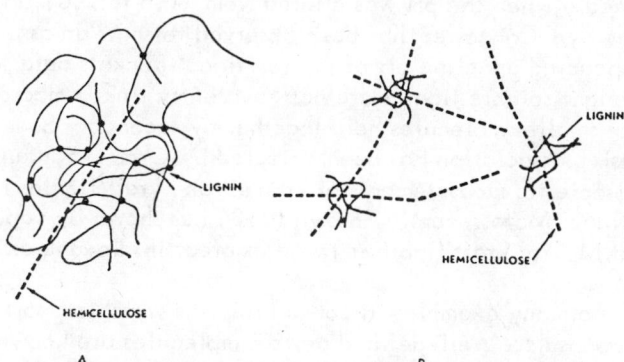

Figure 17.44. Pictorial illustration of possible physical entanglement between gel-like lignin molecules and linear hemicellulose molecules. A: High molecular weight lignin containing a single hemicellulose chain as an impurity. B: Isolated hemicellulose molecules having lignin fragments as impurities.

D. Aggregation by Secondary Bonds. Important evidence for a somewhat different concept of lignin has recently been presented by Benko (125) who found that the molecular weight of a softwood lignosulphonate decreased by a factor of about 12 when measurements were made in methanol instead of aqueous 0.1N KCl. A 30-fold decrease in molecular weight was found when DMSO was used as solvent. Some of his data on which this conclusion is based are shown in Figure 17.45.

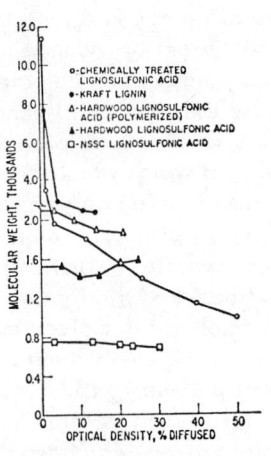

Figure 17.45. Decrease in molecular weights of lignosulphonates during continuous diffusivity measurements in dimethylsulphoxide, according to Benko (125).

Benko (125) also showed that the molecular weight of soluble kraft lignins increased markedly when the pH was altered from 11.5 to 7.0. In addition he notes that the dye Congo red has been observed to show an association of 15 to 250-fold depending upon the pH of the solution. Benko's data suggest that high molecular weight soluble lignins are not covalently linked macromolecules but aggregates of smaller molecules held together by secondary bonds. Similar but not as extensive association has been detected by Gross, Sarkanen and Schuerch (109) from isopiestic measurements of ethanol and kraft lignins in dioxane or tetrahydrofuran. More recently, Brown (178) has shown by vapour pressure osmometry that M_n for kraft lignin is twice as great in dioxane as it is in dimethylsulphoxide.

There are many examples in colloid science which support the plausibility of Benko's proposals. Well-defined protein molecules are known to be composed of several sub-units held together by secondary bonds (51). Recently Némethy and Scheraga (181) have given an elegant thermodynamical treatment of hydrophobic bonding in aqueous solutions of proteins. The free energy of the system is lowered by the intermolecular association of the nonpolar groups on the sidechains with the exclusion of water molecules at the site of the bond. Similar forces produce the well-known micellar properties of long chain fatty acids or sulphonates (190).

Some time ago Gardon and Mason (67) detected discontinuities in the concentration dependence of the electrical conductivity of solutions of low molecular weight lignosulphonates and suggested the possibility that micellar formation was taking place. The electron micrographs of high molecular weight lignosulphonates (72) indicated an aggregated structure, see Figure 17.6. Soda lignins from periodate lignin were found to become irreversibly insoluble on precipitation from alkaline solution with organic solvents (74). This coupled with the extremely high molecular weights of some of the fractions (24) suggests a colloidal aggregation of smaller molecules rather than a covalently linked molecule.

If high molecular lignosulphonates are aggregates, it might be expected that their molecular weights would be temperature dependent. However, careful light scattering measurements have shown (32) that a lignosulphonate having M_w = 105,000 showed very little change in molecular weight when the solution was heated to 120°C and measurements were then made at temperatures of 95°C. Thus if the molecule was made up of smaller sub-units, the secondary bonds between the sub-units were not broken by the higher temperatures.

Recently, the weight average molecular weight of a lignosulphonate fraction was measured in various solvents by the sedimentation equilibrium method (191). Values of M_w in water, dimethylsulphoxide (DMSO) and formamide were 52,000, 43,000, and 117,000, respectively. Apparently, dimerization was occurring with formamide as solvent, but the molecular weights in water and DMSO were essentially equal. Furthermore, prolonged standing in DMSO, for a period of over one month, produced no change in M_w. By gel filtration of a mixture of radioactive and non-radioactive lignosulphonates, Jensen et al. (192) concluded that low molecular weight lignosulphonates do not form aggregates in water. Thus the weight of evidence now indicates that lignosulphonates in water are not markedly aggregated but are true covalent polymers. A similar conclusion has been reached by Ekman and Lindberg (193) for Brauns native lignin and milled wood lignins dissolved in DMSO for which elution patterns from Sephadex gels were unaltered by methylation and acetylation.

E. Heterogeneity in Original Lignin. Thus far in the chapter, the protolignin has been pictured as a uniform type of material. The polydispersity of molecular weight is produced by the cleavage of a network gel or, less probably, by the polymerization of a small reactive molecule. Support for this comes from the chemical similarity of fractions of soluble lignin and from the same wood, particularly a softwood. However, some authors believe that in the wood itself there is, in fact, material which produces soluble lignins of low molecular weight and a different material which dissolves to give high molecular weight lignins. The recent papers of Forss and his coworkers (194-197) emphasize the concept of the heterogeneity of the protolignin. Forss and Fremer fractionated spruce sulphite liquors by gel filtration in a Sephadex column (197). Between

14 and 24% of the aromatics in the wood were recovered as low molecular weight, lignin-like substances which gave ultraviolet spectra in alkaline solution different from the absorption spectrum of a true lignosulphonate (196). Forss and Fremer call this material "hemilignin" and tend to the view that hemilignins are not artefacts produced by cooking but exist as a different but lignin-like species in the wood. The remainder of the aromatic containing materials Forss and Fremer class as lignosulphonates from the true protolignin. The low molecular weight portion of the lignosulphonate was refractionated in a Sephadex column 7.5m. long. A step-wise increase in the ratio of calcium to UV absorption was found which was accompanied by definite changes in the elemental composition of each fraction. In order to explain the results, Forss and Fremer have proposed a comparatively simple model of low molecular weight lignosulphonate in which a series of oligomers are produced by progressive hydrolysis and sulphonation of a molecule contain 16 guaiacylpropane units (eight of which are sulphonated) and two hydroxyphenylpropane units. The authors envisage that lignosulphonates of higher molecular weight are formed by the linking together of the 18-mers (197).

An interesting question is whether or not the high and low molecular weight producing material is located in morphologically different wood tissue. In order to test this, wood was fractionated according to the following tissue differentiations: (a) sapwood and heartwood; (b) early wood, transition wood and latewood; and (c) compression wood, opposite wood and normal wood (198). The fractions of wood were sulphonated and the molecular weights of the purified lignosulphonates were measured. The results shown in Table 17.10 indicate some slight differences in molecular weight, but the effects were small when compared with the wide polydispersity usually expected of soluble lignins. The loss of low molecular weight material by dialysis was also comparable within each of the differentiations studied. Apparently, low and high molecular weight lignin macromolecules were not associated individually with the types of tissues shown in Table 17.10.

Table 17.10. Molecular Weights of Lignosulphonates from Various Differentiations (198).

Sample	M_w	Nondialysable (%)
White spruce		
Sap wood	52,000	51
Heart wood	52,000	57
Western hemlock		
Early wood	20,000	71
Transition wood	28,000	74
Late wood	30,000	65
Black spruce		
Compression wood	27,000	60
Opposite wood	19,000	66
Normal wood	19,000	65

It remains possible that the different molecular weight lignins may arise from different tissue areas at the level of a single cell. We have already noted that in both softwoods and hardwoods the concentration of lignin in the middle lamella is high, but most of the lignin is found in the secondary wall (16, 17). We may thus ask whether the low or the high molecular weight lignins individually find their origins in the middle lamella or secondary wall regions. When wood is delignified, low molecular weight lignin is made soluble first. Therefore, it is tempting to assume that the low molecular weight lignins are produced predominantly in those tissue regions from which the lignin is first extracted (60, 199).

F. <u>Topochemistry of Delignification</u>. Few attempts have been made to discover the topochemical pattern of lignin removal from wood and the results obtained, for the most part, have been inconclusive. By digesting 20μ sections of wood, Bixler (200) concluded that most of the middle lamella was dissolved in kraft pulping before any major dissolution of lignin from the secondary wall; with sulphite pulping he found that lignin was simultaneously removed from the middle lamella and secondary wall regions. Lange (12) observed that the middle lamella of spruce was attacked in the early stages of sulphite pulping. Bucher (201) also observed dissolution of the middle lamella and noted that in sulphite pulp the concentration of residual lignin increased towards the lumen (202). Marth (203) concluded that in the sulphite pulping of aspen, sulphonation of lignin occurred initially in the middle lamella but that lignin was removed preferentially from the cell wall. From the study of the neutral sulphite pulping of eucalypt, Davies (204) suggested that the middle lamella was the site of the original attack by the delignifying chemicals. Jayme and Torgersen (205) have recently claimed that in the partial acid sulphite pulping of spruce, lignin was dissolved from the middle lamella whereas in the case of kraft pulping lignin removal occurred simultaneously in the cell wall and middle lamella.

Wardrop and Davies (206, 207), from their studies on the path of penetration of liquors into hardwoods, found that ray cells and vessels governed the initial flow, and that the pits afforded an intercell penetration path. Moreover, liquor penetration into the cell wall was by diffusion from the lumen such that the order of liquor contact was: secondary wall and then middle lamella. Wardrop (207) concluded that this scheme may be similar to the topochemical pattern of attack on the cell wall.

Recent studies of the topochemistry of delignification of spruce (208) and birch (209) have proved Wardrop's proposal to be substantially correct. Matchstick-size pieces of wood were delignified to various degrees in kraft and sulphite liquors. The specimens were washed, solvent exchanged, embedded in methacrylate, sectioned to 0.5μ and the sections photographed in a Leitz UV microscope (210). The photographs were analysed densitometrically and a quantitative comparison made of the concentration of lignin in the middle lamella with that of the secondary wall.

Shown in Figures 17.46 (208) and 17.47 (209) are series of UV photo-

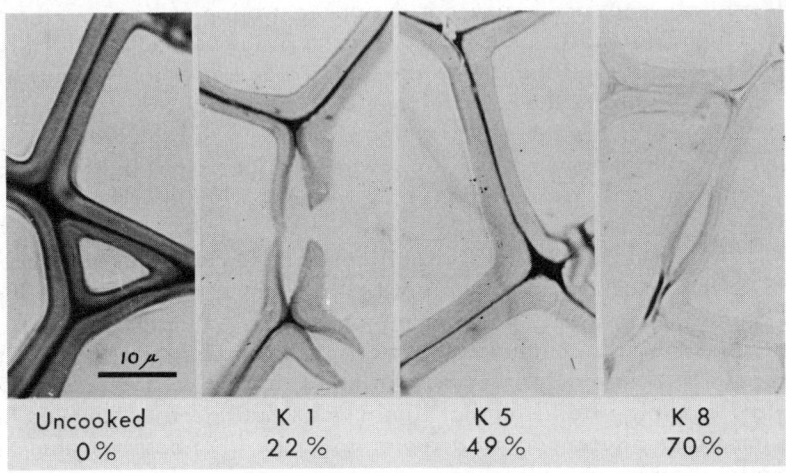

Figure 17.46. UV photomicrographs of fibre walls in spruce earlywood showing the progressive removal of lignin by kraft pulping. The figures give percentage delignification (208).

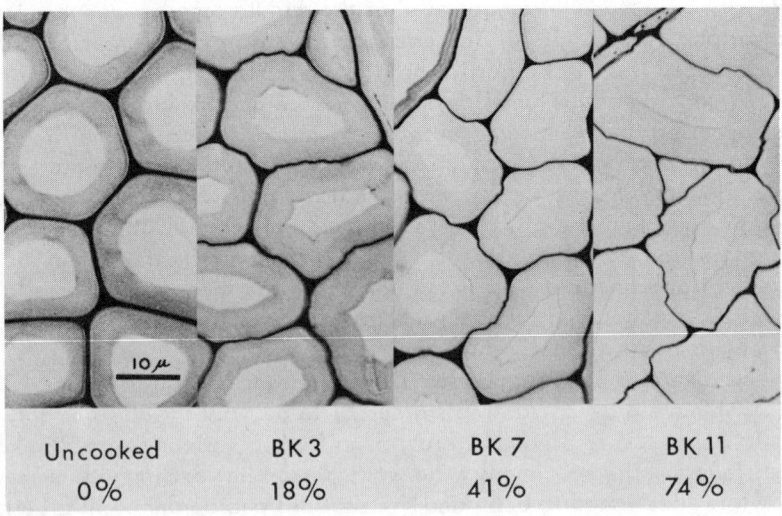

Figure 17.47. UV photomicrographs of fibres in birch earlywood showing the progressive removal of lignin by kraft pulping. The figures give percentage delignification (209).

micrographs taken at various stages of delignification of spruce and birch wood by kraft digestion. It is quite evident that up to about 50% lignin removal, the contrast between the middle lamella and the secondary wall increases. After this point the middle lamella begins to dissolve and in the case of spruce, disappears completely when about 70% of the lignin has been dissolved. It is interesting to note, also, that in the case of birch, Figure 17.47, the fibre walls appeared to swell and considerable distortion of the middle lamella was noted in the later stages of the cook. Such effects were not found in the case of spruce.

Relative lignin concentrations in the secondary wall and the middle lamella of birch fibres during kraft and neutral sulphite pulping are shown in Figure 17.48.

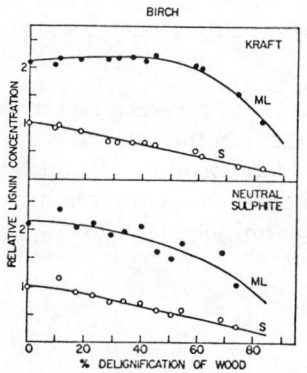

Figure 17.48. Decrease in the relative concentration of lignin in the middle lamella and secondary wall of birch fibres as a function of the lignin removed by kraft and neutral sulphite pulping. Lignin concentration in the uncooked secondary wall is set at unity (209).

The lignin concentration in the middle lamella, C_{ML}, is not much changed in kraft pulping until half way through the cook when C_{ML} begins to decrease rapidly. In contrast, the secondary wall loses lignin right from the start of the reaction. Kraft pulping of spruce yielded somewhat similar trends (208). A less marked effect was noted for the neutral sulphite treatment as shown in Figure 17.48.

If the results are expressed in terms of the percentage of lignin removed from a particular morphological region, we can derive the graphs shown in Figures 17.49 and 17.50. For birch, Figure 17.50, an allowance was made for the change in volume during pulping (209). This correction was not necessary in the case of spruce (208) for which the volume fraction of a given morphological region was assumed to remain constant during the treatment. In the case of birch, the trends shown for kraft and neutral sulphite pulping in Figure 17.50, were derived directly from changes in lignin concentration during delignification (209). When the work on spruce was done (208) topochemical changes during

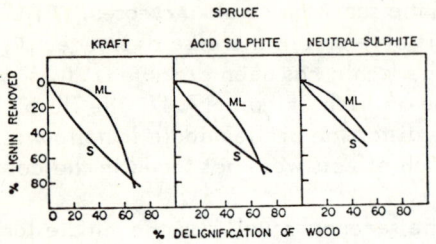

Figure 17.49. Percentage of lignin removed from the secondary wall, S, and middle lamella, ML, of spruce fibres during delignification by kraft, acid sulphite and neutral sulphite (208).

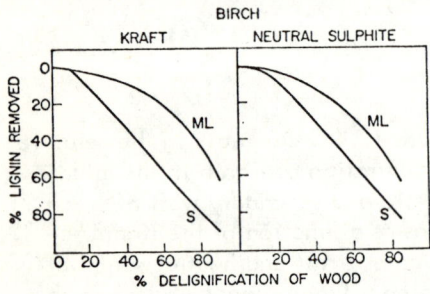

Figure 17.50. Percentage of lignin removed from the secondary wall, S, and middle lamella, ML, of birch fibres during delignification by kraft and neutral sulphite (209).

pulping were expressed in terms of the ratio of the concentration of lignin in the middle lamella to that in the secondary wall; i.e., C_{ML}/C_S. If it is assumed that spruce is composed entirely of the middle lamella and secondary wall of the tracheid and further that the specific volume of all tissue regions is equal and remains constant during delignification, it can be shown that Equation 16 applies

$$\begin{pmatrix} \text{\% of lignin re-} \\ \text{moved from the} \\ \text{secondary wall} \end{pmatrix} = 100 - \frac{V_S + V_{ML}(C_{ML}/C_S)}{V_S + V_{ML}(C_{ML}/C_S)_p} \begin{pmatrix} 100 - \text{\% delignifi-} \\ \text{cation of the} \\ \text{whole wood} \end{pmatrix} \quad (16)$$

in which V_S and V_{ML} are the invarient volume fractions of the secondary wall and middle lamella respectively and $(C_{ML}/C_S)_p$ is the concentration ratio at a particular stage of the treatment. Equation 16 and a similar one for the middle lamella were used to compute the trends for acid and neutral sulphite pulping of spruce wood given in Figure 17.49. The graph for the kraft experiment was derived from relative concentrations determined from the changes in UV absorbance in middle lamella and secondary wall as the treatment proceeded.

The results in Figures 17.49 and 17.50 indicate that in all the pulping processes examined for both spruce and birch wood, lignin is removed first from the secondary wall and later from the middle lamella region. The topochemical preference is greatest in the case of kraft and somewhat less for the sulphite processes. Although it is difficult to generalise about wood, the similarity of the patterns for birch and spruce suggests that preferential removal of lignin from the secondary wall is a characteristic feature of the early stages of most commercial pulping processes.

An immediate and important question is whether or not it is possible to delignify wood in such a way that the lignin in the middle lamella dissolves faster than that in the secondary wall. With such a process, fibres may separate easily at very high yields with little or no mechanical damage, and with substantial amounts of hemicellulose still conserved in the cell wall. Depending on the final product, these fibres could be used as such or could be delignified further with a carbohydrate-conserving process. Alternatively, one might devise a topochemical bleaching reaction whereby lignin is removed only from the outermost layers of the separated fibres, thus obtaining a surface-bleached fibre at high yields.

Returning now to the origin of the low molecular weight material, we can represent the delignification process diagrammatically as shown in Figure 17.51. In the initial stages of lignin removal, low molecular weight material is extracted mainly from the secondary wall. Later on in the cook, the size of the macromolecular hunks increases and lignin is removed from the middle lamella. At the end of the cook, the middle lamella has been almost completely consumed and the residual lignin is in the secondary wall.

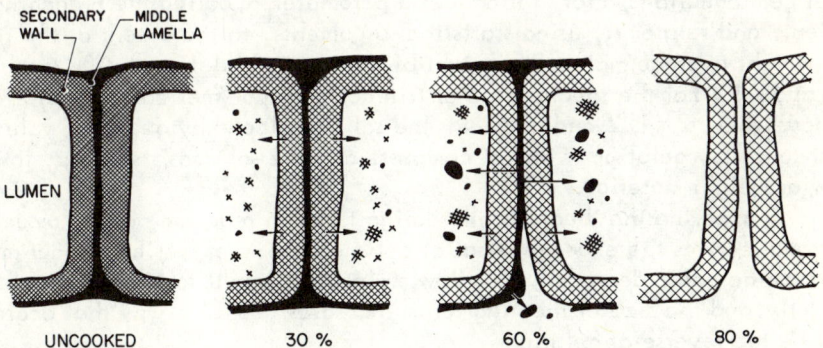

Figure 17.51. Diagrammatic representation of lignin dissolution from the double cell wall. The secondary wall and middle lamella gels are indicated by hashed and black areas respectively. For clarity of presentation, the size of the individual macromolecules diffusing into the lumen has been scaled up.

G. **Sol-Gel Concept of Lignin in Wood.** The above picture of delignification leads to the concept that the wood contains, in addition to an infinite gel fraction, a sol fraction which consists of lignin molecules of a finite low molecular weight. Such a composition is not unexpected if the lignin network is originally formed by the condensation of a reactive monomer. Flory (211) has pointed out that when a gel is formed by polyfunctional condensation there is always a sol fraction formed as well. As the reaction proceeds, the sol is depleted of higher molecular weight species which are selectively combined with the gel fraction. Thus wood could contain, in addition to a lignin network, a residual sol consisting of low molecular weight polymer similar chemically to the material comprising the infinite network. Perhaps, the sol fraction is concentrated in the secondary wall where the polymerization may remain incomplete because of the diluting influence of the polysaccharides. The crucial flaw in this simple argument is the inexplicable insolubility of the "sol" fraction in all neutral solvents.

H. **Gel Degradation Theory of Delignification.** An alternative concept, and one favoured by the present author, is that there can be several types of lignin gel in wood, each with its own pattern of degradation. For softwoods there is likely to be two main types of gel: middle lamella gel and secondary wall gel. (For another view on this, compare Chapter 4, Section II-I.) The secondary wall gel, which of course is rich in hemicellulose also, tends to degrade rapidly in the initial stages of pulping. The middle lamella gel is more resistant chemically to the pulping agent and thus is dissolved later in the cook.

In a recent paper (212), the degradation of the lignin gel is analysed quantitatively by means of Flory's treatment of polyfunctional polymerization (55). The Flory theory predicts, in accord with experiment, that there is always a substantial fraction of low molecular weight material both before and after gel formation. Flory introduces a parameter, α, called the branching coefficient, and relates it, using statistical arguments, to the extent of the reaction, the weight average molecular weight of the system and the relative amounts of gel and sol. For the special case of trifunctional polymerization, gelation commences when $\alpha = 1/2$ and as $\alpha \rightarrow 1$ the sol is replaced by gel. For values of α near to unity, gelation is almost complete and the sol consists of very low molecular weight material.

In applying the theory to the delignification reaction, it was assumed that degradation was the exact reversal of polymerization of a trifunctional monomer. Thus as the gel is degraded, very low molecular weight fragments are obtained initially and the size of the fragments increases in such a way that degradation will be the reverse of gelation.

In order to accommodate the topochemical aspects of the reaction, it was assumed that wood contained two types of lignin gel which degraded simultaneously. The patterns of degradation were described by the time-dependence of the α-values in the equations 17, 18 and 19

$$\frac{da_{ML}}{dt} = -k_{ML}a_{ML} \tag{17}$$

$$\frac{da_S}{dt} = -(k_S a_S + k' a_S^2) \tag{18}$$

$$\underline{\delta} = \frac{(da_S/dt)_{a=1}}{(da_{ML}/dt)_{a=1}} > 1 \tag{19}$$

in which t is time and the subscripts ML and S refer respectively to the middle lamella and secondary wall gels, and k_{ML}, k_S and k' are constants.

As shown later, the trends in a-values set by Equations 17, 18 and 19 were chosen to allow an approximate fit to the topochemical pattern of delignification. The topochemical pattern is set by the choice of the parameter $\underline{\delta}$. For $\delta = 1$, there is no topochemical preference and the middle lamella ans secondary wall gels degrade at equal rates. For $\underline{\delta}$ greater than unity, the initial degradation of the secondary wall gel exceeds that of the middle lamella gel as observed in practice.

The theory permitted the prediction of the weight average molecular weight of the lignin removed from spruce wood (39, 46) in the continuous flow experiments. As shown in Figure 17.52 reasonable agreement was achieved between theory and experiment, particularly when a two-gel calculation was applied with an appropriate value of $\underline{\delta}$. A comparison of the theory with the topochemical data for kraft and acid sulphite pulping of spruce is shown in Figure 17.53. The fit here is not exact but there is good agreement in the trends obtained with the theory correctly predicting a lower topochemical preference for the secondary wall in the case of sulphite pulping.

The treatment described above shows that it is not necessary to postulate that the low molecular weight lignin, produced in the initial stages of pulping, exists as molecules of finite size in the cell wall. Low molecular weight lignin will be produced by all lignin degradation processes simply by the statistical nature of the breakdown of a three-dimensional gel. Even if the middle lamella dissolved first, we would still expect to find low molecular weight lignin produced in the early stages of the delignification reaction.

Of course, even the multi-network theory of the lignin structure in wood is based on a very much oversimplified picture. Many puzzles remain. What is the difference between the lignin of the secondary wall and that of the middle lamella? Do they differ chemically? Are diffusion effects important in delig-

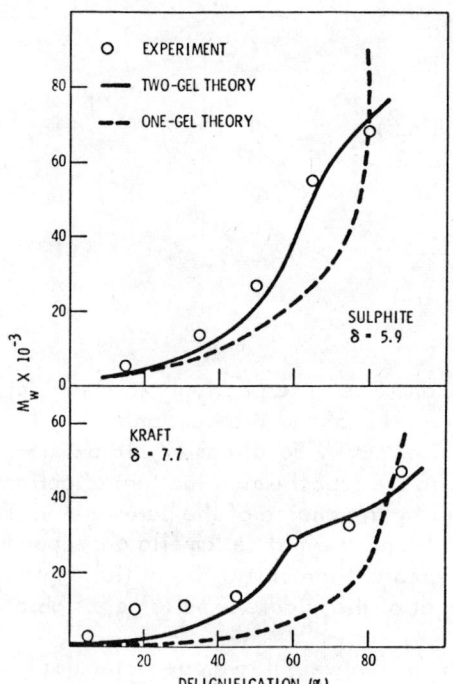

Figure 17.52. Comparison of experimental results with the theoretical predictions for the molecular weights of acid sulphite and kraft lignins made soluble at various stages in the delignification.

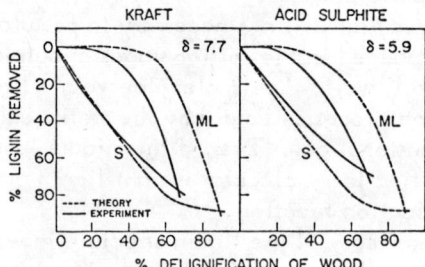

Figure 17.53. A comparison of the theoretically predicted and experimentally determined topochemical pattern of delignification for kraft and acid sulphite pulping of spruce wood.

nification? Why does neutral sulphite pulping show a smaller topochemical effect than kraft? What is the role of hemicelluloses? To answer these questions, knowledge of the detailed structural chemistry of lignin should be combined with knowledge of its polymeric properties. In this respect, it is interesting to report that Brenner and Bolker (213) are refining the gel degradation theory to take into account the rupture of specific chemical bonds known to exist in lignin. Their analysis supports the view that, rather than being formed from the condensation of a polyfunctional monomer, lignin is made up of short linear chains, about 15-20 units long, cross-linked to give a network.

I. <u>Concluding Remarks</u>. In conclusion, the author may be permitted a brief admonition to lignin chemists everywhere. Cease to look at lignin alone. Look at the whole wood. Advances in the elucidation of the lignin problem are now going to depend on deeper understanding of the intimate chemical and morphological relationship of lignin to the other important constituents of the xylem. And we can predict with certainty that such further understanding will lead to a wider and more efficient use of wood as a basic source of fibre and chemical.

REFERENCES

1. D. A. I. Goring and T. E. Timell, Tappi, $\underline{45}$, 454 (1962).
2. A. M. Holtzer, H. Benoit and P. Doty, J. Phys. Chem., $\underline{58}$, 624 (1954).
3. M. M. Huque, D. A. I. Goring and S. G. Mason, Can. J. Chem., $\underline{36}$, 952 (1958).
4. A. B. Wardrop and H. E. Dadswell, Holzforschung, $\underline{11}$, 33 (1957).
5. O. L. Forgacs, Pulp and Paper Mag. Can., $\underline{64}$, T-89 (1963).
6. R. St. J. Manley, Nature, $\underline{204}$, 1155 (1964).
7. R. St. J. Manley and S. Inoue, Polymer Letters, $\underline{3}$, 691 (1965).
8. T. E. Timell, "Wood and Bark Polysaccharides," in <u>Cellular Ultrastructure of Woody Plants</u> (W. A. Côté, Ed.), Syracuse University Press, 1965, pp. 127-156.
9. H. Meier, J. Polymer Sci., $\underline{51}$, 11 (1961).
10. R. H. Marchessault, <u>Chim. Biochim. Lignine, Cellulose et Hemicellulose</u>, Grenoble, 1964, pp. 287-301.
11. A. B. Wardrop and D. E. Bland, "The Process of Lignification in Woody Plants," in <u>Biochemistry of Wood</u>, (K. Kratzl and G. Billek, Eds.) Pergamon Press, London, 1959, pp. 92-116.
12. P. Lange, Svensk Papperstidn., $\underline{57}$, 525, 533 (1954).
13. I. B. Sach, I. R. Clark and J. C. Pew, J. Polymer Sci., $\underline{C2}$, 203 (1963).
14. W. A. Côté, T. E. Timell and R. A. Zabel, Holz Roh und Werkstoff, $\underline{24}$, 432 (1966).
15. G. P. Berlyn and R. E. Mark, Forest Prod. J., $\underline{16}$, 140 (1965).
16. B. J. Fergus, A. R. Procter, J. A. N. Scott and D. A. I. Goring, Wood Science and Technology, in press.

17. B. J. Fergus and D. A. I. Goring, Holzforschung, in press.
18. B. J. Fergus and D. A. I. Goring, ibid., in press.
19. A. Björkman, Svensk Papperstidn., 59, 477 (1956).
20. E. Adler, Ind. Eng. Chem., 49, 1377 (1957).
21. K. Freudenberg, Science, 148, 595 (1965).
22. F. E. Brauns, The Chemistry of Lignin, Academic Press, New York, 1952, pp. 189-197.
23. A. Rezanowich and D. A. I. Goring, J. Colloid Sci., 15, 452 (1960).
24. P. R. Gupta and D. A. I. Goring, Can. J. Chem., 38, 270 (1960).
25. V. F. Felicetta, D. Glennie and J. L. McCarthy, Tappi, 50, 170 (1967).
26. S. W. Schubert, M. G. Anders, C. Ludwig, D. Glennie and J. L. McCarthy, ibid., 50, 186 (1967).
27. S. W. Schubert and J. L. McCarthy, ibid., 50, 202 (1967).
28. J. R. Parrish, Tetrahedron Letters, 11/12, 1964, 555.
29. Q. P. Peniston and J. L. McCarthy, J. Am. Chem. Soc., 70, 1324 (1948).
30. J. L. Gardon and S. G. Mason, Can. J. Chem., 33, 1477 (1955).
31. E. Sjöström, P. Haglund and J. Janson, Svensk Papperstidn., 65, 855 (1962).
32. W. Q. Yean, A. Rezanowich and D. A. I. Goring, in Chim., Biochim., Lignine, Cellulose et Hemicellulose, Grenoble, 1964, p. 327.
33. J. Moacanin, V. F. Felicetta, W. Haller and J. L. McCarthy, J. Am. Chem. Soc., 77, 3470 (1955).
34. V. F. Felicetta, A. Ahola and J. L. McCarthy, ibid., 78, 1899 (1956).
35. P. J. Flory, Principles of Polymer Chemistry, Cornell University Press, New York, 1953, p. 321.
36. K. Schwabe and L. Hasner, Cellulose Chem., 20, 61 (1942).
37. A. Rezanowich, W. Q. Yean and D. A. I. Goring, Svensk Papperstidn., 66, 141 (1963).
38. J. Marton and T. Marton, Tappi, 47, 471 (1964).
39. J. G. McNaughton, W. Q. Yean and D. A. I. Goring, ibid., 50, 548 (1967).
40. J. J. Lindberg, H. Tylli and C. Majani, Paperi Puu, 46, 521 (1964).
41. J. J. Lindberg, K. Penttinen and C. Majani, Suomen Kemistilehti, B38, 95 (1965).
42. C. L. Hess, Tappi, 35, 312 (1952).
43. J. M. Pepper, P. E. T. Baylis and E. Adler, Can. J. Chem., 37, 1241 (1959).
44. R. L. Karpovskaya, V. L. Levdikova, N. M. Dorset and V. M. Reznikov Zh. Prikl. Khim., 6, 1318 (1964); Abstr. Bull. Inst. Paper Chem., 35, 7094 (1965).
45. A. E. Markham, Q. P. Peniston and J. L. McCarthy, J. Am. Chem. Soc 71, 3599 (1949).
46. W. Q. Yean and D. A. I. Goring, Pulp Paper Mag. Can., 65, T-127 (1964).

47. D. A. I. Goring and A. Rezanowich, Can. J. Chem., 36, 1653 (1958).
48. Y. Kyogoku, H. Ida and Y. Hachihama, J. Chem. Soc.Japan, 5, 922 (1961); Abstr. Bull. Inst. Paper Chem., 33, 1590 (1963).
49. K. Hata and K. Nakamura, J. Japan Tappi, 21, 37 (1967); Abstr. Bull. Inst. Paper Chem., 38, 4369 (1967).
50. J. M. Pepper and M. Siddiqueullah, Can. J. Chem., 39, 1454 (1961).
51. K. O. Pederson, Cold Spring Harbour Symposia Quant. Biol, 14, 140 (1950).
52. R. G. Le Bel, D. A. I. Goring and T. E. Timell, J. Polymer Sci., C2, 9 (1963).
53. S. R. Erlander and D. French, ibid., 32, 291 (1958).
54. W. A. J. Bryce, J. M. G. Cowie, C. T. Greenwood and I. G. Jones, J. Chem. Soc, 1958, 3558.
55. Reference 35, Chapter 9.
56. E. Nokihara, M. J. Tuttle, V. F. Felicetta and J. L. McCarthy, J. Am. Chem. Soc., 79, 4495 (1957).
57. Von B. Abrahamson, B. O. Lindgren and E. Hägglund, Svensk Papperstidn., 51, 471 (1948).
58. B. Leopold, Acta Chem. Scand., 6, 64 (1952).
59. W. Q. Yean and D. A. I. Goring, Svensk Papperstidn., 71, 739 (1968).
60. W. Brown, S. I. Falkehag and E. B. Cowling, Nature, 214, 410 (1967).
61. G. Meyerhoff and G. V. Schulz, Makromol. Chem., 7, 294 (1951).
62. A. F. Sirianni, D. J. Worsford and S. Bywater, Trans. Faraday Soc., 55, 2124 (1959).
63. R. G. LeBel and D. A. I. Goring, J. Polymer Sci., C2, 29 (1963).
64. S. Newman, L. Loeb and C. M. Conrad, ibid., 10, 463 (1953).
65. M. Rinaudo and F. Pla, Chim. Anal. (Paris), 49, 320 (1967).
66. Y. Kojima, A. Hayashi and I. Tachi, J. Japan.Tappi, 15, 713 (1961); Abstr. Bull. Inst. Paper Chem., 34, 818 (1964).
67. J. L. Gardon and S. G. Mason, Can. J. Chem., 33, 1491 (1955).
68. Reference 35, pp. 620-629.
69. D. Henley, Arkiv Kemi, 18, 327 (1961).
70. G. Sitaramaiah and D. A. I. Goring, J. Polymer Sci., 58, 1107 (1962).
71. C. Tanford, Physical Chemistry of Macromolecules, John Wiley, New York, 1951, p. 409.
72. A. Rezanowich, W. Q. Yean and D. A. I. Goring, J. Applied Polymer Sci., 81, 1801 (1964).
73. Y. Kojima, A. Nakai and I. Tachi, J. Japan.Tappi, 14, 665 (1960); Abstr. Bull. Inst. Paper Chem., 31, 985 (1961).
74. P. R. Gupta and D. A. I. Goring, Can. J. Chem., 38, 248 (1960).
75. F. Pla, Ph.D. Thesis, Univ. de Grenoble (1967); Abstr. Bull. Inst. Paper Chem., 39, 3784(D) (1968).
76. P. R. Gupta and J. L. McCarthy, Macromolecules, 1, 236 (1968).
77. J. J. Hermans, Ann. Rev. Phys. Chem., 8, 179 (1957).

78. D. A. I. Goring, T. Webb and A. Sehon, Svensk Papperstidn., 61, 1010 (1958).
79. R. E. Davis, E. T. Reaville, Q. P. Peniston and J. L. McCarthy, J.Am. Chem. Soc., 77, 2405 (1955).
80. W. J. Schubert, A. Passannante, G. de Stevens, M. Bier and F. F. Nord, ibid., 75, 1869 (1953).
81. B. O. Lindgren, Acta Chem. Scand., 12, 447 (1958).
82. A. Björkman, Svensk Papperstidn., 60, 243 (1957).
83. H. I. Bolker and P. Y. Wang, Tappi, in press.
84. F. Løsehbrandt and C. V. Wetlesen, Svensk Papperstidn., 61, 656(1958).
85. G. A. Dubey, T. R. McElhinney and A. J. Wiley, Tappi, 48, 95 (1965).
86. M. S. Mintz, R. E. Lacey and E. W. Lang, ibid., 50, 137 (1967).
87. W. Q. Yean and D. A. I. Goring, ibid., 47, 16 (1964).
88. T. Kobayashi and Y. Suhara, Repts. Govt. Chem. Ind. Res. Inst., Tokyo, 52, 255 (1957); Abstr. Bull. Inst. Paper Chem., 28, 524 (1957).
89. I. Petrariu, I. Scondac and M. Dima, Acad. Rep. Pop. Rom., Fil. Iasi Studii Cercet. Stiint, Chim., 13, 81 (1962); Abstr. Bull. Inst. Paper Chem., 33, 1638 (1963).
90. M. H. Khundhar and A. H. Mahmood, Pakistan J. Sci. Ind. Res., 5, 147 (1962); Abstr. Bull. Inst. Paper Chem., 33, 1209 (1963).
91. T. Enkvist, K.-E. Fremer, P. Lehtonen and V. Mölsä, Svensk Papperstidn., 61, 811 (1958).
92. H. B. Marshall and M. Krizsan, J. Am. Leather Chem. Assoc., L, 85 (1955).
93. W. M. Hearon, J. Chem. Ed., 35, 498 (1958).
94. P. R. Wiley, Tappi, 44, 22A (1961).
95. H. L. Hergert, L. E. Van Blaricom, J. C. Steinberg and K. R. Gray, Forest Prod. J., 15, 485 (1965).
96. J. Benko,"Chemistry in Canada," November(1965), p. 50.
97. J. M. Holderby, H. S. Olson and W. H. Wegener, Tappi, 50, 92A (1967).
98. A. Rezanowich, J. F. Jaworzyn and D. A. I. Goring, Pulp and Paper Mag. Can., 62, T-172 (1961).
99. A. Hayashi and D. A. I. Goring, ibid., 66, T154 (1965).
100. H. Ishikawa, Y. Kinoshita, T. Oki and K. Okubo, J. Japan Tappi, 22, 147 (1968); Abstr. Bull. Inst. Paper Chem., 39, 3780 (1968).
101. A. Kobayashi, T. Haga and K. Sato, J. Japan. Wood Res. Soc., 13, 246 (1967); Abstr. Bull. Inst. Paper Chem., 38, 8318 (1968).
102. A. Kobayashi, T. Haga, I. Yamakawa and K. Sato, J. Japan. Wood Res. Soc., 13, 118 (1967); Abstr. Bull. Inst. Paper Chem., 38, 8324 (1968).
103. J. Beeckmans, Can. J. Chem., 40, 265 (1962).
104. A. N. James and P. A. Tice, Tappi, 47, 43 (1964).
105. J. Böeseken, Advances in Carbohydrate Chemistry, 4, 189 (1949).

106. A. Hayashi, J. Japan. Wood Res. Soc., 11, 212, 253 (1965).
107. A. Hayashi and Y. Namura, ibid., 12, 44, 300 (1966
108. A. Hayashi, Y. Namura and T. Uekita, ibid., 13, 194 (1967).
109. S. K. Gross, K. V. Sarkanen and C. Schuerch, Anal. Chem., 30, 518 (1958).
110. L. H. Tung, J. Polymer Sci., 32, 477 (1958).
111. G. R. Quimby and O. Goldschmid, Tappi, 49, 562 (1966).
112. K. Forss, O. Schott and B. Stenlund, Paperi Puu, 49, 525 (1967).
113. N. Gralen, J. Colloid Sci., 1, 453 (1946).
114. P. R. Gupta, R. F. Robertson and D. A. I. Goring, Can. J. Chem., 38, 260 (1960).
115. O. M. Sokolov, B. D. Bogomolov and E. V. Veselova, Lesnoi Zh., 9, 139 (1966); Abstr. Bull. Inst. Paper Chem., 37, 1133 (1966).
116. P. R. Gupta and D. A. I. Goring, J. Chem. Phys., 32, 1890 (1960).
117. H. K. Schachman, Ultracentrifugation in Biochemistry, Academic Press, New York, 1959, p. 216.
118. K. E. Van Holde and R. L. Baldwin, J. Phys. Chem., 62, 734 (1958)
119. L. R. Lawson and J. B. Doughty, Chem. Eng. Data Series, 3, 128 (1958).
120. R. Z. Pen, Lesnoi Zh., 11, 131 (1968); Abstr. Bull. Inst. Paper Chem., 39, 2844 (1968).
121. V. F. Felicetta, A. E. Markham, Q. P. Peniston and J. L. McCarthy, J. Am. Chem. Soc., 71, 2879 (1949).
122. E. Back, V. F. Felicetta and J. L. McCarthy, Anal. Chem., 29, 1903 (1957).
123. J. Moacanin, H. Nelson, E. Back, V. F. Felicetta and J. L. McCarthy, J. Am. Chem. Soc., 81, 2054 (1959).
124. J. Benko, Tappi, 44, 766 (1961).
125. J. Benko, ibid., 47, 508 (1964).
126. A. E. Alexander and P. Johnson, Colloid Science, Oxford Press, 1949, p. 359.
127. P. V. Cheng and K. H. Schachman, J. Polymer Sci., 16, 19 (1955).
128. A. Polson, Biochem. et Biophys. Acta, 21, 185 (1956).
129. A. Polson, Nature, 187, 482 (1960).
130. J. T. Edward, Can. J. Chem., 35, 571 (1957).
131. J. T. Edward, J. Polymer Sci., 25, 483 (1957).
132. A. B. Ivarsson, Svensk Papperstidn., 54, 1 (1951).
133. K. Bárány, F. Guba and G. Tamásovits, Faserforsch. u. Textiltech, 8, 27 (1957); Abstr. Bull. Inst. Paper Chem., 27, 972 (1957).
134. L. Mandelkern and P. J. Flory, J. Chem. Phys., 20, 212 (1952).
135. J. C. Moore, J. Polymer Sci., A2, 835 (1964).
136. J. Porath and P. Flodin, Nature, 183, 1657 (1959).
137. T. C. Laurent and J. Killander, J. Chromatog., 14, 317 (1964).
138. A. N. James, E. Pickard and P. G. Shotton, ibid., 32, 64 (1968).

139. Z. Grubisic, P. Rempp and H. Benoit, Polymer Letters, 5, 753 (1967).
140. T. N. Soundararajan and M. Wayman, J. Polymer Sci., in press.
141. Th. N. Kleinert and K. V. Tayenthal, Angew. Chem., 44, 788 (1931).
142. S. Lagergren, S. Rydholm and L. Stockman, Svensk Papperstidn., 60, 632 (1957).
143. P. I. Baird and S. L. Schwartz, U. S. Forest Products Laboratory Report No. D1928 (1952).
144. A. J. Stamm, U. S. Forest Products Laboratory Report No. 2192 (1960).
145. H. G. Higgins, Australia C. S. I. R. O. Journal, 19, 455 (1946).
146. A. Björkman and B. Person, Svensk Papperstidn., 60, 285 (1957).
147. Th. N. Kleinert, Holzforschung, 18, 139 (1964).
148. F. Luhde, Pulp Paper Mag. Can., 61, T-544 (1960).
149. D. Atack and I. T. Pye, ibid., 65, T-363 (1964).
150. J. E. Stone and A. M. Scallan, ibid., 66, T-440 (1965).
151. D. A. I. Goring, ibid., 64, T-517 (1963).
152. D. A. I. Goring, in Consolidation of the Paper Web (F. Bolam, Ed.), B.P. and BMA, 1966.
153. T. G. Fox and P. J. Flory, J. Appl. Physics, 21, 581 (1950).
154. H. Fujita and A. Kishimoto, J. Polymer Sci., 28, 547 (1958).
155. N. Takamura, J. Japan. Wood Res. Soc., 14, 75 (1968); Abstr. Bull. Inst. Paper Chem., 39, 3893 (1968).
156. A. D. McLaren, T. T. Li, R. Rager and H. J. Mark, J. Polymer Sci., 7, 463 (1951).
157. P. R. Gupta, A. Rexanowich and D. A. I. Goring, Pulp and Paper Mag. Can., 63, T21 (1962).
158. F. Johanson and E. L. Back, Svensk Papperstidn., 69, 199 (1966).
159. W. Q. Yean, T. Higgs, G. Suranyi and D. A. I. Goring, Pulp and Paper Mag.Can., 67, T570 (1966).
160. L. P. Clermont and H. Schwartz, Forest Prod. Res. Soc. Wood Tech. Series, No. 16 (1948).
161. Fibreboard and Particle Board, Food and Agricultural Organization of the United Nations, Geneva (1957).
162. S. H. Baldwin and D. A. I. Goring, Svensk Papperstidn., 71, 646 (1968).
163. A. Björkman, Ph.D. Thesis, Chalmers University, Goteberg (1957).
164. F. Bueche, Physical Properties of Polymers, Interscience, New York, 1962, pp. 99, 100.
165. R. F. Boyer and R. S. Spencer, Advances in Colloid Science, II, 1 (1946).
166. M. V. Ramiah, Ph. D. Thesis, McGill University, 1966, and Reference 167.
167. M. V. Ramiah and D. A. I. Goring, J. Polymer Sci., C11, 27 (1965).
168. M. Wahba and K. Aziz, J. Text. Inst., 55, T291 (1962).
169. E. L. Back, M. T. Htun, M. Jackson and F. Johanson, Tappi, 50, 542 (1967).

170. J. Kubat and C. Pattyranie, Nature, 215, 390 (1967).
171. G. Némethy and H. A. Scheraga, J. Chem. Phys., 36, 3382 (1962).
172. D. A. I. Goring, Pulp and Paper Mag. Can., 67, T519 (1966).
173. J. Nakano and C. Schuerch, J. Am. Chem. Soc., 82, 1677 (1960).
174. C. Schuerch, ibid., 74, 5061 (1952).
175. J. J. Lindberg, Paperi Puu, 42, 193 (1960).
176. K. H. Ekman and J. J. Lindberg, Suomen Kemi, B39, 89 (1966).
177. J. J. Lindberg, ibid., B40, 225 (1967).
178. W. Brown, J. Appl. Polym. Sci., 11, 2381 (1967).
179. G. N. Christensen and K. E. Kelsey, Aust. J. Appl. Sci., 9, 265 (1958).
180. T. S. McKnight and S. G. Mason, Svensk Papperstidn., 61, 383 (1958).
181. G. Némethy and H. A. Scheraga, J. Am. Chem. Soc., 66, 1773 (1962).
182. Th. N. Kleinert and L. M. Marraccini, Tappi, 47, 605 (1964).
183. R. M. Screaton and S. G. Mason, Svensk Papperstidn., 60, 379 (1957).
184. H. W. Giertz, Svensk Papperstidn., 66, 691 (1963).
185. H. J. Kvisgaard, Norsk Skogind, 19, 155 (1965); Abstr. Bull. Inst. Paper Chem., 36, 357 (1965).
186. V. F. Felicetta and J. L. McCarthy, J. Am. Chem. Soc., 79, 4499 (1957).
187. M. J. Hatch, J. A. Dillon and H. B. Smith, Ind. Eng. Chem., 49, 1812 (1957).
188. D. A. I. Goring, Pure and Applied Chemistry, 5, 233 (1962).
189. J. C. Pew and P. Weyna, Tappi, 45, 247 (1962).
190. Reference 126, p. 670.
191. W. Q. Yean and D. A. I. Goring, to be published.
192. W. Jensen, B. C. Fogelberg, K. Forss, K.-E. Fremer and M. Johanson, Holzforschung, 20, 48 (1966).
193. K. H. Ekman and J. J. Lindberg, Paperi Puu, 48, 241 (1966).
194. W. Jensen, K.-E. Fremer and K. Forss, Tappi, 45, 122 (1962).
195. K. Forss and K.-E. Fremer, ibid., 47, 485 (1964).
196. K. Forss and K.-E. Fremer, Paperi Puu, 47, 443 (1965).
197. K. Forss, K.-E. Fremer and B. Stenlund, ibid., 48, 565 (1966).
198. W. Q. Yean and D. A. I. Goring, Svensk Papperstidn., 68, 787 (1965).
199. D. A. I. Goring, Chem. Soc. Special Publication No. 23, p. 115 (1968).
200. A. L. M. Bixler, Tech. Assoc. Pap., 21, 181 (1938).
201. H. Bucher, Das Papier, 14, 542 (1960).
202. H. Bucher, ibid., 22, 390 (1968).
203. D. E. Marth, Tappi, 42, 301 (1959).
204. G. W. Davies, Appita, 19, 95 (1966).
205. G. Jayme and H. F. Torgersen, Holzforschung, 21, 110 (1967).

206. A. B. Wardrop and G. W. Davies, ibid., 15, 129 (1961).
207. A. B. Wardrop, Svensk Papperstidn., 66, 231 (1963).
208. A. R. Procter, W. Q. Yean and D. A. I. Goring, Pulp and Paper Mag. Can., 68, T-445 (1967).
209. B. J. Fergus and D. A. I. Goring, ibid., in press.
210. J. A. N. Scott, A. R. Procter, B. J. Fergus and D. A. I. Goring, Wood Science and Technology, in press.
211. Reference 35, p. 393.
212. A. Szabo and D. A. I. Goring, Tappi, 51, 440 (1968).
213. H. M. Brenner and H. I. Bolker, to be published.

18

MICROBIOLOGICAL DEGRADATION AND THE FORMATION OF HUMUS

R. F. Christman and R. T. Oglesby

I. <u>Introduction</u>. Our understanding of the mechanisms by which lignin is biologically degraded has been significantly advanced as a result of biochemical research within the last decade. Progress in the study of the chemical structure of lignin has stimulated work in this area and has allowed the effective use of model compounds as a research tool. Thus, it is now possible to describe some important features of the metabolic pathways employed by microorganisms in the degradation of lignin. The picture is far from complete but several aspects are well grounded in fact and the trend of future developments can be assayed from them.

The nature of that ubiquitous substance known as humus is less well-defined than lignin. This fact when considered with the respective complex structures of these natural products makes studies of their interrelation in a natural cycle speculative. However, as in the case of lignin decomposition enough facts are known so that several pathways for the formation of humus from lignin and its degradation products can be logically postulated.

In this chapter attention will be focused on recent concepts of the microbial degradation of lignin and the role of this natural product in humus formation.

II. <u>Biological Agents</u>. The chief structural elements of plants—the lignins, celluloses and hemicelluloses—provide the fundamental sources of carbon and energy for the metabolic processes of a varied microbial flora. Gottlieb and Pelczar (1), Still (2), and Cook (3) have published extensive reviews detailing the organisms known to degrade lignin. Chief among these are the white rot fungi belonging to the family Basidiomycetes. Some of the more thoroughly studied species are listed in Tables 18.3 and 18.4, which appear later in the chapter. As a group, the fungi are obligately aerobic, non-chlorophyll bearing plants which are especially adept at utilizing complex organic molecules as food. Those found growing in the soil or on vegetation can withstand a wide variety of environmental conditions and are therefore ideal agents for biotransformations in nature.

Much remains to be learned regarding the enzymology of lignolytic fungi. The extracellular enzyme activity of these microorganisms probably predominates, and, as Schubert (4) has pointed out, this is to be expected considering the size of the lignin macromolecule.

Those specific enzymes most frequently connected with lignin degradation are of a phenoloxidase nature and include tyrosinase (phenolase) and laccase. In addition, a few lignin-degrading species have been reported which are capable of producing a peroxidase similar to horse radish peroxidase. Most white rot fungi contain laccase and phenolase whereas brown rot fungi apparently are devoid of laccase. The properties of these enzymes are compared in Table 18.1. For a more detailed discussion of lignin-degrading enzymes the reader is referred to Schubert (4).

Table 18.1. Properties of Tyrosinase and Laccase

Catalytic Property	Lignolytic Phenoloxidases	
	Tyrosinase (phenolase)	Laccase
o-hydroxylation of monohydric phenols	+	−
oxidation of o-dihydric phenols	+	+
oxidation of p-dihydric phenols	−	+
oxidation of polyhydric phenols	−	+
inhibition by carbon monoxide	+	−

Lyr (5) has proposed that these enzymes may also carry out a second function, that of detoxifying a number of materials in heartwood which would otherwise produce uncoupling of the vital process of oxidative phosphorylation. He has also suggested that peroxidases would be particularly appropriate enzymes for fungi living in the heartwood of trees because they do not compete with intracellular substrate oxidations for oxygen.

The fate of intact lignin molecules or their degradation fragments which reach the soil is a more involved topic. In addition to the white rots a number of common soil fungi belonging to the families Ascomycetes and Fungi imperfecti are known to attack lignin itself [e.g. Henderson (6)] and soil bacteria are undoubtedly active in metabolizing the fragments and probably lignin macromolecules as well [e.g. Higuchi et al. (7), Shibamoto et al. (8) and Sorensen (9)].

In view of current biological knowledge the relationship of lignin degradation to humus formation is a very tenuous one and neither the specific organisms involved nor their various roles in this process have been defined to any significant extent. As can be seen in Table 18.2 the biological organisms in soil are both abundant and varied (10). Biochemical activity is most closely related to surface area and hence bacteria, having by far the largest surface to volume ratio of all members of the soil community, might be expected to exert

Table 18.2. An Example of the Composition and Biomass of Soil Organisms (after Umbreit) (10).

Organisms	Number per gram of soil	Pounds per acre
Bacteria	1,000,000	2
Streptomyces	100,000	5
Protozoa	10,000	15
Algae	10,000	5
Molds	10,000	50
		77
Small worms	----	150
		227

a large proportion of the activity in the lignin to humus pathway.

Many workers have speculated on the probable significance of soil fungi on the formation of humus. Laboratory studies have indicated the probability that fungi are directly responsible for producing at least a portion of the soil humus from more highly oxidized organic precursors. For example, investigations by Konanova and Aleksandrova (11), discussed more fully in Section IVB, have indicated that at least some soil fungi are capable of converting simple carbohydrates into humic-like excretion products.

The possible role of the soil protozoans and metazoans, for example the exceedingly abundant nematode worms, cannot be entirely disregarded, particularly if the postulate is accepted that humus may originate as an excreted metabolic end product. Since nutritionally these groups of organisms subsist by feeding on lower forms of life, they probably exert no direct influence on the decomposition of lignin.

A generalized scheme illustrating the probable relationships of lignin and humus in the natural cycle of organic biosynthesis and decay is shown in Figure 18.1. Lignin and humus finding their way into anaerobic environments are protected from microbial attack and, along with other biosynthesized organics, are the probable precursors of coal.

III. Microbiological Degradation of Lignin.

A. Initial Microbiological Attack. Early investigations of the enzymatic degradation of native and milled wood lignin were generally oriented towards distinguishing the different effects of brown and white rot fungi on the wood substrate. Thus it was generally known from the works of Falck (12) and others that white rot fungi could decompose lignin and cellulose effectively whereas the attack of brown rot fungi was directed primarily at the cellulose content. Little was known, however, of the mechanism of the enzymatic reactions involved in the biological decay of lignin.

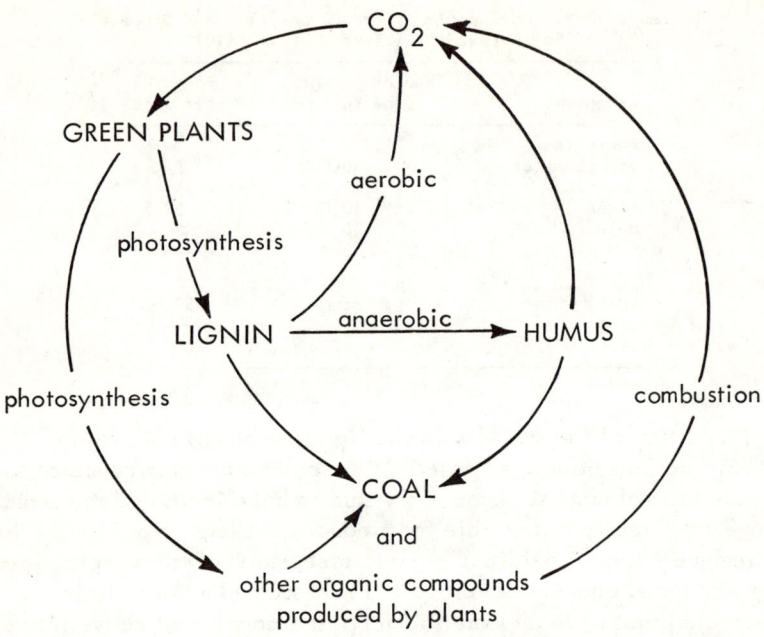

Figure 18.1. Lignin and humus in a probable natural cycle.

In 1955, Higuchi et al. (13) reported the detection of coniferaldehyde, vanillin and syringaldehyde in extracts of beech wood meal that had been decayed by several species of brown and white rot fungi. Their extensive analysis of the chemical properties of the decayed wood confirmed earlier views regarding the modes of action of the two types of fungi. Nitrobenzene oxidation of the white rot decayed wood resulted in significantly reduced yields of vanillin and syringaldehyde, although the syringaldehyde to vanillin molar ratio in the oxidation product mixture was higher than that obtained with fresh beech wood. These facts were interpreted by Higuchi to mean those units of the lignin macromolecule that will yield vanillin or syringaldehyde on subsequent oxidation are preferentially attacked during fungal degradation, and that guaiacyl units are degraded more readily than syringyl units. The identification of the three aromatic aldehydes was a significant step toward an elucidation of the pathway of lignin degradation.

Fukuzumi (14) performed in vivo experiments with Poria subacida (Peck) sacc. grown on Nord's native lignin extracted from a spruce wood that had been decayed by a brown rot fungus. As a result of the biological attack, the residue wood meal showed a higher ketonic carbonyl content as revealed by increased infrared absorption and the absence of a color reaction with phloroglucinol and hydrochloric acid. This worker identified 4-hydroxy-3-methoxyphenylpyruvic acid, 1, in the culture filtrates and an extraction of whole wood meal decayed by the same fungus revealed the presence of the β-O-4-dilignol, 2.

Ishikawa, Schubert and Nord (15) also identified this compound in the culture media of several polyphenoloxidase-poor fungal species that had been adapted to pine and spruce lignins as a sole carbon source. The growth rate of several polyphenoloxidase-rich species was more rapid on the same substrate than was that of other fungal types, and the authors suggest that this would account for the fact that fewer degradation products were isolated from cultures of these organisms and that no trace of the dilignol, 2, was found. The degradation products formed by the growth of these fungi are summarized in Table 18.3.

Table 18.3. Products formed from Pine and Spruce Lignins by White Rot Fungi (Ishikawa, Schubert and Nord) (15).

Fungus[1]	Media Lignin	Vanillic Acid	p-Hydroxy Benzoic Acid	Ferulic Acid	4-Hydroxy-3-methoxy phenylpyruvic acid	p-Hydroxy cinnamic acid	Guaiacyl glycerol	Vanillin	Dehydrodivanillin	Coniferaldehyde	p-Hydroxy cinnamaldehyde	Guaiacyl glycerol -β-O-4-coniferyl alcohol
A	Brauns lignin	+	+			+		+	+	+	+?	
A	Milled wood lignin	+	+			+		+	+?	+		
B	Brauns lignin	+	+	+	+	+	+	+	+	+	+	+
B	Milled wood lignin	+	+	+	+	+	+	+	+	+	+	+
		+				+						

1. A, Polyphenoloxidase rich: Polyporus hirsutus, Polyporus versicolor, Poria subacida J 247
   B, Polyphenoloxidase poor: Poria subacida N 199, Fomes fomentarius, Fomes annosus, Trametes pini

These workers observed an increase in ketonic carbonyl content of the residual lignin as did Fukuzumi and they also found higher contents of phenolic hydroxyl and carboxylic activity. The methoxyl content of the un-utilized portion decreased by as much as one-third of the fresh wood value in some cases and the yield of vanillin after nitrobenzene oxidation was correspondingly lower. The chemical characteristics of the residual lignin are of extreme importance in relation to humus formation and this point will be discussed in Section IV B.

In regard to the initial chemical reactions during lignin degradation, it is attractive to accept the postulates of Higuchi et al. (13) and Ishikawa and Oki (16) which state that those portions of the lignin macromolecule which contain the largest amount of guaiacyl type monomeric units are attacked first, resulting in the liberation of the β-O-4-dilignol $\underline{2}$. Schemes suggested by Fukuzumi (14) for spruce wood decay by Poria subacida (Peck) sacc. and by Ishikawa and Oki (16) for the general case of softwood decay by white rot fungi are shown in Equation 1. Indeed, the identification of dilignol $\underline{2}$ in culture filtrates of fungi grown on lignin substrates suggests that this compound is a primary degradation product of lignin exposed to microorganisms containing polyphenoloxidase enzyme systems and it offers further proof of the existence of β-O-4 linkages in the lignin macromolecule.

$$\text{(1)}$$

B. **Phenylpropanoid Intermediates.** The various phenylpropanoid compounds that have been identified in fungal cultures utilizing lignin as a sole carbon source are listed in Table 18.4. Presumably these compounds are produced from arylglycerol-β-ether units and are intermediate structures in its degradation to benzyl derivatives and ultimately to CO_2 or Krebs cycle compounds such as pyruvic acid. Ishikawa, Schubert and Nord (17) in an excellent series of experiments grew cultures of Polyporus versicolor and Fomes fomentarius on a variety of aromatic substrates and noted the rates of utilization and the nature of the principal degradation products. Both of these organisms were able to convert guaiacylglycerol-β-2-methoxyphenyl ether to guaiacylglycerol and guaiacol thus offering evidence of the susceptibility of the β-O-4 intermonomer ether linkage to fungal activity. It is interesting to note in this connection that Farmer, et al. (18) and Russel et al. (19) found that mats of Polystictus versicolor are able to metabolize the β-aryl ether of guaiacylglycerol only when the phenolic hydroxyl group is not etherified. These results may be related to a fundamental difference in the enzymology of Polystictus versicolor and Polyporu

Table 18.4. Phenylpropanoid Intermediates in the Degradation of Lignin by White Rot Fungi.

Compound Identified	Substrate	Fungi Capable of Producing Compound	Reference
(G)-CH=CH-CHO	beech, pine, spruce	A, B	(13, 15)
(G)-CH=CH-COOH	pine, spruce	C	(15)
(H)-CH=CH-CHO	pine, spruce	C	(15)
(H)-CH=CH-COOH	pine, spruce	B	(15)
(G)-CH$_2$COCOOH	pine, spruce	C, D	(14, 15)
(G)-CHOHCHOHCH$_2$OH	pine, spruce	C	(15)

Explanation of Fungal Groups:

A. <u>Coriolus hirsutus</u>, <u>Fomes fomentarius</u>, <u>Coriolus versicolor</u>, <u>Elfvingia applanata</u>, <u>Tramates saguinea</u>, <u>Fomitopsis pinicola</u>, <u>Poria vaporarid</u>, <u>Merulius lacrymans</u>

B. <u>Polyporus hirsutus</u>, <u>Polyporus versicolor</u>, <u>Poria subacida J247</u>, <u>Poria subacida N199</u>, <u>Fomes annosus</u>, <u>Trametes pini</u>, <u>Fomes fomentarius</u>

C. <u>Poria subacida N199</u>, <u>Fomes fomentarius</u>, <u>Fomes annosus</u>, <u>Trametes pini</u>

D. <u>Poria subacida</u> (Peck) sacc.

versicolor, however, and the existence of a free phenolic hydroxyl may not be a requirement for metabolism by most white rot fungi. Since much of the basic enzymology of the different genera of fungi used by the various investigators is obscure, it is obviously dangerous to generalize.

Referring again to the data of Ishikawa et al. (17), compounds with guaiacyl aromatic configurations were more rapidly metabolized by <u>Polystictus</u> and <u>Fomes</u> than were the p-hydroxyphenyl derivatives, and all phenylpropanoid compounds were utilized faster than phenylethanoid or benzyl derivatives. Some of the compounds studied and their principal degradation products are listed in Table 18.5. All of the aromatic molecules were ultimately converted to vanillic acid 3 through vanillin 4 by <u>Polystictus</u> and <u>Fomes</u>. Coniferaldehyde, 5, was oxidized to vanillin and vanillic acid via ferulic acid, 6;

Table 18.5. Degradation Products Formed from Aromatic Compounds by Fomes fomentarius and Polyporus versicolor

Substrate	Vanillin	Dehydrodivanillin	Vanillic Acid	Ferulic Acid	Coniferaldehyde	3-methoxy-4-hydroxy phenylpyruvic acid
Vanillin		+	+			
Vanillyl alcohol	+	+	+			
Acetovanillone	+	+	+			
Guaiacylmethyl carbinol	+	+	+			
Ferulic acid	+	+	+			
3-methoxy-4-hydroxy phenylpyruvic acid	+	+	+			
Coniferaldehyde	+	+	+	+		
Coniferyl alcohol	+	+	+	+	+	
Isoeugenol	+	+	+	+	+	
Guaiacylglycerol	+		+			+

coniferyl alcohol, 7, and isoeugenol were metabolized via coniferaldehyde. Guaiacylglycerol, 8, was the only compound converted to vanillin and vanillic acid through 3-methoxy-4-hydroxyphenylpyruvic acid, 1, and the latter compound was observed to exist in equilibrium with its enol tautomer 9. It is a logical assumption that guaiacylglycerol is converted to the guaiacylpyruvic acid tautomeric system through β-hydroxyconiferyl alcohol, 10, although the authors did not show the presence of this compound in the culture media. The scheme proposed by Ishikawa et al. for the degradation of the guaiacylglycerol-β-ether units in softwood lignin by white rot fungi is shown in Equation 2.

Fukuzumi (14), however, does not consider coniferaldehyde and vanillin as degradation products of lignin by Poria. This author has hypothesized the conversion of guaiacylglycerol-β-O-4-ether units to coniferyl alcohol and β-oxyconiferyl alcohol. The latter compound is presumed by Fukuzumi to be converted to methoxy homogentisic acid, 11, via guaiacylpyruvic acid, 1, Equation 3.

Although Fukuzumi identified 1 and 2 in extracts of decayed wood meal, it should be pointed out that these compounds are required by the scheme suggested by Ishikawa which is supported by more evidence.

C. Benzyl Derivatives. Although there is good evidence for the conversion of guaiacylglycerol-β-aryl ether units in softwood lignin to compounds such as vanillic acid, the metabolic pathways employed by the white rot fungi in utilizing these benzyl derivatives are almost entirely speculative. In order that microorganisms may utilize these products as carbon sources by whatever pathways are involved, cleavage of the aromatic ring is probably an important feature. Many soil organisms are capable of degrading aromatic substances although bacteria are able to utilize a wider variety of aromatics than yeasts and fungi.

It is of some value for speculative purposes to summarize the general nature of microbial attacks on aromatic nuclei. Excellent reviews in depth of this subject have been published recently by Evans (21) and Towers (22). In spite of the number of different microorganisms that utilize aromatic substrates, it is possible to make general statements regarding the mechanism. Ring cleavage occurs when the compound either contains or can be converted to an ortho or para dihydroxy phenol configuration. The principal products of cleavage are aliphatic acids which are fed into intermediate biochemical pathways which depend on the organism and its environmental conditions. Equation 4 contains an example of the variety of hydroxylation and decarboxylation reactions that organisms carry out on salicylic acid, 12, prior to ring cleavage in an effort to produce the preferred ortho or para dihydroxy configurations such as in gentisic acid, 13; o-pyrocatechuic acid, 14; and catechol, 15.

$$(4)$$

In addition, the enzyme systems of mammalian liver are capable of converting L-tyrosine, 16, to homogentisic acid, 17, via p-hydroxyphenylpyruvic acid, 18, and 2,5-dihydroxyphenylpyruvic acid, 19, as shown in Equation 5 (23).

$$(5)$$

According to Evans, two distinctly different cleavage mechanisms are known for the utilization of o-dihydroxyphenols by microorganisms. The ring may be opened either between or immediately adjacent to the hydroxyl groups. In the case of protocatechuic acid, 20, utilization by various species of Pseudomonas, both mechanisms have been reported, with rupture occurring either between carbons 4 and 5 or between carbons 3 and 4, as shown in Equation 6. p-Dihydroxyphenols are attacked at the bond between the carbon substituted by hydroxyl and the carbon substituted by a side-chain (Evans, Towers, Fukuzumi)

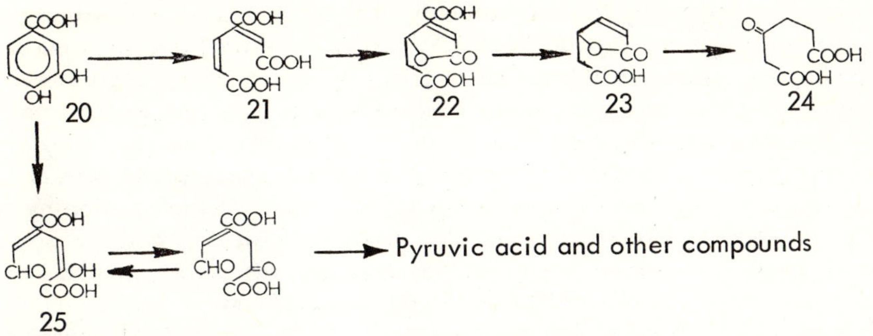

MICROBIOLOGICAL DEGRADATION

$$\text{(6)}$$

20 Protocatechuic acid
21 β-carboxymuconic acid
22 β-carboxymucloactone

23 Muconolactone
24 β-ketoadipic acid
25 α-hydroxy- -carboxymuconic semialdehyde

The cleavage of gentisic, 13, and homogentisic, 17, acids to fumarylpyruvic, 26, and fumarylacetoacetic, 27, acids by this mechanism is shown in Equations 7 and 8, respectively.

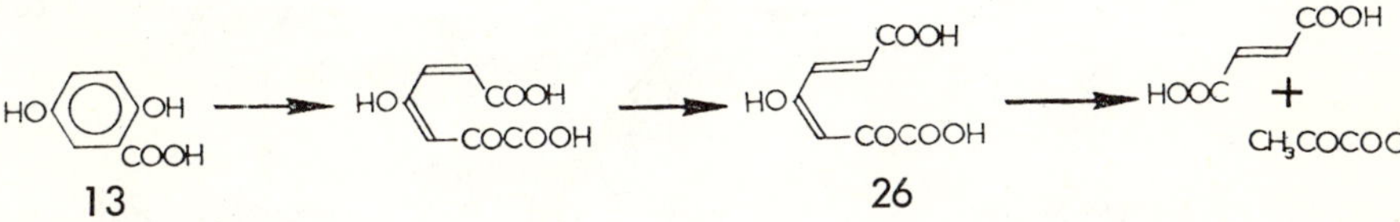

$$\text{(7)}$$

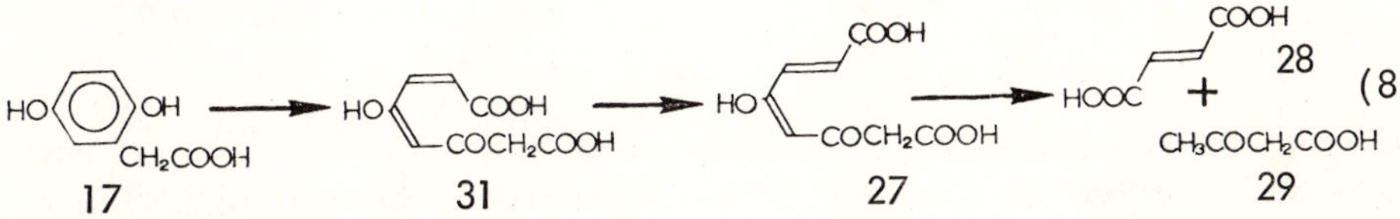

$$\text{(8)}$$

The enzymes responsible for ring cleavage are activated by ferrous ion and the specific effect produced may be substantially altered if this metal is not available. Thus Ichihara (24) found that in the presence of a metal chelate, homogentisic acid is not converted to fumaryl, 28, and acetoacetic, 29, acids by an enzyme solution obtained from rabbit liver, but instead is converted to gentisic acid according to Equation 9.

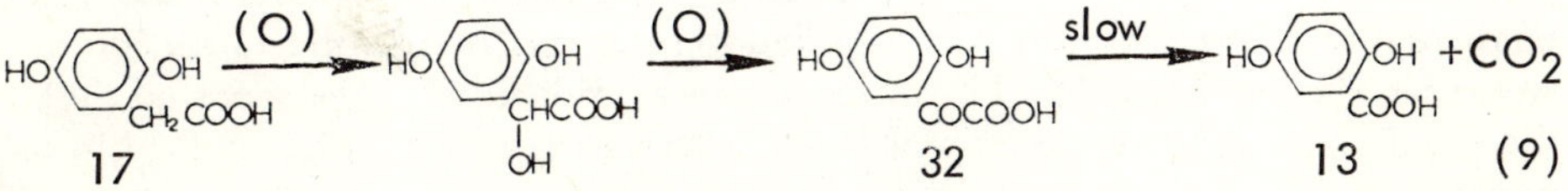

$$\text{(9)}$$

Information regarding the applicability of these relatively well-known pathways to the white rot degradation of the guaiacylglycerol-β-aryl ether units of softwood lignin is meager but considering the uniformity of the degradative mechanisms among the more extensively studied microbial systems it is attractive to assume that they are relevant. It is apparent, of course, from the low yields of degradation products obtained by Ishikawa and the oxygen uptake data obtained by Fukuzumi that the white rot fungi do not terminate their action by the conversion of substrate to guaiacylpyruvic or vanillic acids only and that some unknown reactions are carried out by these microorganisms beyond this point.

It is possible to visualize at least two pathways of benzyl derivative metabolism by white rot fungi that are consistent with the theories of aromatic ring metabolism, Equation 10.

$$ (10) $$

Although the proposed mechanisms differ substantially from each other it must be remembered that the data responsible for each were obtained using different genera of microorganisms. If, as Fukuzumi has suggested, vanillin (and presumably vanillic acid) are not important intermediates, the guaiacylpyruvic acid units must be converted directly to a preferred o- or p-dihydroxy configuration such as occurs in gentisic or homogentisic acids. As was mentioned earlier, such a pathway is known for the conversion of L-tyrosine to homogentisic acid, 17, via p-hydroxyphenylpyruvic acid, 18, Equation 5. There is only limited evidence that such a reaction may be employed by some genera of white rot fungi. Fukuzumi (25) followed the effect of enzyme extracts of Poria on guaiacylpyruvic acid, 1, and observed an oxygen uptake of 1.0 mole/mole substrate with concurrent evolution of carbon dioxide. This would of course be consistent with the reaction shown in Equation 3 resulting in the formation of methoxyhomogentisic acid, 11. In later work, Fukuzumi (26) reported that the reaction of enzyme extracts of Poria with homogentisic acid resulted in an oxygen uptake of approximately one-half of the theoretical requirement

(1.0 mole oxygen/mole substrate) to produce maleylacetoacetic acid, 31, Equation 8. In addition, a phenolic carboxyl compound was found in the reaction media. Since the same enzyme system was capable of reacting with gentisic acid with an oxygen uptake equal to that required to open the ring, Fukuzumi suggested that Poria converts homogentisic acid, 17, to gentisic acid, 13, as in Equation 9, prior to ring cleavage and that the step involving the oxidation of 2,5-dihydroxybenzoyl formic acid, 32, is extremely slow.

This interpretation although intriguing is presumptive since the reaction products in all of these studies were not actually identified and a number of different reactions would exert the same oxygen demand.

The metabolism of guaiacylpyruvic acid, 1, through vanillic acid, Equation 10, as proposed by Ishikawa et al. would require demethylation to convert vanillic acid to an o-dihydroxy phenolic configuration. This could be accomplished by conversion to protocatechuic acid, 20. However, Ishikawa et al. (17) could not detect protocatechuic acid in the culture media of Polyporus or Fomes growing on any of the model compounds listed in Table 18.5. It is possible that these organisms were successful in converting vanillic acid to protocatechuic acid and the latter compound was metabolized too rapidly to be detected. It is known that soil fungi are capable of demethylating and hydroxylating aromatic substrates prior to ring cleavage. Henderson (27) found that Haplographium sp., Hormodendrum sp. and Penicillium sp. converted ortho-, meta-, and para-methoxy benzoic acids to the corresponding monohydroxybenzoic acids and that Hormodendrum and Penicillium further metabolized p-hydroxybenzoic acid via protocatechuic acid 20.

A similar mechanism has been investigated by Sundman (28) for the metabolism of isovanillic acid produced by the degradation of α-conidendrin by lignanolytic agrobacteria. This worker proposed that isovanillic acid is converted to protocatechuic acid via m-hydroxybenzoic acid and although respirometric data did not support the proposed reaction pathway, isovanillic and protocatechuic acids were identified in the test solutions.

Most recently, Kirk, Harkin and Cowling (28a) have studied the degradation of guaiacylglycerol-β-o-methoxyphenyl ether 33 and syringylglycol-β-o-methoxyphenyl ether 34 by Polyporus versicolor and Stereum frustulatum. Interestingly, enzymatic hydrolysis of ether bonds appeared to be absent in both cases, and neither demethylation nor hydrolysis of the β-aryl ether bond could be detected. The main degradative action of both cultures was therefore ascribed to p-diphenyl oxidase which was isolated from cultures of P. versicolor. The identified products and proposed mechanisms are summarized in Equations 11 and 12. The two cultures as well as the isolated p-diphenol oxidase convert guaiacylglycerol-β-o-methoxyphenyl ether mainly to the 5-5 coupled dimer, 35, Equation 11. Syringylglycol-β-o-methoxyphenyl ether yields the corresponding α-carbonyl compound, 36, 2,6-dimethoxy-p-benzoquinone, 37, and 2-(o-methoxyphenoxy)ethanol, 38. The last compound was shown to form via reduction of o-methoxyphenoxyacetaldehyde, 39, Equation 12.

In conformity with the rules of phenoxyradical coupling (see Chapter 4, Section II), a dimer formed through 1-O-4 coupling was envisaged to act as a common intermediate for all isolated products.

The 4-methyl ether of guaiacylglycerol-β-o-methoxyphenyl ether alone was not affected by the two fungi. However, when either wood meal or milled wood lignin was added to the medium, it was converted to the corresponding α-carbonyl compound through the oxidative action of phenoxyradicals generated in lignin.

It is to be expected that a variety of pathways exist in nature for the metabolism of benzyl derivatives among the numerous microorganisms involved in lignin utilization. However, the precise pathway followed by any particular species of soil fungus in oxidizing the C_6C_1 products of lignin degradation has yet to be shown.

IV. Conversion to Humus. The principal objective in the following section is to relate recent theories of lignin degradation to those of humus formation. Waksman (29) and more recently Bremner (30) and Breger (31) have provided detailed reviews of work on many aspects of the classification and properties of humus. No attempt will be made here to review this field other than to introduce those concepts and definitions necessary to discuss the conversion of lignin to humus.

A. <u>The Nature of Humus</u>. The term humus does not refer to a specific chemical compound of fixed composition but rather to the collection of dark colored organic components of soil and other sedimentary accumulations (31). Early workers in soil science were quick to recognize the variable nature of humus and several classification systems have been advanced in attempts to define humus in terms of its physical environment (soil, composts, water) or manner of formation (aerobic or anaerobic). Humus, then, is a generic term and describes an accumulation of relatively refractory organic materials resulting from the microbiological decay of plant and animal tissue.

It should be mentioned that many organic substrates in addition to lignin are subject to humification and therefore any tendency to invariably connect the process of humus formation with that of lignin degradation may be misleading. Even in instances where lignin is a principal organic material being humified (soil humus), a knowledge of the chemical nature of humus leads one to believe that other chemical moities must be present to react with lignin degradation products before a true humus can be formed. The heterogeneous nature of this process and the understandable factual void in the literature permits only general conclusions to be drawn regarding the humification of lignin.

Some confusion has existed in the field of soil science regarding the nomenclature and isolation procedures for the various fractions of humus. The classification system shown in Table 18.6 is at present widely accepted and in the absence of detailed structural information serves as the only definition of these fractions. Attempts to determine significant chemical differences between the components of these fractions have been inconclusive as have been attempts to correlate soil type with a particular percentage distribution of total soil organic matter among the fractions. The four fractions are believed to occur in most if not all soils. The lack of standardized techniques for fractionating humus may in part account for the absence of significant data relating soil type and the percentage of material in each fraction.

Table 18.6. Classification of Soil Organic Matter

Fraction	Solubility in		
	Alkali	Mineral Acid	Ethanol*
Humin or Ulmin	-	-	
Fulvic Acid	+	+	
Hymatomelanic Acid	+	-	+
Humic Acid	+	-	-

*Solubility in ethanol is used only to distinguish between hymatomelamic and humic acids.

In nature, humus is found principally in the upper layers of soil. Some soils have a pronounced litter layer on the surface and possess a transition zone between mineral soil and litter that may be quite rich in humic material. According to Burges (32) as much as 75% of the organic substances in this zone may be humified. In other types of soils as much as 75-80% of the organic content may be isolated as humic acid. Of the four fractions of soil organic matter, humic acid has definitely been the subject of the most intensive investigations. It is generally regarded as the most complex fractions of the four, and it is often inferred that the other fractions are natural precursors of humic acid.

Humus performs several important functions in nature. First, the presence of humus influences the structure and texture of a soil increasing its aeration and moisture holding capacity. Second, humus exhibits a pronounced base exchange capacity and is able to store and gradually release nutrients to the surrounding environment. A principal nutrient provided by humus is carbon dioxide, which is present in the air adjacent to the ground in concentrations several times that found in ambient atmospheres. There is some evidence (11) that humus may participate in a more direct manner in plant nutrition by an actual involvement in plant metabolism and by stimulation of plant enzyme systems. Although the mechanism of this participation is not clear, it is probable that the functions of humus are not limited to providing and mobilizing simple nutrients. As was mentioned earlier, humic acid has received by far the most attention from recent investigators and a discussion of the chemical characteristics of humus must therefore be centered around this fraction.

Humic acid is regarded as a polyelectrolytic macromolecule with an average molecular weight of approximately 25,000 as estimated by Piret et al. (33) from sedimentation and viscosity measurements. Typical values for the elemental and functional group composition of humic acid are listed in Table 18.7. It should be noted that in contrast to lignin the most predominant functional group in humic acid is the carboxyl group and that this material has a significantly lower methoxyl content than lignin. However, humic acid, like lignin, may also be regarded as a polyphenol.

Table 18.7. Elemental and Functional Group Composition of a Typical Humic Acid

Element or Group	Percent	Milliequivalents per gram of dry ash-free solid
Carbon	52-59	
Hydrogen	3-5	
Nitrogen	0.5-5	
Total Acidity		11.5
Carboxyl		8.6
Total Hydroxyl		5.1
Phenolic Hydroxyl		2.9
Alcoholic Hydroxyl		2.2
Methoxyl		0.2
Carbonyl		5.5

Extensive degradation studies have been performed on humic acid although the results have not been as revealing as in the field of lignin chemistry. Benzene polycarboxylic acids and hydroxy acids such as protocatechuic acid were typical of the degradation products isolated in low yields by early investigators using alkaline fusion and other drastic oxidation procedures. Morrison (34) applied the alkaline nitrobenzene oxidation procedure to humic acid and isolated small quantities of vanillin, syringaldehyde and p-hydroxybenzaldehyde. The first evidence that humic acid contains aromatic configurations other than those derivable from lignin was obtained by Steelink and Green (35) using the alkaline CuO oxidation technique. These authors identified m-hydroxybenzoic acid and 3,5-dihydroxybenzoic acid in their oxidation mixtures in addition to the products found by Morrison. Presumably these structures are incorporated into the humic acid macromolecule by reaction with numerous polyphenolic plant constituents containing resorcinol hydroxylation patterns and not by a rearrangement of the aromatic configurations of lignin.

B. <u>Characteristics of Lignin Humification</u>. There is little doubt that lignin is more refractory than cellulosic or proteinaceous material. The actual percentage of lignin degraded in a given period of time is of course a function of the substrate, microorganism and the environmental conditions. However, Cook (3) has cited a number of general studies which indicate that in approximately six months' time 60-75% of the lignin content of natural substrates is destroyed by white rot fungi. These figures are consistent with the more detailed findings of Ichihara et al. (24). The overall conversion of plant carbonaceous material to carbon dioxide and humus is illustrated in Figure 18.2.

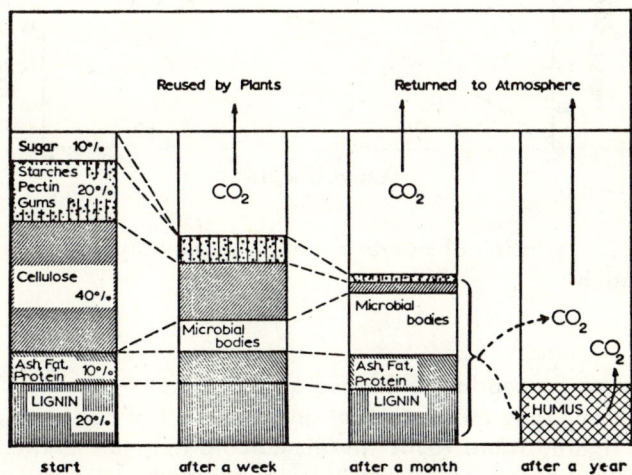

Figure 18.2. The progress of humus formation resulting from the microbial degradation of plants [after Umbreit (10)].

It must not be assumed that the only source of humus is the solid material left after fragmentation of the lignin macromolecule by microorganisms for their various metabolic purposes. From the present state of knowledge at least two other possibilities can be envisaged for the formation of humus from lignin. First, reactive phenylpropanoid or benzyl degradation products as described in Section III may undergo either direct or enzymatic extracellular polymerization with formation of new humus-like macromolecules. Second, humus may be formed as a direct metabolic by-product of lignolytic soil organisms involving intercellular resynthesis of reactive compounds followed by excretion and subsequent polymerization. It is likely that all of these processes are involved in natural humus formation but present knowledge does not allow an estimation of their relative contributions.

Chemical characteristics of the residue left after white rot fungal degradation differ substantially from natural lignin and appear to approach those of humus. Certainly if this material is to become humus it must be significantly decreased in methoxyl content and increased in nitrogen. Ichihara et al. (24) and Flaig (36) have observed such trends in long term lignin degradation studies with white rot fungi (Figure 18.3).

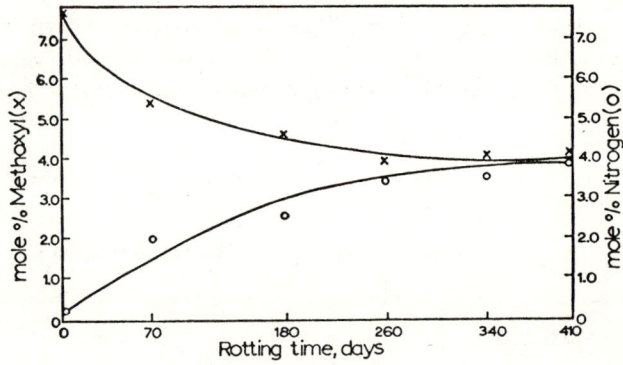

Figure 18.3. Variation of methoxyl and nitrogen contents of straw lignin during degradation.

A general conception of the overall process of lignin humification can be obtained by comparing the functional group content of humic acid with that of gymnosperm lignin, Table 18.8. In this calculation, the functional group data for spruce milled wood lignin (20) in moles per methoxyl group were used. The general process must be oxidative in nature resulting in an increased carboxyl acidity and a decreased alcoholic hydroxyl content, probably due to side-chain

Table 18.8. Comparative Functional Group Analyses of Humic Acid and a Hypothetical Lignin

Group	Group Content, millimoles/gram	
	Spruce milled wood lignin	A Humic Acid
Methoxyl	5.1	0.2
Total Hydroxyl	6.2	5.1
Phenolic Hydroxyl	1.6	2.9
Alcoholic Hydroxyl	4.6	2.2
Carbonyl	1.0	5.5
Carboxyl	Trace	8.6

oxidation. In this comparison it is not possible to account for the magnitude of the increase in carboxyl content solely by side-chain oxidation since such a reaction would result in a maximum value of 6.0 milliequivalents per gram; that is, one equivalent per phenylpropane unit. The additional acidity could be accounted for by ring cleavage of some of the aromatic molecules. Ring oxidation is also indicated by the relatively large carbonyl content of humic acid. The danger in such a hypothetical comparison is obvious; however, these calculations do reveal the general trend of the conversion.

This process must involve more of a chemical change in the residue than a simple removal of guaiacylglycerol-β-aryl ether units by the mechanisms discussed in Section III since such a removal, although resulting in a reduction of methoxyl groups, does not alter the nitrogen content. The process might be visualized as oxidative in nature to account for the increased carboxyl acidity. An oxidation could also produce reactive phenolic and quinonoid intermediates capable of polycondensation reactions with amino acids and with other polyphenols. A general scheme for this conversion is presented in Figure 18.4.

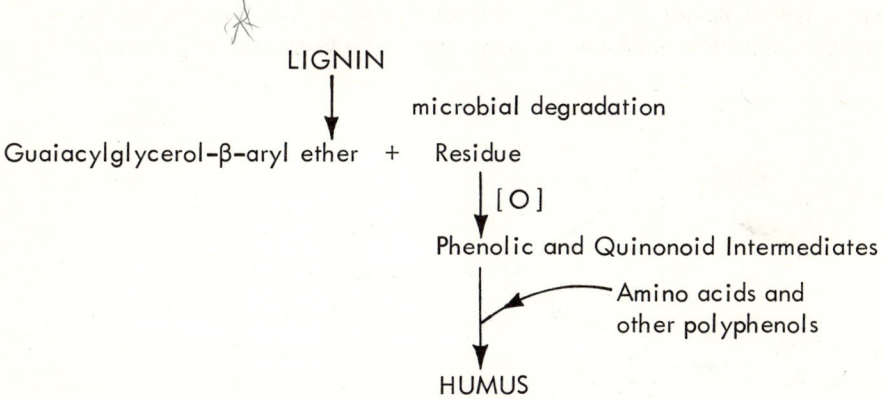

Figure 18.4. Conversion of residual decayed lignin to humus.

Polymerization of reactive intermediates produced by lignin degradation is essential to all of the possible mechanisms of humus formation mentioned above. Evidence exists that such reactions do occur even with the benzyl and phenylpropanoid degradation fragments. Ishikawa and Oki (16) isolated an enzyme from the mycelial pellets of Collybia velutipes N 8 which was catalyzed by H_2O_2. The peroxidase caused dehydrogenative polymerization of the compounds listed in Table 18.9 as well as oxidative cleavage of the alkyl side-chain.

Table 18.9. Effect of Fungal Peroxidase on Model Lignin Degradation Compounds

Model Compound	Principal Degradation Products (%)		
	Vanillin	Dehydrodivanillin	Dehydrodivanillyl-alcohol
Vanillin	--	66.5	
Vanillyl alcohol	47.4	27.0	8.0
Vanillylethyl ether	23.6	13.0	
3-methoxy-4-hydroxy-phenylpyruvic acid	54.2	23.5	
β-hydroxyconiferyl alcohol	29.3	21.1	

This ability to polymerize degradative intermediates is not restricted to C. velutipes since Ishikawa et al. (17) noted the formation of small amounts of dehydrodivanillin with other species of white rot fungi cultured on all of the compounds listed in Table 18.5.

The nature of the polymerization reactions that are presumed to occur during the formation of humus has been studied by Flaig (36). This worker found that in the pH 8-9 region, catechol, 15, and hydroquinone, 40, were non-enzymatically oxidized to humic acid type products, 41, Equation 13.

(13)

Table 18.10. Comparative Elemental and Functional Group Analyses for Natural, Synthetic and Postulated Humic Acids

Sample	Composition %						
	C	H	O	N	OCH_3	glycol	OCH_3 after CH_2 N_2
Humic acid from peat	58.35	4.97	32.2	2.62	1.55	15.6	18.9
Humic acid from p-benzoquinone (Flaig)	59.0	1.64	39.4	--	--	0	0
Humic acid from p-benzoquinone (Murphy and Moore)	59.1	3.7	37.2	--	--	13.3	19.7
Proposed humic acid from p-benzoquinone (Murphy and Moore)	60.7	3.52	35.8	--	--	13.05	39.2
Humic acid from 2-methoxy-1,4 benzoquinone (Murphy and Moore)	63.2	4.9	33.2	--	--	6.1	18.8

Murphy and Moore (37), however, prepared a model humic acid from p-benzoquinone and found Flaig's model formula to be inconsistent with the observed properties of their reaction product (Table 18.10), and proposed the structure 42 for p-benzoquinone humic acid:

42

The percentage composition required by this formulation is in fair agreement with that measured for the p-benzoquinone polymerization product and, with the exception of the methoxyl and nitrogen content, is close to that typical of natural humic acid.

These data suffice only to point out the general nature of the non-enzymatic polymerization of simple phenolic compounds. However, most of the postulated degradation intermediates of lignin are substituted phenols; and since the type of substitution may alter the nature of the reaction, studies on the polymerization of phenols with guaiacyl or syringyl substitution patterns may be interpreted with more confidence. In this regard Murphy and Moore (37) polymerized 4-methyl-1,2-benzoquinone, prepared by sodium iodate oxidation of 4-methylcatechol, at pH 8-9 for 24 hours. A similar polymerization product of 2-methoxy-1,4-benzoquinone had the properties listed in Table 18.10. An inspection of these data reveals that, in many aspects, this product is indeed similar to natural humic acid. The formula 43 was proposed by the authors for the synthetic humic product from 4-methyl-1,2-benzoquinone.

$$\left[\begin{array}{c} \text{Me} \quad \text{Me} \\ \bigcirc - \bigcirc - O \\ \text{OH} \quad O \quad \text{OH} \end{array} \right]_n$$

43

The spectral properties and polymerization reactions of quinones containing different substituents has been investigated by Flaig (36). He showed that 3-hydroxy-1,2-benzoquinones are important intermediates in the oxidative polymerization of lignin-like phenols. Flaig suggested that substituted catechols, as models of guaiacyl structures, and substituted pyrogallols, as models of syringl structures, are converted to these quinones prior to further oxidation and dimerization. It was also shown that these quinones can undergo a variety of additional oxidative polymerizations resulting in a number of complex products. These findings are extremely important and may represent the only available information regarding the actual polymerization pathways involved in humus formation.

The work described above indicates that the fraction of the vanillic acid-like degradation products (Section III) not metabolized by microorganisms are oxidized to a series of quinonoid intermediates probably of the 3-hydroxy-1,2-benzoquinone type prior to humus formation. Demethoxylation reactions must precede any polymerization to account for the low methoxyl content of humic acid. It will be recalled that such reactions are theoretically required for metabolic utilization involving ring fissure. It is conceivable, then, that the benzyl derivatives are demethoxylated prior to either metabolic utilization reactions or to polymerization reactions leading to the formation of humus as shown in Figure 18.5.

It should be pointed out that the foregoing discussion has been concerned exclusively with the concept of lignin humification as an extracellular phenomenon. In contrast to this, evidence exists that certain microorganisms are apparently capable of synthesizing humus precursors within the cell from simple carbohydrate substrates. Thus, Kononova and Aleksandrova (11) cultured Asperigillus niger and Penicillium spp. with glucose as the sole carbon source, and isolated from the culture media a brown organic residue which closely resembled natural soil humic acid, Table 18.11. It is extremely interesting to note that these microbial systems were able to convert inorganic nitrogen available in the nutrient medium to organic nitrogen in the humic residue. Other

MICROBIOLOGICAL DEGRADATION

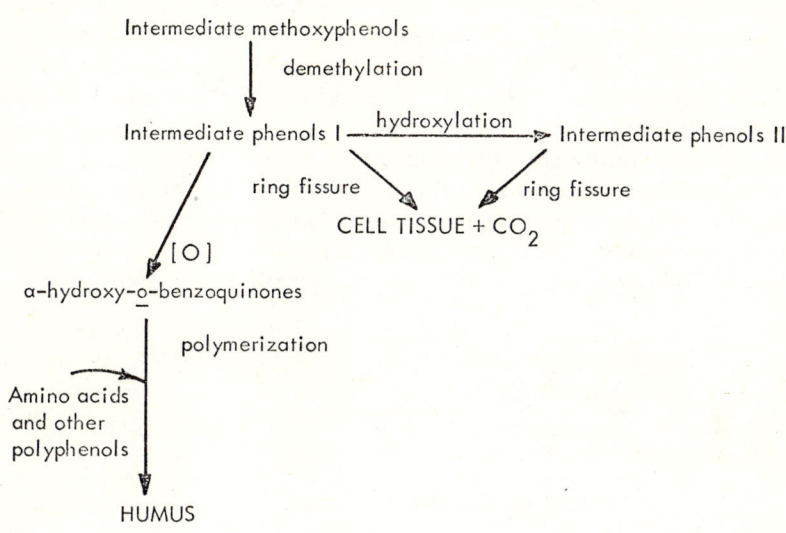

Figure 18.5. Conversion of lignin degradation products to humus or cell tissue.

workers, for example Emanuel (38) and Craigie and McLachlan (39), have similarly noted the production of humic substances by microorganisms growing on simple carbon sources.

These results indicate that humus may be formed as a metabolic by-product of lignolytic soil microorganisms. Considering the size of the humic acid molecule it is unlikely that anything but precursors, probably phenolic or quinonoid compounds, are formed inside the cell wall. Following excretion of these reactive intermediates similar oxidative polymerization reactions of the type described above could result in the formation of the humic acid macromolecule.

Table 18.11. Properties of Natural and Fungal Produced Humic Substances

Humic Substances	% dry-ash free				C/H	pH	Exchange Capacity meg/100g at pH 7
	C	H	O	N			
From Aspergillus cultures	51.0	5.6	40.6	2.8	9.1	3.42	312
From Penicillium cultures	45.2	6.1	45.9	2.8	7.4	3.26	261
Humic Acids from strongly podzolized soil	56.5	3.8	33.8	5.9	14.9	3.3	324

The interesting problem concerning the incorporation of nitrogen into the humic acid polymer is at present unresolved. Certainly a percentage of the nitrogen exists in the form of acid hydrolyzable amino acids, as Kononova and Aleksandrova (11) have identified 16 amino acids in hydrolyzates of humic acids, and Bremner (40) has shown the presence of amino sugars in similar hydrolyzates. In some cases, however, as much as 75% of the nitrogen content exists in an unhydrolyzable state presumably bound in a heterocyclic configuration. It would be difficult to regard this substance as nitrogen-free if such configurations do exist. Reactions have been proposed by Flaig, Equation 14, and by Murphy and Moore, Equation 15, which result in the inclusion of nitrogen in a heterocyclic configuration and in secondary amino linkages.

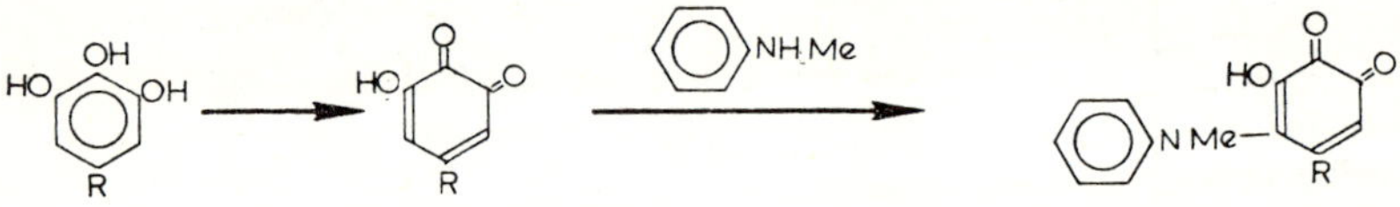

(14)

R = H or COOH

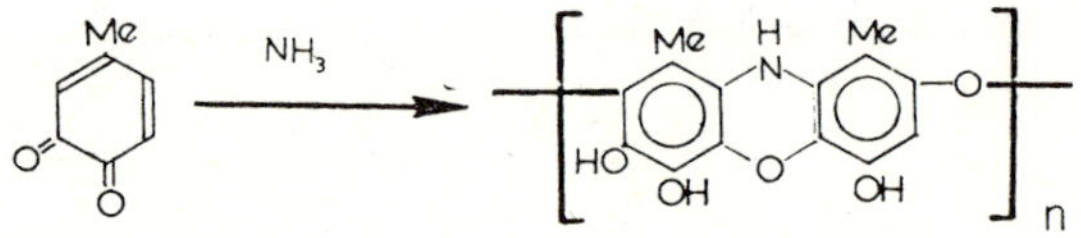

(15)

There are workers, for example Burges (32), who regard humic acid as essentially a nitrogen-free substance and explain the presence of nitrogen in preparations as being due to a mixture with soil protein.

More recently, the excellent work of Martin (41) and Martin and Haider (42, 43) concerned with the synthesis of humic substances by the soil fungi Epicoccum nigrum and Stachybotrys atra has appeared in the literature. During the initial growth phases of these fungi in glucose-asparagine media, these workers observed the appearance of various phenols concurrent with a decrease in culture pH and utilization of glucose and asparagine. Both fungi formed orsellinic acid, 44, which was subsequently converted into various phenols and phenolic acids.

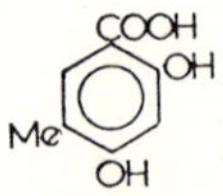

44

In all, twenty-three phenolic constituents of the culture media of E. nigrum were identified by thin-layer chromatography. After two to three weeks of growth, ammonia, peptides and amino acids began to accumulate in the culture solution as the pH increased. The appearance of humic substances in the culture was noted after four to six weeks of growth and this was accompanied by a loss of phenols and an increase in the color of the media.

The nitrogen content of the E. nigrum humic acids obtained by these workers ranged from 6–8.5%, increasing with increasing pH at the time of harvest. Respirometric studies with and without the addition of phenoloxidases conducted in mixtures of culture phenols and added amino acids showed the most readily autoxidizable phenols to be 2,4,5- and 2,3,5-trihydroxytoluene. Mixing these phenols even under aseptic conditions with amino acids, peptides and other phenols resulted in the formation of an acid insoluble-base soluble polymer containing 6–7% nitrogen.

Hydrochloric acid hydrolysis of the humic acids formed by these fungi liberated 40–50% of nitrogen mostly as alpha amino nitrogen. Amino acids identified were the same as those that accumulated in the culture during earlier growth stages. Sodium amalgam reduction of the humic acids following HCl hydrolysis resulted in relatively large yields of ether-soluble degradation products (15–40%). As with the amino acids, many of the phenols detected during earlier growth stages were observable in the degradation mixture, although there was some indication of methyl group oxidation after incorporation into the polymer.

An extremely interesting observation offered by Martin (41) is that polymer formation by soil fungi need not be limited to phenols which a species can synthesize. Lignin-like compounds ferulic acid and p-hydroxycinnamic acid when added, in carbon-labelled forms, to growing E. nigrum cultures were incorporated into the humic polymer after oxidation and demethylation. In these cases, ^{14}C in side-chain and methoxyl groups was recovered as CO_2, but ring carbon-14 was converted into the polymer.

REFERENCES

1. S. Gottlieb and M. J. Pelczar, Jr.., Bacteriol. Revs., 15, 55 (1951).
2. L. R. Lawson and C. N. Still, The Biological Decomposition of Lignin—Literature Survey, Information Services Center, West Virginia Pulp and Paper Company (Westvaco), Charleston, S. C., 1956, 79 pp., 272 ref., Published for private use.
3. W. B. Cook, Tappi, 40, 301 (1957).
4. W. J. Schubert, Lignin Biochemistry, Academic Press, New York, 1965, pp. 77-119.
5. H. Lyr, Nature, 195, 289 (1962).
6. M. E. K. Henderson, J. Gen. Microbiol., 26, 149 (1961).

7. T. Higuchi, I. Kawamura and I. Hayashi, J. Japan.Wood Res. Soc. (Mokuzai Gakkaishi), 2, 31 (1956). English abstract.
8. T. Shibamoto, T. Fukuzumi, H. Mikawa and M. Hayashi, ibid., 6, 135 (1960). English abstract.
9. H. Sorensen, J. Gen. Microbiol., 27, 21 (1962).
10. W. W. Umbreit, Modern Microbiology, W. H. Freeman, San Francisco-London, 1962, p. 331.
11. M. M. Kononova and I. V. Aleksandrova, Soils and Fertilizers, 22, 77 (1959). Translated from Russian from Iz. Akad. Nauk SSSR Seriya biol. No. 1, 79 (1958).
12. R. Falck, Cellulosechemie, 11, 198 (1930).
13. T. Higuchi, I. Kawamura and H. Kawamura, J. Japan. Forestry Soc., 37, 298 (1955).
14. T. Fukuzumi, Bull. Agr. Chem. Soc. Japan, 24, 728 (1960). In English.
15. H. Ishikawa, W. J. Schubert and F. F. Nord, Arch. Biochem. Biophys., 100, 131 (1963).
16. H. Ishikawa and T. Oki, J. Japan.Wood Res. Soc. (Mokuzai Gakkaishi), 10, 207 (1964).
17. H. Ishikawa, W. J. Schubert and F. F. Nord, Arch. Biochem. Biophys., 100, 140 (1963).
18. V. C. Farmer, M. E. K. Henderson and J. D. Russel, Biochem. J., 74, 257 (1960).
19. J. D. Russel, M. E. K. Henderson and V. C. Farmer, Biochem. Biophys. Acta, 52, 565 (1961).
20. K. V. Sarkanen, in The Chemistry of Wood (B. L. Browning, Ed.), John Wiley, New York-London, 1963, Chap. 6.
21. W. C. Evans, J. Gen. Microbiol., 32, 177 (1963).
22. G. H. N. Towers, in Biochemistry of Phenolic Compounds (J. B. Harborne, Ed.), Academic Press, London-New York, 1964, pp. 272-288.
23. J. S. Fruton and S. Simmonds, General Biochemistry, 2nd ed., Wiley, New York, 1958, p. 826.
24. K. Ichihara, S. Ikeda and Y. Sakamoto, J. Biochem., 43, 129 (1956).
25. T. Fukuzumi, J. Japan.Wood Res. Soc. (Mokuzai Gakkaishi), 5, 232 (1959).
26. T. Fukuzumi, Agr. Biol. Chem., 26, 447 (1962).
27. M. E. K. Henderson, J. Gen. Microbiol., 16, 686 (1957).
28. V. Sundman, ibid., 36, 171 (1964).
28a. T. K. Kirk, J. M. Harkin and E. B. Cowling, Biochim. Biophys. Acta, 165, 134, 145 (1968).
29. S. A. Waksman, Humus, 2nd ed., Williams and Wilkins, Baltimore, 1938
30. J. M. Bremner, J. Soil Sci., 2, 67 (1951).
31. I. A. Breger, International Series of Monographs on Earth Sciences, 16, 50 (1963).
32. A. Burges, Scientific Proc. Royal Dublin Soc., Ser.A, Vol.1, p.53 (1960).

33. E. L. Piret, R. G. White, H. C. Walther, Jr., and A. J. Madden, Jr., ibid. p.69.
34. R. I. Morrison, J. Soil Sci., 9, 130 (1958).
35. C. Steelink and G. Green, J. Org. Chem., 27, 170 (1962).
36. W. Flaig, Scientific Proc. Royal Dublin Soc., Ser. A, Vol.1, p.149 (1960).
37. D. Murphy and A. W. Moore, ibid., p. 191.
38. C. F. Emanuel, J. Water Pollution Control Fed., 36, 1229 (1964).
39. J. S. Craigie and J. McLachlan, Can. J. Bot., 42, 23 (1964).
40. J. M. Bremner, J. Agric. Sci., 48, 352 (1956).
41. J. P. Martin, 15th Annual Faculty Research Lecture, University of California, Riverside, April 12, 1967.
42. J. P. Martin, S. J. Richards and K. Kaider, Soil Sci. Soc. Amer. Proc., 31, 657 (1967).
43. K. Haider and J. P. Martin, ibid., 31, 766 (1967).

19

LOW MOLECULAR WEIGHT CHEMICALS

D. W. Goheen

I. Introduction. The discovery, in the latter part of the nineteenth century (1-3), that wood could serve as the raw material for preparation of cellulosic fibers resulted in the formation of the modern pulp and paper industry. This industry is, in a very real sense, a growth industry, and in the United States has operated at over 90% capacity for the past several years. Production is currently increasing at a rate of over 5% a year, nearly three times faster than the population growth. Predictions for the future reflect confidence of continued growth (4). However, one of the toughest problems that confronts the industry is that it still effectively utilizes only about one-half of its raw material.

The wood that is processed contains two of the most abundant naturally occurring organic polymers. Almost full use is made of the most abundant, cellulose, but the story is considerably different for the second polymer, lignin. In plants making pulp by chemical means, the only use for lignin until recent times was to burn it. Otherwise, the liquor has been disposed of as an aqueous solution into streams and waterways. Public resistance to this disposal has increased over the years, and the disposal problem is more acute than ever, particularly for sulfite process mills.

Although kraft process mills do not have the serious disposal problem associated with large scale dumping of spent liquors into waterways, their burning of lignin gives, at best, only a marginal return. The very nature of the pulping operations results in the lignin always being obtained in the form of dilute aqueous solutions. Although the lignins have relatively high calorific values (black liquor solids have a heat content of at least 6000 B.T.U. per pound despite their content of more than 40% inorganic compounds), the heat produced by the burning is only slightly greater than that necessary to evaporate the water from the spent liquors.

Attempts to create an industry based on spent liquors have been only partially successful. Basically, it is not that chemicals are hard to make from lignin, but that it is hard for them to meet the economic competition of the same chemicals produced from other sources. Energy requirements are the key to the problem. During the long ages of the earth's geologic history, natural forces and energies of the earth's crust acted upon deposits of plant remains to convert much of their organic matter into simple molecules found in natural gas and petroleum. To accomplish the same result from lignins, much energy must be used—an expensive step. Chemists and chemical engineers appear to be better able to build up from relatively small chemicals such as are present in petroleum and natural gas than to achieve a controlled breakdown of large polymeric naturally occurring materials such as lignin. Therefore, the petrochemical industry has a built-in advantage over chemicals from lignin. It could be argued that the most profitable way to utilize lignin is to make high yield paper and paperboard out of wood by not removing much of lignin during pulping. It would appear that this is a pessimistic analysis of the potential for chemicals from lignin. It is only pessimistic, however, in comparison with excessive claims which were once made about industries which would arise, based on lignin, in "the coming age of wood." While it is true that a number of commonly produced petrochemicals each produce more revenue than all of the products made from lignin, it is equally true that there have been some substantial commercial successes in the lignin utilization field. With the emphasis on research and development that is being placed by a number of pulp and paper companies, it is extremely likely that more successes will appear in the future.

II. Vanillin and Related Products.

A. Early Studies. A tropical plant belonging to the Orchidaceae family, Vanilla planifolia, was initially used as the sole source of the well-known vanilla extract, which in addition to vanillin contains other odorous principles. Vanillin and its glycosides are actually very widespread in nature, occurring, for example, in materials as different as potato peelings and maple sugar syrup.

Vanillin is the most widely used flavoring agent, due in part to its own very pleasant aroma and in part to its property of enhancing other flavors. In perfumes, it is used to impart sweetness to nearly any kind of odor while its own aroma remains masked. Vanillin is considered as a "safe" flavoring under the U. S. Food, Drug and Cosmetic Act (5), and in ice creams represents the most popular flavor by a wide margin.

B. Vanillin Syntheses. Common syntheses of vanillin are summarized in Figure 19.1. Most of the methods use guaiacol as starting material. The oldest procedure is based on the well-known Reimer-Tiemann reaction (7-9). o-Vanillin is formed as a side-product, as in many other synthetic methods based on guaiacol. Early speculations as to the mechanism for this reaction (10) have been replaced by the concept that $:CCl_2$ acts as the intermediate, method 1, Figure 19.1 (11). The Gatterman synthesis (method 2) (13) has also been used to synthesize vanillin, though the earlier Gatterman-Koch procedure (12) is not applicable. Yields of up to 70% can be achieved by absorbing guaiacol in diatomaceous earth prior to the Gatterman reaction (14). Several patents (15-

LOW MOLECULAR WEIGHT CHEMICALS

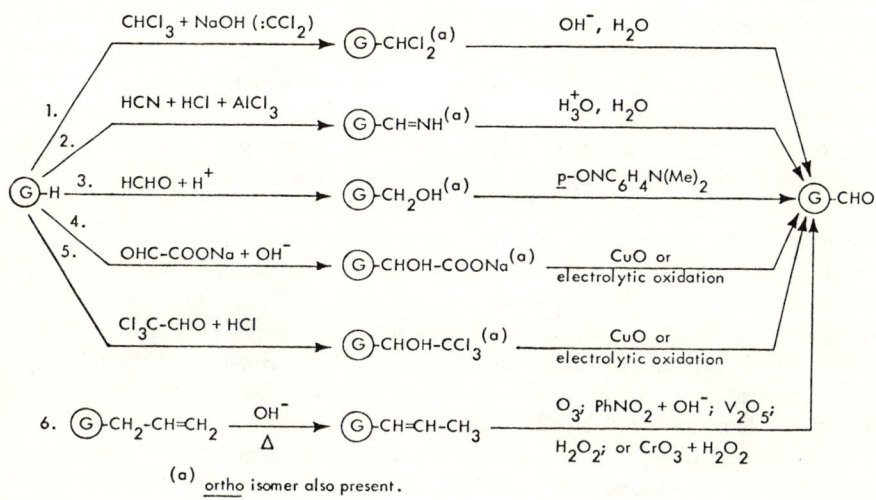

(a) ortho isomer also present.

Figure 19.1. Pathways for synthesis of vanillin from guaiacol and eugenol.

17) describe the acid-catalyzed condensation of guaiacol with formaldehyde in the presence of a nitroso compound (e. g. p-nitrosodimethylaniline) as an oxidant, method 3. Close related synthetic processes are the alkaline condensation with sodium glyoxylate (method 4) (18-20) and the mildly acidic condensation with chloral (method 5), both of which are followed by oxidation using, for example, CuO and alkali or electrolytic procedures. The reported yields for these two methods are 80% (21) and 55% (23), respectively.

For many years synthetic vanillin was obtained almost exclusively by the rearrangement and oxidation of eugenol, a constituent of oil of cloves, method 6. Eugenol is readily rearranged by hot alkali to isoeugenol. The oxidation of the latter to vanillin, first studied by Erlenmeyer (24), has been accomplished by many methods. Of the commercially important ones, oxidations with ozone (25) (especially in the presence of bisulfites (26, 27) or hyposulfites (28) to reduce the ozonide) give yields as high as 90% of theoretical. Nitrobenzene oxidations, patented in several countries at about the same time (29, 30), and oxidations with alkali soluble nitro compounds containing sulfonic acid or carboxyl groups (31) give even more than 90% of the theoretical yield of vanillin. However, these procedures have not been commercialized because of unfavorable economics. Inorganic oxidants give generally lower yields, such as 66% for V_2O_5-H_2O_2 (32) or 58% for CrO_3-H_2O_2 (33). Protection of the aromatic hydroxyl, however, increases yields to above the 90% level. For example, almost quantitative vanillin yields may be obtained from isoeugenol acetate using CrO_3 as the oxidant (34).

C. <u>Vanillin from Lignins</u>. As early as 1898, Pollacsek (6) observed the formation of vanillin in the oxidation of spent sulfite liquor with air and ferric chloride. Several years later, in 1904, Grafe (35) obtained vanillin from an alkaline hydrolysis of sulfite spent liquor solids. Strangely enough, more than twenty years passed before any further work was reported on this reaction. In 1928, Kürschner published a series of papers on alkaline air-oxidation of spent sulfite liquor (36, 37, 38, 39). The reported yields of vanillin (20% of lignin sulfonate) have not been successfully reproduced (40-42). Rather, it has been established that lignin sulfonates yield 6-8.5% vanillin upon alkaline hydrolysis, although highly sulfonated fractions may give somewhat higher yields. The careful studies by Tomlinson and Hibbert (41, 42) established the importance of an α-sulfonic acid function in the hydrolytic formation of vanillin through a reversed aldol condensation mechanism.

Commercial processes based on heating spent liquors with alkali under pressure were developed in the 1930's (43-45), shortly after Tomlinson and Hibbert's work was published. During the period 1936-1938, two commercial plants were established, one in Canada and the other in the United States. The Canadian process of the Howard Smith Paper Mills at Cornwall, Ontario was based largely on the studies made by Tomlinson and Hibbert. Sulfite spent liquor, preferably fermented to remove hexose sugars, was subjected to hydrolysis by sodium hydroxide and the vanillin formed recovered by benzene extraction after neutralization with carbon dioxide.

The United States process, established at the Rothschild, Wisconsin plant of the Marathon Corp. by the Salvo Chemical Co. (now a division of the Sterling Drug Co.), utilized the alkaline hydrolysis of lignin sulfonate materials separated from the spent liquor by the Howard Process (46). In this process, lime is added in two stages. The first yields a precipitate of calcium sulfite. Further lime addition to the filtrate results in precipitation of basic calcium lignin sulfonates. The sugars remain in solution in modified form. The separated calcium lignin sulfonate is hydrolyzed with sodium hydroxide, acidified with SO_2 to precipitate remaining lignin products and to convert vanillin to the water-soluble bisulfite addition product. On acidification, the vanillin is regenerated and can be recovered by solvent extraction. It was soon found (45) that the sodium salt of vanillin is actually extractable directly from the alkaline solution by butyl alcohol. By heating the butyl alcohol extract at 70° with about one-fourth its volume of water containing a small amount of sodium hydroxide, the sodium salt is removed from the alcohol and the vanillin can be recovered by acidification and solvent extraction (47). In both of the above commercial processes, the use of sodium hydroxide as the alkaline reagent was considered necessary (48).

The effect of oxygen and other oxidizing agents on vanillin production was studied by Freudenberg, Lautsch and coworkers (49-51). They showed that yields of vanillin several times higher than those from alkaline hydrolysis alone could be obtained by the use of air, oxygen or nitrobenzene as the oxidizing agent. A yield of 12% vanillin was obtained by heating spent sulfite liquor at 110°C with excess alkali in the presence of finely dispersed oxygen. Alkaline nitrobenzene oxidation at 160°C gave a vanillin yield of 20% and was patented as an industrial process by Schultz (52). The best yields were again obtained from highly sulfonated fractions. In oxidation by air or oxygen, temperatures in excess of 140°C caused extensive degradation of vanillin.

Extensive investigations of the alkaline oxidation of lignins with agents such as cupric hydroxide (53, 58, 59), various nitrobenzene compounds (54, 55), mercuric oxide (56) and silver oxide (57) were carried on by Pearl et al. All of these oxidizing agents provided much better yields of vanillin than straight alkaline hydrolysis. For example, freshly precipitated cupric hydroxide gave 22% vanillin from lignin sulfonates at 160° and 12.5% at 102°. In comparison, the maximum yield of vanillin was 23% in nitrobenzene oxidation under optimum conditions (1 hour at 180°). In addition, reduction products of nitrobenzene are obtained in varying yields. The optimum condition for formation of each product was determined giving the maximum yields, which follow: aniline 67.5%, azoxybenzene 67.5%, azobenzene 65% and sodium p-azobenzenesulfonate 30%. The production of a mixture of reduction products of nitrobenzene complicates the isolation and purification of vanillin, and any commercial process would have to include recovery and sale of these products. So far, there have been no reports of the commercial application of nitrobenzene oxidation.

The use of alkaline mercuric oxide and silver oxide at 100-175° (56,57) resulted in formation of vanillin, major amounts of vanillic acid, and some guaiacol and acetovanillone.

The use of oxygen to increase the vanillin yield during alkaline hydrolysis was applied to the commercial process of the Salvo Chemical Company shortly after World War II (60). An aqueous solution of basic calcium lignin sulfonate was formed by addition of sodium hydroxide, heated to 140-170° and oxygen passed in at such a rate that 25-35 g of oxygen were consumed per 100 g of lignin. This type of oxidation increased the vanillin yield by 80%.

In the period following World War II, two more commercial producers of vanillin from lignin began operations. The Monsanto Chemical Company started up a plant at Seattle, Washington utilizing fermented sulfite spent liquor from the alcohol plant of the Bellingham Division of Georgia-Pacific Corporation. The process has been claimed to form vanillin in at least 10% yield (61). According to available information, spent liquor is made alkaline and oxidized in the presence of copper compounds under carefully regulated oxygen pressure. Steam or nitrogen is used to dilute air admitted to the mixture so that the partial pressure of oxygen is less than 6.6 lb. per sq. in. (62). Vanillin can be recovered from the reaction mixture as the sodium salt by extraction with propyl or isopropyl alcohol (63) and purified by distillation under a moderate vacuum (2-15 mm Hg) at temperatures not exceeding 165°C in the presence of non-reactive, non-solvent aliphatic or chloroaliphatic saturated hydrocarbons (64).

The chief drawback to the use of sodium hydroxide is that this relatively expensive base must be recovered for economical operation. A process based on the use of inexpensive lime, that does not need to be recovered, has been developed and patented (65, 66) by the Ontario Paper Company. Unexpectedly, it was found that the yield of vanillin was highest at low lignin sulfonate concentrations and the yield from the reactor was improved when operated continuously instead of batchwise. This was explained in terms of adsorption of vanillin on the solids (consisting of lime, calcium carbonate, -sulfate and -oxalate, as well as salts of polymeric lignin degradation products) which are present in the process. In more dilute liquors, less vanillin adsorbed. The alkalinity must be kept above pH 12 in the reactor by means of vigorous agitation. The partial pressure of oxygen must be controlled (<20 psi at 170°C) to prevent overoxidation. After the oxidation, the reaction mixture is centrifuged and acidified, first with carbon dioxide and finally, with sulfuric acid. Vanillin (6-8% of lignin) then can be extracted with toluene and purified by distillation at about 2mm Hg pressure followed by crystallization from water.

More recently, the Ontario Paper Company has built a new plant which utilizes an improved process with a higher recovery of vanillin. Additional products such as calcium oxalate, modified lignins and sodium sulfate and/or -carbonate may also be recovered.

Although there are really two methods of operation, they both depend on treatment of the insoluble calcium containing sludges with sodium carbonate to make insoluble calcium carbonate and soluble sodium salts. In the first method of operation, the sludge from the older process is treated at about 160°F with a sodium carbonate solution (67) which liberates an additional 6-8% vanillin along with sodium salts of oxalic acid and lignin residues. The calcium carbonate is filtered, calcined and reused. The filtrate is acidified with sulfuric acid to pH 4 to precipitate lignin. Vanillin and related oxidation products are extracted with an organic solvent and purified. After removal of vanillin, lime is added until calcium oxalate precipitates. Chemical recovery is completed by evaporation to recover the sodium as the sulfate or carbonate.

The alternate method of operation (68) and the one which is undoubtedly used in practice is to conduct the oxidation-hydrolysis of sulfite spent liquor in the presence of a combination of lime and sodium carbonate. This gives a result similar to that from treating the sludge separately with sodium carbonate and, in addition, the oxidation, in the presence of sodium carbonate, improves the filtering properties of calcium carbonate solids so they can be easily separated from the product solution.

The use of this new process made Ontario Paper Company the largest producer of vanillin in the world. At the present time, there are four producers of vanillin from lignin sharing the North American market. At the current price of approximately three dollars per pound, the possibility of finding additional markets for vanillin as an industrial chemical are remote. However, recent devel-

LOW MOLECULAR WEIGHT CHEMICALS

opments in the production of pharmaceutical products from vanillin show some promise. Ways of lowering production costs could logically come from work on processes that do not require elevated temperatures in costly pressure vessels. Furthermore, additional products may be developed from the residual lignin products and other waste liquor constituents. Both the Salvo operation and the Ontario process have made good progress in this direction, and a number of uses for the altered, desulfonated lignin by-product have evolved.

The action of strong oxidizing agents such as ozone and alkaline potassium permanganate has been studied, but no commercial applications have been reported. With these strong agents, lignin has the tendency to become completely degraded to formic, acetic and oxalic acids. A patent (69) claims vanillin production by the ozone oxidation of a suspension of pine wood sawdust in glacial acetic acid. The yield was claimed to be as high as 33%, based on the lignin of the wood. In addition, oxidation of peat and wood meal to vanillin by means of chromic oxide in acetic acid was described. No further reference to this remarkable patent has been found.

While vanillin is the most commonly reported simple compound obtained by alkaline treatment of lignins, there are a number of others that are formed which may eventually prove to be of commercial interest. Most of these products have been obtained in studies aimed at trying to elucidate the structure of lignin. The products are summarized in Table 19.1. (Vanillin is included for reference purposes.) While no mention of the degradation of hardwood lignin has been made in this chapter, it must be remembered that a mixture of vanillin and syringaldehyde is obtained from this source (53, 70-86).

Table 19.1. Compounds Produced by the Alkaline Hydrolysis and Oxidative Alkaline Hydrolysis of Lignins.

Compound	Source	Oxidizing Agent, Base	Yield*	Ref.
vanillin	lignin sulfonates	none, NaOH	7	70
vanillin	lignin sulfonates	$CuSO_4$, NaOH	21.9	53
vanillin	lignin sulfonates	Na_2S, O_2, NaOH	32.4	76
vanillin	sulfate lignin	$Cu(OH)_2$, NaOH	13	77
vanillin	lignin in spruce wood	nitrobenzene, NaOH	24	85
vanillin	lignin in spruce wood	nitrobenzene, NaOH	27.5	86
vanillin	cuproxam lignin	nitrobenzene, NaOH	19	85
vanillin	lignin sulfonates	nitrobenzene, NaOH	20	85
vanillic acid	lignin sulfonates	Ag_2O, NaOH	28	78
vanillic acid	lignin sulfonates	CuO, NaOH	5.2	79
vanillic acid	lignin in spruce wood	nitrobenzene, NaOH	4.8	86
acetoguaiacone	lignin sulfonates	none, NaOH	~0.3	70
acetoguaiacone	lignin sulfonates	O_2, lime	?	80
acetoguaiacone	lignin sulfonates	CuO, NaOH	3.3	79
4,4'-dihydroxy-3,3'dimethoxystilbene	lignin sulfonates	none, NaOH	1	71
4,4'-dihydroxy-3,3'dimethoxystilbene	lignin sulfonates	none, NaOH	4.7	72
p-hydroxybenzoic acid	aspen, lignin sulfonates	none, NaOH	1.5	73
acetaldehyde	lignin sulfonates	none, NaOH	>7	74
formaldehyde	cuproxam spruce lignin	none, NaOH	~3	75
5-carboxyvanillic acid	lignin sulfonates	CuO, NaOH	1.2	79
vanilloylformic acid	lignin sulfonates	none, NaOH	2	82
vanilloylformic acid	lignin sulfonates	CuO, NaOH	0.4	84
5-carboxyvanillin	lignin in spruce wood	nitrobenzene, NaOH	1.2	86
5-carboxyvanillin	lignin sulfonates	O_2, lime	?	80
5-carboxyvanillin	lignin sulfonates	CuO, NaOH	0.8	79

Other compounds found in minor amounts include: 5-formylvanillin (20,86); dehydrodivanillin (20,80,86); dehydrodivanillic acid (80,86); bivanilloyl (82); 4,4'-dihydroxy-3,3'-dimethoxy chalcone (82); 4,4'-dihydroxy-3,3'-dimethoxybenzophenone (82); syringaldehyde (82); p-hydroxybenzaldehyde (80,86); 5-formylvanillic acid (86).

* Yield in percent of lignin.

D. _Purification_. Vanillin, obtained by either synthetic means or from lignin, contains impurities that interfere with its flavoring properties. Consequently, a number of processes for its purification have been devised. Sublimation (87) is not very practical, while distillation under a reduced pressure is a commonly used method (88). A co-distillation with a saturated (or chlorinated) hydrocarbon is used to facilitate the distillation process.

Pearl and Dickey (89) separated vanillin from syringaldehyde quantitatively by adsorption on a combination of magnesol and Celite No. 535 (5:1 ratio by weight). Steam distillation has been used in several processes. In one process (90) crude vanillin extract from the oxidizing reaction is fractionally steam distilled either from a good non-volatile solvent or, alternatively, using a poor solvent for vanillin which is volatile with steam. A Russian process (91) distills crude vanillin at 110-115° C with steam preheated to 180-200° C.

A Japanese patent (92) describes the adsorption of vanillin by active carbon from a 1% solution in 10% sodium hydroxide. The separation of vanillin from acetoguaiacone is claimed in a German patent (93). The mixture is treated with a sodium bisulfite solution and filtered to remove acetoguaiacone as an insoluble bisulfite addition compound. The filtrate is extracted at pH 1.5-3.5 and less than 50° C with organic solvents to remove organic contaminants from the soluble vanillin-bisulfite complex. The complex is then heated above 50° C to decompose the complex yielding vanillin of over 99.5% purity.

A method of recrystallizing vanillin and also recovering some of the lignin oxidation by-products has been described (80). Crude vanillin is purified by two recrystallizations from aqueous methanol with a ratio of water: NaOH of 40:60. The methanol can be distilled and recycled. In addition to pure vanillin, 5-formyl vanillin as its magnesium salt is recovered as is p-hydroxybenzaldehyde.

E. _Properties, Uses and Derivatives_. Vanillin melts at 81-82°, and its vapor pressure over a range of temperature (94), oxidation potential (95), dipole moment and dielectric constant (96) have all been determined. In addition to its flavoring characteristics, it exerts an effective masking effect on various other odors. A patent has been issued (97) for its use to mask the odor of paper coatings made from polymerization of cracked hydrocarbons.

Vanillin also finds use as reductant and as material absorbing ultraviolet light. A patented coating for heat sensitive copy papers consists of organic amine molybdates combined with an organic reducing agent such as vanillin (98). It can be incorporated into various plastics such as regenerated cellulose, polyvinyl ethers and polyamides to protect them against ultraviolet radiation. Ointments for protection against sunburn on the skin can be prepared with vanillin as the active agent (99). Condensed with acetylacetone, acetoacetic ester and similar compounds, vanillin can be used to protect fabrics from light deterioration. Films containing the condensation products protect light-sensitive foods (100). The incorporation of vanillin in ethyl cellulose plastics to protect them from outdoor exposure has been patented (101).

The water-soluble sodium salt of vanillin can be prepared by reacting it in benzene solution with anhydrous sodium carbonate. This product has been used in flavoring and odor masking (102). Adsorbed on starch, vanillin becomes so firmly bound that it cannot be extracted with ether. This may be a result of its becoming packed in amylose spirals of the starch. After the starch is gelatinized by alkali, the vanillin can be extracted (103).

The isonicotinoyl-hydrazone of vanillin is reported to be tuberculostatic in vitro at dilute concentrations (104). Anils are formed in reaction with various amines and anilines (105). Vanillin often can be regenerated from the anils, such as the crystalline anils formed in reactions with p-amino-N, N-dimethylamiline or 4-aminoantipyrine, by treatment with dilute acid (106, 107).

Vanillin oxime has been prepared and converted to vanilloamidine. Various imidazolines and tetrahydropyrimidinediones have been tested for pharmacological properties, such as antihistimine activity (108). Vanillin azine, prepared by reaction of vanillin with serinehydrazide hydrochloride (109) is useful in chemical analysis, and its disodium salt in neutral solutions gives colored complexes with many metal ions (110).

Vanillin reacts with aldehydes in the presence of piperidine to give unsaturated condensation products (111). It also may be condensed with nitromethane by refluxing the two materials in acetic acid in the presence of ammonium acetate (112). A condensation of vanillin with o-aminophenol takes place when the two are refluxed in nitrobenzene to give a benzoxazole (113).

A synthetic tanning agent can be prepared by condensing vanillin with two moles of phenol in 98% sulfuric acid, adding 5% lactic acid and adjusting the pH to 3.5 (114). Vanillin dissolved in sulfuric acid is used for a qualitative test for hydrogen peroxide (115).

Polymeric materials prepared from vanillin and vanillic acid have potential as fibers and films. For example, copolyesters of vanillic and terephthalic acids can be prepared from the corresponding acids or esters using metal salts or alcoholates as catalysts (116). Other reports describe formation of polyesters and polyanhydrides from vanillin and its condensation products (117-120).

Several reports on the comparison of vanillin with "ethyl vanillin" (4-hydroxy-3-ethoxybenzaldehyde) have been published. "Ethyl vanillin" has been estimated to have 3.5 times the flavoring strength of vanillin and is recommended to be used in admixture with vanillin for best flavoring results (121). Taste tests on ethyl vanillin and vanillin in lactose showed that the minimum detectable amounts of the two were in a ratio of about 1:2.6 and in milk, 1:4.25. Ethyl vanillin is more resistant than vanillin to temperature, light, humidity and air (122).

The possibility of converting vanillin to ethyl vanillin has been considered. In a patent issued to Kamlet (123), a process is described for demethylation of vanillin by a complex of aluminum bromide and an aromatic hydrocarbon to aldehyde and methyl bromide. The protocatechuic aldehyde is ethylated to give ethyl vanillin. An alternative method of de-alkylation of vanillin and other alkyl-o-hydroxyphenyl ethers has been described by Lange (124). The active agent is aluminum chloride in pyridine with a methylene dichloride solvent. The process is visualized as proceeding by means of a solvated five-membered cyclic ring 1 which is attacked by pyridine in a nucleophilic displacement reaction resulting in formation of 2 which can be hydrolyzed to protocatechuic aldehyde, Equation 1.

$$\text{(CHO-C}_6\text{H}_3\text{(OMe)(OH)} + \text{AlCl}_3 \longrightarrow \underset{1}{\text{[cyclic intermediate]}} \xrightarrow{C_5H_5N} \underset{2}{\text{[product]}} + \text{MeCl} \quad (1)$$

"Ethyl vanillin" can, of course, be obtained by syntheses analogous to those of vanillin. Its price is, however, considerably greater than that of vanillin and is in the range of seven to eight dollars per pound.

III. Organic Sulfur Chemicals from Lignins.

A. Production of Dimethyl Sulfide (DMS) and Methyl Mercaptan (MM). Depending upon the source of lignin—whether from softwood or hardwood—the methoxyl content of kraft lignin varies from about 14-20% of the lignin. Since the methyl group represents about half of the weight of the methoxyl, it is theoretically possible to recover from 7-10% of the lignin calculated as methyl radicals by reactions which result in cleavage of methoxyl groups. One such reaction is the nucleophilic cleavage of methoxyl by mercaptide or sulfide ions resulting in the formation of dimethyl sulfide (DMS) or methyl mercaptan (MM), respectively. DMS has been known to be present in the blow gases from the digesters of the kraft pulping operation for many years (131). Klason (125) first suggested that DMS results in some manner from the attack by sulfide on the methoxyl group of the lignin during pulping. The amounts of DMS, MM and other sulfur chemicals in the off gases from kraft pulping are quite low, however, being equivalent to about 2 pounds of sulfur per ton of pulp produced (126). A part of DMS and other sulfur compounds formed can be obtained by fractional distillation of the crude kraft turpentine, but there have been only limited attempts to recover DMS from this source (127).

Hägglund and Enkvist (128) found that adding sodium sulfide to kraft black liquor and heating to above 200° C produced a considerable amount of DMS along with MM. This work was studied in detail and confirmed by Cisney et al. (129) who replaced the addition of sulfide to the black liquor by elemental sulfur. The yields of DMS were about 3% based on the liquor solids. MM was obtained in 0.2-0.3% yield.

Following this work, the Crown Zellerbach Corporation obtained the exclusive rights to the Hägglund et al. patent (128) for the United States and

LOW MOLECULAR WEIGHT CHEMICALS

Canada. A pilot plant for production of DMS was built in 1956-7, and the process was studied further. It was found that sugar-free sulfite spent liquor as well as soda lignin can be used as the starting material to obtain DMS and MM in yields comparable with those from black liquor (130). After completed pilot plant studies, a full-scale plant, with a capacity output of ten million pounds per year, was built at Bogalusa, Louisiana, in 1960. The commercial process consists of the following steps: 1, Elemental sulfur is added to concentrated black liquor and the mixture reacted at 200-250°; 2, The DMS formed is flashed from the reactor along with some water, MM, H_2S, etc. and collected by means of a condenser; 3, The crude DMS is scrubbed by passing it through sodium hydroxide and distilled.

B. <u>Mechanism of Dimethyl Sulfide and Methyl Mercaptan Formation.</u> The black liquor from pulping can furnish sulfide ions for formation of DMS without addition of extra sulfide or sulfur. However, the liquor has to be heated to 300° C (at about 1200 psi pressure) in an autoclave in order to obtain a 3% yield, based on the lignin (132). In the commercial process, the addition of sulfur or sulfide in the liquor allows the formation of 3% or more yield at much more moderate temperatures and pressures (133).

The formation of DMS (and MM) is, therefore, dependent on the concentration of sulfide. The reaction of sulfide ions with the methoxyl group of lignin is an example of a nucleophilic displacement reaction shown in Equations 2 and 3 (134).

$$HS^- + MeO-\text{Ar}(C_3)(O-) \longrightarrow HSMe + {}^-O-\text{Ar}(C_3)(O-) \qquad (2)$$

$$MeS^- + MeO-\text{Ar}(C_3)(O-) \longrightarrow MeSMe + {}^-O-\text{Ar}(C_3)(O-) \qquad (3)$$

The initial attacking ion is sulfide; and as a result of its reaction with methoxyl groups, an even more nucleophilic ion, mercaptide, is formed. This attacks another methoxyl, forming DMS. The stepwise nature of the reaction with MM as an intermediate product has been shown in several ways (136, 137). For example, if the reaction is run at about 220-240° C and the gases are vented continuously, a product largely consisting of MM is obtained with a total yield of up to 4% of the weight of the black liquor solids. Due to the rapidity of the reaction of mercaptide ion (Equation 3), there is always a certain amount of DMS formed even though the reaction is run so that mercaptan can flash off soon after being produced. On the other hand, when mercaptans other than methyl mercaptan are heated in basic solution with lignin or with model methylaryl ethers such as guaiacol and conidendrin, corresponding methyalkyl sulfides are readily formed showing that mercaptide ions cleave alkyl aryl ethers to give sulfides.

These two reactions are, then, representative of the nucleophilic reaction of sulfide and mercaptide ions on a tetrahedral carbon atom substrate, the carbon atom of a methoxyl group in lignin. The order of reactivity of a number of nucleophilic agents on a carbon substrate has been determined (135–138) and is shown in Table 19.2. The relative nucleophilities provide an explanation why, in the preparation of DMS, only traces of methanol have been detected. It can be seen that reactivity of mercaptide ions are approximately a thousand-fold of that of hydroxyl ions in nucleophilic displacements, and in the presence of the former ions, only traces of methanol can be formed.

Table 19.2. The Order of Reactivity of Nucleophilic Agents on a Carbon Substrate (135, 138).

$C_4H_9S^-$	680,000
$C_6H_5S^-$	470,000
I^-	3,700
$S_2O_3^=$	3,200
$C_2H_5O^-$	1,000
OH^-	~ 600 – 900
Br^-	500
Cl^-	80
$C_6H_5NMe_2$	20
NO_3^-	1

Again, it may be seen that mercaptide ions have relative rates of reaction 34,000 times that of amines, exemplified by dimethyl aniline. On the other hand, precipitated lignin, which was freed of sulfide ions by the precipitation, does react with ammonia to give a 4% yield of trimethyl amine (139). The ammonia in this case was not competing with sulfide. Formation of only trimethyl amine indicated that mono- and dimethylamines react more rapidly than ammonia. When cyanide ion was used, trimethyl amine was again formed. Apparently, the cyanide was hydrolyzed to ammonia which then reacted as above.

The yield of products by the nucleophilic demethylation reaction has varied from about 50% to 75% of the theoretical value calculated from the methoxyl content of lignin. There are several possible explanations for the relatively low yields. It may be that the mercaptide ion intermediate attacks other positions in the molecule and is thus lost from the product yield. Alternatively, the two successive reactions may not reach completion. The latter explanation is supported by the observation that the yield of DMS from the reaction of black liquor with an excess of added sodium methyl mercaptide approaches 90% of theoretical.

Even though the yield of DMS from black liquor is only about 3% of the liquor solids, the potential production on the basis of black liquor available in the United States is well over a billion pounds per year.

C. Properties and Reactions of Dimethyl Sulfide.

1. Properties. DMS is a colorless liquid that boils at 37°C and is almost insoluble in water but miscible with most organic solvents. It has a characteristic odor; but when pure, this odor is ethereal, and many people do not find it unpleasant. It is a stable molecule in which the hydrogen atoms are quite unreactive, even to butyl lithium (140). Despite its stability, it can undergo reactions in two ways: (1) those replacing hydrogen and (2) those involving the extra electron pairs on the sulfur atom.

2. Halogenation. DMS reacts vigorously with chlorinating agents such as sulfur monochloride (141), thionyl chloride and sulfuryl chloride (142). The chlorination occurs stepwise and can be continued until all six of the hydrogen atoms have been replaced. All of the hydrogens on one methyl group are replaced before any substitution occurs on the other group.

Direct chlorination of DMS at ambient temperature is an extremely rapid reaction in which visible light is generated as well as noise from sudden collapse of chlorine bubbles. Chlorination below about -10°C in contrast yields crystalline dimethylchlorosulfonium chloride in a quiet reaction. The latter product rearranges spontaneously above -10°C to monochlorodimethyl sulfide, releasing hydrogen chloride (143), Equation 4.

$$Me_2S + Cl_2 \longrightarrow [Me_2SCl]^+Cl^- \xrightarrow{-HCl} MeSCH_2Cl \xrightarrow{Cl_2}$$
$$MeSCHCl_2 \xrightarrow{Cl_2} MeSCCl_3 \xrightarrow{Cl_2} ClCH_2SCCl_3 \xrightarrow{Cl_2} CCl_4 + ClSCH_2Cl \tag{4}$$

The direct chlorination can be readily accomplished by mingling vapor streams of DMS and chlorine in stoichiometric amounts in the presence of a diluent gas such as hydrogen chloride (144). The amount of diluent is adjusted so that the reaction zone temperature remains below 300°C and preferably about 200°C.

Chlorination beyond the mono stage is accomplished by passing chlorine gas into the liquid reaction mixture. The direct chlorination continues until tetrachloro-DMS is formed. Chlorination beyond this point requires heating, and the next step results in cleavage of the molecule to give carbon tetrachloride and chloromethyl sulfenyl chloride, Equation 4. Chlorination in the presence of water follows a different course (145). In this case, the chlorine adduct is hydrolyzed to dimethyl sulfoxide, Equation 5.

$$Me_2S \xrightarrow{Cl_2} [Me_2SCl]^+Cl^- \xrightarrow{H_2O} Me_2SO + 2HCl \qquad (5)$$

Addition of another molecule of chlorine forms an adduct with the sulfoxide that in the presence of water reacts in two ways: formation of dimethyl sulfone by oxidation, and formation of methane sulfonyl chloride along with formaldehyde and chlorinated hydrocarbons by rearrangement and hydrolytic cleavage, Equations 6 and 7.

$$Me_2SO \xrightarrow{Cl_2} [Me_2SO\text{-}Cl]^+Cl^- \xrightarrow{H_2O} Me_2SO_2$$

$$\downarrow$$

$$MeSOCH_2Cl \xrightarrow{H_2O} MeSOOH + HCHO$$
$$\xrightarrow{Cl_2} MeSOCl + CH_2Cl_2$$

$$\downarrow Cl_2 \qquad\qquad\qquad\qquad\qquad\qquad\qquad (6)$$

$$MeSOCHCl_2 \xrightarrow{Cl_2} MeSOCl + CHCl_3$$

$$\downarrow Cl_2$$

$$MeSOCCl_3 \xrightarrow{Cl_2} MeSOCl + CCl_4$$

$$MeSOOH \text{ or } MeSOCl \xrightarrow[H_2O]{Cl_2} MeSO_2Cl \qquad (7)$$

The chlorinated derivatives of DMS are reactive α-halothio ethers and react fairly rapidly with water to give various products, depending on the degree of chlorination, Equations 8, 9 and 10.

$$2MeSCH_2Cl + H_2O \longrightarrow MeSCH_2SMe + 2HCl + HCHO \qquad (8)$$

$$\text{MeSCHCl}_2 + 2\text{H}_2\text{O} \longrightarrow \text{MeSH} + 2\text{HCl} + \text{HCOOH} \tag{9}$$

$$\text{MeSCCl}_3 + 2\text{H}_2\text{O} \longrightarrow \text{MeSH} + 3\text{HCl} + \text{CO}_2 \tag{10}$$

3. <u>Reactions Involving Unshared Electron Pairs.</u> Because of its unshared electron pairs, DMS readily forms coordinate bonds with certain metallic ions resulting in the formation of a number of interesting metal salt complexes, Equation 11. The complexes formed from the halide and cyanide salts of the transition metals are usually soluble in DMS, while those of the alkali and alkaline earth metals are insoluble. The crystalline complexes contain one or two molecules of DMS.

$$\text{Me}_2\text{S} + \text{M}^+\text{X}^- \longrightarrow [\text{Me}_2\text{SM}]^+\text{X}^- \tag{11}$$

Alkylation of DMS with benzyl chloride or bromide, alkyl benzyl halides, methyl iodide or sulfate gives sulfonium salts in a formally analogous reaction, Equation 12. These can be converted to the hydroxides with alkaline silver oxide, Equation 12, or by ion exchange.

$$\text{Me}_2\text{S} + \text{RX} \longrightarrow [\text{Me}_2\text{SR}]^+\text{X}^- \xrightarrow{\frac{\text{Ag}_2\text{O}}{\text{H}_2\text{O}}} [\text{Me}_2\text{SR}]^+\text{OH}^- \tag{12}$$

Trimethylsulfonium hydroxide is an organic base nearly as strong as sodium hydroxide.

4. <u>Oxidation.</u> The derivative of DMS which is most useful and best known is the sulfoxide. It is commercially produced at the Bogalusa, Louisiana plant of Crown Zellerbach Corporation, which has an annual capacity of eight to ten million pounds.

Three main processes have been used in pilot or commercial scale to oxidize DMS to DMSO. The first one is based on the oxidation of DMS in the gas phase in the presence of nitrogen oxides. This system was first described by Smedslund (146), and presumably was used by the Stepan Chemical Company until the plant was destroyed by an explosion in 1959.

In a second process described in Hubenett and Reim (147), DMS is oxidized with a solution of NO_2 in DMSO. The oxidant becomes reduced to NO which is separated from the DMSO and unreacted DMS, oxidized back to NO_2 by air or oxygen and returned to the reaction zone in DMSO solution.

A third modification described by Wetterholm and Fossan (148) forms the basis for the commercial plant of Crown Zellerbach. Oxygen is introduced into a solution of DMS in dimethyl sulfoxide containing sufficient oxides of nitrogen to continuously convert the DMS to DMSO. The process operates since the nitric oxide resulting from the oxidation of DMS by nitrogen dioxide is continuously reoxidized by the oxygen bubbling through the system. Excess oxygen must be scrupulously avoided to eliminate the hazard for explosion.

Oxidations with nitrogen dioxide under anhydrous conditions generally terminate at the dimethyl sulfoxide stage. Stronger oxidants such as hydrogen peroxide and nitric acid convert DMS and DMSO by an exothermic reaction to dimethyl sulfone, $DMSO_2$. Oxidation by nitric acid is shown in Equations 13 and 14.

$$3Me_2S + 2HNO_3 \longrightarrow 3Me_2SO + 2NO + 2H_2O \qquad (13)$$

$$Me_2SO \xrightarrow{HNO_3} [Me_2SOH]^+ NO_3^- \xrightarrow{HNO_3} Me_2SO_2 + 2NO_2 + H_2O \qquad (14)$$

The initially formed DMSO is immediately converted to the conjugate acid by the nitric acid. Upon warming to 120–150° C, the conjugate acid reacts with excess nitric acid to form $DMSO_2$, nitrogen dioxide and water (149).

D. Dimethyl Sulfoxide (DMSO). Dimethyl sulfoxide is an almost odorless liquid with a freezing point at 18.5° C. However, even traces of water are sufficient to lower the freezing point considerably. DMSO dissolves readily most aromatic and unsaturated hydrocarbons, organic nitrogen compounds, organic sulfur compounds and many inorganic salts. It is miscible with common organic solvents with the exception of saturated aliphatic hydrocarbons.

DMSO is a reasonably stable chemical substance. Neither hydrogen nor carbon monoxide in the absence of a catalyst will reduce it at 100° C and a pressure of 10 atmospheres. Refluxing it at the boiling point (189° C) for an extended period results in slight decomposition to bis(thiomethyl)methane. The rate of decomposition is accelerated by acids and retarded by many bases, Equation 15.

$$2Me_2SO \xrightarrow{H^+} HCHO + H_2C(SMe)_2 + H_2O \qquad (15)$$

The remarkable properties of DMSO have excited interest in many fields. In addition to its use as solvent for many sparingly soluble materials, it has been used as an outstanding medium promoting organic reactions and as a reactant itself in chemical reactions. Its use in medicine as a carrier for other drugs, as a

LOW MOLECULAR WEIGHT CHEMICALS

drug itself and as a solvent for freeze preservation of body tissues has been given wide attention. The following discussion outlines some of the most important current and potential applications of DMSO.

1. As a Reactant. DMSO is a weak base which can be quantitatively titrated in acetic anhydride using perchloric acid (150, 151), Equation 16. Neither alkyl sulfides nor sulfones interfere with this titration.

$$Me_2SO + HClO_4 \longrightarrow [Me_2SOH]^+ClO_4^- \qquad (16)$$

The nitrate salt of DMSO has been obtained in crystalline form (152). The freezing point depression by DMSO in 100% sulfuric acid suggests the formation of a similar salt with sulfuric acid. Halogen acids react in a more complicated fashion, forming the crystalline di-halide salts of dimethyl sulfide, Equation 17. Both the diiodide and dibromide can be isolated, but the dichloride is too unstable (153).

$$Me_2SO + HX \longrightarrow [Me_2SOH]^+X^- \xrightarrow{HX} [Me_2SX]^+X^- + H_2O \qquad (17)$$

DMSO can also act as a weak acid. Corey and Chaykovsky first prepared a solution of methyl sulfinyl carbanion by reacting either sodium hydride or sodamide with anhydrous DMSO (154). The methyl sulfinyl carbanion is basic enough to convert triphenyl methane to its red colored carbanion. Solutions of the anion react to give the expected products with carbonyl compounds, halides or proton donors. For example, with benzophenone, an 86% yield of methylsulfinylmethyldiphenyl carbinol can be obtained, Equation 18.

$$Ph_2CO + Na^+[MeSOCH_2]^- \longrightarrow Ph_2C(OH)CH_2SOMe \qquad (18)$$

DMSO forms coordination compounds with a number of inorganic ions acting as a donor of electrons by means of the oxygen atom. Some of these complexes are very stable. $SnCl_4 \cdot 2DMSO$ can be prepared in water solution and sublimes unchanged at 180°. $TiF_4 \cdot 2DMSO$ can be recrystallized from water (155). At 160-175°, DMSO can oxidize organic sulfides to their sulfoxides while it is reduced to dimethyl sulfide, Equation 19 (156).

$$R_2S + Me_2SO \longrightarrow R_2SO + Me_2S \qquad (19)$$

Epoxides can also be oxidized with DMSO at 100°, using boron trifluoride as catalyst (157), Equation 20.

$$PhCH\underset{O}{-\!\!-\!\!-}CH_2 + Me_2SO \longrightarrow PhCOCH_2OH + Me_2S \qquad (20)$$

A particularly interesting reaction of DMSO is its ability to oxidize certain active halogen compounds (158), iodides in particular (161), as well as tosylates (159, 160), to the corresponding aldehydes (158). For example, phenacyl bromide can be oxidized in 95% yield to the corresponding aldehyde, Equation 21.

$$PhCOCH_2Br + Me_2SO \longrightarrow PhCOCHO + HBr + Me_2S \qquad (21)$$

An interesting extension of the oxidative ability of DMSO is its use in preparing vanillin from lignin. Lagally (162) has patented a process for producing vanillin from lignin sulfonates using dimethyl sulfoxide as well as sulfoxides and amine oxides as oxidants.

DMSO reacts with acid anhydrides and acid chlorides (163, 164). With acetic anhydride, a good yield of acetoxymethyl sulfide can be obtained, Equation 22.

$$Me_2SO + Ac_2O \longrightarrow MeSCH_2OAc + AcOH \qquad (22)$$

Carefully controlled reaction with thionyl chloride gives monochlorodimethyl sulfide, Equation 23.

$$Me_2SO + SOCl_2 \longrightarrow [Me_2S\text{-}O\text{-}SOCl]^+Cl^- \xrightarrow{-SO_2}$$
$$[Me_2SCl]^+Cl^- \longrightarrow MeSCH_2Cl + HCl \qquad (23)$$

Conversion of DMSO to sulfonium compounds is possible by reaction with aromatic compounds containing groups which promote aromatic substitution. When DMSO and phenol are treated with dry HCl at 45°C, the crystalline sulfonium product is obtained, and this on heating loses methyl chloride to give p-methylthiophenol (165), Equation 24.

$$\text{Me}_2\text{SO} + \text{PhOH} \xrightarrow[-\text{H}_2\text{O}]{\text{HCl}} [\underline{p}\text{-Me}_2\text{S-C}_6\text{H}_4\text{OH}]^+\text{Cl}^- \longrightarrow$$
$$\underline{p}\text{-MeS-C}_6\text{H}_4\text{OH} + \text{MeCl} \tag{24}$$

When DMSO is refluxed in an excess of methyl iodide for several days, the crystalline trimethyl sulfoxonium iodide is gradually deposited. Only the S-methyl compound is obtained with none of the O-methyl derivative (166), Equation 25.

$$\text{Me}_2\text{SO} + \text{MeI} \longrightarrow [\text{Me}_3\text{SO}]^+\text{I}^- \tag{25}$$

The resulting trimethyl sulfoxonium iodide can be converted to other salts by exchanging the anion with various silver salts. Corey and Chaykovsky (167) found that trimethyl sulfoxonium halides are dehydrohalogenated when treated with strong bases in DMSO solution at room temperature to give dimethylsulfoxonium methide, Equation 26,

$$[\text{Me}_3\text{SO}]^+\text{X}^- + \text{NaH} \longrightarrow \underset{\underset{\text{CH}_2}{\|}}{\text{Me}_2\text{SO}} + \text{H}_2 + \text{NaX} \tag{26}$$

which is a reactive material useful in many syntheses. The methylene group adds carbonyl compounds to give epoxides in high yield, Equation 27. With benzalacetophenone, the methylene group adds across the double bond to form cyclopropane derivatives in 95% yield, Equation 27.

$$\underset{\underset{\text{CH}_2}{\|}}{\text{Me}_2\text{SO}} \xrightarrow{\text{R}_2\text{CO}} \text{R}_2\text{C}\underset{\text{O}}{\overset{}{\triangle}}\text{CH}_2 + \text{Me}_2\text{SO}$$

$$\xrightarrow{\text{PhCH=CHCOPh}} \text{PhCH}\underset{\text{CH}_2}{\overset{}{\triangle}}\text{CHCOPh} \tag{27}$$

2. **As a Catalyzing Reaction Solvent.** DMSO exerts a strong catalyzing influence on many displacement reactions. The reaction of methyl iodide with hydroxide and alkoxide ions is thousands of times faster in DMSO than in water or alcohol. Hydrogen-deuterium exchange reactions are also

tremendously accelerated in DMSO. The rate of proton removal from 1-phenyl-1-methoxy ethanol has been found to be 10^7 times faster in DMSO than in t-butanol. The vastly enhanced activity of alkoxide ions in DMSO over their activity in alcohols is attributed to the absende of alkoxide-solvent hydrogen bonds in DMSO (168).

The general conclusion has been reach that in most nucleophilic reactions, DMSO allows a faster reaction at lower temperatures with better yields than in more commonly used solvents such as methanol. In the displacement reactions of an alkyl halide with sodium cyanide to give an alkyl cyanide, the use of DMSO allows the reaction to give a 93% yield in one hour as contrasted with a 30% yield in 11 hours in refluxing methanol (169). Similar results are obtained from the reaction of alkyl halides with sodium nitrite to give organic nitro compounds.

3. As a Solubilizing Reaction Solvent. Since DMSO is an excellent solvent for many carbohydrate materials, it has found application in the preparation of sugar derivatives. Sucroesters can conveniently be prepared in DMSO solution since both the starting materials and the resulting esters are soluble (170).

Due to its property of dissolving large molecules, DMSO has been extensively used as a solvent for polymerization. Polymerization of acrylonitrile gives a solution of polyacrylonitrile which can be spun directly from the DMSO solution (171, 172). It has also been used for polymerization of vinyl acetate (173) and for the reaction of diisocyanates with various materials (174, 175).

4. In Biological Systems. Because of its solvency for various high molecular weight molecules, DMSO has been extensively studied as a solvent for many biological and medical substances. Enzymes, for example trypsin (176), dissolve in DMSO and can be recovered from it with little loss of activity.

DMSO also has found utility as a material to protect cellular tissues from freeze damage. In this case, it has been shown to penetrate the cell walls better than glycerine and thus prevent excessive concentrations of electrolytes in the cells during freezing (177). It has been found to have a very low order of toxicity. The LD-50 in mice was found to be 11g/kg (178). In the same study, the ability of DMSO to protect organisms against damage by short wave radiation was demonstrated. Mice injected with a DMSO dose of 4.5g/kg were given a 70% protection against a lethal dose of X-rays.

DMSO has been used successfully to protect red blood cells against freezing. At a concentration in solution of 15-20% of DMSO, up to 90% of red blood cells are still viable after being subjected to -80° C freezing. With two human patients, bone marrow was preserved using DMSO and then was returned to the patients successfully following treatments which would have destroyed the marrow had it been left in the patients (179).

Widespread interest in DMSO as a solvent for carrying drugs across cell membranes and as a drug itself has been developed. Preliminary reports show that DMSO is effective in relieving the pain associated with musculosketal injuries, and inflammations such as arthritis and bursitis (180). It also is reported to show promise as a carrier of various chemicals throughout plant tissues to combat plant diseases (181).

E. Dimethyl Sulfone. $DMSO_2$ is a white crystalline solid melting at 109°C. It is very stable and quite inert toward most reagents. For this reason, it has been suggested for use as a plasticizer and selective solvent. It is resistant to reduction and requires the use of lithium aluminum hydride to reduce it, in contrast with DMSO which reduces readily with a number of reducing agents.

F. Methyl Mercaptan. Methyl mercaptan (MM) can be made directly from black liquor as described earlier (137a) or by treating DMS with hydrogen sulfide over an alumina catalyst at 400°C.

Methyl mercaptan is a gas at room temperature, boiling at about 6°C. It has the interesting property of forming a solid clathrate compound with water, stable at and below its boiling point. Its most important use is in the production of the amino acid, methionine, used as a supplement in various animal feeds.

MM is readily oxidized to dimethyl disulfide, DMDS. This can easily be accomplished by treatment of MM with oxygen or air using a copper oxide catalyst. DMDS is an excellent solvent for elemental sulfur and holds promise in agricultural uses. DMDS reacts with chlorine in a variety of ways, depending upon the temperature, to give chlorinated products that may also have agricultural interest.

IV. Potential Phenolic Products by High Temperature Sodium Sulfide Treatments. Enkvist and his coworkers in Finland have shown (182-184) that other useful degradation products in addition to dimethyl sulfide may also be prepared by heating black liquors with sulfur. It was demonstrated (182) that regular black liquor contains small quantities of vanillin, guaiacol, acetoguaiacone, vanillic acid, p-hydroxybenzoic acid and 1-methylcyclopenten-2-ol-3-one. When the black liquors were heated with Na_2S at 255°C under pressure, pyrocatechol, p-methyl and p-ethylcatechol, protocatechualdehyde, homoprotocatechualdehyde, protocatechuic acid and vanillin were the phenolic products found in the liquors.

In another paper, Enkvist et al. (183) describe a method of degradation of various spent liquors by heating them with NaOH and Na_2S. The liquors were heated for 1-2 minutes at 95-160 atm. pressure by pumping them through a stainless steel helix heated by means of a salt bath at 330-390°. The reaction products were acidified and extracted with ether. The highest yields of ether soluble, non-volatile degradation products, 25-34% of the organic part of the liquor, were obtained from black liquor heated with Na_2S. Formation of pyrocatechol structures occurred concurrently with formation of methyl mercaptan and dimethyl sulfide.

In later papers (184, 185) more detailed data were presented on the

Table 19.3. By-products from the Kraft Process Before and After Pressure Heating with Na_2S at 250 to 285°C. (185).

Before pressure heating	lb./ton pulp	
Tall oil (100) and sulfate turpentine (35) total	135	
Acetic (80) and formic (80) acids, total	160	
Methanol (in condensate)	10	
Vanillin (4), acetoguaiacone (3), total	7	
Guaiacol (4?) and pyrocatechol (2), total	6	
Guaiacylethanol and -propanol, other phenols	83	
Vanillic acid (4?), homovanillic, 1-guaiacylpropionic, p-hydroxybenzoic, and other phenolcarbonic acids, ca.	90	
Nonvolatiles, ether-soluble acids, and lactic acid	290	
		781 Total
After pressure heating with Na_2S at 250 to 285°C.		
Acetic (120) and formic (120) acids, total	240	
Methyl sulfide (80) and methyl mercaptan (16), total	96	
Pyrocatechol (60), methylcatechol (22), ethylcatechol	100	
Other ether soluble phenols	300	
Homoprotocatechuic, protocatechuic, other phenolcarboxylic acids, perhaps	300	
Ether soluble acids of carbohydrate origin	300	
"Demethylated lignin" with novolac properties	200	
Butanol-soluble lactones, etc.	70	
Total: ether solubles 1336, others 270.		1606 Total

yields of phenolic products obtained in heating in an autoclave, Table 19.3. Under optimum conditions (291°C, 20.4% of organic substances, 3.2% Na_2S and 1.6% NaOH based on the total liquor) a maximum yield of 33% of the organic products was obtained as ether soluble phenols. The yield of pyrocatechol (and its methyl and ethyl homologues) was 5% of the organic substances. The yields at 255°C were lower, consisting of 20% ether soluble phenols and 2.5% pyrocatechol and homologues. These very interesting papers demonstrate the potential of preparing industrially important amounts of phenols from black liquors.

V. Hydrogenolysis of Lignins. The techniques of hydroreforming and hydrocracking, which have been so successful in petroleum technology, have encouraged many investigators to study the hydrogenolysis of lignins in the hope that a substantial portion of the lignin could be broken down into simple chemicals. The very first studies on lignin hydrogenation used noncatalytic methods. Dorée and Hall (186) treated lignin sulfonic acid with zinc in both dilute acetic and hydrochloric acid, accomplishing little more than partial desulfonation. Willstatter and Kalb (187) treated hydrochloric acid spruce lignin with red phosphorous and hydriodic acid, obtaining 60-80% of the lignin as a complex mixture of hydrocarbons. Much more promising results have been obtained by the technique of catalytic hydrogenation of wood using various catalysts and temperatures up to 350°C. In this way, Lindblad (188) obtained heavy oils containing saturated hydrocarbons.

The hydrogenation of isolated methanol lignin was studied by Harris et al. (189, 190) using a copper chromite catalyst in dioxane solution. Substituted monomeric cyclohexyl alcohols and higher boiling products of unknown structure were obtained. Adkins et al. (191) applied similar hydrogenation to a soda lignin, but obtained less low molecular weight products. The same method was applied in systematic structural studies by Hibbert et al. (192-194) on the hydrogenation of both isolated lignins and lignins in wood. Lautsch (195, 196) and Lautsch and Freudenberg (197) carried out extensive investigations of lignin hydrogenation in alkaline media, using various catalysts such as nickel, copper, copper-chromium-magnesium oxides and metal sulfides. Complex mixtures of products including phenols, substituted phenols and hydrogenated phenols were obtained. Lautsch and Piazolo (198) also explored the hydrogenation of lignin and lignin sulfonic acids in an autoclave with no catalyst using caustic and alcohol as reductant. The reduction products were again obtained as extremely complex mixtures. Thus, conventional hydrogenation studies did not encourage, as a whole, the development of commercially important processes.

To avoid excessive hydrogenation of the phenol intermediates to alcohols and hydrocarbons, two separate groups, in the 1950's, applied the techniques of hydrocracking, both in the presence of sulfur resistant catalysts and with no catalysts at all. The first group, the Inventa A. G. fur Forschung und Patent Verwertung in Switzerland, has a number of patents, chiefly in the name of J. Giesen as the inventor (199). By the use of thiomolybdate and thiotungstate catalysts, sulfite spent liquor lignin was converted in 30% yield to an oily distillable product containing 35% phenols. In an alternative procedure, hydrolysis lignin dissolved in xylenol was treated without catalyst at very high pressures of hydrogen (5000-10,000 psi) at temperatures of about 300°C. As much as 40-50% yields of distillable oils, containing 45% phenols, were obtained.

The second group, the Noguchi Institute of Japan, has applied experience from the hydrogenation of coal to the liquefaction of lignin in pilot plant scale. Two novel hydrogenation catalysts form the basis for two patents recently issued in Canada (200 a & b).

The Japanese process consisted of the following steps. Sulfite spent liquor was desulfonated by a two-stage treatment with calcium hydroxide and sulfur dioxide to give desugared, largely desulfonated lignin sulfonate with a lower ash content. This lignin was treated with a pasting oil, generally phenol, to provide a reaction medium in which the lignin could dissolve during the hydrogenation reaction. The mixture together with the catalyst (1-10% of the lignin) was hydrogenated with continuous agitation at 370-430°C for 0.5-4 hours with an initial hydrogen pressure of at least 100 atm. The product mixture was filtered and distilled, yielding about 44% of monophenols together with 20-24% of the heavy oils suitable for use as a recycling pasting oil. The remainder of the lignin was lost as light oils, gas and water. The mono-phenol fraction was reported to be composed of phenol, o-cresol, p-cresol, p-ethyl phenol and p-propyl phenol.

The new hydrogenation catalysts were claimed to have advantageous characteristics of not being deactivated by sulfur, of being inexpensive and of not causing extensive reduction of phenolic materials to saturated alcohols and hydrocarbons. Speculatively, the following sequence of reactions was envisaged for the process, see structure 3.

$$
\begin{array}{c}
H_2COH \\
| \quad a \\
HC \\
| \quad b \\
HC \\
\quad c \\
\text{—Ar(OMe)(OH)} \\
3
\end{array}
$$

The methyl ether linkage (d) was assumed to split first, followed by the homolytic cleavage of side bonds (a), (b) or (c) into various free radicals immediately hydrogenated to give methanol, ethanol and various phenolic derivatives. Obviously, splitting must also occur at the aromatic ether bond of the methoxyl group to give monophenols. The Noguchi catalysts were considered to aid in the splitting of these bonds and stabilizing the radicals formed without promoting the hydrogenation of the aromatic nuclei. This was considered to be the reason for the formation of a relatively small total number of simple phenolic compounds.

Since 1961, the Crown Zellerbach Corporation has expended much effort on the evaluation of the economics and process conditions. Despite many improvements in operating conditions, the process did not turn out to be competitive with the production of phenols from coal and petroleum sources, mainly for the following reasons: The yield of monophenol products was, on the average, 21% instead of the 44% claimed by the Japanese investigators. Instead of p-cresol, a mixture of m- and p-cresol was formed. These isomers are extremely difficult to separate and together have much less value than either one pure. The phenol used as pasting compound entered into the reaction and part of it appeared in the product as p-cresol, thus deceptively boosting the yield. Another part was lost as gas and light oil. Due to the incomplete recovery, the use of phenol for pasting is not economical.

Nevertheless, the improved Noguchi method appears to be the best process for liquefaction of lignin that has yet appeared. The reaction time can be as short as 15 minutes at 450°C and 2500 psi operating pressure. The total yield of distillable products is in the range of 55-65% of lignin, although more than half is obtained in the form of light oils, neutral oils and high boiling fractions with little value. Compared to other hydrogenation processes, the phenolic products were remarkably simple and few in number, consisting of phenol, o-cresol, m- and p-cresol, o-ethylphenol, m- and p-ethylphenol, o-, m- and p-n-propylphenol and 2,4-xylenol. Small amounts of 2,6-xylenol were also ocasionally observed.

If, by means of catalyst improvements, monophenol yield could be increased by, say 50%, and the isolation cost for solid lignin reduced, the process should undoubtedly become economically attractive.

VI. Oxidation of Lignins. Aside from the previously discussed vanillin formation, most oxidative reactions on lignin result in formation of mixtures of simple acids including carbon dioxide. Only a few products have been obtained in sufficient quantities to be commercially interesting. Reed obtained two early patents claiming the formation of oxalic acid by the oxidation of sulfite spent liquor solids. In the first patent (201), spent liquor solids were oxidized with concentrated nitric acid at 95°C. In the second patent (202), dried spent liquor solids were fused with sodium or potassium hydroxide and heated in thin layers. However, Heuser et al. (203) demonstrated that lignin sulfonates alone produced no oxalic acid on fusion and concluded that oxalic acid formation from spent liquor solids was due to the presence of carbohydrates.

In contrast, Willstatter lignin (204) gave a 20% yield of oxalic acid on being heated with potassium hydroxide to 280°. It has been claimed (67) that oxalic acid is a potential product of the new vanillin process of the Ontario Paper Company at Thorold, Ontario.

In more recent times, many additional acids have been found as lignin oxidation products. Some of these are potentially useful industrial products, but no commercial utilization has as yet been reported for any of these chemicals (205-212).

Ozone and hydrogen peroxide have been used repeatedly as oxidizing reagents, but generally, only carbon dioxide and simple acids have been obtained. Other oxidizing agents such as lead tetracetate and potassium ferricyanide have been studied, as has electrolytic oxidation, with no industrially attractive results, however.

Finally, it should be mentioned that oxidation of lignin with hypochlorite or chlorine dioxide is a commercial operation in the bleaching of paper pulp. So far as is known, no attempts to recover the degradation products from the bleach liquors have been made commercially.

VII. Carbonization of Lignins. It is difficult for us in this modern petrochemical age to realize that destructive distillation of wood was once the major industrial chemical source. From this source, much acetic acid, methanol, wood creosote and charcoal were obtained in large quantities. Acetic acid apparently arises mainly from hemicelluloses of the wood, while methanol and wood creosote are largely lignin decomposition products. At one time in the

United States (1910), over 500,000 tons of wood charcoal was produced. This production fell steadily and reached a low of 200,000 tons by 1955. A recent increase in popular interest in home barbecuing has caused production to rise again, and it is now in the range of 300,000 tons per year (213). The recovery of the by-product chemicals from charcoal has steadily declined, however. A principal drawback to the recovery is the tremendously diverse nature of the products. Chemicals such as acetic acid and methanol are commonly collected, but the much larger amounts of tars and oils have no well established general uses and are usually burned in plant boilers for production of steam. This is in marked contrast to the analogous product, coal tar. The latter material, which is also very complex in nature, serves as a source of many industrially important chemicals.

In early studies, Heuser and Skiöldebrand (214) subjected a hydrochloric lignin to pyrolysis, obtaining 51% charcoal and 13% tar along with some acetone, methanol and acetic acid. Carbon monoxide and methane were obtained as non-condensible gases. Similar results have been reported by Fischer and Schrader (215) and, under conditions of reduced pressure, by Tropsch (216). Dry distillation of hydrolysis lignin from Douglas-fir at 400°C has been described by Fletcher and Harris (217). This lignin gave an average of about 60% coke, 17% aqueous distillate, but only 7.5% tar. The aqueous distillate contained acetone, formic and acetic acids. The tar was separated into phenols (36%), neutral substances (32%) and an acidic fraction (8%). The phenolic fraction contained phenol, o- and p-cresol, guaiacol, 2,4-xylenol, 4-methyl guaiacol and 4-ethyl guaiacol along with other compounds.

Freudenberg and Adam (218) improved the dry distillation yields by applying a hydrogen atmosphere in the presence of catalysts. Nickel precipitated onto the lignin gave the best results. Other catalysts studied were copper, mixtures of nickel and copper, cobalt, chromium oxide, aluminum oxide and iron. Up to 60% of various lignins were obtained as ether soluble phenols (35% of lignin), neutrals (7%) and acids (0.3%). The acids consisted of formic, acetic and traces of propionic acid. The composition of the phenolic fraction, Table 19.4, demonstrates that individual compounds are present in too small amounts to make their recovery attractive.

Table 19.4. Phenols Obtained by Dry Distillation of Lignins (218).

phenol	5.5%*
p-ethylphenol	1.1
guaiacol	3.9
p-cresol	7.1
p-ethyl guaiacol	1.6
o-ethyl guaiacol	0.5
isoeugenol	1.3
catechol	2.9
p-propyl catechol	0.5
homocatechol	1.1
higher boiling phenols (135-180°/0.03mµ)	10.0

*% of starting lignin

Lautsch and Piazola (198) used isopropyl alcohol vapor as the hydrogen source and obtained only 26.5% phenols using palladium as catalyst and even less (22%) with a nickel catalyst. Substitution of ethanol or water gas for isopropyl alcohol decreased the yields of phenols to 15 and 10%, respectively. Sergeeva and Surna (219) obtained only 20% tar from a commercial sulfuric acid hydrolysis lignin in a current of hydrogen using nickel and molybdic oxide catalysts.

Much of the recent work on thermal breakdown of lignin has taken place in the U.S.S.R., where utilization of lignin from wood hydrolysis plants is being intensively studied. Panasyuk et al. (220-223) have studied the thermal decomposition of hydrolysis lignin under a great variety of conditions. They concluded that the best results were obtained in the pyrolysis of lignin dispersed in anthracene oil under reduced pressure. Still, their best yields of low boiling phenols amounted only to 12% of the lignin. The distillation residue could be readily converted to activated carbon. Sergeeva and Jaunzems (224) have studied the pyrolysis Klason lignin in high vacuum obtaining a 24% yield of a liquid distillate.

Although these procedures, so far as is known, have not resulted in commercial production of phenols, they hold promise for future improvements. Such techniques as exposing lignin to high temperatures for short periods under vacuum certainly need to be investigated further.

VIII. Future Prospects for Low Molecular Weight Chemicals from Lignins.
There is no denial that at present only a few simple chemicals are being produced from lignin in commercial volume. The reason for this disparity between what might be and what is actually the case is simply the absence of economically competitive production methods for many otherwise attractive chemical products.

Future work on lignin utilization must be carried on with this economic picture firmly in mind. A very important factor, and one that may very well spell the difference between success and failure, is the cost of separation of the lignin from the large volumes of water in pulping liquors. At first glance, this factor may appear trivial because evaporation and drying can be accomplished for approximately 1 cent per pound; however, it must be remembered that the dried product is not pure lignin but still contains the ash, carbohydrates and extraneous material of the spent liquor. Taking into account, furthermore, that a 20% product yield can be considered to be quite good, it is easy to see that the initial lignin cost exerts a considerable leverage on the final chemical cost. For example, at 2 cents per pound for the starting lignin and a 20% yield of, say, a simple monophenol such as o-cresol, the starting material cost for the o-cresol would be 10 cents per pound for the lignin cost alone. When one considers the processing and other costs, it is difficult to see how this can compete with o-cresol from petroleum or coal selling for 16 cents per pound. If the lignin isolation cost could be reduced to less than 0.5 cents per pound, it would be a different story.

Thus, an inexpensive method of separating nearly pure lignin from pulping liquors is needed urgently. If this can be achieved, application of hydrocracking techniques, which are already known, should allow the production of chemicals competitively with those from petrochemicals. In addition, ways can be found to optimize the conversion of the lignin into useful products and achieve a greater yield per unit of lignin. Perhaps this can be achieved by processes and catalysts analogous to those developed in the petrochemical field.

Indeed, many new techniques deserve to be tested for applicability in lignin degradation. For example, various types of radiation to degrade the lignin molecule have hardly been tried. High temperature-short time pyrolysis techniques are just now being studied. The use of extremely high pressures is becoming a commercial possibility and should be applied to lignin degradation.

Over the long haul, lignin, which is a replenishable resource, would appear to have the edge over most other chemical raw materials. This may not, however, become consequential until natural gas is in short supply and petroleum becomes an expensive raw material. For the immediate future, chemicals from lignin can be anticipated to prosper only in rather specialized areas; but to some future generations, lignin will undoubtedly become an essential source of organic chemicals.

REFERENCES

1. C. Watt and H. Burgess, U. S. Patents 1448 and 1449 (1854).
2. B. C. Tilghman, U. S. Patent 70,487 (1866).
3. C. F. Dahl, U. S. Patent 296,935 (1884).
4. Bureau of Census, Washington, D.C., Preliminary Report (1962).
5. Anonymous, Fed. Reg., 26, 3991, May 9, 1961.
6. E. Pollacsek, Austrial Prov., 1898, 1524.
7. K. Reimer, Chem. Ber., 9, 423-4 (1876).
8. F. Tiemann and P. Koppe, ibid., 14, 2021, 2023 (1881).
9. M. C. Traube, ibid., 28R, 524 (1896); Ger. Patent 80,195, April 5, 1894; Frdl., 4, 1287 (1894-97).
10. D. E. Armstrong and D. H. Richardson, J. Chem. Soc., 1933, 496.
11. H. Wnyberg, Chem. Rev., 60, 169 (1960).
12. L. Gatterman and J. A. Koch, Chem. Ber., 30, 1622 (1897).
13. L. Gatterman, Ann. Chem., 357, 313 (1907).
14. A. Roester, Ger. Patent 189,037, Oct. 24, 1906; through Chem. Zentr. 19081, 73.
15. Akt. Ges. für Anilin-Fab., Brit. Patent 145,581, June, 1920; through Chem. Abstr.,14, 3427 (1920).
16. A. Weiss, U. S. Patent 1,345,649, July 6, 1920.
17. A. Gilliard, P. Monnet and E. Cartier, Brit. Patent 164,715, Feb. 25,

1921; through Chem. Abstr., 16, 424 (1922).
18. J. D. Riedel, E. de Haën A.-G., Brit. Patent 401,562, Nov. 18, 1933.
19. M. A. Yakshilevich, Russ. Patent 55,779, Sept. 30, 1939.
20. J. Kamlet, U. S. Patent 2,640,083, May 26, 1953.
21. G. V. Chelintzev and B. N. Rodnevich, J. Applied Chem. (U.S.S.R.), 8, 909 (1935).
22. R. Yu. Shagolova, N. A. Daev, N. I. Gel'perin and M. P. Kuzentsova, Trudy Vsesoyuz. Nauk-Isledovatel. Inst. Sintel. i Natural Dushistykh. Veshchestv., 1958 (4), 34; through Chem. Abstr., 53, 13094f (1959).
23. V. A. Zasosov, E. I. Metel'kova and V. S. Onopienko, Med. Prom. S.S.S.R., 13 (3), 22 (1959); through Chem. Abstr., 54, 17316e, (1960).
24. E. Ehrlenmeyer, Chem. Ber., 9, 273 (1876).
25. B. G. Wood, Chem. and Met. Eng., 28, 806 (1923).
26. F. Hoffman-LaRoche and Co., Swiss Patent 75,449, Aug. 1, 1917.
27. E. C. Spurge, U. S. Patent 829,300, Aug. 21, 1906.
28. Schimmel and Co. A.G., Ger. Patent 513,678, Oct. 18, 1925.
29. J. D. Riedel A.-G., Brit. Patent 298,451, Feb. 17, 1927.
30. R. H. Bots, U. S. Patent 1,643,804, Sept. 27, 1927.
31. F. Hoffman-LaRoche and Co. A.G., Brit. Patent 392,399, May 18, 1933.
32. N. A. Milas, J. Am. Chem. Soc., 59, 2342 (1937).
33. N. A. Milas, U. S. Patent 2,414,385, Jan. 14, 1947.
34. N. T. Farinacci, U. S. Patent 2,794,813, June 4, 1957.
35. V. Grafe, Monatsh., 25, 987 (1904).
36. K. Kürschner, J. Prakt. Chem., 118, 238 (1928).
37. K. Kürschner and W. Schramek, Tech. Chem. Papier-u-Zellstoff Fabr., 28, 65 (1931).
38. K. Kürschner and W. Schremek, ibid., 29, 35 (1932).
39. K. Kürschner, ibid., 30, 1 (1933).
40. M. Honig and W. Ruziczka, Z. Angew, Chem., 44, 845 (1931).
41. P. P. Shorugin and E. K. Smolyaninova, J. Gen. Chem. (U.S.S.R.), 4, 1428 (1934).
42. G. H. Tomlinson and H. Hibbert, J. Am. Chem. Soc., 58, 345 (1936).
43. G. H. Tomlinson and H. Hibbert, U. S. Patent 2,069,185, Jan. 26, 1937.
44. L. T. Sandborn, J. R. Salvesen and G. C. Howard, U. S. Patent 2,057,117, Oct. 13, 1936.
45. L. T. Sandborn, U. S. Patent 2,104,701, Jan. 4, 1938.
46. G. C. Howard, U. S. Patent 1,699,845, Jan. 22, 1929; reissue 18,268 Dec. 1, 1931.
47. R. Servis, U. S. Patent 2,399,607, April 30, 1946.
48. Hoffman LaRoche and Co. Akt. Ges., Swiss Patent 210,834, Oct. 16, 1940.
49. K. Freudenberg, W. Lautsch and K. Engler, Chem. Ber., 73B, 167 (1940).
50. K. Freudenberg and W. Lautsch, Naturwiss., 27, 227 (1939).
51. W. Lautsch, E. Plankenhorn and F. Klink, Angew. Chem., 53, 450 (1940).

52. L. Schultz, U. S. Patent 2,187,366, Jan. 16, 1940.
53. I. Pearl, J. Am. Chem. Soc., 64, 1429 (1942).
54. I. Pearl and H. F. Lewis, Ind. Eng. Chem., 36, 664 (1944).
55. I. A. Pearl, J. Org. Chem., 9, 429 (1944).
56. I. A. Pearl and H. F. Lewis, U. S. Patent 2,433,227, Dec. 23, 1947.
57. I. A. Pearl, U. S. Patent 2,431,419, Nov. 25, 1947.
58. I. A. Pearl and D. L. Beyer, Tappi, 33, 544 (1950).
59. I. A. Pearl, J. Am. Chem. Soc., 72, 2309 (1950).
60. J. R. Salvesen, D. L. Brink, D. G. Diddaus and P. Owzarski, U. S. Patent 2,434,626, Jan. 18, 1948.
61. Monsanto Chemical Co., Brit. Patent 695,301, Aug. 5, 1953.
62. C. C. Bryan, U. S. Patent 2,692,291, Oct. 19, 1954.
63. C. C. Bryan, U. S. Patent 2,721,221, Oct. 18, 1955.
64. C. C. Bryan, U. S. Patent 2,506,540, May 2, 1950.
65. J. H. Fisher and H. B. Marshall, U. S. Patents 2,576,752-3, Nov. 27, 1951.
66. J. H. Fisher and C. A. Sankey, U. S. Patent 2, 576,754, Nov. 27, 1951.
67. D. C. Craig and C. D. Logan, Can. Patent 615,552, Feb. 28, 1961.
68. D. C. Craig and C. D. Logan, Can. Patent 615,553, Feb. 28, 1961.
69. H. Pauly and K. Feuerstein, Ger. Patent 552,887, Sept. 28, 1928.
70. F. Leger and H. Hibbert, J. Am. Chem. Soc., 60, 565 (1938).
71. H. Richtzenhain and C. V. Hofe, Chem. Ber., 72B, 1890 (1939).
72. A. W. Sohn, Ger. Patent 1,002,753, Feb. 21, 1957.
73. I. A. Pearl, U. S. Patent 2,996,540, Aug. 15, 1961.
74. K. Kratzl and I. Keller, Monatsh. Chem., 83, 197 (1952).
75. K. Freudenberg and E. Plankenhorn, Chem. Ber., 80, 149 (1947).
76. N. A. Sorensen and J. Mehlum, U. S. Patent 2,752,394, June 26, 1956.
77. A. A. Sokolova, N. A. Baranova and E. V. Nazarina, Gidroliz i Lesokhim. Prom., 10 (3), 6 (1957).
78. I. A. Pearl, U. S. Patent 2,483,559, Oct. 4, 1949.
79. I. A. Pearl and D. L. Beyer, Tappi, 39, 171 (1956).
80. E. W. Schoeffel, U. S. Patent 3,049,566, Aug. 14, 1962.
81. I. A. Pearl, U. S. Patent 2,602,089, July 1, 1952.
82. I. A. Pearl and E. E. Dickey, J. Am. Chem. Soc., 74, 614 (1952).
83. R. E. Davis, E. T. Reaville, Q. P. Peniston and J. L. McCarthy, J. Am. Chem. Soc., 77, 2405 (1955).
84. D. W. Glennie, H. Techlenberg, E. T. Reaville and J. L. McCarthy, ibid., 77, 2409 (1955).
85. K. Freudenberg, W. Lautsch, K. Engler, Chem. Ber., 73, 167 (1940).
86. B. Leopold, Acta. Chem. Scand., 6, 38 (1952).
87. R. Kempf, Zeit. Anal. Chem., 62, 284 (1923).
88. C. C. Bryan, U. S. Patent 2,506,540, May 2, 1950.
89. I. A. Pearl and E. E. Dickey, J. Am. Chem. Soc., 73, 863 (1951).
90. O. Töppal, U. S. Patent 2,745,796, May 15, 1956.

91. V. G. Voronin, A. P. Antonov and L. P. Zharkova, U.S.S.R. Patent 148,474, filed Aug. 4, 1961.
92. S. Kuriyama and C. Yamashita, Japan. Patent 4268, May 28, 1959; through Chem. Abstr., 53, 20804 (1959).
93. O. Töppel, Ger. Patent 1,119,244, Dec. 14, 1961.
94. V. V. Serpinski, S. A. Voitkevich, N. Ya. Lyuboshits, Zhur. Fiz. Khim., 27, 1032 (1953).
95. D. M. Ritter, J. Am. Chem. Soc., 69, 46 (1947).
96. S. K. K. Jatker and C. M. Despande, J. Indian Chem. Soc., 37, 1 (1960).
97. W. P. Fitzgerald, J. F. Nelson and O. C. Stotterbeck, U. S. Patent 2,857,361, Oct. 21, 1958.
98. R. Owen, U. S. Patent 3,028,255, April 3, 1962.
99. W. Steiner, Ger. Patent 1,021,981, Jan. 2, 1958.
100. D. Lauerer and M. Pesteiner, Ger. Patent 1,087,902, Aug. 25, 1960.
101. W. W. Koch, Brit. Patent 577,875, June 4, 1946.
102. A. Alt, U. S. Patent 2,774,791, Dec. 18, 1956.
103. R. Rutgers, J. Sci. Food Agr., 6, 735 (1955); through Chem. Abstr., 50, 6078 (1956).
104. C. H. Budlann, E. Budlann and A. Town, Rev. Chim. Acad. Rep. Populaire Roumaine, 2, 185 (1957); through Chem. Abstr., 51, 10751 (1957).
105. P. Carre and P. Baranger, Bull. Soc. Chim., 43, 73 (1928).
106. G. Ed. Utzinger and F. A. Regenass, Helv. Chim. Acta, 37, 1901 (1954).
107. O. Manns and S. Pfeifer, Mikrochim. Acta, 1958, 630.
108. K. Kratzl and E. Meisert, Monatsh. Chem., 88, 1056 (1957).
109. K. Oyamoda, Takamine Kenkyujo Nempo, 8, 32 (1956); through Chem. Abstr., 50, 333 (1956).
110. R. Matyja, Chem. Anal. (Warsaw), 8, 437 (1963).
111. J. Kinugawa and M. Ochiai, Yakugaku Zasshi, 78, 98 (1958).
112. L. C. Raiford and D. E. Fox, J. Org. Chem., 9, 170 (1944).
113. V. V. Somayajulu and N. V. Subba Rao, Current Sci. (India), 25, 86 (1956); through Chem. Abstr., 51, 5048 (1957).
114. J. Miglarese, U. S. Patent 2,564,022 Aug. 14, 1951.
115. L. Rosenthaler, Z. Anal. Chem., 183, 193 (1961).
116. Inventa A. G. für Forschung und Patent Verwertung, Brit. Patent 933,448, Aug. 8, 1963.
117. S. Takamuru and Y. Hachihama, Kogyo Kogaku Zasshi, 61, 1097 (1958); through Chem. Abstr., 55, 23425 (1961).
118. I. Tachi and K. Murakami, Nippon Mokuzai Gakkaishi, 5, 101 (1959); through Chem. Abstr., 54, 16001 (1960).
119. K. Murakami and I. Tachi, Nippon Mokuzai Gakkaishi, 6, 16 (1960); through Chem. Abstr., 54, 16001 (1960).

120. S. Kuriyama, Japan. Patent 17,198, Sept. 22, 1961.
121. F. M. Boyles, Am. Perfumer, 25, 243 (1930).
122. W. Diemair and W. Hennig, Z. Lebensm. Untersuch u. Forsch., 109, 150 (1959); through Chem. Abstr., 53, 10587 (1959).
123. J. Kamlet, U. S. Patent 2,975,214, Mar. 14, 1961.
124. R. G. Lange, J. Org. Chem., 27, 2037 (1962).
125. E. Hägglund, Chemistry of Wood, Academic Press, New York, 1951, p. 490.
126. Bialkowsky and DeHass, Tappi, 36, 330 (1953).
127. E. Hägglund, Chemistry of Wood, Academic Press, New York, 1951, p. 491.
128. E. Hägglund and T. Enkvist, U. S. Patent 2,711,430, June 21, 1955; U. S. Patent Reissue 24,293, Mar. 19, 1957.
129. M. E. Cisney and J. D. Wethern, U. S. Patent 2,816,832, Dec. 17, 1957.
130. D. W. Goheen, W. M. Hearon, M. E. Cisney and J. D. Wethern, U. S. Patent 2,914,568, Nov. 24, 1959.
131. H. Falk, Paperfabr., 7, 467 (1909).
132. F. J. Ball and R. Pueschel, U. S. Patent 2,976,273, March 21, 1961.
133. W. M. Hearon, W. S. MacGregor and D. W. Goheen, Tappi, 45, 28A (1962).
134. C. G. Swain and C. B. Scott, J. Am. Chem. Soc., 75, 141 (1953).
135. J. O. Edwards and R. G. Pearson, ibid., 84, 16 (1962).
136. M. E. Cisney and D. W. Goheen, U. S. Patent 2,908,716, Oct. 13, 1959.
137a. D. W. Goheen, U. S. Patent 2,840,614, June 24, 1958.
137b. D. W. Goheen and W. M. Hearon, U. S. Patent 2,914,567, Nov. 24, 1959.
138. A. Streitweiser, Chem. Revs., 56, 571 (1956).
139. D. W. Goheen, Forest Prod. J., 12, 471 (1962).
140. B. A. Dolgoplosk, V. A. Kropacker and N. I. Nikolaer, Doklady Akad. Nauk. U.S.S.R., 110, 789 (1956).
141. H. Richtzenhain and B. Alfredson, Chem. Ber., 86, 142 (1953).
142. W. T. Truce, G. H. Birum and E. T. McBee, J. Am. Chem. Soc., 74, 3594 (1952).
143. F. Boberg, G. Winter and J. Moos, Ann. Chem., 616, 1 (1958).
144. W. S. MacGregor, U. S. Patent 3,069,473, Dec. 18, 1962.
145. C. F. Bennett, D. W. Goheen and W. S. MacGregor, J. Org. Chem., 28, 2485 (1963).
146. T. H. Smedslund, U. S. Patent 2,581,050, Jan. 1, 1952.
147. F. Hübnett and K. Keim, U. S. Patent 2,935,533, May 3, 1960.
148. G. A. Wetterholm and K. R. Fossan, U. S. Patent 2,702,824, Feb. 22, 1935.
149. D. W. Goheen and C. F. Bennett, J. Org. Chem., 26, 1331 (1961).

150. C. A. Streuli, Anal. Chem., 30, 997 (1958).
151. D. C. Winer, ibid., 30, 2060 (1958).
152. P. Nylen, Z. Anorg. Allgem. Chem., 246, 227 (1941).
153. W. Steinkopf and S. Muller, Chem. Ber., 56, 1926-30 (1923).
154. E. J. Corey and M. Chaykovsky, J. Am. Chem. Soc., 84, 866 (1962).
155. E. L. Meutterties, ibid., 82, 1082 (1960).
156. S. Searles, Jr. and H. R. Hayes, J. Org. Chem., 23, 2028 (1958).
157. T. Cohen and T. Tsuji, ibid., 26, 1681 (1961).
158. N. Kornblum, J. W. Powers, G. J. Anderson, W. J. Jones, H. O. Larson, O. Levand and W. M. Weaver, J. Am. Chem. Soc., 79, 6562 (1957).
159. N. Kornblum, W. J. Jones, G. J. Anderson, ibid., 81, 4113 (1959).
160. I. M. Hunsberger and J. M. Tien, Chem. & Ind., 88 (1959).
161. A. P. Johnson and A. Pelter, J. Chem. Soc., 1964, 520.
162. P. Lagally, U. S. Patent 2,547,913, April 3, 1951.
163. L. Horner and P. Kaiser, Ann. Chem., 626, 19 (1959).
164. F. G. Bordwell and B. M. Pitt, J. Am. Chem. Soc., 77, 572 (1955).
165. E. W. Lard and D. P. Claypool, Chem. and Eng. News, 38 (Nov. 21) 46 (1960).
166. S. G. Smith and S. Winstein, Tetrahedron, 3, (3/4), 317 (1958).
167. E. J. Corey and M. Chaykovsky, J. Am. Chem. Soc., 84, 866 (1962).
168. D. J. Cram, B. Rickborn and G. R. Knox, ibid., 82, 6412 (1960).
169. M. P. Cava, R. L. Little, D. R. Napier, ibid., 80, 2260 (1958).
170. D. C. Nelson, U. S. Patent 3,023,183, Feb. 27, 1962.
171. C. W. Davis and F. A. Ehlers, U. S. Patent 2,858,290, Oct. 28, 1958.
172. K. Gutweiler and P. Otto, Ger. Patent 1,141,457, Dec. 20, 1962.
173. Kurashiki Rayon Co. Ltd., Brit. Patent 887,376, Jan. 17, 1962.
174. J. Forago, U. S. Patent 3,004,954, Oct. 17, 1961.
175. M. Katz, U. S. Patent 2,888,438, May 26, 1959.
176. A. D. Rees and S. J. Singer, Archiv. Biochem. Biophys., 63, 144 (1956).
177. J. E. Lovelock and M. W. H. Bishop, Nature, 183, 1394 (1959).
178. M. J. Ashwood-Smith, Int. J. Radiation Biol., 3 (1), 41 (1961).
179. H. M. Pyle and H. F. Boyer, Vox Sanguinis, 6, 199 (1961).
180. E. E. Rosenbaum and S. W. Jacob, Northwest Medicine, 63, 227 (1964).
181. H. L. Keil, Agricultural Chemicals, 20 (4), 23 (1965).
182. T. Enkvist, Paperi Puu, 43, 657 (1961).
183. T. Enkvist, B. Holm, J. Turunen and K. Turunen, Finska Kemistsamfundets Medd., 70, 5 (1961); through Chem. Abstr., 56, 3697 (1962).
184. T. Ashorn and T. Enkvist, Acta. Chem. Scand., 16, 548 (1962).
185. T. Enkvist, J. Turunen and T. Ashorn, Tappi, 45, 128 (1962).
186. C. Dorée and L. Hall, J. Soc. Chem. Ind., 43, 2571 (1924).
187. R. Willstatter and L. Kalb, Chem. Ber., 55, 2637 (1922).
188. A. Lindblad, Swed. Patent 70,795, Dec. 16, 1930; through Chem.

Abstr., 26, 2049 (1932).
189. E. E. Harris and H. Adkins, Paper Trade J., 107 (20), 38 (1938).
190. E. E. Harris, J. D. Tanni and H. Adkins, J. Am. Chem. Soc., 60, 1467 (1938).
191. H. Adkins, R. L. Frank and F. J. Bloom, ibid., 63, 549 (1941).
192. L. M. Cooke, J. L. McCarthy and H. Hibbert, ibid., 63, 3052 (1941).
193. J. R. Bower, J. L. McCarthy and H. Hibbert, ibid., 63, 3066 (1941).
194. C. P. Brewer, L. M. Cooke and H. Hibbert, ibid., 70, 57 (1948).
195. W. Lautsch, Cellulosechem., 19, 69 (1941).
196. W. Lautsch, Brennstoff-Chem., 23, 265 (1941).
197. W. Lautsch and K. Freudenberg, Ger. Patent 741,686, Sept. 30, 1943.
198. W. Lautsch and G. Piazolo, Chem. Ber., 76, 486 (1943).
199a. Inventa A.G. für Forschung und Patent Verwertung, Swiss Pattent 305,712, March 16, 1955.
199b. J. Giesen, Canadian Patent 547,591, Oct. 15, 1957.
199c. J. Giesen, Canadian Patent 555,997, Apr. 15, 1958.
199d. J. Giesen, Canadian Patent 559,006, June 17, 1958.
199e. J. Giesen, Canadian Patent 581,528, Aug. 18, 1959.
199f. J. Giesen, U. S. Patent 2,991,314, July 4, 1961.
200a. M. Oshima, Y. Maeda and K. Kashima, Canadian Patent 700,209, Dec. 22, 1964.
200b. M. Oshima, Y. Maeda and K. Kashima, Canadian Patent 700,210, Dec. 22, 1964.
201. H. C. Reed, U. S. Patent 1,217,218, Feb. 27, 1917.
202. H. C. Reed, U. S. Patent 1,310,713, July 22, 1919.
203. E. Heuser, H. Roesch and L. Gunkel, Cellulosechemie, 2, 13 (1921).
204. E. Heuser and A. Winsvold, Cellulosechemie, 2, 113 (1921).
205. K. Freudenberg, M. Meister and E. Flickinger, Chem. Ber., 70B, 500 (1937).
206. H. Richtzenhain, Acta Chem. Scand., 4, 206 (1950).
207. H. Richtzenhain, Svensk Papperstidn., 53, 644 (1950).
208. A. P. Kresnova, E. A. Porshina, S. I. Sukhanovskii and M. I. Chudako Zhur. Prikdad. Khim., 30, 802 (1957).
209. O. Kleinnickel, Ger. Patent 1,006,412, April 18, 1957.
210. A. Poljak, Ger. Patent 1,013,285, Aug. 8, 1957.
211. D. H. Grengaard, U. S. Patent 2,928,868, March 15, 1960.
212. M. I. Chudakov, U.S.S.R. Patent 134,684, Jan. 10, 1961.
213. L. C. Bratt, Paper Trade J., 147 (14), 40 (1963).
214. E. Heuser and C. Skiöldexbrand, Z. Angew. Chem., 321, 41 (1919); through Chem. Abstr., 13, 2758 (1919).
215. F. Fischer and H. Schrader, Ges. Abh. Kenntnis Kohle, 5, 106 (1921); through Chem. Abstr., 17, 3092 (1923).
216. H. Tropsch, Ges Abhandl. Kenntnis Kohle, 6, 293 (1923).
217. T. L. Fletcher and E. E. Harris, J. Am. Chem. Soc., 69, 3144 (1947)

218. K. Freudenberg and K. Adam, Chem. Ber., 74, 387 (1941).
219. V. N. Sergeeva and Ya. A. Surna, Tr. Inst. Lesokhoz. Probl. i Khim. Drevesiny Akad. Nauk. Latv.SSR, 21, 107 (1960); through Chem. Abstr., 56, 6216 (1962).
220. V. G. Panasyuk, Zhur. Priklad. Khim., 30, 598 (1957).
221. V. G. Panasyuk, ibid., 813 (1957).
222. V. G. Panasyuk, ibid., 1049 (1957).
223. V. G. Panasyuk, L. V. Panasyuk, N. S. Maksimenko and F. S. Lapshin, Gidroliz i Lesokhim. Prom., 12 (7), 16 (1959).
224. V. N. Sergeeva and V. Jaunzems, Tr. Inst. Lesokhoz. Probli i Khim. Akad. Nauk. Latv. SSR, 24, 39 (1962); through Chem. Abstr., 58, 5879 (1963).

20

POLYMERIC PRODUCTS

C. H. Hoyt and D. W. Goheen

I. Introduction.

A. <u>Lignin in Wood</u>. Wood itself can be regarded to be a lignin plastic reinforced by cellulosic fibers. Lignin in wood can be modified, together with other wood components, by plasticizing (liquid ammonia, urea, polyethylene glycol), by acetylation, by reacting with phenolformaldehyde (e.g., use of wood meal in phenol-formaldehyde plastics) and by fiber separation using high pressure steam followed by heat compression to hardboards (Masonite and Asplund processes). These applications are outside of the scope of this book, although better understanding of lignin chemistry will no doubt contribute to the improvement of these products in the future.

Many millions of tons of wood are processed each year by the chemical pulping industry for separation of lignin from cellulosic fibers. The spent liquors from the pulping operation therefore represent the most readily available source of lignin.

B. <u>Available Lignin Products</u>. The total production of pulp in the United States in 1968 can be divided into the following approximate percentages: sulfate (kraft), 65.8%; soda, 0.5%; sulfite, 10.7%; semichemical, 8.2% and groundwood, 14.1% (1). The organic by-products made available by the two principal pulping processes each year are shown in Table 20.1.

Table 20.1. Organic By-products Formed in Pulping Processes in the U.S.A. in 1964.

Kraft Pulping, tons		Sulfite Pulping, tons	
Thiolignin	8,800,000	Lignin sulfonic acid	1,920,000
Fatty and resin acids	405,000	Hexoses	550,000
Formic acid	600,000	Pentoses	87,000
Acetic acid	840,000	Acetic acid	140,000
Acids derived from carbohydrates	4,900,000	Formaldehyde and small amounts of ethyl alcohol, furfural, and acetone	18,000
All other organics (phenols, etc.)	650,000		

The sulfate or kraft process produces by far the largest amount of pulp. Consequently, sulfate lignin is the most plentiful lignin available in spent pulping liquors. In the United States over 27,000,000 tons of kraft black liquor solids were produced in 1968. Of this total, approximately 10,800,000 tons were inorganic salts used principally for the recovery of pulping chemicals. The remainder, approximately 16,200,000 tons, was organic material. Except for the fatty and rosin acids, practically all of this material is burned to provide heat for evaporation of the liquor and to recover the inorganic chemicals by forming sodium carbonate and reducing the sulfur present in various forms to sodium sulfide. These products are recausticized and recycled for subsequent pulping. The burning of the organic material results in formation of approximately 3.2×10^{14} BTU of heat energy, on an annual basis. It is estimated that at least 2.6×10^{14} BTU are necessary for the evaporation of the black liquors so that only a fairly small excess of heat energy is realized from the recovery operation.

The rosins and fats in the wood are generally recovered in the form of tall oil, which is an important and profitable by-product. Attempts to recover organic acids from kraft black liquor have been unsuccessful. The formic and acetic acids are not economically recoverable from the relatively large amount of alkaline black liquor, nor have any of the proposals for recovery of saccharinic and iso-saccharinic acids been commercialized. The small molecular weight lignin degradation products are present as a mixture consisting of such compounds as guaiacol, vanillin and catechol, the sum of which totals only 4% of the black liquor solids.

Kraft lignin (or thiolignin) is the major constituent of kraft black liquor solids. It can be recovered in reasonably high yields by acidifying and filtering the precipitated lignin. To date, however, of the 85 kraft mills in the United States, only the Westvaco mill at Charleston, South Carolina is processing kraft lignin for sale. The production is estimated at about twenty million pounds per year. This is only about a tenth of a percent of the total lignin available in kraft black liquors. Another company, the Crown Zellerbach Corporation, Bogalusa, Louisiana, processes black liquor for the production of dimethyl sulfoxide as described in the preceding chapter.

Although it appears to be a prodigious waste of raw material to burn the organic matter of the black liquor, it must be remembered that fuel value of this material is considerable. If the organic substances are processed for sale, they must be replaced with some other fuel since heat and reduction of oxidized sulfur species are vital in the recovery of pulping chemicals. Consequently, the value of any product from black liquor must be high enough to compensate for the cost of the capital installation and of additional fuel to replace the lignin removed.

In many ways, kraft lignin is a more versatile raw material than are lignin sulfonates from sulfite spent liquor. It can be easily isolated from the spent pulping liquor, is soluble in many organic solvents, possesses thermoplastic

POLYMERIC PRODUCTS

properties, contains less sulfur and is soluble in caustic. Kraft lignin is, of course, a polymeric material consisting of about 65% carbon, 5% hydrogen and 30% oxygen. It is quite reactive and can be etherified, esterified, nitrated, mercurated and halogenated. It reacts with phenols, amines, aldehydes and sulfides. Reaction with sodium sulfite produces a series of sodium lignin sulfonates with varying amounts of sulfonic acid groups. These sulfonated kraft lignins are more expensive to produce than lignin sulfonates, but they have the advantage of being free of carbohydrate constituents.

In contrast with the kraft process, only modest progress toward recovery of pulping chemicals has been made for the sulfite process. In North America, there are about 70 sulfite mills producing nearly 3,000,000 tons per year of spent liquor solids, equivalent to about 4.8×10^{13} BTU of potential heat energy. The more than 27,000,000 tons of water that must be evaporated before the spent liquor solids can be combusted would consume a better part of this energy. There has been an increasing effort for more than forty years to generate utilization for sulfite spent liquor solids, and at present about 28 laboratories are active in this area in North America alone (2). In spite of this intensive effort, only 20% of the spent liquor solids are being sold as by-products, including the spent liquor sprayed on roads. There has been an increase of 27% per year in utilization; and in 1961, the income from lignin sulfonate production was estimated at 10.5 million dollars, 4.95% of the total silvichemical income (3). Trends are shown by results of a survey of thirteen Wisconsin mills, Figure 20.1 (4).

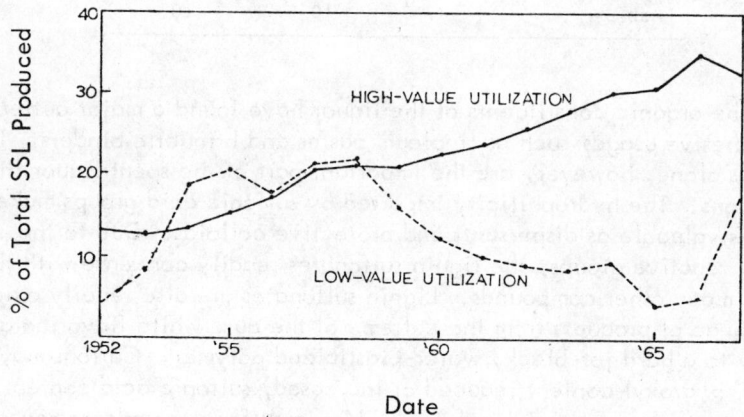

Figure 20.1. Utilization of spent sulfite liquor solids as percent of the total produced by thirteen Wisconsin mills (members of the Sulfite Pulp Manufacturer's Research League) (4).

This represents an annual sale of 210,000,000 pounds of spent liquor solids. By 1963, the income had increased to nearly double that of 1961. The sale of total by-product chemicals from the paper industry overtook that of Naval Stores in 1965. It is important to note that the output of sulfite spent liquor solids is a diminishing rather than increasing one. The peak U.S. sulfite pulp production occurred in 1966, and it dropped 3% in 1967. No increased production facilities for sulfite pulping are now in planning (5). While the sulfite process lends itself especially well for the manufacture of dissolving pulp, it does not equal kraft pulps for papermaking. As a consequence, the trends in future growth in pulp production favor kraft mills. It is possible that before long the increase in sales of sulfite spent liquor solids and by-products will meet production output and water pollution will be less of a problem to much of the industry.

The approximate composition of calcium base spent sulfite liquor is shown in Table 20.2 (6).

Table 20.2. The Approximate Composition of Calcium Base Spent Sulfite Liquor (6).

	Softwood	Hardwood
Lignin sulfonates	55	42
Hexose sugars	14	5
Pentose sugars	6	20
Sugar acids and residues	12	20
Resins and extractives	3	3
Ash	10	10

The organic constituents of the liquor have found a major outlet in the direct adhesive usages such as linoleum pastes and briquette binders. The lignin sulfonates alone, however, are the important part of the spent liquor in many applications. The hydrophilicity imparted by sulfonic acid groups makes lignin sulfonates valuable as dispersants and protective colloids. Due to the presence of highly reactive groups, the lignin sulfonates readily condense with themselves and with many other compounds. Lignin sulfonates are also readily oxidized to a wide range of products from the extreme of the pure white flavoring agent, vanillin, to a hard jet-black, water-plasticized polymer. Cations may be exchanged, hydroxyl content reduced or increased, sulfonic acid content varied, and average molecular weight altered. Lignin sulfonates undergo reactions as phenols, acids, alcohols, ethers, and through carbonyl groups.

A rather impressive number of uses has already been uncovered for the lignin by-products. Many of these uses are the same or similar for kraft and sulfite lignins. In some, kraft lignin or a derivative works better than the lignin sulfonates; in others, lignin sulfonates perform better.

For many years in various parts of the world, plants and pilot plants for the saccharification of wood have been built and studied. Everywhere, these plants have failed to become economically competitive with the production of sugars from agricultural crops and consequently survive only in the Soviet Union where no less than fifteen commercial-scale wood hydrolysis plants are in operation. These plants produce glucose which serves mainly as a substrate for production of alcohol and fodder yeast.

A by-product of the wood saccharification is the insoluble hydrolysis lignin, amounting to 20-30% of the total wood used. These lignins may be modified by methods similar to those applied to kraft lignins. Proposed uses for hydrolysis lignin or its derivatives have included the following: soil conditioners, ingredients and fillers for various resins and rubber, binders in particle- and hardboards, depressants in ore flotation, decay retarding agents in fabrics, briquetted fuels and grinding aids for cement.

II. Preparation of Lignin Products.
    A. Lignin Sulfonates. Lignin sulfonates are produced either as free sulfonic acids or as calcium, magnesium, sodium or ammonium salts. They are commonly sold in the form of spray dried powders or as 40-60% solutions in water. While sodium and ammonium base spent liquors create no particular problems in evaporation, magnesium salts are less soluble and calcium lignin sulfonates particularly generate scaling on heat exchange surfaces consisting of less soluble lignin sulfonates and inorganic calcium salts which interfere with efficient heat transfer. Consequently, specifically constructed evaporators are necessary for the recovery of calcium base lignin sulfonates. This complicating factor plus the acidity of the spent liquor makes concentration a relatively expensive operation (7).

Alternative methods for the concentration of sulfite liquors are being considered. One method showing great promise is the reverse osmosis process (8) developed for the desalination of sea water. To accomplish reverse osmosis, sufficient pressure needs to be placed on the sulfite liquor in contact with a semipermeable membrane to overcome the osmotic pressure of the solution. Under these conditions, only water along with a small amount of sulfurous acid passes through the membrane, leaving a more concentrated solution of spent liquor solids behind. The equipment involved consists of a bank of tubes manifolded in a container. The tubes are made of a permeable resin reinforced by wound fiber. The liquor is circulated under pressure through the tubes on the membrane side and water is collected on the opposite side at atmospheric pressure.

Ammonia base spent liquor can be evaporated without problem of scaling and produces relatively little ash in burning. On the other hand, ammonium lignin sulfonates have the tendency toward self-condensation upon heating. While calcium, sodium and magnesium lignin sulfonates dissolve only in aqueous media, ammonium lignin sulfonates have appreciable solubility in dimethyl sulfoxide, hot formic acid, ethanolamines, acetamide, dimethyl formamide, methanol, glycerine, ethylene glycol, triethylene glycol and propylene glycol. Sodium and magnesium base sulfite processes lend themselves to chemical recovery (9, 10). Solutions of sodium lignin sulfonate are very light in color and are used as additives in detergents or as dispersants for light-colored products. Melms and Schwenzon (11) and Saponitski (12) have published extensive reviews covering the utilization of spent sulfite liquor.

B. Kraft Lignins.

1. Isolation. Black liquors of the conventional kraft (sulfate) pulping process contain lignin, hemicelluloses, acids derived from carbohydrates and small amounts of extractives. Of these, only lignin precipitates upon acidification and can be separated by filtration. In spite of its apparent simplicity, the separation procedure is difficult to conduct in practice. Simple acidification of kraft black liquor results in the formation of a lignin that is almost nonfilterable owing to near-colloidal particle sizes. Consequently, a coagulation step is required prior to the filtration.

For complete precipitation of lignin, the pH of the liquor must be lowered to 3 or less. Since acidification to this level requires an amount of acid equivalent not only to the lignin salts but also to other organic and inorganic sodium salts, a great deal of acid would be wasted if black liquor were simply acidified directly to pH 3. Furthermore, difficulties with frothing due to escape of hydrogen sulfide and carbon dioxide would be encountered. This problem has been overcome by conducting the neutralization in two steps. In the first step, it is sufficient to lower the pH by flue gas carbon dioxide only to about 9, in order to precipitate a major portion of the lignin as sodium salt. By heating, this salt can be coagulated and filtered or, alternatively, if heated to a higher temperature (>90°C), it melts to a viscous tar which can be separated simply by decantation. The sodium salt is then suspended in water and acidified with sulfuric acid to pH 3 or less at 80-90°C. Since the filtrates from both acidifications must be, in the interest of economy, returned to the kraft recovery cycle, the choice is limited to only two acids, carbon dioxide and sulfuric acid. Use of other acids would introduce undesirable anions into the kraft process.

The salient features of the described procedure were contained already in the patent by Tomlinson and Tomlinson in 1946 (13), and practically all later procedures have been variations of the original technique. Gray et al. (14) describe acidification of black liquor with carbon dioxide under pressures of 20 to 165 psi. Hydrogen sulfide is collected for recycle into the recovery operation and thiolignin is filtered. Giesen (15) patented a method of carbonation of weak black liquor in a vertical tube with carbon dioxide at a pressure of 20 atmospheres. N. Nikitin and Obolenskays (16) claimed that carbon dioxide precipitation should be done under a pressure of 3-5 atmospheres, at 80-85°C in 15-20 minutes in order to obtain 80% of the lignin as precipitated sodium salt. Ouchi (17) uses only sulfuric acid for the initial precipitation of the sodium salt. The filtrate is treated with more acid to precipitate additional lignin and some hemicelluloses. The second filtrate is used to precipitate lignin salt from fresh black liquor.

Merewether et al. (18) patented a process in Australia consisting of the acidification of 20-30% solids black liquor to a pH of 8.5-9.5 at 65-75°C. Ball and Vardell (19) have patented a process of heating carbonated black liquor (pH 9.7) to a temperature of 305°F at 150 psi to force the lignin to separate as a viscous lower layer which can be removed by decantation. It can either be used as the salt or further acidified with sulfuric acid. In a second patent (20), a continuous process is described in which black liquor of 30-50% concentration is heated to 202-205°F under pressure prior to carbonation to a pH of 8.2-10.35 in order to achieve adequate coagulation. A thorough review of kraft lignin separation has been published by Merewether (21).

Kennedy and Jernigan (22) have patented a novel electrolytic process in which kraft lignin precipitates at the anode and is collected. Caustic is regenerated at the cathode area. Ball (23) patented a procedure consisting of freezing an acidified black liquor. On thawing, lignin is obtained in a readily filterable form. In another patented procedure (24), kraft lignin is obtained in the form of free-flowing, dust-free, discrete particles. The procedure consists of adding an ammoniacal solution of kraft lignin to a slurry of precipitated and washed kraft lignin and spray drying the resultant mixture. The evaporation of ammonia leaves the lignin particles surrounded with a resistant crust.

2. Properties. Kraft lignin is recovered either as its sodium salt or in the form of free lignin containing very little ash. Both forms when dry are free flowing brown powders.

In the United States the only commercial sulfate lignin, "Indulin" (Westvaco), is produced in three grades: "Indulin C" is the crude sodium salt which contains some occluded black liquor solids and therefore has a relatively high ash content of 22.5%; "Indulin B" is a purified sodium salt which has a lower ash content of about 9%; "Indulin AT" represents an acidified lignin with an ash content of less than 1%.

Free kraft lignin contains about 1.5% of organically bound sulfur. The exact nature of this sulfur is not as yet completely understood. In contrast to lignin sulfonates, kraft lignins have sintering points in the vicinity of 210°C and tend to flow at elevated temperatures. Kraft lignins from softwoods, such as Indulin AT which is derived from pine, have methoxyl contents of about 14%. Hardwood kraft lignins, currently not commercially available, contain about 21% methoxyl. Kraft lignins are soluble in a number of solvents, such as glycols and cellosolves, oxygen containing heterocyclic compounds, and dimethyl sulfoxide. They show limited or no solubility in simple alcohols and ethers and are insoluble in terpenes and other hydrocarbons. Many binary solvent mixtures are better solvents than either component alone. For example, Indulin AT is freely soluble in a mixture of butyl alcohol and nitroethane, but insoluble in butyl alcohol and only slightly soluble in nitroethane (25).

The number average molecular weights ($\overline{M_n}$) for kraft lignins have recently been determined by Marton and Marton (26) by a thermoelectric vapor pressure osmometer. They found a value of 1600 for pine lignin and 1050 for hardwood lignin. When pine lignin was partially demethylated by the method of Ball and Puschel (27), by heating black liquor to a high temperature, the $\overline{M}_n$ value fell to 900. Using weight average molecular weights ($\overline{M}_w$) determined by Goring and Yean for these preparations by the equilibrium sedimentation method (28), the $\overline{M}_w/\overline{M}_n$ ratios could be determined. The pine lignin had a ratio of 2.2, indicating moderate polydispersity. Hardwood lignin had a ratio of 2.8, and the partially demethylated lignin, 3.2, indicating that they are highly polydisperse as a consequence of fragmentation in the secondary treatment.

Kraft lignins contain a variety of functional groups such as carbonyls, carboxyls, aliphatic hydroxyl groups including benzyl alcohols and ketol hydroxyls, phenolic hydroxyls, ethers and mercaptan groups, as well as possibly thioether groups and ethylenic linkages in the side-chains. A review of various functional groups present in kraft lignin has been published (29) and is discussed in detail in Chapter 16.

C. Hydrolysis Lignin. From any wood saccharification process, the lignin residue is recovered by centrifugation or filtration. This hydrolysis lignin is condensed and contains some unhydrolyzed cellulose residues.

Basically two types of hydrolysis are practiced. The first, typified by the commercial Rheinau process, utilizes a two-stage hydrolysis. In the first step, fuming hydrochloric acid is used to hydrolyze the cellulose to a mixture of water-soluble dextrans at near-ambient temperatures (20-30°C). The acid is then diluted and the mixture boiled in a second-stage hydrolysis to give glucose, along with various other sugars and decomposition products, and lignin as the insoluble residue.

The second type of hydrolysis is typified by the Scholler process in which the hydrolysis is completed in one stage with dilute acid at elevated temperatures (160-190°C) and pressures. The yield is lower since more of the glucose is decomposed during the high temperature hydrolysis. Other proposed processes have been variations of the two methods described.

Glucose and its fermentation product, ethyl alcohol, are the principal products of wood saccharification. Unfortunately, they can be produced somewhat more economically from other sources. On the other hand, a substantial return from the by-product lignin could make the wood saccharification process economically attractive.

III. Modification of Lignin Products.
    A. Modifications of Sulfite Lignins.
        1. Removal of Sugars and Other Non-Lignin Components. In many applications of sulfite spent liquor products, it is desirable to remove the sugars, which represent as much as 20% of the liquor solids. Often this is done by fermentation to alcohol or by conversion to yeast (30). At the present time, only one plant in the United States (Georgia Pacific Corporation, Bellingham

Division) produces alcohol from spent sulfite liquor sugars. This plant was built during World War II and now produces 3,500,000 gallons of 95% alcohol per year. The world-wide production of sulfite alcohol reached a peak in 1951 corresponding to about 30,000,000 gallons of 95% alcohol.

In Europe, the production of sulfite pulp has declined since then, as has alcohol production. The yield of 95% alcohol is reported to be 20 gallons per ton of pulp.

Torula yeast production was started in 1947 in the United States. The product is composed of the dried cells of the yeast Candida utilis. It contains a minimum of 50% protein, all of the essential amino acids, and is rich in lysine. It is a source of iron, calcium, phosphorus, and iodine as well as the entire vitamin B complex. Torula yeast has a bland flavor which allows its blending with almost any prepared food combination. In its production, both hexose and pentose sugars contribute to the growth of the yeast and no alcohol is generated. Torula yeast was admitted to the USP XV in 1956 for use as a nutritional supplement. Assuming 15 to 25% sugars in spent sulfite liquor solids, the yeast production amounts to 150-250 pounds per ton of pulp.

The sugars themselves may be separated in the laboratory by using anion exchange resins which retain the anionic lignin sulfonates but not sugars. Dow Chemical Company has developed an ion exclusion process for separation of ionic from nonionic materials, using specially developed ion exchange resins (31). The University of Washington Pulp Mills Research Group has investigated the separation of lignin sulfonates from sugars using this procedure (32). The first effluent from the ion exclusion column contains the lignin sulfonates, followed in later fractions by sugars and other non-ionized materials. By intermittent additions of the spent sulfite liquor to the ion exclusion column, lignin sulfonates and sugars can be separated on a continuous basis, Figure 20.2 (33).

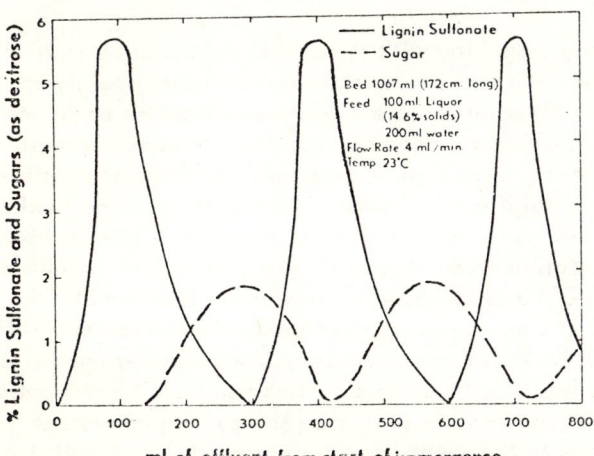

Figure 20.2. Separation of the lignin sulfonates from the sugars of spent sulfite liquor by ion exclusion chromatography (33).

The Finnish Pulp and Paper Research Institute has reported separations of sugars from lignin sulfonates by Sephadex gel chromatography as well as by ion exclusion chromatography (34). As yet, none of these chromatographic procedures has proven to be useful for the production of purified lignin sulfonates on a commercial basis.

The Sulfite Pulp Manufacturers' Research League has explored a novel process of isolating and separating spent liquor sugars by reacting them with acetone in the presence of sulfuric acid (35). The resulting acetone sugars (di-O-isopropylidene derivatives) are soluble in acetone and can be separated from the acetone insoluble lignin sulfonates. Upon concentration of the acetone solution, diacetone mannose (di-O-isopropylidene-D-mannose) precipitates and may be hydrolyzed to free D-mannose. Diacetone pentoses are steam distilled, leaving a residue of diacetone galactose and -glucose. All of these derivatives are hydrolyzable to the corresponding free sugars.

The sugars in spent liquor are chemically altered both by acidic and alkaline treatments. Heating the spent liquor with acid converts hexose sugars to levulinic acid. Concurrently, lignin sulfonates condense to a black insoluble material which at one time was available commercially from Penobscot Chemical Fibre Company for use as a reactive filler in rubber and plastics. Heating in mildly alkaline conditions causes only minor changes in the lignin sulfonates, while the sugars are extensively converted to saccharinic acids. As a consequence, alkaline conditions during the evaporation of the spent liquor result in considerable decrease in the amount of the reducing sugars. Access of air-oxygen during the alkaline heating, as well as the presence of oxidation catalysts accelerates the destruction of sugars; and such oxidative treatments are commonly used to decrease the amount of carbohydrates in spent liquors. Desugared lignin sulfonates find application in uses where the growth of bacteria, mold, slime or other microorganisms is harmful. Sugars also interfere with the surface active properties of the lignin sulfonates and are often removed for this reason.

Aqua-Chem Corporation, working with the Sulfite Pulp Manufacturers' Research League in Appleton, Wisconsin have developed an electrodialysis process for spent liquor. A 50% solution of spent liquor is fed into packs of permeable membranes, each having a cathode and anode. A direct current is applied across the pack, and the spent liquor is separated into three parts: an aqueous solution of inorganic pulping chemicals, lignin sulfonic acids, and a mixture of sugars with low molecular weight organic acids. The process has been studied in pilot plant scale with flow rates up to 15 gallons per minute (36).

Sugar-free basic salts of calcium lignin sulfonates are produced by the Howard Process, developed in connection with the production of vanillin. The process has been in use by Marathon Paper Company, Division of American Can Company, since 1936. The basic calcium lignin sulfonates have found application as dispersants, tanning agents and in plastics. The process consists of addition of calcium hydroxide to the spent liquor in three stages. The first stage recovers calcium sulfite for reuse in pulping liquors. The second stage separates lignin as a basic salt of calcium lignin sulfonate. The third-stage precipitate, which consists of a small additional amount of lignin sulfonate along with excess lime, is added to the lime of the first stage of a succeeding batch. Effluent of the third stage is discarded. For each pound of lignin sulfonate recovered, eleven pounds of liquor are processed. Recoveries of 90-95% of the lignin are generally accomplished (37).

2. Cation Selection. Calcium lignin sulfonates may be easily converted to sulfonates of other bases by the addition of a soluble sulfate of the cation desired, precipitating calcium ions as the insoluble sulfate. Aluminum, potassium, iron, zinc, chromium, and a number of other lignin sulfonates are made in this manner, Figure 20.3. Ammonium lignin sulfonates also can be conveniently exchanged to other bases by the addition of the appropriate hydroxides and by ion exchange procedures (38). Sodium and magnesium lignin sulfonates, on the other hand, require the use of cationic exchange resins for conversion to other bases.

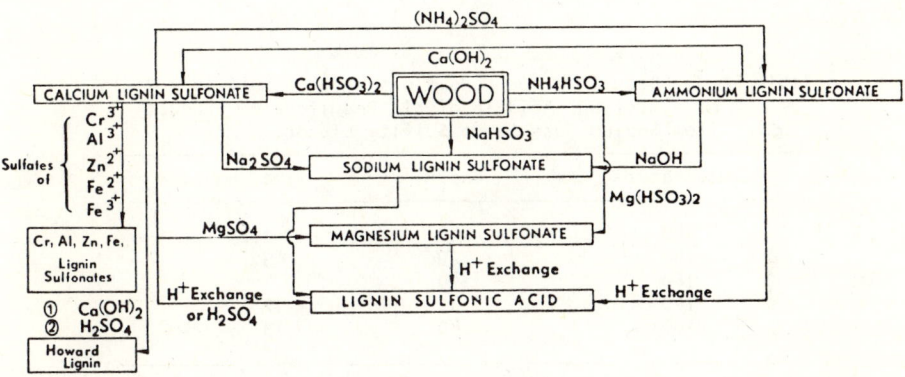

Figure 20.3. Procedures for base exchange of lignin sulfonates.

The Abiperm ion exchange process (Abitibi Power and Paper Company and Pfaudler Permutit Inc.) was developed for the purpose of recovery of chemicals from the sodium bisulfite process of pulping (39). In addition, this process makes lignin sulfonic acid available. The process consists of passing steam-stripped spent liquor through a bed of Permutit Q (a sulfonated polystyrene exchange resin) which removes sodium and other cations from the lignin sulfonate. The cations are released from the resin by a sulfurous acid treatment, and the released bisulfites transferred back to the pulp mill. Pilot plant studies have shown the recovery of over 80% of the base in sodium base spent liquor without deterioration of the resin. Another patent of the same type was issued to L. K. Swenson (40). The Prichard-O.R.F. (Ontario Research Foundation) process (41, 42) provides for removing unwanted multivalent ions prior to exchange of the monovalent base. This process gives a recovery of around 80% of the input base when the effluent contains up to 1% combined sulfur dioxide. However, when higher levels of combined sulfur dioxide are required in the cooking liquor, the recoveries fall off. Only 54% recovery was obtained at 2.25% combined sulfur dioxide. In the Prichard-Fraxon process (43), acetone is added to the sulfurous acid, thus allowing the pH to be lowered and additional sulfur dioxide to be absorbed. In addition, it was found that during the removal of acetone by steam stripping, the multivalent ions were eliminated by precipitation.

3. <u>Desulfonated Products from Lignin Sulfonates</u>. In connection with the vanillin manufacture, partially desulfonated lignin sulfonates are often recovered by precipitation with acidic reagents. Desulfonation also may be accomplished using the conditions shown in Table 20.3, which all result in 40% recovery of the desulfonated product.

Table 20.3. Desulfonation Conditions which Result in a Yield of 40% of an Acid Insoluble Desulfonated Product from Sodium Base Spent Sulfite Liquor.

Temperature, °C.	%NaOH	Time, hours
180	40	2.25
180	20	9
160	40	3
140	40	12
100	40	15

Although desulfonated lignin sulfonates have less effect on surface tension than lignin sulfonates, Figure 20.4, they do reduce the interfacial tension more, Figure 20.5, and are more efficient dispersants for certain materials. In addition, they are more stable than lignin sulfonates in such uses as boiler water treatment and oil well cementing additives. Their increased reactivity with resins and water insolubility make them useful as resin extenders.

The desulfonated product obtained by means of ammonia, called the "AS lignin," has received considerable attention in Czechoslovakia, but is not in production in North America (44). In the process, enough ammonia is added to give a 5-10% solution in the liquor. The desulfonation reaction is carried out above 100°C, under pressure. Excess ammonia and insoluble calcium sulfite are removed, and the liquor acidified to precipitate the AS lignin.

POLYMERIC PRODUCTS 845

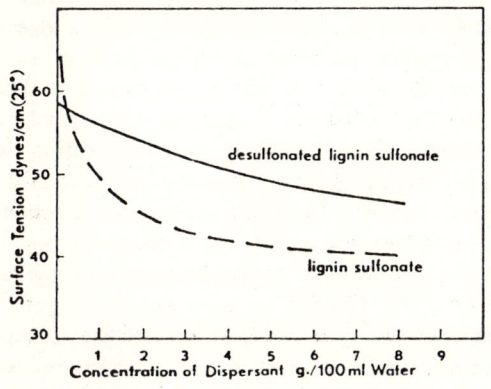

Figure 20.4. Effect of desulfonation of lignin sulfonate on surface tension.

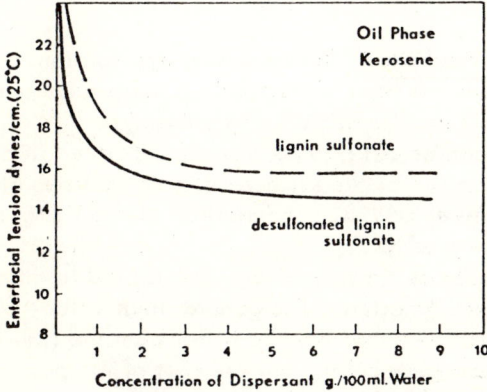

Figure 20.5. Effect of desulfonation of lignin sulfonate on interfacial tension.

B. Modifications of Kraft Lignin.
1. *Chemical Modifications.* Chemical modifications of kraft lignin can drastically change such properties as solublilities in various media, solution viscosities, polyelectrolyte properties, etc. (45). Common modifications consist of the following: a, Formation of ethers by reactions with alkyl aryl halides or with alcohols and acid catalysts. Other ethers can be formed by reactions with acrylonitrile, chloroacetic acid, epichlorohydrin, chloroformates and alkyl chlorides; b, Formation of esters by reactions with carboxylic acids or acyl halides and anhydrides, particularly in the presence of basic catalysts such as pyridine; c, Formation of polymeric condensation products with phenols, aldehydes and amines to produce thermosetting resins. d. Conversion to sparingly

soluble phenolate derivatives in reactions with various salts of heavy metals such as calcium, copper, tin, mercury, zinc and others; e, Sulfonation to water-soluble kraft lignin sulfonates with solutions of alkali sulfites at elevated temperatures and pressures. The degree of sulfonation can be controlled by the selection of appropriate conditions; f, Nitration can be carried out with nitric acid-sulfuric acid mixtures or with other nitrating reagents. Care must be taken to avoid oxidation to oxalic acid and carbon dioxide; g, Chlorination occurs readily by reaction with aqueous chlorine, chlorine in anhydrous organic solvents, thionyl chloride or sulfuryl chloride. As with other lignin preparations, a considerable amount of chlorine can become organically bound (more than 30% by weight). In aqueous solution, chlorination is accompanied by demethylation and other cleavage reactions; h, Oxidation with air or oxygen in alkaline solution produces derivatives with changed solubilities and melting points. At low levels of oxygen uptake, no loss of lignin occurs; but at higher levels, partial degradation to smaller molecules takes place; i, Demethylation of lignin can be accomplished by acidic and alkaline treatments and results in the formation of catechol groups, increasing the reactivity in condensations, e.g., with formaldehyde.

IV. Utilization of Lignin Products. The necessary applications research for successful product development has been discussed in an outstanding manner by Wiley (46), with specific reference to kraft lignin industry.

The uses to which lignins can be put may be broadly separated into three classes: 1, combustion; 2, utilization of the surface active properties of water-soluble salts of the lignin derivative; and 3, condensation of the lignin so that it becomes an integral part of the product.

Burning of spent sulfite liquor as a means of disposal is used in Europe and, to a much smaller extent, in North America. The general heat value, 8000 BTU/lb. of SSL solids, is sufficient to evaporate the water from the liquor and leave some extra calories for steam generation, but the cost of the plant and the operation tends to negate the realizable profit. Several furnace arrangements have been developed which allow the liquors to be burned without the addition of oil or coal and in addition afford practical ash removal (47). In these, the concentrated liquor is conducted into a very hot zone in such a fashion that the water may be driven off without lowering the temperature below the burning point of the liquor solids. This is usually accomplished by some rotary or oscillating system of spraying in order to limit the cooling to only one area of the burning chamber while another area nearby is burning and heating the furnace.

A. Oil Well Drilling Muds. In the drilling of oil wells, water and dispersants are continuously fed into the hole to form, together with suspended fine clay, what is usually called "the drilling mud." The mud serves many purposes. It cools and lubricates the drill, hydraulically cuts, removes cuttings, transports cuttings to the surface, cements or seals the wall of the drill hole, supports the walls of the hole, holds cuttings in suspension during drilling delays, and forms a hydrostatic head which serves as a means of controlling high pressure gas, oil and water flows. The mud must have proper rheological properties, particle size distribution and weight. In addition it must remain functional at temperatures often above 300°F, at pressures of 15-20,000 lbs/sq in. and in the presence of contaminating ions. In early days of rotary drilling for oil, the mud was thinned with water alone. If the fluid loss was too great, the wall cake would stick the drill pipe or slough into the hole. If mud weight was too low, the well blew out. In fact, for years the "blow out" was the method of producing a well. W. F. Rogers in his book "Composition and Properties of Well Drilling Fluids" (48) gives a detailed account of the development and present technology of drilling fluids. In the 1920's, the use of natural tannins, such as quebracho, was introduced and has only recently been supplanted by lignin derivatives as the principal dispersants. Both kraft lignins and lignin sulfonates have been used in drilling, but the preponderance of use has developed with lignin sulfonates, particularly the chrome and ferrochrome derivatives of the latter (49). Addition of the chromium in the form of chromate or dichromate performs a dual purpose. It causes the lignin sulfonates to be oxidized and supplies a multivalent cation. Controlled oxidation is beneficial in the formation of an effective additive. The heavy metal lignin sulfonates are expecially helpful in controlling the rheological properties of mud systems and in maintaining these properties in the presence of natural contaminants.

Some of the requisites of a good drilling fluid appear to an extent to oppose one another. While a good dispersion and hydration of the clay is required, it is undesirable to disperse or hydrate the clay being drilled. Rather, the drilled particles should remain in a condensed form and in discreet chunks so that they may be screened from the mud. The drilling mud must also have sufficient gel to support the cuttings and the weighting agent (commonly natural barium sulfate), yet have a low enough gel for good flow properties and easy release of cuttings. Chrome and ferrochrome lignin sulfonates are remarkably effective in imparting these properties. In addition, some are stable enough to allow working mud temperatures as high as 400°F. Mixtures of chrome or ferrochrome lignin sulfonates with lignites, sulfonated bark extracts, lignin sulfonates, and tannins are often advantageous for drilling of specific shales. Pilot testing of mud samples at the drilling site gives the engineer a guidance as to necessary treatment of dispersant, water loss agents, caustic and weighting agent. Helicopters are coming into use for transporting samples of mud to established laboratories where chromium analysis may be rapidly carried out to determine the content of chrome or ferrochrome lignin sulfonate. The muds may also be tested at high temperatures and under pressure in consistometers to give immediate guidance for field treatment.

Oil-in-water emulsions in a drilling fluid have been in use since the 1940's. Lignin sulfonates are valuable for stabilizing these emulsions which give the mud improved properties in control of water loss, reduction of torque on the drill stem, increase of bit life and general improvement of bore-hole conditions. In 1951, lignin sulfonate stabilized emulsions in saturated brines were introduced and are used as workover fluids (50).

B. Cement and Concrete Additives.

1. Grinding Aid for Portland Cement. In the grinding of the raw materials for portland cement, quarried stone which has been crushed and milled to a satisfactory size is ground either by wet or dry processes. Ground solids are then brought to the kiln where first the water is driven off, organic matter burned and carbon dioxide removed from carbonates. In the hottest part of the kiln a portion of the mass becomes liquid and reaction between the components causes sintering to a clinker which, after cooling, is processed through a further milling operation to give the finished portland cement.

The use of lignin sulfonates and sulfonated kraft lignins as grinding aids in cement and concrete industries represents a major outlet for these derivatives. These aids are often used in the grinding of the clinker and, to a small extent, in the wet initial grinding. Their function is to reduce the agglomeration of the ground particles, to reduce surface tension at the solid-air interface of the particles and to keep the surfaces of the grinding media clean and free. Microscopic examination of cement particles ground with and without grinding aid readily shows the difference in agglomeration (51). This effect has been ascribed to selective adsorption of the lignin on high energy sites of the freshly ground cement particles (52). In addition, the grinding aid may supply ions to satisfy the unbalanced valence forces of the new surfaces generated by grinding. Both effects can be expected to interfere with the reunion of particles. As little as 0.02% of the grinding aid, based on the weight of cement, has been found to be effective. Since approximately ten times more of the lignin sulfonate is required to coat the theoretically calculated total surface area of the particles, only selected surface sites need to be blocked to inhibit agglomeration.

Many grinding aids also make the cement flow more readily by reducing inter-particle attraction. This reduces the problem of caking during storage and transport known as "pack set." It may also result in unexpected overflows in the grinding mills if the operation is not properly adjusted. In the early days of grinding aid usage, the beneficial effects were sometimes overlooked as a consequence of each experience.

Specific quality tests have been developed (ASTM-C-465) for grinding aids used in the finish mills.

2. Air Entraining Agents. Many lignin derivatives, including lignin sulfonates as well as sodium salts and sulfonates of kraft lignins (53), have been used as air entraining agents in concrete. The entrained air gives increased resistance to spalling and cracking of the finished concrete, particularly those resulting from alternate freezing and thawing.

3. Concrete Additives. The dispersive action of certain lignins is useful in mixing and placing concrete. Incorporation of 0.15-0.35 lbs of

lignin sulfonate-concrete admixture per sack of cement will allow water reduction up to 20%. This results in better compressive strength and durability of the concrete, often measurable for several years. In all likelihood, the surfactant properties of lignin sulfonates provide optimum air entrainment to cushion the expansion and contraction during freezing and thawing. Bond strength to steel is increased in lignin treated concrete. For this reason prestressed concrete objects usually have a lignin-concrete admixture included in their formulations.

A set time delay of one hour or more can be achieved by incorporation of an unmodified lignin sulfonate-concrete admixture. It may be desirable to extend the set time to reduce the problem of heat of hydration in large amounts of concrete to permit long hauls, facilitate pumping on difficult operations as tunnel jobs, lessen the chance of cold joints and inhibit premature setting in hot weather or in placing large areas of concrete such as bridge decks. Set times can also be shortened by the addition of calcium chloride to the lignin sulfonate (54). The commercially used concrete admixture must carry the qualification of ASTM specification C-494, types A or D.

4. Oil Well Cementing. Oil well casings are often cemented by a pumpable grout of cement and water under pressure in order to separate oil, gas, and water-bearing strata from each other. The grout must not set until it is all in place. The high temperatures and pressures complicate the picture. Lignin sulfonate additives not only make possible a lower water usage but give an increased set time.

Set time and strength characteristics vary widely with different lignin sulfonate additives. They are compared using specified methods published by the American Petroleum Institute (55). The additives are for the most part calcium lignin sulfonates with specific molecular weight ranges and sulfonate contents. A good additive should delay set time without weakening the strength of the cured grout and the set time should vary uniformly with change of temperature and pressure. For the most part, lignin sulfonate additives do not behave uniformly with changes in temperature and pressure. As a result, extensive programs of testing and development are currently in progress. High pressure, high temperature consistometers or thickening testers are used to characterize the rheological properties of cement mixes (56).

C. Dispersants. Pastes and slurries of solid particles in water are often dramatically thinned when a small amount of a lignin derivative is added. The colloidal lignins coat the surfaces imparting a negative charge to the particles. The similarly charged particles now repel one another so that aggregates are dispersed and the viscosity is lowered.

Lignin sulfonate dispersants usually have a molecular weight range of 10,000 to 40,000 (57). A major outlet for these products is found in the large clay and ceramic industry, which in the United States alone produces 20 million tons of clay products with an annual sales value of 300 million dollars. In addition to common ceramics such as porcelain, refractories and tiles, the products include such varied commodities as absorbants, cosmetics, crayons, medicines, paints, paper coatings, phonograph records and toothpowder. In the manufacture of these products, lignin derivatives are used or have the potential for future use not only as dispersants but also as binders. In the ceramic uses of clay, the reduction of water in the clay mix results in less shrinkage and cracking during firing, and the binding action is needed during shaping of the article. During firing, these organic binders are burned, leaving negligible ash. The lignins are added in amounts ranging from 0.125-1% of the ceramic solids. Lignin sulfonates increase the "green strength" in the production of dry mix floor tiles, refractory bricks, etc. In glazes, they provide good deflocculation and permit finer grinds and faster milling.

Lignin sulfonates are used also in the production of gypsum board. The benefits consist of lowered need of water in the gypsum plaster, increased dryer capacity and higher safe temperature of drying. Normal usage requires 1-3 lbs lignin sulfonates/1000 square feet of 1/2-inch board.

Alkaline salts of kraft lignin, as well as the "Polyfon" sulfonates and lignin sulfonates, are used as "levelling" agents to retard the rate of adsorption of dyes on cotton and rayon fabrics to insure uniform dyeing. The dispersing action is particularly useful with acetate rayon dyes that are insoluble in water (25). In addition, lignin sulfonate imparts smoothness and good "handle" to the yarn.

The dispersing action of sulfonated lignins can be used to decrease the settling rate of suspensions of insoluble insecticide powders such as 50% wettable DDT powders (58). Addition of 2-5% is effective though higher percentages are often applied to bond the insecticide to plants or crops in order to inhibit their removal by wind or rain (59).

Another area of application consists of dispersing carbon black in rubber masterbatching. The dispersion is mixed with latex and the carbon is coprecipitated with the rubber. The presence of lignin sulfonates results in uniform carbon distribution in the rubber. These dispersants also convert carbon pastes to pourable slurries and eliminate thixotropic properties (52).

Scale buildup on the heat transfer surfaces of boiler tubes is a troublesome problem in steam generating plants. Kraft lignin, its sulfonated derivatives, lignin sulfonates and desulfonated lignin sulfonates act as protective colloids and prevent the growth of the dense crystal formations associated with the deposition of inorganic salts at elevated temperatures. A flocculent sludge is formed in the presence of approximately 2% of the lignin conditioners and the sludge is easily removed during blowdown.

The dispersing action of lignin derivatives in suspending fine dirt particles gives them some utility in commercial cleaning formulations. They are used in alkaline soak and spray cleaners to produce foam, in addition to acting as wetting agents and dispersants, and may replace a part of more expensive surfactants in such formulations. Kraft lignin sulfonates are compatible with phosphates, resin acid soaps and wetting agents (52). Some lignin sulfonates are used as detergent and soap extenders (60). Household detergents, however,

D. Ore Flotation. Lignin flotation agents are useful in the refinement of certain ores. They depress such minerals as calcite, barite, talc, sericite, molybdenite, and carbonaceous materials; e.g., graphite. They also disperse slimy ores in tabling and flotation operations. The dispersing action of lignins can be used to improve the filtration properties of lead-zinc systems owing to the lowering of water content in the concentrates.

E. Emulsifiers and Stabilizers. Lignin emulsifiers are quite competitive with other commercial emulsifiers stabilizing oil-in-water emulsions. Efficient stirring is necessary for producing such emulsions, but once formed they remain stable even when frozen or heated above the boiling point of water or in the presence of weak acids, bases, and salts. The effect has been at least partly attributed to the ability of the lignin to prevent the reaction of soaps, which are the primary emulsifying agents, with calcium ions. Stability is imparted even to bitumen emulsions in water without affecting the viscosity (61). Since most emulsions of this type are alkaline, sodium salts of kraft lignin are commonly used.

A lignin amine derivative, obtained by Mannich reaction of kraft lignin, dimethylamine and formaldehyde, has been patented for stabilizing asphalt emulsions (62). In the asphalt emulsion industry, sugar-free lignin sulfonates are preferred to avoid the growth of molds and bacteria (63). Emulsions with large amounts of spent liquor solids present may be spray-dried without coalescence, and the oil droplets remain encapsulated in the lignin sulfonate. On redissolving the dried product in water, the emulsion is reconstituted.

Road oils may be mixed with lignin sulfonates for use in surfacing roads (64). Emulsions of light-colored oils with lignin sulfonate solutions are marketed for use in holding down dust in driveways, school playgrounds and commercial parking areas.

The stabilization of fire fighting foams that result from the interaction of sodium bicarbonate solutions and alum is achieved by addition of sodium salt of kraft lignin. Even better results have been obtained by the use of sulfonated kraft lignin such as "Polyfons" (Westvaco) (65). Kraft lignin with a low degree of sulfonation is most effective for this use (66).

F. Grinding Aids. The use of lignin sulfonates as grinding aids has been discussed in portland cement manufacture. They are also used, however, for controlling dust during grinding and handling of finely ground or mixed materials. A mist of dilute spent sulfite liquor or lignin sulfonate is usually directed into a grinding or blending mill. The particles so coated are smooth and flow easily. The manufacture of garden fertilizers is an example of a successful application of this type.

Nearly all dry and aqueous grinding systems may benefit from the dispersing and protecting action of lignins. Polishing operations in which slurried abrasives are used benefit from the dispersing and suspending action of the lignins.

G. <u>Electrolytic Refining</u>. In electrolytic refining of copper, the addition of ammonium lignin sulfonate helps to form a smooth, ductile, solid cathode and does not add to the calcium deposits of the cell. Similar effect is realized in electrolytic refining of zinc and other metals (67). In alkaline electroplating, kraft lignin has been used successfully as "brightening" agent for the plated surface, preventing pitting and deposition of rough surfaces. If concentrations are kept to 1% or less, it is possible to use the lignin in mildly acidic baths (pH > 4) (25).

H. <u>Binders and Adhesives</u>. From the late 1920's until the present, large quantities of spent sulfite liquor have been sprayed on roads, both for the purpose of making a hard durable surface and for the laying of dust (68). In some instances, spent liquor solids have been bladed into the surface with graders. In areas of low rainfall, spent liquor treatment produces durable roads with less tendency to form chuck holes than those surfaced by an oil spray treatment. In the absence of rain, the lignin sulfonates and spent liquor sugars maintain road surfaces smooth and dust-free for long periods of time. Spent sulfite liquor is also applied as a pretreatment prior to the application of paving materials contributing a hard, water-resistant barrier in the subsurface road bed.

Deep treatment of subsurface road beds and rail beds using pressure injection of spent liquor is practical for preventing frost heave. Well-point type pipes are driven into the ground every few feet along each side of the road or rail bed and liquor is pumped in under pressure until it reappears in the next pipe. The dispersing and binding action of the lignin sulfonate causes the natural clays of the soil to form an impermeable layer, keeping the road bed free of ground water.

The reaction of spent sulfite liquor with chromates to produce gels of water-insoluble oxidized lignin sulfonate has been known for many years (69). The work done for the U. S. Air Force at Cornell University on the solidification of soil received much attention (70). The chemical reactions involved have been studied and are complex, to say the least (71). The solubility of the product is dependent on the cation of the lignin sulfonate, the reaction time and temperature, concentration of dichromate, and the pH in the system. Calcium lignin sulfonate provides the most insoluble gel while ammonium lignin sulfonate, the least. Similar cross-linked gels may be generated by oxidation with permanganate or hydrogen peroxide in the presence of iron salts. These processes require careful control because the oxidation may sometimes provide only a small increase in molecular weight instead of producing a gelled or hard resilient solid. In solid gels, water is a plasticizer and on drying, the strength disappears, leaving a friable solid. These oxidized lignin gels find application in stabilizing soil for roads, tunnel drilling, and foundations of dams. They have also been used in stopping leaks in tunnels and dams (72).

Large tonnages of spent liquor solids are used as binders for pellets and briquettes. Ore binders probably rank first in consumption. Cleveland-Cliffs Iron Company laboratory has reported that 2% of calcium lignin sulfonate in iron ore gives a pellet ten times as hard as that made with bentonite binder costing the same. Cleveland-Cliffs alone could use 120,000 tons of lignin sulfonate a year (73). Tens of thousands of pounds per year are used as a binder for zinc ore. A patent (74) suggests the use of spent liquor for binding smelt materials fed into glass furnaces. The continuous electrodes for refining of aluminum also need a binder, offering a potential use for considerable quantities of lignin sulfonate. At present, natural pitch is generally used for this purpose, but experimental trials with lignin sulfonate materials have demonstrated their practicality (75).

For many years, spent sulfite liquor has been used as a foundry mold and core binder for iron, brass, and other high melting metals (76). Spent liquors and lignin sulfonate preparations impart satisfactory strength to the sand molds, and the time between preparation of the mold and use is no longer critical. Intricate shapes may be handled and release of the mold from the pattern is improved. A combination of ammonium or sodium lignin sulfonate with soda ash allows the clay particles present to give the sand good flow properties without the danger of flocculation. Reaction products of lignin sulfonate with urea or phenol-formaldehyde resins may be used to give a very strong mold with a smooth surface. Also, a patent suggests the use of a mixture of lignin sulfonate with bituminous materials as a mold binder (76), permitting close tolerances for the molded products.

The danger of silicosis among workers handling silica brick can be overcome by treating the bricks with a lignin sulfonate dip or spray. Complete pellet loads of refractory bricks may be carried by fork truck and dipped in a tank of lignin sulfonate solution.

The binding of coal and charcoal briquettes with lignin sulfonates has been practiced to some extent for many years. The incorporation of a small amount of fuel oil gives water resistance to the briquettes, protecting them when stored outdoors. Ammonium lignin sulfonate is preferred owing to its low ash content of 1-1.5%. These sulfonates can be rendered water-insoluble by heating the briquette to 400°F. They are also preferred for the growing industry of pelletizing animal and poultry feeds currently amounting to 25-35 million pounds per year. The binder reduces loss of fines, Figure 20.6, increasing the yield of pellets by as much as 30% and die life by 50%. Less heat is generated during pressing, preserving efficiently the vitamin content of the feed.

Though spent liquor solids are brittle when dry, they form a strong adhesive when plasticized by water. The most familiar adhesive product is the linoleum cement containing principally lignin sulfonates and clay. These cements are also used as adhesives for ceiling tiles. They spread smoothly, have a long useful tack life, bond readily, and are easily washed from tools or the flooring surface. Despite their soluble nature, such adhesives are permanent since most flooring materials protect the adhesive from exposure to moisture. Formulations of similar nature are finding increasing use as temporary adhesives; for example, to hold palletized bags in place during moving and shipping.

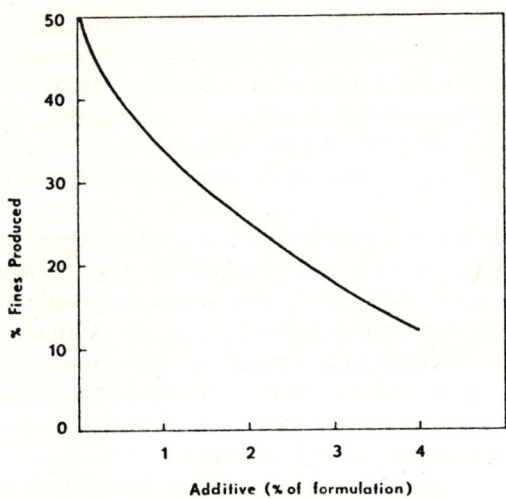

Figure 20.6. Effect of lignin sulfonate pelletizing additive on fines in animal and poultry feed pellets.

Lignin derivatives have been contemplated for use as binding agents in printing inks (77, 78) and claimed to be non-tacky, chemically inactive and non-smudging under printing conditions.

1. Resin Ingredients. Lignin's role in wood as a natural plastic and binding agent has encouraged the use of lignin derivatives as binding agents for fibrous materials and wood particles. The lignin itself has been used with various degrees of success as the sole binding agent in boards and paper laminates. More successful has been the employment of lignin derivatives as phenolic ingredients of more expensive phenol-formaldehyde resins. The effects of lignin on resin structure and properties have been discussed by Soipioni (79). Kratzl et al. (80) have studied the condensation of various lignins with phenols, aldehydes and isocyanates. The condensation of isocyanates with lignin to form thermoplastic molding resins has been patented in the United States (81). The hydrophilic nature of lignin sulfonates detracts from their utility as reactants in resins or waterproof adhesives; and desulfonated lignin sulfonates, therefore, make better additives in resin formulations. Both lignin sulfonates and desulfonated lignin sulfonates are, nevertheless, in commercial usage. Lignin sulfonates in combination with acid catalyzed resins have potential in molding powders as well as in binding agents previously discussed. Ammonium lignin sulfonates impart extended shelf life to the resins through shorter cure time (82, 83). The reactivity of lignin sulfonates with protein is utilized in a thermosetting resin made of lignin sulfonates and animal glue (84).

Combinations of lignin sulfonates with low molecular weight urea- and phenol-formaldehyde resins have received attention as adhesives for plywood, particle- and chip-board. Some lignin sulfonate products are used as resin extenders, but most of the adhesive patents specify desulfonated lignin sulfonates or nonsulfonated lignins (82). One unique lignin sulfonate product is added to particle board pulp suspension and precipitated onto the fibers by means of certain metal ions below pH 6. Improvements in board strength and retention of resins, sizing materials and pigments have been claimed (85).

Many glue formulations for wood or remoistenable tapes incorporate lignin sulfonates with polyvinyl acetate or polyvinyl alcohol or with bone or animal glues (86). Ayers (87) has patented the use of oxidized kraft lignin as an extender in phenol-formaldehyde glues.

Kraft lignin heat-condensed with xylenol or crude phenols ("cresylic acids") has been claimed to be suitable as a wood particle binder by Sokolova (88). Likewise, Rieche and Redinger (89) have condensed lignin with phenol in the presence of sulfuric acid, obtaining a product which can be further condensed with formaldehyde and sodium hydroxide. A thermosetting resin is obtained, claimed to be suitable for laminating fibrous material.

Another way to use lignin in phenolic resins is to substitute up to 50% of the other phenolic materials with kraft lignin when condensing with formaldehyde. This procedure is said to give good "A"-stage resins with a considerable saving of both phenol and formaldehyde (53). Still another way to incorporate kraft lignin into phenol-formaldehyde plastics is to react it with a "resol" resin obtained by alkaline condensation of 1 mole phenol with 1.5-3 moles of formaldehyde. The products are used to make various laminated materials, particularly kraft paper laminates (90).

The reactivity of lignins towards aldehydes can be improved by demethylating guaiacyl groups to catechols. Marton and Adler (91) have patented a procedure of oxidizing kraft lignin with sodium periodate to convert free phenolic units to o-quinone groups which are subsequently reduced to catechols by sulfur dioxide. The lignin so produced is claimed to be suitable for condensation with formaldehyde and phenols to give thermosetting resins.

The "Reax" resins produced by Westvaco (92) by a hitherto undisclosed process have been claimed to be outstanding modifiers for phenolic and cresylic laminating resins. Excellent machining, punching and electrical properties are attributed to such modified resins (93). Mocsny (94) describes a resin mixture containing gilsonite, calcium oxide, magnesium oxide, sugar and lignin which is suitable for making sand molds. Ball et al. (95) have patented lignin resins made by reacting kraft lignin with epoxides.

J. <u>Rubber Additives</u>. The literature contains almost as many references to the use of kraft lignin as an extender, modifier and reinforcing pigment in rubber compounding as it does to resin formation.

Many types of latices such as natural, GR-S, nitrile and neoprene are compatible with lignin either in the acid or sodium salt form, and will tolerate the addition of appreciable quantities without coagulation. The lignin reportedly improves the quality and strength of fibers made from the latex (25, 96). Mixtures of rubber with kraft lignin are also suitable for use as rubber-to-glass adhesives. The lignin after the addition of mineral acid reduces spurting and evaporation (97). Kraft lignin also compares very favorably with other reinforcing agents such as carbon black because of its lower density and opacifying diameter. At high loadings, tough light-weight, light-colored hard rubbers possessing high elongation, tensile and tear strengths are obtained (98). Thus far, however, the use of lignin in automobile tires (99) has not been completely successful (100).

Various treatments of kraft lignin have been suggested for improving its rubber reinforcing properties. By passing oxygen through an alkaline solution of the lignin, a superior reinforcing pigment for butadiene-styrene rubbers (101) can be produced. An alkali hydrolyzed lignin which is at least 90% soluble in acetone has been claimed to function as an improved filler for butadiene-acrylonitrile copolymers (102).

In addition to the use of straight lignin-rubber masterbatches as such, Mills and Haxo (103) have patented the incorporation of an organic polyisocyanate as a coreactant. Finally, Papst in Germany has claimed that almost any lignin can be a satisfactory reinforcing filler if the rubber is a copolymer, graft copolymer or block copolymer obtained by polymerization of natural rubber with vinyl compounds containing tertiary amine groups (104).

K. Protein Precipitants. The ability of lignin to react with amines and proteinaceous materials has been utilized to remove these materials from sugar solutions, beer and similar liquors. The lignin forms a complex with the protein which is generally insoluble in slightly acid media and filterable. Since lignin is relatively non-toxic to cattle and poultry, the protein-lignin complex can be used along with proteinaceous feed supplements (105). Vitamin B-12 can be recovered from antibiotics by similar techniques (106, 107).

L. Tanning Agents. One of the oldest and most worked-over fields of lignosulfonate usage is leather tanning, and probably no single use has resulted in more product and use patents (108). Using the method of the American Leather Chemists Association for determining tanning content, typical lignosulfonates contain the amounts of equivalent tannin corresponding to 36% tannin equivalent for sulfite spent liquor solids and 54-57% for purified lignin sulfonates. The largest use of these lignosulfonate tanning agents is in the retan of chrome upper leathers where the grain is fixed more nearly in its natural state if the lignosulfonate is employed as "first feed." The rapidity of subsequent reaction with vegetable tannins is thus reduced, and a leather with improved "break" results. A replacement of up to 65% of the vegetable tannins may be made; and in addition to the price differential, the lignosulfonates give a more soluble tanning solution which facilitates recovery procedures (109). Calcium lignosulfonates are less desirable since the calcium ions tend to insolubilize the vegetable tannins.

M. Sequestering Agents. A number of metal ions, including Ca,

Ni, Sn, Zn and Al, can be sequestered by lignosulfonates, particularly above pH 7. Copper and iron will not form hydroxides, phosphates, or carbonates in the presence of lignosulfonates. Iron- and trace element-containing lignosulfonates are produced commercially for use in agriculture to combat chlorosis. Iron chlorosis, perhaps the best known of these elemental deficiencies, occurs in Florida, Hawaii and in parts of the western and midwestern sections of the United States. It is due to the unavailability of the soil iron to the plant roots. Direct spraying with iron salts in solution does not, however, result in uniform greening of the leaves, but either brings out spots on the leaves or, if applied at greater concentrations, causes "burning" of the leaves. In contrast, iron as well as zinc, copper, molybdenum and other metals can be successfully applied when their salts are complexed in lignosulfonate solutions because the toxicity of the salts is greatly reduced (110).

Compounds subject to catalytic decomposition in the presence of trace amounts of metals may also be stabilized by the sequestering action of lignosulfonates, and detergents and commercial cleaning compounds containing lignosulfonates are more effective in iron-containing hard water for the same reason.

Solutions of the sodium salt of kraft lignin are also useful as inexpensive sequestering agents for calcium, magnesium and otherpolyvalent metal ions. The sequestering action is most marked near neutral pH values, and the formation of scums and insoluble salts in hard waters can be prevented. Up to seven grams of calcium per 100 grams of kraft lignin can be sequestered (111).

N. Miscellaneous.

1. Storage Battery Plates. In lead-acid storage batteries, the plates consist of a lead-antimony alloy grid filled with a paste composed principally of lead oxide. The plates also contain less than 1.5% of an expander which prevents loss of the spongy character of the plates and gives higher capacity and increased life to the battery (112, 113). Lignins can be used as the expander and as a component of the separators between the positive and negative plates (114).

2. Lime Plaster. The addition of lignin sulfonates to plaster allows the use of less water in the mix and gives improved set properties as a consequence of the dispersant and crystal inhibiting properties of the lignin sulfonates. An easily handled low shrinkage plaster results which is especially suitable for machine application.

3. Crystal Growth Inhibitor. Concentrated commercial aqueous solutions of fertilizers, pesticides or cleaning agents will often crystallize on standing unless a crystal growth inhibitor is added. Spent sulfite liquor solids are effective for this purpose because they are non-toxic and not very chemically reactive at the levels added.

4. Ingot Mold Wash. A coating is used on steel ingot molds to protect the mold and to facilitate release of the ingot. Tars, which have been used for years, give off noxious fumes both during coating and pouring. Ammonium lignin sulfonate is an especially good substitute because mold life and

ingot surface quality are equivalent and the fume problem is greatly reduced.

5. Flame Retardant. Lignin phosphoric esters, prepared by reaction with phosphorus halides and oxyhalides, have been claimed as flame retardants for paper and textiles (115).

O. Liquor Disposal. By far, the largest portion of spent sulfite liquor must be disposed of. No outlet has been defined which could use the vast quantities of liquor available. In a few areas, the spent liquor may be dumped into rivers or the ocean; but probably no mill can look toward this as a long term solution to the disposal problem. The organic materials in the spent liquor use up the oxygen of the water, making it unavailable for fish. In areas where there are tremendous amounts of water, the oxygen content may remain satisfactorily high, but the nutrients of the spent liquor promote the growth of slimes and other unwanted organisms.

Large ponds have been built to allow the liquor to concentrate by evaporation, but these must usually be lined to protect ground water from contamination. The liquor is used to reduce dust on unpaved roads, added to irrigation water (116), or barged to sea. A certain amount may be added to sewage disposal plants with beneficial effect. A most novel method has been used at the Hammermill Paper Company, where wells drilled to a depth of 1600-2300 feet receive the doubly filtered liquor which is pumped under high pressure into the surrounding porous limestone deposits (117).

V. Utilization of Hydrolysis Lignin. In recent years, most of the references to uses for hydrolysis lignin have appeared in journals from the U.S.S.R. This is, of course, due to the fact that the Soviet Union is the only country where industrial plants for the hydrolysis of wood are in operation.

Hydrolysis lignin, like other lignins, has an effect on the mechanical properties of soil, and Yashchenko (118) has suggested its use for soil consolidation.

Hydrolysis lignin, as might be expected, has also been proposed as a filler for resins and rubber (119). It also finds application as a grinding aid for portland cement (120). Heating of hydrolysis lignin with alkali generates acidic groups within the lignin and makes what various Russian authors have called an "activated" hydrolysis lignin. By treatment of the lignin in a vibratory ball mill, Kurdobov and Rubina (121) found that the lignin requires less caustic and the time was shortened for the activation step. The product was suitable as a rubber filling agent, as a reactive ingredient of various resins and as a binding resin itself. Thus, Sokolova (88) describes the use of hydrolysis lignin (40 parts) as the binding agent for wood particles (100 parts) in board manufacture, and Raznikov has studied a combination with phenol-formaldehyde resin for the same application (122).

The "activation" reaction of hydrolysis lignin has been studied by a number of investigators (123, 124). Hosaka in Japan (125) has patented a pressure reaction with sodium hydroxide at 120-210° which affords an active lignin that will condense with aldehydes to give resins.

Hydrolysis lignin can be readily chlorinated, nitrated and sulfonated (126). Thus, Shorygina and Kolotova (127) passed chlorine into a lignin suspension in carbon tetrachloride or water at 20°. Below an uptake of 25-27%, the chlorine was quantitatively absorbed. Excess chlorine cleaved methoxyl groups. Turetskii et al. (128) also chlorinated hydrolysis lignin and reported that the product was useful in the flotation enrichment of iron ore.

Nitrolignins were analogously prepared (129). Both the nitro- and chloro-lignins imparted decay resistance to fabrics (130) after alkaline application and insolubilization with copper sulfate and soluble chromium salts.

Since the wood hydrolysis industry is quite large in the U.S.S.R., it is apparent that much of the lignin produced cannot be utilized by any of the foregoing. It is, therefore, used as a fuel; but before burning, dehydration is necessary since the water content is quite high. This can be accomplished by drying of the lignin with hot flue gases to a moisture content of 12-18%, followed by pressing into briquettes (131). Alternatively (132), the lignin from the hydrolysis reaction can be pressed at 180° and a pressure of 18 kg/cm^2 to a water content of 27-35%.

VI. Concluding Remarks. While it must be admitted that application research in the lignin field has not succeeded in bringing about glorious major breakthroughs, and, because of the very nature of this complex organic polymer and its derivatives, such breakthroughs may not be anticipated in the future, there nevertheless has been sound gradual development towards an increasing number of uses and products. These developments will undoubtedly continue in this challenging field of research and be aided by two independent factors: One is the current emphasis on environmental ecology demanding non-polluting disposals of industrial wastes; the other is provided by the much improved understanding of the chemical structure of lignins and their derivatives, helping the researchers in this area to relate their experimentation to a fundamentally sound framework of ideas.

REFERENCES

1. P. D. Van Derveer, Pulp and Paper, 11 (8), 127 (1969).
2. A. J. Wiley and J. M. Holderby, Pulp Paper Mag. Can., 61, T212 (1960).
3. Anon., Paper Mill News, 86 (14), 10 (1963);
   L. C. Bratt, Paper Trade J., 147 (14), 40 (1963);
   Chem. Eng. News, 43 (10), 30 (1965);
   Tappi, 48, 46A (1965).
4. A. J. Wiley, Pulp Manufacturers' League Annual Report, Appleton, Wis., (1967).

5. Anon., U. S. Dept. of Commerce, Business and Defense Service Administration, Pulp, Paper and Board, Quarterly Industry Report, April, 1968.
6. Anon., Chem. Eng. News, 41, 83 (1963).
   H. F. J. Wenzl, Paper Trade J., 149 (22), 47 (1965).
7. Anon., Chem. Eng. News, 41 (25), 51 (1958);
   Anon., Pulp Paper Mag. Can., 59 (2), 89 (1958);
   G. Edling, Pulp & Paper Industry, 19 (12), 58 (1945).
8. Anon., Chem. Eng. News, 47 (8), 49 (1969).
9. R. S. Hatch, ibid., 19 (11), 38 (1945).
10. R. S. Aries and A. Pollak, Tappi, 35 (12), 142 (1952).
11. F. Melms and K. Schwenzon, Verwertungsgebiete der sulfitablauge, V.E.B. Deutscher Verlag fur Grundstoff Industrie, Leipzig, German Dem. Rep., 1967.
12. S. A. Saponitzki, Verwertung der Sulfitlauge, V.E.B. Fachbuchverlag, Leipzig, 1963 (translation from Russian).
13. G. H. Tomlinson and G. H. Tomlinson II, U. S. Patent 2,406,867 (1946).
14. K. R. Gray, H. L. Crosby and J. C. Steinberg, U. S. Patent 2,772,965 (1956).
15. J. Giesen, Swiss Patent 318,828 (1957); U.S. Patent 2,828,297 (1958).
16. V. M. Nikitin and A. V. Obolenskaya, Trudy Lenningrad Lesotakh Akad. Khim. S. M. Korova, 85, 12 (1960); through Chem. Abstr., 54, 25793 (1960).
17. G. Ouchi, Japan. Patent 8323 (1960).
18. J. W. T. Merewether, J. W. Hallam and H. B. Birch, Australian Patent 231,928 (1961).
19. F. J. Ball and W. G. Vardell, U.S. Patent 2,997,466 (1961).
20. F. J. Ball and W. G. Vardell, U.S. Patent 3,048,576 (1962).
21. J. W. T. Merewether, Tappi, 45, 159 (1962).
22. A. M. Kennedy and J. M. Jernigan, U.S. Patent 2,905,605 (1959).
23. F. J. Ball, Can. Patent 689,565 (1960).
24. F. J. Ball, U.S. Patent 3,017,404 (1962).
25. West Virginia Pulp and Paper Co. (Westvaco), Technical Bulletin L-6 (1951).
26. J. Marton and T. Marton, Tappi, 47, 471 (1964).
27. F. J. Ball and R. Puschel, U. S. Patent 2,967,273 (1961).
28. A. Rezanowich, W. Q. Yesn and D. A. I. Goring, Svensk Papperstidn., 66, 144 (1963).
29. J. Marton and E. Adler, Tappi, 46, 92 (1963).
30. C. A. Watson, Forest Prod. J., 9 (3), 25 (1959);
    E. D. Dahlgren, J. Water Pollution Control Federation, 36, 1543 (1964).
31. R. M. Wheaton and W. C. Bauman, Ind. Eng. Chem., 45, 228 (1953);
    A. H. Seamster and R. M. Wheaton, Chem. Eng., 67 (17), 115 (1960).
32. V. F. Felicetta, M. Lung and J. L. McCarthy, Tappi, 42, 496 (1959).

POLYMERIC PRODUCTS 861

33. C. H. Hoyt, Research Memorandum No. 960-22, Chemical Prod. Div., Crown Zellerbach Corp., Camas, Wash., 1956.
34. W. Jensen, K. E. Fremer and K. Forss, Tappi, $\underline{45}$, 122 (1962).
35. Anon., Chem. Eng. News, 36 (28), 41 (1958).
36. Ibid., $\underline{40}$ (46), 52 (1962);
    G. A. Dubsy, T. R. McElhinney and A. J. Wiley, Tappi, $\underline{48}$, 95 (1965); G. A. Dubsy, U.S. Patent 3,135,710 (1964).
37. G. C. Howard, Ind. Eng. Chem., $\underline{22}$, 1184 (1930); $\underline{26}$, 614 (1934); Anon., Paper Mill Wood Pulp News, $\underline{59}$, 17 (1936); A. J. Bailey, Pacific Paper Industries, $\underline{13}$ (11), 34 (1939).
38. V. F. Felicetta, A. E. Markham and J. L. McCarthy, Tappi, $\underline{37}$, 431 (1954); A. E. Markham and J. L. McCarthy, ibid., $\underline{37}$, 355 (1954).
39. Anon., Can. Chem. Process, $\underline{45}$ (7), $\underline{45}$ (1961); D. G. Manchester and J. P. Termini, Pulp Paper Mag. Can., $\underline{62}$, T415 (1961); L. E. Robinson, Tappi, $\underline{39}$, 182 (1956).
40. L. K. Swenson, U.S. Patent 2,916,355 (1959).
41. W. R. Effer, E. W. Hopper, N. B. Marshall, Pulp Paper Mag. Can., $\underline{62}$, 7447 (1961).
42. W. D. Kerr, ibid., $\underline{62}$, T455 (1961).
43. S. F. Ali, G. G. Wilson and W. H. Whitney, Tappi, $\underline{51}$, 69A (1968).
44. R. Borisek, Czech. Patent 84,217 (1955).
45. J. J. Keilen, Chemurgic Digest, $\underline{14}$ (3), 13 (1955).
46. P. R. Wiley, Tappi, $\underline{44}$, 22A, 24A, 26A (1961).
47. T. Simmons, Svensk Papperstidn., $\underline{56}$, 121 (1953);
    Pulp Paper Mag. Can., $\underline{56}$, 108 (1954);
    G. K. Dickenmen, ibid., $\underline{61}$, T200 (1960).
    H. E. A. Burger, Das Papier., $\underline{18}$, 267 (1964);
    K. A. Kobe and W. McCleave, Paper Trade J., $\underline{160}$ (1), 35 (1938);
    B. Nikander, Finnish Paper Timber J., $\underline{30}$ (7), 53 (1948);
    C. E. Rogers and R. S. Jolley, Paper Trade J., $\underline{131}$ (20), 25 (1950);
    C. E. Rogers, Pulp and Paper, $\underline{25}$, 66 (1951);
    M. V. Moleberry, Pulp Paper Mag. Can., $\underline{62}$, T532 (1961);
    P. J. Wenzl, Paper Trade J., $\underline{149}$ (23), 59 (1965).
48. W. F. Rogers, Compositions and Properties of Oil Well Drilling Fluids, 3rd Ed., Gulf Publishing Co., Houston, Texas, 1963.
49. E. G. King and C. Adolphson, U.S. Patents 2,935,473 and 2,935,504 (1960).
50. W. C. Browning, J. Petrol. Technol., $\underline{7}$ (6), 9 (1955).
51. M. Papadakis, La Revue Des Materiax de Construction, No. 519, December (1958).
52. West Virginia Pulp and Paper Co. (Westvaco), Polychemicals Div., Technical Bull. 300.

53. Ibid., 102.
54. W. J. Halstead, B. Chaiken, Public Roads, 31, 126 (1961).
55. American Petroleum Institute, RP 10B.
56. API Standardization of Oil Well Cements, Chapter 3, "Oil Well Cementing Practices in the United States," American Petroleum Institute, 1959.
57. Anon., Chem. Eng. News, 38 (26), 40 (1960);
    J. Banko, Tappi, 44, 849 (1961).
58. West Virginia Pulp and Paper Co. (Westvaco), Polychemicals Div., Technical Bull. 306B.
59. C. Greenacher and M. Matter, U.S. Patent 2,490,953 (1949);
    R. J. Geary, U.S. Patent 2,858,250 (1958).
60. A. L. Hall, Svensk Papperstidn., 80, 199 (1957);
    J. R. Salvesen and W. C. Browning, Chemical Industries, 61, 232 (1947);
    J. V. Otrhalek, U. S. Patent 2,976,248 (1961).
61. West Virginia Pulp and Paper Co. (Westvaco), Polychemicals Div., Technical Bull. 1010.
62. M. J. Borgfeldt, U.S. Patent 3,123,569 (1964).
63. L. Friedman and C. L. Lindekin, Pacific Pulp Paper Industries, 14 (1), 27 (1940);
    A. J. Wiley, M. F. Kummer, C. R. Faulkender and B. Van Camp, Tappi, 34, 556 (1951);
    D. C. Joseph, U. S. Patent 2,494,708 (1950);
    J. A. A. Lefebve, Can. Patent 617,762 (1961);
    W. D. Stewart, U. S. Patent 2,470,115 (1949).
64. H. Freeman, Can. Patent 535,849 (1957).
65. West Virginia Pulp and Paper Co. (Westvaco), Polychemicals Div., Technical Bull. 301A.
66. W. A. McIntosh, U.S. Patent 2,958,658 (1960).
67. W. C. Cochran, W. P. Kampert, Can. Patent 700,517 (1964).
68. Anon., Paper Industry, 21, 638 (1939);
    Anon., Pacific Pulp and Paper Industry, 18 (11), 48 (1944);
    W. A. Sherman, Paper Trade J., 131, No. 15, 19 (1950).
69. W. Heage, Ger. Patent 228,721 (1908); Chem. Abstr., 5, 2177 (1911).
70. Anon., Paper Trade J., 133 (24), 40 (1951);
    B. K. Hough and J. C. Smith, Industrial Laboratories, 3, 73 (1952);
    Anon., Chem. Eng. News, 30, 74 (1952).
71. A. Hayashi and D. A. I. Goring, Pulp Paper Mag. Can., 66 (C), T154 (1966);
    A. N. James and P. A. Tice, Tappi, 47, 43 (1964).
72. G. Anger, ibid., 38, 242 (1955);
    Das Papier, 8, 542 (1954).
73. Anon., Chem. Week, 87 (14), 64 (1960);
    J. H. Crowe, Can. Patent 571,307 (1959);
    A. Vloeberghs, Can. Patent 474,730 (1951).

POLYMERIC PRODUCTS 863

74. R. Tanberg, U. S. Patent 2,508,629 (1950).
75. R. Tanberg, U. S. Patent 2,495,148 (1950).
76. B. P. Wallace, V. D. Romney and H. H. Becker, U. S. Patent 2,863,781 (1958).
77. C. E. Irion, U. S. Patent 2,449,230 (1948).
78. A. Voet, U. S. Patent 2,525,433 (1959).
79. A. Soipioni, Ann. Chim. Rome, 51, 614 (1961); through Chem. Abstr., 56, 1641 (1962).
80. K. Kratzl, K. Buchtela, J. Gratzl, J. Zauner and O. Ettinghausen, Tappi, 45, 113 (1962).
81. T. R. Santelli and R. T. Wallace, U. S. Patent 3,072,634 (1963).
82. W. C. Goss, U. S. Patents 2,849,314 (1958) and 3,141,873 (1964);
    C. Uschmann, U. S. Patent 2,891,918 (1959);
    H. M. McFarlane, Can. Patent 696,732 (1964).
83. W. C. Goss, U. S. Patent 2,846,431 (1958);
    I. A. Pearl and J. W. Rose, Forest Prod. J., 10 (2), 91 (1960).
84. T. Holzer, U. S. Patent 2,534,908 (1950).
85. Anon., Paper Trade J., 140, 27 (1956).
86. J. E. Fenn, U.S. Patents 2,579,481, 2,579,482, 2,579,483 (1951);
    D. S. Bruce and H. L. Heise, U. S. Patents 2,544,585 (1951) and 2,443,889 (1948);
    E. Bichel, Gelatine Leim Klebstoffe, 7 (7/8), 129 (1939).
87. J. W. Ayers, U. S. Patents 3,093,604, 3,093,605, and 3,093,606 (1963).
88. A. A. Sokolova, Izv. Vysshikh, Uchebn, Zavedenii Lesn. Zh., 5, 147 (1962); Chem. Abstr., 57, 2467 (1962).
89. A. Rieche and L. Redinger, Ger. (East) Patent 24,060 (1962); Chem. Abstr., 58, 14299 (1963).
90. F. J. Ball and J. B. Doughty, U. S. Patent 3,090,700 (1963).
91. J. Marton and E. Adler, U. S. Patent 3,071,570 (1963).
92. C. F. Schulerud and J. B. Doughty, Tappi, 44, 823 (1961).
93. West Virginia Pulp and Paper Co. (Westvaco), Reax Bull., 1960.
94. S. Mocsny, U. S. Patent 3,112,206 (1963).
95. F. J. Ball, W. K. Doughty and H. H. Moorer, Can. Patent 654,728 (1962).
96. West Virginia Pulp and Paper Co. (Westvaco), Polychemicals Div., Technical Bull. 115.
97. T. R. Griffith and J. E. Tyson, U. S. Patent 3,013,094 (1961).
98. Anon., Chemical Week, 74 (26), 62 (1954).
99. Ibid., 89 (11), 106 (1961).
100. L. C. Bratt, Tappi, 48, 46A (1965).
101. R. A. V. Raff and G. H. Tomlinson, Jr., U.S. Patent 2,610,954 (1962).
102. J. B. Doughty, U. S. Patent 2,911,383 (1959).
103. G. S. Mills and H. E. Haxo, Jr., U.S. Patent 2,906,718 (1959).

104. E. Papst, Ger. Patent 1,165,844 (1964); Chem. Abstr., 60, 14718 (1964).
105. West Virginia Pulp and Paper Co. (Westvaco), Polychemicals Div., Technical Bull. 105B.
106. R. P. Papino and J. E. Charlebois, U. S. Patent 2,918,410 (1959).
107. M. Faye and J. B. Doughty, U. S. Patent 2,958,690 (1960).
108. M. Baum, R. Lovin and J. R. Salveson, J. Am. Leather Chem. Assoc., 47, 269 (1952);
K. T. Williams and E. F. Potter, U. S. Patent 2,505,818 (1950);
R. G. Banner and J. H. Pierce, U. S. Patent 2,995,415 (1961);
H. B. Marshall and A. C. Shaw, U. S. Patent 2,952,507 (1960);
G. Manthe and R. Fingado, U. S. Patent 2,857,346 (1958);
J. H. Pierce and R. G. Banner, Brit. Patent 968,332 (1954).
109. M. A. Buchanan and R. M. Lollar, J. Am. Leather Chem. Assoc., 42, 232 (1947);
K. H. Gustavson, Svensk Papperstidn., 44, 193 (1942).
110. J. P. Bennett, U. S. Patent 2,929,700 (1960).
111. West Virginia Pulp and Paper Co. (Westvaco), Polychemicals Div., Technical Bull. 107.
112. A. L. Hindall, Can. Patent 491,345 (1953).
113. West Virginia Pulp and Paper Co., Industrial Chemical Sales Div. Bull. L-5.
114. T. F. Kilroy, U. S. Patent 3,022,366 (1963).
115. J. B. Doughty, U. S. Patent 3,081,293 (1963).
116. R. E. Stephenson, Pulp and Paper Industries, 19 (12), 68 (1945).
117. Anon., Water Wastes Digest, 4 (2), 1 (1964);
C. W. Spalding, D. L. Halke and H. S. Carpenter, Jr., Tappi, 48, 68A (1965);
R. W. Brown and C. W. Spalding, American Paper Industry, 48 (2), 64 (1966).
118. A. V. Yashchanko, Lesnoi Zh., 6 (1), 62 (1963).
119. V. I. Vanina, A. M. Gutman, A. P. Zakoshchikov, S. A. Zakoshchikov and B. M. Rotleider, Gidroliz. i. Lesokhim. Prom., 12 (5), 8 (1959).
120. A. I. Kozlov, E. B. Tsyrkin, Sb. Tr. Gos. Nauk. Issled. Inst. Gidroliz. Sulf. Spirt. Prom., 9, 209 (1961); through Abstr. Bull. Inst. Paper Chem., 33 (5), 710 (1963).
121. Y. F. Kurdubev and S. I. Rubina, Gidroliz i Lesokhim. Prom., 12 (5), 6 (1959).
122. V. M. Raznikov, Struzhechn. Plity i. Svyazuyushchie Materialy, Moscow Sb., 212 (1961); through Chem. Abstr., 58, 7030 (1963).
123. Z. A. Dobrenravova, ibid., 14 (3), 5 (1961).
124. K. E. Kuhl and R. Bahm, Holsforsch., 16 (2), 47 (1962).
125. H. Hosaka, Japan. Patent 8713 (1962).
126. J. Nakano, K. Sasaki, C. Takatsuka and N. Migita, Nippon Mokuzai

Gakkaishi, 9 (3), 107 (1963); through Chem. Abstr., 60, 1920 (1964).
127. N. N. Shorygina and L. I. Kolotova, Izvest. Akad. Nauk. S.S.S.R. Otdel. Khim. Nauk., 562 (1953); through Chem. Abstr., 47, 12806 (1953).
128. Y. M. Turstskii, N. N. Shorygina, T. V. Izumrudova and E. L. Gristan, Gidroliz i. Lesokhim. Prom., 14 (8), 10 (1961).
129. M. I. Chudakov and M. G. Okun, Zh. Prikl. Khim., 35, 1602 (1962).
130. L. I. Kirkina, N. N. Shorygina and T.V. Izumrudova, U.S.S. R. Patent 147,168 (1961).
131. F. S. Lapshin, Gidroliz i. Lesokhim. Prom., 13 (7), 25 (1960).
132. M. Zarubin and O. V. Tishchenko, Izvest. Vysshikh. Ucheb. Zavadenii Lesnoi Zhur., 4 (5), 156 (1961).

AUTHOR INDEX

The numbers in parentheses indicate reference numbers.

Aaltio, E., 222 (281); 552 (408); 614 (99)
Abadie, F. A., 224 (293,294)
Abe, F., 583 (63)
Abe, H., 552 (414)
Abele, E. M., 581, 582 (57)
Abrahamsson, B., 222 (284); 612 (86); 702 (57)
Acerbo, S. N., 96, 105 (4); 102, 109 (19)
Adachi, K., 328, 338 (68); 556, 557 (431)
Adam, K., 577, 579 (38); 822 (218)
Adkins, H., 487 (2); 503 (2,33); 504 (34,35); 514 (28); 552 (399); 819 (189,191)
Adler, E., 2 (5); 6 (17); 8 (30); 119 (106); 150 (189); 151 (190,192); 181 (80,81); 187, 188 (109); 196 (159,171); 198 (159,181,188); 199 (188,192); 200 (195); 201 (80,201,203); 203 (208); 204 (211); 205 (214); 207 (218); 209 (237b); 211 (243); 213 (211,244,245); 226 (188,243,298); 227 (159,171,188,211, 245,298); 230 (171); 245, 250 (41); 246, 251 (42); 252 (42,61); 253 (62,64,65); 271 (40); 281, 287 (66); 284 (40,82); 350 (28,29, 30); 351 (30); 352 (35,36,37,38,40); 353 (37); 355 (35,37,51,53,57); 356 (57); 357 (29,37,40,53); 358 (37,53,61); 360 (68); 361 (70); 362 (70,72); 363 (53,77); 366 (98); 395 (176); 408 (221); 414 (238); 472 (182,183, 185,186); 473 (186,187,188); 488 (11); 512 (8); 514 (31,34,37); 517 (68,71); 518 (77); 521 (108); 528 (147); 531 (173,174); 532 (176); 533 (37); 537 (231); 539 (31,176); 540 (278); 544 (320); 546 (335); 550 (374); 551 (390); 602 (65,67); 603, 604, 605 (65); 618 (117); 626 (129,130,135); 627 (148); 650 (47,48,49,50,54); 651 (50,54,59,60,62); 652 (62); 654 (60,68); 655 (68); 658 (50); 659 (76); 660 (47); 663 (86); 664 (59,68); 671, 672 (139); 673 (54,59,62,68,85,86,144, 145,147,156); 674 (62); 676 (54,62,68,85, 139); 677 (147); 678 (62,86,144,164); 681 (85); 682, 683 (62); 698 (20); 700 (20,43); 840 (29); 855 (91)
Adolphson, C., 847 (49)
Ahmed, Q. A., 525 (120)
Ahola, A., 262 (101); 615, 616 (110); 699, 700, 711 (34)
Aida, K., 82 (113)
Akagane, K., 538 (264)
Akahane, M., 552 (410)
Akhmina, E. I., 577, 579, 583 (32,34), 581 (55, 57); 582 (57)
Akimova, M. P., 469, 470 (168)
Albrecht, A. C., 243 (28)
Aleksandrova, I. V., 771, 784, 790, 792 (11)
Alekseeva, O. P., 541 (294)
Alexander, A. E., 726 (126); 751 (190)
Alfredsson, B., 201 (198); 463, 465 (137); 464, 466 (137,138,139); 552 (401); 623 (122, 123); 624 (123); 641 (6); 660 (80); 667 (102, 103); 809 (141)
Alger, R. S., 327 (53)
Ali, I. F., 844 (43)
Ali, M. Erfan, 380, 418 (71); 525 (120)
Allan, F. J., 555 (430)
Allan, G. G., 6, 7 (22); 55, 61, 68, 69, 70 (49); 73 (83); 175, 178 (43); 248. 249 (56); 257 (56,69); 267, 271, 280, 284, 292 (15,16); 269, 283 (15); 443 (32); 444 (33); 512 (19); 521 (105); 524 (115,116); 528 (149); 530, 531 (162); 538 (264); 539 (149,267); 546 (337); 548 (115); 554 (116,428); 555 (116, 429,430); 590 (117); 591 (118)
Allen, C. E., 27, 37 (35)
Allen, G. A., 614 (100)
Allsop, A., 23 (23)
Alm, B., 201 (198)
Alsleben, H., 601 (37)
Alsup, I. A., 581, 582 (57)
Alt, A., 805 (102)
Althin, B., 201 (199)
Altpeter, J., 548 (352)
Altwicker, E. R., 116 (95); 117 (100); 118, 119 (102); 121 (115)
AMERICAN PETROLEUM INSTITUTE, 849 (55,56)
Anbar, M., 385 (134)
Anders, W., 601 (37)
Anderson, C. B., 457 (114)
Anderson, E., 46, 49, 66 (15)
Anderson, J. J., 814 (158,159)
Anderson, L., 202 (207); 384, 386, 389 (108); 388, 390 (108,164); 419 (164)
Anderzen, O., 469 (157)
Andrews, D. H., 477 (207)
Andrus, M. G., 300, 320, 325 (15); 539 (274); 627 (150); 699 (26)
Anger, G., 852 (72)
Angus, W. R., 381 (89)
ANONYMOUS, 373, 392 (2,4); 376, 414 (2); 639 (1); 686-688 (173); 736 (160); 798 (5, 15); 819 (199a); 835 (3,5,6,7,8); 842 (35); 843 (37,39); 849 (57); 852 (68,70); 853 (73); 855 (85); 856 (98,99); 858 (117)
Antipova, A. V., 421 (278,279); 551 (386)
Anton, J., 282, 285 (72); 288 (108)
Antonov, A. P., 804 (91)
Antonov'skii, S. D., 526 (131,132,134)
Aparicio, F. J. L., 531 (170)
Apitzch, H., 529 (154)
Appel, W. D., 166, 170, 195 (12)
Aprahamian, N. S., 377 (37)
Aries, R. I., 838 (10)
Arima, T., 44-46 (7)

867

AUTHOR INDEX

Arlt, H. G., 489, 497 (16); 490-492 (23); 551 (395)
Armitage, E. R., 194 (149)
Armstrong, D. E., 798 (10)
Arnall, F., 383 (92)
Aronovsky, S., 346 (5)
Arseneau, D. F., 187 (108); 285 (92); 540 (288)
Ashorn, T., 207 (226); 208 (226,228); 364 (79); 365 (85); 512, 526 (13); 587 (85); 589 (85,105,106); 590 (114); 654, 655, 660, 671 (70); 817 (184,185); 818 (185)
Ashwood-Smith, M. J., 816 (178)
Ashworth, R. B., 194 (149)
ASSOCIATION OF OFFICIAL AGR. CHEMISTS., 194 (147)
Atack, D., 730 (149)
Atal, C. K., 306 (27)
Audrieth, L. F., 515 (58)
Augustin, H., 581 (60)
Aulin-Erdtman, G., 132, 143 (154); 201, 226, 227 (197); 217 (262); 241 (6,7); 243 (29); 244 (29,30); 245 (7,31,37,39); 246 (7); 247 (48,49); 248 (30,37,48,49,50,51,52,58); 249 (37,49); 251 (30,49,52); 255 (29,51,52,68); 257 (29,30); 307 (30); 472 (184); 537 (233); 617 (112); 626 (132,133,134); 677 (161)
Awa, T., 542 (303); 543 (303,317)
Ayers, J. W., 855 (87)
Ayroud, A. M., 221, 222 (275); 374, 404, 405, 407, 412, 414, 416 (9); 375 (26)
Aziz, K., 741 (168)

Back, E. L., 725 (122,123); 735, 736 (158); 741 (169)
Baekeland, L. H., 549 (355)
Bahm, R., 858 (124)
Bailey, A. J., 48, 50, 52 (26); 540 (285); 843 (37)
Bailey, C. W., 414 (239); 457, 477 (115); 473-476 (195)
Bailey, I. W., 23 (25); 35 (62)
Bain, J. W., 374, 375, 405, 407, 412 (8); 529 (151)
Bair, R. K., 601 (53)
Baird, J. C., 327 (52)
Baird, P. I., 730 (143)
Bak, B., 300 (11)
Balashova, I. M., 538 (262)
Balatinecz, J. J., 52-54 (43)
Baldwin, R. L., 722 (118)
Baldwin, S. H., 737, 739 (162)
Balkema, D. W., 258 (77)
Ball, C. D., 102 (21); 109 (21,63)
Ball, F. J., 525, 526, 529 (126); 546 (334); 669 (121); 671 (132,133,134); 807 (132); 839 (19,20,23,24,27); 855 (90,95)
Ballersby, A. R., 630 (160)
Bamber, R. K., 29 (46)
Banko, J., 849 (57)
Bannard, R. A. B., 384 (98)
Banner, R. G., 552 (407); 856 (108)
Baranger, P., 805 (105)
Baranova, N. A., 803 (77)
Bárány, K., 726 (133)
Barghoorn, E. S., 21 (8)
Barnes, C. A., 262 (105)

Barnoud, F., 47, 48 (23); 60 (58,59); 61, 70, 73 (59); 95 (1); 104, 105 (35,38); 111 (77); 194 (150); 282, 284 (73,74,75)
Barskaya, E. I., 31 (49)
Barton, D. H. R., 131 (151); 517 (70); 552 (415,416)
Barton, E., 229 (308)
Bassett, K. H., 285 (91)
Battersby, A. R., 2 (2); 116 (94); 123 (121); 128 (132); 552, 554 (419)
Baum, M., 856 (108)
Bauman, W. C., 841 (31)
Baxter, G. F., 590 (117)
Baylis, P. E. T., 187, 188 (109); 358 (61); 700 (43)
Beakbane, A. B., 21 (11)
Becconsall, J. K., 122 (118)
Becker, H. D., 226, 227 (298); 473 (187); 512 (8)
Becker, H. H., 853 (76)
Beckmann, E., 188, 189, 228 (116); 588 (90)
Beekmans, J., 715 (103)
Begliner, E., 576 (27)
Behrman, E. J., 457 (120); 468 (152)
Beijeriuk, K., 110, 112 (72)
Bell, A., 471 (176); 532 (186)
Bell, R. P., 377, 378 (36)
Bellamy, L. J., 268 (19)
Bellamy, W. D., 556 (433)
Belorizky-Perret, N., 308 (32)
Belz, W., 405, 407, 416 (200); 512 (9)
Bender, F., 66 (73); 541 (293); 601 (36)
Benko, J., 601 (48); 713 (96); 725 (124,125); 750 (125)
Bennett, C. F., 810 (145); 812 (149)
Bennett, J. P., 857 (110)
Benoit, H., 695 (2); 729 (139)
Benson, H. K., 262 (103); 405, 407 (206); 418 (252)
Bentley, W. B., 384 (125); 388 (165)
Benton, F. C., 512 (20)
Berchet, G., 515 (54)
Berezkina, S. A., 414 (237)
Bergmann, L., 35 (61)
Bergstrom, C. B., 504 (40)
Berliner, E., 378 (46,56)
Berlyn, G. P., 24 (28a); 52 (41); 173 (34); 697 (15)
Berndt, A., 120 (111)
Berndtson, L., 210, 214 (239); 358 (62)
Bernstein, H. J., 299 (2)
Bersohn, M., 327 (52)
Bethge, P. O., 259 (82)
Bevan, E. J., 373, 376, 392, 414 (2); 405 (204)
Beyer, D. L., 209 (237a); 222 (285,286); 445 (35); 446 (35,40,41,42,44,45,46,47); 447 (41,42); 448 (44,45,46,50); 449 (64); 552 (411); 598 (98,99); 631 (172); 801 (58); 803 (79)
Bezard, A. V., 400, 407 (195); 626 (140)
Bezmosgin, E. S., 577, 579, 583 (34); 581 (55)
Bhacca, N. S., 300 (8,11)
Bhattacharya, A., 490-492 (21); 551 (397,398)
Bialkowsky, 806 (126)
Bichel, E., 855 (86)
Bicho, J. G., 434 (4)
Bier, M., 712 (80)

AUTHOR INDEX

Bigelow, L. A., 377 (38)
Bilberg, E., 670 (123)
Billek, G., 8 (31); 56 (48); 97 (5); 105 (40); 106 (5,40); 198, 200, 210 (187); 207 (220); 539 (276)
⁀irch, A. J., 306, 318 (26); 505, 506 (45); 515 (52,53,59); 516 (52,65,66); 541 (52); 839 (18)
Birum, G. H., 809 (42)
Bishop, M. W. H., 816 (177)
Bittner, F., 110 (73); 551 (391)
Bixler A. L. M., 644 (29); 753 (200)
Björkman, A., 10 (37); 23 (20); 25 (168); 54-56, 68 (46); 165, 167 (5); 166 (5,11); 168 (11); 170 (27); 171 (27,31); 173 (31,36,37); 174 (36); 175 (5,36); 176 (5); 177 (5,31); 178, 201 (36); 195, 226 (11); 220 (31); 245 (31,32); 256 (31); 284 (77); 532 (180); 608 (69,70); 612 (84); 672 (140); 698, 700 (19); 712, 720 (82); 730, 735 (146); 737, 749 (163)
Björkman, C. B., 186 (92); 359 (63)
Björkquist, K. J., 6 (17); 198 (181); 514 (34)
Bjorkvist, J., 626 (129)
Blackall, E. L., 380, 381 (88)
Blair, J. S., 515 (63)
Blanchard, H. S., 119 (103); 125, 126 (124); 552, 555 (417)
Bland, D. E., 20, 31 (7); 25, 26 (33); 47 (21); 50 (37); 59, 60 (54); 66, 73 (78); 70, 71 (81); 72 (78,81); 104 (33); 109 (62); 167 (18,20, 21); 168 (22); 173 (21); 175 (42); 176 (20); 258 (75); 259 (83); 284, 292 (84,85); 285 (90); 291 (130); 300 (17,18); 301 (18); 313 (17,18); 314 (18); 318 (17); 320 (17,18); 322, 324 (18); 434 (9); 454 (91); 539 (276); 696 (11)
Bleckmann, Ch., 384, 399, 408, 415 (105)
Bloom, E. S., 504 (35); 514 (28); 552 (399)
Bloom, J. J., 819 (191)
Bloom, S. M., 218 (265)
Blundell, M. J., 78 (87)
Boberg, F., 809 (143)
Bockman, O. C., 551 (380)
Bocks, S. M., 127 (128); 129 (140,141); 130 (140)
Bodding-Wiger, B., 579, 580 (47)
Böeseken, J., 414 (236); 458, 459 (135b); 717 (105)
Boesenberg, H., 220 (270)
Bogert, M. T., 385, 386 (138)
Bogomolov, B. D., 512 (10,17); 519 (79); 537 (219,237); 541 (294); 720 (115)
Bolker, H. I., 6 (21); 224 (296); 260 (91); 270, 271, 280, 289, 290 (33); 281, 285 (33,62); 287 (94); 288 (33,62,94,102); 384, 385, 393, 397, 418-421 (102); 532 (190); 670 (125); 712 (83); 761 (213)
Bolou, D. A., 117 (96)
Bölsing, F., 673 (142)
Bone, W. A., 469 (165)
Booth, H., 128 (131)
Bordwell, F. G., 380, 381 (74); 601 (43); 814 (164)
Borgfeldt, M. J., 548 (348); 851 (62)
Borisek, R., 538 (259); 548 (350); 844 (44)
Borquez, L. M., 80 (104)

Borra, G., 384 (97)
Bots, R. H., 799 (30)
Bottger, F., 580 (51)
Bourquelot, E., 129 (135)
Bouvier, M., 580 (52)
Bovey, F. A., 300 (6)
Bowden, K., 131 (146)
Bower, J. R., 504 (38); 819 (193)
Boyed, R., 458 (159)
Boyer, H. F., 816 (179)
Boyer, R. F., 740 (165)
Boyles, F. M., 805 (121)
Brabovskii, Y. K., 109 (64)
Brachman, W., 119 (104)
Bradbury, E. M., 300, 326 (12)
Braddon, S. A., 313 (43a); 416 (244)
Branch, S. J., 378 (55)
Brannova, N. A., 448 (49)
Bratt, L. C., 822 (213); 835 (3); 856 (100)
Brauns, D. A., 8 (25); 9 (34); 19 (1); 44, 45 (6); 95 (1); 186 (102); 189, 228 (117); 198 (180); 241, 248 (15); 350 (27); 470 (170); 502, 503 (31); 515, 524, 525, 529, 532, 539, 556 (49); 541 (301); 576, 587, 589 (26); 598, 602, 614, 617 (23); 671 (136)
Brauns, F. E., 8 (25); 9 (34); 11 (39); 19 (1); 44, 45 (6); 83 (126); 95 (1); 165 (4,7); 166 (17); 169, 220 (26); 176, 178 (4,48); 177 (58); 186 (100,102); 189 (119,121); 198 (180); 228 (119); 241 (14,15); 248 (15); 258 (70); 285 (89); 350 (27); 359 (66); 437 (29); 470 (170); 502, 503 (31); 515 (49); 519 (80, 81,85,86); 520 (86); 524 (49,81); 525, 529, 532, 556 (49); 527 (136,140,141); 528 (145); 537 (228); 538 (145,252); 539 (49,275); 540 (145,281); 541 (297,301); 545 (86,328,329); 549 (361); 575 (10); 576, 587, 589 (26); 598 (22,23); 602 (22,23,60); 614, 617 (23); 671 (136); 699 (22)
Breger, I. A., 281, 291 (67); 782, 783 (31)
Bremner, J. G. M., 170 (28); 292 (40); 782 (30)
Brenner, H. M., 761 (213)
Bretschneider, H., 515 (50)
Brewer, C. P., 487, 489-491 (5); 819 (194)
Brickman, W. J., 437 (30)
Bridger, G. L., 575, 577 (22)
Brink, D. L., 434 (4); 801 (60)
Brissaud, L., 379, 418, 419, 420 (63)
Bronovitskii, V. E., 328 (65); 330 (65,66,66a); 337 (65,66); 338 (79)
Brookband, E. G., 189 (121); 519 (85); 527 (141)
Brötz, A., 575 (12)
Brown, B. A., 111 (79)
Brown, B. R., 127 (128,130); 129 (140); 130 (140,143)
Brown, H. C., 385 (139)
Brown, R. W., 858 (117)
Brown, S. A., 2 (1); 9 (32); 12 (43); 95 (1); 99 (17); 104 (17,28,29,30); 105 (28,29,39,43); 106 (28,44,46); 107, 109 (46); 291 (131); 503 (32)
Brown, W., 702, 703, 753 (60); 744, 750 (178)
Brownell, H. H., 166, 168, 170, 171, 221 (15)
Browning, B. L., 88 (149); 190-192 (125); 194 (148); 195 (154); 259, 260 (86); 575 (2)
Browning, W. C., 848 (50); 850 (60)

Brownstein, C. J., 490, 491 (22)
Bruce, D. S., 855 (86)
Brüggem, F., 385 (148)
Brunow, G., 214 (250,251); 470 (169); 551 (384); 650 (92a)
Bruun, H. H., 292 (134,137); 293 (137)
Bryan, C. C., 801 (62,63); 802 (64); 804 (88)
Bryant, M. P., 556 (433)
Bryce, W. A. J., 701, 721 (54)
Bublitz, L. O., 259, 260 (86)
Buchanan, M. A., 83 (126); 176, 178 (48); 258 (70); 477 (201); 527 (141); 545 (328); 856 (109)
Buche, F., 686 (176); 740 (164)
Bucher, H., 753 (201,202)
Buchtela, K., 8 (31). 97, 106 (5); 198, 200, 210 (187); 207 (220); 521, 522 (107); 854 (80)
Buckland, I. K., 366 (94); 541 (297)
Bückman, L., 151 (193); 206 (215)
Budlann, C. H., 805 (104)
Budlann, E., 805 (104)
Budzikiewicz, H., 218 (264)
Bunton, C. A., 380, 383, 392-394 (87); 468 (149); 473 (196)
BUREAU OF CENSUS, WASHINGTON, D.C., Preliminary Report (1962), 797 (4)
Burger, H. E. A., 846 (47)
Burges, A., 784, 792 (32)
Burgess, H., 640 (2); 797 (1)
Burkov, G. L., 284, 285 (78)
Burroway, G. L., 457 (116)
Busch, H., 598 (17)
Busche, L. R., 8 (24); 166, 178, 192 (14); 260 (87); 282, 284 (70)
Butler, J. P., 674 (150)
Byerrum, R. V., 102 (21); 109 (21,63)
Bywater, S., 703, 705 (62)

Caldwell, E., 333 (71)
Calhoun, J. M., 610 (76a)
Callow, H. F., 477 (202,209); 478 (202); 527 (137)
Cambie, R. C., 129 (141)
Campbell, W. G., 191 (132); 259 (85)
Cantor, S. M., 259 (80)
Cardinale, Ct., 214, 216, 217, 227 (254)
Cardinale, G., 150 (185); 365 (82); 452, 454, 455 (78); 660 (78)
Carlander, R., 538 (266)
Carle, A., 192 (139)
Carlsson, G. E., 597 (7)
Carpenter, H. S., 858 (117)
Carpenter, J. S., 418 (252)
Carre, P., 805 (105)
Carroll, C. W., 647 (40)
Carroll, J. H., 546 (333)
Cartier, E., 799 (17)
Cartwright, N. J., 137 (161)
Casey, J. M., 73 (84)
Cason, J., 397 (184)
Caspersson, T., 241 (10)
Casselman, B. W., 321 (41); 446, 448 (48)
Castelfranchi, G., 384 (97)
Cavalieri, L. F., 259 (79,81)
Chaiken, B., 849 (54)
Challis, B. C., 383 (95)

Chang, H.-M., 6, 7 (22); 55, 61, 68-70 (49); 167 (19); 175, 178 (43); 185 (90); 248, 249 (56); 257 (56,69); 267, 271, 280 (14,15,16); 269, 283 (15); 272, 274 (14); 284, 292 (15, 16); 488 (12)
Chang, Y., 82 (120,121); 88 (152)
Chapiro, A., 556, 557 (435)
Charlesbois, J. E., 856 (106)
Chaykovsky, M., 813 (154); 816 (167)
Chédin, J., 380 (73)
Chelintzer, G. V., 799 (21)
Chen, C.-L., 150 (185,188); 197, 200, 205 (176); 211, 218 (242); 214, 216, 227 (254); 217 (242,254); 253 (66); 338 (81); 349 (19); 365 (19,82,83); 452, 454 (77,78); 455 (78, 97); 511 (5); 660 (78)
Cheng, P. V., 726 (127)
Chesters, G., 537 (220)
Chirkin, G., 365 (84); 648, 649 (107); 669 (117); 677 (160)
Choulet, B., 47, 48 (23); 282, 284 (75)
Christensen, G. N., 745 (179)
Christensen, P. K., 600 (28)
Christiansen, C. B., 457 (116)
Christman, F. R., 263 (108)
Chudakov, M. I., 328 (65); 330 (65,66,66a); 331 (69a); 337 (65,66,78); 338 (79); 398 (194); 421 (278,279); 469, 470 (168); 542 (302,307); 551 (385,386,387,388,389); 587 (83); 821 (208,212); 859 (129)
Chuksanova, A. A., 379 (64); 384 (109,110); 388 (109); 400, 401 (109,110); 402 (110); 405, 407 (201); 418 (263); 419 (64,109,267, 268); 420, 421 (64, 267, 268)
Chupka, Z. I., 512 (18)
Cisney, M. E., 806 (129); 807 (130,136)
Claeson, S., 477 (200)
Clark, I. R., 697 (13)
Clark, I. T., 49, 50, 66 (34); 643 (27)
Clark, J. C., 527 (140)
Claugh, S., 122 (118)
Claus, P., 76 (88); 367 (101); 444, 446 (34); 450, 451 (65,66); 625 (126)
Claypool, D. P., 814 (165)
Clayton, D. W., 434 (12)
Clermont, L. P., 66 (73); 287 (100); 541 (292, 293); 601 (36); 736 (160)
Coalson, R. L., 209, 213, 216 (238); 302, 303, 304, 311 (39); 488, 492, 498, 500 (9); 659 (73)
Cochran, W. C., 852 (67)
Cohen, W. E., 6 (13); 70, 71, 72 (81); 190 (128); 192 (133,137,138); 193 (137); 434 (9); 814 (157)
Collias, E. E., 262 (105)
Colpitts, J. H., 170 (28)
Conn, E. C., 98, 99 (12); 104 (37); 108 (55)
Connors, W. J., 122 (117,119); 123, 124 (119); 129, 131, 148 (117); 150 (117,119); 202 (206); 313 (43d)
Conrad, C. M., 703, 705 (64)
Cook, C. D., 120 (108); 121 (116); 122 (118a)
Cook, W. B., 769, 785 (3)
Cooke, L. M., 487, 489-491 (5); 504 (37,38); 819 (192,194)
Cooper, G. D., 117 (96); 125, 126 (124)
Coppinger, G. M., 120 (110); 336 (72)

AUTHOR INDEX

Corey, A. J., 551 (378); 612 (82)
Corey, E. J., 813 (154); 816 (167)
Corner, J. J., 107 (49)
Coscia, C. J., 500, 501 (24)
Cosgrove, S. L., 131 (148)
Côté, W. A., 47 (18,19); 643 (28); 697 (14)
Cousin, H., 132 (152)
Cowie, J. M. G., 701, 721 (54)
Cowling, E. B., 702, 703, 753 (60); 781 (28a)
Cox, E. G., 380 (76)
Craig, D. C., 802 (67,68); 821 (67)
Craigie, J. S., 791 (39)
Cram, D. J., 816 (168)
Crane-Robinson, C., 300, 326 (12)
Creighton, R. H. J., 10 (35); 43, 67 (2,3); 57, 61, 63, 69 (2); 65, 71, 74 (3); 434 (6,7,8); 489 (13)
Crocker, E. C., 27 (36)
Cronyn, M., 111 (78)
Croon, I., 365 (87); 517 (72); 601 (56); 630 (157)
Crosby, H. L., 838 (14)
Cross, C. F., 373, 376, 392, 414 (2); 405 (204)
Crowe, J. H., 853 (73)
Crowell, E. P., 246 (44)
Cunningham, M., 471 (172,179)
Curran, C. E., 48 (27); 404, 405, 417 (196)
Cymerman-Craig, J., 516 (65)
Czepel, H., 577, 578 (35)
Czepel, T. P., 674 (150)

Dadswell, H. E., 23, 20 (22); 24 (30); 27 (30, 34); 37 (34); 48, 49 (25); 66 (75); 180 (76); 192 (138); 695 (4)
Dahl, K., 248 (57)
Dahlgreen, E. D., 840 (30)
Daleski, E. J., 647 (42,43)
Dall, K., 177, 201 (60)
Dambmann, C., 300 (11)
Dandarova-Vasatkova, M., 272, 281, 284 (46)
Daniels, L. C., 539 (268)
Das, N. R., 46 (13); 328, 330 (64)
Datta, N. P., 328, 330 (64)
Daubner-Rettenbacher, H., 541, 542 (299)
Daur, R., 550 (365)
Davidage, H., 458 (126)
Davidson, T. A., 552 (418)
Davis, A., 458 (126)
Davis, B. D., 99 (9,11)
Davis, B. L., 76, 77 (93); 193, 194 (143)
Davis, C. W., 816 (171)
Davis, D. E., 36 (70)
Davis, G. W., 24, 27 (30); 642 (12); 753 (204, 206)
Davis, R. F., 449 (56); 712 (79); 803 (83)
Day, A. C., 47 (18,19)
Day, A. R., 384 (99)
Dean, G. R., 259 (80)
Dean, P. M., 515 (54)
DeBaun, R. M., 199 (190); 267, 280, 282, 285 (9)
deChoudens, Ch., 374, 404, 405, 408, 412, 414, 416, 417 (7)
Deflorin, A. M., 131 (151); 552 (415)
DeHass, 806 (126)
de la Mare, P. B. D., 377 (39,40); 378 (52,53, 54); 379 (53,60); 380 (75,82); 384, 407, 415 (103); 386 (154); 418 (250); 420 (273)
Delin, S., 151 (192); 213, 227 (245); 253 (62); 352 (36); 618 (117)
Dence, C. W., 313 (43a); 373, 375 (1); 384 (106,112); 385, 394, 411, 412 (146); 387 (112,146,157); 388 (157); 390 (112,157); 391 (106,146,157); 392 (1,157); 393 (146, 157); 397 (106,178); 403 (112,218); 404 (112); 407 (218); 409, 410 (106); 413 (106, 218); 414 (106,239); 416 (244); 457, 477 (115); 464, 467 (140); 468 (140,148); 473–476 (195); 512 (11)
Denivelle, L., 385, 391 (147)
Derbyshire, D. H., 378 (50,51)
Derfer, J. M., 579 (45)
Desert, M. D., 555 (429)
Desmet, J., 282, 285 (68,69)
Despande, C. M., 804 (96)
Des Rosiers, P., 477 (207)
de Stevens, G., 11 (40); 177 (65,66); 178 (65, 66,69); 179 (69,75); 180 (66); 258 (71); 267, 280, 282, 285 (11); 712 (80)
Deters, W., 280 (59)
deVries, J. L., 380 (77)
Dewey, L. J., 102, 109 (21)
Dhar, K. L., 306 (27)
Dickenmen, G. K., 846 (47)
Dickey, E. E., 229 (306); 446–448 (39); 803 (82); 804 (89)
Diddaus, D. G., 801 (60)
Diemair, W., 805 (122)
Dietrich, G., 530 (158)
Dietrich, H., 137 (161); 229 (305); 267, 272, 283 (5)
Dillon, J. A., 749 (187)
Dillon, T. E., 512 (20)
Dima, M., 713 (89)
Dimroth, K., 117 (99); 120 (111)
Diwald, J., 374, 405, 407 (17)
Djerassi, C., 517 (69)
Dobrenravova, Z. A., 858 (123)
Dohl, C. F., 797 (3)
Dolgoplask, B. A., 809 (140)
Domansky, R., 581, 583 (59)
Domberg, G. E., 575 (24); 578 (42); 583 (65, 68)
Doree, C., 407, 418, 420, 421 (214); 471 (172, 179); 528, 532 (143); 818 (186)
Dorland, R. M., 471 (177)
Dorset, N. M., 701 (44)
Doty, P., 695 (2)
Doub, L., 241 (16,17); 242 (16,23); 243 (16)
Doucette, E. I., 291 (127)
Doughty, J. B., 523, 524 (114); 530, 531 (159); 546 (334); 725 (119); 855 (90,92,95); 856 (102,107); 858 (115)
Douglas, I. D., 545 (331)
DOW CHEMICAL COMPANY, 523 (113)
Drake, N. L., 384 (101)
Dreher, E., 166 (8)
Dreyfus, C., 545 (330)
Drummond, A. Y., 457 (105)
Drummond, P. E., 288 (115,115); 289 (115); 537 (238); 669 (112)
Dryselius, E., 192 (135); 222 (284)
Dubey, G. A., 601 (46); 712 (85); 842 (36)
Duchaigne, A., 29, 30 (47)

Dulta, T. R., 36 (72)
Durie, R. A., 271, 288, 291 (45)
Dürr, W., 165 (3); 187 (104); 374, 380, 420, 422 (22); 512 (12); 532 (178); 549 (360)
Dzhurinskaya, N. G., 531 (175)

Eargle, Dolan H., 299, 336 (1)
Eberhardt, G., 95, 96, 102 (3)
Ebisawa, K., 59 (51); 415, 416 (240); 538 (245); 548 (345); 618 (113,116); 622, 629 (120); 662, 663, 678 (84); 668, 676 (99)
Edhborg, A., 614 (102)
Edling, G., 837 (7)
Edward, J. T., 726 (130,131)
Edwards, J. O., 457 (120); 468 (152,153,155); 808 (135)
Edwards, O. E., 131 (151); 552 (415)
Effer, W. R., 844 (41)
Ehlers, F. A., 816 (171)
Eholzer, V., 137 (160)
Ehrensvärd, G., 99 (10)
Ehrlenmeyer, E., 799 (24)
Ehrlich, J., 385, 386 (138)
Eibl, J., 61, 63 (61)
Eickner, H. W., 575 (4); 581 (62)
Eisenbraun, E. W., 538 (248)
Eisenhut, W., 272 (56)
Ekenstam, A. af., 166 (8)
Ekman, K. H., 268, 272 (27); 287, 288 (98); 537 (215); 551 (393); 612 (83,85); 664, 673 (87); 676 (154); 687 (179); 743 (176); 751 (193)
El-Basyouni, S. Z., 107 (50)
Elbs, K., 552 (413)
Elder, F. J., 538 (266)
Elias, W., 380, 418-420 (69)
Eliasek, J., 398 (193)
El'khones, N. M., 408 (224)
Ellefsen, O., 175 (41); 222 (287); 271, 273, 285 (41)
Ellis, C., 549 (358)
Ellmer, L. R., 6 (17); 196, 198, 227 (159); 531 (173); 626 (130)
Elofson, R. M., 288, 291 (104)
Emanuel, C. F., 791 (381)
Emerson, H. W., 645 (37)
Endres, G. F., 119 (103); 125 (124,125); 126 (124); 131 (125); 552, 555 (417)
Engleberts, R., 458, 459 (135b)
Engler, K., 186 (95); 434 (2); 452, 453 (73); 515 (60); 803 (85)
Enkvist, T., 201 (198); 207, 208 (226); 272 (52); 364 (79); 365 (85); 368 (104); 408 (223); 512 (13,14); 519, 520 (87); 526 (13); 532 (183); 552 (400,401); 587 (85,87,88); 589 (85,87,100,101,102,103,104,105,106); 590 (107,108,109,110,111); 591 (110); 641 (6); 647 (93); 649 (109); 654, 655 (70); 660 (70,80); 666 (93,104); 668 (95,96,97,111, 115); 669 (120); 671 (70,137); 713 (91); 806 (128); 817 (182,183,184,185); 818 (185)
Enslin, P. R., 176-179 (50); 282 (71)
Enström, B., 259 (83a)
Erdmann, E., 416 (243); 575, 577 (11)
Erdmann, J., 586 (72)
Erdtman, H., 2 (3); 132 (153); 137 (161); 214 (255); 245 (32); 389 (168); 398 (190,191); 455 (95); 538 (243,244,246,256,257); 543 (314); 602 (61,62); 614 (102); 601 (32)
Ericks, K., 380 (77)
Erickson, E., 550 (374)
Ericsson, B., 210, 214 (239); 267, 271, 272, 274, 280 (14); 313 (43); 358 (62); 452 (85); 488 (12); 610, 615 (73)
Eriksoo, E. E., 207 (218); 352, 355 (35)
Erin'sh, P. P., 645 (34,35)
Erlander, S. R., 701 (53)
Erman, W. F., 83 (132)
Ettinghausen, O., 521, 522 (107); 854 (80)
Eustance, J. W., 119 (103); 125, 131 (125)
Evans, T. H., 354 (45)
Evans, W. C., 777 (21)
Evans, W. L., 456 (101)
Evstifeeva, E. B., 581, 582 (57)
Excoffier, G., 306 (29)
Eymery, A., 221, 222 (275)

Fahey, M. D., 83 (137)
Faigle, H., 102, 109 (20)
Falck, R., 771 (12)
Falk, H., 806 (131)
Falkehag, I., 213 (244); 253 (63,64,65); 361 (70,71); 362 (70,71,72); 363 (77); 395 (176); 472 (185); 537 (232); 546 (335); 651 (59,61, 62); 652 (61,62); 654, 655 (68); 663 (86); 664 (59,68); 671, 672 (139); 673 (59,61,62, 68,86); 674, 682, 683 (62); 676 (61,62,68, 139); 678 (62,86); 702, 703, 753 (60)
Farinacci, N. T., 800 (34)
Farmer, V. C., 80 (107,109); 81 (107); 287, 291 (95); 774 (18,19)
Farrar, J. L., 35 (63); 54 (45)
Faulkender, C. R., 851 (63)
Favorskii, A. E., 386 (156)
Faye, M., 856 (107)
Felicetta, V. F., 197 (172); 262 (101,104,105); 601 (47); 609, 610 (72); 615, 616 (110); 619-621 (118); 627 (149); 699 (25,33,34); 700, 711 (34); 702, 748 (56); 705, 719 (33); 725 (33,121,122,123); 748 (186); 841 (32); 843 (38)
Fénéant, S., 380 (73)
Fenn, J. E., 855 (86)
Fergus, B. J., 50, 52 (39,40); 51, 53, 67, 68 (40); 697 (16,17,18); 698 (16); 753 (16,17, 209,210); 754-756 (209)
Fergus, L., 173, 175 (33)
Ferguson, L. N., 242 (21)
Ferguson, W. S., 194 (149)
Fernholz, H., 468 (151)
Ferry, J. D., 553 (422)
Feurstein, K., 803 (69)
Field, B. B., 421 (277)
Field, L., 288 (115,116); 289 (115); 537 (238); 669 (112)
Fieser, L. F., 513 (21)
Fieser, M., 513 (21)
Filar, L. J., 345 (2)
Fingado, R., 856 (108)
Finkbeiner, H., 117 (96); 125, 126 (124); 552, 555 (417)
Finkelstein, A. V., 271 (36,37); 273, 288 (37); 284 (81,83); 285, 291 (81)
Finkle, B. J., 108 (56,57); 109 (57)

AUTHOR INDEX

Fisher, F., 374, 379, 418, 419, 421, 422 (21); 449 (61,62); 450 (61); 469, 470 (164); 575 (13); 822 (215)
Fisher, H. E., 207 (217); 327 (52); 355, 357 (54)
Fisher, J. H., 802 (65,66)
Fitzgerald, W. P., 804 (97)
Fitzpatrick, J. D., 336, 338 (76)
Flaig, W., 123, 129 (122); 291 (129); 398 (189); 786, 788, 790 (36)
Flanagan, H. K., 122 (118a)
Fleck, L. C., 48, 66 (24)
Fletcher, T. L., 377, 407, 416 (34); 577 (29, 30); 578 (30); 822 (217)
Fletscher, C. A., 384 (119)
Flickinger, E., 452, 453 (72,73); 469 (161); 505 (44); 515 (60,61); 516 (61); 618 (115); 821 (205)
Flodin, P., 727 (136)
Flokstra, J. H., 102, 109 (21)
Flory, P. J., 511, 547 (1); 699 (35); 701 (55); 705, 751 (68); 726 (134); 731 (153); 758 (55,211)
Floyd, D. E., 384 (131)
Fogelberg, B. C., 289 (121); 614, 615 (106, 108); 622 (108); 626 (143); 751 (192)
Forago, J., 816 (174)
Forbes, W. F., 288 (110); 669 (113)
Forgacs, O. L., 695, 696 (5)
Forman, L. V., 478 (210)
Forsen, S., 132, 143 (154); 307 (30)
Forss, K., 5 (12); 262 (100); 289 (121); 601 (53); 614, 615 (53,103,104,105,106,107, 108); 617 (104); 622 (108); 626 (103,104, 143); 719 (112); 751 (192,194,195,196,197); 752 (196,197); 842 (34)
Förster, Th., 243 (25)
Fort, R., 385, 391 (147)
Forziati, F. H., 166, 170, 195 (12)
Fossan, K. R., 812 (148)
Foster, D. H., 192 (133); 299, 336 (1)
Fox, A. S., 456 (100)
Fox, D. E., 805 (112)
Fox, T. G., 731 (153)
Francis, A. W., 388, 390, 420 (161)
Frank, R. L., 504 (35); 514 (28); 552 (399); 819 (191)
Frankel, G. S., 21 (10)
Fraser, R. D. B., 245 (36)
Freeman, H. G., 590 (117); 851 (64)
Freeman, J. H., 664 (88)
Fremer, K.-E., 5 (12); 81 (112); 262 (100); 601 (53); 614, 615 (53,103,104,105,106); 617 (104); 626 (103,104,143); 713 (91); 751 (192,194,195,196,197); 752 (196,197); 842 (34)
French, D., 701 (53)
Freudenberg, K., 2 (4); 3 (7,8,9); 4 (8); 5 (8,9); 12, 14 (42); 23 (21); 28 (37,39); 29 (44); 56, 59 (50); 69 (80); 80 (110); 95 (1); 98, 101 (6); 104 (31,32,36); 106 (31,32); 110 (31,65, 66,67,70,73,74); 112 (80,81); 116 (88,89,90, 92,93); 129 (137); 137 (161,162); 138 (88,92, 163,169); 139 (164); 140 (88,164,165,166, 167,168); 141 (171,172,173,174); 142 (173, 174); 143 (80,92,175,176); 146 (93,177, 179); 147 (182); 148 (142,166); 150 (185,

188); 155 (80,92,195,196); 165 (3,6); 168, 175, 190 (24); 171 (30); 176 (44,45); 177 (54,60,63); 180 (44); 185, 191 (87); 186 (95); 187 (63,104,106,107); 195 (156,157); 197 (157,176,178); 200 (176,196); 201 (60, 199,202); 205 (176,212); 210 (240,241); 211 (242); 214 (54,156,248,254); 215 (257,258); 216 (254); 217 (242,254); 218 (242,266); 219 (267,268); 220 (269,270,271); 224 (297); 226 (299,300,301); 227 (202,254); 229 (54,305); 230 (311); 245 (33); 248 (57); 253 (66); 267 (4,5); 272 (5,54,55,56); 281 (4,60,61); 283 (4,5,60,61); 284 (60,61); 328, 330 (67); 338 (81); 349 (19); 359 (64,65,67); 360 (67); 365 (19,82,83); 374 (22); 380 (22, 70); 405, 407, 416 (200); 418 (70); 420 (22, 70); 422 (22); 434 (2); 452 (71,72,73,79,80, 81,82); 454 (77,78); 455 (78,96,97); 469 (161); 471 (175); 505 (44,46); 506 (46); 511 (4,5); 512 (6,9,12); 514 (30,33); 515 (60,61, 62); 516 (61,62); 527 (138); 528 (144); 530 (158); 532 (178); 540 (287); 549 (360,363); 551 (391,392); 577, 579 (38); 588 (95,96); 597 (6); 601 (38); 618 (38,115); 628, 629 (153); 630 (159); 698 (21); 800 (49,50); 803 (75,85); 819 (197); 821 (205); 822 (218)
Frey-Wyssling, A., 22 (16,19); 24 (19); 37 (19, 76,79); 643 (16)
Friedman, L., 851 (63)
Friedmann, M., 141, 142 (173)
Friedrich, A., 374, 405 (17); 377 (32); 407 (17, 213); 449, 450 (61); 469, 470 (164); 531 (171)
Fries, S. L., 458–460, 463 (124)
Friese, H., 374 (20); 380, 420 (67); 418, 419 (20,67)
Frisch, K. C., 521 (106)
Frolov, S. S., 452 (70)
Fronmüller, D., 478 (211)
Frost, P., 35 (65)
Fruton, J. S., 778 (23)
Fuchs, W., 28 (37); 104, 106, 110 (31); 188, 228 (117); 220 (271); 407, 417 (212); 527 (138); 579, 580 (46)
Fugleberg, S., 289 (121); 614, 615, 622 (108)
Fujahara, K., 196 (165); 317 (48); 537 (139)
Fujii, M., 82 (113); 83 (136); 193 (140); 454 (90)
Fujita, F., 530 (93)
Fujita, H., 732, 733 (154)
Fukuhara, S., 177 (59); 281, 282 (64)
Fukui, E., 525 (125); 545 (323)
Fukushimi, T., 527 (139)
Fukuwatari, S., 288 (118); 289 (119)
Fukuzumi, T., 112, 143, 155 (80); 196 (165); 281 (65); 317 (48); 770 (8); 772, 774–776 (14); 780 (25,26)
Fung, T. H., 653, 676 (155)
Furner, A. H., 130, 131 (145)
Fursenko, I. V., 530 (164)
Fürst, H., 374, 418, 419 (20)
Furukawa, J., 520 (92)

Gaber, H., 213 (246); 349 (20)
Gagnaire, D., 301–304, 313, 318, 325 (44)
Galochkin, A. I., 531 (167,169)
Gamborg, D. L., 99, 106 (14)

Gamparin, F., 534 (205)
Ganguli, N. C., 99 (13)
Garanzha, L. P., 417, 421 (245)
Garbisch, E. J., Jr., 380, 381 (74)
Gardner, J. A. F., 177, 182, 186, 206, 207 (56); 355 (49,55,56); 357 (55,56); 611 (78)
Gardner, P. D., 536 (214)
Gardon, J. L., 601 (49); 699, 700, 725 (30); 705, 708, 710 (67)
Garnjobst, L., 99 (10)
Gaslini, F., 548 (349); 674 (152)
Gatterman, L., 798 (12,13)
Gaulis, M., 575, 577 (14)
Geary, R. J., 850 (59)
Geiger, H., 141 (171); 195, 214 (156); 272 (55)
Gelfand, E. D., 512 (10); 519 (79); 537 (219)
Gellerstedt, G., 307 (30b); 365 (81,88); 619, 630 (117a)
Gelles, E., 377, 378 (36)
Georgescu, V. N., 548 (351)
Georgievskaya, G. D., 551 (386)
Gerdjikova, S., 525, 526 (127)
Gess, J., 397, 398, 402 (180)
Gibbs, D., 434 (8)
Gibbs, R. D., 6 (19); 19 (3); 43, 63 (2,4); 57, 61, 67, 69 (2); 65, 72, 74, 79, 80 (4); 489 (13)
Gierer, J., 202 (204,205); 207 (222,223,224, 225); 281, 287 (66); 288, 289 (117); 302, 303, 312 (42); 307 (30b); 350 (28); 360, 261 (69); 362 (69,73,74,75,76); 363 (69,78); 365 (81,88); 387, 391, 399, 415 (158); 518 (75,76); 519 (83); 525 (123,124); 528, 540 (146); 532 (176,177); 537 (240); 539 (176); 551 (375); 619, 630 (117a); 623 (122,123); 624 (123); 627 (145); 649 (108); 651 (57,58, 69); 652 (57,58); 653 (63,64,65,66,67); 654, 662, 664, 676 (69); 656 (69,90,91,92); 657 (90,92); 666 (66); 667 (66,102,103); 673 (146)
Gierisch, W., 405, 416 (207); 407 (210)
Giertz, H. W., 258 (76); 407, 413 (217); 416 (241); 601 (35); 748 (184)
Giesen, J., 819 (199b,199c,199d,199e,199f); 838 (15)
Gilbert, A. R., 117 (96)
Gillespie, R. J., 380 (78,79)
Gillet, A., 575 (25)
Gilliard, A., 799 (17)
Ginsberg, D., 384 (117); 385 (134,135)
Gionola, G., 467, 468 (143)
Gjokic, G., 80 (100)
Glading, R. E., 241 (4)
Glennie, D., 197 (172,173); 300, 320, 325 (15); 311 (35); 539 (274); 602 (58); 613 (98); 626 (137); 627 (149,150); 628 (137,152); 699 (25,26); 803 (152)
Glover, R. L., 529 (151)
Godard, H. P., 487 (3)
Goheen, D. W., 365 (86); 512 (15,16); 575 (17); 577 (33); 583 (69,70); 807 (130,133, 136,137b); 808 (139); 810 (145); 812 (149); 817 (137a)
Gold, V., 381 (90)
Goldschmid, O., 107 (52); 196, 198 (162); 229 (309); 245, 246, 247, 249 (38); 248 (53,55); 258 (77); 262 (105,106); 348 (11); 538 (254); 601 (57); 615, 616 (111); 626 (136);
677 (162); 719 (111)
Goliath, M., 351 (33); 622 (121)
Gömöry, A., 575, 576 (21)
Gondo-Hunwald, K., 378 (47)
Goos, A. W., 575 (1)
Gorbunov, O. F., 537 (237)
Goring, D. A. I., 50 (38,39); 52 (39); 63 (70); 554 (425); 601 (41); 610 (74); 612 (74,87, 92,94); 613 (94,95a); 620 (87); 643 (19,20, 21,22,23); 644 (20); 645 (23,38); 685 (19); 686 (175); 695 (1,3); 697 (16,17,18); 698 (16); 699 (23,24,32); 700 (24,32,37,39); 701 (24,37,39,46,47,52); 702 (37,39,46,47,59); 703 (24,32,37,63); 704 (3,63); 705 (24,32, 37,63,70); 706 (32); 708, 709 (23); 710 (23, 74); 711 (78); 712 (87); 713 (98,99); 754–756 (208,209); 758 (212); 759 (39,46); 840 (28); 852 (71)
Gortner, R., 346 (5)
Goss, M. J., 76, 77 (93); 190 (127); 193, 194 (143,144); 471 (178); 577 (31); 588 (92,93)
Goss, W. C., 854 (82,83); 855 (82)
Gottlieb, S., 769 (1)
Gould, E. S., 408 (226)
Grabovskii, Y. K., 95 (2); 530 (163)
Graf, G., 378 (47)
Grafe, V., 195 (158); 366 (93); 598 (18); 800 (35)
Graham, J., 380 (78)
Gralen, N., 720 (113)
Gran, G., 201 (199); 259 (82)
Granath, M., 398 (191); 490, 492 (19)
Grangaard, D. H., 413, 416 (232); 513 (27); 538 (253); 821 (211)
Gratzl, J., 196 (166); 348, 355, 357 (10); 444, 446 (34); 450, 451 (65,66); 521, 522 (107); 526 (130); 549 (359); 577, 578 (35); 854 (80)
Gray, K. R., 83, 85, 88 (140); 537 (228); 540 (281); 549 (361); 838 (14)
Green, F. O., 455 (98)
Green, G., 785 (35)
Greenacher, C., 850 (59)
Greenwood, C. T., 701, 721 (54)
Greiner, W. J., 538 (265)
Griengl, H., 213 (247)
Griffith, T. R., 856 (97)
Griffoen, K., 29, 37 (41)
Grigorev, G. P., 288 (101); 537 (218)
Grion, G., 138 (169); 219 (267)
Gripenberg, J., 214 (255); 389 (168)
Grison, P. E., 380 (77)
Gristan, E. L., 859 (128)
Grohn, H., 166, 184 (10); 220 (272); 280 (59)
Gross, A. J., 129 (138)
Gross, S. R., 99 (10); 104 (27); 490–492 (23); 551 (395); 719, 750 (109)
Grosskinsky, O. A., 540 (289)
Grovenstein, E., 377 (37); 388 (162)
Gruber, K., 539 (276)
Grubisic, Z., 729 (139)
Grundy, J., 421 (277)
Grushnikiv, O. P., 405, 407, 413, 414 (208); 418 (263)
Guba, F., 726 (133)
Gugnin, Y. A., 547 (342)
Gunkel, L., 421 (274); 586 (78); 821 (203)

AUTHOR INDEX

Gupta, M. K., 464, 467, 468 (140)
Gupta, P. R., 601 (54); 699, 700, 701, 703, 705, 719 (24); 710 (74,76); 711 (76); 720 (76,114); 721 (116); 727, 728 (76); 734, 735, 748 (157); 751 (24,74)
Gustafsson, C., 202 (207); 388, 390, 419 (164)
Gustavson, K. H., 554 (423,424); 856 (109)
Gutman, A. M., 858 (119)
Gutweiler, K., 816 (172)
Gwocock, C. M., 469 (165)

Haag, A., 452, 453 (71)
Haage, W., 552 (405)
Haarmann, W., 28 (38)
Haas, B. R., 88, 89 (150)
Haas, H., 457 (106)
Haber, F., 117 (97)
Habermann, H., 374, 399 (13)
Hachihama, Y., 198 (185); 418, 421, 422 (256); 419 (269,270); 454 (90); 525 (122,125); 534 (204); 542 (306); 545 (322,323,324); 551 (379); 556 (434); 701 (48); 805 (117)
Hafe, C., 631 (170)
Haga, T., 542 (304); 547 (343); 548 (304,343); 715 (101,102)
Hagedorn, I., 137 (160)
Hagerman, D. C., 489, 490 (15)
Hägglund, E., 59 (53); 165, 182 (1); 185 (1,89, 91); 186 (92,96); 189 (122); 201 (198); 241 (3); 359 (63); 379, 407, 418 (62); 532 (183, 187); 533 (201); 537 (226); 543 (318); 575, 576 (19); 586 (76,77); 597 (7,10); 598 (10, 14,15,16,17,21); 602 (10); 612 (86); 645, 646, 662 (83); 666 (83, 104); 669 (120); 702 (57); 806 (125,127,128)
Häggroth, S., 6 (17); 198 (181); 209 (237b); 366 (98); 514 (34,37); 533 (37); 626 (129)
Haglund, P., 261 (97); 699, 701 (31)
Häglund, S. E., 245 (32)
Hahn, O., 384, 400 (121)
Haider, K., 123, 129 (122); 792 (42,43)
Halabisky, D. D., 524, 554, 555 (116); 546 (337)
Halke, D. L., 858 (117)
Hall, A. L., 850 (60)
Hall, E., 556 (433)
Hall, L., 407, 418, 420, 421 (214); 528, 532 (143); 818 (186)
Hallam, J. W., 839 (18)
Haller, W., 699, 705, 719, 725 (33)
Halmekoski, J., 590 (112)
Halstead, W. J., 849 (54)
Halvarson, H., 363 (77); 654, 655, 664, 673, 676 (68)
Hamerstrand, G. E., 532 (192,193)
Hamill, R. L., 109 (63)
Han, S. T., 647 (41)
Hanford, W. E., 526 (133)
Hann, R. M., 384 (126)
Hansen, R. E., 120 (112); 336 (75); 536 (212)
Harada, H., 52, 53 (42)
Harada, T., 198 (184,186); 534, 535, 542 (206, 207)
Harborne, J. B., 107 (49)
Harder, M., 359 (64,65); 588 (95)
Harders-Steinhauser, M., 66 (74)
Hardwell, H., 629, 630 (154)

Harkin, J. M., 3, 4 (10); 28 (39); 80 (110); 95 (1); 98 (8); 110 (67); 112, 143, 155 (80); 116 (91); 138 (169); 150 (187,188); 184, 204 (85); 197 (176,179); 200, 205 (176); 201 (199); 215 (258); 219 (268); 226 (299); 229 (310); 253 (66); 328, 330 (67); 338 (81); 349, 365 (19); 446 (37); 452 (88); 511 (5); 535, 537 (210); 601, 618 (38); 630 (160); 650 (56); 649, 677 (110); 781 (28a)
Harlow, W. M., 540 (284)
Harrington, K. J., 271, 281, 284, 285 (44)
Harris, E. E., 190 (128); 374, 405 (12,16); 377 (12); 407 (12,16,411); 410 (16,411); 487, 503 (2); 504 (34,36,39,40); 577 (29,30); 578 (30); 601 (51); 819 (189,190); 822 (217)
Harris, H. C., 384 (101)
Hart, J. S., 260 (92)
Hart, V. E., 575 (5)
Hartler, N., 601 (39); 640, 641 (4); 642 (9)
Harvey, J. T., 378 (54); 549 (355)
Hasegawa, M., 99 (15); 103 (22); 104, 110 (25)
Hasner, L., 700 (36)
Hastbacka, K., 207, 208 (226); 364 (79); 589 (105); 590 (115); 654, 655, 671 (70); 660 (70,82); 661, 668 (82)
Hata, H., 281, 283–285 (63)
Hata, K., 46, 48 (14); 82 (115,116); 83 (127, 128,130,141); 85 (141,143,144); 86 (115); 87 (115,116,128,130,141,143,144,148); 189 (124); 538 (251); 701 (49)
Hatakeyama, H., 338 (80); 458 (125,129,131, 132); 459 (125,131); 460 (125); 461 (129); 463 (131); 471 (181)
Hatch, M. J., 749 (187)
Hatch, R. S., 837 (9)
Havinga, E., 119 (104)
Hawkins, W. L., 471 (176,177)
Hawley, L. F., 48, 49 (25); 597, 598 (3)
Haworth, R. D., 137 (161)
Haxo, H. E., Jr., 856 (103)
Hay, A. S., 119 (103); 125, 131 (125); 552, 555 (417)
Hayashi, A., 77, 78 (95); 220 (273); 221 (277); 222 (279); 454 (92); 517, 519 (73); 538 (249, 250); 554 (425,426,427); 592 (123); 601 (52); 630 (162,163,164,165,166,167,168); 705 (66); 713, 716 (99); 717 (99, 106, 107); 718 (99,108)
Hayashi, I., 770 (7)
Hayashi, M., 770 (8)
Hayes, H. R., 813 (156)
Haynes, C. G., 130, 131 (145)
Hayward, L. D., 380, 418, 419, 420 (69)
Heage, W., 852 (69)
Hearon, W. M., 228 (302); 365 (86); 512 (16); 583 (70); 713 (93); 807 (130,133,137b)
Heden, S., 597 (9)
Hedler, R., 601 (37)
Hedlund, K., 357, 362 (60)
Heeschen, Jerry P., 299, 336 (1)
Hegbom, L., 244, 257 (30); 248 (30,50,51); 251 (30,52); 255 (51,52); 626 (133)
Heimberger, W., 116 (90); 155 (195)
Heirnberger, W., 267, 281, 283 (4)
Heise, H. L., 855 (86)
Henderson, D. W., 355, 357 (56)
Henderson, G. G., 458 (159)

Henderson, J. T., 583 (69,70)
Henderson, M. E. K., 770 (6); 774 (18,19); 781 (27)
Henderson, U. V., 388 (162)
Henley, D., 704, 705 (69)
Henning, W., 805 (122)
Hentgartner, H., 24 (28); 642 (15)
Hergert, H. L., 82 (117,118,119,124); 83 (117, 118,119,124,133,138,139,140); 84 (118, 124); 85 (117,133,138,139,140,142); 86 (117,124,139); 87 (117,124,139); 88 (140); 107, 176 (52); 177 (55); 229 (309); 242, 243, 246 (22); 267 (13); 272 (13,43); 271, 273, 275, 277, 278, 280, 281, 282, 284, 285–290, 292 (43); 528, 538, 540 (145); 630 (158); 713 (95)
Herissey, H., 132 (152)
Hermann, F., 586 (82)
Hermans, J. J., 711 (77)
Hernestam, S., 177 (61); 181 (80); 201 (80, 201); 414 (238); 472 (182,183); 517 (68); 673, 678 (144)
Herzog, R. O., 241 (1,2)
Hess, C. L., 177, 178 (64); 245 (34); 267, 281, 282, 285 (12); 700 (42)
Hess, D., 108 (58)
Hess, K., 166 (9); 532 (179)
Heumann, K. E., 166 (9); 532 (179)
Heuser, E., 375, 404, 405, 407 (23); 421 (274); 469 (162); 575 (12,15); 576, 579 (15); 586 (79,80,81,82); 588 (80); 821 (203,204); 822 (214)
Hewson, W. B., 206 (216); 626 (142)
Hibbert, H., 10 (35); 43, 67 (2,3); 57, 61, 63, 69 (2); 65, 71, 74 (3); 151 (193); 206 (215, 216); 207 (217); 354 (43,45); 355 (43,50,54); 357 (54,58); 366 (94); 380, 419, 420, 422 (72); 397, 402, 407, 409, 413, 414 (177); 418 (72,177); 434 (6,7,8); 471 (176,177); 487 (3,5); 489 (5,13,14); 490, 491 (5,14); 504 (37,38); 532 (188,189); 537 (227,228, 229); 539 (275); 540 (280,281); 541 (297); 549 (361); 598 (20); 626 (142); 800 (42,43); 803 (70); 819 (192,193,194)
Higashitsuji, K., 601 (52)
Higgins, H. B., 271, 285, 288–290, 292 (34)
Higgins, H. G., 271, 281, 284, 285 (44); 730 (145)
Higgins, P. H., 537 (238)
Higgs, T., 643, 645 (23); 736 (159)
Higuchi, T., 8 (27); 10 (36); 29 (42); 43, 65 (5); 47, 48 (22,23); 59, 66, 72 (22); 60, 70, 73 (59); 61 (5,59); 62 (5,64); 69, 71 (5,82); 75 (86); 76 (86,89,90); 77 (94); 78 (94,97); 79 (90); 82, 87 (114); 95 (1); 99 (16); 103 (22); 104 (25,34,35,38); 105 (35,38,42); 106 (34, 44,46); 107, 109 (46); 108, 115 (59); 110 (25,75); 112 (42,75,82,83,84); 113 (42,82, 85); 114 (42); 168, 188 (23); 177 (53); 186, 187 (101); 189 (123); 194 (150); 196 (167); 207 (221); 221 (276); 222 (278,282,283); 258 (74); 271, 279, 280, 285 (42); 272 (53); 282 (73,74,75); 284 (42,73,74,75,87); 288, 292 (42,103); 434 (11); 532 (194); 770 (7); 772, 774, 775 (13)
Hill, A. J., 388, 390, 420 (161); 546 (338)
Hillmer, A., 241 (1,2); 542 (305)

Hilpert, S., 405, 416, 417 (205)
Hilton, I. C., 377 (40); 379 (60)
Himoe, A., 386 (155)
Hindall, A. L., 857 (112)
Hine, J., 386 (169); 392 (174)
Hirakawa, K., 535 (209)
Hirashitsuji, K., 538 (249)
Hirnyj, B., 258 (75)
Hirst, S., 598 (12)
Hlasiwetz, H., 374, 399 (13)
Hlava, J. B., 437 (29)
Ho, G., 70, 71, 72 (81)
Hodgson, W. G., 340 (85)
Hofbauer, G., 106 (47); 366 (97)
Hofe, C. V., 803 (71)
HOFFMAN-LaROCHE & COMPANY, 799 (26, 31); 800 (48)
Hofman, W., 291 (124,128)
Hofreiter, B. T., 532 (192,193)
Hogan, D., 601 (51)
Hoge, W. H., 627 (144)
Holderby, J. M., 713 (97); 835 (2)
Holladay, P. C., 473 (190)
Holleman, A. F., 386 (150)
Holm, B., 201 (198); 589 (103); 649 (109); 817 (183)
Holmberg, B., 8 (28); 54–56, 63, 64, 80 (47); 189 (120); 224, 227 (295); 345 (1); 351 (31); 469 (157); 532 (181,182); 544 (319); 545 (325,326,327); 588 (94); 597 (8,9)
Holmberg, G. A., 384, 410 (100)
Holmes, G. W., 82, 83 (125)
Holtzer, A. M., 695 (2)
Holzer, T., 854 (84)
Honig, M., 800 (40)
Hopkins, C. Y., 385 (132)
Hoppe-Seyler, F., 586 (73,74)
Hopper, E. W., 844 (41)
Hori, S., 188, 228 (118)
Horner, L., 814 (163)
Horning, E. C., 385 (143)
Hosaka, 858 (125)
Hostomsky, G., 537 (230)
Hough, B. K., 552 (406); 852 (70)
Howard, G. C., 800 (44,46); 843 (37)
Hoyt, C. H., 841 (33)
Hrutfiord, B. F., 262 (105); 312, 317, 318, 321, 323, 324 (40); 365 (84); 488, 490–492, 494–496 (10); 525 (121); 601 (44); 669 (117,118)
Htun, M. T., 741 (169)
Hubbard, J. K., 83 (131); 193 (141)
Huber, H. F., 387, 391, 399, 415 (158)
Hübner, H., 116, 138, 143, 155 (92); 197 (178)
Hübnett, F., 811 (147)
Hückel, W., 515 (50)
Hughes, E. D., 378 (52); 380 (78,83,87,88); 381 (88); 383, 392–394 (87); 468 (149)
Hummel, D. O., 268 (22)
Humphrey, J. D., 83 (131); 193 (141)
Hungate, R. E., 556 (433)
Hunsberger, I. M., 814 (160)
Huque, M. M., 695, 704 (3)
Hurd, C. D., 455 (98)
Husband, R. M., 467, 468 (142)
Hyde, J. S., 336, 338 (76); 337 (77)

AUTHOR INDEX

Ianni, J. D., 504 (34)
Ibne-Rasa, K. M., 383 (94)
Ichihara, K., 779, 785, 786 (24)
Ichino, M., 514 (40,41); 611 (79)
Ida, H., 701 (48)
Ifju, G., 48 (33)
Iizima, R., 288, 289 (107); 671, 672 (138)
Ikeda, S., 779, 785, 786 (24)
Immergut, E. H., 171 (32)
Imoto, M., 525 (122)
Ingersoll, S. C., 458–460, 463 (124)
Ingold, C. K., 379 (58); 380 (78,83,87,88); 381 (88,91); 383, 393, 394 (87); 386 (151,153); 392 (87,175); 418 (249); 468 (149)
Ingruber, O., 611 (77)
Ioannidis, O., 530 (160)
Ionescu, M. V., 548 (351)
Iorio, A. D., 108 (60)
Iosilevich, A. I., 530 (160)
Irion, C. E., 854 (77)
Ishihara, T., 82, 86, 87 (115); 209 (235); 226, 227 (298); 366 (90); 473 (188); 512 (8); 650 (55)
Ishii, T., 301, 315, 318 (19a)
Ishikawa, H., 78 (97); 104, 110 (25); 112, 113 (82); 181 (82); 196 (168); 374, 416 (15); 375, 407 (24); 457, 461 (118); 458 (118,127, 128,133,135a); 459, 460 (127,128); 463 (135a); 469 (160); 473, 474 (194); 520 (91, 93,94,95,98,99); 521 (100,101,102,103,104); 532 (184); 713 (1ง0); 773 (15); 774 (16,17); 775 (15,17); 781 (17); 788 (16,17)
Ishizu, A., 178, 181 (68); 179 (71); 207 (227); 208 (229); 385, 416 (144); 535 (208,209); 537 (223,224,225); 548 (347); 660 (81); 668 (100); 677 (100,163)
Isome, Y., 196 (165); 317 (48); 527 (139)
Israel, G. C., 377 (43,44)
Ito, Y., 8 (27); 10 (36); 75, 76 (86); 82, 87 (114); 168, 188 (23); 258 (74); 284 (87)
Itoh, T., 627 (146)
Ivanenko, A. D., 547 (344)
Ivanov, V. I., 379, 419, 420, 421 (64)
Ivarsson, A. B., 726 (132)
Ivencic, A., 261 (96); 392, 397, 405, 407, 413, 414, 416, 417, 421 (173)
Ivnäs, L., 351 (32); 611 (80)
Iwamida, T., 288, 289 (107); 671, 672 (138)
Izumrudova, T. V., 408 (224); 418, 422 (256); 550 (373); 859 (128,130)

Jackman, L. M., 299 (3)
Jackson, C. L., 385 (149); 397 (183)
Jackson, M., 741 (169)
Jacob, S. W., 817 (180)
Jacobs, D. I. H., 380, 383, 392–394 (87); 468 (149)
Jacobs, T. L., 519 (84)
Jacopian, V., 601 (37)
Jacquiot, C., 60 (57)
Jaeger, C. B., 384 (101)
Jaffe, H. H., 242 (19,20)
Jahn, E. C., 527 (142)
James, A. N., 289, 290 (120); 552, 553 (412); 618, 629, 630 (114); 716 (104); 729 (138); 852 (71)
James, H. N., 537 (216)

Jannzems, V., 417 (247)
Jansen, G. V., 374, 375, 405, 407, 412 (8)
Janson, A., 452, 453 (71); 471 (175)
Janson, J., 699, 701 (31)
Jansons, N., 48, 49 (28)
Jantzen, L., 538 (260); 602 (59)
Jatker, S. K. K., 804 (96)
Jaunzems, V. R., 288, 290 (105); 823 (224)
Javorsky, J. M., 44 (8)
Jaworzyn, J. F., 713–716 (98)
Jayme, G., 66 (74); 241 (13); 263 (107); 292 (136); 644 (30); 645 (36); 753 (205)
Jeffrey, G. A., 380 (76)
Jenkins, S. H., 190 (126)
Jensen, W., 6 (18); 81 (112); 601 (53); 614, 615 (53,106); 626 (143); 751 (192,194); 842 (34)
Jerkeman, P., 364 (80)
Jernigan, J. M., 839 (22)
Jett, A. C., 478 (212)
Jirgensons, B., 511, 558 (3)
Jodai, S., 419 (270)
Johannson, B., 261 (95); 650, 655, 658, 664 (51)
Johanson, F., 735, 736 (158); 741 (169)
Johanson, K., 670 (124)
Johanson, M., 614, 615 (106); 626 (143); 751 (192)
Johnson, A. P., 814 (161)
Johnson, D. B., 260 (89)
Johnson, D. C., 473, 474 (193)
Johnson, L. F., 300 (8)
Johnson, P., 726 (126); 751 (190)
Johnson, T., 189 (122); 533 (201); 543 (318); 545 (331); 598 (15,17)
Johnston, H. W., 612 (88,89)
Jolley, R. S., 846 (47)
Jones, B., 378 (55)
Jones, D. D., 473, 474 (193)
Jones, E. A., 288 (115,116); 289 (115); 537 (238); 669 (112)
Jones, E. J., Jr., 241 (8); 267, 268, 281 (2,3); 282, 283 (3)
Jones, G. W., 477 (201); 519, 520, 545 (86)
Jones, I. G., 701, 721 (54)
Jones, K., 452, 453 (81)
Jones, M. H., 380, 383, 392–394 (87); 468 (149)
Jones, W. J., 814 (158,159)
Jorgensen, L. R., 34 (53)
Joseleau, J. P., 104, 105 (35); 282, 284 (73)
Joseph, D. C., 851 (63)
Joshi, B. S., 120 (110); 336 (73)
Jovanovic, V., 455 (96); 659 (75)
Joyce, J. S., 260 (93); 261 (94)
Jungwirt, A., 398 (193)
Jura, W. H., 340 (85)
Jurd, L., 384 (96)
Juslen, C., 272 (52)
Jyodai, S., 198 (185); 525 (125); 534 (204); 545 (322,323); 556 (434)

Ka, S., 208 (230)
Kaeding, W. W., 131 (147)
Kagino, T., 668 (100); 677 (100,163)
Kahehi, K., 467 (145,146); 468 (146)
Kaiser, P., 814 (163)

Kalb, L., 185 (88); 186 (103); 818 (187)
Kalinskaya, L. L., 338 (79)
Kalk, F., 120 (111)
Kallmes, O., 644 (31)
Kalmykova, T. M., 418, 422 (258)
Kalninsh, T. A., 377, 384, 410 (33); 530 (161); 531 (172); 581, 582 (57)
Kamlet, J., 799 (20); 805 (123)
Kantor, P. I., 542 (307)
Kanyaev, N. P., 377 (41); 378 (48)
Karakozov, N. A., 326 (51)
Karapally, J. C., 312 (41); 446, 448 (48); 487, 489–492, 497 (6)
Karlsson, G., 340 (84)
Karpovskaya, P., 287 (93)
Karpovskaya, R. L., 701 (44)
Karrer, P., 377, 405 (31); 577 (28); 579, 580 (47)
Kashima, K., 531 (168); 819 (200a,200b)
Kashirskii, V. G., 581, 584, 585 (58)
Kato, K., 208 (230)
Katsumi, H., 196 (169)
Katz, M., 816 (175)
Katzen, R., 44 (11); 374, 375, 407, 410 (10)
Kaustinen, O., 457 (108)
Kavanaugh, K. R., 385 (142); 435 (14)
Kawamura, H., 196 (167); 772, 774, 775 (13)
Kawamura, I., 8 (27); 10 (36); 43, 61, 62 (5); 47, 48, 59, 66, 72 (22); 65 (5,76); 69 (5,82); 71 (5,76,82); 75 (86); 76 (86,89,90); 78 (97); 79 (90); 82, 87 (114); 112, 113 (82); 168, 188 (23); 177 (53); 186, 187 (101); 196 (167); 207 (221); 221 (276); 222 (278, 282, 283); 258 (74); 271, 279, 280, 285 (42); 272 (53); 284 (42, 85, 87); 288 (42,103); 292 (42, 85,103); 434 (11); 532 (194); 770 (7); 772, 774, 775 (13)
Kawamura, M., 8 (24); 186, 191, 214 (94)
Kawasaki, M., 542, 543 (303)
Kawvano, T., 542, 543 (303)
Kay, D. J., 532 (193)
Keays, J. L., 260 (92)
Kee, M. L., 384, 393, 418–421 (102, 116); 385, 397 (102); 386, 388, 389, 401–403 (116)
Kefeli, T. Ya., 487 (8); 505 (41,42); 514 (42, 43)
Keilen, J. J., 845 (45)
Keim, K., 811 (147)
Keller, E. L., 627 (146)
Keller, I., 803 (74)
Kelsey, K. E., 745 (179)
Kempf, R., 804 (87)
Kennedy, A. M., 839 (22)
Kennedy, R. W., 35 (63); 44 (8); 52, 53 (43); 54 (43,45)
Kenner, J., 117 (98)
Kenttämaa, J., 272, 273 (49)
Kenyon, J., 458 (126)
Kerr, T., 23 (25)
Kerr, W. D., 599, 626 (26); 844 (42)
Ketley, A. D., 378 (53,57); 379 (53)
Khakhar, M. P., 328, 330 (64)
Kharasch, M. S., 120 (110); 336 (73)
Khovanskaya, A. P., 550 (373)
Khundkar, M. H., 380, 418 (71); 525 (120); 713 (90)
Kiedaisch, W., 397 (179)

Keifer, H. J., 44 (9); 83, 85 (134); 193 (142)
Kilb, R. W., 223 (291)
Killander, J., 727 (137)
Kilroy, T. F., 857 (114)
Kimura, N., 76, 79 (90)
Kin, Z., 601 (42)
King, E. G., 540 (281); 847 (49)
King, F.-L., 397, 398 (185)
Kinney, C. R., 291 (127)
Kinoshita, Y., 458 (128,133); 459, 460 (128); 713 (100)
Kinugawa, J., 805 (11)
Kirby, G. W., 552 (416)
Kirdaisch, W., 122 (118a)
Kirk, T. K., 781 (28a)
Kirkina, L. I., 859 (130)
Kirkman, M. A., 420 (272)
Kirsbaum, I. Z., 578 (42)
Kiselis, O. V., 581, 582 (57)
Kishimoto, A., 732, 733 (154)
Kishimoto, S., 457 (119)
Kispert, L. D., 336, 338 (76)
Kisser, W., 196 (166); 348, 355, 357 (10)
Kitamura, R., 532, 534 (195)
Kitao, K., 578 (41); 610 (75)
Kitaura, S., 198 (183); 407 (220); 408 (225); 533 (203); 542 (303); 543 (303,316,317)
Kitaura, T., 533 (203)
Kjaer, A., 24 (27); 643 (25)
Klason, P., 469 (156); 597 (4); 598 (4,11,13)
Klein, E., 8 (31); 97, 106 (5); 198, 200, 210 (187); 207 (220); 539 (271)
Kleinberg, J., 515 (58)
Kleinert, T. N., 174 (40); 192 (136); 260 (93); 261 (94); 262 (102); 288, 289 (109); 329 (57,58,58a); 330 (59a,59b,59,69); 336 (69); 552 (402,403,404); 612 (90); 642 (8,10,11); 645 (39); 646, 648, 650, 666, 669 (45,46); 730 (141,147); 735 (147); 746 (182)
Kleinnickel, O., 821 (209)
Klemola, A., 198 (189); 283–285 (76); 315, 318, 322–325 (46); 348 (15,16)
Kleppe, P. J., 670 (124)
Klingstedt, F. W., 241 (3)
Klink, F., 449 (63); 505 (44); 515 (60,61); 516 (61); 800 (51)
Knackstedt, W., 187 (115)
Knof, L., 177, 214, 229 (54); 673, 676 (155)
Knopf, E., 452, 453 (71)
Knox, G. R., 816 (168)
Kobayashi, A., 384, 396 (124); 412, 414 (231); 535 (209); 542 (304); 547 (343); 548 (304, 343); 613 (95); 715 (101,102)
Kobayashi, T., 713 (88)
Kobe, K. A., 846 (47)
Koblitz, H., 35 (60,64); 36 (71)
Koch, J. A., 798 (12)
Koch, W. W., 804 (101)
Kochneva, M. N., 6 (16); 19 (2); 61, 80 (60)
Kojima, Y., 538 (249,250); 601 (52); 705 (66); 709 (73)
Kolboe, S., 175 (41); 271, 273, 285 (41)
Kolotova, L. I., 407 (216); 409 (228); 414, 415, 417 (234); 859 (127)
Komshilov, N. F., 531 (175)
Kondo, T., 209 (235,236); 215 (256); 216 (260); 306 (28); 366 (90,91); 458 (123,130);

AUTHOR INDEX

460, 463 (123); 461, 462 (130); 512 (7)
Kondrat'ev, E. V., 80 (102); 540 (286)
König, F., 418, 419 (262); 471 (173)
Konkin, A. A., 540 (282)
Kononova, M. M., 771, 784, 790 (11)
Koppe, P., 798 (8)
Kornblum, N., 814 (158,159)
Koroshilova, T. M., 525 (118)
Korosy, F., 378 (47)
Koshijima, T., 328, 338 (68); 556 (431,432, 436); 557 (431,437,438,439); 558 (439)
Kosikova, B., 272, 281, 284 (46); 287 (97)
Kosyukova, L. V., 578 (43,44)
Kottek, J. F., 27, 37 (34); 180 (76)
Koukol, J., 104 (37)
Kourula, A., 649 (109)
Kovapenko, A. V., 531 (167)
Kozlov, A. I., 858 (120)
Kraft, K., 155 (195)
Kraft, R., 116 (90); 267, 281, 283 (4)
Krajcinovic, M., 537 (221)
Kranzfelder, A. L., 507 (49)
Kratzl, K., 2 (6); 8 (31); 29 (45); 34 (56); 56 (48); 61, 63 (61); 74 (85); 76 (88); 95 (1); 97 (5); 102, 109 (20); 103 (23); 105 (40); 106 (5,40,47,48); 186 (97); 190 (129); 196 (166); 198, 200, 210 (187); 207 (220); 267 (6,7); 285 (6); 339 (82); 346, 354 (8); 348, 355, 357 (10); 366 (95,96,97,99); 367 (100, 101,102); 384, 399, 408, 415 (105); 400, 407 (195); 436 (18,19,20,21,28); 437, 441 (18,19,20,21); 444, 446 (34); 450, 451 (65, 66); 514 (32,35); 521, 522 (107); 532 (199); 539 (271,272,276); 551 (377); 577, 578 (35); 625 (125,126,127); 626 (131,140); 803 (74); 805 (108); 854 (80)
Krauch, H., 547 (340)
Kraus, C. A., 515 (51)
Kreicberga, Z., 165, 182, 185 (2)
Krietsberg, Z. N., 95 (2); 103 (24); 109 (64); 436 (15); 440 (31); 530 (163)
Kremers, R. E., 88, 89 (150); 110 (68,69)
Kresnova, A. P., 821 (208)
Krevelen, W., 127 (127)
Kringstad, K., 222 (287,288)
Krivoyez, I. M., 330 (66a)
Krizsan, M., 713 (92)
Kropacker, V. A., 809 (140)
Krösche, W., 548 (346)
Kroshilova, T. M., 550 (368,372)
Kubat, J., 741 (170)
Kubelka, P., 683 (167,168)
Kubiczek, G., 384 (118)
Kuc, J., 34 (52); 76, 79 (92)
Kudryavtseva, L. G., 414 (237)
Kudzin, S. F., 179, 180 (74); 267, 280, 285 (9, 10); 282 (9)
Kuerschner, K., 537 (230)
Kuhl, K. E., 858 (124)
Kuhn, L. P., 285 (88)
Kuksina, V. L., 449 (59)
Kulish, N. F., 417, 421 (245)
Kulka, M., 355, 357 (54)
Kulkevich, A. I., 581, 582 (57)
Kullgren, C., 601 (31,33)
Kummer, M. F., 851 (63)
Kung, F. L., 384, 385, 393, 397, 418–421 (102)

Kunin, R., 553 (421)
Kunishi, A., 122, 129, 131, 148, 150 (117); 202 (206)
Kunz, W., 547 (340)
Kunze, I., 362 (74); 653 (63)
Kupinskaya, G. V., 377 (42)
KURASHIKI RAYON COMPANY, LTD., 816 (173)
Kurdubev, Y. F., 858 (121)
Kuriyama, A., 583 (64); 804 (92); 805 (120)
Kürschner, K., 374, 419, 421 (18); 418 (18, 264); 422 (280); 577 (39,40); 800 (36,37,38, 39)
Kurth, E. F., 44 (9); 82 (113,125); 83 (125, 126,131,133,134,135,136,137); 85 (133,134, 135); 88 (151); 193 (140,141,142); 267, 272 (13); 375 (27)
Kurz, J. L., 457 (104)
Kushchenko, V. V., 326 (51)
Küster, W., 550 (365)
Kutscha, N. P., 47 (19)
Kvasnicka, E. A., 630 (171)
Kvisgaard, H. J., 600 (29); 748 (185)
Kyoguku, Y., 454 (90); 545 (324); 551 (379); 701 (48)

Lacan, M., 551 (381)
Lacey, R. E., 712 (86)
Lagally, P., 814 (162)
Lagergren, S., 610 (76); 730 (142)
Lai, Y.-Z., 458, 461, 463 (121)
Landmark, P. A., 670 (123,124)
Lane, W. H., 545 (729)
Lang, E. W., 712 (86)
Lang, F. M., 380 (85)
Lange, G., 586 (75)
Lange, P. W., 6 (20); 24 (26,27); 36 (74); 50, 51, 66 (36); 241 (9,11); 643 (24,25,26); 697, 753 (12)
Lange, R. G., 806 (124)
Lapin, V. N., 581 (55)
Lapshin, F. S., 823 (223); 859 (131)
Lard, E. W., 814 (165)
Larson, H. O., 814 (158)
Larson, L. L., 409, 412, 415–417 (227)
Larson, P. R., 48 (29)
Larsson, S. L., 150 (186); 203, 204, 217 (210); 354, 361 (42); 454 (93,94); 455 (94); 652 (71); 660 (79)
Latif, M. A., 6 (15); 60 (55); 62, 81 (62); 166 (16); 167 (16,38); 174 (38); 185 (90)
Latremouille, G., 384 (98)
Laurer, D., 804 (100)
Laurent, T. C., 727 (137)
Lautsch, W., 186 (95); 380, 418, 420 (70); 407 (219); 434 (2); 449 (63); 487 (4); 505, 506 (46); 515, 516 (62); 549 (364); 579 (48); 800 (49,50,51); 803 (85); 819 (195,196,197,198); 823 (198)
Lawson, A. J., 383 (95)
Lawson, L. R., 769 (2); 725 (119)
Lawton, E. J., 556 (433)
Leaf, R. L., 176, 178 (48); 258 (70)
Leary, G. F., 477, 478 (199,204)
Lebach, H., 549 (356)
LeBel, R. G., 701 (52); 703–705 (63)
Leckie, A. H., 381 (89)

Lee, H., 552, 555 (420)
Lee, J. A., 473 (189)
Lee, V. P. F. F., 60 (56)
Lefebve, J. A. A., 851 (63)
Leger, F., 803 (70)
Lehmann, B., 56, 59 (50); 104, 106 (32); 140 (165); 195, 197 (157); 210 (240); 272 (54); 452, 453 (82)
Lehmann, F., 188, 189, 228 (116); 588 (90)
Lehmann, R., 539 (269)
Lehtikoski, D., 467, 468 (144)
Lehtonen, P., 713 (91)
Leith, W. H., 684 (169)
Lemmel, L., 541 (295)
Lemon, H. W., 246, 249 (47)
Lenel, P., 196 (161); 345 (4)
Leney, L., 25 (32)
Lenz, B., 207 (224); 288, 289 (117); 301, 304, 315, 319 (19,47); 302, 303 (47); 306, 318, 320-324 (19); 362 (73,76); 363 (78); 518 (75); 525 (123); 653 (65,66,67); 651, 654, 656, 662, 664, 676 (69); 666, 667 (66); 674, 675, 683 (148)
Leopold, B., 43, 56-58, 67 (1); 216 (259); 434 (10); 435 (13,17); 436 (22,23,24,25,26,27); 438, 439 (22,23,24,25); 440 (22,23,24,25, 26); 446 (17); 457 (107); 538 (255); 601, 612 (50); 614 (102); 659 (74); 702 (58); 803 (86)
Lerch, K., 552 (413)
Lerot, R., 129 (134)
Leslie, R. T., 192 (133)
Letonmyaki, N. M., 531 (175)
Leuce, J. E., 645 (32)
Leuch, H.-U., 673, 676 (155)
Levand, O., 814 (158)
Levdikova, V. L., 701 (44)
Levine, A. A., 385 (145)
Levit, R. M., 581 (55)
Levitt, L. S., 377, 396 (45)
Lewis, C. W., 664 (88)
Lewis, H. F., 83 (126); 477 (201); 478 (211); 519 (85); 527 (141); 801 (54,56)
Ley, K., 119 (105); 120 (109,113); 121 (113); 122 (118a); 397 (179); 398 (187)
Li, T. T., 733 (156)
Liang, C. Y., 285 (91)
Libby, C. E., 598 (24)
Lichty, J. G., 384 (130)
Lieff, M., 532 (188)
Liesche, O., 188, 189, 228 (116); 588 (90)
Lieser, T., 185 (88); 418, 419 (261); 529 (156)
Lindberg, B., 80 (105); 133 (155); 351 (32); 354, 361 (42); 364 (80); 611 (80); 629 (155); 652 (71)
Lindberg, J. J., 268 (23,24,25,27); 272 (23,24, 25,27,49,50,51); 273 (49,50,51); 287 (99); 288 (23,24,25,113,114); 289 (114); 537 (215); 612 (85); 668 (97); 669 (122); 674 (149); 676 (154); 684, 686 (171); 685 (171, 172); 687 (179); 700 (40,41); 703, 705 (40); 727 (41); 743 (175); 744 (177); 751 (193)
Lindblad, A., 487 (1); 819 (188)
Lindekin, C. L., 851 (63)
Lindfors, T., 512 (14); 519, 520 (87); 590 (108, 109)
Lindgren, B. O., 126 (126); 128 (133); 184 (84); 199, 227 (191); 220, 221 (274); 345 (3); 351 (3,33,34); 355 (51); 408 (222); 514 (38,39); 538 (243,247); 544 (320,321); 602, 603 (64,65,68); 604, 605 (64,65); 608 (64, 68); 612 (81,86); 622 (121); 626 (141); 627 (147); 702 (57); 712 (81)
Lindsey, J. B., 597 (4)
Linnell, W. S., 223 (289,290)
Linsbauer, K., 80 (101)
Lipetz, J., 36 (68,69)
Lipsitz, P., 542 (309)
Lisina, Z. I., 577, 579, 583 (32)
Little, R. L., 816 (169)
Liu, Kang-Jen, 300 (10)
Livingston, A., 262 (105)
Ljungren, S., 59 (53)
Lobacheva, N. B., 581, 584, 585 (58)
Loeb, L., 703, 705 (64)
Lofdahl, L. J., 374, 377, 405, 407 (12)
Logan, A., 291 (130); 454 (91)
Logan, C. D., 467, 468 (142); 802 (67,68); 821 (67)
Lollar, R. M., 542 (309); 856 (109)
Lonsky, W., 549 (359)
Lopatin, B. V., 186 (98); 384, 400 (111,114, 115); 386 (114,115); 389, 401 (111,115); 390, 402 (114); 391 (172); 397 (114,186); 419 (111,172,226); 420 (111,172)
Loråas, V., 195 (151)
Löschbrandt, F., 195 (151); 258 (78); 405 (197); 712 (84)
Lovelock, J. E., 816 (177)
Lowe, C., 131 (149)
Lüdecke, W., 380, 418-420 (67)
Ludwig, C. H., 5 (11); 177, 227 (62); 300 (13, 14,15,16); 301-304, 306, 307, 319 (14); 312, 314, 322, 324, 326 (16); 313 (13,16); 318 (13,14,16); 320 (14,15); 325 (15); 496 (26); 539 (274); 627 (150); 699 (26)
Luhde, F., 730 (148)
Lunar, P., 260 (92)
Lung, M., 841 (32)
Lundgren, R., 357 (59)
Lundquist, K., 146 (181); 147 (183); 151 (192); 196 (160,171); 201 (203); 204 (211); 206, 212 (160); 209 (233); 210, 214 (239); 213 (211,245); 227 (171,211,245); 230 (171); 253 (62,67); 352 (36,37,38,39); 353 (37,41); 355 (37,41,48,52); 357 (37,48,60); 358 (37,62); 362 (60); 472, 473 (186); 518 (77); 541 (290); 551 (390); 618 (117); 650, 651, 673, 676 (54); 659 (76)
Lustig, Ernest, 299, 336 (1)
Luttringhaus, A., 137 (160)
Lyness, W. I., 82 (122); 83 (122,132); 190 (130)
Lynch, B. M., 271, 288, 291 (45)
Lyr, H., 770 (5)
Lyuboshits, N. Ya., 804 (94)

Maass, O., 551 (378); 610 (76a); 612 (83,88)
MacDonald, J. A., 384 (129)
Macdonald, P. L., 306, 318 (26)
MacGregor, W. S., 228 (302); 354 (45); 365 (86); 512 (16); 807 (133); 809 (144); 810 (155)
MacInnes, A. S., 354, 355 (43)

AUTHOR INDEX

Maclaren, J. A., 288 (110); 669 (113)
MacLaurin, R. D., 397 (183)
MacLean, H., 355 (49,56); 357 (56)
Madden, A. J., Jr., 784 (33)
Madorsky, S. L., 575 (5,6)
Maeda, Y., 819 (200a,200b)
Maercker, G., 628, 629 (153)
Magnusson, R., 119 (106)
Mahmood, A. H., 713 (90)
Maisaiya, D. D., 288 (101)
Majani, C., 288 (113); 671 (137); 684, 686 (171); 685 (171,172); 700 (40,41); 703, 705 (40); 727 (41)
Maki, A. H., 337 (77)
Makkonen, H., 599, 600 (27)
Maksimenko, N. S., 823 (223)
Maksimov, V. F., 575 (18)
Malm, C. J., 586 (77)
Malström, B. G., 129 (139)
Malström, I. L., 43, 56–58, 67 (1); 434 (10); 436, 438–440 (22)
Maltsev, V. I., 328 (65); 330, 337 (65,66); 331 (69a)
Manchester, D. F., 477 (206)
Manchester, D. G., 843 (39)
Mandelkern, L., 726 (134)
Manley, R. St. J., 695 (6,7)
Mannich, C., 548 (346)
Manns, O., 805 (107)
Manskaya, S. M., 6 (16); 8 (26); 19 (2); 29, 33 (43); 61 (60); 80 (60,103); 110 (71)
Manthe, G., 856 (108)
Maranville, L. F., 248 (55); 262 (106)
Marchadier, M., 129 (135)
Marchessault, R. H., 285 (91); 696 (10)
Marcusson, J., 580 (51)
Marion, L., 380, 418–420, 422 (72)
Mark, H., 171 (32); 733 (156)
Mark, R. E., 24 (28a); 52 (41); 173 (34); 697 (15)
Markert, L., 359 (65); 588 (95)
Markham, A. E., 601 (45); 701 (45); 725 (121, 122,123); 843 (38)
Marraccini, L. M., 287, 288 (94); 434 (12); 642 (8,10,11); 645 (39); 746 (182)
Marshall, H. B., 374, 391, 397, 408, 409, 412–417 (14); 713 (92); 802 (65); 856 (108)
Marshall, N. B., 844 (41)
Martensson, O., 340 (84)
Marth, D. E., 613 (97); 753 (203)
Martin, J. K., 377 (43); 577 (33)
Martin, J. P., 792 (41,42,43); 793 (41)
Martin, R. W., 547 (341)
Martin, S., 575 (9)
Martinsen, H., 380, 383 (84)
Marton, J., 151 (190); 174 (40); 198, 226, 227 (188); 199 (188,192); 213 (244); 245 (41); 246, 251, 252 (42); 250 (41,60); 253 (64, 65); 254 (60); 260 (90); 271 (40); 284 (40, 82); 288 (112); 292 (132); 350, 351 (30); 361 (70); 362 (70,72); 363 (77); 408 (221); 411 (230); 488 (11); 517 (71); 531 (174); 537 (217,231); 540 (278); 546 (335); 651 (59,62); 652 (62); 654, 655 (68); 663 (86); 664 (59,68,89); 668, 680, 688 (101); 671 (101,126,127,139); 672 (101,139); 673 (59, 62,68,85,86,101,145,147,156); 674 (62,105); 676 (62,68,85,139); 677 (101,147); 678 (62, 86,101,164); 679 (101,105); 681 (85); 682 (62,101,126,127); 683 (62); 684 (89,101); 685 (89,105); 687 (89); 700, 701, 719, 720 (38); 839 (26); 840 (29); 855 (91)
Marton, R., 457, 477 (117)
Marton, T., 546 (335); 663, 673, 678 (86); 664, 684, 687 (89); 674, 679 (102); 685 (89,102); 700, 701, 719, 720 (38); 839 (26)
Martteein, J. E., 649 (109)
Maruoka, M., 575 (23)
Mason, H. S., 111 (78)
Mason, R. F., 458 (126)
Mason, S. G., 601 (49); 614 (100); 695, 704 (3); 699, 725 (30); 705, 710 (67); 708 (30, 67); 745 (180); 746 (180,183); 747 (183)
Masri, M. S., 108 (57)
Matasovic, D., 551 (381)
Matell, M., 385, 391 (137)
Matizevic, E., 684 (170)
Matsmoto, K. U., 388 (163)
Matter, M., 850 (59)
Mattila, T., 538 (264); 555 (430)
Matusevich, L. G., 284, 285 (80)
Matyja, R., 805 (110)
Mauranen, P., 444 (33); 554 (428); 555 (429, 430)
Maxwell, J. L., 377 (39)
Mayer, R., 137 (158)
McBee, E. T., 809 (142)
McCalla, D. R., 99, 106 (14)
McCarthy, J. L., 5 (11); 10 (35); 151 (193); 177, 227 (62); 197 (172); 206 (215,216); 262 (101,104,105); 300 (14,15,16); 301–305, 307, 319, 320 (14); 312, 324 (16,40); 313, 314, 322, 326 (16); 317, 321, 323 (40); 318 (14,16,40); 320, 325 (15); 397, 402, 407, 409, 413, 414, 418 (177); 434 (6,7); 449 (56); 487 (3); 488, 490–492, 494, 495 (10); 496 (10,26); 504 (38); 514 (29); 539 (274); 601 (44,45,47,54); 609 (72); 610 (72,73); 615 (73,110); 616 (110); 619–621 (118); 624 (124); 627 (149,150,151); 699 (25,26,27,29, 33,34); 700 (34); 701 (45); 702 (56); 705, 719 (33); 710, 720, 727, 728 (76); 711 (34, 76); 712 (79); 725 (33,121,122,123); 748 (56,186); 803 (83,84); 819 (192,193); 841 (32); 843 (38)
McCleave, W., 846 (47)
McDaniel, D. H., 385 (139)
McDonald, F. R., 300 (9)
McDonald, I. R. C., 191 (132); 259 (85)
McElkinney, T. R., 601 (46); 712 (85); 842 (36)
McElwee, R. L., 46 (17)
McEwen, J. M., 373, 375, 392 (1)
McEwen, R. L., 457 (112)
McFarlane, H. M., 854, 855 (82)
McGinnes, E. Q., 285 (91)
McIlrath, W. J., 36 (72,73)
McInnes, A. S., 540 (280)
McIntosh, W. A., 851 (66)
McIvor, R. A., 384 (127)
McKean, W. T., 669 (118)
McKinney, J. W., 477 (206)
McKnight, T. S., 745, 746 (180)
McLachlan, J., 791 (39)

McLaren, A. D., 733 (156)
McLaughlin, R. R., 630 (171)
McNair, J. J., 527 (142)
McNamee, R. W., 455 (98)
McNaughton, J. G., 612 (92); 643, 685 (19); 700-702, 720, 727, 728, 759 (39)
McPherson, J. A., 222 (280); 601 (35)
Médard, L., 380 (81)
Mehltretter, C. L., 532 (192)
Mehlum, J., 803 (76)
Meier, H., 23 (24); 49 (35); 696 (9)
Meisert, E., 805 (108)
Meister, M., 452, 453 (72); 469 (161); 618 (115); 821 (205)
Melamed, D., 130 (144)
Melander, K. H. A., 587 (86); 598 (19)
Mellor, J. W., 375 (28)
Melms, F., 601 (37); 838 (11)
Menke, A. E., 384 (125)
Menshren, M., 291 (130)
Menshun, M., 109 (62); 167 (18); 175 (42)
Merewether, J. W. T., 176-178, 180 (49); 519, 520 (88,89); 532 (88,89,185); 671 (128,129, 130); 839 (18,21)
Merijan, A., 536 (214)
Meshizuka, G., 458 (132)
Metel'kova, E. I., 799 (23)
Metlesics, W., 218 (264)
Meutterties, E. L., 813 (155)
Meybeck, J., 467, 468 (143)
Meyer, J. A., 385, 387, 391, 393, 394, 411, 412 (146)
Meyer, H., 173 (35)
Meyerhoff, G., 703, 705 (61)
Michell, A. J., 268, 272, 273 (30); 271, 285, (34,44); 281, 284 (44); 288-290, 292 (34)
Middlebrook, M. J., 37 (78)
Mieg, W., 588 (91)
Migita, N., 8 (24); 179 (71); 181, 187 (68); 186, 191, 214 (94); 207 (227); 208 (229, 230); 245 (35); 272, 283 (48); 288, 289 (111); 338 (80); 385, 416 (144); 458 (125, 129,131,132); 459 (125,131); 460 (125); 461 (129); 463 (131); 471 (181); 514 (40,41); 523 (111); 526 (128,129); 535 (208,209); 537 (223,224,225); 548 (347); 611 (79); 660 (81); 668 (100); 677 (100,163); 859 (126)
Miglarese, J., 805 (114)
Mihailov, M., 525, 526 (127)
Mika, T., 530 (157)
Mikailov, N. P., 384, 386, 389, 400, 401 (115); 419 (266)
Mikawa, H., 59 (51); 199, 227 (191); 289, 290 (122); 384 (120,124); 385, 391, 398 (120); 396 (124); 405, 409, 411 (203); 408 (222); 412 (231); 414 (203,231); 415 (203,240); 416 (240); 422 (281); 514 (40); 538 (245); 544 (321); 545 (332); 546 (336); 548 (345); 602 (66); 613 (95); 618 (113,116); 622, 629 (120); 627 (147); 662, 663, 678 (84); 668 (94,99); 676 (99); 770 (8)
Mikhailova, M. A., 526 (131)
Mikkailov, N. P., 550 (373)
Miksche, G. E., 147 (183); 150 (186); 196, 227, 230 (171); 203, 204, 217 (210); 209 (233); 210, 214 (239); 301-304 (21); 352 (37,39); 353, 355, 357 (37); 358 (37,62); 454 (93, 94); 455 (94); 650, 655, 664 (51); 658 (51, 92a); 660 (79)
Milas, N. A., 799 (32,33)
Millen, D. J., 380 (80)
Miller, J. G., 366 (92); 487 (7); 492, 500 (25)
Milligan, B., 515 (59)
Mills, G. S., 856 (103)
Milovanov, A. V., 398 (194); 542 (302)
Minari, H., 630 (168)
Minear, R. A., 263 (108)
Minkoff, E. J., 380, 383, 392-394 (87); 468 (149)
Mintz, M. S., 712 (86)
Mishchenko, K. P., 326 (51); 537 (218)
Misra, P., 23 (23)
Mitchell, L., 357 (58)
Mitchell, R. L., 88 (152); 374, 405, 407, 410 (16); 414, 417 (235)
Mitsukawa, R., 514 (41); 611 (79)
Miyao, S., 288, 289 (111)
Moacanin, J., 699, 705, 719 (33); 725 (33,123)
Mocsny, S., 855 (94)
Mohrberg, W., 645 (36)
Moilanen, M., 552 (401); 641 (6); 668 (95,96)
Moleberry, M. V., 846 (47)
Mollard, A., 104, 105 (35); 282, 284 (73)
Mölsä, V., 713 (91)
Monnberg, R., 580 (53,54); 582 (53)
Monnet, P., 799 (17)
MONSANTO CHEMICAL COMPANY, 801 (61)
Moore, A. W., 789 (37)
Moore, D. R., 529 (155)
Moore, J. C., 727 (135)
Moore, R. G. D., 537 (229)
Moore, W. E., 260 (89)
Moorer, H. H., Jr., 525, 526, 529 (126); 855 (95)
Moos, J., 809 (143)
Morimoto, T., 434 (11); 538 (261)
Morohoshi, N., 347 (9)
Morozov, E. F., 581 (56)
Morozova, V. M., 531 (167)
Morren, H., 539 (269)
Morrison, R. I., 80 (106,107,109); 81 (107); 287, 291 (95); 785 (34)
Morse, B. K., 458-460, 463 (124)
Morton, R. A., 246 (46)
Mosbach, R., 129 (139)
Moschatos, H., 549 (354)
Moss. W. H., 539 (270)
Motherhead, J. S., 197 (173); 311 (35); 613 (98); 626, 628 (137)
Mott, R. A., 170 (29)
Mottier, M., 513 (24)
Mozheiko, L. N., 288, 290 (105); 417 (247); 525 (119)
Mueller, H. G., 359, 360 (67)
Mühlethaler, K., 22 (15)
Müller, A., 133, 143 (156)
Müller, E., 120 (109,113,114); 121 (113,114); 137 (158,159); 397 (179,181); 398 (187); 414 (181)
Müller, G., 122 (118a)
Müller, H. F., 452 (74); 514 (30); 549 (363); 588 (96)
Müller, O., 187 (105); 374, 405, 407, 412, 416 (6); 588 (97)

AUTHOR INDEX

Muller, R. E., 44 (11); 374, 375, 407, 410 (10)
Muller, S., 813 (153)
Munk, F., 683 (167)
Murakami, K., 805 (118,119)
Muraki, E., 328, 338 (68); 556 (431,432,436); 557 (431,437,438)
Murashkevich, T. V., 592 (121)
Murphy, D., 789 (37)
Murphy, E. J., 575 (7)
Murphy, R. A., 467 (145,146); 468 (146)
Musso, H., 118 (101)

Nagar, B. R., 328, 330 (64)
Nagaski, T., 533 (203)
Nagata, M., 526 (128)
Nahum, L. Z., 216 (261); 674 (152)
Nair, P. M., 108 (53,54)
Naito, K., 328, 338 (68); 556, 557 (431)
Nakai, A., 709 (73)
Nakajima, K., 413 (233); 542 (306)
Nakajuma, T., 181 (82); 521 (103,104); 532 (184)
Nakamura, K., 538 (251); 701 (49)
Nakamura, T., 198 (183); 407 (220); 408 (225); 525 (125); 533 (203); 542 (303); 543 (303,316,317); 545 (323)
Nakanishi, K., 268 (20)
Nakano, J., 178, 181 (68); 179 (71); 207 (227); 208 (230); 245 (35); 272, 283 (48); 288, 289 (111); 338 (80); 377 (35); 385, 416 (144); 458 (125,129,131,132); 459 (125,131); 460 (125); 461 (129); 463 (131); 471 (181); 514 (40,41); 523 (111); 526 (128,129); 535 (208, 209); 537 (223,224,225); 548 (347); 611 (79); 660 (81); 668 (100); 677 (100,163); 741, 744, 745 (173); 859 (126)
Nakayama, N., 87 (147); 174, 197 (39); 187 (113,114); 196 (39,113,114); 346 (7); 348 (13,14); 349 (13)
Namura, Y., 454 (92); 517, 519 (73); 554 (427); 630 (162,163,164,166,167); 717 (107); 718 (108)
Nandy, M., 99 (13)
Napier, D. R., 816 (169)
Narayanamurti, D., 46 (13)
Nash, N. G., 122 (118a)
Naveau, H. P., 271, 283 (39)
Naya, K., 138 (163)
Nazareva, E. V., 448 (49); 546 (339)
Nazarina, E. V., 803 (77)
Nečesány, V., 35 (59)
Neill, W. K., 575 (8)
Neish, A. C., 9 (32); 21 (9); 95 (1); 98 (6,7,8); 99 (14,17); 101 (6); 104 (17,28,29,30); 105 (28,29,39,41,43); 106 (14,28,45); 107 (50, 51); 116, 138, 140 (88); 155 (196); 168, 190 (24); 175 (24,44); 180 (44); 200 (196); 201, 227 (202); 210 (241); 214 (248); 218 (266); 220 (269); 226 (301); 503 (32)
Nekrasova, Z. D., 417, 421 (245)
Nelson, D. C., 816 (170)
Nelson, H., 725 (123)
Nelson, J. F., 804 (97)
Nelson, N. A., 515 (56)
Nelson, O. E., 34 (52); 76, 79 (92)
Nelson, P. F., 108 (56); 288 (110); 669 (113)
Nemchenko, A. G., 577, 579, 583 (34)

Némethy, G., 741 (171); 745 (171,180); 746 (180)
Neogi, A. N., 444 (33); 538 (264); 554 (428)
Neubauer, G., 120 (111)
Neville, K., 552, 555 (420)
Newcombe, A. G., 374, 391, 397, 408, 409, 412-417 (14)
Newman, M. S., 386 (152)
Newman, S., 703, 705 (6)
Nichols, R. F., 522 (110)
Nicolaisen, F., 300 (11)
Niederkorn, F., 104, 106 (32); 452, 453 (79, 80); 551 (392)
Niemann, C., 405, 407, 416 (200); 512 (9); 549 (360)
Nieuwland, J. A., 515 (55)
Nifanter, E. E., 530 (164)
Nikander, B., 846 (47)
Nikitin, N. I., 346 (6); 520 (97); 526 (131,134); 528 (148); 532 (191)
Nikitin, V. M., 284, 285 (78); 525 (118); 538 (241); 539 (273); 547 (344); 550 (368,370, 371,372); 838 (16)
Nikitin, Y. V., 538 (241); 539 (273)
Nikolaer, N. I., 809 (140)
Nikuni, Z., 198 (184,186); 534, 535, 542 (206, 207)
Nilsson, E., 6 (14); 80, 81 (108)
Nimz, H., 138 (163); 141 (172); 146 (178,179, 180); 147 (182,184); 148 (184); 150 (188); 154 (194); 176 (46); 180 (78); 196 (164); 197 (175,176,177); 199 (194); 200 (164,176, 194); 205 (175,176,177,194,213); 208 (213, 231,232); 209 (164,213,232); 211 (164,175); 212 (174,175); 213 (175,246); 214 (249,252); 301 (22); 302 (22,23,24,33); 303, 310 (22,24, 33); 304, 305 (22,23,24,25); 308 (22,23,24,25, 33); 309 (25); 319 (24); 338 (81); 349 (17,18, 19,20,21,22,23,24,25,26); 360 (24); 365 (19); 500 (28); 511 (5); 628, 629 (153); 630 (169); 658 (92b)
Nishikori, I., 288 (118)
Nist, B. J., 5 (11); 177, 227 (62); 300 (5,14, 16); 301, 302, 303, 304, 305, 307, 319, 320 (14); 312, 314, 322, 324, 326 (16); 313 (16, 43); 318 (14,16); 452 (85); 496 (26)
Nitkin, V. M., 671 (131)
Nokihara, E., 422 (281); 546 (336); 609, 610 (72); 702, 748 (56)
Nolan, P., 477 (198)
Noll. A., 538 (258)
Nollav, E. H., 539 (268)
Norcross, B. E., 121 (116)
Nord, F. F., 11 (40); 82 (116); 87 (116,148); 95 (1); 96, 105 (4); 102, 109 (19); 108 (61); 176 (47); 177 (65,66); 178 (65,66,69); 179 (69,72,73,74,75); 180 (66,72,74); 196 (167, 168); 199 (190); 258 (71); 267 (8,9,10,11); 280, 285 (8,9,10,11,58); 281 (8); 282 (9,11); 500, 501 (24); 773 (15); 774, 781, 788 (17); 775 (15,17); 712 (80)
Norén, I., 207 (222,223,225); 360, 361 (69); 362 (69,72,75); 363 (69); 651, 652 (57); 653 (64,65)
Norman, A. G., 190 (126)
Norman, R. O. C., 418 (251)
Northcote, D. H., 115 (87)
Nowak, P., 380 (68); 418 (260); 421 (68,260)

Nylen, P., 813 (152)

Obolenskaya, A. V., 512 (18); 671 (131); 838 (16)
Ochiai, M., 805 (111)
Odincovs, P., 165, 182, 185 (2)
Odintsovs, P. N., 186 (93); 436 (15); 440 (31); 645 (34,35)
Ogait, A., 541 (291)
Ogino, T., 548 (347); 677 (163)
Ogiyama, K., 215 (256); 306 (28); 512 (7)
Ohkubo, K., 458 (127,133,135a); 459 (127); 460 (127); 463 (135a)
Ohlsson, K., 259 (82)
Oiwa, K., 531 (168)
Okabe, J., 418, 421, 422 (256)
Okashi, H., 108, 115 (59)
Oki, T., 457, 461 (118); 458 (124,127,128, 133,135a); 459, 460 (124,127,128); 463 (124,135a); 469 (160); 473, 474 (194); 520 (93,94); 713 (100); 774, 788 (16)
Okubo, K., 457, 458, 461 (118); 713 (100)
Okubo, M., 413 (233)
Okuda, A., 188, 228 (118)
Okuma, J., 407 (220)
Okun, M. G., 330, 337 (66); 542 (307); 551 (385); 859 (129)
Olcay, A., 501 (29,30)
Olson, E., 477 (200)
Olson, H. S., 713 (97)
Onishi, M., 419 (269)
Onoe, S., 413 (233)
Onopienko, V. S., 799 (23)
Opdyke, D. L., 82, 83 (122)
Orchin, M., 242 (20)
Orlowa, I. M., 532 (191)
Orsler, R. J., 261 (98)
Oshima, M., 819 (200a,200b)
Östberg, K., 601 (40); 642 (7)
Ostrowski, T., 291 (124,128)
Othmer, D. F., 44 (11); 374, 375, 407, 410 (10)
Otrhalek, J. V., 850 (60)
Otto, P., 816 (172)
Ouchi, G., 838 (17)
Owen, R., 804 (98)
Owzarski, P., 801 (60)
Oya, H., 660 (81)
Oyamoda, K., 805 (109)

Packman, D. F., 261 (98)
Paloheimo, L., 194 (145)
Panasyuk, L. V., 592 (120,122); 823 (223)
Panasyuk, V. G., 592 (120,122); 823 (220,221, 222,223)
Papadakis, M., 848 (51)
Papino, R. P., 856 (106)
Papst, E., 856 (104)
Parker, J. A., 385 (143)
Parker, P. E., 209, 213, 216 (238); 302–304 (39); 311 (38,39); 318 (38); 488, 492, 498, 500 (9); 659 (73)
Parker, W. F., 457 (116)
Parrish, J. R., 197, 205, 212 (174); 311 (36, 37); 626, 628 (138,139); 629 (138); 699 (28)
Parsons, L. G. B., 469 (165)
Paschke, F., 377 (30); 519 (78); 530, 531 (165)

Passannante, A., 712 (80)
Pataky, A. A., 477 (206)
Patterson, R. F., 241 (5); 260 (92)
Pattyranie, C., 741 (170)
Paulson, J., 468 (147)
Pauly, H., 529 (150); 803 (69)
Pavoline, T., 534 (205)
Pearl, I. A., 8 (24); 95 (1); 166, 178, 192 (14); 195 (155); 209 (237a); 222 (285,286); 260 (87); 262 (103); 272, 277, 280 (47); 282, 284 (70); 445 (35,36); 446 (35,36,38,40,41,42, 43,44,45,46,47); 447 (39,41,42,43); 448 (39, 44,45,46,50,51,52,53,54); 449 (51,55,57,64); 473, 474 (192); 552 (411); 589 (98,99); 614, 617 (109); 631 (172); 671 (135); 801 (52,53, 54,55,56,57,58,59); 803 (73,78,79,81,82); 804 (89); 854 (83)
Pearlstein, J., 44 (11); 374, 375, 407, 410 (10)
Pearson, D. A., 610, 615 (73)
Pearson, R. G., 808 (135)
Pedersen, E. J., 300 (11)
Pedersen, N., 597 (2)
Pederson, J. H., 405, 407 (206)
Pederson, K. O., 701, 750 (51)
Peeling, E. R. A., 380 (78)
Peet, C. E., 444 (33); 554 (428)
Peikert, H., 374, 418, 419, 421 (18)
Peill, P. L. D., 477 (205)
Pelczar, M. J., 769 (1)
Peliken, E., 407 (213)
Pellerin, J., 108 (60)
Pelter, A., 306 (26,27); 318 (36); 814 (161)
Pen, R. Z., 271 (36,37); 273, 288 (37); 284 (81,83); 285, 291 (81); 434 (5); 543 (313); 725 (120)
Peniston, Q. P., 397, 402, 407, 409, 413, 414, 418 (177); 449 (56); 514 (29); 601 (45); 624 (124); 699 (29); 701 (45); 712 (79); 725 (121); 803 (83)
Penttinen, K., 700, 727 (41)
Pepper, J. M., 8 (29); 187 (108,109,110,111, 112); 188 (109); 207 (218); 271 (38); 285 (38,92); 312 (41); 319 (50); 352, 355 (35); 358 (61); 384 (127,129); 385 (142); 435 (14); 446, 448 (48); 487, 492, 497 (6); 489 (6,14,15,18); 490 (6,14,15,22); 491 (6,14, 22); 540 (288); 550 (374); 700 (43); 701 (50)
Perez, M., 384 (104)
Pershina, L. A., 449 (58,59); 531 (167,169)
Person, B., 10 (37); 54–56, 68 (46); 166, 168, 195, 226 (11); 173-175, 178, 201 (36); 245, 256 (31); 532 (180); 608 (69); 672 (140); 730, 735 (146)
Persson, K. I., 198, 199, 226, 227 (188); 271, 284 (40); 673 (145)
Pesteiner, M., 804 (100)
Peterson, J. C., 300 (9); 540 (284)
Petrarin, I., 713 (89)
Petrova, A. M., 414 (237); 417 (246); 550 (369)
Petterson, T., 538 (243)
Pew, J. C., 10 (38); 40, 50, 66 (34); 122, 150 (117,119); 123, 124 (119); 129, 131, 148 (117); 166, 169-172, 186, 220, 221 (13); 177, 216 (57); 182 (83); 202 (206); 217, 227 (263); 246 (45); 313 (43d); 435, 438–440 (16); 451 (68); 532, 533 (197); 643 (27); 697 (13); 749 (189)

AUTHOR INDEX 885

Pfannenstiel, A., 397 (182)
Pfau, A. St., 515 (64)
Pfeifer, S., 805 (107)
Philipp, B., 601 (37)
Phillips, E. W. J., 34 (54)
Phillips, G. O., 478 (212)
Phillips, M., 76,77 (93); 190 (127); 193 (143, 144); 194 (143,144,146); 471 (178); 577 (31, 37); 580 (49,50); 588 (92,93)
Piazolo, G., 380, 418, 420 (70); 407 (219); 487 (4); 505, 506 (46); 515, 516 (62); 549 (364); 579 (48); 819, 823 (198)
Pickard, E., 729 (138)
Pickett-Heaps, J. D., 32, 36 (50)
Pictet, A., 575, 577 (14); 580 (52)
Pierce, J. H., 552 (407); 856 (108)
Pierce, J. S., 525 (117); 856 (108)
Pigden, W. J., 78 (96)
Pigman, W. W., 46, 49, 66 (15)
Pilipchuk, Y. S., 268 (26); 271 (36,37); 273, 288 (37); 284 (81,83); 285 (26,81); 291 (81); 551 (385)
Pinchas, S., 270 (32)
Piret, E. L., 784 (33)
Pitt, B. M., 814 (164)
Pla, F., 705 (65); 710 (75)
Plankehorn, E., 187 (106); 449 (63); 514 (33); 800 (51); 803 (75)
Platt, J. R., 243 (26)
Ploetz, Th., 185, 191 (87)
Podgornava, T. A., 577, 579, 583 (32,34)
Pohl, E., 263 (107)
Polcin, J., 272, 281, 284 (46); 287 (97); 375, 410 (25); 405, 415 (25,202); 407 (25,215); 408, 409 (202); 411 (25,229); 416 (202,215, 242); 417 (202,242); 549 (362); 550 (367)
Poljak, A., 457, 462, 463 (109,110,111); 821 (210)
Pollacsek, E., 800 (6)
Pollak, A., 838 (10)
Poller, S., 379, 420 (65)
Polson, A., 726 (128,129)
Polyak, A. B., 421 (278)
Ponurov, G. D., 284 (79)
Pople, J. A., 299 (2)
Porath, J., 727 (136)
Porshina, E. A., 821 (208)
Porto, F., 35 (65)
Possoz, L., 586 (71)
Potter, E. F., 856 (108)
Powell, W. A., 246 (44)
Powell, W. J., 405, 407, 412, 414, 416, 418, 419, 421, 422 (199)
Powers, J. W., 814 (158)
Preston, R. D., 22 (18); 37 (78)
Prett, K., 457 (113)
Preu, E., 550 (366)
Prey, V., 513 (23,26)
Pringsheim, H., 188, 228 (117)
Procter, A. R., 50 (38,39); 52 (39); 241 (12); 612, 613 (94); 643, 644 (20); 697, 698 (16); 753 (16,208,210); 754–756 (208)
Procter, H. R., 598 (12)
Prokshin, G. F., 512 (10,17); 518 (74)
Pryor, W., 583 (66)
Pschorr, R., 384 (107)
Pueschel, R., 807 (132)

Pummerer, R., 130 (144)
Purves, C. B., 11 (41); 181, 182 (79); 186, 214 (99); 385, 420 (133); 437 (30); 467, 468 (142); 469, 470 (166,167); 551 (383)
Puschel, R., 669 (121); 840 (27)
Puttfaicker, H., 130 (144)
Pye, I. T., 730 (149)
Pyle, H. M., 816 (179)
Pyle, J. J., 151 (193); 206 (215)

Quimby, G. R., 538 (254); 601 (57); 719 (111)
Qvist, W., 391 (170)

Raff, R. A. V., 856 (101)
Rager, R., 733 (156)
Raiford, L. C., 384 (130,131); 385 (136); 805 (112)
Raisanen, S., 196, 200, 206, 209 (170)
Ramiah, M. V., 740 (166,167); 741 (167)
Ramsey, J. W., 300 (9)
Rapson, W. H., 466, 467 (141)
Rasenack, D., 140 (167); 220 (270)
Raskin, M. N., 331 (69a); 421 (278); 551 (385)
Rasson, B., 577 (36)
Rassow, B., 374–376, 405, 407 (11); 587, 588 (84)
Ray, P. M., 35 (66)
Raznikov, V. M., 858 (122)
Read, D. E., 186, 214 (99); 469, 470 (166, 167); 551 (383)
Readville, E. T., 449 (56); 712 (79); 803 (83, 84)
Rechert, M., 112, 143, 155 (80)
Redinger, L., 542, 543 (308); 672, 676 (141); 855 (89)
Reece, C. H., 131 (146)
Reed, H. C., 821 (201,202)
Reed, R. I., 380 (83,87); 383, 392–394 (87); 468 (149)
Rees, A. D., 816 (176)
Reeve, K. D., 377 (44)
Reeves, R. H., 473, 474 (192)
Regeness, F. A., 805 (106)
Reiche, A., 542, 543 (308)
Reichert, M., 28 (37); 104, 106, 110 (31)
Reid, T., 330, 332 (61); 536 (211); 677 (158)
Reiff, W. M., 538 (264); 555 (429)
Reilly, C. A., 299, 336 (1)
Rempp, P., 729 (139)
Rendos, F., 581, 583 (59)
Renner, E., 253 (66)
Renner, H., 150 (188); 197, 200, 205 (176); 338 (81); 349, 365 (19); 452, 453 (81); 511 (5)
Renner, K. C., 140, 148 (166); 512, 540 (6)
Repka, V. P., 592 (122)
Rex, R. W., 328, 330, 332 (54)
Rezanowich, A., 610, 612 (74); 614 (100); 699 (23,32); 700, 703, 705, 725 (32,37); 701, 702 (37,47); 706, 722–724 (32); 707 (72); 708–710 (23); 713–716 (98); 719 (23,32); 720 (23,32,37); 726 (37); 734, 735, 748 (157); 751 (32,72)
Reznik, H., 21 (14); 28 (37); 104, 106, 110 (31); 220 (270,271)
Reznikov, V. M., 80, 81 (111); 268 (26); 284 (79,80); 285 (26,80); 291 (126); 701 (44)

Richards, S. J., 792 (42)
Richardson, D. H., 798 (10)
Richter, G. A., 44 (10)
Richtzenhain, H., 116 (89); 129 (136); 184 (86); 186 (96); 192 (135); 203 (209); 209 (234); 214 (253); 222 (284); 366 (89); 452, 453 (75,76,83,84); 463, 465 (137); 464, 466 (137,138,139); 469 (158); 471 (174); 532 (187); 551 (382); 631 (170); 659, 679 (77); 803 (71); 809 (141); 821 (206,207)
Rickborn, B., 816 (168)
Ridd, J. H., 380 (75,82); 386 (154); 418 (250); 420 (273)
Rieche, A., 541 (298); 855 (89)
Riedel, J. D., 799 (18,29)
Rieker, A., 137 (159)
Riggins, P. H., 288, 289 (115)
Rinaman, T. E., 668 (115)
Rinaudo, M., 705 (65)
Ripa, R., 80 (104)
Risnyovszky, E., 367 (100,101)
Rist, C. E., 532 (192,193)
Ritchie, P. F., 11 (41); 181, 182 (79)
Ritter, D. M., 377, 407, 416 (34); 804 (95)
Ritter, F. J., 48, 66 (24); 373, 392 (3); 414, 417 (235)
Robert, A., 47, 48 (23); 95 (1); 282 (68,75); 284 (75); 285 (68); 374 (7,19); 384, 388 (122); 404, 405, 408, 412, 414, 416, 417 (7); 418 (19,254,265); 419 (19,254); 420–422 (19)
Robert, D., 301–304, 313, 318, 325 (44)
Roberts, J. D., 300 (4)
Robertson, P. W., 385 (141)
Robertson, R. F., 720 (114)
Robinson, C. H., 517 (70)
Robinson, L. E., 843 (39)
Robovin, Z. A., 540 (282)
Rodd, E. H., 529 (153)
Rodnevich, B. N., 799 (21)
Roesch, H., 821 (203)
Roester, A., 798 (14)
Rogers, C. E., 846 (47)
Rogers, W. F., 847 (48)
Rohmann, E. M., 292 (136)
Romney, V. D., 853 (76)
Ronssin, S., 379, 418–420 (63)
Ronstrom, P., 601 (39)
Rösch, H., 421 (274); 586 (78)
Roschier, R. H., 222 (281); 552 (408); 614 (99)
Rose, J. W., 854 (83)
Rosenbaum, E. E., 817 (180)
Rosenkranz, K., 384 (104)
Rosenthaler, L., 805 (115)
Rosenwald, R. H., 385 (140)
Roshchin, V. I., 575 (18)
Ross, J. H., 546 (338)
Routala, O., 418 (255)
Rower, J. W., 166, 170, 195 (12)
Rozenberger, N. A., 542 (310)
Rozenkranz, G., 517 (69)
Rubery, P. H., 115 (87)
Rubina, S. I., 858 (121)
Ruch, F., 24 (28); 36 (75); 642 (15)
Ruderman, I. W., 388 (160)
Rudloff, E. V., 229 (307)

Rudneva, T. I., 520 (97); 528 (148)
Runius, S., 224, 227 (295)
Runtso, A. P., 592 (121)
Russel, J. D., 774 (18,19)
Russell, G. A., 327 (52); 451 (69)
Rutgers, R., 805 (103)
Ruzicaka, W., 800 (40)
Ryan, J. F., 515 (57)
Rydholm, S. A., 261 (96); 376 (29); 392, 397, 405, 407, 413, 414, 416, 417, 421 (173); 463, 466, 467 (136); 473 (191); 598, 602, 608, 611, 612, 617, 622, 623 (25); 600 (25, 30); 610 (25,76); 640, 642, 647, 648, 670 (3); 730 (142)

Sach, I. B., 697 (13)
Sacks, I. B., 49, 50, 66 (34); 643 (27)
Sadeh, D., 270 (32)
Sadowski, J., 385, 387, 391, 393, 394, 411, 412 (146)
Saedén, U., 355 (51); 514 (38); 538 (247); 544 (320); 602–605 (65); 626 (141)
Saegusa, T., 520 (92)
Saeman, J., 504 (36,39,40)
Saizeva, A. F., 346 (6)
Sakai, A., 434 (12)
Sakai, K., 457 (119); 458 (123,130); 460, 463 (123); 461, 462 (130)
Sakakibara, A., 87 (147); 129, 148 (142); 174, 197 (39); 187 (113,114); 196 (39,113,114); 301, 315, 318 (19a); 346 (7); 347 (9); 348 (13,14); 349 (13)
Sakamoto, Y., 779, 785, 786 (24)
Sakuma, S., 196 (165); 317 (48); 527 (139)
Salari, R. J., 473 (190)
Saleh, T. M., 25 (32)
Salvesen, J. R., 800 (44); 801 (60); 850 (60); 856 (108)
Salway, A. H., 389 (167)
Samsonova, A. P., 337 (78); 551 (385)
Samuel, D., 270 (32)
Samuelsen, S., 469 (162)
Samuelson, O., 538 (242); 622 (119)
Sandborn, L. T., 800 (44,45)
Sandén, R., 243, 244, 257 (29); 248 (58); 255 (29,68); 626 (134)
Sandermann, W., 477 (197); 529 (151); 581 (60)
Sankey, C. A., 537 (227); 802 (66)
Santelli, T. R., 521 (109); 854 (81)
Sanyer, N., 627 (146)
Sapiro, R. H., 469 (165)
Saponitzki, S. A., 838 (12)
Sarkanen, K. V., 6 (15,22); 7 (22); 25 (32); 55, 61, 68, 69, 70 (49); 55, 61, 68, 69, 70 (49); 62, 81 (62); 73, 74 (83); 133–137, 144, 145, 150 (157); 151 (191); 166 (16); 167 (16,19); 175, 178 (43); 185 (90); 201 (200); 207 (219); 245 (40); 248 (56); 249 (40,56); 257 (56,69); 267, 271, 280 (14,15,16); 269, 283 (15); 272, 274 (14); 284, 292 (15,16); 313 (43); 354, 355 (46); 365 (84); 379, 398 (59); 384, 397, 413, 414 (106); 387, 388, 390, 392 (157); 391 (106,157,171); 393 (157,171); 406, 411, 412, 418 (171); 409, 410 (106, 171); 452 (85); 458, 461, 463 (121, 134); 464 (140); 467 (140,145,146); 468 (140,

AUTHOR INDEX

146,148,150); 488 (12); 520, 524 (96); 523 (112); 540 (279); 602, 617 (63); 669 (117, 118); 674 (151); 719, 750 (109); 776, 786 (20)
Sarkar, P. B., 380, 418, 420 (66); 405, 407, 416, 417 (198); 587 (89)
Sartonetto, P. A., 507 (49)
Sasaki, A., 535 (208)
Sasaki, I., 407 (220); 859 (126)
Sastre, J. A. L., 531 (170)
Sato, H., 525 (125); 545 (323)
Sato, K., 59 (51); 289, 290 (122); 384 (120, 124); 385, 391, 398 (120); 396 (124); 405, 409, 411 (203); 412 (231); 414 (203,231); 415 (203,340); 416 (240); 419 (270); 422 (281); 538 (245); 542 (304); 545 (332); 546 (336); 547 (343); 548 (304,343,345); 613 (95); 618 (116); 622, 629 (120); 662, 663, 678 (84); 668, 676 (99); 715 (101,102)
Sato, L., 601 (55)
Saunders, B. C., 128 (131)
Saunders, J. H., 521 (106)
Sawanobori, T., 523 (111)
Scallan, A. M., 612 (91); 643, 659 (18); 730 (150)
Schaack, W., 418, 419 (261)
Schaarschmidt, A., 380, 421 (68); 418, 421 (260)
Schachman, H. K., 722 (117); 726 (127)
Schafer, E. R., 625 (126)
Schafer, W., 450, 451 (66)
Schech, W., 457 (106)
Scheffer, K., 137 (158)
Schenker, C., 190 (130)
Scheraga, H. A., 741 (171); 745 (171,180); 746 (180)
Scheutz, R. D., 259 (79,81)
Schierbaum, F., 220 (272)
Schildknecht, C. E., 526 (133)
Schilling, P., 122, 124 (120); 450, 451 (66)
SCHIMMEL AND COMPANY, 799 (28)
Schindler, H., 63 (71)
Schlumbom, F., 477 (197)
Schlüter, H., 139, 140 (164); 195, 214 (156); 205 (212); 673 (143)
Schmid, G. H., 304 (21b)
Schmidhuber, W., 120, 121 (114)
Schmidt, V., 592 (119)
Schneider, B., 292 (132)
Schneider, W. G., 299 (2); 545 (330)
Schoeffel, E. W., 803, 804 (80)
Schoning, A. G., 261 (95)
Schorning, P., 601 (34)
Schorning, Z., 241 (2)
Schott, O., 262 (100); 614, 615 (107); 719 (112)
Schotton, P. G., 729 (138)
Schrader, H., 374, 379, 418, 419, 421, 422 (21); 449 (60,61,62); 450 (61); 469, 470 (164); 575 (13); 822 (215)
Schramek, W., 800 (37,38)
Schramm, R. M., 380, 383, 386, 391, 392 (86)
Schraube, H., 137 (162)
Schriesheim, A., 668 (98)
Schroth, D., 259 (84)
Schubert, S. W., 539 (274); 627 (150,151); 699 (26,27)

Schubert, W. J., 82, 87 (116); 95 (1,3); 96 (3, 4); 102 (3,19); 105 (4); 109 (19); 176 (47); 179 (72,73); 180 (72); 196 (168); 267, 281 (8); 280, 285 (8,57,58); 300, 325 (15); 500, 501 (24); 712 (80); 770 (4); 773 (15); 774, 781, 788 (17); 775 (15,17)
Schuerch, C., 201 (200); 207 (219); 209, 213, 216 (238); 245, 249 (40); 302, 303, 304, 311 (39); 354 (44,46); 355 (46); 366 (92); 377 (35); 487 (7); 488, 498 (19); 489, 497 (16, 17); 490 (19,20,21,23); 491 (21,23); 492 (19, 21,23,25); 494 (20); 500 (9,25); 514 (46,47, 48); 520, 524 (96); 537 (234); 540 (227,279); 551 (394,395,396,397,398); 608, 612 (71); 659 (72,73); 674 (151); 687 (177,178); 719, 750 (109); 741, 744, 745 (173); 742, 748 (174)
Schuhmacher, G., 245 (33)
Schulerud, C. F., 855 (92)
Schulz, G. V., 703, 705 (61)
Schulz, L., 434 (3); 801 (52)
Schutz, F., 187 (115); 540 (283)
Schwabe, K., 550 (366); 700 (36)
Schwartz, H., 397, 402, 407, 409, 413, 414, 418 (177); 736 (160)
Schwartz, S. L., 730 (144)
Schweers, W., 56 (48); 539 (272)
Schwenzon, K., 838 (11)
Schwind, V., 529 (156)
Scondac, I., 713 (89)
Scott, A. I., 131 (150); 242 (18); 246 (43); 552 (418)
Scott, C. B., 807 (134)
Scott, G., 122 (118)
Scott, J. A. N., 50, 52 (39); 697, 698 (16); 753 (16,210)
Screaton, R. M., 746, 747 (183)
Scurfield, G., 168 (22); 285 (90)
Seamster, A. H., 841 (31)
Searles, S., Jr., 813 (156)
Seborg, R. M., 414, 417 (235)
Sehon, A., 711 (78)
Seidl, J., 538 (263)
Seifert, K., 532, 534 (198)
Seiler, H., 169, 220 (26); 285 (89)
Sell, L. O., 88 (149)
Semakova, L. A., 546 (339)
Semechkina, A. F., 9 (33); 487 (43); 505 (42, 43); 514 (44,45)
Senju, R., 552 (414)
Senko, I. V., 284, 285 (80)
Sento, A., 668, 677 (100)
Senzyu, R., 520 (91)
Sergeeva, L. L., 279 (64); 384, 400 (109,111, 113,114); 386, 390 (113,114); 388 (109, 113); 389 (111,113); 391 (172); 397 (114, 186). 401 (109,111); 402 (114); 419 (64, 109,111,172,267,268); 420 (64,111,172,267, 268); 421 (64,267,268)
Sergeeva, V. N., 103 (24); 186 (98); 288, 290 (105); 377, 384, 410 (33); 417 (247); 525 (119); 530 (161,163); 531 (172); 575 (24); 578 (42); 583 (65,67,68); 823 (219,224)
Serpinski, V. V., 804 (94)
Servis, R., 800 (47)
Sevón, J., 418 (255)
Shanley, E. S., 468 (154)

Shapiro, I. L., 543 (313)
Shaposhnikov, Y. K., 578 (43,44)
Shaw, A. C., 856 (108)
Shearer, D. A., 490, 491 (22)
Sheldon, F. R., 457 (112)
SHELL CHEMICAL COMPANY, 513 (21)
Sherman, W. A., 852 (68)
Sherrard, E. C., 374, 405, 407, 410, 504 (36)
Shibamoto, T., 196 (165); 281 (65); 317 (48); 527 (139); 770 (8)
Shilov, E. A., 377 (42); 378 (48)
Shimada, M., 8 (27); 82, 87 (114); 108, 115 (59); 112 (83,84); 113 (85); 168, 188 (23); 258 (74); 284 (87)
Shimazono, H., 108 (61)
Shimokoriyama, M., 113 (86)
Shinra, K., 418 (259); 420 (271)
Shoolery, J. N., 300 (8); 327 (52)
Shorygin, P., 517 (67); 800 (41)
Shorygina, N. N., 9 (33); 186 (98); 384, 400 (109,110,111,113,114,115); 386 (113,114, 115); 388 (109,113); 389 (111,113,115); 390 (113,114); 391 (172); 397 (114,186); 401 (109,110,111,115); 402 (110,114); 405 (201, 208); 407 (201,208,216); 408 (224); 409 (228); 413 (208); 414 (208,234); 415, 417 (234); 418 (263); 419 (109,111,172,266,267, 268); 420 (111,172,267,268); 421 (267,268); 487 (8,43); 505 (41,42,43); 514 (42,43,44, 45); 550 (373); 859 (127,128,130)
Shuhmacher, G., 143 (175)
Siddiqueullah, M., 187 (110,111); 701 (50)
Sidhu, G. S., 69 (80); 176 (45)
Sieber, R., 375, 404, 405, 407 (23)
Siebert, W., 116 (90); 267, 283 (4,5); 272 (5); 281 (4)
Siegel, S. M., 28 (40); 30 (48); 34, 37, 38 (57); 35 (40,65); 81 (111a)
Sierila, P., 81 (112)
Siersch, E., 79 (99)
Silbernagel, H., 196 (166); 346, 354 (8); 348, 355, 357 (10); 551 (377)
Silker, R. E., 385 (136)
Simionescu, C., 282, 285 (72); 288 (108)
Simmonds, F. G., 404, 405, 417 (196)
Simmonds, S., 778 (23)
Simpson, W. T., 243 (28)
Simson, B. W., 47 (19)
Sinclair, R. M., 477 (203)
Singer, S. J., 816 (176)
Singh, B., 259 (80)
Singh, R. P., 477 (208)
Sipos, P., 272, 281, 284 (46)
Sirianni, A. F., 703, 705 (62)
Sitaramaiah, G., 705 (70)
Sizer, I. W., 129 (137,138)
Sjöholm, R., 292 (134,137); 293 (137)
Sjöström, E., 259 (83a); 261 (97); 699, 701 (31)
Skachkov, V. M., 284, 285 (78)
Skiöldebrand, C., 575, 576, 579 (15); 822 (214)
Sklar, A. L., 243 (24)
Skoblinskaya, S. A., 517 (67)
Skok, J., 36 (73)
Skurikhin, I. M., 287 (96); 315 (45)
Slabina, M., 292 (132)

Slaytor, M., 516 (65)
Sleziona, J., 385, 391 (147)
Smedman, L. A., 302, 303, 312 (42); 537 (240); 656, 657 (90,92)
Smedslund, T. H., 811 (146)
Smith, B., 395 (176); 472 (185)
Smith, D. C. C., 59 (52); 76 (91); 177 (67); 178 (67,70); 249 (59); 258 (72,73); 268,272, 275, 282 (28,29); 537 (222)
Smith, D. G., 106 (45)
Smith, E., 515 (59)
Smith, H., 515 (53); 749 (187)
Smith, J. E., 83, 85 (135); 552 (406); 852 (71)
Smith, S. G., 815 (166)
Smolyaninova, E. K., 800 (41)
Snyder, J. L., 224 (292)
Sobek, A., 505 (44); 515 (60,61); 516 (61)
Sobolev, I., 384, 391, 392, 397, 420 (123); 489, 497 (16,17); 551 (396); 659 (72)
Sobolevskii, C. A., 436 (15)
Sobue, H., 281, 282 (64)
Söderberg, S., 532 (177); 623, 624 (123); 649 (108); 673 (146)
Sofue, K., 177 (59)
Sogo, M., 82 (115); 83 (127,128,129,130,141); 85 (141,143,144,145); 86 (115,129,130); 87 (115,128,129,130,143,144,141,145); 189 (124)
Sohn, A., 196 (161); 345 (4); 384 (128); 803 (72)
Sohns, F., 471 (175); 549 (360); 597 (6)
Soila, R., 467, 468 (144)
Soipioni, A., 854 (79)
Sokolov, O. M., 720 (115)
Sokolova, A. A., 448 (49); 546 (339); 549 (353); 855, 858 (88)
Sokolova, I. V., 542 (307); 551 (386); 803 (77)
Solovev, L. S., 268, 285 (26); 284 (79)
Soloway, A. H., 458-460, 463 (124)
Somayajulu, V. V., 805 (113)
Somerville, N. G., 6 (21); 260 (91); 270, 271, 280, 281, 285, 288-290 (33); 670 (125)
Sondheimer, E., 490-492 (21); 517 (69); 551 (397)
Soper, F. G., 377 (43); 378 (49)
Sorensen, H., 770 (9)
Sorenson, N. A., 803 (76)
Sorokin, Y. Z., 581 (55)
Sorokina, N. F., 80, 81 (111); 291 (126)
Soundarajan, T. N., 729 (140)
Sowa, F. J., 507 (49)
Spalding, C. W., 858 (117)
Spanagel, H. D., 137 (158,159)
Sparks, H. E., 292 (135)
Speaks, R. N., 515 (59)
Spencer, E. Y., 537 (236)
Spencer, G. C., 384 (126)
Spencer, R. S., 740 (165)
Spona, J., 367 (102); 625 (127)
Sprinson, D. C., 101 (18)
Spurge, E. C., 799 (27)
Squire, G. B., 48, 49 (32)
Srivastava, L. M., 19 (4); 82, 83, 86 (123); 532, 535, 543 (200)
Stafford, H. A., 19 (6); 199 (193)
Stamm, A. J., 575 (3); 730 (144)
Stamvick, A., 226, 227 (298); 512 (8)

AUTHOR INDEX

Stanek, D. A., 196 (163); 348 (12)
Stanik, V., 421 (276); 538 (259)
Stark, B. P., 127 (129)
Starostina, K. M., 408 (224)
Staudinger, H., 166 (8)
Steck, W. F., 487, 489–492, 497 (6); 489 (18)
Steelink, C., 120 (112); 291 (125); 328 (60, 62); 330 (60,61); 332 (61,63); 333 (71); 336 (75,76); 338 (76); 339 (83); 450 (67); 536 (211,212,213); 677 (158,159); 785 (35)
Steinberg, J. C., 83, 85, 88 (140); 838 (14)
Steiner, W., 804 (99)
Steinkopf, W., 813 (153)
Stenemur, B., 203 (208); 528 (147)
Stenlund, B., 292 (100); 614, 615 (105,107); 719 (112); 751, 752 (197)
Stephenson, R. E., 858 (116)
Sternhell, S., 271, 288 (45); 291 (45,130); 300, 313, 320 (17,18); 301, 314, 322, 324 (18); 318 (17)
Stevens, W. P., 457, 477 (117)
Stevenson, F. J., 291 (123); 420 (272)
Stevenson, P. E., 243, 257 (27)
Stewart, C. M., 21 (12,13); 27, 37 (34); 180 (76); 192 (133); 222 (280)
Stewart, R., 456 (99); 457 (102,103)
Stewart, W. D., 851 (63)
Still, C. N., 769 (2)
Stjernstrom, N. E., 398 (192)
Stock, L. M., 386 (155)
Stockman, L., 601 (39,40); 642 (7); 730 (142)
Stoessl, A., 308 (31)
Stoffey, D., 552, 555 (420)
Stone, J. E., 78 (87); 176 (51); 612 (91); 613 (96); 643, 659 (17); 730 (150)
Stone, W. K., 166, 170, 195 (12)
Slotterbeck, O. C., 804 (97)
Strapp, R. K., 260 (92); 599, 626 (26)
Strauss, R. W., 387, 394 (159); 391, 406, 418 (171); 393, 409–412 (159,171); 468 (150)
Straus, S., 575 (5)
Streitweiser, A., 808 (138)
Streuli, C. A., 813 (150)
Strole, U., 457 (106)
Stubbs, A. L., 246 (46)
Stumpf, W., 177, 187 (63); 540 (287,289)
Subba Rao, N. V., 805 (113)
Suchy, J., 272, 281, 284 (46); 287 (97)
Suhara, Y., 713 (88)
Sukhanovskii, S. I., 469, 470 (168); 577, 579, 583 (32,34); 581 (55,57); 582 (57); 587 (83); 592 (122); 821 (205)
Sukhaya, T. V., 284, 285 (80)
Sukhushin, Ya. N., 414 (237)
Sundholm, F., 267, 271, 272 (17,18)
Sundman, V., 781 (28)
Suranyi, G., 643, 645 (23); 736 (159)
Surna, Ya. T., 823 (219)
Suter, C. M., 601 (43)
Sutherland, J. H., 612 (89)
Suzuki, E., 245 (36)
Suzuki, I., 551 (379)
Suzuki, J., 458, 461, 463 (134); 532, 534 (195)
Suzuki, K., 338 (80); 458 (129,132); 461 (129)
Suzuki, U., 313 (43b)
Swain, C. G., 378 (57); 807 (134)
Swan, B., 65 (77); 192 (134); 260 (88); 365 (87); 517 (72); 601 (56); 630 (157)
Swan, E. P., 85, 87, 88 (146); 301, 313, 319, 322, 324 (20)
Swartz, J. N., 397, 402, 407, 409, 413, 414, 418 (177)
Swelim, A. A., 375 (27)
Swenson, H. A., 223 (289,290)
Swenson, L. K., 844 (40)
Swoboda, R., 487, 489–492, 497 (6)
Symons, M. C. R., 327 (52)
Szabo, A., 643 (22); 758 (212)
Szelenyi, G., 575, 576 (21)
Szilagyi, M., 109 (62)

Tachi, I., 77, 78 (95); 220 (273); 222 (279); 538 (249,250); 601 (52); 705 (66); 709 (73); 805 (118,119)
Tahasaki, C., 618 (116); 622, 629 (120); 662, 663, 678 (84); 668, 676(99)
Tai, S., 523 (111); 526 (128,129)
Takahashi, H., 196 (165); 317 (48); 527 (139)
Takahashi, T., 129 (141)
Takamura, N., 733 (155)
Takamuru, S., 805 (117)
Takasaki, C., 59 (51); 288, 289 (111); 538 (245); 548 (345)
Takatsuka, C., 859 (126)
Takeya, G., 301, 315, 318 (19a)
Takeyama, H., 347 (9)
Tamásovits, G., 726 (133)
Tanaka, H., 552 (414)
Tanaka, J., 209 (236); 216 (260); 366 (91); 629 (155)
Tanberg, R., 853 (74,75)
Tanford, C., 511, 558 (2); 705 (71)
Tang, W. K., 575 (8); 581 (61,62)
Tanner, K. G., 78 (87); 176 (51)
Tanni, J. D., 819 (190)
TAPPI STANDARD METHOD T13m, 8 (23)
TAPPI STANDARD METHOD T202, 195 (153)
TAPPI STANDARD METHOD T214, 195 (152)
TAPPI STANDARD METHOD T236, 471 (171)
Tatum, E. L., 99 (10)
Tausend, H., 141, 142 (174); 146 (177)
Tayenthal, K. V., 730 (141)
Taylor, R., 418 (251)
Taylor, W. I., 2 (2); 116 (94); 123 (121); 128 (132); 552, 554 (419); 630 (160)
Techlenberg, H., 803 (84)
Telysheva, G. M., 377, 384, 410 (33); 530 (161); 531 (172)
Terasawa, M., 196 (165)
Terashima, N., 224 (296); 288 (102); 532 (190)
Termini, J. P., 843 (39)
Theander, O., 80 (105); 629 (154,155); 630 (154)
Thimann, K. V., 35 (67)
Thompson, E. C., 21 (11)
Thompson, N. S., 223 (290); 457 (108)
Thompson, S. O., 537 (220)
Thorén, S., 649 (108)
Thorn, G. D., 385, 420 (133)
Thurnher, K., 457 (113)
Thurston, W. H., 336 (74)

Tice, P. A., 289, 290 (120); 537 (216); 552, 553 (412); 618, 629, 630 (114); 716 (104); 852 (71)
Tiemann, F., 28 (38); 798 (8)
Tien, J. M., 814 (160)
Tiers, G. V. D., 300 (7)
Tilghman, B. C., 597 (1); 797 (2)
Timell, T. E., 45 (12); 47 (19,20); 62 (65,66); 63 (70); 66 (72); 180 (77); 224 (292); 643 (28); 695 (18); 697 (14); 701 (52)
Tishchenko, D. V., 543 (312); 592 (120,122); 648, 649 (107); 668 (114); 677 (160); 859 (132)
Tiyama, K., 245 (35)
Todd, A. R., 127 (128,130)
Tokatsuka, C., 208 (229)
Tollens, B., 549 (354); 597 (4)
Tollin, G., 328 (60); 330 (60,61); 332 (61,63); 536 (211,213); 677 (158)
Tominaga, K., 542, 543 (303)
Tomita, K., 196 (165); 317 (48); 527 (139)
Tomita, Y., 132, 143 (154); 307 (30)
Tomlinson, G. H., 366 (94); 598 (20); 800 (42, 43); 838 (13)
Tomlinson, G. H., II, 383 (13); 856 (101)
Tonooka, T., 471 (181)
Toombs, G. L., 262 (105)
Topfmeier, F., 455 (96); 659 (75)
Töppel, O., 292 (133); 804 (90,93)
Torgersen, H. F., 241 (13); 753 (205)
Toribio, F. B., 104 (36)
Torres, Serres, J., 110 (70)
Torrey, H. A., 385 (149)
Tottmar, O., 6 (14); 80, 81 (108)
Towers, G. H. N., 6 (19); 19 (3); 63 (67,68,69); 71, 74, 79, 80 (67); 104 (26); 107 (50); 777 (22)
Town, A., 805 (104)
Tracey, W., 73 (83)
Traube, M. C., 798 (9)
Traynard, P., 111 (77); 221, 222 (275); 374 (19); 375 (26); 384, 385 (122); 418, 419 (19, 254); 420-422 (19)
Treiber, E., 262 (102); 643 (26)
Treibs, W., 449 (62)
Tronov, B. V., 531 (167,169)
Tropsch, H., 530 (166); 575, 577, 579 (20); 822 (216)
Truce, W. T., 809 (142)
Truter, M. R., 380 (76)
Tschamler, H., 186 (97); 267, 285 (6)
Tse, A., 63 (69)
Tsuj, T., 814 (157)
Tsypkina, M. N., 538 (262)
Tsyrkin, E. B., 858 (120)
Tuller, W. N., 614 (101)
Tung, L. H., 719 (110)
Turenen, J., 365 (85); 368 (103,104); 512, 526 (13); 587 (87); 589 (103,106); 590 (87,113, 116); 625 (128); 817 (183,185); 818 (185)
Turstskii, Y. M., 859 (128)
Turunen, K., 589 (103); 817 (183)
Tuttle, M. J., 609, 610 (72); 702, 748 (56)
Twining, R. H., 552 (409)
Tychima, V. D., 405, 407, 413, 414 (208); 550 (373)
Tylli, H., 288 (113); 671 (137); 684, 686 (171);
685 (171,172); 700, 703, 705 (40)
Tyson, J. E., 856 (97)

Uchida, J., 66 (79); 191 (131)
Uekita, T., 592 (123); 630 (167; 718 (108)
Ujiie, M., 284 (86)
Ujioke, Y., 543 (317)
Umbreit, W. W., 770, 785 (10)
Unger, K., 385, 387, 391, 393, 394, 411, 412 (146)
Ungnade, H. E., 513 (25)
Urban, H., 469 (163); 519, 524 (82)
Urbanski, T., 291 (124,128)
Urlings, J., 575 (25)
Uschmann, C., 854, 855 (82)
Usmanov, K. V., 530 (160)
Utzinger, G. Ed., 805 (106)

Vanag, G. Ya., 198 (182); 532, 535 (196)
Van Beckum, W. G., 373, 392 (3)
Van Blaricom, L. E., 83, 85, 88 (140)
Van Buren, J. B., 384, 387, 390, 404 (112); 403 (112,218); 407, 413 (218)
Van Camp, B., 851 (63)
Van den Akker, J. A., 477 (198,201)
Vandenbelt, J. M., 241 (16,17); 242 (16,23); 243 (16)
van der Linden, R., 457 (103)
Van Derveer, P. D., 833 (1)
Vandoni, R., 380 (73)
Van Holde, K. E., 722 (118)
Vanina, V. I., 858 (119)
Vänngård, T., 129 (139)
Vardell, W. G., 546 (334); 671 (133,134); 839 (19,20)
Varma, S., 379 (60)
Varsel, C. J., 246 (44)
Vasatkova, M., 287 (97)
Vasileva, T. M., 537 (218)
Vasileva, V. P., 449 (58,59)
Vaughn, T. H., 515 (55)
Vedernikov, V. G., 575 (18)
Veibel, S., 383 (93)
Vekhov, V. A., 417, 421 (245)
Vener, I. M., 62 (63)
Verbanc, J. J., 507 (49)
Verley, A., 673 (142)
Vernon, C. A., 377 (40); 378 (52,53); 379 (53); 384, 407, 415 (103)
Veselova, E. V., 720 (115)
Viehäuser, M. V. G., 125 (123)
Vincent, T. A., 477 (203)
Vining, L. C., 108 (53,54)
Virkola, N. E., 467, 468 (144)
Visasoro, E., 541 (296)
Vloeberghs, A., 853 (73)
Vodnansky, J., 292 (132)
Voet, A., 854 (78)
Voitkevich, S. A., 804 (94)
von Hofe, C., 209 (234); 366 (89)
von Köppen, A., 644 (30)
Voronin, V. G., 804 (91)
Voykowitsch, A., 313 (43c)
Vroom, K. E., 599, 626 (26)

Wacek, A. V., 110 (76); 213 (247); 259 (84); 313 (43c); 400, 407 (195); 436 (18,19,20,21,

AUTHOR INDEX

28); 437, 441 (18,19,20,21); 514 (35); 541 (299,300); 542 (299); 626 (140)
Wachtmeister, C. A., 455 (95)
Wada, S., 271, 272, 280 (35); 288 (35,106, 107); 289 (107); 671, 672 (138)
Waentig, P., 405, 416 (207); 407 (210)
Wagner, G. H., 291 (123)
Wagner, R. B., 514 (36)
Wahba, M., 741 (168)
Wahlroos, O., 587 (88)
Wainai, T., 313 (43b)
Waksman, S. A., 782 (29)
Walchli, O., 37 (77)
Waldén, I., 673, 678 (144)
Waldichuk, M., 262 (105)
Walker, H. M., 527 (135)
Walker, J. F., 549 (357)
Wallace, B. P., 853 (76)
Wallace, R. T., 521 (109); 854 (81)
Wallace, T. J., 668 (98)
Wallden, I., 201 (201); 472 (183)
Wallin, H. C., 546 (333)
Wallin, N. H., 362 (76); 363 (78); 518 (76); 519 (83); 525 (124); 653, 666, 667 (66); 651, 654, 662, 664, 676 (69); 656 (69,91)
Walling, C., 379 (61)
Wallis, A. F., 133-137, 144, 145, 150 (157); 304, 307 (21a); 309 (34)
Walther, H. C., Jr., 784 (33)
Wang, P. Y., 712 (83)
Wantanabe, S., 457, 458, 461 (118)
Wardrop, A. B., 19, 25, 29, 37 (5); 22 (17); 23, 30 (22); 24 (30); 25 (31,33); 26 (33); 27 (5, 30); 33 (51); 35 (58); 46 (16); 50 (37); 52 (42); 53 (42,44); 54 (44); 66 (75); 78, 79 (98); 612 (93); 642 (12,13,14); 643 (14); 695 (4); 696 (11); 753 (206,207)
Wartiovaara, V., 81 (112)
Watanabe, H., 578 (41)
Waters, W. A., 130 (145); 131 (145,148); 378 (50,51); 433 (1); 457 (105); 458 (122); 471 (180)
Watson, A. J., 27, 37 (34); 180 (76); 192 (133); 271, 288-290, 292 (34)
Watson, C. A., 840 (30)
Watt, C., 640 (2); 797 (1)
Wayman, M., 457 (114; 729 (140)
Weaver, H. E., 327 (52)
Weaver, W. M., 814 (158)
Webb, T., 711 (78)
Wecker, H. K., 201 (199)
Wedekind, E., 187 (105)
Wegener, W. H., 713 (97)
Weinberger, L., 384 (99)
Weinges, K., 228 (303); 229 (304); 230 (311)
Weiss, A., 798, 799 (16)
Wellwood, W., 48 (30,33)
Wennerblom, A., 477 (200)
Wenzl, H. F. J., 640 (5); 835 (6)
Wenzl, P. J., 846 (47)
Werner, H. K., 226 (299); 452 (87,88); 601, 618 (38)
Wershing, H. T., 35 (62)
Wertz, J. E., 327 (52)
Wessely, F., 119 (107); 218 (264)
West, E., 354, 355 (43); 540 (280)
West, K. L., 166, 168, 170, 171, 221 (15)

Westerfield, W. W., 131 (149)
Westheimer, F. H., 380, 383, 386, 391, 392 (86)
Westley, R., 262 (105)
WEST VIRGINIA PULP & PAPER COMPANY, 839 (25); 848 (52,53); 850 (52,58); 851 (61, 65); 855 (53,93); 856 (25,96,106); 857 (111, 113)
Wethern, J. D., 575 (16); 806 (129); 807 (130)
Wetlesen, C. U., 405 (197); 712 (84)
Wetterholm, G. A., 812 (148)
Wexler, A. S., 248 (54)
Weygand, W., 540 (289)
Weyna, P., 749 (189)
Wheaton, R. M., 841 (31)
Whetsel, K., 267 (1)
White, B. B., 539 (270)
White, D. M., 117 (96)
White, E., 63 (68,69); 397, 402, 407, 409, 413, 414, 418 (177)
White, G. F., 515 (51)
White, H., 467, 468 (146)
White, R. G., 784 (33)
Whitney, W. H., 844 (43)
Whittaker, H., 405, 407, 412, 414, 416, 418, 419, 421, 422 (199)
Wiberg, K. B., 300 (5); 456 (100)
Widlund, B., 671 (137)
Widmer, F., 377, 405 (31)
Wiesner, J., 543 (315)
Wilder, H. D., 647 (41,42,43)
Wilds, A. L., 515 (56)
Wiley, A. J., 601 (46); 712 (85); 835 (2,4); 842 (36); 851 (63)
Wiley, P. R., 713 (94); 846 (46)
Wilk, H., 673, 676 (155)
Williams, D. H., 319 (50a)
Williams, D. J., 261 (99)
Williams, K. T., 856 (108)
Willstätter, R., 117 (97); 186 (103); 397 (181, 182); 414 (181); 551 (376); 588 (91); 818 (187)
Wilson, G. G., 844 (43)
Wilson, R. G., 319 (50a)
Wilson, J. W., 48 (30,31,33); 49 (31)
Wilson, W. C., 263 (109)
Wilson, W. J., 378 (49)
Wincor, W., 192 (136); 262 (102)
Windle, J. J., 336 (74)
Winer, D. C., 813 (151)
Wink, W. A., 477 (198)
Winstein, S., 345 (2); 815 (166)
Winsvold, A., 586 (79,80,81); 588 (80); 821 (204)
Winter, G., 809 (143)
Wintzell, T., 588 (94)
Wise, L. E., 540 (284); 597, 598 (3)
Witanowski, M., 291 (124,128)
Wittman, E., 367 (101); 537 (235); 625 (126)
Wold, W. J., 11 (41)
Wolf, J., 417 (248); 421 (276)
Wolfrom, M. L., 259 (79,81)
Wood, B. G., 799 (25)
Wood, D. D. S., 8 (29); 187 (112)
Wooster, C. B., 515 (57)
Worsford, D. J., 703, 705 (62)
Wray, 34 (55)

Wright, D., 104 (29); 105 (29,39)
Wright, G. F., 471 (176); 532 (186,188,189); 537 (236)
Wu, T. T., 48, 49 (31)
Wynberg, H., 798 (11)

Yabe, K., 288, 289 (107); 671, 672 (138)
Yakshilevich, M. A., 799 (19)
Yamakawa, I., 715 (102)
Yamashita, C., 804 (92)
Yamashita, K., 520 (94)
Yano, H., 552 (410)
Yashchanko, A. V., 858 (118)
Yashin, R., 517 (69)
Yasnikov, A. A., 377 (42)
Yates, P., 529 (155)
Yaunzems, V. R., 525 (119); 583 (67)
Yean, W. Q., 50 (38); 601 (41); 612 (89,92,94); 613 (94); 620 (87); 643 (19,20,21,23); 644 (20); 645 (23); 685 (19); 699 (32); 700 (32, 37,39); 701 (37,39,46); 702 (37,39,46,59); 703 (32,37); 705 (32,37); 706 (32); 707 (72); 712 (87); 719 (32); 720 (32,37,39,46,59); 722-724 (32); 725 (32,37); 726 (37); 727, 728 (39); 736 (159); 748 (46); 751 (32,72, 191); 752 (198); 753-756 (208); 759 (39,46)
Yeates, T. E., 532 (192)
Yesn, W. Q., 840 (28)
Yllner, S., 355, 356 (57); 360 (68); 601 (40); 629 (156); 642 (7)
Yokoyama, S., 301, 315, 318 (19a)
Yorston, F. H., 610 (76a)
Yoshida, S., 104 (26); 113 (86)
Yoshikawa, T., 284 (86)
Yoshimura, F., 575 (23)
Yudkevich, Y. D., 577, 579, 583 (34)

Zabel, R. A., 643 (28); 697 (14)
Zakoshchikov, A. P., 858 (119)
Zakoshchikov, S. A., 858 (119)
Zanini, C., 534 (205)
Zank, L. C., 260 (89)
Zarubin, M. Ya., 543 (312); 592 (120,122); 859 (132)
Zasosov, V. A., 799 (23)
Zauner, J., 103 (23); 521, 522 (107); 854 (80)
Zavarin, E., 434 (4)
Zavyalov, A. N., 452 (70)
Zbinden, R., 268 (21)
Zechmeister, L., 187 (107); 551 (376)
Zetsche, W., 380, 421 (68)
Zharkova, L. P., 804 (91)
Zhdanova, R. S., 549 (353)
Zhigalov, Y. V., 668 (114)
Zickmann, P., 374-376, 405, 407 (11); 577 (36); 587, 588 (84)
Ziechmann, W., 269, 291 (31)
Zilch, K. T., 513 (25)
Zincke, Th., 384, 400 (121)
Zobel, B. J., 46 (17)
Zocher, H., 187 (104); 532 (178)
Zollinger, H., 542 (311)
Zook, H. D., 514 (36)
Zoss, A. O., 526 (133)
Zweig, A., 340 (85)

SUBJECT INDEX

Abbreviated formulae, lignins, 15, 16
Abies, 44, 108
 A. amabilis, 49; see also Amabilis fir
 A. balsamea, 47; see also Balsam fir
 A. concolor, 56; see also White fir
Abiperm, ion exchange process, 843
Acacia, Brauns lignin, 282
 A. mellissima, 66
Accessibility, lignins, 608, 741-745
Acer, A. cissifolia, 72
 A. macrophyllum, 68
 A. negundo, 72, 99, 105
 A. plantanoides, 64
 A. rubrum, 72
 A. saccharinum, 71, 318
 see also Maple
Aceraceae, 72
Acetaldehyde, 366, 475, 514, 533, 625, 803
Acetate pathway, 98
Acetic acid formation, from alkaline oxidations, 449, 471, 587, 588, 591
 from kraft liquor, 670, 818
 from lignin, 469, 576, 803
 from wood, 821
Acetic acid lignins, 187
Acetoacetic acid, 779
Acetoguaiacone, 14, 655, 776
 alkaline hydrolysis product, 208, 366-368, 514, 587, 671, 677, 817, 818
 CuO- and nitrobenzene oxidation product, 435, 442, 446, 447, 804
 electrophorosis, 711, 712
 ir, derivatives, 276
 reactions, 388, 390, 463, 469, 474, 502
 uv, 249
Acetolysis lignins, pmr, 317
Acetone, from lignins, 576, 591, 822
 solubility in, of alkali lignins, 685
 sugar derivatives of, 842
Acetophenone, ir, 276
Acetosyringone, 332-336, 446, 447
Acetovanillone, 13, 14; see also Acetoguaiacone
p-Acetoxybenzaldehyde, uv, 242
Acetoxyl protons, 305, 495, 496, 674
Acetylacetone, 534, 535
β-Acetylacrylic acid, 474
Acetylated lignins, ir, 272-274
 pmr, 313-325, 496, 674
Acetylated model compounds, pmr, 301-312
Acetylation, in hydroxyl determination, 673, 674
 lignins, 528, 673-675, 751
Acetyl bromide, 260, 377
Acetylene, 526, 583, 584
Acetyl hypobromite, 377
Acetyl hypochlorite, 377

Acetyl nitrate, 380, 381
Acetylvanillic acid, 275, 460
Acid anhydrides and chlorides, reaction with DMSO, 814
Acidic hydrolysis, see Solvolysis
Acid lignins, 185, 186, 645
 condensed structures, 165, 261, 440, 441
 see also Hydrochloric acid lignins; Klason lignins
Acidolysis, see Solvolysis
Acid soluble lignin, determination by uv, 192, 259, 260
Acrylonitrile, 527, 816, 845
Activated carbon, from lignins, 581
Active alkali, kraft liquors, 640
Active chlorine, determination, 414
"Active insoluble esters," lignification, 107
Acyloin structures, sulfonation, 626
Addition reactions, 225, 382, 398
5-Adenosylmethionine, 102, 103, 109
Adhesive bonding, 731-735, 738, 739
Adhesives, 591, 713, 852
Adipic acid, 587
Aesculus, hippocastanum, 64
Agar gel, 711-713, 725
Agathis alba, 58
 A. australis, 57
A-groups, sulfonation, 602-611, 618
Alcohol lignins, see Alcoholysis lignins
Alcoholysis lignins, 273, 287, 540, 730
Aldehydes, condensation with lignins, 854
 formation by lignification, 122, 200
 by reaction with DMSO, 814
 reaction with vanillin, 805
Aldol condensation, 514, 517, 534, 541
Algae, 21, 771
Aliphatic hydroxyls, lignins, 272-274, 673-675
Aliphatic protons, lignins, pmr, 495, 496, 674
Alkali consumption in pulping, 641, 642, 647
Alkali fusion, 586, 589, 591, 821
Alkali lignins, 188, 470, 670, 685
 dispersing properties, 713-716
 intrinsic viscosities, 703, 705, 706
 macromolecular properties, 684, 707, 733
 molecular weights, 699-701
 solubilities, 687
Alkaline hydrogenolysis, 489-492
Alkaline hydrolysis, lignin sulfonates, 624, 625, 711, 712, 800
 lignins, 275, 360-369, 514, 803
 model compounds, 360-369, 650-658
Alkaline oxidations, 433, 449-452, 685
Alkaline pulping, 639-689
 accessibility effects, 645, 646
 epr studies, 329
 kinetics, 647

893

reactions, 360, 639-689
Alkoxyl groups, effect on aromatic reactivity, 384, 385, 390
Alkyl-aryl ethers, cleavage, 360-366, 391-396, 412, 459, 505-507, 514, 650, 653-658, 674, 680
Alkyl halides, 525, 845
Alkyl substituents, aromatic, 385, 391
Allophanates, 523
6-Allyl-4-hydroxyflavans, nitrobenzene oxidation, 439
Allylic rearrangement, in acidolysis, 353, 355
Alnus glutinosa, 64
  A. rubra, 66, 317, 318, 321
Alpha cellulose, content, woods, 46
Alsophila, 61
  A. mertensina, 62
Aluminum bromides, 805
Aluminum chloride, 513, 806
Aluminum lignin sulfonates, 843
Amabilis fir, 176
Amines, color reactions with lignins, 532, 535
  precipitation, lignin sulfonates, 598, 601
  reactions, 845
Aminoacids, in humic acids, 792, 793
  pools, 103
4-Aminoantipyrine, 805
p-Aminobenzoic acid, 98, 99
p-Amino-N,N-dimethylaniline, 805
Aminoethylation, 523
Aminolignins, 422, 540, 550
o-Aminophenol, 805
3-Amino-1,2,4-triazole, 539
Ammonia, liquid, and sodium, 487, 488
  reactions with lignins, 713, 808, 844
Ammonium base, sulfite pulping, 599
  lignin sulfonates, 843, 852
Amylopectin, 701
Angiosperm lignins, 154, 284, 288; *see also* Hardwood lignins
Aniline, 435, 513, 514, 801
Animal glues, 855
Anion exchange resins, 538, 713, 841
p-Anisic acid, 454
Anisole, 378, 392, 460, 506, 513, 516
Anthranilic acid, diazotized, 550
Antiauxin, 35
Antihistamines, 805
Antimony pentachloride, 530
Apex, lignification, 112
Apocynol, 274, 387, 459, 469, 474, 475, 776
Apple trees, lignification, 34
Arabinogalactans, 696
Araucaria, A. araucana, 58
  A. excelsa, 112
Arborescent forms, 21
Arbutus menziesii, 68
Arctium tomentosum, 64
Aromatic rings in lignins, pmr, 320-325, 495, 496, 674, 675
  oxidative cleavage, 414, 457-474
  substitution reactions, 382, 386
Artificial lignin, *see* Dehydrogenation polymers
Arundo donax, 75
Aryl-alkyl ethers, *see* Alkyl-aryl ethers
α-Aryl ethers, 650
Arylglycerol-β-ethers, 8, 205-208, 488, 650

fungal degradation, 774
reactions in kraft pulping, 666
Aryl glycerols, formation from β-ethers, 356, 362
periodate oxidation, 207
o-Aryloxyphenol structures, in lignin, 209
1,2-Aryloxy shift, 360
o-Arylphenol structures, in lignin, 204
Ascomytes, 770
Ascorbic acid, 29
Ash, lignification, 34
A-storage resins, 855
A S lignin, 844
Asparagus officinalis, 64, 75
Aspen, 64, 66, 68, 71, 177, 178, 192, 196, 256-258, 260, 613
  acid-soluble lignin, 260
  Brauns lignins, 258, 282, 742
  CuO oxidation, 446, 447
  delignification with aq. butanol, 613
  dioxane acidolysis lignin, 701, 730, 731, 734, 735
  hydrolysis with water, 345
  lignins, hydrogenation, 494, 503, 504
  mwl, 248, 256, 257, 284
  neutral sulfite pulping, 613
  nitrobenzene oxidation, 435
  lignin sulfonates, 248, 613, 803
Aspergillus niger, 790
Asphalt emulsions, 851
Asplund process, 733, 735, 736
Athrotaxis selaginoides, 58
Auxins, 34
Avicel, 733
Azobenzene, 436, 801
p-Azobenzenesulfonate, 801
α-Azobisisobutyronitrile, 557
Azoxybenzene, 436, 801

Baccharis halimifolia, 279
Bacteria, biodegradation, 777
Bagasse, 102, 178, 180
  Brauns lignin, 258, 282
Bald cypress, 176
Balsam fir, 47, 699, 700
Bamboo, 64, 177
  Brauns lignin, 282
Barbituric acids, 534
Bark, acidolysis products, 86
  of hardwoods, 85
  lignin content, 88
  nitrobenzene oxidation, 83
  phenolic acid content, 88
Bark lignin, 81-89
  composition, 86, 88
  ir spectra, 250, 291
  isolation, 85
  pmr spectra, 313, 317, 321
Bark phenolic acids, 82, 83, 85, 193
  ir, 83, 84
Bark tannins, 83, 84
Barley, phenylalanine diaminase, 104
Base-catalyzed degradation, 360, 650-658
Basidiomycetes, 769
Basswood, 191
Beech, 30, 64, 68, 177, 191, 196, 208, 221, 222, 256, 257, 279

SUBJECT INDEX

decayed, 772
Beech lignins, hydrolysis by water, 209, 349
   p-hydroxyphenylpropane units, 454
   ir, 282, 287
   lignin sulfonates, 701
   uv spectra, 256, 257
Belliolum haplopus, 71
Benzalacetophenone, 815
Benzaldehyde, 532
Benzene hexacarboxylic acid, 214, 449, 470, 471
Benzene polycarboxylic acids, 469, 551, 785
Benzene-1,3,4,5-tetracarboxylic acid, 470
Benzene sulfonyl halides, 578
Benzilic acid rearrangement, 448
Benzopyran systems, 549
Benzoquinones, 398, 461-463, 518, 532
   chlorinated, alkaline hydrolysis, 417
   from chlorination, 397
   humic acids, 789
   ortho-, 450, 458, 517, 518, 669
   oxidants, 398
   oxidation of, 458-460, 474
   from oxidations, 458-460, 467, 518
   para-, 458-463, 518
   polymerization, 398
Benzoquinone radicals, epr spectra, 332, 335, 339
Benzoxazole, from vanilla, 805
Benzoylacetone, 534
Benzoyl peroxide, 557
Benzyl alcohols, *see* Benzylic hydroxyls
Benzyl derivatives of lignins, fungal attack, 774, 777-782
Benzyl ethers, cleavage, 545
   electrophilic displacement, 390, 512
   ir, 287
   in lignins, 673, 676, 680
   solvolysis, 345
   sulfonation, 618
Benzylic hydroxyls, 325
   displacement by electrophiles, 391
   in lignins, 274, 496, 673, 676, 680
   in phenolic units, 665
   solvolysis, 345
Benzyl thiols, 545, 653
   in kraft pulping, 657, 665
Berberis vulgaris, 64
Betula, 24
   B. lutea, 71
   B. papyrifera, 66-68
   B. verrucosa, 64, 68, 256, 257, 260
   see also Birch
B- and B'-groups, in sulfonation, 602-611, 618, 623
Bibenzyl dimers, from hydrogenolysis, 500
Bicreosol, 549
Binders, from lignin sulfonates, 836, 850, 852, 853
Biodegradation, 769-793
Biogenesis of precursors, 95-116
Biogenetic pathway, lignin formation, 95, 503
Biological function of lignins, 1, 21
Biosynthesis of lignins, 116-155
Biphenyl, *see* Linkages in lignins, 5-1, 5-5, 5-6
Biphenyl model compounds, uv, 140, 216, 254-256, 617; *see also* Lignols, di-, -5-

Birch, 177, 178, 180, 192, 215
   delignification, 753-757
   hydrolysis with water, 348
   lignin-carbohydrate complex, 718
   lignin distribution, 24
   thermal softening, 738
   uv microscopy, 697
Birch lignins, acid soluble, 260
   hydrogenolysis, 500, 501
   hydrotropic, 710
   ir, 284
   Klason lignin, 191
   lignin sulfonates, 701
   periodate, 731
   pmr spectra, 315, 318, 322-324
   uv, 256, 257, 260
Bisepoxylignans, 228
Bisphenol, 526
Bis reagent, 601
Bisulfite pulping, 261, 514, 598
Bisulfite-sulfite pulping, 599
Bjorkman lignin, *see* milled wood lignins
Black gum lignins, pmr, 315, 318, 322-324
   soda lignin, 675
Black liquor, 642, 643, 670, 797, 817, 818, 834, 838, 839
Black spruce, 175, 178, 256, 257
   Brauns lignin, 267
   delignification, 644
   epr spectrum of wood, 328, 329
   lignin sulfonates, 752
   uv microscopy, 697
Black wattle, 177, 178
   sulfonation, 626, 628
Bleaching, of pulps, 407, 670, 671
Boiler water treatment, lignin sulfonates, 844
Bonds, between lignin units, *see* Linkages in lignins
Bordered pits, 643
Borohydride reduction, 247, 531, 547, 627-629, 663
   carbonyl determination, 673-677, 678
   carbonyl groups, 663, 681
   effect on sulfonation, 627
   quinone methides, 650
   spectral changes, epr, 331, 332, ir, 272, 287, uv, 247, 249, 250, 257, 287, 488, 682
Boron tribromide, 512
Branching coefficient, 758, 759
Brauns lignins, 11, 165, 175-180, 258, 267, 282
   aggregation, 751
   electrophoresis, 712
   epr, 328, 332
   ester linkages, 177, 258
   fungal degradation, 771, 773, 774
   ir, 267, 278, 281, 282
   pmr, 314, 322, 323
   uv, 248, 251, 258
Bromopicrine, tracer studies, 96
Bromination, agents, 379
   bromine monoxide, 377
   bromolignins, 374
   N-bromosuccinimide, 377
   of hydroxyethylated lignins, 520
   side-chain displacement, 386, 388
5-Bromovanillic acid, 407

6-Bromovanillin, 407
Brown-rot fungi, 179, 770, 772
Buckwheat, 113
Bulk delignification, 645, 646
"Bulk polymer" coniferyl alcohol, 151, 152
Bunias orientalis, 64
Butadiene-acrylonitrile rubbers, 856
t-Butyl hypochlorite, 377, 385

C_6C_1-components, in plants, 104
C_6C_1-monomers, from hydrogenolysis, 499
C_6/C_1-ratio, in maturing plants, 112
C_6C_2-monomers from hydrogenolysis, 487, 498, 499
C_6C_2-units in kraft lignins, 673
C_6C_3-compounds, biogenesis, 95-112
C_6C_3-monomers, from hydrogenolysis, 487, 498, 499, 504
C_6C_3-units, *see* Phenylpropane units
Cadoxen, solutions of pulp, lignin in, 259
Caffeic acid, 105
Caffeoylshikimic acid, 108
Calamus rotang, 64
Callitris rhomboides, 57
Calluna vulgaris, 64
Callus tissue, lignin, 282
Calorimetry, 740
Cambial sap, glucosides in, 110
Candida utilis, 841
Canizzaro reaction, 436, 437, 448
Carbohydrates, effect on ir spectra, 277
 dissolution in alkaline pulping, 647
 humic products from, 771, 790
Carbon black, 850
Carbonization, 821-823
Carbon-14, labelled compounds, 95-97, 102, 366, 367, 751, 793
Carbon monoxide, from HCl lignin, 822
 inhibition of tyrosinase, 770
 from pyrolysis, 579
Carbonyl groups, bark phenolic acids, 84
 conjugated, 249-251, 257, 275
 determination, 250, 488, 673-678
 DHP and MWL, 281
 effect on sulfonation, 626, 627
 formation, in ethanolysis, 249, 355
  in milling, 256
  in oxidation, 469, 473
  in sulfonation, 626, 627
 ir, 267-269, 274-277, 281, 286, 289
 kraft lignins, 289, 668, 680
 lignin sulfonates, 289, 625-629
 non-conjugated, 200, 250, 251, 488, 673, 677, 678
 reaction with formaldehyde, 663
 reduction, 275-277, 651, 663, 681
α-Carbonyl groups, directive influence, arom. substitution, 384
 effect, on epr spectra, 332-336
  on nitrobenzene oxidation, 436
  on sulfonation, 626
 in lignins, 325, 673-678
 phenolic and non-phenolic, 488
 uv, 247-250, 256, 682
Carboxyl groups, 537, 629, 702
 formation, 413, 477
 humic acids, 787

ir, 275, 278, 281, 289
 kraft lignins, 676
 lignins, 280, 289, 663, 668, 672-676, 680
 lignin sulfonates, 629, 630
Carboxymethylation, 552
β=Carboxymuconic acid, 460, 779
4-Carboxy-2,3',4'-trimethoxydiphenyl ether, 211, 455
5-Carboxyvanillic acid, 803
5-Carboxyvanillin, 435, 439, 440, 448, 803
Carpinus betulus, 64
Carrot, tissue cultures, 35
Casuarina equisetifolia, 279
Cataleptic polymers, 554
Catalysts, hydrogenation, 488-503
Catechin, 83, 85
Catechol groups, in kraft lignins, 669, 673-675, 846
 in lignin sulfonates, 454, 617, 627, 630, 631, 717, 718
 in MWL, 673-675
Catechols, 518, 519, 543, 554, 778, 817, 818
 acetals, 515
 disulfonic acids, 630
 formation in alkali fusion, 540, 586, 588
  in chlorination, 402
  by demethylation, 365, 513, 619, 669, 846
  in oxidation, 458
  in pyrolysis, 578, 822
  in reduction of quinones, 679, 855
 humic acids from, 788
 in kraft liquor, 834
 oxidation, 396
 reaction with formaldehyde, 590, 663
 sulfur containing, 669
Catechol-O-methyltransferases, 108, 109
Cation exchange, 674, 713, 843
Cationic surfactants, 548
Cedrus deodora, 46, 58
Cell corners, delignification, 612, 644
 lignin content, 24
Cell wall, 696, 697
 distribution of lignin, 30, 51, 65-68, 643, 698, 754
 texture and organization, 22-25
Cellophane, 733
Cellulose, polydispersity, 701
 softening point, 645
 solution viscosity, 703, 704
 submicroscopic structure, 22, 695
Cement additives, 715, 848, 849
Cephalodexus drupacea, 57, 58
Chalcones, 312, 446, 447, 533
Charcoal, 821-823, 853
Charge transfer complex, phenoxy radicals, 121
Chamaecyparis, 45
C. nootkatensis, 58
Chelation, lignin sulfonates, 630, 856, 857
Chemical shifts, protons, 299, 300, 310, 312, 494, 495
Chemicals, from lignins, 797-824
Chemotaxonomy, 291
Chestnut, Brauns lignin, 282
Chitin, absorption of lignin sulfonates by, 602
Chloral, 541, 799
Chlorinated model compounds, alkaline hydrolysis, 416

SUBJECT INDEX

pmr, 313
Chlorination, 373-417, 545, 552
   addition reactions, 408
   agents, reactivity, 379, 744, 745
   in chlorine dioxide oxidation, 467
   demethylation, 395, 409-412
   dimethyl sulfide, 809-811
   effect of pH and medium, 376, 391, 392
   effect on uv spectra, 261
   formation of carbonyl and carboxyl, 399, 413, 415
   hydrolysis of ethers, 412
   in hypochlorite treatment, 464
   lignins, 373-417, 520, 846, 859
   model compounds, 393-396
   oxidation reactions, 400, 433, 457
   pulps, 407
   side-chain displacement in, 386-388
   side-chains, 408
   steric effects, demethylation, 394, 409
Chlorine dioxide oxidation, 433, 457, 851
   effect of pH, 466
   electron transfer processes, 467
   non-phenolic units, 467
   reactive species, 466
Chlorine monoxide, 377
Chlorine-sodium sulfite, color reaction, 6, 19, 20
Chlorine-zinc iodine, stain, 37
Chlorite, holocellulose, lignin in, 259
   oxidation, 467
6-Chloro-3-ethoxy-4-hydroxybenzaldehyde, 413
Chloroacetic acid, reaction with lignins, 524, 528, 845
p-Chloroanisole, nitration, 392, 395
Chloroformic acid, reaction with lignin, 845
Chlorolignin, 11, 416
   alkali fusion, 588
   composition, 374, 407, 414
   dehydrohalogenation, 417
   electrophoresis, 712
   hydrolysis, 416, 417
   introduction of functional groups, 417
   methylation, 416
   reaction, with ammonia, 417
      with silver nitrate, 414
Chloromethylated lignins, reactions, 417
Chloromethylphosphonic chloride, 531
Chloromuconic acids, hydrolysis, 417
p-Chlorophenol, 542
o-Chloroquinones, stability, 414
Chlorosis, 857
Chlorous acid, oxidation, 433
Chlorovanillins, 407, 531
Chorismic acid, 98, 100
Chromium complexes, lignin sulfonates, 552, 553, 843, 847, 852
Cinnamaldehyde groups, 195, 249, see also Coniferaldehyde
Cinnamic acid, 98, 105
   end groups, 199
   lignin precursor, 106, 107, 114
Classification of lignins, 43-89, 292
Cleavages, of aromatic rings, 414, 457-474, 777-782
   carbon-carbon bonds, 275, 359, 360, 365, 366, 498, 499, 820
   of ether bonds, 360-366, 391-396, 412, 459, 505-507, 514, 650, 653-658, 674, 680
   homolytic, 368
Clubmosses, 56, 61-63
Coagulants, 548, 555
Coalescence, of lignin in pulping, 645, 736
Coal formation in nature, 771, 772
Cobalt compounds, octacarbonyl catalyst, 502
   oxide, oxidation, 444
   thiocyanate, 535
Cocos nucifera, 64
Co-enzyme A, 107
Cohesive energy density, 687, 742
Coix lachryma, 75
Coleoptiles, wheat, 35
Collagen, 554
Collenchyma, 19, 29, 30
Colloidal behavior, 713-718
Collybia velutibes N8, 788
Color, of lignins, 649, 680, 683
Color reactions, lignins, 6, 19, 20, 31, 43, 63, 74, 82, 86, 277, 531, 532, 535, 543, 773
Compound names, 12
Compression wood, 35, 47, 54, 752
   MWL, 292
Concentrated HCl, color reaction, 531
Concrete additives, 848, 849
Condensation reactions, acid catalyzed, 182, 183, 206, 358, 440, 441
   base catalyzed, 658-664
   benzyl alcohols, 184, 612
   between side-chain carbons, 658, 664
   coniferylaldehyde groups, 184
   with formaldehyde, 545
   in sulfonation, 598, 610
   in water hydrolysis, 346
   uv, 261
Condensed aromatic units, 320, 321, 452, 436, 659, 680
   estimation by pmr, 320, 321, 324
Condensed polylignols, 346
Conductometric titration, 618, 629, 674
Congo-red stain, 37
Conidendrin, 440, 470, 781, 808
Coniferaldehyde, 86, 109, 184, 208, 229, 517, 551, 654, 655, 772, 775, 776,
   determination, 198, 250, 488
   end-groups, in lignins, 250, 286, 488, 531-535, 543, 673, 677, 678
   formation, 109, 139, 140, 196, 348, 357, 773
   ir, 276, 277, 286
   reactions, 141, 366, 499
   sulfonation, 606, 607, 615, 626
   uv, 250-252, 256, 488
Coniferin, 28, 33, 106, 110, 629
Conifers, 45, 284, 292
Coniferyl alcohol, 2, 87, 551, 776
   dehydrogenation and coupling 106, 138-150, 338, 445, 446
   DHP formation, 116, 283
   end groups, 195, 199, 537
   formation, 196, 348
   glucosides of, see Coniferin
   ir, 274
   reactions, 143, 502

898   SUBJECT INDEX

sulfonation, 197, 606, 607, 611, 627, 628
uv, 250-252
Coniferyl aldehyde, see Coniferaldehyde
Continuous pulping, 600, 643, 685
Copernicia australis, 64
Copper chromite catalyst, hydrogenation, 503, 504, 819
Corchorus, 74
Cordyline congerta, 110
Core lignin, 2, 9
Coriolus hirsutus, 775
  C. versicolor, 775
Cork lignins, 712
Corncob lignin, pyrolysis, 577, 580
Cortaderia selloana, 108
Corylus avellana, 64
Cotton hull lignin, 328, 580
Cottonwood lignin, 257
p-Coumaraldehyde, 86, 196, 773, 775
p-Coumaric acid, 105, 106, 275, 773, 793
  ester groups, grasses, 10, 105, 114, 258
  uv, 258
p-Coumaroylquinic and -shikimic acids, 108
p-Coumaryl alcohol, 2, 87, 106, 110, 195
  DHP, 283, 454
  hydrolysis product, 196
Coupling, phenoxy radicals, 117-155, 450, 455, 781, 782
Cracking, 583
Creosol, see Guaiacylmethane
Cresols, 313, 516, 577, 820-822
  p-, 130, 358, 458, 504
Cross-linking, 525, 553, 726
Cryptomeria, 177, 215
  C. japonica, 52, 58
Cubecon, 230
Cuoxam lignins, 186, 187, 505, 514, 529, 803
Cupressaceae, 65
Cupressus macnabiana, 58
  C. torulosa, 46
Cupric oxide, oxidations, 444-448, 531, 799, 803
Cupriethylene diamine, 703, 705
Cuproxam lignin, see Cuoxam lignins
Cyanuric chloride, 555
Cyanide, demethylation by, 808
Cyathea boninsimensis, 61, 62
Cycadales, 44
Cycas revoluta, 56, 61
Cyclohexane derivatives, hydrogenation products, 489, 490, 503, 504, 819
3-Cyclohexyl-1-propanol, 489, 504
Cytisus laburnum, 64

Dacrydium cupressinum, 58
Daedalea quercina, 180
DAHP-synthetase, 100
Dakin reaction, 450, 475
Dawsonia, 81
Dealkylation, see Demethylation
Decayed lignin, spin content, 328
Definition of lignins, 2, 10, 11
Degree of dehydrogenation (DD), of lignins, 224-226
Degree of polymerization (DP), lignin sulfonates, 614, 626
Dehydrodianisic acid, 216

Dehydrodiconiferyl alcohol, 138, 211, 229, 306, 307
Dehydrodi-α-ethylvanillyl alcohol, 123
Dehydrodieugenol, 439
Dehydrodiisoeugenol, 132, 307, 311, 313, 439, 541
Dehydrodipinoresinol, 129
Dehydrodivanillic acid, 216, 435
Dehydrodivanillin, 129, 216, 398, 435, 439, 443, 773, 776, 778
Dehydrodivanillyl alcohol, 518, 788
Dehydrodiveratric acid, 216, 452, 453, 455
Dehydrogenation, free radical intermediates, 338
Dehydrogenation of, coniferyl alcohol 106, 138-150, 338, 445, 446
  p-coumaryl alcohol, 106, 110, 195, 283, 454
  p-cresol, 130
  di-(3,5-di-t-butyl-4-hydroxyphenyl)-methane, 125
  dihydrodehydrodi-isoeugenol, 129
  di- and oligolignols, 148
  2,4-demethylphenol, 131
  3,4-dimethylphenol, 131
  2,4-disubstituted phenols, 128, 129
  p-ethylphenol, 130
  trans-ferulic acid and ester, 137, 138
  guaiacol, 128
  cis-isoeugenol, 134, 135
  trans-isoeugenol, 132-135
  methylphloroacetophenone, 131
  4-phenoxyphenol, 125
  pinoresinol, 129
  podototarin, 129
  p-propenylphenol derivatives, 132, 136, 137, 143, 144
  4-n-propyl-2-methoxyphenol, 131, 132
  4-n-propylphenol, 130
  sinapic acid, 137
  sinapyl alcohol, 106, 110, 143, 155, 195, 283
  2,4,6-trisubstituted phenols, 124, 125, 332-339
  tyramine, 129
  tyrosine, 129
  vanillic acid, 129
  vanillin, 129
Dehydrogenation polymers (DHP), 116, 154, 155
  ir, 267, 280, 283
  methylated, 365
  permanganate oxidation, 454
  spin content, 328
  structure and composition, 149-155, 248
  sulfonate, 629
Dehydrogenative polymerization, 2, 3, 29, 116-155
Dehydroguaiacetic acid, 134
5-Dehydroquinase, 99, 100
5-Dehydroquinate hydrolyase, 112
5-Dehydroquinate synthetase, 100
5-Dehydroshikimic acid, 98-100
Dehydroshikimic acid reductase, 99, 100, 112
Delignification, alkaline, 639, 645-650
  with DMSO-SO$_2$, 601
  kraft, 641, 645-647
  rates, 608, 609, 612, 645-650
  sulfite, 602-614, 620-622
  topochemistry, 612, 735

SUBJECT INDEX

Delphinidin, 85
Delta (δ) values, pmr, 300, 495, 496
Demethoxylation, 280, 497, 504-507, 578, 790
Demethylated lignins, 512, 818
Demethylation, 513, 515, 550, 578, 790
  by base, 365, 589
  by chlorine, 395, 409, 846
  by dichromate, 630, 718, 719
  by hydrosulfide, 512
  of kraft lignin, 669, 680, 840, 846, 855
  in kraft process, 512, 641
  by nitration, 392
  oxidative, 460-463, 467, 469, 471, 476, 478
  sulfite processes, 619, 630
  vanillic acid and vanillin, 781, 805
Dendrodium nobile, 75
Dendroligotrichum, 81
Densitometrical analysis, 753-757
3-Deoxy-arabinoheptulosonic acid phosphate, 100, 101
Destructive distillation, wood, 821
Desugared lignin sulfonates, 842
Desulfonation, lignin sulfonates, 598, 819, 844, 845, 854
  model compounds, 623, 624
Detached side-chain structures, *see* Displaced side-chains
Deuteration, 272, 273, 278, 312, 659
Diacetone sugars, isolation from sulfite liquor, 842
Dialysis, sulfite liquor, 601, 620-621, 699-701
Diamagnetic dimers, phenoxy radicals, 120, 121
Diaryl ethers, 364, 512, 650
Diarylmethane linkages, soda lignins, 665
Diarylmethanes, 536
Diastase, 221
Diazo coupling, 537, 549, 550
Diazomethane methylation, 201, 331, 411, 438, 519, 520, 533, 537, 659, 662, 663, 666, 667, 676
Dibenzoylmethane, 534
4,5-Dibromoguaiacol, 407
4,6-Di-*t*-butylguaiacol, bromination, 397
3,5-Di-*t*-butyl-4-hydroxybenzyl alcohol, 125
Di-(3,5-di-*t*-butyl-4-hydroxyphenyl)-methane, 125
4,6-Di-*t*-butyl-*o*-quinone, 397
Dicarbonyl compounds, reactions with lignin, 532, 534, 548
4,3'-Dicarboxy-2,4',5'-trimethoxydiphenyl ether, 455
2,6-Dichloroanisole, nitration, 395
Dichlorodicyano-*p*-quinone (DDQ), 433, 473
Dichloroprotocatechualdehydes, 2,5-and 5,6-, 396
Dichlorourea, 377
5,6-Dichlorovanillin, 408
4,5-Dichloroveratrole, demethylation, 394
Dichromates, 552, 554
  gelling of lignin sulfonates by, 630, 716-718
  oxidation by, 457, 847
Dicksonia, 61
Dicotyledons, lignin, 284, 436; *see also* Hardwood lignins
Dielectric constants, alkali lignins, 684
  solvent mixtures, 743, 744
Dienone-phenol rearrangement, 218

Diethylaminoethyl cellulose, 601
Diethylene triamine, 526
Difference spectra, ir, 285
  uv, 201, 245, 488, 682
Differential thermal analysis, 581, 684
Diffusion, in molecular weight determinations, 616, 722, 723, 725, 726, 750
  pulping liquors, 643
  spherical microgel, 703-706
Diguaiacylethane, 209, 492
Diguaiacylmethane, 492, 660-662
1,2-Diguaiacyl-1,3-propandiol, 352, 360
Diguaiacylpropanones, 352, 353
Dihydrobenzacridine, 629
Dihydroconiferyl alcohol, *see* 3-guaiacyl-1-propanol
Dihydrodehydrodiconiferyl alcohol, 651
Dihydrodehydrodi-isoeugenol, 129, 439, 651, 652, 667
Dihydroeugenol, *see* Guaiacylpropane
Dihydroferulic acid, 655; *see also* 3-Guaiacyl-1-propanoic acid
Di-(hydroxymethyl)succinic acid dilactone, 215
3,4-Dihydroxyacetophenone, 242, 243
3,5-Dihydroxybenzoic acid, 785
2,5-Dihydroxybenzoylformic acid, 781
4,4'-Dihydroxy-3,3'-dimethoxybenzil, 209, 446, 447
4,4'-Dihydroxy-3,'-dimethoxybenzophenone, 446, 447
4,4'-Dihydroxy-3,3'-dimethoxychalcone, 446, 447
4,4'-Dihydroxy-3,3'-dimethoxydiphenyl, 398
2,4'-Dihydroxy-3,3'-dimethoxy-5-ethylbibenzyl, 307
4,4'-Dihydroxy-3,3'-dimethoxystilbene, 209, 803
Dihydroxydiphenyl methanes, 513, 650
4,5-Dihydroxy-2,7-naphthalene disulfonic acid, 539
2,5-Dihydroxyphenylpyruvic acid, 778
Dihydroxystilbenes, *o*,*p*-, 651, 665, 676
  p, *p*-, 253, 665
Dihydroquercitin, 627
Dihydrosinapyl alcohol, *see* 3-Syringyl-1-propanol
Diisocyanates, 521
Diisoeugenol, 133
Di-O-isopropylidene sugars, 842
α,β-Diketones, 474, 631
Dilatometry, 740, 741
Dilignols, *see* Lignols, di-
Dimeric products, hydrogenolysis, 488, 492, 493, 500
3,4-Dimethoxyacetophenone, 363
Dimethoxybenzenes 459, 460, 515, 516
  1,2-, *see* Veratrole
3,4-Dimethoxy benzoic acid, *see* Veratric acid
2,6-Dimethoxy-*p*-benzoquinone, 122, 155, 335, 339, 397, 393, 422, 781, 782
3,4-Dimethoxycinnamyl alcohol, 199
5,6-Dimethoxyisophthalic acid, *see* Isohemipinic acid
2,6-Dimethoxyphenol, 460
3,4-Dimethoxyphenyl glycerol, 362, 399, 605, 653
  β-(2-methoxyphenyl)-ether, 362, 387, 464,

604, 605
1-(3,4-Dimethoxyphenyl)-1-nitropropane, 402
1-(3,4-Dimethoxyphenyl)-propane-1-sulfonic acid, 399
1-(3,4-Dimethoxyphenyl)-1-propanol, 400-404
1-(3,4-Dimethoxyphenyl)-2-propanol, nitration, 400
4,5-Dimethoxyphthalic acid, 184, 203; see also Metahemipinic acid
2,2-Dimethoxypropane, 288
Dimethylamine, 548
   and formaldehyde, reaction with lignin, 662, 851
4,4'-bis-(p-Dimethylaminodiphenyl)-methane, 601
2,6-Dimethylanisole, 395
Dimethyldisulfide, 817
3,5-Dimethyl-4-hydroxybenzylalcohol, 122
Dimethylphenol, 2,4-, 131, 577
  2,6-, 116, 119, 126, 127, 339
  3,4-, 131
2,4-Dimethylphloroglucinol, 543
Dimethylsulfate, 331, 519, 520, 653
Dimethylsulfide, 365, 512, 589, 591, 641, 669, 806-813
Dimethylsulfone, 810, 812, 817
Dimethylsulfoxide, 287, 601, 705, 810-817
Dimethylsulfoxonium methide, 815
2,3-Di-O-methylxylose, 222
3,5-Dinitrobenzaldehyde, 419
2,6-Dinitro-4-chlorophenol, 395
Dinitrogen pentoxide, 380, 381
Dinitrogen tetroxide, 380
4,6-Dinitroguaiacol, 202, 215, 400, 419, 420
3,7-Dioxabicyclo-[3,3,0]-octane system, pmr, 306
Dioxane acidolysis lignins, 7, 11, 187, 540
  molecular weights, 726
  physical properties, 240, 241, 703-707, 730-735
  reactives, 393, 583
Dioxane hydrochloric acid lignins, see Dioxane acidolysis lignins
Dioxane lignins, see Dioxane acidolysis lignins
Dioxepins, 124, 149
o-Diphenols, see Catechol
2,3-Diphenylbenzofuran, 541
Diphenylethane linkages, see β-1 linkages
Diphenyl ether linkages, 144, 455
Diphenylmethane, derivatives, 514, 547
  linkages, 440, 449, 617, 680
Di-quinonemethides, formation, 134, 137
Dispersants, from lignins, 713-716, 836, 849-851
Displaced side-chains, reactions, 210, 626, 627
Displacement of p-hydroxybenzyl groups, 125
Displacement of side-chains, in coupling, 16, 122, 124, 147, 149
Disposal, of spent liquors, 797, 858
Disproportionation, of phenoxy radicals, 121
Distillable oils, see Hibberts ketones
Disyringylethane, 209
Disyringylmethane, 336
Dithianes, 312, 538, 657, 658, 668
Divanilloyl, 209
Divanillyl, see Diguaiacyl ethane
Divanillyl tetrahydrofuran, 230

Dodecenyl succinic anhydride, 529
Donnan equilibrium, 608, 719
Double bonds, α,β-, 199, 251-254, 256, 668, 673, 676, 677
Douglas fir, 177, 179, 193
  bark, polyphenols, 591
  lignins, 278, 282, 286, 318, 461, 822
  pulping, 627
  see also Pseudotsuga Menziesii
Double layer, lignin sulfonates, 708
Drilling muds, 846-848
Drimys winteri, 71
Dryopteris filixmas, 56
Duroquinone, 580
Durvillea antarctica, 80
Dyes, adsorption rates, 850

ΔE Method, see Ultraviolet spectra, difference spectra
Earlywood, 49, 752
Effective alkali, kraft liquor, 640
Effective hydrodynamic radius and volume, 704, 710, 711
Einstein equation, 703-705, 726
Electrical conductivity, lignin solutions, 331, 751
Electrodialysis, lignin sulfonates, 601, 712, 842
Electron microscopy, 31, 50, 743, 697, 706, 707, 751
Electron nuclear double resonance (endor), 337, 338
Electron paramagnetic resonance (epr), 326-340
  alkali lignins, 537, 676, 677
  peracetic acid, oxidation, 458
  pulping reactions, 648
Electron spin resonance (esr), see Electron paramagnetic resonance
Electrolytic oxidation, 799, 821
Electrophilic reactions, aromatic substitution, 382
  side-chain displacement, 382, 386-391
Electrophoresis, 179, 601, 711-713
Elemental halogens, 377
Elfvingia applanata, 775
β-Elimination, p-tolylsulfones, 364
Elm, 191
Emulsifiers, lignin sulfonates, 847, 851
Endoplastic reticulum, 33
End-wise polymer, from coniferyl alcohol, 151-153
Enol ethers, formation in alkaline pulping, 654, 665
3-Enolpyruvylshikimate synthetase, 100
Enzymatic hydrolysis, wood polysaccharides, 10
Enzyme lignins, 169, 753
Enzyme system, lignification, 112
Enzymically liberated lignins, 11, 179, 180, 267, 280, 285, 712
Enzymic degradation, lignins, 313, 779
Ephedra, 63, 65
  E. gerardiana, 65
  E. procera, 63, 64
  E. trifurca, 65
Epichlorohydrin, 526, 530, 845
Epicoccum nigrum, 792, 793

SUBJECT INDEX

Epipinoresinol, 138
Episulfides, intermediates in kraft process, 364, 654, 657
  pmr, 312
Epoxidation, 525
Epoxide intermediates, in alkaline hydrolysis of ethers, 362, 498, 499, 651-654
Epoxides, oxidation by DMSO, 814
  reaction with kraft lignins, 855
Epr, see Electron paramagnetic resonance
Equisetum, 61-63
Eryngium, 19, 29
  E. vesiculosum, 30
Erythritol, conversion to lignin, 103
Erythrose-4-phosphate, 98-101
Escherichia coli, 99, 102
Ester groups, 2, 63, 75, 76, 177, 258
  effect on uv and ir spectra, 178
Esterification, 477, 527, 530, 552, 845
Ethane, pyrolysis product, 579
Ethanol, from alkali fusion, 591
  hydrogen donor, 823
  from spent sulfite liquor, 840, 841
Ethanol lignin, see Ethanolysis lignin
Ethanolamine, 540
Ethanolysis, 249, 354
Ethanolysis lignin, 11, 453, 454, 743
Ether bonds, cleavage, 354, 361, 505-507, 511, 651-658, 666
  ir, 270, 274, 280
Etherification, 477, 845
1-Ethoxy-1-guaiacyl-2-propanone, see α-Ethoxypropioguaiacone
α-Ethoxypropioguaiacone, 14, 275, 276
α-Ethoxypropiovanillone, see α-Ethoxypropioguaiacone
Ethyl acetoacetate, 534, 541
4-Ethyl-2-methoxyphenol, see Guaiacylethane
Ethylation, lignin sulfonates, 454, 630
4-Ethylcatechol, 590, 817, 818
4-Ethylcyclohexanol, from hydrogenolysis, 489, 504
Ethylene, from pyrolysis, 585
Ethylene glycol, 540
Ethylenediamine tetracetic acid, 553
Ethylene oxide, 520, 523
Ethylenic double bonds, ir, 277, 280
  uv, 252
Ethylenimine, 523, 524
4-Ethylguaiacol, see Guaiacylethane
6-Ethylguaiacol, from pyrolysis, 822
p-Ethylphenol, 130, 822
Ethylphenols, from hydrogenolysis, 820, 821
Ethylvanillin, 805
Ethylvanillyl sulfide, 458
Ethylveratryl alcohol, 389
Eucalypt lignins, acid soluble, 260
  Brauns, 177, 178
  epr, 330
  ir, 282, 288, 290
  kraft, 261, 288, 671
  mwl, 282
  neutral sulfite semichemical, 290
  pmr, 313-316, 319, 322, 323
  species variation, 292
  uv, 258, 261
Eucalyptus botryoides, 66, 72, 73

E. claeophora, 19, 25, 26, 31
E. diversicolor, 71, 72
E. gigantea, 72
E. globulus, 260
E. goniolyx, 66
E. maculata, 71, 72
E. marginata, 72, 193
E. nitens, 66, 103
E. obliqua, 72
E. polyanthemos, 193
E. regnans, 66, 72, 176, 192, 193, 222, 258, 313-323
E. sieberiana, 177, 178
Eugenol, 464, 502, 514, 799, 630
Eugetic acid, 439
Extenders, 713

Fagopyrum tartaricum, 105
Fagus, F. crenata, 279
  F. grandifolia, 66
  F. sylvatica, 30, 64, 68, 256, 257
  see also Beech
Far infrared spectroscopy, 270
Fehling's solution, 517
Fermentation, sugar in sulfite spent liquor, 840, 841
Ferric chloride, 445, 446, 630, 663
Ferricyanides, dehydrogenation, 445, 446
Ferrochrome lignin sulfonates, 847
Ferrous sulfate color reaction, catechol determination, 673-675
Ferulic acid, 229, 775, 776
  biological precursor, 106, 109, 793
  ester groups, in lignins, 10, 258
  from fungal degradation, 773
  reactions of, 137, 138, 143
Feruloylquinic acid, 108
Feruloylshikimic acid, 108
Fiberboard, 736
Fibers, cell wall organization, 23
  reactive dyes, 555
  swelling, influence of lignin, 645
Fir, see Abies
Flame retardants, 530, 531, 583, 858
  Flavan-3,4-diol, 85
Flavanones, 533
Flax, 19
Flory-Huggins theory, 744
Flory theory, polyfunctional polymerization, 758
Fluorene derivatives, 549
Fluorescence, 6, 263, 719
  determination of lignin sulfonates, 263
Fomes annosus, 773, 775
  F. fomentarius, 773-775
  F. pinicola, 329
Fomintopsis pinicola, 775
Formaldehyde, 514, 520, 536, 539, 542, 545, 546, 548, 550
  condensation, 662-664, 678, 680, 713, 799
  formation, in acidolysis, 352, 359, 360
  by base, 363, 365, 366, 351, 654, 665, 803
  in hydrogenolysis, 498, 499
  from lignin sulfonates
Formic acid, from kraft pulping, 670, 803, 818
  from oxidations, 449, 469, 471
  from pyrolysis 577, 591

reaction with phenols, 527
5-Formylvanillic acid, from nitrobenzene oxidation, 435, 439, 440
5-Formylvanilllin, from nitrobenzene oxidation, 435, 439, 440, 443, 804
Fractionation, 262, 282, 287, 601, 700, 702, 710, 711, 727-729
  hydrogenolysis lignins, 490, 494-496
  kraft lignins, 289, 671, 673, 685
  lignin sulfonates, 601, 614, 616, 617, 699, 710, 711, 751, 752, 842
Fraxinus, effect of light on lignification, 34
F. americana, 66, 71
F. excelsior, 64
Free radicals, 536,
  alkali lignin, 677
  epr, 326-340
  grafting, 552
  by oxidation, 332-339, 445, 478
  in pulping, 648
  pyrolysis, 578
  reactions, 457, 552
  scavengers, 648
  sulfonation, 629
  see also Phenoxyradicals
Fremy's salt, 119, 201, 472, 519, 659
Freudenberg structural concept, spruce lignin, 5, 226
Fucus serratus, 80
Fulvic acid, 328, 783
Fumaric acid, 458, 460, 779
Fumarylacetoacetic acid, 779
Fumarylpyruvic acid, 779
Funaria, 80
Fungal degradation, 769, 770, 772-782, 785
Fungicides, 525
Fungi imperfecti, 770
Furfural, 258, 259, 551, 713
Furfuryl alcohol, 534
Fusion, *see* Alkali fusion

Galactans, 696
Galbergin, 134
Gallic acid, 98, 104
Galvinoxyl, 120, 336, 337
Gatterman-Koch synthesis, 798, 799
Gatterman synthesis, 798, 799
Gel filtration, 700, 727-729
  kraft lignins, 685
  lignin sulfonates, 601, 614, 617, 710, 711, 751
Gelling, lignin sulfonates, 630, 713, 716-719
Gel permeation chromatography, 262, 727-729
Gentisic acid, 778, 779, 781
Geotropic response, 35
Gingko biloba, 56
Glass transitions, 645, 729-739
p-Glucocoumaryl alcohol, 28
Glucomannans, 696, 733
Glucose, C-14 labelled, 102
  conversion to humic acid, 790
  from wood saccharification, 840
Glucose-6-phosphate, 99
Glucose-6-phosphate dehydrogenase, 112
β-Glucosidase, 28, 33, 110
Glucosides, of cinnamyl alcohol precursors, 109, 110

Glucuronoxylan, 696
Glutathione, 29
Glyceraldehyde-2-aryl ethers, 358
Glyceraldehyde, displacement product, 399
Glyceric acid, 399
Glycerol, 399, 521, 531
Glycerol-β-aryl ether, 653
Glycidyl groups, 525, 536, 549
Glycol chlorohydrin, 540
Glycolytic pathway, 100
Glycosidase, in lignin isolation, 169
Gnetales, 44, 65
Gnetum indicum, 65
Golgi apparatus, 33
Graft copolymerization, 338, 555, 556
Graminaceae, 76
Grass lignins, 44, 74-79
  ester groups, 258
  mwl, 258, 284
  nitrobenzene oxidation, 436
  reactions, 469, 588
Griffithi excelsa, 55
Grignard reagents, 532
Grinding aids, 848, 851
Groundwood, 833
Guaiacol, 398, 402, 463, 552, 576, 664, 774, 817, 818
  dehydrogenation, 128
  from, acidolysis of models, 355, 357
    hydrogenolysis, 490-493, 503, 54
    kraft and soda process, 363, 368, 671, 834
    oxidations, 435, 447
    pyrolysis, 577, 581, 591, 822
  reactions, 385, 458, 460, 718, 798, 799, 808
Guaiacylacetic acid, from hydrogenolysis, 491
Guaiacylacetone, from water hydrolysis, 348
  hydrogenolysis, 502
  ir, 275
  condensations, 655, 656, 664
Guaiacylethane, from hydrogenolysis, 487, 489, 491, 492, 498-503
  from pyrolysis, 577, 822
1-Guaiacylethanol, *see* Apocynol
2-Guaiacylethanol, 502, 818
Guaiacylethylcarbinol, *see* 1-Guaiacylpropanol
Guaiacylglycerol, 200, 356, 362, 459, 460, 773-776
  hydrogenolysis, 497, 498
  in hydrolysis products, 199
  ir, 266
  optically active, 229
  peracetic acid oxidation, 459, 460
  α-sulfonates, 626
Guaiacylglycerol-β-coniferyl ether, *see* Lignols, di-, β-0-4
Guaiacylglycerol-β-coniferyl alcohol, *see* Lignols, di-, β-0-4
Guaiacylglycerol-β-guaiacyl ether, *see* Guaiacylglycerol-β-(2-methoxyphenyl) ether
Guaiacylglycerol-β-(2-methoxyphenyl)-ether 302-306, 357, 361, 464, 654, 664, 774, 781
Guaiacyl group, definition, 12-14
Guaiacyl lignins, 43-63
  in hardwoods, 68, 494, 613
Guaiacylmethane, 474, 578, 822
  chlorination, 396-398, 660

SUBJECT INDEX

from hydrogenolysis, 503
nitration, 391, 392, 402
α-sulfonic acid, 14, 396, 611
Guaiacylmethylcarbinol, see Apocynol
1-Guaiacylpropane, 12, 277, 506, 515, 577
  dehydrogenation of, 131, 132
  from hydrogenolysis, 490-493, 498-503
  uv, 247, 255, 257
1-Guaiacylpropanediols, 441, 460
1-Guaiacyl-*n*-propanol, 387, 388, 401-404, 505, 506, 518
1-Guaiacyl-2-propanol, 505, 506, 521
3-Guaiacylpropanol, 397, 489, 491, 492, 498-502, 505, 506, 655, 818
Guaiacylpropanones, 289, 442
β-Guaiacylpropionic acid, 491, 818
Guaiacylpyruvic acid, 776, 777, 780, 781
Guaiacyl-syringyl lignins, 43, 44, 63-79, 280, 613
Guaiacyl-syringyl ratio, 69, 280
Gum, kraft lignin, 671-673
Gymnosperm lignins, 4, 5, 43-63, 68, 284, 288, 436, 494, 551, 613, 696-698

Hadromal, 196
Halogenation, agents, 375, 377, 379
  reactions, 398, 400, 409, 809-811
Halogenium ions, 378, 379
Halogens, hydrolysis constants, 375
  radicals, 379
  solvent complexes, 386
Halolignins, 404-407, 416, 417
α-Halothioethers, 810
Haloxylon ammodendron, 64
Halse lignin, 288
Hammett δ-values, *p*-hydroxyl and -methoxyl, 385
Haplographium sp., 781
Hardwood lignins, 697, 698
  ir, 270, 278, 279
  kraft and soda, 664, 674, 685, 700, 701, 839, 840
  molecular weights, 685, 700, 701
  pmr, 313-319, 324
  reactions, 393, 412, 436, 461, 469, 613
  uv, 248
  see also under individual species
Heartwood, 752
Helianthus, 36
  H. tuberosus, 64
Hemicelluloses, 642, 644, 645, 695, 696, 733, 741, 749
  electrodialysis, 712
  impurities in mwl, 284
Hemilignin, 615, 752
Hemlock, 108, 229, 284, 626
  sulfonation, 610, 620, 699-702
Heracleum sibiricum, 64
2,3-*bis*-(Hexahydrobenzyl) butane, 505
Hexahydrophthalic anhydride, 529
Hexalignols, 142
Hexamethylene diisocyanate, 521
Hexamethylene tetramine, 548, 549
Hibberts ketones, 8, 9, 54, 56, 60, 70, 86, 177, 206, 207, 350, 355, 541, 626
  chromatography, 207
  radioactive, 95, 97

Highly shielded protons, pmr, 321, 325, 494, 495, 674
High yield pulps, 600, 640
Hildebrand's solubility parameter, 687, 741-744
Histochemical tests, 6
Holmberg lignins, see Lignin thioglycolic acids
Holocellulose, preparation, 457, 463
Homocatechol, 822
Homogentisic acid, 778-781
Homolytic processes, 368, 648
Homoprotocatechualdehyde, 817, 818
Homoprotocatechuic acid, 590
Homovanillic acid, 655, 818
Homovanillin, 665
Homovanillyl alcohol, 655
Hormodendrum sp., 781
Horse-chestnut, 179
Horseradish peroxidase, 111, 116, 770
Howard process, 800, 842, 843
Huggins plots, 708, 709
Humic acid, 783-792
  ir, 290, 291
  spin content, pmr, 328
Humification, 785-793
Humin, 783
Humus, 771, 772, 781-793
Hydriodic acid, 512, 520, 533
Hydrochloric acid lignins, 11, 170, 186, 288, 328, 449, 453, 454, 470, 822
Hydrocracking, 818, 819
Hydrodynamic radius and specific volume, 726, 727, 729
Hydrofluoric acid, 697
Hydroformylation, 502
Hydrogenation, catalytic, 251, 272, 488, 489, 501, 659, 819
Hydrogen balance, pmr, 320-323
Hydrogen bonding capacity, 687, 741-744
Hydrogen bonds, internal, in lignins, 5
  o,o'-diphenols, uv, 255
Hydrogen cyanide, from lignins, 584
Hydrogenolysis, 487-507, 535, 818-821
  low molecular weight products, 8, 9, 311, 312, 493
Hydrogenolysis lignins, formation and characterization, 11, 487-507
  pmr, 318, 322-324, 494-496
Hydrogen peroxide oxidation, 455, 457, 468, 469, 805, 821
  alkaline, 473, 474, 477
  in bleaching, 477
  of dimethylsulfide, 812
  enzyme-catalyzed, 111-112, 119, 788
  with metal oxides, 799
Hydrogen sulfide ions, 552, 641, 662, 669
Hydrol lignins, see Hydrogenolysis lignins
Hydrolysis, acid, see Solvolysis
  alkaline, see Alkaline hydrolysis
  effect on hydrogenation, 488, 504
  of ether linkages, 651, 652, 674
  of lignin sulfonates, 598, 610, 620, 621, 630
  by steam, products, 348, 349
Hydrolysis lignins, from wood saccharification, 837, 840
  reactions, 819, 823
  Rheinau process, 526, 840
  spin content, 328

utilization, 837, 858, 859
Hydroperoxides, from phenols, 119
Hydroquinones, 458, 788
Hydrosulfite, reduction by, 331
Hydrotropic lignins, 328, 705, 706, 710
Hydroxide ion, nucleophilicity, 808
Hydroxyacetophenone, p-, uv, ir, 243, 276
  m-, uv, 243
p-Hydroxyazobenzene, 436
p-Hydroxybenzaldehyde, 388, 435, 447, 804
p-Hydroxybenzoic acid, 99, 258, 781, 803
  alkali fusion product, 589
  halogenation, 388
  humic acids, 785
  hydrogenolysis product, 491, 503
  ir, 275, 284
  kraft liquor, 817, 818
  lignin decomposition by white rot, 773
p-Hydroxybenzyl alcohol, 388, 540
p-Hydroxybenzyl alcohol units, 202, 531, 603, 612, 662
α-Hydroxy-δ-carboxymuconic semialdehyde, 779
p-Hydroxycinnamaldehyde, see p-Coumaraldehyde
p-Hydroxycinnamic acid, see p-Coumaric acid
β-Hydroxyconiferyl alcohol, 776, 777
  dehydrogenation, 788
  intermediate in acidolysis, 355, 357
  ir, 275, 276
  isolation from bark, 87
3-(4-Hydroxycyclohexyl)-1-propanol, from hydrogenolysis, 503
1-Hydroxycyclopentanecarboxylic acid, 587, 590
1-(4-Hydroxy-3,4-dimethoxyphenyl)-n-propane, see Syringylpropane
4-Hydroxy-3-ethoxybenzaldehyde, 805
5-Hydroxyferulic acid, 106
ω-Hydroxyguaiacylacetone, 206, 350; see also β-Hydroxyconiferyl alcohol
Hydroxylamine, 532
  in carbonyl determination, 673, 677, 678
Hydroxylation, α-, guaiacylpropane, 131
  aromatic, by peracetic acid, 458, 462
  by enzymes, 106, 108, 770, 778
  of phenoxonium ions, 118, 119
Hydroxyl content, 673, 675
  aromatic and aliphatic, pmr, 325, 326, 496, 675
  increase in chlorination, 413
Hydroxyl groups, α-, 440, 650
  ir, 268-270, 273, 280, 287
  phenolic, uv, 248
  protons, delta values, 495, 496
  reactions, 519
Hydroxymatairesinol, 229
2-Hydroxy-6-methoxy-p-benzoquinone, 339
1-(4-Hydroxy-3-methoxyphenyl)-2-propanol, 514; see also 1-Guaiacyl-2-propanol
(4-Hydroxy-3-methoxyphenyl)pyruvic acid, 773
trans-4-Hydroxy-3-methoxystyrene-ω-sulfonic acid, from neutral sulfonation, 364
Hydroxymethylfurfural, 259, 580
Hydroxymethyl groups, 546, 547, 663-665
  displacement as formaldehyde, 122, 123

β-Hydroxymethylmuconic acid lactone, 471
4-Hydroxy-3-nitrobenzaldehyde, 419
4-Hydroxy-3-nitrobenzoic acid, 419
2-Hydroxy-5-methyl-3-nitro-1,4-benzoquinone, 397
p-Hydroxyphenylacetic acid, from hydrogenolysis, 491
p-Hydroxyphenyl groups, uv, 243
p-Hydroxyphenyl-ω-hydroxyacetone, acidolysis product, 357
β-(p-Hydroxyphenyl)lactic acid, lignin precursor, 105
4-Hydroxyphenylpropane, 130, 277, 820, 821
p-Hydroxyphenylpropane units, in beech lignin, 454
p-Hydroxyphenylpyruvic acid, intermediate, 105, 778, 780
α-Hydroxypropioguaiacone, 274, 276, 460
ω-Hydroxypropioguaiacone, hydrogenolysis, 498
p-Hydroxypropiophenone, ir, 276
α-Hydroxypropiovanillone, see α-Hydroxypropioguaiacone
Hydroxyquinones, epr, 332
Hydroxysuccinic acid, oxidation product, 474, 477
Hymatomelanic acid, 783
Hypochlorite, chlorination agent, 385
  oxidation, 433, 457, 463-466, 821
  pulp bleaching, 463, 464
Hypohalous acids, 377, 378
  dissociation constants, 375
  kinetics of halogenation, 378
Hysteresis, sorption isotherm, 745, 746

Imidazolides, 805
Imperata cylindrica, 75
Indican test, for β-glucosidase, 28
Indole, 354
3-Indole acetic acid (IAA), 35
Indophenol reaction, p-hydroxybenzyl alcohol groups, 202
Indophenols, 578
Indoxyl, 534
Indulin, a kraft lignin product, 328, 688, 742, 839
Infrared spectroscopy, 57, 60, 267-293, 523, 537, 552
  bark lignin, 290, 291
  Brauns lignins, 177
  dioxane acidolysis lignins, 286, 494
  humic acids, 290, 291
  kraft lignins, 671, 681, 682
  multiple internal reflectance, 292, 671
  mwl, 69, 70, 671, 681, 682
  near and far, 269, 270
  nitrolignins, 418
  peracetic acid lignins, 461
  polymers related to lignin, 83, 84, 281-283
  wood, 7
Insecticides, 525, 850
Interfacial tension, lignin sulfonates, 844, 845
Interference microscopy, 24
Intrinsic viscosity, 552, 703-705
  effect of ionic strength, 708, 709
  kraft lignins, 686
  lignin sulfonates, 713, 714

SUBJECT INDEX 905

molecular weights, determination from, 723-726
  see also Viscometry
Iodination, 407, 409, 529, 549
5-Iodovanillin, 407, 549
Ion exchange resin, from lignins, 525, 547, 553, 713
Ion exclusion, lignin sulfonates, 601, 842
Ions, sorption by lignins, 746-748
Iron, chelates, 586
  salts of lignin sulfonates, 843, 857
Isoetes, 61, 63
Isoeugenol, *cis*- and *trans*-isomers, 132-136
  difference spectra, uv, 250, 251
  fungal attack, 776
  hydrogenolysis, 502
  methyl ether, reactions, 441, 464
  nitration, 390
  oxidation, 434, 464, 441, 800
  oxidative coupling, 132-136, 445, 446
  δ-sulfonic acid, 14
Isohemipinic acid, 203, 213, 452, 453, 541, 659
  methyl ester, 455
Isolation, of lignins, 165
  of kraft lignins, 670, 671
Isoolivil, nitrobenzene oxidation, 440
Isotope effects, in halogenation 378
  in nitrosation, 383
Isovanillic acid, 781

Jack-in-the-box, retention mechanism, 538
Juglans cinerea, 71
  J. nigra, 71
  J. regia, 64
Juncus bufonius, 64
Juniperus, 45
  J. communis, 56, 58
Jute lignin, alkali fusion, 587

α-Ketoacids, 474
β-Ketoadipic acid, 779
Ketols, condensations, 550
  end-groups, 200
  formation, 665
  sulfonation, 626
Ketones, ir, 275, 281
  permanganate oxidation, 475
  reaction with lignin, 576
  uv, 250-252
Kinase, 100
Kinetin, 35
Kino, 192
Kiri, 178
Klason lignins, 7, 8, 11
  condensation with polyphenols and proteins, 8, 193, 194
  content in herbaceous plants, 74-79, 193
  content in woods, 45-49, 65, 66, 88, 193
  effect of photooxodation, 478
  epr, 328
  ir, 288
  permanganate oxidation, 470
  in pulps, 194
  pyrolysis, 583, 823
  soluble, 259
K-number, 195
Kraft lignins, 11, 639-689, 839
  analytical composition, 289, 671-673
  chromophoric groups, 649
  condensation reactions, 662-664, 713, 855
  condensed units, 324
  demethylation, 411, 855
  diffusion, 750
  epr, 328, 332
  fractionation, 671, 727, 728
  functional groups, 250, 673-679, 840
  gel filtration, 727, 728
  ir, 271, 288, 289, 681, 682
  isolation, 670, 671, 743, 834, 838-840
  from milled wood, 253, 254, 652
  modification, 845, 846, 855
  molecular weights, 643, 659, 680, 727, 728, 744, 839, 840
  optical properties, 683, 684
  oxidation, 803, 855
  pmr, 322-325, 674
  sodium salts, 677, 851
  solvents for, 742, 839
  structural representation, 679-681
  sulfonation, 846
  swelling, 645
  utilization, 834, 835, 848, 850, 855, 857
  uv, 250, 682-684
  viscosity, 685, 686, 703-705
Kraft liquor, 642, 643, 670, 834
Kraft mill odor, 669, 806
Kraft pulping, 639-689
  alkali consumption in, 647
  by-products, 833
  kinetics of, 647, 648
  vs. soda pulping, 642, 665, 666
  topochemistry, 644, 753-757
  yields, softwoods, 642
Kraft pulps, 413, 644
Krebs cycle, 774
Kubelka-Munk reflectance functions, 683
Kullgren acid, 600-602, 624

Lignin building units, 165
Lignin-carbohydrate bonds, 23, 218-224, 749
Lignin-carbohydrate complexes, 168, 220-224, 712
Lignin coke, 575, 576
Lignin content, of bark, 88
  conifers, 45-49, 52, 53
  grasses, 74-79
  of hardwoods, 65, 66
  by ir, 292
  by uv, 258-262, 697
  variation of, in *Poinsettia*, 38
  see also Klason lignin
Lignin distribution, in conifer cell walls, 50-53, 697
  in hardwood cell walls, 66-68
Lignin *in situ*, 11, 267, 748-761; see also Protolignin
Lignin macromolecules, electron microscopy, 706-707
Lignin-phenolformaldehyde resins, 542
Lignin products, 833-846
Lignin skeletons, of cell-walls, 643
Lignin sulfonates, 519, 544, 575-633, 842-845, 850
  α and β, 598, 615, 618

SUBJECT INDEX

alkali fusion of, 587-592, 821
alkaline hydrolysis, 624, 625, 711, 800
analytical composition, 629
chlorination, 261, 405, 406, 409, 417
determination in SSL, 262
diffusion, 643, 722-726, 750
dispersing properties, 713-716, 849, 850
electrical conductivity, 751
electron micrograph of 707
electrophoresis, 711-713
epr, 328, 331
functional groups, 614-631
gel filtration 262, 601-614, 617, 710, 711, 751, 752
gelling, 630, 716-718
by Howard process, 800, 842, 843
insoluble, 602, 613
intrinsic viscosity, 703, 705, 706
ir, 270, 271, 289-290
isolation, 262, 538, 598, 601, 602, 620-625, 699, 712
Kullgren lignin, 602
modifications, 840-845
molecular weights, 609, 613-618, 699-701, 707, 711, 713-716, 719, 750-752
oxidations, 415, 446, 453, 454, 466, 800-806
polydispersity, 701
polyelectrolyte behavior, 708-713
precipitation with amines and salts, 598
products from, 601, 713-716, 835-838, 843, 847-855
pyrolysis, 580, 585
softening temperature, 733
solvents for, 837, 838
structure, 607, 708, 709
uv, 245, 247, 261, 406, 414, 417
vanillin from, 800-806
Lignin sulfonic acid, free, 843
Lignin thioglycolic acids, 194, 543, 544
composition, 56, 61, 64, 74
formation, 351
in lignin identification, 7
pmr, 322-325
Lignites, 847
Lignols, di-,
β-β linked, 138-141, 143, 214, 305, 306; see also Pinoresinol
β-0-4 linked, 14, 87, 138, 140, 197, 200, 205, 304-306, 308, 773, 774
β-1 linked, 146, 147, 209, 351, 352, 365
β-5 linked, 138, 141, 197, 211, 229, 306, 307, 351, 352
1-0-4 linked, 211
4-0-5 linked, 216
5-5 linked, 140, 216
stereochemistry, 144
oligo-, 205, 209
from coniferyl alcohol, 141
formation in hydrolysis, 154
penta-, 141
tetra-, 141, 308, 309
tri-, 141, 146, 308
Lignolytic organisms, 770, 786
Lignosulfonates, see Lignin sulfonates
Linkages in lignins, α-1, from alkaline pulping, 649, 680

α-5, 204, 649, 651, 660, 665, 680
β-β, 144, 150, 213-215, 443, 495, 496, 605, 673, 676, 680
β-0-4, 150, 151, 205-208, 505, 782
acid hydrolysis, 354-359
alkaline hydrolysis, 207, 653-657, 680
cleavage by Na/NH$_3$, 506
hydrogenolysis, 499
oxidation, 442, 443
pmr, 495, 496
β-1, 208-211, 308, 448
acid hydrolysis, 209, 352, 353
alkaline hydrolysis, 209, 253, 666
hydrogenation, 209, 500
oxidation, 443
β-5, 144, 150, 211-213, 443, 495, 496
β-6, 218
1-0-4, 150, 211, 443
4-0-5, 150, 204, 216, 217, 365, 443, 444, 680
5-CH$_2$-5, kraft lignins, 680
1-5, 150, 211
5-5, 150, 204, 216, 217, 308, 443, 447, 500, 502, 680
5-6, 218
Liquefaction of lignins and wood, 487, 504, 819
Liriodendron tulipifera, 71
Lithium aluminum hydride reductions, 254, 287, 675, 681
Lithium hydroxide, 674, 746-748
Lithium metaborate, 674
Loblolly pine, 176, 671-673
Longleaf pine, 176, 278, 286
Low lignin conifers, 44
Low molecular weight chemicals from lignins, 797-824
Low molecular weight LSA, 749
Low-sulfonated lignin, 600-602, 618
Lumina, 642, 643
Lycopodium, 61-63
  L. annoticum, 56
  L. clavatum, 62, 63
  see also Clubmosses
Lysine, 841

Macromolecular properties, 695-761
Magnesium base sulfite pulping, 598, 599
Magnesium lignin sulfonate, 843
Magnoliaceae, 71
Magnolia soulangeana, 64
Maize, mutant lignification, 34
Maleic acid, 458, 460, 462, 474, 477
Maleic anhydride, 529
Maleylacetoacetic acid, 781
Malonic acid, 469, 477, 479
Manganese dioxide, 445, 446, 552
Mannich reactions, 548, 662, 678, 851
Maple, 178, 180, 193
  Brauns lignin, 282, 712
  enzyme-liberated lignin, 712
  hydrogenolysis lignins, 494
  lignin precursors, 99, 105
  sulfonation, 609, 610
  yield of Hibberts ketones, 206
  see also Acer
Marchantia, 80

SUBJECT INDEX

Mark-Houwink equation, 704
Matairesinol, 440
Matteuchia atruthiopteris, 62
Mäule color reaction, 6, 31, 43, 63, 74, 82, 86, 535
Melene, 580
Melting, of lignins, 730
Mercaptide ions, 351, 806-809
Mercuration, 589
Mercuric oxide, 444, 446, 449, 801
Merulius lacrymans, 775
Metahemipinic acid, 203, 214, 218, 453-455, 551
Methane, 515, 579, 585, 822
Methane sulfonic acid, 365, 630, 810
Methanol, by chlorination of models, 394
  by hydrogenolysis, 503
  in kraft liquor, 670, 818
  from lignins, 462, 512, 576, 591, 821
  oxidation by chlorine, 410
Methanolysis, 350, 539, 675
Methionine, 817
α-Methoxyacetophenone, ir, 276
Methoxyanisole, *m* and *p*, 506
Methoxybenzene, *see* Anisole
Methoxybenzene-2,4,6-tricarboxylic acid, 218, 454
Methoxybenzoic acids, *o, m* and *p*, 781
2-Methoxy-1,4-benzoquinone, 460, 474, 790
3-Methoxy-1,2-benzoquinone, 518
m-Methoxycinnamic acid, 106
3-Methoxycyclohex-2-ene-1-carboxylic acid, 516
2-Methoxy-4-ethylcyclohexanol, 504
Methoxyhomogentisic acid, 776, 777, 780
2-Methoxyhydroquinone, 459, 460, 479
3-Methoxy-4-hydroxyphenylpyruvic acid, 776, 788
4-Methoxyisophthalic acid, 218
6-Methoxyisophthalic acid, 454
Methoxyl content, compression wood lignin, 47, 59, 60
  degraded straw lignins, 786
  kraft lignin, 672, 839
  hardwood lignins, 64, 68
  immature conifer xylem lignin, 48
  lignin sulfonates, 620
  normal conifer wood lignin, 55, 56
Methoxyl groups, aromatic, ir, 279; pmr, 310
  cleavage in alkali, 365, 512, 669, 670, 673
  in grafting, 558
  hydrolysis by electrophiles, 393, 420, 461, 512
  methanol and methyl ester formation, 462
  reduction by Na/NH$_3$, 505-507
Methoxyl protons, pmr, 305, 308, 309, 310, 495, 496, 674
o-Methoxymethylanisole, 516
1-Methoxymethyl-7-naphthol, 536
2-Methoxy-4-methylphenol, *see* Guaiacylmethane
Methoxyphenols, 516
o-Methoxy phenols, periodate oxidation, 678
2-(*o*-Methoxyphenoxy)ethanol, 781, 782
o-Methoxyphenoxyacetaldehyde, 781, 782
4-Methoxyphthalic acid, 218
2-Methoxy-4-*n*-propylphenol, *see* 1-Guaiacylpropane
Methoxysuccinic acid, 477
Methoxytrimesic acid, 454
Methylamine, 512
Methylaniline, 808
Methylanisoles, 500, 516
Methylated lignins, 272, 274, 314, 331, 455, 549, 598, 653, 679, 751
Methylated wood, 452-454
Methylation, biological, 106
  with diazomethane, 314, 547, 676
  effect on reactivity, 338, 438, 619
  functional group determinations by, 673, 676
  with methanolic hydrogen chloride, 350, 539, 675
4-Methyl-1,2-benzoquinone, 397, 402, 479, 790
Methyl bromide, from vanillin, 805
4-Methylcatechol, 504, 590, 790, 817, 818
Methyl cellulose, 712
Methyl-*p*-coumarate, uv, 258
4-Methylcyclohexanol, 504
1-Methylcyclopenten-2-ol-3-one, 817, 818
Methyl-3,4-diethoxybenzoate, 454, 554, 630
4-Methyl-1,2-dimethoxybenzene, 462
Methylene bridges, in kraft lignins, 680
Methylene groups, ir, 274
Methylene iodide, 525
Methyl 4-ethoxy-3-methoxybenzoate, 454, 630
Methylfumaric and -maleic acids, 398, 474
Methylglyoxal, 358, 359
Methyl groups, ir, 274
  formation in acidolysis, 207, 354
4-Methylguaiacol, *see* Guaiacylmethane
Methyl *p*-hydroxybenzoate, uv, 258
Methyl mercaptan, 365, 579, 591, 641, 669, 806-809, 817
Methyl methacrylate, 556
Methyl nitrite, 391, 420
Methylol groups, *see* Hydroxymethyl groups
Methylphenols, *see* Cresols
N-Methylpiperidine, 512
5-Methylresorcinol, 532, 542
Methyl sulfide, *see* Dimethyl sulfide
Methylsulfinylmethyldiphenylcarbinol, 813
p-Methylthiophenol, 814
o-Methyl transferase, in bamboos, 115
Michael condensation, 514
Microbiological degradation of lignins, 769-793
Microfibrils, orientation in cell wall, 23
Middle lamella 696
  delignification, 612, 613, 643, 644, 753-758
  lignins from, 23, 24, 51, 613, 643
  sulfonation, 613
Milled wood lignins, (MWL), 10, 54, 68, 167, 173, 226, 683, 698
  borohydride reduction, 250, 652
  condensed units in, 452
  coniferaldehyde units in, 250
  epr, 328
  functional groups, 224-228, 250, 252, 256, 488, 673-679, 786, 787
  fungal decomposition, 771, 773, 774
  hydrogen balance, 322, 323
  hydrogenolysis, 251, 252, 500-502

hydrolysis, 362, 653
  ir, 175, 267, 271, 280, 284-288, 292
  kraft and soda treatment, 666, 667
  methanolysis, 351
  molecular weight, 175, 729
  permanganate oxidation, 454
  pmr, 314, 315, 320-325, 674, 675
  preparation, 166-168
  pyrolysis of, 583
  residual carbohydrates in, 173
  sulfonation, 601, 608, 621-623
  uv, 175, 248-257, 682
  vanillin yields from, 174
Milling, 171-175, 328
Miscanthus condensatus, 75
  M. sacchariflogus, 75
  M. sinensis, 75
  M. tinctonius, 75
Model compounds, halogenation and nitration, 381
  hydrogenolysis, 497-502
  hydrolysis of ether bonds, 346, 650, 653-658
  nitrobenzene oxidations, 434-444
  peracetic acid oxidations of, 458-460
  sulfonation, 603, 604, 611, 612, 617, 619
  see also Lignols, and names of particular model compounds
Molecular weights, determination of, 616, 643, 685, 710, 719-729
  in dichromate gelling, 552, 718, 719
  by diffusion, 616
  distribution of, 685, 698-729, 727-729
  by gel filtration, 727-729
  halolignins, 405
  hydrogenolysis lignins, 494
  kraft and/or soda lignins, 659, 664, 680, 685, 698-701, 750, 839, 840
  lignin sulfonates, 609, 613-618, 713-716, 750-752
  milled wood lignin, 175, 729
  solvent effects, 751
Monimiaceae, 71
Monocotyledon lignins, 284, 436
Monotropa hypopitys, 64
Monterey pine, 26, 29, 193, 261, 313, 314, 319, 322, 323
Mosses, 21, 80, 81, 287
Muconic acid structures, 458, 462, 474, 719
Multiple-effect evaporators, 670
Multiple internal reflectance, ir, 292, 682
Musci, 21; *see also* Mosses
Mutant corn plants, 73

Naphthalene diisocyanate, 521
2-Naphthol, 536, 537
Naphthylamines, 550
β-Naphthyl methyl ether, 458
Native lignin, 11; *see also* Brauns lignin
Network concept of lignins, 698, 701, 702, 741-743, 748
Neurospora, 99
Neutral sulfite pulping, 261, 290, 602, 619, 755, 756
Nickel catalysts, 503, 579, 819, 823
Nitration, 373-422, 545, 859
  agents, 379-381

demethylation in, 392, 395, 419
  of *p*-hydroxybenzyl alcohol groups, 202
  of kraft lignins, 846
  oxidations in, 433, 457, 812, 821
  pulping by nitric acid, 373
  of side-chains, 388-390, 398, 400
m-Nitrobenzaldehyde, 532
Nitrobenzene oxidation, 57-60, 65, 69-71, 73-75, 77, 83, 433-444, 540-543, 551, 557, 799
  aldehyde ratios H/V and S/V, 54
  determination of condensed units, 659
  of humic acids, 785
  of isolated lignins, 177, 185-188, 490, 613
  of 5-5 linked lignols, 216
  optimum conditions for, 434, 435
  of plant tissues, 8, 9, 34, 772
  of radioactive lignins, 95, 96
  in vanillin production, 801
3-Nitrobenzoic acid, 419
2-Nitrobutanol, 540
Nitrodisulfonic acid, oxidation, 433
Nitrogen content, of humic acid, 792
Nitrogen oxides, 380, 811
4-Nitroguaiacol, 420
Nitrolignins, 11, 418-422
Nitromethane, condensation with vanillin, 805
Nitronium ion, 380
Nitrosation, 380, 388-390, 402, 420
  application in lignin determination, 262
Nitrosobenzene, 435
p-Nitrosodimethylaniline, 519, 799
6-Nitrosyringic acid, 419
5-Nitrovanillic acid, 419
5-Nitrovanillin, 419
4-Nitroveratrole, 400
Nomenclature, 10-14
Non-condensed aromatic units, 320
Non-conjugated carbonyl groups, 250-251, 488, 673, 677, 678
Non-dialyzable lignin sulfonates, 620-622, 624, 625
Non-phenolic aryl ethers, alkaline hydrolysis, 362, 651
Non-vascular plants, 80
Normal wood, 47, 752
Norway spruce, 102, 103, 178, 195, 256, 257, 262, 318, 435
Nuclear magnetic reasonance, (nmr), ^{1}H, *see* Proton magnetic resonance
  ^{2}H, ^{14}C, ^{33}S, 326
Nucleophilic displacements, 345, 808
Number average molecular weights, 685, 719
Nyssa aquatica, 86

Oak, 178, 180
  Brauns lignin, 282, 712
  hydrolysis by water, 346
  kraft lignin, 671-673
  lignins, ir, 269, 287; epr, 330
  milled wood lignin, 500, 501
Oat plants, lignin content, 77
Obl gate intermediates, biogenesis, 99
Oil well additives, 844-849
Oligomeric degradation products, 359
Olivil, 230, 440, 441
One-carbon pool, 102 109

SUBJECT INDEX

One-electron transfer oxidants, 117, 332, 444
Onosis repens, 64
Ontario Paper Company, vanillin method, 802
Open phenylcoumarane structures, 213
Opposite wood, 752
Orchis mascula, 64
Orcinol, 532, 542; see also 5-Methylresorcinol
Ore binders and flotation agents, 851-853
"Organosolv" lignins, 540, 541, 719, 725, 742
Orsellinic acid, 792
Ortho-para, ratio, halogenation and nitration, 380, 383-386
Osmometry, 719
Osmundaceae, 62
Oxalacetic acid, 474, 477
Oxalic acid, 398, 401, 460, 462, 469, 471, 474, 477, 586, 591, 803, 821
Oxamine blue stain, 37
Oxidases, in cambium, 112
Oxidation, 433-418, 821, 846, 847
  by air, 116, 119, 275, 685, 800, 801
  aromatic ring cleavage in, 457-474
  catalyzed by phenoloxidases, 111-112, 119, 333-336, 770
  in chlorination, 413, 535
  by chlorine dioxide and chlorites, 463
  of dimethyl sulfide, 811-814
  with Fremy's salt, 119, 201, 472, 519, 659
  in halogenation and nitration, 382, 390, 396
  by hypochlorite, 433, 457, 463-466, 821
  by molecular oxygen, 449-452, 586, 856
  by one electron oxidants, 118, 119, 332, 444
  by ozone, 457, 471, 799, 803, 821
  by peracetic acid, 338, 433, 457-463, 477
  by per benzoic acid, 460, 677
  by periodate, 201, 289, 433, 472, 518, 617, 683
  see also Hydrogen peroxide; Nitrobenzene; and Permanganate oxidation
Oxidation potential, of phenols, 124
Oxidative coupling, 117-155, 554
Oximes, 531, 532, 535
Oxiranes, 518
β-Oxyconiferyl alcohol, see β-Hydroxyconiferyl alcohol
Ozone, 457, 471, 799, 821, 903

Pacific silver fir, 278, 286; see also Amabilis fir
Palladium catalysts, 488, 579, 823
Pandanus tectorius, 75
Pavlownia imperialis, 71
  P. tomentosa, 73, 178, 279
Pearl-Benson method, 262
peat, 291, 789
Penicillum sp., 781, 790, 791
Pentamethylphenol, 580
Pentose phosphate pathway, 100, 112
Pentoses, conversion to lignin, 103
Peracetic acid oxidation, 338, 433, 457-463, 477
Perbenzoic acid oxidation, 460, 677
Periodate lignins, 181, 183, 700
  benzene polycarbolic acids from, 470
  dilatometry, 740, 741
  ir, 288

softening temperature, 731-733
sorption, of vapors and ions, 745-748
Periodate oxidation, 289, 433
  of phenolic groups, 472, 617, 683
  phenolic hyfroxyl determination by, 201, 472, 617
  of side-chain glycols, 518
Permanganate number, of pulps, 195
Permanganate oxidation, 216, 433, 452-457, 469, 551, 554, 803
  in acidic solution, 457, 471
  of ethylated LSA, 630
  of methylated wood and lignins, 203, 469, 659, 679
Peroxidases, 29-36, 111, 122, 332, 770, 788
Persulfate, 119, 445, 446
Pesticides, 525, 578, 850
Phenol, 516, 554
  by hydrogenolysis, 504, 820, 821
  as pasting oil, 819-821
Phenol dehydrogenases, 28
Phenol-fromaldehyde resins, 546, 855
Phenolic end groups, 201-204, 616
Phenolic hydroxyl groups, activating effect on reactivity, 384, 385, 390, 441
  determination, 201, 248, 617, 618, 672, 774
  formation, by acidolysis, 207, 249, 354-358, 360-365
    by alkaline hydrolysis, 362, 417, 624, 625, 650, 653, 817-818
    in halogenation and nitration, 391, 395, 413
  ir, 272-274
  in lignins, 247-249, 598, 613, 616-619, 672, 675, 774
  pmr, 325
  uv, 246-248
Phenolic units, in alkaline pulping, 650, 651, 666
  condensed and uncondensed, 201-204
Phenols, color reactions with lignin, 532
  condensation with lignins, 288, 845, 854
  oxidation potential of, 124
o-Phenolsulfonic acid, 542
Phenoxonium ions, 117, 398
4-Phenoxyphenol, 125
Phenoxy radicals, coupling, 117-155, 450, 455, 781, 782
  disproportionation, 121
  epr, 333-336
  formation, 117-119, 444
  intermediate complex, in coupling, 144
Phenylalanine, as presursor, 34, 104, 105, 113-115
Phenylcoumarans and -elements in lignin, 515, 541, 551, 650, 651
  acidolysis reactions, 350, 360, 514, 618
  alkaline thermal degradation, 361, 368, 369, 652
  as extractive compounds, 229
  hydrogenolysis, 500
  in kraft pulping, 253, 254, 666, 673, 676
  nitration, 398
  open, 204, 213
  phenylcoumarones from, 618, 651
  pmr, 308
  stilbenes from, 618, 651, 680

sulfonation, 514, 605, 618, 623, 624
uv, 253, 256
see also Lignols, -di, β-5; Linkages in lignins, β-5
Phenylenediamines, 523, 677
Phenylglyoxylic acids, 454-456
Phenylhydrazones, 532
Phenylhydroxylamine, 435
Phenyl isocyanate, 521
β-Phenylglyceric acids as precursors, 104
Phenylhydracrylic acid precursor, 104
β-Phenyllactic acid precursor, 104
Phenylosazone, 532
Phenylpropane units, notation of carbons in, 11, 12
Phleum pratense, 19
Phlobaphene, 85
Phloroglucinol, 530, 532, 534, 543
Phloroglucinol-hydrochloric acid, color reaction, 6, 19, 20, 31, 277, 773
Phloroglucinol trimethylether, 533, 543
Phormium tenax, 19
Phosphoenol pyruvate, 98-101
6-Phosphogluconate dehydrogenase, 112
Phosphorus acid derivatives, of lignin, 530, 531
5-Phosphoshikimic acid, 100
Photoxidation, 477, 478
Phragmites communis, 64, 75, 282
Phthalic anhydride, 526, 529
Phylloclauus romboidalis, 58
Phylloglossum drummondi, 63
Phyllostachys heterocycla, 77
   P. nigra, 77
   P. pubescens, 77, 108, 113, 114
   P. reticulata, 108, 113, 114
Picea, 24, 44
   P. abies (P. excelsa), 58, 102, 103, 178, 256, 257, 262, 318; see also Norway spruce
   P. excelsa, 44, 49, 55, 56, 59, 195
   P. glauca, 57
   P. jezoensis Carr., 317
   P. mariana, 49, 50, 52, 55, 178, 256, 257, 328, 329; see also Black spruce
   P. sitchensis, 49
   see also Spruce
Pine, 44, 108, 670, 733
Pine lignins, 324, 330, 773
   alcoholysis, 287
   Brauns lignin, 281
   dioxane acidolysis, 322, 323, 462
   kraft, 254, 322, 323, 662-664, 674, 685, 668, 688, 700
   lignin sulfonates, 701
   milled wood, 215, 284, 501
   soda, 322, 323
   thioglycolic acid, 322, 323
Pinoresinol 389, 515, 652
   alkaline hydrolysis, 652, 667
   condensation, 470
   dehydrogenation, 129
   formation, 138, 213
   methanolysis, 350
   nitration, 214, 390
   nitrobenzene oxidation, 440, 441
   pmr, 305, 306
   sulfonation, 623, 624
   see also Lignols, di, β-β

Pinoresinolide, 141, 214
Pinus, 44, 108
   P. banksiana, 52
   P. densiflora, 46
   P. ponderosa, 55
   P. radiata, 26, 46, 50, 54, 60, 193, 261, 313, 314, 319, 322, 323; see also Monterey pine
   P. resinosa, 104, 229,
   P. strobus, 57, 61, 104, 106, 109,
   P. sylvestris, 35, 55-58, 178, 256, 257; see also Scots pine
   P. taeda, 46
   P. thunbergi, 59
Piperidine, 362, 548, 662
1-(N-Piperidinoacetylamino)-naphthalene, 405
Pisum sativum, 64
Pits, in pulping, 612, 642
Pittosporum crassifolia, 108
Plasmalemma, 31
Podocarpus, 63
   P. acutifolium, 57
   P. amarus, 63, 65
   P. macrophyllus, 57
   P. pedunculatus, 63, 65
   P. spicatus, 58, 63
   P. totara, 57-59, 65
Poinsettia, 38
Polydispersity, of lignins, 620-622, 685, 699-703, 720, 721, 748, 840
Polyelectrolyte behavior, 708-713, 720
Polyethylenimine, 523
Polyfons, 850, 851
Polyleucocyanidins, 83, 290
Polymeric lignin products, 833-859
Polymerization, of lignins, 2, 111
Polymer properties of lignins, 695-761
Polyphenoloxidases, 111, 773
Polyphenylene oxides, 339, 555
Polypodiaceae, 62
Polyporus fomes, 781
   P. hirsutus, 773, 775
   P. versicolor, 130, 773-775, 781
Polysaccharides, 645, 701, 703-706
Polystichum munitum, 62
Polystictus versicolor, 775
Polytrichum commune, 80
Polysulfides, 526, 538, 669
Pomilio process, 373
Poplar, 221, 222, 282, 736; see also *Populus alba*
Populus alba, 35
   P. canadensis, 66
   P. deltoides, 68, 257; see also Cottonwood lignin
   P. euramericana, ir, 279
   P. tremula, 64, 68
   P. tremuloides, 66, 71, 192, 256-258, 260, 613
   see also White poplar
Poria, 780, 781
   P. subacida J247 and N199, 773, 775
   P. subacida (Peck) sacc., 772, 774, 775
   P. vaillanti, 180
   P. vaporarid, 775
Potamogeton natans, 64
Potassium bromide wafers, ir, 268-269

SUBJECT INDEX

Potassium ferricyanide, 118, 821
Potassium lignin sulfonates, 843
Potassium and liquid ammonia, 504, 505, 515
Potassium permanganate, as electron stain, 24, 26, 27
Potato, etiolated shoots, 34
Potentiometric titrations, 248, 674
Precursors of lignins, 25, 95-112
Prehydrolysis, of wood, 346
Prephenic acid, 100, 101
Prichard-O.R.F. process, 844
Primary precursors, 2, 103
Primary wall, 23, 24, 696
Primitive land plants, 19
Propenylbenzene, 441
4-Propenyl-2,6-dimethoxyphenol, dehydrogenation, 136
Propioguaiacone, 208, 277, 399, 654, 655
Propiosyringone, 332-336; see also 3,5-Dimethoxy-4-hydroxypropiophenone
4-Propylcatechol, 504, 822
i-Propylconiferyl alcohol, uv, 252
4-n-Propylcyclohexanediol-1,2, 503
4-Propylcyclohexanol, 489, 503, 504
Propylene oxide, 521
4-n-Propylguaiacol, see 1-Guaiacylpropane
Propylphenol, see 4-Hydroxyphenylpropane
Propylveratrole, 506
Protein, 194, 750, 854, 856
Protocatechualdehyde, 83, 459, 460, 471, 590, 669, 817, 818
Protocatechuic acid, 104, 465, 471, 503, 586, 588, 589, 669, 778, 779, 781, 785, 817, 818
Protolignin, 11, 453, 454, 643, 698, 741, 751-753
Proton magnetic resonance spectroscopy, 88, 299-326, 494, 495, 527, 674, 675, 683
Protozoa, 771
Prunus sp., 110
  P. spinosa, 64
Psalliota campestris, 116, 129
Pseudomonas, 778
Pseudotsuga menziesii, 44, 46, 49, 55, 60
  P. taxifolia, 58
  see also Douglas fir
Psilopsida, 21
Psilotum, 21, 63, 79, 81
  P. nudum, 62
Pteridophyta, 43, 44, 284
Pteris, 61
  P. aquilina, 56
Pulping, 261, 434, 542, 702, 730, 744
  alkaline, 639-689
  sulfite, 597-631
Pulps, 328, 457, 599, 600, 639, 640, 645, 733
Pummerer's ketone, 130
Pyrazolines, 519, 531, 533
Pyrocatechol, 503, 817, 818; see also Catechols
Pyrocatechuic acid, 778
Pyrogallol, 396, 578
Pyrolysis, 575-583, 8210823
Pyromellitic anhydride, 529
Pyrrole, 534
Pyrus communis, 64
Pyruvic acid, 774, 775

Quantitative determinations, of lignin, 190-195; see also Klason lignins; Lignin content
Quaternary ammonium lignosulfonates, 601, 719
Quebracho, 847
Quercitin, 627
Quercus borealis, 71
  Q. robur, 64
Quinhydrones, epr, 332
Quinol ethers, 121, 125, 126, 130
Quinols from phenols, 119
Quinones, see Benzoquinones
Quinonemethide intermediates, 203, 441, 535, 578, 611, 649-651, 665, 677
  in alkaline hydrolysis of models, 652, 658, 662
  in biopolymerization of lignins, 121, 122, 137-142, 147, 151, 153
  of β-0-4 lignols, 138, 654, 657
  in hydrogenolysis, 498, 499
  in oxidations, 444, 445
  in solvolysis, 345
Quinone structures, 556
  in alkali lignins, 677
  in chlorinated jute lignins, 414
  ortho-, 518, 678, 855
  para-, 450, 518
  uv, 682
  see also Benzoquinones
γ-Radiation, 556
Radicals, see Free radicals; Phenoxy radicals
Radioactivity in lignin studies, 25, 33, 95-97, 366, 367, 614, 751
Raney nickel, 489, 490, 658, 659
Rauma process, 580, 625
Ray cells, 612, 613, 642, 753
Reaction wood, 20, 25, 29, 30, 35, 47, 59, 292, 752
Reactive sites, in lignins, 511-558, 602-612, 650, 673
Reax resins, 855
Recovery furnace, 670
Red alder, 322-324, 494, 495, 613, 626, 628; see also Alnus glutinosa, A. rubra
Redox potential, in differentiating xylem, 29
Reducing groups, in lignin sulfonates, 626
Reductases, 109
Reduction difference spectra, carbonyls, 673, 677, 678
Redwoods, 1; see also Sequoia
Reed lignins, ir, 282, 287
Refractive index, 684, 722
Reimer-Tiemann reaction, 798, 799
Residual delignification, 645, 646
Residual lignin, in pulps, 258-262, 292, 293, 644, 682
Resin additives, 844, 854, 855
Resorcinol, 531, 532, 542, 543
Reversed condensation, 360, 364, 367, 368, 436, 499
Reverse osmosis, of SSL, 837
Rhamnus purshiana, 68
Rheinau process, 526, 840
Rheum rhaponticum, 64
Rhitadiadelphia loerious, 81
Riber glossularia, 64
Robinia pseudoacacia, 65, 66, 71, 72, 279

Roe-number of pulps, 195
Rosa wichuriana, 73
Rubber additives, 855, 856
Russula delica, 132
Ruthenium red, as a microscopic stain, 27, 37
Rye, 64

^{31}S, ^{35}S, and ^{37}S, 614
Saccharification of wood, 837, 840; *see also* Hydrolysis lignins
Saccharinic acids, 642, 842
Saccharum officinarum, 102
Safranin-fast green staining, 6
Saguaro wood, spin content, 328
Salicylic acid, 541, 778
Salvia splendens, 105, 106
Salvo Chemical Company, vanillin production, 801
Sapwood lignins, 285, 287, 752
Sassafras, 74
Scholler lignins, 462, 840
Schotten-Baumann reaction, 527, 528
Sciadopitys verticillata, 58
Sclerenchyma, 19, 20
Scots pine, 35, 178-180, 256, 257
Secale cereale, 64
Secondary wall, 23, 24, 49-54, 66-68, 695-697
  delignification, 612, 644, 753-757
Sedimentation, constant, 704, 705, 720-722
  equilibrium, 710, 720-724, 751
  humic acids, 784
Selaginella, 63, 79
Semichemical pulping, 261, 600, 833
Semiconductors, 338, 556
Semiquinones, 332, 536
Sephadex gel fractionation, 601, 710, 711, 727-729, 751, 752, 842
Sequestering agents, 856, 857
Sequoia, 1, 61, 115, 284
  S. sempervirens, 49, 57, 61, 313, 325
Sequoiadendron giganteum, 58
Serinehydrazide, 805
Shikimic acid, 98-102
  in growing bamboos, 113
  occurrence in plants, 99
  pathway, 97-104
Side-chain, biogenesis, 106
  carbons, notation, 12
  displaced structures, 16, 606, 607
Silver oxide, 444-446, 449, 800
Sinapaldehyde, from wood hydrolysis, 196
Sinapic acid, 105, 137
Sinapyl alcohol, 2, 122, 195
  dehydrogenation, 106, 110, 143, 155, 283
  from wood hydrolysis, 197
  sulfonation, 197, 628, 629
Sitka spruce, 176
Sivola process, 600
Slash pine, 176
Smilar china, 75
Soda lignins, 684, 743, 751
  alkali fusion, 588
  hydrogenolysis, 504, 819
  ir, 288, 289
  pmr, 322-324
  soda pulping, 639, 642, 647, 648, 833
Sodamide, 513, 813

Sodium borohydride, *see* Borohydride
Sodium 4,4-dimethyl-4-silapentane-1-sulfonate, pmr standard, 318
Sodium in liquid ammonia, 487, 488, 504, 505, 514
Sodium sulfide, *see* Sulfide
Softening temperatures, 645, 729-739
Softwood lignins, 4, 5, 284, 288, 436, 551, 696-698
  ir, 270, 278-280
  kraft, 839, 940
  lignin sulfonates, 454, 700, 701, 711
  pmr, 313-325
  reactions, 393, 409, 409, 438, 461, 469
  see also individual species
Soil, 770, 771, 783, 858
Solanum dulcamara, 64
  S. tuberosum, 64
Sol-gel concept, lignin in wood, 758
Solubility parameter, Hildebrandt, 745
Solvent effects. molecular weight determinations, 750, 751
  reactivity of wood lignin, 744, 745
Solvolysis, by acids, and bases, 345-369
Sorbus americana, 71
Sorption, of alkali by wood, 642
  of vapors and ions by lignins, 741, 745, 748
Southern pine, kraft lignin, 315, 318, 671-673
Spent sulfite liquor, composition, 836, 837
  dialysis, 601, 712
  disposal, 262, 797
  evaporation, 837
  fractionation, 699, 700
  hydrogenolysis, 487, 819
  oxidation, 449, 800, 821
  products, 800, 836, 837, 852
  removal of sugars, 840-843
  reverse osmosis, 837
Sphagnum species, 80, 81
Spherical microgel, 703, 706
Spin-spin interactions, 299, 308
Spirea ulmaria, 64
Spruce, 191, 222, 229, 329
Spruce lignins, alcoholysis, 287, 504
  alkali, 655, 685, 700
  Brauns, 281, 322, 323
  cuoxam, 803
  dioxane acidolysis, 700, 702, 731, 734, 735
  kraft, 685, 700, 728, 729, 753
  lignin sulfonates, 261, 290, 610, 617, 629, 700, 702
  periodate, 181, 183, 470, 702, 731, 732
Spruce milled wood lignins, condensed units in, 452
  coniferaldehyde groups, 250
  functional groups, 250, 252, 673-679, 786, 787
  hydrogenolysis, 251, 252, 501
  ir, 267, 280, 284
  pmr, 314, 315, 322-324, 674
  uv, 247, 248, 251-254, 256, 257
Spruce wood, delignification, 608, 609, 620, 643-646, 753-757
  fungal attack, 773
  hydrolysis, dioxane-water, 209, 348, 349
  oxidations, 446, 447, 449, 803
  spin content, 328, 329

SUBJECT INDEX

thermal softening, 738
Stachybotrys atra, 792
Steam hydrolysis lignins, 283, 322-325
Steaming of wood, 736-739
Stereum frustulatum, 781
Stilbenequinone, 209
Stilbenes, 537
  by acidolysis of mwl, 353
  from β-1 lignols, 209, 352, 365
  from β-5 lignols, 352, 361, 651
  in kraft lignins, 673, 676, 677, 680
  from lignin sulfonate hydrolyzates, 365, 630
  uv, 253, 256, 682
Stoke's law, 726
Stora process, 599
Streptomycetes, 771
Structural formulations, abbreviated, 14-16, 152
  conifer lignins, 3-5
  kraft lignin, 679
  lignin sulfonates, 606, 607
Structural polyphenols, 81
Structure, of lignins, 195-228
Styrene, 455, 556, 557
Styrene-ω-sulfonic acids, by neutral sulfonation, 619
Succinic acid, 452, 529, 591
Sugar acids, in spent sulfite liquor, 836
Sugar cane, *see* Bagasse
Sugar maple, lignin, pmr, 318
Sugars, removal from spent sulfite liquor, 601, 712, 836, 840-843
Sulfate process, *see* Kraft process
Sulfide, cleavage of methoxyl by, 806-809
  in kraft pulping, 273, 639, 640, 645, 646, 665-670, 806, 817, 818
  oxidation by air in black liquor, 669
  radical scavenger, 648
Sulfidity, definitions, 640, 642
Sulfite pulping, 261, 597-631
  by-products, 833
  classification of precesses, 599, 600
  pulps, 290, 413, 416, 644
  reaction rates, 602-612, 647, 648
  topochemistry of delignification, 753-757
  see also Neutral sulfite pulping
Sulfite spent liquor, *see* Spent sulfite liquor
Sulfitolysis, 611; *see also* Sulfonation
Sulfomethylation, 548, 601
Sulfonamide groups, 539
Sulfonate groups, 270, 415, 620-625, 630, 702, 800
Sulfonation, 186, 597-631, 713, 835, 851, 859
Sulfones, 537
Sulfonium compounds, from DMSO, 811, 814
Sulfur chemicals, organic, 806-817
Sulfur content, kraft lignins, 657, 666, 668, 672, 839
  lignin sulfonates, 620, 622, 623, 626
Sulfur dioxide, 517
  adsorption by pulps and lignin, 645, 746
  in organic media, 541, 618
  reduction of *o*-quinones by, 679, 855
  in sulfite pulping, 598-601
  uv, 261
Sulfuric acid lignins, *see* Klason lignins

Sulfur-methoxyl ratio, lignin sulfonates, 603, 605, 613, 620, 625, 642
Sulfuryl chloride, 377, 531, 846
Sustained release, of pesticides, 528
Svedberg equation, 722
Swelling, 557, 645
Synthetic lignins, *see* Dehydrogenation polymers
Syringaldehyde, 13, 332, 389, 397, 433, 435, 447, 490, 613, 701, 772, 785, 804
Syringaresinol, dehydrogenation product, 143
  from wood hydrolysis, 213
Syringa vulgaris, 73
Syringic acid, 13, 435, 448, 491, 589
  reactions, 122, 389, 397
Syringin, 28, 110
Syringyl alcohol, 13, 123, 389, 397, 499, 500
Syringyl derivatives, from hydrogenolysis, 488, 493, 498-502
  ir, 279
  nmr, 494, 495
  nomenclature, 12-14
  oxidation, 460
  reactivity, 385, 463, 512
  uv, 243, 257, 258
Syringyl ethane, 489-492, 498-502
2-Syringylethanol, 489-492
Syringylglycerol, from wood hydrolysis, 200
  α-sulfonate, 626
Syringylglycerol-β-(2-methoxyphenyl)ether, 781
Syringyl group, nomenclature, 13, 14
Syringyl-guaiacyl ratios, 258, 291, 613
Syringyl lignins, 155-613
Syringylmethane, 490-493
Syringylpropane, dehydrogenation, 122
  epr, 333, 336
  from hydrogenolysis, 489-492, 498-502, 505
  ir, 277
  uv, 257
3-Syringyl-1-propanol, 489, 491, 498-502
β-Syringylpropionic acid, from hydrogenolysis, 491
Syringyl units in lignin, 546, 581, 675
  hydrogenolysis, 496, 497
Syringylvinylcarbinol, 122
Syrinoxyl, 120, 336-338

Tall oil, 526, 833, 834
Tanning agents, 805, 856
Taxodium distichum, 58
Taxonomical studies, 434, 578
Taxus baccata, 57, 58
  T. canadensis, 57
Tension wood, 20, 25, 29, 30, 35, 73
Terephthalic acid, 589
Tetrachloro-*o*-benzoquinone, 397, 416
Tetrachlorocatechol, 397
Tetrachloroguaiacol, 394, 409
Tetrachloroveratrole, 394
Tetraclinis articulata, 63, 65
Thermal coalescence of lignin, 645
Thermal degradation, *see* Pyrolysis
Thermal expansion, 732, 740, 741
Thermal softening, 645, 684, 729-739
Thermogravimetric analysis, 581, 584
Thermosetting resins, 845, 855

SUBJECT INDEX

Thiocompounds, reactivity with lignin, 532, 538, 543, 545
Thioglycolic acid lignins, see Lignin thioglycolic acids
Thiolignins, 666-669
Thionyl chloride, 377, 531, 814, 846
Thiophosphoryl derivatives, 530, 531
Thiosulfate, 622, 623, 669, 808
Thuja, 45
  T. plicata, 49, 55-58, 313, 317, 321; see also Western redcedar lignins
Tilia americana, 71
  T. ulmifolia, 64
Timothy, 19, 20
Tissue cultures, 35, 60, 73, 104, 106, 109-111
Titanium dioxide, 713-716
Tmesipteris, 21
Tolylene di-isocyanate, 522
Topochemistry, of delignification, 612, 639-650 650, 753-757
Torula yeast, 841
Tracer methods, 25, 33, 95-97, 366, 367, 614, 751
Tracheids, 21-23, 643, 697
Trachycarpus excelsa, 75
Trametes saquinea, 775
  T. pini, 773, 775
1,2,4-Triazyl-3-ammonium lignin sulfonate, 539
2,4,6-Tri-*t*-butylphenoxy radical, 120, 121
2,4,6-Trichloroanisole, 394
Trichloropyrogallol, 414
1,2,3-Trichloro-4,5,6-trimethoxybenzene, 394, 412
3,4,5-Trichloroveratrole, 394
Trichoderma viride, 169
Trihydroxytoluenes, autoxidation, 793
Tri-iodobenzoic acid, formation of tension wood by, 35
Trimethoxybenzenes, 211, 365, 393, 397, 460
Trimethoxybenzoic acids, 217, 218, 365, 452, 453
3,4,5-Trimethoxybenzyl alcohol, 387
Trimethylsilyl ethers, pmr, 315
Trimethylsulfoxonium iodide, 815
2,3,4-Tri-O-methylxylose, 221, 222
2,4,6-Triphenylphenoxy radical, 120, 121
Tritiated lignin precursors, 25, 33
Triticum aestivum, 99, 105
Tritylation, 520
Trochodendron sp., 71
  T. aralioides, 279
Tropical hardwood lignins, 44, 71
Trypsin, 816
Tryptophan, 98, 99
Tsuga, 45
  T. canadensis, 56-58
  T. heterophylla, 49, 55, 58, 107, 256, 257
Turpentine, 806
Two-electron transfer processes, 117, 434-436
Type L lignins, see Guaiacyl-syringyl lignins
Type N lignins, see Guaiacyl lignins
Tyramine, dehydrogenation, 129
Tyrase, 105, 111-114, 770
Tyrosinase, see Tyrase
L-Tyrosine, 98-104, 778
  dehydrogenation of, 129

homogentisic acid from, 780
incorporation in maize lignin, 34
lignin precursor, 104, 105
Tyrosine ammonia lyase, see Tyrase

Ulmin, 783
Ulmus americana, 66, 71
  U. campestris, 73
  U. fulva, 71
  U. montana, 64
Ultracentrifuge methods, 720-723
Ultraviolet microscopy, 24-26, 30, 697, 753-757
Ultraviolet spectra, 241-263
  bark lignins, 87
  borohydride reduced lignins, 245, 247-251, 673, 677, 678
  α-carbonyl groups, 247, 673, 677, 678
  in carboxyl group determination, 629
  coniferaldehyde groups, 251
  difference spectra, 245, 248, 251, 257, 616, 617, 629, 673, 677, 678
  halolignins, 261, 405, 406, 417, 466
  hydrogenated lignins, 246, 251, 673
  kraft lignins, 671, 678, 682-684
  lignin sulfonates, 245, 247, 261, 262, 629
  milled wood lignins, 54, 56, 60, 246, 247, 251, 256, 257, 678, 682-684
  peracetic acid lignins, 461
  phenolic structures, 246, 255-256, 616
  quinonemethides, 122, 138
  stilbene structures, 253, 673, 676
Umbelliferae, lignification of collenchyma, 29
Urea, in nitrations, 380, 392
Urea-formaldehyde, resins, 854, 855
Usnic acid hydrate, 131
Utilization of lignin products, 797-858

Vanadium oxides, 449, 535
Vanilla planifolia, 798
Vanillic acid, acidolysis product, 357
  from alkaline oxidations, 435, 438, 440, 446, 448, 464, 469, 589, 801
  from black liquor, 671, 817, 818
  dehydrogenation of, 123, 129
  demethylation of, 506, 515
  ester groups in lignins, 2, 63, 178
  by fungal degradation, 773, 775, 776, 780
  by hydrogenolysis, 491
  ir, 275
  iso-, metabolism of, 781
  Mannich reaction of, 662
  nitration of, 388, 390
  from peracetic acid oxidations, 459, 460, 463
Vanillil, oxidation, 446-448
Vanillin, 515, 542, 543, 551, 557, 558, 613, 625, 718, 780
  acetate of, 96
  by acidolysis, 200, 357
  by alkaline hydrolysis, 365-367, 598
  by alkaline oxidations, 440, 446-449, 591, 668, 814
  from black liquor, 671, 817, 818, 834
  chlorination of, 284, 531
  commercial production, 434, 449, 800-803
  dehydrogenation of, 129, 788
  electrophoresis of, 711, 712

SUBJECT INDEX

from fungal degradation, 772-776
by hydrogen peroxide oxidation, 469
from isoeugenol, 434, 441, 799
as a lignin precursor, 104
nitration of, 96, 388, 390
by nitrobenzene oxidation, 433-435, 438-443, 490, 785
oxime of, 805
by peracetic acid oxidations, 459, 460
properties of, 804-806
radioactive, 96, 102
by reversed aldol condensations, 360, 367, 499
sodium salt of, 805
synthesis of, 798-800
ultraviolet spectroscopy of, 249
uses and derivatives of, 804-806
by water hydrolysis of wood, 348
o-Vanillin, 798, 799
Vanilloylformic acid, 803
Vanillyl alcohol, chlorination, 387
condensations, 660, 662
dehydrogenation, 129, 788
fungal degradation, 776
hydrogenation, 499, 500
ir, 272-274
nitration, 388-390
oxidation, 444, 445, 469, 471, 474, 788
sulfonation, 351, 602, 611, 630
thioglycolic acid derivative of, 351
thiosulfate treatment of, 622, 623
from veratryl alcohol, 459, 460
5-Vanillylcreosol, oxidation of, 649
Vanillyl disulfide, 668
Vanillyl ethers, alkaline hydrolysis, 361, 652, 660
dehydrogenation, 788
hydrogen peroxide oxidation, 469
nitration, 388, 390
Vanillyl mercaptan, 662
Vanillyl sulfide, 660, 662, 668
Vanillyl sulfonic acid, 14
Vapor pressure osmometry, Kraft lignins, 719, 744, 750, 839, 840
Vapors, sorption by lignins, 745, 746
Vascular plants, 21
Veraguensin, racemic, from isoeugenol, 134
Veratraldehyde, 96, 460
Cannizzaro reaction of, 436, 437
from methylated lignin sulfonates, 367, 598
nitration, 389
sulfonation, 365, 630
Veratric acid, 203, 436, 516, 531
bromination, 388
ir, 275
from methylated lignins, 204, 452, 453, 455
Veratrole, 363, 384, 506
Veratryl alcohol, 12, 436, 437
chlorination, 387
nitrosation, 389
ozone oxidation, 471
sulfonation, 351
thiosulfate treatment, 622, 623
Veratrylglycerol, 12; see also 3,4-Dimethoxy-phenylglycerol
Veratryl groups, nomenclature, 12-14
Vesicles, role in lignification, 31, 32
Vessels, 67, 68, 613, 642
Vinyl acetate, 556
Vinyl ether groups, in alkali lignins, 676, 682
4-Vinylphenol, from pyrolyzates, 577
Viscosity, of dichromate-lignin sulfonates, 716-718
effect of ionic strength, 708-710
in molecular weight determinations, 614, 685, 686, 723, 784
spherical microgel, 703-706
Vitamin B-12, recovery by LSA precipitation, 856

Wagner-Meerwein shift, in acidolysis, 353
Water, hydrolysis by, 154, 205, 209, 345-349
sorption, 745, 746
Wattle wood, sulfonates, 628
Weight average molecular weights, 643, 685, 719-723
Welwitschia mirabilis, 65
Western hemlock, 176, 177, 328
lignins, ir, 271, 281, pmr, 322-324, uv, 256, 257
sulfonation, 608, 627, 630, 752
Western redcedar lignins, 176
ir, 278, 286
pmr, 313, 317, 321-323
Wheat lignins, ester groups, 178
ir, 282
lignin-carbohydrate complexes, 221
nitrobenzene oxidation, 78
precursors, 99, 105
White birch, Klason lignin, 191
ultraviolet microscopy, 697
White fir lignins, 180, 191
White liquor, 640-642; see also Kraft pulping; Kraft liquor
White pine lignin, 712
White poplar, tension wood, 35
White rot fungi, 769-782, 785
White spruce, lignin sulfonates, 752
Widdringtonia juniperoides, 58
Wiesner reaction, 6, 82, 532, 543
Willstätter lignins, 11, 185, 186, 462, 551, 659, 821; see also Hydrochloric acid lignins
Winteraceae, 71
Wood, cellular structure, 22-24, 695-698
chlorination, 744, 745
enzymatic degradation of, 179, 703, 774
hydrogenation, 487
hydrolysis, 526, 823, 840
ir, 285
liquefaction of, 487, 504, 819
peracetic acid oxidation, 460
pyrolysis, 581, 821
sodium-liquid ammonia treatment, 505

Xanthation, 529, 530
X groups in sulfonation, 603
Xylan, 22, 284, 292, 696
from birch, 701-704

SUBJECT INDEX

in pulp fibers, 645
softening temperatures, 733
viscosity, 703, 704
Xylem, primary, 19, 20
Xylenols, 821, 822, 855

Yeast, 777

Z-average molecular weight, 723
Zeisel methoxyl determinations, 512, 539

Z groups, in sulfonation, 603, 623
Zinc, lignin sulfonate, 843
   oxide slurries, dispersion of by LSA, 713, 715
   pyrolysis catalyst, 579
   reduction by, 818
Zostera marina, 64
Zulaufverfahren, 146, 150
Zutropfverfahren, 146, 150
Zygogynum viellardi, 71